| Extensive and Varied Exercise Sets | An abundant collection of exercises is included in an exercise set at the end of each section. Exercises are organized within categories. Your instructor will usually provide guidance on which exercises to work. The exercises in the first category, Practice Exercises, follow the same order as the section's worked examples. | The parallel order of the Practice Exercises lets you refer to the worked examples and use them as models for solving these problems. |
| Practice Plus Problems | This category of exercises contains more challenging problems that often require you to combine several skills or concepts. | It is important to dig in and develop your problem-solving skills. Practice Plus Exercises provide you with ample opportunity to do so. |

3 Review for Quizzes and Tests

Feature	Description	Benefit
Mid-Chapter Check Points	At approximately the midway point in the chapter, an integrated set of review exercises allows you to review the skills and concepts you learned separately over several sections.	By combining exercises from the first half of the chapter, the Mid-Chapter Check Points give a comprehensive review before you move on to the material in the remainder of the chapter.
Chapter Review Grids	Each chapter contains a review chart that summarizes the definitions and concepts in every section of the chapter. Examples that illustrate these key concepts are also included in the chart.	Review this chart and you'll know the most important material in the chapter!
Chapter Review Exercises	A comprehensive collection of review exercises for each of the chapter's sections follows the review grid.	Practice makes perfect. These exercises contain the most significant problems for each of the chapter's sections.
Chapter Tests	Each chapter contains a practice test with approximately 25 problems that cover the important concepts in the chapter. Take the practice test, check your answers, and then watch the Chapter Test Prep Video CD to see worked-out solutions for any exercises you miss.	You can use the chapter test to determine whether you have mastered the material covered in the chapter.
Chapter Test Prep Video CD	This video CD found at the back of your text contains worked-out solutions to every exercise in each chapter test.	The video lets you review any exercises you miss on the chapter test.
Cumulative Review Exercises	Beginning with Chapter 2, each chapter concludes with a comprehensive collection of mixed cumulative review exercises. These exercises combine problems from previous chapters and the present chapter, providing an ongoing cumulative review.	Ever forget what you've learned? These exercises ensure that you are not forgetting anything as you move forward.

More Tools for Your Mathematics Success

Student Study Pack

Get math help when YOU need it! The **Student Study Pack** provides you with the ultimate set of support resources to go along with your text. Packaged at no charge with a new textbook, the **Student Study Pack** contains these invaluable tools:

☑ **Student Solutions Manual**

A printed manual containing full solutions to odd-numbered textbook exercises.

☑ **Prentice Hall Math Tutor Center**

Tutors provide one-on-one tutoring for any problem with an answer at the back of the book. You can contact the Tutor Center via a toll-free phone number, fax, or email.

☑ **CD-ROM Lecture Series**

A comprehensive set of textbook-specific CD-ROMs containing short video clips of the textbook objectives being reviewed and key examples being solved.

Tutorial and Homework Options

MYMATHLAB

MyMathLab, packaged at no charge with a new textbook, is a complete online multimedia resource to help you succeed in learning. **MyMathLab** features:

☑ The entire textbook online.

☑ Problem-solving video clips and practice exercises, correlated to the examples and exercises in the text.

☑ Online tutorial exercises with guided examples.

☑ Online homework and tests.

☑ Generates a personalized study plan based on your test results.

☑ Tracks all online homework, tests, and tutorial work you do in **MyMathLab** gradebook.

Essentials of
Intermediate Algebra
for College Students

ANNOTATED INSTRUCTOR'S EDITION

Essentials of Intermediate Algebra for College Students

Robert Blitzer
Miami-Dade College

PEARSON

Prentice
Hall

Upper Saddle River, New Jersey 07458

Senior Acquisitions Editor: *Paul Murphy*
Editor in Chief: *Christine Hoag*
Project Manager: *Liz Covello*
Production Editor: *Barbara Mack*
Assistant Managing Editor: *Bayani Mendoza de Leon*
Senior Managing Editor: *Linda Mihatov Behrens*
Executive Managing Editor: *Kathleen Schiaparelli*
Media Project Manager: *Audra J. Walsh*
Media Production Editor: *Allyson Kloss*
Managing Editor, Digital Supplements: *Nicole M. Jackson*
Assistant Manufacturing Manager/Buyer: *Michael Bell*
Executive Marketing Manager: *Eilish Collins Main*
Marketing Assistant: *Rebecca Alimena*
Director of Marketing: *Patrice Jones*
Development Editor: *Don Gecewicz*
Editor in Chief, Development: *Carol Trueheart*
Editorial Assistant: *Mary Burket*
Assistant Editor/Print Supplements Editor: *Christina Simoneau*
Art Director: *Maureen Eide*
Interior Design: *Running River Design*
Cover Art and Designer: *Koala Bear Design*
Art Editor: *Thomas Benfatti*
Director of Creative Services: *Paul Belfanti*
Director, Image Resource Center: *Melinda Reo*
Manager, Rights and Permissions: *Zina Arabia*
Manager, Visual Research: *Beth Brenzel*
Image Permission Coordinator: *Debbie Hewitson*
Photo Researcher: *Beaura K. Ringrose*
Art Studios: *Scientific Illustrators; Laserwords*
Compositor: *Prepare, Inc.*

© 2006 Pearson Education, Inc.
Pearson Prentice Hall
Pearson Education, Inc.
Upper Saddle River, New Jersey 07458

Pearson Prentice Hall™ is a trademark of Pearson Education, Inc.

Printed in the United States of America
10   9   8   7   6   5   4   3   2   1

ISBN 0-13-186613-3
ISBN 0-13-186559-5 (Student Edition)

Pearson Education LTD., *London*
Pearson Education Australia PTY, Limited, *Sydney*
Pearson Education Singapore, Pte. Ltd
Pearson Education North Asia Ltd, *Hong Kong*
Pearson Education Canada, Ltd., *Toronto*
Pearson Educación de Mexico, S.A. de C.V.
Pearson Education—Japan, *Tokyo*
Pearson Education Malaysia, Pte. Ltd

CONTENTS

A BRIEF GUIDE TO GETTING THE MOST FROM THIS BOOK
Inside front cover

PREFACE   IX
TO THE STUDENT   XVII
ABOUT THE AUTHOR   XIX

CHAPTER 1 ALGEBRA, MATHEMATICAL MODELS, AND PROBLEM SOLVING   1

1.1 Algebraic Expressions and Real Numbers   2
1.2 Operations with Real Numbers and Simplifying Algebraic Expressions   13
1.3 Graphing Equations   28
1.4 Solving Linear Equations   39
      Mid-Chapter Check Point   **51**
1.5 Problem Solving and Using Formulas   51
1.6 Properties of Integral Exponents   66
1.7 Scientific Notation   78
      GROUP PROJECT   85
      SUMMARY   86
      REVIEW EXERCISES   91
      CHAPTER TEST   94

CHAPTER 2 FUNCTIONS AND LINEAR FUNCTIONS   95

2.1 Introduction to Functions   96
2.2 The Algebra of Functions   112
      Mid-Chapter Check Point   **122**
2.3 Linear Functions and Slope   123
2.4 The Point-Slope Form of the Equation of a Line   141
      GROUP PROJECT   152
      SUMMARY   153
      REVIEW EXERCISES   156
      CHAPTER TEST   159
      CUMULATIVE REVIEW EXERCISES (CHAPTERS 1–2)   160

CHAPTER **3** SYSTEMS OF LINEAR EQUATIONS   **162**

3.1 Systems of Linear Equations in Two Variables   163
3.2 Problem Solving and Business Applications Using Systems of Equations   178
3.3 Systems of Linear Equations in Three Variables   193
Mid-Chapter Check Point   **204**
3.4 Matrix Solutions to Linear Systems   205
3.5 Determinants and Cramer's Rule   217
GROUP PROJECT   227
SUMMARY   228
REVIEW EXERCISES   233
CHAPTER TEST   235
CUMULATIVE REVIEW EXERCISES (CHAPTERS 1–3)   236

CHAPTER **4** INEQUALITIES AND PROBLEM SOLVING   **238**

4.1 Solving Linear Inequalities   239
4.2 Compound Inequalities   253
4.3 Equations and Inequalities Involving Absolute Value   262
Mid-Chapter Check Point   **270**
4.4 Linear Inequalities in Two Variables   271
4.5 Linear Programming   282
GROUP PROJECT   290
SUMMARY   290
REVIEW EXERCISES   293
CHAPTER TEST   295
CUMULATIVE REVIEW EXERCISES (CHAPTERS 1–4)   296

CHAPTER **5** POLYNOMIALS, POLYNOMIAL FUNCTIONS, AND FACTORING   **297**

5.1 Introduction to Polynomials and Polynomial Functions   298
5.2 Multiplication of Polynomials   311
5.3 Greatest Common Factors and Factoring by Grouping   324
5.4 Factoring Trinomials   333
Mid-Chapter Check Point   **346**
5.5 Factoring Special Forms   347
5.6 A General Factoring Strategy   356
5.7 Polynomial Equations and Their Applications   362
GROUP PROJECT   375
SUMMARY   376
REVIEW EXERCISES   380
CHAPTER TEST   382
CUMULATIVE REVIEW EXERCISES (CHAPTERS 1–5)   383

CHAPTER **6** RATIONAL EXPRESSIONS, FUNCTIONS, AND EQUATIONS   **384**

6.1 Rational Expressions and Functions: Multiplying and Dividing   385
6.2 Adding and Subtracting Rational Expressions   399
6.3 Complex Rational Expressions   411
6.4 Division of Polynomials   419
   Mid-Chapter Check Point   **427**
6.5 Synthetic Division and the Remainder Theorem   428
6.6 Rational Equations   436
6.7 Formulas and Applications of Rational Equations   445
6.8 Modeling Using Variation   458
   GROUP PROJECT   469
   SUMMARY   470
   REVIEW EXERCISES   474
   CHAPTER TEST   477
   CUMULATIVE REVIEW EXERCISES (CHAPTERS 1–6)   478

CHAPTER **7** RADICALS, RADICAL FUNCTIONS, AND RATIONAL EXPONENTS   **479**

7.1 Radical Expressions and Functions   480
7.2 Rational Exponents   492
7.3 Multiplying and Simplifying Radical Expressions   504
7.4 Adding, Subtracting, and Dividing Radical Expressions   512
   Mid-Chapter Check Point   **519**
7.5 Multiplying with More Than One Term and Rationalizing Denominators   520
7.6 Radical Equations   532
7.7 Complex Numbers   542
   GROUP PROJECT   551
   SUMMARY   551
   REVIEW EXERCISES   555
   CHAPTER TEST   557
   CUMULATIVE REVIEW EXERCISES (CHAPTERS 1–7)   558

CHAPTER **8** QUADRATIC EQUATIONS AND FUNCTIONS   **559**

8.1 The Square Root Property and Completing the Square   560
8.2 The Quadratic Formula   574
8.3 Quadratic Functions and Their Graphs   589
   Mid-Chapter Check Point   **606**
8.4 Equations Quadratic in Form   607
8.5 Polynomial and Rational Inequalities   614
   GROUP PROJECT   627
   SUMMARY   628
   REVIEW EXERCISES   632
   CHAPTER TEST   634
   CUMULATIVE REVIEW EXERCISES (CHAPTERS 1–8)   635

Shaded chapters are available in Blitzer,
Intermediate Algebra for College Students, Fourth Edition.

CHAPTER 9 EXPONENTIAL AND LOGARITHMIC FUNCTIONS 636

9.1 Exponential Functions 637
9.2 Composite and Inverse Functions 650
9.3 Logarithmic Functions 665
9.4 Properties of Logarithms 678
   Mid-Chapter Check Point **688**
9.5 Exponential and Logarithmic Equations 690
9.6 Exponential Growth and Decay; Modeling Data 702
   GROUP PROJECT 716
   SUMMARY 716
   REVIEW EXERCISES 720
   CHAPTER TEST 725
   CUMULATIVE REVIEW EXERCISES (CHAPTERS 1–9) 726

CHAPTER 10 CONIC SECTIONS AND SYSTEMS OF NONLINEAR EQUATIONS 728

10.1 Distance and Midpoint Formulas; Circles 729
10.2 The Ellipse 739
10.3 The Hyperbola 751
   Mid-Chapter Check Point **760**
10.4 The Parabola; Identifying Conic Sections 761
10.5 Systems of Nonlinear Equations in Two Variables 773
   GROUP PROJECT 784
   SUMMARY 785
   REVIEW EXERCISES 787
   CHAPTER TEST 789
   CUMULATIVE REVIEW EXERCISES (CHAPTERS 1–10) 790

CHAPTER 11 SEQUENCES, SERIES, AND THE BINOMIAL THEOREM 791

11.1 Sequences and Summation Notation 792
11.2 Arithmetic Sequences 802
11.3 Geometric Sequences and Series 812
   Mid-Chapter Check Point **827**
11.4 The Binomial Theorem 828
   GROUP PROJECT 835
   SUMMARY 836
   REVIEW EXERCISES 838
   CHAPTER TEST 840
   CUMULATIVE REVIEW EXERCISES (CHAPTERS 1–11) 841

APPENDIX A Distance and Midpoint Formulas; Circles 637
APPENDIX B Summation Notation and the Binomial Theorem 644
Answers to Selected Exercises A1
Graphing Answer Section G1
Applications Index AP1
Subject Index I1
Photo Credits P1

PREFACE

Essentials of Intermediate Algebra for College Students provides comprehensive, in-depth coverage of the topics required in most one-term courses in intermediate algebra. The book is written for college students who have had a course in introductory algebra. I wrote the book to help diverse students, with different backgrounds and career plans, to succeed in intermediate algebra. *Essentials of Intermediate Algebra for College Students* has two primary goals:

1. To help students acquire a solid foundation in the skills and concepts of intermediate algebra.
2. To show students how algebra can model and solve authentic real-world problems.

One major obstacle in the way of achieving these goals is the fact that very few students actually read their textbook. This has been a regular source of frustration for me and for my colleagues in the classroom. Anecdotal evidence gathered over years highlights two basic reasons why students do not take advantage of their textbook:

- "I'll never use this information."
- "I can't follow the explanations."

I've written every page of this textbook with the intent of eliminating these two objections. The ideas and tools I've used to do so are described in the features that follow. These features and their benefits are highlighted for the student in "A Brief Guide to Getting the Most from This Book" that appears inside the front cover.

A Note on the Essentials Version of Intermediate Algebra

Essentials of Intermediate Algebra for College Students is a concise version of the Fourth Edition of *Intermediate Algebra for College Students*. **The essentials version differs from the Fourth Edition only in terms of length**. Chapter 9 (Exponential and Logarithmic Functions), Chapter 10 (Conic Sections and Systems of Nonlinear Equations), and Chapter 11 (Sequences, Series, and the Binomial Theorem) have been eliminated. The omitted chapters and their sections are highlighted for your reference in the Table of Contents. The essentials version contains two appendices (Appendix A: Distance and Midpoint Formulas; Circles and Appendix B: Summation Notation and the Binomial Theorem) for instructors who prefer the lighter, less expensive alternative to the Fourth Edition, but who still want to include these topics.

Features in the Essentials Version of Intermediate Algebra

- **Detailed Worked-Out Examples.** Each worked example is titled, making clear the purpose of the example. Examples are clearly written and provide students with detailed step-by-step solutions. No steps are omitted and each step is thoroughly explained to the right of the mathematics.

- **Check Point Examples.** Each example is followed by a similar matched problem, called a Check Point, offering students the opportunity to test their understanding of the example by working a similar exercise. The answers to the Check Points are provided in the answer section.

- **Explanatory Voice Balloons.** Voice balloons are used in a variety of ways to demystify mathematics. They translate algebraic ideas into everyday English, help clarify problem-solving procedures, present alternative ways of understanding concepts, and connect problem solving to concepts students have already learned.

- **Applications Using Real-World Data.** Interesting and innovative applications from nearly every discipline, supported by up-to-date real-world data, are included in nearly every section.

- **Extensive and Varied Exercise Sets.** An abundant collection of exercises is included in an exercise set at the end of each section. Exercises are organized within seven category types: Practice Exercises, Practice Plus Exercises, Application Exercises, Writing in Mathematics, Technology Exercises, Critical Thinking Exercises, and Review Exercises. This format makes it easy to create well-rounded homework assignments. The order of the practice exercises is exactly the same as the order of the section's worked examples. This parallel order enables students to refer to the titled examples and their detailed explanations to achieve success working the practice exercises.

- **Practice Plus Problems.** This category of exercises contains more challenging practice problems that often require students to combine several skills or concepts. With an average of ten practice plus problems per exercise set, instructors are provided with the option of creating assignments that take practice exercises to a more challenging level.

- **Mid-Chapter Check Points.** At approximately the midway point in each chapter, an integrated set of review exercises allows students to review and assimilate the skills and concepts they learned separately over several sections. The problems that make up the Mid-Chapter Check Points, averaging 23 exercises per check point, are of a mixed nature, requiring students to discriminate which concepts or skills to apply. The Mid-Chapter Check Points should help students bring together the different objectives covered in the first half of the chapter before they move on to the material in the remainder of the chapter.

- **Graphing and Functions.** Graphing is introduced in Chapter 1 and functions are introduced in Chapter 2. Because functions are the core of this course, students are repeatedly shown how functions relate to equations and graphs.

- **Chapter-Opening and Section-Opening Scenarios.** Every chapter and every section open with a scenario presenting a unique application of algebra in students' lives outside the classroom. These scenarios are revisited in the course of the chapter or section in an example, discussion, or exercise.

- **Section Objectives.** Learning objectives are clearly stated at the beginning of each section. These objectives help students recognize and focus on the section's most important ideas. The objectives are restated in the margin at their point of use.

- **Thorough, Yet Optional Technology.** Although the use of graphing utilities is optional, they are utilized in Using Technology boxes to enable students to visualize and gain numerical insight into algebraic concepts. The use of graphing utilities is also reinforced in the technology exercises appearing in the exercise sets for those who want this option. With the book's early introduction to graphing and functions, students can look at the calculator screens in the Using Technology boxes and, with the assistance of the explanatory voice balloons, begin to understand both graphical and numerical approaches to problems even if they are not using a graphing utility in the course.

- **Study Tips.** The book's Study Tip boxes offer suggestions for problem solving, point out common errors to avoid, and provide informal hints and suggestions. These invaluable hints appear throughout the book.

- **Chapter Review Grids.** Each chapter contains a review chart that summarizes the definitions and concepts in every section of the chapter. Examples that illustrate these key concepts are also included in the chart.

- **End-of-Chapter Materials.** A comprehensive collection of review exercises for each of the chapter's sections follows the review grid. This is followed by a chapter test that enables students to test their understanding of the material covered in the chapter. Beginning with Chapter 2, each chapter concludes with a comprehensive collection of mixed cumulative review exercises.

- **Chapter Test Prep Video CD.** Packaged at the back of the text, this video CD provides students with step-by-step solutions for each of the exercises in the book's chapter tests.

- **Enrichment Essays.** These discussions provide historical, interdisciplinary, and otherwise interesting connections to the algebra under study, showing students that math is an interesting and dynamic discipline.

- **Discovery.** Discover for Yourself boxes, found throughout the text, encourage students to further explore algebraic concepts. These explorations are optional and their omission does not interfere with the continuity of the topic under consideration.

- **Chapter Projects.** At the end of each chapter is a collaborative activity that gives students the opportunity to work cooperatively as they think and talk about mathematics. Additional group projects can be found in the *Instructor's Resource Manual*. Many of these exercises should result in interesting group discussions.

Resources for the Instructor

Print

Annotated Instructor's Edition (ISBN: 0-13-186613-3)

- Answers appear in place on the same text page as exercises or in the Graphing Answer Section
- Answers to all exercises in the exercise sets, Mid-Chapter Check Points, Chapter Reviews, Chapter Tests, and Cumulative Reviews

Instructor's Solutions Manual (ISBN: 0-13-133056-X)

- Detailed step-by-step solutions to the even-numbered section exercises
- Solutions to every exercise (odd and even) in the Mid-Chapter Check Points, Chapter Reviews, Chapter Tests, and Cumulative Reviews
- Solution methods reflect those emphasized in the text.

Instructor's Resource Manual with Tests (ISBN: 0-13-133057-8)

- Six test forms per chapter—3 free response, 3 multiple choice
- Two *Cumulative Tests* for all even-numbered chapters
- Two *Final Exams*
- Answers to all test items
- *Mini-Lectures* for each section with brief lectures including key learning objectives, classroom examples, and teaching notes

- Additional *Activities*, two per chapter, providing short group activities in a convenient ready-to-use handout format
- *Skill Builders* providing an enhanced worksheet for each text section, including concept rules, explained examples, and extra problems for students
- Twenty *Additional Exercises* per section for added test exercises or worksheets

Media

Lab Pack CD Lecture Series (ISBN: 0-13-133058-6)

- Organized by section, *Lab Pack CD Lecture Series* contains problem-solving techniques and examples from the textbook.
- Step-by-step solutions to selected exercises from each textbook section marked with a video icon.

TestGen (ISBN: 0-13-133060-8)

- Windows and Macintosh compatible
- Algorithmically driven, text-specific testing program covering all objectives of the text
- Chapter Test file for each chapter provides algorithms specific to exercises in each *Chapter Test* from the text.
- Edit and add your own questions with the built-in question editor, which allows you to create graphs, import graphics, and insert math notation.
- Create a nearly unlimited number of tests and worksheets, as well as assorted reports and summaries.
- Networkable for administering tests and capturing grades online, or on a local area network

MyMathLab
www.mymathlab.com
An all-in-one, online tutorial, homework, assessment, and course management tool with the following features:

- Rich and flexible set of course materials, featuring free-response exercises algorithmically generated for unlimited practice and mastery
- Entire textbook online with links to multimedia resources including video clips, practice exercises, and animations that are correlated to the textbook examples and exercises
- Homework and test managers to select and assign online exercises correlated directly to the text
- A personalized Study Plan generated based on student test results. The Study Plan links directly to unlimited tutorial exercises for the areas students need to study and retest, so they can practice until they have mastered the skills and concepts.
- *MyMathLab Gradebook* allows you to track all of the online homework, tests, and tutorial work while providing grade control.
- Import *TestGen* tests

MathXL®
www.mathxl.com
A powerful online homework, tutorial, and assessment system that allows instructors to:

- Create, edit, and assign online homework and tests using algorithmically generated exercises correlated at the objective level to the textbook
- Track student work in *MathXL*'s online gradebook

Resources for the Student

Student Solutions Manual   CD/Video   PH Math/Tutor Center   MathXL Tutorials on CD   MathXL®   MyMathLab   Interactmath.com

Student Study Pack (ISBN: 0-13-154937-5)

Get math help when YOU need it! Available at no charge when packaged with a new text, *Student Study Pack* provides the ultimate set of support resources to go along with the text. The *Student Study Pack* includes the *Student Solutions Manual*, access to the *Prentice Hall Math Tutor Center*, and the *CD Lecture Videos*.

Print

Student Solutions Manual (ISBN: 0-13-192200-9)

- Solutions to all odd-numbered section exercises
- Solutions to every (even and odd) exercise in the Mid-Chapter Check Points, Chapter Reviews, Chapter Tests, and Cumulative Reviews
- Solution methods reflect those emphasized in the text.

Media

MyMathLab

An all-in-one, online tutorial, homework, assessment, course management tool with the following student features:

- The entire textbook online with links to multimedia resources including video clips, practice exercises, and animations correlated to the textbook examples and exercises
- Online tutorial, homework, and tests
- A personalized Study Plan based on student test results. The Study Plan links directly to unlimited tutorial exercises for the areas students need to study and retest, so they can practice until they have mastered the skills and concepts.

MathXL®

A powerful online homework, tutorial, and assessment system that allows students to:

- Take chapter tests and receive a personalized study plan based on their test results
- See diagnosed weaknesses and link directly to tutorial exercises for the objectives they need to study and retest
- Access supplemental animations and video clips directly from selected exercises

www.InterActMath.com

- Have the power of the *MathXL®* text-specific tutorial exercises available for unlimited practice online, without an access code, and without tracking capabilities

MathXL® Tutorials on CD (ISBN: 0-13-133061-6)

An interactive tutorial that provides:

- Algorithmically generated practice exercises correlated at the objective level
- Practice exercises accompanied by an example and a guided solution
- Tutorial video clips within the exercise to help students visualize concepts
- Easy-to-use tracking of student activity and scores and printed summaries of students' progress

Chapter Test Prep Video CD (ISBN: 0-13-133062-4)

- Provides step-by-step video solutions to each exercise in each Chapter Test in the textbook
- Packaged at no charge with the text, inside the back cover

PH Tutor Center (ISBN: 0-13-064604-0)

- Tutorial support via phone, fax, or email bundled at no charge with a new text, or purchased separately with a used book
- Staffed by developmental math faculty
- Available 5 days a week, 7 hours a day

Acknowledgments

An enormous benefit of authoring a successful series is the broad-based feedback I receive from the students, dedicated users, and reviewers. Every change to this edition is the result of their thoughtful comments and suggestions. I would like to express my appreciation to all the reviewers, whose collective insights form the backbone of this revision. In particular, I would like to thank:

Howard Anderson	*Skagit Valley College*
John Anderson	*Illinois Valley Community College*
Michael H. Andreoli	*Miami Dade College–North Campus*
Hien Bui	*Hillsborough Community College*
Warren J. Burch	*Brevard Community College*
Alice Burstein	*Middlesex Community College*
Edie Carter	*Amarillo College*
Thomas B. Clark	*Trident Technical College*
Sandra Pryor Clarkson	*Hunter College*
Bettyann Daley	*University of Delaware*
Robert A. Davies	*Cuyahoga Community College*
Paige Davis	*Lurleen B. Wallace Community College*
Ben Divers, Jr.	*Ferrum College*
Irene Doo	*Austin Community College*
Charles C. Edgar	*Onondaga Community College*
Rhoderick Fleming	*Wake Technical Community College*
Susan Forman	*Bronx Community College*
Donna Gerken	*Miami-Dade College*
Marion K. Glasby	*Anne Arundel Community College*
Sue Glascoe	*Mesa Community College*
Jay Graening	*University of Arkansas*
Robert B. Hafer	*Brevard Community College*
Mary Lou Hammond	*Spokane Community College*
Donald Herrick	*Northern Illinois University*
Beth Hooper	*Golden West College*
Tracy Hoy	*College of Lake County*
Judy Kasabian	*Lansing Community College*

Gary Kersting	*North Central Michigan College*
Gary Knippenberg	*Lansing Community College*
Mary Kochler	*Cuyahoga Community College*
Kristi Laird	*Jackson Community College*
Jennifer Lempke	*North Central Michigan College*
Sandy Lofstock	*St. Petersburg College*
Hank Martel	*Broward Community College*
Diana Martelly	*Miami-Dade College*
John Robert Martin	*Tarrant County Junior College*
Mikal McDowell	*Cedar Valley College*
Irwin Metviner	*State University of New York at Old Westbury*
Terri Moser	*Austin Community College*
Robert Musselman	*California State University, Fresno*
Kamilia Nemri	*Spokane Community College*
Allen R. Newhart	*Parkersburg Community College*
Steve O'Donnell	*Rogue Community College*
Jeff Parent	*Oakland Community College*
Kate Rozsa	*Mesa Community College*
Scott W. Satake	*Eastern Washington University*
Mike Schramm	*Indian River Community College*
Kathy Shepard	*Monroe County Community College*
Gayle Smith	*Lane Community College*
Linda Smoke	*Central Michigan University*
Dick Spangler	*Tacoma Community College*
Janette Summers	*University of Arkansas*
Robert Thornton	*Loyola University*
Lucy C. Thrower	*Francis Marion College*
Andrew Walker	*North Seattle Community College*
Margaret Williamson	*Milwaukee Area Technical College*
Roberta Yellott	*McNeese State University*
Marilyn Zopp	*McHenry County College*

Special thanks to Professors Doreen Kelly and Kate Rozsa at Mesa Community College for your work on the custom edition that resulted in many of the practice plus exercises and many of the new exercises in the Fourth Edition.

Additional acknowledgments are extended to Paul Murphy, senior acquisitions editor; Liz Covello, project manager; Barbara Mack, production editor; and Bayani Mendoza de Leon, assistant managing editor. Lastly, my thanks to the staff at Prentice Hall for their continuing support: Eilish Main, executive marketing manager; Patrice Jones, director of marketing; Chris Hoag, editor-in-chief; and to the entire Prentice Hall sales force for their confidence and enthusiasm about my books.

I hope that my love for learning, as well as my respect for the diversity of students I have taught and learned from over the years, is apparent throughout this new edition. By connecting algebra to the whole spectrum of learning, it is my intent to show students that their world is profoundly mathematical, and indeed, π is in the sky.

Robert Blitzer

ALGEBRAIC EXPRESSIONS AND REAL NUMBERS

Insatiable killer. That's the reputation the gray wolf acquired in the United States in the nineteenth and early twentieth centuries. Although the label was undeserved, an estimated 2 million wolves were shot, trapped, or poisoned. By 1960, the population was reduced to 800 wolves. In this section, you will learn how the special language of algebra describes your world, including the increasing wolf population in the continental United States following the Endangered Species Act of 1973.

Algebraic Expressions Algebra uses letters, such as x and y, to represent numbers. If a letter is used to represent various numbers, it is called a **variable**. For example, imagine that you are basking in the sun on the beach. We can let x represent the number of minutes that you can stay in the sun without burning with no sunscreen. With a number 6 sunscreen, exposure time without burning is six times as long, or 6 times x. This can be written $6 \cdot x$, but it is usually expressed as $6x$. Placing a number and a letter next to one another indicates multiplication.

Notice that $6x$ combines the number 6 and the variable x using the operation of multiplication. A combination of variables and numbers using the operations of addition, subtraction, multiplication, or division, as well as powers or roots, is called an **algebraic expression**. Here are some examples of algebraic expressions:

$$x + 6, \quad x - 6, \quad 6x, \quad \frac{x}{6}, \quad 3x + 5, \quad x^2 - 3, \quad \sqrt{x} + 7.$$

Is every letter in algebra a variable? No. Some letters stand for a particular number. Such a letter is called a **constant**. For example, let d = the number of days in a week. The letter d represents just one number, namely 7, and is a constant.

Translating English Phrases into Algebraic Expressions Problem solving in algebra involves translating English phrases into algebraic expressions. Here is a list of words and phrases for the four basic operations:

Addition	Subtraction	Multiplication	Division
sum	difference	product	quotient
plus	minus	times	divide
increased by	decreased by	of (used with fractions)	per
more than	less than	twice	ratio

Although the world's attention is often on large endangered mammals from Africa to the oceans, the United States is home to more than 395 endangered animals and close to 600 endangered plants.

This discussion is developed algebraically in Exercises 72 and 75 in the chapter's review exercises.

1

CHAPTER

Algebra, Mathematical Models, and Problem Solving

1.1 Algebraic Expressions and Real Numbers

1.2 Operations with Real Numbers and Simplifying Algebraic Expressions

1.3 Graphing Equations

1.4 Solving Linear Equations

1.5 Problem Solving and Using Formulas

1.6 Properties of Integral Exponents

1.7 Scientific Notation

In the United States, the Department of the Interior classifies species as endangered or threatened. Endangered species are those in danger of extinction. Threatened species are those likely to become endangered species within the foreseeable future. The formula

$$E = 10x + 167$$

approximately describes the number of endangered animal species in the United States, represented by E, x years after 1980. The formula tells of ten new animal species per year classified by the federal government as endangered. Formulas can be used to describe what happened in the past and to make predictions about what might occur in the future. Knowing how to use and create formulas will help you recognize patterns, logic, and order in a world that can appear chaotic to the untrained eye. In many ways, algebra will provide you with a new way of looking at our world.

Essentials of
Intermediate Algebra
for College Students

TO THE STUDENT

I've written this book so that you can learn about the power of algebra and how it relates directly to your life outside the classroom. All concepts are carefully explained, important definitions and procedures are set off in boxes, and worked-out examples that present solutions in a step-by-step manner appear in every section. Each example is followed by a similar matched problem, called a Check Point, for you to try so that you can actively participate in the learning process as you read the book. (Answers to all Check Points appear in the back of the book.) Study Tips offer hints and suggestions and often point out common errors to avoid. A great deal of attention has been given to applying algebra to your life to make your learning experience both interesting and relevant.

As you begin your studies, I would like to offer some specific suggestions for using this book and for being successful in this course:

1. Read the book. Read each section with pen (or pencil) in hand. Move through the worked-out examples with great care. These examples provide a model for doing exercises in the exercise sets. As you proceed through the reading, do not give up if you do not understand every single word. Things will become clearer as you read on and see how various procedures are applied to specific worked-out examples.

2. Work problems every day and check your answers. The way to learn mathematics is by doing mathematics, which means working the Check Points and assigned exercises in the exercise sets. The more exercises you work, the better you will understand the material.

3. Review for quizzes and tests. After completing a chapter, study the chapter review chart, work the exercises in the Chapter Review, and work the exercises in the Chapter Test. Answers to all these exercises are given in the back of the book.

The methods that I've used to help you read the book, work the problems, and review for tests are described in "A Brief Guide to Getting the Most from This Book" that appears inside the front cover. Spend a few minutes reviewing the guide to familiarize yourself with the book's features and their benefits.

4. Use the resources available with this book. Additional resources to aid your study are described following the guide to getting the most from your book. These resources include a Solutions Manual, a Chapter Test Prep Video CD, MyMathLab, an online version of the book with links to multimedia resources, MathXL®, an online homework, tutorial, and assessment system of the text, and tutorial support at no charge at the PH Tutor Center.

5. Attend all lectures. No book is intended to be a substitute for valuable insights and interactions that occur in the classroom. In addition to arriving for lecture on time and being prepared, you will find it useful to read the section before it is covered in lecture. This will give you a clear idea of the new material that will be discussed.

I wrote this book in Point Reyes National Seashore, 40 miles north of San Francisco. The park consists of 75,000 acres with miles of pristine surf-washed beaches, forested ridges, and bays bordered by white cliffs. It was my hope to convey the beauty and excitement of mathematics using nature's unspoiled beauty as a source of inspiration and creativity. Enjoy the pages that follow as you empower yourself with the algebra needed to succeed in college, your career, and in your life.

Regards,
Bob
Robert Blitzer

ABOUT THE AUTHOR

Bob Blitzer is a native of Manhattan and received a Bachelor of Arts degree with dual majors in mathematics and psychology (minor: English literature) from the City College of New York. His unusual combination of academic interests led him toward a Master of Arts in mathematics from the University of Miami and a doctorate in behavioral sciences from Nova University. Bob is most energized by teaching mathematics and has taught a variety of mathematics courses at Miami-Dade College for nearly 30 years. He has received numerous teaching awards, including Innovator of the Year from the League for Innovations in the Community College, and was among the first group of recipients at Miami-Dade College for an endowed chair based on excellence in the classroom. In addition to *Essentials of Intermediate Algebra for College Students*, Bob has written *Introductory Algebra for College Students*, *Intermediate Algebra for College Students*, *Introductory and Intermediate Algebra for College Students*, *Essentials of Introductory and Intermediate Algebra for College Students*, *Algebra for College Students*, *Thinking Mathematically*, *College Algebra*, *Algebra and Trigonometry*, and *Precalculus*, all published by Prentice Hall.

EXAMPLE 1 Translating English Phrases into Algebraic Expressions

Write each English phrase as an algebraic expression. Let x represent the number:

a. Nine less than six times a number.

b. The quotient of five and a number, increased by twice the number.

SOLUTION

a.
Nine less than   six times a number

$$6x - 9$$

b.
The quotient of five and a number,   increased by   twice the number

$$\frac{5}{x} \quad + \quad 2x$$

 CHECK POINT 1 Write each English phrase as an algebraic expression. Let x represent the number:

a. Five more than 8 times a number. $8x + 5$

b. The quotient of a number and 7, decreased by twice the number. $\frac{x}{7} - 2x$

---

2   Evaluate algebraic expressions.

Evaluating Algebraic Expressions   Evaluating an algebraic expression means to find the value of the expression for a given value of the variable.

EXAMPLE 2   Evaluating an Algebraic Expression

The algebraic expression $81 - 0.6x$ approximates the percentage of American adults who smoked cigarettes x years after 1900. Evaluate the expression for $x = 100$. Describe what the answer means in practical terms.

SOLUTION   We begin by substituting 100 for x. Because $x = 100$, we will be finding the percentage of Americans who smoked cigarettes 100 years after 1900, in the year 2000.

$$81 - 0.6x$$

Replace x with 100.

$$= 81 - 0.6(100) = 81 - 60 = 21$$

Thus, in 2000, approximately 21% of American adults smoked cigarettes.

 CHECK POINT 2 Evaluate $81 - 0.6x$ for $x = 50$. Describe what the answer means in practical terms.

Many algebraic expressions involve *exponents*. For example, the algebraic expression

$$0.72x^2 + 9.4x + 783$$

approximates the gray wolf population in the United States x years after 1960. The expression x^2 means $x \cdot x$, and is read "x to the second power" or "x squared." The exponent, 2, indicates that the base, x, appears as a factor two times.

EXPONENTIAL NOTATION If n is a counting number $(1, 2, 3,$ and so on$)$,

Exponent or Power

$$b^n = \underbrace{b \cdot b \cdot b \cdot \ldots \cdot b.}_{\substack{b \text{ appears as a} \\ \text{factor } n \text{ times.}}}$$

Base

b^n is read "the nth power of b" or "b to the nth power." Thus, the nth power of b is defined as the product of n factors of b. The expression b^n is called an **exponential expression**. Furthermore, $b^1 = b$.

USING TECHNOLOGY

You can use a calculator to evaluate exponential expressions. For example, to evaluate 2^4, press the following keys:

Many Scientific Calculators

2 $\boxed{y^x}$ 4 $\boxed{=}$

Many Graphing Calculators

2 $\boxed{\wedge}$ 4 $\boxed{\text{ENTER}}$

Although calculators have special keys to evaluate powers of ten and squaring bases, you can always use one of the sequences shown here.

For example,

$$8^2 = 8 \cdot 8 = 64, \quad 5^3 = 5 \cdot 5 \cdot 5 = 125, \quad \text{and} \quad 2^4 = 2 \cdot 2 \cdot 2 \cdot 2 = 16.$$

When evaluating algebraic expressions, follow these rules for determining the order in which operations should be performed:

ORDER OF OPERATIONS

1. First, perform all operations within grouping symbols.
2. Next, evaluate all exponential expressions.
3. Next, do all multiplications and divisions in the order in which they occur, working from left to right.
4. Finally, do all additions and subtractions in the order in which they occur, working from left to right.

EXAMPLE 3 Evaluating an Algebraic Expression

Evaluate $7 + 5(x - 4)^3$ for $x = 6$.

SOLUTION

$$
\begin{aligned}
7 + 5(x - 4)^3 &= 7 + 5(6 - 4)^3 && \text{Replace x with 6.} \\
&= 7 + 5(2)^3 && \text{First work inside parentheses: } 6 - 4 = 2. \\
&= 7 + 5(8) && \text{Evaluate the exponential expression:} \\
& && 2^3 = 2 \cdot 2 \cdot 2 = 8. \\
&= 7 + 40 && \text{Multiply: } 5(8) = 40. \\
&= 47 && \text{Add.}
\end{aligned}
$$

■

✔ **CHECK POINT 3** Evaluate $8 + 6(x - 3)^2$ for $x = 13$. 608

3 Use mathematical models.

Formulas and Mathematical Models An **equation** is formed when an equal sign is placed between two algebraic expressions. One aim of algebra is to provide a compact, symbolic description of the world. These descriptions involve the use of *formulas*. A **formula** is an equation that uses letters to express a relationship between two or more variables. Here is an example of a formula:

$$C = \frac{5}{9}(F - 32).$$

| Celsius temperature | is | $\frac{5}{9}$ of | the difference between Fahrenheit temperature and 32°. |

The process of finding formulas to describe real-world phenomena is called **mathematical modeling**. Such formulas, together with the meaning assigned to the variables, are called **mathematical models**. We often say that these formulas model, or describe, the relationships among the variables.

Gray Wolf Population (to the Nearest Hundred)

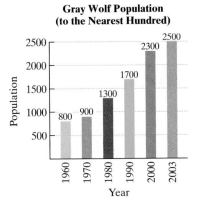

FIGURE 1.1

Source: U.S. Department of the Interior

EXAMPLE 4 Modeling the Gray Wolf Population

The formula

$$P = 0.72x^2 + 9.4x + 783$$

models the gray wolf population, P, in the United States, x years after 1960. Use the formula to find the population in 1990. How well does the formula model the actual data shown in the bar graph in Figure 1.1?

SOLUTION Because 1990 is 30 years after 1960, we substitute 30 for x in the given formula. Then we use the order of operations to find P, the gray wolf population in 1990.

$P = 0.72x^2 + 9.4x + 783$	This is the given mathematical model.
$P = 0.72(30)^2 + 9.4(30) + 783$	Replace each occurrence of x with 30.
$P = 0.72(900) + 9.4(30) + 783$	Evaluate the exponential expression: $30^2 = 30 \cdot 30 = 900$.
$P = 648 + 282 + 783$	Multiply from left to right: $0.72(900) = 648$ and $9.4(30) = 282$.
$P = 1713$	Add.

The formula indicates that in 1990, the gray wolf population in the United States was 1713. The number given in Figure 1.1 is 1700, so the formula models the data quite well. ■

✓ **CHECK POINT 4** Use the formula in Example 4 to find the gray wolf population in 2000. How well does the formula model the data in Figure 1.1?

In 2000, the gray wolf population in the United States was 2311; the formula models the data quite well.

Sometimes a mathematical model gives an estimate that is not a good approximation or is extended to include values of the variable that do not make sense. In these cases, we say that **model breakdown** has occurred.

4 Recognize the sets that make up the real numbers.

The Set of Real Numbers Before we describe the set of real numbers, let's be sure you are familiar with some basic ideas about sets. A **set** is a collection of objects whose contents can be clearly determined. The objects in a set are called the **elements** of the set. For example, the set of numbers used for counting can be represented by

$$\{1, 2, 3, 4, 5, \ldots\}.$$

The braces, { }, indicate that we are representing a set. This form of representation, called the **roster method**, uses commas to separate the elements of the set. The three dots after the 5, called an *ellipsis*, indicate that there is no final element and that the listing goes on forever.

Three common sets of numbers are the *natural numbers*, the *whole numbers*, and the *integers*.

NATURAL NUMBERS, WHOLE NUMBERS, AND INTEGERS

The Set of Natural Numbers

$$\{1, 2, 3, 4, 5, \ldots\}$$

These are the numbers that we use for counting.

The Set of Whole Numbers

$$\{0, 1, 2, 3, 4, 5, \ldots\}$$

The set of whole numbers includes 0 and the natural numbers.

The Set of Integers

$$\{\ldots, -5, -4, -3, -2, -1, 0, 1, 2, 3, 4, 5, \ldots\}$$

The set of integers includes the negatives of the natural number and the whole numbers.

---

5 Use set-builder notation.

A set can also be written in **set-builder notation**. In this notation, the elements of the set are described, but not listed. Here is an example:

$$\{x \,|\, x \text{ is a natural number less than 6}).$$

The set of all x such that x is a natural number less than 6

The same set written using the roster method is

$$\{1, 2, 3, 4, 5\}.$$

6 Use the symbols $\in$ and $\notin$.

The symbol $\in$ is used to indicate that an element is in a particular set. The symbol $\in$ is read "is an element of." Here is an example:

$$7 \in \{1, 2, 3, 4, 5, \ldots\}.$$

7   is an element of   the set of natural numbers.

The symbol $\notin$ is used to indicate that an element is not in a particular set. The symbol $\notin$ is read "is not an element of." Here is an example:

$$\frac{1}{2} \notin \{1, 2, 3, 4, 5, \ldots\}.$$

$\frac{1}{2}$ is not an element of   the set of natural numbers.

EXAMPLE 5 Using the Symbols ∈ and ∉

Determine whether each statement is true or false:

a. $100 \in \{x \mid x \text{ is an integer}\}$ **b.** $20 \notin \{5, 10, 15\}$.

SOLUTION

a. Because 100 is an integer, the statement

$$100 \in \{x \mid x \text{ is an integer}\}$$

is true. The number 100 is an element of the set of integers.

b. Because 20 is not an element of $\{5, 10, 15\}$, the statement $20 \notin \{5, 10, 15\}$ is true. ■

 CHECK POINT 5 Determine whether each statement is true or false:

a. $13 \in \{x \mid x \text{ is an integer}\}$ true   **b.** $6 \notin \{7, 8, 9, 10\}$.   true

Another common set is the set of *rational numbers*. These numbers can be expressed as an integer divided by a nonzero integer.

RATIONAL NUMBERS   The set of **rational numbers** is the set of all numbers that can be expressed as a quotient of two integers, with the denominator not 0.

> This means that b is not equal to zero.

$$\left\{ \frac{a}{b} \,\middle|\, a \text{ and } b \text{ are integers and } b \neq 0 \right\}$$

Three examples of rational numbers are

$$\frac{1}{4} \quad (a = 1, \ b = 4), \qquad \frac{-2}{3} \quad (a = -2, \ b = 3), \quad \text{and } 5 = \frac{5}{1} \quad (a = 5, \ b = 1)$$

Can you see that integers are also rational numbers because they can be written in terms of division by 1?

Rational numbers can be expressed in fraction or decimal notation. To express the fraction $\frac{a}{b}$ as a decimal, divide the denominator, b, into the numerator, a. In decimal notation, rational numbers either terminate (stop) or have a digit, or block of digits, that repeats. For example,

$$\frac{3}{8} = 3 \div 8 = 0.375 \quad \text{and} \quad \frac{7}{11} = 7 \div 11 = 0.6363\ldots = 0.\overline{63}.$$

> The decimal stops: it is a terminating decimal.

> This is a repeating decimal. The bar is written over the repeating part.

Some numbers cannot be expressed as terminating or repeating decimals. An example of such a number is $\sqrt{2}$, the square root of 2. The number $\sqrt{2}$ is a number that can be squared to give 2. No terminating or repeating decimal can be squared to get 2. However, some approximations have squares that come close to 2. We use the symbol $\approx$, which means "is approximately equal to."

- $\sqrt{2} \approx 1.4$ because $(1.4)^2 = (1.4)(1.4) = 1.96$.
- $\sqrt{2} \approx 1.41$ because $(1.41)^2 = (1.41)(1.41) = 1.9881$.
- $\sqrt{2} \approx 1.4142$ because $(1.4142)^2 = 1.99996164$.

$\sqrt{2}$ is an example of an *irrational number*.

> **IRRATIONAL NUMBERS** The set of **irrational numbers** is the set of numbers whose decimal representations neither terminate nor repeat. Irrational numbers cannot be expressed as quotients of integers.

Examples of irrational numbers include

$$\sqrt{3} \approx 1.73205 \qquad \text{and} \qquad \pi(\text{pi}) \approx 3.141593.$$

Not all square roots are irrational. For example, $\sqrt{25} = 5$ because $5^2 = 5 \cdot 5 = 25$. Thus, $\sqrt{25}$ is a natural number, a whole number, an integer, and a rational number $\left(\sqrt{25} = \frac{5}{1}\right)$.

The set of *real numbers* is formed by combining the sets of rational numbers and irrational numbers. Thus, every real number is either rational or irrational, as shown in Figure 1.2.

> **REAL NUMBERS** The set of **real numbers** is the set of numbers that are either rational or irrational:
>
> $$\{x \mid x \text{ is rational or } x \text{ is irrational}\}.$$

FIGURE 1.2 Every real number is either rational or irrational.

The Real Number Line The **real number line** is a graph used to represent the set of real numbers. An arbitrary point, called the **origin**, is labeled 0. Select a point to the right of 0 and label it 1. The distance from 0 to 1 is called the **unit distance**. Numbers to the right of the origin are **positive** and numbers to the left of the origin are **negative**. The real number line is shown in Figure 1.3.

FIGURE 1.3 The real number line

Real numbers are **graphed** on a number line by placing a dot at the correct location for each number. The integers are easiest to locate. In Figure 1.4, we've graphed the integers -3, 0, and 4.

FIGURE 1.4

Every real number corresponds to a point on the number line and every point on the number line corresponds to a real number. We say that there is a **one-to-one correspondence** between all the real numbers and all points on a real number line.

Ordering the Real Numbers On the real number line, the real numbers increase from left to right. The lesser of two real numbers is the one farther to the left on a number line. The greater of two real numbers is the one farther to the right on a number line.

7 Use inequality symbols.

Look at the number line in Figure 1.5. The integers -4 and -1 are graphed.

$$-5 \quad -4 \quad -3 \quad -2 \quad -1 \quad 0 \quad 1 \quad 2 \quad 3 \quad 4 \quad 5$$

FIGURE 1.5

Observe that -4 is to the left of -1 on the number line. This means that -4 is less than -1.

$$-4 < -1$$ — -4 is less than -1 because -4 is to the **left** of -1 on the number line.

In Figure 1.5, we can also observe that -1 is to the right of -4 on the number line. This means that -1 is greater than -4.

$$-1 > -4$$ — -1 is greater than -4 because -1 is to the **right** of -4 on the number line.

The symbols $<$ and $>$ are called **inequality symbols**. These symbols always point to the lesser of the two real numbers when the inequality statement is true.

-4 is less than -1. — $-4 < -1$ The symbol points to -4, the lesser number.

-1 is greater than -4. — $-1 > -4$ The symbol still points to -4, the lesser number.

The symbols $<$ and $>$ may be combined with an equal sign, as shown in the following table:

	Symbols	Meaning	Examples	Explanation
This inequality is true if either the $<$ part or the $=$ part is true.	$a \leq b$	a is less than or equal to b.	$2 \leq 9$ $9 \leq 9$	Because $2 < 9$ Because $9 = 9$
This inequality is true if either the $>$ part or the $=$ part is true.	$b \geq a$	b is greater than or equal to a.	$9 \geq 2$ $2 \geq 2$	Because $9 > 2$ Because $2 = 2$

EXAMPLE 6   Using Inequality Symbols

Write out the meaning of each inequality. Then determine whether the inequality is true or false:

a. $-5 < -1$ **b.** $6 > -2$ **c.** $-6 \leq 3$ **d.** $10 \geq 10$ **e.** $-9 \geq 6$.

SOLUTION   The solution is illustrated by the number line in Figure 1.6.

$$-9 \quad -8 \quad -7 \quad -6 \quad -5 \quad -4 \quad -3 \quad -2 \quad -1 \quad 0 \quad 1 \quad 2 \quad 3 \quad 4 \quad 5 \quad 6 \quad 7 \quad 8 \quad 9 \quad 10$$

FIGURE 1.6

Inequality	Meaning
a. $-5 < -1$	"-5 is less than -1." Because -5 is to the left of -1 on the number line, the inequality is true.
b. $6 > -2$	"6 is greater than -2." Because 6 is to the right of -2 on the number line, the inequality is true.
c. $-6 \leq 3$	"-6 is less than or equal to 3." Because $-6 < 3$ is true (-6 is to the left of 3 on the number line), the inequality is true.
d. $10 \geq 10$	"10 is greater than or equal to 10." Because $10 = 10$ is true, the inequality is true.
e. $-9 \geq 6$	"-9 is greater than or equal to 6." Because neither $-9 > 6$ nor $-9 = 6$ is true, the inequality is false.

✔ CHECK POINT **6** Write out the meaning of each inequality. Then determine whether the inequality is true or false:

a. $-8 < -2$ -8 is less than -2; true

b. $7 > -3$ 7 is greater than -3; true

c. $-1 \leq -4$ -1 is less than or equal to -4; false

d. $5 \geq 5$ 5 is greater than or equal to 5; true

e. $2 \geq -14$. 2 is greater than or equal to -14; true

1.1 EXERCISE SET

Student Solutions Manual   CD/Video   PH Math/Tutor Center   MathXL Tutorials on CD   MathXL®   MyMathLab   Interactmath.com

Practice Exercises

In Exercises 1–14, write each English phrase as an algebraic expression. Let x represent the number.

1. Five more than a number $x + 5$

2. A number increased by six $x + 6$

3. Four less than a number $x - 4$

4. Nine less than a number $x - 9$

5. Four times a number $4x$

6. Twice a number $2x$

7. Ten more than twice a number $2x + 10$

8. Four more than five times a number $5x + 4$

9. The difference of six and half of a number $6 - \frac{1}{2}x$

10. The difference of three and half of a number $3 - \frac{1}{2}x$

11. Two less than the quotient of four and a number $\frac{4}{x} - 2$

12. Three less than the quotient of five and a number $\frac{5}{x} - 3$

13. The quotient of three and the difference of five and a number $\frac{3}{5 - x}$

14. The quotient of six and the difference of ten and a number $\frac{6}{10 - x}$

In Exercises 15–26, evaluate each algebraic expression for the given value or values of the variable(s).

15. $7 + 5x$, for $x = 10$ 57

16. $8 + 6x$, for $x = 5$ 38

17. $6x - y$, for $x = 3$ and $y = 8$ 10

18. $8x - y$, for $x = 3$ and $y = 4$ 20

19. $x^2 + 3x$, for $x = 8$ 88

20. $x^2 + 5x$, for $x = 6$ 66

21. $x^2 - 6x + 3$, for $x = 7$ 10

22. $x^2 - 7x + 4$, for $x = 8$ 12

23. $4 + 5(x - 7)^3$, for $x = 9$ 44

24. $6 + 5(x - 6)^3$, for $x = 8$ 46

25. $x^2 - 3(x - y)$, for $x = 8$ and $y = 2$ 46

26. $x^2 - 4(x - y)$, for $x = 8$ and $y = 3$ 44

In Exercises 27–34, use the roster method to list the elements in each set.

27. $\{x | x \text{ is a natural number less than } 5\}$ $\{1, 2, 3, 4\}$

28. $\{x | x \text{ is a natural number less than } 4\}$ $\{1, 2, 3\}$

29. $\{x | x \text{ is an integer between } -8 \text{ and } -3\}$ $\{-7, -6, -5, -4\}$

30. $\{x | x \text{ is an integer between } -7 \text{ and } -2\}$ $\{-6, -5, -4, -3\}$

31. $\{x | x \text{ is a natural number greater than } 7\}$ $\{8, 9, 10, \dots\}$

32. $\{x | x \text{ is a natural number greater than } 9\}$ $\{10, 11, 12, \dots\}$

33. $\{x | x \text{ is an odd whole number less than } 11\}$ $\{1, 3, 5, 7, 9\}$

34. $\{x | x \text{ is an odd whole number less than } 9\}$ $\{1, 3, 5, 7\}$

In Exercises 35–48, use the meaning of the symbols $\in$ and $\notin$ to determine whether each statement is true or false.

35. $7 \in \{x | x \text{ is an integer}\}$ true

36. $9 \in \{x | x \text{ is an integer}\}$ true

37. $7 \in \{x | x \text{ is a rational number}\}$ true

38. $9 \in \{x | x \text{ is a rational number}\}$ true

39. $7 \in \{x | x \text{ is an irrational number}\}$ false

40. $9 \in \{x | x \text{ is an irrational number}\}$ false

41. $3 \notin \{x | x \text{ is an irrational number}\}$ true

42. $5 \notin \{x \mid x \text{ is an irrational number}\}$ true

43. $\frac{1}{2} \notin \{x \mid x \text{ is a rational number}\}$ false

44. $\frac{1}{4} \notin \{x \mid x \text{ is a rational number}\}$ false

45. $\sqrt{2} \notin \{x \mid x \text{ is a rational number}\}$ true

46. $\pi \notin \{x \mid x \text{ is a rational number}\}$ true

47. $\sqrt{2} \notin \{x \mid x \text{ is a real number}\}$ false

48. $\pi \notin \{x \mid x \text{ is a real number}\}$ false

In Exercises 49–64, write out the meaning of each inequality. Then determine whether the inequality is true or false.

49. $-6 < -2$ -6 is less than -2; true

50. $-7 < -3$ -7 is less than -3; true

51. $5 > -7$ 5 is greater than -7; true

52. $3 > -8$ 3 is greater than -8; true

53. $0 < -4$ 0 is less than -4; false

54. $0 < -5$ 0 is less than -5; false

55. $-4 \le 1$ -4 is less than or equal to 1; true

56. $-5 \le 1$ -5 is less than or equal to 1; true

57. $-2 \le -6$ -2 is less than or equal to -6; false

58. $-3 \le -7$ -3 is less than or equal to -7; false

59. $-2 \le -2$ -2 is less than or equal to -2; true

60. $-3 \le -3$ -3 is less than or equal to -3; true

61. $-2 \ge -2$ -2 is greater than or equal to -2; true

62. $-3 \ge -3$ -3 is greater than or equal to -3; true

63. $2 \le -\dfrac{1}{2}$ 2 is less than or equal to $-\frac{1}{2}$; false

64. $4 \le -\dfrac{1}{2}$ 4 is less than or equal to $-\frac{1}{2}$; false

Practice Plus

By definition, an "and" statement is true only when the statements before and after the "and" connective are both true. Use this definition to determine whether each statement in Exercises 65–74 is true or false.

65. $0.\overline{3} > 0.3$ and $-10 < 4 + 6$.   true

66. $0.6 < 0.\overline{6}$ and $2 \cdot 5 \le 4 + 6$.   true

67. $12 \in \{1, 2, 3, \dots\}$ and $\{3\} \in \{1, 2, 3, 4\}$.   false

68. $17 \in \{1, 2, 3, \dots\}$ and $\{4\} \in \{1, 2, 3, 4, 5\}$.   false

69. $\left(\dfrac{2}{5} + \dfrac{3}{5}\right) \in \{x \mid x \text{ is a natural number}\}$ and the value of $9x^2(x + 11) - 9(x + 11)x^2$, for $x = 100$, is 0.   true

70. $\left(\dfrac{14}{19} + \dfrac{5}{19}\right) \in \{x \mid x \text{ is a natural number}\}$ and the value of $12x^2(x + 10) - 12(x + 10)x^2$, for $x = 50$, is 0.   true

71. $\{x \mid x \text{ is an integer between } -3 \text{ and } 0\} = \{-3, -2, -1, 0\}$ and $-\pi > -3.5$.   false

72. $\{x \mid x \text{ is an integer between } -4 \text{ and } 0\} = \{-4, -3, -2, -1, 0\}$ and $-\dfrac{\pi}{2} > -2.3$.   false

73. Twice the sum of a number and three is represented by $2x + 3$ and $-1,100,000 \in \{x \mid x \text{ is an integer}\}$.   false

74. Three times the sum of a number and five is represented by $3x + 5$ and $-4,500,000 \in \{x \mid x \text{ is an integer}\}$.   false

Application Exercises

The United States spends more money than any developed nation on health care relative to its economy. The nation's 2001 health expenditures were \$5035 per person, up from \$2737 in 1990—and only \$143 in 1960. The bar graph shows per capita expenditures for health care for six selected years from 1980 through 2001. The data in the graph can be modeled by the formula

$$E = 183.6x + 992,$$

where E represents per capita expenditures for health care, in dollars, x years after 1980. Use this formula to solve Exercises 75–76.

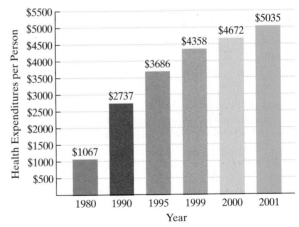

Per Capita Expenditure for Health Care in the U.S.

Source: Department of Health and Human Services

75. According to the formula, how much did each American pay for health care in 2000? Does the formula underestimate or overestimate the actual amount shown in the bar graph? By how much?   \$4664; underestimates by \$8

76. According to the formula, how much did each American pay for health care in 1995? Does the formula underestimate or overestimate the actual amount shown in the bar graph? By how much?   \$3746; overestimates by \$60

The bar graph shows the number of people in the United States living below the poverty level, in millions, from 1997 through 2002. The data can be modeled by the formula

$$N = 0.5x^2 - 3x + 36,$$

where N represents the number of Americans living below the poverty level, in millions, x years after 1997. Use this formula to solve Exercises 77–78.

Millions of Americans Living below the Poverty Level

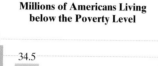

Source: U.S. Census Bureau

77. According to the formula, how many millions of Americans were living below the poverty level in 2001? How well does the formula model the number shown in the bar graph? 32 million; fairly well

78. According to the formula, how many millions of Americans were living below the poverty level in 1999? How well does the formula model the number shown in the bar graph? 32 million; fairly well

The formula

$$C = \frac{5}{9}(F - 32)$$

expresses the relationship between Fahrenheit temperature, F, and Celsius temperature, C. In Exercises 79–80, use the formula to convert the given Fahrenheit temperature to its equivalent temperature on the Celsius scale.

79. 50°F   10°C   **80.** 86°F   30°C

A football is kicked vertically upward from a height of 4 feet with an initial speed of 60 feet per second. The formula

$$h = 4 + 60t - 16t^2$$

describes the ball's height above the ground, h, in feet, t seconds after it was kicked. Use this formula to solve Exercises 81–82.

81. What is the ball's height 2 seconds after it was kicked? 60 feet

82. What is the ball's height 3 seconds after it was kicked? 40 feet

Writing in Mathematics

Writing about mathematics will help you to learn mathematics. For all writing exercises in this book, use complete sentences to respond to the question. Some writing exercises can be answered in

a sentence. Others require a paragraph or two. You can decide how much you need to write as long as your writing clearly and directly answers the question in the exercise. Standard references such as a dictionary and a thesaurus should be helpful.

83. What is a variable? **83–100.** Answers will vary.

84. What is an algebraic expression? Give an example with your explanation.

85. If *n* is a natural number, what does b^n mean? Give an example with your explanation.

86. What does it mean when we say that a formula models real-world phenomena?

87. What is model breakdown?

88. What is a set?

89. Describe the roster method for representing a set.

90. What are the natural numbers?

91. What are the whole numbers?

92. What are the integers?

93. Describe the rational numbers.

94. Describe the difference between a rational number and an irrational number.

95. What are the real numbers?

96. What is set-builder notation?

97. Describe the meanings of the symbols $\in$ and $\notin$. Provide an example showing the correct use of each symbol.

98. What is the real number line?

99. If you are given two real numbers, explain how to determine which one is the lesser.

100. The mathematical model in Exercises 77–78 predicts an increase in the number of Americans living in poverty. What might occur in the future to cause this formula to undergo model breakdown?

Critical Thinking Exercises

101. Which one of the following statements is true? c
   a. Every rational number is an integer.
   b. Some whole numbers are not integers.
   c. Some rational numbers are not positive.
   d. Irrational numbers cannot be negative.

102. A bird lover visited a pet shop where there were twice 4 and 20 parrots. The bird lover purchased $\frac{1}{7}$ of the birds. English, of course, can be ambiguous and "twice 4 and 20" can mean $2(4 + 20)$ or $2 \cdot 4 + 20$. Explain how the conditions of the situation determine if "twice 4 and 20" means $2(4 + 20)$ or $2 \cdot 4 + 20$. $2 \cdot 4 + 20 = 28$ and $\frac{1}{7}$ of 28 is 4, but $2(4 + 20) = 48$ and $\frac{1}{7}$ of 48 is about 6.86. You can't purchase 6.86 birds; the correct translation is $2 \cdot 4 + 20$.

In Exercises 103–104, insert parentheses to make each statement true.

103. $2 \cdot 3 + 3 \cdot 5 = 45$ $(2 \cdot 3 + 3) \cdot 5 = 45$

104. $8 + 2 \cdot 4 - 3 = 10$ $8 + 2 \cdot (4 - 3) = 10$

105. Between which two consecutive integers is $-\sqrt{26}$? Do not use a calculator. -6 and -5

SECTION

1.2

Objectives

1 Find a number's absolute value.

2 Add real numbers.

3 Find opposites.

4 Subtract real numbers.

5 Multiply real numbers.

6 Evaluate exponential expressions.

7 Divide real numbers.

8 Use the order of operations.

9 Use commutative, associative, and distributive properties.

10 Simplify algebraic expressions.

OPERATIONS WITH REAL NUMBERS AND SIMPLIFYING ALGEBRAIC EXPRESSIONS

You fly from Tokyo to Honolulu, about a $6\frac{1}{2}$ hour flight. You leave Tokyo at 9 A.M. and arrive in Honolulu at 8:30 P.M. the day before! This is one of the bizarre travel consequences of the world's time zones, essential to keeping the middle of the day at noon and making today's complex airline schedules possible. In this section, we review operations with real numbers so that you can make sense of the time zone system.

Absolute Value   *Absolute value* is used to describe how to operate with positive and negative numbers.

> **ABSOLUTE VALUE**   The **absolute value** of a real number a, denoted by $|a|$, is the distance from 0 to a on the number line. This distance is always taken to be nonnegative.

1 Find a number's absolute value.

EXAMPLE 1 Finding Absolute Value

Find the absolute value:

a. $|-4|$ **b.** $|3.5|$ **c.** $|0|$.

SOLUTION   The solution is illustrated in Figure 1.7.

a. $|-4| = 4$ The absolute value of -4 is 4 because -4 is 4 units from 0.
b. $|3.5| = 3.5$ The absolute value of 3.5 is 3.5 because 3.5 is 3.5 units from 0.
c. $|0| = 0$ The absolute value of 0 is 0 because 0 is 0 units from itself.

FIGURE 1.7

Can you see that the absolute value of a real number is either positive or zero? Zero is the only real number whose absolute value is 0:

$$|0| = 0.$$

The absolute value of a real number is never negative.

 CHECK POINT 1 Find the absolute value:

    a. $|-6|$ 6   **b.** $|4.5|$ 4.5   **c.** $|0|$.   0

2 Add real numbers.

Adding Real Numbers   Table 1.1 reviews how to add real numbers.

Table 1.1 **Adding Real Numbers**	
Rule	**Example**
To add two real numbers with the same sign, add their absolute values. Use the common sign as the sign of the sum.	$(-7) + (-4) = -(|-7| + |-4|)$ $= -(7 + 4)$ $= -11$
To add two real numbers with different signs, subtract the smaller absolute value from the greater absolute value. Use the sign of the number with the greater absolute value as the sign of the sum.	$7 + (-15) = -(|-15| - |7|)$ $= -(15 - 7)$ $= -8$

EXAMPLE 2   Adding Real Numbers

Add: **a.** $-12 + (-14)$ **b.** $-0.3 + 0.7$ **c.** $-\dfrac{3}{4} + \dfrac{1}{2}$.

SOLUTION

a. $-12 + (-14) = -26$ Add absolute values: $12 + 14 = 26$.

       Use the common sign.

b. $-0.3 + 0.7 = 0.4$ Subtract absolute values: $0.7 - 0.3 = 0.4$.

       Use the sign of the number with the greater absolute value. The sign of the sum is assumed to be positive.

c. $-\dfrac{3}{4} + \dfrac{1}{2} = -\dfrac{1}{4}$ Subtract absolute values: $\frac{3}{4} - \frac{1}{2} = \frac{3}{4} - \frac{2}{4} = \frac{1}{4}$.

       Use the sign of the number with the greater absolute value.

 CHECK POINT 2 Add:

    a. $-10 + (-18)$ -28 **b.** $-0.2 + 0.9$ 0.7   **c.** $-\dfrac{3}{5} + \dfrac{1}{2}$. $-\dfrac{1}{10}$

If one of two numbers being added is zero, the sum is the other number. For example,

$$-3 + 0 = -3 \quad \text{and} \quad 0 + 2 = 2.$$

In general,

$$a + 0 = a \quad \text{and} \quad 0 + a = a.$$

We call 0 the **identity element of addition** or the **additive identity**. Thus, the additive identity can be deleted from a sum.

3 Find opposites.

Numbers with different signs but the same absolute value are called **opposites** or **additive inverses**. For example, 3 and -3 are additive inverses. When additive inverses are added, their sum is 0. For example,

$$3 + (-3) = 0 \quad \text{and} \quad -3 + 3 = 0.$$

INVERSE PROPERTY OF ADDITION The sum of a real number and its additive inverse is 0, the additive identity.

$$a + (-a) = 0 \quad \text{and} \quad (-a) + a = 0$$

The symbol "$-$" is used to name the opposite, or additive inverse, of a. When a is a negative number, $-a$, its opposite, is positive. For example, if a is -4, its opposite is 4. Thus,

$$-(-4) = 4.$$

The opposite of -4 is 4.

In general, if a is any real number,

$$-(-a) = a.$$

EXAMPLE 3 Finding Opposites

Find $-x$ if **a.** $x = -6$ **b.** $x = \dfrac{1}{2}$.

SOLUTION

a. If $x = -6$, then $-x = -(-6) = 6$. The opposite of -6 is 6.

b. If $x = \dfrac{1}{2}$, then $-x = -\dfrac{1}{2}$. The opposite of $\dfrac{1}{2}$ is $-\dfrac{1}{2}$. ■

✔ **CHECK POINT 3** Find $-x$ if **a.** $x = -8$ 8 **b.** $x = \dfrac{1}{3}$. $-\dfrac{1}{3}$

We can define the absolute value of the real number a using opposites, without referring to a number line. The algebraic definition of the absolute value of a is given as follows:

DEFINITION OF ABSOLUTE VALUE

$$|a| = \begin{cases} a & \text{if } a \geq 0 \\ -a & \text{if } a < 0 \end{cases}$$

If a is nonnegative (that is $a \geq 0$), the absolute value of a is the number itself: $|a| = a$. For example,

$$|5| = 5 \qquad |\pi| = \pi \qquad \left|\frac{1}{3}\right| = \frac{1}{3} \qquad |0| = 0$$

Zero is the only number whose absolute value is 0.

If a is a negative number (that is, $a < 0$), the absolute value of a is the opposite of a: $|a| = -a$. This makes the absolute value positive. For example,

$$|-3| = -(-3) = 3 \qquad |-\pi| = -(-\pi) = \pi \qquad \left|-\frac{1}{3}\right| = -\left(-\frac{1}{3}\right) = \frac{1}{3}.$$

This middle step is usually omitted.

4 Subtract real numbers.

Subtracting Real Numbers Subtraction of real numbers is defined in terms of addition.

> **DEFINITION OF SUBTRACTION** If a and b are real numbers,
> $$a - b = a + (-b).$$
> To subtract a real number, add its opposite or additive inverse.

Thus, to subtract real numbers,

1. Change the subtraction to addition.
2. Change the sign of the number being subtracted.
3. Add, using one of the rules for adding numbers with the same sign or different signs.

EXAMPLE 4 Subtracting Real Numbers

Subtract: **a.** $6 - 13$ **b.** $5.1 - (-4.2)$ **c.** $-10 - (-2)$.

SOLUTION

a. $6 - 13 = 6 + (-13) = -7$

Change the subtraction to addition. Replace 13 with its opposite.

b. $5.1 - (-4.2) = 5.1 + 4.2 = 9.3$

Change the subtraction to addition. Replace −4.2 with its opposite.

c. $-10 - (-2) = -10 + 2 = -8$

Change the subtraction to addition. Replace −2 with its opposite.

■

✔ **CHECK POINT 4** Subtract:

a. $7 - 10$ -3 **b.** $4.3 - (-6.2)$ 10.5 **c.** $-9 - (-3)$. -6

5 Multiply real numbers.

Multiplying Real Numbers You can think of multiplication as repeated addition or subtraction that starts at 0. For example,

$$3(-4) = 0 + (-4) + (-4) + (-4) = -12$$

> The numbers have different signs and the product is negative.

and

$$(-3)(-4) = 0 - (-4) - (-4) - (-4) = 0 + 4 + 4 + 4 = 12$$

> The numbers have the same sign and the product is positive.

Table 1.2 reviews how to multiply real numbers.

Table 1.2 **Multiplying Real Numbers**	
Rule	**Example**
The product of two real numbers with different signs is found by multiplying their absolute values. The product is negative.	$7(-5) = -35$
The product of two real numbers with the same sign is found by multiplying their absolute values. The product is positive.	$(-6)(-11) = 66$
The product of 0 and any real number is 0: $a \cdot 0 = 0$ and $0 \cdot a = 0$.	$-17(0) = 0$
If no number is 0, a product with an odd number of negative factors is found by multiplying absolute values. The product is negative.	$-2(-3)(-5) = -30$ Three (odd) negative factors
If no number is 0, a product with an even number of negative factors is found by multiplying absolute values. The product is positive.	$-2(3)(-5) = 30$ Two (even) negative factors

6 Evaluate exponential expressions.

Because exponents indicate repeated multiplication, rules for multiplying real numbers can be used to evaluate exponential expressions.

EXAMPLE 5 Evaluating Exponential Expressions

Evaluate: **a.** $(-6)^2$ **b.** -6^2 **c.** $(-5)^3$ **d.** $(-2)^4$.

SOLUTION

a. $(-6)^2 = (-6)(-6) = 36$

> Base is −6.   Same signs give positive product.

b. $-6^2 = -(6 \cdot 6) = -36$

> Base is 6. The negative is not inside parentheses and is not taken to the second power.

c. $(-5)^3 = (-5)(-5)(-5) = -125$

> An odd number of negative factors gives a negative product.

d. $(-2)^4 = (-2)(-2)(-2)(-2) = 16$

> An even number of negative factors gives a positive product.

✔ **CHECK POINT 5** Evaluate:

a. $(-5)^2$ 25   **b.** -5^2 -25 **c.** $(-4)^3$ -64 **d.** $(-3)^4$.   81

---

7 Divide real numbers.

Dividing Real Numbers If a and b are real numbers and b is not 0, then the quotient of a and b is defined as follows:

$$a \div b = a \cdot \frac{1}{b} \quad \text{or} \quad \frac{a}{b} = a \cdot \frac{1}{b}.$$

Thus, to find the quotient of a and b, we can divide by b or multiply by $\frac{1}{b}$. The nonzero real numbers b and $\frac{1}{b}$ are called **reciprocals**, or **multiplicative inverses**, of one another. When reciprocals are multiplied, their product is 1:

$$b \cdot \frac{1}{b} = 1.$$

Because division is defined in terms of multiplication, the sign rules for dividing numbers are the same as the sign rules for multiplying them.

> **DIVIDING REAL NUMBERS**   The quotient of two numbers with different signs is negative. The quotient of two numbers with the same sign is positive. The quotient is found by dividing absolute values.

EXAMPLE 6   Dividing Real Numbers

Divide: **a.** $\dfrac{20}{-5}$ **b.** $-\dfrac{3}{4} \div \left(-\dfrac{5}{9}\right)$.

SOLUTION

a. $\dfrac{20}{-5} = -4$ Divide absolute values: $\frac{20}{5} = 4$.

> Different signs: negative quotient

b. $-\dfrac{3}{4} \div \left(-\dfrac{5}{9}\right) = \dfrac{27}{20}$ Divide absolute values: $\frac{3}{4} \div \frac{5}{9} = \frac{3}{4} \cdot \frac{9}{5} = \frac{27}{20}$.

> Same sign: positive quotient

✔ **CHECK POINT 6** Divide: **a.** $\dfrac{32}{-4}$ -8 **b.** $-\dfrac{2}{3} \div \left(-\dfrac{5}{4}\right)$. $\frac{8}{15}$

We must be careful with division when 0 is involved. Zero divided by any nonzero real number is 0. For example,

$$\frac{0}{-5} = 0.$$

Can you see why $\frac{0}{-5}$ must be 0? The definition of division tells us that

$$\frac{0}{-5} = 0 \cdot \left(-\frac{1}{5}\right)$$

and the product of 0 and any real number is 0. By contrast, what happens if we divide -5 by 0. The answer must be a number that, when multiplied by 0, gives -5. However, any number multiplied by 0 is 0. Thus, we cannot divide -5, or any other real number, by 0.

> **DIVISION BY ZERO** Division by zero is not allowed; it is undefined. A real number can never have a denominator of 0.

8 | Use the order of operations.

Order of Operations The rules for order of operations can be applied to positive and negative real numbers. Recall that if no grouping symbols are present, we

- Evaluate exponential expressions.
- Multiply and divide, from left to right.
- Add and subtract, from left to right.

EXAMPLE 7 Using the Order of Operations

Simplify: $4 - 7^2 + 8 \div 2(-3)^2$.

SOLUTION

$$4 - 7^2 + 8 \div 2(-3)^2$$

$= 4 - 49 + 8 \div 2(9)$ Evaluate exponential expressions: $7^2 = 7 \cdot 7 = 49$ and $(-3)^2 = (-3)(-3) = 9$.

$= 4 - 49 + 4(9)$ Divide: $8 \div 2 = 4$.

$= 4 - 49 + 36$ Multiply: $4(9) = 36$.

$= -45 + 36$ Subtract: $4 - 49 = 4 + (-49) = -45$.

$= -9$ Add. ∎

✔ **CHECK POINT 7** Simplify: $3 - 5^2 + 12 \div 2(-4)^2$. 74

If an expression contains grouping symbols, we perform operations within these symbols first. Common grouping symbols are parentheses, brackets, and braces. Other grouping symbols include fraction bars, absolute value symbols, and radical symbols, such as square root signs ($\sqrt{}$).

EXAMPLE 8 Using the Order of Operations

Simplify: $\dfrac{13 - 3(-2)^4}{3 - (6 - 10)}$.

SOLUTION Simplify the numerator and the denominator separately. Then divide.

$$\frac{13 - 3(-2)^4}{3 - (6 - 10)}$$

$$= \frac{13 - 3(16)}{3 - (-4)}$$

Evaluate the exponential expression in the numerator: $(-2)^4 = (-2)(-2)(-2)(-2) = 16$.
Subtract inside parentheses in the denominator: $6 - 10 = 6 + (-10) = -4$.

$$= \frac{13 - 48}{7}$$

Multiply in the numerator: $3(16) = 48$.
Subtract in the denominator: $3 - (-4) = 3 + 4 = 7$.

$$= \frac{-35}{7}$$

Subtract in the numerator: $13 - 48 = 13 + (-48) = -35$.

$$= -5$$

Divide. ∎

 CHECK POINT 8 Simplify: $\dfrac{4 + 3(-2)^3}{2 - (6 - 9)}$. -4

---

9 Use commutative, associative, and distributive properties.

The Commutative, Associative, and Distributive Properties Basic algebraic properties enable us to write *equivalent algebraic expressions*. Two algebraic expressions that have the same value for all replacements are called **equivalent algebraic expressions**. In Section 1.4, you will use such expressions to solve equations.

In arithmetic, when two numbers are added or multiplied, the order in which the numbers are written does not affect the answer. These facts are called **commutative properties**.

> **THE COMMUTATIVE PROPERTIES** Let a and b represent real numbers, variables, or algebraic expressions.
>
> $$\text{Addition:}\quad a + b = b + a$$
> $$\text{Multiplication:}\quad ab = ba$$
>
> Changing order when adding or multiplying does not affect a sum or product.

EXAMPLE 9 Using the Commutative Properties

Write an algebraic expression equivalent to $3x + 7$ using each of the commutative properties.

SOLUTION

Commutative of Addition
$$3x + 7 = 7 + 3x$$
Change the order of the addition.

Commutative of Multiplication
$$3x + 7 = x \cdot 3 + 7$$
Change the order of the multiplication.

∎

 CHECK POINT 9 Write an algebraic expression equivalent to $4x + 9$ using each of the commutative properties.

commutative of addition: $9 + 4x$; commutative of multiplication: $x \cdot 4 + 9$

The **associative properties** enable us to form equivalent expressions by regrouping.

THE ASSOCIATIVE PROPERTIES   Let a, b, and c represent real numbers, variables, or algebraic expressions.

$$\text{Addition:} \quad (a + b) + c = a + (b + c)$$

$$\text{Multiplication:} \quad (ab)c = a(bc)$$

Changing grouping when adding or multiplying does not affect a sum or product.

EXAMPLE 10   Using the Associative Properties

Use an associative property to write an equivalent expression and simplify:

a. $7 + (3 + x)$ **b.** $-6(5x)$.

SOLUTION

a. $7 + (3 + x) = (7 + 3) + x = 10 + x$

b. $-6(5x) = (-6 \cdot 5)x = -30x$ ■

 CHECK POINT 10   Use an associative property to write an equivalent expression and simplify:

a. $6 + (12 + x)$ $(6 + 12) + x = 18 + x$ **b.** $-7(4x)$. $(-7 \cdot 4)x = -28x$

ENRICHMENT ESSAY

The Associative Property and the English Language

In the English language, phrases and sentences can take on different meanings depending on the way the words are associated with commas. Here are two examples.

- What's the latest dope?
  What's the latest, dope?

- Population of Amsterdam broken
  down by age and sex
  Population of Amsterdam, broken
  down by age and sex

The **distributive property** allows us to rewrite the product of a number and a sum as the sum of two products.

THE DISTRIBUTIVE PROPERTY   Let a, b, and c represent real numbers, variables, or algebraic expressions.

$$a(\overset{\frown}{b + c}) = ab + ac$$

Multiplication distributes over addition.

EXAMPLE 11 Using the Distributive Property

Use the distributive property to write an equivalent expression:

$$-2(3x + 5).$$

SOLUTION

$$-2(3x + 5) = -2 \cdot 3x + (-2) \cdot 5 = -6x + (-10) = -6x - 10 \qquad \blacksquare$$

 CHECK POINT 11 Use the distributive property to write an equivalent expression: $-4(7x + 2)$. $-28x - 8$

Table 1.3 shows a number of other forms of the distributive property.

Table 1.3 **Other Forms of the Distributive Property**

Property	Meaning	Example
$a(b - c) = ab - ac$	Multiplication distributes over subtraction.	$6(4x - 5) = 6 \cdot 4x - 6 \cdot 5$ $= 24x - 30$
$a(b + c + d) = ab + ac + ad$	Multiplication distributes over three or more terms in parentheses.	$5(x + 4 + 7y)$ $= 5x + 5 \cdot 4 + 5 \cdot 7y$ $= 5x + 20 + 35y$
$(b + c)a = ba + ca$	Multiplication on the right distributes over addition (or subtraction).	$(x + 10)8 = x \cdot 8 + 10 \cdot 8$ $= 8x + 80$

10 Simplify algebraic expressions.

Combining Like Terms and Simplifying Algebraic Expressions The **terms** of an algebraic expression are those parts that are separated by addition. For example, the algebraic expression $7x + 3$ contains two terms, namely $7x$ and 3. Notice that a term is a number, a variable, or a number multiplied by one or more variables.

The numerical part of a term is called its **coefficient**. In the term $7x$, the 7 is the coefficient. If a term containing one or more variables is written without a coefficient, the coefficient is understood to be 1. Thus, x means $1x$ and ab means $1ab$. If a term is a constant, its coefficient is that constant. Thus, the coefficient of the constant term 3 is 3.

The parts of each term that are multiplied are called the **factors** of the term. The factors of the term $7x$ are 7 and x.

Like terms are terms that have exactly the same variable factors. For example, $3x$ and $7x$ are like terms. The distributive property in the form

$$ba + ca = (b + c)a$$

enables us to add or subtract like terms. For example,

$$3x + 7x = (3 + 7)x = 10x$$
$$7y^2 - y^2 = 7y^2 - 1y^2 = (7 - 1)y^2 = 6y^2.$$

This process is called **combining like terms**.

An algebraic expression is **simplified** when grouping symbols have been removed and like terms have been combined.

EXAMPLE 12 Simplifying an Algebraic Expression

Simplify: $7x + 12x^2 + 3x + x^2$.

SOLUTION

$$7x + 12x^2 + 3x + x^2$$

$$= (7x + 3x) + (12x^2 + x^2)$$
Rearrange terms and group like terms using commutative and associative properties. This step is often done mentally.

 $x^2 = 1x^2$

$$= (7 + 3)x + (12 + 1)x^2$$
Apply the distributive property.

$$= 10x + 13x^2$$
Simplify. Because $10x$ and $13x^2$ are not like terms, this is the final answer.

Using the commutative property of addition, we can write this simplified expression as $13x^2 + 10x$. ■

✔ CHECK POINT **12** Simplify: $3x + 14x^2 + 11x + x^2$. $14x + 15x^2$ or $15x^2 + 14x$

EXAMPLE 13 Simplifying an Algebraic Expression

Simplify: $4(7x - 3) - 10x$.

SOLUTION

$$4(\overbrace{7x - 3}) - 10x$$

$$= 4 \cdot 7x - 4 \cdot 3 - 10x$$
Use the distributive property to remove the parentheses.

$$= 28x - 12 - 10x$$
Multiply.

$$= (28x - 10x) - 12$$
Group like terms.

$$= (28 - 10)x - 12$$
Apply the distributive property.

$$= 18x - 12$$
Simplify. ■

✔ CHECK POINT **13** Simplify: $8(2x - 5) - 4x$. $12x - 40$

It is not uncommon to see algebraic expressions with parentheses preceded by a negative sign or subtraction. An expression of the form $-(b + c)$ can be simplified as follows:

$$-(b + c) = -1(\overbrace{b + c}) = (-1)b + (-1)c = -b + (-c) = -b - c.$$

Do you see a fast way to obtain the simplified expression on the right? **If a negative sign or a subtraction symbol appears outside parentheses, drop the parentheses and change the sign of every term within the parentheses.** For example,

$$-(3x^2 - 7x - 4) = -3x^2 + 7x + 4.$$

EXAMPLE 14 Simplifying an Algebraic Expression

Simplify: $8x + 2[5 - (x - 3)]$.

SOLUTION

$$8x + 2[5 - (x - 3)]$$
$$= 8x + 2[5 - x + 3]$$ Drop parentheses and change the sign of each term in parentheses: $-(x - 3) = -x + 3$.

$$= 8x + 2[8 - x]$$ Simplify inside brackets: $5 + 3 = 8$.
$$= 8x + 16 - 2x$$ Apply the distributive property:

$$2[8 - x] = 2 \cdot 8 - 2x = 16 - 2x.$$

$$= (8x - 2x) + 16$$ Group like terms.
$$= (8 - 2)x + 16$$ Apply the distributive property.
$$= 6x + 16$$ Simplify.

✔ **CHECK POINT 14** Simplify: $6 + 4[7 - (x - 2)]$. $42 - 4x$

1.2 EXERCISE SET

 Student Solutions Manual   CD/Video   PH Math/Tutor Center   MathXL Tutorials on CD   MathXL®   MyMathLab   Interactmath.com

Practice Exercises

In Exercises 1–12, find each absolute value.

1. $|-7|$ 7
2. $|-10|$ 10
3. $|4|$ 4
4. $|13|$ 13
5. $|-7.6|$ 7.6
6. $|-8.3|$ 8.3
7. $\left|\dfrac{\pi}{2}\right|$ $\dfrac{\pi}{2}$
8. $\left|\dfrac{\pi}{3}\right|$ $\dfrac{\pi}{3}$
9. $|-\sqrt{2}|$ $\sqrt{2}$
10. $|-\sqrt{3}|$ $\sqrt{3}$
11. $-\left|-\dfrac{2}{5}\right|$ $-\dfrac{2}{5}$
12. $-\left|-\dfrac{7}{10}\right|$ $-\dfrac{7}{10}$

In Exercises 13–28, add as indicated.

13. $-3 + (-8)$ -11
14. $-5 + (-10)$ -15
15. $-14 + 10$ -4
16. $-15 + 6$ -9
17. $-6.8 + 2.3$ -4.5
18. $-7.9 + 2.4$ -5.5
19. $\dfrac{11}{15} + \left(-\dfrac{3}{5}\right)$ $\dfrac{2}{15}$
20. $\dfrac{7}{10} + \left(-\dfrac{4}{5}\right)$ $-\dfrac{1}{10}$
21. $-\dfrac{2}{9} - \dfrac{3}{4}$ $-\dfrac{35}{36}$
22. $-\dfrac{3}{5} - \dfrac{4}{7}$ $-\dfrac{41}{35}$
23. $-3.7 + (-4.5)$ -8.2
24. $-6.2 + (-5.9)$ -12.1
25. $0 + (-12.4)$ -12.4
26. $0 + (-15.3)$ -15.3
27. $12.4 + (-12.4)$ 0
28. $15.3 + (-15.3)$ 0

In Exercises 29–34, find $-x$ for the given value of x.

29. $x = 11$ -11
30. $x = 13$ -13
31. $x = -5$ 5
32. $x = -9$ 9
33. $x = 0$ 0
34. $x = -\sqrt{2}$ $\sqrt{2}$

In Exercises 35–46, subtract as indicated.

35. $3 - 15$ -12
36. $4 - 20$ -16
37. $8 - (-10)$ 18
38. $7 - (-13)$ 20
39. $-20 - (-5)$ -15
40. $-30 - (-10)$ -20
41. $\dfrac{1}{4} - \dfrac{1}{2}$ $-\dfrac{1}{4}$
42. $\dfrac{1}{10} - \dfrac{2}{5}$ $-\dfrac{3}{10}$
43. $-2.3 - (-7.8)$ 5.5
44. $-4.3 - (-8.7)$ 4.4
45. $0 - (-\sqrt{2})$ $\sqrt{2}$
46. $0 - (-\sqrt{3})$ $\sqrt{3}$

In Exercises 47–58, multiply as indicated.

47. $9(-10)$ -90
48. $8(-10)$ -80
49. $(-3)(-11)$ 33
50. $(-7)(-11)$ 77
51. $\dfrac{15}{13}(-1)$ $-\dfrac{15}{13}$
52. $\dfrac{11}{13}(-1)$ $-\dfrac{11}{13}$
53. $-\sqrt{2} \cdot 0$ 0
54. $-\sqrt{3} \cdot 0$ 0
55. $(-4)(-2)(-1)$ -8
56. $(-5)(-3)(-2)$ -30
57. $2(-3)(-1)(-2)(-4)$ 48
58. $3(-2)(-1)(-5)(-3)$ 90

In Exercises 59–70, evaluate each exponential expression.

59. $(-10)^2$ 100

60. $(-8)^2$ 64

61. -10^2 -100

62. -8^2 -64

63. $(-2)^3$ -8

64. $(-3)^3$ -27

65. $(-1)^4$ 1

66. $(-4)^4$ 256

67. $(-1)^{33}$ -1

68. $(-1)^{35}$ -1

69. $-\left(-\dfrac{1}{2}\right)^3$ $\dfrac{1}{8}$

70. $-\left(-\dfrac{1}{4}\right)^3$ $\dfrac{1}{64}$

In Exercises 71–82, divide as indicated or state that the division is undefined.

71. $\dfrac{12}{-4}$ -3

72. $\dfrac{30}{-5}$ -6

73. $\dfrac{-90}{-2}$ 45

74. $\dfrac{-55}{-5}$ 11

75. $\dfrac{0}{-4.6}$ 0

76. $\dfrac{0}{-5.3}$ 0

77. $-\dfrac{4.6}{0}$ undefined

78. $-\dfrac{5.3}{0}$ undefined

79. $-\dfrac{1}{2} \div \left(-\dfrac{7}{9}\right)$ $\dfrac{9}{14}$

80. $-\dfrac{1}{2} \div \left(-\dfrac{3}{5}\right)$ $\dfrac{5}{6}$

81. $6 \div \left(-\dfrac{2}{5}\right)$ -15

82. $8 \div \left(-\dfrac{2}{9}\right)$ -36

In Exercises 83–100, use the order of operations to simplify each expression.

83. $4(-5) - 6(-3)$ -2

84. $8(-3) - 5(-6)$ 6

85. $3(-2)^2 - 4(-3)^2$ -24

86. $5(-3)^2 - 2(-2)^2$ 37

87. $8^2 - 16 \div 2^2 \cdot 4 - 3$ 45

88. $10^2 - 100 \div 5^2 \cdot 2 - 3$ 89

89. $\dfrac{5 \cdot 2 - 3^2}{[3^2 - (-2)]^2}$ $\dfrac{1}{121}$

90. $\dfrac{10 \div 2 + 3 \cdot 4}{(12 - 3 \cdot 2)^2}$ $\dfrac{17}{36}$

91. $8 - 3[-2(2 - 5) - 4(8 - 6)]$ 14

92. $8 - 3[-2(5 - 7) - 5(4 - 2)]$ 26

93. $\dfrac{2(-2) - 4(-3)}{5 - 8}$ $-\dfrac{8}{3}$

94. $\dfrac{6(-4) - 5(-3)}{9 - 10}$ 9

95. $\dfrac{(5 - 6)^2 - 2|3 - 7|}{89 - 3 \cdot 5^2}$ $-\dfrac{1}{2}$

96. $\dfrac{12 \div 3 \cdot 5|2^2 + 3^2|}{7 + 3 - 6^2}$ -10

97. $15 - \sqrt{3 - (-1)} + 12 \div 2 \cdot 3$ 31

98. $17 - |5 - (-2)| + 12 \div 2 \cdot 3$ 28

99. $20 + 1 - \sqrt{10^2 - (5 + 1)^2}(-2)$ 37

100. $24 \div \sqrt{3 \cdot (5 - 2)} \div [-1 - (-3)]^2$ 2

In Exercises 101–104, write an algebraic expression equivalent to the given expression using each of the commutative properties.

101. $4x + 10$ commutative of addition: $10 + 4x$; commutative of multiplication: $x \cdot 4 + 10$

102. $5x + 30$ commutative of addition: $30 + 5x$; commutative of multiplication: $x \cdot 5 + 30$

103. $7x - 5$ commutative of addition: $-5 + 7x$; commutative of multiplication: $x \cdot 7 - 5$

104. $3x - 7$ commutative of addition: $-7 + 3x$; commutative of multiplication: $x \cdot 3 - 7$

In Exercises 105–110, use an associative property to write an algebraic expression equivalent to each expression and simplify.

105. $4 + (6 + x)$ $(4 + 6) + x = 10 + x$

106. $12 + (3 + x)$ $(12 + 3) + x = 15 + x$

107. $-7(3x)$ $(-7 \cdot 3)x = -21x$

108. $-10(5x)$ $(-10 \cdot 5)x = -50x$

109. $-\dfrac{1}{3}(-3y)$ $\left(-\dfrac{1}{3} \cdot -3\right)y = y$

110. $-\dfrac{1}{4}(-4y)$ $\left(-\dfrac{1}{4} \cdot -4\right)y = y$

In Exercises 111–116, use the distributive property to write an equivalent expression.

111. $3(2x + 5)$ $6x + 15$

112. $5(4x + 7)$ $20x + 35$

113. $-7(2x + 3)$ $-14x - 21$

114. $-9(3x + 2)$ $-27x - 18$

115. $-(3x - 6)$ $-3x + 6$

116. $-(6x - 3)$ $-6x + 3$

In Exercises 117–130, simplify each algebraic expression.

117. $7x + 5x$ $12x$

118. $8x + 10x$ $18x$

119. $6x^2 - x^2$ $5x^2$

120. $9x^2 - x^2$ $8x^2$

121. $6x + 10x^2 + 4x + 2x^2$ $10x + 12x^2$

122. $9x + 5x^2 + 3x + 4x^2$ $12x + 9x^2$

123. $8(3x - 5) - 6x$ $18x - 40$

124. $7(4x - 5) - 8x$ $20x - 35$

125. $5(3y - 2) - (7y + 2)$ $8y - 12$

126. $4(5y - 3) - (6y + 3)$ $14y - 15$

127. $7 - 4[3 - (4y - 5)]$ $16y - 25$

128. $6 - 5[8 - (2y - 4)]$ $10y - 54$

129. $18x^2 + 4 - [6(x^2 - 2) + 5]$ $12x^2 + 11$

130. $14x^2 + 5 - [7(x^2 - 2) + 4]$ $7x^2 + 15$

Practice Plus

In Exercises 131–138, write each English phrase as an algebraic expression. Then simplify the expression. Let x represent the number.

131. A number decreased by the sum of the number and four.
$x - (x + 4); -4$

132. A number decreased by the difference between eight and the number. $x - (8 - x); 2x - 8$

133. Six times the product of negative five and a number.
$6(-5x); -30x$

134. Ten times the product of negative four and a number.
$10(-4x); -40x$

135. The difference between the product of five times a number and twice the number. $5x - 2x; 3x$

136. The difference between the product of six and a number and negative two times the number. $6x - (-2x); 8x$

137. The difference between eight times a number and six more than three times the number. $8x - (3x + 6); 5x - 6$

138. Eight decreased by three times the sum of a number and six. $8 - 3(x + 6); -3x - 10$

Application Exercises

The world is divided into 24 time zones, measured in relation to 12 noon Greenwich Mean Time, on the Greenwich Meridian (0°). Time increases by one hour for every 15° longitude east of Greenwich and decreases by one hour for every 15° west, although the system is adjusted in line with national and state boundaries. The numbers on the bottom of the map show the number of hours that must be added to or subtracted from Greenwich Mean Time to calculate the time in each zone. Use these numbers to solve Exercises 139–142. Use the word "ahead" or "behind" in each of your answers.

The World's 24 Time Zones

139. What is the time difference between Hong Kong and Los Angeles?   Hong Kong is 16 hours ahead.

140. What is the time difference between Hong Kong and New York?   Hong Kong is 13 hours ahead.

141. What is the time difference between Los Angeles and New York?   Los Angeles is 3 hours behind.

142. What is the time difference between New York and Rio de Janeiro?   New York is 2 hours behind.

In 2003, the mean, or average, SAT verbal score was 507. The mean SAT math score was 519. The bar graph on the next pages shows the deviation in points from the mean score by the highest level of educational attainment of a test-taker's parent, indicated by voice balloons. Use the numbers displayed by the bars to solve Exercises 143–146.

Average SAT Scores by Parental Education, 2003

Points below or above the Mean Score

Source: The College Board

143. For the group whose highest level of parental education is an associate degree, what is the point difference between the verbal score and the math score?   8 points

144. For the group whose highest level of parental education is a high school diploma, what is the point difference between the verbal score and the math score?   7 points

145. What is the point difference in math between the group whose highest level of parental education is a graduate degree and the group whose parent does not have a high school diploma?   126 points

146. What is the point difference in the verbal score between the group whose highest level of parental education is a graduate degree and the group whose parent does not have a high school diploma?   146 points

147. You had $10,000 to invest. You put x dollars in a safe, government-insured certificate of deposit paying 5% per year. You invested the remainder of the money in noninsured corporate bonds paying 12% per year. Your total interest earned at the end of the year is given by the algebraic expression

$$0.05x + 0.12(10,000 - x).$$

   a. Simplify the algebraic expression. $1200 - 0.07x$

   b. Use each form of the algebraic expression to determine your total interest earned at the end of the year if you invested $6000 in the safe, government-insured certificate of deposit. $780

148. It takes you 50 minutes to get to campus. You spend t minutes walking to the bus stop and the rest of the time riding the bus. Your walking rate is 0.06 miles per minute and the bus travels at a rate of 0.5 miles per minute. The total distance walking and traveling by bus is given by the algebraic expression

$$0.06t + 0.5(50 - t).$$

   a. Simplify the algebraic expression. $25 - 0.44t$

   b. Use each form of the algebraic expression to determine the total distance that you travel if you spend 20 minutes walking to the bus stop.   16.2 miles

Writing in Mathematics

149. What is the meaning of $|a|$ in terms of a number line?
149–166. Answers will vary.

150. Explain how to add two numbers with the same sign. Give an example with your explanation.

151. Explain how to add two numbers with different signs. Give an example with your explanation.

152. What are opposites, or additive inverses? What happens when finding the sum of a number and its opposite?

153. Explain how to subtract real numbers.

154. Explain how to multiply two numbers with different signs. Give an example with your explanation.

155. Explain how to multiply two numbers with the same sign. Give an example with your explanation.

156. Explain how to determine the sign of a product that involves more than two numbers.

157. Explain how to divide real numbers.

158. Why is $\dfrac{0}{4} = 0$, although $\dfrac{4}{0}$ is undefined?

159. What are equivalent algebraic expressions?

160. State a commutative property and give an example of how it is used to write equivalent algebraic expressions.

161. State an associative property and give an example of how it is used to write equivalent algebraic expressions.

162. State a distributive property and give an example of how it is used to write equivalent algebraic expressions.

163. What are the terms of an algebraic expression? How can you tell if terms are like terms?

164. What does it mean to simplify an algebraic expression?

165. If a negative sign appears outside parentheses, explain how to simplify the expression. Give an example.

166. What explanations can you offer for the trend in average SAT Scores, by parental education, shown in the bar graph in Exercises 143–146?

Critical Thinking Exercises

167. Which one of the following is true?   d

   a. $16 \div 4 \cdot 2 = 16 \div 8 = 2$

   b. $6 - 2(4 + 3) = 4(4 + 3) = 4(7) = 28$

   c. $3 + 7x = 10x$

   d. None of the above is true.

In Exercises 168–169, insert parentheses to make each statement true.

168. $8 - 2 \cdot 3 - 4 = 14$ $(8 - 2) \cdot 3 - 4 = 14$

169. $2 \cdot 5 - \dfrac{1}{2} \cdot 10 \cdot 9 = 45$ $\left(2 \cdot 5 - \dfrac{1}{2} \cdot 10\right) \cdot 9 = 45$

170. Simplify: $\dfrac{9[4 - (1 + 6)] - (3 - 9)^2}{5 + \dfrac{12}{5 - \dfrac{6}{2 + 1}}}.$ -7

Review Exercises

From here on, each exercise set will contain three review exercises. It is important to review previously covered topics to improve your understanding of the topics and to help maintain your mastery of the material. If you are not certain how to solve a review exercise, turn to the section and the worked-out example given in parentheses at the end of each exercise.

171. Write the following English phrase as an algebraic expression: "The quotient of ten and a number, decreased by four times the number." Let x represent the number. (Section 1.1, Example 1) $\dfrac{10}{x} - 4x$

172. Evaluate $10 + 2(x - 5)^4$ for $x = 7$. (Section 1.1, Example 3) 42

173. Determine whether the following statement is true or false: $\frac{1}{2} \notin \{x \mid x \text{ is an irrational number}\}$. (Section 1.1, Example 5) true

SECTION 1.3

GRAPHING EQUATIONS

Objectives

1 Plot points in the rectangular coordinate system.

2 Graph equations in the rectangular coordinate system.

3 Use the rectangular coordinate system to visualize relationships between variables.

4 Interpret information about a graphing utility's viewing rectangle or table.

Boxer Oscar De La Hoya (1973–) began his career by winning the lightweight gold medal at the 1992 Olympics.

Singers Ricky Martin and Jennifer Lopez, television journalist Soledad O'Brien, boxer Oscar De La Hoya, Internet entrepreneur Carlos Cardona, actor John Leguizamo, and political organizer Luigi Crespo are members of a generation of young Hispanics that is changing the way America looks, feels, thinks, eats, dances, and votes. There are 38 million Hispanics in the United States, pumping $350 billion a year into the economy. By 2050, the Hispanic population is expected to reach 96 million—an increase of more than 150 percent. Hispanics are also younger than the rest of the nation: One third are under 18. The graph in Figure 1.8 shows the average age of whites, African Americans, and Americans of Hispanic origin.

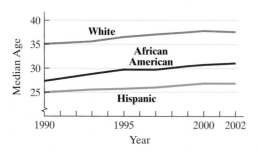

Median Age of U.S. Whites, African Americans, and Hispanics

FIGURE 1.8

Source: U.S. Census Bureau

Magazines and newspapers often display information using **line graphs** like those in Figure 1.8. Line graphs are often used to illustrate trends over time. Some measure of time, such as months or years, frequently appears on the horizontal axis. Amounts are generally listed on the vertical axis.

A line graph displays information based on an idea developed by the French philosopher and mathematician René Descartes (1596–1650). Descartes developed a new branch of mathematics that brought together algebra and geometry in a unified way—a way that visualized numbers as points on a graph, equations as geometric figures, and geometric figures as equations. We begin this section by looking at Descartes's deceptively simple idea, called the **rectangular coordinate system** or (in his honor) the **Cartesian coordinate system**.

1 Plot points in the rectangular coordinate system.

Points and Ordered Pairs Descartes used two number lines that intersect at right angles at their zero points, as shown in Figure 1.9. The horizontal number line is the **x-axis**. The vertical number line is the **y-axis**. The point of intersection of these axes is their zero points, called the **origin**. Positive numbers are shown to the right and above the origin. Negative numbers are shown to the left and below the origin. The axes divide the plane into four quarters, called **quadrants**. The points located on the axes are not in any quadrant.

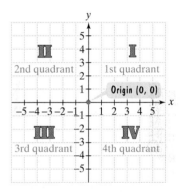

FIGURE 1.9 The rectangular coordinate system

Each point in the rectangular coordinate system corresponds to an **ordered pair** of real numbers, (x, y). Examples of such pairs are $(-5, 3)$ and $(3, -5)$. The first number in each pair, called the **x-coordinate**, denotes the distance and direction from the origin along the x-axis. The second number, called the **y-coordinate**, denotes vertical distance and direction along a line parallel to the y-axis or along the y-axis itself.

Figure 1.10 shows how we **plot**, or locate, the points corresponding to the ordered pairs $(-5, 3)$ and $(3, -5)$. We plot $(-5, 3)$ by going 5 units from 0 to the left along the x-axis. Then we go 3 units up parallel to the y-axis. We plot $(3, -5)$ by going 3 units from 0 to the right along the x-axis and 5 units down parallel to the y-axis. The phrase "the points corresponding to the ordered pairs $(-5, 3)$ and $(3, -5)$" are often abbreviated as "the points $(-5, 3)$ and $(3, -5)$."

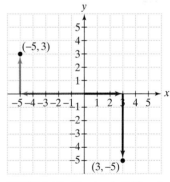

FIGURE 1.10 Plotting $(-5, 3)$ and $(3, -5)$

STUDY TIP

The phrase *ordered pair* is used because order is important. The order in which coordinates appear makes a difference in a point's location. This is illustrated in Figure 1.10.

EXAMPLE 1 Plotting Points in the Rectangular Coordinate System

Plot the points: $A(-4, 5)$, $B(3, -4)$, $C(-5, 0)$, $D(-4, -2)$, $E(0, 3.5)$, and $F(0, 0)$.

SOLUTION See Figure 1.11. We move from the origin and plot the points in the following way:

$A(-4, 5)$: 4 units left, 5 units up

$B(3, -4)$: 3 units right, 4 units down

$C(-5, 0)$: 5 units left, 0 units up or down

$D(-4, -2)$: 4 units left, 2 units down

$E(0, 3.5)$: 0 units right or left, 3.5 units up

$F(0, 0)$: 0 units right or left,

        0 units up or down

Notice that the origin is represented by (0, 0).

FIGURE 1.11 Plotting points

 CHECK POINT 1 Plot the points:

$A(2, 5)$, $B(-1, 3)$, $C(-1.5, -4.5)$, and $D(0, -2)$. See graphing answer section.

---

2 Graph equations in the rectangular coordinate system.

Graphs of Equations A relationship between two quantities can be expressed as an **equation in two variables**, such as

$$y = 4 - x^2.$$

A **solution of an equation in two variables**, x and y, is an ordered pair of real numbers with the following property: When the x-coordinate is substituted for x and the y-coordinate is substituted for y in the equation, we obtain a true statement. For example, consider the equation $y = 4 - x^2$ and the ordered pair $(3, -5)$. When 3 is substituted for x and -5 is substituted for y, we obtain the statement $-5 = 4 - 3^2$, or $-5 = 4 - 9$, or $-5 = -5$. Because this statement is true, the ordered pair $(3, -5)$ is a solution of the equation $y = 4 - x^2$. We also say that $(3, -5)$ **satisfies** the equation.

We can generate as many ordered-pair solutions as desired to $y = 4 - x^2$ by substituting numbers for x and then finding the corresponding values for y. For example, suppose we let $x = 3$:

Start with x.	Compute y.	Form the ordered pair (x, y).
x	$y = 4 - x^2$	**Ordered Pair (x, y)**
3	$y = 4 - 3^2 = 4 - 9 = -5$	$(3, -5)$
Let $x = 3$.		$(3, -5)$ is a solution of $y = 4 - x^2$.

The **graph of an equation in two variables** is the set of all points whose coordinates satisfy the equation. One method for graphing such equations is the **point-plotting method**. First, we find several ordered pairs that are solutions of the equation. Next, we plot these ordered pairs as points in the rectangular coordinate system. Finally, we connect the points with a smooth curve or line. This often gives us a picture of all ordered pairs that satisfy the equation.

EXAMPLE 2   Graphing an Equation Using the Point-Plotting Method

Graph $y = 4 - x^2$. Select integers for x, starting with -3 and ending with 3.

SOLUTION   For each value of x we find the corresponding value for y.

| | Start with x. | Compute y. | Form the ordered pair (x, y). |

x	$y = 4 - x^2$	Ordered Pair (x, y)
-3	$y = 4 -(-3)^2 = 4 - 9 = -5$	$(-3, -5)$
-2	$y = 4 -(-2)^2 = 4 - 4 = 0$	$(-2, 0)$
-1	$y = 4 -(-1)^2 = 4 - 1 = 3$	$(-1, 3)$
0	$y = 4 - 0^2 = 4 - 0 = 4$	$(0, 4)$
1	$y = 4 - 1^2 = 4 - 1 = 3$	$(1, 3)$
2	$y = 4 - 2^2 = 4 - 4 = 0$	$(2, 0)$
3	$y = 4 - 3^2 = 4 - 9 = -5$	$(3, -5)$

We selected integers from -3 to 3, inclusive, to include three negative numbers, 0, and three positive numbers. We also wanted to keep the resulting computations for y relatively simple.

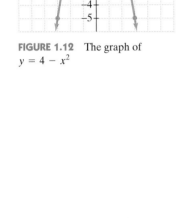

FIGURE 1.12   The graph of $y = 4 - x^2$

Now we plot the seven points and join them with a smooth curve, as shown in Figure 1.12. The graph of $y = 4 - x^2$ is a curve where the part of the graph to the right of the y-axis is a reflection of the part to the left of it and vice versa. The arrows on the left and the right of the curve indicate that it extends indefinitely in both directions. ■

 CHECK POINT **2** Graph $y = 1 - x^2$. Select integers for x, starting with -3 and ending with 3.   See graphing answer section.

EXAMPLE 3   Graphing an Equation Using the Point-Plotting Method

Graph $y = |x|$. Select integers for x, starting with -3 and ending with 3.

SOLUTION   For each value of x we find the corresponding value for y.

x	$y = |x|$	Ordered Pair (x, y)
-3	$y = |-3| = 3$	$(-3, 3)$
-2	$y = |-2| = 2$	$(-2, 2)$
-1	$y = |-1| = 1$	$(-1, 1)$
0	$y = |0| = 0$	$(0, 0)$
1	$y = |1| = 1$	$(1, 1)$
2	$y = |2| = 2$	$(2, 2)$
3	$y = |3| = 3$	$(3, 3)$

FIGURE 1.13   The graph of $y = |x|$

We plot the points and connect them, resulting in the graph shown in Figure 1.13. The graph is V-shaped and centered at the origin. For every point (x, y) on the graph, the point $(-x, y)$ is also on the graph. This shows that the absolute value of a positive number is the same as the absolute value of its opposite. ■

 CHECK POINT **3** Graph $y = |x + 1|$. Select integers for x, starting with -4 and ending with 2.   See graphing answer section.

3 Use the rectangular coordinate system to visualize relationships between variables.

FIGURE 1.14 Body temperature from 8 A. M. through 3 P.M.

Applications The rectangular coordinate system allows us to visualize relationships between two variables by associating any equation in two variables with a graph. Graphs in the rectangular coordinate system can also be used to tell a story.

EXAMPLE 4 Telling a Story with a Graph

Too late for that flu shot now! It's only 8 A.M. and you're feeling lousy. Fascinated by the way that algebra models the world (your author is projecting a bit here), you construct a graph showing your body temperature from 8 A.M. through 3 P.M. You decide to let x represent the number of hours after 8 A.M. and y your body temperature at time x. The graph is shown in Figure 1.14. The symbol $\frac{7}{7}$ on the y-axis shows that there is a break in values between 0 and 98. Thus, the first tick mark on the y-axis represents a temperature of 98°F.

a. What is your temperature at 8 A.M.?

b. During which period of time is your temperature decreasing?

c. Estimate your minimum temperature during the time period shown. How many hours after 8 A.M. does this occur? At what time does this occur?

d. During which period of time is your temperature increasing?

e. Part of the graph is shown as a horizontal line segment. What does this mean about your temperature and when does this occur?

SOLUTION

a. Because x is the number of hours after 8 A.M., your temperature at 8 A.M. corresponds to $x = 0$. Locate 0 on the horizontal axis and look at the point on the graph above 0. The first figure on the right shows that your temperature at 8 A.M. is 101°F.

b. Your temperature is decreasing when the graph falls from left to right. This occurs between $x = 0$ and $x = 3$, also shown in the first figure on the right. Because x represents the number of hours after 8 A.M., your temperature is decreasing between 8 A.M. and 11 A.M.

c. Your minimum temperature can be found by locating the lowest point on the graph. This point lies above 3 on the horizontal axis, shown in the second figure on the right. The y-coordinate of this point falls more than midway between 98 and 99, at approximately 98.6. The lowest point on the graph, (3, 98.6), shows that your minimum temperature, 98.6°F, occurs 3 hours after 8 A.M., at 11 A.M.

d. Your temperature is increasing when the graph rises from left to right. This occurs between $x = 3$ and $x = 5$, shown in the second figure. Because x represents the number of hours after 8 A.M., your temperature is increasing between 11 A.M. and 1 P.M.

e. The horizontal line segment shown in the figure on the right indicates that your temperature is neither increasing nor decreasing. Your temperature remains the same, 100°F, between $x = 5$ and $x = 7$. Thus, your temperature is at a constant 100°F between 1 P.M. and 3 P.M.

Hours after 8 A.M.

✔ CHECK POINT 4 When a person receives a drug injected into a muscle, the concentration of the drug in the body, measured in milligrams per 100 milliliters, depends on the time elapsed after the injection, measured in hours. The figure shows the graph of drug concentration over time, where x represents hours after the injection and y represents the drug concentration at time x.

a. During which period of time is the drug concentration increasing? 0 to 3 hours

b. During which period of time is the drug concentration decreasing? 3 to 13 hours

c. What is the drug's maximum concentration and when does this occur? 0.05 milligrams per 100 milliliters; after 3 hours.

d. What happens by the end of 13 hours? None of the drug is left in the body.

Graphing Equations and Creating Tables Using a Graphing Utility Graphing calculators or graphing software packages for computers are referred to as **graphing utilities** or **graphers**. A graphing utility is a powerful tool that quickly generates the graph of an equation in two variables. Figures 1.15(a) and 1.15(b) show two such graphs for the equations in Examples 2 and 3.

4 Interpret information about a graphing utility's viewing rectangle or table.

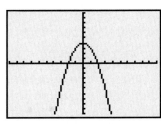

FIGURE 1.15(a) The graph of $y = 4 - x^2$

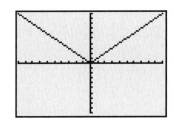

FIGURE 1.15(b) The graph of $y = |x|$

What differences do you notice between these graphs and the graphs that we drew by hand? They do seem a bit "jittery." Arrows do not appear on the left and right ends of the graphs. Furthermore, numbers are not given along the axes. For both graphs in Figure 1.15, the x-axis extends from -10 to 10 and the y-axis also extends from -10 to 10. The distance represented by each consecutive tick mark is one unit. We say that the **viewing rectangle**, or the **viewing window**, is $[-10, 10, 1]$ by $[-10, 10, 1]$.

To graph an equation in x and y using a graphing utility, enter the equation and specify the size of the viewing rectangle. The size of the viewing rectangle sets minimum and maximum values for both the x- and y-axes. Enter these values, as well as the values between consecutive tick marks, on the respective axes. The $[-10, 10, 1]$ by $[-10, 10, 1]$ viewing rectangle used in Figure 1.15 is called the **standard viewing rectangle**.

EXAMPLE 5 Understanding the Viewing Rectangle

What is the meaning of a $[-2, 3, 0.5]$ by $[-10, 20, 5]$ viewing rectangle?

SOLUTION We begin with $[-2, 3, 0.5]$, which describes the x-axis. The minimum x-value is -2 and the maximum x-value is 3. The distance between consecutive tick marks is 0.5.

Next, consider $[-10, 20, 5]$, which describes the y-axis. The minimum y-value is -10 and the maximum y-value is 20. The distance between consecutive tick marks is 5.

Figure 1.16 illustrates a $[-2, 3, 0.5]$ by $[-10, 20, 5]$ viewing rectangle. To make things clearer, we've placed numbers by each tick mark. These numbers do not appear on the axes when you use a graphing utility to graph an equation. ∎

 CHECK POINT 5 What is the meaning of a $[-100, 100, 50]$ by $[-100, 100, 10]$ viewing rectangle? Create a figure like the one in Figure 1.16 that illustrates this viewing rectangle.

FIGURE 1.16 A $[-2, 3, 0.5]$ by $[-10, 20, 5]$ viewing rectangle

The minimum x-value is -100, the maximum x-value is 100, and the distance between consecutive tick marks is 50. The minimum y-value is -100, the maximum y-value is 100, and the distance between consecutive tick marks is 10. See graphing answer section.

On most graphing utilities, the display screen is two-thirds as high as it is wide. By using a square setting, you can equally space the x and y tick marks. (This does not occur in the standard viewing rectangle.) Graphing utilities can also *zoom in* and *zoom out*. When you zoom in, you see a smaller portion of the graph, but you do so in greater detail. When you zoom out, you see a larger portion of the graph. Thus, zooming out may help you to develop a better understanding of the overall character of the graph. With practice, you will become more comfortable with graphing equations in two variables using your graphing utility. You will also develop a better sense of the size of the viewing rectangle that will reveal needed information about a particular graph.

Graphing utilities can also be used to create tables showing solutions of equations in two variables. Use the Table Setup function to choose the starting value of x and to input the increment, or change, between the consecutive x-values. The corresponding y-values are calculated based on the equation(s) in two variables in the $\boxed{Y=}$ screen. In Figure 1.17, we used a TI-83 Plus to create a table for $y = 4 - x^2$ and $y = |x|$, the equations in Examples 2 and 3.

We entered two equations:
$y_1 = 4 - x^2$ and $y_2 = |x|$.

We entered -3 for the starting x-value and 1 as increment between x-values.

The x-values are in the first column and the corresponding values of $y_1 = 4 - x^2$ and $y_2 = |x|$ are in the second and third columns, respectively. Arrow keys permit scrolling through the table to find other x-values and corresponding y-values.

FIGURE 1.17 Creating a table for $y_1 = 4 - x^2$ and $y_2 = |x|$.

1.3 EXERCISE SET

Student Solutions Manual CD/Video PH Math/Tutor Center MathXL Tutorials on CD MathXL® MyMathLab Interactmath.com

Practice Exercises

In Exercises 1–10, plot the given point in a rectangular coordinate system.   **1–10.** See graphing answer section.

1. $(1, 4)$ **2.** $(2, 5)$

3. $(-2, 3)$ **4.** $(-1, 4)$

5. $(-3, -5)$ **6.** $(-4, -2)$

7. $(4, -1)$ **8.** $(3, -2)$

9. $(-4, 0)$ **10.** $(0, -3)$

Graph each equation in Exercises 11–26. Let $x = -3, -2, -1, 0,$ 1, 2, and 3.   **11–26.** See graphing answer section.

11. $y = x^2 - 4$ **12.** $y = x^2 - 9$

13. $y = x - 2$ **14.** $y = x + 2$

15. $y = 2x + 1$ **16.** $y = 2x - 4$

17. $y = -\dfrac{1}{2}x$ **18.** $y = -\dfrac{1}{2}x + 2$

19. $y = |x| + 1$ **20.** $y = |x| - 1$

21. $y = 2|x|$ **22.** $y = -2|x|$

23. $y = -x^2$ **24.** $y = -\dfrac{1}{2}x^2$

25. $y = x^3$ **26.** $y = x^3 - 1$

In Exercises 27–30, match the viewing rectangle with the correct figure. Then label the tick marks in the figure to illustrate this viewing rectangle.

27. $[-5, 5, 1]$ by $[-5, 5, 1]$ c

28. $[-10, 10, 2]$ by $[-4, 4, 2]$ d

29. $[-20, 80, 10]$ by $[-30, 70, 10]$ b

30. $[-40, 40, 20]$ by $[-1000, 1000, 100]$ a

a.   **b.**

c.   **d.**

The table of values was generated by a graphing utility with a TABLE feature. Use the table to solve Exercises 31–38.

31. Which equation corresponds to Y_2 in the table?   c
     a. $y_2 = x + 8$ **b.** $y_2 = x - 2$
     c. $y_2 = 2 - x$ **d.** $y_1 = 1 - 2x$

32. Which equation corresponds to Y_1 in the table?   b
     a. $y_1 = -3x$ **b.** $y_1 = x^2$
     c. $y_1 = -x^2$ **d.** $y_1 = 2 - x$

33. Does the graph of Y_2 pass through the origin?   no

34. Does the graph of Y_1 pass through the origin?   yes

35. At which point does the graph of Y_2 cross the x-axis? $(2, 0)$

36. At which point does the graph of Y_2 cross the y-axis? $(0, 2)$

37. At which points do the graphs of Y_1 and Y_2 intersect? $(-2, 4)$ and $(1, 1)$

38. For which values of x is $Y_1 = Y_2$? -2 and 1

Practice Plus

In Exercises 39–42, write each English sentence as an equation in two variables. Then graph the equation.   39–42. See graphing answer section.

39. The y-value is four more than twice the x-value.
$y = 2x + 4$

40. The y-value is the difference between four and twice the x-value. $y = 4 - 2x$

41. The y-value is three decreased by the square of the x-value.
$y = 3 - x^2$

42. The y-value is two more than the square of the x-value.
$y = x^2 + 2$

In Exercises 43–46, graph each equation.   43–46. See graphing answer section.

43. $y = 5$ (Let $x = -3, -2, -1, 0, 1, 2,$ and 3.)

44. $y = -1$ (Let $x = -3, -2, -1, 0, 1, 2,$ and 3.)

45. $y = \dfrac{1}{x}$ (Let $x = -2, -1, -\dfrac{1}{2}, -\dfrac{1}{3}, \dfrac{1}{3}, \dfrac{1}{2}, 1,$ and 2.)

46. $y = -\dfrac{1}{x}$ (Let $x = -2, -1, -\dfrac{1}{2}, -\dfrac{1}{3}, \dfrac{1}{3}, \dfrac{1}{2}, 1,$ and 2.)

Application Exercises

The graph shows the percent distribution of divorces in the United States by number of years of marriage. Use the graph to solve Exercises 47–50.

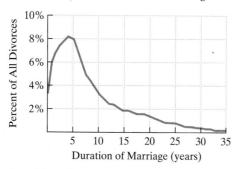

Percent Distribution of Divorces by Number of Years of Marriage

Source: Divorce Center

47. During which years of marriage is the chance of divorce increasing?   0 to 4 years

48. During which years of marriage is the chance of divorce decreasing?   4 to 35 years

49. During which year of marriage is the chance of divorce the highest? Estimate, to the nearest tenth of a percent, the percentage of divorces that occur during this year.   4th year; about 8.3%

50. During which year of marriage is the chance of divorce the lowest? Estimate, to the nearest tenth of a percent, the percentage of divorces that occur during this year.   35th year; about 0.2%

The graph shows the percentage of high school seniors in the United States who regularly used alcohol, cigarettes, and marijuana, from 1975 through 2001. Use the graph to solve Exercises 51–54. Round all estimates to the nearest percent.

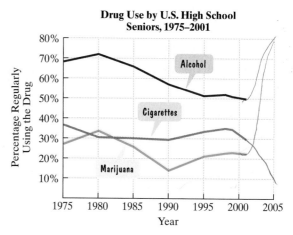

Drug Use by U.S. High School Seniors, 1975–2001

Source: Department of Health and Human Services

51. During which year did the greatest percentage of seniors regularly use alcohol? What percentage used the drug in that year?   1980; about 72%

52. During which year did the least percentage of seniors regularly use marijuana? What percentage used the drug in that year?   1990; about 14%

53. Find the difference between the percentage of seniors who regularly used marijuana and those who regularly used cigarettes in 1980.   about 3%

54. Find the difference between the percentage of seniors who regularly used alcohol and those who regularly used cigarettes in 2001.   about 20%

In Exercises 55–58, match the story with the correct figure. The figures are labeled (a), (b), (c), and (d).

c.

d.

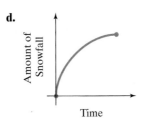

A portion of the rectangular coordinate system can be used to show bands of healthy weights for different heights. The figure shows such a weight chart for men between the ages of 20 and 65. Use the information shown to solve Exercises 61–64. Round all answers to the nearest pound.

55. As the blizzard got worse, the snow fell harder and harder.   a

56. The snow fell more and more softly.   d

57. It snowed hard, but then it stopped. After a short time, the snow started falling softly.   b

58. It snowed softly, and then it stopped. After a short time, the snow started falling hard.   c

In Exercises 59–60, select the graph that best illustrates each story.

59. An airplane flew from Miami to San Francisco.   b

a.

b.

Weight Chart for Men Ages 20 Through 65

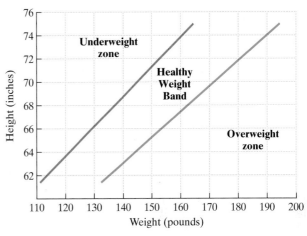

Source: United States Department of Agriculture

61. What is the maximum healthy weight for a man who is 6 feet (72 inches) tall?   about 181 pounds

62. What is the minimum healthy weight for a man who is 6 feet (72 inches) tall?   about 153 pounds

63. A man who is 70 inches tall needs to gain ten pounds. Approximately how much does this man weigh?   about 135 pounds

64. A man who is 70 inches tall needs to lose ten pounds. Approximately how much does this man weigh?   about 182 pounds

c.

d.

60. At noon, you begin to breathe in.   a

a.

b.

Writing in Mathematics

65. What is the rectangular coordinate system?
65–72. Answers will vary.

66. Explain how to plot a point in the rectangular coordinate system. Give an example with your explanation.

67. Explain why $(5, -2)$ and $(-2, 5)$ do not represent the same point.

c.

d.

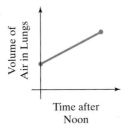

68. Explain how to graph an equation in the rectangular coordinate system.

69. What does a $[-20, 2, 1]$ by $[-4, 5, 0.5]$ viewing rectangle mean?

In Exercises 70–71, write a story, or description, to match each graph.

70.   **71.**

72. The graph for Exercises 47–50 shows that the chance of divorce is greatest during the fourth year of marriage. What explanation can you offer for this trend?

Technology Exercises

73. Use a graphing utility to verify each of your hand-drawn graphs in Exercises 11–26. Experiment with the viewing rectangle to make the graph displayed by the graphing utility resemble your hand-drawn graph as much as possible.   Answers will vary.

74. The stated intent of the 1994 "don't ask, don't tell" policy was to reduce the number of discharges of gay men and lesbians from the military. The formula

$$y = 45.48x^2 - 334.35x + 1237.9$$

models the number of gay service members, y, discharged from the military for homosexuality x years after 1990. Graph the equation in a $[0, 10, 1]$ by $[0, 2200, 100]$ viewing rectangle. Then describe something about the relationship between x and y that is revealed by looking at the graph that is not obvious from the equation. What does the graph reveal about the success or lack of success of "don't ask, don't tell"?   See graphing answer section. The number of gay service members being discharged was decreasing before the policy was implemented, but began to increase after its implementation. Answers will vary.

A graph of an equation is a complete graph if it shows all of the important features of the graph. Use a graphing utility to graph the equations in Exercises 75–78 in each of the given viewing rectangles. Then choose which viewing rectangle gives a complete graph.

75. $y = x^2 + 10$ c
 a. $[-5, 5, 1]$ by $[-5, 5, 1]$
 b. $[-10, 10, 1]$ by $[-10, 10, 1]$
 c. $[-10, 10, 1]$ by $[-50, 50, 1]$

76. $y = 0.1x^4 - x^3 + 2x^2$ c
 a. $[-5, 5, 1]$ by $[-8, 2, 1]$
 b. $[-10, 10, 1]$ by $[-10, 10, 1]$
 c. $[-8, 16, 1]$ by $[-16, 8, 1]$

77. $y = \sqrt{x} + 18$ c
 a. $[-10, 10, 1]$ by $[-10, 10, 1]$
 b. $[-50, 50, 10]$ by $[-10, 10, 1]$
 c. $[-10, 10, 1]$ by $[-50, 50, 10]$

78. $y = x^3 - 30x + 20$ c
 a. $[-10, 10, 1]$ by $[-10, 10, 1]$
 b. $[-10, 10, 1]$ by $[-50, 50, 10]$
 c. $[-10, 10, 1]$ by $[-50, 100, 10]$

Critical Thinking Exercises

79. Which one of the following is true?   d
 a. If the product of a point's coordinates is positive, the point must be in quadrant I.
 b. If a point is on the x-axis, it is neither up nor down, so $x = 0$.
 c. A graphing utility displays numbers along the x- and y-axes.
 d. None of the above is true.

The graph shows the height, y, in feet, of a ball dropped from the top of the Empire State Building at various times during its fall to the ground. Use the graph to solve Exercises 80–81.

80. What is a reasonable estimate of the height of the Empire State Building?   1250 feet

81. What is a reasonable estimate, to the nearest tenth of a second, of how long it takes for the ball to hit the ground?   8.8 seconds

Review Exercises

82. Find the absolute value: $|-14.3|$. (Section 1.2, Example 1)
 14.3

83. Simplify: $[12 - (13 - 17)] - [9 - (6 - 10)]$.
 (Section 1.2, Examples 7 and 8)   3

84. Simplify: $6x - 5(4x + 3) - 10$. (Section 1.2, Example 13)
 $-14x - 25$

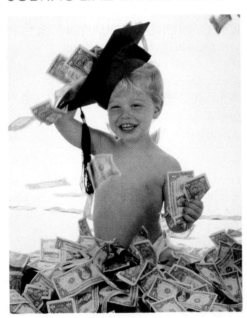

<table>
<tr><td>

SECTION 1.4

Objectives

1 Solve linear equations.

2 Recognize identities, conditional equations, and inconsistent equations.

3 Solve applied problems using mathematical models.

</td></tr>
</table>

SOLVING LINEAR EQUATIONS

The bar graph in Figure 1.18 confirms what you already know: The cost of a college education is skyrocketing. Tuition and fees in 2002–2003 were nearly three times what they were as recently as 1985–1986 and nearly ten times what they were in 1971–1972.

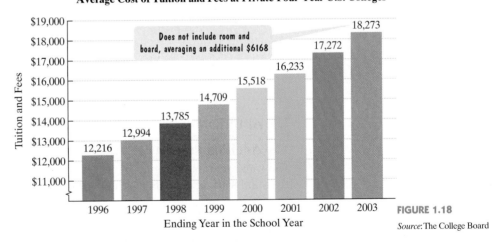

Average Cost of Tuition and Fees at Private Four-Year U.S. Colleges

Does not include room and board, averaging an additional $6168

FIGURE 1.18

Source: The College Board

The data in the bar graph can be modeled by the formula

$$T = 857x + 12{,}127,$$

where T represents the average cost of tuition and fees at private U.S. colleges for the school year ending x years after 1996. So, when will tuition and fees average $24,125? Substitute 24,125 for T in the formula:

$$24{,}125 = 857x + 12{,}127.$$

Notice that the exponent on the variable in this equation is 1. In this section, we will study how to determine the value of x in such equations. With this skill, you will make predictions about future costs of private and public colleges.

1 Solve linear equations.

Solving Linear Equations in One Variable   We begin by stating the definition of a linear equation in one variable.

> **DEFINITION OF A LINEAR EQUATION**   A **linear equation in one variable** x is an equation that can be written in the form
>
> $$ax + b = 0,$$
>
> where a and b are real numbers, and $a \neq 0$ (a is not equal to 0).

An example of a linear equation in one variable is

$$4x + 12 = 0.$$

Solving an equation in x involves determining all values of x that result in a true statement when substituted into the equation. Such values are **solutions**, **roots**, or **zeros** of the equation. For example, substitute -3 for x in $4x + 12 = 0$. We obtain

$$4(-3) + 12 = 0, \quad \text{or} \quad -12 + 12 = 0.$$

This simplifies to the true statement $0 = 0$. Thus, -3 is a solution of the equation $4x + 12 = 0$. We also say that -3 **satisfies** the equation $4x + 12 = 0$, because when we substitute -3 for x, a true statement results. The set of all such solutions is called the equation's **solution set**. For example, the solution set of the equation $4x + 12 = 0$ is $\{-3\}$.

Two or more equations that have the same solution set are called **equivalent equations**. For example, the equations

$$4x + 12 = 0 \quad \text{and} \quad 4x = -12 \quad \text{and} \quad x = -3$$

are equivalent equations because the solution set for each is $\{-3\}$. To solve a linear equation in x, we transform the equation into an equivalent equation one or more times. Our final equivalent equation should be of the form

$$x = \text{ a number.}$$

The solution set of this equation is the set consisting of the number.

To generate equivalent equations, we will use the following properties:

> **THE ADDITION AND MULTIPLICATION PROPERTIES OF EQUALITY**
>
> **The Addition Property of Equality**
>
> The same real number or algebraic expression may be added to both sides of an equation without changing the equation's solution set.
>
> $$a = b \text{ and } a + c = b + c \text{ are equivalent equations.}$$
>
> **The Multiplication Property of Equality**
>
> The same nonzero real number may multiply both sides of an equation without changing the equation's solution set.
>
> $$a = b \text{ and } ac = bc \text{ are equivalent equations as long as } c \neq 0.$$

Because subtraction is defined in terms of addition, the addition property also lets us subtract the same number from both sides of an equation without changing the equation's solution set. Similarly, because division is defined in terms of multiplication, the multiplication property of equality can be used to divide both sides of an equation by the same nonzero number to obtain an equivalent equation.

Table 1.4 illustrates how these properties are used to isolate x to obtain an equation of the form $x = $ a number.

STUDY TIP

Your final equivalent equation should not be of the form

$$-x = \text{ a number.}$$

We're not finished. A negative sign should not precede the variable.

Isolate x by multiplying or dividing both sides of this equation by -1.

Table 1.4 **Using Properties of Equality to Solve Linear Equations**

Equation	How to Isolate x	Solving the Equation	The Equation's Solution Set
$x - 3 = 8$	Add 3 to both sides.	$x - 3 + 3 = 8 + 3$ $x = 11$	$\{11\}$
$x + 7 = -15$	Subtract 7 from both sides.	$x + 7 - 7 = -15 - 7$ $x = -22$	$\{-22\}$
$6x = 30$	Divide both sides by 6 (or multiply both sides by $\frac{1}{6}$).	$\dfrac{6x}{6} = \dfrac{30}{6}$ $x = 5$	$\{5\}$
$\dfrac{x}{5} = 9$	Multiply both sides by 5.	$5 \cdot \dfrac{x}{5} = 5 \cdot 9$ $x = 45$	$\{45\}$

These equations are solved using the Addition Property of Equality.

These equations are solved using the Multiplication Property of Equality.

Sadmep

EXAMPLE 1 Solving a Linear Equation

Solve and check: $2x + 3 = 17$.

SOLUTION Our goal is to obtain an equivalent equation with x isolated on one side and a number on the other side.

$$2x + 3 = 17 \quad \text{This is the given equation.}$$

$$2x + 3 - 3 = 17 - 3 \quad \text{Subtract 3 from both sides.}$$

$$2x = 14 \quad \text{Simplify.}$$

$$\frac{2x}{2} = \frac{14}{2} \quad \text{Divide both sides by 2.}$$

$$x = 7 \quad \text{Simplify.}$$

Now we check the proposed solution, 7, by replacing x with 7 in the original equation.

$$2x + 3 = 17 \quad \text{This is the original equation.}$$

$$2 \cdot 7 + 3 \overset{?}{=} 17 \quad \text{Substitute 7 for x. The question mark indicates that we do not yet know if the two sides are equal.}$$

$$14 + 3 \overset{?}{=} 17 \quad \text{Multiply: } 2 \cdot 7 = 14.$$

$$17 = 17 \quad \text{Add: } 14 + 3 = 17.$$

This statement is true.

Because the check results in a true statement, we conclude that the solution of the given equation is 7, or the solution set is $\{7\}$. ■

CHECK POINT 1 Solve and check: $4x + 5 = 29$.   6 or $\{6\}$

Here is a step-by-step procedure for solving a linear equation in one variable. Not all of these steps are necessary to solve every equation.

SOLVING A LINEAR EQUATION

1. Simplify the algebraic expression on each side.

2. Collect all the variable terms on one side and all the numbers, or constant terms, on the other side.

3. Isolate the variable and solve.

4. Check the proposed solution in the original equation.

EXAMPLE 2 Solving a Linear Equation

Solve and check: $2x - 7 + x = 3x + 1 + 2x$.

SOLUTION

Step 1. **Simplify the algebraic expression on each side.**

$$2x - 7 + x = 3x + 1 + 2x \qquad \text{This is the given equation.}$$

$$3x - 7 = 5x + 1 \qquad \text{Combine like terms:}$$
$$\text{2x + x = 3x and 3x + 2x = 5x.}$$

DISCOVER FOR YOURSELF

Solve the equation in Example 2 by collecting terms with the variable on the right and constant terms on the left. What do you observe?

Step 2. **Collect variable terms on one side and constant terms on the other side.** We will collect variable terms on the left by subtracting $5x$ from both sides. We will collect the numbers on the right by adding 7 to both sides.

$$3x - 5x - 7 = 5x - 5x + 1 \qquad \text{Subtract 5x from both sides.}$$

$$-2x - 7 = 1 \qquad \text{Simplify.}$$

$$-2x - 7 + 7 = 1 + 7 \qquad \text{Add 7 to both sides.}$$

$$-2x = 8 \qquad \text{Simplify.}$$

Step 3. **Isolate the variable and solve.** We isolate x by dividing both sides by -2.

$$\frac{-2x}{-2} = \frac{8}{-2} \qquad \text{Divide both sides by } -2.$$

$$x = -4 \qquad \text{Simplify.}$$

Step 4. **Check the proposed solution in the original equation.** Substitute -4 for x in the original equation.

$$2x - 7 + x = 3x + 1 + 2x \qquad \text{This is the original equation.}$$

$$2(-4) - 7 + (-4) \stackrel{?}{=} 3(-4) + 1 + 2(-4) \qquad \text{Substitute } -4 \text{ for x.}$$

$$-8 - 7 + (-4) \stackrel{?}{=} -12 + 1 + (-8) \qquad \text{Multiply: } 2(-4) = -8,$$
$$3(-4) = -12, \text{ and } 2(-4) = -8.$$

$$-15 + (-4) \stackrel{?}{=} -11 + (-8) \qquad \text{Add or subtract from left to right:}$$
$$-8 - 7 = -15 \text{ and } -12 + 1 = -11.$$

$$-19 = -19 \qquad \text{Add.}$$

The true statement $-19 = -19$ verifies that -4 is the solution, or the solution set is $\{-4\}$. ∎

 CHECK POINT 2 Solve and check: $2x - 12 + x = 6x - 4 + 5x$. -1 or $\{-1\}$

EXAMPLE 3 Solving a Linear Equation

Solve and check: $4(2x + 1) - 29 = 3(2x - 5)$.

SOLUTION

Step 1.   **Simplify the algebraic expression on each side.**

$$4(2x + 1) - 29 = 3(2x - 5)$$ This is the given equation.

$$8x + 4 - 29 = 6x - 15$$ Use the distributive property.

$$8x - 25 = 6x - 15$$ Simplify.

Step 2.   **Collect variable terms on one side and constant terms on the other side.** We will collect the variable terms on the left by subtracting 6x from both sides. We will collect the numbers on the right by adding 25 to both sides.

$$8x - 6x - 25 = 6x - 6x - 15$$ Subtract 6x from both sides.

$$2x - 25 = -15$$ Simplify.

$$2x - 25 + 25 = -15 + 25$$ Add 25 to both sides.

$$2x = 10$$ Simplify.

Step 3.   **Isolate the variable and solve.** We isolate x by dividing both sides by 2.

$$\frac{2x}{2} = \frac{10}{2}$$ Divide both sides by 2.

$$x = 5$$ Simplify.

Step 4.   **Check the proposed solution in the original equation.** Substitute 5 for x in the original equation.

$$4(2x + 1) - 29 = 3(2x - 5)$$ This is the original equation.

$$4(2 \cdot 5 + 1) - 29 \overset{?}{=} 3(2 \cdot 5 - 5)$$ Substitute 5 for x.

$$4(11) - 29 \overset{?}{=} 3(5)$$ Simplify inside parentheses: $2 \cdot 5 + 1 = 10 + 1 = 11$ and $2 \cdot 5 - 5 = 10 - 5 = 5$.

$$44 - 29 \overset{?}{=} 15$$ Multiply: $4(11) = 44$ and $3(5) = 15$.

$$15 = 15$$ Subtract.

The true statement $15 = 15$ verifies that 5 is the solution, or the solution set is $\{5\}$. ∎

CHECK POINT **3** Solve and check: $2(x - 3) - 17 = 13 - 3(x + 2)$.   6 or $\{6\}$

USING TECHNOLOGY

You can use a graphing utility to check the solution of a linear equation. Enter each side of the equation separately under y_1 and y_2. Then use the table or the graphs to locate the x-value for which the y-values are the same. This x-value is the solution.

Let's verify our work in Example 3 and show that 5 is the solution of

$$4(2x + 1) - 29 = 3(2x - 5).$$

Enter $y_1 = 4(2x + 1) - 29$ in the $\boxed{y=}$ screen.

Enter $y_2 = 3(2x - 5)$ in the $\boxed{y=}$ screen.

Numeric Check

Display a table for y_1 and y_2.

Graphic Check

Display graphs for y_1 and y_2 and use the intersection feature. The solution is the x-coordinate of the intersection point.

When $x = 5$, y_1 and y_2 have the same value, namely, 15. This verifies 5 is the solution of $4(2x + 1) - 29 = 3(2x - 5)$.

$y_1 = 4(2x + 1) - 29 \qquad y_2 = 3(2x - 5)$

X	Y1	Y2
1	-17	-9
2	-9	-3
3	-1	3
4	7	9
5	15	15
6	23	21
7	31	27

X=5

Graphs intersect at (5, 15). When $x = 5$, y_1 and y_2 have the same value, namely, 15. This graphically verifies 5 is the solution of $4(2x + 1) - 29 = 3(2x - 5)$.

$y_2 = 3(2x - 5)$

$y_1 = 4(2x + 1) - 29$

Intersection
X=5 — Y=15

$[0, 7, 1]$ by $[-1, 20, 1]$

Choose a large enough viewing rectangle so that you can see the intersection point.

Linear Equations with Fractions Equations are easier to solve when they do not contain fractions. How do we remove fractions from an equation? We begin by multiplying both sides of the equation by the least common denominator (LCD) of all fractions in the equation. The least common denominator is the smallest number that all denominators will divide into. Multiplying every term on both sides of the equation by the least common denominator will eliminate the fractions in the equation. Example 4 shows how we "clear an equation of fractions."

EXAMPLE 4 Solving a Linear Equation Involving Fractions

Solve: $\dfrac{2x + 5}{5} + \dfrac{x - 7}{2} = \dfrac{3x + 1}{2}$.

SOLUTION The denominators are 5, 2, and 2. The smallest number that is divisible by 5, 2, and 2 is 10. We begin by multiplying both sides of the equation by 10, the least common denominator.

$$\dfrac{2x + 5}{5} + \dfrac{x - 7}{2} = \dfrac{3x + 1}{2}$$

This is the given equation.

$$10\left(\dfrac{2x + 5}{5} + \dfrac{x - 7}{2}\right) = 10\left(\dfrac{3x + 1}{2}\right)$$

Multiply both sides by 10.

$$\dfrac{10}{1} \cdot \left(\dfrac{2x + 5}{5}\right) + \dfrac{10}{1} \cdot \left(\dfrac{x - 7}{2}\right) = \dfrac{10}{1} \cdot \left(\dfrac{3x + 1}{2}\right)$$

Use the distributive property and multiply each term by 10.

$$\frac{\overset{2}{\cancel{10}}}{1} \cdot \left(\frac{2x + 5}{\underset{1}{\cancel{5}}}\right) + \frac{\overset{5}{\cancel{10}}}{1} \cdot \left(\frac{x - 7}{\underset{1}{\cancel{2}}}\right) = \frac{\overset{5}{\cancel{10}}}{1} \cdot \left(\frac{3x + 1}{\underset{1}{\cancel{2}}}\right)$$ Divide out common factors in each multiplication.

$$2(2x + 5) + 5(x - 7) = 5(3x + 1)$$ The fractions are now cleared.

At this point, we have an equation similar to those we have previously solved. Use the distributive property to begin simplifying each side.

$$4x + 10 + 5x - 35 = 15x + 5$$ Use the distributive property.

$$9x - 25 = 15x + 5$$ Combine like terms on the left side: $4x + 5x = 9x$ and $10 - 35 = -25$.

For variety, let's collect variable terms on the right and constant terms on the left.

$$9x - 9x - 25 = 15x - 9x + 5$$ Subtract $9x$ from both sides.

$$-25 = 6x + 5$$ Simplify.

$$-25 - 5 = 6x + 5 - 5$$ Subtract 5 from both sides.

$$-30 = 6x$$ Simplify.

Isolate x on the right side by dividing both sides by 6.

$$\frac{-30}{6} = \frac{6x}{6}$$ Divide both sides by 6.

$$-5 = x$$ Simplify.

Check the proposed solution in the original equation. Substitute -5 for x in the original equation. You should obtain $-7 = -7$. This true statement verifies that -5 is the solution, or the solution set is $\{-5\}$.   ■

✔ **CHECK POINT 4** Solve: $\dfrac{x + 5}{7} + \dfrac{x - 3}{4} = \dfrac{5}{14}$.   1 or {1}

Types of Equations   We tend to place things in categories, allowing us to order and structure the world. For example, you can categorize yourself by your age group, your ethnicity, your academic major, or your gender. Equations can be placed into categories that depend on their solution sets.

An equation that is true for all real numbers for which both sides are defined is called an **identity**. An example of an identity is

$$x + 3 = x + 2 + 1.$$

Every number plus 3 is equal to that number plus 2 plus 1. Therefore, the solution set to this equation is the set of all real numbers. This set is written either as

$$\{x \mid x \text{ is a real number}\} \quad \text{or} \quad \mathbb{R}.$$

An equation that is not an identity, but that is true for at least one real number, is called a **conditional equation**. The equation $2x + 3 = 17$ is an example of a conditional equation. The equation is not an identity and is true only if x is 7.

An **inconsistent equation**, also called a **contradiction**, is an equation that is not true for even one real number. An example of an inconsistent equation is

$$x = x + 7.$$

---

2   Recognize identities, conditional equations, and inconsistent equations.

STUDY TIP

If you are concerned by the vocabulary of equation types, keep in mind that there are three possible situations. We can state these situations informally as follows:

1. $x = $ a real number

  conditional equation

2. $x = $ all real numbers

  identity

3. $x = $ no real numbers.

  inconsistent equation

There is no number that is equal to itself plus 7. The equation $x = x + 7$ has no solution. Its solution set is written either as

$$\{ \quad \} \quad \text{or} \quad \varnothing.$$

These symbols stand for the empty set, a set with no elements.

If you attempt to solve an identity or an inconsistent equation, you will eliminate the variable. A true statement such as $6 = 6$ or a false statement such as $2 = 3$ will be the result. **If a true statement results, the equation is an identity that is true for all real numbers. If a false statement results, the equation is an inconsistent equation with no solution.**

EXAMPLE 5 Categorizing an Equation

Solve and determine whether the equation

$$2(x + 1) = 2x + 3$$

is an identity, a conditional equation, or an inconsistent equation.

SOLUTION Begin by applying the distributive property on the left side. We obtain

$$2x + 2 = 2x + 3.$$

Does something look strange? Can doubling a number and increasing the product by 2 give the same result as doubling the same number and increasing the product by 3? No. Let's continue solving the equation by subtracting $2x$ from both sides.

$$2x - 2x + 2 = 2x - 2x + 3$$

Keep reading. 2 = 3 is not the solution.

$$2 = 3$$

The original equation is equivalent to the statement $2 = 3$, which is false for every value of x. The equation is inconsistent and has no solution. You can express this by writing "no solution" or using one of the symbols for the empty set, $\{ \ \}$ or $\varnothing$. ∎

USING TECHNOLOGY

The graphs of $y_1 = 2(x + 1)$ and $y_2 = 2x + 3$ appear to be parallel lines with no intersection point. This verifies that the equation

$$2(x + 1) = 2x + 3$$

has no solution and is inconsistent.

$y_2 = 2x + 3$

$y_1 = 2(x + 1)$

$[-5, 2, 1]$ by $[-5, 5, 1]$

 CHECK POINT 5 Solve and determine whether the equation

$$4x - 7 = 4(x - 1) + 3$$

is an identity, a conditional equation, or an inconsistent equation.

no solution; inconsistent equation

EXAMPLE 6	Categorizing an Equation

Solve and determine whether the equation

$$4x + 6 = 6(x + 1) - 2x$$

is an identity, a conditional equation, or an inconsistent equation.

SOLUTION

$$4x + 6 = 6(x + 1) - 2x \qquad \text{This is the given equation.}$$

$$4x + 6 = 6x + 6 - 2x \qquad \text{Apply the distributive property on the right side.}$$

$$4x + 6 = 4x + 6 \qquad \text{Combine like terms on the right side: } 6x - 2x = 4x.$$

Can you see that the equation $4x + 6 = 4x + 6$ is true for every value of x? Let's continue solving the equation by subtracting $4x$ from both sides.

$$4x - 4x + 6 = 4x - 4x + 6$$

Keep reading. $6 = 6$ is not the solution.

$$6 = 6$$

The original equation is equivalent to the statement $6 = 6$, which is true for every value of x. The equation is an identity, and all real numbers are solutions. You can express this by writing "all real numbers" or using one of the following notations:

$$\{x \mid x \text{ is a real number}\} \quad \text{or} \quad \mathbb{R}.$$

USING TECHNOLOGY

A graphing utility's ⌐TABLE⌐ feature can be used to numerically verify that the solution set of

$$4x + 6 = 6(x + 1) - 2x$$

is the set of all real numbers.

No matter how far up or down you scroll, y_1 and y_2 have the same corresponding value for each x-value. The expressions $4x + 6$ and $6(x + 1) - 2x$ are always equal.

 CHECK POINT 6 Solve and determine whether the equation

$$7x + 9 = 9(x + 1) - 2x$$

is an identity, a conditional equation, or an inconsistent equation.

all real numbers; identity

3 Solve applied problems using mathematical models.

Applications Our next example shows how the procedure for solving linear equations can be used to find the value of a variable in a mathematical model.

EXAMPLE 7 The Cost of a College Education

The formula

$$T = 857x + 12{,}127$$

models the average cost of tuition and fees, T, at private U.S. colleges for the school year ending x years after 1996. When will tuition and fees average $24,125?

SOLUTION We are interested in when tuition and fees will average $24,125, so substitute 24,125 for T in the formula and solve for x, the number of years after 1996.

$T = 857x + 12{,}127$	This is the given formula.
$24{,}125 = 857x + 12{,}127$	Replace T with 24,125.
$24{,}125 - 12{,}127 = 857x + 12{,}127 - 12{,}127$	Subtract 12,127 from both sides.
$11{,}998 = 857x$	Simplify.
$\dfrac{11{,}998}{857} = \dfrac{857x}{857}$	Divide both sides by 857.
$14 = x$	Simplify.

The formula indicates that for the school year ending 14 years after 1996, or in 2010, tuition and fees at private U.S. colleges will average $24,125. ■

 CHECK POINT 7 Use the formula in Example 7 to find when tuition and fees will average $21,554.   the school year ending in 2007

1.4 EXERCISE SET

Student Solutions Manual   CD/Video   PH Math/Tutor Center   MathXL Tutorials on CD   MathXL®   MyMathLab   Interactmath.com

Practice Exercises

In Exercises 1–24, solve and check each linear equation.

1. $5x + 3 = 18$ 3
2. $3x + 8 = 50$ 14
3. $6x - 3 = 63$ 11
4. $5x - 8 = 72$ 16
5. $14 - 5x = -41$ 11
6. $25 - 6x = -83$ 18
7. $11x - (6x - 5) = 40$ 7
8. $5x - (2x - 8) = 35$ 9
9. $2x - 7 = 6 + x$ 13
10. $3x + 5 = 2x + 13$ 8
11. $7x + 4 = x + 16$ 2
12. $8x + 1 = x + 43$ 6
13. $8y - 3 = 11y + 9$ −4
14. $5y - 2 = 9y + 2$ −1
15. $3(x - 2) + 7 = 2(x + 5)$ 9
16. $2(x - 1) + 3 = x - 3(x + 1)$ −1
17. $3(x - 4) - 4(x - 3) = x + 3 - (x - 2)$ −5
18. $2 - (7x + 5) = 13 - 3x$ −4
19. $16 = 3(x - 1) - (x - 7)$ 6

20. $5x - (2x + 2) = x + (3x - 5)$ 3
21. $7(x + 1) = 4[x - (3 - x)]$ 19
22. $2[3x - (4x - 6)] = 5(x - 6)$ 6
23. $\dfrac{1}{2}(4z + 8) - 16 = -\dfrac{2}{3}(9z - 12)$ $\dfrac{5}{2}$
24. $\dfrac{3}{4}(24 - 8z) - 16 = -\dfrac{2}{3}(6z - 9)$ −2

In Exercises 25–38, solve each equation.

25. $\dfrac{x}{3} = \dfrac{x}{2} - 2$ 12
26. $\dfrac{x}{5} = \dfrac{x}{6} + 1$ 30
27. $20 - \dfrac{x}{3} = \dfrac{x}{2}$ 24
28. $\dfrac{x}{5} - \dfrac{1}{2} = \dfrac{x}{6}$ 15
29. $\dfrac{3x}{5} = \dfrac{2x}{3} + 1$ −15
30. $\dfrac{x}{2} = \dfrac{3x}{4} + 5$ −20
31. $\dfrac{3x}{5} - x = \dfrac{x}{10} - \dfrac{5}{2}$ 5
32. $2x - \dfrac{2x}{7} = \dfrac{x}{2} + \dfrac{17}{2}$ 7

33. $\dfrac{x + 3}{6} = \dfrac{2}{3} + \dfrac{x - 5}{4}$ 13   **34.** $\dfrac{x + 1}{4} = \dfrac{1}{6} + \dfrac{2 - x}{3}$ 1

35. $\dfrac{x}{4} = 2 + \dfrac{x - 3}{3}$ -12 **36.** $5 + \dfrac{x - 2}{3} = \dfrac{x + 3}{8}$ -19

37. $\dfrac{x + 1}{3} = 5 - \dfrac{x + 2}{7}$ $\dfrac{46}{5}$ **38.** $\dfrac{3x}{5} - \dfrac{x - 3}{2} = \dfrac{x + 2}{3}$ $\dfrac{25}{7}$

In Exercises 39–50, solve each equation. Then state whether the equation is an identity, a conditional equation, or an inconsistent equation.

39. $5x + 9 = 9(x + 1) - 4x$ all real numbers; identity

40. $4x + 7 = 7(x + 1) - 3x$ all real numbers; identity

41. $3(y + 2) = 7 + 3y$ no solution; inconsistent equation

42. $4(y + 5) = 21 + 4y$ no solution; inconsistent equation

43. $10x + 3 = 8x + 3$ 0; conditional equation

44. $5x + 7 = 2x + 7$ 0; conditional equation

45. $\dfrac{1}{2}(6z + 20) - 8 = 2(z - 4)$ -10; conditional equation

46. $\dfrac{1}{3}(6z + 12) = \dfrac{1}{5}(20z + 30) - 8$ 3; conditional equation

47. $-4x - 3(2 - 2x) = 7 + 2x$ no solution; inconsistent equation

48. $3x - 3(2 - x) = 6(x - 1)$ all real numbers; identity

49. $y + 3(4y + 2) = 6(y + 1) + 5y$ 0; conditional equation

50. $9y - 3(6 - 5y) = y - 2(3y + 9)$ 0; conditional equation

In Exercises 51–54, use the $\boxed{Y=}$ *screen to write the equation being solved. Then use the table to solve the equation.*

51.

$3(x - 4) = 3(2 - 2x); 2$

52.

$3(2x - 5) = 5x + 2; 17$

53.

$-3(x - 3) = 5(2 - x); 0.5$

54.

$2x - 5 = 4(3x + 1) - 2; -0.7$

Practice Plus

55. Evaluate $x^2 - x$ for the value of x satisfying $4(x - 2) + 2 = 4x - 2(2 - x)$.   2

56. Evaluate $x^2 - x$ for the value of x satisfying $2(x - 6) = 3x + 2(2x - 1)$.   6

57. Evaluate $x^2 - (xy - y)$ for x satisfying $\dfrac{3(x + 3)}{5} = 2x + 6$ and y satisfying $-2y - 10 = 5y + 18$. -7

58. Evaluate $x^2 - (xy - y)$ for x satisfying $\dfrac{13x - 6}{4} = 5x + 2$ and y satisfying $5 - y = 7(y + 4) + 1$. -5

In Exercises 59–66, solve each equation.

59. $[(3 + 6)^2 \div 3] \cdot 4 = -54x$ -2

60. $2^3 - [4(5 - 3)^3] = -8x$ 3

61. $5 - 12x = 8 - 7x - [6 \div 3(2 + 5^3) + 5x]$ $\varnothing$; no solution

62. $2(5x + 58) = 10x + 4(21 \div 3.5 - 11)$ $\varnothing$; no solution

63. $0.7x + 0.4(20) = 0.5(x + 20)$ 10

64. $0.5(x + 2) = 0.1 + 3(0.1x + 0.3)$ 0

65. $4x + 13 - \{2x - [4(x - 3) - 5]\} = 2(x - 6)$ -2

66. $-2\{7 - [4 - 2(1 - x) + 3]\} = 10 - [4x - 2(x - 3)]$ $\dfrac{4}{3}$

Application Exercises

The bar graph on the next page shows the average cost of tuition and fees at public four-year colleges in the United States. The data can be modeled by the formula

$$T = 165x + 2771,$$

where T represents the average cost of tuition and fees for the school year ending x years after 1996. Use the formula to solve Exercises 67–68.

Average Cost of Tuition and Fees at Public Four-Year U.S. Colleges

Does not include room and board, averaging $4931

Source: The College Board

(Refer to the formula on page 49 when solving Exercises 67–68.)

67. When will tuition and fees at public U.S. colleges average $4421?   the school year ending 2006

68. When will tuition and fees at public U.S. colleges average $4751?   the school year ending 2008

Medical researchers have found that the desirable heart rate, R, in beats per minute, for beneficial exercise is approximated by the formulas

$$R = 143 - 0.65A \quad \text{for women}$$
$$R = 165 - 0.75A \quad \text{for men,}$$

where A is the person's age. Use these formulas to solve Exercises 69–70.

69. If the desirable heart rate for a woman is 117 beats per minute, how old is she? How is the solution shown on the line graph?
40 years old; approximately (40, 117) on female graph

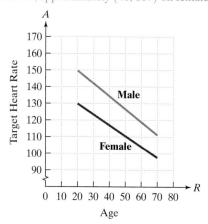

70. If the desirable heart rate for a man is 147 beats per minute, how old is he? How is the solution shown on the line graph?
24 years old; approximately (24, 147) on male graph

Writing in Mathematics

71. What is a linear equation in one variable? Give an example of this type of equation.   71–81. Answers will vary.

72. What does it mean to solve an equation?

73. How do you determine if a number is a solution of an equation?

74. What are equivalent equations? Give an example.

75. What is the addition property of equality?

76. What is the multiplication property of equality?

77. Explain how to clear an equation of fractions.

78. What is an identity? Give an example.

79. What is a conditional equation? Give an example.

80. What is an inconsistent equation? Give an example.

81. Despite low rates of inflation, the cost of a college education continues to skyrocket. This is a departure from the trend during the 1970s: In constant dollars (which negate the effect of inflation), the cost of college actually decreased several times. What explanations can you offer for the increasing cost of a college education?

Technology Exercises

In Exercises 82–85, use your graphing utility to enter each side of the equation separately under y_1 *and* y_2. *Then use the utility's* TABLE *or* GRAPH *feature to solve the equation.*   82–85. See graphing answer section.

82. $5x + 2(x - 1) = 3x + 10$ 3

83. $2x + 3(x - 4) = 4x - 7$ 5

84. $3(2x + 11) = 3(5 + x)$ -6

85. $\dfrac{2x - 1}{3} - \dfrac{x - 5}{6} = \dfrac{x - 3}{4}$ -5

Critical Thinking Exercises

86. Which one of the following is true?   c

   a. The equation $-7x = x$ has no solution.

   b. The equations $\dfrac{x}{x - 4} = \dfrac{4}{x - 4}$ and $x = 4$ are equivalent.

   c. The equations $3y - 1 = 11$ and $3y - 7 = 5$ are equivalent.

   d. If a and b are any real numbers, then $ax + b = 0$ always has only one number in its solution set.

87. Solve for x: $ax + b = c$. $x = \dfrac{c - b}{a}$

88. Write three equations that are equivalent to $x = 5$.   Answers will vary.

89. If x represents a number, write an English sentence about the number that results in an inconsistent equation.   Answers will vary.

90. Find b such that $\dfrac{7x + 4}{b} + 13 = x$ will have a solution set given by $\{-6\}$.   2

Review Exercises

In Exercises 91–93, perform the indicated operations.

91. $-\dfrac{1}{5} - \left(-\dfrac{1}{2}\right)$ (Section 1.2, Example 4) $\dfrac{3}{10}$

92. $4(-3)(-1)(-5)$ (Section 1.2, Examples in Table 1.2) -60

93. Graph $y = x^2 - 4$. Let $x = -3, -2, -1, 0, 1, 2,$ and 3. (Section 1.3, Example 2)   See graphing answer section.

✔ MID-CHAPTER CHECK POINT

CHAPTER 1

What You Know: We reviewed a number of topics from introductory algebra, including the real numbers and their representations on number lines. We performed operations with real numbers and applied the order-of-operations agreement to expressions containing more than one operation. We used commutative, associative, and distributive properties to simplify algebraic expressions. We used the rectangular coordinate system to represent ordered pairs of real numbers and graph equations in two variables. Finally, we solved linear equations, including equations with fractions. We saw that some linear equations have no solution, whereas others have all real numbers as solutions.

In Exercises 1–14, simplify the expression or solve the equation, whichever is appropriate.

1. $-5 + 3(x + 5)$ $3x + 10$

2. $-5 + 3(x + 5) = 2(3x - 4)$ 6

3. $3[7 - 4(5 - 2)]$ -15

4. $\dfrac{x - 3}{5} - 1 = \dfrac{x - 5}{4}$ -7

5. $\dfrac{-2^4 + (-2)^2}{-4 - (2 - 2)}$ 3

6. $7x - [8 - 3(2x - 5)]$ $13x - 23$

7. $3(2x - 5) - 2(4x + 1) = -5(x + 3) - 2$ 0

8. $3(2x - 5) - 2(4x + 1) - 5(x + 3) - 2$ $-7x - 34$

9. $-4^2 \div 2 + (-3)(-5)$ 7

10. $3x + 1 - (x - 5) = 2x - 4$ no solution or $\varnothing$ or inconsistent equation

11. $\dfrac{3x}{4} - \dfrac{x}{3} + 1 = \dfrac{4x}{5} - \dfrac{3}{20}$ 3

12. $(6 - 9)(8 - 12) \div \dfrac{5^2 + 4 \div 2}{8^2 - 9^2 + 8}$ -4

13. $4x - 2(1 - x) = 3(2x + 1) - 5$ $\{x \mid x \text{ is a real number}\}$ or $\mathbb{R}$ or identity

14. $\dfrac{3[4 - 3(-2)^2]}{2^2 - 2^4}$ 2

In Exercises 15–17, graph each equation in a rectangular coordinate system.   **15–17.** See graphing answer section.

15. $y = 2x - 1$ **16.** $y = 1 - |x|$ **17.** $y = x^2 + 2$

In Exercises 18–21, determine whether each statement is true or false.

18. $-\left| -\dfrac{\sqrt{3}}{5} \right| = -\dfrac{\sqrt{3}}{5}$ true

19. $\{x \mid x \text{ is a negative integer greater than } -4\} = \{-4, -3, -2, -1\}$ false

20. $-17 \notin \{x \mid x \text{ is a rational number}\}$ false

21. $-128 \div (2 \cdot 4) > (-128 \div 2) \cdot 4$ true

SECTION 1.5

Objectives

1 Solve algebraic word problems using linear equations.

2 Solve a formula for a variable.

PROBLEM SOLVING AND USING FORMULAS

The human race is undeniably becoming a faster race. Since the beginning of the past century, track-and-field records have fallen in everything from sprints to miles to marathons. The performance arc is clearly rising, but no one knows how much higher it can climb. At some point, even the best-trained body simply has to up and quit. The question is, just where is that point, and is it possible for athletes, trainers, and genetic engineers to push it higher? In this section, you will learn a problem-solving strategy that uses linear equations to determine if anyone will ever run a 3-minute mile.

Step 4. **Solve the equation and answer the question.**

$$(x + 11) + (x + 3) + x = 23 \qquad \text{This is the equation for the problem's conditions.}$$

$$3x + 14 = 23 \qquad \text{Remove parentheses, regroup, and combine like terms.}$$

$$3x = 9 \qquad \text{Subtract 14 from both sides.}$$

$$x = 3 \qquad \text{Divide both sides by 3.}$$

Thus,

the number of countries in the 1990s $= x = 3$.

the number of countries in the 1980s $= x + 3 = 3 + 3 = 6$.

the number of countries in the 1970s $= x + 11 = 3 + 11 = 14$.

There were 3 countries granting women the right to vote in the 1990s, 6 in the 1980s, and 14 in the 1970s.

Step 5. **Check the proposed solution in the original wording of the problem.** The problem states that from 1970 through 1999, 23 countries granted women the right to vote. By adding 3, 6, and 14, the numbers that we found for the 1990s, the 1980s, and the 1970s, respectively, we obtain

$$3 + 6 + 14, \quad \text{or} \quad 23,$$

as specified by the problem's conditions. ∎

✓ CHECK POINT **1** English, French, and Arabic are the official languages spoken in the greatest number of countries. There are 113 countries in which one of these languages is the official language. English is spoken in 34 more countries than Arabic and 24 more countries than French. Determine the number of countries in which each of these three languages is the official language.

English: 57; French: 33; Arabic: 23

EXAMPLE 2 Will Anyone Ever Run a Three-Minute Mile?

Mile Records			
1886	4:12.3	1958	3:54.5
1923	4:10.4	1966	3:51.3
1933	4:07.6	1979	3:48.9
1945	4:01.3	1985	3:46.3
1954	3:59.4	1999	3:43.1

Source: U.S.A. Track and Field

One yardstick for measuring how steadily—if slowly—athletic performance has improved is the mile run. In 1923, the record for the mile was a comparatively sleepy 4 minutes, 10.4 seconds. In 1954, Roger Bannister of Britain cracked the 4-minute mark, coming in at 3 minutes, 59.4 seconds. In the half-century since, about 0.3 second per year has been shaved off Bannister's record. If this trend continues, by which year will someone run a 3-minute mile?

SOLUTION In solving this problem, we will express time for the mile run in seconds. Our interest is in a time of 3 minutes, or 180 seconds.

Step 1. **Let x represent one of the quantities.** Here is the critical information in the problem:

- In 1954, the record was 3 minutes, 59.4 seconds, or 239.4 seconds.
- The record has decreased by 0.3 second per year since then.

We are interested in when the record will be 180 seconds. Let

$x =$ the number of years after 1954 when someone will run a 3-minute mile.

Step 2. **Represent other quantities in terms of x.** There are no other unknown quantities to find, so we can skip this step.

Step 3. **Write an equation in x that describes the conditions.**

The 1954 record time	decreased by	0.3 second per year for x years	equals	the 3-minute, or 180-second, mile.
239.4	−	0.3x	=	180

Step 4. **Solve the equation and answer the question.**

$$239.4 - 0.3x = 180$$ This is the equation for the problem's conditions.

$$239.4 - 239.4 - 0.3x = 180 - 239.4$$ Subtract 239.4 from both sides.

$$-0.3x = -59.4$$ Simplify.

$$\frac{-0.3x}{-0.3} = \frac{-59.4}{-0.3}$$ Divide both sides by -0.3.

$$x = 198$$ Simplify.

Using current trends, by 198 years (gasp!) after 1954, or in 2152, someone will run a 3-minute mile.

Step 5. **Check the proposed solution in the original wording of the problem.** The problem states that the record time should be 180 seconds. Do we obtain 180 seconds if we decrease the 1954 record time, 239.4 seconds, by 0.3 second per year for 198 years, our proposed solution?

$$239.4 - 0.3(198) = 239.4 - 59.4 = 180$$

This verifies that, using current trends, the 3-minute mile will be run 198 years after 1954. ∎

 CHECK POINT 2 In 2002, the population of the United States was approximately 288 million, increasing at a rate of 2.4 million people per year. If this trend continues, when will the population be 312 million? 2012

EXAMPLE 3 Selecting a Long-Distance Carrier

You are choosing between two long-distance telephone plans. Plan A has a monthly fee of $20 with a charge of $0.05 per minute for all long-distance calls. Plan B has a monthly fee of $5 with a charge of $0.10 per minute for all long-distance calls. For how many minutes of long-distance calls will the costs for the two plans be the same?

SOLUTION

Step 1. **Let x represent one of the quantities.** Let

$x =$ the number of minutes of long-distance calls
for which the two plans cost the same.

Step 2. **Represent other unknown quantities in terms of x.** There are no other unknown quantities, so we can skip this step.

Step 3. **Write an equation in x that describes the conditions.** The monthly cost for plan A is the monthly fee, $20, plus the per minute charge, $0.05, times the number of minutes of long-distance calls, x. The monthly cost for plan B is the monthly fee, $5, plus the per-minute charge, $0.10, times the number of minutes of long-distance calls, x.

The monthly cost for plan A	must equal	the monthly cost for plan B.
$20 + 0.05x$	$=$	$5 + 0.10x$

Step 4. **Solve the equation and answer the question.**

$$20 + 0.05x = 5 + 0.10x \qquad \text{This is the equation for the problem's conditions.}$$
$$20 = 5 + 0.05x \qquad \text{Subtract } 0.05x \text{ from both sides.}$$
$$15 = 0.05x \qquad \text{Subtract 5 from both sides.}$$
$$\frac{15}{0.05} = \frac{0.05x}{0.05} \qquad \text{Divide both sides by 0.05.}$$
$$300 = x \qquad \text{Simplify.}$$

Because x represents the number of minutes of long-distance calls for which the two plans cost the same, the costs will be the same for 300 minutes of long-distance calls.

Step 5. **Check the proposed solution in the original wording of the problem.** The problem states that the costs for the two plans should be the same. Let's see if they are with 300 minutes of long distance calls:

$$\text{Cost for plan A} = \$20 + \$0.05(300) = \$20 + \$15 = \$35$$

Monthly fee	Per-minute charge

$$\text{Cost for plan B} = \$5 + \$0.10(300) = \$5 + \$30 = \$35.$$

With 300 minutes, or 5 hours, of long-distance chatting, both plans cost $35 for the month. Thus, the proposed solution, 300 minutes, satisfies the problem's conditions.

USING TECHNOLOGY

We can use a graphing utility to numerically or graphically verify our work in Example 3.

The monthly cost for plan A	must equal	the monthly cost for plan B.
$20 + 0.05x$	$=$	$5 + 0.10x$
Enter $y_1 = 20 + .05x$		Enter $y_2 = 5 + .10x$

Numeric Check

Display a table for y_1 and y_2.

When $x = 300$, y_1 and y_2 have the same value, 35. With 300 minutes of calls, costs are the same, $35, for both plans.

X	Y₁	Y₂
100	25	15
150	27.5	20
200	30	25
250	32.5	30
300	35	35
350	37.5	40
400	40	45

X=300

Graphic Check

Display graphs for y_1 and y_2. Use the intersection feature.

$y_1 = 20 + 0.05x$

$y_2 = 5 + 0.10x$

Intersection
X=300   Y=35

[0, 500, 100] by [0, 50, 5]

Graphs intersect at (300, 35). With 300 minutes of calls, costs are the same, $35, for both plans.

> ✔ CHECK POINT **3** You are choosing between two long-distance telephone plans. Plan A has a monthly fee of $15 with a charge of $0.08 per minute for all long-distance calls. Plan B has a monthly fee of $3 with a charge of $0.12 per minute for all long-distance calls. For how many minutes of long-distance calls will the costs for the two plans be the same?   300 minutes

EXAMPLE 4 A Price Reduction on a Digital Camera

Your local computer store is having a terrific sale on digital cameras. After a 40% price reduction, you purchase a digital camera for $276. What was the camera's price before the reduction?

SOLUTION

Step 1. **Let x represent one of the quantities.** We will let

$$x = \text{the original price of the digital camera}$$
$$\text{prior to the reduction.}$$

Step 2. **Represent other quantities in terms of x.** There are no other unknown quantities to find, so we can skip this step.

Step 3. **Write an equation in x that describes the conditions.** The camera's original price minus the 40% reduction is the reduced price, $276.

Step 4. **Solve the equation and answer the question.**

$x - 0.4x = 276$	This is the equation for the problem's conditions.
$0.6x = 276$	Combine like terms: $x - 0.4x = 1x - 0.4x = 0.6x$.
$\dfrac{0.6x}{0.6} = \dfrac{276}{0.6}$	Divide both sides by 0.6.
$x = 460$	Simplify: $0.6\overline{)276.0}$ $460.$

The digital camera's price before the reduction was $460.

Step 5. **Check the proposed solution in the original wording of the problem.** The price before the reduction, $460, minus the 40% reduction should equal the reduced price given in the original wording, $276:

$$460 - 40\% \text{ of } 460 = 460 - 0.4(460) = 460 - 184 = 276.$$

This verifies that the digital camera's price before the reduction was $460. ■

> ✔ CHECK POINT **4** After a 30% price reduction, you purchase a new computer for $840. What was the computer's price before the reduction? $1200

STUDY TIP

Observe that the original price, x, reduced by 40% is $x - 0.4x$ and *not* $x - 0.4$.

Solving geometry problems usually requires a knowledge of basic geometric ideas and formulas. Formulas for area, perimeter, and volume are given in Table 1.5.

Table 1.5   Common Formulas for Area, Perimeter, and Volume

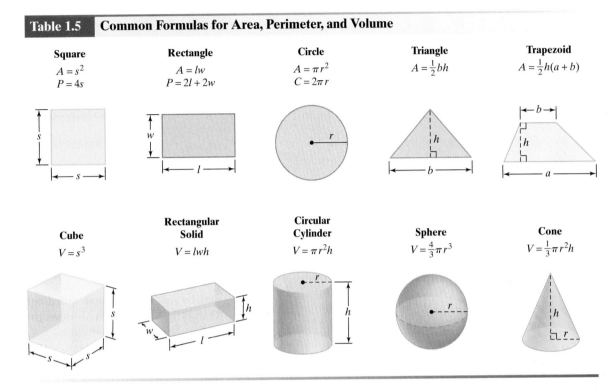

Square	Rectangle	Circle	Triangle	Trapezoid
$A = s^2$	$A = lw$	$A = \pi r^2$	$A = \frac{1}{2}bh$	$A = \frac{1}{2}h(a+b)$
$P = 4s$	$P = 2l + 2w$	$C = 2\pi r$		

Cube	Rectangular Solid	Circular Cylinder	Sphere	Cone
$V = s^3$	$V = lwh$	$V = \pi r^2 h$	$V = \frac{4}{3}\pi r^3$	$V = \frac{1}{3}\pi r^2 h$

We will be using the formula for the perimeter of a rectangle, $P = 2l + 2w$, in our next example. The formula states that a rectangle's perimeter is the sum of twice its length and twice its width.

EXAMPLE 5   Finding the Dimensions of an American Football Field

The length of an American football field is 200 feet more than the width. If the perimeter of the field is 1040 feet, what are its dimensions?

SOLUTION

Step 1.   Let x represent one of the quantities. We know something about the length; the length is 200 feet more than the width. We will let

$$x = \text{the width.}$$

Step 2.   Represent other quantities in terms of x. Because the length is 200 feet more than the width, let

$$x + 200 = \text{the length.}$$

Figure 1.19 illustrates an American football field and its dimensions.

Step 3.   Write an equation in x that describes the conditions. Because the perimeter of the field is 1040 feet,

Twice the length	plus	twice the width	is	the perimeter.
$2(x + 200)$	$+$	$2x$	$=$	$1040.$

FIGURE 1.19   An American football field

$x + 200$
Length

x
Width

FIGURE 1.19 (repeated) An American football field

Step 4. Solve the equation and answer the question.

$$2(x + 200) + 2x = 1040 \quad \text{This is the equation for the problem's conditions.}$$

$$2x + 400 + 2x = 1040 \quad \text{Apply the distributive property.}$$

$$4x + 400 = 1040 \quad \text{Combine like terms: } 2x + 2x = 4x.$$

$$4x = 640 \quad \text{Subtract 400 from both sides.}$$

$$x = 160 \quad \text{Divide both sides by 4.}$$

Thus,

$$\text{width} = x = 160.$$

$$\text{length} = x + 200 = 160 + 200 = 360.$$

The dimensions of an American football field are 160 feet by 360 feet. (The 360-foot length is usually described as 120 yards.)

Step 5. Check the proposed solution in the original wording of the problem. The perimeter of the football field using the dimensions that we found is

$$2(160 \text{ feet}) + 2(360 \text{ feet}) = 320 \text{ feet} + 720 \text{ feet} = 1040 \text{ feet}.$$

Because the problem's wording tells us that the perimeter is 1040 feet, our dimensions are correct. ■

 CHECK POINT 5 The length of a rectangular basketball court is 44 feet more than the width. If the perimeter of the basketball court is 288 feet, what are its dimensions?   50 feet by 94 feet

2 Solve a formula for a variable.

Solving a Formula for One of Its Variables   We know that solving an equation is the process of finding the number (or numbers) that make the equation a true statement. All of the equations we have solved contained only one letter, such as x or y.

By contrast, the formulas we have seen contain two or more letters, representing two or more variables. An example is the formula for the perimeter of a rectangle:

$$2l + 2w = P.$$

We say that this formula is solved for the variable P because P is alone on one side of the equation and the other side does not contain a P.

Solving a formula for a variable means using the addition and multiplication properties of equality to rewrite the formula so that the variable is isolated on one side of the equation. It does not mean obtaining a numerical value for that variable.

To solve a formula for one of its variables, treat that variable as if it were the only variable in the equation. Think of the other variables as if they were numbers. Use the addition property of equality to isolate all terms with the specified variable on one side of the equation and all terms without the specified variable on the other side. Then use the multiplication property of equality to get the specified variable alone. The next example shows how to do this.

EXAMPLE 6 Solving a Formula for a Variable

Solve the formula $2l + 2w = P$ for l.

SOLUTION First, isolate $2l$ on the left by subtracting $2w$ from both sides. Then solve for l by dividing both sides by 2.

We need to isolate l.

$$2l + 2w = P \qquad \text{This is the given formula.}$$

$$2l + 2w - 2w = P - 2w \qquad \text{Isolate } 2l \text{ by subtracting } 2w \text{ from both sides.}$$

$$2l = P - 2w \qquad \text{Simplify.}$$

$$\frac{2l}{2} = \frac{P - 2w}{2} \qquad \text{Solve for } l \text{ by dividing both sides by 2.}$$

$$l = \frac{P - 2w}{2} \qquad \text{Simplify.} \qquad \blacksquare$$

 CHECK POINT 6 Solve the formula $2l + 2w = P$ for w. $\quad w = \dfrac{P - 2l}{2}$

EXAMPLE 7 Solving a Formula for a Variable

Table 1.5 shows that the volume of a circular cylinder is given by the formula
$$V = \pi r^2 h,$$
where r is the radius of the circle at either end and h is the height. Solve this formula for h.

SOLUTION Our goal is to get h by itself on one side of the formula. There is only one term with h, $\pi r^2 h$, and it is already isolated on the right side. We isolate h on the right by dividing both sides by πr^2.

We need to isolate h.

$$V = \pi r^2 h \qquad \text{This is the given formula.}$$

$$\frac{V}{\pi r^2} = \frac{\pi r^2 h}{\pi r^2} \qquad \text{Isolate } h \text{ by dividing both sides by } \pi r^2.$$

$$\frac{V}{\pi r^2} = h \qquad \text{Simplify: } \frac{\pi r^2 h}{\pi r^2} = \frac{\pi r^2}{\pi r^2} \cdot h = 1h = h.$$

Equivalently,

$$h = \frac{V}{\pi r^2}. \qquad \blacksquare$$

 CHECK POINT 7 The volume of a rectangular solid is the product of its length, width, and height:
$$V = lwh.$$

Solve this formula for h. $\quad h = \dfrac{V}{lw}$

Circular Cylinder

$V = \pi r^2 h$

The Celsius scale is on the left and the Fahrenheit scale is on the right.

You'll be leaving the cold of winter for a vacation to Hawaii. CNN International reports a temperature in Hawaii of 30°C. Should you pack a winter coat? You can convert from Celsius temperature, C, to Fahrenheit temperature, F, using the formula

$$F = \frac{9}{5}C + 32.$$

A temperature of 30°C corresponds to a Fahrenheit temperature of

$$F = \frac{9}{5} \cdot 30 + 32 = \frac{9}{\underset{1}{5}} \cdot \frac{\overset{6}{30}}{1} + 32 = 54 + 32 = 86,$$

or a balmy 86°F. (Don't pack the coat.)

Visitors to the United States are more likely to be familiar with the Celsius temperature scale. For them, a useful formula is one that can be used to convert from Fahrenheit to Celsius. In Example 8, you will see how to obtain such a formula.

EXAMPLE 8 Solving a Formula for a Variable

Solve the formula

$$F = \frac{9}{5}C + 32$$

for C.

SOLUTION We begin by multiplying both sides of the formula by 5 to clear the fraction. Then we isolate the variable C.

$$F = \frac{9}{5}C + 32 \qquad \text{This is the given formula.}$$

$$5F = 5\left(\frac{9}{5}C + 32\right) \qquad \text{Multiply both sides by 5.}$$

$$5F = 5 \cdot \frac{9}{5}C + 5 \cdot 32 \qquad \text{Apply the distributive property.}$$

We need to isolate C.

$$5F = 9C + 160 \qquad \text{Simplify.}$$

$$5F - 160 = 9C + 160 - 160 \qquad \text{Subtract 160 from both sides.}$$

$$5F - 160 = 9C \qquad \text{Simplify.}$$

$$\frac{5F - 160}{9} = \frac{9C}{9} \qquad \text{Divide both sides by 9.}$$

$$\frac{5F - 160}{9} = C \qquad \text{Simplify.}$$

Using the distributive property, we can express $5F - 160$ as $5(F - 32)$. Thus,

$$C = \frac{5F - 160}{9} = \frac{5(F - 32)}{9}.$$

This formula, used to convert from Fahrenheit to Celsius, is usually given as

$$C = \frac{5}{9}(F - 32).$$

✔ CHECK POINT **8** The formula

$$\frac{W}{2} - 3H = 53$$

models the recommended weight, W, in pounds, for a male, where H represents his height, in inches, over 5 feet. Solve this formula for W. $W = 106 + 6H$

EXAMPLE 9   Solving a Formula for a Variable That Occurs Twice

The formula

$$A = P + Prt$$

describes the amount, A, that a principal of P dollars is worth after t years when invested at a simple annual interest rate, r. Solve this formula for P.

SOLUTION   Notice that all the terms with P already occur on the right side of the formula.

We need to isolate P.

$$A = P + Prt$$

STUDY TIP

You cannot solve $A = P + Prt$ for P by subtracting Prt from both sides and writing

$$A - Prt = P.$$

When a formula is solved for a specified variable, that variable must be isolated on one side. The variable P occurs on both sides of

$$A - Prt = P.$$

We can use the distributive property in the form $ab + ac = a(b + c)$ to convert the two occurrences of P into one.

$$A = P + Prt \qquad \text{This is the given formula.}$$

$$A = P(1 + rt) \qquad \text{Use the distributive property to obtain a single occurrence of P.}$$

$$\frac{A}{1 + rt} = \frac{P(1 + rt)}{1 + rt} \qquad \text{Divide both sides by 1 + rt.}$$

$$\frac{A}{1 + rt} = P \qquad \text{Simplify: } \frac{P(1 + rt)}{1(1 + rt)} = \frac{P}{1} = P.$$

Equivalently,

$$P = \frac{A}{1 + rt}.$$

✔ CHECK POINT **9** Solve the formula $P = C + MC$ for C. $C = \dfrac{P}{1 + M}$

ENRICHMENT ESSAY

Einstein's Famous Formula: $E = mc^2$

One of the most famous formulas in the world is $E = mc^2$ formulated by Albert Einstein. Einstein showed that any form of energy has mass and that mass itself is a form of energy. In this formula, E represents energy, in ergs, m represents mass, in grams, and c represents the speed of light. Because light travels at 30 billion centimeters per second, the formula indicates that 1 gram of mass will produce 900 billion ergs of energy.

Einstein's formula implies that the mass of a golf ball could provide the daily energy needs of the metropolitan Boston area. Mass and energy are equivalent, and the transformation of even a tiny amount of mass releases an enormous amount of energy. If this energy is released suddenly, a destructive force is unleashed, as in an atom bomb. When the release is gradual and controlled, the energy can be used to generate power.

The theoretical results implied by Einstein's formula $E = mc^2$ have not been realized because scientists have not yet developed a way of converting a mass completely to energy.

1.5 EXERCISE SET

Student Solutions Manual   CD/Video   PH Math/Tutor Center   MathXL Tutorials on CD   MathXL®   MyMathLab   Interactmath.com

Practice Exercises

Use the five-step strategy for solving word problems to find the number or numbers described in Exercises 1–10.

1. When five times a number is decreased by 4, the result is 26. What is the number?   6

2. When two times a number is decreased by 3, the result is 11. What is the number?   7

3. When a number is decreased by 20% of itself, the result is 20. What is the number?   25

4. When a number is decreased by 30% of itself, the result is 28. What is the number?   40

5. When 60% of a number is added to the number, the result is 192. What is the number?   120

6. When 80% of a number is added to the number, the result is 252. What is the number?   140

7. 70% of what number is 224?   320

8. 70% of what number is 252?   360

9. One number exceeds another by 26. The sum of the numbers is 64. What are the numbers?   19 and 45

10. One number exceeds another by 24. The sum of the numbers is 58. What are the numbers?   17 and 41

Application Exercises

In Exercises 11–44, use the five-step strategy for solving word problems.

11. The bar graph shows the number of Internet users, in millions, for the countries with the most users. The number of Internet users in Japan exceeds China by 10 million, and the number of Internet users in the United States exceeds China by 123 million. There are a total of 271 million Internet users in the United States, Japan, and China. Determine the number of users, in millions, in each country.   U.S.: 169 million; Japan: 56 million; China: 46 million

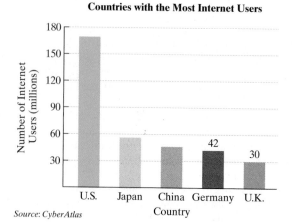

Countries with the Most Internet Users

Source: CyberAtlas

12. The bar graph shows the defense budget, in billions of dollars, for the countries with the largest defense budgets. The defense budget of Russia exceeds Japan's by $4 billion, and the defense budget of the United States exceeds Japan's by $251 billion. The combined defense budgets of the United States, Russia, and Japan is $375 billion. Determine the defense budget, in billions of dollars, for each country.   U.S.: $291 billion; Russia: $44 billion; Japan: $40 billion

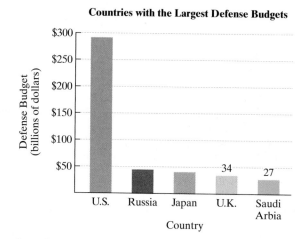

Countries with the Largest Defense Budgets

Source: Russell Ash, *The Top 10 of Everything 2004*

Solve Exercises 13–16 using the fact that the sum of the measures of the three angles of a triangle is 180°.

13. In a triangle, the measure of the first angle is twice the measure of the second angle. The measure of the third angle is 8° less than the measure of the second angle. What is the measure of each angle?   94°, 47°, 39°

14. In a triangle, the measure of the first angle is three times the measure of the second angle. The measure of the third angle is 35° less than the measure of the second angle. What is the measure of each angle?   129°, 43°, 8°

15. In a triangle, the measures of the three angles are consecutive integers. What is the measure of each angle?   59°, 60°, 61°

16. In a triangle, the measures of the three angles are consecutive even integers. What is the measure of each angle?   58°, 60°, 62°

According to one mathematical model, the average life expectancy for American men born in 1900 was 55 years. Life expectancy has increased by about 0.2 year for each birth year after 1900. Use this information to solve Exercises 17–18.

17. If this trend continues, for which birth year will the average life expectancy be 85 years?   2050

18. If this trend continues, for which birth year will the average life expectancy be 91 years?   2080

According to the National Center for Health Statistics, in 1990, 28% of babies in the United States were born to parents who were not married. Throughout the 1990s, this increased by approximately 0.6% per year. Use this information to solve Exercises 19–20.

19. If this trend continues, in which year will 40% of babies be born out of wedlock?   2010

20. If this trend continues, in which year will 46% of babies be born out of wedlock?   2020

In 2003, the price of a BMW 7 Series was approximately $80,500 with a depreciation of $8705 per year. Use this information to solve Exercises 21–22.

21. After how many years will the car's value be $19,565?   after 7 years

22. After how many years will the car's value be $36,975?   after 5 years

23. You are choosing between two health clubs. Club A offers membership for a fee of $40 plus a monthly fee of $25. Club B offers membership for a fee of $15 plus a monthly fee of $30. After how many months will the total cost at each health club be the same? What will be the total cost for each club?   after 5 months; $165

24. Video Store A charges $9 to rent a video game for one week. Although only members can rent from the store, membership is free. Video Store B charges only $4 to rent a video game for one week. Only members can rent from the store and membership is $50 per year. After how many video-game rentals will the total amount spent at each store be the same? What will be the total amount spent at each store?   10 rentals; $90

25. The bus fare in a city is $1.25. People who use the bus have the option of purchasing a monthly coupon book for $15.00. With the coupon book, the fare is reduced to $0.75. Determine the number of times in a month the bus must be used so that the total monthly cost without the coupon book is the same as the total monthly cost with the coupon book.   30 times

26. A coupon book for a bridge costs $30 per month. The toll for the bridge is normally $5.00, but it is reduced to $3.50 for people who have purchased the coupon book. Determine the number of times in a month the bridge must be crossed so that the total monthly cost without the coupon book is the same as the total monthly cost with the coupon book.   20 times

27. In 2005, there were 13,300 students at college A, with a projected enrollment increase of 1000 students per year. In the same year, there were 26,800 students at college B, with a projected enrollment decline of 500 students per year.

   a. According to these projections, when will the colleges have the same enrollment? What will be the enrollment in each college at that time?   2014; 22,300 students

b. Use the following table to numerically check your work in part (a). What equations were entered for Y_1 and Y_2 to obtain this table? $y_1 = 13300 + 1000x$; $y_2 = 26800 - 500x$

28. In 2000, the population of Greece was 10,600,000, with projections of a population decrease of 28,000 people per year. In the same year, the population of Belgium was 10,200,000, with projections of a population decrease of 12,000 people per year. (*Source*: United Nations) According to these projections, when will the two countries have the same population? What will be the population at that time?   2025; 9,900,000

29. After a 20% reduction, you purchase a television for $336. What was the television's price before the reduction? $420

30. After a 30% reduction, you purchase a dictionary for $30.80. What was the dictionary's price before the reduction? $44

31. Including 8% sales tax, an inn charges $162 per night. Find the inn's nightly cost before the tax is added. $150

32. Including 5% sales tax, an inn charges $252 per night. Find the inn's nightly cost before the tax is added. $240

The graph shows average yearly earnings in the United States by highest educational attainment. Use the relevant information shown in the graph to solve Exercises 33–34.

Average Earnings of Full-Time Workers in the U.S., by Highest Educational Attainment

Source: U.S. Census Bureau

33. The annual salary for men with some college is an increase of 22% over the annual salary for men whose highest educational attainment is a high school degree. What is the annual salary, to the nearest thousand dollars, for men whose highest educational attainment is a high school degree? $36,000

34. (Be sure to refer to the graph on page 63.) The annual salary for women with an associate degree is an increase of 26% over the annual salary for women whose highest educational attainment is a high school degree. What is the annual salary, to the nearest thousand dollars, for women whose highest educational attainment is a high school degree? $27 thousand

Exercises 35–36, involve markup, the amount added to the dealer's cost of an item to arrive at the selling price of that item.

35. The selling price of a refrigerator is $584. If the markup is 25% of the dealer's cost, what is the dealer's cost of the refrigerator? $467.20

36. The selling price of a scientific calculator is $15. If the markup is 25% of the dealer's cost, what is the dealer's cost of the calculator? $12

37. A rectangular soccer field is twice as long as it is wide. If the perimeter of the soccer field is 300 yards, what are its dimensions?   50 yards by 100 yards

38. A rectangular swimming pool is three times as long as it is wide. If the perimeter of the pool is 320 feet, what are its dimensions?   40 feet by 120 feet

39. The length of the rectangular tennis court at Wimbledon is 6 feet longer than twice the width. If the court's perimeter is 228 feet, what are the court's dimensions?   36 feet by 78 feet

40. The length of a rectangular pool is 6 meters less than twice the width. If the pool's perimeter is 126 meters, what are its dimensions?   23 meters by 40 meters

41. The rectangular painting in the figure shown measures 12 inches by 16 inches and contains a frame of uniform width around the four edges. The perimeter of the rectangle formed by the painting and its frame is 72 inches. Determine the width of the frame.   2 inches

42. The rectangular swimming pool in the figure shown measures 40 feet by 60 feet and contains a path of uniform width around the four edges. The perimeter of the rectangle formed by the pool and the surrounding path is 248 feet. Determine the width of the path.   6 feet

43. For a long-distance person-to-person telephone call, a telephone company charges $0.43 for the first minute, $0.32 for each additional minute, and a $2.10 service charge. If the cost of a call is $5.73, how long did the person talk?   11 minutes

44. A job pays an annual salary of $33,150, which includes a holiday bonus of $750. If paychecks are issued twice a month, what is the gross amount for each paycheck? $1350

In Exercises 45–70, solve each formula for the specified variable.

45. $A = lw$ for l $l = \dfrac{A}{w}$

46. $A = lw$ for w $w = \dfrac{A}{l}$

47. $A = \dfrac{1}{2}bh$ for b $b = \dfrac{2A}{h}$

48. $A = \dfrac{1}{2}bh$ for h $h = \dfrac{2A}{b}$

49. $I = Prt$ for P $P = \dfrac{I}{rt}$

50. $I = Prt$ for t $t = \dfrac{I}{Pr}$

51. $T = D + pm$ for p $p = \dfrac{T - D}{m}$

52. $P = C + MC$ for M $M = \dfrac{P - C}{C}$

53. $A = \dfrac{1}{2}h(a + b)$ for a $a = \dfrac{2A - hb}{h}$ or $a = \dfrac{2A}{h} - b$

54. $A = \dfrac{1}{2}h(a + b)$ for b $b = \dfrac{2A - ha}{h}$ or $b = \dfrac{2A}{h} - a$

55. $V = \dfrac{1}{3}\pi r^2 h$ for h $h = \dfrac{3V}{\pi r^2}$

56. $V = \dfrac{1}{3}\pi r^2 h$ for r^2 $r^2 = \dfrac{3V}{\pi h}$

57. $y - y_1 = m(x - x_1)$ for m $m = \dfrac{y - y_1}{x - x_1}$

58. $y_2 - y_1 = m(x_2 - x_1)$ for m $m = \dfrac{y_2 - y_1}{x_2 - x_1}$

59. $V = \dfrac{d_1 - d_2}{t}$ for d_1 $d_1 = Vt + d_2$

60. $z = \dfrac{x - u}{s}$ for x $x = zs + u$

61. $Ax + By = C$ for x $x = \dfrac{C - By}{A}$

62. $Ax + By = C$ for y $y = \dfrac{C - Ax}{B}$

63. $s = \dfrac{1}{2}at^2 + vt$ for v $\quad v = \dfrac{2s - at^2}{2t}$

64. $s = \dfrac{1}{2}at^2 + vt$ for a $\quad a = \dfrac{2s - 2vt}{t^2}$

65. $L = a + (n - 1)d$ for n $\quad n = \dfrac{L - a}{d} + 1$

66. $L = a + (n - 1)d$ for d $\quad d = \dfrac{L - a}{n - 1}$

67. $A = 2lw + 2lh + 2wh$ for l $\quad l = \dfrac{A - 2wh}{2w + 2h}$

68. $A = 2lw + 2lh + 2wh$ for h $\quad h = \dfrac{A - 2lw}{2l + 2w}$

69. $IR + Ir = E$ for I $\quad I = \dfrac{E}{R + r}$

70. $A = \dfrac{x_1 + x_2 + x_3}{n}$ for n $\quad n = \dfrac{x_1 + x_2 + x_3}{A}$

Writing in Mathematics

71. In your own words, describe a step-by-step approach for solving algebraic word problems. **71–76.** Answers will vary.

72. Write an original word problem that can be solved using a linear equation. Then solve the problem.

73. Explain what it means to solve a formula for a variable.

74. Did you have difficulties solving some of the problems that were assigned in this exercise set? Discuss what you did if this happened to you. Did your course of action enhance your ability to solve algebraic word problems?

75. The mile records in Example 2 on page 53 are a yardstick for measuring how athletes are getting better and better. Do you think that there is a limit to human performance? Explain your answer. If so, when might we reach it?

76. The bar graph in Exercises 33–34 shows average earnings of U.S. men and women, by highest educational attainment. Describe the trend shown by the graph. Discuss any aspects of the data that surprised you.

Technology Exercises

77. Use a graphing utility to numerically or graphically verify your work in any one exercise from Exercises 23–26. For assistance on how to do this, refer to the Using Technology box on page 55. Answers will vary.

78. The formula $y = 55 + 0.2x$ models the average life expectancy, y, of American men born x years after 1900. Graph the formula in a $[0, 200, 20]$ by $[0, 100, 10]$ viewing rectangle. Then use the TRACE or ZOOM feature to verify your answer in Exercise 17 or 18. See graphing answer section.

79. The formula $y = 28 + 0.6x$ models the percentage, y, of U.S. babies born out of wedlock x years after 1990.

Graph the formula in a $[0, 30, 1]$ by $[0, 50, 1]$ viewing rectangle. Then use the TRACE or ZOOM feature to verify your answer in Exercise 19 or 20. See graphing answer section.

Critical Thinking Exercises

80. Which one of the following is true? **b**

a. If $I = prt$, then $t = I - pr$.

b. If $y = \dfrac{kx}{z}$, then $z = \dfrac{kx}{y}$.

c. It is not necessary to use the distributive property to solve $P = C + MC$ for C.

d. None of the above is true.

81. The price of a dress is reduced by 40%. When the dress still does not sell, it is reduced by 40% of the reduced price. If the price of the dress after both reductions is $72, what was the original price? $200

82. Suppose that we agree to pay you 8¢ for every problem in this chapter that you solve correctly and fine you 5¢ for every problem done incorrectly. If at the end of 26 problems we do not owe each other any money, how many problems did you solve correctly? 10 problems

83. It was wartime when the Ricardos found out Mrs. Ricardo was pregnant. Ricky Ricardo was drafted and made out a will, deciding that $14,000 in a savings account was to be divided between his wife and his child-to-be. Rather strangely, and certainly with gender bias, Ricky stipulated that if the child were a boy, he would get twice the amount of the mother's portion. If it were a girl, the mother would get twice the amount the girl was to receive. We'll never know what Ricky was thinking, for (as fate would have it) he did not return from war. Mrs. Ricardo gave birth to twins—a boy and a girl. How was the money divided? Mrs. Ricardo received $4000; boy received $8000; girl received $2000.

84. A thief steals a number of rare plants from a nursery. On the way out, the thief meets three security guards, one after another. To each security guard, the thief is forced to give one-half the plants that he still has, plus 2 more. Finally, the thief leaves the nursery with 1 lone palm. How many plants were originally stolen? 36 plants

85. Solve for C: $V = C - \dfrac{C - S}{L}N$. $\quad C = \dfrac{LV - SN}{L - N}$

Review Exercises

86. What does $-6 \le -6$ mean? Is the statement true or false? (Section 1.1, Example 6) $\quad -6$ is less than or equal to -6; true

87. Simplify: $\dfrac{(2 + 4)^2 + (-1)^5}{12 \div 2 \cdot 3 - 3}$. (Section 1.2, Example 8) $\quad \dfrac{7}{3}$

88. Solve: $\dfrac{2x}{3} - \dfrac{8}{3} = x$. (Section 1.4, Example 4) $\quad -8$

SECTION

1.6

Objectives

1 Use the product rule.

2 Use the quotient rule.

3 Use the zero-exponent rule.

4 Use the negative-exponent rule.

5 Use the power rule.

6 Find the power of a product.

7 Find the power of a quotient.

8 Simplify exponential expressions.

PROPERTIES OF INTEGRAL EXPONENTS

Horses were born with eight deformed legs, pigs with no eyes, and eggs contained several yolks. This was part of the grotesque aftermath of the 1986 explosion at the Chernobyl nuclear power plant in the former Soviet Union. Nearby cities were abandoned. Because of dangerous radiation levels for at least one hundred years, an exclusion zone was formed around the area. The spread of radioactive fallout from the world's worst nuclear accident is expected to result in at least 100,000 additional deaths from cancer in the Northern Hemisphere. In this section's exercise set, you will work with a formula that sadly explains why 335,000 people in cities surrounding Chernobyl were permanently displaced from their homes. The formula involves exponents, our topic for this section.

The Product and Quotient Rules We have seen that exponents are used to indicate repeated multiplication. Now consider the multiplication of two exponential expressions, such as $b^4 \cdot b^3$. We are multiplying 4 factors of b and 3 factors of b. We have a total of 7 factors of b:

4 factors of b 3 factors of b

$$b^4 \cdot b^3 = (b \cdot b \cdot b \cdot b)(b \cdot b \cdot b) = b^7.$$

Total: 7 factors of b

The product is exactly the same if we add the exponents:

$$b^4 \cdot b^3 = b^{4+3} = b^7.$$

This suggests the following rule:

THE PRODUCT RULE

$$b^m \cdot b^n = b^{m+n}$$

When multiplying exponential expressions with the same base, add the exponents. Use this sum as the exponent of the common base.

1 Use the product rule.

EXAMPLE 1   Using the Product Rule

Multiply each expression using the product rule:

  a. $b^8 \cdot b^{10}$ **b.** $(6x^4y^3)(5x^2y^7)$.

SOLUTION

a. $b^8 \cdot b^{10} = b^{8+10} = b^{18}$

b. $(6x^4y^3)(5x^2y^7)$

$= 6 \cdot 5 \cdot x^4 \cdot x^2 \cdot y^3 \cdot y^7$ Use the associative and commutative properties. This step can be done mentally.

$= 30x^{4+2}y^{3+7}$

$= 30x^6y^{10}$ ■

✔ **CHECK POINT 1** Multiply each expression using the product rule:

a. $b^6 \cdot b^5$ b^{11} **b.** $(4x^3y^4)(10x^2y^6)$. $40x^5y^{10}$

2 Use the quotient rule.

Now, consider the division of two exponential expressions, such as the quotient of b^7 and b^3. We are dividing 7 factors of b by 3 factors of b.

$$\frac{b^7}{b^3} = \frac{b \cdot b \cdot b \cdot b \cdot b \cdot b \cdot b}{b \cdot b \cdot b} = \boxed{\frac{b \cdot b \cdot b}{b \cdot b \cdot b}} \cdot b \cdot b \cdot b \cdot b = 1 \cdot b \cdot b \cdot b \cdot b = b^4$$

This factor is equal to 1.

The quotient is exactly the same if we subtract the exponents:

$$\frac{b^7}{b^3} = b^{7-3} = b^4.$$

This suggests the following rule:

THE QUOTIENT RULE

$$\frac{b^m}{b^n} = b^{m-n}, \quad b \neq 0$$

When dividing exponential expressions with the same nonzero base, subtract the exponent in the denominator from the exponent in the numerator. Use this difference as the exponent of the common base.

EXAMPLE 2

Divide each expression using the quotient rule:

a. $\dfrac{b^{20}}{b^5}$ **b.** $\dfrac{30x^{12}y^9}{5x^3y^7}$.

SOLUTION

a. $\dfrac{b^{20}}{b^5} = b^{20-5} = b^{15}$

b. $\dfrac{30x^{12}y^9}{5x^3y^7} = \dfrac{30}{5} \cdot \dfrac{x^{12}}{x^3} \cdot \dfrac{y^9}{y^7} = 6x^{12-3}y^{9-7} = 6x^9y^2$. ■

✔ **CHECK POINT 2** Divide each expression using the quotient rule:

a. $\dfrac{b^{15}}{b^3}$ b^{12} **b.** $\dfrac{27x^{14}y^8}{3x^3y^5}$. $9x^{11}y^3$

3 Use the zero-exponent rule.

Zero as an Exponent A nonzero base can be raised to the 0 power. The quotient rule can be used to help determine what zero as an exponent should mean. Consider the quotient of b^4 and b^4, where b is not zero. We can determine this quotient in two ways.

$$\frac{b^4}{b^4} = 1 \qquad\qquad \frac{b^4}{b^4} = b^{4-4} = b^0$$

| Any nonzero expression divided by itself is 1. | Use the quotient rule and subtract exponents. |

This means that b^0 must equal 1.

> **THE ZERO–EXPONENT RULE** If b is any real number other than 0,
> $$b^0 = 1.$$

EXAMPLE 3 Using the Zero-Exponent Rule

Use the zero-exponent rule to simplify each expression:

a. 8^0 **b.** $(-6)^0$ **c.** -6^0 **d.** $5x^0$ **e.** $(5x)^0$.

SOLUTION

a. $8^0 = 1$ Any nonzero number raised to the 0 power is 1.

b. $(-6)^0 = 1$

c. $-6^0 = -(6^0) = -1$

| Only 6 is raised to the 0 power. |

d. $5x^0 = 5 \cdot 1 = 5$ Only x is raised to the 0 power.

e. $(5x)^0 = 1$ The entire expression, 5x, is raised to the 0 power.

 CHECK POINT 3 Use the zero-exponent rule to simplify each expression:

a. 7^0 1   **b.** $(-5)^0$ 1   **c.** -5^0 -1   **d.** $10x^0$ 10   **e.** $(10x)^0$.   1

4 Use the negative-exponent rule.

Negative Integers as Exponents A nonzero base can be raised to a negative power. The quotient rule can be used to help determine what a negative integer as an exponent should mean. Consider the quotient of b^3 and b^5, where b is not zero. We can determine this quotient in two ways.

$$\frac{b^3}{b^5} = \frac{\cancel{b} \cdot \cancel{b} \cdot \cancel{b}}{\cancel{b} \cdot \cancel{b} \cdot \cancel{b} \cdot b \cdot b} = \frac{1}{b^2} \qquad\qquad \frac{b^3}{b^5} = b^{3-5} = b^{-2}$$

| After dividing common factors, we have two factors of b in the denominator. | Use the quotient rule and subtract exponents. |

Notice that $\dfrac{b^3}{b^5}$ equals both b^{-2} and $\dfrac{1}{b^2}$. This means that b^{-2} must equal $\dfrac{1}{b^2}$. This example is a special case of the **negative-exponent rule**.

THE NEGATIVE–EXPONENT RULE If b is any real number other than 0 and n is a natural number, then

$$b^{-n} = \frac{1}{b^n}.$$

EXAMPLE 4 Using the Negative-Exponent Rule

Use the negative-exponent rule to write each expression with a positive exponent. Simplify, if possible:

    a. 9^{-2} **b.** $(-2)^{-5}$ **c.** $\dfrac{1}{6^{-2}}$ **d.** $7x^{-5}y^2$.

SOLUTION

    a. $9^{-2} = \dfrac{1}{9^2} = \dfrac{1}{81}$

    b. $(-2)^{-5} = \dfrac{1}{(-2)^5} = \dfrac{1}{(-2)(-2)(-2)(-2)(-2)} = \dfrac{1}{-32} = -\dfrac{1}{32}$

> Only the sign of the exponent, −5, changes. The base, −2, does not change sign.

    c. $\dfrac{1}{6^{-2}} = \dfrac{1}{\frac{1}{6^2}} = 1 \cdot \dfrac{6^2}{1} = 6^2 = 36$

    d. $7x^{-5}y^2 = 7 \cdot \dfrac{1}{x^5} \cdot y^2 = \dfrac{7y^2}{x^5}$ ■

 CHECK POINT 4 Use the negative-exponent rule to write each expression with a positive exponent. Simplify, if possible:

    a. 5^{-2} $\frac{1}{25}$ **b.** $(-3)^{-3}$ $-\frac{1}{27}$ **c.** $\dfrac{1}{4^{-2}}$ 16 **d.** $3x^{-6}y^4$. $\frac{3y^4}{x^6}$

In Example 4 and Check Point 4, did you notice that

$$\frac{1}{6^{-2}} = 6^2 \quad \text{and} \quad \frac{1}{4^{-2}} = 4^2?$$

In general, if a negative exponent appears in a denominator, an expression can be written with a positive exponent using

$$\frac{1}{b^{-n}} = b^n.$$

NEGATIVE EXPONENTS IN NUMERATORS AND DENOMINATORS If b is any real number other than 0 and n is a natural number, then

$$b^{-n} = \frac{1}{b^n} \quad \text{and} \quad \frac{1}{b^{-n}} = b^n.$$

When a negative number appears as an exponent, switch the position of the base (from numerator to denominator or from denominator to numerator) and make the exponent positive. The sign of the base does not change.

EXAMPLE 5 Using Negative Exponents

Write each expression with positive exponents only. Then simplify, if possible:

a. $\dfrac{5^{-3}}{4^{-2}}$ **b.** $\dfrac{1}{6x^{-4}}$.

SOLUTION

a. $\dfrac{5^{-3}}{4^{-2}} = \dfrac{4^2}{5^3} = \dfrac{4 \cdot 4}{5 \cdot 5 \cdot 5} = \dfrac{16}{125}$ Switch the position of each base to the other side of the fraction bar and change the sign of the exponent.

b. $\dfrac{1}{6x^{-4}} = \dfrac{x^4}{6}$ Switch the position of x to the other side of the fraction bar and change −4 to 4.

> Don't switch the position of 6. It is not affected by a negative exponent.

 CHECK POINT 5 Write each expression with positive exponents only. Then simplify, if possible:

a. $\dfrac{7^{-2}}{4^{-3}} \quad \dfrac{4^3}{7^2} = \dfrac{64}{49}$ **b.** $\dfrac{1}{5x^{-2}} \cdot \dfrac{x^2}{5}$

---

5 Use the power rule.

The Power Rule for Exponents (Powers to Powers) The next property of exponents applies when an exponential expression is raised to a power. Here is an example:

$$(b^2)^4.$$

> The exponential expression b^2 is raised to the fourth power.

There are 4 factors of b^2. Thus,

$$(b^2)^4 = b^2 \cdot b^2 \cdot b^2 \cdot b^2 = b^{2+2+2+2} = b^8$$

> Add exponents when multiplying with the same base.

We can obtain the answer, b^8, by multiplying the exponents:

$$(b^2)^4 = b^{2 \cdot 4} = b^8.$$

This suggests the following rule:

THE POWER RULE (POWERS TO POWERS)

$$(b^m)^n = b^{mn}$$

When an exponential expression is raised to a power, multiply the exponents. Place the product of the exponents on the base and remove the parentheses.

EXAMPLE 6 Using the Power Rule (Powers to Powers)

Simplify each expression using the power rule:

a. $(x^6)^4$ **b.** $(y^5)^{-3}$ **c.** $(b^{-4})^{-2}$.

SOLUTION

a. $(x^6)^4 = x^{6 \cdot 4} = x^{24}$ **b.** $(y^5)^{-3} = y^{5(-3)} = y^{-15} = \dfrac{1}{y^{15}}$

c. $(b^{-4})^{-2} = b^{(-4)(-2)} = b^8$

 CHECK POINT 6 Simplify each expression using the power rule:

a. $(x^5)^3$ $\quad x^{15}$ **b.** $(y^7)^{-2}$ $\quad \dfrac{1}{y^{14}}$ **c.** $(b^{-3})^{-4}$. $\quad b^{12}$

6 Find the power of a product.

The Products-to-Powers Rule for Exponents The next property of exponents applies when we are raising a product to a power. Here is an example:

$$(2x)^4.$$

The product $2x$ is raised to the fourth power.

There are four factors of $2x$. Thus,

$$(2x)^4 = 2x \cdot 2x \cdot 2x \cdot 2x = 2 \cdot 2 \cdot 2 \cdot 2 \cdot x \cdot x \cdot x \cdot x = 2^4 x^4.$$

We can obtain the answer, $2^4 x^4$, by raising each factor within the parentheses to the fourth power:

$$(2x)^4 = 2^4 x^4.$$

This suggests the following rule:

PRODUCTS TO POWERS

$$(ab)^n = a^n b^n$$

When a product is raised to a power, raise each factor to that power.

EXAMPLE 7 Using the Products-to-Powers Rule

Simplify each expression using the products-to-powers rule:

a. $(6x)^3$ **b.** $(-2y^2)^4$ **c.** $(-3x^{-1}y^3)^{-2}$.

SOLUTION

a. $\quad (6x)^3 = 6^3 x^3$ Raise each factor to the third power.
         $= 216x^3$ Simplify: $6^3 = 6 \cdot 6 \cdot 6 = 216$.

b. $(-2y^2)^4 = (-2)^4(y^2)^4$ Raise each factor to the fourth power.
         $= (-2)^4 y^{2 \cdot 4}$ To raise an exponential expression to a power, multiply exponents: $(b^m)^n = b^{mn}$.
         $= 16y^8$ Simplify: $(-2)^4 = (-2)(-2)(-2)(-2) = 16$.

c. $(-3x^{-1}y^3)^{-2} = (-3)^{-2}(x^{-1})^{-2}(y^3)^{-2}$ Raise each factor to the -2 power.

$= (-3)^{-2}x^{(-1)(-2)}y^{3(-2)}$ Use $(b^m)^n = b^{mn}$ on the second and third factors.

$= (-3)^{-2}x^2 y^{-6}$ Simplify.

$= \dfrac{1}{(-3)^2} \cdot x^2 \cdot \dfrac{1}{y^6}$ Apply $b^{-n} = \dfrac{1}{b^n}$ to the first and last factors.

$= \dfrac{x^2}{9y^6}$ Simplify: $(-3)^2 = (-3)(-3) = 9$. ∎

 CHECK POINT 7 Simplify each expression using the products-to-powers rule:

a. $(2x)^4$ $16x^4$ **b.** $(-3y^2)^3$ $-27y^6$ **c.** $(-4x^5y^{-1})^{-2}$. $\dfrac{y^2}{16x^{10}}$

The Quotients-to-Powers Rule for Exponents   The following rule is used to raise a quotient to a power:

7 Find the power of a quotient.

> **QUOTIENTS TO POWERS**   If b is a nonzero real number, then
>
> $$\left(\frac{a}{b}\right)^n = \frac{a^n}{b^n}.$$
>
> When a quotient is raised to a power, raise the numerator to the power and divide by the denominator to the power.

EXAMPLE 8   Using the Quotients-to-Powers Rule

Simplify each expression using the quotients-to-powers rule:

a. $\left(\dfrac{x^2}{4}\right)^3$ **b.** $\left(\dfrac{2x^3}{y^{-4}}\right)^5$ **c.** $\left(\dfrac{x^3}{y^2}\right)^{-4}$.

SOLUTION

a. $\left(\dfrac{x^2}{4}\right)^3 = \dfrac{(x^2)^3}{4^3} = \dfrac{x^{2 \cdot 3}}{4 \cdot 4 \cdot 4} = \dfrac{x^6}{64}$ Cube the numerator and the denominator.

b. $\left(\dfrac{2x^3}{y^{-4}}\right)^5 = \dfrac{(2x^3)^5}{(y^{-4})^5}$ Raise the numerator and the denominator to the fifth power.

$= \dfrac{2^5(x^3)^5}{(y^{-4})^5}$ Raise each factor in the numerator to the fifth power.

$= \dfrac{2^5 \cdot x^{3 \cdot 5}}{y^{(-4)(5)}}$ Multiply exponents in both powers-to-powers expressions: $(b^m)^n = b^{mn}$.

$= \dfrac{32x^{15}}{y^{-20}}$ Simplify.

$= 32x^{15}y^{20}$ Move y to the other side of the fraction bar and change -20 to 20: $\dfrac{1}{b^{-n}} = b^n$.

STUDY TIP

When simplifying exponential expressions, the first step should be to simplify inside parentheses. In Example 8, the expressions inside parentheses have different bases and cannot be simplified. This is why we begin with the quotients-to-powers rule.

c. $\left(\dfrac{x^3}{y^2}\right)^{-4} = \dfrac{(x^3)^{-4}}{(y^2)^{-4}}$ Raise the numerator and the denominator to the -4 power.

$= \dfrac{x^{3(-4)}}{y^{2(-4)}}$ Multiply exponents in both powers-to-powers expressions: $(b^m)^n = b^{mn}$.

$= \dfrac{x^{-12}}{y^{-8}}$ Simplify.

$= \dfrac{y^8}{x^{12}}$ Move each base to the other side of the fraction bar and make each exponent positive. ■

✔ **CHECK POINT 8** Simplify each expression using the quotients-to-powers rule:

a. $\left(\dfrac{x^5}{4}\right)^3$ $\dfrac{x^{15}}{64}$ **b.** $\left(\dfrac{2x^{-3}}{y^2}\right)^4$ $\dfrac{16}{x^{12}y^8}$ **c.** $\left(\dfrac{x^{-3}}{y^4}\right)^{-5}$ $x^{15}y^{20}$

---

8 Simplify exponential expressions.

Simplifying Exponential Expressions Properties of exponents are used to simplify exponential expressions. An exponential expression is **simplified** when

- No parentheses appear.
- No powers are raised to powers.
- Each base occurs only once.
- No negative or zero exponents appear.

SIMPLIFYING EXPONENTIAL EXPRESSIONS

1. If necessary, remove parentheses by using   **Example**

$(ab)^n = a^n b^n$ or $\left(\dfrac{a}{b}\right)^n = \dfrac{a^n}{b^n}$. $(xy)^3 = x^3 y^3$

2. If necessary, simplify powers to powers by using

$(b^m)^n = b^{mn}$. $(x^4)^3 = x^{4\cdot3} = x^{12}$

3. If necessary, be sure that each base appears only once by using

$b^m \cdot b^n = b^{m+n}$ or $\dfrac{b^m}{b^n} = b^{m-n}$. $x^4 \cdot x^3 = x^{4+3} = x^7$

4. If necessary, rewrite exponential expressions with zero powers as 1 ($b^0 = 1$). Furthermore, write the answer with positive exponents by using

$b^{-n} = \dfrac{1}{b^n}$ or $\dfrac{1}{b^{-n}} = b^n$. $\dfrac{x^5}{x^8} = x^{5-8} = x^{-3} = \dfrac{1}{x^3}$

The following example shows how to simplify exponential expressions. Throughout the example, assume that no variable in a denominator is equal to zero.

EXAMPLE 9   Simplifying Exponential Expressions

Simplify:

a. $(-2xy^{-14})(-3x^4y^5)^3$ **b.** $\left(\dfrac{25x^2y^4}{-5x^6y^{-8}}\right)^2$ **c.** $\left(\dfrac{x^{-4}y^7}{2}\right)^{-5}$.

SOLUTION

a. $(-2xy^{-14})(-3x^4y^5)^3$

$= (-2xy^{-14})(-3)^3(x^4)^3(y^5)^3$ Cube each factor in the second parentheses.

$= (-2xy^{-14})(-27)x^{12}y^{15}$ Multiply the exponents when raising a power to a power: $(x^4)^3 = x^{4\cdot3} = x^{12}$ and $(y^5)^3 = y^{5\cdot3} = y^{15}$.

$= (-2)(-27)x^{1+12}y^{-14+15}$ Mentally rearrange factors and multiply like bases by adding the exponents.

$= 54x^{13}y$ Simplify.

b. $\left(\dfrac{25x^2y^4}{-5x^6y^{-8}}\right)^2$ The expression inside parentheses contains the same bases and can be simplified. Begin with this simplification.

$= (-5x^{2-6}y^{4-(-8)})^2$ Simplify inside the parentheses. Subtract the exponents when dividing.

$= (-5x^{-4}y^{12})^2$ Simplify.

$= (-5)^2(x^{-4})^2(y^{12})^2$ Square each factor in parentheses.

$= 25x^{-8}y^{24}$ Multiply the exponents when raising a power to a power: $(x^{-4})^2 = x^{-4(2)} = x^{-8}$ and $(y^{12})^2 = y^{12\cdot2} = y^{24}$.

$= \dfrac{25y^{24}}{x^8}$ Simplify x^{-8} using $b^{-n} = \dfrac{1}{b^n}$.

c. $\left(\dfrac{x^{-4}y^7}{2}\right)^{-5}$

$= \dfrac{(x^{-4}y^7)^{-5}}{2^{-5}}$ Raise the numerator and the denominator to the -5 power.

$= \dfrac{(x^{-4})^{-5}(y^7)^{-5}}{2^{-5}}$ Raise each factor in the numerator to the -5 power.

$= \dfrac{x^{20}y^{-35}}{2^{-5}}$ Multiply the exponents when raising a power to a power: $(x^{-4})^{-5} = x^{-4(-5)} = x^{20}$ and $(y^7)^{-5} = y^{7(-5)} = y^{-35}$.

$= \dfrac{2^5x^{20}}{y^{35}}$ Move each base with a negative exponent to the other side of the fraction bar and make each negative exponent positive.

$= \dfrac{32x^{20}}{y^{35}}$ Simplify: $2^5 = 2\cdot2\cdot2\cdot2\cdot2 = 32$.

STUDY TIP

Try to avoid the following common errors that can occur when simplifying exponential expressions.

Correct	Incorrect	Description of Error
$b^3 \cdot b^4 = b^7$	$b^3 \cdot b^4 = b^{12}$	The exponents should be added, not multiplied.
$3^2 \cdot 3^4 = 3^6$	$3^2 \cdot 3^4 = 9^6$	The common base should be retained, not multiplied.
$\dfrac{5^{16}}{5^4} = 5^{12}$	$\dfrac{5^{16}}{5^4} = 5^4$	The exponents should be subtracted, not divided.
$(4a)^3 = 64a^3$	$(4a)^3 = 4a^3$	Both factors should be cubed.
$b^{-n} = \dfrac{1}{b^n}$	$b^{-n} \neq -\dfrac{1}{b^n}$	Only the exponent should change sign.
$(a+b)^{-1} = \dfrac{1}{a+b}$	$(a+b)^{-1} = \dfrac{1}{a}+\dfrac{1}{b}$	The exponent applies to the entire expression $a+b$.

✓ **CHECK POINT 9** Simplify:

a. $(-3x^{-6}y)(-2x^3y^4)^2$ $\quad -12y^9$ **b.** $\left(\dfrac{10x^3y^5}{5x^6y^{-2}}\right)^2$ $\dfrac{4y^{14}}{x^6}$ **c.** $\left(\dfrac{x^3y^5}{4}\right)^{-3}$ $\cdot$ $\dfrac{64}{x^9y^{15}}$

ENRICHMENT ESSAY

Exponents and Large Numbers

- Although people do a great deal of talking, the total output since the beginning of gabble to the present day, including all baby talk, love songs, and congressional debates, only amounts to about 10^{16} (10 million billion) words.

- A googol is 10^{100}. The number of raindrops falling on Miami in a century is much less than a googol.

- A googol times a googol is $10^{100} \cdot 10^{100}$, or 10^{200}, which is much less than a googolplex! A googolplex is 10 to the googol power, or $10^{10^{100}}$, or 1 with a googol of zeros! If you tried to write a googolplex without using exponential notation, there would not be enough room to write it if you went to the farthest star, touring all the nebulae, and putting down zeros every inch of the way.

- The number of grains of sand on the beach at Coney Island, New York, is about 10^{20}.

- The number of snow crystals necessary to form the Ice Age was approximately 10^{30}.

- The number 10^{80} is greater than the total number of atoms in our entire universe.

1.6 EXERCISE SET

Student Solutions Manual   CD/Video   PH Math/Tutor Center   MathXL Tutorials on CD   MathXL®   MyMathLab   Interactmath.com

Practice Exercises

In Exercises 1–14, multiply using the product rule.

1. $b^4 \cdot b^7$ $\quad b^{11}$ **2.** $b^5 \cdot b^9$ $\quad b^{14}$

3. $x \cdot x^3$ $\quad x^4$ **4.** $x \cdot x^4$ $\quad x^5$

5. $2^3 \cdot 2^2$ $\quad 32$ **6.** $2^4 \cdot 2^2$ $\quad 64$

7. $3x^4 \cdot 2x^2$ $\quad 6x^6$ **8.** $5x^3 \cdot 3x^2$ $\quad 15x^5$

9. $(-2y^{10})(-10y^2)$ $\quad 20y^{12}$ **10.** $(-4y^8)(-8y^4)$ $\quad 32y^{12}$

11. $(5x^3y^4)(20x^7y^8)$ $\quad 100x^{10}y^{12}$

12. $(4x^5y^6)(20x^7y^4)$ $\quad 80x^{12}y^{10}$

13. $(-3x^4y^0z)(-7xyz^3)$ $\quad 21x^5yz^4$

14. $(-9x^3yz^4)(-5xy^0z^2)$ $\quad 45x^4yz^6$

In Exercises 15–24, divide using the quotient rule.

15. $\dfrac{b^{12}}{b^3}$ $\quad b^9$ **16.** $\dfrac{b^{25}}{b^5}$ $\quad b^{20}$

17. $\dfrac{15x^9}{3x^4}$ $\quad 5x^5$ **18.** $\dfrac{18x^{11}}{3x^4}$ $\quad 6x^7$

19. $\dfrac{x^9y^7}{x^4y^2}$ $\quad x^5y^5$ **20.** $\dfrac{x^9y^{12}}{x^2y^6}$ $\quad x^7y^6$

21. $\dfrac{50x^2y^7}{5xy^4}$ $\quad 10xy^3$ **22.** $\dfrac{36x^{12}y^4}{4xy^2}$ $\quad 9x^{11}y^2$

23. $\dfrac{-56a^{12}b^{10}c^8}{7ab^2c^4}$ $\quad -8a^{11}b^8c^4$ **24.** $\dfrac{-66a^9b^7c^6}{6a^3bc^2}$ $\quad -11a^6b^6c^4$

In Exercises 25–34, use the zero-exponent rule to simplify each expression.

25. 6^0 $\quad 1$ **26.** 9^0 $\quad 1$

27. $(-4)^0$ $\quad 1$ **28.** $(-2)^0$ $\quad 1$

29. -4^0 $\quad -1$ **30.** -2^0 $\quad -1$

31. $13y^0$ $\quad 13$ **32.** $17y^0$ $\quad 17$

33. $(13y)^0$ $\quad 1$ **34.** $(17y)^0$ $\quad 1$

In Exercises 35–52, write each expression with positive exponents only. Then simplify, if possible.

35. 3^{-2} $\quad \dfrac{1}{3^2} = \dfrac{1}{9}$ **36.** 4^{-2} $\quad \dfrac{1}{4^2} = \dfrac{1}{16}$

37. $(-5)^{-2}$ $\quad \dfrac{1}{(-5)^2} = \dfrac{1}{25}$ **38.** $(-7)^{-2}$ $\quad \dfrac{1}{(-7)^2} = \dfrac{1}{49}$

39. -5^{-2} $\quad -\dfrac{1}{5^2} = -\dfrac{1}{25}$ **40.** -7^{-2} $\quad -\dfrac{1}{7^2} = -\dfrac{1}{49}$

41. x^2y^{-3} $\quad \dfrac{x^2}{y^3}$ **42.** x^3y^{-4} $\quad \dfrac{x^3}{y^4}$

43. $8x^{-7}y^3$ $\quad \dfrac{8y^3}{x^7}$ **44.** $9x^{-8}y^4$ $\quad \dfrac{9y^4}{x^8}$

45. $\dfrac{1}{5^{-3}}$ $\quad 5^3 = 125$ **46.** $\dfrac{1}{2^{-5}}$ $\quad 2^5 = 32$

47. $\dfrac{1}{(-3)^{-4}}$ $\quad (-3)^4 = 81$ **48.** $\dfrac{1}{(-2)^{-4}}$ $\quad (-2)^4 = 16$

49. $\dfrac{x^{-2}}{y^{-5}} \cdot \dfrac{y^5}{x^2}$ $\dfrac{y^{10}}{x^4}$

50. $\dfrac{x^{-3}}{y^{-7}} \cdot \dfrac{y^7}{x^3}$ $\dfrac{y^{14}}{x^6}$

51. $\dfrac{a^{-4}b^7}{c^{-3}} \cdot \dfrac{b^7c^3}{a^4}$ $\dfrac{b^{14}c^6}{a^8}$

52. $\dfrac{a^{-3}b^8}{c^{-2}} \cdot \dfrac{b^8c^2}{a^3}$ $\dfrac{b^{16}c^4}{a^6}$

In Exercises 53–58, simplify each expression using the power rule.

53. $(x^6)^{10}$ x^{60}

54. $(x^3)^2$ x^6

55. $(b^4)^{-3}$ $\dfrac{1}{b^{12}}$

56. $(b^8)^{-3}$ $\dfrac{1}{b^{24}}$

57. $(7^{-4})^{-5}$ 7^{20}

58. $(9^{-4})^{-5}$ 9^{20}

In Exercises 59–72, simplify each expression using the products-to-powers rule.

59. $(4x)^3$ $64x^3$

60. $(2x)^5$ $32x^5$

61. $(-3x^7)^2$ $9x^{14}$

62. $(-4x^9)^2$ $16x^{18}$

63. $(2xy^2)^3$ $8x^3y^6$

64. $(3x^2y)^4$ $81x^8y^4$

65. $(-3x^2y^5)^2$ $9x^4y^{10}$

66. $(-3x^4y^6)^2$ $9x^8y^{12}$

67. $(-3x^{-2})^{-3}$ $-\dfrac{x^6}{27}$

68. $(-2x^{-4})^{-3}$ $-\dfrac{x^{12}}{8}$

69. $(5x^3y^{-4})^{-2}$ $\dfrac{y^8}{25x^6}$

70. $(7x^2y^{-5})^{-2}$ $\dfrac{y^{10}}{49x^4}$

71. $(-2x^{-5}y^4z^2)^{-4}$ $\dfrac{x^{20}}{16y^{16}z^8}$

72. $(-2x^{-4}y^5z^3)^{-4}$ $\dfrac{x^{16}}{16y^{20}z^{12}}$

In Exercises 73–84, simplify each expression using the quotients-to-powers rule.

73. $\left(\dfrac{2}{x}\right)^4$ $\dfrac{16}{x^4}$

74. $\left(\dfrac{y}{2}\right)^5$ $\dfrac{y^5}{32}$

75. $\left(\dfrac{x^3}{5}\right)^2$ $\dfrac{x^6}{25}$

76. $\left(\dfrac{x^4}{6}\right)^2$ $\dfrac{x^8}{36}$

77. $\left(-\dfrac{3x}{y}\right)^4$ $\dfrac{81x^4}{y^4}$

78. $\left(-\dfrac{2x}{y}\right)^5$ $-\dfrac{32x^5}{y^5}$

79. $\left(\dfrac{x^4}{y^2}\right)^6$ $\dfrac{x^{24}}{y^{12}}$

80. $\left(\dfrac{x^5}{y^3}\right)^6$ $\dfrac{x^{30}}{y^{18}}$

81. $\left(\dfrac{x^3}{y^{-4}}\right)^3$ x^9y^{12}

82. $\left(\dfrac{x^4}{y^{-2}}\right)^3$ $x^{12}y^6$

83. $\left(\dfrac{a^{-2}}{b^3}\right)^{-4}$ a^8b^{12}

84. $\left(\dfrac{a^{-3}}{b^5}\right)^{-4}$ $a^{12}b^{20}$

In Exercises 85–116, simplify each exponential expression.

85. $\dfrac{x^3}{x^9}$ $\dfrac{1}{x^6}$

86. $\dfrac{x^6}{x^{10}}$ $\dfrac{1}{x^4}$

87. $\dfrac{20x^3}{-5x^4}$ $-\dfrac{4}{x}$

88. $\dfrac{10x^5}{-2x^6}$ $-\dfrac{5}{x}$

89. $\dfrac{16x^3}{8x^{10}}$ $\dfrac{2}{x^7}$

90. $\dfrac{15x^2}{3x^{11}}$ $\dfrac{5}{x^9}$

91. $\dfrac{20a^3b^8}{2ab^{13}}$ $\dfrac{10a^2}{b^5}$

92. $\dfrac{72a^5b^{11}}{9ab^{17}}$ $\dfrac{8a^4}{b^6}$

93. $x^3 \cdot x^{-12}$ $\dfrac{1}{x^9}$

94. $x^4 \cdot x^{-12}$ $\dfrac{1}{x^8}$

95. $(2a^5)(-3a^{-7})$ $-\dfrac{6}{a^2}$

96. $(4a^2)(-2a^{-5})$ $-\dfrac{8}{a^3}$

97. $\left(-\dfrac{1}{4}x^{-4}y^5z^{-1}\right)(-12x^{-3}y^{-1}z^4)$ $\dfrac{3y^4z^3}{x^7}$

98. $\left(-\dfrac{1}{3}x^{-5}y^4z^6\right)(-18x^{-2}y^{-1}z^{-7})$ $\dfrac{6y^3}{x^7z}$

99. $\dfrac{6x^2}{2x^{-8}}$ $3x^{10}$

100. $\dfrac{12x^5}{3x^{-10}}$ $4x^{15}$

101. $\dfrac{x^{-7}}{x^3}$ $\dfrac{1}{x^{10}}$

102. $\dfrac{x^{-10}}{x^4}$ $\dfrac{1}{x^{14}}$

103. $\dfrac{30x^2y^5}{-6x^8y^{-3}}$ $-\dfrac{5y^8}{x^6}$

104. $\dfrac{24x^2y^{13}}{-2x^5y^{-2}}$ $-\dfrac{12y^{15}}{x^3}$

105. $\dfrac{-24a^3b^{-5}c^5}{-3a^{-6}b^{-4}c^{-7}}$ $\dfrac{8a^9c^{12}}{b}$

106. $\dfrac{-24a^2b^{-2}c^8}{-8a^{-5}b^{-1}c^{-3}}$ $\dfrac{3a^7c^{11}}{b}$

107. $\left(\dfrac{x^3}{x^{-5}}\right)^2$ x^{16}

108. $\left(\dfrac{x^4}{x^{-11}}\right)^3$ x^{45}

109. $\left(\dfrac{-15a^4b^2}{5a^{10}b^{-3}}\right)^3$ $-\dfrac{27b^{15}}{a^{18}}$

110. $\left(\dfrac{-30a^{14}b^8}{10a^{17}b^{-2}}\right)^3$ $-\dfrac{27b^{30}}{a^9}$

111. $\left(\dfrac{3a^{-5}b^2}{12a^3b^{-4}}\right)^0$ 1

112. $\left(\dfrac{4a^{-5}b^3}{12a^3b^{-5}}\right)^0$ 1

113. $\left(\dfrac{x^{-5}y^8}{3}\right)^{-4}$ $\dfrac{81x^{20}}{y^{32}}$

114. $\left(\dfrac{x^6y^{-7}}{2}\right)^{-3}$ $\dfrac{8y^{21}}{x^{18}}$

115. $\left(\dfrac{20a^{-3}b^4c^5}{-2a^{-5}b^{-2}c}\right)^{-2}$ $\dfrac{1}{100a^4b^{12}c^8}$

116. $\left(\dfrac{-2a^{-4}b^3c^{-1}}{3a^{-2}b^{-5}c^{-2}}\right)^{-4}$ $\dfrac{81a^8}{16b^{32}c^4}$

Practice Plus

In Exercises 117–124, simplify each exponential expression.

117. $\dfrac{9y^4}{x^{-2}} + \left(\dfrac{x^{-1}}{y^2}\right)^{-2}$ $10x^2y^4$

118. $\dfrac{7x^3}{y^{-9}} + \left(\dfrac{x^{-1}}{y^3}\right)^{-3}$ $8x^3y^9$

119. $\left(\dfrac{3x^4}{y^{-4}}\right)^{-1}\left(\dfrac{2x}{y^2}\right)^3$ $\dfrac{8}{3xy^{10}}$

120. $\left(\dfrac{2^{-1}x^{-2}y}{x^4y^{-1}}\right)^{-2}\left(\dfrac{xy^{-3}}{x^{-3}y}\right)^3$ $\dfrac{4x^{24}}{y^{16}}$

121. $(-4x^3y^{-5})^{-2}(2x^{-8}y^{-5})$ $\dfrac{y^5}{8x^{14}}$

122. $(-4x^{-4}y^5)^{-2}(-2x^5y^{-6})$ $-\dfrac{x^{13}}{8y^{16}}$

123. $\dfrac{(2x^2y^4)^{-1}(4xy^3)^{-3}}{(x^2y)^{-5}(x^3y^2)^4}$ $\dfrac{1}{128x^7y^{16}}$

124. $\dfrac{(3x^3y^2)^{-1}(2x^2y)^{-2}}{(xy^2)^{-5}(x^2y^3)^3}$ $\dfrac{1}{12x^8y^3}$

Application Exercises

The 1986 explosion at the Chernobyl nuclear power plant in the former Soviet Union sent about 1000 kilograms of radioactive cesium-137 into the atmosphere. The formula

$$y = 1000\left(\frac{1}{2}\right)^{x/30}$$

models the amount, y, in kilograms, of cesium-137 remaining in Chernobyl x years after 1986. If even 100 kilograms of cesium-137 remain in Chernobyl's atmosphere, the area is considered unsafe for human habitation. Use this information to solve Exercises 125–126.

125. Substitute 90 for x in the formula and determine if Chernobyl will be safe for human habitation 90 years after 1986, by 2076.   125 kilograms; not safe

126. Substitute 60 for x in the formula and determine if Chernobyl will be safe for human habitation 60 years after 1986, by 2046.   250 kilograms; not safe

The astronomical unit (AU) is often used to measure distances within the solar system. One AU is equal to the average distance between Earth and the sun, or 92,955,630 miles. The distance, d, of the nth planet from the sun is modeled by the formula

$$d = \frac{3(2^{n-2}) + 4}{10}$$

where d is measured in astronomical units. Use this formula to solve Exercises 127–130.

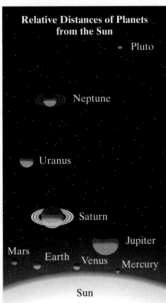

Relative Distances of Planets from the Sun

127. Substitute 1 for n and find the distance between Mercury and the sun.   0.55 astronomical unit

128. Substitute 2 for n and find the distance between Venus and the sun.   0.7 astronomical unit

129. Substitute 3 for n and evaluate the exponential expression. Describe what this means in practical terms.   1; Earth is 1 astronomical unit from the sun.

130. Substitute 5 for n and evaluate the exponential expression. Describe what this means in practical terms.   2.8; Jupiter is 2.8 astronomical units from the sun.

Writing in Mathematics

131. Explain the product rule for exponents. Use $b^2 \cdot b^3$ in your explanation.   131–138. Answers will vary.

132. Explain the quotient rule for exponents. Use $\dfrac{b^8}{b^2}$ in your explanation.

133. Explain how to find any nonzero number to the 0 power.

134. Explain the negative-exponent rule and give an example.

135. Explain the power rule for exponents. Use $(b^2)^3$ in your explanation.

136. Explain how to simplify an expression that involves a product raised to a power. Give an example.

137. Explain how to simplify an expression that involves a quotient raised to a power. Give an example.

138. How do you know if an exponential expression is simplified?

Technology Exercise

139. Graph $y = 1000(0.5)^{x/30}$, the Chernobyl model described in Exercises 125–126, in a [0, 110, 10] by [0, 1000, 100] viewing rectangle. Enter the equation using

$$y_1 = 1000 \boxed{\times} .5 \boxed{\wedge} \boxed{(} (x \boxed{\div} 30 \boxed{)} .$$

Then $\boxed{\text{TRACE}}$ along the graph and determine when y is less than 100 kilograms of cesium-137. Approximately how many years after 1986 will Chernobyl be safe for human habitation?   about 100 years or 2086   See graphing answer section.

Critical Thinking Exercises

140. Which one of the following is true?   d

    a. $2^2 \cdot 2^4 = 2^8$ **b.** $5^6 \cdot 5^2 = 25^8$

    c. $2^3 \cdot 3^2 = 6^5$ **d.** None of these is true.

141. Which one of the following is true?   d

    a. $\dfrac{1}{(-2)^3} = 2^{-3}$ **b.** $\dfrac{2^8}{2^{-3}} = 2^5$

    c. $2^4 + 2^5 = 2^9$ **d.** None of these is true.

In Exercises 142–145, simplify the expression. Assume that all variables used as exponents represent integers and that all other variables represent nonzero real numbers.

142. $x^{n-1} \cdot x^{3n+4}$ x^{4n+3} **143.** $(x^{-4n} \cdot x^n)^{-3}$ x^{9n}

144. $\left(\dfrac{x^{3-n}}{x^{6-n}}\right)^{-2}$ x^6 **145.** $\left(\dfrac{x^n y^{3n+1}}{y^n}\right)^3$ $x^{3n}y^{6n+3}$

Review Exercises

146. Graph $y = 2x - 1$ in a rectangular coordinate system. Let $x = -3, -2, -1, 0, 1, 2,$ and 3. (Section 1.3, Example 2)   See graphing answer section.

147. Solve $Ax + By = C$ for y. (Section 1.5, Example 6) $y = \dfrac{C - Ax}{B}$

148. The length of a rectangular playing field is 5 meters less than twice its width. If 230 meters of fencing enclose the field, what are its dimensions? (Section 1.5, Example 5)   40 meters by 75 meters

SECTION

1.7

SCIENTIFIC NOTATION

Objectives

1 Convert from scientific to decimal notation.

2 Convert from decimal to scientific notation.

3 Perform computations with scientific notation.

4 Use scientific notation to solve problems.

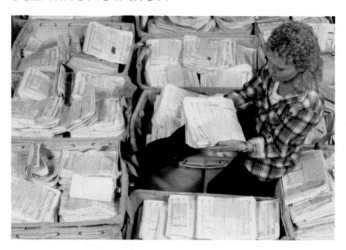

People who complain about paying their income tax can be divided into two types: men and women. Perhaps we can quantify the complaining by examining the data in Figure 1.20. The bar graphs show the U.S. population, in millions, and the total amount we paid in federal taxes, in trillions of dollars, for six selected years.

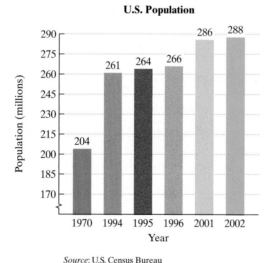

U.S. Population

Source: U.S. Census Bureau

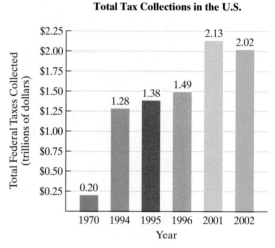

Total Tax Collections in the U.S.

Source: Internal Revenue Service

FIGURE 1.20   Population and total tax collections in the U.S.

The bar graph on the right shows that in 2002, total tax collections were $2.02 trillion. How can we place this amount in the proper perspective? If the total tax collections were evenly divided among all Americans, how much would each citizen pay in taxes?

In this section, you will learn to use exponents to provide a way of putting large and small numbers in perspective. Using this skill, we will explore the per capita tax for some of the years shown in Figure 1.20.

1 Convert from scientific to decimal notation.

Table 1.6

Names of Large Numbers

10^2	hundred
10^3	thousand
10^6	million
10^9	billion
10^{12}	trillion
10^{15}	quadrillion
10^{18}	quintillion
10^{21}	sextillion
10^{24}	septillion
10^{27}	octillion
10^{30}	nonillion
10^{100}	googol

Scientific Notation We have seen that in 2002, total tax collections were \$2.02 trillion. Because a trillion is 10^{12} (see Table 1.6), this amount can be expressed as

$$2.02 \times 10^{12}.$$

The number 2.02×10^{12} is written in a form called *scientific notation*.

> **SCIENTIFIC NOTATION** A number is written in **scientific notation** when it is expressed in the form
>
> $$a \times 10^n,$$
>
> where the absolute value of a is greater than or equal to 1 and less than 10 $(1 \le |a| < 10)$, and n is an integer.

It is customary to use the multiplication symbol, $\times$, rather than a dot, when writing a number in scientific notation.

Converting from Scientific to Decimal Notation Here are two examples of numbers in scientific notation:

$$6.4 \times 10^5 \quad \text{means} \quad 640{,}000.$$
$$2.17 \times 10^{-3} \quad \text{means} \quad 0.00217.$$

Do you see that the number with the positive exponent is relatively large and the number with the negative exponent is relatively small?

We can use n, the exponent on the 10 in $a \times 10^n$, to change a number in scientific notation to decimal notation. If n is **positive**, move the decimal point in a to the **right** n places. If n is **negative**, move the decimal point in a to the **left** $|n|$ places.

EXAMPLE 1 Converting from Scientific to Decimal Notation

Write each number in decimal notation:

a. 6.2×10^7 **b.** -6.2×10^7 **c.** 2.019×10^{-3} **d.** -2.019×10^{-3}

SOLUTION In each case, we use the exponent on the 10 to move the decimal point. In parts (a) and (b), the exponent is positive, so we move the decimal point to the right. In parts (c) and (d), the exponent is negative, so we move the decimal point to the left.

a. $6.2 \times 10^7 = 62{,}000{,}000$

$n = 7$ Move the decimal point 7 places to the right.

b. $-6.2 \times 10^7 = -62{,}000{,}000$

$n = 7$ Move the decimal point 7 places to the right.

c. $2.019 \times 10^{-3} = 0.002019$

$n = -3$ Move the decimal point $|-3|$ places, or 3 places, to the left.

d. $-2.019 \times 10^{-3} = -0.002019$

$n = -3$ Move the decimal point $|-3|$ places, or 3 places, to the left.

✔ **CHECK POINT 1** Write each number in decimal notation:

a. -2.6×10^9 $-2{,}600{,}000{,}000$ **b.** 3.017×10^{-6}. 0.000003017

2 Convert from decimal to scientific notation.

Converting from Decimal to Scientific Notation To convert from decimal notation to scientific notation, we reverse the procedure of Example 1.

> **CONVERTING FROM DECIMAL TO SCIENTIFIC NOTATION** Write the number in the form $a \times 10^n$.
>
> • Determine a, the numerical factor. Move the decimal point in the given number to obtain a number whose absolute value is between 1 and 10, including 1.
> • Determine n, the exponent on 10^n. The absolute value of n is the number of places the decimal point was moved. The exponent n is positive if the decimal point was moved to the left, negative if the decimal point was moved to the right, and 0 if the decimal point was not moved.

EXAMPLE 2 Converting from Decimal Notation to Scientific Notation

Write each number in scientific notation:

a. 34,970,000,000,000
b. −34,970,000,000,000
c. 0.0000000000802
d. −0.0000000000802.

SOLUTION

a. $34{,}970{,}000{,}000{,}000 = 3.497 \times 10^{13}$

Move the decimal point to get a number whose absolute value is between 1 and 10.

The decimal point was moved 13 places to the left, so $n = 13$.

b. $-34{,}970{,}000{,}000{,}000 = -3.497 \times 10^{13}$

c. $0.0000000000802 = 8.02 \times 10^{-11}$

Move the decimal point to get a number whose absolute value is between 1 and 10.

The decimal point was moved 11 places to the right, so $n = -11$.

d. $-0.0000000000802 = -8.02 \times 10^{-11}$

USING TECHNOLOGY

You can use your calculator's EE (enter exponent) or EXP key to convert from decimal to scientific notation. Here is how it's done for 0.0000000000802.

Many Scientific Calculators

Keystrokes

.0000000000802 EE =

Display

8.02 − 11

Many Graphing Calculators

Use the mode setting for scientific notation.

Keystrokes

.0000000000802 ENTER

Display

8.02 E −11

✔ **CHECK POINT 2** Write each number in scientific notation:

a. 5,210,000,000 5.21×10^9 **b.** −0.00000006893 -6.893×10^{-8}

EXAMPLE 3 Expressing the U.S. Population in Scientific Notation

In 2002, the population of the United States was approximately 288 million. Express the population in scientific notation.

SOLUTION Because one million is 10^6 (see Table 1.6 on page 79), the 2002 population can be expressed as

$$288 \times 10^6.$$

This factor is not between 1 and 10, so the number is not in scientific notation.

The voice balloon indicates that we need to convert 288 to scientific notation.

$$288 \times 10^6 = (2.88 \times 10^2) \times 10^6 = 2.88 \times 10^{2+6} = 2.88 \times 10^8$$

$$288 = 2.88 \times 10^2$$

In scientific notation, the population is 2.88×10^8. ■

✔ CHECK POINT **3** Express 410×10^7 in scientific notation. $\quad 4.1 \times 10^9$

3 Perform computations with scientific notation.

Computations with Scientific Notation Properties of exponents are used to perform computations with numbers that are expressed in scientific notation.

COMPUTATIONS WITH NUMBERS IN SCIENTIFIC NOTATION

Multiplication

$$(a \times 10^n)(b \times 10^m) = (a \times b) \times 10^{n+m}$$

> Add the exponents on 10 and multiply the other parts of the numbers separately.

Division

$$\frac{a \times 10^n}{b \times 10^m} = \left(\frac{a}{b}\right) \times 10^{n-m}$$

> Subtract the exponents on 10 and divide the other parts of the numbers separately.

After the computation is completed, the answer may require an adjustment before it is back in scientific notation.

EXAMPLE 4 Computations with Scientific Notation

Perform the indicated computations, writing the answers in scientific notation:

a. $(6.1 \times 10^5)(4 \times 10^{-9})$

b. $\dfrac{1.8 \times 10^4}{3 \times 10^{-2}}$.

SOLUTION

a. $(6.1 \times 10^5)(4 \times 10^{-9})$

$= (6.1 \times 4) \times (10^5 \times 10^{-9})$ Regroup factors.

$= 24.4 \times 10^{5+(-9)}$ Add the exponents on 10 and multiply the other parts.

$= 24.4 \times 10^{-4}$ Simplify.

$= (2.44 \times 10^1) \times 10^{-4}$ Convert 24.4 to scientific notation: $24.4 = 2.44 \times 10^1$

$= 2.44 \times 10^{-3}$ $10^1 \times 10^{-4} = 10^{1+(-4)} = 10^{-3}$

USING TECHNOLOGY

$(6.1 \times 10^5)(4 \times 10^{-9})$
On a Calculator:

Many Scientific Calculators

6.1 $\boxed{\text{EE}}$ 5 $\boxed{\times}$ 4 $\boxed{\text{EE}}$ 9 $\boxed{+/-}$ $\boxed{=}$

Display

$2.44 - 03$

Many Graphing Calculators

6.1 $\boxed{\text{EE}}$ 5 $\boxed{\times}$ 4 $\boxed{\text{EE}}$ $\boxed{(-)}$ 9 $\boxed{\text{ENTER}}$

Display (in scientific notation mode)

$2.44\text{E} - 3$

b. $\dfrac{1.8 \times 10^4}{3 \times 10^{-2}} = \left(\dfrac{1.8}{3}\right) \times \left(\dfrac{10^4}{10^{-2}}\right)$ Regroup factors.

$= 0.6 \times 10^{4-(-2)}$ Subtract the exponents on 10 and divide the other parts.

$= 0.6 \times 10^6$ Simplify: $4 - (-2) = 4 + 2 = 6$.

$= (6 \times 10^{-1}) \times 10^6$ Convert 0.6 to scientific notation: $0.6 = 6 \times 10^{-1}$.

$= 6 \times 10^5$ $10^{-1} \times 10^6 = 10^{-1+6} = 10^5$ ■

✔ **CHECK POINT 4** Perform the indicated computations, writing the answers in scientific notation:

a. $(7.1 \times 10^5)(5 \times 10^{-7})$ 3.55×10^{-1} **b.** $\dfrac{1.2 \times 10^6}{3 \times 10^{-3}}.$ 4×10^8

---

4   Use scientific notation to solve problems.

Applications: Putting Numbers in Perspective   We have seen that in 2002, the U.S. government collected $2.02 trillion in taxes. Example 5 shows how we can use scientific notation to comprehend the meaning of a number such as 2.02 trillion.

EXAMPLE 5   Tax per Capita

In 2002, the U.S. government collected 2.02×10^{12} dollars in taxes. At that time, the U.S. population was approximately 288 million, or 2.88×10^8. If the total tax collections were evenly divided among all Americans, how much would each citizen pay? Express the answer in decimal notation, rounded to the nearest dollar.

SOLUTION   The amount that we would each pay, or the tax per capita, is the total amount collected, 2.02×10^{12}, divided by the number of Americans, 2.88×10^8.

$$\frac{2.02 \times 10^{12}}{2.88 \times 10^8} = \left(\frac{2.02}{2.88}\right) \times \left(\frac{10^{12}}{10^8}\right) \approx 0.7014 \times 10^{12-8} = 0.7014 \times 10^4 = 7014$$

To obtain an answer in decimal notation, it is not necessary to express this number in scientific notation.

Move the decimal point 4 places to the right.

If total tax collections were evenly divided, we would each pay approximately $7014 in taxes. ■

✔ **CHECK POINT 5** In 2001, the U.S. government collected 2.13×10^{12} dollars in taxes. At that time, the U.S. population was approximately 286 million, or 2.86×10^8. Find the per capita tax, rounded to the nearest dollar, in 2001. $7448

Many problems in algebra involve motion. Suppose that you ride your bike at an average speed of 12 miles per hour. What distance do you cover in 2 hours? Your distance is the product of your speed and the time that you travel:

$$\frac{12 \text{ miles}}{\text{hour}} \times 2 \text{ hours} = 24 \text{ miles}.$$

Your distance is 24 miles. Notice how the hour units cancel. The distance is expressed in miles.

In general, the distance covered by any moving body is the product of its average speed, or rate, and its time in motion.

A FORMULA FOR MOTION

$$d = rt$$

Distance equals rate times time.

EXAMPLE 6   Using the Motion Formula

Light travels at a rate of approximately 1.86×10^5 miles per second. It takes light 5×10^2 seconds to travel from the sun to Earth. What is the distance between Earth and the sun?

SOLUTION

$d = rt$	Use the motion formula.
$d = (1.86 \times 10^5) \times (5 \times 10^2)$	Substitute the given values.
$d = (1.86 \times 5) \times (10^5 \times 10^2)$	Rearrange factors.
$d = 9.3 \times 10^7$	Add the exponents on 10 and multiply the other parts.

The distance between Earth and the sun is approximately 9.3×10^7 miles, or 93 million miles.

CHECK POINT 6 A futuristic spacecraft traveling at 1.55×10^3 miles per hour takes 20,000 hours (about 833 days) to travel from Venus to Mercury. What is the distance from Venus to Mercury? 3.1×10^7 miles, or 31 million miles

1.7 EXERCISE SET

Student Solutions Manual   CD/Video   PH Math/Tutor Center   MathXL Tutorials on CD   MathXL®   MyMathLab   Interactmath.com

Practice Exercises

In Exercises 1–14, write each number in decimal notation without the use of exponents.

1. 3.8×10^2 380
2. 9.2×10^2 920
3. 6×10^{-4} 0.0006
4. 7×10^{-5} 0.00007
5. -7.16×10^6 $-7,160,000$
6. -8.17×10^6 $-8,170,000$
7. 1.4×10^0 1.4
8. 2.4×10^0 2.4
9. 7.9×10^{-1} 0.79
10. 6.8×10^{-1} 0.68
11. -4.15×10^{-3} -0.00415
12. -3.14×10^{-3} -0.00314
13. -6.00001×10^{10} $-60,000,100,000$
14. -7.00001×10^{10} $-70,000,100,000$

In Exercises 15–30, write each number in scientific notation.

15. 32,000 3.2×10^4
16. 64,000 6.4×10^4
17. 638,000,000,000,000,000 6.38×10^{17}
18. 579,000,000,000,000,000 5.79×10^{17}
19. -317 -3.17×10^2
20. -326 -3.26×10^2

21. -5716 -5.716×10^3
22. -3829 -3.829×10^3
23. 0.0027 2.7×10^{-3}
24. 0.0083 8.3×10^{-3}
25. -0.00000000504 -5.04×10^{-9}
26. -0.00000000405 -4.05×10^{-9}
27. 0.007 7×10^{-3}
28. 0.005 5×10^{-3}
29. 3.14159 3.14159×10^0
30. 2.71828 2.71828×10^0

In Exercises 31–50, perform the indicated computations. Write the answers in scientific notation. If necessary, round the decimal factor in your scientific notation answer to two decimal places.

31. $(3 \times 10^4)(2.1 \times 10^3)$ 6.3×10^7
32. $(2 \times 10^4)(4.1 \times 10^3)$ 8.2×10^7
33. $(1.6 \times 10^{15})(4 \times 10^{-11})$ 6.4×10^4
34. $(1.4 \times 10^{15})(3 \times 10^{-11})$ 4.2×10^4
35. $(6.1 \times 10^{-8})(2 \times 10^{-4})$ 1.22×10^{-11}
36. $(5.1 \times 10^{-8})(3 \times 10^{-4})$ 1.53×10^{-11}
37. $(4.3 \times 10^8)(6.2 \times 10^4)$ 2.67×10^{13}

38. $(8.2 \times 10^8)(4.6 \times 10^4)$ 3.77×10^{13}

39. $\dfrac{8.4 \times 10^8}{4 \times 10^5}$ 2.1×10^3 **40.** $\dfrac{6.9 \times 10^8}{3 \times 10^5}$ 2.3×10^3

41. $\dfrac{3.6 \times 10^4}{9 \times 10^{-2}}$ 4×10^5 **42.** $\dfrac{1.2 \times 10^4}{2 \times 10^{-2}}$ 6×10^5

43. $\dfrac{4.8 \times 10^{-2}}{2.4 \times 10^6}$ 2×10^{-8} **44.** $\dfrac{7.5 \times 10^{-2}}{2.5 \times 10^6}$ 3×10^{-8}

45. $\dfrac{2.4 \times 10^{-2}}{4.8 \times 10^{-6}}$ 5×10^3 **46.** $\dfrac{1.5 \times 10^{-2}}{3 \times 10^{-6}}$ 5×10^3

47. $\dfrac{480,000,000,000}{0.00012}$ 4×10^{15} **48.** $\dfrac{282,000,000,000}{0.00141}$ 2×10^{14}

49. $\dfrac{0.00072 \times 0.003}{0.00024}$ 9×10^{-3} **50.** $\dfrac{66,000 \times 0.001}{0.003 \times 0.002}$ 1.1×10^7

Practice Plus

In Exercises 51–58, solve each equation. Express the solution in scientific notation.

51. $(2 \times 10^{-5})x = 1.2 \times 10^9$ 6×10^{13}

52. $(3 \times 10^{-2})x = 1.2 \times 10^4$ 4×10^5

53. $\dfrac{x}{2 \times 10^8} = -3.1 \times 10^{-5}$ -6.2×10^3

54. $\dfrac{x}{5 \times 10^{11}} = -2.9 \times 10^{-3}$ -1.45×10^9

55. $x - (7.2 \times 10^{18}) = 9.1 \times 10^{18}$ 1.63×10^{19}

56. $x - (5.3 \times 10^{-16}) = 8.4 \times 10^{-16}$ 1.37×10^{-15}

57. $(-1.2 \times 10^{-3})x = (1.8 \times 10^{-4})(2.4 \times 10^6)$ -3.6×10^5

58. $(-7.8 \times 10^{-4})x = (3.9 \times 10^{-7})(6.8 \times 10^5)$ -3.4×10^2

Application Exercises

The graph shows the number of people in the United States ages 65 and over for the year 2000 and projections beyond. Use 10^6 for one million and the figures shown to solve Exercises 59–62. Express all answers in scientific notation.

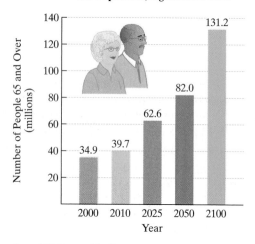

U.S Population, Ages 65 and Over

Source: U.S. Bureau of the Census

59. How many people 65 and over will there be in 2025? 6.26×10^7 people

60. How many people 65 and over will there be in 2050? 8.2×10^7 people

61. How many more people 65 and over will there be in 2100 than in 2000? 9.63×10^7 people

62. How many more people 65 and over will there be in 2050 than in 2000? 4.71×10^7 people

Our ancient ancestors hunted for their meat and expended a great deal of energy chasing it down. Today, our animal protein is raised in cages and on feedlots, delivered in great abundance nearly to our door. Use the numbers shown below to solve Exercises 63–66. Use 10^6 for one million and 10^9 for one billion.

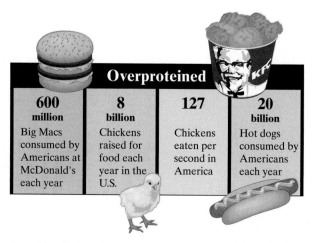

Overproteined

600 million	**8** billion	**127**	**20** billion
Big Macs consumed by Americans at McDonald's each year	Chickens raised for food each year in the U.S.	Chickens eaten per second in America	Hot dogs consumed by Americans each year

Source: Time, October 20, 2003

In Exercises 63–64, use 288 million, or 2.88×10^8, for the U.S. population. Express answers in decimal notation, rounded to the nearest whole number.

63. Find the number of hot dogs consumed by each American in a year.   approximately 69 hot dogs per person

64. If the consumption of Big Macs was divided evenly among all Americans, how many Big Macs would we each consume in a year?   approximately 2 Big Macs per person

In Exercises 65–66, use the fact that there are approximately 3.2×10^7 seconds in a year.

65. How many chickens are raised for food each second in the United States? Express the answer in scientific and decimal notations. $2.5 \times 10^2 = 250$ chickens

66. How many chickens are eaten per year in the United States? Express the answer in scientific notation. 4.064×10^9

67. Approximately 2×10^4 people run in the New York City Marathon each year. Each runner runs a distance of 26 miles. Write the total distance covered by all the runners (assuming that each person completes the marathon) in scientific notation. 5.2×10^5 miles

68. The area of Alaska is approximately 3.66×10^8 acres. The state was purchased in 1867 from Russia for $7.2 million. What price per acre, to the nearest cent, did the United States pay Russia?   Approximately $0.02, or 2¢, per acre

69. The mass of one oxygen molecule is 5.3×10^{-23} gram. Find the mass of 20,000 molecules of oxygen. Express the answer in scientific notation. 1.06×10^{-18} gram

70. The mass of one hydrogen atom is 1.67×10^{-24} gram. Find the mass of 80,000 hydrogen atoms. Express the answer in scientific notation. 1.336×10^{-19} gram

71. In Exercises 65–66, we used 3.2×10^7 as an approximation for the number of seconds in a year. Convert 365 days (one year) to hours, to minutes, and, finally, to seconds, to determine precisely how many seconds there are in a year. Express the answer in scientific notation. 3.1536×10^7

Writing in Mathematics

72. How do you know if a number is written in scientific notation?   72–75. Answers will vary.

73. Explain how to convert from scientific to decimal notation and give an example.

74. Explain how to convert from decimal to scientific notation and give an example.

75. Describe one advantage of expressing a number in scientific notation over decimal notation.

Technology Exercises

76. Use a calculator to check any three of your answers in Exercises 1–14.   76–78. Answers will vary.

77. Use a calculator to check any three of your answers in Exercises 15–30.

78. Use a calculator with an $\boxed{\text{EE}}$ or $\boxed{\text{EXP}}$ key to check any four of your computations in Exercises 31–50. Display the result of the computation in scientific notation.

Critical Thinking Exercises

79. Which one of the following is true?   d

   a. $534.7 = 5.347 \times 10^3$

   b. $\dfrac{8 \times 10^{30}}{4 \times 10^{-5}} = 2 \times 10^{25}$

   c. $(7 \times 10^5) + (2 \times 10^{-3}) = 9 \times 10^2$

   d. $(4 \times 10^3) + (3 \times 10^2) = 43 \times 10^2$

In Exercises 80–81, perform the indicated additions. Write the answers in scientific notation.

80. $5.6 \times 10^{13} + 3.1 \times 10^{13}$ 8.7×10^{13}

81. $8.2 \times 10^{-16} + 4.3 \times 10^{-16}$ 1.25×10^{-15}

82. Our hearts beat approximately 70 times per minute. Express in scientific notation how many times the heart beats over a lifetime of 80 years. Round the decimal factor in your scientific notation answer to two decimal places.   about 2.94×10^9 times

83. Give an example of a number where there is no advantage in using scientific notation over decimal notation.   Answers will vary.

Review Exercises

84. Simplify: $9(10x - 4) - (5x - 10)$. (Section 1.2, Example 14) $85x - 26$

85. Solve: $\dfrac{4x - 1}{10} = \dfrac{5x + 2}{4} - 4$. (Section 1.4, Example 4)   4

86. Simplify: $(8x^4y^{-3})^{-2}$. (Section 1.6, Example 7) $\dfrac{y^6}{64x^8}$

GROUP PROJECT

CHAPTER 1

Additional group projects can be found in the *Instructor's Resource Manual.*

One of the best ways to learn how to *solve* a word problem in algebra is to *design* word problems of your own. Creating a word problem makes you very aware of precisely how much information is needed to solve the problem. You must also focus on the best way to present information to a reader and on how much information to give. As you write your problem, you gain skills that will help you solve problems created by others.

The group should design five different word problems that can be solved using an algebraic equation. All of the problems should be on different topics. For example, the group should not have more than one problem on a price reduction. The group should turn in both the problems and their algebraic solutions.

Definitions and Concepts	Examples

Section 1.1 Algebraic Expressions and Real Numbers

Definitions and Concepts	Examples
Letters that represent numbers are called variables. An algebraic expression is a combination of variables, numbers, and operation symbols. English phrases can be translated into algebraic expressions: • Addition: sum, plus, increased by, more than • Subtraction: difference, minus, decreased by, less than • Multiplication: product, times, of, twice • Division: quotient, divide, per, ratio	Translate: Six less than the product of a number and five. 6 less than — the product of a number and 5 $$5x - 6$$ or $x \cdot 5 - 6$
Many algebraic expressions contain exponents. If b is a natural number, b^n, the nth power of b, is the product of n factors of b. Furthermore, $b^1 = b$. Evaluating an algebraic expression means to find the value of the expression for a given value of the variable.	Evaluate $6 + 5(x - 10)^3$ for $x = 12$. $$= 6 + 5(12 - 10)^3$$ $$= 6 + 5 \cdot 2^3$$ $$= 6 + 5 \cdot 8$$ $$= 6 + 40 = 46$$
An equation is a statement that two expressions are equal. Formulas are equations that express relationships among two or more variables. Mathematical modeling is the process of finding formulas to describe real-world phenomena. Such formulas, together with the meaning assigned to the variables, are called mathematical models. The formulas are said to model, or describe, the relationships among the variables.	The formula $$h = -16t^2 + 200t + 4$$ models the height, h, in feet, of fireworks t seconds after launch. What is the height after 2 seconds? $$h = -16(2)^2 + 200(2) + 4$$ $$= -16(4) + 200(2) + 4$$ $$= -64 + 400 + 4 = 340$$ The height is 340 feet.
A set is a collection of objects, called elements, enclosed in braces. The roster method uses commas to separate the elements of the set. Set-builder notation describes the elements of a set, but does not list them. The symbol $\in$ means that an element is in a set; $\notin$ means that an element is not in a set. The set of real numbers is the set of all numbers that can be represented by points on the number line. Sets that make up the real numbers include Natural numbers: $\{1, 2, 3, 4, \dots\}$ Whole numbers: $\{0, 1, 2, 3, 4, \dots\}$ Integers: $\{\dots, -4, -3, -2, -1, 0, 1, 2, 3, 4, \dots\}$ Rational numbers: $\{\frac{a}{b} \mid a \text{ and } b \text{ are integers and } b \neq 0\}$ Irrational numbers: $\{x \mid x \text{ is a real number and } x \text{ is not a rational number}\}$. In decimal form, rational numbers terminate or repeat. In decimal form, irrational numbers do neither.	• Use the roster method to list the elements of $$\{x \mid x \text{ is a natural number less than 6}\}.$$ **Solution** $$\{1, 2, 3, 4, 5\}$$ • True or false: $$\sqrt{2} \notin \{x \mid x \text{ is a rational number}\}.$$ **Solution** The statement is true: $$\sqrt{2} \text{ is not a rational number.}$$ The decimal form of $\sqrt{2}$ neither terminates nor repeats. Thus, $\sqrt{2}$ is an irrational number.

Definitions and Concepts	Examples

Section 1.1 Algebraic Expressions and Real Numbers (continued)

For any two real numbers, a and b, a is less than b if a is to the left of b on the number line.

Inequality Symbols
$<$: is less than $>$: is greater than
$\leq$: is less than or equal to
$\geq$: is greater than or equal to

- $-1 < 5$, or -1 is less than 5, is true because -1 is to the left of 5 on a number line.

- $-3 \geq 7$, -3 is greater than or equal to 7, is false. Neither $-3 > 7$ nor $-3 = 7$ is true.

Section 1.2 Operations with Real Numbers and Simplifying Algebraic Expressions

Absolute Value
$$|a| = \begin{cases} a & \text{if} \quad a \geq 0 \\ -a & \text{if} \quad a < 0 \end{cases}$$
The opposite, or additive inverse, of a is $-a$. When a is a negative number, $-a$ is positive.

- $|6.03| = 6.03$
- $|0| = 0$
- $|-4.9| = -(-4.9) = 4.9$

Adding Real Numbers
To add two numbers with the same sign, add their absolute values and use their common sign. To add two numbers with different signs, subtract the smaller absolute value from the greater absolute value and use the sign of the number with the greater absolute value.

$$-4.1 + (-6.2) = -10.3$$
$$-30 + 25 = -5$$
$$12 + (-8) = 4$$

Subtracting Real Numbers
$$a - b = a + (-b)$$

$$-\frac{3}{4} - \left(-\frac{1}{2}\right) = -\frac{3}{4} + \frac{1}{2} = -\frac{3}{4} + \frac{2}{4} = -\frac{1}{4}$$

Multiplying and Dividing Real Numbers
The product or quotient of two numbers with the same sign is positive and with different signs is negative. If no number is 0, a product with an even number of negative factors is positive and a product with an odd number of negative factors is negative. Division by 0 is undefined.

- $$2(-6)(-1)(-5) = -60$$

  > Three (odd) negative factors give a negative product.

- $(-2)^3 = (-2)(-2)(-2) = -8$
- $-\dfrac{1}{3}\left(-\dfrac{2}{5}\right) = \dfrac{2}{15}$
- $\dfrac{-14}{2} = -7$

Order of Operations
1. Perform operations within grouping symbols, starting with the innermost grouping symbols. Grouping symbols include parentheses, brackets, fraction bars, absolute value symbols, and square root signs.
2. Evaluate exponential expressions.
3. Multiply and divide, from left to right.
4. Add and subtract, from left to right.

Simplify: $\dfrac{6(8 - 10)^3 + (-2)}{(-5)^2(-2)}$.

$$= \frac{6(-2)^3 + (-2)}{(-5)^2(-2)} = \frac{6(-8) + (-2)}{25(-2)}$$

$$= \frac{-48 + (-2)}{-50} = \frac{-50}{-50} = 1$$

Basic Algebraic Properties

Commutative: $a + b = b + a$ and $ab = ba$
Associative: $(a + b) + c = a + (b + c)$
$(ab)c = a(bc)$
Distributive: $a(b + c) = ab + ac$
$a(b - c) = ab - ac$
$(b + c)a = ba + ca$

- Commutative
$$3x + 5 = 5 + 3x = 5 + x \cdot 3$$
- Associative
$$-4(6x) = (-4 \cdot 6)x = -24x$$
- Distributive
$$-4(9x + 3) = -4(9x) + (-4) \cdot 3$$
$$= -36x + (-12)$$
$$= -36x - 12$$

Definitions and Concepts	Examples

Section 1.2 Operations with Real Numbers and Simplifying Algebraic Expressions (continued)

Simplifying Algebraic Expressions
Terms are separated by addition. Like terms have the same variable factors and are combined using the distributive property. An algebraic expression is simplified when grouping symbols have been removed and like terms have been combined.

Simplify: $7(3x - 4) - (10x - 5)$.

$$= 21x - 28 - 10x + 5$$
$$= 21x - 10x - 28 + 5$$
$$= 11x - 23$$

Section 1.3 Graphing Equations

The rectangular coordinate system consists of a horizontal number line, the x-axis, and a vertical number line, the y-axis, intersecting at their zero points, the origin. Each point in the system corresponds to an ordered pair of real numbers (x, y). The first number in the pair is the x-coordinate; the second number is the y-coordinate.

Plot: $(4, 2)$, $(-3, 4)$, $(-5, -4)$, and $(4, -3)$.

An ordered pair is a solution of an equation in two variables if replacing the variables by the corresponding coordinates results in a true statement. The ordered pair is said to satisfy the equation. The graph of the equation is the set of all points whose coordinates satisfy the equation. One method for graphing an equation is to plot ordered-pair solutions and connect them with a smooth curve or line.

Graph: $y = x^2 - 1$.

x	$y = x^2 - 1$
-2	$(-2)^2 - 1 = 3$
-1	$(-1)^2 - 1 = 0$
0	$0^2 - 1 = -1$
1	$1^2 - 1 = 0$
2	$2^2 - 1 = 3$

Section 1.4 Solving Linear Equations

A linear equation in one variable can be written in the form $ax + b = 0, a \neq 0$. A solution is a value of the variable that makes the equation a true statement. The set of all such solutions is the equation's solution set. Equivalent equations have the same solution set. To solve a linear equation,

1. Simplify each side.

2. Collect variable terms on one side and constant terms on the other side.

3. Isolate the variable and solve.

4. Check the proposed solution in the original equation.

Solve: $4(x - 5) = 2x - 14$.

$$4x - 20 = 2x - 14$$
$$4x - 2x - 20 = 2x - 2x - 14$$
$$2x - 20 = -14$$
$$2x - 20 + 20 = -14 + 20$$
$$2x = 6$$
$$\frac{2x}{2} = \frac{6}{2}$$
$$x = 3$$

Checking gives $-8 = -8$, so 3 is the solution, or $\{3\}$ is the solution set.

Definitions and Concepts	Examples

Section 1.4 Solving Linear Equations (continued)

Equations Containing Fractions

Multiply both sides (all terms) by the least common denominator. This clears the equation of fractions.

Solve: $\dfrac{x-2}{5} + \dfrac{x+2}{2} = \dfrac{x+4}{3}$.

$$30\left(\dfrac{x-2}{5} + \dfrac{x+2}{2}\right) = 30\left(\dfrac{x+4}{3}\right)$$

$$6(x-2) + 15(x+2) = 10(x+4)$$

$$6x - 12 + 15x + 30 = 10x + 40$$

$$21x + 18 = 10x + 40$$

$$11x = 22$$

$$x = 2$$

Checking gives $2 = 2$, so 2 is the solution, or $\{2\}$ is the solution set.

Types of Equations

An equation that is true for all real numbers, $\mathbb{R}$, is called an identity. When solving an identity, the variable is eliminated and a true statement, such as $3 = 3$, results. An equation that is not true for even one real number is called an inconsistent equation. A false statement, such as $3 = 7$, results when solving such an equation, whose solution set is $\varnothing$, the empty set. A conditional equation is not an identity, but is true for at least one real number.

Solve: $4x + 5 = 4(x + 2)$.

$$4x + 5 = 4x + 8$$

$$5 = 8, \quad \text{false}$$

The inconsistent equation has no solution: $\varnothing$.

Solve: $5x - 4 = 5(x + 1) - 9$.

$$5x - 4 = 5x + 5 - 9$$

$$5x - 4 = 5x - 4$$

$$-4 = -4, \quad \text{true}$$

All real numbers satisfy the identity: $\mathbb{R}$.

Section 1.5 Problem Solving and Using Formulas

Strategy for Solving Algebraic Word Problems

1. Let x represent one of the quantities.
2. Represent other quantities in terms of x.
3. Write an equation that describes the conditions.
4. Solve the equation and answer the question.
5. Check the proposed solution in the original wording of the problem.

After a 60% reduction, a suit sold for $32. What was the original price?

Let x = the original price.

Original price	minus	60% reduction	=	reduced price
x	−	$0.6x$	=	32

$$0.4x = 32$$

$$\dfrac{0.4x}{0.4} = \dfrac{32}{0.4}$$

$$x = 80$$

The original price was $80. Check this using the first sentence in the problem's conditions.

To solve a formula for a variable, use the steps for solving a linear equation and isolate that variable on one side of the equation.

Solve for r:

$$E = I(R + r).$$

We need to isolate r.

$$E = IR + Ir$$

$$E - IR = Ir$$

$$\dfrac{E - IR}{I} = r$$

Definitions and Concepts	Examples

Section 1.6 Properties of Integral Exponents

The Product Rule

$$b^m \cdot b^n = b^{m+n}$$

$$(-3x^{10})(5x^{20}) = -3 \cdot 5x^{10+20}$$
$$= -15x^{30}$$

The Quotient Rule

$$\frac{b^m}{b^n} = b^{m-n}, b \neq 0$$

$$\frac{5x^{20}}{10x^{10}} = \frac{5}{10} \cdot x^{20-10} = \frac{x^{10}}{2}$$

Zero and Negative Exponents

$$b^0 = 1, b \neq 0$$

$$b^{-n} = \frac{1}{b^n} \quad \text{and} \quad \frac{1}{b^{-n}} = b^n$$

- $(3x)^0 = 1$
- $3x^0 = 3 \cdot 1 = 3$
- $\frac{2^{-3}}{4^{-2}} = \frac{4^2}{2^3} = \frac{16}{8} = 2$

Power Rule

$$(b^m)^n = b^{mn}$$

$$(x^5)^{-4} = x^{5(-4)} = x^{-20} = \frac{1}{x^{20}}$$

Products to Powers

$$(ab)^n = a^n b^n$$

$$(5x^3 y^{-4})^{-2} = 5^{-2} \cdot (x^3)^{-2} \cdot (y^{-4})^{-2}$$
$$= 5^{-2} x^{-6} y^8$$
$$= \frac{y^8}{5^2 x^6} = \frac{y^8}{25x^6}$$

Quotients to Powers

$$\left(\frac{a}{b}\right)^n = \frac{a^n}{b^n}$$

$$\left(\frac{2}{x^3}\right)^{-4} = \frac{2^{-4}}{(x^3)^{-4}} = \frac{2^{-4}}{x^{-12}}$$
$$= \frac{x^{12}}{2^4} = \frac{x^{12}}{16}$$

An exponential expression is simplified when

- No parentheses appear.
- No powers are raised to powers.
- Each base occurs only once.
- No negative or zero exponents appear.

Simplify: $\dfrac{-5x^{-3} y^2}{-20x^2 y^{-6}}$.

$$= \frac{-5}{-20} \cdot x^{-3-2} \cdot y^{2-(-6)}$$
$$= \frac{1}{4} x^{-5} y^8 = \frac{y^8}{4x^5}$$

Section 1.7 Scientific Notation

A number in scientific notation is expressed in the form

$$a \times 10^n,$$

where $|a|$ is greater than or equal to 1 and less than 10, and n is an integer.

Write in decimal notation: 3.8×10^{-3}.

$$3.8 \times 10^{-3} = .0038 = 0.0038$$

Write in scientific notation: 26,000.

$$26,000 = 2.6 \times 10^4$$

Computations with Numbers in Scientific Notation

$$(a \times 10^n)(b \times 10^m) = (a \times b) \times 10^{n+m}$$

$$\frac{a \times 10^n}{b \times 10^m} = \left(\frac{a}{b}\right) \times 10^{n-m}$$

$$(8 \times 10^3)(5 \times 10^{-8})$$
$$= 8 \cdot 5 \times 10^{3+(-8)}$$
$$= 40 \times 10^{-5}$$
$$= (4 \times 10^1) \times 10^{-5} = 4 \times 10^{-4}$$

CHAPTER 1   REVIEW EXERCISES

1.1 *In Exercises 1–3, write each English phrase as an algebraic expression. Let x represent the number.*

1. Ten less than twice a number $2x - 10$

2. Four more than the product of six and a number $6x + 4$

3. The quotient of nine and a number, increased by half of the number $\frac{9}{x} + \frac{1}{2}x$

In Exercises 4–6, evaluate each algebraic expression for the given value or values of the variable.

4. $x^2 - 7x + 4$, for $x = 10$ 34

5. $6 + 2(x - 8)^3$, for $x = 11$ 60

6. $x^4 - (x - y)$, for $x = 2$ and $y = 1$ 15

In Exercises 7–8, use the roster method to list the elements in each set.

7. $\{x \mid x$ is a natural number less than $3\}$ $\{1, 2\}$

8. $\{x \mid x$ is an integer greater than -4 and less than $2\}$
$\{-3, -2, -1, 0, 1\}$

In Exercises 9–11, determine whether each statement is true or false.

9. $0 \in \{x \mid x$ is a natural number$\}$ false

10. $-2 \in \{x \mid x$ is a rational number$\}$ true

11. $\frac{1}{3} \notin \{x \mid x$ is an irrational number$\}$ true

In Exercises 12–14, write out the meaning of each inequality. Then determine whether the inequality is true or false.

12. $-5 < 2$ -5 is less than 2; true

13. $-7 \geq -3$ -7 is greater than or equal to -3; false

14. $-7 \leq -7$ -7 is less than or equal to -7; true

15. You are riding along an expressway traveling x miles per hour. The formula
$$S = 0.015x^2 + x + 10$$
models the recommended safe distance, S, in feet, between your car and other cars on the expressway. What is the recommended safe distance when your speed is 60 miles per hour?   124 feet

1.2 *In Exercises 16–18, find each absolute value.*

16. $|-9.7|$ 9.7   **17.** $|5.003|$ 5.003   **18.** $|0|$ 0

In Exercises 19–30, perform the indicated operation.

19. $-2.4 + (-5.2)$ -7.6 **20.** $-6.8 + 2.4$ -4.4

21. $-7 - (-20)$ 13   **22.** $(-3)(-20)$ 60

23. $-\frac{3}{5} - \left(-\frac{1}{2}\right)$ $-\frac{1}{10}$ **24.** $\left(\frac{2}{7}\right)\left(-\frac{3}{10}\right)$ $-\frac{3}{35}$

25. $4(-3)(-2)(-10)$ -240 **26.** $(-2)^4$ 16

27. -2^5 -32 **28.** $-\frac{2}{3} \div \frac{8}{5}$ $-\frac{5}{12}$

29. $\frac{-35}{-5}$ 7   **30.** $\frac{54.6}{-6}$ -9.1

31. Find $-x$ if $x = -7$.   7

In Exercises 32–38, simplify each expression.

32. $-11 - [-17 + (-3)]$ 9   **33.** $\left(-\frac{1}{2}\right)^3 \cdot 2^4$ -2

34. $-3[4 - (6 - 8)]$ -18 **35.** $8^2 - 36 \div 3^2 \cdot 4 - (-7)$ 55

36. $\dfrac{(-2)^4 + (-3)^2}{2^2 - (-21)}$ 1   **37.** $\dfrac{(7 - 9)^3 - (-4)^2}{2 + 2(8) \div 4}$ -4

38. $4 - (3 - 8)^2 + 3 \div 6 \cdot 4^2$ -13

In Exercises 39–43, simplify each algebraic expression.

39. $5(2x - 3) + 7x$ $17x - 15$

40. $5x + 7x^2 - 4x + 2x^2$ $9x^2 + x$

41. $3(4y - 5) - (7y + 2)$ $5y - 17$

42. $8 - 2[3 - (5x - 1)]$ $10x$

43. $6(2x - 3) - 5(3x - 2)$ $-3x - 8$

1.3 *In Exercises 44–46, plot the given point in a rectangular coordinate system.*   **44–46.** See graphing answer section.

44. $(-1, 3)$ **45.** $(2, -5)$ **46.** $(0, -6)$

In Exercises 47–50, graph each equation. Let $x = -3, -2, -1, 0, 1, 2,$ and 3.   **47–50.** See graphing answer section.

47. $y = 2x - 2$ **48.** $y = x^2 - 3$

49. $y = x$ **50.** $y = |x| - 2$

51. What does a $[-20, 40, 10]$ by $[-5, 5, 1]$ viewing rectangle mean? Draw axes with tick marks and label the tick marks to illustrate this viewing rectangle.   The minimum x-value is -20 and the maximum x-value is 40. The distance between tick marks is 10. The minimum y-value is -5 and the maximum y-value is 5. The distance between tick marks is 1. See graphing answer section.

The caseload of Alzheimer's disease in the United States is expected to explode as baby boomers head into their later years. The graph shows the percentage of Americans with the disease, by age. Use the graph to solve Exercises 52–54.

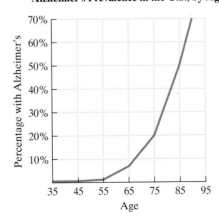

Alzheimer's Prevalence in the U.S., by Age

Source: Centers for Disease Control

(In Exercises 52–54, be sure to refer to the graph on the previous page.)

52. What percentage of Americans who are 75 have Alzheimer's disease?   20%

53. What age represents 50% prevalence of Alzheimer's disease?   85 years old

54. Describe the trend shown by the graph.   Answers will vary.

55. Select the graph that best illustrates the following description: A train pulls into a station and lets off its passengers.   c

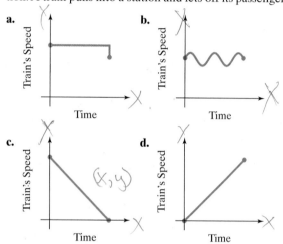

a.

b.

c.

d.

1.4 *In Exercises 56–61, solve and check each linear equation.*

56. $2x - 5 = 7$ 6

57. $5x + 20 = 3x$ −10

58. $7(x - 4) = x + 2$ 5

59. $1 - 2(6 - x) = 3x + 2$ −13

60. $2(x - 4) + 3(x + 5) = 2x - 2$ −3

61. $2x - 4(5x + 1) = 3x + 17$ −1

In Exercises 62–66, solve each equation.

62. $\dfrac{2x}{3} = \dfrac{x}{6} + 1$ 2

63. $\dfrac{x}{2} - \dfrac{1}{10} = \dfrac{x}{5} + \dfrac{1}{2}$ 2

64. $\dfrac{2x}{3} = 6 - \dfrac{x}{4}$ $\dfrac{72}{11}$

65. $\dfrac{x}{4} = 2 + \dfrac{x - 3}{3}$ −12

66. $\dfrac{3x + 1}{3} - \dfrac{13}{2} = \dfrac{1 - x}{4}$ $\dfrac{77}{15}$

In Exercises 67–71, solve each equation. Then state whether the equation is an identity, a conditional equation, or an inconsistent equation.

67. $7x + 5 = 5(x + 3) + 2x$ no solution; inconsistent equation

68. $7x + 13 = 4x - 10 + 3x + 23$ all real numbers; identity

69. $7x + 13 = 3x - 10 + 2x + 23$ 0; conditional equation

70. $4(x - 3) + 5 = x + 5(x - 2)$ $\dfrac{3}{2}$; conditional equation

71. $(2x - 3)2 - 3(x + 1) = (x - 2)4 - 3(x + 5)$ no solution; inconsistent equation

72. The bar graph shows the number of endangered animal species in the United States for six selected years. The data can be modeled by the formula $E = 10x + 167$, in which E represents the number of endangered species x years after 1980. If the trend indicated by this model continues, in which year will there be 437 endangered animal species in the United States?   2007

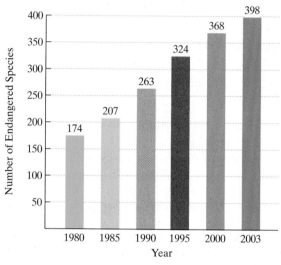

Endangered Animal Species in the U.S.

Source: U.S. Fish and Wildlife Service

1.5 *In Exercises 73–79, use the five-step strategy for solving word problems.*

73. The fast-food chains may be touting their "new and improved" salads, but how do they measure up in terms of calories?

Burger King
Chicken Caesar

Taco Bell
Express Taco Salad

Wendy's
Mandarin Chicken Salad

Number of calories exceeds the Chicken Caesar by 125.

Number of calories exceeds the Chicken Caesar by 95.

Source: Newsweek

Combined, the three salads contain 1705 calories. Determine the number of calories in each salad.   Chicken Caesar: 495; Express Taco: 620; Mandarin Chicken: 590

74. One angle of a triangle measures 10° more than the second angle. The measure of the third angle is twice the sum of the measures of the first two angles. Determine the measure of each angle.   25°, 35°, 120°

75. The bar graph shows the number of endangered plant species in the United States from 1998 through 2003. In 1998, there were 567 endangered species. For the period shown, the number of endangered plants increased at an average rate of 6.4 species per year. If this trend continues, in which year will there be 663 endangered plant species in the United States?   2013

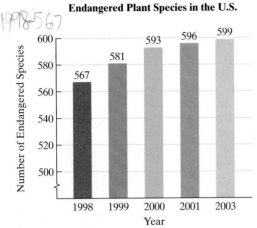

Endangered Plant Species in the U.S.

Source: U.S. Fish and Wildlife Service

76. You are choosing between two long-distance telephone plans. One plan has a monthly fee of $15 with a charge of $0.05 per minute. The other plan has a monthly fee of $5 with a charge of $0.07 per minute. For how many minutes of long-distance calls will the costs for the two plans be the same?   500 minutes

77. After a 20% price reduction, a cordless phone sold for $48. What was the phone's price before the reduction? $60

78. A salesperson earns $300 per week plus 5% commission of sales. How much must be sold to earn $800 in a week? $10,000 in sales

79. The length of a rectangular field is 6 yards less than triple the width. If the perimeter of the field is 340 yards, what are its dimensions?   44 yards by 126 yards

80. In 2005, there were 14,100 students at college A, with a projected enrollment increase of 1500 students per year. In the same year, there were 41,700 students at college B, with a projected enrollment decline of 800 students per year.

   a. Let x represent the number of years after 2005. Write, but do not solve, an equation that can be used to find how many years after 2005 the colleges will have the same enrollment. $14{,}100 + 1500x = 41{,}700 - 800x$

   b. The following table is based on your equation in part (a). Y_1 represents one side of the equation and Y_2 represents the other side of the equation. Use the table to answer the following questions: In which year will the colleges have the same enrollment? What will be the enrollment in each college at that time?   2017; 32,100

X	Y1	Y2
7	24600	36100
8	26100	35300
9	27600	34500
10	29100	33700
11	30600	32900
12	32100	32100
13	33600	31300

X=7

In Exercises 81–86, solve each formula for the specified variable.

81. $V = \dfrac{1}{3}Bh$ for h $h = \dfrac{3V}{B}$

82. $y - y_1 = m(x - x_1)$ for x $x = \dfrac{y - y_1}{m} + x_1$

83. $E = I(R + r)$ for R $R = \dfrac{E}{I} - r$ or $R = \dfrac{E - Ir}{I}$

84. $C = \dfrac{5F - 160}{9}$ for F $F = \dfrac{9C + 160}{5}$ or $F = \dfrac{9}{5}C + 32$

85. $s = vt + gt^2$ for g $g = \dfrac{s - vt}{t^2}$

86. $T = gr + gvt$ for g $g = \dfrac{T}{r + vt}$

1.6 *In Exercises 87–101, simplify each exponential expression. Assume that no denominators are 0.*

87. $(-3x^7)(-5x^6)$ $15x^{13}$

88. x^2y^{-5} $\dfrac{x^2}{y^5}$

89. $\dfrac{3^{-2}x^4}{y^{-7}}$ $\dfrac{x^4y^7}{9}$

90. $(x^3)^{-6}$ $\dfrac{1}{x^{18}}$

91. $(7x^3y)^2$ $49x^6y^2$

92. $\dfrac{16y^3}{-2y^{10}}$ $-\dfrac{8}{y^7}$

93. $(-3x^4)(4x^{-11})$ $-\dfrac{12}{x^7}$

94. $\dfrac{12x^7}{4x^{-3}}$ $3x^{10}$

95. $\dfrac{-10a^5b^6}{20a^{-3}b^{11}}$ $-\dfrac{a^8}{2b^5}$

96. $(-3xy^4)(2x^2)^3$ $-24x^7y^4$ **97.** $2^{-2} + \dfrac{1}{2}x^0$ $\dfrac{3}{4}$

98. $(5x^2y^{-4})^{-3}$ $\dfrac{y^{12}}{125x^6}$

99. $(3x^4y^{-2})(-2x^5y^{-3})$ $-\dfrac{6x^9}{y^5}$

100. $\left(\dfrac{3xy^3}{5x^{-3}y^{-4}}\right)^2$ $\dfrac{9x^8y^{14}}{25}$ **101.** $\left(\dfrac{-20x^{-2}y^3}{10x^5y^{-6}}\right)^{-3}$ $-\dfrac{x^{21}}{8y^{27}}$

1.7 *In Exercises 102–103, write each number in decimal notation.*

102. 7.16×10^6 7,160,000 **103.** 1.07×10^{-4} 0.000107

In Exercises 104–105, write each number in scientific notation.

104. $-41{,}000{,}000{,}000{,}000$ -4.1×10^{13}

105. 0.00809 8.09×10^{-3}

In Exercises 106–107, perform the indicated computations. Write the answers in scientific notation.

106. $(4.2 \times 10^{13})(3 \times 10^{-6})$ 1.26×10^8

107. $\dfrac{5 \times 10^{-6}}{20 \times 10^{-8}}$ 2.5×10^1

108. If the population of the United States is 2.9×10^8 and each person spends about $150 per year going to the movies (or renting movies), express the total annual spending on movies in scientific notation. 4.35×10^{10}

CHAPTER 1 TEST Remember to use your Chapter Test Prep Video CD to see the worked-out solutions to the test questions you want to review.

1. Write the following English phrase as an algebraic expression:
   Five less than the product of a number and four.
   Let x represent the number. $4x - 5$

2. Evaluate $8 + 2(x - 7)^4$, for $x = 10$.   170

3. Use the roster method to list the elements in the set:
   $\{x \mid x$ is a negative integer greater than $-5\}$.
   $\{-4, -3, -2, -1\}$

4. Determine whether the following statement is true or false:
   $\frac{1}{4} \notin \{x \mid x$ is a natural number$\}$.   true

5. Write out the meaning of the inequality $-3 > -1$. Then determine whether the inequality is true or false.
   -3 is greater than -1: false

6. The formula
   $$F = 24t^2 - 260t + 816$$
   models the number of police officers in the United States, F, convicted of felonies t years after 1990. How many convictions of police officers were there in 2000? (*Source*: FBI)
   616 convictions

7. Find the absolute value: $|-17.9|$.   17.9

In Exercises 8–12, perform the indicated operation.

8. $-10.8 + 3.2$ -7.6 9. $-\frac{1}{4} - \left(-\frac{1}{2}\right)$ $\frac{1}{4}$

10. $2(-3)(-1)(-10)$ -60 11. $-\frac{1}{4}\left(-\frac{1}{2}\right)$ $\frac{1}{8}$

12. $\frac{-27.9}{-9}$ 3.1

In Exercises 13–18, simplify each expression.

13. $24 - 36 \div 4 \cdot 3$ -3

14. $(5^2 - 2^4) + [9 \div (-3)]$ 6

15. $\frac{(8 - 10)^3 - (-4)^2}{2 + 8(2) \div 4}$ -4

16. $7x - 4(3x + 2) - 10$ $-5x - 18$

17. $5(2y - 6) - (4y - 3)$ $6y - 27$

18. $9x - [10 - 4(2x - 3)]$ $17x - 22$

19. Plot $(-2, -4)$ in a rectangular coordinate system.
    See graphing answer section.

20. Graph $y = x^2 - 4$ in a rectangular coordinate system.
    See graphing answer section.

In Exercises 21–23, solve each equation. If the solution set is $\emptyset$ or $\mathbb{R}$, classify the equation as an inconsistent equation or an identity.

21. $3(2x - 4) = 9 - 3(x + 1)$ 2

22. $\frac{2x - 3}{4} = \frac{x - 4}{2} - \frac{x + 1}{4}$ -6

23. $3(x - 4) + x = 2(6 + 2x)$ no solution; inconsistent equation

In Exercises 24–28, use the five-step strategy for solving word problems.

24. Find two numbers such that the second number is 3 more than twice the first number and the sum of the two numbers is 72.   23 and 49

25. You bought a new car for $13,805. Its value is decreasing by $1820 per year. After how many years will its value be $4705?   5 years

26. Photo Shop A charges $1.60 to develop a roll of film plus $0.11 for each print. Photo Shop B charges $1.20 to develop a roll of film plus $0.13 per print. For how many prints will the amount spent at each photo shop be the same? What will be that amount?   20 prints; $3.80

27. After a 60% reduction, a jacket sold for $20. What was the jacket's price before the reduction? $50

28. The length of a rectangular field exceeds the width by 260 yards. If the perimeter of the field is 1000 yards, what are its dimensions?   120 yards by 380 yards

In Exercises 29–30, solve each formula for the specified variable.

29. $V = \frac{1}{3}lwh$ for h $h = \frac{3V}{lw}$

30. $Ax + By = C$ for y $y = \frac{C - Ax}{B}$

In Exercises 31–35, simplify each exponential expression.

31. $(-2x^5)(7x^{-10})$ $-\frac{14}{x^5}$

32. $(-8x^{-5}y^{-3})(-5x^2y^{-5})$ $\frac{40}{x^3y^8}$

33. $\frac{-10x^4y^3}{-40x^{-2}y^6}$ $\frac{x^6}{4y^3}$

34. $(4x^{-5}y^2)^{-3}$ $\frac{x^{15}}{64y^6}$

35. $\left(\frac{-6x^{-5}y}{2x^3y^{-4}}\right)^{-2}$ $\frac{x^{16}}{9y^{10}}$

36. Write in decimal notation: 3.8×10^{-6}.   0.0000038

37. Write in scientific notation: $407{,}000{,}000{,}000$. 4.07×10^{11}

38. Divide and write the answer in scientific notation:
    $$\frac{4 \times 10^{-3}}{8 \times 10^{-7}}.$$ 5×10^3

39. In 2003, world population was approximately 6.3×10^9. By some projections, world population will double by 2040. Express the population at that time in scientific notation. 1.26×10^{10}

Note to Instructor: The Chapter 1 Test file in the TestGen program provides algorithms specifically matched to these problems so you can easily replicate this chapter test for additional practice or assessment purposes.

In the dramatic arts, ours is the era of the movies. As individuals and as a nation, we've grown up with them. Our images of love, war, family, country—even of things that terrify us—owe much to what we've seen on screen.

Using mathematics in a way that is similar to making a movie is discussed in the enrichment essay on page 137.

2

Take a rapid sequence of still photographs of a moving scene and project them onto a screen at thirty shots a second or faster. Our eyes see the result as continuous motion. The small difference between one frame and the next cannot be detected by the human visual system. In this chapter, you will see how the notion of a function and the steepness of a line can analyze motion and change by "freezing the frame" of a continuously changing process, instant by instant. Understanding the concept of a function will give you a new perspective on many situations, including using mathematics in a way that is similar to making a movie.

Functions and Linear Functions

2.1 Introduction to Functions

2.2 The Algebra of Functions

2.3 Linear Functions and Slope

2.4 The Point-Slope Form of the Equation of a Line

SECTION

2.1

Objectives

1 Find the domain and range of a relation.

2 Determine whether a relation is a function.

3 Evaluate a function.

4 Graph functions by plotting points.

5 Use the vertical line test to identify functions.

6 Obtain information about a function from its graph.

7 Identify the domain and range of a function from its graph.

INTRODUCTION TO FUNCTIONS

Jerry Orbach	$34,000
Charles Shaughnessy	$31,800
Andy Richter	$29,400
Norman Schwarzkopf	$28,000
Jon Stewart	$28,000

The answer: See the above list. The question: Who are *Celebrity Jeopardy's* five all-time highest earners? The list indicates a correspondence between the five all-time highest earners and their winnings. We can write this correspondence using a set of ordered pairs:

{(Orbach, $34,000), (Shaughnessy, $31,800), (Richter, $29,400), (Schwarzkopf, $28,000), (Stewart, $28,000)}.

The mathematical term for a set of ordered pairs is a *relation*.

> **DEFINITION OF A RELATION** A **relation** is any set of ordered pairs. The set of all first components of the ordered pairs is called the **domain** of the relation and the set of all second components is called the **range** of the relation.

---

1 Find the domain and range of a relation.

EXAMPLE 1 Finding the Domain and Range of a Relation

Find the domain and range of the relation:

{(Orbach, $34,000), (Shaughnessy, $31,800), (Richter, $29,400), (Schwarzkopf, $28,000), (Stewart, $28,000)}.

SOLUTION The domain is the set of all first components. Thus, the domain is

{Orbach, Shaughnessy, Richter, Schwarzkopf, Stewart}.

The range is the set of all second components. Thus, the range is

{$34,000, $31,800, $29,400, $28,000}.

> Although Schwarzkopf and Stewart both won $28,000, it is not necessary to list $28,000 twice.

 CHECK POINT 1 Find the domain and the range of the relation:

{(5, 12.8), (10, 16.2), (15, 18.9), (20, 20.7), (25, 21.8)}.

domain: {5, 10, 15, 20, 25}; range: {12.8, 16.2, 18.9, 20.7, 21.8}

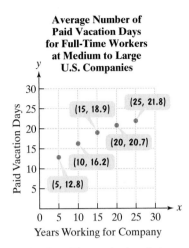

Average Number of Paid Vacation Days for Full-Time Workers at Medium to Large U.S. Companies

(15, 18.9) (25, 21.8) (20, 20.7) (10, 16.2) (5, 12.8)

Years Working for Company

FIGURE 2.1 The graph of a relation showing a correspondence between years with a company and paid vacation days

Source: Bureau of Labor Statistics

As you worked Check Point 1, did you wonder if there was a rule that assigned the "inputs" in the domain to the "outputs" in the range? For example, for the ordered pair (15, 18.9), how does the output 18.9 depend on the input 15? Think paid vacation days! The first number in each ordered pair is the number of years that a full-time employee has been employed by a medium to large U.S. company. The second number is the average number of paid vacation days each year. Consider, for example, the ordered pair (15, 18.9).

(15, 18.9)

After 15 years, workers average 18.9 paid vacation days per year.

The relation in the vacation-days example can be pictured as follows:

```
5  ──────→ 12.8
10 ──────→ 16.2
15 ──────→ 18.9
20 ──────→ 20.7
25 ──────→ 21.8
Domain      Range
```

The five points in Figure 2.1 are another way to visually represent the relation.

2 Determine whether a relation is a function.

Jerry Orbach	$34,000
Charles Shaughnessy	$31,800
Andy Richter	$29,400
Norman Schwarzkopf	$28,000
Jon Stewart	$28,000

Functions Shown, again, in the margin are *Celebrity Jeopardy's* five all-time highest winners and their winnings. We've used this information to define two relations. Figure 2.2(a) shows a correspondence between winners and their winnings. Figure 2.2(b) shows a correspondence between winnings and winners.

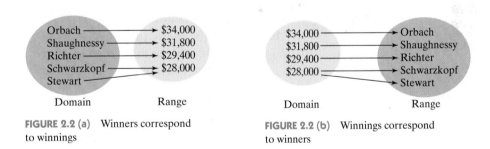

FIGURE 2.2 (a) Winners correspond to winnings

FIGURE 2.2 (b) Winnings correspond to winners

A relation in which each member of the domain corresponds to exactly one member of the range is a **function**. Can you see that the relation in Figure 2.2(a) is a function? Each winner in the domain corresponds to exactly one winning amount in the range. If we know the winner, we can be sure of the amount won. Notice that more than one element in the domain can correspond to the same element in the range. (Schwarzkopf and Stewart both won $28,000.)

Is the relation in Figure 2.2(b) a function? Does each member of the domain correspond to precisely one member of the range? This relation is not a function because there is a member of the domain that corresponds to two different members of the range:

($28,000, Schwarzkopf)   ($28,000, Stewart).

The member of the domain, $28,000, corresponds to both Schwarzkopf and Stewart in the range. If we know the amount won, $28,000, we cannot be sure of the winner. Because **a function is a relation in which no two ordered pairs have the same**

first component and different second components, the ordered pairs ($28,000, Schwarzkopf) and ($28,000, Stewart) are not ordered pairs of a function.

Same first component

($28,000, Schwarzkopf)   ($28,000, Stewart)

Different second components

> **DEFINITION OF A FUNCTION**   A **function** is a correspondence from a first set, called the **domain**, to a second set, called the **range**, such that each element in the domain corresponds to *exactly one* element in the range.

Example 2 illustrates that not every correspondence between sets is a function.

EXAMPLE 2   Determining Whether a Relation Is a Function

Determine whether each relation is a function:

    a. $\{(1, 5), (2, 5), (3, 7), (4, 8)\}$ **b.** $\{(5, 1), (5, 2), (7, 3), (8, 4)\}$.

SOLUTION   We begin by making a figure for each relation that shows the domain and the range (Figure 2.3).

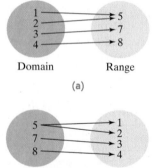

Domain   Range

(a)

Domain   Range

(b)

FIGURE 2.3

    a. Figure 2.3(a) shows that every element in the domain corresponds to exactly one element in the range. The element 1 in the domain corresponds to the element 5 in the range. Furthermore, 2 corresponds to 5, 3 corresponds to 7, and 4 corresponds to 8. No two ordered pairs in the given relation have the same first component and different second components. Thus, the relation is a function.

    b. Figure 2.3(b) shows that 5 corresponds to both 1 and 2. If any element in the domain corresponds to more than one element in the range, the relation is not a function. This relation is not a function because two ordered pairs have the same first component and different second components.

Same first component

(5, 1)   (5, 2)

Different second components

Look at Figure 2.3(a) again. The fact that 1 and 2 in the domain correspond to the same number, 5, in the range does not violate the definition of a function. **A function can have two different first components with the same second component**. By contrast, a relation is not a function when two different ordered pairs have the same first component and different second components. Thus, the relation in Example 2(b) is not a function.

 CHECK POINT 2 Determine whether each relation is a function:

    a. $\{(1, 2), (3, 4), (5, 6), (5, 7)\}$ not a function

    b. $\{(1, 2), (3, 4), (6, 5), (7, 5)\}$.   function

Functions as Equations and Function Notation Functions are usually given in terms of equations rather than as sets of ordered pairs. For example, here is an equation that models paid vacation days each year as a function of years working for a company:

$$y = -0.016x^2 + 0.93x + 8.5.$$

The variable x represents years working for a company. The variable y represents the average number of vacation days each year. The variable y is a function of the variable x. For each value of x, there is one and only one value of y. The variable x is called the **independent variable** because it can be assigned any value from the domain. Thus, x can be assigned any positive integer representing years working for a company. The variable y is called the **dependent variable** because its value depends on x. Paid vacation days depend on years working for a company. The value of the dependent variable, y, is calculated after selecting a value for the independent variable, x.

If an equation in x and y gives one and only one value of y for each value of x, then the variable y is a function of the variable x. When an equation represents a function, the function is often named by a letter such as $f, g, h, F, G,$ or H. Any letter can be used to name a function. Suppose that f names a function. Think of the domain as the set of the function's inputs and the range as the set of the function's outputs. As shown in Figure 2.4, the input is represented by x and the output by $f(x)$. The special notation $f(x)$, read "f of x" or "f at x," represents the **value of the function at the number x**.

Let's make this clearer by considering a specific example. We know that the equation

$$y = -0.016x^2 + 0.93x + 8.5$$

defines y as a function of x. We'll name the function f. Now, we can apply our new function notation.

Input x

FIGURE 2.4 A "function machine" with inputs and outputs

Output $f(x)$

3 Evaluate a function.

We read this equation as "f of x equals $-0.016x^2 + 0.93x + 8.5$."

Input	Output	Equation
x	$f(x)$	$f(x) = -0.016x^2 + 0.93x + 8.5$

STUDY TIP

The notation $f(x)$ does *not* mean "f times x." The notation describes the value of the function at x.

Suppose we are interested in finding $f(10)$, the function's output when the input is 10. To find the value of the function at 10, we substitute 10 for x. We are **evaluating the function** at 10.

$$f(x) = -0.016x^2 + 0.93x + 8.5 \qquad \text{This is the given function.}$$
$$f(10) = -0.016(10)^2 + 0.93(10) + 8.5 \qquad \text{Replace each occurrence of x with 10.}$$
$$= -0.016(100) + 0.93(10) + 8.5 \qquad \text{Evaluate the exponential expression:}$$
$$\text{} \qquad 10^2 = 100.$$
$$= -1.6 + 9.3 + 8.5 \qquad \text{Perform the multiplications.}$$
$$= 16.2 \qquad \text{Add from left to right.}$$

The statement $f(10) = 16.2$, read "f of 10 equals 16.2," tells us that the value of the function at 10 is 16.2. When the function's input is 10, its output is 16.2. (After 10 years, workers average 16.2 vacation days each year.) To find other function values, such as $f(15)$, $f(20)$, or $f(23)$, substitute the specified input values for x into the function's equation.

If a function is named f and x represents the independent variable, the notation $f(x)$ corresponds to the y-value for a given x. Thus,

$$f(x) = -0.016x^2 + 0.93x + 8.5 \quad \text{and} \quad y = -0.016x^2 + 0.93x + 8.5$$

define the same function. This function may be written as

$$y = f(x) = -0.016x^2 + 0.93x + 8.5.$$

EXAMPLE 3   Using Function Notation

Find the indicated function value:

a. $f(4)$ for $f(x) = 2x + 3$
b. $g(-2)$ for $g(x) = 2x^2 - 1$
c. $h(-5)$ for $h(r) = r^3 - 2r^2 + 5$
d. $F(a + h)$ for $F(x) = 5x + 7$.

SOLUTION

a. $f(x) = 2x + 3$ This is the given function.
$\quad f(4) = 2 \cdot 4 + 3$ To find f of 4, replace x with 4.
$\quad\quad = 8 + 3$ Multiply: $2 \cdot 4 = 8$.
$\quad\quad = 11$ *f of 4 is 11.*   Add.

b. $\quad g(x) = 2x^2 - 1$ This is the given function.
$\quad g(-2) = 2(-2)^2 - 1$ To find g of −2, replace x with −2.
$\quad\quad = 2(4) - 1$ Evaluate the exponential expression: $(-2)^2 = 4$.
$\quad\quad = 8 - 1$ Multiply: $2(4) = 8$.
$\quad\quad = 7$ *g of −2 is 7.*   Subtract.

c. $\quad h(r) = r^3 - 2r^2 + 5$ The function's name is h and r represents the independent variable.
$\quad h(-5) = (-5)^3 - 2(-5)^2 + 5$ To find h of −5, replace each occurrence of r with −5.
$\quad\quad = -125 - 2(25) + 5$ Evaluate exponential expressions.
$\quad\quad = -125 - 50 + 5$ Multiply.
$\quad\quad = -170$ *h of −5 is −170.* $-125 - 50 = -175$ and $-175 + 5 = -170$.

d. $\quad F(x) = 5x + 7$ This is the given function.
$\quad F(a + h) = 5(a + h) + 7$ Replace x with $a + h$.
$\quad\quad = 5a + 5h + 7$ Apply the distributive property.   ∎

F of a + h is 5a + 5h + 7.

 CHECK POINT 3 Find the indicated function value:

a. $f(6)$ for $f(x) = 4x + 5$ 29
b. $g(-5)$ for $g(x) = 3x^2 - 10$ 65
c. $h(-4)$ for $h(r) = r^2 - 7r + 2$ 46
d. $F(a + h)$ for $F(x) = 6x + 9$. $6a + 6h + 9$

---

4 Graph functions by plotting points.

Graphs of Functions and the Vertical Line Test   The **graph of a function** is the graph of its ordered pairs. For example, the graph of $f(x) = 2x$ is the set of points (x, y) in the rectangular coordinate system satisfying $y = 2x$. Similarly, the graph of $g(x) = 2x + 4$ is the set of points (x, y) in the rectangular coordinate system satisfying the equation $y = 2x + 4$. In the next example, we graph both of these functions in the same rectangular coordinate system.

EXAMPLE 4 Graphing Functions

Graph the functions $f(x) = 2x$ and $g(x) = 2x + 4$ in the same rectangular coordinate system. Select integers for x, starting with -2 and ending with 2.

SOLUTION We begin by setting up a partial table of coordinates for each function. Then, we plot the five points in each table and connect them, as shown in Figure 2.5. The graph of each function is a straight line. Do you see a relationship between the two graphs? The graph of g is the graph of f shifted vertically up by 4 units.

x	$f(x) = 2x$	(x, y) or $(x, f(x))$
-2	$f(-2) = 2(-2) = -4$	$(-2, -4)$
-1	$f(-1) = 2(-1) = -2$	$(-1, -2)$
0	$f(0) = 2 \cdot 0 = 0$	$(0, 0)$
1	$f(1) = 2 \cdot 1 = 2$	$(1, 2)$
2	$f(2) = 2 \cdot 2 = 4$	$(2, 4)$

x	$g(x) = 2x + 4$	(x, y) or $(x, g(x))$
-2	$g(-2) = 2(-2) + 4 = 0$	$(-2, 0)$
-1	$g(-1) = 2(-1) + 4 = 2$	$(-1, 2)$
0	$g(0) = 2 \cdot 0 + 4 = 4$	$(0, 4)$
1	$g(1) = 2 \cdot 1 + 4 = 6$	$(1, 6)$
2	$g(2) = 2 \cdot 2 + 4 = 8$	$(2, 8)$

Choose x. | Compute $f(x)$ by evaluating f at x. | Form the ordered pair. | Choose x. | Compute $g(x)$ by evaluating g at x. | Form the ordered pair.

FIGURE 2.5

The graphs in Example 4 are straight lines. All functions with equations of the form $f(x) = mx + b$ graph as straight lines. Such functions, called **linear functions**, will be discussed in detail in Sections 2.3–2.4.

USING TECHNOLOGY

We can use a graphing utility to check the tables and the graphs in Example 4 for the functions

$$f(x) = 2x \quad \text{and} \quad g(x) = 2x + 4.$$

Enter $y_1 = 2x$ in the $\boxed{y=}$ screen.

Enter $y_2 = 2x + 4$ in the $\boxed{y=}$ screen.

We entered -2 for the starting x-value and 1 as an increment between x-values to check our tables in Example 4.

Checking Tables

X	Y1	Y2
-2	-4	0
-1	-2	2
0	0	4
1	2	6
2	4	8
3	6	10
4	8	12

X= -2

Use the first five ordered pairs (x, y_1) to check the first table.

Use the first five ordered pairs (x, y_2) to check the second table.

Checking Graphs

$[-6, 6, 1]$ by $[-6, 9, 1]$

We selected this viewing rectangle, or window, to match Figure 2.5.

✔ **CHECK POINT 4** Graph the functions $f(x) = 2x$ and $g(x) = 2x - 3$ in the same rectangular coordinate system. Select integers for x, starting with -2 and ending with 2. How is the graph of g related to the graph of f?

See graphing answer section; The graph of g is the graph of f shifted down by 3 units.

5 Use the vertical line test to identify functions.

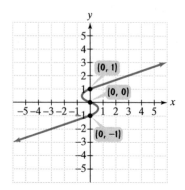

FIGURE 2.6 *y* is not a function of *x* because 0 is paired with three values of *y*, namely, 1, 0, and −1.

Not every graph in the rectangular coordinate system is the graph of a function. The definition of a function specifies that no value of *x* can be paired with two or more different values of *y*. Consequently, if a graph contains two or more different points with the same first coordinate, the graph cannot represent a function. This is illustrated in Figure 2.6. Observe that points sharing a common first coordinate are vertically above or below each other.

This observation is the basis of a useful test for determining whether a graph defines *y* as a function of *x*. The test is called the **vertical line test**.

THE VERTICAL LINE TEST FOR FUNCTIONS If any vertical line intersects a graph in more than one point, the graph does not define *y* as a function of *x*.

EXAMPLE 5 Using the Vertical Line Test

Use the vertical line test to identify graphs in which *y* is a function of *x*.

a.

b.

c.

d.
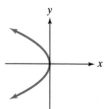

SOLUTION *y* is a function of *x* for the graphs in (b) and (c).

a.
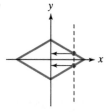

y is not a function of *x*. Two values of *y* correspond to one *x*-value.

b.

y is a function of *x*.

c.

y is a function of *x*.

d.

y is not a function of *x*. Two values of *y* correspond to one *x*-value.

 CHECK POINT 5 Use the vertical line test to identify graphs in which *y* is a function of *x*.

a. function   **b.** function   **c.** not a function

6 Obtain information about a function from its graph.

Obtaining Information from Graphs You can obtain information about a function from its graph. At the right or left of a graph, you will often find closed dots, open dots, or arrows.

- A closed dot indicates that the graph does not extend beyond this point and the point belongs to the graph.
- An open dot indicates that the graph does not extend beyond this point and the point does not belong to the graph.
- An arrow indicates that the graph extends indefinitely in the direction in which the arrow points.

EXAMPLE 6 Analyzing the Graph of a Function

The function

$$f(x) = -0.016x^2 + 0.93x + 8.5$$

models the average number of paid vacation days each year, $f(x)$, for full-time workers at medium to large U.S. companies after x years. The graph of f is shown in Figure 2.7.

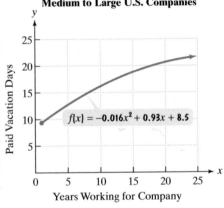

Average Number of Paid Vacation Days for Full-Time Workers at Medium to Large U.S. Companies

$f(x) = -0.016x^2 + 0.93x + 8.5$

Paid Vacation Days

Years Working for Company

FIGURE 2.7

Source: Bureau of Labor Statistics.

a. Explain why f represents the graph of a function.

b. Use the graph to find a reasonable estimate of $f(5)$.

c. For what value of x is $f(x) = 20$?

d. Describe the general trend shown by the graph.

SOLUTION

a. No vertical line intersects the graph of f more than once. By the vertical line test, f represents the graph of a function.

b. To find $f(5)$, or f of 5, we locate 5 on the x-axis. The figure shows the point on the graph of f for which 5 is the first coordinate. From this point, we look to the y-axis to find the corresponding y-coordinate. A reasonable estimate of the y-coordinate is 13. Thus, $f(5) \approx 13$. After 5 years, a worker can expect approximately 13 paid vacation days.

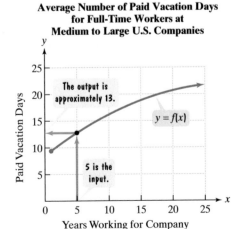

Average Number of Paid Vacation Days for Full-Time Workers at Medium to Large U.S. Companies

The output is approximately 13.

$y = f(x)$

5 is the input.

Paid Vacation Days

Years Working for Company

c. To find the value of x for which $f(x) = 20$, we locate 20 on the y-axis. The figure shows that there is one point on the graph of f for which 20 is the second coordinate. From this point, we look to the x-axis to find the corresponding x-coordinate. A reasonable estimate of the x-coordinate is 18. Thus, $f(x) = 20$ for $x \approx 18$. A worker with 20 paid vacation days has been with the company approximately 18 years.

d. The graph of f is rising from left to right. This shows that paid vacation days increase as time with the company increases. However, the rate of increase is slowing down as the graph moves to the right. This means that the increase in paid vacation days takes place more slowly the longer an employee is with the company. ∎

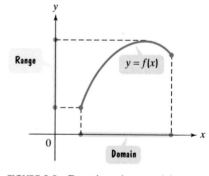

Average Number of Paid Vacation Days for Full-Time Workers at Medium to Large U.S. Companies

✓ CHECK POINT 6

a. Use the graph of f in Figure 2.7 on page 103 to find a reasonable estimate of $f(10)$. $f(10) \approx 16$

b. For what value of x is $f(x) = 15$? Round to the nearest whole number.
$x \approx 8$

---

7 Identify the domain and range of a function from its graph.

Identifying Domain and Range from a Function's Graph   Figure 2.8 illustrates how the graph of a function is used to determine the function's domain and its range.

Domain: set of inputs

Found on the x-axis

Range: set of outputs

Found on the y-axis

FIGURE 2.8   Domain and range of f

Let's apply these ideas to the graph of the function shown in Figure 2.9. To find the domain, look for all the inputs on the x-axis that correspond to points on the graph. Can you see that they extend from -4 to 2, inclusive? Using set-builder notation, the function's domain can be represented as follows:

$$\{ x \mid -4 \le x \le 2 \}.$$

The set of all x — such that — x is greater than or equal to -4 and less than or equal to 2.

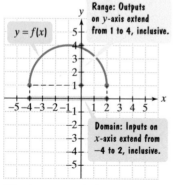

FIGURE 2.9   Domain and range of f

To find the range, look for all the outputs on the y-axis that correspond to points on the graph. They extend from 1 to 4, inclusive. Using set-builder notation, the function's range can be represented as follows:

$$\{\, y \mid 1 \leq y \leq 4 \,\}$$

| The set of all y | such that | y is greater than or equal to 1 and less than or equal to 4. |

EXAMPLE 7 Identifying the Domain and Range of a Function from Its Graph

Use the graph of each function to identify its domain and its range.

a.

b.

c.

d.

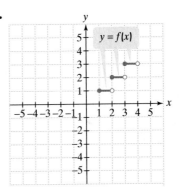

SOLUTION For the graph of each function, the domain is highlighted in blue on the x-axis and the range is highlighted in green on the y-axis.

a.

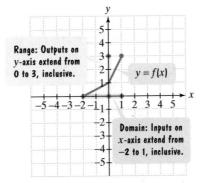

Range: Outputs on y-axis extend from 0 to 3, inclusive.

Domain: Inputs on x-axis extend from -2 to 1, inclusive.

Domain = $\{x \mid -2 \leq x \leq 1\}$
Range = $\{y \mid 0 \leq y \leq 3\}$

b.

Range: Outputs on y-axis extend from 1, excluding 1, to 2, including 2.

Domain: Inputs on x-axis extend from -3, excluding -3, to 2, including 2.

Domain = $\{x \mid -3 < x \leq 2\}$
Range = $\{y \mid 1 < y \leq 2\}$

c.

Range: Outputs on y-axis include real numbers greater than or equal to 0.

$y = f(x)$

Domain: Inputs on x-axis include real numbers less than or equal to 4.

Domain $= \{x | x \le 4\}$
Range $= \{y | y \ge 0\}$

d.

$y = f(x)$

Range: Outputs on y-axis "step" from 1 to 2 to 3.

Domain: Inputs on x-axis extend from 1, including 1, to 4, excluding 4.

Domain $= \{x | 1 \le x < 4\}$
Range $= \{y | y = 1, 2, 3\}$

CHECK POINT **7** Use the graph of each function to identify its domain and its range.

a.

$y = f(x)$

Domain $= \{x | -2 \le x \le 1\}$
Range $= \{y | 0 \le y \le 3\}$

b.

$y = f(x)$

Domain $= \{x | -2 < x \le 1\}$
Range $= \{y | -1 \le y < 2\}$

c.

$y = f(x)$

Domain $= \{x | -3 \le x < 0\}$
Range $= \{y | y = -3, -2, -1\}$

2.1 EXERCISE SET

Student Solutions Manual CD/Video PH Math/Tutor Center MathXL Tutorials on CD MathXL® MyMathLab Interactmath.com

Practice Exercises

In Exercises 1–8, determine whether each relation is a function. Give the domain and range for each relation.

1. $\{(1, 2), (3, 4), (5, 5)\}$ function; domain: $\{1, 3, 5\}$; range: $\{2, 4, 5\}$

2. $\{(4, 5), (6, 7), (8, 8)\}$ function; domain: $\{4, 6, 8\}$; range: $\{5, 7, 8\}$

3. $\{(3, 4), (3, 5), (4, 4), (4, 5)\}$ not a function; domain: $\{3, 4\}$; range: $\{4, 5\}$

4. $\{(5, 6), (5, 7), (6, 6), (6, 7)\}$ not a function; domain: $\{5, 6\}$; range: $\{6, 7\}$

5. $\{(-3, -3), (-2, -2), (-1, -1), (0, 0)\}$ function; domain: $\{-3, -2, -1, 0\}$; range: $\{-3, -2, -1, 0\}$

6. $\{(-7, -7), (-5, -5), (-3, -3), (0, 0)\}$ function; domain: $\{-7, -5, -3, 0\}$; range: $\{-7, -5, -3, 0\}$

7. $\{(1, 4), (1, 5), (1, 6)\}$ not a function; domain: $\{1\}$; range: $\{4, 5, 6\}$

8. $\{(4, 1), (5, 1), (6, 1)\}$ function; domain: $\{4, 5, 6\}$; range: $\{1\}$

In Exercises 9–18, find the indicated function values.

9. $f(x) = x + 1$

   a. $f(0)$ 1 **b.** $f(5)$ 6 **c.** $f(-8)$ −7

   d. $f(2a)$ $2a + 1$ **e.** $f(a + 2)$ $a + 3$

10. $f(x) = x + 3$

   a. $f(0)$ 3   **b.** $f(5)$ 8   **c.** $f(-8)$ −5

   d. $f(2a)$ $2a + 3$ **e.** $f(a + 2)$ $a + 5$

11. $g(x) = 3x - 2$

   a. $g(0)$ −2   **b.** $g(-5)$ −17   **c.** $g\left(\dfrac{2}{3}\right)$ 0

   d. $g(4b)$ $12b - 2$ **e.** $g(b + 4)$ $3b + 10$

12. $g(x) = 4x - 3$

   a. $g(0)$ −3   **b.** $g(-5)$ −23   **c.** $g\left(\dfrac{3}{4}\right)$ 0

   d. $g(5b)$ $20b - 3$ **e.** $g(b + 5)$ $4b + 17$

13. $h(x) = 3x^2 + 5$

   a. $h(0)$ 5   **b.** $h(-1)$ 8   **c.** $h(4)$ 53

   d. $h(-3)$ 32   **e.** $h(4b)$ $48b^2 + 5$

14. $h(x) = 2x^2 - 4$

   a. $h(0)$ −4   **b.** $h(-1)$ −2   **c.** $h(5)$ 46

   d. $h(-3)$ 14   **e.** $h(5b)$ $50b^2 - 4$

15. $f(x) = 2x^2 + 3x - 1$

   a. $f(0)$ −1   **b.** $f(3)$ 26   **c.** $f(-4)$ 19

   d. $f(b)$ $2b^2 + 3b - 1$

   e. $f(5a)$ $50a^2 + 15a - 1$

16. $f(x) = 3x^2 + 4x - 2$

   a. $f(0)$ −2   **b.** $f(3)$ 37   **c.** $f(-5)$ 53

   d. $f(b)$ $3b^2 + 4b - 2$

   e. $f(5a)$ $75a^2 + 20a - 2$

17. $f(x) = \dfrac{2x - 3}{x - 4}$

   a. $f(0)$ $\dfrac{3}{4}$ **b.** $f(3)$ −3   **c.** $f(-4)$ $\dfrac{11}{8}$

   d. $f(-5)$ $\dfrac{13}{9}$ **e.** $f(a + h)$ $\dfrac{2a + 2h - 3}{a + h - 4}$

   f. Why must 4 be excluded from the domain of f?
     Denominator would be zero.

18. $f(x) = \dfrac{3x - 1}{x - 5}$

   a. $f(0)$ $\dfrac{1}{5}$ **b.** $f(3)$ −4   **c.** $f(-3)$ $\dfrac{5}{4}$

   d. $f(10)$ $\dfrac{29}{5}$ **e.** $f(a + h)$ $\dfrac{3a + 3h - 1}{a + h - 5}$

   f. Why must 5 be excluded from the domain of f?
     Denominator would be zero.

In Exercises 19–28, graph the given functions, f and g, in the same rectangular coordinate system. Select integers for x, starting with −2 and ending with 2. Once you have obtained your graphs, describe how the graph of g is related to the graph of f.
19–28. See graphing answer section.

19. $f(x) = x, g(x) = x + 3$ The graph of g is the graph of f shifted up by 3 units.

20. $f(x) = x, g(x) = x - 4$ The graph of g is the graph of f shifted down by 4 units.

21. $f(x) = -2x, g(x) = -2x - 1$ The graph of g is the graph of f shifted down by 1 unit.

22. $f(x) = -2x, g(x) = -2x + 3$ The graph of g is the graph of f shifted up by 3 units.

23. $f(x) = x^2, g(x) = x^2 + 1$ The graph of g is the graph of f shifted up by 1 unit.

24. $f(x) = x^2, g(x) = x^2 - 2$ The graph of g is the graph of f shifted down by 2 units.

25. $f(x) = |x|, g(x) = |x| - 2$ The graph of g is the graph of f shifted down by 2 units.

26. $f(x) = |x|, g(x) = |x| + 1$ The graph of g is the graph of f shifted up by 1 unit.

27. $f(x) = x^3, g(x) = x^3 + 2$ The graph of g is the graph of f shifted up by 2 units.

28. $f(x) = x^3, g(x) = x^3 - 1$ The graph of g is the graph of f shifted down by 1 unit.

In Exercises 29–36, use the vertical line test to identify graphs in which y is a function of x.

29.

function

30.

function

31.

not a function

32.

not a function

33.

function

34.

not a function

35.

not a function

36.

function

In Exercises 37–42, use the graph of f to find each indicated function value.

37. $f(-2)$ -4

38. $f(2)$ -4

39. $f(4)$ 4

40. $f(-4)$ 4

41. $f(-3)$ 0

42. $f(-1)$ 0

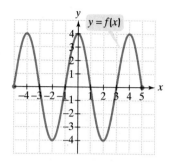

Use the graph of g to solve Exercises 43–48.

43. Find $g(-4)$. 2

44. Find $g(2)$. -2

45. Find $g(-10)$. 2

46. Find $g(10)$. -2

47. For what value of x is $g(x) = 1$? -2

48. For what value of x is $g(x) = -1$? 1

In Exercises 49–58, use the graph of each function to identify its domain and its range.

49.

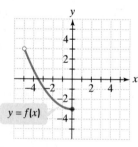

domain $= \{x \mid 0 \le x < 5\}$;
range $= \{y \mid -1 \le y < 5\}$

50.

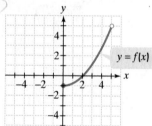

domain $= \{x \mid -5 < x \le 0\}$;
range $= \{y \mid -3 \le y < 3\}$

51.

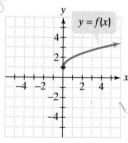

domain $= \{x \mid x \ge 0\}$;
range $= \{y \mid y \ge 1\}$

52.

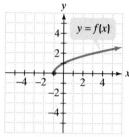

domain $= \{x \mid x \ge -1\}$;
range $= \{y \mid y \ge 0\}$

53.

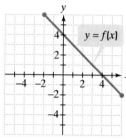

domain $= \{x \mid -2 \le x \le 6\}$;
range $= \{y \mid -2 \le y \le 6\}$

54.

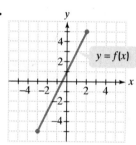

domain $= \{x \mid -3 \le x \le 2\}$;
range $= \{y \mid -5 \le y \le 5\}$

55.

domain $= \{x \mid x \text{ is a real number}\}$;
range $= \{y \mid y \le -2\}$

56.

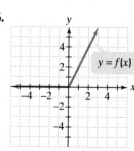

domain $= \{x \mid x \text{ is a real number}\}$;
range $= \{y \mid y \ge 0\}$

57.

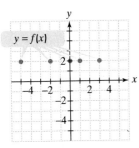

domain = $\{x \mid x = -5, -2, 0, 1, 3\}$;
range = $\{y \mid y = 2\}$

58.

domain = $\{x \mid x = -5, -2, 0, 1, 4\}$;
range = $\{y \mid y = -2\}$

Practice Plus

In Exercises 59–60, let $f(x) = x^2 - x + 4$ *and* $g(x) = 3x - 5$.

59. Find $g(1)$ and $f(g(1))$. $-2; 10$

60. Find $g(-1)$ and $f(g(-1))$. $-8; 76$

In Exercises 61–62, let f and g be defined by the following table:

x	$f(x)$	$g(x)$
-2	6	0
-1	3	4
0	-1	1
1	-4	-3
2	0	-6

61. Find $\sqrt{f(-1) - f(0)} - [g(2)]^2 + f(-2) \div g(2) \cdot g(-1)$ -38

62. Find $|f(1) - f(0)| - [g(1)]^2 + g(1) \div f(-1) \cdot g(2)$ 0

In Exercises 63–64, find $f(-x) - f(x)$ *for the given function f. Then simplify the expression.*

63. $f(x) = x^3 + x - 5$ $-2x^3 - 2x$

64. $f(x) = x^2 - 3x + 7$ $6x$

In Exercises 65–66, each function is defined by two equations. The equation in the first row gives the output for negative numbers in the domain. The equation in the second row gives the output for non-negative numbers in the domain. Find the indicated function values.

65. $f(x) = \begin{cases} 3x + 5 & \text{if } x < 0 \\ 4x + 7 & \text{if } x \geq 0 \end{cases}$

a. $f(-2)$ -1 **b.** $f(0)$ 7

c. $f(3)$ 19   **d.** $f(-100) + f(100)$ 112

66. $f(x) = \begin{cases} 6x - 1 & \text{if } x < 0 \\ 7x + 3 & \text{if } x \geq 0 \end{cases}$

a. $f(-3)$ -19 **b.** $f(0)$ 3

c. $f(4)$ 31   **d.** $f(-100) + f(100)$ 102

In Exercises 67–68, use the graph of each function to identify its domain and its range.

67.

Graph approaches but never touches the dashed vertical line.

On both sides, graph never touches the x-axis.

domain = $\{x \mid x \neq 1\}$; range = $\{y \mid y \neq 0\}$

68.

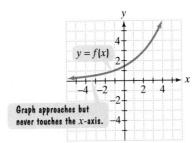

Graph approaches but never touches the x-axis.

domain = $\{x \mid x \text{ is a real number}\}$; range = $\{y \mid y > 0\}$

Application Exercises

69. The bar graph shows the breakdown of political ideologies in the United States.

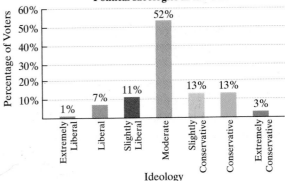

Political Ideologies in the U.S.

Source: Center for Political Studies, University of Michigan

a. Write a set of seven ordered pairs in which political ideologies correspond to percentages. Each ordered pair should be in the form

(ideology, percent).

Use EL, L, SL, M, SC, C, and EC to represent the respective ideologies from left to right. $\{$(EL, 1%), (L, 7%), (SL, 11%), (M, 52%), (SC, 13%), (C, 13%), (EC, 3%)$\}$

b. Is the relation in part (a) a function? Explain your answer.   Yes; each ideology corresponds to exactly one percentage.

c. Write a set of seven ordered pairs in which percentages correspond to political ideologies. Each ordered pair should be in the form

(percent, ideology).

{(1%, EL), (7%, L), (11%, SL), (52%, M), (13%, SC), (13%, C), (3%, EC)}

d. Is the relation in part (c) a function? Explain your answer.   No; 13% in the domain corresponds to two ideologies in the range, SC and C.

70. The bar graph shows the percentage of teens in the United States who say they really need various technologies.

Percentage of U.S. Teens Saying They Really Need Various Technologies

Source: Gallup Organization

a. Write a set of six ordered pairs in which technologies correspond to percentages. Each ordered pair should be in the form

(technology, percent).

{(Computer, 77%), (TV, 71%), (Calculator, 71%), (Stereo, 69%), (Ph. Ans. M.,62%), (VCR, 51%)}

b. Is the relation in part (a) a function? Explain your answer.   Yes; each technology corresponds to exactly one percentage.

c. Write a set of six ordered pairs in which percentages correspond to technologies. Each ordered pair should be in the form

(percent, technology).

{(77%, Computer), (71%, TV), (71%, Calculator), (69%, Stereo), (62%, Ph. Ans. M.), (51%, VCR)}

d. Is the relation in part (c) a function? Explain your answer.   No; 71% in the domain corresponds to two technologies in the range, TV and calculator.

The male minority? The graphs show enrollment in U.S. colleges, with projections through 2009. The trend indicated by the graphs is among the hottest topics of debate among college-admissions officers. Some private liberal arts colleges have quietly begun special efforts to recruit men—including admissions preferences for them.

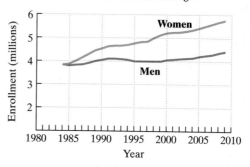

Enrollment in U.S. Colleges

Source: Department of Education

The function

$$W(x) = 0.07x + 4.1$$

models the number of women, $W(x)$, in millions, enrolled in U.S. colleges x years after 1984. The function

$$M(x) = 0.01x + 3.9$$

models the number of men, $M(x)$, in millions, enrolled in U.S. colleges x years after 1984. Use these functions to solve Exercises 71–74.

71. Find and interpret $W(16)$. Identify this information as a point on the graph for women.   5.22; In 2000, there were 5.22 million women enrolled in U.S. colleges; (2000, 5.22).

72. Find and interpret $M(16)$. Identify this information as a point on the graph for men.   4.06; In 2000, there were 4.06 million men enrolled in U.S. colleges; (2000, 4.06).

73. Find and interpret $W(20) - M(20)$.   1.4; In 2004, there were 1.4 million more women than men enrolled in U.S. colleges.

74. Find and interpret $W(25) - M(25)$.   1.7; In 2009, there will be 1.7 million more women than men enrolled in U.S. colleges.

The function

$$f(x) = 0.4x^2 - 36x + 1000$$

models the number of accidents, $f(x)$, per 50 million miles driven as a function of a driver's age, x, in years, where x includes drivers from ages 16 through 74, inclusive. The graph of f is shown. Use the equation for f to solve Exercises 75–78.

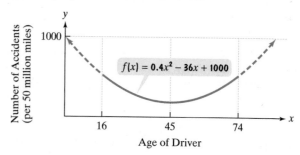

75. Find and interpret $f(20)$. Identify this information as a point on the graph of f.   440; For 20-year-old drivers, there are 440 accidents per 50 million miles driven; (20, 440)

76. Find and interpret $f(50)$. Identify this information as a point on the graph of f.   200; For 50-year-old drivers, there are 200 accidents per 50 million miles driven; (50, 200)

77. For what value of x does the graph reach its lowest point? Use the equation for f to find the minimum value of y. Describe the practical significance of this minimum value.
$x = 45$; $y = 190$; The minimum number of accidents is 190 per 50 million miles driven and is attributed to 45-year-old drivers.

78. Use the graph to identify two different ages for which drivers have the same number of accidents. Use the equation for f to find the number of accidents for drivers at each of these ages.
Answers will vary; an example is 16 and 74 years old. For these ages, the number of accidents is 526.4 per 50 million miles.

The figure shows the percentage of Jewish Americans in the U.S. population, $f(x)$, x years after 1900. Use the graph to solve Exercises 79–86.   79–84. Answers will vary. Example are given.

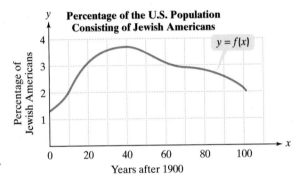

Source: American Jewish Yearbook

79. Use the graph to find a reasonable estimate of $f(60)$. What does this mean in terms of the variables in this situation?
$f(60) \approx 3.1$; In 1960, Jewish Americans made up about 3.1% of the U.S. population.

80. Use the graph to find a reasonable estimate of $f(100)$. What does this mean in terms of the variables in this situation?
$f(100) \approx 2.2$; In 2000, Jewish Americans made up about 2.2% of the U.S. population.

81. For what value or values of x is $f(x) = 3$? Round to the nearest year. What does this mean in terms of the variables in this situation? $x \approx 19$ and $x \approx 64$; In 1919 and in 1964, Jewish Americans made up about 3% of the U.S. population.

82. For what value or values of x is $f(x) = 2.5$? Round to the nearest year. What does this mean in terms of the variables in this situation? $x \approx 14$ and $x \approx 90$; In 1914 and in 1990, Jewish Americans made up 2.5% of the U.S. population.

83. In which year did the percentage of Jewish Americans in the U.S. population reach a maximum? What is a reasonable estimate of the percentage for that year?   1940; 3.7%

84. In which year was the percentage of Jewish Americans in the U.S. population at a minimum? What is a reasonable estimate of the percentage for that year?   1900; 1.4%

85. Explain why f represents the graph of a function.   Each year corresponds to only one percentage.

86. Describe the general trend shown by the graph.   As a percentage of the U.S. population, the Jewish-American population was increasing until 1940 and has been decreasing since then.

The figure shows the cost of mailing a first-class letter, $f(x)$, as a function of its weight, x, in ounces. Use the graph to solve Exercises 87–90.

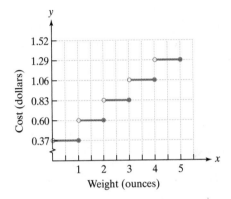

87. Find $f(3)$. What does this mean in terms of the variables in this situation?   0.83; It costs $0.83 to mail a 3-ounce first-class letter.

88. Find $f(4)$. What does this mean in terms of the variables in this situation?   1.06; It costs $1.06 to mail a 4-ounce first-class letter.

89. What is the cost of mailing a letter that weighs 1.5 ounces? $0.60

90. What is the cost of mailing a letter that weighs 1.8 ounces? $0.60

Writing in Mathematics

91. What is a relation? Describe what is meant by its domain and its range.   91–98. Answers will vary.

92. Explain how to determine whether a relation is a function. What is a function?

93. Does $f(x)$ mean f times x when referring to function f? If not, what does $f(x)$ mean? Provide an example with your explanation.

94. What is the graph of a function?

95. Explain how the vertical line test is used to determine whether a graph represents a function.

96. Explain how to identify the domain and range of a function from its graph.

97. For people filing a single return, federal income tax is a function of adjusted gross income because for each value of adjusted gross income there is a specific tax to be paid. By contrast, the price of a house is not a function of the lot size on which the house sits because houses on same-sized lots can sell for many different prices.
  a. Describe an everyday situation between variables that is a function.
  b. Describe an everyday situation between variables that is not a function.

98. Do you believe that the trend shown by the graphs for Exercises 71–74 should be reversed by providing admissions preferences for men? Explain your position on this issue.

Technology Exercises

99. Use a graphing utility to verify the pairs of graphs that you drew by hand in Exercises 19–28.   Answers will vary.

100. The function

$$f(x) = -0.00002x^3 + 0.008x^2 - 0.3x + 6.95$$

  models the number of annual physician visits, $f(x)$, by a person of age x. Graph the function in a $[0, 100, 5]$ by $[0, 40, 2]$ viewing rectangle. What does the shape of the graph indicate about the relationship between one's age and the number of annual physician visits? Use the TRACE or minimum function capability to find the coordinates of the minimum point on the graph of the function. What does this mean?   See graphing answer section; The number of annual visits decreases until about age 20 and then increases; (20.3, 4.0); The minimum is at about age 20 when there are about 4 annual physician visits.

Critical Thinking Exercises

101. Which one of the following is true?   d
  a. All relations are functions.
  b. No two ordered pairs of a function can have the same second component and different first components.
  c. The graph of every line is a function.
  d. A horizontal line can intersect the graph of a function in more than one point.

102. If $f(x) = 3x + 7$, find $\dfrac{f(a + h) - f(a)}{h}$.   3

103. Give an example of a relation with the following characteristics: The relation is a function containing two ordered pairs. Reversing the components in each ordered pair results in a relation that is not a function.   Answers will vary; an example is $\{(1, 1), (2, 1)\}$.

104. If $f(x + y) = f(x) + f(y)$ and $f(1) = 3$, find $f(2)$, $f(3)$, and $f(4)$. Is $f(x + y) = f(x) + f(y)$ for all functions? $f(2) = 6$; $f(3) = 9$; $f(4) = 12$; no

Review Exercises

105. Simplify: $24 \div 4[2 - (5 - 2)]^2 - 6$. (Section 1.2, Example 7)   0

106. Simplify: $\left(\dfrac{3x^2 y^{-2}}{y^3}\right)^{-2}$. (Section 1.6, Example 9) $\dfrac{y^{10}}{9x^4}$

107. Solve: $\dfrac{x}{3} = \dfrac{3x}{5} + 4$. (Section 1.4, Example 4) -15

SECTION 2.2

THE ALGEBRA OF FUNCTIONS

Objectives

1 Find the domain of a function.

2 Find the sum of two functions.

3 Use the algebra of functions to combine functions and determine domains.

2006 Pontiac Solstice, the first two-seater from Pontiac in more than 20 years. Starting price: $30,000

America's big three automakers, GM, Ford, and Chrysler, know that consumers have always been willing to pay more for cool design and a hot car. In 2004, Detroit gave consumers that opportunity, unleashing 40 new cars or updated models. The Big Three's discovery of the automobile in 2004 might seem odd, but from 1990 through 2003, Detroit focused much of its energy and money on SUVs and trucks, which commanded high prices and high profits, and saw less competition from imports. The line graphs in Figure 2.10 show the number, in millions, of cars and SUVs sold by the Big Three from 1990 through 2003. In this section, we will look at these data from the perspective of functions. By considering total sales of cars and SUVs, you will see that functions can be combined using procedures that will remind you of combining algebraic expressions.

Cars in Reverse: Millions of Cars and SUVs Sold

FIGURE 2.10 Sales by America's Big Three automakers

1 Find the domain of a function.

The Domain of a Function We begin with two functions that model the data in Figure 2.10. (In the next section of the chapter, you will learn how to obtain these functions from the data.)

$$C(x) = -0.14x + 9 \qquad S(x) = 0.22x + 1$$

Car sales, $C(x)$, in millions, x years after 1990

SUV sales, $S(x)$, in millions, x years after 1990

How far beyond 1990 should we extend these models? The trend in Figure 2.10 shows decreasing car sales and increasing SUV sales. Because the three automakers focused on SUVs from 1990 through 2003 and launched a fleet of new cars in 2004, the trend shown by the data changed in 2004. Thus, we should not extend the models beyond 2003. Because x represents the number of years after 1990,

$$\text{Domain of } C = \{x | x = 0, 1, 2, 3, \ldots, 13\}$$

and

$$\text{Domain of } S = \{x | x = 0, 1, 2, 3, \ldots, 13\}.$$

Functions that model data often have their domains explicitly given with the function's equation. However, for most functions, only an equation is given, and the domain is not specified. In cases like this, the domain of a function f is the largest set of real numbers for which the value of $f(x)$ is a real number. For example, consider the function

$$f(x) = \frac{1}{x - 3}.$$

Because division by 0 is undefined (and not a real number), the denominator, $x - 3$, cannot be 0. Thus, x cannot equal 3. The domain of the function consists of all real numbers other than 3, represented by

$$\text{Domain of } f = \{x | x \text{ is a real number and } x \neq 3\}.$$

In Chapter 7, we will be studying square root functions such as

$$g(x) = \sqrt{x}.$$

The equation tells us to take the square root of x. Because only nonnegative numbers have square roots that are real numbers, the expression under the square root sign, x, must be nonnegative. Thus,

$$\text{Domain of } g = \{x \mid x \text{ is a nonnegative real number}\}.$$

Equivalently,

$$\text{Domain of } g = \{x \mid x \geq 0\}.$$

FINDING A FUNCTION'S DOMAIN If a function f does not model data or verbal conditions, its domain is the largest set of real numbers for which the value of $f(x)$ is a real number. Exclude from a function's domain real numbers that cause division by zero and real numbers that result in a square root of a negative number.

EXAMPLE 1 Finding the Domain of a Function

Find the domain of each function:

a. $f(x) = 3x + 2$ **b.** $g(x) = \dfrac{3x + 2}{x + 1}$.

SOLUTION

a. The function $f(x) = 3x + 2$ contains neither division nor a square root. For every real number, x, the algebraic expression $3x + 2$ is a real number. Thus, the domain of f is the set of all real numbers.

$$\text{Domain of } f = \{x \mid x \text{ is a real number}\}$$

b. The function $g(x) = \dfrac{3x + 2}{x + 1}$ contains division. Because division by 0 is undefined, we must exclude from the domain the value of x that causes $x + 1$ to be 0. Thus, x cannot equal -1.

$$\text{Domain of } g = \{x \mid x \text{ is a real number and } x \neq -1\} \qquad \blacksquare$$

 CHECK POINT 1 Find the domain of each function:

a. $f(x) = \dfrac{1}{2}x + 3$ Domain of $f = \{x \mid x \text{ is a real number}\}$

b. $g(x) = \dfrac{7x + 4}{x + 5}$.   Domain of $g = \{x \mid x \text{ is a real number and } x \neq -5\}$

The Algebra of Functions We return to the functions that model millions of car and SUV sales from 1990 through 2003:

$$C(x) = -0.14x + 9 \qquad S(x) = 0.22x + 1.$$

Car sales, $C(x)$, in millions, x years after 1990

SUV sales, $S(x)$, in millions, x years after 1990

How can we use these functions to find total sales of cars and SUVs in 2003? Because 2003 is 13 years after 1990 and 13 is in the domain of each function, we need to find the sum of two function values:

$$C(13) + S(13).$$

Here is how it's done:

$$C(13) = -0.14(13) + 9 = 7.18 \qquad S(13) = 0.22(13) + 1 = 3.86$$

Substitute 13 for x in $C(x) = -0.14x + 9$. / 7.18 million cars were sold in 2003.

Substitute 13 for x in $S(x) = 0.22x + 1$. / 3.86 million SUVs were sold in 2003.

$$C(13) + S(13) = 7.18 + 3.86 = 11.04.$$

Thus, a total of 11.04 million cars and SUVs were sold in 2003.

There is a second way we can obtain this number. We can first add the functions C and S to obtain a new function, $C + S$. To do so, we add the terms to the right of the equal sign for $C(x)$ to the terms to the right of the equal sign for $S(x)$:

$$\begin{aligned}(C + S)(x) &= C(x) + S(x) \\ &= (-0.14x + 9) + (0.22x + 1) \qquad \text{Add terms for } C(x) \text{ and } S(x). \\ &= 0.08x + 10. \qquad \text{Combine like terms.}\end{aligned}$$

Thus,

$$(C + S)(x) = 0.08x + 10$$

Total car and SUV sales, in millions, x years after 1990

Do you see how we can use this new function to find total car and SUV sales in 2003? Substitute 13 for x in the equation for $C + S$:

$$(C + S)(13) = 0.08(13) + 10 = 11.04$$

Substitute 13 for x in $(C + S)(x) = 0.08x + 10$.

As we found above, a total of 11.04 million cars and SUVs were sold in 2003.

The domain of the new function, $C + S$, consists of the numbers x that are in the domain of C and in the domain of S. Because both functions model data from 1990 through 2003,

$$\text{Domain of } C + S = \{x \mid x = 0, 1, 2, 3, \ldots, 13\}.$$

Here is a general definition for function addition:

> **THE SUM OF FUNCTIONS** Let f and g be two functions. The **sum $f + g$** is the function defined by
> $$(f + g)(x) = f(x) + g(x).$$
> The domain of $f + g$ is the set of all real numbers that are common to the domain of f and the domain of g.

EXAMPLE 2 Finding the Sum of Two Functions

Let $f(x) = x^2 - 3$ and $g(x) = 4x + 5$. Find each at the following:

a. $(f + g)(x)$ **b.** $(f + g)(3)$.

SOLUTION

a. $(f + g)(x) = f(x) + g(x) = (x^2 - 3) + (4x + 5) = x^2 + 4x + 2$

Thus,

$$(f + g)(x) = x^2 + 4x + 2.$$

b. We find $(f + g)(3)$ by substituting 3 for x in the equation for $f + g$.

$$(f + g)(x) = x^2 + 4x + 2 \qquad \text{This is the equation for } f + g.$$

Substitute 3 for x.

$$(f + g)(3) = 3^2 + 4 \cdot 3 + 2 = 9 + 12 + 2 = 23 \qquad \blacksquare$$

 CHECK POINT **2** Let $f(x) = 3x^2 + 4x - 1$ and $g(x) = 2x + 7$. Find each of the following:

a. $(f + g)(x)$ $\quad 3x^2 + 6x + 6$ **b.** $(f + g)(4)$. $\quad 78$

EXAMPLE 3   Adding Functions and Determining the Domain

Let $f(x) = \dfrac{4}{x}$ and $g(x) = \dfrac{3}{x + 2}$. Find each of the following:

a. $(f + g)(x)$ **b.** the domain of $f + g$.

SOLUTION

a. $(f + g)(x) = f(x) + g(x) = \dfrac{4}{x} + \dfrac{3}{x + 2}$

(In Chapter 6, we will discuss how to combine these algebraic fractions. Perhaps you remember how to do so from your work in introductory algebra. For now, we will leave these fractions in the form shown.)

b. The domain of $f + g$ is the set of all real numbers that are common to the domain of f and the domain of g. Thus, we must find the domains of f and g. We will do so for f first.

Note that $f(x) = \dfrac{4}{x}$ is a function involving division. Because division by 0 is undefined, x cannot equal 0.

$$\text{Domain of } f = \{x \mid x \text{ is a real number and } x \neq 0\}$$

The function $g(x) = \dfrac{3}{x + 2}$ is also a function involving division. Because division by 0 is undefined, x cannot equal -2.

$$\text{Domain of } g = \{x \mid x \text{ is a real number and } x \neq -2\}$$

To find $f(x) + g(x)$, x must be in both domains listed. Thus,

$$\text{Domain of } f + g = \{x \mid x \text{ is a real number and } x \neq 0 \text{ and } x \neq -2\}. \qquad \blacksquare$$

 CHECK POINT **3** Let $f(x) = \dfrac{5}{x}$ and $g(x) = \dfrac{7}{x - 8}$. Find each of the following:

a. $(f + g)(x)$ $\quad \dfrac{5}{x} + \dfrac{7}{x - 8}$ **b.** the domain of $f + g$.

Domain of $f + g = \{x \mid x$ is a real number and $x \neq 0$ and $x \neq 8\}$

3   Use the algebra of functions to combine functions and determine domains.

We can also combine functions using subtraction, multiplication, and division by performing operations with the algebraic expressions that appear on the right side of the equations. For example, the functions $f(x) = 2x$ and $g(x) = x - 1$ can be combined to form the difference, product, and quotient of f and g. Here's how it's done:

Difference: $f - g$

$(f - g)(x) = f(x) - g(x)$

$= 2x - (x - 1) = 2x - x + 1 = x + 1$

For each function,
$f(x) = 2x$ and
$g(x) = x - 1.$

Product: fg

$(fg)(x) = f(x) \cdot g(x)$

$= 2x(x - 1) = 2x^2 - 2x$

Quotient: $\dfrac{f}{g}$

$\left(\dfrac{f}{g}\right)(x) = \dfrac{f(x)}{g(x)} = \dfrac{2x}{x - 1}, x \neq 1.$

Just like the domain for $f + g$, the domain for each of these functions consists of all real numbers that are common to the domains of f and g. In the case of the quotient function $\dfrac{f(x)}{g(x)}$, we must remember not to divide by 0, so we add the further restriction that $g(x) \neq 0$.

The following definitions summarize our discussion:

THE ALGEBRA OF FUNCTIONS: SUM, DIFFERENCE, PRODUCT, AND QUOTIENT OF FUNCTIONS Let f and g be two functions. The **sum** $f + g$, the **difference** $f - g$, the **product** fg, and the **quotient** $\dfrac{f}{g}$ are functions whose domains are the set of all real numbers common to the domains of f and g, defined as follows:

1. **Sum:** $(f + g)(x) = f(x) + g(x)$
2. **Difference:** $(f - g)(x) = f(x) - g(x)$
3. **Product:** $(fg)(x) = f(x) \cdot g(x)$
4. **Quotient:** $\left(\dfrac{f}{g}\right)(x) = \dfrac{f(x)}{g(x)}$, provided $g(x) \neq 0$.

EXAMPLE 4   Using the Algebra of Functions

Let $f(x) = x^2 + x$ and $g(x) = x - 5$. Find each of the following:

a. $(f + g)(4)$

b. $(f - g)(x)$ and $(f - g)(-3)$

c. $\left(\dfrac{f}{g}\right)(x)$ and $\left(\dfrac{f}{g}\right)(7)$

d. $(fg)(-2)$.

SOLUTION

a. We can find $(f + g)(4)$ using $f(4)$ and $g(4)$.

$$f(x) = x^2 + x \qquad g(x) = x - 5$$
$$f(4) = 4^2 + 4 = 20 \quad g(4) = 4 - 5 = -1$$

Thus,

$$(f + g)(4) = f(4) + g(4) = 20 + (-1) = 19.$$

We can also find $(f + g)(4)$ by first finding $(f + g)(x)$ and then substituting 4 for x:

$(f + g)(x) = f(x) + g(x)$ This is the definition of the sum $f + g$.

$= (x^2 + x) + (x - 5)$ Substitute the given functions.

$= x^2 + 2x - 5.$ Simplify.

Using $(f + g)(x) = x^2 + 2x - 5$. we have

$$(f + g)(4) = 4^2 + 2 \cdot 4 - 5 = 16 + 8 - 5 = 19.$$

$f(x) = x^2 + x$
$g(x) = x - 5$
The given functions,
f and g, repeated

b. $(f - g)(x) = f(x) - g(x)$ This is the definition of the difference $f - g$.

$$= (x^2 + x) - (x - 5)$$ Substitute the given functions.

$$= x^2 + x - x + 5$$ Remove parentheses and change the sign of each term in the second set of parentheses.

$$= x^2 + 5$$ Simplify.

Using $(f - g)(x) = x^2 + 5$, we have

$$(f - g)(-3) = (-3)^2 + 5 = 9 + 5 = 14.$$

c. $\left(\dfrac{f}{g}\right)(x) = \dfrac{f(x)}{g(x)}$ This is the definition of the quotient $\dfrac{f}{g}$.

$$= \dfrac{x^2 + x}{x - 5}$$ Substitute the given functions.

$\{x | x \text{ is real } \# \}$

Using $\left(\dfrac{f}{g}\right)(x) = \dfrac{x^2 + x}{x - 5}$, we have

$$\left(\dfrac{f}{g}\right)(7) = \dfrac{7^2 + 7}{7 - 5} = \dfrac{56}{2} = 28.$$

d. We can find $(fg)(-2)$ using the fact that

$$(fg)(-2) = f(-2) \cdot g(-2).$$

$f(x) = x^2 + x$ $g(x) = x - 5$

$f(-2) = (-2)^2 + (-2) = 4 - 2 = 2$ $g(-2) = -2 - 5 = -7$

Thus,

$$(fg)(-2) = f(-2) \cdot g(-2) = 2(-7) = -14.$$

We could also have found $(fg)(-2)$ by multiplying $f(x) \cdot g(x)$ and then substituting -2 into the product. We will discuss how to multiply expressions such as $x^2 + x$ and $x - 5$ in Chapter 5. ■

 CHECK POINT 4 Let $f(x) = x^2 - 2x$ and $g(x) = x + 3$. Find each of the following:

a. $(f + g)(5)$ 23

b. $(f - g)(x)$ and $(f - g)(-1)$ $x^2 - 3x - 3; 1$

c. $\left(\dfrac{f}{g}\right)(x)$ and $\left(\dfrac{f}{g}\right)(7)$ $\dfrac{x^2 - 2x}{x + 3}; \dfrac{7}{2}$

d. $(fg)(-4)$. -24

2.2 EXERCISE SET

Student Solutions Manual   CD/Video   PH Math/Tutor Center   MathXL Tutorials on CD   MathXL®   MyMathLab   Interactmath.com

Practice Exercises

In Exercises 1–10, find the domain of each function.

1. $f(x) = 3x + 5$ Domain of $f = \{x \mid x \text{ is a real number}\}$

2. $f(x) = 4x + 7$ Domain of $f = \{x \mid x \text{ is a real number}\}$

3. $g(x) = \dfrac{1}{x + 4}$
Domain of $g = \{x \mid x \text{ is a real number and } x \neq -4\}$

4. $g(x) = \dfrac{1}{x + 5}$
Domain of $g = \{x \mid x \text{ is a real number and } x \neq -5\}$

5. $f(x) = \dfrac{2x}{x - 3}$
Domain of $f = \{x \mid x \text{ is a real number and } x \neq 3\}$

6. $f(x) = \dfrac{4x}{x - 2}$
Domain of $f = \{x \mid x \text{ is a real number and } x \neq 2\}$

7. $g(x) = x + \dfrac{3}{5 - x}$
Domain of $g = \{x \mid x \text{ is a real number and } x \neq 5\}$

8. $g(x) = x + \dfrac{7}{6 - x}$
Domain of $g = \{x \mid x \text{ is a real number and } x \neq 6\}$

9. $f(x) = \dfrac{1}{x + 7} + \dfrac{3}{x - 9}$ Domain of
$f = \{x \mid x \text{ is a real number and } x \neq -7 \text{ and } x \neq 9\}$

10. $f(x) = \dfrac{1}{x + 8} + \dfrac{3}{x - 10}$ Domain of
$f = \{x \mid x \text{ is a real number and } x \neq -8 \text{ and } x \neq 10\}$

In Exercises 11–16, find **a.** $(f + g)(x)$ **b.** $(f + g)(5)$.

11. $f(x) = 3x + 1, g(x) = 2x - 6$
$(f + g)(x) = 5x - 5; (f + g)(5) = 20$

12. $f(x) = 4x + 2, g(x) = 2x - 9$
$(f + g)(x) = 6x - 7; (f + g)(5) = 23$

13. $f(x) = x - 5, g(x) = 3x^2$
$(f + g)(x) = 3x^2 + x - 5; (f + g)(5) = 75$

14. $f(x) = x - 6, g(x) = 2x^2$
$(f + g)(x) = 2x^2 + x - 6; (f + g)(5) = 49$

15. $f(x) = 2x^2 - x - 3, g(x) = x + 1$
$(f + g)(x) = 2x^2 - 2; (f + g)(5) = 48$

16. $f(x) = 4x^2 - x - 3, g(x) = x + 1$
$(f + g)(x) = 4x^2 - 2; (f + g)(5) = 98$

In Exercises 17–28, for each pair of functions, f and g, determine the domain of f + g.

17. $f(x) = 3x + 7, g(x) = 9x + 10$
Domain of $f + g = \{x \mid x \text{ is a real number}\}$

18. $f(x) = 7x + 4, g(x) = 5x - 2$
Domain of $f + g = \{x \mid x \text{ is a real number}\}$

19. $f(x) = 3x + 7, g(x) = \dfrac{2}{x - 5}$
Domain of $f + g = \{x \mid x \text{ is a real number and } x \neq 5\}$

20. $f(x) = 7x + 4, g(x) = \dfrac{2}{x - 6}$
Domain of $f + g = \{x \mid x \text{ is a real number and } x \neq 6\}$

21. $f(x) = \dfrac{1}{x}, g(x) = \dfrac{2}{x - 5}$ Domain of
$f + g = \{x \mid x \text{ is a real number and } x \neq 0 \text{ and } x \neq 5\}$

22. $f(x) = \dfrac{1}{x}, g(x) = \dfrac{2}{x - 6}$ Domain of
$f + g = \{x \mid x \text{ is a real number and } x \neq 0 \text{ and } x \neq 6\}$

23. $f(x) = \dfrac{8x}{x - 2}, g(x) = \dfrac{6}{x + 3}$ Domain of
$f + g = \{x \mid x \text{ is a real number and } x \neq 2 \text{ and } x \neq -3\}$

24. $f(x) = \dfrac{9x}{x - 4}, g(x) = \dfrac{7}{x + 8}$ Domain of
$f + g = \{x \mid x \text{ is a real number and } x \neq 4 \text{ and } x \neq -8\}$

25. $f(x) = \dfrac{8x}{x - 2}, g(x) = \dfrac{6}{2 - x}$
Domain of $f + g = \{x \mid x \text{ is a real number and } x \neq 2\}$

26. $f(x) = \dfrac{9x}{x - 4}, g(x) = \dfrac{7}{4 - x}$
Domain of $f + g = \{x \mid x \text{ is a real number and } x \neq 4\}$

27. $f(x) = x^2, g(x) = x^3$
Domain of $f + g = \{x \mid x \text{ is a real number}\}$

28. $f(x) = x^2 + 1, g(x) = x^3 - 1$
Domain of $f + g = \{x \mid x \text{ is a real number}\}$

In Exercises 29–48, let
$$f(x) = x^2 + 4x \quad \text{and} \quad g(x) = 2 - x.$$

Find each of the following.

29. $(f + g)(x)$ and $(f + g)(3)$
$(f + g)(x) = x^2 + 3x + 2; (f + g)(3) = 20$

30. $(f + g)(x)$ and $(f + g)(4)$
$(f + g)(x) = x^2 + 3x + 2; (f + g)(4) = 30$

31. $f(-2) + g(-2)$ 0

32. $f(-3) + g(-3)$ 2

33. $(f - g)(x)$ and $(f - g)(5)$
$(f - g)(x) = x^2 + 5x - 2; (f - g)(5) = 48$

34. $(f - g)(x)$ and $(f - g)(6)$
$(f - g)(x) = x^2 + 5x - 2; (f - g)(6) = 64$

35. $f(-2) - g(-2)$ −8   **36.** $f(-3) - g(-3)$ −8

37. $(fg)(-2)$ -16 **38.** $(fg)(-3)$ -15

39. $(fg)(5)$ -135 **40.** $(fg)(6)$ -240

41. $\left(\dfrac{f}{g}\right)(x)$ and $\left(\dfrac{f}{g}\right)(1)$ $\left(\dfrac{f}{g}\right)(x) = \dfrac{x^2 + 4x}{2 - x}; \left(\dfrac{f}{g}\right)(1) = 5$

42. $\left(\dfrac{f}{g}\right)(x)$ and $\left(\dfrac{f}{g}\right)(3)$ $\left(\dfrac{f}{g}\right)(x) = \dfrac{x^2 + 4x}{2 - x}; \left(\dfrac{f}{g}\right)(3) = -21$

43. $\left(\dfrac{f}{g}\right)(-1)$ -1 **44.** $\left(\dfrac{f}{g}\right)(0)$ 0

45. The domain of $f + g$
Domain of $f + g = \{x | x$ is a real number$\}$

46. The domain of $f - g$
Domain of $f - g = \{x | x$ is a real number$\}$

47. The domain of $\dfrac{f}{g}$

Domain of $\dfrac{f}{g} = \{x | x$ is a real number and $x \neq 2\}$

48. The domain of fg Domain of $fg = \{x | x$ is a real number$\}$

Practice Plus

Use the graphs of f and g to solve Exercises 49–56.

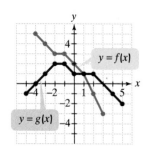

49. Find $(f + g)(-3)$.   5   **50.** Find $(g - f)(-2)$. -1

51. Find $(fg)(2)$. -1 **52.** Find $\left(\dfrac{g}{f}\right)(3)$.   0

53. Find the domain of $f + g$. $\{x | -4 \leq x \leq 3\}$

54. Find the domain of $\dfrac{f}{g}$. $\{x | -4 < x < 3\}$

55. Graph $f + g$.   55–56. See graphing answer section.

56. Graph $f - g$.

Use the table defining f and g to solve Exercises 57–60.

x	$f(x)$	$g(x)$
-2	5	0
-1	3	-2
0	-2	4
1	-6	-3
2	0	1

57. Find $(f + g)(1) - (g - f)(-1)$. -4

58. Find $(f + g)(-1) - (g - f)(0)$. -5

59. Find $(fg)(-2) - \left[\left(\dfrac{f}{g}\right)(1)\right]^2$. -4

60. Find $(fg)(2) - \left[\left(\dfrac{g}{f}\right)(0)\right]^2$. -4

Application Exercises

The table shows the total number of births and the total number of deaths in the United States from 1995 through 2002.

Births and Deaths in the U.S.

Year	Births	Deaths
1995	3,899,589	2,312,132
1996	3,891,494	2,314,690
1997	3,880,894	2,314,245
1998	3,941,553	2,337,256
1999	3,959,417	2,391,399
2000	4,058,814	2,403,351
2001	4,025,933	2,416,425
2002	4,022,000	2,436,000

Source: Department of Health and Human Services

The data can be modeled by the following functions:

Number of births $B(x) = 24{,}770x + 3{,}873{,}266$

Number of deaths $D(x) = 20{,}205x + 2{,}294{,}970.$

In each function, x represents the number of years after 1995. Assume that the functions apply only to the years shown in the table. Use these functions to solve Exercises 61–64.

61. Find the domain of B. $\{x | x = 0, 1, 2, \ldots, 7\}$

62. Find the domain of D. $\{x | x = 0, 1, 2, \ldots, 7\}$

63. a. Find $(B - D)(x)$. What does this function represent?
$(B - D)(x) = 4565x + 1{,}578{,}296$; change in U.S. population

   b. Use the function in part (a) to find $(B - D)(6)$. What does this mean in terms of the U.S. population and to which year does this apply? $(B - D)(6) = 1{,}605{,}686$; In 2001, U.S. population increased by 1,605,686.

   c. Use the data shown in the table to find $(B - D)(6)$. How well does the difference of functions used in part (b) model this number?   1,609,508; fairly well

64. a. Find $(B - D)(x)$. What does this function represent?
$(B - D)(x) = 4565x + 1{,}578{,}296$; change in U.S. population

   b. Use the function in part (a) to find $(B - D)(2)$. What does this mean in terms of the U.S. population and to which year does this apply? $(B - D)(2) = 1{,}587{,}426$; In 1997, U.S. population increased by 1,587,426.

c. Use the data shown in the table to find $(B - D)(2)$. How well does the difference of functions used in part (b) model this number?   1,566,649; fairly well, although answers may vary

Consider the following functions:

$f(x)$ = population of the world's more-developed

regions in year x

$g(x)$ = the population of the world's less-developed

regions in year x

$h(x)$ = total world population in year x.

Use these functions and the graphs shown to answer Exercises 65–68.

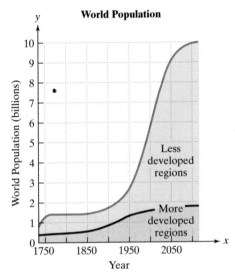

World Population

Source: Population Reference Bureau

65. What does the function $f + g$ represent? $h(x)$ or total world population in year x

66. What does the function $h - g$ represent? $f(x)$ or population of the world's more-developed regions in year x

67. Use the graph to estimate $(f + g)(2000)$.
$(f + g)(2000) \approx 5.9$ billion

68. Use the graph to estimate $(h - g)(2000)$.
$(h - g)(2000) \approx 1.6$ billion

69. A company that sells radios has yearly fixed costs of $600,000. It costs the company $45 to produce each radio. Each radio will sell for $65. The company's costs and revenue are modeled by the following functions:

$C(x) = 600,000 + 45x$ This function models the company's costs.

$R(x) = 65x$ This function models the company's revenue.

Find and interpret $(R - C)(20,000)$, $(R - C)(30,000)$, and $(R - C)(40,000)$. $(R - C)(20,000) = -200,000$: If the company produces and sells 20,000 radios, it will lose $200,000; $(R - C)(30,000) = 0$: If the company produces and sells 30,000 radios, it will break even with its costs equal to its revenue; $(R - C)(40,000) = 200,000$: If the company produces and sells 40,000 radios, it will make a profit of $200,000.

70. The function $f(t) = -0.14t^2 + 0.51t + 31.6$ models the U.S. population, $f(t)$, in millions, ages 65 and older, t years after 1990. The function $g(t) = 0.54t^2 + 12.64t + 107.1$ models the total yearly cost of Medicare, $g(t)$, in billions of dollars, t years after 1990.

a. What does the function $\dfrac{g}{f}$ represent?

the yearly cost of Medicare per person aged 65 and older t years after 1990 (in thousands of dollars)

b. Find and interpret $\dfrac{g}{f}(10)$.

In 2000, the yearly cost of Medicare per person aged 65 and older was about $12.7 thousand, or $12,700.

Writing in Mathematics

71. If a function is defined by an equation, explain how to find its domain.   71–74. Answers will vary.

72. If equations for functions f and g are given, explain how to find $f + g$.

73. If the equations of two functions are given, explain how to obtain the quotient function and its domain.

74. If equations for functions f and g are given, describe two ways to find $(f - g)(3)$.

Technology Exercises

75. Graph the function $\dfrac{g}{f}$ from Exercise 70 in a $[0, 15, 1]$ by $[0, 60, 1]$ viewing rectangle. What does the shape of the graph indicate about the per capita costs of Medicare for the U.S. population ages 65 and over with increasing time?   Per capita costs of Medicare are increasing more and more rapidly over time.

In Exercises 76–79, graph each of the three functions in the same $[-10, 10, 1]$ by $[-10, 10, 1]$ viewing rectangle.   76–79. See graphing answer section.

76. $y_1 = 2x + 3$
$y_2 = 2 - 2x$
$y_3 = y_1 + y_2$

77. $y_1 = x - 4$
$y_2 = 2x$
$y_3 = y_1 - y_2$

78. $y_1 = x$
$y_2 = x - 4$
$y_3 = y_1 \cdot y_2$

79. $y_1 = x^2 - 2x$
$y_2 = x$
$y_3 = \dfrac{y_1}{y_2}$

80. In Exercise 79, use the TRACE feature to trace along y_3. What happens at $x = 0$? Explain why this occurs.   No y-value is displayed; y_3 is undefined at $x = 0$.

Critical Thinking Exercises

81. Which one of the following is false?   d

    a. If $(f + g)(a) = 0$, then $f(a)$ and $g(a)$ must be opposites, or additive inverses.

    b. If $(f - g)(a) = 0$, then $f(a)$ and $g(a)$ must be equal.

    c. If $\left(\dfrac{f}{g}\right)(a) = 0$, then $f(a)$ must be 0.

    d. If $(fg)(a) = 0$, then $f(a)$ must be 0.

82. Use the graphs given in Exercises 65–68 to create a graph that shows the population, in billions, of less-developed regions from 1950 through 2050.   See graphing answer section.

Review Exercises

83. Solve for b: $R = 3(a + b)$. (Section 1.5, Example 6)
$$b = \frac{R - 3a}{3} \text{ or } b = \frac{R}{3} - a$$

84. Solve: $3(6 - x) = 3 - 2(x - 4)$. (Section 1.4, Example 3)   7

85. If $f(x) = 6x - 4$, find $f(b + 2)$. (Section 2.1, Example 3)
$6b + 8$

✓ MID-CHAPTER CHECK POINT

CHAPTER 2

What You Know: We learned that a function is a relation in which no two ordered pairs have the same first component and different second components. We represented functions as equations and used function notation. We graphed functions and applied the vertical line test to identify graphs of functions. We determined the domain and range of a function from its graph, using inputs on the *x*-axis for the domain and outputs on the *y*-axis for the range. Finally, we developed an algebra of functions to combine functions and determine their domains.

In Exercises 1–6, determine whether each relation is a function. Give the domain and range for each relation.

1. $\{(2, 6), (1, 4), (2, -6)\}$
not a function; domain: $\{1, 2\}$; range: $\{-6, 4, 6\}$

2. $\{(0, 1), (2, 1), (3, 4)\}$
function; domain: $\{0, 2, 3\}$; range: $\{1, 4\}$

3.

function; domain:
$\{x \mid -2 \le x < 2\}$;
range: $\{y \mid 0 \le y \le 3\}$

4.

not a function; domain:
$\{x \mid -3 < x \le 4\}$; range:
$\{y \mid -1 \le y \le 2\}$

5.

not a function; domain:
$\{-2, -1, 0, 1, 2\}$;
range: $\{-2, -1, 1, 3\}$

6.

function; domain:
$\{x \mid x \le 1\}$;
range: $\{y \mid y \ge -1\}$

Use the graph of f to solve Exercises 7–12.

7. Explain why f represents the graph of a function.
No vertical line intersects the graph of f more than once.

8. Use the graph to find $f(-4)$.   3

9. For what value or values of x is $f(x) = 4$? -2

10. For what value or values of x is $f(x) = 0$? -6 and 2

11. Find the domain of f. $\{x \mid x \text{ is a real number}\}$

12. Find the range of f. $\{y \mid y \le 4\}$

In Exercises 13–14, find the domain of each function.

13. $f(x) = (x + 2)(x - 2)$ $\{x | x \text{ is a real number}\}$

14. $g(x) = \dfrac{1}{(x + 2)(x - 2)}$

$\{x | x \text{ is a real number and } x \neq -2 \text{ and } x \neq 2\}$

In Exercises 15–22, let

$$f(x) = x^2 - 3x + 8 \text{ and } g(x) = -2x - 5.$$

Find each of the following.

15. $f(0) + g(-10)$ 23   **16.** $f(-1) - g(3)$ 23

17. $f(a) + g(a + 3)$ $a^2 - 5a - 3$

18. $(f + g)(x)$ and $(f + g)(-2)$ $x^2 - 5x + 3; 17$

19. $(f - g)(x)$ and $(f - g)(5)$ $x^2 - x + 13; 33$

20. $(fg)(-1)$ -36

21. $\left(\dfrac{f}{g}\right)(x)$ and $\left(\dfrac{f}{g}\right)(-4)$ $\dfrac{x^2 - 3x + 8}{-2x - 5}; 12$

22. The domain of $\dfrac{f}{g}$ $\left\{x | x \text{ is a real number and } x \neq -\dfrac{5}{2}\right\}$

SECTION 2.3

Objectives

1 Use intercepts to graph a linear function in standard form.

2 Compute a line's slope.

3 Find a line's slope and *y*-intercept from its equation.

4 Graph linear functions in slope-intercept form.

5 Graph horizontal or vertical lines.

6 Calculate average rates of change in applied situations.

7 Use slope and *y*-intercept to model data.

LINEAR FUNCTIONS AND SLOPE

Is there a relationship between literacy and child mortality? As the percentage of adult females who are literate increases, does the mortality of children under age five decrease? Figure 2.11, based on data from the United Nations, indicates that this is, indeed, the case. Each point in the figure represents one country.

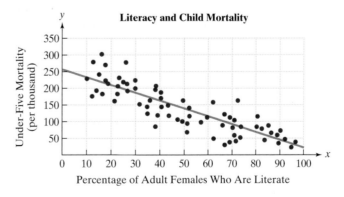

FIGURE 2.11

Source: United Nations

Data presented in a visual form as a set of points is called a **scatter plot**. Also shown in Figure 2.11 is a line that passes through or near the points. A line that best fits the data points in a scatter plot is called a **regression line**. By writing the equation of this line, we can obtain a model of the data and make predictions about child mortality based on the percentage of literate adult females in a country.

Data often fall on or near a line. In the remainder of this chapter, we will use equations to model such data and make predictions. We begin with a discussion of graphing linear functions using intercepts.

1 Use intercepts to graph a linear function in standard form.

Graphing Using Intercepts   The equation of the regression line in Figure 2.11 is

$$y = -2.39x + 254.47.$$

The variable x represents the percentage of adult females in a country who are literate. The variable y represents child mortality, per thousand, for children under five in that country. Using function notation, we can rewrite the equation as

$$f(x) = -2.39x + 254.47.$$

A function such as this, whose graph is a straight line, is called a **linear function**. There is another way that we can write the function's equation

$$y = -2.39x + 254.47.$$

We will collect the x- and y-terms on the left side. This is done by adding $2.39x$ to both sides:

$$2.39x + y = 254.47.$$

The form of this equation is $Ax + By = C$.

$$2.39x \quad + \quad y = 254.47$$

| A, the coefficient of x, is 2.39. | B, the coefficient of y, is 1. | C, the constant on the right, is 254.47. |

All equations of the form $Ax + By = C$ are straight lines when graphed, as long as A and B are not both zero. Such an equation is called the **standard form of the equation of a line**. To graph equations of this form, we will use two important points: the **intercepts**.

An **x-intercept** of a graph is the x-coordinate of a point where the graph intersects the x-axis. For example, look at the graph of $2x - 4y = 8$ in Figure 2.12. The graph crosses the x-axis at $(4, 0)$. Thus, the x-intercept is 4. **The y-coordinate corresponding to an x-intercept is always zero.**

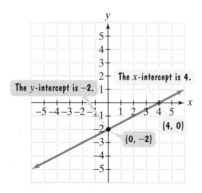

The y-intercept is -2.   The x-intercept is 4.

$(4, 0)$

$(0, -2)$

FIGURE 2.12   The graph of $2x - 4y = 8$

A **y-intercept** of a graph is the y-coordinate of a point where the graph intersects the y-axis. The graph of $2x - 4y = 8$ in Figure 2.12 shows that the graph crosses the y-axis at $(0, -2)$. Thus, the y-intercept is -2. **The x-coordinate corresponding to a y-intercept is always zero**.

STUDY TIP

Mathematicians tend to use two ways to describe intercepts. Did you notice that we are using single numbers? If a graph's x-intercept is a, it passes through the point $(a, 0)$. If a graph's y-intercept is b, it passes through the point $(0, b)$.

Some books state that the x-intercept is the *point* $(a, 0)$ and the x-intercept is *at a* on the x-axis. Similarly, the y-intercept is the *point* $(0, b)$ and the y-intercept is *at b* on the y-axis. In these descriptions, the intercepts are the actual points where a graph crosses the axes.

Although we'll describe intercepts as single numbers, we'll immediately state the point on the x- or y-axis that the graph passes through. Here's the important thing to keep in mind:

x-intercept: The corresponding y coordinate is 0.

y-intercept: The corresponding x coordinate is 0.

When graphing using intercepts, it is a good idea to use a third point, a checkpoint, before drawing the line. A checkpoint can be obtained by selecting a value for either variable, other than 0, and finding the corresponding value for the other variable. The checkpoint should lie on the same line as the x- and y-intercepts. If it does not, recheck your work and find the error.

USING INTERCEPTS TO GRAPH $Ax + By = C$

1. Find the x-intercept. Let $y = 0$ and solve for x.

2. Find the y-intercept. Let $x = 0$ and solve for y.

3. Find a checkpoint, a third ordered-pair solution.

4. Graph the equation by drawing a line through the three points.

EXAMPLE 1   Using Intercepts to Graph a Linear Equation

Graph: $4x - 3y = 6$.

SOLUTION

Step 1.   **Find the x-intercept.** Let $y = 0$ and solve for x.

$$4x - 3 \cdot 0 = 6$$
$$4x = 6$$
$$x = \frac{6}{4} = \frac{3}{2}$$

The x-intercept is $\frac{3}{2}$, so the line passes through $\left(\frac{3}{2}, 0\right)$ or $(1.5, 0)$.

Step 2.   **Find the y-intercept.** Let $x = 0$ and solve for y.

$$4 \cdot 0 - 3y = 6$$
$$-3y = 6$$
$$y = -2$$

The y-intercept is -2, so the line passes through $(0, -2)$.

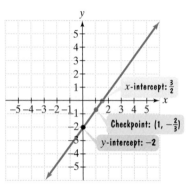

FIGURE 2.13 The graph of $4x - 3y = 6$

Step 3. **Find a checkpoint, a third ordered-pair solution.** For our checkpoint, we will let $x = 1$ and find the corresponding value for y.

$$4x - 3y = 6 \qquad \text{This is the given equation.}$$
$$4 \cdot 1 - 3y = 6 \qquad \text{Substitute 1 for x.}$$
$$4 - 3y = 6 \qquad \text{Simplify.}$$
$$-3y = 2 \qquad \text{Subtract 4 from both sides.}$$
$$y = -\frac{2}{3} \qquad \text{Divide both sides by } -3.$$

The checkpoint is the ordered pair $\left(1, -\frac{2}{3}\right)$.

Step 4. **Graph the equation by drawing a line through the three points.** The three points in Figure 2.13 lie along the same line. Drawing a line through the three points results in the graph of $4x - 3y = 6$. ∎

USING TECHNOLOGY

You can use a graphing utility to graph equations of the form $Ax + By = C$. Begin by solving the equation for y. For example, to graph $4x - 3y = 6$, solve the equation for y.

$$4x - 3y = 6 \qquad \text{This is the equation to be graphed.}$$
$$4x - 4x - 3y = -4x + 6 \qquad \text{Add } -4x \text{ to both sides.}$$
$$-3y = -4x + 6 \qquad \text{Simplify.}$$
$$\frac{-3y}{-3} = \frac{-4x + 6}{-3} \qquad \text{Divide both sides by } -3.$$
$$y = \frac{4}{3}x - 2 \qquad \text{Simplify.}$$

This is the equation to enter in your graphing utility. The graph of $y = \frac{4}{3}x - 2$, or, equivalently, $4x - 3y = 6$ is shown above in a $[-6, 6, 1]$ by $[-6, 6, 1]$ viewing rectangle.

✔ **CHECK POINT 1** Graph: $3x - 2y = 6$.   See graphing answer section.

2 Compute a line's slope.

The Slope of a Line Mathematicians have developed a useful measure of the steepness of a line, called the **slope** of the line. Slope compares the vertical change (the **rise**) to the horizontal change (the **run**) when moving from one fixed point to another along the line. To calculate the slope of a line, we use a ratio that compares the change in y (the rise) to the change in x (the run).

DEFINITION OF SLOPE The **slope** of the line through the distinct points (x_1, y_1) and (x_2, y_2) is

$$\frac{\text{Change in } y}{\text{Change in } x} = \frac{\text{Rise}}{\text{Run}}$$
$$= \frac{y_2 - y_1}{x_2 - x_1}$$

where $x_2 - x_1 \neq 0$.

It is common notation to let the letter m represent the slope of a line. The letter m is used because it is the first letter of the French verb *monter*, meaning to rise, or to ascend.

EXAMPLE 2 Using the Definition of Slope

Find the slope of the line passing through each pair of points:

    a. $(-3, -4)$ and $(-1, 6)$ **b.** $(-1, 3)$ and $(-4, 5)$.

SOLUTION

    a. Let $(x_1, y_1) = (-3, -4)$ and $(x_2, y_2) = (-1, 6)$. The slope is obtained as follows:

$$m = \frac{\text{Change in } y}{\text{Change in } x} = \frac{y_2 - y_1}{x_2 - x_1} = \frac{6 - (-4)}{-1 - (-3)} = \frac{6 + 4}{-1 + 3} = \frac{10}{2} = 5.$$

The situation is illustrated in Figure 2.14. The slope of the line is 5, or $\frac{10}{2}$. For every vertical change, or rise, of 10 units, there is a corresponding horizontal change, or run, of 2 units. The slope is positive, and the line rises from left to right.

STUDY TIP

When computing slope, it makes no difference which point you call (x_1, y_1) and which point you call (x_2, y_2). If we let $(x_1, y_1) = (-1, 6)$ and $(x_2, y_2) = (-3, -4)$, the slope is still 5:

$$m = \frac{y_2 - y_1}{x_2 - x_1} = \frac{-4 - 6}{-3 - (-1)} = \frac{-10}{-2} = 5.$$

However, you should not subtract in one order in the numerator $(y_2 - y_1)$ and then in a different order in the denominator $(x_1 - x_2)$.

$$\frac{-4 - 6}{-1 - (-3)} = \frac{-10}{2} = -5. \quad \text{Incorrect! The slope is not } -5.$$

    b. We can let $(x_1, y_1) = (-1, 3)$ and $(x_2, y_2) = (-4, 5)$. The slope is computed as follows:

$$m = \frac{\text{Change in } y}{\text{Change in } x} = \frac{y_2 - y_1}{x_2 - x_1} = \frac{5 - 3}{-4 - (-1)} = \frac{2}{-3} = -\frac{2}{3}.$$

The situation is illustrated in Figure 2.15. The slope of the line is $-\frac{2}{3}$. For every vertical change of -2 units (2 units down), there is a corresponding horizontal change of 3 units. The slope is negative and the line falls from left to right.

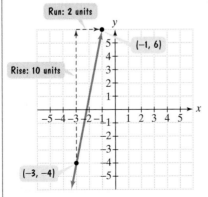

FIGURE 2.14 Visualizing a slope of 5

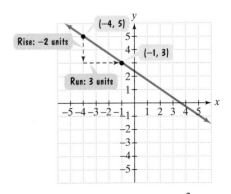

FIGURE 2.15 Visualizing a slope of $-\frac{2}{3}$

✔ **CHECK POINT 2** Find the slope of the line passing through each pair of points:

    a. $(-3, 4)$ and $(-4, -2)$ $m = 6$ **b.** $(4, -2)$ and $(-1, 5)$. $m = -\dfrac{7}{5}$

Example 2 illustrates that a line with a positive slope is rising from left to right and a line with a negative slope is falling from left to right. By contrast, a horizontal line neither rises nor falls and has a slope of zero. A vertical line has no horizontal change, so $x_2 - x_1 = 0$ in the formula for slope. Because we cannot divide by zero, the slope of a vertical line is undefined. This discussion is summarized in Table 2.1.

Table 2.1   Possibilities for a Line's Slope

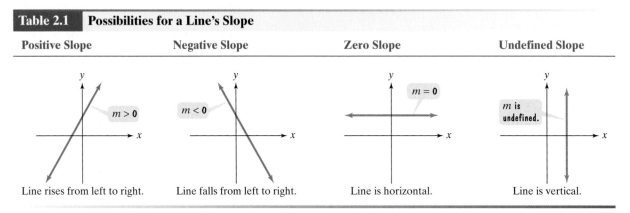

Positive Slope	Negative Slope	Zero Slope	Undefined Slope
$m > 0$	$m < 0$	$m = 0$	m is undefined.
Line rises from left to right.	Line falls from left to right.	Line is horizontal.	Line is vertical.

The Slope-Intercept Form of the Equation of a Line   We opened this section with a linear function that modeled child mortality as a function of literacy. The function's equation can be expressed as

$$y = -2.39x + 254.47 \quad \text{or} \quad f(x) = -2.39x + 254.47.$$

What is the significance of -2.39, the x-coefficient, or of 254.47, the constant term? To answer this question, let's look at an equation in the same form with simpler numbers. In particular, consider the equation $y = 2x + 4$.

Figure 2.16 shows the graph of $y = 2x + 4$. Verify that the x-intercept is -2 by setting y equal to 0 and solving for x. Similarly, verify that the y-intercept is 4 by setting x equal to 0 and solving for y.

Now that we have two points on the line, we can calculate the slope of the graph of $y = 2x + 4$.

$$\text{Slope} = \frac{\text{Change in } y}{\text{Change in } x}$$

$$= \frac{4 - 0}{0 - (-2)} = \frac{4}{2} = 2$$

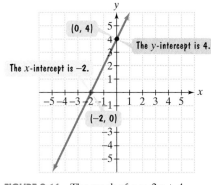

FIGURE 2.16   The graph of $y = 2x + 4$

We see that the slope of the line is 2, the same as the coefficient of x in the equation $y = 2x + 4$. The y-intercept is 4, the same as the constant in the equation $y = 2x + 4$.

$$y = 2x + 4$$

The slope is 2.   The y-intercept is 4.

It is not merely a coincidence that the x-coefficient is the line's slope and the constant term is the y-intercept. Let's find the equation of any nonvertical line with slope m and y-intercept b. Because the y-intercept is b, the point $(0, b)$ lies on the line. Now, let (x, y) represent any other point on the line, shown in Figure 2.17. Keep in mind that the point (x, y) is arbitrary and is not in one fixed position. By contrast, the point $(0, b)$ is fixed.

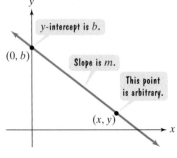

FIGURE 2.17   A line with slope m and y-intercept b

Regardless of where the point (x, y) is located, the steepness of the line in Figure 2.17 remains the same. Thus, the ratio for slope stays a constant m. This means that for all points along the line,

$$m = \frac{\text{Change in } y}{\text{Change in } x} = \frac{y - b}{x - 0} = \frac{y - b}{x}.$$

We can clear the fraction by multiplying both sides by x, the least common denominator.

$$m = \frac{y - b}{x} \qquad \text{This is the slope of the line in Figure 2.17.}$$

$$mx = \frac{y - b}{x} \cdot x \qquad \text{Multiply both sides by } x.$$

$$mx = y - b \qquad \text{Simplify: } \frac{y - b}{\cancel{x}} \cdot \cancel{x} = y - b.$$

$$mx + b = y - b + b \qquad \text{Add } b \text{ to both sides and solve for } y.$$

$$mx + b = y \qquad \text{Simplify.}$$

Now, if we reverse the two sides, we obtain the slope-intercept form of the equation of a line.

SLOPE-INTERCEPT FORM OF THE EQUATION OF A LINE The **slope-intercept equation** of a nonvertical line with slope m and y-intercept b is

$$y = mx + b.$$

The slope-intercept equation, $y = mx + b$, can be expressed in function notation by replacing y with $f(x)$:

$$f(x) = mx + b.$$

We have seen that functions in this form are called **linear functions**. Thus, in the equation of a linear function, the x-coefficient is the line's slope and the constant term is the y-intercept. Here are two examples:

$$y = 2x - 4 \qquad\qquad f(x) = \frac{1}{2}x + 2$$

The slope is 2. The y-intercept is -4. The slope is $\frac{1}{2}$. The y-intercept is 2.

If a linear function's equation is in standard form, $Ax + By = C$, do you see how we can identify the line's slope and y-intercept? Solve the equation for y and convert to slope-intercept form.

3 Find a line's slope and y-intercept from its equation.

STUDY TIP

The variables in $y = mx + b$ vary in different ways. The values for slope, m, and y-intercept, b, vary from one line's equation to another. However, they remain constant in the equation of a single line. By contrast, the variables x and y represent the infinitely many points, (x, y), on a single line. Thus, these variables vary in both the equation of a single line, as well as from one equation to another.

EXAMPLE 3 Converting from Standard Form to Slope-Intercept Form

Give the slope and the y-intercept for the line whose equation is

$$5x + 3y = -12.$$

SOLUTION We convert to slope-intercept form by solving $5x + 3y = 3y = -12$ for y. In this form, the coefficient of x is the line's slope and the constant term is the y-intercept.

$$5x + 3y = -12$$

This is the given equation in standard form, $Ax + By = C$.

Our goal is to isolate y.

$$5x - 5x + 3y = -5x - 12$$

Add $-5x$ to both sides.

$$3y = -5x - 12$$

Simplify.

$$\frac{3y}{3} = \frac{-5x - 12}{3}$$

Divide both sides by 3.

$$y = -\frac{5}{3}x - 4$$

Divide each term in the numerator by 3.

The slope is $-\frac{5}{3}$.　　The y-intercept is -4.

■

✔ **CHECK POINT 3** Give the slope and the y-intercept for the line whose equation is $8x - 4y = 20$.　Slope is 2; y-intercept is -5.

If a linear function's equation is in slope-intercept form, we can use the y-intercept and the slope to obtain its graph.

4 Graph linear functions in slope-intercept form.

GRAPHING $y = mx + b$ **USING THE SLOPE AND** y**-INTERCEPT**

1. Plot the point containing the y-intercept on the y-axis. This is the point $(0, b)$.
2. Obtain a second point using the slope, m. Write m as a fraction, and use rise over run, starting at the point containing the y-intercept, to plot this point.
3. Use a straightedge to draw a line through the two points. Draw arrowheads at the ends of the line to show that the line continues indefinitely in both directions.

EXAMPLE 4 Graphing by Using the Slope and y-Intercept

Graph the line whose equation is $y = 3x - 4$.

SOLUTION The equation $y = 3x - 4$ is in the form $y = mx + b$.

$$y = 3x + (-4)$$

The slope is 3.　The y-intercept is -4.

Now that we have identified the slope and the y-intercept, we use the three steps in the box to graph the equation.

Step 1. **Plot the point containing the y-intercept on the y-axis.** The y-intercept is -4. We plot the point $(0, -4)$, shown in Figure 2.18(a).

We can use $y = mx + b$, the slope-intercept form of a line's equation, to obtain an equation that models the federal minimum wage, y, in current dollars, in year x, where x is between 1997 and 2003, inclusive. The blue line segment in Figure 2.20 provides the values for m and b:

$$y = mx + b.$$

| Because the line segment is horizontal, $m = 0$. | The y-intercept is shown as \$5.15, so $b = 5.15$. |

Thus, an equation that models the federal minimum wage between 1997 and 2003, inclusive, is

$$y = 0x + 5.15, \quad \text{or } y = 5.15.$$

The federal minimum wage remained constant at \$5.15 per hour. Using function notation, we can write

$$f(x) = 5.15.$$

| For all year inputs, x, in the domain from 1997 through 2003, inclusive, | the output is a constant 5.15. |

In general, if a line is horizontal, its slope is zero: $m = 0$. Thus, the equation $y = mx + b$ becomes $y = b$, where b is the y-intercept. For example, the graph of $y = -4$ is a horizontal line with a y-intercept of -4. The graph is shown in Figure 2.21. Three of the points along the line are shown and labeled.

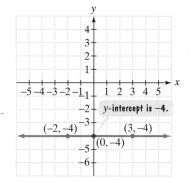

FIGURE 2.21 The graph of $y = -4$ or $f(x) = -4$

x	$y = -4$	(x, y)
-2	-4	$(-2, -4)$
0	-4	$(0, -4)$
3	-4	$(3, -4)$

| For all choices of x, | y is a constant -4. |

No matter what the x-coordinate is, the corresponding y-coordinate for every point on the line in Figure 2.21 is -4.

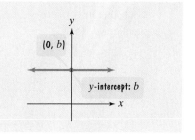

EQUATION OF A HORIZONTAL LINE A horizontal line is given by an equation of the form

$$y = b$$

where b is the y-intercept.

Because any vertical line can intersect the graph of a horizontal line $y = b$ only once, a horizontal line is the graph of a function. Thus, we can express the equation $y = b$ as $f(x) = b$. This linear function is often called a **constant function**. The function modeling the federal minimum wage from 1997 through 2003, namely $f(x) = 5.15$, is an example of a constant function.

Next, let's see what we can discover about the graph of an equation of the form $x = a$ by looking at an example.

EXAMPLE 6 Graphing a Vertical Line

Graph the linear equation: $x = 2$.

SOLUTION All ordered pairs that are solutions of $x = 2$ have a value of x that is always 2. Any value can be used for y. Let's select three of the possible values for y: -2, 0, and 3:

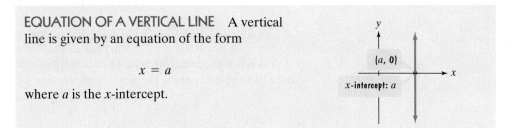

For all choices of y,

$x = 2$	y	(x, y)
2	-2	$(2, -2)$
2	0	$(2, 0)$
2	3	$(2, 3)$

x is always **2**.

The table shows that three ordered pairs that are solutions of $x = 2$ are $(2, -2)$, $(2, 0)$, and $(2, 3)$. Drawing a line that passes through the three points gives the vertical line shown in Figure 2.22.

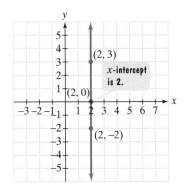

FIGURE 2.22 The graph of $x = 2$

EQUATION OF A VERTICAL LINE A vertical line is given by an equation of the form

$$x = a$$

where a is the x-intercept.

$(a, 0)$
x-intercept: a

Does a vertical line represent the graph of a linear function? No. Look at the graph of $x = 2$ in Figure 2.22. A vertical line drawn through $(2, 0)$ intersects the graph infinitely many times. This shows that infinitely many outputs are associated with the input 2. **No vertical line is a linear function**.

See graphing answer section.

✔ **CHECK POINT 6** Graph the linear equation: $x = -3$.

6 Calculate average rates of change in applied situations.

SLOPE AS RATE OF CHANGE

Slope is defined as the ratio of a change in y to a corresponding change in x. It describes how fast y is changing with respect to x. For a linear function, slope may be interpreted as the rate of change of the dependent variable per unit change in the independent variable. This is illustrated in Table 2.2.

Table 2.2   Interpreting Slope

Linear Function	What the Function Models	Interpretation of Slope
$W = 0.3x + 46.9$ *Slope*	Average number of hours, W, Americans worked per week x years after 1980	The rate of change of hours worked per week is 0.3 hour per year. Each year, hours worked per week increased by 0.3.
$y = -2.39x + 254.47$ *Slope*	Child mortality, y, per thousand, for children under five in a country where x percent of adult females are literate	The rate of change in child mortality is -2.39 per unit change in literacy. For each 1% increase in literacy, child mortality decreases by 2.39 per thousand.

If the graph of a function is not a straight line, the **average rate of change** between any two points is the slope of the line containing the two points. For example, Figure 2.23 shows the graph of a particular man's height, in inches, as a function of his age, in years. Two points on the graph are labeled (13, 57) and (18, 76). At age 13, this man was 57 inches tall, and at age 18, he was 76 inches tall.

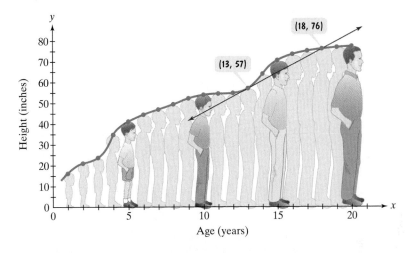

FIGURE 2.23 Height as a function of age

The man's average growth rate between ages 13 and 18 is the slope of the line containing (13, 57) and (18, 76):

$$m = \frac{\text{Change in } y}{\text{Change in } x} = \frac{76 - 57}{18 - 13} = \frac{19}{5} = 3\frac{4}{5}.$$

This man's average rate of change, or average growth rate, from age 13 to age 18 was $3\frac{4}{5}$, or 3.8, inches per year.

For any function, $y = f(x)$, the slope of the line between any two points is the **average change in *y* per unit change in *x*.**

STUDY TIP

Units used to describe *x* and *y* tend to "pile up" when expressing the rate of change of *y* with respect to *x*. The unit used to express the rate of change of *y* with respect to *x* is

the unit used to describe *y*	**per**	the unit used to describe *x*.
In Figure **2.23**, *y*, or height, is described in inches.		In Figure **2.23**, *x*, or age, is described in years.

In Figure 2.23, the rate of change is described in terms of inches per year.

EXAMPLE 7 Slope as Average Rate of Change

When a person receives a drug injected into a muscle, the concentration of the drug in the body, measured in milligrams per 100 milliliters, is a function of the time elapsed after the injection, measured in hours. Figure 2.24 on the next page shows the graph of such a function, where *x* represents hours after the injection and $f(x)$ is the drug's concentration at time *x*. Find the average rate of change in the drug's concentration between 3 and 7 hours.

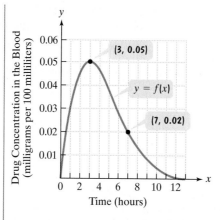

FIGURE 2.24 Concentration of a drug as a function of time

SOLUTION At 3 hours, the drug's concentration is 0.05 and at 7 hours, the concentration is 0.02. The average rate of change in its concentration between 3 and 7 hours is the slope of the line connecting the points (3, 0.05) and (7, 0.02).

$$m = \frac{\text{Change in } y}{\text{Change in } x} = \frac{0.02 - 0.05}{7 - 3} = \frac{-0.03}{4} = -0.0075$$

The average rate of change is −0.0075. This means that the drug's concentration is decreasing at an average rate of 0.0075 milligrams per 100 milliliters per hour. ■

CHECK POINT 7 Use Figure 2.24 to find the average rate of change in the drug's concentration between 1 hour and 3 hours. 0.01 milligrams per 100 milliliters per hour

 Use slope and y-intercept to model data.

Modeling Data with the Slope-Intercept Form of the Equation of a Line Linear functions are useful for modeling data that fall on or near a line. In the previous section, we looked at data showing the number, in millions, of cars and SUVs sold by America's big three automakers from 1990 through 2003, repeated in Figure 2.25. Example 8 and Check Point 8 illustrate how we can use the equation $y = mx + b$ to develop models for the data. The red portions of the figure show additions to the original display so that we can obtain these models.

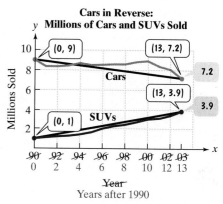

FIGURE 2.25 Obtaining models from data

EXAMPLE 8 Modeling with the Slope-Intercept Equation

Find a function in the form $C(x) = mx + b$ that models car sales, $C(x)$, in millions, x years after 1990.

SOLUTION We will use the line segment passing through the points $(0, 9)$ and $(13, 7.2)$ to obtain a model. We need values for m, the slope, and b, the y-intercept.

$$C(x) = mx + b$$

$$m = \frac{\text{Change in } y}{\text{Change in } x}$$

$$= \frac{7.2 - 9}{13 - 0} \approx -0.14$$

The point $(0, 9)$ lies on the line segment, so the y-intercept is 9: $b = 9$.

Car sales, $C(x)$, in millions, x years after 1990 can be modeled by the linear function

$$C(x) = -0.14x + 9.$$

The slope, approximately -0.14, indicates a decrease in sales of about 0.14 million cars per year from 1990 through 2003. ■

 CHECK POINT 8 Use Figure 2.25 to find a function in the form $S(x) = mx + b$ that models SUV sales, $S(x)$, in millions, x years after 1990. $\quad S(x) = 0.22x + 1$

ENRICHMENT ESSAY

How Calculus Studies Change

Calculus allows motion and change to be analyzed by "freezing the frame" of a continuous changing process, instant by instant. For example, Figure 2.26 shows a male's changing height over intervals of time. Over the period of time from P to D, his average rate of growth is his change in height—that is, his height at time D minus his height at time P—divided by the change in time from P to D. This is the slope of line PD.

FIGURE 2.26 Analyzing continuous growth over intervals of time and at an instant in time

The lines PD, PC, PB, and PA shown in Figure 2.26 have slopes that show average growth rates for successively shorter periods of time. Calculus makes these time frames so small that they approach a single point—that is, a single instant in time. This point is shown as point P in Figure 2.26. The slope of the line that touches the graph at P gives the male's growth rate at one instant in time, P.

2.3 EXERCISE SET

Student Solutions Manual   CD/Video   PH Math/Tutor Center   MathXL Tutorials on CD   MathXL®   MyMathLab   Interactmath.com

Practice Exercises

In Exercises 1–14, use intercepts and a checkpoint to graph each linear function.   1–14. See graphing answer section.

1. $x + y = 4$ **2.** $x + y = 2$ **3.** $x + 3y = 6$

4. $2x + y = 4$ **5.** $6x - 2y = 12$ **6.** $6x - 9y = 18$

7. $3x - y = 6$ **8.** $x - 4y = 8$ **9.** $x - 3y = 9$

10. $2x - y = 5$ **11.** $2x = 3y + 6$ **12.** $3x = 5y - 15$

13. $6x - 3y = 15$ **14.** $8x - 2y = 12$

In Exercises 15–24, find the slope of the line passing through each pair of points or state that the slope is undefined. Then indicate whether the line through the points rises, falls, is horizontal, or is vertical.

15. $(2, 4)$ and $(3, 8)$ $m = 4$; rises

16. $(3, 1)$ and $(5, 4)$ $m = \frac{3}{2}$; rises

17. $(-1, 4)$ and $(2, 5)$ $m = \frac{1}{3}$; rises

18. $(-3, -2)$ and $(2, 5)$ $m = \frac{7}{5}$; rises

19. $(2, 5)$ and $(-1, 5)$ $m = 0$; horizontal

20. $(-6, -3)$ and $(4, -3)$ $m = 0$; horizontal

21. $(-7, 1)$ and $(-4, -3)$ $m = -\frac{4}{3}$; falls

22. $(2, -1)$ and $(-6, 3)$ $m = -\frac{1}{2}$; falls

23. $\left(\frac{7}{2}, -2\right)$ and $\left(\frac{7}{2}, \frac{1}{4}\right)$ undefined slope; vertical

24. $\left(\frac{3}{2}, -6\right)$ and $\left(\frac{3}{2}, \frac{1}{6}\right)$ undefined slope; vertical

In Exercises 25–26, find the slope of each line.

25.

26.

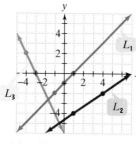

L_1: $\frac{2}{3}$; L_2: -2; L_3: $-\frac{1}{2}$

L_1: 1; L_2: $\frac{2}{3}$; L_3: -2

In Exercises 27–36, give the slope and y-intercept of each line whose equation is given. Then graph the linear function.
27–36. See graphing answer section.

27. $y = 2x + 1$ $m = 2; b = 1$

28. $y = 3x + 2$ $m = 3; b = 2$

29. $y = -2x + 1$ $m = -2; b = 1$

30. $y = -3x + 2$ $m = -3; b = 2$

31. $f(x) = \frac{3}{4}x - 2$ $m = \frac{3}{4}; b = -2$

32. $f(x) = \frac{3}{4}x - 3$ $m = \frac{3}{4}; b = -3$

33. $f(x) = -\frac{3}{5}x + 7$ $m = -\frac{3}{5}; b = 7$

34. $f(x) = -\frac{2}{5}x + 6$ $m = -\frac{2}{5}; b = 6$

35. $y = -\frac{1}{2}x$ $m = -\frac{1}{2}; b = 0$

36. $y = -\frac{1}{3}x$ $m = -\frac{1}{3}; b = 0$

In Exercises 37–44,

a. *Rewrite the given equation in slope-intercept form by solving for y.*

b. *Give the slope and y-intercept.*

c. *Use the slope and y-intercept to graph the linear function.*
37c–44c. See graphing answer section.

37. $2x + y = 0$ **a.** $y = -2x$ **b.** $m = -2; b = 0$

38. $3x + y = 0$ **a.** $y = -3x$ **b.** $m = -3; b = 0$

39. $5y = 4x$ **a.** $y = \frac{4}{5}x$ **b.** $m = \frac{4}{5}; b = 0$

40. $4y = 3x$ **a.** $y = \frac{3}{4}x$ **b.** $m = \frac{3}{4}; b = 0$

41. $3x + y = 2$ **a.** $y = -3x + 2$ **b.** $m = -3; b = 2$

42. $2x + y = 4$ **a.** $y = -2x + 4$ **b.** $m = -2; b = 4$

43. $5x + 3y = 15$ **a.** $y = -\frac{5}{3}x + 5$ **b.** $m = -\frac{5}{3}; b = 5$

44. $7x + 2y = 14$ **a.** $y = -\frac{7}{2}x + 7$ **b.** $m = -\frac{7}{2}; b = 7$

In Exercises 45–58, graph each equation in a rectangular coordinate system.   45–58. See graphing answer section.

45. $y = 3$ **46.** $y = 5$ **47.** $f(x) = -2$

48. $f(x) = -4$ **49.** $3y = 18$ **50.** $5y = -30$

51. $f(x) = 2$ **52.** $f(x) = 1$ **53.** $x = 5$

54. $x = 4$ **55.** $3x = -12$ **56.** $4x = -12$

57. $x = 0$ **58.** $y = 0$

Practice Plus   Little Lady

In Exercises 59–62, find the slope of the line passing through each pair of points or state that the slope is undefined. Assume that all variables represent positive real numbers. Then indicate whether the line through the points rises, falls, is horizontal, or is vertical.

59. $(0, a)$ and $(b, 0)$ $m = -\frac{a}{b}$; falls

60. $(-a, 0)$ and $(0, -b)$ $m = -\frac{b}{a}$; falls

61. (a, b) and $(a, b + c)$ undefined slope; vertical

62. $(a - b, c)$ and $(a, a + c)$ $m = \frac{a}{b}$; rises

In Exercises 63–64, give the slope and y-intercept of each line whose equation is given. Assume that $B \neq 0$.

63. $Ax + By = C$ $m = -\frac{A}{B}; b = \frac{C}{B}$

64. $Ax = By - C$ $m = \frac{A}{B}; b = \frac{C}{B}$

In Exercises 65–66, find the value of y if the line through the two given points is to have the indicated slope.

65. $(3, y)$ and $(1, 4)$, $m = -3$ $y = -2$

66. $(-2, y)$ and $(4, -4)$, $m = \frac{1}{3}$ $y = -6$

In Exercises 67–68, graph each linear function.   67–68. See graphing answer section.

67. $3x - 4f(x) = 6$ **68.** $6x - 5f(x) = 20$

69. If one point on a line is $(3, -1)$ and the line's slope is -2, find the y-intercept.   5

70. If one point on a line is $(2, -6)$ and the line's slope is $-\frac{3}{2}$, find the y-intercept. -3

Use the figure to make the lists in Exercises 71–72.

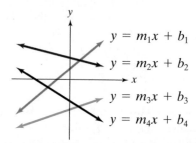

71. Use the figure on page 138 and list the slopes m_1, m_2, m_3, and m_4 in order of decreasing size. $\quad m_1, m_3, m_2, m_4$

72. Use the figure on page 138 and list the y-intercepts b_1, b_2, b_3, and b_4 in order of decreasing size. $\quad b_2, b_1, b_4, b_3$

Application Exercises

In Exercises 73–76, a linear function that models data is described. Find the slope of each model. Then describe what this means in terms of the rate of change of the dependent variable per unit change in the independent variable.

73. The linear function $y = -0.4x + 38$ models the percentage of U.S. men, y, smoking cigarettes x years after 1980. $m = -0.4$; The percentage of U.S. men smoking has decreased at a rate of 0.4% per year after 1980.

74. The linear function $y = 2x + 24$ models the average cost in dollars of a retail drug prescription in the United States, y, x years after 1991. $m = 2$; The average cost of a retail drug prescription has increased at a rate of $2 per year since 1991.

75. The linear function $f(x) = 0.01x + 57.7$ models the global average temperature of Earth, $f(x)$, in degrees Fahrenheit, x years after 1995. $m = 0.01$; The global average temperature of Earth has increased at a rate of 0.01°F per year since 1995.

76. The linear function $f(x) = -0.7x + 60$ models the percentage of U.S. adults, $f(x)$, who read a newspaper x years after 1995. $m = -0.7$; The percentage of U.S. adults who read a newspaper has decreased at a rate of 0.7% per year since 1995.

The graph shows the percentage of sales of recorded music in the United States for rock and rap/hip-hop from 1997 through 2002. Use the information shown to solve Exercises 77–78.

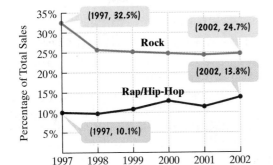

Sales of Recorded Music in the U.S.

Source: RIAA

77. Find the average rate of change in the percentage of total sales of rock from 1997 through 2002. $m = -1.56\%$; Average decrease of 1.56% of total sales per year

78. Find the average rate of change in the percentage of total sales of rap/hip-hop from 1997 through 2002. $m = 0.74\%$; Average increase of 0.74% of total sales per year

79. The graph shows the number of global HIV/AIDS cases, in millions, from 1999 through 2003.

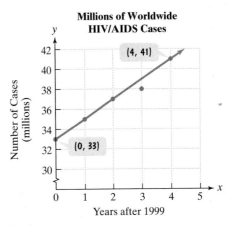

Millions of Worldwide HIV/AIDS Cases

Source: UNAIDS

a. Use the points labeled by the voice balloons to find a function in the form $A(x) = mx + b$ that models the number of HIV/AIDS cases worldwide, $A(x)$, in millions, x years after 1999. $\quad A(x) = 2x + 33$

b. Use the linear function from part (a) to predict the number of HIV/AIDS cases worldwide, in millions, in 2010. 55 million

80. Thanks to low mortgage rates and investors' shifting assets from stocks to real estate, real-estate sales and prices keep rising. The graph shows the median, or middlemost, sales prices of existing single-family houses in the United States from 1998 through 2002.

Median Sales Prices of Existing Single-Family Houses in the U.S.

Source: National Association of Realtors

a. Use the points labeled by the voice balloons to find a function in the form $S(x) = mx + b$ that models the median sales price of a single-family house x years after 1998. $\quad S(x) = 5375x + 128{,}400$

b. Use the linear function from part (a) to predict the median sales price of a single-family house in 2010. $192,900

In Exercises 81–86, find a linear function, in slope-intercept form or a constant function that models the given description. Describe what each variable in your model represents. Then use the model to make a prediction. **81–86.** Predictions will vary. Examples are given.

81. In 1995, the average salary for a college professor was $49,100 and has increased at a rate of $1700 per year since then. $S(x) = 1700x + 49{,}100$; average salary, $S(x)$, x years after 1995

82. In 1998, there were 84 million Internet users in the United States and this number has increased at a rate of 21 million users per year since then. $I(x) = 21x + 84$; number of U.S. Internet users, $I(x)$, in millions, x years after 1998

83. In 1996, 1.7% of U.S. taxpayers were audited by the IRS and this percentage has decreased at a rate of 0.28% per year since then. $A(x) = -0.28x + 1.7$; percentage of audited taxpayers, $A(x)$, x years after 1996

84. A new minivan that was purchased for $21,300 is depreciating at a rate of $1900 per year. $V(x) = -1900x + 21{,}300$; value of the minivan, $V(x)$, x years after its purchase

85. For each year from 1980 through the present, approximately 98% of households in the United States had televisions. $f(x) = 98$; percentage of U.S. households with televisions, $f(x)$, x years after 1980

86. For each year from 1990 through the present, approximately 2.8% of U.S. families had four or more children. $f(x) = 2.8$; percentage of U.S. families with four or more children, $f(x)$, x years after 1990

In Exercises 87–88, draw a graph that illustrates each data description. **87–88.** Answers will vary; See graphing answer section for examples.

87. From 1971 through 1980, the percentage of Americans who were obese held constant at 15%. After 1980, this percentage increased at a rate of 0.8% per year. (*Source:* National Center for Health Statistics)

88. In 1970, the daily calories in the U.S. food supply was 3300 per person. From 1970 through 1975, daily calories in the food supply decreased at a rate of 20 calories per person per year. From 1975 through the present, this number, increased at a rate of 28 calories per person per year. (*Source:* USDA, Economic Research Service)

Writing in Mathematics

89. What is a scatter plot? **89–104.** Answers will vary.

90. What is a regression line?

91. What is the standard form of the equation of a line?

92. What is an x-intercept of a graph?

93. What is a y-intercept of a graph?

94. If you are given the standard form of the equation of a line, explain how to find the x-intercept.

95. If you are given the standard form of the equation of a line, explain how to find the y-intercept.

96. What is the slope of a line?

97. Describe how to calculate the slope of a line passing through two points.

98. What does it mean if the slope of a line is zero?

99. What does it mean if the slope of a line is undefined?

100. Describe how to find the slope of a line whose equation is given.

101. Describe how to graph a line using the slope and y-intercept. Provide an original example with your description.

102. Describe the graph of $y = b$.

103. Describe the graph of $x = a$.

104. We saw that the federal minimum wage remained constant from 1997 through 2003. Exercise 85 showed that the percentage of U.S. households with televisions remained constant over time. Give another example of a real-world phenomenon that has remained relatively constant. Try writing an equation that models this phenomenon.

Technology Exercises

105. Use a graphing utility to verify any three of your hand-drawn graphs in Exercises 1–14. Solve the equation for y before entering it. Answers will vary.

In Exercises 106–109, use a graphing utility to graph each linear function. Then use the TRACE *feature to trace along the line and find the coordinates of two points. Use these points to compute the line's slope. Check your result by using the coefficient of x in the line's equation.* **106–109.** See graphing answer section; Answers will vary.

106. $y = 2x + 4$ $m = 2$

107. $y = -3x + 6$ $m = -3$

108. $f(x) = -\dfrac{1}{2}x - 5$ $m = -\dfrac{1}{2}$

109. $f(x) = \dfrac{3}{4}x - 2$ $m = \dfrac{3}{4}$

Critical Thinking Exercises

110. Which one of the following is true?   c

a. A linear function with nonnegative slope has a graph that rises from left to right.

b. Every line in the rectangular coordinate system has an equation that can be expressed in slope-intercept form.

c. The graph of the linear function $5x + 6y = 30$ is a line passing through the point $(6, 0)$ with slope $-\dfrac{5}{6}$.

d. The graph of $x = 7$ in the rectangular coordinate system is the single point $(7, 0)$.

In Exercises 111–112, find the coefficients that must be placed in each shaded area so that the function's graph will be a line satisfying the specified conditions.

111. ▓ $x +$ ▓ $y = 12$; x-intercept $= -2$; y-intercept $= 4$
$-6x + 3y = 12$; Coefficients are -6 and 3.

112. ▓ $x +$ ▓ $y = 12$; y-intercept $= -6$; slope $= \dfrac{1}{2}$
$x - 2y = 12$; Coefficients are 1 and -2.

113. For the linear function

$$f(x) = mx + b,$$

a. Find $f(x_1 + x_2)$ $mx_1 + mx_2 + b$

b. Find $f(x_1) + f(x_2)$. $mx_1 + mx_2 + 2b$

c. Is $f(x_1 + x_2) = f(x_1) + f(x_2)$?   no

Review Exercises

114. Simplify: $\left(\dfrac{4x^2}{y^{-3}}\right)^2$. (Section 1.6, Example 9) $16x^4y^6$

115. Multiply and write the answer in scientific notation:

$$(8 \times 10^{-7})(4 \times 10^3).$$

(Section 1.7, Example 4) 3.2×10^{-3}

116. Simplify: $5 - [3(x - 4) - 6x]$. (Section 1.2, Example 14)
$3x + 17$

THE POINT-SLOPE FORM OF THE EQUATION OF A LINE

If present trends continue, is it possible that our descendants could live to be 200 years of age? To answer this question, we need to develop a function that models life expectancy by birth year. In this section, you will learn to use another form of a line's equation to obtain functions that model data.

Point-Slope Form   We can use the slope of a line to obtain another useful form of the line's equation. Consider a nonvertical line that has slope m and contains the point (x_1, y_1). Now, let (x, y) represent any other point on the line, shown in Figure 2.27. Keep in mind that the point (x, y) is arbitrary and is not in one fixed position. By contrast, the point (x_1, y_1) is fixed.

Regardless of where the point (x, y) is located, the steepness of the line in Figure 2.27 remains the same. Thus, the ratio for slope stays a constant m. This means that for all points along the line

$$m = \frac{\text{Change in } y}{\text{Change in } x} = \frac{y - y_1}{x - x_1}.$$

We can clear the fraction by multiplying both sides by $x - x_1$, the least common denominator.

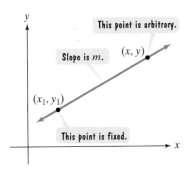

FIGURE 2.27   A line passing through (x_1, y_1) with slope m

$$m = \frac{y - y_1}{x - x_1} \qquad \text{This is the slope of the line in Figure 2.27.}$$

$$m(x - x_1) = \frac{y - y_1}{x - x_1} \cdot x - x_1 \qquad \text{Multiply both sides by } x - x_1.$$

$$m(x - x_1) = y - y_1 \qquad \text{Simplify: } \frac{y - y_1}{x - x_1} \cdot x - x_1 = y - y_1.$$

Now, if we reverse the two sides, we obtain the **point-slope form** of the equation of a line.

POINT-SLOPE FORM OF THE EQUATION OF A LINE The **point-slope equation** of a nonvertical line with slope m that passes through the point (x_1, y_1) is

$$y - y_1 = m(x - x_1).$$

For example, the point-slope equation of the line passing through $(1, 5)$ with slope 2 ($m = 2$) is

$$y - 5 = 2(x - 1).$$

Using the Point-Slope Form to Write a Line's Equation If we know the slope of a line and a point not containing the y-intercept through which the line passes, the point-slope form is the equation that we should use. Once we have obtained this equation, it is customary to solve for y and write the equation in slope-intercept form. Examples 1 and 2 illustrate these ideas.

| 1 | Use the point-slope form to write equations of a line. |

EXAMPLE 1 Writing the Point-Slope Form and the Slope-Intercept Form

Write the point-slope form and the slope-intercept form of the equation of the line with slope 7 that passes through the point $(-4, 5)$.

SOLUTION We begin with the point-slope equation of a line with $m = 7$, $x_1 = -4$, and $y_1 = 5$.

$$y - y_1 = m(x - x_1) \qquad \text{This is the point-slope form of the equation.}$$
$$y - 5 = 7[x - (-4)] \qquad \text{Substitute the given values.}$$
$$y - 5 = 7(x + 4) \qquad \text{We now have the point-slope form of the equation for the given line.}$$

Now we solve this equation for y and write an equivalent equation in slope-intercept form ($y = mx + b$).

$$y - 5 = 7(x + 4) \qquad \text{This is the point-slope equation.}$$
$$y - 5 = 7x + 28 \qquad \text{Use the distributive property.}$$
$$y = 7x + 33 \qquad \text{Add 5 to both sides.}$$

The slope-intercept form of the line's equation is $y = 7x + 33$. Using function notation, the equation is $f(x) = 7x + 33$. ∎

 CHECK POINT 1 Write the point-slope form and the slope-intercept form of the equation of the line with slope -2 that passes through the point $(4, -3)$.

$y + 3 = -2(x - 4); y = -2x + 5$ or $f(x) = -2x + 5$

EXAMPLE 2 Writing the Point-Slope Form and the Slope-Intercept Form

A line passes through the points $(1, -3)$ and $(-2, 4)$. (See Figure 2.28 at the top of the next page.) Find an equation of the line

  a. in point-slope form.

  b. in slope-intercept form.

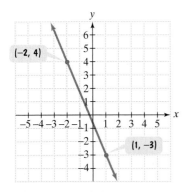

FIGURE 2.28 Line passing through $(1, -3)$ and $(-2, 4)$

SOLUTION

a. To use the point-slope form, we need to find the slope. The slope is the change in the y-coordinates divided by the corresponding change in the x-coordinates.

$$m = \frac{4 - (-3)}{-2 - 1} = \frac{7}{-3} = -\frac{7}{3}$$

This is the definition of slope using $(1, -3)$ and $(-2, 4)$.

We can take either point on the line to be (x_1, y_1). Let's use $(x_1, y_1) = (1, -3)$. Now, we are ready to write the point-slope equation.

$$y - y_1 = m(x - x_1)$$ *This is the point-slope form of the equation.*

$$y - (-3) = -\frac{7}{3}(x - 1)$$ *Substitute: $(x_1, y_1) = (1, -3)$ and $m = -\frac{7}{3}$.*

$$y + 3 = -\frac{7}{3}(x - 1)$$ *Simplify.*

This equation is the point-slope form of the equation of the line shown in Figure 2.28.

b. Now, we solve this equation for y and write an equivalent equation in slope-intercept form $(y = mx + b)$.

$$y + 3 = -\frac{7}{3}(x - 1)$$ *This is the point-slope equation.*

$$y + 3 = -\frac{7}{3}x + \frac{7}{3}$$ *Use the distributive property.*

$$y = -\frac{7}{3}x - \frac{2}{3}$$ *Subtract 3 from both sides:*
$$\frac{7}{3} - 3 = \frac{7}{3} - \frac{9}{3} = -\frac{2}{3}.$$

This equation is the slope-intercept form of the equation of the line shown in Figure 2.28. Using function notation, the equation is $f(x) = -\frac{7}{3}x - \frac{2}{3}$. ■

DISCOVER FOR YOURSELF

If you are given two points on a line, you can use either point for (x_1, y_1) when you write its point-slope equation. Rework Example 2 using $(-2, 4)$ for (x_1, y_1). Once you solve for y, you should obtain the same slope-intercept equation as the one shown in the last line of the solution to Example 2.

 CHECK POINT **2** A line passes through the points $(6, -3)$ and $(2, 5)$. Find an equation of the line

a. in point-slope form. $y + 3 = -2(x - 6)$ or $y - 5 = -2(x - 2)$

b. in slope-intercept form. $y = -2x + 9$ or $f(x) = -2x + 9$

Here is a summary of the various forms for equations of lines:

EQUATIONS OF LINES

1. Standard form: $Ax + By = C$

2. Slope-intercept form: $y = mx + b$ or $f(x) = mx + b$

3. Horizontal line: $y = b$

4. Vertical line: $x = a$

5. Point-slope form: $y - y_1 = m(x - x_1)$

In Examples 1 and 2, we eventually wrote a line's equation in slope-intercept form, or in function notation. But where do we start our work?

Starting with $y = mx + b$	Starting with $y - y_1 = m(x - x_1)$
Begin with the slope-intercept form if you know	Begin with the point-slope form if you know
1. The slope of the line and the y-intercept.	1. The slope of the line and a point on the line not containing the y-intercept or
	2. Two points on the line, neither of which contains the y-intercept.

2 Model data with linear functions and make predictions.

Applications   We have seen that linear functions are useful for modeling data that fall on or near a line. For example, the bar graph in Figure 2.29(a) gives the life expectancy for Americans born in the indicated year. The data are displayed as a set of seven points in the scatter plot in Figure 2.29(b).

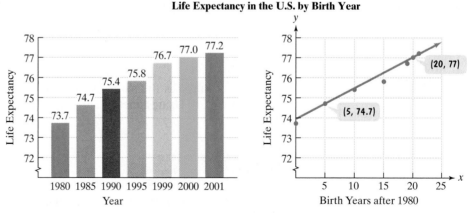

Life Expectancy in the U.S. by Birth Year

Source: National Center for Health Statistics

FIGURE 2.29 (a)   FIGURE 2.29 (b)

Also shown on the scatter plot in Figure 2.29(b) is a line that passes through or near the seven points. By writing the equation of this line, we can obtain a model for life expectancy and make predictions about how long Americans will live in the future.

EXAMPLE 3   Modeling Life Expectancy

Write the slope-intercept equation of the line shown in Figure 2.29(b). Use the equation to predict the life expectancy of an American born in 2020.

You can use a graphing utility to obtain a model for a scatter plot in which the data points fall on or near a straight line. After entering the data in Figure 2.29, a graphing utility displays a scatter plot of the data and the regression line, that is, the line that best fits the data.

[0, 25, 5] by [72, 78, 1]

Also displayed is the regression line's equation.

SOLUTION The line in Figure 2.29(b) passes through (5, 74.7) and (20, 77). We start by finding its slope.

$$m = \frac{\text{Change in } y}{\text{Change in } x} = \frac{77 - 74.7}{20 - 5} = \frac{2.3}{15} = 0.15\overline{3} \approx 0.15$$

The slope indicates that for each subsequent birth year, life expectancy is increasing by approximately 0.15 years.

Now we write the line's slope-intercept equation.

$y - y_1 = m(x - x_1)$ Begin with the point-slope form.

$y - 77 = 0.15(x - 20)$ Either ordered pair can be (x_1, y_1). Let $(x_1, y_1) = (20, 77)$. From above, $m \approx 0.15$.

$y - 77 = 0.15x - 3$ Apply the distributive property on the right side.

$y = 0.15x + 74$ Add 77 to both sides and solve for y.

A linear function that models life expectancy for Americans born x years after 1980 is

$$f(x) = 0.15x + 74.$$

Now let's use this function to predict the life expectancy of an American born in 2020. Because 2020 is 40 years after 1980, substitute 40 for x and evaluate the function at 40.

$$f(40) = 0.15(40) + 74 = 80$$

Our model predicts that Americans born in 2020 will have a life expectancy of 80 years. ∎

✓ **CHECK POINT 3** The six points in the scatter plot in Figure 2.30 show the population of the United States, in millions, for six selected years. Also shown is a line that passes through or near the six points. Use the data points labeled by the voice balloons to write the slope-intercept equation of this line. Then use the linear function to predict U.S. population in 2010.

$y = 2.35x + 179.5$ or

$f(x) = 2.35x + 179.5$; 297 million

FIGURE 2.30

Parallel and Perpendicular Lines Two nonintersecting lines that lie in the same plane are **parallel**. If two lines do not intersect, the ratio of the vertical change to the horizontal change is the same for each line. Because two parallel lines have the same "steepness," they must have the same slope.

3 Find slopes and equations of parallel and perpendicular lines.

SLOPE AND PARALLEL LINES

1. If two nonvertical lines are parallel, then they have the same slope.
2. If two distinct nonvertical lines have the same slope, then they are parallel.
3. Two distinct vertical lines, both with undefined slopes, are parallel.

EXAMPLE 4 Writing Equations of a Line Parallel to a Given Line

Write an equation of the line passing through $(-3, 1)$ and parallel to the line whose equation is $y = 2x + 1$. Express the equation in point-slope form and slope-intercept form.

SOLUTION The situation is illustrated in Figure 2.31. We are looking for the equation of the red line shown on the left. How do we obtain this equation? Notice that the line passes through the point $(-3, 1)$. Using the point-slope form of the line's equation, we have $x_1 = -3$ and $y_1 = 1$.

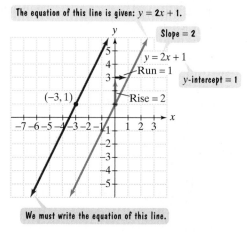

The equation of this line is given: $y = 2x + 1$.

Slope = 2

$y = 2x + 1$

Run = 1

y-intercept = 1

$(-3, 1)$

Rise = 2

We must write the equation of this line.

FIGURE 2.31

$$y - y_1 = m(x - x_1)$$

$y_1 = 1$ $x_1 = -3$

Now the only thing missing from the equation is m, the slope of the red line. Do we know anything about the slope of either line in Figure 2.31? The answer is yes; we know the slope of the dark blue line on the right, whose equation is given.

$$y = 2x + 1$$

The slope of the blue line on the right in Figure 2.31 is 2.

Parallel lines have the same slope. Because the slope of the blue line is 2, the slope of the red line, the line whose equation we must write, is also 2: $m = 2$. We now have values for x_1, y_1, and m for the red line.

$$y - y_1 = m(x - x_1)$$

$y_1 = 1$ $m = 2$ $x_1 = -3$

The point-slope form of the red line's equation is

$$y - 1 = 2[x - (-3)] \text{ or}$$
$$y - 1 = 2(x + 3).$$

Solving for y, we obtain the slope-intercept form of the equation.

$$y - 1 = 2x + 6 \qquad \text{Apply the distributive property.}$$

$$y = 2x + 7 \qquad \text{Add 1 to both sides. This is the slope-intercept form, } y = mx + b \text{, of the equation. Using function notation, the equation is } f(x) = 2x + 7. \blacksquare$$

 CHECK POINT 4 Write an equation of the line passing through $(-2, 5)$ and parallel to the line whose equation is $y = 3x + 1$. Express the equation in point-slope form and slope-intercept form. $y - 5 = 3(x + 2); y = 3x + 11 \text{ or } f(x) = 3x + 11$

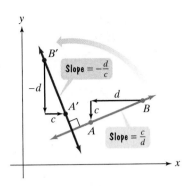

FIGURE 2.32 Slopes of perpendicular lines

Two lines that intersect at a right angle (90°) are said to be **perpendicular**, shown in Figure 2.32. The relationship between the slopes of perpendicular lines is not as obvious as the relationship between parallel lines. Figure 2.32 shows line AB, with slope $\frac{c}{d}$. Rotate line AB through 90° to the left to obtain line $A'B'$, perpendicular to line AB. The figure indicates that the rise and the run of the new line are reversed from the original line, but the rise is now negative. This means that the slope of the new line is $-\frac{d}{c}$. Notice that the product of the slopes of the two perpendicular lines is -1:

$$\left(\frac{c}{d}\right)\left(-\frac{d}{c}\right) = -1.$$

This relationship holds for all perpendicular lines and is summarized in the following box:

SLOPE AND PERPENDICULAR LINES

1. If two nonvertical lines are perpendicular, then the product of their slopes is -1.

2. If the product of the slopes of two lines is -1, then the lines are perpendicular.

3. A horizontal line having zero slope is perpendicular to a vertical line having undefined slope.

An equivalent way of stating this relationship is to say that one line is perpendicular to another line if its slope is the *negative reciprocal* of the slope of the other. For example, if a line has slope 5, any line having slope $-\frac{1}{5}$ is perpendicular to it. Similarly, if a line has slope $-\frac{3}{4}$, any line having slope $\frac{4}{3}$ is perpendicular to it.

EXAMPLE 5 Finding the Slope of a Line Perpendicular to a Given Line

Find the slope of any line that is perpendicular to the line whose equation is $x + 4y - 8 = 0$.

SOLUTION We begin by writing the equation of the given line in slope-intercept form. Solve for y.

$$x + 4y - 8 = 0 \qquad \text{This is the given equation.}$$

$$4y = -x + 8 \qquad \text{To isolate the } y\text{-term, subtract } x \text{ and add 8 on both sides.}$$

$$y = -\frac{1}{4}x + 2 \qquad \text{Divide both sides by 4.}$$

Slope is $-\frac{1}{4}$.

The given line has slope $-\frac{1}{4}$. Any line perpendicular to this line has a slope that is the negative reciprocal of $-\frac{1}{4}$. Thus, the slope of any perpendicular line is 4. ∎

✔ **CHECK POINT 5** Find the slope of any line that is perpendicular to the line whose equation is $x + 3y - 12 = 0$. *m = 3*

2.4 EXERCISE SET

Student Solutions Manual   CD/Video   PH Math/Tutor Center   MathXL Tutorials on CD   MathXL®   MyMathLab   Interactmath.com

Practice Exercises

Write the point-slope form of the line satisfying each of the conditions in Exercises 1–28. Then use the point-slope form of the equation to write the <u>*slope-intercept form*</u> *of the equation in function notation.*

1. Slope $= 3$, passing through $(2, 5)$
$y + 5 = 3(x - 2); f(x) = 3x - 1$

2. Slope $= 4$, passing through $(3, 1)$
$y - 1 = 4(x - 3); f(x) = 4x - 11$

3. Slope $= 5$, passing through $(-2, 6)$
$y - 6 = 5(x + 2); f(x) = 5x + 16$

4. Slope $= 8$, passing through $(-4, 1)$
$y - 1 = 8(x + 4); f(x) = 8x + 33$

5. Slope $= -4$, passing through $(-3, -2)$
$y + 2 = -4(x + 3); f(x) = -4x - 14$

6. Slope $= -6$, passing through $(-2, -4)$
$y + 4 = -6(x + 2); f(x) = -6x - 16$

7. Slope $= -5$, passing through $(-2, 0)$
$y - 0 = -5(x + 2); f(x) = -5x - 10$

8. Slope $= -4$, passing through $(0, -3)$
$y + 3 = -4(x - 0); f(x) = -4x - 3$

9. Slope $= -1$, passing through $\left(-2, -\frac{1}{2}\right)$
$y + \frac{1}{2} = -1(x + 2); f(x) = -x - \frac{5}{2}$

10. Slope $= -1$, passing through $\left(-\frac{1}{4}, -4\right)$
$y + 4 = -1\left(x + \frac{1}{4}\right); f(x) = -x - \frac{17}{4}$

11. Slope $= \frac{1}{4}$, passing through the origin
$y - 0 = \frac{1}{4}(x - 0); f(x) = \frac{1}{4}x$

12. Slope $= \frac{1}{5}$, passing through the origin
$y - 0 = \frac{1}{5}(x - 0); f(x) = \frac{1}{5}x$

13. Slope $= -\frac{2}{3}$, passing through $(6, -4)$
$y + 4 = -\frac{2}{3}(x - 6); f(x) = -\frac{2}{3}x$

14. Slope $= -\frac{2}{5}$, passing through $(15, -4)$
$y + 4 = -\frac{2}{5}(x - 15); f(x) = -\frac{2}{5}x + 2$

15. Passing through $(6, 3)$ and $(5, 2)$
$y - 3 = 1(x - 6)$ or $y - 2 = 1(x - 5); f(x) = x - 3$

16. Passing through $(1, 3)$ and $(2, 4)$
$y - 3 = 1(x - 1)$ or $y - 4 = 1(x - 2); f(x) = x + 2$

17. Passing through $(-2, 0)$ and $(0, 4)$
$y - 0 = 2(x + 2)$ or $y - 4 = 2(x - 0); f(x) = 2x + 4$

18. Passing through $(2, 0)$ and $(0, -1)$
$y - 0 = \frac{1}{2}(x - 2)$ or $y + 1 = \frac{1}{2}(x - 0); f(x) = \frac{1}{2}x - 1$

19. Passing through $(-6, 13)$ and $(-2, 5)$
$y - 13 = -2(x + 6)$ or $y - 5 = -2(x + 2); f(x) = -2x + 1$

20. Passing through $(-3, 2)$ and $(2, -8)$
$y - 2 = -2(x + 3)$ or $y + 8 = -2(x - 2); f(x) = -2x - 4$

21. Passing through $(1, 9)$ and $(4, -2)$
$y - 9 = -\frac{11}{3}(x - 1)$ or $y + 2 = -\frac{11}{3}(x - 4); f(x) = -\frac{11}{3}x + \frac{38}{3}$

22. Passing through $(4, -8)$ and $(8, -3)$
$y + 8 = \frac{5}{4}(x - 4)$ or $y + 3 = \frac{5}{4}(x - 8); f(x) = \frac{5}{4}x - 13$

23. Passing through $(-2, -5)$ and $(3, -5)$
$y + 5 = 0(x + 2)$ or $y + 5 = 0(x - 3); f(x) = -5$

24. Passing through $(-1, -4)$ and $(3, -4)$
$y + 4 = 0(x + 1)$ or $y + 4 = 0(x - 3); f(x) = -4$

25. Passing through $(7, 8)$ with x-intercept $= 3$
$y - 8 = 2(x - 7)$ or $y - 0 = 2(x - 3); f(x) = 2x - 6$

26. Passing through $(-4, 5)$ with y-intercept $= -3$
$y - 5 = -2(x + 4)$ or $y + 3 = -2(x - 0); f(x) = -2x - 3$

27. x-intercept $= 2$ and y-intercept $= -1$
$y - 0 = \frac{1}{2}(x - 2)$ or $y + 1 = \frac{1}{2}(x - 0); f(x) = \frac{1}{2}x - 1$

28. x-intercept $= -2$ and y-intercept $= 4$
$y - 0 = 2(x + 2)$ or $y - 4 = 2(x - 0); f(x) = 2x + 4$

In Exercises 29–44, the equation of a line is given. Find the slope of a line that is **a.** *parallel to the line with the given equation; and* **b.** *perpendicular to the line with the given equation.*

29. $y = 5x$ **a.** $m = 5$ **b.** $m = -\frac{1}{5}$

30. $y = 3x$ **a.** $m = 3$ **b.** $m = -\frac{1}{3}$

31. $y = -7x$ **a.** $m = -7$ **b.** $m = \frac{1}{7}$

32. $y = -9x$ **a.** $m = -9$ **b.** $m = \frac{1}{9}$

33. $y = \frac{1}{2}x + 3$ **a.** $m = \frac{1}{2}$ **b.** $m = -2$

34. $y = \frac{1}{4}x - 5$ **a.** $m = \frac{1}{4}$ **b.** $m = -4$

35. $y = -\frac{2}{5}x - 1$ **a.** $m = -\frac{2}{5}$ **b.** $m = \frac{5}{2}$

36. $y = -\frac{3}{7}x - 2$ **a.** $m = -\frac{3}{7}$ **b.** $m = \frac{7}{3}$

37. $4x + y = 7$ **a.** $m = -4$ **b.** $m = \frac{1}{4}$

38. $8x + y = 11$ **a.** $m = -8$ **b.** $m = \frac{1}{8}$

39. $2x + 4y - 8 = 0$ **a.** $m = -\frac{1}{2}$ **b.** $m = 2$

40. $3x + 2y - 6 = 0$ **a.** $m = -\frac{3}{2}$ **b.** $m = \frac{2}{3}$

41. $2x - 3y - 5 = 0$ **a.** $m = \frac{2}{3}$ **b.** $m = -\frac{3}{2}$

42. $3x - 4y + 7 = 0$ **a.** $m = \frac{3}{4}$ **b.** $m = -\frac{4}{3}$

43. $x = 6$ **a.** m is undefined.   **b.** $m = 0$

44. $y = 9$ **a.** $m = 0$ **b.** m is undefined.

In Exercises 45–48, write an equation for line L in point-slope form and slope-intercept form.

45.

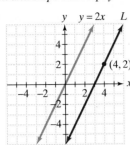

$y - 2 = 2(x - 4);$
$y = 2x - 6$ or
$f(x) = 2x - 6$

L is parallel to $y = 2x$.

46.

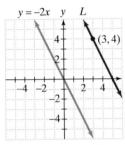

$y - 4 = -2(x - 3);$
$y = -2x + 10$ or
$f(x) = -2x + 10$

L is parallel to $y = -2x$.

47.

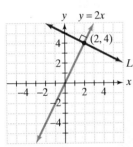

$y - 4 = -\frac{1}{2}(x - 2);$

$y = -\frac{1}{2}x + 5$

or $f(x) = -\frac{1}{2}x + 5$

L is perpendicular to $y = 2x$.

48.

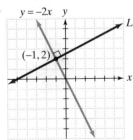

$y - 2 = \frac{1}{2}(x + 1);$

$y = \frac{1}{2}x + \frac{5}{2}$

or $f(x) = \frac{1}{2}x + \frac{5}{2}$

L is perpendicular to $y = -2x$.

In Exercises 49–56, use the given conditions to write an equation for each line in point-slope form and slope-intercept form.

49. Passing through $(-8, -10)$ and parallel to the line whose equation is $y = -4x + 3$
$y + 10 = -4(x + 8); y = -4x - 42$ or $f(x) = -4x - 42$

50. Passing through $(-2, -7)$ and parallel to the line whose equation is $y = -5x + 4$
$y + 7 = -5(x + 2); y = -5x - 17$ or $f(x) = -5x - 17$

51. Passing through $(2, -3)$ and perpendicular to the line whose equation is $y = \frac{1}{5}x + 6$
$y + 3 = -5(x - 2); y = -5x + 7$ or $f(x) = -5x + 7$

52. Passing through $(-4, 2)$ and perpendicular to the line whose equation is $y = \frac{1}{3}x + 7$
$y - 2 = -3(x + 4); y = -3x - 10$ or $f(x) = -3x - 10$

53. Passing through $(-2, 2)$ and parallel to the line whose equation is $2x - 3y - 7 = 0$
$y - 2 = \frac{2}{3}(x + 2); y = \frac{2}{3}x + \frac{10}{3}$ or $f(x) = \frac{2}{3}x + \frac{10}{3}$

54. Passing through $(-1, 3)$ and parallel to the line whose equation is $3x - 2y - 5 = 0$
$y - 3 = \frac{3}{2}(x + 1); y = \frac{3}{2}x + \frac{9}{2}$ or $f(x) = \frac{3}{2}x + \frac{9}{2}$

55. Passing through $(4, -7)$ and perpendicular to the line whose equation is $x - 2y - 3 = 0$
$y + 7 = -2(x - 4); y = -2x + 1$ or $f(x) = -2x + 1$

56. Passing through $(5, -9)$ and perpendicular to the line whose equation is $x + 7y - 12 = 0$
$y + 9 = 7(x - 5); y = 7x - 44$ or $f(x) = 7x - 44$

Practice Plus

In Exercises 57–64, write the slope-intercept equation of a function f whose graph satisfies the given conditions.

57. The graph of f passes through $(-1, 5)$ and is perpendicular to the line whose equation is $x = 6$. $f(x) = 5$

58. The graph of f passes through $(-2, 6)$ and is perpendicular to the line whose equation is $x = -4$. $f(x) = 6$

59. The graph of f passes through $(-6, 4)$ and is perpendicular to the line that has an x-intercept of 2 and a y-intercept of -4. $f(x) = -\frac{1}{2}x + 1$

60. The graph of f passes through $(-5, 6)$ and is perpendicular to the line that has an x-intercept of 3 and a y-intercept of -9. $f(x) = -\frac{1}{3}x + \frac{13}{3}$

61. The graph of f is perpendicular to the line whose equation is $3x - 2y = 4$ and has the same y-intercept as this line.
$f(x) = -\frac{2}{3}x - 2$

62. The graph of f is perpendicular to the line whose equation is $4x - y = 6$ and has the same y-intercept as this line.
$f(x) = -\frac{1}{4}x - 6$

63. The graph of f is the graph of $g(x) = 4x - 3$ shifted down 2 units. $f(x) = 4x - 5$

64. The graph of f is the graph of $g(x) = 2x - 5$ shifted up 3 units. $f(x) = 2x - 2$

65. What is the slope of a line that is parallel to the line whose equation is $Ax + By = C, B \neq 0$? $-\frac{A}{B}$

66. What is the slope of a line that is perpendicular to the line whose equation is $Ax + By = C, A \neq 0$ and $B \neq 0$? $\frac{B}{A}$

Application Exercises

Though increasing numbers of Americans are obese, fewer are trimming down by watching what they eat. The bar graph shows the percentage of American adults on weight-loss diets for four selected years. The data are displayed as two sets of four points each, one scatter plot for the percentage of dieting women and one for the percentage of dieting men Also shown in each scatter plot is a line that passes through or near the four points. Use these lines to solve Exercises 67–68.

Percentage of American Adults on Weight-Loss Diets

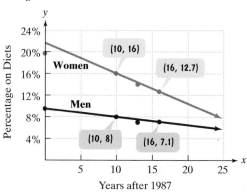

Source: Mediamark Research, American Demographics

67. In this exercise, you will use the blue line for the women shown on the scatter plot to develop a model for the percentage of dieting American women.

a. Use the two points whose coordinates are shown by the voice balloons to find the point-slope equation of the line that models the percentage of adult women on diets, y, x years after 1987.

$$y - 16 = -0.55(x - 10) \text{ or } y - 12.7 = -0.55(x - 16)$$

b. Write the equation in part (a) in slope-intercept form. Use function notation. $f(x) = -0.55x + 21.5$

c. Use the linear function to predict the percentage of adult women on weight-loss diets in 2007.   10.5%

68. In this exercise, you will use the red line for men shown on the scatter plot to develop a model for the percentage of dieting American men.

a. Use the two points whose coordinates are shown by the voice balloons to find the point-slope equation of the line that models the percentage of adult men on diets, y, x years after 1987.

$$y - 8 = -0.15(x - 10) \text{ or } y - 7.1 = -0.15(x - 16)$$

b. Write the equation in part (a) in slope-intercept form. Use function notation. $f(x) = -0.15x + 9.5$

c. Use the linear function to predict the percentage of adult men on weight-loss diets in 2007.   6.5%

69. The bar graph shows the percentage of U.S. students who dropped out of high school in seven selected years.

Percentage of U.S. Students Dropping Out of High School

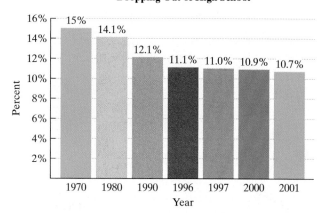

Source: Digest of Education Statistics 2002

a. Let x represent the number of years after 1970 and let y represent the percentage of students who dropped out of high school. Create a scatter plot that displays the data as a set of seven points in a rectangular coordinate system. See graphing answer section.

b. Draw a line through the two points that show the dropout percents for 1990 and 2000. Use the coordinates of these points to write the line's point-slope and slope-intercept equations. See graphing answer section; $y - 12.1 = -0.12(x - 20)$ or $y - 10.9 = -0.12(x - 30)$; $f(x) = -0.12x + 14.5$

c. Use the slope-intercept equation in part (b) to predict the percentage of U.S. students who will drop out of high school in 2010.   9.7%

70. The bar graph shows the percentage of U.S. homes with wireless telephones for seven selected years.

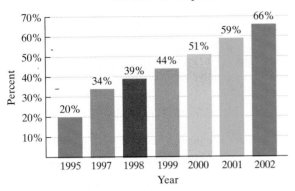

Percentage of U.S. Homes with Wireless Telephones

Source: Telecommunications Association

a. Let x represent the number of years after 1995 and let y represent the percentage of homes with wireless telephones. Create a scatter plot that displays the data as a set of seven points in a rectangular coordinate system. See graphing answer section.

b. Draw a line through the two points that show the percents for 1998 and 2002. Use the coordinates of these points to write the line's point-slope and slope-intercept equations. See graphing answer section; $y - 39 = 6.75(x - 3)$ or $y - 66 = 6.75(x - 7)$; $f(x) = 6.75x + 18.75$

c. Use the slope-intercept equation in part (b) to predict the percentage of U.S. homes that will have wireless telephones in 2006. 93%

71. Drugs are an increasingly common way to treat depressed and hyperactive kids. The line graphs represent models that show users per 1000 U.S. children, ages 9 through 17.

U.S. Children, Ages 9 through 17, Using Drugs to Treat Mental Disorders

Source: Newsweek

a. Find the slope of the blue line segment for children using stimulants. Describe what this means in terms of rate of change. $m = 1.7$; Stimulant use has increased at a rate of 1.7 per thousand children per year.

b. Find the slope of the red line segment for children using antidepressants. Describe what this means in terms of rate of change. $m = 1.4$; Antidepressant use has increased at a rate of 1.4 per thousand children per year.

c. Do the blue and red line segments lie on parallel lines? What does this mean in terms of the rate of change for children using stimulants and children using antidepressants? no; stimulant use has increased at a faster rate than the use of antidepressants.

72. Shown, again, is the scatter plot that indicates a relationship between the percentage of adult females in a country who are literate and the mortality of children under five. Also shown is a line that passes through or near the points. Find a function that models the data by finding the slope-intercept equation of the line. Use the function to make a prediction about child mortality based on the percentage of adult females in a country who are literate. Answers will vary; an example is $(10, 230)$ and $(60, 110)$; $f(x) = -2.4x + 254$; If 50% of the adult women in a country are literate, the mortality rate of children under 5 will be 134 per thousand.

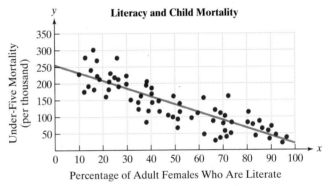

Literacy and Child Mortality

Source: United Nations

Writing in Mathematics

73. Describe how to write the equation of a line if its slope and a point along the line are known. 73–79. Answers will vary.

74. Describe how to write the equation of a line if two points along the line are known.

75. If two lines are parallel, describe the relationship between their slopes.

76. If two lines are perpendicular, describe the relationship between their slopes.

77. If you know a point on a line and you know the equation of a line parallel to this line, explain how to write the line's equation.

78. In Example 3 on page 144, we developed a model that predicted Americans born in 2020 will have a life expectancy of 80 years. Describe something that might occur that would make this prediction inaccurate.

79. Take another look at the scatter plot in Exercise 72. Although there is a relationship between literacy and child mortality, we cannot conclude that increased literacy causes child mortality to decrease. Offer two or more possible explanations for the data in the scatter plot.

Technology Exercises

80. The lines whose equations are $y = \frac{1}{3}x + 1$ and

$y = -3x - 2$ are perpendicular because the product of their slopes, $\frac{1}{3}$ and -3, respectively, is -1.

a. Use a graphing utility to graph the equations in a $[-10, 10, 1]$ by $[-10, 10, 1]$ viewing rectangle. Do the lines appear to be perpendicular? See graphing answer section. Answers will vary.

b. Now use the zoom square feature of your utility. Describe what happens to the graphs. Explain why this is so. Answers will vary.

81. a. Use the statistical menu of your graphing utility to enter the seven data points shown in the scatter plot that you drew in Exercise 69(a). 81–82. See graphing answer section. Answers will vary.

b. Use the DRAW menu and the scatter plot capability to draw a scatter plot of the data points.

c. Select the linear regression option. Use your utility to obtain values for a and b for the equation of the regression line, $y = ax + b$. Compare this equation to the one that you obtained by hand in Exercise 69. You may also be given a **correlation coefficient**, r. Values of r close to 1 indicate that the points can be modeled by a linear function and the regression line has a positive slope. Values of r close to -1 indicate that the points can be modeled by a linear function and the regression line has a negative slope. Values of r close to 0 indicate no linear relationship between the variables. In this case, a linear model does not accurately describe the data.

d. Use the appropriate sequence (consult your manual) to graph the regression equation on top of the points in the scatter plot.

82. Repeat Exercise 81 using the seven data points shown in the scatter plot that you drew in Exercise 70(a).

Critical Thinking Exercises

83. Determine the value of B so that the line whose equation is $By = 8x - 1$ has slope -2. -4

84. Determine the value of A so that the line whose equation is $Ax + y = 2$ is perpendicular to the line containing the points $(1, -3)$ and $(-2, 4)$. $-\frac{3}{7}$

85. Consider a line whose x-intercept is -3 and whose y-intercept is -6. Provide the missing coordinate for the following two points that lie on this line:

$$(-40, \quad) \text{ and } (\quad, -200).$$

$(-40, 74)$ and $(97, -200)$

86. Prove that the equation of a line passing through $(a, 0)$ and $(0, b)$ $(a \neq 0, b \neq 0)$ can be written in the form $\frac{x}{a} + \frac{y}{b} = 1$. Why is this called the *intercept form* of a line?

$$y - 0 = -\frac{b}{a}(x - a) \quad \leftarrow \text{Point-Slope Form}$$

$$y = -\frac{b}{a}x + b \quad \leftarrow \text{Slope-Intercept Form}$$

$$\frac{y}{b} = -\frac{x}{a} + 1 \quad \leftarrow \text{Divide by } b.$$

$$\frac{x}{a} + \frac{y}{b} = 1 \quad \leftarrow \text{Intercept Form}$$

Review Exercises

87. If $f(x) = 3x^2 - 8x + 5$, find $f(-2)$. (Section 2.1, Example 3) 33

88. If $f(x) = x^2 - 3x + 4$ and $g(x) = 2x - 5$, find $(fg)(-1)$. (Section 2.2, Example 4) -56

89. The sum of the angles of a triangle is $180°$. Find the three angles of a triangle if one angle is $20°$ greater than the smallest angle and the third angle is twice the smallest angle. (Section 1.5, Example 1) $40°, 60°,$ and $80°$

GROUP PROJECT

CHAPTER 2

Additional group projects can be found in the *Instructor's Resource Manual.*

In Example 3 on page 144, we used the data in Figure 2.29 to develop a linear function that modeled life expectancy. For this group exercise, you might find it helpful to pattern your work after Figure 2.29 and the solution to Example 3. Group members should begin by consulting an almanac, newspaper, magazine, or the Internet to find data that appear to lie approximately on or near a line. Working by hand or using a graphing utility, group members should construct scatter plots for the data that were assembled. If working by hand, draw a line that approximately fits the data in each scatter plot and then write its equation as a function in slope-intercept form. If using a graphing utility, obtain the equation of each regression line. Then use each linear function's equation to make predictions about what might occur in the future. Are there circumstances that might affect the accuracy of the prediction? List some of these circumstances.

CHAPTER 2 SUMMARY

Definitions and Concepts	Examples

Section 2.1 Introduction to Functions

A relation is any set of ordered pairs. The set of first components of the ordered pairs is the domain and the set of second components is the range. A function is a relation in which each member of the domain corresponds to exactly one member of the range. No two ordered pairs of a function can have the same first component and different second components.

The domain of the relation $\{(1, 2), (3, 4), (3, 7)\}$ is $\{1, 3\}$. The range is $\{2, 4, 7\}$. The relation is not a function: 3, in the domain, corresponds to both 4 and 7 in the range.

If a function is defined by an equation, the notation $f(x)$, read "f of x" or "f at x," describes the value of the function at the number, or input, x.

If $f(x) = 7x - 5$, then
$$f(a + 2) = 7(a + 2) - 5$$
$$= 7a + 14 - 5$$
$$= 7a + 9.$$

The graph of a function is the graph of its ordered pairs.

The Vertical Line Test for Functions
If any vertical line intersects a graph in more than one point, the graph does not define y as a function of x.

At the left or right of a function's graph, you will often find closed dots, open dots, or arrows. A closed dot shows that the graph ends and the point belongs to the graph. An open dot shows that the graph ends and the point does not belong to the graph. An arrow indicates that the graph extends indefinitely.

The graph of a function can be used to determine the function's domain and its range. To find the domain, look for all the inputs on the x-axis that correspond to points on the graph. To find the range, look for all the outputs on the y-axis that correspond to points on the graph.

Not the graph of a function

The graph of a function

(0, 2) belongs to the graph of f; (0, 0) does not.

$y = f(x)$

To find $f(2)$, locate 2 on the x-axis. The graph shows $f(2) = 4$.
Domain of $f = \{x \mid x \text{ is a real number}\}$
Range of $f = \{y \mid y > 0\}$

Section 2.2 The Algebra of Functions

A Function's Domain
If a function f does not model data or verbal conditions, its domain is the largest set of real numbers for which the value of $f(x)$ is a real number. Exclude from a function's domain real numbers that cause division by zero and real numbers that result in a square root of a negative number.

$$f(x) = 7x + 13$$

Domain of $f = \{x \mid x \text{ is a real number}\}$

$$g(x) = \frac{7x}{12 - x}$$

Domain of $g = \{x \mid x \text{ is a real number and } x \neq 12\}$

Definitions and Concepts	Examples

Section 2.2 The Algebra of Functions (continued)

The Algebra of Functions

Let f and g be two functions. The sum $f + g$, the difference $f - g$, the product fg, and the quotient $\dfrac{f}{g}$ are functions whose domains are the set of all real numbers common to the domains of f and g, defined as follows:

1. Sum: $\quad (f + g)(x) = f(x) + g(x)$

2. Difference: $\quad (f - g)(x) = f(x) - g(x)$

3. Product: $\quad (fg)(x) = f(x) \cdot g(x)$

4. Quotient: $\quad \left(\dfrac{f}{g}\right)(x) = \dfrac{f(x)}{g(x)}, g(x) \neq 0.$

Let $f(x) = x^2 + 2x$ and $g(x) = 4 - x$.

- $(f + g)(x) = (x^2 + 2x) + (4 - x) = x^2 + x + 4$
  $(f + g)(-2) = (-2)^2 + (-2) + 4 = 4 - 2 + 4 = 6$
- $(f - g)(x) = (x^2 + 2x) - (4 - x) = x^2 + 2x - 4 + x$
  $\qquad\qquad\qquad = x^2 + 3x - 4$
  $(f - g)(5) = 5^2 + 3 \cdot 5 - 4 = 25 + 15 - 4 = 36$
- $(fg)(1) = f(1) \cdot g(1) = (1^2 + 2 \cdot 1)(4 - 1)$
  $\qquad\qquad = 3(3) = 9$
- $\left(\dfrac{f}{g}\right)(x) = \dfrac{x^2 + 2x}{4 - x}, x \neq 4$
  $\left(\dfrac{f}{g}\right)(3) = \dfrac{3^2 + 2 \cdot 3}{4 - 3} = \dfrac{9 + 6}{1} = 15$

Section 2.3 Linear Functions and Slope

Data presented in a visual form as a set of points is called a scatter plot. A line that best fits the data points is called a regression line. A function whose graph is a straight line is called a linear function. All linear functions can be written in the form $f(x) = mx + b$.

$f(x) = 3x + 10$ is a linear function.

$g(x) = 3x^2 + 10$ is not a linear function.

If a graph intersects the x-axis at $(a, 0)$, then a is an x-intercept. If a graph intersects the y-axis at $(0, b)$, then b is a y-intercept. The standard form of the equation of a line,

$$Ax + By = C,$$

can be graphed using intercepts and a checkpoint.

Graph using intercepts: $4x + 3y = 12$.

x-intercept: $\qquad 4x = 12$ Line passes
(Set $y = 0$.) $\qquad\ x = 3$ through (3, 0).

y-intercept: $\qquad 3y = 12$ Line passes
(Set $x = 0$.) $\qquad\ y = 4$ through (0, 4).

checkpoint: $\qquad$ Let $x = 2$.

$\qquad\qquad 4 \cdot 2 + 3y = 12$

$\qquad\qquad\ 8 + 3y = 12$

$\qquad\qquad\qquad 3y = 4$

$\qquad\qquad\qquad\ y = \dfrac{4}{3}$

y-intercept is 4.
Checkpoint: $\left(2, \frac{4}{3}\right)$
x-intercept is 3.

Definitions and Concepts	Examples

Section 2.3 Linear Functions and Slope (continued)

The slope, m, of the line through the points (x_1, y_1) and (x_2, y_2) is

$$m = \frac{y_2 - y_1}{x_2 - x_1}, \quad x_2 - x_1 \neq 0.$$

If the slope is positive, the line rises from left to right. If the slope is negative, the line falls from left to right. The slope of a horizontal line is 0. The slope of a vertical line is undefined.

For points $(-7, 2)$ and $(3, -4)$, the slope of the line through the points is

$$m = \frac{\text{Change in } y}{\text{Change in } x} = \frac{-4 - 2}{3 - (-7)} = \frac{-6}{10} = -\frac{3}{5}.$$

Slope is negative, so the line falls.

For points $(2, -5)$ and $(2, 16)$, the slope of the line through the points is:

$$m = \frac{\text{Change in } y}{\text{Change in } x} = \frac{16 - (-5)}{2 - 2} = \frac{21}{0}$$

undefined

Slope is undefined, so the line is vertical.

The slope-intercept equation of a nonvertical line with slope m and y-intercept b is

$$y = mx + b.$$

Using function notation, the equation is

$$f(x) = mx + b.$$

Graph: $f(x) = -\dfrac{3}{4}x + 1$

Slope is $-\frac{3}{4}$. y-intercept is 1.

y-intercept is 1.

Rise $= -3$

Run $= 4$

Horizontal and Vertical Lines

The graph of $y = b$, or $f(x) = b$, is a horizontal line. The y-intercept is b. The linear function $f(x) = b$ is called a constant function.

The graph of $x = a$ is a vertical line. The x-intercept is a. A vertical line is not a linear function.

If the graph of a function is not a straight line, the average rate of change between any two points is the slope of the line containing the two points.

For a linear function, slope is the rate of change of the dependent variable per unit change of the independent variable.

The function

$$p(t) = -0.59t + 80.75$$

slope

models the percentage, $p(t)$, of Americans smoking cigarettes t years after 1900. Slope, -0.59, shows that the percentage of smokers is decreasing by 0.59% per year.

Definitions and Concepts	Examples

Section 2.4 The Point-Slope Form of the Equation of a Line

The point-slope equation of a nonvertical line with slope m that passes through the point (x_1, y_1) is $$y - y_1 = m(x - x_1).$$	Slope $= -4$, passing through $(-1, 5)$ $$m = -4 \qquad x_1 = -1 \quad y_1 = 5$$ The line's point-slope equation is $$y - 5 = -4[x - (-1)].$$ Simplify: $$y - 5 = -4(x + 1).$$
To write the point-slope form of the line passing through two points, begin by using the points to compute the slope, m. Use either given point as (x_1, y_1) and write the point-slope equation: $$y - y_1 = m(x - x_1).$$ Solving this equation for y gives the slope-intercept form of the line's equation.	Write an equation in point-slope form and slope-intercept form of the line passing through $(4, 1)$ and $(3, -2)$. $$m = \frac{-2 - 1}{3 - 4} = \frac{-3}{-1} = 3$$ Using $(4, 1)$ as (x_1, y_1), the point-slope equation is $$y - 1 = 3(x - 4).$$ Solve for y to obtain the slope-intercept form. $$y - 1 = 3x - 12$$ $$y = 3x - 11$$ In function notation, $$f(x) = 3x - 11.$$
Nonvertical parallel lines have the same slope. If the product of the slopes of two lines is -1, then the lines are perpendicular. One line is perpendicular to another line if its slope is the negative reciprocal of the slope of the other.	Write point-slope and slope-intercept equations of the line passing through $(2, -1)$ $$x_1 \qquad y_1$$ and perpendicular to $y = -\frac{1}{5}x + 6$. $$\text{slope}$$ Slope, m, of perpendicular line is 5, the negative reciprocal of $-\frac{1}{5}$. $$y - (-1) = 5(x - 2) \qquad \text{Point-slope equation}$$ $$y + 1 = 5(x - 2)$$ $$y + 1 = 5x - 10$$ $$y = 5x - 11 \text{ or } f(x) = 5x - 11$$ Slope intercept equation

CHAPTER 2 REVIEW EXERCISES

2.1 *In Exercises 1–3, determine whether each relation is a function. Give the domain and range for each relation.*

1. $\{(3, 10), (4, 10), (5, 10)\}$
function; domain: $\{3, 4, 5\}$; range: $\{10\}$

2. $\{(1, 12), (2, 100), (3, \pi), (4, -6)\}$
function; domain: $\{1, 2, 3, 4\}$; range: $\{12, 100, \pi, -6\}$

3. $\{(13, 14), (15, 16), (13, 17)\}$
not a function; domain: $\{13, 15\}$; range: $\{14, 16, 17\}$

In Exercises 4–5, find the indicated function values.

4. $f(x) = 7x - 5$
a. $f(0)$ -5 **b.** $f(3)$ 16 **c.** $f(-10)$ -75
d. $f(2a)$ $14a - 5$ **e.** $f(a + 2)$ $7a + 9$

5. $g(x) = 3x^2 - 5x + 2$

   a. $g(0)$ 2   **b.** $g(5)$ 52   **c.** $g(-4)$ 70

   d. $g(b)$ $3b^2 - 5b + 2$ **e.** $g(4a)$ $48a^2 - 20a + 2$

In Exercises 6–7, graph the given functions, f and g, in the same rectangular coordinate system. Select integers for x, starting with −2 and ending with 2. Once you have obtained your graphs, describe how the graph of g is related to the graph of f.   6–7. See graphing answer section.

6. $f(x) = x^2$, $g(x) = x^2 - 1$

   The graph of g is the graph of f shifted down by 1 unit.

7. $f(x) = |x|$, $g(x) = |x| + 2$

   The graph of g is the graph of f shifted up by 2 units.

In Exercises 8–13, use the vertical line test to identify graphs in which y is a function of x.

8.

not a function

9.

function

10.

function

11.

not a function

12.

not a function

13.

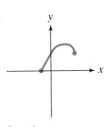

function

Use the graph of f to solve Exercises 14–18.

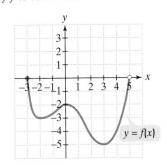

14. Find $f(-2)$.   −3   **15.** Find $f(0)$.   −2

16. For what value of x is $f(x) = -5$?   3

17. Find the domain of f. $\{x \mid -3 \le x < 5\}$

18. Find the range of f. $\{y \mid -5 \le y \le 0\}$

19. The graph shows the height, in meters, of a vulture in terms of its time, in seconds, in flight.

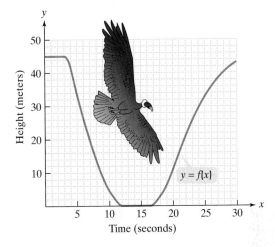

   a. Use the graph to explain why the vulture's height is a function of its time in flight.   For each time, there is only one height.

   b. Find $f(15)$. Describe what this means in practical terms.   0; The vulture was on the ground after 15 seconds.

   c. What is a reasonable estimate of the vulture's maximum height?   45 meters

   d. For what values of x is $f(x) = 20$? Describe what this means in practical terms.   7 and 22; After 7 and 22 seconds, the vulture's height is 20 meters.

   e. Use the graph of the function to write a description of the vulture's flight.   Answers will vary.

2.2 *In Exercises 20–22, find the domain of each function.*

20. $f(x) = 7x - 3$ Domain of $f = \{x \mid x \text{ is a real number}\}$

21. $g(x) = \dfrac{1}{x + 8}$

   Domain of $g = \{x \mid x \text{ is a real number and } x \neq -8\}$

22. $f(x) = x + \dfrac{3x}{x - 5}$

   Domain of $f = \{x \mid x \text{ is a real number and } x \neq 5\}$

In Exercises 23–24, find **a.** $(f + g)(x)$ *and* **b.** $(f + g)(3)$.

23. $f(x) = 4x - 5$, $g(x) = 2x + 1$ **a.** $6x - 4$ **b.** 14

24. $f(x) = 5x^2 - x + 4$, $g(x) = x - 3$ **a.** $5x^2 + 1$ **b.** 46

In Exercises 25–26, for each pair of functions, f and g, determine the domain of f + g.

25. $f(x) = 3x + 4$, $g(x) = \dfrac{5}{4 - x}$

   Domain of $f + g = \{x \mid x \text{ is a real number and } x \neq 4\}$

26. $f(x) = \dfrac{7x}{x + 6}$, $g(x) = \dfrac{4}{x + 1}$ Domain of

   $f + g = \{x \mid x \text{ is a real number and } x \neq -6 \text{ and } x \neq -1\}$

In Exercises 27–34, let

$$f(x) = x^2 - 2x \quad \text{and} \quad g(x) = x - 5.$$

Find each of the following.

27. $(f + g)(x)$ and $(f + g)(-2)$ $x^2 - x - 5; 1$

28. $f(3) + g(3)$ 1

29. $(f - g)(x)$ and $(f - g)(1)$ $x^2 - 3x + 5; 3$

30. $f(4) - g(4)$ 9

31. $(fg)(-3)$ -120

32. $\left(\dfrac{f}{g}\right)(x)$ and $\left(\dfrac{f}{g}\right)(4)$ $\dfrac{x^2 - 2x}{x - 5}; -8$

33. The domain of $f - g$
Domain of $f - g = \{x \mid x \text{ is a real number}\}$

34. The domain of $\dfrac{f}{g}$

Domain of $\dfrac{f}{g} = \{x \mid x \text{ is a real number and } x \neq 5\}$

2.3 In Exercises 35–37, use intercepts and a checkpoint to graph each linear function.   **35–37.** See graphing answer section.

35. $x + 2y = 4$ **36.** $2x - 3y = 12$ **37.** $4x = 8 - 2y$

In Exercises 38–41, find the slope of the line passing through each pair of points or state that the slope is undefined. Then indicate whether the line through the points rises, falls, is horizontal, or is vertical.

38. $(5, 2)$ and $(2, -4)$ $m = 2$; rises

39. $(-2, 3)$ and $(7, -3)$ $m = -\frac{2}{3}$; falls

40. $(3, 2)$ and $(3, -1)$ m is undefined; vertical

41. $(-3, 4)$ and $(-1, 4)$ $m = 0$; horizontal

In Exercises 42–44, give the slope and y-intercept of each line whose equation is given. Then graph the linear function.
42–44. See graphing answer section.

42. $y = 2x - 1$ $m = 2; b = -1$

43. $f(x) = -\dfrac{1}{2}x + 4$ $m = -\frac{1}{2}; b = 4$

44. $y = \dfrac{2}{3}x$ $m = \frac{2}{3}; b = 0$

In Exercises 45–47, rewrite the equation in slope-intercept form. Give the slope and y-intercept.

45. $2x + y = 4$ $y = -2x + 4; m = -2; b = 4$

46. $-3y = 5x$ $y = -\frac{5}{3}x; m = -\frac{5}{3}; b = 0$

47. $5x + 3y = 6$ $y = -\frac{5}{3}x + 2; m = -\frac{5}{3}; b = 2$

In Exercises 48–52, graph each equation in a rectangular coordinate system.   **48–52.** See graphing answer section.

48. $y = 2$ **49.** $7y = -21$ **50.** $f(x) = -4$

51. $x = 3$ **52.** $2x = -10$

53. The function $f(t) = -0.27t + 70.45$ models record time, $f(t)$, in seconds, for the women's 400-meter run t years after 1900. What is the slope of this model? Describe what this means in terms of rate of change. $m = -0.27$, Record time has been decreasing at a rate of 0.27 second per year since 1900.

54. The graph shows U.S. military spending, in inflation-adjusted billions of dollars. Find the average rate of change in military spending per year, to the nearest tenth of a billion dollars, from 1946 through 2004. $m \approx -3.4$; average decrease of approximately \$3.4 billion per year

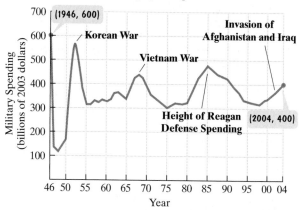

U.S. Military Spending: 1946–2004

Source: Center for Defense Information

55. The graph shows that a linear function describes the relationship between Fahrenheit temperature, F, and Celsius temperature, C.

a. Use the points labeled by the voice balloons to find a function in the form $F = mC + b$ that expresses Fahrenheit temperature, F, in terms of Celsius temperature, C. $F = \frac{9}{5}C + 32$

b. Use the function from part (a) to find the Fahrenheit temperature when the Celsius temperature is 30°. $86°$

2.4 In Exercises 56–59, use the given conditions to write an equation for each line in point-slope form and slope-intercept form.

56. Passing through $(-3, 2)$ with slope -6
$y - 2 = -6(x + 3); y = -6x - 16$ or $f(x) = -6x - 16$

57. Passing through $(1, 6)$ and $(-1, 2)$ $y - 6 = 2(x - 1)$ or $y - 2 = 2(x + 1); y = 2x + 4$ or $f(x) = 2x + 4$

58. Passing through $(4, -7)$ and parallel to the line whose equation is $3x + y - 9 = 0$

$y + 7 = -3(x - 4); y = -3x + 5$ or $f(x) = -3x + 5$

59. Passing through $(-2, 6)$ and perpendicular to the line whose equation is $y = \frac{1}{3}x + 4$

$y - 6 = -3(x + 2); y = -3x$ or $f(x) = -3x$

60. When adjusted for inflation, the federal minimum wage from 1997 through 2003 continued to decrease. The points in the scatter plot show the minimum hourly inflation-adjusted wage for this period. Also shown is a line that passes through or near the points.

a. Use the two points whose coordinates are shown by the voice balloons to find the point-slope equation of the line that models the minimum hourly inflation-adjusted wage, y, x years after 1997.

$y - 4.85 = -0.12(x - 2)$ or $y - 4.49 = -0.12(x - 5)$

b. Write the equation in part (a) in slope-intercept form. Use function notation. $f(x) = -0.12x + 5.09$

Federal Minimum Wage Adjusted for Inflation

Source: www.dul.gov/esa/pudha/minwage

c. Use the linear function to predict the minimum hourly inflation-adjusted wage in 2007. $3.89

CHAPTER 2 TEST

Remember to use your Chapter Test Prep Video CD to see the worked-out solutions to the test questions you want to review.

In Exercises 1–2, determine whether each relation is a function. Give the domain and range for each relation.

1. $\{(1, 2), (3, 4), (5, 6), (6, 6)\}$

function; domain: $\{1, 3, 5, 6\}$; range: $\{2, 4, 6\}$

2. $\{(2, 1), (4, 3), (6, 5), (6, 6)\}$

not a function; domain: $\{2, 4, 6\}$; range: $\{1, 3, 5, 6\}$

3. If $f(x) = 3x - 2$, find $f(a + 4)$. $3a + 10$

4. If $f(x) = 4x^2 - 3x + 6$, find $f(-2)$. 28

5. Graph $f(x) = x^2 - 1$ and $g(x) = x^2 + 1$ in the same rectangular coordinate system. Select integers for x, starting with -2 and ending with 2. Once you have obtained your graphs, describe how the graph of g is related to the graph of f. See graphing answer section. The graph of g is the graph of f shifted up by 2 units.

In Exercises 6–7, identify the graph or graphs in which y is a function of x.

6.

function

7.

not a function

Use the graph of f to solve Exercises 8–11.

8. Find $f(6)$. -3

9. List two values of x for which $f(x) = 0$. -2 and 3

10. Find the domain of f. $\{x \mid x \text{ is a real number}\}$

11. Find the range of f. $\{y \mid y \leq 3\}$

12. Find the domain of $f(x) = \dfrac{6}{10 - x}$.

Domain of $f = \{x \mid x \text{ is a real number and } x \neq 10\}$

In Exercises 13–17, let

$$f(x) = x^2 + 4x \quad and \quad g(x) = x + 2.$$

Find each of the following.

13. $(f + g)(x)$ and $(f + g)(3)$ $x^2 + 5x + 2; 26$

14. $(f - g)(x)$ and $(f - g)(-1)$ $x^2 + 3x - 2; -4$

15. $(fg)(-5)$ -15

16. $\left(\dfrac{f}{g}\right)(x)$ and $\left(\dfrac{f}{g}\right)(2)$ $\dfrac{x^2 + 4x}{x + 2}$; 3

17. The domain of $\dfrac{f}{g}$

Domain of $\dfrac{f}{g} = \{x \mid x$ is a real number and $x \neq -2\}$

In Exercises 18–20, graph each linear function.   **18–20.** See graphing answer section.

18. $4x - 3y = 12$ **19.** $f(x) = -\dfrac{1}{3}x + 2$

20. $f(x) = 4$

In Exercises 21–22, find the slope of the line passing through each pair of points or state that the slope is undefined. Then indicate whether the line through the points rises, falls, is horizontal, or is vertical.

21. $(5, 2)$ and $(1, 4)$ $m = -\dfrac{1}{2}$; falls

22. $(4, 5)$ and $(4, -5)$ m is undefined; vertical

The function $V(t) = 3.6t + 140$ models the number of Super Bowl viewers, $V(t)$, in millions, t years after 1995. Use the model to solve Exercises 23–24.

23. Find $V(10)$. Describe what this means in terms of the variables in the model.   176; In 2005, the number of Super Bowl viewers was 176 million.

24. What is the slope of this model? Describe what this means in terms of rate of change. $m = 3.6$; The number of Super Bowl viewers is increasing at a rate of 3.6 million per year.

In Exercises 25–27, use the given conditions to write an equation for each line in point-slope form and slope-intercept form.

25. Passing through $(-1, -3)$ and $(4, 2)$ $y + 3 = 1(x + 1)$ or $y - 2 = 1(x - 4)$; $y = x - 2$ or $f(x) = x - 2$

26. Passing through $(-2, 3)$ and perpendicular to the line whose equation is $y = -\dfrac{1}{2}x - 4$
$y - 3 = 2(x + 2)$; $y = 2x + 7$ or $f(x) = 2x + 7$

27. Passing through $(6, -4)$ and parallel to the line whose equation is $x + 2y = 5$
$y + 4 = -\dfrac{1}{2}(x - 6)$; $y = -\dfrac{1}{2}x - 1$ or $f(x) = -\dfrac{1}{2}x - 1$

28. The scatter plot shows the annual number of adult and child deaths worldwide due to AIDS from 1999 through 2003. Also shown is a line that models the data.

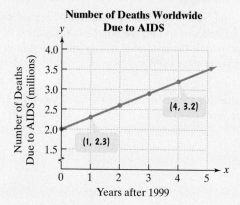

Number of Deaths Worldwide Due to AIDS

Source: World Health Organization

a. Use the two points whose coordinates are shown by the voice balloons to find the point-slope equation of the line that models the number of worldwide deaths due to AIDS, y, in millions, x years after 1999.
$y - 2.3 = 0.3(x - 1)$ or $y - 3.2 = 0.3(x - 4)$

b. Write the equation in part (a) in slope-intercept form. Use function notation. $f(x) = 0.3x + 2$

c. Use the linear function to predict the number of worldwide deaths due to AIDS, in millions, in 2007.
4.4 million

Note to Instructor: The Chapter 2 Test file in the TestGen program provides algorithms specifically matched to these problems so you can easily replicate this chapter test for additional practice or assessment purposes.

CUMULATIVE REVIEW EXERCISES (CHAPTERS 1–2)

1. Use the roster method to list the elements in the set:
$\{x \mid x$ is a whole number less than 4$\}$. $\{0, 1, 2, 3\}$

2. Determine whether the following statement is true or false:
$\pi \notin \{x \mid x$ is an irrational number$\}$.   false

In Exercises 3–4, use the order of operations to simplify each expression.

3. $\dfrac{8 - 3^2 \div 9}{|-5| - [5 - (18 \div 6)]^2}$ 7

4. $4 - (2 - 9)^0 + 3^2 \div 1 + 3$ 15

5. Simplify: $3 - [2(x - 2) - 5x]$. $7 + 3x$ or $3x + 7$

In Exercises 6–8, solve each equation. If the solution set is $\varnothing$ or $\mathbb{R}$, classify the equation as an inconsistent equation or an identity.

6. $2 + 3x - 4 = 2(x - 3)$ -4

7. $4x + 12 - 8x = -6(x - 2) + 2x$ all real numbers; identity

8. $\dfrac{x - 2}{4} = \dfrac{2x + 6}{3}$ -6

9. After a 20% reduction, a computer sold for $1800. What was the computer's price before the reduction? $2250

10. Solve for t: $A = p + prt$. $t = \dfrac{A - p}{pr}$

In Exercises 11–12, simplify each exponential expression.

11. $(3x^4y^{-5})^{-2}$ $\dfrac{y^{10}}{9x^8}$

12. $\left(\dfrac{3x^2y^{-4}}{x^{-3}y^2}\right)^2$ $\dfrac{9x^{10}}{y^{12}}$

13. Multiply and write the answer in scientific notation:

$$(7 \times 10^{-8})(3 \times 10^2). \quad 2.1 \times 10^{-5}$$

14. Is $\{(1,5),(2,5),(3,5),(4,5),(6,5)\}$ a function? Give the relation's domain and range.   function; domain: $\{1,2,3,4,6\}$; range: $\{5\}$

15. Graph $f(x) = |x| - 1$ and $g(x) = |x| + 2$ in the same rectangular coordinate system. Select integers for x, starting with -2 and ending with 2. Once you have obtained your graphs, describe how the graph of g is related to the graph of f.

See graphing answer section. The graph of g is the graph of f shifted up by 3 units.

16. Find the domain of $f(x) = \dfrac{1}{15 - x}$.

Domain of $f = \{x \mid x$ is a real number and $x \neq 15\}$

17. If $f(x) = 3x^2 - 4x + 2$ and $g(x) = x^2 - 5x - 3$, find $(f - g)(x)$ and $(f - g)(-1)$. $2x^2 + x + 5; 6$

In Exercises 18–19, graph each linear function.   18–19. See graphing answer section.

18. $f(x) = -2x + 4$.

19. $x - 2y = 6$

20. Write equations in point-slope form and slope-intercept form for the line passing through $(3, -5)$ and parallel to the line whose equation is $y = 4x + 7$.

$y + 5 = 4(x - 3); y = 4x - 17$ or $f(x) = 4x - 17$

According to the *American Journal of Public Health*, smoking on-screen has surpassed its 1950s level of 11 incidents per hour. Although we still may see celebrities smoking in movies, in music videos, and on television, there has been a remarkable decline in the percentage of cigarette smokers in the United States.

The declining use of cigarettes is developed using pairs of linear equations in Exercises 93–94 in Exercise Set 3.1.

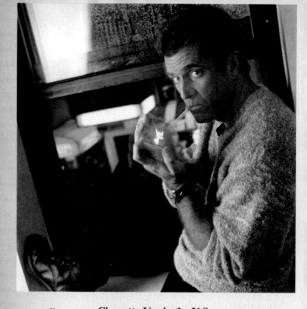

Systems of Linear Equations

3.1 Systems of Linear Equations in Two Variables

3.2 Problem Solving and Business Applications Using Systems of Equations

3.3 Systems of Linear Equations in Three Variables

3.4 Matrix Solutions to Linear Systems

3.5 Determinants and Cramer's Rule

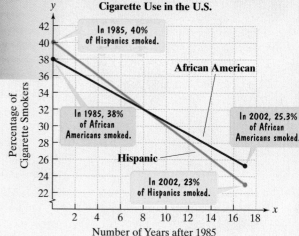

Cigarette Use in the U.S.

In 1985, 40% of Hispanics smoked.

African American

In 1985, 38% of African Americans smoked.

In 2002, 25.3% of African Americans smoked.

Hispanic

In 2002, 23% of Hispanics smoked.

Percentage of Cigarette Smokers

Number of Years after 1985

Source: Dept. of Health and Human Services

Many situations in the real world require more than one equation to model the relationships between two or more variables. The decline in cigarette smoking among African Americans and Hispanics can be analyzed using a pair of linear models in two variables. In this chapter, you will learn to model your world with two equations in two variables and three equations in three variables. The methods you learn for solving these systems provide the foundation for solving complex problems involving thousands of equations containing thousands of variables.

SECTION 3.1

Objectives

1 Determine whether an ordered pair is a solution of a system of linear equations.

2 Solve systems of linear equations by graphing.

3 Solve systems of linear equations by substitution.

4 Solve systems of linear equations by addition.

5 Select the most efficient method for solving a system of linear equations.

6 Identify systems that do not have exactly one ordered-pair solution.

SYSTEMS OF LINEAR EQUATIONS IN TWO VARIABLES

Women now attain higher levels of education than ever before and a majority of women choose careers in the labor force rather than homemaking. One of the most important developments in the work force has been the dramatic increase in the number of women, at approximately $\frac{1}{2}$% per year. By contrast, the percentage of men is decreasing by $\frac{1}{4}$% per year. These changes are illustrated in Figure 3.1. If current trends continue, the graphs will intersect. The intersection point reveals when the percentage of women and men in the labor force will be equal.

Projections frequently involve graphs that intersect. In this section, you will learn three methods for finding intersection points. With these methods, you can use mathematical models to determine when variable quantities, such as the percentage of women and men in the work force, will be equal.

Percentage of U.S. Men and Women in the Civilian Labor Force

FIGURE 3.1
Source: U.S. Department of Labor

1 Determine whether an ordered pair is a solution of a system of linear equations.

Systems of Linear Equations and Their Solutions We have seen that all equations in the form $Ax + By = C$ are straight lines when graphed. Two such equations are called a **system of linear equations**. A **solution of a system of linear equations** is an ordered pair that satisfies all equations in the system. For example, $(3, 4)$ satisfies the system

$$x + y = 7 \quad \text{(3 + 4 is, indeed, 7.)}$$
$$x - y = -1. \quad \text{(3 - 4 is, indeed, -1.)}$$

Thus, $(3, 4)$ satisfies both equations and is a solution of the system. The solution can be described by saying that $x = 3$ and $y = 4$. The solution can also be described using set notation. The solution set of the system is $\{(3, 4)\}$—that is, the set consisting of the ordered pair $(3, 4)$.

A system of linear equations can have exactly one solution, no solution, or infinitely many solutions. We begin with systems with exactly one solution.

EXAMPLE 1 Determining Whether an Ordered Pair Is a Solution of a System

Determine whether $(-5, -6)$ is a solution of the system

$$2x - y = -4$$
$$3x - 5y = 15.$$

SOLUTION Because -5 is the x-coordinate and -6 is the y-coordinate of $(-5, -6)$, we replace x with -5 and y with -6.

$$2x - y = -4 \qquad\qquad 3x - 5y = 15$$

$$2(-5) - (-6) \stackrel{?}{=} -4 \qquad 3(-5) - 5(-6) \stackrel{?}{=} 15$$

$$-10 + 6 \stackrel{?}{=} -4 \qquad\qquad -15 + 30 \stackrel{?}{=} 15$$

$$-4 = -4, \quad \text{true} \qquad\qquad 15 = 15, \quad \text{true}$$

The pair $(-5, -6)$ satisfies both equations: It makes each equation true. Thus, $(-5, -6)$ is a solution of the system. ∎

 CHECK POINT 1 Determine whether $(-2, -3)$ is a solution of the system

$$2x - 4y = 8$$
$$2x - y = -1. \quad \text{solution}$$

Solving Linear Systems by Graphing The solution of a system of two linear equations in two variables can be found by graphing both of the equations in the same rectangular coordinate system. For a system with one solution, **the coordinates of the point of intersection give the system's solution.**

STUDY TIP

When solving linear systems by graphing, neatly drawn graphs are essential for determining points of intersection.

• Use rectangular coordinate graph paper.

• Use a ruler or straightedge.

• Use a pencil with a sharp point.

SOLVING SYSTEMS OF TWO LINEAR EQUATIONS IN TWO VARIABLES, x AND y, BY GRAPHING

1. Graph the first equation.
2. Graph the second equation on the same set of axes.
3. If the lines representing the two graphs intersect at a point, determine the coordinates of this point of intersection. The ordered pair is the solution to the system.
4. Check the solution in both equations.

EXAMPLE 2 Solving a Linear System by Graphing

Solve by graphing:

$$y = -x - 1$$
$$4x - 3y = 24.$$

SOLUTION

Step 1. **Graph the first equation.** We use the y-intercept and slope to graph $y = -x - 1$.

$$y = -x - 1$$

The slope is -1.　The y-intercept is -1.

The graph of the linear function is shown as a blue line in Figure 3.2.

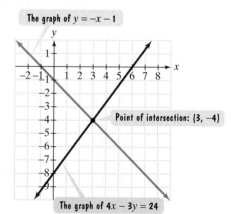

The graph of $y = -x - 1$

Point of intersection: $(3, -4)$

The graph of $4x - 3y = 24$

FIGURE 3.2

Step 2. **Graph the second equation on the same axes.** We use intercepts to graph $4x - 3y = 24$.

***x*-intercept (Set $y = 0$.)**	***y*-intercept (Set $x = 0$.)**
$4x - 3 \cdot 0 = 24$	$4 \cdot 0 - 3y = 24$
$4x = 24$	$-3y = 24$
$x = 6$	$y = -8$

The x-intercept is 6, so the line passes through $(6, 0)$. The y-intercept is -8, so the line passes through $(0, -8)$. The graph of $4x - 3y = 24$ is shown as a red line in Figure 3.2.

Step 3. **Determine the coordinates of the intersection point. This ordered pair is the system's solution.** Using Figure 3.2, it appears that the lines intersect at $(3, -4)$. The "apparent" solution of the system is $(3, -4)$.

Step 4. **Check the solution in both equations.**

Check $(3, -4)$ in $y = -x - 1$:	**Check $(3, -4)$ in** $4x - 3y = 24$:
$-4 \overset{?}{=} -3 - 1$	$4(3) - 3(-4) \overset{?}{=} 24$
$-4 = -4,$ true	$12 + 12 \overset{?}{=} 24$
	$24 = 24,$ true

Because both equations are satisfied, $(3, -4)$ is the solution and $\{(3, -4)\}$ is the solution set. ∎

✔ CHECK POINT **2** Solve by graphing:
$$y = -2x + 6$$
$$2x - y = -2. \qquad (1, 4) \text{ or } \{(1, 4)\}$$

3 Solve systems of linear equations by substitution.

Eliminating a Variable Using the Substitution Method Finding the solution to a linear system by graphing equations may not be easy to do. For example, a solution of $\left(-\frac{2}{3}, \frac{157}{29}\right)$ would be difficult to "see" as an intersection point on a graph.

Let's consider a method that does not depend on finding a system's solution visually: the substitution method. This method involves converting the system to one equation in one variable by an appropriate substitution.

EXAMPLE 3 Solving a System by Substitution

Solve by the substitution method:
$$y = -2x + 4$$
$$7x - 2y = 3.$$

SOLUTION

Step 1. **Solve either of the equations for one variable in terms of the other.** This step has already been done for us. The first equation, $y = -2x + 4$, has y solved in terms of x.

Step 2. **Substitute the expression from step 1 into the other equation.** We substitute the expression $-2x + 4$ for y in the other equation:

$$y = \boxed{-2x + 4} \qquad 7x - 2\boxed{y} = 3. \quad \text{Substitute } -2x + 4 \text{ for } y.$$

This gives us an equation in one variable, namely
$$7x - 2(-2x + 4) = 3.$$
The variable y has been eliminated.

Step 3. **Solve the resulting equation containing one variable.**

$$7x - 2(-2x + 4) = 3 \qquad \text{This is the equation containing one variable.}$$

$$7x + 4x - 8 = 3 \qquad \text{Apply the distributive property.}$$

$$11x - 8 = 3 \qquad \text{Combine like terms.}$$

$$11x = 11 \qquad \text{Add 8 to both sides.}$$

$$x = 1 \qquad \text{Divide both sides by 11.}$$

Step 4. **Back-substitute the obtained value into one of the original equations.** We now know that the x-coordinate of the solution is 1. To find the y-coordinate, we back-substitute the x-value into either original equation. We will use

$$y = -2x + 4.$$

Substitute 1 for x.

$$y = -2 \cdot 1 + 4 = -2 + 4 = 2$$

With $x = 1$ and $y = 2$, the proposed solution is $(1, 2)$.

Step 5. **Check the proposed solution in both of the system's given equations.** Replace x with 1 and y with 2.

$$y = -2x + 4 \qquad\qquad 7x - 2y = 3$$

$$2 \overset{?}{=} -2 \cdot 1 + 4 \qquad\qquad 7(1) - 2(2) \overset{?}{=} 3$$

$$2 \overset{?}{=} -2 + 4 \qquad\qquad 7 - 4 \overset{?}{=} 3$$

$$2 = 2, \ \text{true} \qquad\qquad 3 = 3, \ \text{true}$$

The pair $(1, 2)$ satisfies both equations. The solution is $(1, 2)$ and the system's solution set is $\{(1, 2)\}$. ■

 CHECK POINT 3 Solve by the substitution method:

$$y = 3x - 7$$

$$5x - 2y = 8. \qquad (6, 11) \text{ or } \{(6, 11)\}$$

Before considering additional examples, let's summarize the steps used in the substitution method.

SOLVING LINEAR SYSTEMS BY SUBSTITUTION

1. Solve either of the equations for one variable in terms of the other. (If one of the equations is already in this form, you can skip this step.)

2. Substitute the expression found in step 1 into the other equation. This will result in an equation in one variable.

3. Solve the equation containing one variable.

4. Back-substitute the value found in step 3 into one of the original equations. Simplify and find the value of the remaining variable.

5. Check the proposed solution in both of the system's given equations.

EXAMPLE 4 Solving a System by Substitution

Solve by the substitution method:

$$5x + 2y = 1$$
$$x - 3y = 7.$$

SOLUTION

Step 1. **Solve either of the equations for one variable in terms of the other.** We begin by isolating one of the variables in either of the equations. By solving for x in the second equation, which has a coefficient of 1, we can avoid fractions.

$x - 3y = 7$	This is the second equation in the given system.
$x = 3y + 7$	Solve for x by adding 3y to both sides.

Step 2. **Substitute the expression from step 1 into the other equation.** We substitute $3y + 7$ for x in the first equation.

$$x = \boxed{3y + 7} \qquad 5\boxed{x} + 2y = 1$$

This gives us an equation in one variable, namely

$$5(3y + 7) + 2y = 1.$$

The variable x has been eliminated.

Step 3. **Solve the resulting equation containing one variable.**

$5(3y + 7) + 2y = 1$	This is the equation containing one variable.
$15y + 35 + 2y = 1$	Apply the distributive property.
$17y + 35 = 1$	Combine like terms.
$17y = -34$	Subtract 35 from both sides.
$y = -2$	Divide both sides by 17.

Step 4. **Back-substitute the obtained value into one of the original equations.** We back-substitute -2 for y into one of the original equations to find x. Let's use both equations to show that we obtain the same value for x in either case.

Using the first equation:	**Using the second equation:**
$5x + 2y = 1$	$x - 3y = 7$
$5x + 2(-2) = 1$	$x - 3(-2) = 7$
$5x - 4 = 1$	$x + 6 = 7$
$5x = 5$	$x = 1$
$x = 1$	

With $x = 1$ and $y = -2$, the proposed solution is $(1, -2)$.

Step 5. **Check.** Take a moment to show that $(1, -2)$ satisfies both given equations. The solution is $(1, -2)$ and the solution set is $\{(1, -2)\}$. ∎

 CHECK POINT 4 Solve by the substitution method:

$$3x + 2y = 4$$
$$2x + y = 1. \quad (-2, 5) \text{ or } \{(-2, 5)\}$$

4 Solve systems of linear equations by addition.

Eliminating a Variable Using the Addition Method The substitution method is most useful if one of the given equations has an isolated variable. A third method for solving a linear system is the addition method. Like the substitution method, the addition method involves eliminating a variable and ultimately solving an equation containing only one variable. However, this time we eliminate a variable by adding the equations.

For example, consider the following equations:

$$3x - 4y = 11$$
$$-3x + 2y = -7.$$

When we add these two equations, the x-terms are eliminated. This occurs because the coefficients of the x-terms, 3 and -3, are opposites (additive inverses) of each other:

$$
\begin{array}{rl}
3x - 4y = & 11 \\
-3x + 2y = & -7 \\
\hline
\text{Add: } 0x - 2y = & 4 \\
-2y = & 4 \\
y = & -2.
\end{array}
$$

The sum is an equation in one variable.

Divide both sides by -2 and solve for y.

Now we can back-substitute -2 for y into one of the original equations to find x. It does not matter which equation you use; you will obtain the same value for x in either case. If we use either equation, we can show that $x = 1$ and the solution $(1, -2)$ satisfies both equations in the system.

When we use the addition method, we want to obtain two equations whose sum is an equation containing only one variable. The key step is to **obtain, for one of the variables, coefficients that differ only in sign**. To do this, we may need to multiply one or both equations by some nonzero number so that the coefficients of one of the variables, x or y, become opposites. Then when the two equations are added, this variable is eliminated.

EXAMPLE 5 Solving a System by the Addition Method

Solve by the addition method:

$$3x + 4y = -10$$
$$5x - 2y = 18.$$

STUDY TIP

Although the addition method is also known as the elimination method, variables are eliminated when using both the substitution and addition methods. The name *addition method* specifically tells us that the elimination of a variable is accomplished by adding two equations.

SOLUTION We must rewrite one or both equations in equivalent forms so that the coefficients of the same variable (either x or y) are opposites of each other. Consider the terms in y in each equation, that is, $4y$ and $-2y$. To eliminate y, we can multiply each term of the second equation by 2 and then add equations.

$$
\begin{array}{rl}
3x + 4y = -10 & \xrightarrow{\text{No change}} \quad 3x + 4y = -10 \\
5x - 2y = 18 & \xrightarrow{\text{Multiply by 2.}} \quad 10x - 4y = 36 \\
\hline
& \text{Add: } 13x + 0y = 26 \\
& \quad\quad\quad 13x = 26 \\
& \quad\quad\quad\quad x = 2
\end{array}
$$

Divide both sides by 13 and solve for x.

Thus, $x = 2$. We back-substitute this value into either one of the given equations. We'll use the first one.

$$3x + 4y = -10 \qquad \text{This is the first equation in the given system.}$$

$$3(2) + 4y = -10 \qquad \text{Substitute 2 for x.}$$

$$6 + 4y = -10 \qquad \text{Multiply.}$$

$$4y = -16 \qquad \text{Subtract 6 from both sides.}$$

$$y = -4 \qquad \text{Divide both sides by 4.}$$

We see that $x = 2$ and $y = -4$. The ordered pair $(2, -4)$ can be shown to satisfy both equations in the system. Consequently, the solution is $(2, -4)$ and the solution set is $\{(2, -4)\}$.

SOLVING LINEAR SYSTEMS BY ADDITION

1. If necessary, rewrite both equations in the form $Ax + By = C$.
2. If necessary, multiply either equation or both equations by appropriate nonzero numbers so that the sum of the x-coefficients or the sum of the y-coefficients is 0.
3. Add the equations in step 2. The sum is an equation in one variable.
4. Solve the equation in one variable.
5. Back-substitute the value obtained in step 4 into either of the given equations and solve for the other variable.
6. Check the solution in both of the original equations.

 CHECK POINT **5** Solve by the addition method:

$$4x - 7y = -16$$
$$2x + 5y = \quad 9. \qquad \left(-\tfrac{1}{2}, 2\right) \quad \text{or} \quad \left\{\left(-\tfrac{1}{2}, 2\right)\right\}$$

EXAMPLE 6   Solving a System by the Addition Method

Solve by the addition method:

$$7x = 5 - 2y$$
$$3y = 16 - 2x.$$

SOLUTION

Step 1.   **Rewrite both equations in the form $Ax + By = C$.** We first arrange the system so that variable terms appear on the left and constants appear on the right. We obtain

$$7x + 2y = \quad 5 \qquad \text{Add 2y to both sides of the first equation.}$$
$$2x + 3y = 16. \qquad \text{Add 2x to both sides of the second equation.}$$

Step 2. **If necessary, multiply either equation or both equations by appropriate numbers so that the sum of the *x*-coefficients or the sum of the *y*-coefficients is 0.** We can eliminate *x* or *y*. Let's eliminate *y* by multiplying the first equation by 3 and the second equation by −2.

$$7x + 2y = 5 \xrightarrow{\text{Multiply by 3.}} 21x + 6y = 15$$
$$2x + 3y = 16 \xrightarrow{\text{Multiply by } -2.} -4x - 6y = -32$$

Step 3. **Add the equations.**

$$\text{Add: } 17x + 0y = -17$$
$$17x = -17$$

Step 4. **Solve the equation in one variable.** We solve $17x = -17$ by dividing both sides by 17.

$$\frac{17x}{17} = \frac{-17}{17} \qquad \text{Divide both sides by 17.}$$

$$x = -1 \qquad \text{Simplify.}$$

Step 5. **Back-substitute and find the value for the other variable.** We can back-substitute −1 for *x* into either one of the given equations. We'll use the second one.

$$3y = 16 - 2x \qquad \text{This is the second equation in the given system.}$$

$$3y = 16 - 2(-1) \qquad \text{Substitute } -1 \text{ for x.}$$

$$3y = 16 + 2 \qquad \text{Multiply.}$$

$$3y = 18 \qquad \text{Add.}$$

$$y = 6 \qquad \text{Divide both sides by 3.}$$

With $x = -1$ and $y = 6$, the proposed solution is $(-1, 6)$.

Step 6. **Check.** Take a moment to show that $(-1, 6)$ satisfies both given equations. The solution is $(-1, 6)$ and the solution set is $\{(-1, 6)\}$. ■

 CHECK POINT 6 Solve by the addition method:

$$3x = 2 - 4y$$
$$5y = -1 - 2x. \qquad (2, -1) \text{ or } \{(2, -1)\}$$

Some linear systems have solutions that are not integers. If the value of one variable turns out to be a "messy" fraction, back-substitution might lead to cumbersome arithmetic. If this happens, you can return to the original system and use addition to find the value of the other variable.

EXAMPLE 7 Solving a System by the Addition Method

Solve by the addition method:

$$\frac{x}{2} - 5y = 32$$

$$\frac{3x}{2} - 7y = 45.$$

SOLUTION

Step 1. **Rewrite both equations in the form $Ax + By = C$.** Although each equation is already in this form, the coefficients of *x* are not integers. There is less chance for error if the coefficients for x and y in $Ax + By = C$ are integers. Consequently, we begin by clearing fractions. Multiply both sides of each equation by 2.

$$\frac{x}{2} - 5y = 32 \xrightarrow{\text{Multiply by 2.}} x - 10y = 64$$

$$\frac{3x}{2} - 7y = 45 \xrightarrow{\text{Multiply by 2.}} 3x - 14y = 90$$

Step 2. **If necessary, multiply either equation or both equations by appropriate numbers so that the sum of the x-coefficients or the sum of the y-coefficients is 0.** We will eliminate x. Multiply the first equation with integral coefficients by -3 and leave the second equation unchanged.

$$x - 10y = 64 \xrightarrow{\text{Multiply by } -3.} -3x + 30y = -192$$
$$3x - 14y = 90 \xrightarrow{\text{No change}} \underline{3x - 14y = 90}$$

Step 3. **Add the equations.**
$$\text{Add:} \quad 0x + 16y = -102$$
$$16y = -102$$

Step 4. **Solve the equation in one variable.** We solve $16y = -102$ by dividing both sides by 16.

$$\frac{16y}{16} = \frac{-102}{16} \qquad \text{Divide both sides by 16.}$$

$$y = -\frac{102}{16} = -\frac{51}{8} \qquad \text{Simplify.}$$

Step 5. **Back-substitute and find the value of the other variable.** Back-substitution of $-\frac{51}{8}$ for y into either of the given equations results in cumbersome arithmetic. Instead, let's use the addition method on the system with integral coefficients to find the value of x. Thus, we eliminate y by multiplying the first equation by -7 and the second equation by 5.

$$x - 10y = 64 \xrightarrow{\text{Multiply by } -7.} -7x + 70y = -448$$
$$3x - 14y = 90 \xrightarrow{\text{Multiply by 5.}} \underline{15x - 70y = 450}$$
$$\text{Add:} \quad 8x = 2$$
$$x = \frac{2}{8} = \frac{1}{4}$$

With $x = \frac{1}{4}$ and $y = -\frac{51}{8}$, the proposed solution is $\left(\frac{1}{4}, -\frac{51}{8}\right)$.

Step 6. **Check.** For this system, a calculator is helpful in showing that $\left(\frac{1}{4}, -\frac{51}{8}\right)$ satisfies both of the original equations of the system. The solution is $\left(\frac{1}{4}, -\frac{51}{8}\right)$ and the solution set is $\left\{\left(\frac{1}{4}, -\frac{51}{8}\right)\right\}$. ∎

 CHECK POINT 7 Solve by the addition method:

$$\frac{3x}{2} - 2y = \frac{5}{2}$$

$$x - \frac{5y}{2} = -\frac{3}{2}. \qquad \left(\frac{37}{7}, \frac{19}{7}\right) \text{ or } \left\{\left(\frac{37}{7}, \frac{19}{7}\right)\right\}$$

5 Select the most efficient method for solving a system of linear equations.

Comparing the Three Solution Methods The following chart compares the graphing, substitution, and addition methods for solving systems of linear equations in two variables. With increased practice, you will find it easier to select the best method for solving a particular linear system.

Comparing Solution Methods

Method	Advantages	Disadvantages
Graphing	You can see the solutions.	If the solutions do not involve integers or are too large to be seen on the graph, it's impossible to tell exactly what the solutions are.
Substitution	Gives exact solutions. Easy to use if a variable is on one side by itself.	Solutions cannot be seen. Introduces extensive work with fractions when no variable has a coefficient of 1 or -1.
Addition	Gives exact solutions. Easy to use if no variable has a coefficient of 1 or -1.	Solutions cannot be seen.

6 Identify systems that do not have exactly one ordered-pair solution.

Linear Systems Having No Solution or Infinitely Many Solutions We have seen that a system of linear equations in two variables represents a pair of lines. The lines either intersect at one point, are parallel, or are identical. Thus, there are three possibilities for the number of solutions to a system of two linear equations.

THE NUMBER OF SOLUTIONS TO A SYSTEM OF TWO LINEAR EQUATIONS
The number of solutions to a system of two linear equations in two variables is given by one of the following. (See Figure 3.3.)

Number of Solutions	What This Means Graphically
Exactly one ordered-pair solution	The two lines intersect at one point.
No solution	The two lines are parallel.
Infinitely many solutions	The two lines are identical.

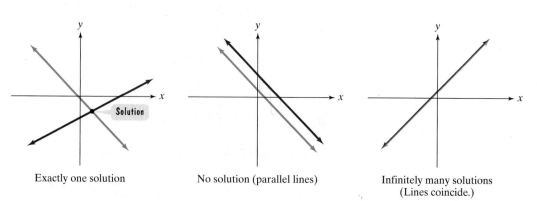

Exactly one solution | No solution (parallel lines) | Infinitely many solutions (Lines coincide.)

FIGURE 3.3 Possible graphs for a system of two linear equations in two variables

A linear system with no solution is called an **inconsistent system**. If you attempt to solve such a system by substitution or addition, you will eliminate both variables. A false statement, such as $0 = 6$, will be the result.

EXAMPLE 8   A System with No Solution

Solve the system:

$$3x - 2y = 6$$
$$6x - 4y = 18.$$

SOLUTION   Because no variable is isolated, we will use the addition method. To obtain coefficients of x that differ only in sign, we multiply the first equation by -2.

$$
\begin{array}{llll}
3x - 2y = 6 & \xrightarrow{\text{Multiply by } -2.} & -6x + 4y & = -12 \\
6x - 4y = 18 & \xrightarrow{\text{No change}} & 6x - 4y & = 18 \\
& & \text{Add:} \quad 0 & = 6
\end{array}
$$

> There are no values of x and y for which $0 = 6$.

The false statement $0 = 6$ indicates that the system is inconsistent and has no solution. The solution set is the empty set, $\varnothing$. ∎

The lines corresponding to the two equations in Example 8 are shown in Figure 3.4. The lines are parallel and have no point of intersection.

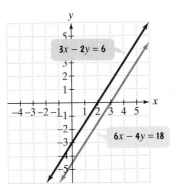

$3x - 2y = 6$

$6x - 4y = 18$

FIGURE 3.4   The graph of an inconsistent system

DISCOVER FOR YOURSELF

Show that the graphs of $3x - 2y = 6$ and $6x - 4y = 18$ must be parallel lines by solving each equation for y. What is the slope and y-intercept for each line? What does this mean? If a linear system is inconsistent, what must be true about the slopes and y-intercepts for the system's graphs?

✔ **CHECK POINT 8**   Solve the system:

$$5x - 2y = 4$$
$$-10x + 4y = 7 \qquad \text{no solution or } \varnothing$$

A linear system that has at least one solution is called a **consistent system**. Lines that intersect and lines that coincide both represent consistent systems. If the lines coincide, then the consistent system has infinitely many solutions, represented by every point on the coinciding lines.

The equations in a linear system with infinitely many solutions are called **dependent**. If you attempt to solve such a system by substitution or addition, you will eliminate both variables. However, a true statement, such as $10 = 10$, will be the result.

EXAMPLE 9   A System with Infinitely Many Solutions

Solve the system:

$$y = 3x - 2$$
$$15x - 5y = 10.$$

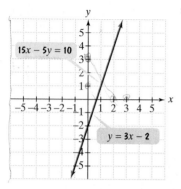

FIGURE 3.5 The graph of a system with infinitely many solutions

SOLUTION Because the variable y is isolated in $y = 3x - 2$, the first equation, we can use the substitution method. We substitute the expression for y into the second equation.

$$y = \boxed{3x - 2} \qquad 15x - 5\boxed{y} = 10 \quad \text{Substitute } 3x - 2 \text{ for } y.$$

$$15x - 5(3x - 2) = 10 \quad \text{The substitution results in an equation in one variable.}$$

$$15x - 15x + 10 = 10 \quad \text{Apply the distributive property.}$$

This statement is true for all values of x and y. $\quad 10 = 10 \quad$ Simplify.

In our final step, both variables have been eliminated and the resulting statement, $10 = 10$, is true. This true statement indicates that the system has infinitely many solutions. The solution set consists of all points (x, y) lying on either of the coinciding lines, $y = 3x - 2$ or $15x - 5y = 10$, as shown in Figure 3.5.

We express the solution set for the system in one of two equivalent ways:

$$\{(x, y) \mid y = 3x - 2\} \qquad \text{or} \qquad \{(x, y) \mid 15x - 5y = 10\}.$$

The set of all ordered pairs (x, y) such that $y = 3x - 2$ *The set of all ordered pairs (x, y) such that $15x - 5y = 10$*

✓ **CHECK POINT 9** Solve the system:

$$x = 4y - 8$$
$$5x - 20y = -40.$$

$\{(x, y) \mid x = 4y - 8\}$ or $\{(x, y) \mid 5x - 20y = -40\}$

3.1 EXERCISE SET

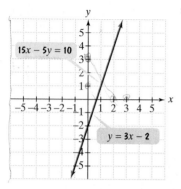 Student Solutions Manual   CD/Video   PH Math/Tutor Center   MathXL Tutorials on CD   MathXL®   MyMathLab   Interactmath.com

Practice Exercises

In Exercises 1–6, determine whether the given ordered pair is a solution of the system.

1. $(7, -5)$
$x - y = 12$
$x + y = 2$ solution

2. $(-3, 1)$
$x - y = -4$
$2x + 10y = 4$ solution

3. $(2, -1)$
$3x + 4y = 2$
$2x + 5y = 1$
not a solution

4. $(4, 2)$
$2x - 5y = -2$
$3x + 4y = 18$
not a solution

5. $(5, -3)$
$y = 2x - 13$
$4x + 9y = -7$ solution

6. $(-3, -4)$
$y = 3x + 5$
$5x - 2y = -7$ solution

In Exercises 7–24, solve each system by graphing. Identify systems with no solution and systems with infinitely many solutions, using set notation to express their solution sets.

7. $x + y = 4$
$x - y = 2$ $(3, 1)$

8. $x + y = 6$
$x - y = -4$ $(1, 5)$

9. $2x + y = 4$
$y = 4x + 1$ $\left(\frac{1}{2}, 3\right)$

10. $x + 2y = 4$
$y = -2x - 1$ $(-2, 3)$

11. $3x - 2y = 6$
$x - 4y = -8$ $(4, 3)$

12. $4x + y = 4$
$3x - y = 3$ $(1, 0)$

13. $2x + 3y = 6$
$4x = -6y + 12$
$\{(x, y) \mid 2x + 3y = 6\}$ or $\{(x, y) \mid 4x = -6y + 12\}$

14. $3x - 3y = 6$
$2x = 2y + 4$
$\{(x, y) \mid 3x - 3y = 6\}$ or $\{(x, y) \mid 2x = 2y + 4\}$

15. $y = 2x - 2$
$y = -5x + 5$ $(1, 0)$

16. $y = -x + 1$
$y = 3x + 5$ $(-1, 2)$

17. $3x - y = 4$
$6x - 2y = 4$ $\varnothing$

18. $2x - y = -4$
$4x - 2y = 6$ $\varnothing$

19. $2x + y = 4$
$4x + 3y = 10$ $(1, 2)$

20. $4x - y = 9$
$x - 3y = 16$ $(1, -5)$

21. $x - y = 2$
$\quad\quad y = 1 \quad (3, 1)$

22. $x + 2y = 1$
$\quad\quad\quad x = 3 \quad (3, -1)$

23. $3x + y = 3$
$\quad 6x + 2y = 12 \quad \varnothing$

24. $2x - 3y = 6$
$\quad 4x - 6y = 24 \quad \varnothing$

In Exercises 25–42, solve each system by the substitution method. Identify inconsistent systems and systems with dependent equations, using set notation to express their solution sets.

25. $x + y = 6$
$\quad\quad y = 2x \quad (2, 4)$

26. $x + y = 10$
$\quad\quad y = 4x \quad\quad (2, 8)$

27. $2x + 3y = 9$
$\quad\quad x = y + 2 \quad (3, 1)$

28. $3x - 4y = 18$
$\quad\quad y = 1 - 2x \quad (2, -3)$

29. $y = -3x + 7$
$\quad 5x - 2y = 8 \quad (2, 1)$

30. $x = 3y + 8$
$\quad 2x - y = 6 \quad (2, -2)$

31. $4x + y = 5$
$\quad 2x - 3y = 13 \quad (2, -3)$

32. $x - 3y = 3$
$\quad 3x + 5y = -19 \quad (-3, -2)$

33. $x - 2y = 4$
$\quad 2x - 4y = 5 \quad \varnothing$

34. $x - 3y = 6$
$\quad 2x - 6y = 5 \quad \varnothing$

35. $2x + 5y = -4$
$\quad 3x - y = 11 \quad (3, -2)$

36. $2x + 5y = 1$
$\quad -x + 6y = 8 \quad (-2, 1)$

37. $2(x - 1) - y = -3$
$\quad\quad y = 2x + 3 \quad\quad \varnothing$

38. $x + y - 1 = 2(y - x)$
$\quad\quad y = 3x - 1$
$\{(x, y) | x + y - 1 = 2(y - x)\}$ or $\{(x, y) | y = 3x - 1\}$

39. $\dfrac{x}{4} - \dfrac{y}{4} = -1$
$\quad x + 4y = -9 \quad (-5, -1)$

40. $\dfrac{x}{6} - \dfrac{y}{2} = \dfrac{1}{3}$
$\quad x + 2y = -3 \quad (-1, -1)$

41. $y = \dfrac{2}{5}x - 2$
$\quad 2x - 5y = 10$
$\left\{(x, y) | y = \dfrac{2}{5}x - 2\right\}$ or $\{(x, y) | 2x - 5y = 10\}$

42. $y = \dfrac{1}{3}x + 4$
$\quad 3y = x + 12$
$\left\{(x, y) | y = \dfrac{1}{3}x + 4\right\}$ or $\{(x, y) | 3y = x + 12\}$

In Exercises 43–58, solve each system by the addition method. Identify inconsistent systems and systems with dependent equations, using set notation to express their solution sets.

43. $x + y = 7$
$\quad x - y = 3 \quad (5, 2)$

44. $2x + y = 3$
$\quad x - y = 3 \quad (2, -1)$

45. $12x + 3y = 15$
$\quad 2x - 3y = 13 \quad (2, -3)$

46. $4x + 2y = 12$
$\quad 3x - 2y = 16 \quad (4, -2)$

47. $x + 3y = 2$
$\quad 4x + 5y = 1 \quad (-1, 1)$

48. $x + 2y = -1$
$\quad 2x - y = 3 \quad (1, -1)$

49. $6x - y = -5$
$\quad 4x - 2y = 6 \quad (-2, -7)$

50. $x - 2y = 5$
$\quad 5x - y = -2 \quad (-1, -3)$

51. $3x - 5y = 11$
$\quad 2x - 6y = 2 \quad (7, 2)$

52. $4x - 3y = 12$
$\quad 3x - 4y = 2 \quad (6, 4)$

53. $2x - 5y = 13$
$\quad 5x + 3y = 17 \quad (4, -1)$

54. $4x + 5y = -9$
$\quad 6x - 3y = -3 \quad (-1, -1)$

55. $2x + 6y = 8$
$\quad 3x + 9y = 12$
$\{(x, y) | 2x + 6y = 8\}$ or $\{(x, y) | 3x + 9y = 12\}$

56. $x - 3y = -6$
$\quad 3x - 9y = 9 \quad \varnothing$

57. $2x - 3y = 4$
$\quad 4x + 5y = 3 \quad \left(\dfrac{29}{22}, -\dfrac{5}{11}\right)$

58. $4x - 3y = 8$
$\quad 2x - 5y = -14 \quad \left(\dfrac{41}{7}, \dfrac{36}{7}\right)$

In Exercises 59–82, solve each system by the method of your choice. Identify inconsistent systems and systems with dependent equations, using set notation to express solution sets.

59. $3x - 7y = 1$
$\quad 2x - 3y = -1 \quad (-2, -1)$

60. $2x - 3y = 2$
$\quad 5x + 4y = 51 \quad (7, 4)$

61. $x = y + 4$
$\quad 3x + 7y = -18 \quad (1, -3)$

62. $y = 3x + 5$
$\quad 5x - 2y = -7 \quad (-3, -4)$

63. $9x + \dfrac{4y}{3} = 5$
$\quad 4x - \dfrac{y}{3} = 5 \quad (1, -3)$

64. $\dfrac{x}{6} - \dfrac{y}{5} = -4$
$\quad \dfrac{x}{4} - \dfrac{y}{6} = -2 \quad (12, 30)$

65. $\dfrac{1}{4}x - \dfrac{1}{9}y = \dfrac{2}{3}$
$\quad \dfrac{1}{2}x - \dfrac{1}{3}y = 1 \quad (4, 3)$

66. $\dfrac{1}{16}x - \dfrac{3}{4}y = -1$
$\quad \dfrac{3}{4}x + \dfrac{5}{2}y = 11 \quad (8, 2)$

67. $x = 3y - 1$
$\quad 2x - 6y = -2$
$\{(x, y) | x = 3y - 1\}$ or $\{(x, y) | 2x - 6y = -2\}$

68. $x = 4y - 1$
$\quad 2x - 8y = -2$
$\{(x, y) | x = 4y - 1\}$ or $\{(x, y) | 2x - 8y = -2\}$

69. $y = 2x + 1$
$\quad y = 2x - 3 \quad \varnothing$

70. $y = 2x + 4$
$\quad y = 2x - 1 \quad \varnothing$

71. $0.4x + 0.3y = 2.3$
$\quad 0.2x - 0.5y = 0.5 \quad (5, 1)$

72. $0.2x - y = -1.4$
$\quad 0.7x - 0.2y = -1.6 \quad (-2, 1)$

73. $5x - 40 = 6y$
$\quad 2y = 8 - 3x \quad \left(\dfrac{32}{7}, -\dfrac{20}{7}\right)$

74. $4x - 24 = 3y$
$\quad 9y = 3x - 1 \quad \left(\dfrac{71}{9}, \dfrac{68}{27}\right)$

75. $3(x + y) = 6$
$\quad 3(x - y) = -36 \quad (-5, 7)$

76. $4(x - y) = -12$
$\quad 4(x + y) = -20 \quad (-4, -1)$

77. $3(x - 3) - 2y = 0$
$\quad 2(x - y) = -x - 3 \quad \varnothing$

78. $5x + 2y = -5$
$\quad 4(x + y) = 6(2 - x) \quad \varnothing$

79. $x + 2y - 3 = 0$
$\quad 12 = 8y + 4x$
$\{(x, y) | x + 2y - 3 = 0\}$ or $\{(x, y) | 12 = 8y + 4x\}$

80. $2x - y - 5 = 0$
$10 = 4x - 2y$
$\{(x, y) | 2x - y - 5 = 0\}$ or $\{(x, y) | 10 = 4x - 2y\}$

81. $3x + 4y = 0$
$7x = 3y$ $(0, 0)$

82. $5x + 8y = 20$
$4y = -5x$ $(-4, 5)$

Practice Plus

In Exercises 83–84, solve each system by the method of your choice.

83. $\dfrac{x + 2}{2} - \dfrac{y + 4}{3} = 3$

$\dfrac{x + y}{5} = \dfrac{x - y}{2} - \dfrac{5}{2}$ $(6, -1)$

84. $\dfrac{x - y}{3} = \dfrac{x + y}{2} - \dfrac{1}{2}$

$\dfrac{x + 2}{2} - 4 = \dfrac{y + 4}{3}$ $(8, -1)$

In Exercises 85–86, solve each system for x and y, expressing either value in terms of a or b, if necessary. Assume that a ≠ 0 and b ≠ 0.

85. $5ax + 4y = 17$
$ax + 7y = 22$ $(\frac{1}{a}, 3)$

86. $4ax + by = 3$
$6ax + 5by = 8$ $(\frac{1}{2a}, \frac{1}{b})$

87. For the linear function $f(x) = mx + b$, $f(-2) = 11$ and $f(3) = -9$. Find m and b. $m = -4, b = 3$

88. For the linear function $f(x) = mx + b$, $f(-3) = 23$ and $f(2) = -7$. Find m and b. $m = -6, b = 5$

Use the graphs of the linear functions to solve Exercises 89–90.

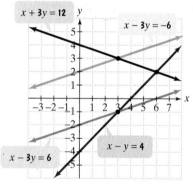

89. Write the linear system whose solution set is $\{(6, 2)\}$. Express each equation in the system in slope-intercept form.
$y = x - 4, \ y = -\frac{1}{3}x + 4$

90. Write the linear system whose solution set is ∅. Express each equation in the system in slope-intercept form.
$y = \frac{1}{3}x + 2, \ y = \frac{1}{3}x - 2$

Application Exercises

91. Life expectancies for U.S. women have been longer than life expectancies for U.S. men for many years. However, the bar graph at the top of the next column shows that life expectancies for men are increasing at a faster rate than those for women.

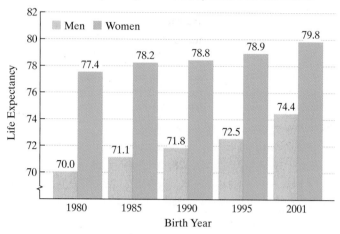

Life Expectancy in the U.S., by Birth Year and Gender

Source: National Center for Health Statistics

Life expectancy for men can be modeled by $E = 0.20x + 70.0$ and for women by $E = 0.11x + 77.5$. In both models, x is the number of birth years after 1980 and E is life expectancy, in years. Use these models to project when life expectancies for women and men will be the same. Round to the nearest year. What are the life expectancies for women and men at that time? Do you feel confident about this prediction? Explain why or why not. 2063; 87 years; Answers will vary.

92. The formula $y = 0.52x + 35.7$ models the percentage, y, of U.S. women in the work force x years after 1955. The formula $0.25x + y = 85.4$ models the percentage, y, of U.S. men in the work force x years after 1955. Use these models to determine when the percentage of women in the work force will be the same as the percentage of men in the work force. Round to the nearest year. What percentage of women and what percentage of men will be in the work force at that time? 2020; about 69.3%

The bar graph shows the percentage of Americans who used cigarettes, by ethnicity, in 1985 and 2002. For each of the groups shown, cigarette use has been linearly decreasing. Use this information to solve Exercises 93–94.

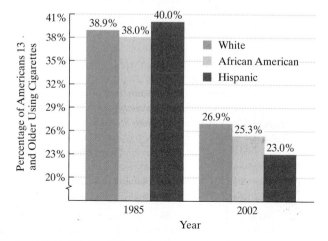

Cigarette Use in the U.S.

Source: Dept. of Health and Human Services

93. In this exercise, let x represent the number of years after 1985 and let y represent the percentage of Americans in one of the groups shown who used cigarettes.

a. Use the data points $(0, 38)$ and $(17, 25.3)$ to find the slope-intercept equation of the line that models the percentage of African Americans who used cigarettes, y, x years after 1985. Round the value of m to two decimal places. $y = -0.75x + 38$

b. Use the data points $(0, 40)$ and $(17, 23)$ to find the slope-intercept equation of the line that models the percentage of Hispanics who used cigarettes, y, x years after 1985. $y = -x + 40$

c. Use the models from parts (a) and (b) to find the year during which cigarette use was the same for African Americans and Hispanics. What percentage of each group used cigarettes during that year?   1993; 32%

94. In this exercise, let x represent the number of years after 1985 and let y represent the percentage of Americans in one of the groups shown who used cigarettes.

a. Use the data points $(0, 38.9)$ and $(17, 26.9)$ to find the slope-intercept equation of the line that models the percentage of whites who used cigarettes, y, x years after 1985. Round the value of m to two decimal places. $y = -0.71x + 38.9$

b. Use the data points $(0, 40)$ and $(17, 23)$ to find the slope-intercept equation of the line that models the percentage of Hispanics who used cigarettes, y, x years after 1985. $y = -x + 40$

c. Use the models from parts (a) and (b) to find the year, to the nearest whole year, during which cigarette use was the same for whites and Hispanics. What percentage of each group, to the nearest percent, used cigarettes during that year?   1989; 36%

An important application of systems of equations arises in connection with supply and demand. As the price of a product increases, the demand for that product decreases. However, at higher prices, suppliers are willing to produce greater quantities of the product. Exercises 95–96 involve supply and demand.

95. A chain of electronics stores sells hand-held color televisions. The weekly demand and supply models are given as follows:

Number sold per week → **Demand model** $N = -5p + 750$

Price of television

Number supplied to the chain per week → $N = 2.5p.$ **Supply model**

a. How many hand-held color televisions can be sold and supplied at $120 per television?   150 sold; 300 supplied

b. Find the price at which supply and demand are equal. At this price, how many televisions can be supplied and sold each week? $100; 250

96. At a price of p dollars per ticket, the number of tickets to a rock concert that can be sold is given by the demand model $N = -25p + 7800$. At a price of p dollars per ticket, the number of tickets that the concert's promoters are willing to make available is given by the supply model $N = 5p + 6000$.

a. How many tickets can be sold and supplied for $50 per ticket?   6550 sold; 6250 supplied

b. Find the ticket price at which supply and demand are equal. At this price, how many tickets will be supplied and sold? $60; 6300 tickets

Writing in Mathematics

97. What is a system of linear equations? Provide an example with your description.   97–105. Answers will vary.

98. What is a solution of a system of linear equations?

99. Explain how to determine if an ordered pair is a solution of a system of linear equations.

100. Explain how to solve a system of linear equations by graphing.

101. Explain how to solve a system of equations using the substitution method. Use $y = 3 - 3x$ and $3x + 4y = 6$ to illustrate your explanation.

102. Explain how to solve a system of equations using the addition method. Use $5x + 8y = -1$ and $3x + y = 7$ to illustrate your explanation.

103. When is it easier to use the addition method rather than the substitution method to solve a system of equations?

104. When using the addition or substitution method, how can you tell if a system of linear equations has no solution? What is the relationship between the graphs of the two equations?

105. When using the addition or substitution method, how can you tell if a system of linear equations has infinitely many solutions? What is the relationship between the graphs of the two equations?

106. The formula $y = 0.94x + 5.64$ models annual U.S. consumption of chicken, y, in pounds per person, x years after 1950. The formula $0.74x + y = 146.76$ models annual U.S. consumption of red meat, y, in pounds per person, x years after 1950. What is the most efficient method for solving this system? What does the solution mean in terms of the variables in the formulas? (It is not necessary to solve the system.)   substitution; The x-coordinate represents the number of years after 1950 that per capita consumption of chicken and red meat will be the same; the y-coordinate represents the per capita consumption of each, in pounds, at that time.

107. In Exercise 106, find the slope of each model. Describe what this means in terms of the rate of change of chicken consumption and the rate of change of red meat consumption. Why must the graphs have an intersection point? What happens to the right of the intersection point?   chicken: 0.94; red meat: -0.74; Chicken consumption is increasing; red meat consumption is decreasing; The lines are not parallel; Chicken consumption is greater than red meat consumption.

Technology Exercises

108. Verify your solutions to any five exercises from Exercises 7–24 by using a graphing utility to graph the two equations in the system in the same viewing rectangle. Then use the intersection feature to display the solution. Answers will vary.

Critical Thinking Exercises

109. Which one of the following is true? d

    a. The addition method cannot be used to eliminate either variable in a system of two equations in two variables.

    b. The solution set of the system
$$5x - y = 1$$
$$10x - 2y = 2$$
    is $\{(2, 9)\}$.

    c. A system of linear equations can have a solution set consisting of precisely two ordered pairs.

    d. The solution set of the system
$$y = 4x - 3$$
$$y = 4x + 5$$
    is the empty set.

110. Determine a and b so that $(2, 1)$ is a solution of this system:
$$ax - by = 4$$
$$bx + ay = 7. \qquad a = 3, b = 2$$

111. Write a system of equations having $\{(-2, 7)\}$ as a solution set. (More than one system is possible.) Answers will vary.

112. Solve the system for x and y in terms of $a_1, b_1, c_1, a_2, b_2,$ and c_2:
$$a_1 x + b_1 y = c_1$$
$$a_2 x + b_2 y = c_2.$$
$$\left\{ \left(\frac{b_2 c_1 - b_1 c_2}{a_1 b_2 - a_2 b_1}, \frac{a_1 c_2 - a_2 c_1}{a_1 b_2 - a_2 b_1} \right) \right\}$$

Review Exercises

113. Solve: $6x = 10 + 5(3x - 4)$. (Section 1.4, Example 3) $\frac{10}{9}$

114. Simplify: $(4x^2 y^4)^2 (-2x^5 y^0)^3$. (Section 1.6, Example 9) $-128x^{19} y^8$

115. If $f(x) = x^2 - 3x + 7$, find $f(-1)$. (Section 2.1, Example 3) 11

SECTION

3.2

Objectives

1 Solve problems using systems of equations.

2 Use functions to model revenue, cost, and profit.

3 Perform a break-even analysis.

PROBLEM SOLVING AND BUSINESS APPLICATIONS USING SYSTEMS OF EQUATIONS

Driving through your neighborhood, you see kids selling lemonade. Would it surprise you to know that this activity can be analyzed using functions and systems of equations? By doing so, you will view profit and loss in the business world in a new way. In this section, we use systems of equations to solve problems and model business ventures.

1 Solve problems using systems of equations.

A Strategy for Solving Word Problems Using Systems of Equations When we solved problems in Chapter 1, we let x represent a quantity that was unknown. Problems in this section involve two unknown quantities. We will let x and y represent these quantities. We then translate from the verbal conditions of the problem into a *system* of linear equations.

EXAMPLE 1 Solving a Problem Involving Energy Efficiency of Building Materials

A heat-loss survey by an electric company indicated that a wall of a house containing 40 square feet of glass and 60 square feet of plaster lost 1920 Btu (British thermal units) of heat. A second wall containing 10 square feet of glass and 100 square feet of plaster lost 1160 Btu of heat. Determine the heat lost per square foot for the glass and for the plaster.

SOLUTION

Step 1. **Use variables to represent unknown quantities.**
Let x = the heat lost per square foot for the glass.
Let y = the heat lost per square foot for the plaster.

Step 2. **Write a system of equations describing the problem's conditions.** The heat loss for each wall is the heat lost by the glass plus the heat lost by the plaster. One wall containing 40 square feet of glass and 60 square feet of plaster lost 1920 Btu of heat.

| Heat lost by the glass | + | heat lost by the plaster | = | total heat lost. |

| $\left(\begin{array}{c}\text{Number}\\\text{of ft}^2\end{array}\right) \cdot \left(\begin{array}{c}\text{heat lost}\\\text{per ft}^2\end{array}\right)$ | + | $\left(\begin{array}{c}\text{number}\\\text{of ft}^2\end{array}\right) \cdot \left(\begin{array}{c}\text{heat lost}\\\text{per ft}^2\end{array}\right)$ | = | total heat lost. |

$$40 \cdot x + 60 \cdot y = 1920$$

A second wall containing 10 square feet of glass and 100 square feet of plaster lost 1160 Btu of heat.

| $\left(\begin{array}{c}\text{Number}\\\text{of ft}^2\end{array}\right) \cdot \left(\begin{array}{c}\text{heat lost}\\\text{per ft}^2\end{array}\right)$ | + | $\left(\begin{array}{c}\text{number}\\\text{of ft}^2\end{array}\right) \cdot \left(\begin{array}{c}\text{heat lost}\\\text{per ft}^2\end{array}\right)$ | = | total heat lost. |

$$10 \cdot x + 100 \cdot y = 1160$$

Step 3. **Solve the system and answer the problem's question.** The system

$$40x + 60y = 1920$$
$$10x + 100y = 1160$$

can be solved by addition. We'll multiply the second equation by -4 and then add equations to eliminate x.

$$40x + 60y = 1920 \xrightarrow{\text{No change}} 40x + 60y = 1920$$
$$10x + 100y = 1160 \xrightarrow{\text{Multiply by } -4.} -40x - 400y = -4640$$
$$\text{Add:} \qquad -340y = -2720$$
$$y = \frac{-2720}{-340} = 8$$

Now we can find the value of x by back-substituting 8 for y in either of the system's equations.

$$10x + 100y = 1160 \qquad \text{We'll use the second equation.}$$
$$10x + 100(8) = 1160 \qquad \text{Back-substitute 8 for } y.$$
$$10x + 800 = 1160 \qquad \text{Multiply.}$$
$$10x = 360 \qquad \text{Subtract 800 from both sides.}$$
$$x = 36 \qquad \text{Divide both sides by 10.}$$

We see that $x = 36$ and $y = 8$. Because x represents heat lost per square foot for the glass and y for the plaster, the glass lost 36 Btu of heat per square foot and the plaster lost 8 Btu per square foot.

Step 4. **Check the proposed solution in the original wording of the problem.** The problem states that the wall with 40 square feet of glass and 60 square feet of plaster lost 1920 Btu.

$$40(36) + 60(8) = 1440 + 480 = 1920 \text{ Btu of heat}$$

Proposed solution is 36 Btu per ft² for glass and 8 Btu per ft² for plaster.

Our proposed solution checks with the first statement. The problem also states that the wall with 10 square feet of glass and 100 square feet of plaster lost 1160 Btu.

$$10(36) + 100(8) = 360 + 800 = 1160 \text{ Btu of heat}$$

Our proposed solution also checks with the second statement. ∎

CHECK POINT 1 Two Snickers bars and one Think Thin bar contain 70 grams of carbohydrates. One Snickers bar and two Think Thin bars contain 35 grams of carbohydrates. Find the carbohydrate content in each item.

Snickers: 35 grams; Think Thin: 0 grams

(*Source*: *Newsweek*)

Next, we will solve problems involving investments, mixtures, and motion with systems of equations. We will continue using our four-step problem-solving strategy. We will also use tables to help organize the information in the problems.

Dual Investments with Simple Interest Simple interest involves interest calculated only on the amount of money that we invest, called the **principal**. The formula $I = Pr$ is used to find the simple interest, I, earned for one year when the principal, P, is invested at an annual interest rate, r. Dual investment problems involve different amounts of money in two or more investments, each paying a different rate.

EXAMPLE 2 Solving a Dual Investment Problem

Your grandmother needs your help. She has $50,000 to invest. Part of this money is to be invested in noninsured bonds paying 15% annual interest. The rest of this money is to be invested in a government-insured certificate of deposit paying 7% annual interest. She told you that she requires $6000 per year in extra income from both of these investments. How much money should be placed in each investment?

SOLUTION

Step 1. **Use variables to represent unknown quantities.**
Let x = the amount invested in the 15% noninsured bonds.
Let y = the amount invested in the 7% certificate of deposit.

Step 2. **Write a system of equations describing the problem's conditions.** Because Grandma has $50,000 to invest,

The amount invested at 15%	plus	the amount invested at 7%	equals	$50,000.
x	$+$	y	$=$	50,000

Furthermore, Grandma requires $6000 in total interest. We can use a table to organize the information in the problem and obtain a second equation.

	Principal (amount invested)	×	Interest rate	=	Interest earned
15% Investment	x		0.15		$0.15x$
7% Investment	y		0.07		$0.07y$

The interest for the two investments combined must be $6000.

Interest from the 15% investment	plus	interest from the 7% investment	is	$6000.

$$0.15x \quad + \quad 0.07y \quad = \quad 6000$$

Step 3. **Solve the system and answer the problem's question.** The system

$$x + y = 50{,}000$$
$$0.15x + 0.07y = 6000$$

can be solved by substitution or addition. Substitution works well because both variables in the first equation have coefficients of 1. Addition also works well; if we multiply the first equation by -0.15 or -0.07, adding equations will eliminate a variable. We will use addition.

$$
\begin{array}{ll}
x + y = 50{,}000 & \xrightarrow{\text{Multiply by } -0.07.} \quad -0.07x - 0.07y = -3500 \\
0.15x + 0.07y = 6000 & \xrightarrow{\text{No change}} \quad\;\; 0.15x + 0.07y = \underline{\;\;6000} \\
& \qquad\qquad\text{Add:} \quad 0.08x \qquad\quad = 2500 \\
& \qquad\qquad\qquad\qquad\qquad x = \dfrac{2500}{0.08} \\
& \qquad\qquad\qquad\qquad\qquad x = 31{,}250
\end{array}
$$

Because x represents the amount that should be invested at 15%, Grandma should place $31,250 in 15% noninsured bonds. Now we can find y, the amount that she should place in the 7% certificate of deposit. We do so by back-substituting 31,250 for x in either of the system's equations.

$$x + y = 50{,}000 \qquad \text{We'll use the first equation.}$$
$$31{,}250 + y = 50{,}000 \qquad \text{Back-substitute 31,250 for x.}$$
$$y = 18{,}750 \qquad \text{Subtract 31,250 from both sides.}$$

Because $x = 31{,}250$ and $y = 18{,}750$, Grandma should invest $31,250 at 15% and $18,750 at 7%.

Step 4. **Check the proposed answers in the original wording of the problem.** Has Grandma invested $50,000?

$$\$31{,}250 + \$18{,}750 = \$50{,}000$$

Yes, all her money was placed in the dual investments. Can she count on $6000 interest? The interest earned on $31,250 at 15% is ($31,250)(0.15), or $4687.50. The interest earned on $18,750 at 7% is ($18,750)(0.07), or $1312.50. The total interest is $4687.50 + $1312.50, or $6000, exactly as it should be. You've made your grandmother happy. (Now if you would just visit her more often ...)

> ✓ **CHECK POINT 2** You inherited $5000 with the stipulation that for the first year the money had to be invested in two funds paying 9% and 11% annual interest. How much did you invest at each rate if the total interest earned for the year was $487? $3150 at 9%; $1850 at 11%

Problems Involving Mixtures   Chemists and pharmacists often have to change the concentration of solutions and other mixtures. In these situations, the amount of a particular ingredient in the solution or mixture is expressed as a percentage of the total.

EXAMPLE 3 Solving a Mixture Problem

A chemist working on a flu vaccine needs to mix a 10% sodium-iodine solution with a 60% sodium-iodine solution to obtain 50 milliliters of a 30% sodium-iodine solution. How many milliliters of the 10% solution and of the 60% solution should be mixed?

SOLUTION

Step 1. **Use variables to represent unknown quantities.**
Let x = the number of milliliters of the 10% solution to be used in the mixture.
Let y = the number of milliliters of the 60% solution to be used in the mixture.

Step 2. **Write a system of equations describing the problem's conditions.** The situation is illustrated in Figure 3.6.

| Volume | x | y | 50 |
| Strength | 10% solution | 60% solution | 30% solution |

FIGURE 3.6

The chemist needs 50 milliliters of a 30% sodium-iodine solution. We form a table that shows the amount of sodium-iodine in each of the three solutions.

Solution	Number of milliliters	×	Percent of Sodium-Iodine	=	Amount of Sodium-Iodine
10% Solution	x		10% = 0.1		$0.1x$
60% Solution	y		60% = 0.6		$0.6y$
30% Mixture	50		30% = 0.3		$0.3(50) = 15$

The chemist needs to obtain a 50-milliliter mixture.

$$x \quad + \quad y \quad = \quad 50$$

The 50-milliliter mixture must be 30% sodium-iodine. The amount of sodium-iodine must be 30% of 50, or $(0.3)(50) = 15$ milliliters.

$$0.1x \quad + \quad 0.6y \quad = \quad 15$$

Step 3. **Solve the system and answer the problem's question.** The system

$$x + y = 50$$

$$0.1x + 0.6y = 15$$

can be solved by substitution or addition. Let's use substitution. The first equation can easily be solved for x or y. Solving for y, we obtain $y = 50 - x$.

$$y = \boxed{50 - x} \qquad 0.1x + 0.6\boxed{y} = 15$$

We substitute $50 - x$ for y in the second equation. This gives us an equation in one variable.

$0.1x + 0.6(50 - x) = 15$	This equation contains one variable, x.
$0.1x + 30 - 0.6x = 15$	Apply the distributive property.
$-0.5x + 30 = 15$	Combine like terms.
$-0.5x = -15$	Subtract 30 from both sides.
$x = \dfrac{-15}{-0.5} = 30$	Divide both sides by -0.5.

Back-substituting 30 for x in either of the system's equations ($x + y = 50$ is easier to use) gives $y = 20$. Because x represents the number of milliliters of the 10% solution and y the number of milliliters of the 60% solution, the chemist should mix 30 milliliters of the 10% solution with 20 milliliters of the 60% solution.

Step 4. **Check the proposed solution in the original wording of the problem.** The problem states that the chemist needs 50 milliliters of a 30% sodium-iodine solution. The amount of sodium-iodine in this mixture is 0.3(50), or 15 milliliters. The amount of sodium-iodine in 30 milliliters of the 10% solution is 0.1(30), or 3 milliliters. The amount of sodium-iodine in 20 milliliters of the 60% solution is 0.6(20) = 12 milliliters. The amount of sodium-iodine in the two solutions used in the mixture is 3 milliliters + 12 milliliters, or 15 milliliters, exactly as it should be.

Problems involving dual investments and problems involving mixtures are both based on the same idea: The total amount times the rate gives the amount.

total money amount	rate of interest	amount of interest

Dual Investment Problems: principal · rate = interest

Percents are expressed as decimals in these equations.

Mixture Problems: solution · concentration = ingredient

total solution amount	percent of ingredient, or rate of concentration	amount of ingredient

Our dual investment problem involved mixing two investments. Our mixture problem involved mixing two liquids. The equations in these problems are obtained from similar conditions:

Dual Investment Problems

| Interest from investment 1 | + | interest from investment 2 | = | amount of interest from mixed investments. |

Mixture Problems

| Ingredient amount in solution 1 | + | ingredient amount in solution 2 | = | amount of ingredient in mixture. |

Being aware of the similarities between dual investment and mixture problems should make you a better problem solver in a variety of situations that involve mixtures.

 CHECK POINT 3 A chemist needs to mix a 12% acid solution with a 20% acid solution to obtain 160 ounces of a 15% acid solution. How many ounces of each of the acid solutions must be used? 12% solution: 100 ounces; 20% solution: 60 ounces

Problems Involving Motion We have seen that the distance, d, covered by any moving body is the product of its average rate, r, and its time in motion, t:

$$d = rt \qquad \text{Distance equals rate times time.}$$

Wind and water current have the effect of increasing or decreasing a traveler's rate.

EXAMPLE 4 Solving a Motion Problem

When a small airplane flies with the wind, it can travel 450 miles in 3 hours. When the same airplane flies in the opposite direction against the wind, it takes 5 hours to fly the same distance. Find the average rate of the plane in still air and the average rate of the wind.

SOLUTION

Step 1. **Use variables to represent unknown quantities.**
Let x = the average rate of the plane in still air.
Let y = the average rate of the wind.

Step 2. **Write a system of equations describing the problem's conditions.** As it travels with the wind, the plane's rate is increased. The net rate is its rate in still air, x, plus the rate of the wind, y, given by the expression $x + y$. As it travels against the wind, the plane's rate is decreased. The net rate is its rate in still air, x, minus the rate of the wind, y, given by the expression $x - y$. Here is a chart that summarizes the problem's information and includes the increased and decreased rates.

	Rate	$\times$	Time	$=$	Distance
Trip with the Wind	$x + y$		3		$3(x + y)$
Trip against the Wind	$x - y$		5		$5(x - y)$

The problem states that the distance in each direction is 450 miles. We use this information to write our system of equations.

The distance of the trip with the wind	is	450 miles.

$$3(x + y) \quad = \quad 450$$

The distance of the trip against the wind	is	450 miles.

$$5(x - y) \quad = \quad 450$$

Step 3. **Solve the system and answer the problem's question.** We can simplify the system by dividing both sides of the equations by 3 and 5, respectively.

$$3(x + y) = 450 \xrightarrow[\text{Divide by 5.}]{\text{Divide by 3.}} \begin{array}{l} x + y = 150 \\ x - y = 90 \end{array}$$
$$5(x - y) = 450$$

Solve the system on the right by the addition method.

$$\begin{array}{r} x + y = 150 \\ \underline{x - y = 90} \\ \text{Add:} \quad 2x \phantom{{}- y} = 240 \\ x = 120 \quad \text{Divide both sides by 2.} \end{array}$$

Back-substituting 120 for x in either of the system's equations gives $y = 30$. Because $x = 120$ and $y = 30$, the average rate of the plane in still air is 120 miles per hour and the average rate of the wind is 30 miles per hour.

Step 4. **Check the proposed solution in the original wording of the problem.** The problem states that the distance in each direction is 450 miles. The average rate of the plane with the wind is $120 + 30 = 150$ miles per hour. In 3 hours, it travels $150 \cdot 3$, or 450 miles, which checks with the stated condition. Furthermore, the average rate of the plane against the wind is $120 - 30 = 90$ miles per hour. In 5 hours, it travels $90 \cdot 5 = 450$ miles, which is the stated distance. ∎

CHECK POINT 4 With the current, a motorboat can travel 84 miles in 2 hours. Against the current, the same trip takes 3 hours. Find the average rate of the boat in still water and the average rate of the current.

Boat in still water: 35 miles per hour; Current: 7 miles per hour

2 Use functions to model revenue, cost, and profit.

Functions of Business: Break-Even Analysis   As a young entrepreneur, did you ever try selling lemonade in your front yard? Suppose that you charged 55 cents for each cup and you sold 45 cups. Your **revenue** is your income from selling these 45 units, or $0.55(45) = 24.75. Your *revenue function* from selling x cups is

$$R(x) = 0.55x.$$

> This is the unit price: 55¢ for each cup.   This is the number of units sold.

For any business, the **revenue function**, $R(x)$, is the money generated by selling x units of the product:

$$R(x) = px.$$

> Price per unit x units sold

Back to selling lemonade and energizing the neighborhood with white sugar: Is your revenue for the afternoon also your profit? No. We need to consider the cost of the business. You estimate that the lemons, white sugar, and bottled water cost 5 cents per cup. Furthermore, mommy dearest is charging you a $10 rental fee for use of your (her?) front yard. Your *cost function* for selling x cups of lemonade is

$$C(x) = 10 + 0.05x.$$

> This is your $10 fixed cost.   This is your variable cost: 5¢ for each cup produced.

For any business, the **cost function**, $C(x)$, is the cost of producing x units of the product:

$$C(x) = (\text{fixed cost}) + cx.$$

> Cost per unit x units produced

The term on the right, cx, represents **variable cost**, because it varies based on the number of units produced. Thus, the cost function is the sum of the fixed cost and the variable cost.

REVENUE AND COST FUNCTIONS   A company produces and sells x units of a product.

Revenue Function

$$R(x) = (\text{price per unit sold})x$$

Cost Function

$$C(x) = \text{fixed cost} + (\text{cost per unit produced})x$$

Figure 3.7 shows the graphs of the revenue and cost functions for the lemonade business. Similar graphs and models apply no matter how small or large a business venture may be.

$$R(x) = 0.55x \qquad C(x) = 10 + 0.05x$$

Revenue is 55¢ times the number of cups sold.

Cost is $10 plus 5¢ times the number of cups produced.

$R(x) = 0.55x$

$C(x) = 10 + 0.05x$

Gain

Loss

Break-even point: (20, $11)

Units Produced and Sold

FIGURE 3.7

The lines intersect at the point (20, 11). This means that when 20 cups of lemonade are produced and sold, both cost and revenue are $11. In business, this point of intersection is called the **break-even point**. At the break-even point, the money coming in is equal to the money going out. Can you see what happens for x-values less than 20? The red cost graph is above the blue revenue graph. The cost is greater than the revenue and the business is losing money. Thus, if you sell fewer than 20 cups of lemonade, the result is a *loss*. By contrast, look at what happens for x-values greater than 20. The blue revenue graph is above the red cost graph. The revenue is greater than the cost and the business is making money. Thus, if you sell more than 20 cups of lemonade, the result is a *gain*.

Because the break-even point is the point of intersection of the graphs of revenue and cost functions, we can interpret each coordinate of this point. The x-coordinate of the point reveals the number of units that a company must produce and sell so that money coming in, the revenue, is equal to money going out, the cost. The y-coordinate of the break-even point gives the amount of money coming in and going out. Example 5 illustrates the use of the substitution method in determining a company's break-even point.

3 Perform a break-even analysis.

EXAMPLE 5 Finding a Break-Even Point

Technology is now promising to bring light, fast, and beautiful wheelchairs to millions of disabled people. A company is planning to manufacture these radically different wheelchairs. Fixed cost will be $500,000 and it will cost $400 to produce each wheelchair. Each wheelchair will be sold for $600.

   a. Write the cost function, C, of producing x wheelchairs.
   b. Write the revenue function, R, from the sale of x wheelchairs.
   c. Determine the break-even point. Describe what this means.

SOLUTION

   a. The cost function is the sum of the fixed cost and variable cost.

Fixed cost of $500,000

plus

Variable cost: $400 for each chair produced

$$C(x) = 500,000 + 400x$$

b. The revenue function is the money generated from the sale of x wheelchairs.

Revenue per chair, $600, times   the number of chairs sold

$$R(x) = 600x$$

c. The break-even point occurs where the graphs of C and R intersect. Thus, we find this point by solving the system

$$
\begin{aligned}
C(x) &= 500{,}000 + 400x \\
R(x) &= 600x
\end{aligned}
\quad \text{or} \quad
\begin{aligned}
y &= 500{,}000 + 400x \\
y &= 600x.
\end{aligned}
$$

Using substitution, we can substitute $600x$ for y in the first equation.

$600x = 500{,}000 + 400x$ Substitute $600x$ for y in $y = 500{,}000 + 400x$.

$200x = 500{,}000$ Subtract $400x$ from both sides.

$x = 2500$ Divide both sides by 200.

Back-substituting 2500 for x in either of the system's equations (or functions), we obtain

$$R(2500) = 600(2500) = 1{,}500{,}000.$$

We used $R(x) = 600x$.

The break-even point is $(2500, 1{,}500{,}000)$. This means that the company will break even if it produces and sells 2500 wheelchairs. At this level, the money coming in is equal to the money going out: $1,500,000. ∎

> ✓ **CHECK POINT 5** A company that manufactures running shoes has a fixed cost of $300,000. Additionally, it costs $30 to produce each pair of shoes. They are sold at $80 per pair.
>
> **a.** Write the cost function, C, of producing x pairs of running shoes.
> $C(x) = 300{,}000 + 30x$
>
> **b.** Write the revenue function, R, from the sale of x pairs of running shoes.
> $R(x) = 80x$
>
> **c.** Determine the break-even point. Describe what this means.
>
> $(6000, 480{,}000)$; The company will break even when it produces and sells 6000 pairs of shoes. At this level, both revenue and cost are $480,000.

What does every entrepreneur, from a kid selling lemonade to Donald Trump, want to do? Generate profit, of course. The *profit* made is the money taken in, or the revenue, minus the money spent, or the cost. This relationship between revenue and cost allows us to define the *profit function*, $P(x)$.

> **THE PROFIT FUNCTION** The profit, $P(x)$, generated after producing and selling x units of a product is given by the **profit function**
>
> $$P(x) = R(x) - C(x),$$
>
> where R and C are the revenue and cost functions, respectively.

EXAMPLE 6 Writing a Profit Function

Use the revenue and cost functions for the lemonade business
$$R(x) = 0.55x \quad \text{and} \quad C(x) = 10 + 0.05x$$
to write the profit function for producing and selling x cups of lemonade.

SOLUTION The profit function is the difference between the revenue function and the cost function.

$$
\begin{aligned}
P(x) &= R(x) - C(x) && \text{This is the definition of the profit function.}\\
&= 0.55x - (10 + 0.05x) && \text{Substitute the given functions.}\\
&= 0.55x - 10 - 0.05x && \text{Distribute } -1 \text{ to each term in parentheses.}\\
&= 0.50x - 10 && \text{Simplify.}
\end{aligned}
$$

The profit function is $P(x) = 0.50x - 10$.

The graph of the profit function, which can be expressed as $y = 0.50x - 10$, is shown in Figure 3.8. The red portion of the graph lies below the x-axis and shows a loss when fewer than 20 units are sold. The lemonade business is "in the red." The black portion of the graph lies above the x-axis and shows a gain when more than 20 units are sold. The lemonade business is "in the black."

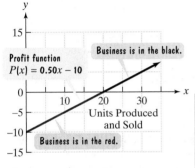

FIGURE 3.8 In the red and in the black

CHECK POINT 6 Use the revenue and cost functions for the wheelchair business in Example 5

$$R(x) = 600x$$
$$C(x) = 500,000 + 400x$$

to write the profit function for producing and selling x wheelchairs.

$$P(x) = 200x - 500,000$$

3.2 EXERCISE SET

 Student Solutions Manual   CD/Video   PH Math/Tutor Center   MathXL Tutorials on CD   MathXL®   MyMathLab   Interactmath.com

Practice Exercises

In Exercises 1–4, let x represent one number and let y represent the other number. Use the given conditions to write a system of equations. Solve the system and find the numbers.

1. The sum of two numbers is 7. If one number is subtracted from the other, the result is −1. Find the numbers.   3 and 4

2. The sum of two numbers is 2. If one number is subtracted from the other, the result is 8. Find the numbers.   −3 and 5

3. Three times a first number decreased by a second number is 1. The first number increased by twice the second number is 12. Find the numbers.   first number: 2; second number: 5

   $3x + 2y = 8$

4. The sum of three times a first number and twice a second number is 8. If the second number is subtracted from twice the first number, the result is 3. Find the numbers.   first number: 2; second number: 1

In Exercises 5–8, cost and revenue functions for producing and selling x units of a product are given. Cost and revenue are expressed in dollars.

a. Find the number of units that must be produced and sold to break even. At this level, what is the dollar amount coming in and going out?

b. Write the profit function from producing and selling x units of the product.

5. $C(x) = 25,500 + 15x$ a. 1500 units; $48,000
   $R(x) = 32x$ b. $P(x) = 17x - 25,500$

6. $C(x) = 15,000 + 12x$ a. 750 units; $24,000
   $R(x) = 32x$ b. $P(x) = 20x - 15,000$

7. $C(x) = 105x + 70,000$ a. 500 units; $122,500
   $R(x) = 245x$ b. $P(x) = 140x - 70,000$

8. $C(x) = 1.2x + 1500$ a. 3000 units; $5100
   $R(x) = 1.7x$ b. $P(x) = 0.5x - 1500$

Application Exercises

In Exercises 9–40, use the four-step strategy to solve each problem.

9. The graph shows the education level of Americans by their early thirties. Thirty-six percent have college degrees. The difference between the percentage with bachelor's degrees or higher and the percentage with associate degrees is 16%. Find the percentage of Americans with bachelor's degrees or higher and with associate degrees by their early thirties. Bachelor's or higher: 26%; Associate: 10%

Education Level of Americans by Their Early Thirties

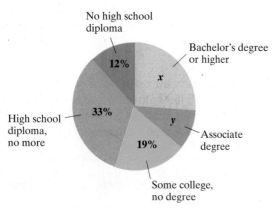

Source: U.S. Census Bureau

10. The graph shows the percentage of Americans in their early thirties who said they were happy, satisfied, or unhappy with their lives. For this age group, 58% said they were happy. The difference between the percentage who said they were very happy and the percentage who said they were extremely happy was 24%. Find the percentage of Americans in their early thirties who said they were very happy and the percentage who said they were extremely happy. very happy: 41%; extremely happy: 17%

Percentage of Americans in Their Early Thirties Who Said They were Happy, Satisfied, or Unhappy

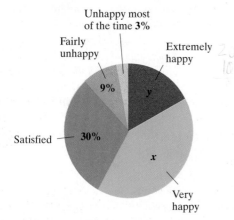

Source: Albertson et al., *Book of Ages 30*

11. One week, a computer store sold a total of 36 computers and hard drives. The revenue from these sales was $27,710. If computers sold for $1180 per unit and hard drives for $125 per unit, how many of each did the store sell? 22 computers and 14 hard drives

12. There were 180 people at a civic club fundraiser. Members paid $4.50 per ticket and nonmembers paid $8.25 per ticket. If total receipts amounted to $1222.50, how many members and how many nonmembers attended the fundraiser? 70 members and 110 nonmembers

13. You invested $7000 in two accounts paying 6% and 8% annual interest, respectively. If the total interest earned for the year was $520, how much was invested at each rate? $2000 at 6% and $5000 at 8%

14. You invested $11,000 in stocks and bonds, paying 5% and 8% annual interest, respectively. If the total interest earned for the year was $730, how much was invested in stocks and how much was invested in bonds? stocks: $5000; bonds: $6000

15. You invested money in two funds. Last year, the first fund paid a dividend of 9% and the second a dividend of 3%, and you received a total of $900 in interest. This year, the first fund paid a 10% dividend and the second only 1%, and you received a total of $860 in interest. How much money did you invest in each fund? first fund: $8000; second fund: $6000

16. You invested money in two funds. Last year, the first fund paid a dividend of 8% and the second a dividend of 5%, and you received a total of $1330 in interest. This year, the first fund paid a 12% dividend and the second only 2%, and you received a total of $1500 in interest. How much money did you invest in each fund? first fund: $11,000; second fund: $9000

17. Things did not go quite as planned. You invested $20,000, part of it in a stock that paid 12% annual interest. However, the rest of the money suffered a 5% loss. If the total annual income from both investments was $1890, how much was invested at each rate? $17,000 was invested at 12% annual interest and $3000 was invested at a 5% loss.

18. Things did not go quite as planned. You invested $30,000, part of it in a stock that paid 14% annual interest. However, the rest of the money suffered a 6% loss. If the total annual income from both investments was $200, how much was invested at each rate? $10,000 was invested at 14% annual interest and $20,000 was invested at a 6% loss.

19. A wine company needs to blend a California wine with a 5% alcohol content and a French wine with a 9% alcohol content to obtain 200 gallons of wine with a 7% alcohol content. How many gallons of each kind of wine must be used? California: 100 gallons; French: 100 gallons

20. A jeweler needs to mix an alloy with a 16% gold content and an alloy with a 28% gold content to obtain 32 ounces of a new alloy with a 25% gold content. How many ounces of each of the original alloys must be used? 16% gold content: 8 ounces; 28% gold content: 24 ounces

21. For thousands of years, gold has been considered one of Earth's most precious metals. One hundred percent pure gold is 24-karat gold, which is too soft to be made into jewelry. In the United States, most gold jewelry is 14-karat gold, approximately 58% gold. If 18-karat gold is 75% gold and 12-karat gold is 50% gold, how much of each should be used to make a 14-karat gold bracelet weighing 300 grams? 96 grams of 18-karat gold and 204 grams of 12-karat gold

22. In the "Peanuts" cartoon shown, solve the problem that is sending Peppermint Patty into an agitated state. How much cream and how much milk, to the nearest thousandth of a gallon, must be mixed together to obtain 50 gallons of cream that contains 12.5% butterfat?

© 1978 United Media/United Feature Syndicate, Inc.

cream: 20.93 gallons; milk: 29.07 gallons

$X(161) + Y(162) = 75$ $1.6x + 2.1y = 1.9(75)$

23. The manager of a candystand at a large multiplex cinema has a popular candy that sells for $1.60 per pound. The manager notices a different candy worth $2.10 per pound that is not selling well. The manager decides to form a mixture of both types of candy to help clear the inventory of the more expensive type. How many pounds of each kind of candy should be used to create a 75-pound mixture selling for $1.90 per pound?   30 pounds of the cheaper candy and 45 pounds of the more expensive candy

24. A grocer needs to mix raisins at $2.00 per pound with granola at $3.25 per pound to obtain 10 pounds of a mixture that costs $2.50 per pound. How many pounds of raisins and how many pound of granola must be used?   6 pounds of raisins and 4 pounds of granola

25. A coin purse contains a mixture of 15 coins in nickels and dimes. The coins have a total value of $1.10. Determine the number of nickels and the number of dimes in the purse.   8 nickels and 7 dimes

26. A coin purse contains a mixture of 15 coins in dimes and quarters. The coins have a total value of $3.30. Determine the number of dimes and the number of quarters in the purse.   3 dimes and 12 quarters

27. When a small plane flies with the wind, it can travel 800 miles in 5 hours. When the plane flies in the opposite direction, against the wind, it takes 8 hours to fly the same distance. Find the rate of the plane in still air and the rate of the wind.   The plane's rate in still air is 130 miles per hour and the rate of the wind is 30 miles per hour.

28. When a plane flies with the wind, it can travel 4200 miles in 6 hours. When the plane flies in the opposite direction, against the wind, it takes 7 hours to fly the same distance. Find the rate of the plane in still air and the rate of the wind.   The plane's rate in still air is 650 miles per hour and the rate of the wind is 50 miles per hour.

29. A boat's crew rowed 16 kilometers downstream, with the current, in 2 hours. The return trip upstream, against the current, covered the same distance, but took 4 hours. Find the crew's rowing rate in still water and the rate of the current.   The crew's rowing rate in still water is 6 kilometers per hour and the rate of the current is 2 kilometers per hour.

30. A motorboat traveled 36 miles downstream, with the current, in 1.5 hours. The return trip upstream, against the current, covered the same distance, but took 2 hours. Find the boat's rate in still water and the rate of the current.   The boat's rate in still water is 21 miles per hour and the rate of the current is 3 miles per hour.

31. With the current, you can canoe 24 miles in 4 hours. Against the same current, you can canoe only $\frac{3}{4}$ of this distance in 6 hours. Find your rate in still water and the rate of the current.   Rate in still water is 4.5 miles per hour and the rate of the current is 1.5 miles per hour.

32. With the current, you can row 24 miles in 3 hours. Against the same current, you can row only $\frac{2}{3}$ of this distance in 4 hours. Find your rowing rate in still water and the rate of the current.   Rate in still water is 6 miles per hour and the rate of the current is 2 miles per hour.

33. A student has two test scores. The difference between the scores is 12 and the mean, or average, of the scores is 80. What are the two test scores?   86 and 74

34. A student has two test scores. The difference between the scores is 8 and the mean, or average, of the scores is 88. What are the two test scores?   92 and 84

In Exercises 35–36, an isosceles triangle containing two angles with equal measure is shown. The degree measure of each triangle's three interior angles and an exterior angle is represented with variables. Find the measure of the three interior angles.

35.

80°; 50°; 50°

36.

30°; 75°; 75°

37. A rectangular lot whose perimeter is 220 feet is fenced along three sides. An expensive fencing along the lot's length costs $20 per foot, and an inexpensive fencing along the two side widths costs only $8 per foot. The total cost of the fencing along the three sides comes to $2040. What are the lot's dimensions?   70 feet by 40 feet

38. A rectangular lot whose perimeter is 260 feet is fenced along three sides. An expensive fencing along the lot's length costs $16 per foot, and an inexpensive fencing along the two side widths costs only $5 per foot. The total cost of the fencing along the three sides comes to $1780. What are the lot's dimensions?   80 feet by 50 feet

39. A new restaurant is to contain two-seat tables and four-seat tables. Fire codes limit the restaurant's maximum occupancy to 56 customers. If the owners have hired enough servers to handle 17 tables of customers, how many of each kind of table should they purchase?   6 two-seat tables and 11 four-seat tables

40. A hotel has 200 rooms. Those with kitchen facilities rent for $100 per night and those without kitchen facilities rent for $80 per night. On a night when the hotel was completely occupied, revenues were $17,000. How many of each type of room does the hotel have?   The hotel has 50 rooms with kitchen facilities and 150 rooms without kitchen facilities.

The figure shows the graphs of the cost and revenue functions for a company that manufactures and sells small radios. Use the information in the figure to solve Exercises 41–46.

Radios Produced and Sold

41. How many radios must be produced and sold for the company to break even?   500 radios must be produced and sold to break even.

42. More than how many radios must be produced and sold for the company to have a profit?   More than 500 radios must be produced and sold to have a profit.

43. Use the formulas shown in the voice balloons to find $R(200) - C(200)$. Describe what this means for the company.   −6000; When the company produces and sells 200 radios, the loss is $6000.

44. Use the formulas shown in the voice balloons to find $R(300) - C(300)$. Describe what this means for the company.   −4000; When the company produces and sells 300 radios, the loss is $4000.

45. a. Use the formulas shown in the voice balloons to write the company's profit function, P, from producing and selling x radios. $P(x) = 20x - 10,000$

b. Find the company's profit if 10,000 radios are produced and sold. $190,000

46. a. Use the formulas shown in the voice balloons to write the company's profit function, P, from producing and selling x radios. $P(x) = 20x - 10,000$

b. Find the company's profit if 20,000 radios are produced and sold. $390,000

Exercises 47–50 describe a number of business ventures. For each exercise,

a. Write the cost function, C.
b. Write the revenue function, R.
c. Determine the break-even point. Describe what this means.

47. A company that manufactures small canoes has a fixed cost of $18,000. It costs $20 to produce each canoe. The selling price is $80 per canoe. (In solving this exercise, let x represent the number of canoes produced and sold.)   **a.** $C(x) = 18,000 + 20x$ **b.** $R(x) = 80x$ **c.** (300, 24,000); When 300 canoes are produced and sold, both cost and revenue are $24,000.

48. A company that manufactures bicycles has a fixed cost of $100,000. It costs $100 to produce each bicycle. The selling price is $300 per bike. (In solving this exercise, let x represent the number of bicycles produced and sold.)   **a.** $C(x) = 100,000 + 100x$ **b.** $R(x) = 300x$ **c.** (500, 150,000); When 500 bicycles are produced and sold, both cost and revenue are $150,000.

49. You invest in a new play. The cost includes an overhead of $30,000, plus production costs of $2500 per performance. A sold-out performance brings in $3125. (In solving this exercise, let x represent the number of sold-out performances.)   **a.** $C(x) = 30,000 + 2500x$ **b.** $R(x) = 3125x$ **c.** (48, 150,000); For 48 sold-out performances, both cost and revenue are $150,000.

50. You invested $30,000 and started a business writing greeting cards. Supplies cost 2¢ per card and you are selling each card for 50¢. (In solving this exercise, let x represent the number of cards produced and sold.)   **a.** $C(x) = 30,000 + 0.02x$ **b.** $R(x) = 0.5x$ **c.** (62,500, 31,250); For 62,500 cards, both cost and revenue are $31,250.

Writing in Mathematics

51. Describe the conditions in a problem that enable it to be solved using a system of linear equations.   51–57. Answers will vary.

52. Write a word problem that can be solved by translating to a system of linear equations. Then solve the problem.

53. Describe a revenue function for a business venture.

54. Describe a cost function for a business venture. What are the two kinds of costs that are modeled by this function?

55. What is the profit function for a business venture and how is it determined?

56. Describe the break-even point for a business.

57. The law of supply and demand states that, in a free market economy, a commodity tends to be sold at its equilibrium price. At this price, the amount that the seller will supply is the same amount that the consumer will buy. Explain how graphs can be used to determine the equilibrium price.

Technology Exercises

In Exercises 58–59, graph the revenue and cost functions in the same viewing rectangle. Then use the intersection feature to determine the break-even point.   58–59. See graphing answer section.

58. $R(x) = 50x$, $C(x) = 20x + 180$ (6, 300)

59. $R(x) = 92.5x$, $C(x) = 52x + 1782$ (44, 4070)

60. Use the procedure in Exercises 58–59 to verify your work for any one of the break-even points that you found in Exercises 47–50.   Answers will vary.

Critical Thinking Exercises

61. A marching band has 52 members, and there are 24 in the pom-pom squad. They wish to form several hexagons and squares like those diagrammed below. Can it be done with no people left over?

B   B
              P   P

B   P   B   B   B = Band Member

                              P = Pom-pom Person

B   B   P   P

              yes; 8 hexagons and 4 squares

62. The radiator in your car contains 4 gallons of antifreeze and water. The mixture is 45% antifreeze. How much of this mixture should be drained and replaced with pure antifreeze in order to have a 60% antifreeze solution? Round to the nearest tenth of a gallon.   1.1 gallons

63. When entering your test score into a computer, your professor accidently reversed the two digits. This error reduced your score by 36 points. Your professor told you that the sum of the digits of your actual score was 14, correct the error, and agreed to give you extra credit if you could determine the actual score without looking back at the test. What was your actual test score? (*Hint*: Let t = the tens-place digit of your actual score and let u = the units-place digit of your actual score. Thus, $10t + u$ represents your actual test score.)   95

64. A boy has as many brothers as he has sisters. Each of his sisters has twice as many brothers as she has sisters. How many boys and girls are in this family?   4 boys and 3 girls

65. A dealer paid a total of $67 for mangos and avocados. The mangos were sold at a profit of 20% on the dealer's cost, but the avocados started to spoil, resulting in a selling price of a 2% loss on the dealer's cost. The dealer made a profit of $8.56 on the total transaction. How much did the dealer pay for the mangos and for the the avocados? $45 for mangos and $22 for avocados

Review Exercises

In Exercises 66–67, use the given conditions to write an equation for each line in point-slope form and slope-intercept form.

66. Passing through $(-2, 5)$ and $(-6, 13)$
(Section 2.4, Example 2)
$y - 5 = -2(x + 2)$ or $y - 13 = -2(x + 6)$;
$y = -2x + 1$ or $f(x) = -2x + 1$

67. Passing through $(-3, 0)$ and parallel to the line whose equation is $-x + y = 7$
(Section 2.4, Example 4)
$y - 0 = 1(x + 3)$; $y = x + 3$ or $f(x) = x + 3$

68. Find the domain of $g(x) = \dfrac{x - 2}{3 - x}$.
(Section 2.2, Example 1)
Domain of $g = \{x \mid x$ is a real number and $x \neq 3\}$

SECTION

3.3

Objectives

1 Verify the solution of a system of linear equations in three variables.

2 Solve systems of linear equations in three variables.

3 Identify inconsistent and dependent systems.

4 Solve problems using systems in three variables.

SYSTEMS OF LINEAR EQUATIONS IN THREE VARIABLES

All animals sleep, but the length of time they sleep varies widely: Cattle sleep for only a few minutes at a time. We humans seem to need more sleep than other animals, up to eight hours a day. Without enough sleep, we have difficulty concentrating, make mistakes in routine tasks, lose energy, and feel bad-tempered. There is a relationship between hours of sleep and death rate per year per 100,000 people. How many hours of sleep will put you in the group with the minimum death rate? In this section, you will learn how to solve systems of linear equations with more than two variables to answer this question.

1 Verify the solution of a system of linear equations in three variables.

Point of intersection

FIGURE 3.9

Systems of Linear Equations in Three Variables and Their Solutions An equation such as $x + 2y - 3z = 9$ is called a *linear equation in three variables.* In general, any equation of the form

$$Ax + By + Cz = D,$$

where A, B, C, and D are real numbers such that A, B, and C are not all 0, is a **linear equation in three variables: x, y, and z.** The graph of this linear equation in three variables is a plane in three-dimensional space.

The process of solving a system of three linear equations in three variables is geometrically equivalent to finding the point of intersection (assuming that there is one) of three planes in space (see Figure 3.9). A **solution** of a system of linear equations in three variables is an ordered triple of real numbers that satisfies all equations in the system. The **solution set** of the system is the set of all its solutions.

EXAMPLE 1 Determining Whether an Ordered Triple Satisfies a System

Show that the ordered triple $(-1, 2, -2)$ is a solution of the system:

$$x + 2y - 3z = 9$$
$$2x - y + 2z = -8$$
$$-x + 3y - 4z = 15.$$

SOLUTION Because -1 is the x-coordinate, 2 is the y-coordinate, and -2 is the z-coordinate of $(-1, 2, -2)$, we replace x with -1, y with 2, and z with -2 in each of the three equations.

$$x + 2y - 3z = 9 \qquad\qquad 2x - y + 2z = -8 \qquad\qquad -x + 3y - 4z = 15$$
$$-1 + 2(2) - 3(-2) \stackrel{?}{=} 9 \qquad 2(-1) - 2 + 2(-2) \stackrel{?}{=} -8 \qquad -(-1) + 3(2) - 4(-2) \stackrel{?}{=} 15$$
$$-1 + 4 + 6 \stackrel{?}{=} 9 \qquad\qquad -2 - 2 - 4 \stackrel{?}{=} -8 \qquad\qquad 1 + 6 + 8 \stackrel{?}{=} 15$$
$$9 = 9, \text{ true} \qquad\qquad -8 = -8, \text{ true} \qquad\qquad 15 = 15, \text{ true}$$

The ordered triple $(-1, 2, -2)$ satisfies the three equations: It makes each equation true. Thus, the ordered triple is a solution of the system. ∎

 CHECK POINT 1 Show that the ordered triple $(-1, -4, 5)$ is a solution of the system:

$$x - 2y + 3z = 22 \qquad (-1) - 2(-4) + 3(5) = 22;$$
$$2x - 3y - z = 5 \qquad 2(-1) - 3(-4) - 5 = 5;$$
$$3x + y - 5z = -32. \qquad 3(-1) + (-4) - 5(5) = -32$$

---

2 Solve systems of linear equations in three variables.

Solving Systems of Linear Equations in Three Variables by Eliminating Variables The method for solving a system of linear equations in three variables is similar to that used on systems of linear equations in two variables. We use addition to eliminate any variable, reducing the system to two equations in two variables. Once we obtain a system of two equations in two variables, we use addition or substitution to eliminate another variable. The result is a single equation in one variable. We solve this equation to get the value of the remaining variable. Other variable values are found by back-substitution.

STUDY TIP

It does not matter which variable you eliminate, as long as you do it in two different pairs of equations.

SOLVING LINEAR SYSTEMS IN THREE VARIABLES BY ELIMINATING VARIABLES

1. Reduce the system to two equations in two variables. This is usually accomplished by taking two different pairs of equations and using the addition method to eliminate the same variable from both pairs.

2. Solve the resulting system of two equations in two variables using addition or substitution. The result is an equation in one variable that gives the value of that variable.

3. Back-substitute the value of the variable found in step 2 into either of the equations in two variables to find the value of the second variable.

4. Use the values of the two variables from steps 2 and 3 to find the value of the third variable by back-substituting into one of the original equations.

5. Check the proposed solution in each of the original equations.

EXAMPLE 2   Solving a System in Three Variables

Solve the system:

$$5x - 2y - 4z = 3 \qquad \text{Equation 1}$$
$$3x + 3y + 2z = -3 \qquad \text{Equation 2}$$
$$-2x + 5y + 3z = 3. \qquad \text{Equation 3}$$

SOLUTION   There are many ways to proceed. Because our initial goal is to reduce the system to two equations in two variables, **the central idea is to take two different pairs of equations and eliminate the same variable from both pairs**.

Step 1.   **Reduce the system to two equations in two variables.** We choose any two equations and use the addition method to eliminate a variable. Let's eliminate z from Equations 1 and 2. We do so by multiplying Equation 2 by 2. Then we add equations.

(Equation 1) $5x - 2y - 4z = 3 \xrightarrow{\text{No change}} 5x - 2y - 4z = 3$

(Equation 2) $3x + 3y + 2z = -3 \xrightarrow{\text{Multiply by 2.}} \underline{6x + 6y + 4z = -6}$

Add: $11x + 4y \qquad = -3$ (Equation 4)

Now we must eliminate the *same* variable from another pair of equations. We can eliminate z from Equations 2 and 3. First, we multiply Equation 2 by -3. Next, we multiply Equation 3 by 2. Finally, we add equations.

(Equation 2) $3x + 3y + 2z = -3 \xrightarrow{\text{Multiply by } -3.} -9x - 9y - 6z = 9$

(Equation 3) $-2x + 5y + 3z = 3 \xrightarrow{\text{Multiply by 2.}} \underline{-4x + 10y + 6z = 6}$

Add: $-13x + y \qquad = 15$ (Equation 5)

Equations 4 and 5 give us a system of two equations in two variables.

Step 2.   **Solve the resulting system of two equations in two variables.** We will use the addition method to solve Equations 4 and 5 for x and y. To do so, we multiply Equation 5 by -4 and add this to Equation 4.

(Equation 4) $11x + 4y = -3 \xrightarrow{\text{No change}} 11x + 4y = -3$

(Equation 5) $-13x + y = 15 \xrightarrow{\text{Multiply by } -4.} \underline{52x - 4y = -60}$

Add: $63x \qquad = -63$

$x \qquad = -1$ Divide both sides by 63.

Step 3. **Use back-substitution in one of the equations in two variables to find the value of the second variable.** We back-substitute -1 for x in either Equation 4 or 5 to find the value of y. We will use Equation 5.

$$-13x + y = 15 \qquad \text{Equation 5}$$
$$-13(-1) + y = 15 \qquad \text{Substitute } -1 \text{ for } x.$$
$$13 + y = 15 \qquad \text{Multiply.}$$
$$y = 2 \qquad \text{Subtract 13 from both sides.}$$

Step 4. **Back-substitute the values found for two variables into one of the original equations to find the value of the third variable.** We can now use any one of the original equations and back-substitute the values of x and y to find the value for z. We will use Equation 2.

$$3x + 3y + 2z = -3 \qquad \text{Equation 2}$$
$$3(-1) + 3(2) + 2z = -3 \qquad \text{Substitute } -1 \text{ for } x \text{ and 2 for } y.$$
$$3 + 2z = -3 \qquad \text{Multiply and then add:}$$
$$\qquad\qquad\qquad\qquad\qquad 3(-1) + 3(2) = -3 + 6 = 3.$$
$$2z = -6 \qquad \text{Subtract 3 from both sides.}$$
$$z = -3 \qquad \text{Divide both sides by 2.}$$

With $x = -1$, $y = 2$, and $z = -3$, the proposed solution is the ordered triple $(-1, 2, -3)$.

Step 5. **Check.** Check the proposed solution, $(-1, 2, -3)$, by substituting the values for x, y, and z into each of the three original equations. These substitutions yield three true statements. Thus, the solution is $(-1, 2, -3)$ and the solution set is $\{(-1, 2, -3)\}$. ■

 CHECK POINT **2** Solve the system:

$$x + 4y - z = 20$$
$$3x + 2y + z = 8$$
$$2x - 3y + 2z = -16. \qquad (1, 4, -3) \text{ or } \{(1, 4, -3)\}$$

In some examples, one of the variables is already eliminated from a given equation. In this case, the same variable should be eliminated from the other two equations, thereby making it possible to omit one of the elimination steps. We illustrate this idea in Example 3.

EXAMPLE 3   Solving a System of Equations with a Missing Term

Solve the system:

$$x + \qquad z = 8 \qquad \text{Equation 1}$$
$$x + y + 2z = 17 \qquad \text{Equation 2}$$
$$x + 2y + z = 16. \qquad \text{Equation 3}$$

SOLUTION

Step 1. **Reduce the system to two equations in two variables.** Because Equation 1 contains only x and z, we could eliminate y from Equations 2 and 3. This will give us two equations in x and z. To eliminate y from Equations 2 and 3, we multiply Equation 2 by -2 and add Equation 3.

(Equation 2) $\quad x + \;\; y + 2z = 17 \xrightarrow{\text{Multiply by } -2.} \quad -2x - 2y - 4z = -34$

(Equation 3) $\quad x + 2y + \;\; z = 16 \xrightarrow{\text{No change}} \qquad\quad\; \underline{x + 2y + \;\; z = \;\;\; 16}$

$$\text{Add: } -x \qquad\quad - 3z = -18 \quad \text{(Equation 4)}$$

Equation 4 and the given Equation 1 provide us with a system of two equations in two variables.

Step 2. **Solve the resulting system of two equations in two variables.** We will solve Equations 1 and 4 for x and z.

$$x + \;\; z = \qquad 8 \quad \text{Equation 1}$$
$$\underline{-x - 3z = -18} \quad \text{Equation 4}$$
$$\text{Add: } \qquad -2z = -10$$
$$z = \qquad 5 \quad \text{Divide both sides by } -2.$$

Step 3. **Use back-substitution in one of the equations in two variables to find the value of the second variable.** To find x, we back-substitute 5 for z in either Equation 1 or 4. We will use Equation 1.

$$x + z = 8 \qquad \text{Equation 1}$$
$$x + 5 = 8 \qquad \text{Substitute 5 for } z.$$
$$x = 3 \qquad \text{Subtract 5 from both sides.}$$

Step 4. **Back-substitute the values found for two variables into one of the original equations to find the value of the third variable.** To find y, we back-substitute 3 for x and 5 for z into Equation 2 or 3. We can't use Equation 1 because y is missing in this equation. We will use Equation 2.

$$x + y + 2z = 17 \qquad \text{Equation 2}$$
$$3 + y + 2(5) = 17 \qquad \text{Substitute 3 for } x \text{ and 5 for } z.$$
$$y + 13 = 17 \qquad \text{Multiply and add.}$$
$$y = 4 \qquad \text{Subtract 13 from both sides.}$$

We found that $z = 5$, $x = 3$, and $y = 4$. Thus, the proposed solution is the ordered triple $(3, 4, 5)$.

Step 5. Check. Substituting 3 for x, 4 for y, and 5 for z into each of the three original equations yields three true statements. Consequently, the solution is $(3, 4, 5)$ and the solution set is $\{(3, 4, 5)\}$. ∎

 CHECK POINT 3 Solve the system:

$$2y - z = \;\; 7$$
$$x + 2y + z = \;\; 17$$
$$2x - 3y + 2z = -1. \qquad (4, 5, 3) \text{ or } \{(4, 5, 3)\}$$

3 Identify inconsistent and dependent systems.

Inconsistent and Dependent Systems A system of linear equations in three variables represents three planes. The three planes need not intersect at one point. The planes may have no common point of intersection and represent an **inconsistent system** with no solution. Figure 3.10 illustrates some of the geometric possibilities for inconsistent systems.

Three planes are parallel with no common intersection point.

Two planes are parallel with no common intersection point.

Planes intersect two at a time. There is no intersection point common to all three planes.

FIGURE 3.10 Three planes may have no common point of intersection.

If you attempt to solve an inconsistent system algebraically, at some point in the solution process you will eliminate all three variables. A false statement, such as $0 = 17$, will be the result. For example, consider the system

$$2x + 5y + \;\; z = \;\; 12 \qquad \text{Equation 1}$$
$$x - 2y + \;\; 4z = -10 \qquad \text{Equation 2}$$
$$-3x + 6y - 12z = \;\; 20. \qquad \text{Equation 3}$$

Suppose we reduce the system to two equations in two variables by eliminating x. To eliminate x in Equations 2 and 3, we multiply Equation 2 by 3 and add Equation 3:

$$
\begin{array}{ll}
x - 2y + \;\; 4z = -10 & \xrightarrow{\;\text{Multiply by 3.}\;} \quad 3x - 6y + 12z = -30 \\
-3x + 6y - 12z = \;\; 20 & \xrightarrow{\;\text{No change}\;} \quad \underline{-3x + 6y - 12z = \;\; 20} \\
& \qquad\qquad\qquad\qquad \text{Add:} \qquad\qquad 0 = -10
\end{array}
$$

There are no values of $x, y,$ and z for which $0 = -10$. The false statement $0 = -10$ indicates that the system is inconsistent and has no solution. The solution set is the empty set, $\varnothing$.

We have seen that a linear system that has at least one solution is called a **consistent system**. Planes that intersect at one point and planes that intersect at infinitely many points both represent consistent systems. Figure 3.11 illustrates planes that intersect at infinitely many points. The equations in these linear systems with infinitely many solutions are called **dependent**.

FIGURE 3.11 Three planes may intersect at infinitely many points.

The planes intersect along a common line.

The planes coincide.

If you attempt to solve a system with dependent equations algebraically, at some point in the solution process you will eliminate all three variables. A true statement, such as $0 = 0$, will be the result. If this occurs as you are solving a linear system, simply state that the equations are dependent.

4 Solve problems using systems in three variables.

Applications Systems of equations may allow us to find models for data without using a graphing utility. Three data points that do not lie on or near a line determine the graph of a function of the form

$$y = ax^2 + bx + c, a \neq 0.$$

Such a function is called a **quadratic function**. Its graph has a cuplike U-shape, making it ideal for modeling situations in which values of y are decreasing and then increasing. In Chapter 8, we'll have lots of interesting things to tell you about quadratic functions and their graphs.

The process of determining a function whose graph contains given points is called **curve fitting**. In our next example, we fit the curve whose equation is $y = ax^2 + bx + c$ to three data points. Using a system of equations, we find values for a, b, and c.

EXAMPLE 4 Modeling Data Relating Sleep and Death Rate

In a study relating sleep and death rate, the following data were obtained. Use the function $y = ax^2 + bx + c$ to model the data.

x (Average Number of Hours of Sleep)	y (Death Rate per Year per 100,000 Males)
4	1682
7	626
9	967

SOLUTION We need to find values for a, b, and c in $y = ax^2 + bx + c$. We can do so by solving a system of three linear equations in a, b, and c. We obtain the three equations by using the values of x and y from the data as follows:

$y = ax^2 + bx + c$ *Use the quadratic function to model the data.*

When $x = 4, y = 1682$: $1682 = a \cdot 4^2 + b \cdot 4 + c$ or $16a + 4b + c = 1682$

When $x = 7, y = 626$: $626 = a \cdot 7^2 + b \cdot 7 + c$ or $49a + 7b + c = 626$

When $x = 9, y = 967$: $967 = a \cdot 9^2 + b \cdot 9 + c$ or $81a + 9b + c = 967.$

The easiest way to solve this system is to eliminate c from two pairs of equations, obtaining two equations in a and b. Solving this system gives $a = 104.5$, $b = -1501.5$, and $c = 6016$. We now substitute the values for a, b, and c into $y = ax^2 + bx + c$. The function that models the given data is

$$y = 104.5x^2 - 1501.5x + 6016.$$ ∎

We can use the model that we obtained in Example 4 to find the death rate of males who average, say, 6 hours of sleep. First, write the model in function notation:

$$f(x) = 104.5x^2 - 1501.5x + 6016.$$

Substitute 6 for x:

$$f(6) = 104.5(6)^2 - 1501.5(6) + 6016 = 769.$$

According to the model, the death rate for males who average 6 hours of sleep is 769 deaths per 100,000 males.

USING TECHNOLOGY

The graph of

$$y = 104.5x^2 - 1501.5x + 6016$$

is displayed in a $[3, 12, 1]$ by $[500, 2000, 100]$ viewing rectangle. The minimum function feature shows that the lowest point on the graph is approximately $(7.2, 622.5)$. Men who average 7.2 hours of sleep are in the group with the lowest death rate, approximately 622.5 deaths per 100,000 males.

CHECK POINT 4 Find the quadratic function $y = ax^2 + bx + c$ whose graph passes through the points $(1, 4)$, $(2, 1)$, and $(3, 4)$.

$y = 3x^2 - 12x + 13$ or $f(x) = 3x^2 - 12x + 13$

Problems involving three unknowns can be solved using the same strategy for solving problems with two unknown quantities. You can let x, y, and z represent the unknown quantities. We then translate from the verbal conditions of the problem to a system of three equations in three variables. Problems of this type are included in the exercise set that follows.

3.3 EXERCISE SET

Student Solutions Manual CD/Video PH Math/Tutor Center MathXL Tutorials on CD MathXL® MyMathLab Interactmath.com

Practice Exercises

In Exercises 1–4, determine if the given ordered triple is a solution of the system.

1. $(2, -1, 3)$
$$x + y + z = 4$$
$$x - 2y - z = 1 \quad \text{not a solution}$$
$$2x - y - z = -1$$

2. $(5, -3, -2)$
$$x + y + z = 0$$
$$x + 2y - 3z = 5 \quad \text{solution}$$
$$3x + 4y + 2z = -1$$

3. $(4, 1, 2)$
$$x - 2y = 2$$
$$2x + 3y = 11 \quad \text{solution}$$
$$y - 4z = -7$$

4. $(-1, 3, 2)$
$$x - 2z = -5$$
$$y - 3z = -3 \quad \text{solution}$$
$$2x - z = -4$$

Solve each system in Exercises 5–22. If there is no solution or if there are infinitely many solutions and a system's equations are dependent, so state.

5. $x + y + 2z = 11$
$x + y + 3z = 14 \quad (2, 3, 3)$
$x + 2y - z = 5$

6. $2x + y - 2z = -1$
$3x - 3y - z = 5 \quad (1, -1, 1)$
$x - 2y + 3z = 6$

7. $4x - y + 2z = 11$
$x + 2y - z = -1 \quad (2, -1, 1)$
$2x + 2y - 3z = -1$

8. $x - y + 3z = 8$
$3x + y - 2z = -2 \quad (1, -1, 2)$
$2x + 4y + z = 0$

9. $3x + 2y - 3z = -2$
$2x - 5y + 2z = -2 \quad (1, 2, 3)$
$4x - 3y + 4z = 10$

10. $2x + 3y + 7z = 13$
$3x + 2y - 5z = -22 \quad (-1, -2, 3)$
$5x + 7y - 3z = -28$

11. $2x - 4y + 3z = 17$
$x + 2y - z = 0 \quad (3, 1, 5)$
$4x - y - z = 6$

12. $x + z = 3$
$x + 2y - z = 1 \quad (1, 1, 2)$
$2x - y + z = 3$

13. $2x + y = 2$
$x + y - z = 4 \quad (1, 0, -3)$
$3x + 2y + z = 0$

14. $x + 3y + 5z = 20$
$y - 4z = -16 \quad (0, 0, 4)$
$3x - 2y + 9z = 36$

15. $x + y = -4$
$y - z = 1 \quad (1, -5, -6)$
$2x + y + 3z = -21$

16. $x + y = 4$
$x + z = 4 \quad (2, 2, 2)$
$y + z = 4$

17. $2x + y + 2z = 1$
$3x - y + z = 2 \quad \text{no solution or } \emptyset$
$x - 2y - z = 0$

18. $3x + 4y + 5z = 8$
$x - 2y + 3z = -6 \quad \text{no solution or } \emptyset$
$2x - 4y + 6z = 8$

19. $5x - 2y - 5z = 1$
$10x - 4y - 10z = 2$ *infinitely many solutions;*
$15x - 6y - 15z = 3$ *dependent equations*

20. $x + 2y + z = 4$
$3x - 4y + z = 4$ *infinitely many solutions;*
$6x - 8y + 2z = 8$ *dependent equations*

21. $3(2x + y) + 5z = -1$
$2(x - 3y + 4z) = -9$ $\left(\frac{1}{2}, \frac{1}{3}, -1\right)$
$4(1 + x) = -3(z - 3y)$

22. $7z - 3 = 2(x - 3y)$
$5y + 3z - 7 = 4x$ $\left(\frac{1}{2}, 3, -2\right)$
$4 + 5z = 3(2x - y)$

*In Exercises 23–26, find the quadratic function $y = ax^2 + bx + c$
whose graph passes through the given points.*

23. $(-1, 6), (1, 4) (2, 9)$ $y = 2x^2 - x + 3$

24. $(-2, 7), (1, -2), (2, 3)$ $y = 2x^2 - x - 3$

25. $(-1, -4), (1, -2), (2, 5)$ $y = 2x^2 + x - 5$

26. $(1, 3), (3, -1), (4, 0)$ $y = x^2 - 6x + 8$

*In Exercises 27–28, let x represent the first number, y the second
number, and z the third number. Use the given conditions to write
a system of equations. Solve the system and find the numbers.*

27. The sum of three numbers is 16. The sum of twice the first
number, 3 times the second number, and 4 times the third
number is 46. The difference between 5 times the first number and the second number is 31. Find the three numbers.
first number: 7; second number: 4; third number: 5

28. The following is known about three numbers: Three times
the first number plus the second number plus twice the third
number is 5. If 3 times the second number is subtracted from
the sum of the first number and 3 times the third number, the
result is 2. If the third number is subtracted from the sum of 2
times the first number and 3 times the second number, the
result is 1. Find the numbers. *first number: −1; second
number: 2; third number: 3*

Practice Plus

Solve each system in Exercises 29–30.

29. $\dfrac{x + 2}{6} - \dfrac{y + 4}{3} + \dfrac{z}{2} = 0$

$\dfrac{x + 1}{2} + \dfrac{y - 1}{2} - \dfrac{z}{4} = \dfrac{9}{2}$ $(4, 8, 6)$

$\dfrac{x - 5}{4} + \dfrac{y + 1}{3} + \dfrac{z - 2}{2} = \dfrac{19}{4}$

30. $\dfrac{x + 3}{2} - \dfrac{y - 1}{2} + \dfrac{z + 2}{4} = \dfrac{3}{2}$

$\dfrac{x - 5}{2} + \dfrac{y + 1}{3} - \dfrac{z}{4} = -\dfrac{25}{6}$ $(-3, 0, 2)$

$\dfrac{x - 3}{4} - \dfrac{y + 1}{2} + \dfrac{z - 3}{2} = -\dfrac{5}{2}$

*In Exercises 31–32, find the equation of the quadratic function
$y = ax^2 + bx + c$ whose graph is shown. Select three points
whose coordinates appear to be integers.*

31.

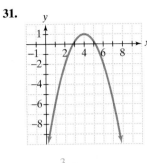

$y = -\dfrac{3}{4}x^2 + 6x - 11$

32.

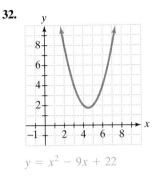

$y = x^2 - 9x + 22$

*In Exercises 33–34, solve each system for (x, y, z) in terms of the
nonzero constants a, b, and c.*

33. $ax - by - 2cz = 21$
$ax + by + cz = 0$
$2ax - by + cz = 14$

$\left(\dfrac{8}{a}, -\dfrac{3}{b}, -\dfrac{5}{c}\right)$

34. $ax - by + 2cz = -4$
$ax + 3by - cz = 1$
$2ax + by + 3cz = 2$

$\left(-\dfrac{9}{a}, \dfrac{5}{b}, \dfrac{5}{c}\right)$

Application Exercises

35. The bar graph shows the percentage of the U.S. population
that was foreign-born from 1900 through 2002.

**Percentage of U.S. Population
That was Foreign-Born, 1900-2002**

Source: U.S. Census Bureau

  a. Write the data for 1900, 1970, and 2002 as ordered pairs
(x, y), where x is the number of years after 1900 and y is
the percentage of the U.S. population that was foreign-
born in that year. $(0, 13.6), (70, 4.7), (102, 11.5)$

  b. The three data points in part (a) can be modeled by the
quadratic function $y = ax^2 + bx + c$. Substitute each
ordered pair into this function, one ordered pair at a time,
and write a system of linear equations in three variables
that can be used to find values for a, b, and c. It is not
necessary to solve the system. $0a + 0b + c = 13.6$,
$4900a + 70b + c = 4.7, 10{,}404a + 102b + c = 11.5$

36. The bar graph shows that the U.S. divorce rate increased from 1970 to 1985 and then decreased from 1985 to 2001.

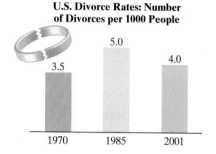

U.S. Divorce Rates: Number of Divorces per 1000 People

3.5 — 1970
5.0 — 1985
4.0 — 2001

Source: U.S. Census Bureau

a. Write the data for 1970, 1985, and 2001 as ordered pairs (x, y), where x is the number of years after 1970 and y is that year's divorce rate. $(0, 3.5), (15, 5.0), (31, 4.0)$

b. The three data points in part (a) can be modeled by the quadratic function $y = ax^2 + bx + c$. Substitute each ordered pair into this function, one ordered pair at a time, and write a system of linear equations in three variables that can be used to find values for a, b, and c. It is not necessary to solve the system. $0a + 0b + c = 3.5,$
$225a + 15b + c = 5.0,\ 961a + 31b + c = 4.0$

37. You throw a ball straight up from a rooftop. The ball misses the rooftop on its way down and eventually strikes the ground. A mathematical model can be used to describe the relationship for the ball's height above the ground, y, after x seconds. Consider the following data.

x, seconds after the ball is thrown	y, ball's height, in feet, above the ground
1	224
3	176
4	104

a. Find the quadratic function $y = ax^2 + bx + c$ whose graph passes through the given points.
$y = -16x^2 + 40x + 200$

b. Use the function in part (a) to find the value for y when $x = 5$. Describe what this means.   0; After 5 seconds, the ball hits the ground.

38. A mathematical model can be used to describe the relationship between the number of feet a car travels once the brakes are applied, y, and the number of seconds the car is in motion after the brakes are applied, x. A research firm collects the data shown at the top of the next column.

x, seconds in motion after brakes are applied	y, feet car travels once the brakes are applied
1	46
2	84
3	114

a. Find the quadratic function $y = ax^2 + bx + c$ whose graph passes through the given points. $y = -4x^2 + 50x$

b. Use the function in part (a) to find the value for y when $x = 6$. Describe what this means.   156; During the 6 seconds after the brakes are applied, the car has traveled 156 feet.

In Exercises 39–46, use the four-step strategy to solve each problem. Use x, y, and z to represent unknown quantities. Then translate from the verbal conditions of the problem to a system of three equations in three variables.

39. In current U.S. dollars, John D. Rockefeller's 1913 fortune of $900 million would be worth about $189 billion. The bar graph shows that Rockefeller is the wealthiest among the world's five richest people of all time. The combined estimated wealth, in current billions of U.S. dollars, of Andrew Carnegie, Cornelius Vanderbilt, and Bill Gates is $256 billion. The difference between Carnegie's estimated wealth and Vanderbilt's is $4 billion. The difference between Vanderbilt's estimated wealth and Gates's is $36 billion. Find the estimated wealth, in current billions of U.S. dollars, of Carnegie, Vanderbilt, and Gates.   Carnegie: $100 billion; Vanderbilt: $96 billion; Gates: $60 billion

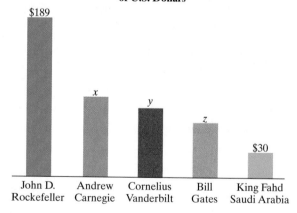

The Richest People of All Time
Estimated Wealth, in Current Billions of U.S. Dollars

$189 — John D. Rockefeller
x — Andrew Carnegie
y — Cornelius Vanderbilt
z — Bill Gates
$30 — King Fahd Saudi Arabia

Source: Scholastic Book of World Records

40. The circle graph at the top of the next page indicates computers in use for the United States and the rest of the world. The percentage of the world's computers in Europe and Japan combined is 13% less than the percentage of the world's computers in the United States. If the percentage of the world's computers in Europe is doubled, it is only 3% more than the percentage of the world's computers in the United States. Find the percentage of the world's computers in the United States, Europe, and Japan. United States: 43%; Europe: 23%; Japan: 7%

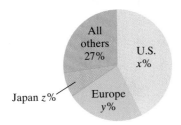

Percentage of the World's Computers: U.S. and the World

Source: Jupiter Communications

41. A person invested $6700 for one year, part at 8%, part at 10%, and the remainder at 12%. The total annual income from these investments was $716. The amount of money invested at 12% was $300 more than the amount invested at 8% and 10% combined. Find the amount invested at each rate. $1200 was invested at 8%, $2000 was invested at 10%, and $3500 was invested at 12%.

42. A person invested $17,000 for one year, part at 10%, part at 12%, and the remainder at 15%. The total annual income from these investments was $2110. The amount of money invested at 12% was $1000 less than the amount invested at 10% and 15% combined. Find the amount invested at each rate. $4000 was invested at 10%, $8000 was invested at 12%, and $5000 was invested at 15%.

43. At a college production of *Streetcar Named Desire*, 400 tickets were sold. The ticket prices were $8, $10, and $12, and the total income from ticket sales was $3700. How many tickets of each type were sold if the combined number of $8 and $10 tickets sold was 7 times the number of $12 tickets sold?   200 $8 tickets, 150 $10 tickets and 50 $12 tickets were sold.

44. A certain brand of razor blades comes in packages of 6, 12, and 24 blades, costing $2, $3, and $4 per package, respectively. A store sold 12 packages containing a total of 162 razor blades and took in $35. How many packages of each type were sold?   5 $2 packages, 3 $3 packages and 4 $4 packages were sold.

45. Three foods have the following nutritional content per ounce.

	Calories	Protein (in grams)	Vitamin C (in milligrams)
Food A	40	5	30
Food B	200	2	10
Food C	400	4	300

If a meal consisting of the three foods allows exactly 660 calories, 25 grams of protein, and 425 milligrams of vitamin C, how many ounces of each kind of food should be used?   4 ounces of food A, 0.5 ounce of food B and 1 ounce of food C should be used.

46. A furniture company produces three types of desks: a children's model, an office model, and a deluxe model. Each desk is manufactured in three stages: cutting, construction, and finishing. The time requirements for each model and manufacturing stage are given in the following table.

	Children's model	Office model	Deluxe model
Cutting	2 hr	3 hr	2 hr
Construction	2 hr	1 hr	3 hr
Finishing	1 hr	1 hr	2 hr

Each week the company has available a maximum of 100 hours for cutting, 100 hours for construction, and 65 hours for finishing. If all available time must be used, how many of each type of desk should be produced each week?   15 children's models, 10 office models, and 20 deluxe models, should be produced each week.

Writing in Mathematics

47. What is a system of linear equations in three variables?   47–52. Answers will vary.

48. How do you determine whether a given ordered triple is a solution of a system in three variables?

49. Describe in general terms how to solve a system in three variables.

50. Describe what happens when using algebraic methods to solve an inconsistent system.

51. Describe what happens when using algebraic methods to solve a system with dependent equations.

52. AIDS is taking a deadly toll on southern Africa. Describe how to use the techniques that you learned in this section to obtain a model for African life span using projections with AIDS. Let x represent the number of years after 1985 and let y represent African life span in that year.

African Life Span

Source: United Nations

Technology Exercises

53. Does your graphing utility have a feature that allows you to solve linear systems by entering coefficients and constant terms? If so, use this feature to verify the solutions to any five exercises that you worked by hand from Exercises 5–16.   53–54. Answers will vary.

54. Verify your results in Exercises 23–26 by using a graphing utility to graph the quadratic function. Trace along the curve and convince yourself that the three points given in the exercise lie on the function's graph.

Critical Thinking Exercises

55. Which one of the following is true?   c

   a. The ordered triple $(2, 15, 14)$ is the only solution of the equation $x + y - z = 3$.

   b. The equation $x - y - z = -6$ is satisfied by $(2, -3, 5)$.

   c. If two equations in a system are $x + y - z = 5$ and $2x + 2y - 2z = 7$, then the system must be inconsistent.

   d. An equation with four variables, such as $x + 2y - 3z + 5w = 2$, cannot be satisfied by real numbers.

56. A modernistic painting consists of triangles, rectangles, and pentagons, all drawn so as to not overlap or share sides. Within each rectangle are drawn 2 red roses, and each pentagon contains 5 carnations. How many triangles, rectangles, and pentagons appear in the painting if the painting contains a total of 40 geometric figures, 153 sides of geometric figures, and 72 flowers?   13 triangles, 21 rectangles, and 6 pentagons

57. In the following triangle, the degree measures of the three interior angles and two of the exterior angles are represented with variables. Find the measure of each interior angle.

$60°, 55°, 65°$

58. Two blocks of wood having the same length and width are placed on the top and bottom of a table, as shown in (a). Length A measures 32 centimeters. The blocks are rearranged as shown in (b). Length B measures 28 centimeters. Determine the height of the table.   30 centimeters

(a)   **(b)**

Review Exercises

In Exercises 59–61, graph each linear function.

59. $f(x) = -\dfrac{3}{4}x + 3$ (Section 2.3, Example 5)   59–61. See graphing answer section.

60. $-2x + y = 6$ (Section 2.3, Example 1)

61. $f(x) = -5$ (Section 2.3, Figure 2.21)

✓ MID-CHAPTER CHECK POINT

CHAPTER 3

What You Know: We learned to solve systems of linear equations. We solved systems in two variables by graphing, by the substitution method, and by the addition method. We solved systems in three variables by eliminating a variable, reducing the system to two equations in two variables. We saw that some systems, called inconsistent systems, have no solution, whereas other systems, called dependent systems, have infinitely many solutions. We used systems of linear equations to solve a variety of applied problems, including dual investment problems, mixture problems, motion problems, and business problems.

In Exercises 1–8, solve each system by the method of your choice.

1. $x = 3y - 7$
$4x + 3y = 2$ $(-1, 2)$

2. $3x + 4y = -5$
$2x - 3y = 8$ $(1, -2)$

3. $\dfrac{2x}{3} + \dfrac{y}{5} = 6$
$\dfrac{x}{6} - \dfrac{y}{2} = -4$ $(6, 10)$

4. $y = 4x - 5$
$8x - 2y = 10$
$\{(x, y) \mid y = 4x - 5\}$ or $\{(x, y) \mid 8x - 2y = 10\}$

5. $2x + 5y = 3$
$3x - 2y = 1$ $\left(\dfrac{11}{19}, \dfrac{7}{19}\right)$

6. $\dfrac{x}{12} - y = \dfrac{1}{4}$
$4x - 48y = 16$ $\varnothing$

7. $2x - y + 2z = -8$
$x + 2y - 3z = 9$
$3x - y - 4z = 3$
$(-1, 2, -2)$

8. $x \quad\quad - 3z = -5$
$2x - y + 2z = 16$
$7x - 3y - 5z = 19$
$(4, -2, 3)$

In Exercises 9–10, solve each system by graphing.

9. $2x - y = 4$
$x + y = 5$ $(3, 2)$

10. $y = x + 3$
$y = -\dfrac{1}{2}x$ $(-2, 1)$

11. A company is planning to manufacture PDAs (personal digital assistants). The fixed cost will be $400,000 and it will cost $20 to produce each PDA. Each PDA will be sold for $100.

    a. Write the cost function, C, of producing x PDAs. $C(x) = 400,000 + 20x$

    b. Write the revenue function, R, from the sale of x PDAs. $R(x) = 100x$

    c. Write the profit function, P, from producing and selling x PDAs. $P(x) = 80x - 400,000$

    d. Determine the break-even point. Describe what this means. $(5000, 500,000)$; The company will break even when it produces and sells 5000 PDAs. At this level, both revenue and cost are $500,000.

In Exercises 12–18, solve each problem.

12. Roses sell for $3 each and carnations for $1.50 each. If a mixed bouquet of 20 flowers consisting of roses and carnations costs $39, how many of each type of flower is in the bouquet?   6 roses and 14 carnations

13. You invested $15,000 in two funds paying 5% and 6% annual interest, respectively. At the end of the year, the total interest from these investments was $837. How much was invested at each rate? $6300 at 5% and $8700 at 6%

14. The manager of a gardening center needs to mix a plant food that is 13% nitrogen with one that is 18% nitrogen to obtain 50 gallons of a plant food that is 16% nitrogen. How many gallons of each of the plant foods must be used?   20 gallons of 13% nitrogen and 30 gallons of 18% nitrogen

15. With the current, you can row 9 miles in 2 hours. Against the current, your return trip takes 6 hours. Find your average rowing rate in still water and the average rate of the current.   rowing rate in still water: 3 miles per hour; current: 1.5 miles per hour

16. You invested $8000 in two funds paying 2% and 5% annual interest, respectively. At the end of the year, the interest from the 5% investment exceeded the interest from the 2% investment by $85. How much money was invested at each rate? $4500 at 2% and $3500 at 5%

17. Find the quadratic function $y = ax^2 + bx + c$ whose graph passes through the points $(-1, 0)$, $(1, 4)$, and $(2, 3)$. $y = -x^2 + 2x + 3$

18. A coin collection contains a mixture of 26 coins in nickels, dimes, and quarters. The coins have a total value of $4.00. The number of quarters is 2 less than the number of nickels and dimes combined. Determine the number of nickels, the number of dimes, and the number of quarters in the collection.   8 nickels, 6 dimes, and 12 quarters

SECTION

3.4

Objectives

1 Write the augmented matrix for a linear system.

2 Perform matrix row operations.

3 Use matrices to solve systems.

4 Use matrices to identify inconsistent and dependent systems.

MATRIX SOLUTIONS TO LINEAR SYSTEMS

Yes, we overindulged, but it was delicious. Anyway, a few hours of moderate activity and we'll just burn off those extra calories. The chart on the next page should help. It shows that the number of calories burned per hour depends on our weight. Four hours of tennis and we'll be as good as new!

How Fast You Burn Off Calories

Activity	Weight (pounds)					
	110	132	154	176	187	209
	Calories Burned per Hour					
Housework	175	210	245	285	300	320
Cycling	190	215	245	270	280	295
Tennis	335	380	425	470	495	520
Watching TV	60	70	80	85	90	95

The 24 numbers inside the red brackets are arranged in four rows and six columns. This rectangular array of 24 numbers, arranged in rows and columns and placed in brackets, is an example of a **matrix** (plural: **matrices**). The numbers inside the brackets are called **elements** of the matrix. Matrices are used to display information and to solve systems of linear equations.

1 Write the augmented matrix for a linear system.

Solving Linear Systems by Using Matrices A matrix gives us a shortened way of writing a system of equations. The first step in solving a system of linear equations using matrices is to write the *augmented matrix*. An **augmented matrix** has a vertical bar separating the columns of the matrix into two groups. The coefficients of each variable are placed to the left of the vertical line and the constants are placed to the right. If any variable is missing, its coefficient is 0. Here are two examples:

System of Linear Equations	Augmented Matrix
$\begin{aligned} x + 3y &= 5 \\ 2x - y &= -4 \end{aligned}$	$\left[\begin{array}{cc|c} 1 & 3 & 5 \\ 2 & -1 & -4 \end{array}\right]$
$\begin{aligned} x + 2y - 5z &= -19 \\ y + 3z &= 9 \\ z &= 4 \end{aligned}$	$\left[\begin{array}{ccc|c} 1 & 2 & -5 & -19 \\ 0 & 1 & 3 & 9 \\ 0 & 0 & 1 & 4 \end{array}\right]$

Notice how the second augmented matrix contains 1s down the diagonal from upper left to lower right and 0s below the 1s. This arrangement makes it easy to find the solution of the system of equations, as Example 1 shows.

EXAMPLE 1 Solving a System Using a Matrix

Write the solution for a system of equations represented by the matrix

$$\left[\begin{array}{ccc|c} 1 & 2 & -5 & -19 \\ 0 & 1 & 3 & 9 \\ 0 & 0 & 1 & 4 \end{array}\right].$$

SOLUTION The system represented by the given matrix is

$$\left[\begin{array}{ccc|c} 1 & 2 & -5 & -19 \\ 0 & 1 & 3 & 9 \\ 0 & 0 & 1 & 4 \end{array}\right] \rightarrow \begin{aligned} 1x + 2y - 5z &= -19 \\ 0x + 1y + 3z &= 9 \\ 0x + 0y + 1z &= 4 \end{aligned}.$$

This system can be simplified as follows.

$$\begin{aligned} x + 2y - 5z &= -19 && \text{Equation 1} \\ y + 3z &= 9 && \text{Equation 2} \\ z &= 4 && \text{Equation 3} \end{aligned}$$

The value of z is known. We can find y by back-substitution.

$$
\begin{array}{ll}
y + 3z = 9 & \text{Equation 2} \\
y + 3(4) = 9 & \text{Substitute 4 for } z. \\
y + 12 = 9 & \text{Multiply.} \\
y = -3 & \text{Subtract 12 from both sides.}
\end{array}
$$

With values for y and z, we can now use back-substitution to find x.

$$
\begin{array}{ll}
x + 2y - 5z = -19 & \text{Equation 1} \\
x + 2(-3) - 5(4) = -19 & \text{Substitute } -3 \text{ for } y \text{ and 4 for } z. \\
x - 6 - 20 = -19 & \text{Multiply.} \\
x - 26 = -19 & \text{Add.} \\
x = 7 & \text{Add 26 to both sides.}
\end{array}
$$

We see that $x = 7$, $y = -3$, and $z = 4$. The solution is $(7, -3, 4)$ and the solution set for the system is $\{(7, -3, 4)\}$. ■

 CHECK POINT 1 Write the solution for a system of equations represented by the matrix

$$
\left[
\begin{array}{ccc|c}
1 & -1 & 1 & 8 \\
0 & 1 & -12 & -15 \\
0 & 0 & 1 & 1
\end{array}
\right]. \quad (4, -3, 1) \text{ or } \{(4, -3, 1)\}
$$

Our goal in solving a linear system using matrices is to produce a matrix with 1s down the diagonal from upper left to lower right, called the **main diagonal**, and 0s below the 1s. In general, the matrix will be one of the following forms.

This is the desired form for systems with two equations.
$$
\left[
\begin{array}{cc|c}
1 & a & b \\
0 & 1 & c
\end{array}
\right]
$$

$$
\left[
\begin{array}{ccc|c}
1 & a & b & c \\
0 & 1 & d & e \\
0 & 0 & 1 & f
\end{array}
\right]
$$
This is the desired form for systems with three equations.

The last row of these matrices gives us the value of one variable. The other variables can then be found by back-substitution.

A matrix with 1s down the main diagonal and 0s below the 1s is said to be in **row-echelon form**. How do we produce a matrix in this form? We use **row operations** on the augmented matrix. These row operations are just like what you did when solving a linear system by the addition method. The difference is that we no longer write the variables, usually represented by $x, y,$ and z.

2 Perform matrix row operations.

MATRIX ROW OPERATIONS The following row operations produce matrices that represent systems with the same solution:

1. Two rows of a matrix may be interchanged. This is the same as interchanging two equations in the linear system.
2. The elements in any row may be multiplied by a nonzero number. This is the same as multiplying both sides of an equation by a nonzero number.
3. The elements in any row may be multiplied by a nonzero number, and these products may be added to the corresponding elements in any other row. This is the same as multiplying an equation by a nonzero number and then adding equations to eliminate a variable.

Two matrices are **row equivalent** if one can be obtained from the other by a sequence of row operations.

Each matrix row operation in the preceding box can be expressed symbolically as follows:

1. Interchange the elements in the ith and jth rows: $R_i \leftrightarrow R_j$.

2. Multiply each element in the ith row by k: kR_i.

3. Add k times the elements in row i to the corresponding elements in row j: $kR_i + R_j$.

EXAMPLE 2 Performing Matrix Row Operations

Use the matrix

$$\begin{bmatrix} 3 & 18 & -12 & 21 \\ 1 & 2 & -3 & 5 \\ -2 & -3 & 4 & -6 \end{bmatrix}$$

and perform each indicated row operation:

a. $R_1 \leftrightarrow R_2$ **b.** $\dfrac{1}{3}R_1$ **c.** $2R_2 + R_3$.

SOLUTION

a. The notation $R_1 \leftrightarrow R_2$ means to interchange the elements in row 1 and row 2. This results in the row-equivalent matrix

$$\begin{bmatrix} 1 & 2 & -3 & 5 \\ 3 & 18 & -12 & 21 \\ -2 & -3 & 4 & -6 \end{bmatrix}.$$

This was row 2; now it's row 1.

This was row 1; now it's row 2.

b. The notation $\frac{1}{3}R_1$ means to multiply each element in row 1 by $\frac{1}{3}$. This results in the row-equivalent matrix

$$\begin{bmatrix} \frac{1}{3}(3) & \frac{1}{3}(18) & \frac{1}{3}(-12) & \frac{1}{3}(21) \\ 1 & 2 & -3 & 5 \\ -2 & -3 & 4 & -6 \end{bmatrix} = \begin{bmatrix} 1 & 6 & -4 & 7 \\ 1 & 2 & -3 & 5 \\ -2 & -3 & 4 & -6 \end{bmatrix}.$$

c. The notation $2R_2 + R_3$ means to add 2 times the elements in row 2 to the corresponding elements in row 3. Replace the elements in row 3 by these sums. First, we find 2 times the elements in row 2, namely, $1, 2, -3$ and 5:

$$2(1) \text{ or } 2, \qquad 2(2) \text{ or } 4, \qquad 2(-3) \text{ or } -6, \qquad 2(5) \text{ or } 10.$$

Now we add these products to the corresponding elements in row 3. Although we use row 2 to find the products, row 2 does not change. It is the elements in row 3 that change, resulting in the row-equivalent matrix

$$\begin{bmatrix} 3 & 18 & -12 & 21 \\ 1 & 2 & -3 & 5 \\ -2 + 2 = 0 & -3 + 4 = 1 & 4 + (-6) = -2 & -6 + 10 = 4 \end{bmatrix} = \begin{bmatrix} 3 & 18 & -12 & 21 \\ 1 & 2 & -3 & 5 \\ 0 & 1 & -2 & 4 \end{bmatrix}.$$

✔ CHECK POINT **2** Use the matrix

$$\begin{bmatrix} 4 & 12 & -20 & | & 8 \\ 1 & 6 & -3 & | & 7 \\ -3 & -2 & 1 & | & -9 \end{bmatrix}$$

and perform each indicated row operation:

a. $R_1 \leftrightarrow R_2$ $\begin{bmatrix} 1 & 6 & -3 & | & 7 \\ 4 & 12 & -20 & | & 8 \\ -3 & -2 & 1 & | & -9 \end{bmatrix}$

b. $\dfrac{1}{4}R_1$ $\begin{bmatrix} 1 & 3 & -5 & | & 2 \\ 1 & 6 & -3 & | & 7 \\ -3 & -2 & 1 & | & -9 \end{bmatrix}$

c. $3R_2 + R_3$. $\begin{bmatrix} 4 & 12 & -20 & | & 8 \\ 1 & 6 & -3 & | & 7 \\ 0 & 16 & -8 & | & 12 \end{bmatrix}$

3 Use matrices to solve systems.

The process that we use to solve linear systems using matrix row operations is often called **Gaussian elimination**, after the German mathematician Carl Friedrich Gauss (1777–1855). Here are the steps used in solving systems with matrices:

SOLVING LINEAR SYSTEMS USING MATRICES

1. Write the augmented matrix for the system.

2. Use matrix row operations to simplify the matrix to one with 1s down the main diagonal from upper left to lower right, and 0s below the 1s.

3. Write the system of linear equations corresponding to the matrix in step 2 and use back-substitution to find the system's solution.

EXAMPLE 3 Using Matrices to Solve a Linear System

Use matrices to solve the system:

$$4x - 3y = -15$$
$$x + 2y = -1.$$

SOLUTION

Step 1. **Write the augmented matrix for the system.**

Linear System	**Augmented Matrix**
$4x - 3y = -15$	$\begin{bmatrix} 4 & -3 & | & -15 \\ 1 & 2 & | & -1 \end{bmatrix}$
$x + 2y = -1$	

Step 2. **Use matrix row operations to simplify the matrix to one with 1s down the diagonal from upper left to lower right, and 0s below the 1s.** Our goal is to obtain a matrix of the form

$$\begin{bmatrix} 1 & a & | & b \\ 0 & 1 & | & c \end{bmatrix}.$$

Our first step in achieving this goal is to get 1 in the top position of the first column.

We want 1 in this position. $\begin{bmatrix} 4 & -3 & | & -15 \\ 1 & 2 & | & -1 \end{bmatrix}$

To get 1 in this position, we interchange rows 1 and 2: $R_1 \leftrightarrow R_2$.

$\begin{bmatrix} 1 & 2 & | & -1 \\ 4 & -3 & | & -15 \end{bmatrix}$ This was row 2; now it's row 1.

This was row 1; now it's row 2.

Now we want 0 below the 1 in the first column.

We want 0 in this position. $\begin{bmatrix} 1 & 2 & | & -1 \\ 4 & -3 & | & -15 \end{bmatrix}$

Let's get a 0 where there is now a 4. If we multiply the top row of numbers by -4 and add these products to the second row of numbers, we will get 0 in this position: $-4R_1 + R_2$. *We change only row 2.*

$-4R_1 + R_2$ $\begin{bmatrix} 1 & 2 & | & -1 \\ -4(1) + 4 & -4(2) + (-3) & | & -4(-1) + (-15) \end{bmatrix} = \begin{bmatrix} 1 & 2 & | & -1 \\ 0 & -11 & | & -11 \end{bmatrix}$

row 1 element · row 2 element · row 1 element · row 2 element · row 1 element · row 2 element

We move on to the second column. We want 1 in the second row, second column.

We want 1 in this position. $\begin{bmatrix} 1 & 2 & | & -1 \\ 0 & -11 & | & -11 \end{bmatrix}$

To get 1 in the desired position, we multiply -11 by its multiplicative inverse, or reciprocal, $-\frac{1}{11}$. Therefore, we multiply all the numbers in the second row by $-\frac{1}{11}$: $-\frac{1}{11}R_2$.

$-\frac{1}{11}R_2$ $\begin{bmatrix} 1 & 2 & | & -1 \\ -\frac{1}{11}(0) & -\frac{1}{11}(-11) & | & -\frac{1}{11}(-11) \end{bmatrix} = \begin{bmatrix} 1 & 2 & | & -1 \\ 0 & 1 & | & 1 \end{bmatrix}$

We now have the desired matrix with 1s down the main diagonal and 0 below the 1.

Step 3. **Write the system of linear equations corresponding to the matrix in step 2 and use back-substitution to find the system's solution.** The system represented by the matrix in step 2 is

$$\begin{bmatrix} 1 & 2 & | & -1 \\ 0 & 1 & | & 1 \end{bmatrix} \rightarrow \begin{array}{l} 1x + 2y = -1 \\ 0x + 1y = 1 \end{array} \quad \text{or} \quad \begin{array}{l} x + 2y = -1 \\ y = 1. \end{array}$$

We immediately see that the value for y is 1. To find x, we back-substitute 1 for y in the first equation.

$$\begin{array}{ll} x + 2y = -1 & \text{Equation 1} \\ x + 2 \cdot 1 = -1 & \text{Substitute 1 for } y. \\ x = -3 & \text{Solve for } x. \end{array}$$

With $x = -3$ and $y = 1$, the proposed solution is $(-3, 1)$. Take a moment to show that $(-3, 1)$ satisfies both equations. The solution is $(-3, 1)$ and the solution set is $\{(-3, 1)\}$.

CHECK POINT 3 Use matrices to solve the system:

$$2x - y = -4$$
$$x + 3y = 5. \quad (-1, 2) \text{ or } \{(-1, 2)\}$$

When solving a system of linear equations in three variables, use matrix row operations in the following order to obtain a matrix with 1s down the main diagonal and 0s below the 1s:

Get 1 in the upper left-hand corner.   Use the 1 in the first column to get 0s below it.   Get 1 in the second row, second column position.   Use the 1 in the second column to get 0 below it.   Get 1 in the third row, third column position.

EXAMPLE 4   Using Matrices to Solve a Linear System

Use matrices to solve the system:

$$3x + y + 2z = 31$$
$$x + y + 2z = 19$$
$$x + 3y + 2z = 25.$$

SOLUTION

Step 1. **Write the augmented matrix for the system.**

Linear System	**Augmented Matrix**
$3x + y + 2z = 31$	$\begin{bmatrix} 3 & 1 & 2 & 31 \\ 1 & 1 & 2 & 19 \\ 1 & 3 & 2 & 25 \end{bmatrix}$
$x + y + 2z = 19$	
$x + 3y + 2z = 25$	

Step 2. **Use matrix row operations to simplify the matrix to one with 1s down the diagonal from upper left to lower right, and 0s below the 1s.** Our first step in achieving this goal is to get 1 in the top position of the first column.

We want 1 in this position. $\begin{bmatrix} 3 & 1 & 2 & 31 \\ 1 & 1 & 2 & 19 \\ 1 & 3 & 2 & 25 \end{bmatrix}$

To get 1 in this position, we interchange rows 1 and 2: $R_1 \leftrightarrow R_2$. (We could also interchange rows 1 and 3 to attain our goal.)

$\begin{bmatrix} 1 & 1 & 2 & 19 \\ 3 & 1 & 2 & 31 \\ 1 & 3 & 2 & 25 \end{bmatrix}$ This was row 2; now it's row 1.

This was row 1; now it's row 2.

Now we want to get 0s below the 1 in the first column.

We want 0 in these positions. $\begin{bmatrix} 1 & 1 & 2 & 19 \\ 3 & 1 & 2 & 31 \\ 1 & 3 & 2 & 25 \end{bmatrix}$

To get a 0 where there is now a 3, multiply the top row of numbers by -3 and add these products to the second row of numbers: $-3R_1 + R_2$. To get a 0 where there is now a 1, multiply the top row of numbers by -1 and add these products to the third row of numbers: $-1R_1 + R_3$. Although we are using row 1 to find the products, the numbers in row 1 do not change.

$$\begin{array}{c} \\ -3R_1 + R_2 \\ \\ -1R_1 + R_3 \end{array} \left[\begin{array}{ccc|c} 1 & 1 & 2 & 19 \\ -3(1)+3 & -3(1)+1 & -3(2)+2 & -3(19)+31 \\ -1(1)+1 & -1(1)+3 & -1(2)+2 & -1(19)+25 \end{array}\right]$$

$$= \left[\begin{array}{ccc|c} 1 & 1 & 2 & 19 \\ 0 & -2 & -4 & -26 \\ 0 & 2 & 0 & 6 \end{array}\right]$$

We move on to the second column. We want 1 in the second row, second column.

We want 1 in this position.
$$\left[\begin{array}{ccc|c} 1 & 1 & 2 & 19 \\ 0 & -2 & -4 & -26 \\ 0 & 2 & 0 & 6 \end{array}\right]$$

To get 1 in the desired position, we multiply -2 by its reciprocal, $-\frac{1}{2}$. Therefore, we multiply all the numbers in the second row by $-\frac{1}{2}$: $-\frac{1}{2}R_2$.

$$\begin{array}{c} \\ -\frac{1}{2}R_2 \\ \\ \end{array} \left[\begin{array}{ccc|c} 1 & 1 & 2 & 19 \\ -\frac{1}{2}(0) & -\frac{1}{2}(-2) & -\frac{1}{2}(-4) & -\frac{1}{2}(-26) \\ 0 & 2 & 0 & 6 \end{array}\right] = \left[\begin{array}{ccc|c} 1 & 1 & 2 & 19 \\ 0 & 1 & 2 & 13 \\ 0 & 2 & 0 & 6 \end{array}\right].$$

We want 0 in this position.

We are not yet done with the second column. The voice balloon shows that we want to get a 0 where there is now a 2. If we multiply the second row of numbers by -2 and add these products to the third row of numbers, we will get 0 in this position: $-2R_2 + R_3$. Although we are using the numbers in row 2 to find the products, the numbers in row 2 do not change.

$$\begin{array}{c} \\ \\ -2R_2 + R_3 \end{array} \left[\begin{array}{ccc|c} 1 & 1 & 2 & 19 \\ 0 & 1 & 2 & 13 \\ -2(0)+0 & -2(1)+2 & -2(2)+0 & -2(13)+6 \end{array}\right] = \left[\begin{array}{ccc|c} 1 & 1 & 2 & 19 \\ 0 & 1 & 2 & 13 \\ 0 & 0 & -4 & -20 \end{array}\right]$$

We move on to the third column. We want 1 in the third row, third column.

We want 1 in this position.
$$\left[\begin{array}{ccc|c} 1 & 1 & 2 & 19 \\ 0 & 1 & 2 & 13 \\ 0 & 0 & -4 & -20 \end{array}\right]$$

To get 1 in the desired position, we multiply -4 by its reciprocal, $-\frac{1}{4}$. Therefore, we multiply all the numbers in the third row by $-\frac{1}{4}$: $-\frac{1}{4}R_3$.

$$\begin{array}{c} \\ \\ -\frac{1}{4}R_3 \end{array} \left[\begin{array}{ccc|c} 1 & 1 & 2 & 19 \\ 0 & 1 & 2 & 13 \\ -\frac{1}{4}(0) & -\frac{1}{4}(0) & -\frac{1}{4}(-4) & -\frac{1}{4}(-20) \end{array}\right] = \left[\begin{array}{ccc|c} 1 & 1 & 2 & 19 \\ 0 & 1 & 2 & 13 \\ 0 & 0 & 1 & 5 \end{array}\right].$$

We now have the desired matrix with 1s down the main diagonal and 0s below the 1s.

Most graphing utilities can convert an augmented matrix to one with 1s down the diagonal from upper left to lower right, and 0s below the 1s. Recall that this is called row-echelon form. However, row-echelon form is not unique. Your graphing utility might give a row-echelon form different from the one you obtained by hand. However, all row-echelon forms for a given system's augmented matrix produce the same solution to the system. Enter the augmented matrix and name it A. Then use the $\boxed{\text{REF}}$ (row-echelon form) command on matrix A.

This is the augmented matrix in Example 4

This matrix is equivalent to the row-echelon form we obtained by hand. It produces the same solution: $z = 5$, $y = 3$, and $x = 6$.

4 Use matrices to identify inconsistent and dependent systems.

Step 3. **Write the system of linear equations corresponding to the matrix in step 2 and use back-substitution to find the system's solution.** The system represented by the matrix in step 2 is

$$\begin{bmatrix} 1 & 1 & 2 & | & 19 \\ 0 & 1 & 2 & | & 13 \\ 0 & 0 & 1 & | & 5 \end{bmatrix} \rightarrow \begin{matrix} 1x + 1y + 2z = 19 \\ 0x + 1y + 2z = 13 \\ 0x + 0y + 1z = 5 \end{matrix} \text{ or } \begin{matrix} x + y + 2z = 19 & (1) \\ y + 2z = 13. & (2) \\ z = 5 & (3) \end{matrix}$$

We immediately see from equation (3) that the value for z is 5. To find y, we back-substitute 5 for z in the second equation.

$$y + 2z = 13 \quad \text{Equation (2)}$$
$$y + 2(5) = 13 \quad \text{Substitute 5 for } z.$$
$$y = 3 \quad \text{Solve for } y.$$

Finally, back-substitute 3 for y and 5 for z in the first equation.

$$x + y + 2z = 19 \quad \text{Equation (1)}$$
$$x + 3 + 2(5) = 19 \quad \text{Substitute 3 for } y \text{ and 5 for } z.$$
$$x + 13 = 19 \quad \text{Multiply and add.}$$
$$x = 6 \quad \text{Subtract 13 both sides.}$$

The solution of the original system is $(6, 3, 5)$ and the solution set is $\{(6, 3, 5)\}$. Check to see that the solution satisfies all three equations in the given system. ∎

✔ **CHECK POINT 4** Use matrices to solve the system:

$$\begin{matrix} 2x + & y + 2z = 18 \\ x - & y + 2z = 9 \\ x + 2y - & z = 6. \end{matrix} \quad (5, 2, 3) \text{ or } \{(5, 2, 3)\}$$

Modern supercomputers are capable of solving systems with more than 600,000 variables. The augmented matrices for such systems are huge, but the solution using matrices is exactly like what we did in Example 4. Work with the augmented matrix, one column at a time. First, get 1 in the desired position. Then get 0s below the 1.

Inconsistent Systems and Systems with Dependent Equations When solving a system using matrices, you might obtain a matrix with a row in which the numbers to the left of the vertical bar are all zeros, but a nonzero number number appears on the right. In such a case, the system is inconsistent and has no solution. For example, a system of equations that yields the following matrix is an inconsistent system:

$$\begin{bmatrix} 1 & -2 & | & 3 \\ 0 & 0 & | & -4 \end{bmatrix}.$$

The second row of the matrix represents the equation $0x + 0y = -4$, which is false for all values of x and y.

If you obtain a matrix in which a 0 appears across an entire row, the system contains dependent equations and has infinitely many solutions. This row of zeros represents $0x + 0y = 0$ or $0x + 0y + 0z = 0$. These equations are satisfied by infinitely many ordered pairs or triples.

3.4 EXERCISE SET

Student Solutions Manual   CD/Video   PH Math/Tutor Center   MathXL Tutorials on CD   MathXL®   MyMathLab   Interactmath.com

Practice Exercises

In Exercises 1–8, write the system of linear equations represented by the augmented matrix. Use x, y, and, if necessary, z, for the variables. Once the system is written, use back-substitution to find its solution.

1. $\begin{bmatrix} 1 & -3 & | & 11 \\ 0 & 1 & | & -3 \end{bmatrix}$

$x - 3y = 11$
$y = -3$
$(2, -3)$

2. $\begin{bmatrix} 1 & 3 & | & 5 \\ 0 & 1 & | & 2 \end{bmatrix}$

$x + 3y = 5$
$y = 2$
$(-1, 2)$

3. $\begin{bmatrix} 1 & -3 & | & 1 \\ 0 & 1 & | & -1 \end{bmatrix}$

$x - 3y = 1$
$y = -1$
$(-2, -1)$

4. $\begin{bmatrix} 1 & 2 & | & 13 \\ 0 & 1 & | & 4 \end{bmatrix}$

$x + 2y = 13$
$y = 4$
$(5, 4)$

5. $\begin{bmatrix} 1 & 0 & -4 & | & 5 \\ 0 & 1 & -12 & | & 13 \\ 0 & 0 & 1 & | & -\frac{1}{2} \end{bmatrix}$

$x - 4z = 5$
$y - 12z = 13$
$z = -\frac{1}{2}$
$\left(3, 7, -\frac{1}{2}\right)$

6. $\begin{bmatrix} 1 & 2 & 1 & | & 0 \\ 0 & 1 & 0 & | & -2 \\ 0 & 0 & 1 & | & 3 \end{bmatrix}$

$x + 2y + z = 0$
$y = -2$
$z = 3$
$(1, -2, 3)$

7. $\begin{bmatrix} 1 & \frac{1}{2} & 1 & | & \frac{11}{2} \\ 0 & 1 & \frac{3}{2} & | & 7 \\ 0 & 0 & 1 & | & 4 \end{bmatrix}$

$x + \frac{1}{2}y + z = \frac{11}{2}$
$y + \frac{3}{2}z = 7$
$z = 4$
$(1, 1, 4)$

8. $\begin{bmatrix} 1 & 1 & 0 & | & 3 \\ 0 & 1 & \frac{3}{2} & | & -2 \\ 0 & 0 & 1 & | & 0 \end{bmatrix}$

$x + y = 3$
$y + \frac{3}{2}z = -2$
$z = 0$
$(5, -2, 0)$

In Exercises 9–22, perform each matrix row operation and write the new matrix.

9. $\begin{bmatrix} 2 & 2 & | & 5 \\ 1 & -\frac{3}{2} & | & 5 \end{bmatrix} R_1 \leftrightarrow R_2$ $\begin{bmatrix} 1 & -\frac{3}{2} & | & 5 \\ 2 & 2 & | & 5 \end{bmatrix}$

10. $\begin{bmatrix} -6 & 9 & | & 4 \\ 1 & -\frac{3}{2} & | & 4 \end{bmatrix} R_1 \leftrightarrow R_2$ $\begin{bmatrix} 1 & -\frac{3}{2} & | & 4 \\ -6 & 9 & | & 4 \end{bmatrix}$

11. $\begin{bmatrix} -6 & 8 & | & -12 \\ 3 & 5 & | & -2 \end{bmatrix} -\frac{1}{6}R_1$ $\begin{bmatrix} 1 & -\frac{4}{3} & | & 2 \\ 3 & 5 & | & -2 \end{bmatrix}$

12. $\begin{bmatrix} -2 & 3 & | & -10 \\ 4 & 2 & | & 5 \end{bmatrix} -\frac{1}{2}R_1$ $\begin{bmatrix} 1 & -\frac{3}{2} & | & 5 \\ 4 & 2 & | & 5 \end{bmatrix}$

13. $\begin{bmatrix} 1 & -3 & | & 5 \\ 2 & 6 & | & 4 \end{bmatrix} -2R_1 + R_2$ $\begin{bmatrix} 1 & -3 & | & 5 \\ 0 & 12 & | & -6 \end{bmatrix}$

14. $\begin{bmatrix} 1 & -3 & | & 1 \\ 2 & 1 & | & -5 \end{bmatrix} -2R_1 + R_2$ $\begin{bmatrix} 1 & -3 & | & 1 \\ 0 & 7 & | & -7 \end{bmatrix}$

15. $\begin{bmatrix} 1 & -\frac{3}{2} & | & \frac{7}{2} \\ 3 & 4 & | & 2 \end{bmatrix} -3R_1 + R_2$ $\begin{bmatrix} 1 & -\frac{3}{2} & | & \frac{7}{2} \\ 0 & \frac{17}{2} & | & -\frac{17}{2} \end{bmatrix}$

16. $\begin{bmatrix} 1 & -\frac{2}{5} & | & \frac{3}{4} \\ 4 & 2 & | & -1 \end{bmatrix} -4R_1 + R_2$ $\begin{bmatrix} 1 & -\frac{2}{5} & | & \frac{3}{4} \\ 0 & \frac{18}{5} & | & -4 \end{bmatrix}$

17. $\begin{bmatrix} 2 & -6 & 4 & | & 10 \\ 1 & 5 & -5 & | & 0 \\ 3 & 0 & 4 & | & 7 \end{bmatrix} \frac{1}{2}R_1$ $\begin{bmatrix} 1 & -3 & 2 & | & 5 \\ 1 & 5 & -5 & | & 0 \\ 3 & 0 & 4 & | & 7 \end{bmatrix}$

18. $\begin{bmatrix} 3 & -12 & 6 & | & 9 \\ 1 & -4 & 4 & | & 0 \\ 2 & 0 & 7 & | & 4 \end{bmatrix} \frac{1}{3}R_1$ $\begin{bmatrix} 1 & -4 & 2 & | & 3 \\ 1 & -4 & 4 & | & 0 \\ 2 & 0 & 7 & | & 4 \end{bmatrix}$

19. $\begin{bmatrix} 1 & -3 & 2 & | & 0 \\ 3 & 1 & -1 & | & 7 \\ 2 & -2 & 1 & | & 3 \end{bmatrix} -3R_1 + R_2$ $\begin{bmatrix} 1 & -3 & 2 & | & 0 \\ 0 & 10 & -7 & | & 7 \\ 2 & -2 & 1 & | & 3 \end{bmatrix}$

20. $\begin{bmatrix} 1 & -1 & 5 & | & -6 \\ 3 & 3 & -1 & | & 10 \\ 1 & 3 & 2 & | & 5 \end{bmatrix} -3R_1 + R_2$ $\begin{bmatrix} 1 & -1 & 5 & | & -6 \\ 0 & 6 & -16 & | & 28 \\ 1 & 3 & 2 & | & 5 \end{bmatrix}$

21. $\begin{bmatrix} 1 & 1 & -1 & | & 6 \\ 2 & -1 & 1 & | & -3 \\ 3 & -1 & -1 & | & 4 \end{bmatrix} \begin{matrix} -2R_1 + R_2 \\ \text{and} \\ -3R_1 + R_3 \end{matrix}$ $\begin{bmatrix} 1 & 1 & -1 & | & 6 \\ 0 & -3 & 3 & | & -15 \\ 0 & -4 & 2 & | & -14 \end{bmatrix}$

22. $\begin{bmatrix} 1 & 2 & 1 & | & 2 \\ -2 & -1 & 2 & | & 5 \\ 1 & 3 & -2 & | & -8 \end{bmatrix} \begin{matrix} 2R_1 + R_2 \\ \text{and} \\ -1R_1 + R_3 \end{matrix}$ $\begin{bmatrix} 1 & 2 & 1 & | & 2 \\ 0 & 3 & 4 & | & 9 \\ 0 & 1 & -3 & | & -10 \end{bmatrix}$

In Exercises 23–46, solve each system using matrices. If there is no solution or if there are infinitely many solutions and a system's equations are dependent, so state.

23. $x + y = 6$
$x - y = 2$ $(4, 2)$

24. $x + 2y = 11$
$x - y = -1$ $(3, 4)$

25. $2x + y = 3$
$x - 3y = 12$ $(3, -3)$

26. $3x - 5y = 7$
$x - y = 1$ $(-1, -2)$

27. $5x + 7y = -25$
$11x + 6y = -8$ $(2, -5)$

28. $3x - 5y = 22$
$4x - 2y = 20$ $(4, -2)$

29. $4x - 2y = 5$
$-2x + y = 6$ no solution or $\emptyset$

30. $-3x + 4y = 12$
$6x - 8y = 16$ no solution or $\emptyset$

31. $x - 2y = 1$ infinitely many solutions;
$-2x + 4y = -2$ dependent equations

32. $3x - 6y = 1$
$2x - 4y = \dfrac{2}{3}$ infinitely many solutions; dependent equations

33. $x + y - z = -2$
$2x - y + z = 5$ $(1, -1, 2)$
$-x + 2y + 2z = 1$

34. $x - 2y - z = 2$
$2x - y + z = 4$ $(1, -1, 1)$
$-x + y - 2z = -4$

35. $x + 3y = 0$
$x + y + z = 1$ $(3, -1, -1)$
$3x - y - z = 11$

36. $3y - z = -1$
$x + 5y - z = -4$ $(-3, 0, 1)$
$-3x + 6y + 2z = 11$

37. $2x + 2y + 7z = -1$
$2x + y + 2z = 2$ $(1, 2, -1)$
$4x + 6y + z = 15$

38. $3x + 2y + 3z = 3$
$4x - 5y + 7z = 1$ $(2, 0, -1)$
$2x + 3y - 2z = 6$

39. $x + y + z = 6$
$x - z = -2$ $(1, 2, 3)$
$y + 3z = 11$

40. $x + y + z = 3$
$-y + 2z = 1$ $(1, 1, 1)$
$-x + z = 0$

41. $x - y + 3z = 4$
$2x - 2y + 6z = 7$ no solution or $\emptyset$
$3x - y + 5z = 14$

42. $3x - y + 2z = 4$
$-6x + 2y - 4z = 1$ no solution or $\emptyset$
$5x - 3y + 8z = 0$

43. $x - 2y + z = 4$
$5x - 10y + 5z = 20$ infinitely many solutions; dependent equations
$-2x + 4y - 2z = -8$

44. $x - 3y + z = 2$
$4x - 12y + 4z = 8$ infinitely many solutions; dependent equations
$-2x + 6y - 2z = -4$

45. $x + y = 1$
$y + 2z = -2$ $(-1, 2, -2)$
$2x - z = 0$

46. $x + 3y = 3$
$y + 2z = -8$ $(3, 0, -4)$
$x - z = 7$

Practice Plus

In Exercises 47–48, write the system of linear equations represented by the augmented matrix. Use w, x, y, and z, for the variables. Once the system is written, use back-substitution to find its solution, (w, x, y, z).

47. $\begin{bmatrix} 1 & -1 & 1 & 1 & | & 3 \\ 0 & 1 & -2 & -1 & | & 0 \\ 0 & 0 & 1 & 6 & | & 17 \\ 0 & 0 & 0 & 1 & | & 3 \end{bmatrix}$ $w - x + y + z = 3; (2, 1, -1, 3)$
$x - 2y - z = 0$
$y + 6z = 17$
$z = 3$

48. $\begin{bmatrix} 1 & 2 & -1 & 0 & | & 2 \\ 0 & 1 & 1 & -2 & | & -3 \\ 0 & 0 & 1 & -1 & | & -2 \\ 0 & 0 & 0 & 1 & | & 3 \end{bmatrix}$ $w + 2x - y = 2; (-1, 2, 1, 3)$
$x + y - 2z = -3$
$y - z = -2$
$z = 3$

In Exercises 49–50, perform each matrix row operation and write the new matrix.

49. $\begin{bmatrix} 1 & -1 & 1 & 1 & | & 3 \\ 0 & 1 & -2 & -1 & | & 0 \\ 2 & 0 & 3 & 4 & | & 11 \\ 5 & 1 & 2 & 4 & | & 6 \end{bmatrix}$ $\begin{matrix} \\ \\ -2R_1 + R_3 \\ -5R_1 + R_4 \end{matrix}$ $\begin{bmatrix} 1 & -1 & 1 & 1 & | & 3 \\ 0 & 1 & -2 & -1 & | & 0 \\ 0 & 2 & 1 & 2 & | & 5 \\ 0 & 6 & -3 & -1 & | & -9 \end{bmatrix}$

50. $\begin{bmatrix} 1 & -5 & 2 & -2 & | & 4 \\ 0 & 1 & -3 & -1 & | & 0 \\ 3 & 0 & 2 & -1 & | & 6 \\ -4 & 1 & 4 & 2 & | & -3 \end{bmatrix}$ $\begin{matrix} \\ \\ -3R_1 + R_3 \\ 4R_1 + R_4 \end{matrix}$ $\begin{bmatrix} 1 & -5 & 2 & -2 & | & 4 \\ 0 & 1 & -3 & -1 & | & 0 \\ 0 & 15 & -4 & 5 & | & -6 \\ 0 & -19 & 12 & -6 & | & 13 \end{bmatrix}$

In Exercises 51–52, solve each system using matrices. You will need to use matrix row operations to obtain matrices like those in Exercises 47 and 48, with 1s down the main diagonal and 0s below the 1s. Express the solution as (w, x, y, z).

51. $w + x + y + z = 4$
$2w + x - 2y - z = 0$
$w - 2x - y - 2z = -2$
$3w + 2x + y + 3z = 4$ $(1, 2, 3, -2)$

52. $w + x + y + z = 5$
$w + 2x - y - 2z = -1$
$w - 3x - 3y - z = -1$
$2w - x + 2y - z = -2$ $(2, 1, -1, 3)$

Application Exercises

53. The table shows the number of inmates in federal and state prisons in the United States for three selected years.

x (Number of Years after 1980)	1	5	10
y (Number of Inmates, in thousands)	344	480	740

a. Use the quadratic function $y = ax^2 + bx + c$ to model the data. Solve the system of linear equations involving a, b, and c using matrices. $y = 2x^2 + 22x + 320$

b. Predict the number of inmates in the year 2010.   In 2010, there will be 2780 thousand, or 2,780,000, inmates.

c. List one factor that would change the accuracy of this model for the year 2010.   Answers will vary.

54. A football is kicked straight upward. The graph shows the ball's height, y, in feet, after x seconds.

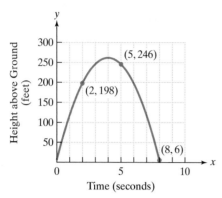

a. Find the quadratic function

$$y = ax^2 + bx + c$$

whose graph passes through the three points labeled on the graph. Solve the system of linear equations involving $a, b,$ and c using matrices. $y = -16x^2 + 128x + 6$

b. Use the function in part (a) to find the value for y when $x = 7$. Describe what this means. 118; After 7 seconds, the ball's height will be 118 feet.

Write a system of linear equations in three variables to solve Exercises 55–56. Then use matrices to solve the system. Exercises 55–56 are based on a Time/CNN telephone poll that included never-married single women between the ages of 18 and 49 and never-married men between the ages of 18 and 49. The circle graphs show the results for one of the questions in the poll.

If You Couldn't Find the Perfect Mate, Would You Marry Someone Else?

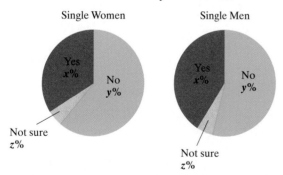

55. For single women in the poll, the percentage who said no exceeded the combined percentages for those who said yes and those who said not sure by 22%. If the percentage who said yes is doubled, it is 7% more than the percentage who said no. Find the percentage of single women who responded yes, no, and not sure. yes: 34%; no: 61%; not sure: 5%

56. For single men in the poll, the percentage who said no exceeded the combined percentages for those who said yes and those who said not sure by 8%. If the percentage who said yes is doubled, it is 28% more than the percentage who said no. Find the percentage of single men who responded yes, no, and not sure. yes: 41%; no: 54%; not sure: 5%

Writing in Mathematics

57. What is a matrix? 57–64. Answers will vary.
58. Describe what is meant by the augmented matrix of a system of linear equations.
59. In your own words, describe each of the three matrix row operations. Give an example of each of the operations.
60. Describe how to use matrices and row operations to solve a system of linear equations.
61. When solving a system using matrices, how do you know if the system has no solution?
62. When solving a system using matrices, how do you know if the system has infinitely many solutions?

Technology Exercises

63. Most graphing utilities can perform row operations on matrices. Consult the owner's manual for your graphing utility to learn proper keystroke sequences for performing these operations. Then duplicate the row operations of any three exercises that you solved from Exercises 11–20.
64. If your graphing utility has a REF (row-echelon form) command, use this feature to verify your work with any five systems from Exercises 23–46.

Critical Thinking Exercises

65. Which one of the following is true? d

a. A matrix row operation such as $-\frac{4}{5}R_1 + R_2$ is not permitted because of the negative fraction.
b. The augmented matrix for the system

$$\begin{aligned} x - 3y &= 5 \\ y - 2z &= 7 \\ 2x + z &= 4 \end{aligned} \quad \text{is} \quad \begin{bmatrix} 1 & -3 & 5 \\ 1 & -2 & 7 \\ 2 & 1 & 4 \end{bmatrix}.$$

c. In solving a linear system of three equations in three variables, we begin with the augmented matrix and use row operations to obtain a row-equivalent matrix with 0s down the diagonal from left to right and 1s below each 0.
d. None of the above is true.

66. The vitamin content per ounce for three foods is given in the following table.

	Milligrams per Ounce		
	Thiamin	**Riboflavin**	**Niacin**
Food A	3	7	1
Food B	1	5	3
Food C	3	8	2

a. Use matrices to show that no combination of these foods can provide exactly 14 milligrams of thiamin, 32 milligrams of riboflavin, and 9 milligrams of niacin. Matrices show that the system of equations for this problem is inconsistent.
b. Use matrices to describe in practical terms what happens if the riboflavin requirement is increased by 5 mg and the other requirements stay the same. Matrices show that the system has infinitely many solutions. There are many combinations of foods that satisfy the new requirements.

Review Exercises

67. If $f(x) = -3x + 10$, find $f(2a - 1)$. (Section 2.1, Example 3) $-6a + 13$

68. If $f(x) = 3x$ and $g(x) = 2x - 3$, find $(fg)(-1)$. (Section 2.2, Example 4) 15

69. Simplify: $\dfrac{-4x^8 y^{-12}}{12x^{-3} y^{24}}$. (Section 1.6, Example 9) $-\dfrac{x^{11}}{3y^{36}}$

SECTION 3.5

Objectives

1 Evaluate a second-order determinant.

2 Solve a system of linear equations in two variables using Cramer's rule.

3 Evaluate a third-order determinant.

4 Solve a system of linear equations in three variables using Cramer's rule.

5 Use determinants to identify inconsistent and dependent systems.

DETERMINANTS AND CRAMER'S RULE

A portion of Charles Babbage's unrealized difference engine

As cyberspace absorbs more and more of our work, play, shopping, and socializing, where will it all end? Which activities will still be offline in 2025?

Our technologically transformed lives can be traced back to the English inventor Charles Babbage (1791–1871). Babbage knew of a method for solving linear systems called *Cramer's rule*, in honor of the Swiss geometer Gabriel Cramer (1704–1752). Cramer's rule was simple, but involved numerous multiplications for large systems. Babbage designed a machine, called the "difference engine," that consisted of toothed wheels on shafts for performing these multiplications. Despite the fact that only one-seventh of the functions ever worked, Babbage's invention demonstrated how complex calculations could be handled mechanically. In 1944, scientists at IBM used the lessons of the difference engine to create the world's first computer.

Those who invented computers hoped to relegate the drudgery of repeated computation to a machine. In this section, we look at a method for solving linear systems that played a critical role in this process. The method uses arrays of numbers, called *determinants*. As with matrix methods, solutions are obtained by writing down the coefficients and constants of a linear system and performing operations with them.

The Determinant of a 2 × 2 Matrix   A matrix of **order** $m \times n$ has m rows and n columns. If $m = n$, a matrix has the same number of rows as columns and is called a **square matrix**. Associated with every square matrix is a real number, called its **determinant**. The determinant for a 2×2 square matrix is defined as follows:

1 Evaluate a second-order determinant.

STUDY TIP

To evaluate a second-order determinant, find the difference of the product of the two diagonals.

$$\begin{vmatrix} a_1 & b_1 \\ a_2 & b_2 \end{vmatrix} = a_1 b_2 - a_2 b_1$$

DEFINITION OF THE DETERMINANT OF A 2 × 2 MATRIX   The determinant of the matrix $\begin{bmatrix} a_1 & b_1 \\ a_2 & b_2 \end{bmatrix}$ is denoted by $\begin{vmatrix} a_1 & b_1 \\ a_2 & b_2 \end{vmatrix}$ and is defined by

$$\begin{vmatrix} a_1 & b_1 \\ a_2 & b_2 \end{vmatrix} = a_1 b_2 - a_2 b_1.$$

We also say that the **value** of the **second-order determinant** $\begin{vmatrix} a_1 & b_1 \\ a_2 & b_2 \end{vmatrix}$ is $a_1 b_2 - a_2 b_1$.

Example 1 illustrates that the determinant of a matrix may be positive or negative. The determinant can also have 0 as its value.

EXAMPLE 1 Evaluating the Determinant of a 2 × 2 Matrix

Evaluate the determinant of each of the following matrices:

a. $\begin{bmatrix} 5 & 6 \\ 7 & 3 \end{bmatrix}$ **b.** $\begin{bmatrix} 2 & 4 \\ -3 & -5 \end{bmatrix}$.

SOLUTION   We multiply and subtract as indicated.

DISCOVER FOR YOURSELF

Write and then evaluate three determinants, one whose value is positive, one whose value is negative, and one whose value is 0.

a. $\begin{vmatrix} 5 & 6 \\ 7 & 3 \end{vmatrix} = 5 \cdot 3 - 7 \cdot 6 = 15 - 42 = -27$ *The value of the second-order determinant is −27.*

b. $\begin{vmatrix} 2 & 4 \\ -3 & -5 \end{vmatrix} = 2(-5) - (-3)(4) = -10 + 12 = 2$ *The value of the second-order determinant is 2.*

  CHECK POINT 1 Evaluate the determinant of each of the following matrices:

a. $\begin{bmatrix} 10 & 9 \\ 6 & 5 \end{bmatrix}$ −4   **b.** $\begin{bmatrix} 4 & 3 \\ -5 & -8 \end{bmatrix}$. −17

2 Solve a system of linear equations in two variables using Cramer's rule.

Solving Systems of Linear Equations in Two Variables Using Determinants

Determinants can be used to solve a linear system in two variables. In general, such a system appears as

$$a_1 x + b_1 y = c_1$$
$$a_2 x + b_2 y = c_2.$$

Let's first solve this system for x using the addition method. We can solve for x by eliminating y from the equations. Multiply the first equation by b_2 and the second equation by $-b_1$. Then add the two equations:

$$a_1 x + b_1 y = c_1 \xrightarrow[\text{Multiply by } -b_1.]{\text{Multiply by } b_2.} a_1 b_2 x + b_1 b_2 y = c_1 b_2$$
$$a_2 x + b_2 y = c_2 \xrightarrow{\hspace{2cm}} -a_2 b_1 x - b_1 b_2 y = -c_2 b_1$$
$$\text{Add: } (a_1 b_2 - a_2 b_1)x = c_1 b_2 - c_2 b_1$$
$$x = \frac{c_1 b_2 - c_2 b_1}{a_1 b_2 - a_2 b_1}$$

Because

$$\begin{vmatrix} c_1 & b_1 \\ c_2 & b_2 \end{vmatrix} = c_1 b_2 - c_2 b_1 \quad \text{and} \quad \begin{vmatrix} a_1 & b_1 \\ a_2 & b_2 \end{vmatrix} = a_1 b_2 - a_2 b_1$$

we can express our answer for x as the quotient of two determinants:

$$x = \frac{\begin{vmatrix} c_1 & b_1 \\ c_2 & b_2 \end{vmatrix}}{\begin{vmatrix} a_1 & b_1 \\ a_2 & b_2 \end{vmatrix}}.$$

Similarly, we could use the addition method to solve our system for y, again expressing y as the quotient of two determinants. This method of using determinants to solve the linear system, called **Cramer's rule**, is summarized in the box.

SOLVING A LINEAR SYSTEM IN TWO VARIABLES USING DETERMINANTS

Cramer's Rule

If

$$a_1 x + b_1 y = c_1$$
$$a_2 x + b_2 y = c_2$$

then

$$x = \frac{\begin{vmatrix} c_1 & b_1 \\ c_2 & b_2 \end{vmatrix}}{\begin{vmatrix} a_1 & b_1 \\ a_2 & b_2 \end{vmatrix}} \quad \text{and} \quad y = \frac{\begin{vmatrix} a_1 & c_1 \\ a_2 & c_2 \end{vmatrix}}{\begin{vmatrix} a_1 & b_1 \\ a_2 & b_2 \end{vmatrix}},$$

where

$$\begin{vmatrix} a_1 & b_1 \\ a_2 & b_2 \end{vmatrix} \neq 0.$$

Here are some helpful tips when solving

$$a_1 x + b_1 y = c_1$$
$$a_2 x + b_2 y = c_2$$

using determinants:

1. Three different determinants are used to find x and y. The determinants in the denominators for x and y are identical. The determinants in the numerators for x and y differ. In abbreviated notation, we write

$$x = \frac{D_x}{D} \quad \text{and} \quad y = \frac{D_y}{D}, \text{where } D \neq 0.$$

2. The elements of D, the determinant in the denominator, are the coefficients of the variables in the system.

$$D = \begin{vmatrix} a_1 & b_1 \\ a_2 & b_2 \end{vmatrix}$$

3. D_x, the determinant in the numerator of x, is obtained by replacing the x-coefficients in D, a_1 and a_2, with the constants on the right sides of the equations, c_1 and c_2.

$$D = \begin{vmatrix} a_1 & b_1 \\ a_2 & b_2 \end{vmatrix} \quad \text{and} \quad D_x = \begin{vmatrix} c_1 & b_1 \\ c_2 & b_2 \end{vmatrix} \quad \begin{array}{l} \text{Replace the column with } a_1 \text{ and } a_2 \text{ with the} \\ \text{constants } c_1 \text{ and } c_2 \text{ to get } D_x. \end{array}$$

4. D_y, the determinant in the numerator for y, is obtained by replacing the y-coefficients in D, b_1 and b_2, with the constants on the right sides of the equations, c_1 and c_2.

$$D = \begin{vmatrix} a_1 & b_1 \\ a_2 & b_2 \end{vmatrix} \quad \text{and} \quad D_y = \begin{vmatrix} a_1 & c_1 \\ a_2 & c_2 \end{vmatrix} \quad \begin{array}{l} \text{Replace the column with } b_1 \text{ and } b_2 \text{ with the} \\ \text{constants } c_1 \text{ and } c_2 \text{ to get } D_y. \end{array}$$

EXAMPLE 2   Using Cramer's Rule to Solve a Linear System

Use Cramer's rule to solve the system:

$$5x - 4y = 2$$
$$6x - 5y = 1.$$

SOLUTION   Because

$$x = \frac{D_x}{D} \quad \text{and} \quad y = \frac{D_y}{D},$$

we will set up and evaluate the three determinants D, D_x, and D_y.

1. D, the determinant in both denominators, consists of the x- and y-coefficients.

$$D = \begin{vmatrix} 5 & -4 \\ 6 & -5 \end{vmatrix} = (5)(-5) - (6)(-4) = -25 + 24 = -1$$

Because this determinant is not zero, we continue to use Cramer's rule to solve the system.

2. D_x, the determinant in the numerator for x, is obtained by replacing the x-coefficients in D, 5 and 6, by the constants on the right sides of the equations, 2 and 1.

$$D_x = \begin{vmatrix} 2 & -4 \\ 1 & -5 \end{vmatrix} = (2)(-5) - (1)(-4) = -10 + 4 = -6$$

3. D_y, the determinant in the numerator for y, is obtained by replacing the y-coefficients in D, -4 and -5, by the constants on the right sides of the equations, 2 and 1.

$$D_y = \begin{vmatrix} 5 & 2 \\ 6 & 1 \end{vmatrix} = (5)(1) - (6)(2) = 5 - 12 = -7$$

4. Thus,

$$x = \frac{D_x}{D} = \frac{-6}{-1} = 6 \quad \text{and} \quad y = \frac{D_y}{D} = \frac{-7}{-1} = 7.$$

As always, the ordered pair $(6, 7)$ should be checked by substituting these values into the original equations. The solution is $(6, 7)$ and the solution set is $\{(6, 7)\}$. ∎

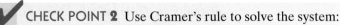 **CHECK POINT 2** Use Cramer's rule to solve the system:

$$5x + 4y = 12$$
$$3x - 6y = 24. \qquad (4, -2) \text{ or } \{(4, -2)\}$$

3 Evaluate a third-order determinant.

The Determinant of a 3 × 3 Matrix The determinant for a 3 × 3 matrix is defined in terms of second-order determinants:

DEFINITION OF THE DETERMINANT OF A 3 × 3 MATRIX A third-order determinant is defined by

Subtract. Add.

$$
\begin{vmatrix} a_1 & b_1 & c_1 \\ a_2 & b_2 & c_2 \\ a_3 & b_3 & c_3 \end{vmatrix} = a_1 \begin{vmatrix} b_2 & c_2 \\ b_3 & c_3 \end{vmatrix} - a_2 \begin{vmatrix} b_1 & c_1 \\ b_3 & c_3 \end{vmatrix} + a_3 \begin{vmatrix} b_1 & c_1 \\ b_2 & c_2 \end{vmatrix}
$$

Each a on the right comes from the first column.

Here are some tips that should be helpful when evaluating the determinant of a 3 × 3 matrix:

EVALUATING THE DETERMINANT OF A 3 × 3 MATRIX

1. Each of the three terms in the definition contains two factors—a numerical factor and a second-order determinant.
2. The numerical factor in each term is an element from the first column of the third-order determinant.
3. The minus sign precedes the second term.
4. The second-order determinant that appears in each term is obtained by crossing out the row and the column containing the numerical factor.

$$
a_1 \begin{vmatrix} b_2 & c_2 \\ b_3 & c_3 \end{vmatrix} - a_2 \begin{vmatrix} b_1 & c_1 \\ b_3 & c_3 \end{vmatrix} + a_3 \begin{vmatrix} b_1 & c_1 \\ b_2 & c_2 \end{vmatrix}
$$

$$
\begin{vmatrix} a_1 & b_1 & c_1 \\ a_2 & b_2 & c_2 \\ a_3 & b_3 & c_3 \end{vmatrix} \quad \begin{vmatrix} a_1 & b_1 & c_1 \\ a_2 & b_2 & c_2 \\ a_3 & b_3 & c_3 \end{vmatrix} \quad \begin{vmatrix} a_1 & b_1 & c_1 \\ a_2 & b_2 & c_2 \\ a_3 & b_3 & c_3 \end{vmatrix}
$$

The **minor** of an element is the determinant that remains after deleting the row and column of that element. For this reason, we call this method **expansion by minors**.

EXAMPLE 3 Evaluating the Determinant of a 3 × 3 Matrix

Evaluate the determinant of the following matrix:

$$
\begin{bmatrix} 4 & 1 & 0 \\ -9 & 3 & 4 \\ -3 & 8 & 1 \end{bmatrix}.
$$

SOLUTION We know that each of the three terms in the determinant contains a numerical factor and a second-order determinant. The numerical factors are from the first column of the given matrix. They are shown in red in the following matrix:

$$
\begin{bmatrix} 4 & 1 & 0 \\ -9 & 3 & 4 \\ -3 & 8 & 1 \end{bmatrix}.
$$

We find the minor for each numerical factor by deleting the row and column of that element:

$$\begin{bmatrix} 4 & 1 & 0 \\ -9 & 3 & 4 \\ -3 & 8 & 1 \end{bmatrix} \quad \begin{bmatrix} 4 & 1 & 0 \\ -9 & 3 & 4 \\ -3 & 8 & 1 \end{bmatrix} \quad \begin{bmatrix} 4 & 1 & 0 \\ -9 & 3 & 4 \\ -3 & 8 & 1 \end{bmatrix}$$

The minor for 4 is $\begin{vmatrix} 3 & 4 \\ 8 & 1 \end{vmatrix}$.

The minor for -9 is $\begin{vmatrix} 1 & 0 \\ 8 & 1 \end{vmatrix}$.

The minor for -3 is $\begin{vmatrix} 1 & 0 \\ 3 & 4 \end{vmatrix}$.

Now we have three numerical factors, 4, -9, and -3, and three second-order determinants. We multiply each numerical factor by its second-order determinant to find the three terms of the third-order determinant:

$$4\begin{vmatrix} 3 & 4 \\ 8 & 1 \end{vmatrix}, \quad -9\begin{vmatrix} 1 & 0 \\ 8 & 1 \end{vmatrix}, \quad -3\begin{vmatrix} 1 & 0 \\ 3 & 4 \end{vmatrix}.$$

Based on the preceding definition, we subtract the second term from the first term and add the third term:

Don't forget to supply the minus sign.

$$\begin{vmatrix} 4 & 1 & 0 \\ -9 & 3 & 4 \\ -3 & 8 & 1 \end{vmatrix} = 4\begin{vmatrix} 3 & 4 \\ 8 & 1 \end{vmatrix} - (-9)\begin{vmatrix} 1 & 0 \\ 8 & 1 \end{vmatrix} - 3\begin{vmatrix} 1 & 0 \\ 3 & 4 \end{vmatrix}$$

$= 4(3 \cdot 1 - 8 \cdot 4) + 9(1 \cdot 1 - 8 \cdot 0) - 3(1 \cdot 4 - 3 \cdot 0)$ Begin evaluating the three second-order determinants.

$= 4(3 - 32) + 9(1 - 0) - 3(4 - 0)$ Multiply within parentheses.

$= 4(-29) + 9(1) - 3(4)$ Subtract within parentheses.

$= -116 + 9 - 12$ Multiply.

$= -119$ Add and subtract as indicated. ∎

 CHECK POINT 3 Evaluate the determinant of the following matrix:

$$\begin{bmatrix} 2 & 1 & 7 \\ -5 & 6 & 0 \\ -4 & 3 & 1 \end{bmatrix}. \quad 80$$

4 Solve a system of linear equations in three variables using Cramer's rule.

Solving Systems of Linear Equations in Three Variables Using Determinants
Cramer's rule can be applied to solving systems of linear equations in three variables. The determinants in the numerator and denominator of all variables are third-order determinants.

SOLVING THREE EQUATIONS IN THREE VARIABLES USING DETERMINANTS

Cramer's Rule

If

$$a_1x + b_1y + c_1z = d_1$$
$$a_2x + b_2y + c_2z = d_2$$
$$a_3x + b_3y + c_3z = d_3$$

then

$$x = \frac{D_x}{D}, \quad y = \frac{D_y}{D}, \quad \text{and} \quad z = \frac{D_z}{D}.$$

These four third-order determinants are given by

$$D = \begin{vmatrix} a_1 & b_1 & c_1 \\ a_2 & b_2 & c_2 \\ a_3 & b_3 & c_3 \end{vmatrix}$$

These are the coefficients of the variables x, y, and z, $D \neq 0$.

$$D_x = \begin{vmatrix} d_1 & b_1 & c_1 \\ d_2 & b_2 & c_2 \\ d_3 & b_3 & c_3 \end{vmatrix}$$

Replace x-coefficients in D with the constants on the right of the three equations.

$$D_y = \begin{vmatrix} a_1 & d_1 & c_1 \\ a_2 & d_2 & c_2 \\ a_3 & d_3 & c_3 \end{vmatrix}$$

Replace y-coefficients in D with the constants on the right of the three equations.

$$D_z = \begin{vmatrix} a_1 & b_1 & d_1 \\ a_2 & b_2 & d_2 \\ a_3 & b_3 & d_3 \end{vmatrix}.$$

Replace z-coefficients in D with the constants on the right of the three equations.

EXAMPLE 4 Using Cramer's Rule to Solve a Linear System in Three Variables

Use Cramer's rule to solve:

$$x + 2y - z = -4$$
$$x + 4y - 2z = -6$$
$$2x + 3y + z = 3.$$

SOLUTION Because

$$x = \frac{D_x}{D}, \quad y = \frac{D_y}{D}, \quad \text{and} \quad z = \frac{D_z}{D},$$

we need to set up and evaluate four determinants.

Step 1. **Set up the determinants.**

1. D, the determinant in all three denominators, consists of the x-, y-, and z-coefficients.

$$D = \begin{vmatrix} 1 & 2 & -1 \\ 1 & 4 & -2 \\ 2 & 3 & 1 \end{vmatrix}$$

$$x + 2y - z = -4$$
$$x + 4y - 2z = -6$$
$$2x + 3y + z = 3$$

The equation we are solving,
(repeated)

2. D_x, the determinant in the numerator for x, is obtained by replacing the x-coefficients in D, 1, 1, and 2, with the constants on the right sides of the equations, -4, -6, and 3.

$$D_x = \begin{vmatrix} -4 & 2 & -1 \\ -6 & 4 & -2 \\ 3 & 3 & 1 \end{vmatrix}$$

3. D_y, the determinant in the numerator for y, is obtained by replacing the y-coefficients in D, 2, 4, and 3, with the constants on the right sides of the equations, -4, -6, and 3.

$$D_y = \begin{vmatrix} 1 & -4 & -1 \\ 1 & -6 & -2 \\ 2 & 3 & 1 \end{vmatrix}$$

4. D_z, the determinant in the numerator for z, is obtained by replacing the z-coefficients in D, -1, -2, and 1, with the constants on the right sides of the equations, -4, -6, and 3.

$$D_z = \begin{vmatrix} 1 & 2 & -4 \\ 1 & 4 & -6 \\ 2 & 3 & 3 \end{vmatrix}$$

Step 2. **Evaluate the four determinants.**

$$D = \begin{vmatrix} 1 & 2 & -1 \\ 1 & 4 & -2 \\ 2 & 3 & 1 \end{vmatrix} = 1\begin{vmatrix} 4 & -2 \\ 3 & 1 \end{vmatrix} - 1\begin{vmatrix} 2 & -1 \\ 3 & 1 \end{vmatrix} + 2\begin{vmatrix} 2 & -1 \\ 4 & -2 \end{vmatrix}$$

$$= 1(4 + 6) - 1(2 + 3) + 2(-4 + 4)$$

$$= 1(10) - 1(5) + 2(0) = 5$$

Using the same technique to evaluate each determinant, we obtain

$$D_x = -10, \quad D_y = 5, \quad \text{and} \quad D_z = 20.$$

Step 3. **Substitute these four values and solve the system.**

$$x = \frac{D_x}{D} = \frac{-10}{5} = -2$$

$$y = \frac{D_y}{D} = \frac{5}{5} = 1$$

$$z = \frac{D_z}{D} = \frac{20}{5} = 4$$

The ordered triple $(-2, 1, 4)$ can be checked by substitution into the original three equations. The solution is $(-2, 1, 4)$ and the solution set is $\{(-2, 1, 4)\}$. ∎

STUDY TIP

To find D_x, D_y, and D_z, you'll need to apply the evaluation process for a 3×3 determinant three times. The values of D_x, D_y, and D_z cannot be obtained from the numbers that occur in the computation of D.

✔ CHECK POINT **4** Use Cramer's rule to solve the system:

$$3x - 2y + z = 16$$
$$2x + 3y - z = -9 \qquad (2, -3, 4) \text{ or } \{(2, -3, 4)\}$$
$$x + 4y + 3z = 2.$$

5 Use determinants to identify inconsistent and dependent systems.

Cramer's Rule with Inconsistent and Dependent Systems If D, the determinant in the denominator, is 0, the variables described by the quotient of determinants are not real numbers. However, when $D = 0$, this indicates that the system is inconsistent or contains dependent equations. This gives rise to the following two situations:

DETERMINANTS: INCONSISTENT AND DEPENDENT SYSTEMS

1. If $D = 0$ and at least one of the determinants in the numerator is not 0, then the system is inconsistent. The solution set is Ø.

2. If $D = 0$ and all the determinants in the numerators are 0, then the equations in the system are dependent. The system has infinitely many solutions.

Although we have focused on applying determinants to solve linear systems, they have other applications, some of which we consider in the exercise set that follows.

3.5 EXERCISE SET

Student Solutions Manual   CD/Video   PH Math/Tutor Center   MathXL Tutorials on CD   MathXL®   MyMathLab   Interactmath.com

Practice Exercises

Evaluate each determinant in Exercises 1–10.

1. $\begin{vmatrix} 5 & 7 \\ 2 & 3 \end{vmatrix}$ 1

2. $\begin{vmatrix} 4 & 8 \\ 5 & 6 \end{vmatrix}$ -16

3. $\begin{vmatrix} -4 & 1 \\ 5 & 6 \end{vmatrix}$ -29

4. $\begin{vmatrix} 7 & 9 \\ -2 & -5 \end{vmatrix}$ -17

5. $\begin{vmatrix} -7 & 14 \\ 2 & -4 \end{vmatrix}$ 0

6. $\begin{vmatrix} 1 & -3 \\ -8 & 2 \end{vmatrix}$ -22

7. $\begin{vmatrix} -5 & -1 \\ -2 & -7 \end{vmatrix}$ 33

8. $\begin{vmatrix} \frac{1}{5} & \frac{1}{6} \\ -6 & 5 \end{vmatrix}$ 2

9. $\begin{vmatrix} \frac{1}{2} & \frac{1}{2} \\ \frac{1}{8} & -\frac{3}{4} \end{vmatrix}$ $-\frac{7}{16}$

10. $\begin{vmatrix} \frac{2}{3} & \frac{1}{3} \\ -\frac{1}{2} & \frac{3}{4} \end{vmatrix}$ $\frac{2}{3}$

For Exercises 11–26, use Cramer's rule to solve each system or to determine that the system is inconsistent or contains dependent equations.

11. $x + y = 7$
$x - y = 3$ $(5, 2)$

12. $2x + y = 3$
$x - y = 3$ $(2, -1)$

13. $12x + 3y = 15$
$2x - 3y = 13$ $(2, -3)$

14. $x - 2y = 5$
$5x - y = -2$ $(-1, -3)$

15. $4x - 5y = 17$
$2x + 3y = 3$ $(3, -1)$

16. $3x + 2y = 2$
$2x + 2y = 3$ $\left(-1, \frac{5}{2}\right)$

17. $x - 3y = 4$
$3x - 4y = 12$ $(4, 0)$

18. $2x - 9y = 5$
$3x - 3y = 11$ $\left(4, \frac{1}{3}\right)$

19. $3x - 4y = 4$
$2x + 2y = 12$ $(4, 2)$

20. $3x = 7y + 1$
$2x = 3y - 1$ $(-2, -1)$

21. $2x = 3y + 2$
$5x = 51 - 4y$ $(7, 4)$

22. $y = -4x + 2$
$2x = 3y + 8$ $(1, -2)$

23. $3x = 2 - 3y$
$2y = 3 - 2x$
inconsistent;
no solution or Ø

24. $x + 2y - 3 = 0$
$12 = 8y + 4x$
dependent equations;
infinitely many solutions

25. $4y = 16 - 3x$
$6x = 32 - 8y$
dependent equations;
infinitely many solutions

26. $2x = 7 + 3y$
$4x - 6y = 3$
inconsistent;
no solution or Ø

Evaluate each determinant in Exercises 27–32.

27. $\begin{vmatrix} 3 & 0 & 0 \\ 2 & 1 & -5 \\ 2 & 5 & -1 \end{vmatrix}$ 72

28. $\begin{vmatrix} 4 & 0 & 0 \\ 3 & -1 & 4 \\ 2 & -3 & 5 \end{vmatrix}$ 28

29. $\begin{vmatrix} 3 & 1 & 0 \\ -3 & 4 & 0 \\ -1 & 3 & -5 \end{vmatrix}$ -75

30. $\begin{vmatrix} 2 & -4 & 2 \\ -1 & 0 & 5 \\ 3 & 0 & 4 \end{vmatrix}$ -76

31. $\begin{vmatrix} 1 & 1 & 1 \\ 2 & 2 & 2 \\ -3 & 4 & -5 \end{vmatrix}$ 0

32. $\begin{vmatrix} 1 & 2 & 3 \\ 2 & 2 & -3 \\ 3 & 2 & 1 \end{vmatrix}$ -20

In Exercises 33–40, use Cramer's rule to solve each system.

33. $x + y + z = 0$
$2x - y + z = -1$
$-x + 3y - z = -8$
$(-5, -2, 7)$

34. $x - y + 2z = 3$
$2x + 3y + z = 9$
$-x - y + 3z = 11$
$(-2, 3, 4)$

35. $4x - 5y - 6z = -1$
$x - 2y - 5z = -12$
$2x - y = 7$
$(2, -3, 4)$

36. $x - 3y + z = -2$
$x + 2y = 8$
$2x - y = 1$
$(2, 3, 5)$

37. $x + y + z = 4$
$x - 2y + z = 7$
$x + 3y + 2z = 4$
$(3, -1, 2)$

38. $2x + 2y + 3z = 10$
$4x - y + z = -5$
$5x - 2y + 6z = 1$
$(-1, 3, 2)$

39. $x + 2z = 4$
$2y - z = 5$
$2x + 3y = 13$
$(2, 3, 1)$

40. $3x + 2z = 4$
$5x - y = -4$
$4y + 3z = 22$
$(0, 4, 2)$

Practice Plus

In Exercises 41–42, evaluate each determinant.

41. $\begin{vmatrix} 3 & 1 \\ -2 & 3 \end{vmatrix} \begin{vmatrix} 7 & 0 \\ 1 & 5 \end{vmatrix}$

$\begin{vmatrix} 3 & 0 \\ 0 & 7 \end{vmatrix} \begin{vmatrix} 9 & -6 \\ 3 & 5 \end{vmatrix}$

-42

42. $\begin{vmatrix} 5 & 0 \\ 4 & -3 \end{vmatrix} \begin{vmatrix} -1 & 0 \\ 0 & -1 \end{vmatrix}$

$\begin{vmatrix} 7 & -5 \\ 4 & 6 \end{vmatrix} \begin{vmatrix} 4 & 1 \\ -3 & 5 \end{vmatrix}$

-407

In Exercises 43–44, write the system of linear equations for which Cramer's rule yields the given determinants.

43. $D = \begin{vmatrix} 2 & -4 \\ 3 & 5 \end{vmatrix}, \quad D_x = \begin{vmatrix} 8 & -4 \\ -10 & 5 \end{vmatrix}$ $\quad 2x - 4y = 8$
$\quad 3x + 5y = -10$

44. $D = \begin{vmatrix} 2 & -3 \\ 5 & 6 \end{vmatrix}, \quad D_x = \begin{vmatrix} 8 & -3 \\ 11 & 6 \end{vmatrix}$ $\quad 2x - 3y = 8$
$\quad 5x + 6y = 11$

In Exercises 45–48, solve each equation for x.

45. $\begin{vmatrix} -2 & x \\ 4 & 6 \end{vmatrix} = 32 \quad -11$

46. $\begin{vmatrix} x + 3 & -6 \\ x - 2 & -4 \end{vmatrix} = 28 \quad 26$

47. $\begin{vmatrix} 1 & x & -2 \\ 3 & 1 & 1 \\ 0 & -2 & 2 \end{vmatrix} = -8 \quad 4$

48. $\begin{vmatrix} 2 & x & 1 \\ -3 & 1 & 0 \\ 2 & 1 & 4 \end{vmatrix} = 39 \quad 3$

Application Exercises

Determinants are used to find the area of a triangle whose vertices are given by three points in a rectangular coordinate system. The area of a triangle with vertices (x_1, y_1), (x_2, y_2), and (x_3, y_3) is

$$\text{Area} = \pm\frac{1}{2}\begin{vmatrix} x_1 & y_1 & 1 \\ x_2 & y_2 & 1 \\ x_3 & y_3 & 1 \end{vmatrix}.$$

where the $\pm$ symbol indicates that the appropriate sign should be chosen to yield a positive area. Use this information to work Exercises 49–50.

49. Use determinants to find the area of the triangle whose vertices are $(3, -5)$, $(2, 6)$, and $(-3, 5)$. 28 square units

50. Use determinants to find the area of the triangle whose vertices are $(1, 1)$, $(-2, -3)$, and $(11, -3)$. 26 square units

Determinants are used to show that three points lie on the same line (are collinear). If

$$\begin{vmatrix} x_1 & y_1 & 1 \\ x_2 & y_2 & 1 \\ x_3 & y_3 & 1 \end{vmatrix} = 0,$$

then the points (x_1, y_1), (x_2, y_2), and (x_3, y_3) are collinear. If the determinant does not equal 0, then the points are not collinear. Use this information to work Exercises 51–52.

51. Are the points $(3, -1)$, $(0, -3)$, and $(12, 5)$ collinear? Yes, the points are collinear.

52. Are the points $(-4, -6)$, $(1, 0)$, and $(11, 12)$. collinear? Yes, the points are collinear.

Determinants are used to write an equation of a line passing through two points. An equation of the line passing through the distinct points (x_1, y_1) and (x_2, y_2) is given by

$$\begin{vmatrix} x & y & 1 \\ x_1 & y_1 & 1 \\ x_2 & y_2 & 1 \end{vmatrix} = 0.$$

Use this information to work Exercises 53–54.

53. Use the determinant to write an equation for the line passing through $(3, -5)$ and $(-2, 6)$. Then expand the determinant, expressing the line's equation in slope-intercept form.

$\begin{vmatrix} x & y & 1 \\ 3 & -5 & 1 \\ -2 & 6 & 1 \end{vmatrix} = 0; y = -\frac{11}{5}x + \frac{8}{5}$

54. Use the determinant to write an equation for the line passing through $(-1, 3)$ and $(2, 4)$. Then expand the determinant, expressing the line's equation in slope-intercept form.

$\begin{vmatrix} x & y & 1 \\ -1 & 3 & 1 \\ 2 & 4 & 1 \end{vmatrix} = 0; y = \frac{1}{3}x + \frac{10}{3}$

Writing in Mathematics

55. Explain how to evaluate a second-order determinant. 55–62. Answers will vary.

56. Describe the determinants D, D_x, and D_y in terms of the coefficients and constants in a system of two equations in two variables.

57. Explain how to evaluate a third-order determinant.

58. When expanding a determinant by minors, when is it necessary to supply a minus sign?

59. Without going into too much detail, describe how to solve a linear system in three variables using Cramer's rule.

60. In applying Cramer's rule, what does it mean if $D = 0$?

61. The process of solving a linear system in three variables using Cramer's rule can involve tedious computation. Is there a way of speeding up this process, perhaps using Cramer's rule to find the value for only one of the variables? Describe how this process might work, presenting a specific example with your description. Remember that your goal is still to find the value for each variable in the system.

62. If you could use only one method to solve linear systems in three variables, which method would you select? Explain why this is so.

Technology Exercises

63. Use the feature of your graphing utility that evaluates the determinant of a square matrix to verify any five of the determinants that you evaluated by hand in Exercises 1–10 or 27–32.   Answers will vary.

64. What is the fastest method for solving a linear system with your graphing utility?   Answers will vary.

Critical Thinking Exercises

65. Which one of the following is true?   d

  a. If $D = 0$, then every variable has a value of 0.

  b. Using Cramer's rule, we use $\dfrac{D}{D_y}$ to get the value of y.

  c. Because there are different determinants in the numerators of x and y, if a system is solved using Cramer's rule, x and y cannot have the same value.

  d. Only one 2×2 determinant is needed to evaluate

$$\begin{vmatrix} 2 & 3 & -2 \\ 0 & 1 & 3 \\ 0 & 4 & -1 \end{vmatrix}.$$

66. What happens to the value of a second-order determinant if the two columns are interchanged?   The value is multiplied by -1.

67. Consider the system

$$a_1x + b_1y = c_1$$
$$a_2x + b_2y = c_2.$$

Use Cramer's rule to prove that if the first equation of the system is replaced by the sum of the two equations, the resulting system has the same solution as the original system.   67–68. Answers will vary.

68. Show that the equation of a line through (x_1, y_1) and (x_2, y_2) is given by the determinant

$$\begin{vmatrix} x & y & 1 \\ x_1 & y_1 & 1 \\ x_2 & y_2 & 1 \end{vmatrix} = 0.$$

Review Exercises

69. Solve: $6x - 4 = 2 + 6(x - 1)$. (Section 1.4, Example 6)
all real numbers or $\{x \mid x \text{ is a real number}\}$

70. Solve for y: $-2x + 3y = 7$. (Section 1.5, Example 6)
$y = \dfrac{2x + 7}{3}$

71. Solve: $\dfrac{4x + 1}{3} = \dfrac{x - 3}{6} + \dfrac{x + 5}{6}$. (Section 1.4, Example 4)
0 or $\{0\}$

GROUP PROJECT

CHAPTER 3

Additional group projects can be found in the *Instructor's Resource Manual.*

The group is going into business for the next year. Your first task is to determine the product that you plan to manufacture and sell. Choose something unique, but realistic, reflecting the abilities and interests of group members. How much will it cost to produce each unit of your product? What will be the selling price for each unit? What is a reasonable estimate of your fixed cost for the entire year? Be sure to include utilities, labor, materials, marketing, and anything else that is relevant to your business venture. Once you have determined the product and these three figures,

  a. Write the cost function, C, of producing x units of your product.

  b. Write the revenue function, R, from the sale of x units of your product.

  c. Determine the break-even point. Does this seem realistic in terms of your product and its target market?

  d. Graph the cost function and the revenue function in the same rectangular coordinate system. Label the figure just like Figure 3.7 on page 187. Indicate the regions that show the loss and gain for your business venture.

  e. Write the profit function, P, from producing and selling x units of your product. Graph the profit function, in the same style as Figure 3.8 on page 189. Indicate where the business is in the red and where it is in the black.

  f. MTV's *The Real World* has offered to pay the fixed cost for the business venture, videotaping all group interactions for its forthcoming *The Real World: Profit and Loss.* Group members need to determine whether or not to accept the offer.

CHAPTER 3 SUMMARY

Definitions and Concepts	Examples

Section 3.1 Systems of Linear Equations in Two Variables

A system of linear equations in two variables, x and y, consists of two equations that can be written in the form $Ax + By = C$. The solution set is the set of all ordered pairs that satisfy both equations. Using the graphing method, a solution of a linear system is a point common to the graphs of both equations in the system.

Solve by graphing: $2x - y = 6$
$x + y = 6$.

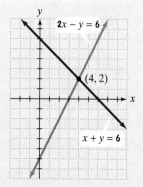

The intersection point gives the solution: $(4, 2)$. The solution set is $\{(4, 2)\}$.

To solve a linear system by the substitution method,

1. Solve either equation for one variable in terms of the other.
2. Substitute the expression for that variable into the other equation.
3. Solve the equation in one variable.
4. Back-substitute the value of the variable into one of the original equations and find the value of the other variable.
5. Check the proposed solution in both equations.

Solve by the substitution method:

$$y = 2x - 3$$
$$4x - 3y = 5.$$

Substitute $2x - 3$ for y in the second equation.

$$4x - 3(2x - 3) = 5$$
$$4x - 6x + 9 = 5$$
$$-2x + 9 = 5$$
$$-2x = -4$$
$$x = 2$$

Find y. Substitute 2 for x in $y = 2x - 3$.

$$y = 2(2) - 3 = 4 - 3 = 1$$

The ordered pair $(2, 1)$ checks. The solution is $(2, 1)$ and $\{(2, 1)\}$ is the solution set.

To solve a linear system by the addition method,

1. Write equations in $Ax + By = C$ form.
2. Multiply one or both equations by nonzero numbers so that coefficients of one variable are opposites.
3. Add equations.
4. Solve the resulting equation for the variable.
5. Back-substitute the value of the variable into either original equation and find the value of the remaining variable.
6. Check the proposed solution in both equations.

Solve by the addition method:

$$2x + y = 10$$
$$3x + 4y = 25.$$

Eliminate y. Multiply the first equation by -4.

$$-8x - 4y = -40$$
$$\underline{3x + 4y = \quad 25}$$
Add: $-5x \qquad = -15$
$$x = 3$$

Find y. Back-substitute 3 for x. Use the first equation, $2x + y = 10$.

$$2(3) + y = 10$$
$$6 + y = 10$$
$$y = 4$$

The ordered pair $(3, 4)$ checks. The solution is $(3, 4)$ and $\{(3, 4)\}$ is the solution set.

Definitions and Concepts	Examples

Section 3.1 Systems of Linear Equations in Two Variables (continued)

A linear system with at least one solution is a consistent system. A system that has no solution, with Ø as its solution set, is an inconsistent system. A linear system with infinitely many solutions has dependent equations. Solving inconsistent systems by substitution or addition leads to a false statement, such as $0 = 3$. Solving systems with dependent equations leads to a true statement, such as $7 = 7$.

One solution: consistent

No solution: inconsistent

Infinitely many solutions: dependent and consistent

Section 3.2 Problem Solving and Business Applications Using Systems of Equations

A Problem-Solving Strategy

1. Use variables, usually x and y, to represent unknown quantities.
2. Write a system of equations describing the problem's conditions.
3. Solve the system and answer the problem's question.
4. Check proposed answers in the problem's wording.

You invested $14,000 in two stocks paying 7% and 9% interest. Total year-end interest was $1180. How much was invested at each rate?

$\quad$ Let x = amount invested at 7% and
$\quad\quad\quad$ y = amount invested at 9%.

$\quad$ amount invested at 7% $\quad$ amount invested at 9%

$$x \quad + \quad y \quad = \quad 14,000$$

$\quad$ interest from 7% investment $\quad$ interest from 9% investment

$$0.07x \quad + \quad 0.09y \quad = \quad 1180$$

Solving by substitution or addition, $x = 4000$ and $y = 10,000$. Thus, 4000 was invested at 7% and $10,000$ at 9%.

Functions of Business
A company produces and sells x units of a product.

Revenue Function

$$R(x) = (\text{price per unit sold})x$$

Cost Function

$$C(x) = \text{fixed cost} + (\text{cost per unit produced})x$$

Profit Function

$$P(x) = R(x) - C(x)$$

The point of intersection of the graphs of R and C is the break-even point. The x-coordinate of the point reveals the number of units that a company must produce and sell so that the money coming in, the revenue, is equal to the money going out, the cost. The y-coordinate gives the amount of money coming in and going out.

A company that manufactures lamps has a fixed cost of $80,000 and it costs $20 to produce each lamp. Lamps are sold for $70.

a. Write the cost function.

$$C(x) = 80,000 + 20x$$

$\quad$ Fixed cost $\quad$ Variable cost: $20 per lamp.

b. Write the revenue function.

$$R(x) = 70x$$

$\quad$ Revenue per lamp, $70, times number of lamps sold

c. Find the break-even point.

$\quad$ Solve $\quad\quad\quad y = 80,000 + 20x$
$\quad\quad\quad\quad\quad\quad\quad y = 70x$

by substitution. Solving

$$70x = 80,000 + 20x$$

yields $x = 1600$. Back-substituting, $y = 112,000$. The break-even point is $(1600, 112,000)$: The company breaks even if it sells 1600 lamps. At this level, money coming in equals money going out: $112,000.

Definitions and Concepts	Examples

Section 3.3 Systems of Linear Equations in Three Variables

A system of linear equations in three variables, x, y, and z, consists of three equations of the form $Ax + By + Cz = D$. The solution set is the set consisting of the ordered triple that satisfies all three equations. The solution represents the point of intersection of three planes in space.

Is $(2, -1, 3)$ a solution of

$$3x + 5y - 2z = -5$$
$$2x + 3y - z = -2$$
$$2x + 4y + 6z = 18?$$

Replace x with 2, y with -1, and z with 3. Using the first equation, we obtain:

$$3 \cdot 2 + 5(-1) - 2(3) \overset{?}{=} -5$$
$$6 - 5 - 6 \overset{?}{=} -5$$
$$-5 = -5, \quad \text{true}$$

The ordered triple $(2, -1, 3)$ satisfies the first equation. In a similar manner, it satisfies the other two equations and is a solution.

To solve a linear system in three variables by eliminating variables,

1. Reduce the system to two equations in two variables.
2. Solve the resulting system of two equations in two variables.
3. Use back-substitution in one of the equations in two variables to find the value of the second variable.
4. Back-substitute the values for two variables into one of the original equations to find the value of the third variable.
5. Check.

If all variables are eliminated and a false statement results, the system is inconsistent and has no solution. If a true statement results, the system contains dependent equations and has infinitely many solutions.

Solve:
$$2x + 3y - 2z = 0 \quad \boxed{1}$$
$$x + 2y - z = 1 \quad \boxed{2}$$
$$3x - y + z = -15. \quad \boxed{3}$$

Add equations $\boxed{2}$ and $\boxed{3}$ to eliminate z.

$$4x + y = -14 \quad \boxed{4}$$

Eliminate z again. Multiply equation $\boxed{3}$ by 2 and add to equation $\boxed{1}$.

$$8x + y = -30 \quad \boxed{5}$$

Multiply equation $\boxed{4}$ by -1 and add to equation $\boxed{5}$.

$$-4x - y = 14$$
$$\underline{8x + y = -30}$$
$$\text{Add:} \quad 4x \qquad = -16$$
$$x = -4$$

Substitute -4 for x in $\boxed{4}$.

$$4(-4) + y = -14$$
$$y = 2$$

Substitute -4 for x and 2 for y in $\boxed{3}$.

$$3(-4) - 2 + z = -15$$
$$-14 + z = -15$$
$$z = -1$$

Checking verifies that $(-4, 2, -1)$ is the solution and $\{(-4, 2, -1)\}$ is the solution set.

Definitions and Concepts	Examples

Section 3.3 Systems of Linear Equations in Three Variables (continued)

Curve Fitting
Curve fitting is determining a function whose graph contains given points. Three points that do not lie on a line determine the graph of a quadratic function

$$y = ax^2 + bx + c.$$

Use the three given points to create a system of three equations. Solve the system to find $a, b,$ and c.

Find the quadratic function $y = ax^2 + bx + c$ whose graph passes through the points $(-1, 2), (1, 8),$ and $(2, 14)$.
Use $y = ax^2 + bx + c$.

When $x = -1, y = 2$: $2 = a(-1)^2 + b(-1) + c$
When $x = 1, y = 8$: $8 = a \cdot 1^2 + b \cdot 1 + c$
When $x = 2, y = 14$: $14 = a \cdot 2^2 + b \cdot 2 + c$

Solving,
$$a - b + c = 2$$
$$a + b + c = 8$$
$$4a + 2b + c = 14,$$

$a = 1, b = 3,$ and $c = 4$. The quadratic function, $y = ax^2 + bx + c,$ is $y = x^2 + 3x + 4$.

Section 3.4 Matrix Solutions to Linear Systems

A matrix is a rectangular array of numbers. The augmented matrix of a linear system is obtained by writing the coefficients of each variable, a vertical bar, and the constants of the system.	$$x + 4y = 9$$ $$3x + y = 5$$ The augmented matrix is $\begin{bmatrix} 1 & 4 & \mid & 9 \\ 3 & 1 & \mid & 5 \end{bmatrix}$.

The following row operations produce matrices that represent systems with the same solution. Two matrices are row equivalent if one can be obtained from the other by a sequence of these row operations.

1. Interchange the elements in the ith and jth rows: $R_i \leftrightarrow R_j$.
2. Multiply each element in the ith row by k: kR_i.
3. Add k times the elements in row i to the corresponding elements in row j: $kR_i + R_j$.

Find $-4R_1 + R_2$:
$$\begin{bmatrix} 1 & 0 & -2 & \mid & 5 \\ 4 & -1 & 2 & \mid & 6 \\ 3 & -7 & 9 & \mid & 10 \end{bmatrix}.$$

Add -4 times the elements in row 1 to the corresponding elements in row 2.

$$\begin{bmatrix} 1 & 0 & -2 & \mid & 5 \\ -4(1)+4 & -4(0)+(-1) & -4(-2)+2 & \mid & -4(5)+6 \\ 3 & -7 & 9 & \mid & 10 \end{bmatrix}$$

$$= \begin{bmatrix} 1 & 0 & -2 & \mid & 5 \\ 0 & -1 & 10 & \mid & -14 \\ 3 & -7 & 9 & \mid & 10 \end{bmatrix}$$

Solving Linear Systems Using Matrices

1. Write the augmented matrix for the system.
2. Use matrix row operations to simplify the matrix to one with 1s down the main diagonal from upper left to lower right, and 0s below the 1s.
3. Write the system of linear equations corresponding to the matrix in step 2 and use back-substitution to find the system's solution.

If you obtain a matrix with a row containing 0s to the left of the vertical bar and a nonzero number on the right, the system is inconsistent. If 0s appear across an entire row, the system contains dependent equations.

Solve using matrices:
$$3x + y = 5$$
$$x + 4y = 9.$$

$$\begin{bmatrix} 3 & 1 & \mid & 5 \\ 1 & 4 & \mid & 9 \end{bmatrix} \xrightarrow{R_1 \leftrightarrow R_2} \begin{bmatrix} 1 & 4 & \mid & 9 \\ 3 & 1 & \mid & 5 \end{bmatrix}$$

$$\xrightarrow{-3R_1 + R_2} \begin{bmatrix} 1 & 4 & \mid & 9 \\ 0 & -11 & \mid & -22 \end{bmatrix} \xrightarrow{-\frac{1}{11}R_2} \begin{bmatrix} 1 & 4 & \mid & 9 \\ 0 & 1 & \mid & 2 \end{bmatrix}$$

$$\rightarrow x + 4y = 9$$
$$y = 2.$$

When $y = 2, x + 4 \cdot 2 = 9,$ so $x = 1$. The solution is $(1, 2)$ and the solution set is $\{(1, 2)\}$.

Definitions and Concepts	Examples

Section 3.5 Determinants and Cramer's Rule

A square matrix has the same number of rows as columns. A determinant is a real number associated with a square matrix. The determinant is denoted by placing vertical bars about the array of numbers. The value of a second-order determinant is

$$\begin{vmatrix} a_1 & b_1 \\ a_2 & b_2 \end{vmatrix} = a_1 b_2 - a_2 b_1.$$

Evaluate:

$$\begin{vmatrix} 2 & -1 \\ 3 & 4 \end{vmatrix} = 2(4) - 3(-1) = 8 + 3 = 11.$$

Cramer's Rule for Two Linear Equations in Two Variables

If

$$a_1 x + b_1 y = c_1$$
$$a_2 x + b_2 y = c_2,$$

then

$$x = \frac{\begin{vmatrix} c_1 & b_1 \\ c_2 & b_2 \end{vmatrix}}{\begin{vmatrix} a_1 & b_1 \\ a_2 & b_2 \end{vmatrix}} = \frac{D_x}{D} \quad \text{and} \quad y = \frac{\begin{vmatrix} a_1 & c_1 \\ a_2 & c_2 \end{vmatrix}}{\begin{vmatrix} a_1 & b_1 \\ a_2 & b_2 \end{vmatrix}} = \frac{D_y}{D}, \quad D \neq 0.$$

If $D = 0$ and any numerator is not zero, the system is inconsistent and has no solution. If all determinants are 0, the system contains dependent equations and has infinitely many solutions.

Solve by Cramer's rule:

$$5x + 3y = 7$$
$$-x + 2y = 9.$$

$$D = \begin{vmatrix} 5 & 3 \\ -1 & 2 \end{vmatrix} = 5(2) - (-1)(3) = 10 + 3 = 13$$

$$D_x = \begin{vmatrix} 7 & 3 \\ 9 & 2 \end{vmatrix} = 7 \cdot 2 - 9 \cdot 3 = 14 - 27 = -13$$

$$D_y = \begin{vmatrix} 5 & 7 \\ -1 & 9 \end{vmatrix} = 5(9) - (-1)(7) = 45 + 7 = 52$$

$$x = \frac{D_x}{D} = \frac{-13}{13} = -1, \, y = \frac{D_y}{D} = \frac{52}{13} = 4$$

The solution is $(-1, 4)$ and the solution set is $\{(-1, 4)\}$.

The value of a third-order determinant is

$$\begin{vmatrix} a_1 & b_1 & c_1 \\ a_2 & b_2 & c_2 \\ a_3 & b_3 & c_3 \end{vmatrix}$$

$$= a_1 \begin{vmatrix} b_2 & c_2 \\ b_3 & c_3 \end{vmatrix} - a_2 \begin{vmatrix} b_1 & c_1 \\ b_3 & c_3 \end{vmatrix} + a_3 \begin{vmatrix} b_1 & c_1 \\ b_2 & c_2 \end{vmatrix}.$$

Each second-order determinant is called a minor.

Evaluate:

$$\begin{vmatrix} 1 & -2 & 1 \\ 3 & 1 & -2 \\ 5 & 5 & 3 \end{vmatrix}$$

$$= 1 \begin{vmatrix} 1 & -2 \\ 5 & 3 \end{vmatrix} - 3 \begin{vmatrix} -2 & 1 \\ 5 & 3 \end{vmatrix} + 5 \begin{vmatrix} -2 & 1 \\ 1 & -2 \end{vmatrix}$$

$$= 1(3 - (-10)) - 3(-6 - 5) + 5(4 - 1)$$

$$= 1(13) - 3(-11) + 5(3)$$

$$= 13 + 33 + 15 = 61.$$

Cramer's Rule for Three Linear Equations in Three Variables

If

$$a_1 x + b_1 y + c_1 z = d_1$$
$$a_2 x + b_2 y + c_2 z = d_2$$
$$a_3 x + b_3 y + c_3 z = d_3,$$

Solve by Cramer's rule:

$$x - 2y + z = 4$$
$$3x + y - 2z = 3$$
$$5x + 5y + 3z = -8.$$

Definitions and Concepts

Examples

Section 3.5 Determinants and Cramer's Rule (continued)

then

$$x = \frac{D_x}{D}, \qquad y = \frac{D_y}{D}, \qquad z = \frac{D_z}{D}.$$

$$D = \begin{vmatrix} a_1 & b_1 & c_1 \\ a_2 & b_2 & c_2 \\ a_3 & b_3 & c_3 \end{vmatrix} \neq 0, \quad D_x = \begin{vmatrix} d_1 & b_1 & c_1 \\ d_2 & b_2 & c_2 \\ d_3 & b_3 & c_3 \end{vmatrix}$$

$$D_y = \begin{vmatrix} a_1 & d_1 & c_1 \\ a_2 & d_2 & c_2 \\ a_3 & d_3 & c_3 \end{vmatrix}, \qquad D_z = \begin{vmatrix} a_1 & b_1 & d_1 \\ a_2 & b_2 & d_2 \\ a_3 & b_3 & d_3 \end{vmatrix}$$

If $D = 0$ and any numerator is not zero, the system is inconsistent. If all determinants are 0, the system contains dependent equations.

$$D = \begin{vmatrix} 1 & -2 & 1 \\ 3 & 1 & -2 \\ 5 & 5 & 3 \end{vmatrix} = 61 \qquad \textit{shown on page 232}$$

$$D_x = \begin{vmatrix} 4 & -2 & 1 \\ 3 & 1 & -2 \\ -8 & 5 & 3 \end{vmatrix} = 61$$

$$D_y = \begin{vmatrix} 1 & 4 & 1 \\ 3 & 3 & -2 \\ 5 & -8 & 3 \end{vmatrix} = -122$$

$$D_z = \begin{vmatrix} 1 & -2 & 4 \\ 3 & 1 & 3 \\ 5 & 5 & -8 \end{vmatrix} = -61$$

$$x = \frac{D_x}{D} = \frac{61}{61} = 1, \quad y = \frac{D_y}{D} = \frac{-122}{61} = -2,$$

$$z = \frac{D_z}{D} = \frac{-61}{61} = -1.$$

The solution is $(1, -2, -1)$ and the solution set is $\{(1, -2, -1)\}$.

CHAPTER 3   REVIEW EXERCISES

3.1 In Exercises 1–2, determine whether the given ordered pair is a solution of the system.

1. $(4, 2)$
$2x - 5y = -2$
$3x + 4y = 4$
not a solution

2. $(-5, 3)$
$-x + 2y = 11$
$y = -\dfrac{x}{3} + \dfrac{4}{3}$
solution

In Exercises 3–6, solve each system by graphing. Identify systems with no solution and systems with infinitely many solutions, using set notation to express their solution sets.

3. $x + y = 5$
$3x - y = 3$ $(2, 3)$

4. $3x - 2y = 6$
$6x - 4y = 12$
$\{(x, y) | 3x - 2y = 6\}$ or $\{(x, y) | 6x - 4y = 12\}$

5. $y = \dfrac{3}{5}x - 3$
$2x - y = -4$
$(-5, -6)$

6. $y = -x + 4$
$3x + 3y = -6$
$\varnothing$

In Exercises 7–13, solve each system by the substitution method or the addition method. Identify systems with no solution and systems with infinitely many solutions, using set notation to express their solution sets.

7. $2x - y = 2$
$x + 2y = 11$
$(3, 4)$

8. $y = -2x + 3$
$3x + 2y = -17$
$(23, -43)$

9. $3x + 2y = -8$
$2x + 5y = 2$
$(-4, 2)$

10. $5x - 2y = 14$
$3x + 4y = 11$ $\left(3, \dfrac{1}{2}\right)$

11. $y = 4 - x$
$3x + 3y = 12$
$\{(x, y) | y = 4 - x\}$ or $\{(x, y) | 3x + 3y = 12\}$

12. $\dfrac{x}{8} + \dfrac{3y}{4} = \dfrac{19}{8}$
$-\dfrac{x}{2} + \dfrac{3y}{4} = \dfrac{1}{2}$ $\left(3, \dfrac{8}{3}\right)$

13. $x - 2y + 3 = 0$
$2x - 4y + 7 = 0$ $\varnothing$

3.2 *In Exercises 14–18, use the four-step strategy to solve each problem.*

14. An appliance store is having a sale on small TVs and stereos. One day a salesperson sells 3 of the TVs and 4 stereos for $2530. The next day the salesperson sells 4 of the same TVs and 3 of the same stereos for $2510. What are the prices of a TV and a stereo?   A TV is $350 and a stereo is $370.

15. You invested $9000 in two funds paying 4% and 7% annual interest, respectively. At the end of the year, the total interest from these investments was $555. How much was invested at each rate? $2500 was invested at 4% and $6500 at 7%.

16. A chemist needs to mix a solution that is 34% silver nitrate with one that is 4% silver nitrate to obtain 100 milliliters of a mixture that is 7% silver nitrate. How many milliliters of each of the solutions must be used?   10 milliliters of 34% and 90 milliliters of 4% silver nitrate must be used.

17. When a plane flies with the wind, it can travel 2160 miles in 3 hours. When the plane flies in the opposite direction, against the wind, it takes 4 hours to fly the same distance. Find the rate of the plane in still air and the rate of the wind.   The rate of the plane in still air is 630 miles per hour and the rate of the wind is 90 miles per hour.

18. The perimeter of a rectangular table top is 34 feet. The difference between 4 times the length and 3 times the width is 33 feet. Find the dimensions.   12 feet by 5 feet

The cost and revenue functions for producing and selling x units of a new graphing calculator are

$$C(x) = 22{,}500 + 40x \quad \text{and} \quad R(x) = 85x.$$

Use these functions to solve Exercises 19–21.

19. Find the loss or the gain from selling 400 graphing calculators.   loss of $4500

20. Determine the break-even point. Describe what this means.   (500, 42,500); When 500 calculators are produced and sold, both the cost and revenue are $42,500.

21. Write the profit function, P, from producing and selling x graphing calculators. $P(x) = 45x - 22{,}500$

22. A company is planning to manufacture computer desks. The fixed cost will be $60,000 and it will cost $200 to produce each desk. Each desk will be sold for $450.
a. Write the cost function, C, of producing x desks. $C(x) = 60{,}000 + 200x$
b. Write the revenue function, R, from the sale of x desks. $R(x) = 450x$
c. Determine the break-even point. Describe what this means.   (240, 108,000); When 240 desks are produced and sold, both cost and revenue are $108,000.

3.3

23. Is $(-3, -2, 5)$ a solution of the system

$$x + y + z = 0$$
$$2x - 3y + z = 5$$
$$4x + 2y + 4z = 3?$$ no, not a solution

Solve each system in Exercises 24–26 by eliminating variables using the addition method. If there is no solution or if there are infinitely many solutions and a system's equations are dependent, so state.

24. $2x - y + z = 1$
$3x - 3y + 4z = 5$
$4x - 2y + 3z = 4$ $(0, 1, 2)$

25. $x + 2y - z = 5$
$2x - y + 3z = 0$
$2y + z = 1$ $(2, 1, -1)$

26. $3x - 4y + 4z = 7$
$x - y - 2z = 2$
$2x - 3y + 6z = 5$ infinitely many solutions; dependent equations

27. Find the quadratic function $y = ax^2 + bx + c$ whose graph passes through the points $(1, 4)$, $(3, 20)$, and $(-2, 25)$. $y = 3x^2 - 4x + 5$

28. The graph shows the average yearly cost for veterinary care, per animal, in the United States. The combined yearly cost for a dog, a horse, and a cat is $307. The difference between the cost for a dog and a horse is $32, and the difference between the cost for a horse and a cat is $16. Find the average yearly cost for veterinary care per dog, per horse, and per cat.   per dog: $129; per horse: $97; per cat: $81

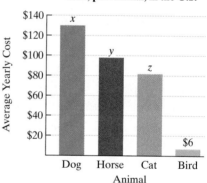

Average Yearly Cost for Veterinary Care, per Animal, in the U.S.

Source: American Veterinary Medical Association

3.4 *In Exercises 29–30, write the system of linear equations represented by the augmented matrix. Use x, y, and, if necessary, z, for the variables. Once the system is written, use back-substitution to find its solution.*

29. $\begin{bmatrix} 2 & 3 & | & -10 \\ 0 & 1 & | & -6 \end{bmatrix}$ $2x + 3y = -10$
$y = -6; (4, -6)$

30. $\begin{bmatrix} 1 & 1 & 3 & | & 12 \\ 0 & 1 & -2 & | & -4 \\ 0 & 0 & 1 & | & 3 \end{bmatrix}$ $x + y + 3z = 12$
$y - 2z = -4$
$z = 3; (1, 2, 3)$

In Exercises 31–34, perform each matrix row operation and write the new matrix.

31. $\begin{bmatrix} 1 & -8 & | & 3 \\ 0 & 7 & | & -14 \end{bmatrix} \frac{1}{7}R_2$ $\begin{bmatrix} 1 & -8 & | & 3 \\ 0 & 1 & | & -2 \end{bmatrix}$

32. $\begin{bmatrix} 1 & -3 & | & 1 \\ 2 & 1 & | & -5 \end{bmatrix} \xrightarrow{-2R_1 + R_2} \begin{bmatrix} 1 & -3 & | & 1 \\ 0 & 7 & | & -7 \end{bmatrix}$

33. $\begin{bmatrix} 2 & -2 & 1 & | & -1 \\ 1 & 2 & -1 & | & 2 \\ 6 & 4 & 3 & | & 5 \end{bmatrix} \xrightarrow{\frac{1}{2}R_1} \begin{bmatrix} 1 & -1 & \frac{1}{2} & | & -\frac{1}{2} \\ 1 & 2 & -1 & | & 2 \\ 6 & 4 & 3 & | & 5 \end{bmatrix}$

34. $\begin{bmatrix} 1 & 2 & 2 & | & 2 \\ 0 & 1 & -1 & | & 2 \\ 0 & 5 & 4 & | & 1 \end{bmatrix} \xrightarrow{-5R_2 + R_3} \begin{bmatrix} 1 & 2 & 2 & | & 2 \\ 0 & 1 & -1 & | & 2 \\ 0 & 0 & 9 & | & -9 \end{bmatrix}$

In Exercises 35–38, solve each system using matrices. If there is no solution or if a system's equations are dependent, so state.

35. $x + 4y = 7$
$3x + 5y = 0$ $(-5, 3)$

36. $2x - 3y = 8$
$-6x + 9y = 4$ inconsistent; no solution or $\varnothing$

37. $x + 2y + 3z = -5$
$2x + y + z = 1$
$x + y - z = 8$ $(1, 3, -4)$

38. $x - 2y + z = 0$
$y - 3z = -1$
$2y + 5z = -2$ $(-2, -1, 0)$

3.5 *In Exercises 39–42, evaluate each determinant.*

39. $\begin{vmatrix} 3 & 2 \\ -1 & 5 \end{vmatrix}$ 17

40. $\begin{vmatrix} -2 & -3 \\ -4 & -8 \end{vmatrix}$ 4

41. $\begin{vmatrix} 2 & 4 & -3 \\ 1 & -1 & 5 \\ -2 & 4 & 0 \end{vmatrix}$ −86

42. $\begin{vmatrix} 4 & 7 & 0 \\ -5 & 6 & 0 \\ 3 & 2 & -4 \end{vmatrix}$ −236

In Exercises 43–46, use Cramer's rule to solve each system. If there is no solution or if a system's equations are dependent, so state.

43. $x - 2y = 8$
$3x + 2y = -1$ $\left(\dfrac{7}{4}, -\dfrac{25}{8}\right)$

44. $7x + 2y = 0$
$2x + y = -3$ $(2, -7)$

45. $x + 2y + 2z = 5$
$2x + 4y + 7z = 19$
$-2x - 5y - 2z = 8$ $(23, -12, 3)$

46. $2x + y = -4$
$y - 2z = 0$
$3x - 2z = -11$ $(-3, 2, 1)$

47. Use the quadratic function $y = ax^2 + bx + c$ to model the following data:

x (Age of a Driver)	y (Average Numbers of Automobile Accidents per Day in the United States)
20	400
40	150
60	400

Use Cramer's rule to determine values for a, b, and c. Then use the model to write a statement about the average number of automobile accidents in which 30-year-old drivers and 50-year-old drivers are involved daily. $y = \frac{5}{8}x^2 - 50x + 1150$; 30-year-old drivers are involved in 212.5 accidents daily and 50-year-old drivers are involved in 212.5 accidents daily.

CHAPTER 3   TEST

Remember to use your Chapter Test Prep Video CD to see the worked-out solutions to the test questions you want to review.

1. Solve by graphing:

$x + y = 6$
$4x - y = 4.$ $(2, 4)$

In Exercises 2–4, solve each system by the substitution method or the addition method. Identify systems with no solution and systems with infinitely many solutions, using set notation to express their solution sets.

2. $5x + 4y = 10$
$3x + 5y = -7$ $(6, -5)$

3. $x = y + 4$
$3x + 7y = -18$ $(1, -3)$

4. $4x = 2y + 6$ $\{(x, y) | 4x = 2y + 6\}$ or
$y = 2x - 3$ $\{(x, y) | y = 2x - 3\}$

In Exercises 5–8, solve each problem.

5. In a new development, 50 one- and two-bedroom condominiums were sold. Each one-bedroom condominium sold for $120 thousand and each two-bedroom condominium sold for $150 thousand. If sales totaled $7050 thousand, how many of each type of unit was sold?   15 one-bedroom and 35 two-bedroom condominium units were sold.

6. You invested $9000 in two funds paying 6% and 7% annual interest, respectively. At the end of the year, the total interest from these investments was $610. How much was invested at each rate? $2000 was invested at 6% and $7000 at 7%.

7. You need to mix a 6% peroxide solution with a 9% peroxide solution to obtain 36 ounces of an 8% peroxide solution. How many ounces of each of the solutions must be used? 12 ounces of a 6% peroxide solution and 24 ounces of a 9% peroxide solution must be used.

8. A paddleboat on the Mississippi River travels 48 miles downstream, with the current, in 3 hours. The return trip, against the current, takes the paddleboat 4 hours. Find the boat's rate in still water and the rate of the current. rate in still water: 14 miles per hour; rate of current: 2 miles per hour

Use this information to solve Exercises 9–11: A company is planning to produce and sell a new line of computers. The fixed cost will be $360,000 and it will cost $850 to produce each computer. Each computer will be sold for $1150.

9. Write the cost function, C, of producing x computers. $C(x) = 360,000 + 850x$

10. Write the revenue function, R, from the sale of x computers. $R(x) = 1150x$

11. Determine the break-even point. Describe what this means. $(1200, 1,380,000)$; When 1200 computers are produced and sold, both cost and revenue are $1,380,000.

12. The cost and revenue functions for producing and selling x units of a toaster oven are

$$C(x) = 40x + 350,000 \quad \text{and} \quad R(x) = 125x.$$

Write the profit function, P, from producing and selling x toaster ovens. $P(x) = 85x - 350,000$

13. Solve by eliminating variables using the addition method:

$$x + y + z = 6$$
$$3x + 4y - 7z = 1$$
$$2x - y + 3z = 5. \quad (1, 3, 2) \text{ or } \{(1, 3, 2)\}$$

14. Perform the indicated matrix row operation and write the new matrix.

$$\begin{bmatrix} 1 & 0 & -4 & | & 5 \\ 6 & -1 & 2 & | & 10 \\ 2 & -1 & 4 & | & -3 \end{bmatrix} -6R_1 + R_2 \quad \begin{bmatrix} 1 & 0 & -4 & | & 5 \\ 0 & -1 & 26 & | & -20 \\ 2 & -1 & 4 & | & -3 \end{bmatrix}$$

In Exercises 15–16, solve each system using matrices.

15. $2x + y = 6$

$3x - 2y = 16 \quad (4, -2)$

16. $x - 4y + 4z = -1$

$2x - y + 5z = 6$

$-x + 3y - z = 5 \quad (-1, 2, 2)$

In Exercises 17–18, evaluate each determinant.

17. $\begin{vmatrix} -1 & -3 \\ 7 & 4 \end{vmatrix} \quad 17$

18. $\begin{vmatrix} 3 & 4 & 0 \\ -1 & 0 & -3 \\ 4 & 2 & 5 \end{vmatrix} \quad -10$

In Exercises 19–20, use Cramer's rule to solve each system.

19. $4x - 3y = 14$

$3x - y = 3 \quad (-1, -6)$

20. $2x + 3y + z = 2$

$3x + 3y - z = 0$

$x - 2y - 3z = 1 \quad (4, -3, 3)$

Note to Instructor: The Chapter 3 Test file in the TestGen program provides algorithms specifically matched to these problems so you can easily replicate this chapter test for additional practice or assessment purposes.

CUMULATIVE REVIEW EXERCISES (CHAPTERS 1–3)

1. Simplify: $\dfrac{6(8 - 10)^3 + (-2)}{(-5)^2(-2)}$. 1

2. Simplify: $7x - [5 - 2(4x - 1)]$. $15x - 7$

In Exercises 3–5, solve each equation.

3. $5 - 2(3 - x) = 2(2x + 5) + 1 \quad -6$

4. $\dfrac{3x}{5} + 4 = \dfrac{x}{3} \quad -15$

5. $3x - 4 = 2(3x + 2) - 3x \quad$ no solution or Ø

6. For a summer sales job, you are choosing between two pay arrangements: a weekly salary of $200 plus 5% commission on sales, or a straight 15% commission. For how many dollars of sales will the earnings be the same regardless of the pay arrangement? For $2000 of sales, the earnings will be the same for both pay arrangements.

7. Simplify: $\dfrac{-5x^6y^{-10}}{20x^{-2}y^{20}}$. $-\dfrac{x^8}{4y^{30}}$

8. If $f(x) = -4x + 5$, find $f(a + 2)$. $-4a - 3$

9. Find the domain of $f(x) = \dfrac{4}{x + 3}$.

$\{x | x \text{ is a real number and } x \neq -3\}$

10. If $f(x) = 2x^2 - 5x + 2$ and $g(x) = x^2 - 2x + 3$, find $(f - g)(x)$ and $(f - g)(3)$.
$(f - g)(x) = x^2 - 3x - 1; -1$

In Exercises 11–12, graph each linear function.

11. $f(x) = -\dfrac{2}{3}x + 2$ **11–12.** See graphing answer section.

12. $2x - y = 6$

In Exercises 13–14, use the given conditions to write an equation for each line in point-slope form and slope-intercept form.

13. Passing through $(2, 4)$ and $(4, -2)$
$y - 4 = -3(x - 2)$ or $y + 2 = -3(x - 4)$;
$y = -3x + 10$ or $f(x) = -3x + 10$

14. Passing through $(-1, 0)$ and parallel to the line whose equation is $3x + y = 6$ $y - 0 = -3(x + 1); y = -3x - 3$
or $f(x) = -3x - 3$

In Exercises 15–16, solve each system by eliminating variables using the addition method.

15. $3x + 12y = 25$
$2x - 6y = 12$ $\left(7, \dfrac{1}{3}\right)$

16. $x + 3y - z = 5$
$-x + 2y + 3z = 13$
$2x - 5y - z = -8$ $(3, 2, 4)$

17. If two pads and 19 pens are sold for \$5.40 and 7 of the same pads and 4 of the same pens sell for \$6.40, find the cost of one pad and one pen. One pad costs \$0.80 and one pen costs \$0.20.

18. Evaluate: $\begin{vmatrix} 0 & 1 & -2 \\ -7 & 0 & -4 \\ 3 & 0 & 5 \end{vmatrix}$. 23

19. Solve using matrices:
$$2x + 3y - z = -1$$
$$x + 2y + 3z = 2$$
$$3x + 5y - 2z = -3.$$ $(3, -2, 1)$

20. Solve using Cramer's rule (determinants):
$$3x + 4y = -1$$
$$-2x + y = 8.$$ $(-3, 2)$

Had a good workout lately? If so, could you tell if you were overdoing it or not pushing yourself hard enough?

Problems involving target heart rate ranges for various exercise goals appear as Exercises 59–62 in Exercise Set 4.4.

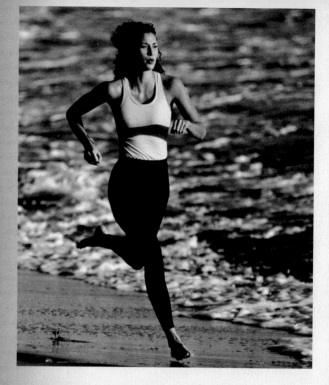

Inequalities and Problem Solving

4.1 Solving Linear Inequalities

4.2 Compound Inequalities

4.3 Equations and Inequalities Involving Absolute Value

4.4 Linear Inequalities in Two Variables

4.5 Linear Programming

Experts suggest a method for determining the heart rate you should maintain during aerobic activities, such as running, biking, or swimming. Subtract your age from 220; find 70% of that figure for the low end of the range and 80% for the high end. Your heart rate, in beats per minute, should stay between these two figures if you want to get the most cardiovascular benefit from your training.

The phrase "should stay between" indicates an algebraic inequality, a symbolic statement containing an inequality symbol such as $<$ (is less than). In this chapter, you will learn methods for modeling your world with inequalities. You will even see how inequalities are used to describe some of the most magnificent places in our nation's landscape.

SECTION 4.1

Objectives

1 Graph an inequality's solution set.

2 Use set-builder and interval notations.

3 Solve linear inequalities.

4 Recognize inequalities with no solution or all real numbers as solutions.

5 Solve applied problems using linear inequalities.

SOLVING LINEAR INEQUALITIES

You can go online and obtain a list of telecommunication companies that provide residential long-distance phone service. The list contains the monthly fee, the monthly minimum, and the rate per minute for each service provider. You've chosen a plan that has a monthly fee of $15 with a charge of $0.08 per minute for all long-distance calls. Suppose you are limited by how much money you can spend for the month: You can spend at most $35. If we let x represent the number of minutes of long-distance calls in a month, we can write an inequality that describes the given conditions:

The monthly fee of $15	plus	the charge of $0.08 per minute for x minutes	must be less than or equal to	$35.
15	+	0.08x	≤	35.

Using the commutative property of addition, we can express this inequality as

$$0.08x + 15 \leq 35.$$

The form of this inequality is $ax + b \leq c$, with $a = 0.08$, $b = 15$, and $c = 35$. Any inequality in this form is called **a linear inequality in one variable**. The greatest exponent on the variable in such an inequality is 1. The symbol between $ax + b$ and c can be ≤ (is less than or equal to), < (is less than), ≥ (is greater than or equal to), or > (is greater than).

   In this section, we will study how to solve linear inequalities such as $0.08x + 15 \leq 35$. **Solving an inequality** is the process of finding the set of numbers that make the inequality a true statement. These numbers are called the **solutions** of the inequality and we say that they **satisfy** the inequality. The set of all solutions is called the **solution set** of the inequality. We begin by discussing how to graph and how to represent these solution sets.

1 Graph an inequality's solution set.

Graphs of Inequalities There are infinitely many solutions of the inequality $x > -4$, namely all real numbers that are greater than -4. Although we cannot list all the solutions, we can make a drawing on a number line that represents these solutions. Such a drawing is called the **graph of the inequality**.

   Graphs of solutions to linear inequalities are shown on a number line by shading all points representing numbers that are solutions. For example, the solutions of $x > 4$ are all real numbers that are greater than 4. They are graphed on a number line by shading all points to the right of 4. The open dot at 4 indicates that 4 is not a solution:

In this book, a convenient notation, called *interval notation*, will be used to represent solution sets of inequalities. To help understand this notation, a different graphing notation will be used. Parentheses, rather than open dots, indicate endpoints that are not solutions. Thus, we graph the solutions of $x > 4$ as

Square brackets, rather than closed dots, indicate endpoints that are solutions. Thus, we graph the solutions of $x \geq 4$ as

EXAMPLE 1 Graphing Inequalities

Graph the solutions of each inequality:

   a. $x < 2$ **b.** $x \geq -3$ **c.** $-2 < x \leq 4$.

SOLUTION

   a. The solutions of $x < 2$ are all real numbers that are less than 2. They are graphed on a number line by shading all points to the left of 2. The parenthesis at 2 indicates that 2 is not a solution, but numbers such as 1.9999 and 1.6 are. The arrow shows that the graph extends indefinitely to the left.

$$\begin{array}{cccccccccccc} & -5 & -4 & -3 & -2 & -1 & 0 & 1 & 2 & 3 & 4 & 5 \end{array}$$

   b. The solutions of $x \geq -3$ are all real numbers that are greater than or equal to -3. We shade all points to the right of -3 and the point for -3 itself. The bracket at -3 shows that -3 is a solution of the given inequality. The arrow shows that the graph extends indefinitely to the right.

$$\begin{array}{cccccccccccc} & -5 & -4 & -3 & -2 & -1 & 0 & 1 & 2 & 3 & 4 & 5 \end{array}$$

   c. The inequality $-2 < x \leq 4$ is read "-2 is less than x *and* x is less than or equal to 4," or "x is greater than -2 *and* less than or equal to 4." The solutions of $-2 < x \leq 4$ are all real numbers between -2 and 4, not including -2 but including 4. The parenthesis at -2 indicates that -2 is not a solution. By contrast, the bracket at 4 shows that 4 is a solution. Shading indicates the other solutions.

$$\begin{array}{cccccccccccc} & -5 & -4 & -3 & -2 & -1 & 0 & 1 & 2 & 3 & 4 & 5 \end{array}$$

   CHECK POINT 1 Graph the solutions of each inequality:

   a. $x \leq 3$ **b.** $x > -5$ **c.** $-4 \leq x < 1$.   See graphing answer section.

② Use set-builder and interval notations.

Set-Builder and Interval Notations Now that we know how to graph the solution set of an inequality such as $x > -4$, let's see how to represent the solution set. One method is with **set-builder notation**. Using this method, the solution set of $x > -4$ can be expressed as

$$\{x \mid x > -4\}.$$

We read this as "the set of all real numbers x such that x is greater than -4."

A second method used to represent solution sets of inequalities is **interval notation**. Using this notation, the solution set of $x > -4$ is expressed as $(-4, \infty)$. The parenthesis at -4 indicates that -4 is not included in the interval. The infinity symbol, ∞, does not represent a real number. It indicates that the interval extends indefinitely to the right.

Table 4.1 lists nine possible types of intervals used to describe sets of real numbers.

Table 4.1	Intervals on the Real Number Line	

Let a and b be real numbers such that $a < b$.

Interval Notation	Set-Builder Notation	Graph
(a, b)	$\{x \mid a < x < b\}$	a b
$[a, b]$	$\{x \mid a \leq x \leq b\}$	a b
$[a, b)$	$\{x \mid a \leq x < b\}$	a b
$(a, b]$	$\{x \mid a < x \leq b\}$	a b
(a, ∞)	$\{x \mid x > a\}$	a
$[a, \infty)$	$\{x \mid x \geq a\}$	a
$(-\infty, b)$	$\{x \mid x < b\}$	b
$(-\infty, b]$	$\{x \mid x \leq b\}$	b
$(-\infty, \infty)$	$\{x \mid x \text{ is a real number}\}$ or $\mathbb{R}$ (set of all real numbers)	

EXAMPLE 2 Intervals and Inequalities

Express each interval in set-builder notation and graph:

   a. $(-1, 4]$ **b.** $[2.5, 4]$ **c.** $(-3, \infty)$.

SOLUTION

   a. $(-1, 4] = \{x \mid -1 < x \leq 4\}$

   b. $[2.5, 4] = \{x \mid 2.5 \leq x \leq 4\}$

   c. $(-3, \infty) = \{x \mid x > -3\}$

See graphing answer section.

✔ **CHECK POINT 2** Express each interval in set-builder notation and graph:

   a. $[-2, 5)$ **b.** $[1, 3.5]$ **c.** $(-\infty, -1)$.

    $\{x \mid -2 \leq x < 5\}$ $\{x \mid 1 \leq x \leq 3.5\}$ $\{x \mid x < -1\}$

3 Solve linear inequalities.

Solving Linear Inequalities Let's return to the question that opened this section: How many minutes of long-distance calls can you make in a month if you can spend at most $35? We answer the question by solving the linear inequality

$$0.08x + 15 \le 35$$

for x. The solution procedure is nearly identical to that for solving the equation

$$0.08x + 15 = 35.$$

Our goal is to get x by itself on the left side. We do this by first subtracting 15 from both sides to isolate $0.08x$:

$$0.08x + 15 \le 35$$

$$0.08x + 15 - 15 \le 35 - 15$$

$$0.08x \le 20.$$

Finally, we isolate x from $0.08x$ by dividing both sides of the inequality by 0.08:

$$\frac{0.08x}{0.08} \le \frac{20}{0.08}$$

$$x \le 250.$$

With at most $35 per month to spend, you can make no more than 250 minutes of long-distance calls each month.

We started with the inequality $0.08x + 15 \le 35$ and obtained the inequality $x \le 250$ in the final step. Both of these inequalities have the same solution set, namely $\{x \mid x \le 250\}$. Inequalities such as these, with the same solution set, are said to be **equivalent**.

We isolated x from $0.08x$ by dividing both sides of $0.08x \le 20$ by 0.08, a positive number. Let's see what happens if we divide both sides of an inequality by a negative number. Consider the inequality $10 < 14$. Divide 10 and 14 by -2:

$$\frac{10}{-2} = -5 \quad \text{and} \quad \frac{14}{-2} = -7.$$

Because -5 lies to the right of -7 on the number line, -5 is greater than -7:

$$-5 > -7.$$

Notice that the direction of the inequality symbol is reversed:

$$10 < 14$$
$$\downarrow$$
$$-5 > -7.$$

Dividing by -2 changes the direction of the inequality symbol.

In general, **when we multiply or divide both sides of an inequality by a negative number, the direction of the inequality symbol is reversed**. When we reverse the direction of the inequality symbol, we say that we change the *sense* of the inequality.

We can isolate a variable in a linear inequality the same way we can isolate a variable in a linear equation. The following properties are used to create equivalent inequalities:

Properties of Inequalities

Property	The Property in Words	Example
The Addition Property of Inequality If $a < b$, then $a + c < b + c$. If $a < b$, then $a - c < b - c$.	If the same quantity is added to or subtracted from both sides of an inequality, the resulting inequality is equivalent to the original one.	$2x + 3 < 7$ Subtract 3: $2x + 3 - 3 < 7 - 3$. Simplify: $2x < 4$.
The Positive Multiplication Property of Inequality If $a < b$ and c is positive, then $ac < bc$. If $a < b$ and c is positive, then $\dfrac{a}{c} < \dfrac{b}{c}$.	If we multiply or divide both sides of an inequality by the same positive quantity, the resulting inequality is equivalent to the original one.	$2x < 4$ Divide by 2: $\dfrac{2x}{2} < \dfrac{4}{2}$. Simplify: $x < 2$.
The Negative Multiplication Property of Inequality If $a < b$ and c is negative, then $ac > bc$. If $a < b$ and c is negative, then $\dfrac{a}{c} > \dfrac{b}{c}$.	If we multiply or divide both sides of an inequality by the same negative quantity and reverse the direction of the inequality symbol, the resulting inequality is equivalent to the original one.	$-4x < 20$ Divide by -4 and reverse the sense of the inequality: $\dfrac{-4x}{-4} > \dfrac{20}{-4}$. Simplify: $x > -5$.

If an inequality does not contain fractions, it can be solved using the following procedure. (In Example 5, we will see how to clear fractions.) Notice, again, how similar this procedure is to the procedure for solving an equation.

SOLVING A LINEAR INEQUALITY

1. Simplify the algebraic expression on each side.
2. Use the addition property of inequality to collect all the variable terms on one side and all the constant terms on the other side.
3. Use the multiplication property of inequality to isolate the variable and solve. Reverse the sense of the inequality when multiplying or dividing both sides by a negative number.
4. Express the solution set in set-builder or interval notation and graph the solution set on a number line.

EXAMPLE 3 Solving a Linear Inequality

Solve and graph the solution set on a number line:

$$3x - 5 > -17.$$

SOLUTION

Step 1. **Simplify each side.** Because each side is already simplified, we can skip this step.

Step 2. **Collect variable terms on one side and constant terms on the other side.** The variable term, $3x$, is already on the left side of $3x - 5 > -17$. We will collect constant terms on the right side by adding 5 to both sides.

$$3x - 5 > -17 \qquad \text{This is the given inequality.}$$
$$3x - 5 + 5 > -17 + 5 \qquad \text{Add 5 to both sides.}$$
$$3x > -12 \qquad \text{Simplify.}$$

Step 3. **Isolate the variable and solve.** We isolate the variable, x, by dividing both sides by 3. Because we are dividing by a positive number, we do not reverse the inequality symbol.

$$\frac{3x}{3} > \frac{-12}{3} \qquad \text{Divide both sides by 3.}$$
$$x > -4 \qquad \text{Simplify.}$$

Step 4. **Express the solution set in set-builder or interval notation and graph the set on a number line.** The solution set consists of all real numbers that are greater than -4, expressed as $\{x \mid x > -4\}$ in set-builder notation. The interval notation for this solution set is $(-4, \infty)$. The graph of the solution set is shown as follows:

CHECK POINT 3 Solve and graph the solution set on a number line:

$4x - 3 > -23$. $\{x \mid x > -5\}$ or $(-5, \infty)$; See graphing answer section.

EXAMPLE 4 Solving a Linear Inequality

Solve and graph the solution set on a number line:

$$-2x - 4 > x + 5.$$

SOLUTION

Step 1. **Simplify each side.** Because each side is already simplified, we can skip this step.

Step 2. **Collect variable terms on one side and constant terms on the other side.** We will collect variable terms on the left and constant terms on the right.

$$-2x - 4 > x + 5 \qquad \text{This is the given inequality.}$$
$$-2x - 4 - x > x + 5 - x \qquad \text{Subtract x from both sides.}$$
$$-3x - 4 > 5 \qquad \text{Simplify.}$$
$$-3x - 4 + 4 > 5 + 4 \qquad \text{Add 4 to both sides.}$$
$$-3x > 9 \qquad \text{Simplify.}$$

Step 3. **Isolate the variable and solve.** We isolate the variable, x, by dividing both sides by -3. Because we are dividing by a negative number, we must reverse the inequality symbol.

$$\frac{-3x}{-3} < \frac{9}{-3} \qquad \text{Divide both sides by } -3 \text{ and reverse the sense of the inequality.}$$
$$x < -3 \qquad \text{Simplify.}$$

Step 4. **Express the solution set in set-builder or interval notation and graph the set on a number line.** The solution set consists of all real numbers that are less than -3, expressed in set-builder notation as $\{x \mid x < -3\}$. The interval notation for this solution set is $(-\infty, -3)$. The graph of the solution set is shown as follows:

USING TECHNOLOGY

You can use a graphing utility to check the solution set of a linear inequality. Enter each side of the inequality separately under y_1 and y_2. Then use the table or the graphs. To use the table, first locate the x-value for which the y-values are the same. Then scroll up or down to locate x values for which y_1 is greater than y_2 or for which y_1 is less than y_2. To use the graphs, locate the intersection point and then find the x-values for which the graph of y_1 lies above the graph of y_2 ($y_1 > y_2$) or for which the graph of y_1 lies below the graph of y_2 ($y_1 < y_2$).

Let's verify our work in Example 4 and show that $(-\infty, -3)$ is the solution set of

$$-2x - 4 > x + 5.$$

Enter $y_1 = -2x - 4$
in the $\boxed{y=}$ screen.

Enter $y_2 = x + 5$
in the $\boxed{y=}$ screen.

We are looking for values of x for which y_1 is greater than y_2.

Numeric Check

Scrolling through the table shows that $y_1 > y_2$ for values of x that are less than -3 (when $x = -3$, $y_1 = y_2$). This verifies $(-\infty, -3)$ is the solution set of $-2x - 4 > x + 5$.

Graphic Check

Display the graphs for y_1 and y_2. Use the intersection feature. The solution set is the set of x-values for which the graph of y_1 lies above the graph of y_2.

Graphs intersect at $(-3, 2)$. When x is less than -3, the graph of y_1 lies above the graph of y_2. This graphically verifies $(-\infty, -3)$ is the solution set of $-2x - 4 > x + 5$.

$[-10, 10, 1]$ by $[-10, 10, 1]$

✔ **CHECK POINT 4** Solve and graph the solution set: $3x + 1 > 7x - 15$.
$\{x \mid x < 4\}$ or $(-\infty, 4)$; See graphing answer section.

If an inequality contains fractions, begin by multiplying both sides by the least common denominator. This will clear the inequality of fractions.

EXAMPLE 5   Solving a Linear Inequality Containing Fractions

Solve and graph the solution set on a number line:

$$\frac{x + 3}{4} \geq \frac{x - 2}{3} + \frac{1}{4}.$$

SOLUTION   The denominators are 4, 3, and 4. The least common denominator is 12. We begin by multiplying both sides of the inequality by 12.

$$\frac{x + 3}{4} \geq \frac{x - 2}{3} + \frac{1}{4}$$ This is the given inequality.

$$12\left(\frac{x + 3}{4}\right) \geq 12\left(\frac{x - 2}{3} + \frac{1}{4}\right)$$ Multiply both sides by 12. Multiplying by a positive number preserves the sense of the inequality.

$$\frac{12}{1} \cdot \frac{x + 3}{4} \geq \frac{12}{1} \cdot \frac{x - 2}{3} + \frac{12}{1} \cdot \frac{1}{4}$$ Multiply each term by 12. Use the distributive property on the right side.

$$\frac{\overset{3}{\cancel{12}}}{1} \cdot \frac{x + 3}{\underset{1}{\cancel{4}}} \geq \frac{\overset{4}{\cancel{12}}}{1} \cdot \frac{x - 2}{\underset{1}{\cancel{3}}} + \frac{\overset{3}{\cancel{12}}}{1} \cdot \frac{1}{\underset{1}{\cancel{4}}}$$ Divide out common factors in each multiplication.

$$3(x + 3) \geq 4(x - 2) + 3$$ The fractions are now cleared.

Now that fractions are cleared, we follow the four steps that we used in the previous examples.

Step 1.   Simplify each side.

$$3(x + 3) \geq 4(x - 2) + 3$$ This is the inequality with the fractions cleared.

$$3x + 9 \geq 4x - 8 + 3$$ Use the distributive property.

$$3x + 9 \geq 4x - 5$$ Simplify.

Step 2.   Collect variable terms on one side and constant terms on the other side. We will collect variable terms on the left and constant terms on the right.

$$3x + 9 - 4x \geq 4x - 5 - 4x$$ Subtract 4x from both sides.

$$-x + 9 \geq -5$$ Simplify.

$$-x + 9 - 9 \geq -5 - 9$$ Subtract 9 from both sides.

$$-x \geq -14$$ Simplify.

Step 3.   Isolate the variable and solve. To isolate x, we must eliminate the negative sign in front of the x. Because $-x$ means $-1x$, we can do this by multiplying (or dividing) both sides of the inequality by -1. We are multiplying by a negative number. Thus, we must reverse the inequality symbol.

$$(-1)(-x) \leq (-1)(-14)$$ Multiply both sides by -1 and reverse the sense of the inequality.

$$x \leq 14$$ Simplify.

Step 4.   Express the solution set in set-builder or interval notation and graph the set on a number line. The solution set consists of all real numbers that are less than or equal to 14, expressed in set-builder notation as $\{x | x \leq 14\}$.

The interval notation for this solution set is $(-\infty, 14]$. The graph of the solution set is shown as follows:

$$\xleftarrow{\hspace{1cm}} \begin{array}{cccccccccc} 5 & 6 & 7 & 8 & 9 & 10 & 11 & 12 & 13 & 14 & 15 \end{array} \xrightarrow{\hspace{1cm}}$$

CHECK POINT 5 Solve and graph the solution set on a number line:

$$\frac{x-4}{2} \geq \frac{x-2}{3} + \frac{5}{6}. \quad \{x | x \geq 13\} \text{ or } [13, \infty); \text{ See graphing answer section.}$$

---

4 Recognize inequalities with no solution or all real numbers as solutions.

Inequalities with Unusual Solution Sets

We have seen that some equations have no solution. This is also true for some inequalities. An example of such an inequality is

$$x > x + 1.$$

There is no number that is greater than itself plus 1. This inequality has no solution and its solution set is $\varnothing$, the empty set.

By contrast, some inequalities are true for all real numbers. An example of such an inequality is

$$x < x + 1.$$

Every real number is less than itself plus 1. The solution set is $\{x | x \text{ is a real number}\}$ or $\mathbb{R}$. In interval notation, the solution set is $(-\infty, \infty)$.

If you attempt to solve an inequality that has no solution, you will eliminate the variable and obtain a false statement, such as $0 > 1$. If you attempt to solve an inequality that is true for all real numbers, you will eliminate the variable and obtain a true statement, such as $0 < 1$.

EXAMPLE 6   Solving Linear Inequalities

Solve each inequality:

a. $2(x + 4) > 2x + 3$ **b.** $x + 7 \leq x - 2.$

SOLUTION

a.

$2(x + 4) > 2x + 3$	This is the given inequality.
$2x + 8 > 2x + 3$	Apply the distributive property.
$2x + 8 - 2x > 2x + 3 - 2x$	Subtract $2x$ from both sides.
$8 > 3$	Simplify. The statement $8 > 3$ is true.

The inequality $8 > 3$ is true for all values of x. Because this inequality is equivalent to the original inequality, the original inequality is true for all real numbers. The solution set is

$$\{x | x \text{ is a real number}\} \text{ or } \mathbb{R} \text{ or } (-\infty, \infty).$$

b.

$x + 7 \leq x - 2$	This is the given inequality.
$x + 7 - x \leq x - 2 - x$	Subtract x from both sides.
$7 \leq -2$	Simplify. The statement $7 \leq -2$ is false.

The inequality $7 \leq -2$ is false for all values of x. Because this inequality is equivalent to the original inequality, the original inequality has no solution. The solution set is $\varnothing$.

CHECK POINT **6** Solve each inequality:

a. $3(x + 1) > 3x + 2$ $\{x \mid x \text{ is a real number}\}$ or $\mathbb{R}$ or $(-\infty, \infty)$

b. $x + 1 \leq x - 1$. $\varnothing$

5 Solve applied problems using linear inequalities.

Applications Commonly used English phrases such as "at least" and "at most" indicate inequalities. Table 4.2 lists sentences containing these phrases and their algebraic translations into inequalities.

Table 4.2	English Sentences and Inequalities
English Sentence	**Inequality**
x is at least 5.	$x \geq 5$
x is at most 5.	$x \leq 5$
x is between 5 and 7.	$5 < x < 7$
x is no more than 5.	$x \leq 5$
x is no less than 5.	$x \geq 5$

Our next example shows how to use an inequality to select the better deal between two pricing options. We use our strategy for solving word problems, translating from the verbal conditions of the problem to a linear inequality.

EXAMPLE 7 Selecting the Better Deal

Acme Car rental agency charges $4 a day plus $0.15 per mile, whereas Interstate rental agency charges $20 a day and $0.05 per mile. How many miles must be driven to make the daily cost of an Acme rental a better deal than an Interstate rental?

SOLUTION

Step 1. **Let x represent one of the quantities.** We are looking for the number of miles that must be driven in a day to make Acme the better deal. Thus,

let x = the number of miles driven in a day.

Step 2. **Represent other quantities in terms of x.** We are not asked to find another quantity, so we can skip this step.

Step 3. **Write an inequality in x that describes the conditions.** Acme is a better deal than Interstate if the daily cost of Acme is less than the daily cost of Interstate.

Step 4. **Solve the inequality and answer the question.**

$4 + 0.15x < 20 + 0.05x$	This is the inequality that describes the verbal conditions.
$4 + 0.15x - 0.05x < 20 + 0.05x - 0.05x$	Subtract 0.05x from both sides.
$4 + 0.1x < 20$	Simplify.
$4 + 0.1x - 4 < 20 - 4$	Subtract 4 from both sides.
$0.1x < 16$	Simplify.
$\dfrac{0.1x}{0.1} < \dfrac{16}{0.1}$	Divide both sides by 0.1.
$x < 160$	Simplify.

Thus, driving fewer than 160 miles per day makes Acme the better deal.

Step 5. **Check the proposed solution in the original wording of the problem.** One way to do this is to take a mileage less than 160 miles per day to see if Acme is the better deal. Suppose that 150 miles are driven in a day.

$$\text{Cost for Acme} = 4 + 0.15(150) = 26.50$$
$$\text{Cost for Interstate} = 20 + 0.05(150) = 27.50$$

Acme has a lower daily cost, making Acme the better deal. ∎

✔ **CHECK POINT 7** A car can be rented from Basic Rental for $260 per week with no extra charge for mileage. Continental charges $80 per week plus 25 cents for each mile driven to rent the same car. How many miles must be driven in a week to make the rental cost for Basic Rental a better deal than Continental's?
Driving more than 720 miles per week makes Basic Rental a better deal.

4.1 EXERCISE SET

Student Solutions Manual   CD/Video   PH Math/Tutor Center   MathXL Tutorials on CD   MathXL®   MyMathLab   Interactmath.com

Practice Exercises

In Exercises 1–12, graph the solutions of each inequality on a number line. 1–12. See graphing answer section.

1. $x > 5$

2. $x > -4$

3. $x < -2$

4. $x < 0$

5. $x \geq -2.5$

6. $x \geq -4.5$

7. $x \leq 3$

8. $x \leq 6$

9. $-3 < x \leq 5$

10. $-2 \leq x < 7$

11. $-1 < x < 3$

12. $-6 \leq x \leq 0$

In Exercises 13–26, express each interval in set-builder notation and graph the interval on a number line. 13–26. See graphing answer section.

13. $(1, 6]$ $\{x | 1 < x \leq 6\}$

14. $(-2, 4]$ $\{x | -2 < x \leq 4\}$

15. $[-5, 2)$ $\{x | -5 \leq x < 2\}$

16. $[-4, 3)$ $\{x | -4 \leq x < 3\}$

17. $[-3, 1]$ $\{x | -3 \leq x \leq 1\}$

18. $[-2, 5]$ $\{x | -2 \leq x \leq 5\}$

19. $(2, \infty)$ $\{x | x > 2\}$

20. $(3, \infty)$ $\{x | x > 3\}$

21. $[-3, \infty)$ $\{x | x \geq -3\}$

22. $[-5, \infty)$ $\{x | x \geq -5\}$

23. $(-\infty, 3)$ $\{x | x < 3\}$

24. $(-\infty, 2)$ $\{x | x < 2\}$

25. $(-\infty, 5.5)$ $\{x | x < 5.5\}$

26. $(-\infty, 3.5]$ $\{x | x \leq 3.5\}$

In Exercises 27–58, solve each linear inequality. Other than $\varnothing$, graph the solution set on a number line. 27–58. See graphing answer section.

27. $5x + 11 < 26$
$\{x | x < 3\}$ or $(-\infty, 3)$

28. $2x + 5 < 17$
$\{x | x < 6\}$ or $(-\infty, 6)$

29. $3x - 8 \geq 13$
$\{x | x \geq 7\}$ or $[7, \infty)$

30. $8x - 2 \geq 14$
$\{x | x \geq 2\}$ or $[2, \infty)$

31. $-9x \geq 36$
$\{x | x \leq -4\}$ or $(-\infty, -4]$

32. $-5x \leq 30$
$\{x | x \geq -6\}$ or $[-6, \infty)$

33. $8x - 11 \leq 3x - 13$
$\{x | x \leq -\frac{2}{5}\}$ or $\left(-\infty, -\frac{2}{5}\right]$

34. $18x + 45 \leq 12x - 8$
$\{x | x \leq -\frac{53}{6}\}$ or $\left(-\infty, -\frac{53}{6}\right]$

35. $4(x + 1) + 2 \geq 3x + 6$ $\{x | x \geq 0\}$ or $[0, \infty)$

36. $8x + 3 > 3(2x + 1) + x + 5$ $\{x | x > 5\}$ or $(5, \infty)$

37. $2x - 11 < -3(x + 2)$ $\{x | x < 1\}$ or $(-\infty, 1)$

38. $-4(x + 2) > 3x + 20$ $\{x | x < -4\}$ or $(-\infty, -4)$

39. $1 - (x + 3) \geq 4 - 2x$ $\{x | x \geq 6\}$ or $[6, \infty)$

40. $5(3 - x) \leq 3x - 1$ $\{x | x \geq 2\}$ or $[2, \infty)$

41. $\dfrac{x}{4} - \dfrac{1}{2} \leq \dfrac{x}{2} + 1$ $\{x | x \geq -6\}$ or $[-6, \infty)$

42. $\dfrac{3x}{10} + 1 \geq \dfrac{1}{5} - \dfrac{x}{10}$ $\{x | x \geq -2\}$ or $[-2, \infty)$

43. $1 - \dfrac{x}{2} > 4$ $\{x | x < -6\}$ or $(-\infty, -6)$

44. $7 - \dfrac{4}{5}x < \dfrac{3}{5}$ $\{x | x > 8\}$ or $(8, \infty)$

45. $\dfrac{x - 4}{6} \geq \dfrac{x - 2}{9} + \dfrac{5}{18}$ $\{x | x \geq 13\}$ or $[13, \infty)$

46. $\dfrac{4x - 3}{6} + 2 \geq \dfrac{2x - 1}{12}$ $\{x | x \geq -\frac{19}{6}\}$ or $[-\frac{19}{6}, \infty)$

47. $4(3x - 2) - 3x < 3(1 + 3x) - 7$
$\{x | x \text{ is a real number}\}$ or $\mathbb{R}$ or $(-\infty, \infty)$

48. $3(x - 8) - 2(10 - x) < 5(x - 1)$
$\{x | x \text{ is a real number}\}$ or $\mathbb{R}$ or $(-\infty, \infty)$

49. $8(x + 1) \leq 7(x + 5) + x$
$\{x | x \text{ is a real number}\}$ or $\mathbb{R}$ or $(-\infty, \infty)$

50. $4(x - 1) \geq 3(x - 2) + x$
$\{x | x \text{ is a real number}\}$ or $\mathbb{R}$ or $(-\infty, \infty)$

51. $3x < 3(x - 2)$ $\varnothing$

52. $5x < 5(x - 3)$ $\varnothing$

53. $7(y + 4) - 13 < 12 + 13(3 + y)$ $\{y | y > -6\}$ or $(-6, \infty)$

54. $-3[7y - (2y - 3)] > -2(y + 1)$ $\{y | y < -\frac{7}{13}\}$ or $(-\infty, -\frac{7}{13})$

55. $6 - \dfrac{2}{3}(3x - 12) \leq \dfrac{2}{5}(10x + 50)$ $\{x | x \geq -1\}$ or $[-1, \infty)$

56. $\dfrac{2}{7}(7 - 21x) - 4 > 10 - \dfrac{3}{11}(11x - 11)$
$\{x | x < -5\}$ or $(-\infty, -5)$

57. $3[3(y + 5) + 8y + 7] + 5[3(y - 6)$
$- 2(3y - 5)] < 2(4y + 3)$ $\{y | y < -2\}$ or $(-\infty, -2)$

58. $5[3(2 - 3y) - 2(5 - y)] - 6[5(y - 2)$
$- 2(4y - 3)] < 3y + 19$ $\{y | y > -\frac{3}{4}\}$ or $(-\frac{3}{4}, \infty)$

59. Let $f(x) = 3x + 2$ and $g(x) = 5x - 8$. Find all values of x for which $f(x) > g(x)$. $\{x | x < 5\}$ or $(-\infty, 5)$

60. Let $f(x) = 2x - 9$ and $g(x) = 5x + 4$. Find all values of x for which $f(x) > g(x)$. $\{x | x < -\frac{13}{3}\}$ or $(-\infty, -\frac{13}{3})$

61. Let $f(x) = \dfrac{2}{5}(10x + 15)$ and $g(x) = \dfrac{1}{4}(8 - 12x)$. Find all values of x for which $g(x) \leq f(x)$. $\{x | x \geq -\frac{4}{7}\}$ or $[-\frac{4}{7}, \infty)$

62. Let $f(x) = \dfrac{3}{5}(10x - 15) + 9$ and $g(x) = \dfrac{3}{8}(16 - 8x) - 7$. Find all values of x for which $g(x) \leq f(x)$. $\{x | x \geq -\frac{1}{9}\}$ or $[-\frac{1}{9}, \infty)$

Practice Plus

In Exercises 63–64, solve each linear inequality and graph the solution set on a number line.   **63–64.** See graphing answer section.

63. $2(x + 3) > 6 - \{4[x - (3x - 4) - x] + 4\}$
$\{x | x < 2\}$ or $(-\infty, 2)$

64. $3(4x - 6) < 4 - \{5x - [6x - (4x - (3x + 2))]\}$
$\{x | x < 2\}$ or $(-\infty, 2)$

In Exercises 65–66, write an inequality with x isolated on the left side that is equivalent to the given inequality.

65. $ax + b > c$; Assume $a < 0$. $x < \dfrac{c - b}{a}$

66. $\dfrac{ax + b}{c} > b$; Assume $a > 0$ and $c < 0$. $x < \dfrac{bc - b}{a}$

In Exercises 67–68, use the graphs of y_1 and y_2 to solve each inequality.

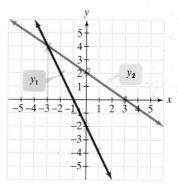

67. $y_1 \geq y_2$ $\{x | x \leq -3\}$ or $(-\infty, -3]$

68. $y_1 \leq y_2$ $\{x | x \geq -3\}$ or $[-3, \infty)$

In Exercises 69–70, use the table of values for the linear functions y_1 and y_2 to solve each inequality.

X	Y1	Y2
-1.8	9.4	7.4
-1.7	9.1	7.6
-1.6	8.8	7.8
-1.5	8.5	8
-1.4	8.2	8.2
-1.3	7.9	8.4
-1.2	7.6	8.6

X=-1.8

69. $y_1 < y_2$ $\{x | x > -1.4\}$ or $(-1.4, \infty)$

70. $y_1 > y_2$ $\{x | x < -1.4\}$ or $(-\infty, -1.4)$

Application Exercises

The graphs show that the three components of love, namely passion, intimacy, and commitment, progress differently over time. Passion peaks early in a relationship and then declines. By contrast, intimacy and commitment build gradually. Use the graphs to solve Exercises 71–78.

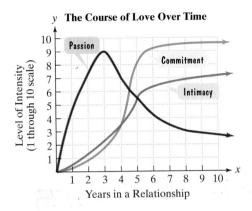

The Course of Love Over Time

Source: R. J. Sternberg, A Triangular Theory of Love, *Psychological Review,* 93, 119–135.

71. Use interval notation to write an inequality that expresses for which years in a relationship intimacy is greater that commitment.   (0, 4)

72. Use interval notation to write an inequality that expresses for which years in a relationship passion is greater than or equal to intimacy.   [0, 5]

73. What is the relationship between passion and intimacy on the interval [5, 7)?   intimacy ≥ passion or passion ≤ intimacy

74. What is the relationship between intimacy and commitment on the interval [4, 7)?   commitment ≥ intimacy or intimacy ≤ commitment

75. What is the relationship between passion and commitment for {x|6 < x < 8}?   commitment > passion or passion < commitment

76. What is the relationship between passion and commitment for {x|7 < x < 9}?   commitment > passion or passion < commitment

77. What is the maximum level of intensity for passion? After how many years in a relationship does this occur?   9; after 3 years

78. After approximately how many years do levels of intensity for commitment exceed the maximum level of intensity for passion?   after approximately $5\frac{1}{2}$ years

79. The percentage, *P*, of U.S. voters who used electronic voting systems, such as optical scans, in national elections can be modeled by the formula

$$P = 3.1x + 25.8,$$

where *x* is the number of years after 1994. In which years will more than 63% of U.S. voters use electronic systems?   voting years after 2006

80. The percentage, *P*, of U.S. voters who used punch cards or lever machines in national elections can be modeled by the formula

$$P = -2.5x + 63.1,$$

where *x* is the number of years after 1994. In which years will fewer than 38.1% of U.S. voters use punch cards or lever machines?   voting years after 2004

The Olympic 500-meter speed skating times have generally been decreasing over time. The formulas

$$W = -0.19t + 57 \quad \text{and} \quad M = -0.15t + 50$$

model the winning times, in seconds, for women, W, and men, M, t years after 1900. Use these models to solve Exercises 81–82.

81. Find values of *t* such that *W* < *M*. Describe what this means in terms of winning times.   {t|t > 175} or (175, ∞); The women's speed skating times will be less than the men's after the year 2075.

82. Find values of *t* such that *W* > *M*. Describe what this means in terms of winning times.   {t|t < 175} or (−∞, 175); The women's speed skating times will be greater than the men's until the year 2075.

In Exercises 83–90, use the strategy for solving word problems, translating from the verbal conditions of the problem to a linear inequality.

83. A truck can be rented from Basic Rental for $50 a day plus $0.20 per mile. Continental charges $20 per day plus $0.50 per mile to rent the same truck. How many miles must be driven in a day to make the rental cost for Basic Rental a better deal than Continental's?   Basic rental is a better deal when driving more than 100 miles per day.

84. You are choosing between two long-distance telephone plans. Plan A has a monthly fee of $15 with a charge of $0.08 per minute for all long-distance calls. Plan B has a monthly fee of $3 with a charge of $0.12 per minute for all long-distance calls. How many minutes of long-distance calls in a month make plan A the better deal?   More than 300 minutes make plan A the better deal.

85. A city commission has proposed two tax bills. The first bill requires that a homeowner pay $1800 plus 3% of the assessed home value in taxes, The second bill requires taxes of $200 plus 8% of the assessed home value. What price range of home assessment would make the first bill a better deal?   A home assessment of greater than $32,000 would make the first bill a better deal.

86. A local bank charges $8 per month plus 5¢ per check. The credit union charges $2 per month plus 8¢ per check. How many checks should be written each month to make the credit union a better deal?   Less than 200 checks should be written per month to make the credit union a better deal.

87. A company manufactures and sells blank audiocassette tapes. The weekly fixed cost is $10,000 and it costs $0.40 to produce each tape. The selling price is $2.00 per tape. How many tapes must be produced and sold each week for the company to have a profit?   More than 6250 tapes must be produced and sold each week for a profit.

88. A company manufactures and sells personalized stationery. The weekly fixed cost is $3000 and it costs $3.00 to produce each package of stationery. The selling price is $5.50 per package. How many packages of stationery must be produced and sold each week for the company to have a profit?   More than 1200 packages of stationery must be produced and sold each week for a profit.

89. An elevator at a construction site has a maximum capacity of 3000 pounds. If the elevator operator weighs 200 pounds and each cement bag weighs 70 pounds, how many bags of cement can be safely lifted on the elevator in one trip?   40 bags or fewer of cement can be safely lifted on the elevator in one trip.

90. An elevator at a construction site has a maximum capacity of 2500 pounds. If the elevator operator weighs 160 pounds and each cement bag weighs 60 pounds, how many bags of cement can be safely lifted on the elevator in one trip?   39 bags or fewer of cement can be safely lifted on the elevator in one trip.

Writing in Mathematics

91. When graphing the solutions of an inequality, what does a parenthesis signify? What does a bracket signify?
91–97. Answers will vary.

92. When solving an inequality, when is it necessary to change the sense of the inequality? Give an example.

93. Describe ways in which solving a linear inequality is similar to solving a linear equation.

94. Describe ways in which solving a linear inequality is different from solving a linear equation.

95. When solving a linear inequality, describe what happens if the solution set is $(-\infty, \infty)$.

96. When solving a linear inequality, describe what happens if the solution set is $\varnothing$.

97. What is the slope of each model in Exercises 79–80? What does this mean in terms of the percentage of U.S. voters using electronic voting systems and more traditional methods, such as punch cards or lever machines? What explanations can you offer for these changes in vote-counting systems?

Technology Exercises

In Exercises 98–99, solve each inequality using a graphing utility. Graph each side separately. Then determine the values of x for which the graph on the left side lies above the graph on the right side.

98. $-3(x - 6) > 2x - 2$ $\{x \mid x < 4\}$ or $(-\infty, 4)$
98–99. See graphing answer section.

99. $-2(x + 4) > 6x + 16$ $\{x \mid x < -3\}$ or $(-\infty, -3)$

100. Use a graphing utility's $\boxed{\text{TABLE}}$ feature to verify your work in Exercises 98–99.   See graphing answer section.

Use the same technique employed in Exercises 98–99 to solve each inequality in Exercises 101–102. In each case, what conclusion can you draw? What happens if you try solving the inequalities algebraically?

101. $12x - 10 > 2(x - 4) + 10x$ $\varnothing$

102. $2x + 3 > 3(2x - 4) - 4x$
$\{x \mid x$ is a real number$\}$ or $\mathbb{R}$ or $(-\infty, \infty)$

103. A bank offers two checking account plans. Plan A has a base service charge of $4.00 per month plus 10¢ per check. Plan B charges a base service charge of $2.00 per month plus 15¢ per check.

   a. Write models for the total monthly costs for each plan if x checks are written.   plan A: $4 + 0.10x$; plan B: $2 + 0.15x$

   b. Use a graphing utility to graph the models in the same $[0, 50, 1]$ by $[0, 10, 1]$ viewing rectangle.   See graphing answer section.

   c. Use the graphs (and the intersection feature) to determine for what number of checks per month plan A will be better than plan B.   c–d. more than 40 checks per month

   d. Verify the result of part (c) algebraically by solving an inequality.

Critical Thinking Exercises

104. Which one of the following is true?   d

   a. The inequality $3x > 6$ is equivalent to $2 > x$.

   b. The smallest real number in the solution set of $2x > 6$ is 4.

   c. If x is at least 7, then $x > 7$.

   d. The inequality $-3x > 6$ is equivalent to $-2 > x$.

105. What's wrong with this argument? Suppose x and y represent two real numbers, where $x > y$.

$2 > 1$	This is a true statement.
$2(y - x) > 1(y - x)$	Multiply both sides by $y - x$.
$2y - 2x > y - x$	Use the distributive property.
$y - 2x > -x$	Subtract y from both sides.
$y > x$	Add $2x$ to both sides.

The final inequality, $y > x$, is impossible because we were initially given $x > y$.   Since $x > y$, $y - x < 0$. Thus, when both sides were multiplied by $y - x$, the sense of the inequality should have been changed.

106. The figure shows a square with four identical equilateral triangles attached. If the perimeter of the figure is at least 24 centimeters, what are the possible values for k?

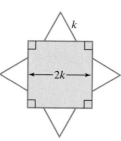

2 centimeters or greater

Review Exercises

107. If $f(x) = x^2 - 2x + 5$, find $f(-4)$. (Section 2.1, Example 3) 29

108. Solve the system:

$$2x - y - z = -3$$
$$3x - 2y - 2z = -5$$
$$-x + y + 2z = 4.$$

(Section 3.3, Example 2) $(-1, -1, 2)$

109. Simplify: $\left(\dfrac{2x^4 y^{-2}}{4xy^3}\right)^3$ (Section 1.6, Example 9) $\dfrac{x^9}{8y^{15}}$

SECTION 4.2

COMPOUND INEQUALITIES

Objectives

1 Find the intersection of two sets.

2 Solve compound inequalities involving *and*.

3 Find the union of two sets.

4 Solve compound inequalities involving *or*.

Boys and girls differ in their toy preferences. These differences are functions of our gender stereotypes—that is, our widely shared beliefs about males' and females' abilities, personality traits, and social behavior. Generally, boys have less leeway to play with "feminine" toys than girls do with "masculine" toys.

Which toys are requested by more than 40% of boys *and* more than 10% of girls? Which toys are requested by more than 40% of boys *or* more than 10% of girls? These questions are not the same. One involves inequalities joined by the word "and"; the other involves inequalities joined by the word "or." A **compound inequality** is formed by joining two inequalities with the word *and* or the word *or*.

Examples of Compound Inequalities

- $x - 3 < 5$ and $2x + 4 < 14$
- $3x - 5 \le 13$ or $5x + 2 > -3$

In this section, you will learn to solve compound inequalities. With this skill, you will be able to analyze childrens' toy preferences in the exercise set.

Compound Inequalities Involving *And* You need to determine whether there is sufficient support on campus to have a blood drive. You take a survey to obtain information, asking students two questions:

1 Find the intersection of two sets.

Would you be willing to donate blood?

Would you be willing to help serve a free breakfast to blood donors?

Part of your report involves the number of students willing to both donate blood *and* serve breakfast. You must focus on the set containing all students who are common to both the set of donors and the set of breakfast servers. This set is called the *intersection* of the two sets.

> **DEFINITION OF THE INTERSECTION OF SETS** The **intersection** of sets A and B, written $A \cap B$, is the set of elements common to both set A **and** set B. This definition can be expressed in set-builder notation as follows:
>
> $$A \cap B = \{x \mid x \in A \text{ AND } x \in B\}.$$

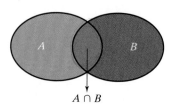

FIGURE 4.1 Picturing the intersection of two sets

Figure 4.1 shows a useful way of picturing the intersection of sets A and B. The figure indicates that $A \cap B$ contains those elements that belong to both A and B at the same time.

EXAMPLE 1 Finding the Intersection of Two Sets

Find the intersection: $\{7, 8, 9, 10, 11\} \cap \{6, 8, 10, 12\}$.

SOLUTION The elements common to $\{7, 8, 9, 10, 11\}$ and $\{6, 8, 10, 12\}$ are 8 and 10. Thus,

$$\{7, 8, 9, 10, 11\} \cap \{6, 8, 10, 12\} = \{8, 10\}.$$ ■

✔ **CHECK POINT 1** Find the intersection: $\{3, 4, 5, 6, 7\} \cap \{3, 7, 8, 9\}$. $\{3, 7\}$

---

2 Solve compound inequalities involving *and*.

A number is a **solution of a compound inequality formed by the word *and*** if it is a solution of both inequalities. For example, the solution set of the compound inequality

$$x \le 6 \quad \text{and} \quad x \ge 2$$

is the set of values of x that satisfy $x \le 6$ and $x \ge 2$. Thus, the solution set is the intersection of the solution sets of the two inequalities.

What are the numbers that satisfy both $x \le 6$ and $x \ge 2$? These numbers are easier to see if we graph the solution set to each inequality on a number line. These graphs are shown in Figure 4.2. The intersection is shown in the third graph.

FIGURE 4.2 Numbers satisfying both $x \le 6$ and $x \ge 2$

The numbers common to both sets are those that are less than or equal to 6 and greater than or equal to 2. This set is $\{x \mid 2 \le x \le 6\}$, or, in interval notation, $[2, 6]$.

Here is a procedure for finding the solution set of a compound inequality containing the word *and*.

SOLVING COMPOUND INEQUALITIES INVOLVING *AND*

1. Solve each inequality separately.
2. Graph the solution set to each inequality on a number line and take the intersection of these solution sets.

> Keep in mind that this intersection appears as the portion of the number line that the two graphs have in common.

EXAMPLE 2 Solving a Compound Inequality with *And*

Solve: $x - 3 < 5$ and $2x + 4 < 14$.

SOLUTION

Step 1. **Solve each inequality separately.**

$$x - 3 < 5 \quad \text{and} \quad 2x + 4 < 14$$
$$x < 8 \qquad\qquad 2x < 10$$
$$\qquad\qquad\qquad x < 5$$

Step 2. **Take the intersection of the solution sets of the two inequalities.** We graph the solution sets of $x < 8$ and $x < 5$. The intersection is shown in the third graph.

The numbers common to both sets are those that are less than 5. The solution set is $\{x \mid x < 5\}$, or, in interval notation, $(-\infty, 5)$. Take a moment to check that any number in $(-\infty, 5)$ satisfies both of the original inequalities. ■

✔ **CHECK POINT 2** Solve: $x + 2 < 5$ and $2x - 4 < -2$. $\{x \mid x < 1\}$ or $(-\infty, 1)$

EXAMPLE 3 Solving a Compound Inequality with *And*

Solve: $2x - 7 > 3$ and $5x - 4 < 6$.

SOLUTION

Step 1. **Solve each inequality separately.**

$$2x - 7 > 3 \quad \text{and} \quad 5x - 4 < 6$$
$$2x > 10 \qquad\qquad 5x < 10$$
$$x > 5 \qquad\qquad\quad x < 2$$

Step 2. **Take the intersection of the solution sets of the two inequalities.** We graph the solution sets of $x > 5$ and $x < 2$. We use these graphs to find their intersection.

$\{x \mid x > 5\}$ $(5, \infty)$

$\{x \mid x < 2\}$ $(-\infty, 2)$

$\{x \mid x > 5\} \cap \{x \mid x < 2\}$
$= \emptyset$ $\emptyset$

There is no number that is both greater than 5 and at the same time less than 2. Thus, the solution set is the empty set, $\emptyset$.

CHECK POINT 3 Solve: $4x - 5 > 7$ and $5x - 2 < 3$. $\emptyset$

If $a < b$, the compound inequality

$$a < x \text{ and } x < b$$

can be written in the shorter form

$$a < x < b.$$

For example, the compound inequality

$$-3 < 2x + 1 \text{ and } 2x + 1 < 3$$

can be abbreviated

$$-3 < 2x + 1 < 3.$$

The word "and" does not appear when the inequality is written in the shorter form, although it is implied. The shorter form enables us to solve both inequalities at once. By performing the same operations on all three parts of the inequality, our goal is to **isolate x in the middle.**

EXAMPLE 4 Solving a Compound Inequality

Solve and graph the solution set:

$$-3 < 2x + 1 \leq 3.$$

SOLUTION We would like to isolate x in the middle. We can do this by first subtracting 1 from all three parts of the compound inequality. Then we isolate x from $2x$ by dividing all three parts of the inequality by 2.

$$-3 < 2x + 1 \leq 3 \qquad \text{This is the given inequality.}$$
$$-3 - 1 < 2x + 1 - 1 \leq 3 - 1 \qquad \text{Subtract 1 from all three parts.}$$
$$-4 < 2x \leq 2 \qquad \text{Simplify.}$$
$$\frac{-4}{2} < \frac{2x}{2} \leq \frac{2}{2} \qquad \text{Divide each part by 2.}$$
$$-2 < x \leq 1 \qquad \text{Simplify.}$$

The solution set consists of all real numbers greater than -2 and less than or equal to 1, represented by $\{x \mid -2 < x \leq 1\}$ in set-builder notation and $(-2, 1]$ in interval notation. The graph is shown as follows:

✔ CHECK POINT 4 Solve and graph the solution set: $1 \le 2x + 3 < 11$.
$\{x|-1 \le x < 4\}$ or $[-1, 4)$; See graphing answer section.

Compound Inequalities Involving *Or* Earlier, we considered a survey asking students if they were willing to donate blood or serve breakfast to donors. You continue to sort the results. Part of your report involves the number of students willing to donate blood *or* serve breakfast *or* do both. You must focus on the set containing donors or breakfast servers or both. This set is called the *union* of two sets.

3 Find the union of two sets.

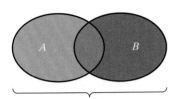

$A \cup B$

FIGURE 4.3 Picturing the union of two sets

> **DEFINITION OF THE UNION OF SETS** The **union** of sets A and B, written $A \cup B$, is the set of elements that are members of set A **or** of set B or of both sets. This definition can be expressed in set-builder notation as follows:
>
> $$A \cup B = \{x | x \in A \text{ OR } x \in B\}.$$

Figure 4.3 shows a useful way of picturing the union of sets A and B. The figure indicates that $A \cup B$ is formed by joining the sets together.

We can find the union of set A and set B by listing the elements of set A. Then, we include any elements of set B that have not already been listed. Enclose all elements that are listed with braces. This shows that the union of two sets is also a set.

EXAMPLE 5 Finding the Union of Two Sets

Find the union: $\{7, 8, 9, 10, 11\} \cup \{6, 8, 10, 12\}$.

SOLUTION To find $\{7, 8, 9, 10, 11\} \cup \{6, 8, 10, 12\}$, start by listing all the elements from the first set, namely 7, 8, 9, 10, and 11. Now list all the elements from the second set that are not in the first set, namely 6 and 12. The union is the set consisting of all these elements. Thus,

$$\{7, 8, 9, 10, 11\} \cup \{6, 8, 10, 12\} = \{6, 7, 8, 9, 10, 11, 12\}. \qquad ■$$

✔ CHECK POINT 5 Find the union: $\{3, 4, 5, 6, 7\} \cup \{3, 7, 8, 9\}$. $\{3, 4, 5, 6, 7, 8, 9\}$

4 Solve compound inequalities involving *or*.

A number is a **solution of a compound inequality formed by the word** *or* if it is a solution of either inequality. Thus, the solution set of a compound inequality formed by the word *or* is the union of the solution sets of the two inequalities.

> **SOLVING COMPOUND INEQUALITIES INVOLVING** *OR*
>
> **1.** Solve each inequality separately.
> **2.** Graph the solution set to each inequality on a number line and take the union of these solution sets.
>
> > Keep in mind that this union appears as the portion of the number line representing the total collection of numbers in the two graphs.

EXAMPLE 6 Solving a Compound Inequality with *Or*

Solve: $2x - 3 < 7$ or $35 - 4x \leq 3$.

SOLUTION

Step 1. **Solve each inequality separately.**

$$2x - 3 < 7 \quad \text{or} \quad 35 - 4x \leq 3$$
$$2x < 10 \qquad\qquad -4x \leq -32$$
$$x < 5 \qquad\qquad\quad x \geq 8$$

Step 2. **Take the union of the solution sets of the two inequalities.** We graph the solution sets of $x < 5$ and $x \geq 8$. We use these graphs to find their union.

The solution set consists of all numbers that are less than 5 or greater than or equal to 8. The solution set is $\{x \mid x < 5 \text{ or } x \geq 8\}$, or, in interval notation, $(-\infty, 5) \cup [8, \infty)$. There is no shortcut way to express this union when interval notation is used. ■

✔ **CHECK POINT 6** Solve: $3x - 5 \leq -2$ or $10 - 2x < 4$. $\{x \mid x \leq 1 \text{ or } x > 3\}$ or $(-\infty, 1] \cup (3, \infty)$

EXAMPLE 7 Solving a Compound Inequality with *Or*

Solve: $3x - 5 \leq 13$ or $5x + 2 > -3$.

SOLUTION

Step 1. **Solve each inequality separately.**

$$3x - 5 \leq 13 \quad \text{or} \quad 5x + 2 > -3$$
$$3x \leq 18 \qquad\qquad 5x > -5$$
$$x \leq 6 \qquad\qquad\quad x > -1$$

Step 2. **Take the union of the solution sets of the two inequalities.** We graph the solution sets of $x \leq 6$ and $x > -1$. We use these graphs to find their union.

$\{x \mid x \leq 6\}$ ⟵─┼─┼─┼─┼─┼─┼─┼─┼─┤─┼─┼─⟶ $(-\infty, 6]$
　　　　　-2　-1　0　1　2　3　4　5　6　7　8

$\{x \mid x > -1\}$ ⟵─┼─(─┼─┼─┼─┼─┼─┼─┼─┼─⟶ $(-1, \infty)$
　　　　　-2　-1　0　1　2　3　4　5　6　7　8

$\{x \mid x \leq 6\} \cup \{x \mid x > -1\}$ ⟵─┼─┼─┼─┼─┼─┼─┼─┼─┼─┼─⟶ $(-\infty, \infty)$
$= \{x \mid x \text{ is a real number}\}$ 　-2　-1　0　1　2　3　4　5　6　7　8

Because all real numbers are less than or equal to 6 or greater than -1, the union of the two sets fills the entire number line. Thus, the solution set is $\{x \mid x \text{ is a real number}\}$, or $\mathbb{R}$. The solution set in interval notation is $(-\infty, \infty)$. Any real number that you select will satisfy either or both of the original inequalities. ∎

CHECK POINT 7 Solve: $2x + 5 \geq 3$ or $2x + 3 < 3$. $\{x \mid x \text{ is a real number}\}$ or $\mathbb{R}$ or $(-\infty, \infty)$

4.2 EXERCISE SET

Student Solutions Manual   CD/Video   PH Math/Tutor Center   MathXL Tutorials on CD   MathXL®   MyMathLab   Interactmath.com

Practice Exercises

In Exercises 1–6, find the intersection of the sets.

1. $\{1, 2, 3, 4\} \cap \{2, 4, 5\}$ $\{2, 4\}$

2. $\{1, 3, 7\} \cap \{2, 3, 8\}$ $\{3\}$

3. $\{1, 3, 5, 7\} \cap \{2, 4, 6, 8, 10\}$ $\varnothing$

4. $\{0, 1, 3, 5\} \cap \{-5, -3, -1\}$ $\varnothing$

5. $\{a, b, c, d\} \cap \varnothing$ $\varnothing$

6. $\{w, y, z\} \cap \varnothing$ $\varnothing$

In Exercises 7–24, solve each compound inequality. Use graphs to show the solution set to each of the two given inequalities, as well as a third graph that shows the solution set of the compound inequality. Except for the empty set, express the solution set in both set-builder and interval notations. 7–24. See graphing answer section.

7. $x > 3$ and $x > 6$ $\{x \mid x > 6\}$; $(6, \infty)$

8. $x > 2$ and $x > 4$ $\{x \mid x > 4\}$; $(4, \infty)$

9. $x \leq 5$ and $x \leq 1$ $\{x \mid x \leq 1\}$; $(-\infty, 1]$

10. $x \leq 6$ and $x \leq 2$ $\{x \mid x \leq 2\}$; $(-\infty, 2]$

11. $x < 2$ and $x \geq -1$ $\{x \mid -1 \leq x < 2\}$; $[-1, 2)$

12. $x < 3$ and $x \geq -1$ $\{x \mid -1 \leq x < 3\}$; $[-1, 3)$

13. $x > 2$ and $x < -1$ $\varnothing$

14. $x > 3$ and $x < -1$ $\varnothing$

15. $5x < -20$ and $3x > -18$ $\{x \mid -6 < x < -4\}$; $(-6, -4)$

16. $3x \leq 15$ and $2x > -6$ $\{x \mid -3 < x \leq 5\}$; $(-3, 5]$

17. $x - 4 \leq 2$ and $3x + 1 > -8$ $\{x \mid -3 < x \leq 6\}$; $(-3, 6]$

18. $3x + 2 > -4$ and $2x - 1 < 5$ $\{x \mid -2 < x < 3\}$; $(-2, 3)$

19. $2x > 5x - 15$ and $7x > 2x + 10$ $\{x \mid 2 < x < 5\}$; $(2, 5)$

20. $6 - 5x > 1 - 3x$ and $4x - 3 > x - 9$ $\{x \mid -2 < x < \frac{5}{2}\}$; $(-2, \frac{5}{2})$

21. $4(1 - x) < -6$ and $\dfrac{x - 7}{5} \leq -2$ $\varnothing$

22. $5(x - 2) > 15$ and $\dfrac{x - 6}{4} \leq -2$ $\varnothing$

23. $x - 1 \leq 7x - 1$ and $4x - 7 < 3 - x$ $\{x \mid 0 \leq x < 2\}$; $[0, 2)$

24. $2x + 1 > 4x - 3$ and $x - 1 \geq 3x + 5$ $\{x \mid x \leq -3\}$; $(-\infty, -3]$

In Exercises 25–32, solve each inequality and graph the solution set on a number line. Express the solution set in both set-builder and interval notations. 25–32. See graphing answer section.

25. $6 < x + 3 < 8$ $\{x \mid 3 < x < 5\}$; $(3, 5)$

26. $7 < x + 5 < 11$ $\{x \mid 2 < x < 6\}$; $(2, 6)$

27. $-3 \leq x - 2 < 1$ $\{x \mid -1 \leq x < 3\}$; $[-1, 3)$

28. $-6 < x - 4 \leq 1$ $\{x \mid -2 < x \leq 5\}$; $(-2, 5]$

29. $-11 < 2x - 1 \leq -5$ $\{x \mid -5 < x \leq -2\}$; $(-5, -2]$

30. $3 \leq 4x - 3 < 19$ $\{x \mid \frac{3}{2} \leq x < \frac{11}{2}\}$; $[\frac{3}{2}, \frac{11}{2})$

31. $-3 \leq \dfrac{2x}{3} - 5 < -1$ $\{x \mid 3 \leq x < 6\}$; $[3, 6)$

32. $-6 \leq \dfrac{x}{2} - 4 < -3$ $\{x \mid -4 \leq x < 2\}$; $[-4, 2)$

In Exercises 33–38, find the union of the sets.

33. $\{1, 2, 3, 4\} \cup \{2, 4, 5\}$ $\{1, 2, 3, 4, 5\}$

34. $\{1, 3, 7, 8\} \cup \{2, 3, 8\}$ $\{1, 2, 3, 7, 8\}$

35. $\{1, 3, 5, 7\} \cup \{2, 4, 6, 8, 10\}$ $\{1, 2, 3, 4, 5, 6, 7, 8, 10\}$

36. $\{0, 1, 3, 5\} \cup \{2, 4, 6\}$ $\{0, 1, 2, 3, 4, 5, 6\}$

37. $\{a, e, i, o, u\} \cup \varnothing$ $\{a, e, i, o, u\}$

38. $\{e, m, p, t, y\} \cup \varnothing$ $\{e, m, p, t, y\}$

In Exercises 39–54, solve each compound inequality. Use graphs to show the solution set to each of the two given inequalities, as well as a third graph that shows the solution set of the compound inequality. Express the solution set in both set-builder and interval notations. 39–54. See graphing answer section.

39. $x > 3$ or $x > 6$ $\{x \mid x > 3\}$; $(3, \infty)$

40. $x > 2$ or $x > 4$ $\{x \mid x > 2\}$; $(2, \infty)$

41. $x \leq 5$ or $x \leq 1$ $\{x \mid x \leq 5\}$; $(-\infty, 5]$

42. $x \leq 6$ or $x \leq 2$ $\{x|x \leq 6\}; (-\infty, 6]$

43. $x < 2$ or $x \geq -1$ $\{x|x \text{ is a real number}\}; (-\infty, \infty)$

44. $x < 3$ or $x \geq -1$ $\{x|x \text{ is a real number}\}; (-\infty, \infty)$

45. $x \geq 2$ or $x < -1$ $\{x|x < -1 \text{ or } x \geq 2\}; (-\infty, -1) \cup [2, \infty)$

46. $x \geq 3$ or $x < -1$ $\{x|x < -1 \text{ or } x \geq 3\}; (-\infty, -1) \cup [3, \infty)$

47. $3x > 12$ or $2x < -6$ $\{x|x < -3 \text{ or } x > 4\}; (-\infty, -3) \cup (4, \infty)$

48. $3x < 3$ or $2x > 10$ $\{x|x < 1 \text{ or } x > 5\}; (-\infty, 1) \cup (5, \infty)$

49. $3x + 2 \leq 5$ or $5x - 7 \geq 8$
$\{x|x \leq 1 \text{ or } x \geq 3\}; (-\infty, 1] \cup [3, \infty)$

50. $2x - 5 \leq -11$ or $5x + 1 \geq 6$
$\{x|x \leq -3 \text{ or } x \geq 1\}; (-\infty, -3] \cup [1, \infty)$

51. $4x + 3 < -1$ or $2x - 3 \geq -11$
$\{x|x \text{ is a real number}\}; (-\infty, \infty)$

52. $2x + 1 < 15$ or $3x - 4 \geq -1$
$\{x|x \text{ is a real number}\}; (-\infty, \infty)$

53. $-2x + 5 > 7$ or $-3x + 10 > 2x$ $\{x|x < 2\}; (-\infty, 2)$

54. $16 - 3x \geq -8$ or $13 - x > 4x + 3$ $\{x|x \leq 8\}; (-\infty, 8]$

55. Let $f(x) = 2x + 3$ and $g(x) = 3x - 1$. Find all values of x for which $f(x) \geq 5$ and $g(x) > 11$. $\{x|x > 4\}$ or $(4, \infty)$

56. Let $f(x) = 4x + 5$ and $g(x) = 3x - 4$. Find all values of x for which $f(x) \geq 5$ and $g(x) \leq 2$. $\{x|0 \leq x \leq 2\}$ or $[0, 2]$

57. Let $f(x) = 3x - 1$ and $g(x) = 4 - x$. Find all values of x for which $f(x) < -1$ or $g(x) < -2$.
$\{x|x < 0 \text{ or } x > 6\}$ or $(-\infty, 0) \cup (6, \infty)$

58. Let $f(x) = 2x - 5$ and $g(x) = 3 - x$. Find all values of x for which $f(x) \geq 3$ or $g(x) < 0$. $\{x|x > 3\}$ or $(3, \infty)$

Practice Plus

In Exercises 59–60, write an inequality with x isolated in the middle that is equivalent to the given inequality. Assume $a > 0, b > 0,$ and $c > 0$.

59. $-c < ax - b < c$ $\dfrac{b - c}{a} < x < \dfrac{b + c}{a}$

60. $-2 < \dfrac{ax - b}{c} < 2$ $\dfrac{b - 2c}{a} < x < \dfrac{b + 2c}{a}$

In Exercises 61–62, use the graphs of $y_1, y_2,$ and $y_3,$ to solve each compound inequality.

61. $-3 \leq 2x - 1 \leq 5$

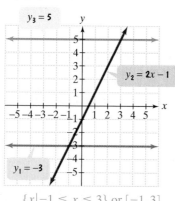

$\{x|-1 \leq x \leq 3\}$ or $[-1, 3]$

62. $x - 2 < 2x - 1 < x + 2$

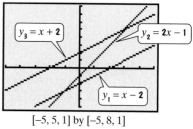

$[-5, 5, 1]$ by $[-5, 8, 1]$

$\{x|-1 < x < 3\}$ or $(-1, 3)$

63. Solve $x - 2 < 2x - 1 < x + 2$, the inequality in Exercise 62, using algebraic methods. (*Hint*: Rewrite the inequality as $2x - 1 > x - 2$ and $2x - 1 < x + 2$.) $\{x|-1 < x < 3\}$ or $(-1, 3)$

64. Use the hint given in Exercise 63 to solve $x \leq 3x - 10 \leq 2x$. $\{x|5 \leq x \leq 10\}$ or $[5, 10]$

In Exercises 65–66, use the table to solve each inequality.

65. $-2 \leq 5x + 3 < 13$

$\{x|-1 \leq x < 2\}$ or $[-1, 2)$

66. $-3 < 2x - 5 \leq 3$

$\{x|1 < x \leq 4\}$ or $(-1, 4]$

In Exercises 67–68, use the roster method to find the set of negative integers that are solutions of each inequality.

67. $5 - 4x \geq 1$ and $3 - 7x < 31$ $\{-3, -2, -1\}$

68. $-5 < 3x + 4 \leq 16$ $\{-2, -1\}$

Application Exercises

As a result of cultural expectations about what is appropriate behavior for each gender, boys and girls differ substantially in their toy preferences. The graph on the next page shows the percentage of boys and girls asking for various types of toys in letters to Santa Claus. Use the information in the graph to solve Exercises 69–74.

Toys Requested by Children

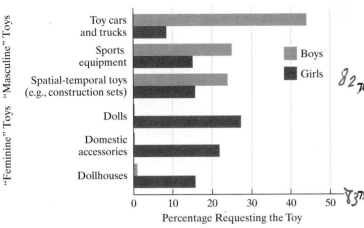

Source: Richardson, J. G., & Simpson, C. H. (1982). Children, gender and social structure: An analysis of the contents of letters to Santa Claus. *Child Development, 53*, 429–436.

69–79 is Printed

69. Which toys were requested by more than 10% of the boys *and* less than 20% of the girls?   spatial-temporal toys, sports equipment, and toy cars and trucks

70. Which toys were requested by fewer than 5% of the boys *and* fewer than 20% of the girls?   dollhouses

71. Which toys were requested by more than 10% of the boys *or* less than 20% of the girls?   dollhouses, spatial-temporal toys, sports equipment, and toy cars and trucks

72. Which toys were requested by fewer than 5% of the boys *or* fewer than 20% of the girls?   dollhouses, domestic accessories, dolls, spatial-temporal toys, sports equipment, and toy cars and trucks

73. Which toys were requested by more than 40% of the boys *and* more than 10% of the girls?   none of the toys

74. Which toys were requested by more than 40% of the boys *or* more than 10% of the girls?   dollhouses, domestic accessories, dolls, spatial-temporal toys, sports equipment, and toy cars and trucks

75. A basic cellular phone plan costs $20 per month for 60 calling minutes. Additional time costs $0.40 per minute. The formula

$$C = 20 + 0.40(x - 60)$$

gives the monthly cost for this plan, C, for x calling minutes, where $x > 60$. How many calling minutes are possible for a monthly cost of at least $28 and at most $40?   between 80 and 110 minutes, inclusive

76. The formula for converting Fahrenheit temperature, F, to Celsius temperature, C, is

$$C = \frac{5}{9}(F - 32).$$

If Celsius temperature ranges from 15° to 35°, inclusive, what is the range for the Fahrenheit temperature? Use interval notation to express this range.   [59°F, 95°F]

77. On the first of four exams, your grades are 70, 75, 87, and 92. There is still one more exam, and you are hoping to earn a B in the course. This will occur if the average of your five exam grades is greater than or equal to 80 and less than 90. What range of grades on the fifth exam will result in earning a B? Use interval notation to express this range.   [76, 126); If the highest grade is 100, then [76, 100].

78. On the first of four exams, your grades are 82, 75, 80, and 90. There is still a final exam, and it counts as two grades. You are hoping to earn a B in the course. This will occur if the average of your six exam grades is greater than or equal to 80 and less than 90. What range of grades on the final exam will result in earning a B? Use interval notation to express this range.   [76.5, 106.5); If the highest grade is 100, then [76.5, 100].

79. The toll to a bridge is $3.00. A three-month pass costs $7.50 and reduces the toll to $0.50. A six-month pass costs $30 and permits crossing the bridge for no additional fee. How many crossings per three-month period does it take for the three-month pass to be the best deal?   more than 3 and less than 15 crossings per three-month period

80. Parts for an automobile repair cost $175. The mechanic charges $34 per hour. If you receive an estimate for at least $226 and at most $294 for fixing the car, what is the time interval that the mechanic will be working on the job?   The mechanic will be working on the job from 1.5 to 3.5 hours, inclusive.

Writing in Mathematics

81. Describe what is meant by the intersection of two sets. Give an example.   81–87. Answers will vary.

82. Explain how to solve a compound inequality involving "and."

83. Why is $1 < 2x + 3 < 9$ a compound inequality? Where are the two inequalities and what is the word that joins them?

84. Explain how to solve $1 < 2x + 3 < 9$.

85. Describe what is meant by the union of two sets. Give an example.

86. Explain how to solve a compound inequality involving "or."

87. How many Christmas trees are sold each holiday season? The function $T(x) = 0.9x + 32$ models the number of trees sold, $T(x)$, in millions, x years after 1990. Use the function to write a problem that can be solved using a compound inequality. Then solve the problem.

Technology Exercises

In Exercises 88–91, solve each inequality using a graphing utility. Graph each of the three parts of the inequality separately in the same viewing rectangle. The solution set consists of all values of x for which the graph of the linear function in the middle lies between the graphs of the constant functions on the left and the right.
88–91. See graphing answer section.

88. $1 < x + 3 < 9$ $\{x | -2 < x < 6\}$ or $(-2, 6)$

89. $-1 < \frac{x + 4}{2} < 3$ $\{x | -6 < x < 2\}$ or $(-6, 2)$

90. $1 \leq 4x - 7 \leq 3$ $\left\{x \mid 2 \leq x \leq \frac{5}{2}\right\}$ or $\left[2, \frac{5}{2}\right]$

91. $2 \leq 4 - x \leq 7$ $\{x \mid -3 \leq x \leq 2\}$ or $[-3, 2]$

Critical Thinking Exercises

92. Which one of the following is true?   *True or false*

a. $(-\infty, -1] \cap [-4, \infty) = [-4, -1]$

b. $(-\infty, 3) \cup (-\infty, -2) = (-\infty, -2)$

c. The union of two sets can never give the same result as the intersection of those sets.

d. The solution set of $x < a$ and $x > a$ is the set of all real numbers excluding a.

93. Solve and express the solution set in interval notation:

$$-7 \leq 8 - 3x \leq 20 \text{ and } -7 < 6x - 1 < 41. (-1, 5]$$

The graphs of $f(x) = \sqrt{4 - x}$ and $g(x) = \sqrt{x + 1}$ are shown in a $[-3, 10, 1]$ by $[-2, 5, 1]$ viewing rectangle.

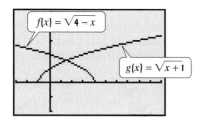

In Exercises 94–97, use the graphs and interval notation to express the domain of the given function.

94. The domain of f $(-\infty, 4]$

95. The domain of g $[-1, \infty)$

96. The domain of $f + g$ $[-1, -4]$

97. The domain of $\dfrac{f}{g}$ $(-1, 4]$

98. At the end of the day, the change machine at a laundrette contained at least \$3.20 and at most \$5.45 in nickels, dimes, and quarters. There were 3 fewer dimes than twice the number of nickels and 2 more quarters than twice the number of nickels. What was the least possible number and the greatest possible number of nickels? The least possible number of nickels was 4 and the greatest possible number of nickels was 7.

Review Exercises

99. If $f(x) = x^2 - 3x + 4$ and $g(x) = 2x - 5$,   find $(g - f)(x)$ and $(g - f)(-1)$. (Section 2.2, Example 4) $(g - f)(x) = -x^2 + 5x - 9; -15$

100. Use function notation to write the equation of the line passing through $(4, 2)$ and perpendicular to the line whose equation is $4x - 2y = 8$. (Section 2.4, Examples 4 and 5) $f(x) = -\frac{1}{2}x + 4$

101. Simplify: $4 - [2(x - 4) - 5]$. (Section 1.2, Example 14) $17 - 2x$

EQUATIONS AND INEQUALITIES INVOLVING ABSOLUTE VALUE

Objectives

1 Solve equations involving absolute value.

2 Solve inequalities involving absolute value.

3 Recognize absolute value inequalities with no solution or all real numbers as solutions.

4 Solve problems using absolute value inequalities.

$M*A*S*H$ was set in the early 1950s during the Korean War. By the final episode, the show had lasted four times as long as the Korean War.

At the end of the twentieth century, there were 94 million households in the United States with television sets. The television program viewed by the greatest percentage of such households in that century was the final episode of $M*A*S*H$. Over 50 million American households watched this program.

Numerical information, such as the number of households watching a television program, is often given with a margin of error. Inequalities involving absolute value are used to describe errors in polling, as well as errors of measurement in manufacturing, engineering, science, and other fields. In this section, you will learn to solve equations and inequalities containing absolute value. With these skills, you will be able to analyze the percentage of households that watched the final episode of $M*A*S*H$.

Equations Involving Absolute Value We have seen that the absolute value of a, denoted $|a|$, is the distance from 0 to a on a number line. Now consider **absolute value equations**, such as

$$|x| = 2.$$

This means that we must determine real numbers whose distance from the origin on a number line is 2. Figure 4.4 shows that there are two numbers such that $|x| = 2$, namely, 2 and -2. We write $x = 2$ or $x = -2$. This observation can be generalized as follows:

1 Solve equations involving absolute value.

FIGURE 4.4 If $|x| = 2$, then $x = 2$ or $x = -2$.

> **REWRITING AN ABSOLUTE VALUE EQUATION WITHOUT ABSOLUTE VALUE BARS** If c is a positive real number and X represents any algebraic expression, then $|X| = c$ is equivalent to $X = c$ or $X = -c$.

EXAMPLE 1 Solving an Equation Involving Absolute Value

Solve: $|2x - 3| = 11$.

SOLUTION

$|2x - 3| = 11$	This is the given equation.
$2x - 3 = 11$ or $2x - 3 = -11$	Rewrite the equation without absolute value bars.
$2x = 14 \qquad\qquad 2x = -8$	Add 3 to both sides of each equation.
$x = 7 \qquad\qquad x = -4$	Divide both sides of each equation by 2.

Check 7:

$|2x - 3| = 11$

$|2(7) - 3| \stackrel{?}{=} 11$

$|14 - 3| \stackrel{?}{=} 11$

$|11| \stackrel{?}{=} 11$

$11 = 11$, true

Check -4:

$|2x - 3| = 11$

$|2(-4) - 3| \stackrel{?}{=} 11$

$|-8 - 3| \stackrel{?}{=} 11$

$|-11| \stackrel{?}{=} 11$

$11 = 11$, true

This is the original equation.

Substitute the proposed solutions.

Perform operations inside the absolute value bars.

These true statements indicate that 7 and -4 are solutions.

The solutions are -4 and 7. We can also say that the solution set is $\{-4, 7\}$. ■

✔ **CHECK POINT 1** Solve: $|2x - 1| = 5$. -2 and 3, or $\{-2, 3\}$

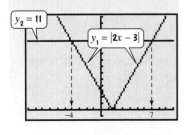
The absolute value of a number is never negative. Thus, if X is an algebraic expression and c is a negative number, then $|X| = c$ has no solution. For example, the equation $|3x - 6| = -2$ has no solution because $|3x - 6|$ cannot be negative. The solution set is $\varnothing$, the empty set.

The absolute value of 0 is 0. Thus, if X is an algebraic expression and $|X| = 0$, the solution is found by solving $X = 0$. For example, the solution of $|x - 2| = 0$ is obtained by solving $x - 2 = 0$. The solution is 2 and the solution set is $\{2\}$.

To solve some absolute value equations, it is necessary to first isolate the expression containing the absolute value symbols. For example, consider the equation

$$3|2x - 3| - 8 = 25.$$

> We need to isolate $|2x - 3|$.

How can we isolate $|2x - 3|$? Add 8 to both sides of the equation and then divide both sides by 3.

$3\|2x - 3\| - 8 = 25$	This is the given equation.
$3\|2x - 3\| = 33$	Add 8 to both sides.
$\|2x - 3\| = 11$	Divide both sides by 3.

This results in the equation we solved in Example 1.

Some equations have two absolute value expressions, such as

$$|3x - 1| = |x + 5|.$$

These absolute value expressions are equal when the expressions inside the absolute value bars are equal to or opposites of each other.

> **REWRITING AN ABSOLUTE VALUE EQUATION WITH TWO ABSOLUTE VALUES WITHOUT ABSOLUTE VALUE BARS** If $|X_1| = |X_2|$, then $X_1 = X_2$ or $X_1 = -X_2$.

EXAMPLE 2 Solving an Absolute Value Equation with Two Absolute Values

Solve: $|3x - 1| = |x + 5|$.

SOLUTION We rewrite the equation without absolute value bars.

$$|X_1| = |X_2| \quad \text{means} \quad X_1 = X_2 \quad \text{or} \quad X_1 = -X_2$$

$$|3x - 1| = |x + 5| \quad \text{means} \quad 3x - 1 = x + 5 \quad \text{or} \quad 3x - 1 = -(x + 5).$$

We now solve the two equations that do not contain absolute value bars.

$$
\begin{aligned}
3x - 1 &= x + 5 &\text{or}\quad 3x - 1 &= -(x + 5) \\
2x - 1 &= 5 & 3x - 1 &= -x - 5 \\
2x &= 6 & 4x - 1 &= -5 \\
x &= 3 & 4x &= -4 \\
& & x &= -1
\end{aligned}
$$

Take a moment to complete the solution process by checking the two proposed solutions in the original equation. The solutions are -1 and 3, and the solution set is $\{-1, 3\}$. ∎

✔ **CHECK POINT 2** Solve: $|2x - 7| = |x + 3|$. $\frac{4}{3}$ and 10, or $\left\{\frac{4}{3}, 10\right\}$

Inequalities Involving Absolute Value Absolute value can also arise in inequalities. Consider, for example,

$$|x| < 2.$$

This means that the distance of x from 0 is *less than* 2, as shown in Figure 4.5. The interval shows values of x that lie less than 2 units from 0. Thus, x can lie between -2 and 2. That is, x is greater than -2 and less than 2: $-2 < x < 2$.

2 Solve inequalities involving absolute value.

FIGURE 4.5 $|x| < 2$, so $-2 < x < 2$.

FIGURE 4.6 $|x| > 2$, so $x < -2$ or $x > 2$.

Some absolute value inequalities use the "greater than" symbol. For example, $|x| > 2$ means that the distance of x from 0 is *greater than* 2, as shown in Figure 4.6. Thus, x can be less than -2 *or* greater than 2: $x < -2$ or $x > 2$.

These observations suggest the following principles for solving inequalities with absolute value.

SOLVING AN ABSOLUTE VALUE INEQUALITY If X is an algebraic expression and c is a positive number,

  1. The solutions of $|X| < c$ are the numbers that satisfy $-c < X < c$.
  2. The solutions of $|X| > c$ are the numbers that satisfy $X < -c$ or $X > c$.

These rules are valid if $<$ is replaced by $\leq$ and $>$ is replaced by $\geq$.

EXAMPLE 3 Solving an Absolute Value Inequality with $<$

Solve and graph the solution set on a number line: $|x - 4| < 3$.

SOLUTION We rewrite the inequality without absolute value bars.

$$|X| < c \text{ means } -c < X < c.$$

$$|x - 4| < 3 \quad \text{means} \quad -3 < x - 4 < 3.$$

We solve the compound inequality by adding 4 to all three parts.

$$-3 < x - 4 < 3$$
$$-3 + 4 < x - 4 + 4 < 3 + 4$$
$$1 < x < 7$$

The solution set is all real numbers greater than 1 and less than 7, denoted by $\{x \mid 1 < x < 7\}$ or $(1, 7)$. The graph of the solution set is shown as follows:

We can use the rectangular coordinate system to visualize the solution set of

$$|x - 4| < 3.$$

Figure 4.7 shows the graphs of $f(x) = |x - 4|$ and $g(x) = 3$. The solution set of $|x - 4| < 3$ consists of all values of x for which the blue graph of f lies below the red graph of g. These x-values make up the interval $(1, 7)$, which is the solution set.

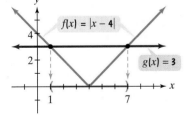

FIGURE 4.7 The solution set of $|x - 4| < 3$ is $(1, 7)$.

✓ **CHECK POINT 3** Solve and graph the solution set on a number line: $|x - 2| < 5$.
$\{x \mid -3 < x < 7\}$ or $(-3, 7)$; See graphing answer section.

EXAMPLE 4 Solving an Absolute Value Inequality with $\geq$

Solve and graph the solution set on a number line: $|2x + 3| \geq 5$.

SOLUTION We rewrite the inequality without absolute value bars.

$$|X| \geq c \quad \text{means} \quad X \leq -c \quad \text{or} \quad X \geq c.$$

$$|2x + 3| \geq 5 \quad \text{means} \quad 2x + 3 \leq -5 \quad \text{or} \quad 2x + 3 \geq 5.$$

We solve this compound inequality by solving each of these inequalities separately. Then we take the union of their solution sets.

$$2x + 3 \leq -5 \quad \text{or} \quad 2x + 3 \geq 5 \qquad \text{These are the inequalities without absolute value bars.}$$

$$2x \leq -8 \qquad\qquad 2x \geq 2 \qquad \text{Subtract 3 from both sides.}$$

$$x \leq -4 \qquad\qquad\; x \geq 1 \qquad \text{Divide both sides by 2.}$$

The solution set consists of all numbers that are less than or equal to -4 or greater than or equal to 1. The solution set is $\{x | x \leq -4 \text{ or } x \geq 1\}$, or, in interval notation, $(-\infty, -4] \cup [1, \infty)$. The graph of the solution set is shown as follows:

$$\begin{array}{ccccccccccc} \text{-7} & \text{-6} & \text{-5} & \text{-4} & \text{-3} & \text{-2} & \text{-1} & 0 & 1 & 2 & 3 \end{array}$$

✔ **CHECK POINT 4** Solve and graph the solution set on a number line: $|2x - 5| \geq 3$. $\{x | x \leq 1 \text{ or } x \geq 4\}$ or $(-\infty, 1] \cup [4, \infty)$; See graphing answer section.

3 Recognize absolute value inequalities with no solution or all real numbers as solutions.

Absolute Value Inequalities with Unusual Solution Sets If c is a positive number, the solutions of $|X| < c$ comprise a single interval. By contrast, the solutions of $|X| > c$ make up two intervals.

The solutions of $|X| < c$ comprise the single interval $-c < X < c$.

The solutions of $|X| > c$ comprise two intervals: $X < -c$ or $X > c$.

Now let's see what happens to these inequalities if c is a negative number. Consider, for example, $|x| < -2$. Because $|x|$ always has a value that is greater than or equal to 0, there is no number whose absolute value is less than -2. The inequality $|x| < -2$ has no solution. The solution set is $\varnothing$.

Now consider the inequality $|x| > -2$. Because $|x|$ is never negative, all numbers have an absolute value that is greater than -2. All real numbers satisfy the inequality $|x| > -2$. The solution set is $(-\infty, \infty)$.

ABSOLUTE VALUE INEQUALITIES WITH UNUSUAL SOLUTION SETS If X is an algebraic expression and c is a negative number,

1. the inequality $|X| < c$ has no solution.
2. the inequality $|X| > c$ is true for all real numbers for which X is defined.

Applications When you were between the ages of 6 and 14, how would you have responded to this question:

What is bad about being a kid?

In a random sample of 1172 children ages 6 to 14, 17% of the children responded, "Getting bossed around." The problem is that this is a single random sample. Do 17% of kids in the entire population of children ages 6 to 14 think that getting bossed around is a bad thing?

If you look at the results of a poll like the one in Table 4.3, you will observe that a **margin of error** is reported. The margin of error is $\pm 2.9\%$. This means that the actual percentage of children who feel getting bossed around is a bad thing is at most 2.9% greater than or less than 17%. If x represents the percentage of children in the population who think that getting bossed around is a bad thing, then the poll's margin of error can be expressed as an absolute value inequality:

$$|x - 17| \leq 2.9.$$

Table 4.3

What Is Bad About Being a Kid?

Kids Say	
Getting bossed around	17%
School, homework	15%
Can't do everything I want	11%
Chores	9%
Being grounded	9%

Source: Penn, Schoen, and Berland using 1172 interviews with children ages 6 to 14 from May 14 to June 1, 1999. Margin of error: ±2.9%

Note the margin of error.

4 Solve problems using absolute value inequalities.

EXAMPLE 5 Analyzing a Poll's Margin of Error

The inequality

$$|x - 9| \le 2.9$$

describes the percentage of children in the population who think that being grounded is a bad thing about being a kid. (See Table 4.3.) Solve the inequality and interpret the solution.

SOLUTION We rewrite the inequality without absolute value bars.

$$|X| \le c \quad \text{means} \quad -c \le X \le c.$$

$$|x - 9| \le 2.9 \quad \text{means} \quad -2.9 \le x - 9 \le 2.9.$$

We solve the compound inequality by adding 9 to all three parts.

$$-2.9 \le x - 9 \le 2.9$$
$$-2.9 + 9 \le x - 9 + 9 \le 2.9 + 9$$
$$6.1 \le x \le 11.9$$

The percentage of children in the population who think that being grounded is a bad thing is somewhere between a low of 6.1% and a high of 11.9%. Notice that these percents are always within 2.9% above and below the given 9%, and that 2.9% is the poll's margin of error. ■

$\{x | 8.1 \le x \le 13.9\}$ or $[8.1, 13.9]$; The percentage of children in the population who think that not being able to do everything they want is a bad thing is between a low of 8.1% and a high of 13.9%.

✔ **CHECK POINT 5** Solve the inequality:

$$|x - 11| \le 2.9.$$

Interpret the solution in terms of the information in Table 4.3 on page 266.

4.3 EXERCISE SET

Student Solutions Manual · CD/Video · PH Math/Tutor Center · MathXL Tutorials on CD · MathXL® · MyMathLab · Interactmath.com

Practice Exercises

In Exercises 1–38, find the solution set for each equation.

1. $|x| = 8$ $\{-8, 8\}$
2. $|x| = 6$ $\{-6, 6\}$
3. $|x - 2| = 7$ $\{-5, 9\}$
4. $|x + 1| = 5$ $\{-6, 4\}$
5. $|2x - 1| = 7$ $\{-3, 4\}$
6. $|2x - 3| = 11$ $\{-4, 7\}$
7. $\left|\dfrac{4x - 2}{3}\right| = 2$ $\{-1, 2\}$
8. $\left|\dfrac{3x - 1}{5}\right| = 1$ $\left\{-\dfrac{4}{3}, 2\right\}$
9. $|x| = -8$ $\varnothing$
10. $|x| = -6$ $\varnothing$
11. $|x + 3| = 0$ $\{-3\}$
12. $|x + 2| = 0$ $\{-2\}$
13. $2|y + 6| = 10$ $\{-11, -1\}$
14. $3|y + 5| = 12$ $\{-9, -1\}$

15. $3|2x - 1| = 21$ $\{-3, 4\}$
16. $2|3x - 2| = 14$ $\left\{-\dfrac{5}{3}, 3\right\}$
17. $|6y - 2| + 4 = 32$ $\left\{-\dfrac{13}{3}, 5\right\}$
18. $|3y - 1| + 10 = 25$ $\left\{-\dfrac{14}{3}, \dfrac{16}{3}\right\}$
19. $7|5x| + 2 = 16$ $\left\{-\dfrac{2}{5}, \dfrac{2}{5}\right\}$
20. $7|3x| + 2 = 16$ $\left\{-\dfrac{2}{3}, \dfrac{2}{3}\right\}$
21. $|x + 1| + 5 = 3$ $\varnothing$
22. $|x + 1| + 6 = 2$ $\varnothing$
23. $|4y + 1| + 10 = 4$ $\varnothing$
24. $|3y - 2| + 8 = 1$ $\varnothing$
25. $|2x - 1| + 3 = 3$ $\left\{\dfrac{1}{2}\right\}$
26. $|3x - 2| + 4 = 4$ $\left\{\dfrac{2}{3}\right\}$
27. $|5x - 8| = |3x + 2|$ $\left\{\dfrac{3}{4}, 5\right\}$
28. $|4x - 9| = |2x + 1|$ $\left\{\dfrac{4}{3}, 5\right\}$
29. $|2x - 4| = |x - 1|$ $\left\{\dfrac{5}{3}, 3\right\}$
30. $|6x| = |3x - 9|$ $\{-3, 1\}$

(handwritten: 74) $9 \leq |4x + 7|$

31. $|2x - 5| = |2x + 5|$ $\{0\}$ **32.** $|3x - 5| = |3x + 5|$ $\{0\}$

33. $|x - 3| = |5 - x|$ $\{4\}$ **34.** $|x - 3| = |6 - x|$ $\left\{\frac{9}{2}\right\}$

35. $|2y - 6| = |10 - 2y|$ $\{4\}$ **36.** $|4y + 3| = |4y + 5|$ $\{-1\}$

37. $\left|\frac{2x}{3} - 2\right| = \left|\frac{x}{3} + 3\right|$ $\{-1, 15\}$

38. $\left|\frac{x}{2} - 2\right| = \left|x - \frac{1}{2}\right|$ $\left\{-3, \frac{5}{3}\right\}$

In Exercises 39–70, solve and graph the solution set on a number line.
39–70. See graphing answer section.

39. $|x| < 3$ $\{x | -3 < x < 3\}$ or $(-3, 3)$

40. $|x| < 5$ $\{x | -5 < x < 5\}$ or $(-5, 5)$

41. $|x - 2| < 1$ $\{x | 1 < x < 3\}$ or $(1, 3)$

42. $|x - 1| < 5$ $\{x | -4 < x < 6\}$ or $(-4, 6)$

43. $|x + 2| \le 1$ $\{x | -3 \le x \le -1\}$ or $[-3, -1]$

44. $|x + 1| \le 5$ $\{x | -6 \le x \le 4\}$ or $[-6, 4]$

45. $|2x - 6| < 8$ $\{x | -1 < x < 7\}$ or $(-1, 7)$

46. $|3x + 5| < 17$ $\{x | -\frac{22}{3} < x < 4\}$ or $\left(-\frac{22}{3}, 4\right)$

47. $|x| > 3$ $\{x | x < -3$ or $x > 3\}$ or $(-\infty, -3) \cup (3, \infty)$

48. $|x| > 5$ $\{x | x < -5$ or $x > 5\}$ or $(-\infty, -5) \cup (5, \infty)$

49. $|x + 3| > 1$ $\{x | x < -4$ or $x > -2\}$ or $(-\infty, -4) \cup (-2, \infty)$

50. $|x - 2| > 5$ $\{x | x < -3$ or $x > 7\}$ or $(-\infty, -3) \cup (7, \infty)$

51. $|x - 4| \ge 2$ $\{x | x \le 2$ or $x \ge 6\}$ or $(-\infty, 2] \cup [6, \infty)$

52. $|x - 3| \ge 4$ $\{x | x \le -1$ or $x \ge 7\}$ or $(-\infty, -1] \cup [7, \infty)$

53. $|3x - 8| > 7$ $\{x | x < \frac{1}{3}$ or $x > 5\}$ or $\left(-\infty, \frac{1}{3}\right) \cup (5, \infty)$

54. $|5x - 2| > 13$ $\{x | x < -\frac{11}{5}$ or $x > 3\}$ or $\left(-\infty, -\frac{11}{5}\right) \cup (3, \infty)$

55. $|2(x - 1) + 4| \le 8$ $\{x | -5 \le x \le 3\}$ or $[-5, 3]$

56. $|3(x - 1) + 2| \le 20$ $\{x | -\frac{19}{3} \le x \le 7\}$ or $\left[-\frac{19}{3}, 7\right]$

57. $\left|\frac{2x + 6}{3}\right| < 2$ $\{x | -6 < x < 0\}$ or $(-6, 0)$

58. $\left|\frac{3x - 3}{4}\right| < 6$ $\{x | -7 < x < 9\}$ or $(-7, 9)$

59. $\left|\frac{2x + 2}{4}\right| \ge 2$ $\{x | x \le -5$ or $x \ge 3\}$ or $(-\infty, -5] \cup [3, \infty)$

60. $\left|\frac{3x - 3}{9}\right| \ge 1$ $\{x | x \le -2$ or $x \ge 4\}$ or $(-\infty, -2] \cup [4, \infty)$

61. $\left|3 - \frac{2x}{3}\right| > 5$ $\{x | x < -3$ or $x > 12\}$ or $(-\infty, -3) \cup (12, \infty)$

62. $\left|3 - \frac{3x}{4}\right| > 9$ $\{x | x < -8$ or $x > 16\}$ or $(-\infty, -8) \cup (16, \infty)$

63. $|x - 2| < -1$ $\varnothing$ **64.** $|x - 3| < -2$ $\varnothing$

65. $|x + 6| > -10$ $\{x | x$ is a real number$\}$ or $(-\infty, \infty)$

66. $|x + 4| > -12$ $\{x | x$ is a real number$\}$ or $(-\infty, \infty)$

67. $|x + 2| + 9 \le 16$ $\{x | -9 \le x \le 5\}$ or $[-9, 5]$

68. $|x - 2| + 4 \le 5$ $\{x | 1 \le x \le 3\}$ or $[1, 3]$

69. $2|2x - 3| + 10 > 12$
$\{x | x < 1$ or $x > 2\}$ or $(-\infty, 1) \cup (2, \infty)$

70. $3|2x - 1| + 2 > 8$
$\{x | x < -\frac{1}{2}$ or $x > \frac{3}{2}\}$ or $\left(-\infty, -\frac{1}{2}\right) \cup \left(\frac{3}{2}, \infty\right)$

(handwritten: 75)

71. Let $f(x) = |5 - 4x|$. Find all values of x for which $f(x) = 11$. $-\frac{3}{2}$ and 4

72. Let $f(x) = |2 - 3x|$. Find all values of x for which $f(x) = 13$. $-\frac{11}{3}$ and 5

73. Let $f(x) = |3 - x|$ and $g(x) = |3x + 11|$. Find all values of x for which $f(x) = g(x)$. -7 and -2

74. Let $f(x) = |3x + 1|$ and $g(x) = |6x - 2|$. Find all values of x for which $f(x) = g(x)$. $\frac{1}{9}$ and 1

75. Let $g(x) = |-1 + 3(x + 1)|$. Find all values of x for which $g(x) \le 5$. $\{x | -\frac{7}{3} \le x \le 1\}$ or $\left[-\frac{7}{3}, 1\right]$

76. Let $g(x) = |-3 + 4(x + 1)|$. Find all values of x for which $g(x) \le 3$. $\{x | -1 \le x \le \frac{1}{2}\}$ or $\left[-1, \frac{1}{2}\right]$

77. Let $h(x) = |2x - 3| + 1$. Find all values of x for which $h(x) > 6$. $\{x | x < -1$ or $x > 4\}$ or $(-\infty, -1) \cup (4, \infty)$

(handwritten: 82L)

78. Let $h(x) = |2x - 4| - 6$. Find all values of x for which $h(x) > 18$. $\{x | x < -10$ or $x > 14\}$ or $(-\infty, -10) \cup (14, \infty)$

Practice Plus

79. When 3 times a number is subtracted from 4, the absolute value of the difference is at least 5. Use interval notation to express the set of all numbers that satisfy this condition. $\left(-\infty, -\frac{1}{3}\right] \cup [3, \infty)$

80. When 4 times a number is subtracted from 5, the absolute value of the difference is at most 13. Use interval notation to express the set of all numbers that satisfy this condition. $\left[-2, \frac{9}{2}\right]$

In Exercises 81–82, solve each inequality. Assume that $a > 0$ and $c > 0$. Use set-builder notation to express each solution set.

81. $|ax + b| < c$ $\left\{x \middle| \frac{-c - b}{a} < x < \frac{c - b}{a}\right\}$

82. $|ax + b| \ge c$ $\left\{x \middle| x \le \frac{-c - b}{a}$ or $x \ge \frac{c - b}{a}\right\}$

In Exercises 83–84, use the graph of $f(x) = |4 - x|$ to solve each equation or inequality.

83. $|4 - x| = 1$ 3 and 5

84. $|4 - x| < 5$ $\{x | -1 < x < 9\}$ or $(-1, 9)$

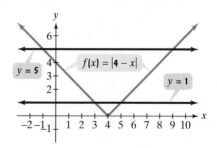

In Exercises 85–86, use the table to solve each inequality.

$y_1 = |2x + 1|$

X	Y1
-4	7
-3	5
-2	3
-1	1
0	1
1	3
2	5

X = -4

85. $|2x + 1| \leq 3$ $\{x | -2 \leq x \leq 1\}$ or $[-2, 1]$

86. $|2x + 1| \geq 3$ $\{x | x \leq -2 \text{ or } x \geq 1\}$ or $(-\infty, -2] \cup [1, \infty)$

Application Exercises

The three television programs viewed by the greatest percentage of U.S. households in the twentieth century are shown in the table. The data are from a random survey of 4000 TV households by Nielsen Media Research. In Exercises 87–88, let x represent the actual viewing percentage in the U.S. population.

TV Programs with the Greatest U.S. Audience Viewing Percentage of the Twentieth Century

Program	Viewing Percentage in Survey
1. "M*A*S*H" Feb. 28, 1983	60.2%
2. "Dallas" Nov. 21, 1980	53.3%
3. "Roots" Part 8 Jan. 30, 1977	51.1%

Source: Nielsen Media Research

87. Solve the inequality: $|x - 60.2| \leq 1.6$. Interpret the solution in terms of the information in the table. What is the margin of error? $\{x | 58.6 \leq x \leq 61.8\}$ or $[58.6, 61.8]$; The percentage of the U.S. population that watched M*A*S*H is between a low of 58.6% and a high of 61.8%. The margin of error is 1.6%.

88. Solve the inequality: $|x - 51.1| \leq 1.6$. Interpret the solution in terms of the information in the table. What is the margin of error? $\{x | 49.5 \leq x \leq 52.7\}$ or $[49.5, 52.7]$; The percentage of the U.S. population that watched "Roots" is between a low of 49.5% and a high of 52.7%. The margin of error is 1.6%.

89. The inequality $|T - 57| \leq 7$ describes the range of monthly average temperature, T, in degrees Fahrenheit, for San Francisco, California. Solve the inequality and interpret the solution. $\{T | 50 \leq T \leq 64\}$ or $[50, 64]$; The monthly average temperature for San Francisco, California is between a low of 50°F and a high of 64°F.

90. The inequality $|T - 50| \leq 22$ describes the range of monthly average temperature, T, in degrees Fahrenheit, for Albany, New York. Solve the inequality and interpret the solution. $\{T | 28 \leq T \leq 72\}$ or $[28, 72]$; The monthly average temperature for Albany, New York is between a low of 28°F and a high of 72°F.

The specifications for machine parts are given with tolerance limits that describe a range of measurements for which the part is acceptable. In Exercises 91–92, x represents the length of a machine part, in centimeters. The tolerance limit is 0.01 centimeter.

91. Solve: $|x - 8.6| \leq 0.01$. If the length of the machine part is supposed to be 8.6 centimeters, interpret the solution. $\{x | 8.59 \leq x \leq 8.61\}$ or $[8.59, 8.61]$; A machine part that is supposed to be 8.6 centimeters is acceptable between a low of 8.59 and a high of 8.61 centimeters.

92. Solve: $|x - 9.4| \leq 0.01$. If the length of the machine part is supposed to be 9.4 centimeters, interpret the solution. $\{x | 9.39 \leq x \leq 9.41\}$ or $[9.39, 9.41]$; A machine part that is supposed to be 9.4 centimeters is acceptable between a low of 9.39 and a high of 9.41 centimeters.

93. If a coin is tossed 100 times, we would expect approximately 50 of the outcomes to be heads. It can be demonstrated that a coin is unfair if h, the number of outcomes that result in heads, satisfies $\left|\dfrac{h - 50}{5}\right| \geq 1.645$. Describe the number of outcomes that determine an unfair coin that is tossed 100 times. If the number of outcomes that result in heads is 41 or less or 59 or more, then the coin is unfair.

Writing in Mathematics

94. Explain how to solve an equation containing one absolute value expression.   94–101. Answers will vary.

95. Explain why the procedure that you described in Exercise 94 does not apply to the equation $|x - 5| = -3$. What is the solution set of this equation?

96. Describe how to solve an absolute value equation with two absolute values.

97. Describe how to solve an absolute value inequality involving the symbol $<$. Give an example.

98. Explain why the procedure that you described in Exercise 97 does not apply to the inequality $|x - 5| < -3$. What is the solution set of this inequality?

99. Describe how to solve an absolute value inequality involving the symbol $>$. Give an example.

100. Explain why the procedure that you described in Exercise 99 does not apply to the inequality $|x - 5| > -3$. What is the solution set of this inequality?

101. The final episode of $M*A*S*H$ was viewed by more than 58% of U.S. television households. Is it likely that a popular television series in the twenty-first century will achieve a 58% market share? Explain your answer.

Technology Exercises

In Exercises 102–104, solve each equation using a graphing utility. Graph each side separately in the same viewing rectangle. The solutions are the x-coordinates of the intersection points.

102. $|x + 1| = 5$ −6 and 4   **102–110.** See graphing answer section.

103. $|3(x + 4)| = 12$ −8 and 0

104. $|2x − 3| = |9 − 4x|$ 2 and 3

In Exercises 105–107, solve each inequality using a graphing utility. Graph each side separately in the same viewing rectangle. The solution set consists of all values of x for which the the graph of the left side lies below the graph of the right side.

105. $|2x + 3| < 5$ $\{x | −4 < x < 1\}$ or $(−4, 1)$

106. $\left|\dfrac{2x − 1}{3}\right| < \dfrac{5}{3}$ $\{x | −2 < x < 3\}$ or $(−2, 3)$

107. $|x + 4| < −1$ $\varnothing$

In Exercises 108–110, solve each inequality using a graphing utility. Graph each side separately in the same viewing rectangle. The solution set consists of all values of x for which the graph of the left side lies above the graph of the right side.

108. $|2x − 1| > 7$ $\{x | x < −3 \text{ or } x > 4\}$ or $(−\infty, −3) \cup (4, \infty)$

109. $|0.1x − 0.4| + 0.4 > 0.6$ $\{x | x < 2 \text{ or } x > 6\}$ or $(−\infty, 2) \cup (6, \infty)$

110. $|x + 4| > −1$ $\{x | x \text{ is a real number}\}$ or $(−\infty, \infty)$

111. Use a graphing utility to verify the solution sets for any five equations or inequalities that you solved by hand in Exercises 1–70.   Answers will vary.

Critical Thinking Exercises

112. Which one of the following is true?   d

   a. All absolute value equations have two solutions.

   b. The equation $|x| = −6$ is equivalent to $x = 6$ or $x = −6$.

   c. We can rewrite the inequality $x > 5$ or $x < −5$ more compactly as $−5 < x < 5$.

   d. Absolute value inequalities in the form $|ax + b| < c$ translate into *and* compound statements, which may be written as three-part inequalities.

113. Write an absolute value inequality for which the interval shown is the solution.

   a. $|x − 4| < 3$ **b.** $|x − 4| \geq 3$

114. The percentage, p, of defective products manufactured by a company is given as $|p − 0.3\%| \leq 0.2\%$. If 100,000 products are manufactured and the company offers a $5 refund for each defective product, describe the company's cost for refunds.   The cost for refunds will be between a low of $500 and a high of $2500.

115. Solve: $|2x + 5| = 3x + 4.$ 1

Review Exercises

116–118. See graphing answer section.

In Exercises 116–118, graph each linear function.

116. $3x − 5y = 15$ (Section 2.3, Example 1)

117. $f(x) = −\dfrac{2}{3}x$ (Section 2.3, Example 5)

118. $f(x) = −2$ (Section 2.3, Figure 2.21)

✔ MID-CHAPTER CHECK POINT

CHAPTER 4

What You Know: We learned to solve linear inequalities, expressing solution sets in set-builder and interval notations. We know that it is necessary to reverse the sense of an inequality when multiplying or dividing both sides by a negative number. We solved compound inequalities with *and* by finding the intersection of solution sets and with *or* by finding the union of solution sets. Finally, we solved equations and inequalities involving absolute value by carefully rewriting the given equation or inequality without absolute value bars. For positive values of c, we wrote $|X| = c$ as $X = c$ or $X = −c$, we wrote $|X| < c$ as $−c < X < c$, and we wrote $|X| > c$ as $X < −c$ or $X > c$.

In Exercises 1–18, solve each inequality or equation.

1. $4 − 3x \geq 12 − x$ $\{x | x \leq −4\}; (−\infty, −4]$

2. $5 \leq 2x − 1 < 9$ $\{x | 3 \leq x < 5\}; [3, 5)$

3. $|4x − 7| = 5$ $\left\{\frac{1}{2}, 3\right\}$

4. $−10 − 3(2x + 1) > 8x + 1$ $\{x | x < −1\}; (−\infty, −1)$

5. $2x + 7 < -11$ or $-3x - 2 < 13$
$\{x | x < -9 \text{ or } x > -5\}; (-\infty, -9) \cup (-5, \infty)$

6. $|3x - 2| \le 4$ $\quad \{x | -\frac{2}{3} \le x \le 2\}; [-\frac{2}{3}, 2]$

7. $|x + 5| = |5x - 8|$ $\quad \{\frac{1}{2}, \frac{13}{4}\}$

8. $5 - 2x \ge 9$ and $5x + 3 > -17$
$\{x | -4 < x \le -2\}$ or $(-4, -2]$

9. $3x - 2 > -8$ or $2x + 1 < 9$
$\{x | x \text{ is a real number}\}; (-\infty, \infty)$

10. $\dfrac{x}{2} + 3 \le \dfrac{x}{3} + \dfrac{5}{2}$ $\quad \{x | x \le -3\}$ or $(-\infty, -3]$

11. $\dfrac{2}{3}(6x - 9) + 4 > 5x + 1$ $\quad \{x | x < -3\}$ or $(-\infty, -3)$

12. $|5x + 3| > 2$ $\quad \{x | x < -1 \text{ or } x > -\frac{1}{5}\}; (-\infty, -1) \cup (-\frac{1}{5}, \infty)$

13. $7 - \left| \dfrac{x}{2} + 2 \right| \le 4$ $\quad \{x | x \le -10 \text{ or } x \ge 2\}; (-\infty, -10] \cup [2, \infty)$

14. $5(x - 2) - 3(x + 4) \ge 2x - 20$ $\quad \varnothing$

15. $\dfrac{x + 3}{4} < \dfrac{1}{3}$ $\quad \{x | x < -\frac{5}{3}\}; (-\infty, -\frac{5}{3})$

16. $5x + 1 \ge 4x - 2$ and $2x - 3 > 5$
$\{x | x > 4\}; (4, \infty)$

17. $3 - |2x - 5| = -6$ $\quad \{-2, 7\}$

18. $3 + |2x - 5| = -6$ $\quad \varnothing$

In Exercises 19–22, solve each problem.

19. A car rental agency rents a certain car for $40 per day with unlimited mileage or $24 per day plus $0.20 per mile. How far can a customer drive this car per day for the $24 option to cost no more than the unlimited mileage option? no more than 80 miles per day

20. To receive a B in a course, you must have an average of at least 80% but less than 90% on five exams. Your grades on the first four exams were 95%, 79%, 91%, and 86%. What range of grades on the fifth exam will result in a B for the course? [49%, 99%)

21. A retiree requires an annual income of at least $9000 from an investment paying 7.5% annual interest. How much should the retiree invest to achieve the desired return? at least $120,000

22. A company that manufactures compact discs has fixed monthly overhead costs of $60,000. Each disc costs $0.18 to produce and sells for $0.30. How many discs should be produced and sold each month for the company to have a profit of at least $30,000? at least 750,000 discs each month

SECTION

4.4

LINEAR INEQUALITIES IN TWO VARIABLES

Objectives

1 Graph a linear inequality in two variables.

2 Graph a system of linear inequalities.

3 Solve applied problems involving systems of inequalities.

This book was written in Point Reyes National Seashore, 40 miles north of San Francisco. The park consists of 75,000 acres with miles of pristine surf-washed beaches, forested ridges, and bays bordered by white cliffs.

Like your author, many people are kept inspired and energized surrounded by nature's unspoiled beauty. In this section, you will see how systems of inequalities model whether a region's natural beauty manifests itself in forests, grasslands, or deserts.

Linear Inequalities in Two Variables and Their Solutions We have seen that the graphs of equations in the form $Ax + By = C$ are straight lines. If we change the $=$ sign to $>, <, \geq,$ or $\leq$, we obtain a **linear inequality in two variables**. Some examples of linear inequalities in two variables are $x + y > 2, 3x - 5y \leq 15,$ and $2x - y < 4$.

A **solution of an inequality in two variables**, x and y, is an ordered pair of real numbers with the following property: When the x-coordinate is substituted for x and the y-coordinate is substituted for y in the inequality, we obtain a true statement. For example, $(3, 2)$ is a solution of the inequality $x + y > 1$. When 3 is substituted for x and 2 is substituted for y, we obtain the true statement $3 + 2 > 1$, or $5 > 1$. Because there are infinitely many pairs of numbers that have a sum greater than 1, the inequality $x + y > 1$ has infinitely many solutions. Each ordered-pair solution is said to **satisfy** the inequality. Thus, $(3, 2)$ satisfies the inequality $x + y > 1$.

1 Graph a linear inequality in two variables.

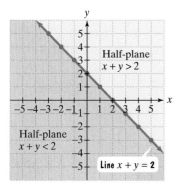

FIGURE 4.8

The Graph of a Linear Inequality in Two Variables We know that the graph of an equation in two variables is the set of all points whose coordinates satisfy the equation. Similarly, the **graph of an inequality in two variables** is the set of all points whose coordinates satisfy the inequality.

Let's use Figure 4.8 to get an idea of what the graph of a linear inequality in two variables looks like. Part of the figure shows the graph of the linear equation $x + y = 2$. The line divides the points in the rectangular coordinate system into three sets. First, there is the set of points along the line, satisfying $x + y = 2$. Next, there is the set of points in the green region above the line. Points in the green region satisfy the linear inequality $x + y > 2$. Finally, there is the set of points in the purple region below the line. Points in the purple region satisfy the linear inequality $x + y < 2$.

A **half-plane** is the set of all the points on one side of a line. In Figure 4.8, the green region is a half-plane. The purple region is also a half-plane. A half-plane is the graph of a linear inequality that involves $>$ or $<$. The graph of an inequality that involves $\geq$ or $\leq$ is a half-plane and a line. A solid line is used to show that a line is part of a graph. A dashed line is used to show that a line is not part of a graph.

GRAPHING A LINEAR INEQUALITY IN TWO VARIABLES

1. Replace the inequality symbol with an equal sign and graph the corresponding linear equation. Draw a solid line if the original inequality contains a $\leq$ or $\geq$ symbol. Draw a dashed line if the original inequality contains a $<$ or $>$ symbol.

2. Choose a test point in one of the half-planes that is not on the line. Substitute the coordinates of the test point into the inequality.

3. If a true statement results, shade the half-plane containing this test point. If a false statement results, shade the half-plane not containing this test point.

EXAMPLE 1　Graphing a Linear Inequality in Two Variables

Graph:　$2x - 3y \geq 6$.

SOLUTION

Step 1.　Replace the inequality symbol by = and graph the linear equation. We need to graph $2x - 3y = 6$. We can use intercepts to graph this line.

We set $y = 0$ to find the x-intercept:	We set $x = 0$ to find the y-intercept:
$2x - 3y = 6$	$2x - 3y = 6$
$2x - 3 \cdot 0 = 6$	$2 \cdot 0 - 3y = 6$
$2x = 6$	$-3y = 6$
$x = 3.$	$y = -2.$

The x-intercept is 3, so the line passes through $(3, 0)$. The y-intercept is -2, so the line passes through $(0, -2)$. Using the intercepts, the line is shown in Figure 4.9 as a solid line. This is because the inequality $2x - 3y \geq 6$ contains a $\geq$ symbol, in which equality is included.

Step 2.　Choose a test point in one of the half-planes that is not on the line. Substitute its coordinates into the inequality. The line $2x - 3y = 6$ divides the plane into three parts—the line itself and two half-planes. The points in one half-plane satisfy $2x - 3y > 6$. The points in the other half-plane satisfy $2x - 3y < 6$. We need to find which half-plane belongs to the solution of $2x - 3y \geq 6$. To do so, we test a point from either half-plane. The origin, $(0, 0)$, is the easiest point to test.

$$2x - 3y \geq 6 \qquad \text{This is the given inequality.}$$
$$2 \cdot 0 - 3 \cdot 0 \stackrel{?}{\geq} 6 \qquad \text{Test } (0, 0) \text{ by substituting 0 for x and 0 for y.}$$
$$0 - 0 \stackrel{?}{\geq} 6 \qquad \text{Multiply.}$$
$$0 \geq 6 \qquad \text{This statement is false.}$$

Step 3.　If a false statement results, shade the half-plane not containing the test point. Because 0 is not greater than or equal to 6, the test point, $(0, 0)$, is not part of the solution set. Thus, the half-plane below the solid line $2x - 3y = 6$ is part of the solution set. The solution set is the line and the half-plane that does not contain the point $(0, 0)$, indicated by shading this half-plane. The graph is shown using green shading and a blue line in Figure 4.10. ■

✔ **CHECK POINT 1** Graph:　$4x - 2y \geq 8$.　See graphing answer section.

When graphing a linear inequality, test a point that lies in one of the half-planes and *not on the line dividing the half-planes*. The test point $(0, 0)$ is convenient because it is easy to calculate when 0 is substituted for each variable. However, if $(0, 0)$ lies on the dividing line and not in a half-plane, a different test point must be selected.

EXAMPLE 2　Graphing a Linear Inequality in Two Variables

Graph:　$y > -\dfrac{2}{3}x$.

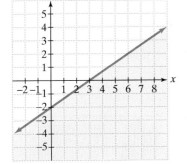

FIGURE 4.9 Preparing to graph $2x - 3y \geq 6$

FIGURE 4.10 The graph of $2x - 3y \geq 6$

SOLUTION

Step 1. **Replace the inequality symbol by = and graph the linear equation.** Because we are interested in graphing $y > -\frac{2}{3}x$, we begin by graphing $y = -\frac{2}{3}x$. We can use the slope and the y-intercept to graph this linear function.

$$y = -\frac{2}{3}x + 0$$

Slope $= \dfrac{-2}{3} = \dfrac{\text{rise}}{\text{run}}$ y-intercept $= 0$

The y-intercept is 0, so the line passes through $(0, 0)$. Using the y-intercept and the slope, the line is shown in Figure 4.11 as a dashed line. This is because the inequality $y > -\frac{2}{3}x$ contains a $>$ symbol, in which equality is not included.

Step 2. **Choose a test point in one of the half-planes that is not on the line. Substitute its coordinates into the inequality.** We cannot use $(0, 0)$ as a test point because it lies on the line and not in a half-plane. Let's use $(1, 1)$, which lies in the half-plane above the line.

$y > -\dfrac{2}{3}x$ This is the given inequality.

$1 \overset{?}{>} -\dfrac{2}{3} \cdot 1$ Test $(1, 1)$ by substituting 1 for x and 1 for y.

$1 > -\dfrac{2}{3}$ This statement is true.

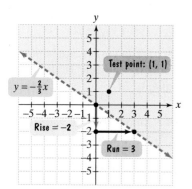

FIGURE 4.11 The graph of $y > -\frac{2}{3}x$

Step 3. **If a true statement results, shade the half-plane containing the test point.** Because 1 is greater than $-\frac{2}{3}$, the test point $(1, 1)$ is part of the solution set. All the points on the same side of the line $y = -\frac{2}{3}x$ as the point $(1, 1)$ are members of the solution set. The solution set is the half-plane that contains the point $(1, 1)$, indicated by shading this half-plane. The graph is shown using green shading and a dashed blue line in Figure 4.11. ∎

USING TECHNOLOGY

Most graphing utilities can graph inequalities in two variables with the $\boxed{\text{SHADE}}$ feature. The procedure varies by model, so consult your manual. For most graphing utilities, you must first solve for y if it is not already isolated. The figure shows the graph of $y > -\frac{2}{3}x$. Most displays do not distinguish between dashed and solid boundary lines.

 CHECK POINT 2 Graph: $y > -\dfrac{3}{4}x$.   See graphing answer section.

Graphing Linear Inequalities without Using Test Points You can graph inequalities in the form $y > mx + b$ or $y < mx + b$ without using test points. The inequality symbol indicates which half-plane to shade.

- If $y > mx + b$, shade the half-plane above the line $y = mx + b$.
- If $y < mx + b$, shade the half-plane below the line $y = mx + b$.

It is also not necessary to use test points when graphing inequalities involving half-planes on one side of a vertical or a horizontal line.

For the Vertical Line $x = a$:
- If $x > a$, shade the half-plane to the right of $x = a$.
- If $x < a$, shade the half-plane to the left of $x = a$.

For the Horizontal Line $y = b$:
- If $y > b$, shade the half-plane above $y = b$.
- If $y < b$, shade the half-plane below $y = b$.

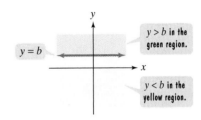

EXAMPLE 3 Graphing Inequalities without Using Test Points

Graph each inequality in a rectangular coordinate system:

a. $y \leq -3$ **b.** $x > 2$.

SOLUTION

a. $y \leq -3$

Graph $y = -3$, a horizontal line with y-intercept -3. The line is solid because equality is included in $y \leq -3$. Because of the less than part of $\leq$, shade the half-plane below the horizontal line.

b. $x > 2$

Graph $x = 2$, a vertical line with x-intercept 2. The line is dashed because equality is not included in $x > 2$. Because of $>$, the greater than symbol, shade the half-plane to the right of the vertical line.

 CHECK POINT 3 Graph each inequality in a rectangular coordinate system:

a. $y > 1$ **b.** $x \leq -2$.   See graphing answer section.

2 Graph a system of linear inequalities.

Systems of Linear Inequalities Just as two linear equations make up a system of linear equations, two or more linear inequalities make up a **system of linear inequalities**. Here is an example of a system of linear inequalities:

$$x - y < 1$$
$$2x + 3y \geq 12.$$

A **solution of a system of linear inequalities** in two variables is an ordered pair that satisfies each inequality in the system. The set of all such ordered pairs is the **solution set** of the system. Thus, to graph a system of inequalities in two variables, begin by graphing each individual inequality in the same rectangular coordinate system. Then find the region, if there is one, that is common to every graph in the system. This region of intersection gives a picture of the system's solution set.

EXAMPLE 4 Graphing a System of Linear Inequalities

Graph the solution set of the system:

$$x - y < 1$$
$$2x + 3y \geq 12.$$

SOLUTION Replacing each inequality symbol with an equal sign indicates that we need to graph $x - y = 1$ and $2x + 3y = 12$. We can use intercepts to graph these lines.

$x - y = 1$		$2x + 3y = 12$

x-intercept: $x - 0 = 1$ [Set $y = 0$ in each equation.] x-intercept: $2x + 3 \cdot 0 = 12$
$x = 1$ $2x = 12$
The line passes through $(1, 0)$. $x = 6$
The line passes through $(6, 0)$.

y-intercept: $0 - y = 1$ [Set $x = 0$ in each equation.] y-intercept: $2 \cdot 0 + 3y = 12$
$-y = 1$ $3y = 12$
$y = -1$ $y = 4$
The line passes through $(0, -1)$ The line passes through $(0, 4)$.

Now we are ready to graph the solution set of the system of linear inequalities.

Graph $x - y < 1$. The blue line, $x - y = 1$, is dashed: Equality is not included in $x - y < 1$. Because $(0, 0)$ makes the inequality true $(0 - 0 < 1$, or $0 < 1$, is true), shade the half-plane containing $(0, 0)$ in yellow.

Add the graph of $2x + 3y \geq 12$. The red line, $2x + 3y = 12$, is solid: Equality is included in $2x + 3y \geq 12$. Because $(0, 0)$ makes the inequality false $(2 \cdot 0 + 3 \cdot 0 \geq 12$, or $0 \geq 12$, is false), shade the half-plane not containing $(0, 0)$ using green vertical shading.

The solution set of the system is graphed as the intersection (the overlap) of the two half-planes. This is the region in which the yellow shading and the green vertical shading overlap.

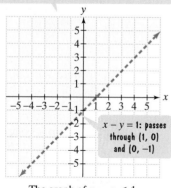

$x - y = 1$: passes through $(1, 0)$ and $(0, -1)$

The graph of $x - y < 1$

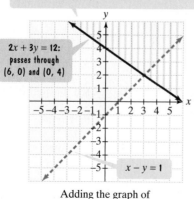

$2x + 3y = 12$: passes through $(6, 0)$ and $(0, 4)$

$x - y = 1$

Adding the graph of $2x + 3y \geq 12$

This open dot shows $(3, 2)$ is not in the solution set.

The graph of $x - y < 1$ and $2x + 3y \geq 12$

CHECK POINT 4 Graph the solution set of the system: See graphing answer section.

$$x - 3y < 6$$
$$2x + 3y \geq -6.$$

A system of inequalities has no solution if there are no points in the rectangular coordinate system that simultaneously satisfy each inequality in the system. For example, the system

$$2x + 3y \geq 6$$
$$2x + 3y \leq 0$$

whose separate graphs are shown in Figure 4.12 has no overlapping region. Thus, the system has no solution. The solution set is ∅, the empty set.

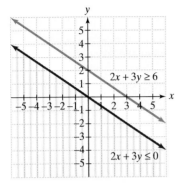

FIGURE 4.12 A system of inequalities with no solution

EXAMPLE 5   Graphing a System of Inequalities

Graph the solution set:

$$x - y < 2$$
$$-2 \leq x < 4$$
$$y < 3.$$

SOLUTION   We begin by graphing $x - y < 2$, the first given inequality. The line $x - y = 2$ has an x-intercept of 2 and a y-intercept of -2. The test point $(0, 0)$ makes the inequality $x - y < 2$ true, and its graph is shown in Figure 4.13.

Now, let's consider the second given inequality, $-2 \leq x < 4$. Replacing the inequality symbols by =, we obtain $x = -2$ and $x = 4$, graphed as red vertical lines in Figure 4.14. The line of $x = 4$ is not included. Using $(0, 0)$ as a test point and substituting the x-coordinate, 0, into $-2 \leq x < 4$, we obtain the true statement $-2 \leq 0 < 4$. We therefore shade the region between the vertical lines. We must intersect this region with the yellow region in Figure 4.13. The resulting region is shown in yellow and green vertical shading in Figure 4.14.

Finally, let's consider the third given inequality, $y < 3$. Replacing the inequality symbol by =, we obtain $y = 3$, which graphs as a horizontal line. Because of the less than symbol in $y < 3$, the graph consists of the half-plane below the line $y = 3$. We must intersect this half-plane with the region in Figure 4.14. The resulting region is shown in yellow and green vertical shading in Figure 4.15. This region represents the graph of the solution set of the given system.

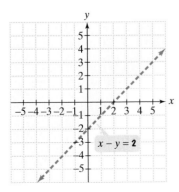

FIGURE 4.13   The graph of $x - y < 2$

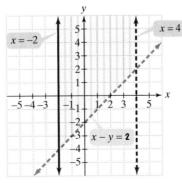

FIGURE 4.14   The graph of $x - y < 2$ and $-2 \leq x < 4$

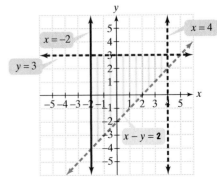

FIGURE 4.15   The graph of $x - y < 2$ and $-2 \leq x < 4$ and $y < 3$

> ✓ **CHECK POINT 5** Graph the solution set:
>
> $$x + y < 2$$
> $$-2 \leq x < 1$$
> $$y > -3. \qquad \text{See graphing answer section.}$$

3 Solve applied problems involving systems of inequalities.

Applications Temperature and precipitation affect whether or not trees and forests can grow. At certain levels of precipitation and temperature, only grasslands and deserts will exist. Figure 4.16 shows three kinds of regions—deserts, grasslands, and forests—that result from various ranges of temperature, T, and precipitation, P.

Systems of inequalities can be used to describe where forests, grasslands, and deserts occur. Because these regions occur when the average annual temperature, T, is 35°F or greater, each system contains the inequality $T \geq 35$.

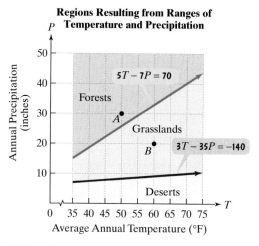

Regions Resulting from Ranges of Temperature and Precipitation

FIGURE 4.16

Source: A. Miller and J. Thompson, *Elements of Meteorology*

Forests occur if	Grasslands occur if	Deserts occur if
$T \geq 35$	$T \geq 35$	$T \geq 35$
$5T - 7P < 70.$	$5T - 7P \geq 70$	$3T - 35P > -140.$
	$3T - 35P \leq -140.$	

EXAMPLE 6 Forests and Systems of Inequalities

Show that point A in Figure 4.16 is a solution of the system of inequalities that describes where forests occur.

SOLUTION Point A has coordinates $(50, 30)$. This means that if a region has an average annual temperature of 50°F and an average annual precipitation of 30 inches, a forest occurs. We can show that $(50, 30)$ satisfies the system of inequalities for forests by substituting 50 for T and 30 for P in each inequality in the system.

$$
\begin{array}{ll}
T \geq 35 & 5T - 7P < 70 \\
50 \geq 35, \;\; \text{true} & 5 \cdot 50 - 7 \cdot 30 \overset{?}{<} 70 \\
& 250 - 210 \overset{?}{<} 70 \\
& 40 < 70, \;\; \text{true}
\end{array}
$$

The coordinates $(50, 30)$ make each inequality true. Thus, $(50, 30)$ satisfies the system for forests. ∎

> ✓ **CHECK POINT 6** Show that point B in Figure 4.16 is a solution of the system of inequalities that describes where grasslands occur. Point $B = (60, 20)$; Using $T = 60$ and $P = 20$, each of the three inequalities for grasslands is true: $60 \geq 35$, true; $5(60) - 7(20) \geq 70$, true; $3(60) - 35(20) \leq -140$, true.

4.4 EXERCISE SET

 Student Solutions Manual   CD/Video   PH Math/Tutor Center   MathXL Tutorials on CD Math XL MathXL® MyMathLab MyMathLab   Interactmath.com

Practice Exercises

In Exercises 1–22, graph each inequality.
1–46. See graphing answer section.

1. $x + y \geq 3$

2. $x + y \geq 2$

3. $x - y < 5$

4. $x - y < 6$

5. $x + 2y > 4$

6. $2x + y > 6$

7. $3x - y \leq 6$

8. $x - 3y \leq 6$

9. $\dfrac{x}{2} + \dfrac{y}{3} < 1$

10. $\dfrac{x}{4} + \dfrac{y}{2} < 1$

11. $y > \dfrac{1}{3}x$

12. $y > \dfrac{1}{4}x$

13. $y \leq 3x + 2$

14. $y \leq 2x - 1$

15. $y < -\dfrac{1}{4}x$

16. $y < -\dfrac{1}{3}x$

17. $x \leq 2$

18. $x \leq -4$

19. $y > -4$

20. $y > -2$

21. $y \geq 0$

22. $x \leq 0$

In Exercises 23–46 graph the solution set of each system of inequalities or indicate that the system has no solution.

23. $3x + 6y \leq 6$
$\quad 2x + y \leq 8$

24. $x - y \geq 4$
$\quad x + y \leq 6$

25. $2x - 5y \leq 10$
$\quad 3x - 2y > 6$

26. $2x - y \leq 4$
$\quad 3x + 2y > -6$

27. $y > 2x - 3$
$\quad y < -x + 6$

28. $y < -2x + 4$
$\quad y < x - 4$

29. $x + 2y \leq 4$
$\quad y \geq x - 3$

30. $x + y \leq 4$
$\quad y \geq 2x - 4$

31. $x \leq 2$
$\quad y \geq -1$

32. $x \leq 3$
$\quad y \leq -1$

33. $-2 \leq x < 5$

34. $-2 < y \leq 5$

35. $x - y \leq 1$
$\quad x \geq 2$

36. $4x - 5y \geq -20$
$\quad x \geq -3$

37. $x + y > 4$
$\quad x + y < -1 \quad \varnothing$

38. $x + y > 3$
$\quad x + y < -2 \quad \varnothing$

39. $x + y > 4$
$\quad x + y > -1$

40. $x + y > 3$
$\quad x + y > -2$

41. $x - y \leq 2$
$\quad x \geq -2$
$\quad y \leq 3$

42. $3x + y \leq 6$
$\quad x \geq -2$
$\quad y \leq 4$

43. $x \geq 0$
$\quad y \geq 0$
$\quad 2x + 5y \leq 10$
$\quad 3x + 4y \leq 12$

44. $x \geq 0$
$\quad y \geq 0$
$\quad 2x + y \leq 4$
$\quad 2x - 3y \leq 6$

45. $3x + y \leq 6$
$\quad 2x - y \leq -1$
$\quad x \geq -2$
$\quad y \leq 4$

46. $2x + y \leq 6$
$\quad x + y \geq 2$
$\quad 1 \leq x \leq 2$
$\quad y \leq 3$

Practice Plus

In Exercises 47–48, write each sentence as a linear inequality in two variables. Then graph the inequality.
47–50. See graphing answer section.

47. The *y*-variable is at least 4 more than the product of -2 and the *x*-variable. $\quad y \geq -2x + 4$

48. The *y*-variable is at least 2 more than the product of -3 and the *x*-variable. $\quad y \geq -3x + 2$

In Exercises 49–50, write the given sentences as a system of linear inequalities in two variables. Then graph the system.

49. The sum of the *x*-variable and the *y*-variable is at most 4. The *y*-variable added to the product of 3 and the *x*-variable does not exceed 6. $\quad x + y \leq 4$ and $3x + y \leq 6$

50. The sum of the *x*-variable and the *y*-variable is at most 3. The *y*-variable added to the product of 4 and the *x*-variable does not exceed 6. $\quad x + y \leq 3$ and $4x + y \leq 6$

In Exercises 51–52, rewrite each inequality in the system without absolute value bars. Then graph the rewritten system in rectangular coordinates. **51–52.** See graphing answer section.

51. $|x| \leq 2 \quad -2 \leq x \leq 2$
$\quad |y| \leq 3 \quad -3 \leq y \leq 3$

52. $|x| \leq 1 \quad -1 \leq x \leq 1$
$\quad |y| \leq 2 \quad -2 \leq y \leq 2$

*The graphs of solution sets of systems of inequalities involve finding the intersection of the solution sets of two or more inequalities. By contrast, in Exercises 53–54 you will be graphing the **union** of the solution sets of two inequalities.* **53–54.** See graphing answer section.

53. Graph the union of $y > \dfrac{3}{2}x - 2$ and $y < 4$.

54. Graph the union of $x - y \geq -1$ and $5x - 2y \leq 10$.

Without graphing, in Exercises 55–58, determine if each system has no solution or infinitely many solutions.

55. $3x + y < 9$
$\quad 3x + y > 9$
$\quad$ no solution

56. $6x - y \leq 24$
$\quad 6x - y > 24$
$\quad$ no solution

57. $3x + y \leq 9$
$\quad 3x + y \geq 9$
$\quad$ infinitely many solutions

58. $6x - y \leq 24$
$\quad 6x - y \geq 24$
$\quad$ infinitely many solutions

Application Exercises

Maximum heart rate, H, in beats per minute is a function of age, a, modeled by the formula

$$H = 220 - a,$$

where $10 \leq a \leq 70$. The bar graph shows the target heart rate ranges for four types of exercise goals.

Target Heart Rate Ranges for Exercise Goals

Exercise Goal

Boost performance as a competitive athlete

Improve cardiovascular conditioning

Lose weight

Improve overall health and reduce risk of heart attack

40% 50% 60% 70% 80% 90% 100%
Percentage of Maximum Heart Rate

Source: Vitality

In Exercises 59–62, systems of inequalities will be used to model three of the target heart rate ranges shown in the bar graph. We begin with the target heart rate range for cardiovascular conditioning, modeled by the following system of inequalities:

$10 \leq a \leq 70$ — Heart rate ranges apply to ages 10 through 70, inclusive.

$H \geq 0.7(220 - a)$ — Target heart rate range is greater than or equal to 70% of maximum heart rate

$H \leq 0.8(220 - a)$ — and less than or equal to 80% of maximum heart rate.

The graph of this system is shown in the figure. Use the graph to solve Exercises 59–60.

Target Heart Rate Range for Cardiovascular Conditioning

H

$a = 10$

$H = 0.8(220 - a)$

• A

• B

$a = 70$

$H = 0.7(220 - a)$

Heart Rate (beats per minute)

10 20 30 40 50 60 70 80 90 a

Age

59. a. What are the coordinates of point A and what does this mean in terms of age and heart rate? $A = (20, 150)$; A 20-year-old with a heart rate of 150 beats per minute is within the target range.

b. Show that point A is a solution of the system of inequalities. $10 \leq 20 \leq 70$, true; $150 \geq 0.7(220 - 20)$, true; $150 \leq 0.8(220 - 20)$, true

60. a. What are the coordinates of point B and what does this mean in terms of age and heart rate? $B = (40, 130)$; A 40-year-old with a heart rate of 130 beats per minute is within the target range.

b. Show that point B is a solution of the system of inequalities. $10 \leq 40 \leq 70$, true; $130 \geq 0.7(220 - 40)$, true; $130 \leq 0.8(220 - 40)$, true

61. Write a system of inequalities that models the target heart rate range for the goal of losing weight. $10 \leq a \leq 70$, $H \geq 0.6(220 - a)$, $H \leq 0.7(220 - a)$

62. Write a system of inequalities that models the target heart rate range for improving overall health. $10 \leq a \leq 70$, $H \geq 0.5(220 - a)$, $H \leq 0.6(220 - a)$

63. On your next vacation, you will divide lodging between large resorts and small inns. Let x represent the number of nights spent in large resorts. Let y represent the number of nights spent in small inns.

a. Write a system of inequalities that models the following conditions:

> You want to stay at least 5 nights. At least one night should be spent at a large resort. Large resorts average $200 per night and small inns average $100 per night. Your budget permits no more than $700 for lodging. $y \geq 0, x + y \geq 5, x \geq 1, 200x + 100y \leq 700$

b. Graph the solution set of the system of inequalities in part (a).   See graphing answer section.

c. Based on your graph in part (b), how many nights could you spend at a large resort and still stay within your budget?   2 nights

64. a. An elevator can hold no more than 2000 pounds. If children average 80 pounds and adults average 160 pounds, write a system of inequalities that models when the elevator holding x children and y adults is overloaded. $x \geq 0, y \geq 0, 80x + 160y > 2000$

b. Graph the solution set of the system of inequalities in part (a).   See graphing answer section.

Writing in Mathematics

65. What is a linear inequality in two variables? Provide an example with your description.   **65–74.** Answers will vary.

66. How do you determine if an ordered pair is a solution of an inequality in two variables, x and y?

67. What is a half-plane?

68. What does a solid line mean in the graph of an inequality?

69. What does a dashed line mean in the graph of an inequality?

70. Explain how to graph $x - 2y < 4$.

71. What is a system of linear inequalities?

72. What is a solution of a system of linear inequalities?

73. Explain how to graph the solution set of a system of inequalities.

74. What does it mean if a system of linear inequalities has no solution?

Technology Exercises

Graphing utilities can be used to shade regions in the rectangular coordinate system, thereby graphing an inequality in two variables. Read the section of the user's manual for your graphing utility that describes how to shade a region. Then use your graphing utility to graph the inequalities in Exercises 75–78. 75–78. See graphing answer section.

75. $y \le 4x + 4$

76. $y \ge \dfrac{2}{3}x - 2$

77. $2x + y \le 6$

78. $3x - 2y \ge 6$

79. Does your graphing utility have any limitations in terms of graphing inequalities? If so, what are they? 79–81. Answers will vary.

80. Use a graphing utility with a $\boxed{\text{SHADE}}$ feature to verify any five of the graphs that you drew by hand in Exercises 1–22.

81. Use a graphing utility with a $\boxed{\text{SHADE}}$ feature to verify any five of the graphs that you drew by hand for the systems in Exercises 23–46.

Critical Thinking Exercises

82. Write a linear inequality in two variables whose graph is shown. $x - y < 2$ or $y > x - 2$

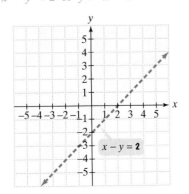

In Exercises 83–84, write a system of inequalities for each graph.

83. $x \ge -2, y > -1$

84. $y > x - 3, y \le x$

85. Write a system of inequalities whose solution set includes every point in the rectangular coordinate system.   Answers will vary.

86. Sketch the graph of the solution set for the following system of inequalities: See graphing answer section.

$$y \ge nx + b\ (n < 0, b > 0)$$
$$y \le mx + b\ (m > 0, b > 0).$$

Review Exercises

87. Solve using matrices:

$$3x - \ y = \ \ 8$$
$$x - 5y = -2.$$

(Section 3.4, Example 3) $(3, 1)$ or $\{(3, 1)\}$

88. Solve by graphing:

$$y = \ \ 3x - 2$$
$$y = -2x + 8.$$

(Section 3.1, Example 2) $(2, 4)$ or $\{(2, 4)\}$

89. Evaluate:

$$\begin{vmatrix} 8 & 2 & -1 \\ 3 & 0 & 5 \\ 6 & -3 & 4 \end{vmatrix}.$$

(Section 3.5, Example 3)   165

SECTION 4.5

Objectives

1 Write an objective function describing a quantity that must be maximized or minimized.

2 Use inequalities to describe limitations in a situation.

3 Use linear programming to solve problems.

LINEAR PROGRAMMING

West Berlin children at Tempelhof airport watch fleets of U.S. airplanes bringing in supplies to circumvent the Soviet blockade. The airlift began June 28, 1948 and continued for 15 months.

The Berlin Airlift (1948–1949) was an operation by the United States and Great Britain in response to military action by the former Soviet Union: Soviet troops closed all roads and rail lines between West Germany and Berlin, cutting off supply routes to the city. The Allies used a mathematical technique developed during World War II to maximize the amount of supplies transported. During the 15-month airlift, 278,228 flights provided basic necessities to blockaded Berlin, saving one of the world's great cities.

In this section, we will look at an important application of systems of linear inequalities. Such systems arise in **linear programming**, a method for solving problems in which a particular quantity that must be maximized or minimized is limited by other factors. Linear programming is one of the most widely used tools in management science. It helps businesses allocate resources to manufacture products in a way that will maximize profit. Linear programming accounts for more than 50% and perhaps as much as 90% of all computing time used for management decisions in business. The Allies used linear programming to save Berlin.

1 Write an objective function describing a quantity that must be maximized or minimized.

Objective Functions in Linear Programming Many problems involve quantities that must be maximized or minimized. Businesses are interested in maximizing profit. An operation in which bottled water and medical kits are shipped to earthquake victims needs to maximize the number of victims helped by this shipment. An **objective function** is an algebraic expression in two or more variables describing a quantity that must be maximized or minimized.

EXAMPLE 1 Writing an Objective Function

Bottled water and medical supplies are to be shipped to victims of an earthquake by plane. Each container of bottled water will serve 10 people and each medical kit will aid 6 people. If x represents the number of bottles of water to be shipped and y represents the number of medical kits, write the objective function that describes the number of people that can be helped.

SOLUTION Because each bottle of water serves 10 people and each medical kit aids 6 people, we have

The number of people helped	is	10 times the number of bottles of water	plus	6 times the number of medical kits.
=		$10x$	+	$6y.$

Using z to represent the number of people helped, the objective function is

$$z = 10x + 6y.$$

Unlike the functions that we have seen so far, the objective function is an equation in three variables. For a value of x and a value of y, there is one and only one value of z. Thus, z is a function of x and y. ■

✔ **CHECK POINT 1** A company manufactures bookshelves and desks for computers. Let x represent the number of bookshelves manufactured daily and y the number of desks manufactured daily. The company's profits are \$25 per bookshelf and \$55 per desk. Write the objective function that describes the company's total daily profit, z, from x bookshelves and y desks. (Check Points 2 through 4 are related to this situation, so keep track of your answers.) $z = 25x + 55y$

2 Use inequalities to describe limitations in a situation.

Constraints in Linear Programming Ideally, the number of earthquake victims helped in Example 1 should increase without restriction so that every victim receives water and medical kits. However, the planes that ship these supplies are subject to weight and volume restrictions. In linear programming problems, such restrictions are called **constraints**. Each constraint is expressed as a linear inequality. The list of constraints forms a system of linear inequalities.

EXAMPLE 2 Writing a Constraint

Each plane can carry no more than 80,000 pounds. The bottled water weighs 20 pounds per container and each medical kit weighs 10 pounds. Let x represent the number of bottles of water to be shipped and y the number of medical kits. Write an inequality that describes this constraint.

SOLUTION Because each plane can carry no more than 80,000 pounds, we have

The total weight of the water bottles	plus	the total weight of the medical kits	must be less than or equal to	80,000 pounds.
$20x$	$+$	$10y$	$\leq$	$80{,}000.$

Each bottle weighs 20 pounds. Each kit weighs 10 pounds.

The plane's weight constraint is described by the inequality

$$20x + 10y \leq 80{,}000.$$ ■

✔ **CHECK POINT 2** To maintain high quality, the company in Check Point 1 should not manufacture more than a total of 80 bookshelves and desks per day. Write an inequality that describes this constraint. $x + y \leq 80$

In addition to a weight constraint on its cargo, each plane has a limited amount of space in which to carry supplies. Example 3 demonstrates how to express this constraint.

EXAMPLE 3 Writing a Constraint

Planes can carry a total volume for supplies that does not exceed 6000 cubic feet. Each water bottle is 1 cubic foot and each medical kit also has a volume of 1 cubic foot. With x still representing the number of water bottles and y the number of medical kits, write an inequality that describes this second constraint.

SOLUTION   Because each plane can carry a volume of supplies that does not exceed 6000 cubic feet, we have

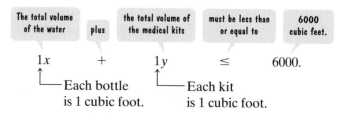

The total volume of the water	plus	the total volume of the medical kits	must be less than or equal to	6000 cubic feet.

$$1x \quad + \quad 1y \quad \leq \quad 6000.$$

└─ Each bottle is 1 cubic foot.   └─ Each kit is 1 cubic foot.

The plane's volume constraint is described by the inequality $x + y \leq 6000$.

In summary, here's what we have described so far in this aid-to-earthquake-victims situation:

$$z = 10x + 6y$$ This is the objective function describing the number of people helped with x bottles of water and y medical kits.

$$20x + 10y \leq 80{,}000$$
$$x + y \leq 6000.$$ These are the constraints based on each plane's weight and volume limitations.

$30 \leq x \leq 80$
$10 \leq y \leq 30;$
$z = 25x + 55y$
$x + y \leq 80$
$30 \leq x \leq 80$
$10 \leq y \leq 30$

✔ **CHECK POINT 3** To meet customer demand, the company in Check Point 1 must manufacture between 30 and 80 bookshelves per day, inclusive. Furthermore, the company must manufacture at least 10 and no more than 30 desks per day. Write an inequality that describes each of these sentences. Then summarize what you have described about this company by writing the objective function for its profits, and the three constraints.

3 Use linear programming to solve problems.

Solving Problems with Linear Programming   The problem in the earthquake situation described previously is to maximize the number of victims who can be helped, subject to the planes' weight and volume constraints. The process of solving this problem is called *linear programming*, based on a theorem that was proven during World War II.

SOLVING A LINEAR PROGRAMMING PROBLEM   Let $z = ax + by$ be an objective function that depends on x and y. Furthermore, z is subject to a number of constraints on x and y. If a maximum or minimum value of z exists, it can be determined as follows:

1. Graph the system of inequalities representing the constraints.
2. Find the value of the objective function at each corner, or **vertex**, of the graphed region. The maximum and minimum of the objective function occur at one or more of the corner points.

EXAMPLE 4   Solving a Linear Programming Problem

Determine how many bottles of water and how many medical kits should be sent on each plane to maximize the number of earthquake victims who can be helped.

SOLUTION   We must maximize $z = 10x + 6y$ subject to the following constraints:

$$20x + 10y \leq 80{,}000$$
$$x + y \leq 6000.$$

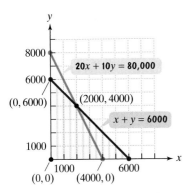

FIGURE 4.17 The region in quadrant I representing the constraints $20x + 10y \leq 80,000$ and $x + y \leq 6000$

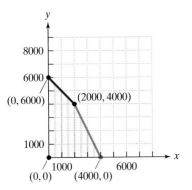

FIGURE 4.18

Step 1. **Graph the system of inequalities representing the constraints.** Because x (the number of bottles of water per plane) and y (the number of medical kits per plane) must be nonnegative, we need to graph the system of inequalities in quadrant I and its boundary only. To graph the inequality $20x + 10y \leq 80,000$, we graph the equation $20x + 10y = 80,000$ as a solid blue line (Figure 4.17). Setting $y = 0$, the x-intercept is 4000 and setting $x = 0$, the y-intercept is 8000. Using $(0, 0)$ as a test point, the inequality is satisfied, so we shade below the blue line, as shown in yellow in Figure 4.17. Now we graph $x + y \leq 6000$ by first graphing $x + y = 6000$ as a solid red line. Setting $y = 0$, the x-intercept is 6000. Setting $x = 0$, the y-intercept is 6000. Using $(0, 0)$ as a test point, the inequality is satisfied, so we shade below the red line, as shown using green vertical shading in Figure 4.17.

We use the addition method to find where the lines $20x + 10y = 80,000$ and $x + y = 6000$ intersect.

$$20x + 10y = 80,000 \xrightarrow{\text{No change}} 20x + 10y = 80,000$$
$$x + y = 6000 \xrightarrow{\text{Multiply by } -10.} -10x - 10y = -60,000$$
$$\text{Add:} \quad 10x \qquad = 20,000$$
$$x \quad = 2000$$

Back-substituting 2000 for x in $x + y = 6000$, we find $y = 4000$, so the intersection point is $(2000, 4000)$.

The system of inequalities representing the constraints is shown by the region in which the yellow shading and the green vertical shading overlap in Figure 4.17. The graph of the system of inequalities is shown again in Figure 4.18. The red and blue line segments are included in the graph.

Step 2. **Find the value of the objective function at each corner of the graphed region. The maximum and minimum of the objective function occur at one or more of the corner points.** We must evaluate the objective function, $z = 10x + 6y$, at the four corners, or vertices, of the region in Figure 4.18.

Corner (x, y)	Objective Function $z = 10x + 6y$
$(0, 0)$	$z = 10(0) + 6(0) = 0$
$(4000, 0)$	$z = 10(4000) + 6(0) = 40,000$
$(2000, 4000)$	$z = 10(2000) + 6(4000) = 44,000$ ← maximum
$(0, 6000)$	$z = 10(0) + 6(6000) = 36,000$

Thus, the maximum value of z is 44,000 and this occurs when $x = 2000$ and $y = 4000$. In practical terms, this means that the maximum number of earthquake victims who can be helped with each plane shipment is 44,000. This can be accomplished by sending 2000 water bottles and 4000 medical kits per plane. ∎

 CHECK POINT 4 For the company in Check Points 1–3, how many bookshelves and how many desks should be manufactured per day to obtain maximum profit? What is the maximum daily profit? To maximize profit, 50 bookshelves and 30 desks should be manufactured per day. The maximum daily profit is $2900.

EXAMPLE 5 Solving a Linear Programming Problem

Find the maximum value of the objective function

$$z = 2x + y$$

subject to the following constraints:

$$x \geq 0, y \geq 0$$

$$x + 2y \leq 5$$

$$x - y \leq 2.$$

SOLUTION We begin by graphing the region in quadrant I ($x \geq 0, y \geq 0$) formed by the constraints. The graph is shown in Figure 4.19.

Now we evaluate the objective function at the four vertices of this region.

Objective function: $z = 2x + y$

At $(0, 0)$: $z = 2 \cdot 0 + 0 = 0$

At $(2, 0)$: $z = 2 \cdot 2 + 0 = 4$

At $(3, 1)$: $z = 2 \cdot 3 + 1 = 7$ **Maximum value of** z

At $(0, 2.5)$: $z = 2 \cdot 0 + 2.5 = 2.5$

Thus, the maximum value of z is 7, and this occurs when $x = 3$ and $y = 1$.

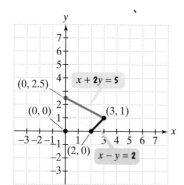

FIGURE 4.19 The graph of $x + 2y \leq 5$ and $x - y \leq 2$ in quadrant I

We can see why the objective function in Example 5 has a maximum value that occurs at a vertex by solving the equation for y.

$$z = 2x + y \qquad \text{This is the objective function of Example 5.}$$

$$y = -2x + z \qquad \text{Solve for } y. \text{ Recall that the slope-intercept form of a line is } y = mx + b.$$

Slope $= -2$ y-intercept $= z$

In this form, z represents the y-intercept of the objective function. The equation describes infinitely many parallel lines, each with slope -2. The process in linear programming involves finding the maximum z-value for all lines that intersect the region determined by the constraints. Of all the lines whose slope is -2, we're looking for the one with the greatest y-intercept that intersects the given region. As we see in Figure 4.20, such a line will pass through one (or possibly more) of the vertices of the region.

FIGURE 4.20 The line with slope -2 with the greatest y-intercept that intersects the shaded region passes through one of its vertices.

✓ **CHECK POINT 5** Find the maximum value of the objective function $z = 3x + 5y$ subject to the constraints $x \geq 0, y \geq 0, x + y \geq 1, x + y \leq 6$.   30

4.5 EXERCISE SET

Student Solutions Manual CD/Video PH Math/Tutor Center MathXL Tutorials on CD MathXL® MyMathLab Interactmath.com

Practice Exercises

In Exercises 1–4, find the value of the objective function at each corner of the graphed region. What is the maximum value of the objective function? What is the minimum value of the objective function?

1. Objective Function
$z = 5x + 6y$

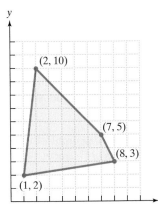

$(1, 2): 17; (2, 10): 70;$
$(7, 5): 65; (8, 3): 58;$
maximum: 70;
minimum: 17

2. Objective Function
$z = 3x + 2y$

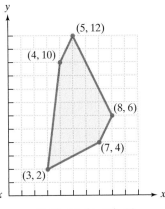

$(3, 2): 13; (4, 10): 32;$
$(5, 12): 39; (8, 6): 36; (7, 4): 29;$
maximum: 39;
minimum: 13

3. Objective Function
$z = 40x + 50y$

$(0, 0): 0; (0, 8): 400;$
$(4, 9): 610; (8, 0): 320;$
maximum: 610;
minimum: 0

4. Objective Function
$z = 30x + 45y$

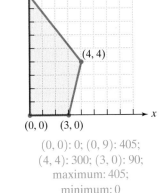

$(0, 0): 0; (0, 9): 405;$
$(4, 4): 300; (3, 0): 90;$
maximum: 405;
minimum: 0

In Exercises 5–14, an objective function and a system of linear inequalities representing constraints are given.

a. *Graph the system of inequalities representing the constraints.*
   5–14a. See graphing answer section.

b. *Find the value of the objective function at each corner of the graphed region.*

c. *Use the values in part (b) to determine the maximum value of the objective function and the values of x and y for which the maximum occurs.*

5. Objective Function $z = 3x + 2y$ **b.** $(0, 4): 8;$
Constraints $x \geq 0, y \geq 0$ $(0, 8): 16;$
  $2x + y \leq 8$ $(4, 0): 12$
  $x + y \geq 4$
   c. maximum: 16; at $(0, 8)$

6. Objective Function $z = 2x + 3y$ **b.** $(0, 0): 0;$
Constraints $x \geq 0, y \geq 0$ $(0, 4): 12;$
  $2x + y \leq 8$ $(3, 2): 12;$
  $2x + 3y \leq 12$ $(4, 0): 8;$
   c. maximum: 12; at $(0, 4)$ and $(3, 2)$

7. Objective Function $z = 4x + y$ **b.** $(0, 3): 3;$
Constraints $x \geq 0, y \geq 0$ $(0, 4): 4;$
  $2x + 3y \leq 12$ $(6, 0): 24;$
  $x + y \geq 3$ $(3, 0): 12$
   c. maximum: 24; at $(6, 0)$

8. Objective Function $z = x + 6y$ **b.** $(0, 0): 0;$
Constraints $x \geq 0, y \geq 0$ $(0, 5): 30;$
  $2x + y \leq 10$ $(2, 6): 38;$
  $x - 2y \geq -10$ $(5, 0): 5$
   c. maximum: 38; at $(2, 6)$

9. Objective Function $z = 3x - 2y$ **b.** $(1, 2): -1;$
Constraints $1 \leq x \leq 5$ $(1, 4): -5;$
  $y \geq 2$ $(5, 8): -1;$
  $x - y \geq -3$ $(5, 2): 11;$
   c. maximum: 11; at $(5, 2)$

10. Objective Function $z = 5x - 2y$ **b.** $(0, 2): -4;$
Constraints $0 \leq x \leq 5$ $(0, 3): -6;$
  $0 \leq y \leq 3$ $(5, 3): 19;$
  $x + y \geq 2$ $(5, 0): 25;$
   $(2, 0): 10$
   c. maximum: 25; at $(5, 0)$

11. Objective Function $z = 4x + 2y$ **b.** $(0, 2): 4;$
Constraints $x \geq 0, y \geq 0$ $(0, 4): 8;$
  $2x + 3y \leq 12$ $\left(\frac{12}{5}, \frac{12}{5}\right): \frac{72}{5} = 14.4;$
  $3x + 2y \leq 12$ $(4, 0): 16;$
  $x + y \geq 2$ $(2, 0): 8$
   c. maximum: 16; at $(4, 0)$

12. Objective Function $z = 2x + 4y$ **b.** $\left(\frac{3}{2}, \frac{3}{2}\right): 9;$
Constraints $x \geq 0, y \geq 0$ $(0, 3): 12;$
  $x + 3y \geq 6$ $(0, 9): 36;$
  $x + y \geq 3$ $(9, 0): 18;$
  $x + y \leq 9$ $(6, 0): 12$
   c. maximum: 36; at $(0, 9)$

13. Objective Function $z = 10x + 12y$ b. $(0, 0): 0;$
     Constraints $x \geq 0, y \geq 0$ $(0, 6): 72;$
               $x + y \leq 7$ $(3, 4): 78;$
               $2x + y \leq 10$ $(5, 0): 50$
               $2x + 3y \leq 18$

   c. maximum: 78; at $(3, 4)$

14. Objective Function $z = 5x + 6y$ b. $\left(\frac{10}{3}, \frac{10}{3}\right): \frac{110}{3} = 36\frac{2}{3};$
     Constraints $x \geq 0, y \geq 0$ $(0, 10): 60;$
               $2x + y \geq 10$ $(10, 0): 50$
               $x + 2y \geq 10$
               $x + y \leq 10$

   c. maximum: 60; at $(0, 10)$

Application Exercises

15. A television manufacturer makes console and wide-screen televisions. The profit per unit is $125 for the console televisions and $200 for the wide-screen televisions.

   a. Let x = the number of consoles manufactured in a month and y = the number of wide-screens manufactured in a month. Write the objective function that describes the total monthly profit. $z = 125x + 200y$

   b. The manufacturer is bound by the following constraints:

     • Equipment in the factory allows for making at most 450 console televisions in one month.

     • Equipment in the factory allows for making at most 200 wide-screen televisions in one month.

     • The cost to the manufacturer per unit is $600 for the console televisions and $900 for the wide-screen televisions. Total monthly costs cannot exceed $360,000.

     Write a system of three inequalities that describes these constraints. $x \leq 450; y \leq 200; 600x + 900y \leq 360,000$

   c. Graph the system of inequalities in part (b). Use only the first quadrant and its boundary, because x and y must both be nonnegative.   See graphing answer section.

   d. Evaluate the objective function for total monthly profit at each of the five vertices of the graphed region. [The vertices should occur at $(0, 0)$, $(0, 200)$, $(300, 200)$, $(450, 100)$, and $(450, 0)$.]   0; 40,000; 77,500; 76,250; 56,250

   e. Complete the missing portions of this statement: The television manufacturer will make the greatest profit by manufacturing 300 console televisions each month and 200 wide-screen televisions each month. The maximum monthly profit is $77,500.

16. a. A student earns $10 per hour for tutoring and $7 per hour as a teacher's aid. Let x = the number of hours each week spent tutoring and y = the number of hours each week spent as a teacher's aid. Write the objective function that describes total weekly earnings. $z = 10x + 7y$

   b. The student is bound by the following constraints:

     • To have enough time for studies, the student can work no more than 20 hours a week.

     • The tutoring center requires that each tutor spend at least three hours a week tutoring.

     • The tutoring center requires that each tutor spend no more than eight hours a week tutoring.

     Write a system of three inequalities that describes these constraints. $x + y \leq 20; x \geq 3; x \leq 8$

   c. Graph the system of inequalities in part (b). Use only the first quadrant and its boundary, because x and y are nonnegative.   See graphing answer section.

   d. Evaluate the objective function for total weekly earnings at each of the four vertices of the graphed region. [The vertices should occur at $(3, 0)$, $(8, 0)$, $(3, 17)$, and $(8, 12)$.]   30; 80; 149; 164

   e. Complete the missing portions of this statement: The student can earn the maximum amount per week by tutoring for 8 hours per week and working as a teacher's aid for 12 hours per week. The maximum amount that the student can earn each week is $164.

Use the two steps for solving a linear programming problem, given in the box on page 284, to solve the problems in Exercises 17–23.

17. A manufacturer produces two models of mountain bicycles. The times (in hours) required for assembling and painting each model are given in the following table:

	Model A	Model B
Assembling	5	4
Painting	2	3

The maximum total weekly hours available in the assembly department and the paint department are 200 hours and 108 hours, respectively. The profits per unit are $25 for model A and $15 for model B. How many of each type should be produced to maximize profit?   40 of model A and 0 of model B

18. A large institution is preparing lunch menus containing foods A and B. The specifications for the two foods are given in the following table:

Food	Units of Fat per Ounce	Units of Carbohydrates per Ounce	Units of Protein per Ounce
A	1	2	1
B	1	1	1

Each lunch must provide at least 6 units of fat per serving, no more than 7 units of protein, and at least 10 units of carbohydrates. The institution can purchase food A for $0.12 per ounce and food B for $0.08 per ounce. How many ounces of each food should a serving contain to meet the dietary requirements at the least cost?   4 ounces of A and 2 ounces of B

19. Food and clothing are shipped to victims of a hurricane. Each carton of food will feed 12 people, while each carton of clothing will help 5 people. Each 20-cubic-foot box of food weighs 50 pounds and each 10-cubic-foot box of clothing weighs 20 pounds. The commercial carriers transporting food and clothing are bound by the following constraints:

- The total weight per carrier cannot exceed 19,000 pounds.
- The total volume must be less than 8000 cubic feet.

How many cartons of food and how many cartons of clothing should be sent with each plane shipment to maximize the number of people who can be helped?   300 boxes of food and 200 boxes of clothing

20. On June 24, 1948, the former Soviet Union blocked all land and water routes through East Germany to Berlin. A gigantic airlift was organized using American and British planes to bring food, clothing, and other supplies to the more than 2 million people in West Berlin. The cargo capacity was 30,000 cubic feet for an American plane and 20,000 cubic feet for a British plane. To break the Soviet blockade, the Western Allies had to maximize cargo capacity, but were subject to the following restrictions:

- No more than 44 planes could be used.
- The larger American planes required 16 personnel per flight, double that of the requirement for the British planes. The total number of personnel available could not exceed 512.
- The cost of an American flight was $9000 and the cost of a British flight was $5000. Total weekly costs could not exceed $300,000.

Find the number of American planes and the number of British planes that were used to maximize cargo capacity.   20 American planes and 24 British planes

21. A theater is presenting a program on drinking and driving for students and their parents. The proceeds will be donated to a local alcohol information center. Admission is $2.00 for parents and $1.00 for students. However, the situation has two constraints: The theater can hold no more than 150 people and every two parents must bring at least one student. How many parents and students should attend to raise the maximum amount of money?   100 parents and 50 students

22. You are about to take a test that contains computation problems worth 6 points each and word problems worth 10 points each. You can do a computation problem in 2 minutes and a word problem in 4 minutes. You have 40 minutes to take the test and may answer no more than 12 problems. Assuming you answer all the problems attempted correctly, how many of each type of problem must you do to maximize your score? What is the maximum score?   4 computation and 8 word problems; a score of 104

23. In 1978, a ruling by the Civil Aeronautics Board allowed Federal Express to purchase larger aircraft. Federal Express's options included 20 Boeing 727s that United Airlines was retiring and/or the French-built Dassault Fanjet Falcon 20. To aid in their decision, executives at Federal Express analyzed the data at the top of the next column.

	Boeing 727	**Falcon 20**
Direct Operating Cost	$1400 per hour	$500 per hour
Payload	42,000 pounds	6000 pounds

Federal Express was faced with the following constraints:
- Hourly operating cost was limited to $35,000.
- Total payload had to be at least 672,000 pounds.
- Only twenty 727s were available.

Given the constraints, how many of each kind of aircraft should Federal Express have purchased to maximize the number of aircraft?   10 Boeing 727s and 42 Falcon 20s

Writing in Mathematics

24. What kinds of problems are solved using the linear programming method?   24–28. Answers will vary.

25. What is an objective function in a linear programming problem?

26. What is a constraint in a linear programming problem? How is a constraint represented?

27. In your own words, describe how to solve a linear programming problem.

28. Describe a situation in your life in which you would like to maximize something, but are limited by at least two constraints. Can linear programming be used in this situation? Explain your answer.

Critical Thinking Exercises

29. Suppose that you inherit $10,000. The will states how you must invest the money. Some (or all) of the money must be invested in stocks and bonds. The requirements are that at least $3000 be invested in bonds, with expected returns of $0.08 per dollar, and at least $2000 be invested in stocks, with expected returns of $0.12 per dollar. Because the stocks are medium risk, the final stipulation requires that the investment in bonds should never be less than the investment in stocks. How should the money be invested so as to maximize your expected returns? $5000 in stocks and $5000 in bonds

30. Consider the objective function $z = Ax + By$ ($A > 0$ and $B > 0$) subject to the following constraints: $2x + 3y \le 9$, $x - y \le 2$, $x \ge 0$, and $y \ge 0$. Prove that the objective function will have the same maximum value at the vertices $(3, 1)$ and $(0, 3)$ if $A = \frac{2}{3}B$.   If $A = \frac{2}{3}B$, then $z = \frac{2}{3}Bx + By$. Thus, $z = \frac{2}{3}B(3) + B(1) = 3B$ at the vertex $(3, 1)$, and $z = \frac{2}{3}B(0) + B(3) = 3B$ at the vertex $(0, 3)$. At $(0, 0)$, $z = 0$, and at $(2, 0)$, $z = \frac{4}{3}B$.

Review Exercises

31. Simplify: $(2x^4y^3)(3xy^4)^3$.
(Section 1.6, Example 9) $54x^7y^{15}$

32. Solve for L: $3P = \dfrac{2L - W}{4}$.
(Section 1.5, Example 8) $L = \dfrac{12P + W}{2}$

33. If $f(x) = x^3 + 2x^2 - 5x + 4$, find $f(-1)$.
(Section 2.1, Example 3)   10

CHAPTER 4

Additional group projects can be found in the *Instructor's Resource Manual.*

Each group member should research one situation that provides two different pricing options. These can involve areas such as public transportation options (with or without coupon books) or long-distance telephone plans or anything of interest. Be sure to bring in all the details for each option. At the group meeting, select the two pricing situations that are most interesting and relevant. Using each situation, write a word problem about selecting the better of the two options. The word problem should be one that can be solved using a linear inequality. The group should turn in the two problems and their solutions.

CHAPTER 4 SUMMARY

Definitions and Concepts	Examples

Section 4.1 Solving Linear Inequalities

A linear inequality in one variable can be written in the form $ax + b < c, ax + b \leq c, ax + b > c$, or $ax + b \geq c$. The set of all numbers that make the inequality a true statement is its solution set. Graphs of solution sets are shown on a number line by shading all points representing numbers that are solutions. Parentheses indicate endpoints that are not solutions. Square brackets indicate endpoints that are solutions.

- $(-2, 1] = \{x \mid -2 < x \leq 1\}$

- $[-2, \infty) = \{x \mid x \geq -2\}$

Solving a Linear Inequality

1. Simplify each side.

2. Collect variable terms on one side and constant terms on the other side.

3. Isolate the variable and solve.

If an inequality is multiplied or divided by a negative number, the inequality symbol must be reversed.

Solve:

$$2(x + 3) - 5x \leq 15.$$
$$2x + 6 - 5x \leq 15$$
$$-3x + 6 \leq 15$$
$$-3x \leq 9$$
$$\frac{-3x}{-3} \geq \frac{9}{-3}$$
$$x \geq -3$$

Solution set: $\{x \mid x \geq -3\}$ or $[-3, \infty)$

Section 4.2 Compound Inequalities

Intersection (∩) and Union (∪)
$A \cap B$ is the set of elements common to both set A and set B.
$A \cup B$ is the set of elements that are members of set A or set B or of both sets.

$\{1, 3, 5, 7\} \cap \{5, 7, 9, 11\} = \{5, 7\}$
$\{1, 3, 5, 7\} \cup \{5, 7, 9, 11\} = \{1, 3, 5, 7, 9, 11\}$

A compound inequality is formed by joining two inequalities with the word *and* or *or*. When the connecting word is *and*, graph each inequality separately and take the intersection of their solution sets.

Solve: $x + 1 > 3$ and $x + 4 \leq 8$.
$x > 2$ and $x \leq 4$

Solution set: $\{x \mid 2 < x \leq 4\}$ or $(2, 4]$

Definitions and Concepts	Examples

Section 4.2 Compound Inequalities (continued)

The compound inequality $a < x < b$ means $a < x$ and $x < b$. Solve by isolating the variable in the middle.

Solve: $-1 < \dfrac{2x + 1}{3} \le 2$.

$-3 < 2x + 1 \le 6$ Multiply by 3.

$-4 < 2x \le 5$ Subtract 1.

$-2 < x \le \dfrac{5}{2}$ Divide by 2.

Solution set: $\left\{ x \mid -2 < x \le \dfrac{5}{2} \right\}$ or $\left(-2, \dfrac{5}{2} \right]$

$$\xleftarrow{\hspace{1cm}} \underset{-4\ -3\ -2\ -1\ \ 0\ \ 1\ \ 2\ \ 3\ \ 4}{\hspace{0.3cm}(\hspace{1cm}]\hspace{0.3cm}} \xrightarrow{\hspace{1cm}}$$

When the connecting word in a compound inequality is *or*, graph each inequality separately and take the union of their solution sets.

Solve: $x - 2 > -3$ or $2x \le -6$.

$\qquad\quad x > -1$ or $\quad x \le -3$

$$\xleftarrow{\hspace{1cm}} \underset{-6\ -5\ -4\ -3\ -2\ -1\ \ 0\ \ 1\ \ 2}{\hspace{0.3cm}]\hspace{1cm}(\hspace{0.3cm}} \xrightarrow{\hspace{1cm}}$$

Solution set: $\{ x \mid x \le -3 \text{ or } x > -1 \}$ or $(\infty, -3] \cup (-1, \infty)$

Section 4.3 Equations and Inequalities Involving Absolute Value

Absolute Value Equations

1. If $c > 0$, then $|X| = c$ means $X = c$ or $X = -c$.
2. If $c < 0$, then $|X| = c$ has no solution.
3. If $c = 0$, then $|X| = 0$ means $X = 0$.

Solve: $|2x - 7| = 3$.

$2x - 7 = 3$ or $2x - 7 = -3$

$\quad 2x = 10 \qquad\qquad 2x = 4$

$\qquad x = 5 \qquad\qquad\quad x = 2$

The solution set is $\{2, 5\}$.

Absolute Value Equations with Two Absolute Value Bars

If $|X_1| = |X_2|$, then $X_1 = X_2$ or $X_1 = -X_2$.

Solve: $|x - 6| = |2x + 1|$.

$x - 6 = 2x + 1$ or $x - 6 = -(2x + 1)$

$-x - 6 = 1 \qquad\qquad x - 6 = -2x - 1$

$\quad -x = 7 \qquad\qquad\ 3x - 6 = -1$

$\qquad x = -7 \qquad\qquad\quad 3x = 5$

$\qquad\qquad\qquad\qquad\qquad\ x = \dfrac{5}{3}$

The solutions are -7 and $\frac{5}{3}$, and the solution set is $\left\{ -7, \frac{5}{3} \right\}$.

If c is a positive number, then to solve $|X| < c$, solve the compound inequality $-c < X < c$. If c is negative, $|X| < c$ has no solution.

Solve: $|x - 4| < 3$.

$-3 < x - 4 < 3$

$1 < x < 7$

The solution set is $\{ x \mid 1 < x < 7 \}$ or $(1, 7)$.

$$\xleftarrow{\hspace{1cm}} \underset{0\ \ 1\ \ 2\ \ 3\ \ 4\ \ 5\ \ 6\ \ 7\ \ 8}{\hspace{0.3cm}(\hspace{1.5cm})\hspace{0.3cm}} \xrightarrow{\hspace{1cm}}$$

Definitions and Concepts	Examples

Section 4.3 Equations and Inequalities Involving Absolute Value (continued)

If c is a positive number, then to solve $|X| > c$, solve the compound inequality $X < -c$ or $X > c$. If c is negative, $|X| > c$ is true for all real numbers for which X is defined.

Solve: $\left|\dfrac{x}{3} - 1\right| \geq 2$.

$$\dfrac{x}{3} - 1 \leq -2 \quad \text{or} \quad \dfrac{x}{3} - 1 \geq 2.$$

$$x - 3 \leq -6 \quad \text{or} \quad x - 3 \geq 6 \qquad \text{Multiply by 3.}$$

$$x \leq -3 \quad \text{or} \qquad\quad x \geq 9 \qquad \text{Add 3.}$$

The solution set is $\{x \mid x \leq -3 \text{ or } x \geq 9\}$ or $(-\infty, -3] \cup [9, \infty)$.

$$\xleftarrow{\quad} \begin{array}{ccccccccc} &] & & & & & & & [\\ \hline -6 & -4 & -2 & 0 & 2 & 4 & 6 & 8 & 10 \end{array} \xrightarrow{\quad}$$

Section 4.4 Linear Inequalities in Two Variables

If the equal sign in $Ax + By = C$ is replaced with an inequality symbol, the result is a linear inequality in two variables. Its graph is the set of all points whose coordinates satisfy the inequality. To obtain the graph,

1. Replace the inequality symbol with an equal sign and graph the boundary line. Use a solid line for $\leq$ or $\geq$ and a dashed line for $<$ or $>$.

2. Choose a test point not on the line and substitute its coordinates into the inequality.

3. If a true statement results, shade the half-plane containing the test point. If a false statement results, shade the half-plane not containing the test point.

Graph: $x - 2y \leq 4$.

1. Graph $x - 2y = 4$. Use a solid line because the inequality symbol is $\leq$.

2. Test $(0, 0)$.

$$x - 2y \leq 4$$

$$0 - 2 \cdot 0 \overset{?}{\leq} 4$$

$$0 \leq 4, \qquad \text{true}$$

3. The inequality is true. Shade the half-plane containing $(0, 0)$.

Two or more linear inequalities make up a system of linear inequalities. A solution is an ordered pair satisfying all inequalities in the system. To graph a system of inequalities, graph each inequality in the system. The overlapping region represents the solutions of the system.

Graph the solutions of the system:

$$y \leq -2x$$

$$x - y \geq 3.$$

Definitions and Concepts	Examples

Section 4.5 Linear Programming

Linear programming is a method for solving problems in which a particular quantity that must be maximized or minimized is limited. An objective function is an algebraic expression in three variables describing a quantity that must be maximized or minimized. Constraints are restrictions, expressed as linear inequalities.

Solving a Linear Programming Problem

1. Graph the system of inequalities representing the constraints.

2. Find the value of the objective function at each corner, or vertex, of the graphed region. The maximum and minimum of the objective function occur at one or more vertices.

Find the maximum value of the objective function $z = 3x + 2y$ subject to the following constraints: $x \geq 0, y \geq 0, 2x + 3y \leq 18, 2x + y \leq 10$.

1. Graph the system of inequalities representing the constraints.

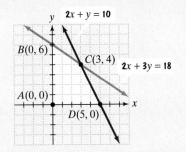

2. Evaluate the objective function at each vertex.

Vertex	$z = 3x + 2y$
$A(0,0)$	$z = 3(0) + 2(0) = 0$
$B(0,6)$	$z = 3(0) + 2(6) = 12$
$C(3,4)$	$z = 3(3) + 2(4) = 17$
$D(5,0)$	$z = 3(5) + 2(0) = 15$

The maximum value of the objective function is 17.

CHAPTER 4   REVIEW EXERCISES

4.1 *In Exercises 1–3, graph the solutions of each inequality on a number line.*   **1–3.** See graphing answer section.

1. $x > 5$ 2. $x \leq 1$ 3. $-3 \leq x < 0$

In Exercises 4–6, express each interval in set-builder notation and graph the interval on a number line.   **4–6.** See graphing answer section.

4. $(-2, 3]$ $\{x | -2 < x \leq 3\}$

5. $[-1.5, 2]$ $\{x | -1.5 \leq x \leq 2\}$

6. $[-1, \infty)$ $\{x | x > -1\}$

In Exercises 7–12, solve each linear inequality. Other than Ø, graph the solution set on a number line. Express the solution set in both set-builder and interval notations.   **7–11.** See graphing answer section.

7. $-6x + 3 \leq 15$ $\{x | x \geq -2\}; [-2, \infty)$

8. $6x - 9 \geq -4x - 3$ $\{x | x \geq \frac{3}{5}\}; [\frac{3}{5}, \infty)$

9. $\frac{x}{3} - \frac{3}{4} - 1 > \frac{x}{2}$ $\{x | x < -\frac{21}{2}\}; (-\infty, -\frac{21}{2})$

10. $6x + 5 > -2(x - 3) - 25$ $\{x | x > -3\}; (-3, \infty)$

11. $3(2x - 1) - 2(x - 4) \geq 7 + 2(3 + 4x)$
$\{x | x \leq -2\}; (-\infty, -2]$

12. $2x + 7 \leq 5x - 6 - 3x$ Ø

13. A person can choose between two charges on a checking account. The first method involves a fixed cost of $11 per month plus 6¢ for each check written. The second method involves a fixed cost of $4 per month plus 20¢ for each check written. How many checks should be written to make the first method a better deal?   more than 50 checks

14. A salesperson earns $500 per month plus a commission of 20% of sales. Describe the sales needed to receive a total income that exceeds $3200 per month.   more than $13,500 in sales

4.2 *In Exercises 15–18, let $A = \{a, b, c\}, B = \{a, c, d, e\}$ and $C = \{a, d, f, g\}$. Find the indicated set.*

15. $A \cap B$ $\{a, c\}$ 16. $A \cap C$ $\{a\}$

17. $A \cup B$ $\{a, b, c, d, e\}$ 18. $A \cup C$ $\{a, b, c, d, f, g\}$

In Exercises 19–29, solve each compound inequality. Except for the empty set, express the solution set in both set-builder and interval notations. Graph the solution set on a number line.
19–29. See graphing answer section.

19. $x \leq 3$ and $x < 6$ $\{x | x \leq 3\}; (-\infty, 3]$

20. $x \leq 3$ or $x < 6$ $\{x | x < 6\}; (-\infty, 6)$

21. $-2x < -12$ and $x - 3 < 5$ $\{x | 6 < x < 8\}; (6, 8)$

22. $5x + 3 \leq 18$ and $2x - 7 \leq -5$ $\{x | x \leq 1\}; (-\infty, 1]$

23. $2x - 5 > -1$ and $3x < 3$ $\varnothing$

24. $2x - 5 > -1$ or $3x < 3$
$\{x | x < 1 \text{ or } x > 2\}; (-\infty, 1) \cup (2, \infty)$

25. $x + 1 \leq -3$ or $-4x + 3 < -5$
$\{x | x \leq -4 \text{ or } x > 2\}; (-\infty, -4] \cup (2, \infty)$

26. $5x - 2 \leq -22$ or $-3x - 2 > 4$ $\{x | x < -2\}; (-\infty, -2)$

27. $5x + 4 \geq -11$ or $1 - 4x \geq 9$
$\{x | x \text{ is a real number}\}; (-\infty, \infty)$

28. $-3 < x + 2 \leq 4$ $\{x | -5 < x \leq 2\}; (-5, 2]$

29. $-1 \leq 4x + 2 \leq 6$ $\{x | -\frac{3}{4} \leq x \leq 1\}; \left[-\frac{3}{4}, 1\right]$

30. On the first of four exams, your grades are $72, 73, 94$, and 80. There is still one more exam, and you are hoping to earn a B in the course. This will occur if the average of your five exam grades is greater than or equal to 80 and less than 90. What range of grades on the fifth exam will result in receiving a B? Use interval notation to express this range.   [81, 131); If the highest grade is 100, then [81, 100].

4.3 *In Exercises 31–34, find the solution set for each equation.*

31. $|2x + 1| = 7$ $\{-4, 3\}$

32. $|3x + 2| = -5$ $\varnothing$

33. $2|x - 3| - 7 = 10$ $\left\{-\frac{11}{2}, \frac{23}{2}\right\}$

34. $|4x - 3| = |7x + 9|$ $\left\{-4, -\frac{6}{11}\right\}$

In Exercises 35–38, solve and graph the solution set on a number line. Except for the empty set, express the solution set in both set-builder and interval notations.   35–37. See graphing answer section.

35. $|2x + 3| \leq 15$ $\{x | -9 \leq x \leq 6\}; [-9, 6]$

36. $\left|\dfrac{2x + 6}{3}\right| > 2$ $\{x | x < -6 \text{ or } x > 0\}; (-\infty, -6) \cup (0, \infty)$

37. $|2x + 5| - 7 < -6$ $\{x | -3 < x < -2\}; (-3, -2)$

38. $|2x - 3| + 4 \leq -10$ $\varnothing$

39. Approximately 90% of the population sleeps h hours daily, where h is modeled by the inequality $|h - 6.5| \leq 1$. Write a sentence describing the range for the number of hours that most people sleep. Do *not* use the phrase "absolute value" in your description.   Approximately 90% of the population sleeps between 5.5 hours and 7.5 hours daily, inclusive.

4.4 *In Exercises 40–45, graph each inequality in a rectangular coordinate system.*   40–45. See graphing answer section.

40. $3x - 4y > 12$

41. $x - 3y \leq 6$

42. $y \leq -\dfrac{1}{2}x + 2$

43. $y > \dfrac{3}{5}x$

44. $x \leq 2$

45. $y > -3$

In Exercises 46–54, graph the solution set of each system of inequalities or indicate that the system has no solution.
46–53. See graphing answer section.

46. $2x - y \leq 4$
$x + y \geq 5$

47. $y < -x + 4$
$y > x - 4$

48. $-3 \leq x < 5$

49. $-2 < y \leq 6$

50. $x \geq 3$
$y \leq 0$

51. $2x - y > -4$
$x \geq 0$

52. $x + y \leq 6$
$y \geq 2x - 3$

53. $3x + 2y \geq 4$
$x - y \leq 3$
$x \geq 0, y \geq 0$

54. $2x - y > 2$
$2x - y < -2$ $\varnothing$

4.5

55. Find the value of the objective function $z = 2x + 3y$ at each corner of the graphed region shown. What is the maximum value of the objective function? What is the minimum value of the objective function? $\left(\frac{1}{2}, \frac{1}{2}\right): \frac{5}{2}; (2, 2): 10; (4, 0): 8;$ $(1, 0): 2;$ maximum: 10; minimum: 2

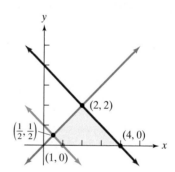

In Exercises 56–58, graph the region determined by the constraints. Then find the maximum value of the given objective function, subject to the constraints.   56–58. See graphing answer section.

56. Objective Function $z = 2x + 3y$
   Constraints $x \geq 0, y \geq 0$
   $x + y \leq 8$
   maximum: 24 $3x + 2y \geq 6$

57. Objective Function $z = x + 4y$
   Constraints $0 \leq x \leq 5, 0 \leq y \leq 7$
   maximum: 33 $x + y \geq 3$

58. Objective Function $z = 5x + 6y$
   Constraints $x \geq 0, y \geq 0$
   $y \leq x$
   $2x + y \leq 12$
   maximum: 44 $2x + 3y \geq 6$

59. A paper manufacturing company converts wood pulp to writing paper and newsprint. The profit on a unit of writing paper is $500 and the profit on a unit of newsprint is $350.

a. Let x represent the number of units of writing paper produced daily. Let y represent the number of units of newsprint produced daily. Write the objective function that models total daily profit. $z = 500x + 350y$

b. The manufacturer is bound by the following constraints:

- Equipment in the factory allows for making at most 200 units of paper (writing paper and newsprint) in a day.

- Regular customers require at least 10 units of writing paper and at least 80 units of newsprint daily.

Write a system of inequalities that models these constraints. $x + y \le 200; x \ge 10; y \ge 80$

c. Graph the inequalities in part (b). Use only the first quadrant, because x and y must both be positive. (*Suggestion*: Let each unit along the x- and y-axes represent 20.)   See graphing answer section.

d. Evaluate the objective profit function at each of the three vertices of the graphed region. $(10, 80)$: 33,000; $(10, 190)$: 71,500; $(120, 80)$: 88,000

e. Complete the missing portions of this statement: The company will make the greatest profit by producing <u>120</u> units of writing paper and <u>80</u> units of newsprint each day. The maximum daily profit is $<u>88,000</u>.

60. A manufacturer of lightweight tents makes two models whose specifications are given in the following table.

	Cutting Time per Tent	Assembly Time per Tent
Model A	0.9 hour	0.8 hour
Model B	1.8 hours	1.2 hours

Each month, the manufacturer has no more than 864 hours of labor available in the cutting department and at most 672 hours in the assembly division. The profits come to $25 per tent for model A and $40 per tent for model B. How many of each should be manufactured monthly to maximize the profit?   480 of model A and 240 of model B

CHAPTER 4   TEST Remember to use your Chapter Test Prep Video CD to see the worked-out solutions to the test questions you want to review.

In Exercises 1–2, express each interval in set-builder notation and graph the interval on a number line.   **1–2.** See graphing answer section.

1. $[-3, 2)$ $\{x \mid -3 \le x < 2\}$

2. $(-\infty, -1]$ $\{x \mid x \le -1\}$

In Exercises 3–4, solve and graph the solution set on a number line. Express the solution set in both set-builder and interval notations.   **3–4.** See graphing answer section.

3. $3(x + 4) \ge 5x - 12$ $\{x \mid x \le 12\}; (-\infty, 12]$

4. $\dfrac{x}{6} + \dfrac{1}{8} \le \dfrac{x}{2} - \dfrac{3}{4}$ $\left\{x \mid x \ge \dfrac{21}{8}\right\}; \left[\dfrac{21}{8}, \infty\right)$

5. You are choosing between two telephone plans for local calls. Plan A charges $25 per month for unlimited calls. Plan B has a monthly fee of $13 with a charge of $0.06 per local call. How many local telephone calls in a month make plan A the better deal?   more than 200 calls

6. Find the intersection: $\{2, 4, 6, 8, 10\} \cap \{4, 6, 12, 14\}$. $\{4, 6\}$

7. Find the union: $\{2, 4, 6, 8, 10\} \cup \{4, 6, 12, 14\}$. $\{2, 4, 6, 8, 10, 12, 14\}$

In Exercises 8–12, solve each compound inequality. Except for the empty set, express the solution set in both set-builder and interval notations. Graph the solution set on a number line.   **8–12.** See graphing answer section.

8. $2x + 4 < 2$ and $x - 3 > -5$ $\{x \mid -2 < x < -1\}; (-2, -1)$

9. $x + 6 \ge 4$ and $2x + 3 \ge -2$ $\{x \mid x \ge -2\}; [-2, \infty)$

10. $2x - 3 < 5$ or $3x - 6 \le 4$ $\{x \mid x < 4\}; (-\infty, 4)$

11. $x + 3 \le -1$ or $-4x + 3 < -5$ $\{x \mid x \le -4 \text{ or } x > 2\}; (-\infty, -4] \cup (2, \infty)$

12. $-3 \le \dfrac{2x + 5}{3} < 6$ $\left\{x \mid -7 \le x < \dfrac{13}{2}\right\}; \left[-7, \dfrac{13}{2}\right)$

In Exercises 13–14, find the solution set for each equation.

13. $|5x + 3| = 7$ $\left\{-2, \dfrac{4}{5}\right\}$

14. $|6x + 1| = |4x + 15|$ $\left\{-\dfrac{8}{5}, 7\right\}$

In Exercises 15–16, solve and graph the solution set on a number line. Express the solution set in both set-builder and interval notations.   **15–16.** See graphing answer section.

15. $|2x - 1| < 7$ $\{x \mid -3 < x < 4\}; (-3, 4)$

16. $|2x - 3| \ge 5$ $\{x \mid x \le -1 \text{ or } x \ge 4\}; (-\infty, -1] \cup [4, \infty)$

17. The inequality $|b - 98.6| > 8$ describes a person's body temperature, b, in degrees Fahrenheit, when hyperthermia (extremely high body temperature) or hypothermia (extremely low body temperature) occurs. Solve the inequality and interpret the solution. $\{b \mid b < 90.6 \text{ or } b > 106.6\}$ or $(-\infty, 90.6) \cup (106.6, \infty)$; Hypothermia: Body temperature below 90.6°F; Hyperthermia: Body temperature above 106.6°F.

In Exercises 18–20, graph each inequality in a rectangular coordinate system.   **18–20.** See graphing answer section.

18. $3x - 2y < 6$ **19.** $y \ge \dfrac{1}{2}x - 1$ **20.** $y \le -1$

In Exercises 21–23, graph the solution set of each system of inequalities. **21–23.** See graphing answer section.

21. $x + y \geq 2$
$\quad\;\; x - y \geq 4$

22. $3x + y \leq 9$
$\quad\;\; 2x + 3y \geq 6$
$\quad\;\; x \geq 0, y \geq 0$

23. $-2 < x \leq 4$

24. Find the maximum value of the objective function $z = 3x + 5y$ subject to the following constraints: $x \geq 0, y \geq 0, x + y \leq 6, x \geq 2.$ maximum: 26

Note to Instructor: The Chapter 4 Test file in the TestGen program provides algorithms specifically matched to these problems so you can easily replicate this chapter test for additional practice or assessment purposes.

25. A manufacturer makes two types of jet skis, regular and deluxe. The profit on a regular jet ski is $200 and the profit on the deluxe model is $250. To meet customer demand, the company must manufacture at least 50 regular jet skis per week and at least 75 deluxe models. To maintain high quality, the total number of both models of jet skis manufactured by the company should not exceed 150 per week. How many jet skis of each type should be manufactured per week to obtain maximum profit? What is the maximum weekly profit? 50 regular and 100 deluxe; $35,000

CUMULATIVE REVIEW EXERCISES (CHAPTERS 1–4)

In Exercises 1–2, solve each equation.

1. $5(x + 1) + 2 = x - 3(2x + 1)$ $\quad -1$

2. $\dfrac{2(x + 6)}{3} = 1 + \dfrac{4x - 7}{3}$ $\quad 8$

3. Simplify: $\dfrac{-10x^2y^4}{15x^7y^{-3}} \cdot \dfrac{2y^7}{3x^5}$ $\quad -\dfrac{4}{9}$

4. If $f(x) = x^2 - 3x + 4$, find $f(-3)$ and $f(2a)$.
22; $\quad 4a^2 - 6a + 4$

5. If $f(x) = 3x^2 - 4x + 1$ and $g(x) = x^2 - 5x - 1$, find $(f - g)(x)$ and $(f - g)(2).$ $\quad 2x^2 + x + 2; 12$

6. Use function notation to write the equation of the line passing through $(2, 3)$ and perpendicular to the line whose equation is $y = 2x - 3.$ $\quad f(x) = -\frac{1}{2}x + 4$

In Exercises 7–10, graph each equation or inequality in a rectangular coordinate system. **7–10.** See graphing answer section.

7. $f(x) = 2x + 1$

8. $y > 2x$

9. $2x - y \geq 6$

10. $f(x) = -1$

11. Solve the system:
$$3x - y + z = -15$$
$$x + 2y - z = 1.$$
$$2x + 3y - 2z = 0 \quad (-4, 2, -1)$$

12. Solve using matrices:
$$2x - y = -4$$
$$x + 3y = 5. \quad (-1, 2)$$

13. Evaluate: $\begin{vmatrix} 4 & 3 \\ -1 & -5 \end{vmatrix}.$ $\quad -17$

14. A motel with 60 rooms charges $90 per night for rooms with kitchen facilities and $80 per night for rooms without kitchen facilities. When all rooms are occupied, the nightly revenue is $5260. How many rooms of each kind are there? 46 rooms with kitchen facilities and 14 without kitchen facilities

15. Which of the following are functions?

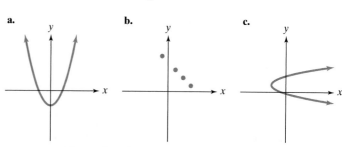

a. **b.** **c.**

a. and **b.** are functions.

In Exercises 16–20, solve and graph the solution set on a number line. Express the solution set in both set-builder and interval notations. **16–20.** See graphing answer section.

16. $\dfrac{x}{4} - \dfrac{3}{4} - 1 \leq \dfrac{x}{2}$ $\quad \{x | x \geq -7\}; [-7, \infty)$

17. $2x + 5 \leq 11$ and $-3x > 18$ $\quad \{x | x < -6\}; (-\infty, -6)$

18. $x - 4 \geq 1$ or $-3x + 1 \geq -5 - x$
$\{x | x \leq 3 \text{ or } x \geq 5\}; (-\infty, 3] \cup [5, \infty)$

19. $|2x + 3| \leq 17$ $\quad \{x | -10 \leq x \leq 7\}; [-10, 7]$

20. $|3x - 8| > 7$ $\quad \{x | x < \frac{1}{3} \text{ or } x > 5\}; \left(\infty, \frac{1}{3}\right) \cup (5, \infty)$

Coffee, our culture's most thoroughly approved drug, has become relevant and contemporary. Coffee bars, providing everything from personal computers so that customers can surf the Net to match-making services, have made coffee an important part of social gathering places. There is a wide variety of coffee offerings— from size, flavor, roast, preparation, and toppings— and plenty of gourmet shops to provide them.

Irritability for coffee drinkers who do not receive the beverage at expected times is explored algebraically and graphically in the enrichment essay on page 368.

Polynomials, Polynomial Functions, and Factoring

5.1 Introduction to Polynomials and Polynomial Functions

5.2 Multiplication of Polynomials

5.3 Greatest Common Factors and Factoring by Grouping

5.4 Factoring Trinomials

5.5 Factoring Special Forms

5.6 A General Factoring Strategy

5.7 Polynomial Equations and Their Applications

Many of the 110 million daily coffee drinkers in the United States are surprised to learn that caffeine is a classically addictive drug. Many regular users cannot think clearly in the morning until they have had their first cup. Coffee is most notorious for causing insomnia, irritability, and nervousness. There is a function that models the percentage of coffee drinkers, $P(x)$, who are x years old who become irritable if they do not have coffee at their regular time:

$$P(x) = -0.05x^2 + 4.2x - 26.$$

The algebraic expression on the right side contains variables to powers that are whole numbers and is an example of a polynomial. Much of what we do in algebra involves polynomials and polynomial functions. In this chapter, we study polynomials, polynomial functions, and their applications.

INTRODUCTION TO POLYNOMIALS AND POLYNOMIAL FUNCTIONS

Objectives

1 Use the vocabulary of polynomials.

2 Evaluate polynomial functions.

3 Determine end behavior.

4 Add polynomials.

5 Subtract polynomials.

Magnified 6000 times, this color-scanned image shows a T-lymphocyte blood cell (green) infected with the HIV virus (red). Depletion of the number of T-cells causes destruction of the immune system.

In 1980, U.S. doctors diagnosed 41 cases of a rare form of cancer, Kaposi's sarcoma, that involved skin lesions, pneumonia, and severe immunological deficiencies. All cases involved gay men ranging in age from 26 to 51. By the end of 2001, approximately 820,000 Americans, straight and gay, male and female, old and young, were infected with the HIV virus.

Modeling AIDS-related data and making predictions about the epidemic's havoc is serious business. Figure 5.1 shows the number of AIDS cases diagnosed in the United States from 1983 through 2001.

AIDS Cases Diagnosed in the U.S., 1983–2001

FIGURE 5.1 AIDS Cases Diagnosed in the U.S., 1983–2001

Source: Department of Health and Human Services

Changing circumstances and unforeseen events can result in models for AIDS-related data that are not particularly useful over long periods of time. For example, the function

$$f(x) = -49x^3 + 806x^2 + 3776x + 2503$$

$f(x) = -49x^3 + 806x^2 + 3776x + 2503$

Cases Diagnosed

60,000

5000

0 1 2 3 4 5 6 7 8

Years after 1983

[0, 8, 1] by [0, 60,000, 5000]

FIGURE 5.2 The graph of a function modeling the number of AIDS cases from 1983 through 1991

1 Use the vocabulary of polynomials.

STUDY TIP

We can express 0 in many ways, including $0x$, $0x^2$, and $0x^3$. It is impossible to assign a single exponent on the variable. This is why 0 has no defined degree.

models the number of AIDS cases diagnosed in the United States x years after 1983. The model was obtained using a portion of the data shown in Figure 5.1, namely cases diagnosed from 1983 through 1991, inclusive. Figure 5.2 shows the graph of f from 1983 through 1991.

The voice balloon in Figure 5.2 displays the algebraic expression used to define f. This algebraic expression is an example of a *polynomial*. A **polynomial** is a single term or the sum of two or more terms containing variables in the numerator with whole-number exponents. Functions containing polynomials are used in such diverse areas as science, business, medicine, psychology, and sociology. In this section, we present basic ideas about polynomials and polynomial functions. We will use our knowledge of combining like terms to find sums and differences of polynomials.

How We Describe Polynomials   Consider the polynomial

$$-49x^3 + 806x^2 + 3776x + 2503.$$

This polynomial contains four terms. It is customary to write the terms in the order of descending powers of the variable. This is the **standard form** of a polynomial.

Some polynomials contain only one variable. Each term of such a polynomial in x is of the form ax^n. If $a \neq 0$, the **degree** of ax^n is n. For example, the degree of the term $-49x^3$ is 3.

THE DEGREE OF ax^n   If $a \neq 0$, the degree of ax^n is n. The degree of a nonzero constant is 0. The constant 0 has no defined degree.

Here is an example of a polynomial and the degree of each of its four terms:

$$-49x^3 + 806x^2 + 3776x + 2503$$

degree 3   degree 2   degree 1   degree of nonzero constant: 0

Notice that the exponent on x for the term $3776x$ is understood to be 1: $3776x^1$. For this reason, the degree of $3776x$ is 1. You can think of 2503 as $2503x^0$; thus, its degree is 0.

A polynomial with exactly one term is called a **monomial**. A **binomial** is a polynomial that has two terms, each with a different exponent. A **trinomial** is a polynomial with three terms, each with a different exponent. Polynomials with four or more terms have no special names.

Some polynomials contain two or more variables. Here is an example of a polynomial in two variables, x and y:

$$7x^2y^3 - 17x^4y^2 + xy - 6y^2 + 9.$$

A polynomial in two variables, x and y, contains the sum of one or more monomials of the form ax^ny^m. The constant a is the **coefficient**. The exponents, n and m, represent whole numbers. The **degree of the term** ax^ny^m is the sum of the exponents of the variables, $n + m$.

The **degree of a polynomial** is the greatest of the degrees of its terms. If there is precisely one term of the greatest degree, it is called the **leading term**. Its coefficient is called the **leading coefficient**.

EXAMPLE 1 Using the Vocabulary of Polynomials

Determine the coefficient of each term, the degree of each term, the degree of the polynomial, the leading term, and the leading coefficient of the polynomial

$$7x^2y^3 - 17x^4y^2 + xy - 6y^2 + 9.$$

SOLUTION

Term	Coefficient	Degree (Sum of Exponents on the Variables)
$7x^2y^3$	7	$2 + 3 = 5$
$-17x^4y^2$	-17	$4 + 2 = 6$
xy	1	$1 + 1 = 2$
$-6y^2$	-6	$0 + 2 = 2$
9	9	$0 + 0 = 0$

Think of xy as $1x^1y^1$.
Think of $-6y^2$ as $-6x^0y^2$.
Think of 9 as $9x^0y^0$.

The degree of the polynomial is the greatest degree of all its terms, which is 6. The leading term is the term of the greatest degree, which is $-17x^4y^2$. Its coefficient, -17, is the leading coefficient.

✔ **CHECK POINT 1** Determine the coefficient of each term, the degree of each term, the degree of the polynomial, the leading term, and the leading coefficient of the polynomial

$$8x^4y^5 - 7x^3y^2 - x^2y - 5x + 11.$$

Term	Coefficient	Degree
$8x^4y^5$	8	9
$-7x^3y^2$	-7	5
$-x^2y$	-1	3
$-5x$	-5	1
11	11	0

The degree of the polynomial is 9, the leading term is $8x^4y^5$, and the leading coefficient is 8.

If a polynomial contains three or more variables, the degree of a term is the sum of the exponents of all the variables. Here is an example of a polynomial in three variables, x, y, and z:

The coefficients are $\frac{1}{4}$, -2, 6 and 5.

$$\frac{1}{4}xy^2z^4 - 2xyz + 6x^2 + 5.$$

Degree: $1 + 2 + 4 = 7$ Degree: $1 + 1 + 1 = 3$ Degree: 2   Degree: 0

The degree of this polynomial is the greatest degree of all its terms, which is 7.

2 Evaluate polynomial functions.

Polynomial Functions The expression $4x^3 - 5x^2 + 3$ is a polynomial. If we write

$$f(x) = 4x^3 - 5x^2 + 3$$

then we have a **polynomial function**. In a polynomial function, the expression that defines the function is a polynomial. How do we evaluate a polynomial function? Use substitution, just as we did to evaluate other functions in Chapter 2.

EXAMPLE 2 Evaluating a Polynomial Function

The polynomial function

$$f(x) = -49x^3 + 806x^2 + 3776x + 2503$$

models the number of AIDS cases diagnosed in the United States, $f(x)$, x years after 1983, where $0 \leq x \leq 8$. Find $f(6)$ and describe what this means in practical terms.

SOLUTION To find $f(6)$, or f of 6, we replace each occurrence of x in the function's formula with 6.

$$f(x) = -49x^3 + 806x^2 + 3776x + 2503 \qquad \text{This is the given function.}$$

$$f(6) = -49(6)^3 + 806(6)^2 + 3776(6) + 2503 \qquad \text{Replace each occurrence of } x \text{ with 6.}$$

$$= -49(216) + 806(36) + 3776(6) + 2503 \qquad \text{Evaluate exponential expressions.}$$

$$= -10{,}584 + 29{,}016 + 22{,}656 + 2503 \qquad \text{Multiply.}$$

$$= 43{,}591 \qquad \text{Add.}$$

Thus, $f(6) = 43{,}591$. According to the model, this means that 6 years after 1983, in 1989, there were 43,591 AIDS cases diagnosed in the United States. (The actual number, shown in Figure 5.1, is 43,499.) ∎

USING TECHNOLOGY

Once each occurrence of x in $f(x) = -49x^3 + 806x^2 + 3776x + 2503$ is replaced with 6, the resulting computation can be performed using a scientific calculator or a graphing calculator.

$$-49(6)^3 + 806(6)^2 + 3776(6) + 2503$$

Many Scientific Calculators

49 [+/−] [×] 6 [y^x] 3 [+] 806 [×] 6 [y^x] 2 [+] 3776 [×] 6 [+] 2503 [=]

Many Graphing Calculators

[(−)] 49 [×] 6 [^] 3 [+] 806 [×] 6 [^] 2 [+] 3776 [×] 6 [+] 2503 [ENTER]

The display should be 43591. This number can also be obtained by using a graphing utility's feature that evaluates a function or by using its table feature.

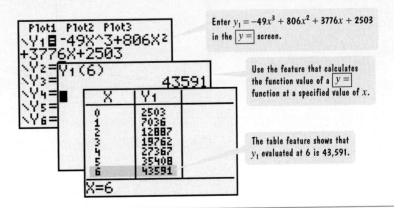

Enter $y_1 = -49x^3 + 806x^2 + 3776x + 2503$ in the [y =] screen.

Use the feature that calculates the function value of a [y =] function at a specified value of x.

The table feature shows that y_1 evaluated at 6 is 43,591.

 CHECK POINT **2** For the polynomial function

$$f(x) = 4x^3 - 3x^2 - 5x + 6,$$

find $f(2)$. $f(2) = 16$

Smooth, Continuous Graphs Polynomial functions of degree 2 or higher have graphs that are *smooth* and *continuous*. By **smooth**, we mean that the graph contains only rounded curves with no sharp corners. By **continuous**, we mean that the graph has no breaks and can be drawn without lifting your pencil from the rectangular coordinate system. These ideas are illustrated in Figure 5.3.

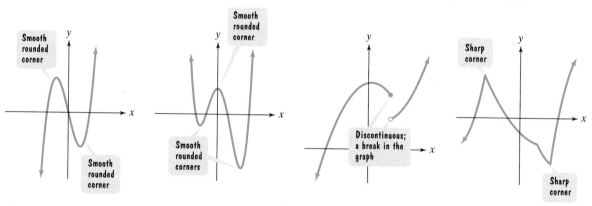

FIGURE 5.3 Recognizing graphs of polynomial functions

---

3 Determine end behavior.

End Behavior of Polynomial Functions Figure 5.4 shows the graph of the function

$$f(x) = -49x^3 + 806x^2 + 3776x + 2503,$$

which models U.S. AIDS cases from 1983 through 1991. Look what happens to the graph when we extend the year up through 2005. By year 21 (2004), the values of y are negative and the function no longer models AIDS cases. We've added an arrow to the graph at the far right to emphasize that it continues to decrease without bound. It is this far-right *end behavior* of the graph that makes it inappropriate for modeling AIDS cases into the future.

Years after 1983

[0, 22, 1] by [−10,000, 85,000, 5000]

FIGURE 5.4 By extending the viewing rectangle, we see that y is eventually negative and the function no longer models the number of AIDS cases. Model breakdown occurs by 2004.

The behavior of a graph of a function to the far left or the far right is called its **end behavior**. Although the graph of a polynomial function may have intervals where it increases or decreases, the graph will eventually rise or fall without bound as it moves far to the left or far to the right.

How can you determine whether the graph of a polynomial function goes up or down at each end? **The end behavior depends upon the leading term**. In particular, the sign of the leading coefficient and the degree of the polynomial reveal the graph's end behavior. With regard to end behavior, only the leading term—that is, the term of the greatest degree—counts, as summarized by the **Leading Coefficient Test**.

THE LEADING COEFFICIENT TEST As x increases or decreases without bound, the graph of a polynomial function eventually rises or falls. In particular,

1. For odd-degree polynomials:

If the leading coefficient is positive, the graph falls to the left and rises to the right.	If the leading coefficient is negative, the graph rises to the left and falls to the right.

2. For even-degree polynomials:

If the leading coefficient is positive, the graph rises to the left and to the right.	If the leading coefficient is negative, the graph falls to the left and to the right.

FIGURE 5.5(a)

FIGURE 5.5(b)

FIGURE 5.5(c)

FIGURE 5.5(d)

STUDY TIP

Odd-degree polynomial functions have graphs with opposite behavior at each end. Even-degree polynomial functions have graphs with the same behavior at each end.

EXAMPLE 3 Using the Leading Coefficient Test

Use the Leading Coefficient Test to determine the end behavior of the graph of

$$f(x) = x^3 + 3x^2 - x - 3.$$

SOLUTION We begin by identifying the sign of the leading coefficient and the degree of the polynomial.

$$f(x) = x^3 + 3x^2 - x - 3$$

The leading coefficient, 1, is positive.   The degree of the polynomial, 3, is odd.

The degree of the function f is 3, which is odd. Odd-degree polynomial functions have graphs with opposite behavior at each end. The leading coefficient, 1, is positive. Thus, the graph falls to the left and rises to the right. This behavior is shown in Figure 5.5(a) in the blue box above. The graph of f is shown in Figure 5.6.   ■

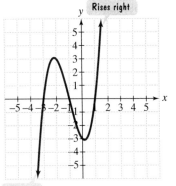

FIGURE 5.6 The graph of $f(x) = x^3 + 3x^2 - x - 3$

CHECK POINT **3** Use the Leading Coefficient Test to determine the end behavior of the graph of $f(x) = x^4 - 4x^2$.   The graph rises to the left and to the right.

EXAMPLE 4   Using the Leading Coefficient Test

Use end behavior to explain why

$$f(x) = -49x^3 + 806x^2 + 3776x + 2503$$

is only an appropriate model for AIDS cases for a limited time period.

SOLUTION   We begin by identifying the sign of the leading coefficient and the degree of the polynomial.

$$f(x) = -49x^3 + 806x^2 + 3776x + 2503$$

> The leading coefficient, −49, is negative.

> The degree of the polynomial, 3, is odd.

The degree of f is 3, which is odd. Odd-degree polynomial functions have graphs with opposite behavior at each end. The leading coefficient, −49, is negative. Thus, the graph rises to the left and falls to the right. This behavior is shown in Figure 5.5(b) in the box on page 303 that describes the Leading Coefficient Test. The fact that the graph falls to the right indicates that at some point the number of AIDS cases will be negative, an impossibility. If a function has a graph that decreases without bound over time, it will not be capable of modeling nonnegative phenomena over long time periods. Model breakdown will eventually occur. ∎

 CHECK POINT 4   The polynomial function

$$f(x) = -0.27x^3 + 9.2x^2 - 102.9x + 400$$

models the ratio of students to computers in U.S. public schools x years after 1980. Use end behavior to determine whether this function could be an appropriate model for computers in the classroom well into the twenty-first century. Explain your answer.   This would not be appropriate over long time periods. Because the graph falls to the right, at some point the ratio would be negative, which is not possible.

If you use a graphing utility to graph a polynomial function, it is important to select a viewing rectangle that accurately reveals the graph's end behavior. If the viewing rectangle, or window, is too small, it may not accurately show the end behavior.

EXAMPLE 5   Using the Leading Coefficient Test

The graph of $f(x) = -x^4 + 8x^3 + 4x^2 + 2$ was obtained with a graphing utility using a $[-8, 8, 1]$ by $[-10, 10, 1]$ viewing rectangle. The graph is shown in Figure 5.7(a). Does the graph show the end behavior of the function?

FIGURE 5.7   **(a)** $[-8, 8, 1]$ by $[-10, 10, 1]$ **(b)** $[-10, 10, 1]$ by $[-1000, 750, 250]$

SOLUTION   We begin by identifying the sign of the leading coefficient and the degree of the polynomial.

$$f(x) = -x^4 + 8x^3 + 4x^2 + 2$$

The leading coefficient, −1, is negative.

The degree of the polynomial, 4, is even.

The degree of f is 4, which is even. Even-degree polynomial functions have graphs with the same behavior at each end. The leading coefficient, −1, is negative. Thus, the graph should fall to the left and fall to the right. [See Figure 5.5(d) in the box on page 303 on the Leading Coefficient Test.] The graph in Figure 5.7(a) is falling to the left, but it is not falling to the right. Therefore, the graph is not complete enough to show end behavior. A more complete graph of the function is shown in a larger viewing rectangle in Figure 5.7(b). ∎

FIGURE 5.8

✔ CHECK POINT **5** The graph of $f(x) = x^3 + 13x^2 + 10x - 4$ is shown in a standard viewing rectangle in Figure 5.8. Use the Leading Coefficient Test to determine whether the graph shows the end behavior of the function. Explain your answer.   The graph does not show the end behavior of the function. The graph should fall to the left.

4 Add polynomials.

Adding Polynomials   Polynomials are added by combining like terms. Here are two examples that illustrate the use of the distributive property in adding monomials and combining like terms:

$$-9x^3 + 13x^3 = (-9 + 13)x^3 = 4x^3$$

Add coefficients and keep the same variable factor(s).

$$-7x^3y^2 + 4x^3y^2 = (-7 + 4)x^3y^2 = -3x^3y^2.$$

EXAMPLE 6   Adding Polynomials

Add: $(-6x^3 + 5x^2 + 4) + (2x^3 + 7x^2 - 10)$.

SOLUTION

$$(-6x^3 + 5x^2 + 4) + (2x^3 + 7x^2 - 10)$$

$$= -6x^3 + 5x^2 + 4 + 2x^3 + 7x^2 - 10 \qquad \text{Remove the parentheses.}$$

$$= \underbrace{-6x^3 + 2x^3}_{} + \underbrace{5x^2 + 7x^2}_{} + \underbrace{4 - 10}_{} \qquad \text{Rearrange the terms so that like terms are adjacent.}$$

$$= \qquad -4x^3 \qquad\qquad +12x^2 \qquad -6 \qquad \text{Combine like terms.} \quad ∎$$

✔ CHECK POINT **6** Add: $(-7x^3 + 4x^2 + 3) + (4x^3 + 6x^2 - 13)$. $-3x^3 + 10x^2 - 10$

EXAMPLE 7   Adding Polynomials

Add: $(5x^3y - 4x^2y - 7y) + (2x^3y + 6x^2y - 4y - 5)$.

SOLUTION

$(5x^3y - 4x^2y - 7y) + (2x^3y + 6x^2y - 4y - 5)$ *The given problem involves adding polynomials in two variables.*

$= 5x^3y - 4x^2y - 7y + 2x^3y + 6x^2y - 4y - 5$ *Remove the parentheses.*

$= \underbrace{5x^3y + 2x^3y} \quad \underbrace{- 4x^2y + 6x^2y} \quad \underbrace{- 7y - 4y} \quad \underbrace{- 5}$ *Rearrange the terms so that like terms are adjacent.*

$= \quad 7x^3y \qquad +2x^2y \qquad -11y \qquad -5$ *Combine like terms.* ∎

Polynomials can be added by arranging like terms in columns. Then combine like terms, column by column. Here's the solution to Example 7 using columns and a vertical format:

$$5x^3y - 4x^2y - 7y$$
$$\underline{2x^3y + 6x^2y - 4y - 5}.$$
$$7x^3y + 2x^2y - 11y - 5$$

 CHECK POINT 7 Add: $(7xy^3 - 5xy^2 - 3y) + (2xy^3 + 8xy^2 - 12y - 9)$.
$9xy^3 + 3xy^2 - 15y - 9$

---

5 Subtract polynomials.

Subtracting Polynomials   We subtract real numbers by adding the opposite, or additive inverse, of the number being subtracted. For example,

$$8 - 3 = 8 + (-3) = 5.$$

Similarly, we subtract one polynomial from another by adding the opposite of the polynomial being subtracted.

> **SUBTRACTING POLYNOMIALS**   To subtract two polynomials, change the sign of every term of the second polynomial. Add this result to the first polynomial.

EXAMPLE 8   Subtracting Polynomials

Subtract: $(7x^3 - 8x^2 + 9x - 6) - (2x^3 - 6x^2 - 3x + 9)$.

SOLUTION

$(7x^3 - 8x^2 + 9x - 6) - (2x^3 - 6x^2 - 3x + 9)$

$= (7x^3 - 8x^2 + 9x - 6) + (-2x^3 + 6x^2 + 3x - 9)$ *Change the sign of each term of the second polynomial and add the two polynomials.*

$= \underbrace{7x^3 - 2x^3} \quad \underbrace{- 8x^2 + 6x^2} \quad \underbrace{+ 9x + 3x} \quad \underbrace{- 6 - 9}$ *Rearrange terms.*

$= \quad 5x^3 \qquad -2x^2 \qquad +12x \qquad -15$ *Combine like terms.* ∎

STUDY TIP

You can also subtract polynomials using a vertical format. Here's the solution to Example 8 using a vertical format:

$$7x^3 - 8x^2 + 9x - 6$$
$$\underline{-(2x^3 - 6x^2 - 3x + 9)}$$

$$7x^3 - 8x^2 + 9x - 6$$
$$\underline{+ -2x^3 + 6x^2 + 3x - 9}.$$
$$5x^3 - 2x^2 + 12x - 15$$

 CHECK POINT 8 Subtract: $(14x^3 - 5x^2 + x - 9) - (4x^3 - 3x^2 - 7x + 1)$.
$10x^3 - 2x^2 + 8x - 10$

EXAMPLE 9   Subtracting Polynomials

Subtract $-2x^5y^2 - 3x^3y + 7$ from $3x^5y^2 - 4x^3y - 3$.

SOLUTION

$(3x^5y^2 - 4x^3y - 3) - (-2x^5y^2 - 3x^3y + 7)$

$= 3x^5y^2 - 4x^3y - 3 + 2x^5y^2 + 3x^3y - 7$ Change subtraction to addition and change the sign of every term of the polynomial in parentheses.

$= \underbrace{3x^5y^2 + 2x^5y^2} \quad \underbrace{-4x^3y + 3x^3y} \quad \underbrace{-3 - 7}$ Rearrange terms.

$= \qquad 5x^5y^2 \qquad\qquad -x^3y \qquad\quad -10$ Combine like terms. ∎

✔ **CHECK POINT 9** Subtract $-7x^2y^5 - 4xy^3 + 2$ from $6x^2y^5 - 2xy^3 - 8$.
$13x^2y^5 + 2xy^3 - 10$

5.1 EXERCISE SET

 Student Solutions Manual   CD/Video   PH Math/Tutor Center   MathXL Tutorials on CD   MathXL®   MyMathLab   Interactmath.com

Practice Exercises

In Exercises 1–10, determine the coefficient of each term, the degree of each term, the degree of the polynomial, the leading term, and the leading coefficient of the polynomial.

1. $-x^4 + x^2$

Term	Coefficient	Degree
$-x^4$	-1	4
x^2	1	2

The degree of the polynomial is 4, the leading term is $-x^4$, and the leading coefficient is -1.

2. $x^3 - 4x^2$

Term	Coefficient	Degree
x^3	1	3
$-4x^2$	-4	2

The degree of the polynomial is 3, the leading term is x^3, and the leading coefficient is 1.

3. $5x^3 + 7x^2 - x + 9$

Term	Coefficient	Degree
$5x^3$	5	3
$7x^2$	7	2
$-x$	-1	1
9	9	0

The degree of the polynomial is 3, the leading term is $5x^3$, and the leading coefficient is 5.

4. $11x^3 - 6x^2 + x + 3$

Term	Coefficient	Degree
$11x^3$	11	3
$-6x^2$	-6	2
x	1	1
3	3	0

The degree of the polynomial is 3, the leading term is $11x^3$, and the leading coefficient is 11.

5. $3x^2 - 7x^4 - x + 6$

Term	Coefficient	Degree
$3x^2$	3	2
$-7x^4$	-7	4
$-x$	-1	1
6	6	0

The degree of the polynomial is 4, the leading term is $-7x^4$, and the leading coefficient is -7.

6. $2x^2 - 9x^4 - x + 5$

Term	Coefficient	Degree
$2x^2$	2	2
$-9x^4$	-9	4
$-x$	-1	1
5	5	0

The degree of the polynomial is 4, the leading term is $-9x^4$, and the leading coefficient is -9.

7. $x^3y^2 - 5x^2y^7 + 6y^2 - 3$

Term	Coefficient	Degree
x^3y^2	1	5
$-5x^2y^7$	-5	9
$6y^2$	6	2
-3	-3	0

The degree of the polynomial is 9, the leading term is $-5x^2y^7$, and the leading coefficient is -5.

8. $12x^4y - 5x^3y^7 - x^2 + 4$

Term	Coefficient	Degree
$12x^4y$	12	5
$-5x^3y^7$	-5	10
$-x^2$	-1	2
4	4	0

The degree of the polynomial is 10, the leading term is $-5x^3y^7$, and the leading coefficient is -5.

9. $x^5 + 3x^2y^4 + 7xy + 9x - 2$

Term	Coefficient	Degree
x^5	1	5
$3x^2y^4$	3	6
$7xy$	7	2
$9x$	9	1
-2	-2	0

The degree of the polynomial is 6, the leading term is $3x^2y^4$, and the leading coefficient is 3.

10. $3x^6 + 4x^4y^4 - x^3y + 4x^2 - 5$

Term	Coefficient	Degree
$3x^6$	3	6
$4x^4y^4$	4	8
$-x^3y$	-1	4
$4x^2$	4	2
-5	-5	0

The degree of the polynomial is 8, the leading term is $4x^4y^4$, and the leading coefficient is 4.

In Exercises 11–20, let

$$f(x) = x^2 - 5x + 6 \quad \text{and} \quad g(x) = 2x^3 - x^2 + 4x - 1.$$

Find the indicated function values.

11. $f(3)$ 0

12. $f(4)$ 2

13. $f(-1)$ 12

14. $f(-2)$ 20

15. $g(3)$ 56

16. $g(2)$ 19

17. $g(-2)$ -29

18. $g(-3)$ -76

19. $g(0)$ -1

20. $f(0)$ 6

In Exercises 21–24, identify which graphs are not those of polynomial functions.

21.

22.

23.

24.

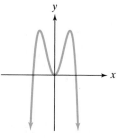

21–24. Graphs # 22 and # 23 are not those of polynomial functions.

In Exercises 25–28, use the Leading Coefficient Test to determine the end behavior of the graph of the given polynomial function. Then use this end behavior to match the polynomial function with its graph. [The graphs are in the next column and are labeled (a) through (d).]

25. $f(x) = -x^4 + x^2$ falls to the left and falls to the right; graph **(b)**

26. $f(x) = x^3 - 4x^2$ falls to the left and rises to the right; graph **(c)**

27. $f(x) = x^2 - 6x + 9$ rises to the left and rises to the right; graph **(a)**

28. $f(x) = -x^3 - x^2 + 5x - 3$ rises to the left and falls to the right; graph **(d)**

In Exercises 29–40, add the polynomials. Assume that all variable exponents represent whole numbers.

29. $(-6x^3 + 5x^2 - 8x + 9) + (17x^3 + 2x^2 - 4x - 13)$
$11x^3 + 7x^2 - 12x - 4$

30. $(-7x^3 + 6x^2 - 11x + 13) + (19x^3 - 11x^2 + 7x - 17)$
$12x^3 - 5x^2 - 4x - 4$

31. $\left(\dfrac{2}{5}x^4 + \dfrac{2}{3}x^3 + \dfrac{5}{8}x^2 + 7\right) + \left(-\dfrac{4}{5}x^4 + \dfrac{1}{3}x^3 - \dfrac{1}{4}x^2 - 7\right)$
$-\dfrac{2}{5}x^4 + x^3 + \dfrac{3}{8}x^2$

32. $\left(\dfrac{1}{5}x^4 + \dfrac{1}{3}x^3 + \dfrac{3}{8}x^2 + 6\right) + \left(-\dfrac{3}{5}x^4 + \dfrac{2}{3}x^3 - \dfrac{1}{2}x^2 - 6\right)$
$-\dfrac{2}{5}x^4 + x^3 - \dfrac{1}{8}x^2$

33. $(7x^2y - 5xy) + (2x^2y - xy)$ $9x^2y - 6xy$

34. $(-4x^2y + xy) + (7x^2y + 8xy)$ $3x^2y + 9xy$

35. $(5x^2y + 9xy + 12) + (-3x^2y + 6xy + 3)$ $2x^2y + 15xy + 15$

36. $(8x^2y + 12xy + 14) + (-2x^2y + 7xy + 4)$ $6x^2y + 19xy + 18$

37. $(9x^4y^2 - 6x^2y^2 + 3xy) + (-18x^4y^2 - 5x^2y - xy)$
$-9x^4y^2 - 6x^2y^2 - 5x^2y + 2xy$

38. $(10x^4y^2 - 3x^2y^2 + 2xy) + (-16x^4y^2 - 4x^2y - xy)$
$-6x^4y^2 - 3x^2y^2 - 4x^2y + xy$

39. $(x^{2n} + 5x^n - 8) + (4x^{2n} - 7x^n + 2)$ $5x^{2n} - 2x^n - 6$

40. $(6y^{2n} + y^n + 5) + (3y^{2n} - 4y^n - 15)$ $9y^{2n} - 3y^n - 10$

In Exercises 41–50, subtract the polynomials. Assume that all variable exponents represent whole numbers.

41. $(17x^3 - 5x^2 + 4x - 3) - (5x^3 - 9x^2 - 8x + 11)$
$12x^3 + 4x^2 + 12x - 14$

42. $(18x^3 - 2x^2 - 7x + 8) - (9x^3 - 6x^2 - 5x + 7)$
$9x^3 + 4x^2 - 2x + 1$

43. $(13y^5 + 9y^4 - 5y^2 + 3y + 6) - (-9y^5 - 7y^3 + 8y^2 + 11)$
$22y^5 + 9y^4 + 7y^3 - 13y^2 + 3y - 5$

44. $(12y^5 + 7y^4 - 3y^2 + 6y + 7) - (-10y^5 - 8y^3 + 3y^2 + 14)$
$22y^5 + 7y^4 + 8y^3 - 6y^2 + 6y - 7$

45. $(x^3 + 7xy - 5y^2) - (6x^3 - xy + 4y^2)$ $\quad -5x^3 + 8xy - 9y^2$

46. $(x^4 - 7xy - 5y^3) - (6x^4 - 3xy + 4y^3)$ $\quad -5x^4 - 4xy - 9y^3$

47. $(3x^4y^2 + 5x^3y - 3y) - (2x^4y^2 - 3x^3y - 4y + 6x)$
$x^4y^2 + 8x^3y + y - 6x$

48. $(5x^4y^2 + 6x^3y - 7y) - (3x^4y^2 - 5x^3y - 6y + 8x)$
$2x^4y^2 + 11x^3y - y - 8x$

49. $(7y^{2n} + y^n - 4) - (6y^{2n} - y^n - 1)$ $\quad y^{2n} + 2y^n - 3$

50. $(8x^{2n} + x^n - 4) - (9x^{2n} - x^n - 2)$ $\quad -x^{2n} + 2x^n - 2$

51. Subtract $-5a^2b^4 - 8ab^2 - ab$ from $3a^2b^4 - 5ab^2 + 7ab$.
$8a^2b^4 + 3ab^2 + 8ab$

52. Subtract $-7a^2b^4 - 8ab^2 - ab$ from $13a^2b^4 - 17ab^2 + ab$.
$20a^2b^4 - 9ab^2 + 2ab$

53. Subtract $-4x^3 - x^2y + xy^2 + 3y^3$ from $x^3 + 2x^2y - y^3$.
$5x^3 + 3x^2y - xy^2 - 4y^3$

54. Subtract $-6x^3 + x^2y - xy^2 + 2y^3$ from $x^3 + 2xy^2 - y^3$.
$7x^3 - x^2y + 3xy^2 - 3y^3$

Practice Plus

55. Add $6x^4 - 5x^3 + 2x$ to the difference between $4x^3 + 3x^2 - 1$
and $x^4 - 2x^2 + 7x - 3$. $\quad 5x^4 - x^3 + 5x^2 - 5x + 2$

56. Add $5x^4 - 2x^3 + 7x$ to the difference between $2x^3 + 5x^2 - 3$
and $-x^4 - x^2 - x - 1$. $\quad 6x^4 + 6x^2 + 8x - 2$

57. Subtract $9x^2y^2 - 3x^2 - 5$ from the sum of $-6x^2y^2 - x^2 - 1$
and $5x^2y^2 + 2x^2 - 1$. $\quad -10x^2y^2 + 4x^2 + 3$

58. Subtract $6x^2y^3 - 2x^2 - 7$ from the sum of $-5x^2y^3 + 3x^2 - 4$
and $4x^2y^3 - 2x^2 - 6$. $\quad -7x^2y^3 + 3x^2 - 3$

In Exercises 59–64, let
$$f(x) = -3x^3 - 2x^2 - x + 4$$
$$g(x) = x^3 - x^2 - 5x - 4$$
$$h(x) = -2x^3 + 5x^2 - 4x + 1.$$

Find the indicated function, function value, or polynomial.

59. $(f - g)(x)$ and $(f - g)(-1)$
$(f - g)(x) = -4x^3 - x^2 + 4x + 8; (f - g)(-1) = 7$

60. $(g - h)(x)$ and $(g - h)(-1)$
$(g - h)(x) = 3x^3 - 6x^2 - x - 5; (g - h)(-1) = -13$

61. $(f + g - h)(x)$ and $(f + g - h)(-2)$
$(f + g - h)(x) = -8x^2 - 2x - 1; (f + g - h)(-2) = -29$

62. $(g + h - f)(x)$ and $(g + h - f)(-2)$
$(g + h - f)(x) = 2x^3 + 6x^2 - 8x - 7; (g + h - f)(-2) = 17$

63. $2f(x) - 3g(x)$ $\quad -9x^3 - x^2 + 13x + 20$

64. $-2g(x) - 3h(x)$ $\quad 4x^3 - 13x^2 + 22x + 5$

Application Exercises

The polynomial function
$$f(x) = -2212x^2 + 57{,}575x + 107{,}896$$

models the cumulative number of deaths from AIDS in the United States, $f(x)$, x years after 1990. Use this function to solve Exercises 65–66.

65. Find and interpret $f(10)$. $\quad f(10) = 462{,}446$; In 2000, the cumulative number of AIDS deaths in the U.S. was 462,446.

66. Find and interpret $f(8)$. $\quad f(8) = 426{,}928$; In 1998, the cumulative number of AIDS deaths in the U.S. was 426,928.

The graph of the polynomial function in Exercises 65–66 is shown in the figure. Use the graph to solve Exercises 67–68.

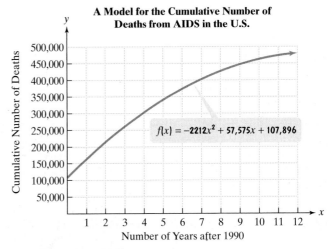

A Model for the Cumulative Number of Deaths from AIDS in the U.S.

$f(x) = -2212x^2 + 57{,}575x + 107{,}896$

67. Identify your answer from Exercise 65 as a point on the graph. $(10, 462{,}446)$

68. Identify your answer from Exercise 66 as a point on the graph. $(8, 426{,}928)$

The bar graph shows the actual data for the cumulative number of deaths from AIDS in the United States from 1990 through 2002.

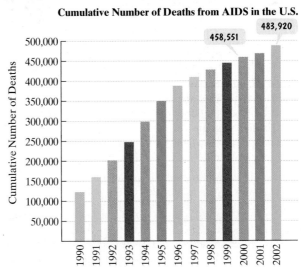

Cumulative Number of Deaths from AIDS in the U.S.

483,920
458,551

Source: Centers for Disease Control

The data in the bar graph on page 309 can be modeled by the following second- and third-degree polynomial functions:

Cumulative number of AIDS deaths x years after 1990.

$$f(x) = -2212x^2 + 57,575x + 107,896$$
$$g(x) = -84x^3 - 702x^2 + 50,609x + 113,435.$$

Use these functions to solve Exercises 69–72.

69. Use both functions to find the cumulative number of AIDS deaths in 2000. Which function provides a better description for the actual number shown in the bar graph on page 309? $f(10) = 462,446$; $g(10) = 465,325$; f

70. Use both functions to find the cumulative number of AIDS deaths in 2002. Which function provides a better description for the actual number shown in the bar graph on page 309? $f(12) = 480,268$; $g(12) = 474,503$; f

71. Use the Leading Coefficient Test to determine the end behavior to the right for the graph of f. Will this function be useful in modeling the cumulative number of AIDS deaths over an extended period of time? Explain your answer. Falls to the right; no; cumulative number of deaths cannot decrease.

72. Use the Leading Coefficient Test to determine the end behavior to the right for the graph of g. Will this function be useful in modeling the cumulative number of AIDS deaths over an extended period of time? Explain your answer. Falls to the right; no; cumulative number of deaths cannot decrease.

73. The common cold is caused by a rhinovirus. After x days of invasion by the viral particles, the number of particles in our bodies, $f(x)$, in billions, can be modeled by the polynomial function

$$f(x) = -0.75x^4 + 3x^3 + 5.$$

Use the Leading Coefficient Test to determine the graph's end behavior to the right. What does this mean about the number of viral particles in our bodies over time? Falls to the right; the viral particles eventually decrease as the days increase.

74. The polynomial function

$$f(x) = -0.87x^3 + 0.35x^2 + 81.62x + 7684.94$$

models the number of thefts, $f(x)$, in thousands, in the United States x years after 1987. Will this function be useful in modeling the number of thefts over an extended period of time? Explain your answer. No; the graph falls to the right, so eventually there would be a negative number of thefts, which is not possible.

75. A herd of 100 elk is introduced to a small island. The number of elk, $f(x)$, after x years is modeled by the polynomial function

$$f(x) = -x^4 + 21x^2 + 100.$$

Use the Leading Coefficient Test to determine the graph's end behavior to the right. What does this mean about what will eventually happen to the elk population? Falls to the right; eventually the elk population will be extinct.

Writing in Mathematics

76. What is a polynomial? 76–87. Answers will vary.
77. Explain how to determine the degree of each term of a polynomial.
78. Explain how to determine the degree of a polynomial.
79. Explain how to determine the leading coefficient of a polynomial.
80. What is a polynomial function?
81. What do we mean when we describe the graph of a polynomial function as smooth and continuous?
82. What is meant by the end behavior of a polynomial function?
83. Explain how to use the Leading Coefficient Test to determine the end behavior of a polynomial function.
84. Why is a polynomial function of degree 3 with a negative leading coefficient not appropriate for modeling nonnegative real-world phenomena over a long period of time?
85. Explain how to add polynomials.
86. Explain how to subtract polynomials.
87. In a favorable habitat and without natural predators, a population of reindeer is introduced to an island preserve. The reindeer population t years after their introduction is modeled by the polynomial function

$$f(t) = -0.125t^5 + 3.125t^4 + 4000.$$

Discuss the growth and decline of the reindeer population. Describe the factors that might contribute to this population model.

Technology Exercises

Write a polynomial function that imitates the end behavior of each graph in Exercises 88–91. The dashed portions of the graphs indicate that you should focus only on imitating the left and right end behavior of the graph and can be flexible about what occurs between the left and right ends. Then use your graphing utility to graph the polynomial function and verify that you imitated the end behavior shown in the given graph.

88. 89.

88. Answers will vary. An example is $f(x) = x^2 - x + 1$.
89. Answers will vary. An example is $f(x) = -x^2 - x + 1$.

90. 91.

90. Answers will vary. An example is $f(x) = -x^3 + x + 1$.
91. Answers will vary. An example is $f(x) = x^3 + x + 1$.

In Exercises 92–95, use a graphing utility with a viewing rectangle large enough to show end behavior to graph each polynomial function.   **92–95.** See graphing answer section.

92. $f(x) = x^3 + 13x^2 + 10x - 4$

93. $f(x) = -2x^3 + 6x^2 + 3x - 1$

94. $f(x) = -x^4 + 8x^3 + 4x^2 + 2$

95. $f(x) = -x^5 + 5x^4 - 6x^3 + 2x + 20$

In Exercises 96–97, use a graphing utility to graph f and g in the same viewing rectangle. Then use the $\boxed{\text{ZOOM OUT}}$ feature to show that f and g have identical end behavior.
96–97. See graphing answer section.

96. $f(x) = x^3 - 6x + 1, \quad g(x) = x^3$

97. $f(x) = -x^4 + 2x^3 - 6x, \quad g(x) = -x^4$

Critical Thinking Exercises

98. Which one of the following is true?   d

a. $4x^3 + 7x^2 - 5x + \dfrac{2}{x}$ is a polynomial containing four terms.

b. If two polynomials of degree 2 are added, their sum must be a polynomial of degree 2.

c. $(x^2 - 7x) - (x^2 - 4x) = -11x$ for all values of x.

d. All terms of a polynomial are monomials.

In Exercises 99–100, perform the indicated operations. Assume that exponents represent whole numbers.

99. $(x^{2n} - 3x^n + 5) + (4x^{2n} - 3x^n - 4) - (2x^{2n} - 5x^n - 3)$
$3x^{2n} - x^n + 4$

100. $(y^{3n} - 7y^{2n} + 3) - (-3y^{3n} - 2y^{2n} - 1) + (6y^{3n} - y^{2n} + 1)$
$10y^{3n} - 6y^{2n} + 5$

101. From what polynomial must $4x^2 + 2x - 3$ be subtracted to obtain $5x^2 - 5x + 8$? $9x^2 - 3x + 5$

Review Exercises

102. Solve: $9(x - 1) = 1 + 3(x - 2)$. (Section 1.4, Example 3) $\frac{2}{3}$

103. Graph: $2x - 3y < -6$. (Section 4.4, Example 1)   See graphing answer section.

104. Write the point-slope and slope-intercept equations of a line passing through the point $(-2, 5)$ and parallel to the line whose equation is $3x - y = 9$. (Section 2.4, Example 4) $y - 5 = 3(x + 2); \; y = 3x + 11 \text{ or } f(x) = 3x + 11$

SECTION
5.2

Objectives

1 Multiply monomials.

2 Multiply a monomial and a polynomial.

3 Multiply polynomials when neither is a monomial.

4 Use FOIL in polynomial multiplication.

5 Square binomials.

6 Multiply the sum and difference of two terms.

7 Find the product of functions.

8 Use polynomial multiplication to evaluate functions.

MULTIPLICATION OF POLYNOMIALS

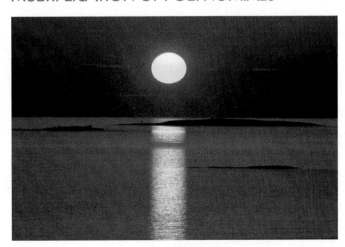

The rising and setting of the sun suggest the obvious: Things change over time. Calculus is the study of rates of change, allowing the motion of the rising sun to be measured by "freezing the frame" at one instant in time. Calculus reveals a function's rate of change at any "frozen" instant. In this section, you will encounter an algebraic expression that plays an important role in this process. You will be able to work with this expression by learning how polynomials are multiplied.

Multiplying Monomials   To multiply monomials, begin by multiplying the coefficients. Then multiply the variables. Use the product rule for exponents to multiply the variables: Retain the variable and add the exponents.

1 Multiply monomials.

EXAMPLE 1 Multiplying Monomials

Multiply:

    a. $(5x^3y^4)(-6x^7y^8)$ **b.** $(4x^3y^2z^5)(2x^5y^2z^4)$.

SOLUTION

   a. $(5x^3y^4)(-6x^7y^8) = 5(-6)x^3 \cdot x^7 \cdot y^4 \cdot y^8$ Rearrange factors. This step is usually done mentally.

                      $= -30x^{3+7}y^{4+8}$ Multiply coefficients and add exponents.

                      $= -30x^{10}y^{12}$ Simplify.

   b. $(4x^3y^2z^5)(2x^5y^2z^4) = 4 \cdot 2 \cdot x^3 \cdot x^5 \cdot y^2 \cdot y^2 \cdot z^5 \cdot z^4$ Rearrange factors.

                      $= 8x^{3+5}y^{2+2}z^{5+4}$ Multiply coefficients and add exponents.

                      $= 8x^8y^4z^9$ Simplify. ∎

✓ **CHECK POINT 1** Multiply:

    a. $(6x^5y^7)(-3x^2y^4)$ $-18x^7y^{11}$ **b.** $(10x^4y^3z^6)(3x^6y^3z^2)$. $30x^{10}y^6z^8$

2 Multiply a monomial and a polynomial.

Multiplying a Monomial and a Polynomial That Is Not a Monomial We use the distributive property to multiply a monomial and a polynomial that is not a monomial. For example,

$$3x^2(2x^3 + 5x) = 3x^2 \cdot 2x^3 + 3x^2 \cdot 5x = 3 \cdot 2x^{2+3} + 3 \cdot 5x^{2+1} = 6x^5 + 15x^3.$$

Monomial   Binomial   Multiply coefficients and add exponents.

To multiply a monomial and a polynomial, multiply each term of the polynomial by the monomial. Once the monomial factor is distributed, we multiply the resulting monomials using the procedure shown in Example 1.

EXAMPLE 2 Multiplying a Monomial and a Trinomial

Multiply:

    a. $4x^3(6x^5 - 2x^2 + 3)$ **b.** $5x^3y^4(2x^7y - 6x^4y^3 - 3)$.

SOLUTION

   a. $4x^3(6x^5 - 2x^2 + 3) = 4x^3 \cdot 6x^5 - 4x^3 \cdot 2x^2 + 4x^3 \cdot 3$ Use the distributive property.

                         $= 24x^8 - 8x^5 + 12x^3$ Multiply coefficients and add exponents.

   b. $5x^3y^4(2x^7y - 6x^4y^3 - 3)$

          $= 5x^3y^4 \cdot 2x^7y - 5x^3y^4 \cdot 6x^4y^3 - 5x^3y^4 \cdot 3$ Use the distributive property.

          $= 10x^{10}y^5 - 30x^7y^7 - 15x^3y^4$ Multiply coefficients and add exponents.

CHECK POINT **2** Multiply:

a. $6x^4(2x^5 - 3x^2 + 4)$ $12x^9 - 18x^6 + 24x^4$

b. $2x^4y^3(5xy^6 - 4x^3y^4 - 5)$. $10x^5y^9 - 8x^7y^7 - 10x^4y^3$

3 Multiply polynomials when neither is a monomial.

Multiplying Polynomials when Neither Is a Monomial   How do we multiply two polynomials if neither is a monomial? For example, consider

$$(3x + 7)(x^2 + 4x + 5).$$

Binomial   Trinomial

One way to perform this multiplication is to distribute $3x$ throughout the trinomial

$$3x(x^2 + 4x + 5)$$

and 7 throughout the trinomial

$$7(x^2 + 4x + 5).$$

Then combine the like terms that result. In general, the product of two polynomials is the polynomial obtained by multiplying each term of one polynomial by each term of the other polynomial and then combining like terms.

EXAMPLE 3   Multiplying a Binomial and a Trinomial

Multiply: $(3x + 7)(x^2 + 4x + 5)$.

SOLUTION

$(3x + 7)(x^2 + 4x + 5)$

$= 3x(x^2 + 4x + 5) + 7(x^2 + 4x + 5)$ Multiply the trinomial by each term of the binomial.

$= 3x \cdot x^2 + 3x \cdot 4x + 3x \cdot 5 + 7x^2 + 7 \cdot 4x + 7 \cdot 5$ Use the distributive property.

$= 3x^3 + 12x^2 + 15x + 7x^2 + 28x + 35$ Multiply monomials: Multiply coefficients and add exponents.

$= 3x^3 + 19x^2 + 43x + 35$ Combine like terms: $12x^2 + 7x^2 = 19x^2$ and $15x + 28x = 43x$.   ■

CHECK POINT **3** Multiply: $(3x + 2)(2x^2 - 2x + 1)$. $6x^3 - 2x^2 - x + 2$

Another method for solving Example 3 is to use a vertical format similar to that used for multiplying whole numbers.

$$
\begin{array}{r}
x^2 + 4x + 5 \\
3x + 7 \\
\hline
7x^2 + 28x + 35 \\
3x^3 + 12x^2 + 15x \\
\hline
3x^3 + 19x^2 + 43x + 35
\end{array}
$$

Write like terms in the same column.

$7(x^2 + 4x + 5)$

$3x(x^2 + 4x + 5)$

Combine like terms.

EXAMPLE 4 Multiplying a Binomial and a Trinomial

Multiply: $(2x^2y + 3y)(5x^4y - 4x^2y + y)$.

SOLUTION

$(2x^2y + 3y)(5x^4y - 4x^2y + y)$

$= 2x^2y(5x^4y - 4x^2y + y) + 3y(5x^4y - 4x^2y + y)$ Multiply the trinomial by each term of the binomial.

$= 2x^2y \cdot 5x^4y - 2x^2y \cdot 4x^2y + 2x^2y \cdot y + 3y \cdot 5x^4y - 3y \cdot 4x^2y + 3y \cdot y$

            Use the distributive property.

$= 10x^6y^2 - 8x^4y^2 + 2x^2y^2 + 15x^4y^2 - 12x^2y^2 + 3y^2$ Multiply coefficients and add exponents.

$= 10x^6y^2 + 7x^4y^2 - 10x^2y^2 + 3y^2$ Combine like terms:
$-8x^4y^2 + 15x^4y^2 = 7x^4y^2$
$2x^2y^2 - 12x^2y^2 = -10x^2y^2$. ∎

✔ **CHECK POINT 4** Multiply: $(4xy^2 + 2y)(3xy^4 - 2xy^2 + y)$.
$12x^2y^6 - 8x^2y^4 + 6xy^5 + 2y^2$

---

4 Use FOIL in polynomial multiplication.

The Product of Two Binomials: FOIL Frequently we need to find the product of two binomials. One way to perform this multiplication is to distribute each term in the first binomial throughout the second binomial. For example, we can find the product of the binomials $7x + 2$ and $4x + 5$ as follows:

$$(7x + 2)(4x + 5) = 7x(4x + 5) + 2(4x + 5)$$
$$= 7x(4x) + 7x(5) + 2(4x) + 2(5)$$
$$= 28x^2 + 35x + 8x + 10.$$

Distribute $7x$ over $4x + 5$.   Distribute 2 over $4x + 5$.

We can also find the product of $7x + 2$ and $4x + 5$ using a method called FOIL, which is based on our work shown above. Any two binomials can be quickly multiplied using the FOIL method, in which F represents the product of the **first** terms in each binomial, O represents the product of the **outside** terms, I represents the product of the two **inside** terms, and L represents the product of the **last**, or second, terms in each binomial. For example, we can use the FOIL method to find the product of the binomials $7x + 2$ and $4x + 5$ as follows:

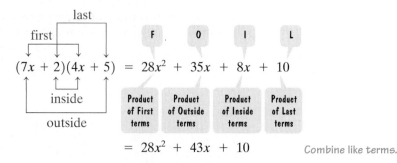

last
first
$(7x + 2)(4x + 5) = 28x^2 + 35x + 8x + 10$
inside
outside

F   O   I   L

Product of First terms   Product of Outside terms   Product of Inside terms   Product of Last terms

$= 28x^2 + 43x + 10$ Combine like terms.

In general, here's how to use the FOIL method to find the product of $ax + b$ and $cx + d$:

USING THE FOIL METHOD TO MULTIPLY BINOMIALS

Visualizing Polynomial Multiplication

Rectangles and their areas make it possible to visualize certain polynomial multiplications. What does the FOIL product

$$(x + 3)(x + 2) = x^2 + 5x + 6$$

look like? Use a rectangle with dimensions $x + 2$ and $x + 3$.

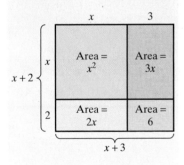

The area of the large rectangle is

$$(x + 3)(x + 2).$$

The sum of the areas of the four smaller rectangles is

$$x^2 + 3x + 2x + 6$$
$$= x^2 + 5x + 6.$$

The area of the large rectangle equals the sum of the areas of the four smaller rectangles:

$$(x + 3)(x + 2) = x^2 + 5x + 6.$$

---

a. $x^2 + 8x + 15$

b. $14x^2 + xy - 4y^2$

c. $4x^6 - 12x^4 - 5x^3 + 15x$

EXAMPLE 5 Using the FOIL Method

Multiply:

a. $(x + 3)(x + 2)$ **b.** $(3x + 5y)(x - 2y)$ **c.** $(5x^3 - 6)(4x^3 - x)$.

SOLUTION

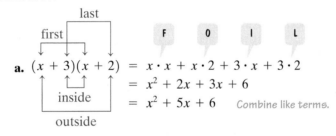

a. $(x + 3)(x + 2) = x \cdot x + x \cdot 2 + 3 \cdot x + 3 \cdot 2$
$$= x^2 + 2x + 3x + 6$$
$$= x^2 + 5x + 6 \quad \text{Combine like terms.}$$

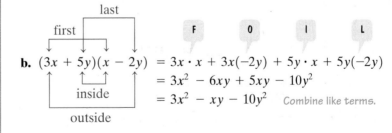

b. $(3x + 5y)(x - 2y) = 3x \cdot x + 3x(-2y) + 5y \cdot x + 5y(-2y)$
$$= 3x^2 - 6xy + 5xy - 10y^2$$
$$= 3x^2 - xy - 10y^2 \quad \text{Combine like terms.}$$

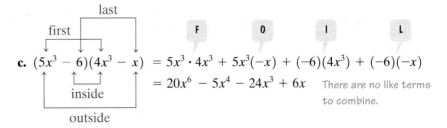

c. $(5x^3 - 6)(4x^3 - x) = 5x^3 \cdot 4x^3 + 5x^3(-x) + (-6)(4x^3) + (-6)(-x)$
$$= 20x^6 - 5x^4 - 24x^3 + 6x \quad \text{There are no like terms to combine.}$$

✔ **CHECK POINT 5** Multiply:

a. $(x + 5)(x + 3)$ **b.** $(7x + 4y)(2x - y)$ **c.** $(4x^3 - 5)(x^3 - 3x)$.

5 Square binomials.

The Square of a Binomial Let us find $(A + B)^2$, the square of a binomial sum. To do so, we begin with the FOIL method and look for a general rule.

$$(A + B)^2 = (A + B)(A + B) = \overset{F}{A \cdot A} + \overset{O}{A \cdot B} + \overset{I}{A \cdot B} + \overset{L}{B \cdot B}$$
$$= A^2 + 2AB + B^2$$

This result implies the following rule, which is often called a **special-product formula**:

THE SQUARE OF A BINOMIAL SUM

$$(A + B)^2 \;=\; A^2 \;+\; 2AB \;+\; B^2$$

| The square of a binomial sum | is | first term squared | plus | 2 times the product of the terms | plus | last term squared. |

STUDY TIP

Caution! The square of a sum is *not* the sum of the squares.

$$(A + B)^2 \neq A^2 + B^2$$

> The middle term $2AB$ is missing.

$$(x + 5)^2 \neq x^2 + 25$$

> Incorrect

Show that $(x + 5)^2$ and $x^2 + 25$ are not equal by substituting 3 for x in each expression and simplifying.

EXAMPLE 6 Finding the Square of a Binomial Sum

Multiply:

a. $(x + 5)^2$ **b.** $(3x + 2y)^2$.

SOLUTION   Use the special-product formula shown.

$$(A + B)^2 = A^2 + 2AB + B^2$$

	(First Term)²	+	2 · Product of the Terms	+	(Last Term)²	= Product
a. $(x + 5)^2 =$	x^2	+	$2 \cdot x \cdot 5$	+	5^2	$= x^2 + 10x + 25$
b. $(3x + 2y)^2 =$	$(3x)^2$	+	$2 \cdot 3x \cdot 2y$	+	$(2y)^2$	$= 9x^2 + 12xy + 4y^2$

■

✔ **CHECK POINT 6** Multiply:

a. $(x + 8)^2$ $x^2 + 16x + 64$ **b.** $(4x + 5y)^2$. $16x^2 + 40xy + 25y^2$

ENRICHMENT ESSAY

Visualizing a Special-Product Formula

The formula
$$(A + B)^2 = A^2 + 2AB + B^2$$
can be interpreted geometrically.

The area of the large rectangle is
$$(A + B)(A + B) = (A + B)^2.$$

The sum of the areas of the four smaller rectangles is
$$A^2 + AB + AB + B^2$$
$$= A^2 + 2AB + B^2.$$

The area of the large rectangle equals the sum of the areas of the four smaller rectangles:
$$(A + B)^2 = A^2 + 2AB + B^2.$$

Using the FOIL method on $(A - B)^2$, the square of a binomial difference, we obtain the following rule:

THE SQUARE OF A BINOMIAL DIFFERENCE

$$(A - B)^2 \;=\; A^2 \;-\; 2AB \;+\; B^2$$

| The square of a binomial difference | is | first term squared | minus | 2 times the product of the terms | plus | last term squared. |

EXAMPLE 7 Finding the Square of a Binomial Difference

Multiply:

   a. $(x - 8)^2$ **b.** $\left(\frac{1}{2}x - 4y^3\right)^2$.

SOLUTION Use the special-product formula shown.

$$(A - B)^2 = \quad A^2 \quad - \quad 2AB \quad + \quad B^2$$

	(First Term)^2	$-$	2 · Product of the Terms	$+$	(Last Term)^2	= Product
a. $(x - 8)^2 =$	x^2	$-$	$2 \cdot x \cdot 8$	$+$	8^2	$= x^2 - 16x + 64$
b. $\left(\frac{1}{2}x - 4y^3\right)^2 =$	$\left(\frac{1}{2}x\right)^2$	$-$	$2 \cdot \frac{1}{2}x \cdot 4y^3$	$+$	$(4y^3)^2$	$= \frac{1}{4}x^2 - 4xy^3 + 16y^6$

✔ **CHECK POINT 7** Multiply:

   a. $(x - 5)^2$ $x^2 - 10x + 25$ **b.** $(2x - 6y^4)^2$. $4x^2 - 24xy^4 + 36y^8$

6 Multiply the sum and difference of two terms.

Multiplying the Sum and Difference of Two Terms We can use the FOIL method to multiply $A + B$ and $A - B$ as follows:

$$\overset{F\ \ \ \ O\ \ \ \ I\ \ \ \ L}{(A + B)(A - B)} = A^2 - AB + AB - B^2 = A^2 - B^2.$$

Notice that the outside and inside products have a sum of 0 and the terms cancel. The FOIL multiplication provides us with a quick rule for multiplying the sum and difference of two terms, which is another example of a special-product formula.

THE PRODUCT OF THE SUM AND DIFFERENCE OF TWO TERMS

$$(A + B)(A - B) = A^2 - B^2$$

| The product of the sum and the difference of the same two terms | is | the square of the first term minus the square of the second term. |

EXAMPLE 8 Finding the Product of the Sum and Difference of Two Terms

Multiply:

   a. $(x + 8)(x - 8)$ **b.** $(9x + 5y)(9x - 5y)$ **c.** $(6a^2b - 3b)(6a^2b + 3b)$.

SOLUTION   Use the special-product formula shown.

$$(A + B)(A - B) \quad = \quad A^2 \quad - \quad B^2$$

First term squared $-$ Second term squared $=$ Product

a. $(x + 8)(x - 8) \quad = \quad x^2 \quad - \quad 8^2 \quad = \quad x^2 - 64$

b. $(9x + 5y)(9x - 5y) \quad = (9x)^2 \quad - (5y)^2 \quad = \quad 81x^2 - 25y^2$

c. $(6a^2b - 3b)(6a^2b + 3b) = (6a^2b)^2 - (3b)^2 = 36a^4b^2 - 9b^2$ ∎

✔ **CHECK POINT 8** Multiply:

a. $(x + 3)(x - 3)$ $x^2 - 9$

b. $(5x + 7y)(5x - 7y)$ $25x^2 - 49y^2$

c. $(5ab^2 - 4a)(5ab^2 + 4a)$. $25a^2b^4 - 16a^2$

Special products can sometimes be used to find the products of certain trinomials, as illustrated in Example 9.

EXAMPLE 9   Using the Special Products

Multiply:

a. $(7x + 5 + 4y)(7x + 5 - 4y)$ **b.** $(3x + y + 1)^2$.

SOLUTION

a. By grouping the first two terms within each set of parentheses, we can find the product using the form for the sum and difference of two terms.

$$(A \quad + \quad B) \quad \cdot \quad (A \quad - \quad B) \quad = \quad A^2 \quad - \quad B^2$$

$$[(7x + 5) + 4y] \cdot [(7x + 5) - 4y] = (7x + 5)^2 - (4y)^2$$

$$= (7x)^2 + 2 \cdot 7x \cdot 5 + 5^2 - (4y)^2$$

$$= 49x^2 + 70x + 25 - 16y^2$$

b. We can group the terms so that the formula for the square of a binomial can be applied.

$$(A \quad + \quad B)^2 \quad = \quad A^2 \quad + \quad 2 \quad \cdot \quad A \quad \cdot \quad B + B^2$$

$$[(3x + y) + 1]^2 = (3x + y)^2 + 2 \cdot (3x + y) \cdot 1 + 1^2$$

$$= 9x^2 + 6xy + y^2 + 6x + 2y + 1$$ ∎

✔ **CHECK POINT 9** Multiply:

a. $(3x + 2 + 5y)(3x + 2 - 5y)$ $9x^2 + 12x + 4 - 25y^2$

b. $(2x + y + 3)^2$. $4x^2 + 4xy + y^2 + 12x + 6y + 9$

7 Find the product of functions.

Multiplication of Polynomial Functions In Chapter 2, we developed an algebra of functions, defining the product of functions f and g as follows:

$$(fg)(x) = f(x) \cdot g(x).$$

Now that we know how to multiply polynomials, we can find the product of functions.

EXAMPLE 10 Using the Algebra of Functions

Let $f(x) = x - 5$ and $g(x) = x - 2$. Find:

a. $(fg)(x)$ **b.** $(fg)(1)$.

SOLUTION

a. $(fg)(x) = f(x) \cdot g(x)$ This is the definition of the product function, fg.

$= (x - 5)(x - 2)$ Substitute the given functions.

$\overbrace{\quad}^{F} \overbrace{\quad}^{O} \overbrace{\quad}^{I} \overbrace{\quad}^{L}$

$= x^2 - 2x - 5x + 10$ Multiply by the FOIL method.

$= x^2 - 7x + 10$ Combine like terms.

Thus,

$$(fg)(x) = x^2 - 7x + 10.$$

b. We use the product function to find $(fg)(1)$—that is, the value of the function fg at 1. Replace x with 1.

$$(fg)(1) = 1^2 - 7 \cdot 1 + 10 = 4$$

Curious about what all this looks like? Figure 5.9 shows the graphs of the three functions $f(x) = x - 5$, $g(x) = x - 2$, and $(fg)(x) = x^2 - 7x + 10$. The three points indicated by the voice balloons show that

$$f(1) = -4, g(1) = -1, \text{ and } (fg)(1) = 4.$$

The y-coordinate of every point on the product function, fg, can be obtained by multiplying the y-coordinates of functions f and g.

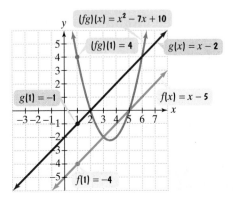

FIGURE 5.9

In Chapter 3, we introduced quadratic functions, functions of the form

$$f(x) = ax^2 + bx + c \quad \text{or} \quad y = ax^2 + bx + c, a \neq 0.$$

Functions f and g in Figure 5.9 on page 319 are linear functions of the form $f(x) = mx + b$. By contrast, the product function is a quadratic function of the form $f(x) = ax^2 + bx + c$. Take a moment to turn back a page and notice the cuplike shape of the blue graph of the quadratic function.

Are linear functions and quadratic functions also polynomial functions? The answer is yes. A linear function is a first-degree polynomial function. A quadratic function is a second-degree polynomial function.

Did you notice that the blue graph of the product function in Figure 5.9 exhibits the end behavior expected of a polynomial function? The degree is 2, which is even. Even-degree polynomial functions have graphs with the same behavior at each end. The leading coefficient, 1, is positive. Thus, the graph should rise to the left and rise to the right, and it does.

CHECK POINT 10 Let $f(x) = x - 3$ and $g(x) = x - 7$. Find:

a. $(fg)(x)$ $(fg)(x) = x^2 - 10x + 21$

b. $(fg)(2)$. $(fg)(2) = 5$

8 Use polynomial multiplication to evaluate functions.

If you are given a function, f, calculus can reveal how it is changing at any instant in time. The algebraic expression

$$\frac{f(a + h) - f(a)}{h}$$

plays an important role in this process. Our work with polynomial multiplication can be used to evaluate the numerator of this expression.

EXAMPLE 11 Using Polynomial Multiplication to Evaluate Functions

Given $f(x) = x^2 - 7x + 3$, find and simplify each of the following:

a. $f(a + 4)$ **b.** $f(a + h) - f(a)$.

SOLUTION

a. We find $f(a + 4)$, read "f at a plus 4," by replacing x with $a + 4$ each time that x appears in the polynomial.

$$f(x) \quad = \quad x^2 \quad - 7x \quad + 3$$

| Replace x with $a + 4$. | Replace x with $a + 4$. | Replace x with $a + 4$. | Copy the 3. There is no x in this term. |

$$f(a + 4) = (a + 4)^2 - 7(a + 4) + 3$$
$$= a^2 + 8a + 16 - 7a - 28 + 3 \quad \text{Multiply as indicated.}$$
$$= a^2 + a - 9 \quad \text{Combine like terms:}$$
$$8a - 7a = a \text{ and}$$
$$16 - 28 + 3 = -9.$$

b. To find $f(a + h) - f(a)$, we first replace each occurrence of x with $a + h$ and then replace each occurrence of x with a. Then we perform the resulting operations and simplify.

$$\overbrace{f(a+h)}^{} \underset{\downarrow}{-} \overbrace{f(a)}^{}$$

$$= (a+h)^2 - 7(a+h) + 3 - (a^2 - 7a + 3)$$

Use $f(x) = x^2 - 7x + 3$, replacing x with $a + h$ and a, respectively.

$$= a^2 + 2ah + h^2 - 7a - 7h + 3 - a^2 + 7a - 3$$

Multiply as indicated.

$$= 2ah + h^2 - 7h$$

Combine like terms.
Observe that
$a^2 - a^2$, $-7a + 7a$, and
$3 - 3$ all have zero sums. ■

 CHECK POINT 11 Given $f(x) = x^2 - 5x + 4$, find and simplify each of the following:

a. $f(a+3)$ $a^2 + a - 2$

b. $f(a+h) - f(a)$. $2ah + h^2 - 5h$

5.2 EXERCISE SET

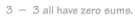
Student Solutions Manual   CD/Video   PH Math/Tutor Center   MathXL Tutorials on CD   MathXL®   MyMathLab   Interactmath.com

Practice Exercises

Throughout the practice exercises, assume that any variable exponents represent whole numbers.

In Exercises 1–8, multiply the monomials.

1. $(3x^2)(5x^4)$ $15x^6$ **2.** $(4x^2)(6x^4)$ $24x^6$

3. $(3x^2y^4)(5xy^7)$ $15x^3y^{11}$ **4.** $(6x^4y^2)(3x^7y)$ $18x^{11}y^3$

5. $(-3xy^2z^5)(2xy^7z^4)$ $-6x^2y^9z^9$

6. $(11x^2yz^4)(-3xy^5z^6)$ $-33x^3y^6z^{10}$

7. $(-8x^{2n}y^{n-5})\left(-\dfrac{1}{4}x^ny^3\right)$ $2x^{3n}y^{n-2}$

8. $(-9x^{3n}y^{n-3})\left(-\dfrac{1}{3}x^ny^2\right)$ $3x^{4n}y^{n-1}$

In Exercises 9–22, multiply the monomial and the polynomial.

9. $4x^2(3x + 2)$ $12x^3 + 8x^2$ **10.** $5x^2(6x + 7)$ $30x^3 + 35x^2$

11. $2y(y^2 - 5y)$ $2y^3 - 10y^2$ **12.** $3y(y^2 - 4y)$ $3y^3 - 12y^2$

13. $5x^3(2x^5 - 4x^2 + 9)$ $10x^8 - 20x^5 + 45x^3$

14. $6x^3(3x^5 - 5x^2 + 7)$ $18x^8 - 30x^5 + 42x^3$

15. $4xy(7x + 3y)$ $28x^2y + 12xy^2$

16. $5xy(8x + 3y)$ $40x^2y + 15xy^2$

17. $3ab^2(6a^2b^3 + 5ab)$ $18a^3b^5 + 15a^2b^3$

18. $5ab^2(10a^2b^3 + 7ab)$ $50a^3b^5 + 35a^2b^3$

19. $-4x^2y(3x^4y^2 - 7xy^3 + 6)$ $-12x^6y^3 + 28x^3y^4 - 24x^2y$

20. $-3x^2y(10x^2y^4 - 2xy^3 + 7)$ $-30x^4y^5 + 6x^3y^4 - 21x^2y$

21. $-4x^n\left(3x^{2n} - 5x^n + \dfrac{1}{2}x\right)$ $-12x^{3n} + 20x^{2n} - 2x^{n+1}$

22. $-10x^n\left(4x^{2n} - 3x^n + \dfrac{1}{5}x\right)$ $-40x^{3n} + 30x^{2n} - 2x^{n+1}$

In Exercises 23–34, find each product using either a horizontal or a vertical format.

23. $(x - 3)(x^2 + 2x + 5)$ $x^3 - x^2 - x - 15$

24. $(x + 4)(x^2 - 5x + 8)$ $x^3 - x^2 - 12x + 32$

25. $(x - 1)(x^2 + x + 1)$ $x^3 - 1$

26. $(x - 2)(x^2 + 2x + 4)$ $x^3 - 8$

27. $(a - b)(a^2 + ab + b^2)$ $a^3 - b^3$

28. $(a + b)(a^2 - ab + b^2)$ $a^3 + b^3$

29. $(x^2 + 2x - 1)(x^2 + 3x - 4)$ $x^4 + 5x^3 + x^2 - 11x + 4$

30. $(x^2 - 2x + 3)(x^2 + x + 1)$ $x^4 - x^3 + 2x^2 + x + 3$

31. $(x - y)(x^2 - 3xy + y^2)$ $x^3 - 4x^2y + 4xy^2 - y^3$

32. $(x - y)(x^2 - 4xy + y^2)$ $x^3 - 5x^2y + 5xy^2 - y^3$

33. $(xy + 2)(x^2y^2 - 2xy + 4)$ $x^3y^3 + 8$

34. $(xy + 3)(x^2y^2 - 2xy + 5)$ $x^3y^3 + x^2y^2 - xy + 15$

In Exercises 35–54, use the FOIL method to multiply the binomials.

35. $(x + 4)(x + 7)$ $x^2 + 11x + 28$

36. $(x + 5)(x + 8)$ $x^2 + 13x + 40$

37. $(y + 5)(y - 6)$ $y^2 - y - 30$

38. $(y + 5)(y - 8)$ $y^2 - 3y - 40$

39. $(5x + 3)(2x + 1)$ $10x^2 + 11x + 3$

40. $(4x + 3)(5x + 1)$ $20x^2 + 19x + 3$

41. $(3y - 4)(2y - 1)$ $6y^2 - 11y + 4$

42. $(5y - 2)(3y - 1)$ $15y^2 - 11y + 2$

43. $(3x - 2)(5x - 4)$ $\quad 15x^2 - 22x + 8$

44. $(2x - 3)(4x - 5)$ $\quad 8x^2 - 22x + 15$

45. $(x - 3y)(2x + 7y)$ $\quad 2x^2 + xy - 21y^2$

46. $(3x - y)(2x + 5y)$ $\quad 6x^2 + 13xy - 5y^2$

47. $(7xy + 1)(2xy - 3)$ $\quad 14x^2y^2 - 19xy - 3$

48. $(3xy - 1)(5xy + 2)$ $\quad 15x^2y^2 + xy - 2$

49. $(x - 4)(x^2 - 5)$ $\quad x^3 - 4x^2 - 5x + 20$

50. $(x - 5)(x^2 - 3)$ $\quad x^3 - 5x^2 - 3x + 15$

51. $(8x^3 + 3)(x^2 - 5)$ $\quad 8x^5 - 40x^3 + 3x^2 - 15$

52. $(7x^3 + 5)(x^2 - 2)$ $\quad 7x^5 - 14x^3 + 5x^2 - 10$

53. $(3x^n - y^n)(x^n + 2y^n)$ $\quad 3x^{2n} + 5x^ny^n - 2y^{2n}$

54. $(5x^n - y^n)(x^n + 4y^n)$ $\quad 5x^{2n} + 19x^ny^n - 4y^{2n}$

In Exercises 55–68, multiply using one of the rules for the square of a binomial.

55. $(x + 3)^2$ $\quad x^2 + 6x + 9$ **56.** $(x + 4)^2$ $\quad x^2 + 8x + 16$

57. $(y - 5)^2$ $\quad y^2 - 10y + 25$ **58.** $(y - 6)^2$ $\quad y^2 - 12y + 36$

59. $(2x + y)^2$ $\quad 4x^2 + 4xy + y^2$

60. $(4x + y)^2$ $\quad 16x^2 + 8xy + y^2$

61. $(5x - 3y)^2$ $\quad 25x^2 - 30xy + 9y^2$

62. $(3x - 4y)^2$ $\quad 9x^2 - 24xy + 16y^2$

63. $(2x^2 + 3y)^2$ $\quad 4x^4 + 12x^2y + 9y^2$

64. $(4x^2 + 5y)^2$ $\quad 16x^4 + 40x^2y + 25y^2$

65. $(4xy^2 - xy)^2$ $\quad 16x^2y^4 - 8x^2y^3 + x^2y^2$

66. $(5xy^2 - xy)^2$ $\quad 25x^2y^4 - 10x^2y^3 + x^2y^2$

67. $(a^n + 4b^n)^2$ $\quad a^{2n} + 8a^nb^n + 16b^{2n}$

68. $(3a^n - b^n)^2$ $\quad 9a^{2n} - 6a^nb^n + b^{2n}$

In Exercises 69–82, multiply using the rule for the product of the sum and difference of two terms.

69. $(x + 4)(x - 4)$ $\quad x^2 - 16$ **70.** $(x + 5)(x - 5)$ $\quad x^2 - 25$

71. $(5x + 3)(5x - 3)$ $\quad 25x^2 - 9$

72. $(3x + 2)(3x - 2)$ $\quad 9x^2 - 4$

73. $(4x + 7y)(4x - 7y)$ $\quad 16x^2 - 49y^2$

74. $(8x + 7y)(8x - 7y)$ $\quad 64x^2 - 49y^2$

75. $(y^3 + 2)(y^3 - 2)$ $\quad y^6 - 4$

76. $(y^3 + 3)(y^3 - 3)$ $\quad y^6 - 9$

77. $(1 - y^5)(1 + y^5)$ $\quad 1 - y^{10}$

78. $(2 - y^5)(2 + y^5)$ $\quad 4 - y^{10}$

79. $(7xy^2 - 10y)(7xy^2 + 10y)$ $\quad 49x^2y^4 - 100y^2$

80. $(3xy^2 - 4y)(3xy^2 + 4y)$ $\quad 9x^2y^4 - 16y^2$

81. $(5a^n - 7)(5a^n + 7)$ $\quad 25a^{2n} - 49$

82. $(10b^n - 3)(10b^n + 3)$ $\quad 100b^{2n} - 9$

In Exercises 83–94, find each product.

83. $[(2x + 3) + 4y][(2x + 3) - 4y]$ $\quad 4x^2 + 12x + 9 - 16y^2$

84. $[(3x + 2) + 5y][(3x + 2) - 5y]$ $\quad 9x^2 + 12x + 4 - 25y^2$

85. $(x + y + 3)(x + y - 3)$ $\quad x^2 + 2xy + y^2 - 9$

86. $(x + y + 4)(x + y - 4)$ $\quad x^2 + 2xy + y^2 - 16$

87. $(5x + 7y - 2)(5x + 7y + 2)$ $\quad 25x^2 + 70xy + 49y^2 - 4$

88. $(7x + 5y - 2)(7x + 5y + 2)$ $\quad 49x^2 + 70xy + 25y^2 - 4$

89. $[5y + (2x + 3)][5y - (2x + 3)]$ $\quad 25y^2 - 4x^2 - 12x - 9$

90. $[8y + (3x + 2)][8y - (3x + 2)]$ $\quad 64y^2 - 9x^2 - 12x - 4$

91. $(x + y + 1)^2$ $\quad x^2 + 2xy + y^2 + 2x + 2y + 1$

92. $(x + y + 2)^2$ $\quad x^2 + 2xy + y^2 + 4x + 4y + 4$

93. $(x + 1)(x - 1)(x^2 + 1)$ $\quad x^4 - 1$

94. $(x + 2)(x - 2)(x^2 + 4)$ $\quad x^4 - 16$

95. Let $f(x) = x - 2$ and $g(x) = x + 6$. Find each of the following.

   a. $(fg)(x)$ $\quad x^2 + 4x - 12$

   b. $(fg)(-1)$ $\quad -15$

   c. $(fg)(0)$ $\quad -12$

96. Let $f(x) = x - 4$ and $g(x) = x + 10$. Find each of the following.

   a. $(fg)(x)$ $\quad x^2 + 6x - 40$

   b. $(fg)(-1)$ $\quad -45$

   c. $(fg)(0)$ $\quad -40$

97. Let $f(x) = x - 3$ and $g(x) = x^2 + 3x + 9$. Find each of the following.

   a. $(fg)(x)$ $\quad x^3 - 27$

   b. $(fg)(-2)$ $\quad -35$

   c. $(fg)(0)$ $\quad -27$

98. Let $f(x) = x + 3$ and $g(x) = x^2 - 3x + 9$. Find each of the following.

   a. $(fg)(x)$ $\quad x^3 + 27$

   b. $(fg)(-2)$ $\quad 19$

   c. $(fg)(0)$ $\quad 27$

In Exercises 99–102, find each of the following and simplify:

   a. $f(a + 2)$ **b.** $f(a + h) - f(a)$.

99. $f(x) = x^2 - 3x + 7$ **a.** $a^2 + a + 5$ **b.** $2ah + h^2 - 3h$

100. $f(x) = x^2 - 4x + 9$ **a.** $a^2 + 5$ **b.** $2ah + h^2 - 4h$

101. $f(x) = 3x^2 + 2x - 1$ **a.** $3a^2 + 14a + 15$ **b.** $6ah + 3h^2 + 2h$

102. $f(x) = 4x^2 + 5x - 1$ **a.** $4a^2 + 21a + 25$ **b.** $8ah + 4h^2 + 5h$

Practice Plus

In Exercises 103–112, perform the indicated operation or operations.

103. $(3x + 4y)^2 - (3x - 4y)^2$ $\quad 48xy$

104. $(5x + 2y)^2 - (5x - 2y)^2$ $\quad 40xy$

105. $(5x - 7)(3x - 2) - (4x - 5)(6x - 1)$ $\quad -9x^2 + 3x + 9$

106. $(3x + 5)(2x - 9) - (7x - 2)(x - 1)$ $\quad -x^2 - 8x - 47$

107. $(2x + 5)(2x - 5)(4x^2 + 25)$ $\quad 16x^4 - 625$

108. $(3x + 4)(3x - 4)(9x^2 + 16)$ $\quad 81x^4 - 256$

109. $(x - 1)^3$ $\quad x^3 - 3x^2 + 3x - 1$

110. $(x - 2)^3$ `$x^3 - 6x^2 + 12x - 8$`

111. $\dfrac{(2x - 7)^5}{(2x - 7)^3}$ $4x^2 - 28x + 49$

112. $\dfrac{(5x - 3)^6}{(5x - 3)^4}$ $25x^2 - 30x + 9$

Application Exercises

In Exercises 113–114, find the area of the large rectangle in two ways:

a. *Find the sum of the areas of the four smaller rectangles.*

b. *Multiply the length and the width of the large rectangle using the FOIL method. Compare this product with your answer to part (a).*

113.
a. $x^2 + 6x + 4x + 24$
b. $(x + 6)(x + 4) = x^2 + 10x + 24$

114.
a. $x^2 + 4x + 3x + 12$
b. $(x + 4)(x + 3) = x^2 + 7x + 12$

In Exercises 115–116, express each polynomial in standard form— that is, in descending powers of x.

a. *Write a polynomial that represents the area of the large rectangle.*
b. *Write a polynomial that represents the area of the small, unshaded rectangle.*
c. *Write a polynomial that represents the area of the shaded region.*

115.
a. $x^2 + 12x + 27$
b. $x^2 + 6x + 5$
c. $6x + 22$

116.
a. $x^2 + 7x + 12$
b. $x^2 + 3x + 2$
c. $4x + 10$

In Exercises 117–118, express each polynomial in standard form.

a. *Write a polynomial that represents the area of of the rectangular base of the open box.*
b. *Write a polynomial that represents the volume of the open box.*

117.
a. $4x^2 - 36x + 80$
b. $4x^3 - 36x^2 + 80x$

118.
a. $4x^2 - 26x + 40$
b. $4x^3 - 26x^2 + 40x$

Writing in Mathematics   119–127. Answers will vary.

119. Explain how to multiply monomials. Give an example.

120. Explain how to multiply a monomial and a polynomial that is not a monomial. Give an example.

121. Explain how to multiply a binomial and a trinomial.

122. What is the FOIL method and when is it used? Give an example of the method.

123. Explain how to square a binomial sum. Give an example.

124. Explain how to square a binomial difference. Give an example.

125. Explain how to find the product of the sum and difference of two terms. Give an example with your explanation.

126. How can the graph of function fg be obtained from the graphs of functions f and g?

127. Explain how to find $f(a + h) - f(a)$ for a given function f.

Technology Exercises

In Exercises 128–131, use a graphing utility to graph the functions y_1 and y_2. Select a viewing rectangle that is large enough to show the end behavior of y_2. What can you conclude? Verify your conclusions using polynomial multiplication.
128–131. See graphing answer section.

128. $y_1 = (x - 2)^2$
$y_2 = x^2 - 4x + 4$ conclusion: $y_1 = y_2$

129. $y_1 = (x - 4)(x^2 - 3x + 2)$
$y_2 = x^3 - 7x^2 + 14x - 8$ conclusion: $y_1 = y_2$

130. $y_1 = (x - 1)(x^2 + x + 1)$
$y_2 = x^3 - 1$ conclusion: $y_1 = y_2$

131. $y_1 = (x + 1.5)(x - 1.5)$
$y_2 = x^2 - 2.25$ conclusion: $y_1 = y_2$

132. Graph $f(x) = x + 4$, $g(x) = x - 2$, and the product function, fg, in a $[-6, 6, 1]$ by $[-10, 10, 1]$ viewing rectangle. Trace along the curves and show that $(fg)(1) = f(1) \cdot g(1)$. See graphing answer section.

Critical Thinking Exercises

133. Which one of the following is true?   **d**

a. If f is a polynomial function, then
$f(a + h) - f(a) = f(a) + f(h) - f(a) = f(h)$.

b. $(x - 5)^2 = x^2 - 5x + 25$

c. $(x + 1)^2 = x^2 + 1$

d. Suppose a square garden has an area represented by $9x^2$ square feet. If one side is made 7 feet longer and the other side is made 2 feet shorter, then the trinomial that represents the area of the larger garden is $9x^2 + 15x - 14$ square feet.

134. Express the area of the plane figure shown as a polynomial in standard form. $x^2 + 2x$

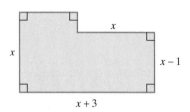

In Exercises 135–136, represent the volume of each figure as a polynomial in standard form.

135.

$x^3 + 7x^2 - 3x$

136.

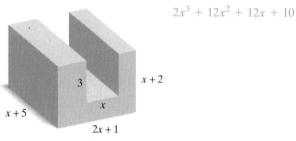

$2x^3 + 12x^2 + 12x + 10$

137. Simplify: $(y^n + 2)(y^n - 2) - (y^n - 3)^2$. $6y^n - 13$

138. The product of two consecutive odd integers is 22 less than the square of the greater integer. Find the integers. 9 and 11

Review Exercises

139. Solve: $|3x + 4| \geq 10$. (Section 4.3, Example 4)
$\left\{x \mid x \leq -\frac{14}{3} \text{ or } x \geq 2\right\}$ or $\left(-\infty, -\frac{14}{3}\right] \cup [2, \infty)$

140. Solve: $2 - 6x \leq 20$. (Section 4.1, Example 3)
$\{x \mid x \geq -3\}$ or $[-3, \infty)$

141. Write in scientific notation: 8,034,000,000. (Section 1.7, Example 2) 8.034×10^9

SECTION 5.3

GREATEST COMMON FACTORS AND FACTORING BY GROUPING

Objectives

1 Factor out the greatest common factor of a polynomial.

2 Factor out a common factor with a negative coefficient.

3 Factor by grouping.

The inability to understand numbers and their meanings is called *innumeracy* by mathematics professor John Allen Paulos. Paulos has written a book about mathematical illiteracy. Entitled *Innumeracy*, the book seeks to explain why so many people are numerically inept and to show how the problem can be corrected.

Jasper Johns *0 Through 9*, 1961. Oil on canvas, 137 × 105 cm. The Saatchi Collection, Courtesy of the Leo Castelli Gallery. © Jasper Johns/Licensed by VAGA, New York, NY.

Did you know that one of the most common ways that you are given numerical information is with percents? Unfortunately, many people are innumerate when it comes to this topic. For example, a computer whose price has been reduced by 40% and then another 40% is not selling at 20% of its original price.

To cure this bout of innumeracy (see Exercise 81 in Exercise Set 5.3), we turn to a process that reverses polynomial multiplication. For example, we can multiply polynomials and show that

$$7x(3x + 4) = 21x^2 + 28x.$$

We can also reverse this process and express the resulting polynomial as

$$21x^2 + 28x = 7x(3x + 4).$$

Factoring a polynomial containing the sum of monomials means finding an equivalent expression that is a product.

Factoring $21x^2 + 28x$

Sum of monomials	Equivalent expression that is a product

$$21x^2 + 28x = 7x(3x + 4).$$

The factors of $21x^2 + 28x$ are $7x$ and $3x + 4$.

In this chapter, we will be factoring over the set of integers, meaning that the coefficients in the factors are integers. Polynomials that cannot be factored using integer coefficients are called **prime polynomials over the set of integers**.

1 Factor out the greatest common factor of a polynomial.

Factoring Out the Greatest Common Factor In any factoring problem, the first step is to look for the *greatest common factor*. The **greatest common factor**, abbreviated GCF, is an expression of the highest degree that divides each term of the polynomial. Can you see that $7x$ is the greatest common factor of $21x^2 + 28x$? 7 is the greatest integer that divides both 21 and 28. Furthermore, x is the greatest expression that divides x^2 and x.

The variable part of the greatest common factor always contains the smallest power of a variable that appears in all terms of the polynomial. For example, consider the polynomial

$$21x^2 + 28x.$$

x^1, or x, is the variable raised to the smallest exponent.

We see that x is the variable part of the greatest common factor, $7x$.

When factoring a monomial from a polynomial, determine the greatest common factor of all terms in the polynomial. Sometimes there may not be a GCF other than 1. When a GCF other than 1 exists, we use the following procedure:

FACTORING A MONOMIAL FROM A POLYNOMIAL

1. Determine the greatest common factor of all terms in the polynomial.

2. Express each term as the product of the GCF and its other factor.

3. Use the distributive property to factor out the GCF.

EXAMPLE 1 Factoring Out the Greatest Common Factor

Factor: $21x^2 + 28x$.

SOLUTION The GCF of the two terms of the polynomial is $7x$.

$$21x^2 + 28x$$

$$= 7x(3x) + 7x(4) \qquad \text{Express each term as the product of the GCF and its other factor.}$$

$$= 7x(3x + 4) \qquad \text{Factor out the GCF.}$$

We can check this factorization by multiplying $7x$ and $3x + 4$, obtaining the original polynomial as the answer. ∎

✔ **CHECK POINT 1** Factor: $20x^2 + 30x$. $10x(2x + 3)$

EXAMPLE 2 Factoring Out the Greatest Common Factor

Factor:

   a. $9x^5 + 15x^3$ **b.** $16x^2y^3 - 24x^3y^4$ **c.** $12x^5y^4 - 4x^4y^3 + 2x^3y^2$.

SOLUTION

   a. First, determine the greatest common factor.

> 3 is the greatest integer that divides 9 and 15.

$$9x^5 + 15x^3$$

> x^3 is the variable raised to the smallest exponent.

The GCF of the two terms of the polynomial is $3x^3$.

$$9x^5 + 15x^3$$

$$= 3x^3 \cdot 3x^2 + 3x^3 \cdot 5 \qquad \text{Express each term as the product of the GCF and its other factor.}$$

$$= 3x^3(3x^2 + 5) \qquad \text{Factor out the GCF.}$$

   b. Begin by determining the greatest common factor.

> 8 is the greatest integer that divides 16 and 24.

$$16x^2y^3 - 24x^3y^4$$

> The variables raised to the smallest exponents are x^2 and y^3.

The GCF of the two terms of the polynomial is $8x^2y^3$.

$$16x^2y^3 - 24x^3y^4$$

$$= 8x^2y^3 \cdot 2 - 8x^2y^3 \cdot 3xy \qquad \text{Express each term as the product of the GCF and its other factor.}$$

$$= 8x^2y^3(2 - 3xy) \qquad \text{Factor out the GCF.}$$

c. First, determine the greatest common factor of the three terms.

> 2 is the greatest integer that divides 12, 4, and 2.

$$12x^5y^4 - 4x^4y^3 + 2x^3y^2$$

> The variables raised to the smallest exponents are x^3 and y^2.

The GCF is $2x^3y^2$.

$$12x^5y^4 - 4x^4y^3 + 2x^3y^2$$
$$= 2x^3y^2 \cdot 6x^2y^2 - 2x^3y^2 \cdot 2xy + 2x^3y^2 \cdot 1 \qquad \text{Express each term as the product of the GCF and its other factor.}$$

$$= 2x^3y^2(6x^2y^2 - 2xy + 1) \qquad \text{Factor out the GCF.}$$

Because factoring reverses the process of multiplication, all factoring results can be checked by multiplying. Take a few minutes to check each of the three factorizations in Example 2. Use the distributive property to multiply the factors. This should give the original polynomial.

> ✓ CHECK POINT **2** Factor:
>
> **a.** $9x^4 + 21x^2$ $3x^2(3x^2 + 7)$
> **b.** $15x^3y^2 - 25x^4y^3$ $5x^3y^2(3 - 5xy)$
> **c.** $16x^4y^5 - 8x^3y^4 + 4x^2y^3$. $4x^2y^3(4x^2y^2 - 2xy + 1)$

2   Factor out a common factor with a negative coefficient.

When the leading coefficient of a polynomial is negative, it is often desirable to factor out a common factor with a negative coefficient. The common factor is the GCF preceded by a negative sign.

EXAMPLE 3   Using a Common Factor with a Negative Coefficient

Factor: $-3x^3 + 12x^2 - 15x$.

SOLUTION   The GCF is $3x$. Because the leading coefficient, -3, is negative, we factor out a common factor with a negative coefficient. We will factor out the negative of the GCF, or $-3x$.

$$-3x^3 + 12x^2 - 15x$$
$$= -3x(x^2) - 3x(-4x) - 3x(5) \qquad \text{Express each term as the product of the common factor and its other factor.}$$

$$= -3x(x^2 - 4x + 5) \qquad \text{Factor out the GCF.}$$

> ✓ CHECK POINT **3** Factor out a common factor with a negative coefficient:
> $-2x^3 + 10x^2 - 6x$. $-2x(x^2 - 5x + 3)$

Factoring by Grouping   Up to now, we have factored a monomial from a polynomial. By contrast, in our next example, the greatest common factor of the polynomial is a binomial.

EXAMPLE 4 Factoring Out the Greatest Common Binomial Factor

Factor:

a. $2(x - 7) + 9a(x - 7)$ **b.** $5y(a - b) - (a - b)$.

SOLUTION Let's identify the common binomial factor in each part of the problem.

$$2(x - 7) + 9a(x - 7) \qquad 5y(a - b) - (a - b)$$

The GCF, a binomial, is $x - 7$.   The GCF, a binomial, is $a - b$.

We factor out each common binomial factor as follows.

a. $2(x - 7) + 9a(x - 7)$

$= (x - 7)2 + (x - 7)9a$ This step, usually omitted, shows each term as the product of the GCF and its other factor, in that order.

$= (x - 7)(2 + 9a)$ Factor out the GCF.

b. $5y(a - b) - (a - b)$

$= 5y(a - b) - 1(a - b)$ Write $-(a - b)$ as $-1(a - b)$ to aid in the factoring.

$= (a - b)(5y - 1)$ Factor out the GCF.   ■

CHECK POINT 4 Factor:

a. $3(x - 4) + 7a(x - 4)$ $(x - 4)(3 + 7a)$

b. $7x(a + b) - (a + b)$. $(a + b)(7x - 1)$

③ Factor by grouping.

Some polynomials have only a greatest common factor of 1. However, by a suitable grouping of the terms, it still may be possible to factor. This process, called **factoring by grouping**, is illustrated in Example 5.

EXAMPLE 5 Factoring by Grouping

Factor: $x^3 - 5x^2 + 3x - 15$.

SOLUTION There is no factor other than 1 common to all terms. However, we can group terms that have a common factor:

$$\boxed{x^3 - 5x^2} + \boxed{3x - 15}$$

Common factor is x^2.   Common factor is 3.

DISCOVER FOR YOURSELF

In Example 5, group the terms as follows:

$$x^2 + 3x - 5x^2 - 15.$$

Factor out the common factor from each group and complete the factoring process. Describe what happens. What can you conclude?

We now factor the given polynomial as follows:

$x^3 - 5x^2 + 3x - 15$

$= (x^3 - 5x^2) + (3x - 15)$ Group terms with common factors.

$= x^2(x - 5) + 3(x - 5)$ Factor out the greatest common factor from the grouped terms. The remaining two terms have $x - 5$ as a common binomial factor.

$= (x - 5)(x^2 + 3)$.   Factor out the GCF.

Thus, $x^3 - 5x^2 + 3x - 15 = (x - 5)(x^2 + 3)$. Check the factorization by multiplying the right side of the equation using the FOIL method. Because the factorization is correct, you should obtain the original polynomial. ■

 CHECK POINT 5 Factor: $x^3 - 4x^2 + 5x - 20$. $(x - 4)(x^2 + 5)$

FACTORING BY GROUPING

1. Group terms that have a common monomial factor. There will usually be two groups. Sometimes the terms must be rearranged.

2. Factor out the common monomial factor from each group.

3. Factor out the remaining common binomial factor (if one exists).

EXAMPLE 6 Factoring by Grouping

Factor: $3x^2 + 12x - 2xy - 8y$.

SOLUTION There is no factor other than 1 common to all terms. However, we can group terms that have a common factor:

$$\boxed{3x^2 + 12x} \; + \; \boxed{-2xy - 8y}\;.$$

Common factor is $3x$:
$3x^2 + 12x = 3x(x + 4)$.

Use $-2y$, rather than $2y$, as the common factor:
$-2xy - 8y = -2y(x + 4)$. In this way, the common binomial factor, $x + 4$, appears.

The voice balloons illustrate that it is sometimes necessary to use a factor with a negative coefficient to obtain a common binomial factor for the two groupings. We now factor the given polynomial as follows:

$$
\begin{aligned}
3x^2 + 12x &- 2xy - 8y \\
&= (3x^2 + 12x) + (-2xy - 8y) \quad &&\text{Group terms with common factors.} \\
&= 3x(x + 4) - 2y(x + 4) \quad &&\text{Factor out the common factors from the grouped terms.} \\
&= (x + 4)(3x - 2y). \quad &&\text{Factor out the GCF.}
\end{aligned}
$$

Thus, $3x^2 + 12x - 2xy - 8y = (x + 4)(3x - 2y)$. Using the commutative property of multiplication, the factorization can also be expressed as $(3x - 2y)(x + 4)$. Verify the factorization by showing that, regardless of the order, FOIL multiplication gives the original polynomial. ■

 CHECK POINT 6 Factor: $4x^2 + 20x - 3xy - 15y$. $(x + 5)(4x - 3y)$

Factoring by grouping sometimes requires that the terms be rearranged before the groupings are made. For example, consider the polynomial

$$3x^2 - 8y + 12x - 2xy.$$

The first two terms have no common factor other than 1. We must rearrange the terms and try a different grouping. Example 6 showed one such rearrangement of two groupings.

5.3 EXERCISE SET

Student Solutions Manual   CD/Video   PH Math/Tutor Center   MathXL Tutorials on CD   MathXL®   MyMathLab   Interactmath.com

Practice Exercises

Throughout the practice exercises, assume that any variable exponents represent whole numbers.

In Exercises 1–22, factor the greatest common factor from each polynomial.

1. $10x^2 + 4x$ $2x(5x + 2)$
2. $12x^2 + 9x$ $3x(4x + 3)$
3. $y^2 - 4y$ $y(y - 4)$
4. $y^2 - 7y$ $y(y - 7)$
5. $x^3 + 5x^2$ $x^2(x + 5)$
6. $x^3 + 7x^2$ $x^2(x + 7)$
7. $12x^4 - 8x^2$ $4x^2(3x^2 - 2)$
8. $20x^4 - 8x^2$ $4x^2(5x^2 - 2)$
9. $32x^4 + 2x^3 + 8x^2$ $2x^2(16x^2 + x + 4)$
10. $9x^4 + 18x^3 + 6x^2$ $3x^2(3x^2 + 6x + 2)$
11. $4x^2y^3 + 6xy$ $2xy(2xy^2 + 3)$
12. $6x^3y^2 + 9xy$ $3xy(2x^2y + 3)$
13. $30x^2y^3 - 10xy^2$ $10xy^2(3xy - 1)$
14. $27x^2y^3 - 18xy^2$ $9xy^2(3xy - 2)$
15. $12xy - 6xz + 4xw$ $2x(6y - 3z + 2w)$
16. $14xy - 10xz + 8xw$ $2x(7y - 5z + 4w)$
17. $15x^3y^6 - 9x^4y^4 + 12x^2y^5$ $3x^2y^4(5xy^2 - 3x^2 + 4y)$
18. $15x^4y^6 - 3x^3y^5 + 12x^4y^4$ $3x^3y^4(5xy^2 - y + 4x)$
19. $25x^3y^6z^2 - 15x^4y^4z^4 + 25x^2y^5z^3$ $5x^2y^4z^2(5xy^2 - 3x^2z^2 + 5yz)$
20. $49x^4y^3z^5 - 70x^3y^5z^4 + 35x^4y^4z^3$ $7x^3y^3z^3(7xz^2 - 10y^2z + 5xy)$
21. $15x^{2n} - 25x^n$ $5x^n(3x^n - 5)$
22. $12x^{3n} - 9x^{2n}$ $3x^{2n}(4x^n - 3)$

In Exercises 23–34, factor out the negative of the greatest common factor.

23. $-4x + 12$ $-4(x - 3)$
24. $-5x + 20$ $-5(x - 4)$
25. $-8x - 48$ $-8(x + 6)$
26. $-7x - 63$ $-7(x + 9)$
27. $-2x^2 + 6x - 14$ $-2(x^2 - 3x + 7)$
28. $-2x^2 + 8x - 12$ $-2(x^2 - 4x + 6)$
29. $-5y^2 + 40x$ $-5(y^2 - 8x)$
30. $-9y^2 + 45x$ $-9(y^2 - 5x)$
31. $-4x^3 + 32x^2 - 20x$ $-4x(x^2 - 8x + 5)$
32. $-5x^3 + 50x^2 - 10x$ $-5x(x^2 - 10x + 2)$
33. $-x^2 - 7x + 5$ $-1(x^2 + 7x - 5)$
34. $-x^2 - 8x + 8$ $-1(x^2 + 8x - 8)$

In Exercises 35–44, factor the greatest common binomial factor from each polynomial.

35. $4(x + 3) + a(x + 3)$ $(x + 3)(4 + a)$
36. $5(x + 4) + a(x + 4)$ $(x + 4)(5 + a)$
37. $x(y - 6) - 7(y - 6)$ $(y - 6)(x - 7)$
38. $x(y - 9) - 5(y - 9)$ $(y - 9)(x - 5)$
39. $3x(x + y) - (x + y)$ $(x + y)(3x - 1)$
40. $7x(x + y) - (x + y)$ $(x + y)(7x - 1)$
41. $4x^2(3x - 1) + 3x - 1$ $(3x - 1)(4x^2 + 1)$
42. $6x^2(5x - 1) + 5x - 1$ $(5x - 1)(6x^2 + 1)$
43. $(x + 2)(x + 3) + (x - 1)(x + 3)$ $(x + 3)(2x + 1)$
44. $(x + 4)(x + 5) + (x - 1)(x + 5)$ $(x + 5)(2x + 3)$

In Exercises 45–68, factor by grouping.

45. $x^2 + 3x + 5x + 15$ $(x + 3)(x + 5)$
46. $x^2 + 2x + 4x + 8$ $(x + 2)(x + 4)$
47. $x^2 + 7x - 4x - 28$ $(x + 7)(x - 4)$
48. $x^2 + 3x - 5x - 15$ $(x + 3)(x - 5)$
49. $x^3 - 3x^2 + 4x - 12$ $(x - 3)(x^2 + 4)$
50. $x^3 - 2x^2 + 5x - 10$ $(x - 2)(x^2 + 5)$
51. $xy - 6x + 2y - 12$ $(y - 6)(x + 2)$
52. $xy - 5x + 9y - 45$ $(y - 5)(x + 9)$
53. $xy + x - 7y - 7$ $(y + 1)(x - 7)$
54. $xy + x - 5y - 5$ $(y + 1)(x - 5)$
55. $10x^2 - 12xy + 35xy - 42y^2$ $(5x - 6y)(2x + 7y)$
56. $3x^2 - 6xy + 5xy - 10y^2$ $(x - 2y)(3x + 5y)$
57. $4x^3 - x^2 - 12x + 3$ $(4x - 1)(x^2 - 3)$
58. $3x^3 - 2x^2 - 6x + 4$ $(3x - 2)(x^2 - 2)$
59. $x^2 - ax - bx + ab$ $(x - a)(x - b)$
60. $x^2 + ax - bx - ab$ $(x + a)(x - b)$
61. $x^3 - 12 - 3x^2 + 4x$ $(x - 3)(x^2 + 4)$

62. $2x^3 - 10 + 4x^2 - 5x$ $(x + 2)(2x^2 - 5)$

63. $ay - by + bx - ax$ $(a - b)(y - x)$·

64. $cx - dx + dy - cy$ $(c - d)(x - y)$

65. $ay^2 + 2by^2 - 3ax - 6bx$ $(a + 2b)(y^2 - 3x)$

66. $3a^2x + 6a^2y - 2bx - 4by$ $(3a^2 - 2b)(x + 2y)$

67. $x^n y^n + 3x^n + y^n + 3$ $(x^n + 1)(y^n + 3)$

68. $x^n y^n - x^n + 2y^n - 2$ $(x^n + 2)(y^n - 1)$

Practice Plus

In Exercises 69–78, factor each polynomial.

69. $ab - c - ac + b$ $(a + 1)(b - c)$

70. $ab - 3c - ac + 3b$ $(a + 3)(b - c)$

71. $x^3 - 5 + 4x^3y - 20y$ $(x^3 - 5)(1 + 4y)$

72. $x^3 - 2 + 3x^3y - 6y$ $(x^3 - 2)(1 + 3y)$

73. $2y^7(3x - 1)^5 - 7y^6(3x - 1)^4$ $y^6(3x - 1)^4(6xy - 2y - 7)$

74. $3y^9(3x - 2)^7 - 5y^8(3x - 2)^6$ $y^8(3x - 2)^6(9xy - 6y - 5)$

75. $ax^2 + 5ax - 2a + bx^2 + 5bx - 2b$ $(x^2 + 5x - 2)(a + b)$

76. $ax^2 + 3ax - 11a + bx^2 + 3bx - 11b$ $(x^2 + 3x - 11)(a + b)$

77. $ax + ay + az - bx - by - bz + cx + cy + cz$
$(x + y + z)(a - b + c)$

78. $ax^2 + ay^2 - az^2 + bx^2 + by^2 - bz^2 + cx^2 + cy^2 - cz^2$
$(x^2 + y^2 - z^2)(a + b + c)$

Application Exercises

79. A ball is thrown straight upward. The function

$$f(t) = -16t^2 + 40t$$

describes the ball's height above the ground, $f(t)$, in feet, t seconds after it is thrown.

   a. Find and interpret $f(2)$.   16; The ball is 16 feet above the ground after 2 seconds.

   b. Find and interpret $f(2.5)$.   0; The ball is on the ground after 2.5 seconds.

   c. Factor the polynomial $-16t^2 + 40t$ and write the function in factored form. $f(t) = -8t(2t - 5)$

   d. Use the factored form of the function to find $f(2)$ and $f(2.5)$. Do you get the same answers as you did in parts (a) and (b)? If so, does this prove that your factorization is correct? Explain.   16; 0; yes; no; Answers will vary.

80. An explosion causes debris to rise vertically. The function

$$f(t) = -16t^2 + 72t$$

describes the height of the debris above the ground, $f(t)$, in feet, t seconds after the explosion.

   a. Find and interpret $f(2)$.   80; The debris is 80 feet above the ground after 2 seconds.

   b. Find and interpret $f(4.5)$.   0; The debris is on the ground after 4.5 seconds.

   c. Factor the polynomial $-16t^2 + 72t$ and write the function in factored form. $f(t) = -8t(2t - 9)$

   d. Use the factored form of the function to find $f(2)$ and $f(4.5)$. Do you get the same answers as you did in parts (a) and (b)? If so, does this prove that your factorization is correct? Explain.   80; 0; yes; no; Answers will vary.

81. Your computer store is having an incredible sale. The price on one model is reduced by 40%. Then the sale price is reduced by another 40%. If x is the computer's original price, the sale price can be represented by

$$(x - 0.4x) - 0.4(x - 0.4x).$$

   a. Factor out $(x - 0.4x)$ from each term. Then simplify the resulting expression. $(x - 0.4x)(1 - 0.4) = (0.6x)(0.6) = 0.36x$

   b. Use the simplified expression from part (a) to answer these questions. With a 40% reduction followed by a 40% reduction, is the computer selling at 20% of its original price? If not, at what percentage of the original price is it selling?   no; The computer is selling at 36% of its original price.

82. Your local electronics store is having an end-of-the-year sale. The price on a large-screen television had been reduced by 30%. Now the sale price is reduced by another 30%. If x is the television's original price, the sale price can be represented by

$$(x - 0.3x) - 0.3(x - 0.3x).$$

   a. Factor out $(x - 0.3x)$ from each term. Then simplify the resulting expression. $(x - 0.3x)(1 - 0.3) = (0.7x)(0.7) = 0.49x$

   b. Use the simplified expression from part (a) to answer these questions. With a 30% reduction followed by a 30% reduction, is the television selling at 40% of its original price? If not, at what percentage of the original price is it selling?   no; The television is selling at 49% of its original price.

Exercises 83–84 involve compound interest. **Compound interest** *is interest computed on your original savings as well as on any accumulated interest.*

83. After 2 years, the balance, A, in an account with principal P and interest rate r compounded annually is given by the formula

$$A = P + Pr + (P + Pr)r.$$

Use factoring by grouping to express the formula as $A = P(1 + r)^2$. $A = P + Pr + (P + Pr)r$
$= P(1 + r) + Pr(1 + r) = (P + Pr)(1 + r)$
$= P(1 + r)(1 + r) = P(1 + r)^2$

84. After 3 years, the balance, A, in an account with principal P and interest rate r compounded annually is given by the formula

$$A = P(1 + r)^2 + P(1 + r)^2 r.$$

Use factoring by grouping to express the formula as $A = P(1 + r)^3$. $A = P(1 + r)^2 + P(1 + r)^2 r$
$= P(1 + r)^2(1 + r) = P(1 + r)^3$

85. The area of the skating rink with semicircular ends shown is $A = \pi r^2 + 2rl$. Express the area, A, in factored form. $A = r(\pi r + 2l)$

86. The amount of sheet metal needed to manufacture a cylindrical tin can, that is, its surface area, S, is $S = 2\pi r^2 + 2\pi rh$. Express the surface area, S, in factored form. $S = 2\pi r(r + h)$

Writing in Mathematics

87. What is factoring? 87–92. Answers will vary.

88. If a polynomial has a greatest common factor other than 1, explain how to find its GCF.

89. Using an example, explain how to factor out the greatest common factor of a polynomial.

90. Suppose that a polynomial contains four terms and can be factored by grouping. Explain how to obtain the factorization.

91. Use two different groupings to factor

$$ac - ad + bd - bc$$

in two ways. Then explain why the two factorizations are the same.

92. Write a sentence that uses the word "factor" as a noun. Then write a sentence that uses the word "factor" as a verb.

Technology Exercises

In Exercises 93–96, use a graphing utility to graph the function on each side of the equation in the same viewing rectangle. Use end behavior to show a complete picture of the polynomial function on the left side. Do the graphs coincide? If so, this means that the polynomial on the left side has been factored correctly. If not, factor the polynomial correctly and then use your graphing utility to verify the factorization. 93–96. See graphing answer section.

93. $x^2 - 4x = x(x - 4)$ Graphs coincide; factored correctly.

94. $x^2 - 2x + 5x - 10 = (x - 2)(x - 5)$ Graphs do not coincide; $x^2 - 2x + 5x - 10 = (x - 2)(x + 5)$.

95. $x^2 + 2x + x + 2 = x(x + 2) + 1$ Graphs do not coincide; $x^2 + 2x + x + 2 = (x + 2)(x + 1)$.

96. $x^3 - 3x^2 + 4x - 12 = (x^2 + 4)(x - 3)$ Graphs coincide; factored correctly.

Critical Thinking Exercises

97. Which one of the following is true? c

   a. If all terms of a polynomial contain the same variable raised to different powers, the exponent on the variable of the GCF is the highest power that appears in all the terms.

   b. Because the GCF of $9x^3 + 6x^2 + 3x$ is $3x$, it is not necessary to write the 1 when $3x$ is factored from the last term.

   c. Some polynomials with four terms, such as $x^3 + x^2 + 4x - 4$, cannot be factored by grouping.

   d. None of the above is true.

In Exercises 98–100, factor each polynomial. Assume that all variable exponents represent whole numbers.

98. $x^{4n} + x^{2n} + x^{3n}$ $x^{2n}(x^{2n} + 1 + x^n)$

99. $3x^{3m}y^m - 6x^{2m}y^{2m}$ $3x^{2m}y^m(x^m - 2y^m)$

100. $8y^{2n+4} + 16y^{2n+3} - 12y^{2n}$ $4y^{2n}(2y^4 + 4y^3 - 3)$

In Exercises 101–102, write a polynomial that fits the given description. Do not use a polynomial that appeared in this section or in the exercise set.

101. The polynomial has three terms and can be factored using a greatest common factor that has both a negative coefficient and a variable. Answers will vary; an example is $-4x^4 + 8x^3 - 16x^2$.

102. The polynomial has four terms and can be factored by grouping. Answers will vary; an example is $6x^2 - 4x + 9x - 6$.

Review Exercises

103. Solve by Cramer's rule:

$$3x - 2y = 8$$
$$2x - 5y = 10.$$

(Section 3.5, Example 2) $\left(\frac{20}{11}, -\frac{14}{11}\right)$ or $\left\{\left(\frac{20}{11}, -\frac{14}{11}\right)\right\}$

104. Determine whether each relation is a function.

   a. $\{(0, 5), (3, -5), (5, 5), (7, -5)\}$ function

   b. $\{(1, 2), (3, 4), (5, 5), (5, 6)\}$ (Section 2.1, Example 2) not a function

105. The length of a rectangle is 2 feet greater than twice its width. If the rectangle's perimeter is 22 feet, find the length and width. (Section 1.5, Example 5) The length is 8 feet and the width is 3 feet.

SECTION
5.4

FACTORING TRINOMIALS

Eadweard Muybridge (English 1830–1904), "Annie G. Cantering, Saddled," 1887. Collotype print, size: sheet: 19 in. × 24 in., image: $7\frac{1}{2}$ in. × $16\frac{1}{4}$ in. Philadelphia Museum of Art: City of Philadelphia, Trade Convention Center, Dept. of Commerce (Commercial Museum). 1962–135–280.

Objectives

1 Factor a trinomial whose leading coefficient is 1.

2 Factor using a substitution.

3 Factor a trinomial whose leading coefficient is not 1.

4 Factor trinomials by grouping.

Motion and change are the very essence of life. Moving air brushes against our faces; rain falls on our heads; birds fly past us; plants spring from the earth, grow, and then die; and rocks thrown upward reach a maximum height before falling to the ground.

How can we use mathematical models to analyze the where and when of moving objects? The first step is to learn techniques for factoring trinomials. We will use these techniques in Section 5.7 to model and visualize motion.

1 Factor a trinomial whose leading coefficient is 1.

Factoring a Trinomial Whose Leading Coefficient Is 1 In Section 5.2, we used the FOIL method to multiply two binomials. The product was often a trinomial. The following are some examples:

Factored Form	F	O	I	L	Trinomial Form
$(x + 3)(x + 4)$	=	x^2	$+ 4x$	$+ 3x$	$+ 12$ = $x^2 + 7x + 12$
$(x - 3)(x - 4)$	=	x^2	$- 4x$	$- 3x$	$+ 12$ = $x^2 - 7x + 12$
$(x + 3)(x - 5)$	=	x^2	$- 5x$	$+ 3x$	$- 15$ = $x^2 - 2x - 15$

Observe that each trinomial is of the form $x^2 + bx + c$, where the coefficient of the squared term is 1. Our goal in the first part of this section is to start with the trinomial form and, assuming that it is factorable, return to the factored form.

The first FOIL multiplication shown in our list indicates that $(x + 3)(x + 4) = x^2 + 7x + 12$. Let's reverse the sides of this equation:

$$x^2 + 7x + 12 = (x + 3)(x + 4).$$

We can make several important observations about the factors on the right side.

These observations provide us with a procedure for factoring $x^2 + bx + c$.

> ### A STRATEGY FOR FACTORING $x^2 + bx + c$
>
> **1.** Enter x as the first term of each factor.
>
> $$(x \qquad)(x \qquad) = x^2 + bx + c$$
>
> **2.** List pairs of factors of the constant c.
>
> **3.** Try various combinations of these factors. Select the combination in which the sum of the Outside and Inside products is equal to bx.
>
> $$(x + \square)(x + \square) = x^2 + bx + c$$
>
> I
>
> O
>
> Sum of O+I
>
> **4.** Check your work by multiplying the factors using the FOIL method. You should obtain the original trinomial.

If none of the possible combinations yield an Outside product and an Inside product whose sum is equal to bx, the trinomial cannot be factored using integers and is called **prime** over the set of integers.

EXAMPLE 1 Factoring a Trinomial Whose Leading Coefficient Is 1

Factor: $x^2 + 5x + 6$.

SOLUTION

Step 1. **Enter x as the first term of each factor.**

$$x^2 + 5x + 6 = (x \qquad)(x \qquad)$$

To find the second term of each factor, we must find two integers whose product is 6 and whose sum is 5.

Step 2. **List pairs of factors of the constant, 6.**

Factors of 6	6, 1	3, 2	−6, −1	−3, −2

Step 3. **Try various combinations of these factors.** The correct factorization of $x^2 + 5x + 6$ is the one in which the sum of the Outside and Inside products is equal to $5x$. Here is a list of the possible factorizations.

Possible Factorizations of $x^2 + 5x + 6$	Sum of Outside and Inside Products (Should Equal $5x$)
$(x + 6)(x + 1)$	$x + 6x = 7x$
$(x + 3)(x + 2)$	$2x + 3x = 5x$
$(x - 6)(x - 1)$	$-x - 6x = -7x$
$(x - 3)(x - 2)$	$-2x - 3x = -5x$

This is the required middle term.

Thus,

$$x^2 + 5x + 6 = (x + 3)(x + 2).$$

Check this result by multiplying the right side using the FOIL method. You should obtain the original trinomial. Because of the commutative property, the factorization can also be expressed as

$$x^2 + 5x + 6 = (x + 2)(x + 3).$$

In factoring a trinomial of the form $x^2 + bx + c$, you can speed things up by listing the factors of c and then finding their sums. We are interested in a sum of b. For example, in factoring $x^2 + 5x + 6$, we are interested in the factors of 6 whose sum is 5.

Factors of 6	6, 1	3, 2	−6, −1	−3, −2
Sum of Factors	7	5	−7	−5

This is the desired sum.

Thus, $x^2 + 5x + 6 = (x + 3)(x + 2).$

USING TECHNOLOGY

If a polynomial contains one variable, a graphing utility can be used to check its factorization. For example, the factorization in Example 1 can be checked graphically or numerically.

Given polynomial   The factorization we are checking

$$x^2 + 5x + 6 = (x + 3)(x + 2)$$

Enter $y_1 = x^2 + 5x + 6$ in the $\boxed{y=}$ screen.   Enter $y_2 = (x + 3)(x + 2)$ in the $\boxed{y=}$ screen.

Numeric Check

Use the $\boxed{\text{TABLE}}$ feature.

Graphic Check

Use the $\boxed{\text{GRAPH}}$ feature to display graphs for y_1 and y_2.

Scrolling through the table shows $y_1 = y_2$ for all displayed values of x. This verifies that $x^2 + 5x + 6 = (x + 3)(x + 2)$.

Rises left   Rises right

The graphs are identical, showing that $y_1 = y_2$ for all values of x. This verifies that $x^2 + 5x + 6 = (x + 3)(x + 2)$.

$[-7, 5, 1]$ by $[-1, 10, 1]$

Notice the cuplike shape of the graph of the quadratic function. The graph of the even-degree quadratic function exhibits the same behavior at each end, rising to the left and rising to the right.

✔ **CHECK POINT 1** Factor: $x^2 + 6x + 8$. $(x + 4)(x + 2)$ or $(x + 2)(x + 4)$

EXAMPLE 2 Factoring a Trinomial Whose Leading Coefficient Is 1

Factor: $x^2 - 14x + 24$.

SOLUTION

Step 1. **Enter x as the first term of each factor.**

$$x^2 - 14x + 24 = (x \qquad)(x \qquad)$$

To find the second term of each factor, we must find two integers whose product is 24 and whose sum is -14.

Step 2. **List pairs of factors of the constant, 24.** Because the desired sum, -14, is negative, we will list only the negative pairs of factors of 24.

| **Negative Factors of 24** | $-24, -1$ | $-12, -2$ | $-8, -3$ | $-6, -4$ |

Step 3. **Try various combinations of these factors.** We are interested in the factors whose sum is -14.

Negative Factors of 24	$-24, -1$	$-12, -2$	$-8, -3$	$-6, -4$
Sum of Factors	-25	-14	-11	-10

This is the desired sum.

Thus, $x^2 - 14x + 24 = (x - 12)(x - 2)$. ∎

STUDY TIP

To factor $x^2 + bx + c$ when c is positive, find two numbers with the same sign as the middle term.

$$x^2 + 5x + 6 = (x + 3)(x + 2)$$

Same signs

$$x^2 - 14x + 24 = (x - 12)(x - 2)$$

Same signs

✔ **CHECK POINT 2** Factor: $x^2 - 9x + 20$. $(x - 5)(x - 4)$

EXAMPLE 3 Factoring a Trinomial Whose Leading Coefficient Is 1

Factor: $y^2 + 7y - 60$

SOLUTION

Step 1. **Enter y as the first term of each factor.**

$$y^2 + 7y - 60 = (y \qquad)(y \qquad)$$

To find the second term of each factor, we must find two integers whose product is -60 and whose sum is 7.

Steps 2 and 3. List pairs of factors of the constant, -60, and try various combinations of these factors. Because the desired sum, 7, is positive, the positive factor of -60 must be farther from 0 than the negative factor is. Thus, we will only list pairs of factors of -60 in which the positive factor has the larger absolute value.

Some Factors of -60	60, -1	30, -2	20, -3	15, -4	12, -5	10, -6
Sum of Factors	59	28	17	11	7	4

This is the desired sum.

Thus, $y^2 + 7y - 60 = (y + 12)(y - 5)$.

STUDY TIP

To factor $x^2 + bx + c$ when c is negative, find two numbers with opposite signs whose sum is the coefficient of the middle term.

$$y^2 + 7y - 60 = (y + 12)(y - 5)$$

Negative   Opposite signs

 CHECK POINT 3 Factor: $y^2 + 19y - 66$. $(y + 22)(y - 3)$

EXAMPLE 4 Factoring a Trinomial in Two Variables

Factor: $x^2 - 4xy - 21y^2$.

SOLUTION

Step 1. Enter x as the first term of each factor. Because the last term of the trinomial contains y^2, the second term of each factor must contain y.

$$x^2 - 4xy - 21y^2 = (x \ ?y)(x \ ?y)$$

The question marks indicate that we are looking for the coefficients of y in each factor. To find these coefficients, we must find two integers whose product is -21 and whose sum is -4.

Steps 2 and 3. List pairs of factors of the coefficient of the last term, -21, and try various combinations of these factors. We are interested in the factors whose sum is -4.

Factors of -21	1, -21	3, -7	-1, 21	-3, 7
Sum of Factors	-20	-4	20	4

This is the desired sum.

Thus, $x^2 - 4xy - 21y^2 = (x + 3y)(x - 7y)$ or $(x - 7y)(x + 3y)$.

Step 4. Verify the factorization using the FOIL method.

$$(x + 3y)(x - 7y) = x^2 - 7xy + 3xy - 21y^2 = x^2 - 4xy - 21y^2$$

Because the product of the factors is the original polynomial, the factorization is correct.

✔ **CHECK POINT 4** Factor: $x^2 - 5xy + 6y^2$. $(x - 3y)(x - 2y)$

Can every trinomial be factored? The answer is no. For example, consider

$$x^2 + x - 5 = (x \quad)(x \quad).$$

To find the second term of each factor, we must find two integers whose product is -5 and whose sum is 1. Because no such integers exist, $x^2 + x - 5$ cannot be factored. This trinomial is prime.

To factor some polynomials, more than one technique must be used. **Always begin by trying to factor out the greatest common factor**. A polynomial is **factored completely** when it is written as the product of prime polynomials.

EXAMPLE 5 Factoring Completely

Factor: $8x^3 - 40x^2 - 48x$.

SOLUTION The GCF of the three terms of the polynomial is $8x$. We begin by factoring out $8x$. Then we factor the remaining trinomial.

$$8x^3 - 40x^2 - 48x$$
$$= 8x(x^2 - 5x - 6) \qquad \text{Factor out the GCF.}$$
$$= 8x(x \quad)(x \quad) \qquad \text{Begin factoring } x^2 - 5x - 6. \text{ Find two integers whose product is } -6 \text{ and whose sum is } -5.$$
$$= 8x(x - 6)(x + 1) \qquad \text{The integers are } -6 \text{ and } 1.$$

Thus,

$$8x^3 - 40x^2 - 48x = 8x(x - 6)(x + 1).$$

Be sure to include the GCF in the factorization.

You can check this factorization by multiplying the binomials using the FOIL method. Then use the distributive property and multiply each term in this product by $8x$. Try doing this now. Because the factorization is correct, you should obtain the original polynomial. ■

✔ **CHECK POINT 5** Factor: $3x^3 - 15x^2 - 42x$. $3x(x - 7)(x + 2)$

Some trinomials, such as $-x^2 + 5x + 6$, have a leading coefficient of -1. Because it is easier to factor a trinomial with a positive leading coefficient, begin by factoring out -1. For example,

$$-x^2 + 5x + 6 = -1(x^2 - 5x - 6) = -(x - 6)(x + 1).$$

In some trinomials, the highest power is greater than 2, and the exponent in one of the terms is half that of the other term. By letting t equal the variable to the smaller power, the trinomial can be written in a form that makes its possible factorization more obvious. Here are some examples:

2 Factor using a substitution.

Given Trinomial	Substitution	New Trinomial
$x^6 - 8x^3 + 15$ or $(x^3)^2 - 8x^3 + 15$	$t = x^3$	$t^2 - 8t + 15$
$x^4 - 8x^2 - 9$ or $(x^2)^2 - 8x^2 - 9$	$t = x^2$	$t^2 - 8t - 9$

In each case, we factor the given trinomial by working with the new trinomial on the right. If a factorization is found, we replace all occurrences of t with the substitution shown in the middle column.

EXAMPLE 6   Factoring by Substitution

Factor: $x^6 - 8x^3 + 15$.

SOLUTION   Notice that the exponent on x^3 is half that of the exponent on x^6. We will let t equal the variable to the power that is half of 6. Thus, let $t = x^3$.

$$(x^3)^2 - 8x^3 + 15 \qquad \text{This is the given polynomial, with } x^6 \text{ written as } (x^3)^2.$$
$$\downarrow \qquad \downarrow$$
$$= t^2 - 8t + 15 \qquad \text{Let } t = x^3. \text{ Rewrite the trinomial in terms of } t.$$
$$= (t - 5)(t - 3) \qquad \text{Factor.}$$
$$= (x^3 - 5)(x^3 - 3) \qquad \text{Now substitute } x^3 \text{ for } t.$$

Thus, the given trinomial can be factored as

$$x^6 - 8x^3 + 15 = (x^3 - 5)(x^3 - 3).$$

Check this result using FOIL multiplication on the right.   ∎

✔ **CHECK POINT 6** Factor: $x^6 - 7x^3 + 10$. $(x^3 - 5)(x^3 - 2)$

③ Factor a trinomial whose leading coefficient is not 1.

Factoring Trinomials Whose Leading Coefficient Is Not 1   How do we factor a trinomial such as $5x^2 - 14x + 8$? Notice that the leading coefficient is 5. We must find two binomials whose product is $5x^2 - 14x + 8$. The product of the First terms must be $5x^2$:

$$(5x \quad)(x \quad).$$

From this point on, the factoring strategy is exactly the same as the one we use to factor trinomials whose leading coefficient is 1.

A STRATEGY FOR FACTORING $ax^2 + bx + c$ Assume, for the moment, that there is no greatest common factor.

1. Find two **F**irst terms whose product is ax^2:

$$(\square x + \quad)(\square x + \quad) = ax^2 + bx + c.$$

2. Find two **L**ast terms whose product is c:

$$(\square x + \square)(\square x + \square) = ax^2 + bx + c.$$

3. By trial and error, perform steps 1 and 2 until the sum of the **O**utside product and **I**nside product is bx:

$$(\square x + \square)(\square x + \square) = ax^2 + bx + c.$$

Sum of O + I

If no such combinations exist, the polynomial is prime.

EXAMPLE 7 Factoring a Trinomial Whose Leading Coefficient Is Not 1

Factor: $5x^2 - 14x + 8$.

SOLUTION

Step 1. **Find two First terms whose product is $5x^2$.**

$$5x^2 - 14x + 8 = (5x \quad)(x \quad)$$

Step 2. **Find two Last terms whose product is 8.** The number 8 has pairs of factors that are either both positive or both negative. Because the middle term, $-14x$, is negative, both factors must be negative. The negative factorizations of 8 are $-1(-8)$ and $-2(-4)$.

Step 3. **Try various combinations of these factors.** The correct factorization of $5x^2 - 14x + 8$ is the one in which the sum of the Outside and Inside products is equal to $-14x$. Here is a list of the possible factorizations:

Possible Factorizations of $5x^2 - 14x + 8$	Sum of Outside and Inside Products (Should Equal $-14x$)
$(5x - 1)(x - 8)$	$-40x - x = -41x$
$(5x - 8)(x - 1)$	$-5x - 8x = -13x$
$(5x - 2)(x - 4)$	$-20x - 2x = -22x$
$(5x - 4)(x - 2)$	$-10x - 4x = -14x$

This is the required middle term.

Thus,

$$5x^2 - 14x + 8 = (5x - 4)(x - 2).$$

Show that this factorization is correct by multiplying the factors using the FOIL method. You should obtain the original trinomial. ∎

✔ **CHECK POINT 7** Factor: $3x^2 - 20x + 28$. $\quad (3x - 14)(x - 2)$

EXAMPLE 8 Factoring a Trinomial Whose Leading Coefficient Is Not 1

Factor: $8x^6 - 10x^5 - 3x^4$.

SOLUTION The GCF of the three terms of the polynomial is x^4. We begin by factoring out x^4.

$$8x^6 - 10x^5 - 3x^4 = x^4(8x^2 - 10x - 3)$$

Now we factor the remaining trinomial, $8x^2 - 10x - 3$.

Step 1. **Find two First terms whose product is $8x^2$.**

$$8x^2 - 10x - 3 \overset{?}{=} (8x \quad)(x \quad)$$
$$8x^2 - 10x - 3 \overset{?}{=} (4x \quad)(2x \quad)$$

Step 2. **Find two Last terms whose product is -3.** The possible factorizations are $1(-3)$ and $-1(3)$.

Step 3. **Try various combinations of these factors.** The correct factorization of $8x^2 - 10x - 3$ is the one in which the sum of the Outside and Inside products is equal to $-10x$. Here is a list of the possible factorizations:

Possible Factorizations of $8x^2 - 10x - 3$	Sum of Outside and Inside Products (Should Equal $-10x$)
$(8x + 1)(x - 3)$	$-24x + x = -23x$
$(8x - 3)(x + 1)$	$8x - 3x = 5x$
$(8x - 1)(x + 3)$	$24x - x = 23x$
$(8x + 3)(x - 1)$	$-8x + 3x = -5x$
$(4x + 1)(2x - 3)$	$-12x + 2x = -10x$
$(4x - 3)(2x + 1)$	$4x - 6x = -2x$
$(4x - 1)(2x + 3)$	$12x - 2x = 10x$
$(4x + 3)(2x - 1)$	$-4x + 6x = 2x$

This is the required middle term.

The factorization of $8x^2 - 10x - 3$ is $(4x + 1)(2x - 3)$. Now we include the GCF in the complete factorization of the given polynomial. Thus,

$$8x^6 - 10x^5 - 3x^4 = x^4(8x^2 - 10x - 3) = x^4(4x + 1)(2x - 3).$$

This is the complete factorization with the GCF, x^4, included. ■

 CHECK POINT 8 Factor: $6x^6 + 19x^5 - 7x^4$. $x^4(3x - 1)(2x + 7)$

We have seen that not every trinomial can be factored. For example, consider

$$6x^2 + 14x + 7 = (6x + \square)(x + \square)$$

$$6x^2 + 14x + 7 = (3x + \square)(2x + \square).$$

The possible factors for the last term are 1 and 7. However, regardless of how these factors are placed in the boxes shown, the sum of the Outside and Inside products is not equal to $14x$. Thus, the trinomial $6x^2 + 14x + 7$ cannot be factored and is prime.

EXAMPLE 9 Factoring a Trinomial in Two Variables

Factor: $3x^2 - 13xy + 4y^2$.

SOLUTION

Step 1. **Find two First terms whose product is $3x^2$.**

$$3x^2 - 13xy + 4y^2 = (3x \quad ?y)(x \quad ?y)$$

The question marks indicate that we are looking for the coefficients of y in each factor.

Steps 2 and 3. **List pairs of factors of the coefficient of the last term, 4, and try various combinations of these factors.** The correct factorization is the one in which the sum of the Outside and Inside products is equal to $-13xy$. Because of the negative coefficient, -13, we will consider only the negative pairs of factors of 4. The possible factorizations are $-1(-4)$ and $-2(-2)$.

Possible Factorizations of $3x^2 - 13xy + 4y^2$	Sum of Outside and Inside Products (Should Equal $-13xy$)
$(3x - y)(x - 4y)$	$-12xy - xy = -13xy$
$(3x - 4y)(x - y)$	$-3xy - 4xy = -7xy$
$(3x - 2y)(x - 2y)$	$-6xy - 2xy = -8xy$

This is the required middle term.

Thus,
$$3x^2 - 13xy + 4y^2 = (3x - y)(x - 4y).$$ ■

 CHECK POINT 9 Factor: $2x^2 - 7xy + 3y^2$. $(2x - y)(x - 3y)$

EXAMPLE 10 Factoring by Substitution

Factor: $6y^4 + 13y^2 + 6$.

SOLUTION Notice that the exponent on y^2 is half that of the exponent on y^4. We will let t equal the variable to the power that is half of 4. Thus, let $t = y^2$.

$$6(y^2)^2 + 13y^2 + 6$$
This is the given polynomial, with y^4 written as $(y^2)^2$.

$$= 6t^2 + 13t + 6$$
Let $t = y^2$. Rewrite the trinomial in terms of t.

$$= (3t + 2)(2t + 3)$$
Factor the trinomial.

$$= (3y^2 + 2)(2y^2 + 3)$$
Now substitute y^2 for t.

Therefore, $6y^4 + 13y^2 + 6 = (3y^2 + 2)(2y^2 + 3)$. Check using FOIL multiplication. ■

 CHECK POINT 10 Factor: $3y^4 + 10y^2 - 8$. $(3y^2 - 2)(y^2 + 4)$

---

4 Factor trinomials by grouping.

Factoring Trinomials by Grouping A second method for factoring $ax^2 + bx + c$, $a \neq 0$, is called the **grouping method**. The method involves both trial and error, as well as grouping. The trial and error in factoring $ax^2 + bx + c$ depends upon finding two numbers, p and q, for which $p + q = b$. Then we factor $ax^2 + px + qx + c$ using grouping.

Let's see how this works by looking at a particular factorization:
$$15x^2 - 7x - 2 = (3x - 2)(5x + 1).$$

If we multiply using FOIL on the right, we obtain
$$(3x - 2)(5x + 1) = 15x^2 + 3x - 10x - 2.$$

In this case, the desired numbers, p and q, are $p = 3$ and $q = -10$. Compare these numbers to ac and b in the given polynomial.

$ac = 15(-2) = -30$

$$15x^2 - 7x - 2$$

$a = 15$ $b = -7$ $c = -2$

Can you see that p and q, 3 and -10, are factors of ac, or -30? Furthermore, p and q have a sum of b, namely -7. By expressing the middle term, $-7x$, in terms of p and q, we can factor by grouping as follows:

$$15x^2 - 7x - 2$$

$$= 15x^2 + (3x - 10x) - 2 \qquad \text{Rewrite } -7x \text{ as } 3x - 10x.$$

$$= (15x^2 + 3x) + (-10x - 2) \qquad \text{Group terms.}$$

$$= 3x(5x + 1) - 2(5x + 1) \qquad \text{Factor from each group.}$$

$$= (5x + 1)(3x - 2). \qquad \text{Factor out } 5x + 1, \text{ the common binomial factor.}$$

FACTORING $ax^2 + bx + c$ USING GROUPING ($a \neq 1$)

1. Multiply the leading coefficient, a, and the constant, c.
2. Find the factors of ac whose sum is b.
3. Rewrite the middle term, bx, as a sum or difference using the factors from step 2.
4. Factor by grouping.

EXAMPLE 11 Factoring a Trinomial by Grouping

Factor by grouping: $12x^2 - 5x - 2$.

SOLUTION The trinomial is of the form $ax^2 + bx + c$.

$$12x^2 - 5x - 2$$

$$\boxed{a = 12} \quad \boxed{b = -5} \quad \boxed{c = -2}$$

DISCOVER FOR YOURSELF

In step 2, we found that the desired numbers were -8 and 3. We wrote $-5x$ as $-8x + 3x$. What happens if we write $-5x$ as $3x - 8x$? Use factoring by grouping on

$$12x^2 - 5x - 2$$
$$= 12x^2 + 3x - 8x - 12.$$

Describe what happens.

Step 1. **Multiply the leading coefficient, a, and the constant, c.** Using $a = 12$ and $c = -2$,

$$ac = 12(-2) = -24.$$

Step 2. **Find the factors of ac whose sum is b.** We want the factors of -24 whose sum is b, or -5. The factors of -24 whose sum is -5 are -8 and 3.

Step 3. **Rewrite the middle term, $-5x$, as a sum or difference using the factors from step 2, -8 and 3.**

$$12x^2 - 5x - 2 = 12x^2 - 8x + 3x - 2$$

Step 4. **Factor by grouping.**

$$= (12x^2 - 8x) + (3x - 2) \qquad \text{Group terms.}$$

$$= 4x(3x - 2) + 1(3x - 2) \qquad \text{Factor from each group.}$$

$$= (3x - 2)(4x + 1) \qquad \text{Factor out } 3x - 2, \text{ the common binomial factor.}$$

Thus,

$$12x^2 - 5x - 2 = (3x - 2)(4x + 1).$$

 CHECK POINT 11 Factor by grouping: $8x^2 - 22x + 5$. $(2x - 5)(4x - 1)$

5.4 EXERCISE SET

Student Solutions Manual   CD/Video   PH Math/Tutor Center   MathXL Tutorials on CD   MathXL®   MyMathLab   Interactmath.com

Practice Exercises

In Exercises 1–30, factor each trinomial, or state that the trinomial is prime. Check each factorization using FOIL multiplication.

1. $x^2 + 5x + 6$ $(x + 3)(x + 2)$
2. $x^2 + 10x + 9$ $(x + 9)(x + 1)$
3. $x^2 + 8x + 12$ $(x + 6)(x + 2)$
4. $x^2 + 8x + 15$ $(x + 5)(x + 3)$
5. $x^2 + 9x + 20$ $(x + 5)(x + 4)$
6. $x^2 + 11x + 24$ $(x + 8)(x + 3)$
7. $y^2 + 10y + 16$ $(y + 8)(y + 2)$
8. $y^2 + 9y + 18$ $(y + 6)(y + 3)$
9. $x^2 - 8x + 15$ $(x - 3)(x - 5)$
10. $x^2 - 5x + 6$ $(x - 2)(x - 3)$
11. $y^2 - 12y + 20$ $(y - 2)(y - 10)$
12. $y^2 - 25y + 24$ $(y - 1)(y - 24)$
13. $a^2 + 5a - 14$ $(a + 7)(a - 2)$
14. $a^2 + a - 12$ $(a + 4)(a - 3)$
15. $x^2 + x - 30$ $(x + 6)(x - 5)$
16. $x^2 + 14x - 32$ $(x + 16)(x - 2)$
17. $x^2 - 3x - 28$ $(x + 4)(x - 7)$
18. $x^2 - 4x - 21$ $(x + 3)(x - 7)$
19. $y^2 - 5y - 36$ $(y + 4)(y - 9)$
20. $y^2 - 3y - 40$ $(y + 5)(y - 8)$
21. $x^2 - x + 7$ prime
22. $x^2 + 3x + 8$ prime
23. $x^2 - 9xy + 14y^2$ $(x - 2y)(x - 7y)$
24. $x^2 - 8xy + 15y^2$ $(x - 3y)(x - 5y)$
25. $x^2 - xy - 30y^2$ $(x + 5y)(x - 6y)$
26. $x^2 - 3xy - 18y^2$ $(x + 3y)(x - 6y)$
27. $x^2 + xy + y^2$ prime
28. $x^2 - xy + y^2$ prime
29. $a^2 - 18ab + 80b^2$ $(a - 10b)(a - 8b)$
30. $a^2 - 18ab + 45b^2$ $(a - 15b)(a - 3b)$

In Exercises 31–38, factor completely.

31. $3x^2 + 3x - 18$ $3(x + 3)(x - 2)$
32. $4x^2 - 4x - 8$ $4(x + 1)(x - 2)$
33. $2x^3 - 14x^2 + 24x$ $2x(x - 3)(x - 4)$
34. $2x^3 + 6x^2 + 4x$ $2x(x + 2)(x + 1)$
35. $3y^3 - 15y^2 + 18y$ $3y(y - 2)(y - 3)$
36. $4y^3 + 12y^2 - 72y$ $4y(y + 6)(y - 3)$
37. $2x^4 - 26x^3 - 96x^2$ $2x^2(x + 3)(x - 16)$
38. $3x^4 + 54x^3 + 135x^2$ $3x^2(x + 15)(x + 3)$

In Exercises 39–44, factor by introducing an appropriate substitution.

39. $x^6 - x^3 - 6$ $(x^3 + 2)(x^3 - 3)$
40. $x^6 + x^3 - 6$ $(x^3 + 3)(x^3 - 2)$
41. $x^4 - 5x^2 - 6$ $(x^2 - 6)(x^2 + 1)$
42. $x^4 - 4x^2 - 5$ $(x^2 - 5)(x^2 + 1)$
43. $(x + 1)^2 + 6(x + 1) + 5$ (Let $t = x + 1$.) $(x + 6)(x + 2)$
44. $(x + 1)^2 + 8(x + 1) + 7$ (Let $t = x + 1$.) $(x + 8)(x + 2)$

In Exercises 45–68, use the method of your choice to factor each trinomial, or state that the trinomial is prime. Check each factorization using FOIL multiplication.

45. $3x^2 + 8x + 5$ $(3x + 5)(x + 1)$
46. $2x^2 + 9x + 7$ $(2x + 7)(x + 1)$
47. $5x^2 + 56x + 11$ $(x + 11)(5x + 1)$
48. $5x^2 - 16x + 3$ $(5x - 1)(x - 3)$
49. $3y^2 + 22y - 16$ $(y + 8)(3y - 2)$
50. $5y^2 + 33y - 14$ $(y + 7)(5y - 2)$
51. $4y^2 + 9y + 2$ $(y + 2)(4y + 1)$
52. $8y^2 + 10y + 3$ $(4y + 3)(2y + 1)$
53. $10x^2 + 19x + 6$ $(2x + 3)(5x + 2)$
54. $6x^2 + 19x + 15$ $(3x + 5)(2x + 3)$
55. $8x^2 - 18x + 9$ $(4x - 3)(2x - 3)$
56. $4x^2 - 27x + 18$ $(4x - 3)(x - 6)$
57. $6y^2 - 23y + 15$ $(y - 3)(6y - 5)$
58. $16y^2 - 6y - 27$ $(8y + 9)(2y - 3)$
59. $6y^2 + 14y + 3$ prime
60. $4y^2 + 22y - 5$ prime
61. $3x^2 + 4xy + y^2$ $(x + y)(3x + y)$
62. $2x^2 + 3xy + y^2$ $(x + y)(2x + y)$
63. $6x^2 - 7xy - 5y^2$ $(2x + y)(3x - 5y)$
64. $6x^2 - 5xy - 6y^2$ $(3x + 2y)(2x - 3y)$
65. $15x^2 - 31xy + 10y^2$ $(5x - 2y)(3x - 5y)$
66. $15x^2 + 11xy - 14y^2$ $(5x + 7y)(3x - 2y)$
67. $3a^2 - ab - 14b^2$ $(3a - 7b)(a + 2b)$
68. $15a^2 - ab - 6b^2$ $(5a + 3b)(3a - 2b)$

In Exercises 69–82, factor completely.

69. $15x^3 - 25x^2 + 10x$ $5x(3x - 2)(x - 1)$
70. $10x^3 + 24x^2 + 14x$ $2x(5x + 7)(x + 1)$
71. $24x^4 + 10x^3 - 4x^2$ $2x^2(3x + 2)(4x - 1)$
72. $15x^4 - 39x^3 + 18x^2$ $3x^2(5x - 3)(x - 2)$
73. $15y^5 - 2y^4 - y^3$ $y^3(5y + 1)(3y - 1)$
74. $10y^5 - 17y^4 + 3y^3$ $y^3(5y - 1)(2y - 3)$
75. $24x^2 + 3xy - 27y^2$ $3(8x + 9y)(x - y)$

76. $12x^2 + 10xy - 8y^2$ $2(3x + 4y)(2x - y)$

77. $6a^2b - 2ab - 60b$ $2b(a + 3)(3a - 10)$

78. $8a^2b + 34ab - 84b$ $2b(a + 6)(4a - 7)$

79. $12x^2y - 34xy^2 + 14y^3$ $2y(2x - y)(3x - 7y)$

80. $12x^2y - 46xy^2 + 14y^3$ $2y(3x - y)(2x - 7y)$

81. $13x^3y^3 + 39x^3y^2 - 52x^3y$ $13x^3y(y + 4)(y - 1)$

82. $4x^3y^5 + 24x^2y^5 - 64xy^5$ $4xy^5(x + 8)(x - 2)$

In Exercises 83–92, factor by introducing an appropriate substitution.

83. $2x^4 - x^2 - 3$ $(2x^2 - 3)(x^2 + 1)$

84. $5x^4 + 2x^2 - 3$ $(5x^2 - 3)(x^2 + 1)$

85. $2x^6 + 11x^3 + 15$ $(2x^3 + 5)(x^3 + 3)$

86. $2x^6 + 13x^3 + 15$ $(2x^3 + 3)(x^3 + 5)$

87. $2y^{10} + 7y^5 + 3$ $(2y^5 + 1)(y^5 + 3)$

88. $5y^{10} + 29y^5 - 42$ $(5y^5 - 6)(y^5 + 7)$

89. $5(x + 1)^2 + 12(x + 1) + 7$ (Let $t = x + 1$.) $(5x + 12)(x + 2)$

90. $3(x + 1)^2 - 5(x + 1) + 2$ (Let $t = x + 1$.) $x(3x + 1)$

91. $2(x - 3)^2 - 5(x - 3) - 7$ $(2x - 13)(x - 2)$

92. $3(x - 2)^2 - 5(x - 2) - 2$ $(3x - 5)(x - 4)$

Practice Plus

In Exercises 93–100, factor completely.

93. $x^2 - 0.5x + 0.06$ $(x - 0.3)(x - 0.2)$

94. $x^2 + 0.3x - 0.04$ $(x - 0.1)(x + 0.4)$

95. $x^2 - \dfrac{3}{49} + \dfrac{2}{7}x$ $\left(x + \dfrac{3}{7}\right)\left(x - \dfrac{1}{7}\right)$

96. $x^2 - \dfrac{6}{25} + \dfrac{1}{5}x$ $\left(x + \dfrac{3}{5}\right)\left(x - \dfrac{2}{5}\right)$

97. $acx^2 - bcx + adx - bd$ $(ax - b)(cx + d)$

98. $acx^2 - bcx - adx + bd$ $(ax - b)(cx - d)$

99. $-4x^5y^2 + 7x^4y^3 - 3x^3y^4$ $-x^3y^2(4x - 3y)(x - y)$

100. $-5x^4y^3 + 7x^3y^4 - 2x^2y^5$ $-x^2y^3(5x - 2y)(x - y)$

101. If $(fg)(x) = 3x^2 - 22x + 39$, find f and g. $f(x) = 3x - 13$ and $g(x) = x - 3$, or vice versa

102. If $(fg)(x) = 4x^2 - x - 5$, find f and g. $f(x) = 4x - 5$ and $g(x) = x + 1$, or vice versa

In Exercises 103–104, a large rectangle formed by a number of smaller rectangles is shown. Factor the sum of the areas of the smaller rectangles to determine the dimensions of the large rectangle.

103.

x^2	x^2	x
x	x	1
x	x	1
x	x	1

$2x + 1$ by $x + 3$

104.

$x + 4$ by $x + 1$

Application Exercises

105. A diver jumps directly upward from a board that is 32 feet high. The function

$$f(t) = -16t^2 + 16t + 32$$

describes the diver's height above the water, $f(t)$, in feet, after t seconds.

  a. Find and interpret $f(1)$.   32; The diver is 32 feet above the water after 1 second.

  b. Find and interpret $f(2)$.   0; The diver hits the water after 2 seconds.

  c. Factor the expression for $f(t)$ and write the function in completely factored form. $f(t) = -16(t - 2)(t + 1)$

  d. Use the factored form of the function to find $f(1)$ and $f(2)$. $f(1) = 32$ and $f(2) = 0$

106. The function $V(x) = 3x^3 - 2x^2 - 8x$ describes the volume, $V(x)$, in cubic inches, of the box shown whose height is x inches.

  a. Find and interpret $V(4)$.   128; The volume of a box whose height is 4 inches is 128 cubic inches.

  b. Factor the expression for $V(x)$ and write the function in completely factored form. $V(x) = x(3x + 4)(x - 2)$

  c. Use the factored form of the function to find $V(4)$ and $V(5)$. $V(4) = 128$ and $V(5) = 285$

107. Find the area of the large rectangle shown below in two ways:

  a. Find the sum of the areas of the six smaller rectangles and squares. $x^2 + x + x + x + 1 + 1 = x^2 + 3x + 2$

  b. Express the area of the large rectangle as the product of its length and width. $(x + 2)(x + 1)$

  c. Explain how the figure serves as a geometric model for the factorization of the sum that you wrote in part (a).   Answers will vary.

108. If x represents a positive integer, factor $x^3 + 3x^2 + 2x$ to show that the trinomial represents the product of three consecutive integers. $x^3 + 3x^2 + 2x = x(x^2 + 3x + 2) = x(x + 1)(x + 2)$

Writing in Mathematics

109. Explain how to factor $x^2 + 8x + 15$. 109–116. Answers will vary.

110. Give two helpful suggestions for factoring $x^2 - 5x + 6$.

111. In factoring $x^2 + bx + c$, describe how the last term in each binomial factor is related to b and c.

112. Describe the first thing that you should try doing when factoring a polynomial.

113. What does it mean to factor completely?

114. Explain how to factor $x^6 - 7x^3 + 10$ by substitution.

115. Is it possible to factor $x^6 - 7x^3 + 10$ without using substitution? How might this be done?

116. Explain how to factor $2x^2 - x - 1$.

Technology Exercises

In Exercises 117–120, use a graphing utility to graph the function on each side of the equation in the same viewing rectangle. Use end behavior to show a complete picture of the polynomial function on the left side. Do the graphs coincide? If so, this means that the polynomial on the left side has been factored correctly. If not, factor the polynomial correctly and then use your graphing utility to verify the factorization. 117–120. See graphing answer section.

117. $x^2 + 7x + 12 = (x + 4)(x + 3)$ Graphs coincide; factored correctly.

118. $x^2 - 7x + 6 = (x - 2)(x - 3)$ Graphs do not coincide; $x^2 - 7x + 6 = (x - 6)(x - 1)$.

119. $6x^3 + 5x^2 - 4x = x(3x + 4)(2x - 1)$ Graphs coincide; factored correctly.

120. $x^4 - x^2 - 20 = (x^2 + 5)(x^2 - 4)$ Graphs do not coincide; $x^4 - x^2 - 20 = (x^2 - 5)(x^2 + 4)$.

121. Use the TABLE feature of a graphing utility to verify any two of your factorizations in Exercises 39–44. Answers will vary.

Critical Thinking Exercises

122. Which one of the following is true? b
   a. Once a GCF is factored from $6y^6 - 19y^5 + 10y^4$, the remaining trinomial factor is prime.
   b. One factor of $8y^2 - 51y + 18$ is $8y - 3$.
   c. We can immediately tell that $6x^2 - 11xy - 10y^2$ is prime because 11 is a prime number and the polynomial contains two variables.
   d. A factor of $12x^2 - 19xy + 5y$ is $4x - y$.

In Exercises 123–124, find all integers b so that the trinomial can be factored.

123. $4x^2 + bx - 1$ $-3, 0, 3$ **124.** $3x^2 + bx + 5$ $-16, -8, 8, 16$

In Exercises 125–130, factor each polynomial. Assume that all variable exponents represent whole numbers.

125. $9x^{2n} + x^n - 8$ $(9x^n - 8)(x^n + 1)$

126. $4x^{2n} - 9x^n + 5$ $(4x^n - 5)(x^n - 1)$

127. $a^{2n+2} - a^{n+2} - 6a^2$ $a^2(a^n - 3)(a^n + 2)$

128. $b^{2n+2} + 3b^{n+2} - 10b^2$ $b^2(b^n - 2)(b^n + 5)$

129. $3c^{n+2} - 10c^{n+1} + 3c^n$ $c^n(3c - 1)(c - 3)$

130. $2d^{n+2} - 5d^{n+1} + 3d^n$ $d^n(2d - 3)(d - 1)$

Review Exercises

131. Solve: $-2x \le 6$ and $-2x + 3 < -7$. (Section 4.2, Example 2) $\{x | x > 5\}$ or $(5, \infty)$

132. Solve the system:
$$2x - y - 2z = -1$$
$$x - 2y - z = 1$$
$$x + y + z = 4.$$
(Section 3.3, Example 2) $(2, -1, 3)$

133. Graph the solution set of the system:
$$x - y \ge -4$$
$$x + 2y \ge 2.$$
(Section 4.4, Example 4) See graphing answer section.

✔ **MID-CHAPTER CHECK POINT**

CHAPTER 5

What You Know: We learned the vocabulary of polynomials and observed the smooth, continuous graphs of polynomial functions. We used the Leading Coefficient Test to describe the end behavior of these graphs. We learned to add, subtract, and multiply polynomials. We used a number of fast methods for finding products of polynomials, including the FOIL method for multiplying binomials, special-product formulas for squaring binomials $[(A + B)^2 = A^2 + 2AB + B^2; (A - B)^2 = A^2 - 2AB + B^2]$, and a special-product formula for the product of the sum and difference of two terms $[(A + B)(A - B) = A^2 - B^2]$. We learned to factor out a polynomial's greatest common factor and to use grouping to factor polynomials with more than three terms. We factored polynomials with three terms, beginning with trinomials with leading coefficient 1 and moving on to $ax^2 + bx + c$, with $a \ne 1$. We saw that the factoring process should begin by looking for a GCF and, if there is one, factoring it out first.

In Exercises 1–18, perform the indicated operations.

1. $(-8x^3 + 6x^2 - x + 5) - (-7x^3 + 2x^2 - 7x - 12)$
   $-x^3 + 4x^2 + 6x + 17$

2. $(6x^2yz^4)\left(-\dfrac{1}{3}x^5y^2z\right)$ $-2x^7y^3z^5$

3. $5x^2y\left(6x^3y^2 - 7xy - \dfrac{2}{5}\right)$ $30x^5y^3 - 35x^3y^2 - 2x^2y$

4. $(3x - 5)(x^2 + 3x - 8)$ $3x^3 + 4x^2 - 39x + 40$

5. $(x^2 - 2x + 1)(2x^2 + 3x - 4)$
   $2x^4 - x^3 - 8x^2 + 11x - 4$

6. $(x^2 - 2x + 1) - (2x^2 + 3x - 4)$ $-x^2 - 5x + 5$

7. $(6x^3y - 11x^2y - 4y) + (-11x^3y + 5x^2y - y - 6)$
   $- (-x^3y + 2y - 1)$ $-4x^3y - 6x^2y - 7y - 5$

8. $(2x + 5)(4x - 1)$ $8x^2 + 18x - 5$

9. $(2xy - 3)(5xy + 2)$ $10x^2y^2 - 11xy - 6$

10. $(3x - 2y)(3x + 2y)$ $9x^2 - 4y^2$

11. $(3xy + 1)(2x^2 - 3y)$ $6x^3y - 9xy^2 + 2x^2 - 3y$

12. $(7x^3y + 5x)(7x^3y - 5x)$ $49x^6y^2 - 25x^2$

13. $3(x + h)^2 - 2(x + h) + 5 - (3x^2 - 2x + 5)$
    $6xh + 3h^2 - 2h$

14. $(x^2 - 3)^2$ $x^4 - 6x^2 + 9$

15. $(x^2 - 3)(x^3 + 5x + 2)$ $x^5 + 2x^3 + 2x^2 - 15x - 6$

16. $(2x + 5y)^2$ $4x^2 + 20xy + 25y^2$

17. $(x + 6 + 3y)(x + 6 - 3y)$ $x^2 + 12x + 36 - 9y^2$

18. $(x + y + 5)^2$ $x^2 + 2xy + y^2 + 10x + 10y + 25$

In Exercises 19–30, factor completely or state that the polynomial is prime.

19. $x^2 - 5x - 24$ $(x - 8)(x + 3)$

20. $15xy + 5x + 6y + 2$ $(5x + 2)(3y + 1)$

21. $5x^2 + 8x - 4$ $(5x - 2)(x + 2)$

22. $35x^2 + 10x - 50$ $5(7x^2 + 2x - 10)$

23. $9x^2 - 9x - 18$ $9(x - 2)(x + 1)$

24. $10x^3y^2 - 20x^2y^2 + 35x^2y$ $5x^2y(2xy - 4y + 7)$

25. $18x^2 + 21x + 5$ $(3x + 1)(6x + 5)$

26. $12x^2 - 9xy - 16x + 12y$ $(4x - 3y)(3x - 4)$

27. $9x^2 - 15x + 4$ $(3x - 4)(3x - 1)$

28. $3x^6 + 11x^3 + 10$ $(3x^3 + 5)(x^3 + 2)$

29. $25x^3 + 25x^2 - 14x$ $x(5x - 2)(5x + 7)$

30. $2x^4 - 6x - x^3y + 3y$ $(x^3 - 3)(2x - y)$

SECTION 5.5

Objectives

1 Factor the difference of two squares.

2 Factor perfect square trinomials.

3 Use grouping to obtain the difference of two squares.

4 Factor the sum and difference of two cubes.

FACTORING SPECIAL FORMS

Bees use honeycombs to store honey and house larvae. They construct honey storage cells from wax. Each cell has the shape of a six-sided figure whose sides are all the same length and whose angles all have the same measure, called a regular hexagon. The cells fit together perfectly, preventing dirt or predators from entering. Squares or equilateral triangles would fit equally well, but regular hexagons provide the largest storage space for the amount of wax used.

In this section, we develop factoring techniques by reversing the formulas for special products discussed in Section 5.2. Like the construction of honeycombs, these factorizations can be visualized by perfectly fitting together "cells" of squares and rectangles to form larger rectangles.

1 Factor the difference of two squares.

Factoring the Difference of Two Squares A method for factoring the difference of two squares is obtained by reversing the special product for the sum and difference of two terms.

> **THE DIFFERENCE OF TWO SQUARES** If A and B are real numbers, variables, or algebraic expressions, then
>
> $$A^2 - B^2 = (A + B)(A - B).$$
>
> In words: The difference of the squares of two terms factors as the product of a sum and a difference of those terms.

EXAMPLE 1 Factoring the Difference of Two Squares

Factor:

    a. $9x^2 - 100$ **b.** $36y^6 - 49x^4$.

SOLUTION We must express each term as the square of some monomial. Then we use the formula for factoring $A^2 - B^2$.

    a. $9x^2 - 100 = (3x)^2 - 10^2 = (3x + 10)(3x - 10)$

$$A^2 \ - \ B^2 \ = \ (A \ + \ B) \ (A \ - \ B)$$

    b. $36y^6 - 49x^4 = (6y^3)^2 - (7x^2)^2 = (6y^3 + 7x^2)(6y^3 - 7x^2)$ ∎

✔ CHECK POINT 1 Factor:

    a. $16x^2 - 25$ $(4x + 5)(4x - 5)$ **b.** $100y^6 - 9x^4$. $(10y^3 + 3x^2)(10y^3 - 3x^2)$

    Can $x^2 - 5$ be factored using integers and the formula for factoring $A^2 - B^2$? No. The number 5 is not the square of an integer. Thus, $x^2 - 5$ is prime over the set of integers.

    When factoring, always check first for common factors. If there are common factors, factor out the GCF and then factor the resulting polynomial.

EXAMPLE 2 Factoring Out the GCF and Then Factoring the Difference of Two Squares

Factor: $3y - 3x^6y^5$.

SOLUTION The GCF of the two terms of the polynomial is $3y$. We begin by factoring out $3y$.

$$3y - 3x^6y^5 = 3y(1 - x^6y^4) = 3y[1^2 - (x^3y^2)^2] = 3y(1 + x^3y^2)(1 - x^3y^2)$$

✔ CHECK POINT 2 Factor: $6y - 6x^2y^7$. $6y(1 + xy^3)(1 - xy^3)$

    We have seen that a polynomial is factored completely when it is written as the product of prime polynomials. To be sure that you have factored completely, check to see whether any factors with more than one term in the factored polynomial can be factored further. If so, continue factoring.

EXAMPLE 3 A Repeated Factorization

Factor completely: $81x^4 - 16$.

SOLUTION

$$
\begin{aligned}
81x^4 - 16 &= (9x^2)^2 - 4^2 && \text{Express as the difference of two squares.} \\
&= (9x^2 + 4)(9x^2 - 4) && \text{The factors are the sum and difference of the} \\
&&& \text{expressions being squared.} \\
&= (9x^2 + 4)[(3x)^2 - 2^2] && \text{The factor } 9x^2 - 4 \text{ is the difference of two} \\
&&& \text{squares and can be factored.} \\
&= (9x^2 + 4)(3x + 2)(3x - 2) && \text{The factors of } 9x^2 - 4 \text{ are the sum and} \\
&&& \text{difference of the expressions being squared.} \quad\blacksquare
\end{aligned}
$$

> **STUDY TIP**
>
> Factoring $81x^4 - 16$ as
>
> $$(9x^2 + 4)(9x^2 - 4)$$
>
> is not a complete factorization. The second factor, $9x^2 - 4$, is itself a difference of two squares and can be factored.

Are you tempted to further factor $9x^2 + 4$, the sum of two squares, in Example 3? Resist the temptation! **The sum of two squares, $A^2 + B^2$, with no common factor other than 1 is a prime polynomial**.

✔ CHECK POINT 3 Factor completely: $16x^4 - 81$. $(4x^2 + 9)(2x + 3)(2x - 3)$

In our next example, we begin with factoring by grouping. We can then factor further using the difference of two squares.

EXAMPLE 4 Factoring Completely

Factor completely: $x^3 + 5x^2 - 9x - 45$.

SOLUTION

$$
\begin{aligned}
x^3 + 5x^2 - 9x - 45 \\
= (x^3 + 5x^2) + (-9x - 45) && \text{Group terms with common factors.} \\
= x^2(x + 5) - 9(x + 5) && \text{Factor out the common factor from each group.} \\
= (x + 5)(x^2 - 9) && \text{Factor out } x + 5, \text{ the common binomial factor,} \\
&& \text{from both terms.} \\
= (x + 5)(x + 3)(x - 3) && \text{Factor } x^2 - 3^2, \text{ the difference of two squares.} \quad\blacksquare
\end{aligned}
$$

✔ CHECK POINT 4 Factor completely: $x^3 + 7x^2 - 4x - 28$. $(x + 7)(x + 2)(x - 2)$

> **2** Factor perfect square trinomials.

Factoring Perfect Square Trinomials Our next factoring technique is obtained by reversing the special products for squaring binomials. The trinomials that are factored using this technique are called **perfect square trinomials**.

FACTORING PERFECT SQUARE TRINOMIALS Let A and B be real numbers, variables, or algebraic expressions.

1. $A^2 + 2AB + B^2 = (A + B)^2$

Same sign

2. $A^2 - 2AB + B^2 = (A - B)^2$

Same sign

The two items in the box on the previous page show that perfect square trinomials, $A^2 + 2AB + B^2$ and $A^2 - 2AB + B^2$, come in two forms: one in which the coefficient of the middle term is positive and one in which the coefficient of the middle term is negative. Here's how to recognize a perfect square trinomial:

1. The first and last terms are squares of monomials or integers.
2. The middle term is twice the product of the expressions being squared in the first and last terms.

EXAMPLE 5 | Factoring Perfect Square Trinomials

Factor:

a. $x^2 + 14x + 49$ **b.** $4x^2 + 12xy + 9y^2$ **c.** $9y^4 - 12y^2 + 4$.

SOLUTION

a. $x^2 + 14x + 49 = x^2 + 2 \cdot x \cdot 7 + 7^2 = (x + 7)^2$ *The middle term has a positive sign.*

$$A^2 + 2AB + B^2 = (A + B)^2$$

b. We suspect that $4x^2 + 12xy + 9y^2$ is a perfect square trinomial because $4x^2 = (2x)^2$ and $9y^2 = (3y)^2$. The middle term can be expressed as twice the product of $2x$ and $3y$.

$$4x^2 + 12xy + 9y^2 = (2x)^2 + 2 \cdot 2x \cdot 3y + (3y)^2 = (2x + 3y)^2$$

$$A^2 + 2AB + B^2 = (A + B)^2$$

c. $9y^4 - 12y^2 + 4 = (3y^2)^2 - 2 \cdot 3y^2 \cdot 2 + 2^2 = (3y^2 - 2)^2$ *The middle term has a negative sign.*

$$A^2 - 2AB + B^2 = (A - B)^2$$

✔ **CHECK POINT 5** Factor:

a. $x^2 + 6x + 9$ $(x + 3)^2$ **b.** $16x^2 + 40xy + 25y^2$ $(4x + 5y)^2$
c. $4y^4 - 20y^2 + 25$. $(2y^2 - 5)^2$

---

3   Use grouping to obtain the difference of two squares.

Using Special Forms When Factoring by Grouping   If a polynomial contains four terms, try factoring by grouping. In the next example, we group the terms to obtain the difference of two squares. One of the squares is a perfect square trinomial.

EXAMPLE 6 | Using Grouping to Obtain the Difference of Two Squares

Factor: $x^2 - 8x + 16 - y^2$.

SOLUTION

$$x^2 - 8x + 16 - y^2$$
$$= (x^2 - 8x + 16) - y^2$$ *Group as a perfect square trinomial minus y^2 to obtain a difference of two squares.*
$$= (x - 4)^2 - y^2$$ *Factor the perfect square trinomial.*
$$= (x - 4 + y)(x - 4 - y)$$ *Factor the difference of two squares. The factors are the sum and difference of the expressions being squared.*

✔ **CHECK POINT 6** Factor: $x^2 + 10x + 25 - y^2$. $(x + 5 + y)(x + 5 - y)$

EXAMPLE 7 Using Grouping to Obtain the Difference of Two Squares

Factor: $a^2 - b^2 + 10b - 25$.

SOLUTION Grouping into two groups of two terms does not result in a common binomial factor. Let's look for a perfect square trinomial. Can you see that the perfect square trinomial is the expression being subtracted from a^2?

$$a^2 - b^2 + 10b - 25$$
$$= a^2 - (b^2 - 10b + 25)$$
Factor out -1 and group as $a^2 - (\text{perfect square trinomial})$ to obtain a difference of two squares.

$$= a^2 - (b - 5)^2$$
Factor the perfect square trinomial.

$$= [a + (b - 5)][a - (b - 5)]$$
Factor the difference of squares. The factors are the sum and difference of the expressions being squared.

$$= (a + b - 5)(a - b + 5)$$
Simplify. ∎

✔ **CHECK POINT 7** Factor: $a^2 - b^2 + 4b - 4$. $(a + b - 2)(a - b + 2)$

---

4 | Factor the sum and difference of two cubes.

Factoring the Sum and Difference of Two Cubes Here are two multiplications that lead to factoring formulas for the sum of two cubes and the difference of two cubes:

$$(A + B)(A^2 - AB + B^2) = A(A^2 - AB + B^2) + B(A^2 - AB + B^2)$$
$$= A^3 - A^2B + AB^2 + A^2B - AB^2 + B^3$$
$$= A^3 + B^3$$

Combine like terms: $-A^2B^2 + A^2B^2 = 0$ and $AB^2 - AB^2 = 0$.

> The product results in the sum of two cubes.

◆

and

$$(A - B)(A^2 + AB + B^2) = A(A^2 + AB + B^2) - B(A^2 + AB + B^2)$$
$$= A^3 + A^2B + AB^2 - A^2B - AB^2 - B^3$$
$$= A^3 - B^3.$$

Combine like terms: $A^2B - A^2B = 0$ and $AB^2 - AB^2 = 0$.

> The product results in the difference of two cubes.

By reversing the two sides of these equations, we obtain formulas that allow us to factor a sum or difference of two cubes. These formulas should be memorized.

FACTORING THE SUM AND DIFFERENCE OF TWO CUBES

1. Factoring the Sum of Two Cubes

$$A^3 + B^3 = (A + B)(A^2 - AB + B^2)$$

Same signs   Opposite signs

2. Factoring the Difference of Two Cubes

$$A^3 - B^3 = (A - B)(A^2 + AB + B^2)$$

Same signs   Opposite signs

EXAMPLE 8 Factoring the Sum of Two Cubes

Factor:

  a. $x^3 + 125$ **b.** $x^6 + 64y^3$.

SOLUTION   We must express each term as the cube of some monomial. Then we use the formula for factoring $A^3 + B^3$.

  a. $x^3 + 125 = x^3 + 5^3 = (x + 5)(x^2 - x \cdot 5 + 5^2) = (x + 5)(x^2 - 5x + 25)$

$$A^3 + B^3 = (A + B)(A^2 - AB + B^2)$$

  b. $x^6 + 64y^3 = (x^2)^3 + (4y)^3 = (x^2 + 4y)[(x^2)^2 - x^2 \cdot 4y + (4y)^2]$

$$A^3 + B^3 = (A + B)(A^2 - AB + B^2)$$

$$= (x^2 + 4y)(x^4 - 4x^2y + 16y^2)$$ ■

✔ **CHECK POINT 8** Factor:

  a. $x^3 + 27$ $(x + 3)(x^2 - 3x + 9)$
  b. $x^6 + 1000y^3$. $(x^2 + 10y)(x^4 - 10x^2y + 100y^2)$

EXAMPLE 9 Factoring the Difference of Two Cubes

Factor:

  a. $x^3 - 216$ **b.** $8 - 125x^3y^3$.

SOLUTION   We must express each term as the cube of some monomial. Then we use the formula for factoring $A^3 - B^3$.

  a. $x^3 - 216 = x^3 - 6^3 = (x - 6)(x^2 + x \cdot 6 + 6^2) = (x - 6)(x^2 + 6x + 36)$

$$A^3 - B^3 = (A - B)(A^2 + AB + B^2)$$

  b. $8 - 125x^3y^3 = 2^3 - (5xy)^3 = (2 - 5xy)[2^2 + 2 \cdot 5xy + (5xy)^2]$

$$A^3 - B^3 = (A - B)(A^2 + AB + B^2)$$

$$= (2 - 5xy)(4 + 10xy + 25x^2y^2)$$ ■

✔ **CHECK POINT 9** Factor:

  a. $x^3 - 8$ $(x - 2)(x^2 + 2x + 4)$
  b. $1 - 27x^3y^3$. $(1 - 3xy)(1 + 3xy + 9x^2y^2)$

5.5 EXERCISE SET

Student Solutions Manual   CD/Video   PH Math/Tutor Center   MathXL Tutorials on CD   MathXL®   MyMathLab   Interactmath.com

Practice Exercises

In Exercises 1–22, factor each difference of two squares. Assume that any variable exponents represent whole numbers.

1. $x^2 - 4$ $(x + 2)(x - 2)$

2. $x^2 - 16$ $(x + 4)(x - 4)$

3. $9x^2 - 25$ $(3x + 5)(3x - 5)$

4. $4x^2 - 9$ $(2x + 3)(2x - 3)$

5. $9 - 25y^2$ $(3 + 5y)(3 - 5y)$

6. $16 - 49y^2$ $(4 + 7y)(4 - 7y)$

7. $36x^2 - 49y^2$ $(6x + 7y)(6x - 7y)$

8. $64x^2 - 25y^2$ $(8x + 5y)(8x - 5y)$

9. $x^2y^2 - 1$ $(xy + 1)(xy - 1)$

10. $x^2y^2 - 100$ $(xy + 10)(xy - 10)$

11. $9x^4 - 25y^6$ $(3x^2 + 5y^3)(3x^2 - 5y^3)$

12. $25x^4 - 9y^6$ $(5x^2 + 3y^3)(5x^2 - 3y^3)$

13. $x^{14} - y^4$ $(x^7 + y^2)(x^7 - y^2)$

14. $x^4 - y^{10}$ $(x^2 + y^5)(x^2 - y^5)$

15. $(x - 3)^2 - y^2$ $(x - 3 + y)(x - 3 - y)$

16. $(x - 6)^2 - y^2$ $(x - 6 + y)(x - 6 - y)$

17. $a^2 - (b - 2)^2$ $(a + b - 2)(a - b + 2)$

18. $a^2 - (b - 3)^2$ $(a + b - 3)(a - b + 3)$

19. $x^{2n} - 25$ $(x^n + 5)(x^n - 5)$

20. $x^{2n} - 36$ $(x^n + 6)(x^n - 6)$

21. $1 - a^{2n}$ $(1 + a^n)(1 - a^n)$

22. $4 - b^{2n}$ $(2 + b^n)(2 - b^n)$

In Exercises 23–48, factor completely, or state that the polynomial is prime.

23. $2x^3 - 8x$ $2x(x + 2)(x - 2)$

24. $2x^3 - 72x$ $2x(x + 6)(x - 6)$

25. $50 - 2y^2$ $2(5 + y)(5 - y)$

26. $72 - 2y^2$ $2(6 + y)(6 - y)$

27. $8x^2 - 8y^2$ $8(x + y)(x - y)$

28. $6x^2 - 6y^2$ $6(x + y)(x - y)$

29. $2x^3y - 18xy$ $2xy(x + 3)(x - 3)$

30. $2x^3y - 32xy$ $2xy(x + 4)(x - 4)$

31. $a^3b^2 - 49ac^2$ $a(ab + 7c)(ab - 7c)$

32. $4a^3c^2 - 16ax^2y^2$ $4a(ac + 2xy)(ac - 2xy)$

33. $5y - 5x^2y^7$ $5y(1 + xy^3)(1 - xy^3)$

34. $2y - 2x^6y^3$ $2y(1 + x^3y)(1 - x^3y)$

35. $8x^2 + 8y^2$ $8(x^2 + y^2)$

36. $6x^2 + 6y^2$ $6(x^2 + y^2)$

37. $x^2 + 25y^2$ prime

38. $x^2 + 36y^2$ prime

39. $x^4 - 16$ $(x^2 + 4)(x + 2)(x - 2)$

40. $x^4 - 1$ $(x^2 + 1)(x + 1)(x - 1)$

41. $81x^4 - 1$ $(9x^2 + 1)(3x + 1)(3x - 1)$

42. $1 - 81x^4$ $(1 + 9x^2)(1 + 3x)(1 - 3x)$

43. $2x^5 - 2xy^4$ $2x(x^2 + y^2)(x + y)(x - y)$

44. $3x^5 - 3xy^4$ $3x(x^2 + y^2)(x + y)(x - y)$

45. $x^3 + 3x^2 - 4x - 12$ $(x + 3)(x + 2)(x - 2)$

46. $x^3 + 3x^2 - 9x - 27$ $(x + 3)^2(x - 3)$

47. $x^3 - 7x^2 - x + 7$ $(x - 7)(x + 1)(x - 1)$

48. $x^3 - 6x^2 - x + 6$ $(x - 6)(x + 1)(x - 1)$

In Exercises 49–64, factor any perfect square trinomials, or state that the polynomial is prime.

49. $x^2 + 4x + 4$ $(x + 2)^2$

50. $x^2 + 2x + 1$ $(x + 1)^2$

51. $x^2 - 10x + 25$ $(x - 5)^2$

52. $x^2 - 14x + 49$ $(x - 7)^2$

53. $x^4 - 4x^2 + 4$ $(x^2 - 2)^2$

54. $x^4 - 6x^2 + 9$ $(x^2 - 3)^2$

55. $9y^2 + 6y + 1$ $(3y + 1)^2$

56. $4y^2 + 4y + 1$ $(2y + 1)^2$

57. $64y^2 - 16y + 1$ $(8y - 1)^2$

58. $25y^2 - 10y + 1$ $(5y - 1)^2$

59. $x^2 - 12xy + 36y^2$ $(x - 6y)^2$

60. $x^2 + 16xy + 64y^2$ $(x + 8y)^2$

61. $x^2 - 8xy + 64y^2$ prime

62. $x^2 - 9xy + 81y^2$ prime

63. $9x^2 + 48xy + 64y^2$ $(3x + 8y)^2$

64. $16x^2 - 40xy + 25y^2$ $(4x - 5y)^2$

In Exercises 65–74, factor by grouping to obtain the difference of two squares.

65. $x^2 - 6x + 9 - y^2$ $(x - 3 + y)(x - 3 - y)$

66. $x^2 - 12x + 36 - y^2$ $(x - 6 + y)(x - 6 - y)$

67. $x^2 + 20x + 100 - x^4$ $(x + 10 + x^2)(x + 10 - x^2)$

68. $x^2 + 16x + 64 - x^4$ $(x + 8 + x^2)(x + 8 - x^2)$

69. $9x^2 - 30x + 25 - 36y^2$ $(3x - 5 + 6y)(3x - 5 - 6y)$

70. $25x^2 - 20x + 4 - 81y^2$ $(5x - 2 + 9y)(5x - 2 - 9y)$

71. $x^4 - x^2 - 2x - 1$ $(x^2 + x + 1)(x^2 - x - 1)$

72. $x^4 - x^2 - 6x - 9$ $(x^2 + x + 3)(x^2 - x - 3)$

73. $z^2 - x^2 + 4xy - 4y^2$ $(z + x - 2y)(z - x + 2y)$

74. $z^2 - x^2 + 10xy - 25y^2$ $(z + x - 5y)(z - x + 5y)$

In Exercises 75–94, factor using the formula for the sum or difference of two cubes.

75. $x^3 + 64$ $(x + 4)(x^2 - 4x + 16)$

76. $x^3 + 1$ $(x + 1)(x^2 - x + 1)$

77. $x^3 - 27$ $(x - 3)(x^2 + 3x + 9)$

78. $x^3 - 1000$ $(x - 10)(x^2 + 10x + 100)$

79. $8y^3 + 1$ $(2y + 1)(4y^2 - 2y + 1)$

80. $27y^3 + 1$ $(3y + 1)(9y^2 - 3y + 1)$

81. $125x^3 - 8$ $(5x - 2)(25x^2 + 10x + 4)$

82. $27x^3 - 8$ $(3x - 2)(9x^2 + 6x + 4)$

83. $x^3y^3 + 27$ $(xy + 3)(x^2y^2 - 3xy + 9)$

84. $x^3y^3 + 64$ $(xy + 4)(x^2y^2 - 4xy + 16)$

85. $64x - x^4$ $x(4 - x)(16 + 4x + x^2)$

86. $216x - x^4$ $x(6 - x)(36 + 6x + x^2)$

87. $x^6 + 27y^3$ $(x^2 + 3y)(x^4 - 3x^2y + 9y^2)$

88. $x^6 + 8y^3$ $(x^2 + 2y)(x^4 - 2x^2y + 4y^2)$

89. $125x^6 - 64y^6$ $(5x^2 - 4y^2)(25x^4 + 20x^2y^2 + 16y^4)$

90. $125x^6 - y^6$ $(5x^2 - y^2)(25x^4 + 5x^2y^2 + y^4)$

91. $x^9 + 1$ $(x + 1)(x^2 - x + 1)(x^6 - x^3 + 1)$

92. $x^9 - 1$ $(x - 1)(x^2 + x + 1)(x^6 + x^3 + 1)$

93. $(x - y)^3 - y^3$ $(x - 2y)(x^2 - xy + y^2)$

94. $x^3 + (x + y)^3$ $(2x + y)(x^2 + xy + y^2)$

Practice Plus

In Exercises 95–104, factor completely.

95. $0.04x^2 + 0.12x + 0.09$ $(0.2x + 0.3)^2$ or $\dfrac{1}{100}(2x + 3)^2$

96. $0.09x^2 - 0.12x + 0.04$ $(0.3x - 0.2)^2$ or $\dfrac{1}{100}(3x - 2)^2$

97. $8x^4 - \dfrac{x}{8}$ $x\left(2x - \dfrac{1}{2}\right)\left(4x^2 + x + \dfrac{1}{4}\right)$

98. $27x^4 + \dfrac{x}{27}$ $x\left(3x + \dfrac{1}{3}\right)\left(9x^2 - x + \dfrac{1}{9}\right)$

99. $x^6 - 9x^3 + 8$ $(x - 1)(x^2 + x + 1)(x - 2)(x^2 + 2x + 4)$

100. $x^6 + 9x^3 + 8$ $(x + 1)(x^2 - x + 1)(x + 2)(x^2 - 2x + 4)$

101. $x^8 - 15x^4 - 16$ $(x^4 + 1)(x^2 + 4)(x + 2)(x - 2)$

102. $x^8 + 15x^4 - 16$ $(x^4 + 16)(x^2 + 1)(x + 1)(x - 1)$

103. $x^5 - x^3 - 8x^2 + 8$ $(x + 1)(x - 1)(x - 2)(x^2 + 2x + 4)$

104. $x^5 - x^3 + 27x^2 - 27$ $(x + 1)(x - 1)(x + 3)(x^2 - 3x + 9)$

Application Exercises

In Exercises 105–108, find the formula for the area of the shaded region and express it in factored form.

105.

$x^2 - 16 = (x + 4)(x - 4)$

106.

$25x^2 - 9 = (5x + 3)(5x - 3)$

107.

$9x^2 - 16 = (3x + 4)(3x - 4)$

108.

$49x^2 - 36 = (7x + 6)(7x - 6)$

In Exercises 109–110, find the formula for the volume of the region outside the smaller rectangular solid and inside the larger rectangular solid. Then express the volume in factored form.

109.

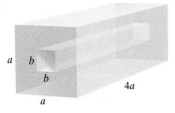

$4a^3 - 4ab^2 = 4a(a + b)(a - b)$

110.

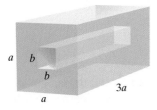

$3a^3 - 3ab^2 = 3a(a + b)(a - b)$

Writing in Mathematics

111. Explain how to factor the difference of two squares. Provide an example with your explanation. **111–114.** Answers will vary.

112. What is a perfect square trinomial and how is it factored?

113. Explain how to factor $x^2 - y^2 + 8x - 16$. Should the expression be grouped into two groups of two terms? If not, why not, and what sort of grouping should be used?

114. Explain how to factor $x^3 + 1$.

Technology Exercises

In Exercises 115–122, use a graphing utility to graph the function on each side of the equation in the same viewing rectangle. Use end behavior to show a complete picture of the polynomial function on the left side. Do the graphs coincide? If so, this means that the polynomial on the left side has been factored correctly. If not, factor the polynomial correctly and then use your graphing utility to verify the factorization.   **115–122.** See graphing answer section.

115. $9x^2 - 4 = (3x + 2)(3x - 2)$ Graphs coincide; factored correctly.

116. $x^2 + 4x + 4 = (x + 4)^2$ Graphs do not coincide; $x^2 + 4x + 4 = (x + 2)^2$.

117. $9x^2 + 12x + 4 = (3x + 2)^2$ Graphs coincide; factored correctly.

118. $25 - (x^2 + 4x + 4) = (x + 7)(x - 3)$ Graphs do not coincide; $25 - (x^2 + 4x + 4) = (7 + x)(3 - x)$.

119. $(2x + 3)^2 - 9 = 4x(x + 3)$ Graphs coincide; factored correctly.

120. $(x - 3)^2 + 8(x - 3) + 16 = (x - 1)^2$ Graphs do not coincide; $(x - 3)^2 + 8(x - 3) + 16 = (x + 1)^2$.

121. $x^3 - 1 = (x - 1)(x^2 - x + 1)$ Graphs do not coincide; $x^3 - 1 = (x - 1)(x^2 + x + 1)$.

122. $(x + 1)^3 + 1 = (x + 1)(x^2 + x + 1)$ Graphs do not coincide; $(x + 1)^3 + 1 = (x + 2)(x^2 + x + 1)$.

123. Use the $\boxed{\text{TABLE}}$ feature of a graphing utility to verify any two of your factorizations in Exercises 67–68 or 77–78. Answers will vary.

Critical Thinking Exercises

124. Which one of the following is true?   d

a. $9x^2 + 15x + 25 = (3x + 5)^2$

b. $x^3 - 27 = (x - 3)(x^2 + 6x + 9)$

c. $x^3 - 64 = (x - 4)^3$

d. None of the above is true.

In Exercises 125–128, factor each polynomial completely. Assume that any variable exponents represent whole numbers.

125. $y^3 + x + x^3 + y$ $(y + x)(y^2 - xy + x^2 + 1)$

126. $36x^{2n} - y^{2n}$ $(6x^n + y^n)(6x^n - y^n)$

127. $x^{3n} + y^{12n}$ $(x^n + y^{4n})(x^{2n} - x^n y^{4n} + y^{8n})$

128. $4x^{2n} + 20x^n y^m + 25y^{2m}$ $(2x^n + 5y^m)^2$

129. Factor $x^6 - y^6$ first as the difference of squares and then as the difference of cubes. From these two factorizations, determine a factorization for $x^4 + x^2y^2 + y^4$.

$x^6 - y^6 = (x^3 + y^3)(x^3 - y^3)$
$= (x + y)(x^2 - xy + y^2)(x - y)(x^2 + xy + y^2);$
$x^6 - y^6 = (x^2 - y^2)(x^4 + x^2y^2 + y^4)$
$= (x + y)(x - y)(x^4 + x^2y^2 + y^4);$
$x^4 + x^2y^2 + y^4 = (x^2 - xy + y^2)(x^2 + xy + y^2)$

In Exercises 130–131, find all integers k so that the trinomial is a perfect square trinomial.

130. $kx^2 + 8xy + y^2$ 16

131. $64x^2 - 16x + k$ 1

Review Exercises

132. Solve: $2x + 2 \geq 12$ and $\dfrac{2x - 1}{3} \leq 7$.   (Section 4.2, Example 2) $\{x | 5 \leq x \leq 11\}$ or $[5, 11]$

133. Solve using matrices:

$$3x - 2y = -8$$
$$x + 6y = 4.$$

(Section 3.4, Example 3) $(-2, 1)$

134. Simplify: $\dfrac{4^{-2}x^{-3}y^1}{x^2y^{-5}}$.

(Section 1.6, Example 9) $\dfrac{y^6}{16x^5}$

5.6

Objectives

1 Use the appropriate method for factoring a polynomial.

2 Use a general strategy for factoring polynomials.

A GENERAL FACTORING STRATEGY

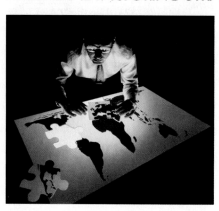

Successful problem solving involves understanding the problem, devising a plan for solving it, and then carrying out the plan. In this section, you will learn a step-by-step strategy that provides a plan and direction for solving factoring problems.

---

1 Use the appropriate method for factoring a polynomial.

A Strategy for Factoring Polynomials It is important to practice factoring a wide variety of polynomials so that you can quickly select the appropriate technique. The polynomial is factored completely when all its polynomial factors, except possibly for monomial factors, are prime. Because of the commutative property, the order of the factors does not matter.

Here is a general strategy for factoring polynomials:

A STRATEGY FOR FACTORING A POLYNOMIAL

1. If there is a common factor, factor out the GCF or factor out a common factor with a negative coefficient.

2. Determine the number of terms in the polynomial and try factoring as follows:

   a. If there are two terms, can the binomial be factored by using one of the following special forms?

   Difference of two squares: $A^2 - B^2 = (A + B)(A - B)$

   Sum of two cubes: $A^3 + B^3 = (A + B)(A^2 - AB + B^2)$

   Difference of two cubes: $A^3 - B^3 = (A - B)(A^2 + AB + B^2)$

   b. If there are three terms, is the trinomial a perfect square trinomial? If so, factor by using one of the following special forms:

   $$A^2 + 2AB + B^2 = (A + B)^2$$
   $$A^2 - 2AB + B^2 = (A - B)^2.$$

   If the trinomial is not a perfect square trinomial, try factoring by trial and error or grouping.

   c. If there are four or more terms, try factoring by grouping.

3. Check to see if any factors with more than one term in the factored polynomial can be factored further. If so, factor completely.

Remember to check the factored form by multiplying or by using the $\boxed{\text{TABLE}}$ or $\boxed{\text{GRAPH}}$ feature of a graphing utility.

2 Use a general strategy for factoring polynomials.

The following examples and those in the exercise set are similar to the previous factoring problems. However, these factorizations are not all of the same type. They are intentionally mixed to promote the development of a general factoring strategy.

USING TECHNOLOGY

The polynomial functions

$$y_1 = 2x^3 + 8x^2 + 8x$$

and

$$y_2 = 2x(x + 2)^2$$

have identical graphs. This verifies that

$$2x^3 + 8x^2 + 8x = 2x(x + 2)^2.$$

$[-4, 2, 1]$ by $[-10, 10, 1]$

The degree of y_1 is 3, which is odd. Odd-degree polynomial functions have graphs with opposite behavior at each end. The leading coefficient, 2, is positive. The graph should fall to the left and rise to the right. The viewing rectangle used is complete enough to show this end behavior.

EXAMPLE 1 Factoring a Polynomial

Factor: $2x^3 + 8x^2 + 8x$.

SOLUTION

Step 1. **If there is a common factor, factor out the GCF.** Because $2x$ is common to all terms, we factor it out.

$$2x^3 + 8x^2 + 8x = 2x(x^2 + 4x + 4) \qquad \text{Factor out the GCF.}$$

Step 2. **Determine the number of terms and factor accordingly.** The factor $x^2 + 4x + 4$ has three terms and is a perfect square trinomial. We factor using $A^2 + 2AB + B^2 = (A + B)^2$.

$$2x^3 + 8x^2 + 8x = 2x(x^2 + 4x + 4)$$

$$= 2x(x^2 + 2 \cdot x \cdot 2 + 2^2)$$

$$\underbrace{A^2} + \underbrace{2AB} + \underbrace{B^2}$$

$$= 2x(x + 2)^2 \qquad A^2 + 2AB + B^2 = (A + B)^2$$

Step 3. **Check to see if factors can be factored further.** In this problem, they cannot. Thus,

$$2x^3 + 8x^2 + 8x = 2x(x + 2)^2.$$ ∎

✔ **CHECK POINT 1** Factor: $3x^3 - 30x^2 + 75x$. $3x(x - 5)^2$

EXAMPLE 2 Factoring a Polynomial

Factor: $4x^2y - 16xy - 20y$.

SOLUTION

Step 1. **If there is a common factor, factor out the GCF.** Because $4y$ is common to all terms, we factor it out.

$$4x^2y - 16xy - 20y = 4y(x^2 - 4x - 5) \qquad \text{Factor out the GCF.}$$

Step 2. **Determine the number of terms and factor accordingly.** The factor $x^2 - 4x - 5$ has three terms, but it is not a perfect square trinomial. We factor it using trial and error.

$$4x^2y - 16xy - 20y = 4y(x^2 - 4x - 5) = 4y(x + 1)(x - 5)$$

Step 3. **Check to see if factors can be factored further.** In this case, they cannot, so we have factored completely. ∎

✔ **CHECK POINT 2** Factor: $3x^2y - 12xy - 36y$. $3y(x + 2)(x - 6)$

EXAMPLE 3 Factoring a Polynomial

Factor: $9b^2x - 16y - 16x + 9b^2y$.

SOLUTION

Step 1. **If there is a common factor, factor out the GCF.** Other than 1 or -1, there is no common factor.

Step 2. **Determine the number of terms and factor accordingly.** There are four terms. We try factoring by grouping. Notice that the first and last terms have a common factor of $9b^2$ and the two middle terms have a common factor of -16. Thus, we begin by rearranging the terms.

$9b^2x - 16y - 16x + 9b^2y$

$= (9b^2x + 9b^2y) + (-16x - 16y)$ Rearrange terms and group terms with common factors.

$= 9b^2(x + y) - 16(x + y)$ Factor from each group.

$= (x + y)(9b^2 - 16)$ Factor out the common binomial factor, $x + y$.

Step 3. **Check to see if factors can be factored further.** We note that $9b^2 - 16$ is the difference of two squares, $(3b)^2 - 4^2$, so we continue factoring.

$9b^2x - 16y - 16x + 9b^2y$

$= (x + y)[(3b)^2 - 4^2]$ Express $9b^2 - 16$ as the difference of squares.

$= (x + y)(3b + 4)(3b - 4)$ The factors of $9b^2 - 16$ are the sum and difference of the expressions being squared. ∎

✔ **CHECK POINT 3** Factor: $16a^2x - 25y - 25x + 16a^2y$. $(x + y)(4a + 5)(4a - 5)$

EXAMPLE 4 Factoring a Polynomial

Factor: $x^2 - 25a^2 + 8x + 16$.

SOLUTION

Step 1. **If there is a common factor, factor out the GCF.** Other than 1 or -1, there is no common factor.

Step 2. **Determine the number of terms and factor accordingly.** There are four terms. We try factoring by grouping. Grouping into two groups of two terms does not result in a common binomial factor. Let's try grouping as a difference of squares.

$x^2 - 25a^2 + 8x + 16$

$= (x^2 + 8x + 16) - 25a^2$ Rearrange terms and group as a perfect square trinomial minus $25a^2$ to obtain a difference of squares.

$= (x + 4)^2 - (5a)^2$ Factor the perfect square trinomial.

$= (x + 4 + 5a)(x + 4 - 5a)$ Factor the difference of squares. The factors are the sum and difference of the expressions being squared.

Step 3. **Check to see if factors can be factored further.** In this case, they cannot, so we have factored completely. ∎

✔ **CHECK POINT 4** Factor: $x^2 - 36a^2 + 20x + 100$. $(x + 10 + 6a)(x + 10 - 6a)$

EXAMPLE 5 Factoring a Polynomial

Factor: $3x^{10} + 3x$.

SOLUTION

Step 1. **If there is a common factor, factor out the GCF.** Because $3x$ is common to both terms, we factor it out.

$$3x^{10} + 3x = 3x(x^9 + 1) \qquad \text{Factor out the GCF.}$$

Step 2. **Determine the number of terms and factor accordingly.** The factor $x^9 + 1$ has two terms. This binomial can be expressed as $(x^3)^3 + 1^3$, so it can be factored as the sum of two cubes.

$$3x^{10} + 3x = 3x(x^9 + 1)$$
$$= 3x[(x^3)^3 + 1^3] = 3x(x^3 + 1)[(x^3)^2 - x^3 \cdot 1 + 1^2]$$

$$A^3 + B^3 = (A + B)(A^2 - AB + B^2)$$

$$= 3x(x^3 + 1)(x^6 - x^3 + 1) \qquad \text{Simplify.}$$

Step 3. **Check to see if factors can be factored further.** We note that $x^3 + 1$ is the sum of two cubes, $x^3 + 1^3$, so we continue factoring.

$$3x^{10} + 3x$$

$$= 3x(x^3 + 1)(x^6 - x^3 + 1) \qquad \text{This is our factorization in the previous step.}$$

$$A^3 + B^3 = (A + B)(A^2 - AB + B^2)$$

$$= 3x(x + 1)(x^2 - x + 1)(x^6 - x^3 + 1) \qquad \text{Factor completely by factoring } x^3 + 1^3, \text{ the sum of cubes.} ■$$

✔ **CHECK POINT 5** Factor: $x^{10} + 512x$. *Hint:* $512 = 8^3$.
  $x(x + 2)(x^2 - 2x + 4)(x^6 - 8x^3 + 64)$

5.6 EXERCISE SET

 Student Solutions Manual   CD/Video   PH Math/Tutor Center   MathXL Tutorials on CD   MathXL®   MyMathLab   Interactmath.com

Practice Exercises

In Exercises 1–68, factor completely, or state that the polynomial is prime.

1. $x^3 - 16x$ $x(x + 4)(x - 4)$

2. $x^3 - x$ $x(x + 1)(x - 1)$

3. $3x^2 + 18x + 27$ $3(x + 3)^2$

4. $8x^2 + 40x + 50$ $2(2x + 5)^2$

5. $81x^3 - 3$ $3(3x - 1)(9x^2 + 3x + 1)$

6. $24x^3 - 3$ $3(2x - 1)(4x^2 + 2x + 1)$

7. $x^2y - 16y + 32 - 2x^2$ $(x + 4)(x - 4)(y - 2)$

8. $12x^2y - 27y - 4x^2 + 9$ $(2x + 3)(2x - 3)(3y - 1)$

9. $4a^2b - 2ab - 30b$ $2b(2a + 5)(a - 3)$

10. $32y^2 - 48y + 18$ $2(4y - 3)^2$

11. $ay^2 - 4a - 4y^2 + 16$ $(y + 2)(y - 2)(a - 4)$

12. $ax^2 - 16a - 2x^2 + 32$ $(x + 4)(x - 4)(a - 2)$

13. $11x^5 - 11xy^2$ $11x(x^2 + y)(x^2 - y)$

14. $4x^9 - 400x$ $4x(x^4 + 10)(x^4 - 10)$

15. $4x^5 - 64x$ $4x(x^2 + 4)(x + 2)(x - 2)$

16. $7x^5 - 7x$ $7x(x^2 + 1)(x + 1)(x - 1)$

17. $x^3 - 4x^2 - 9x + 36$ $(x - 4)(x + 3)(x - 3)$

18. $x^3 - 5x^2 - 4x + 20$ $(x - 5)(x + 2)(x - 2)$

19. $2x^5 + 54x^2$ $2x^2(x + 3)(x^2 - 3x + 9)$

20. $3x^5 + 24x^2$ $3x^2(x + 2)(x^2 - 2x + 4)$

21. $3x^4y - 48y^5$ $3y(x^2 + 4y^2)(x + 2y)(x - 2y)$

22. $32x^4y - 2y^5$ $2y(4x^2 + y^2)(2x + y)(2x - y)$

23. $12x^3 + 36x^2y + 27xy^2$ $3x(2x + 3y)^2$

24. $18x^3 + 48x^2y + 32xy^2$ $2x(3x + 4y)^2$

25. $x^2 - 12x + 36 - 49y^2$ $(x - 6 + 7y)(x - 6 - 7y)$

26. $x^2 - 10x + 25 - 36y^2$ $(x - 5 + 6y)(x - 5 - 6y)$

27. $4x^2 + 25y^2$ prime

28. $16x^2 + 49y^2$ prime

29. $12x^3y - 12xy^3$ $12xy(x + y)(x - y)$

30. $9x^2y^2 - 36y^2$ $9y^2(x + 2)(x - 2)$

31. $6bx^2 + 6by^2$ $6b(x^2 + y^2)$

32. $6x^2 - 66$ $6(x^2 - 11)$

33. $x^4 - xy^3 + x^3y - y^4$ $(x + y)(x - y)(x^2 + xy + y^2)$

34. $x^3 - xy^2 + x^2y - y^3$ $(x + y)^2(x - y)$

35. $x^2 - 4a^2 + 12x + 36$ $(x + 6 + 2a)(x + 6 - 2a)$

36. $x^2 - 49a^2 + 14x + 49$ $(x + 7 + 7a)(x + 7 - 7a)$

37. $5x^3 + x^6 - 14$ $(x^3 - 2)(x^3 + 7)$

38. $6x^3 + x^6 - 16$ $(x + 2)(x^2 - 2x + 4)(x^3 - 2)$

39. $4x - 14 + 2x^3 - 7x^2$ $(2x - 7)(2 + x^2)$

40. $3x^3 + 8x + 9x^2 + 24$ $(3x^2 + 8)(x + 3)$

41. $54x^3 - 16y^3$ $2(3x - 2y)(9x^2 + 6xy + 4y^2)$

42. $54x^3 - 250y^3$ $2(3x - 5y)(9x^2 + 15xy + 25y^2)$

43. $x^2 + 10x - y^2 + 25$ $(x + 5 + y)(x + 5 - y)$

44. $x^2 + 6x - y^2 + 9$ $(x + 3 + y)(x + 3 - y)$

45. $x^8 - y^8$ $(x^4 + y^4)(x^2 + y^2)(x + y)(x - y)$

46. $x^8 - 1$ $(x^4 + 1)(x^2 + 1)(x + 1)(x - 1)$

47. $x^3y - 16xy^3$ $xy(x + 4y)(x - 4y)$

48. $x^3y - 100xy^3$ $xy(x + 10y)(x - 10y)$

49. $x + 8x^4$ $x(1 + 2x)(1 - 2x + 4x^2)$

50. $x + 27x^4$ $x(1 + 3x)(1 - 3x + 9x^2)$

51. $16y^2 - 4y - 2$ $2(4y + 1)(2y - 1)$

52. $32y^2 + 4y - 6$ $2(2y + 1)(8y - 3)$

53. $14y^3 + 7y^2 - 10y$ $y(14y^2 + 7y - 10)$

54. $5y^3 - 45y^2 + 70y$ $5y(y - 2)(y - 7)$

55. $27x^2 + 36xy + 12y^2$ $3(3x + 2y)^2$

56. $125x^2 + 50xy + 5y^2$ $5(5x + y)^2$

57. $12x^3 + 3xy^2$ $3x(4x^2 + y^2)$

58. $3x^4 + 27x^2$ $3x^2(x^2 + 9)$

59. $x^6y^6 - x^3y^3$ $x^3y^3(xy - 1)(x^2y^2 + xy + 1)$

60. $x^3 - 2x^2 - x + 2$ $(x - 2)(x + 1)(x - 1)$

61. $(x + 5)(x - 3) + (x + 5)(x - 7)$ $2(x + 5)(x - 5)$

62. $(x + 4)(x - 9) + (x + 4)(2x - 3)$ $3(x + 4)(x - 4)$

63. $a^2(x - y) + 4(y - x)$ $(x - y)(a + 2)(a - 2)$

64. $b^2(x - 3) + c^2(3 - x)$ $(x - 3)(b + c)(b - c)$

65. $(c + d)^4 - 1$ $[(c + d)^2 + 1](c + d + 1)(c + d - 1)$ or $(c^2 + 2cd + d^2 + 1)(c + d + 1)(c + d - 1)$

66. $(c + d)^4 - 16$ $[(c + d)^2 + 4](c + d + 2)(c + d - 2)$ or $(c^2 + 2cd + d^2 + 4)(c + d + 2)(c + d - 2)$

67. $p^3 - pq^2 + p^2q - q^3$ $(p + q)^2(p - q)$

68. $p^3 - pq^2 - p^2q + q^3$ $(p - q)^2(p + q)$

Practice Plus

In Exercises 69–80, factor completely.

69. $x^4 - 5x^2y^2 + 4y^4$ $(x + 2y)(x - 2y)(x + y)(x - y)$

70. $x^4 - 10x^2y^2 + 9y^4$ $(x + 3y)(x - 3y)(x + y)(x - y)$

71. $(x + y)^2 + 6(x + y) + 9$ $(x + y + 3)^2$

72. $(x - y)^2 - 8(x - y) + 16$ $(x - y - 4)^2$

73. $(x - y)^4 - 4(x - y)^2$ $(x - y)^2(x - y + 2)(x - y - 2)$

74. $(x + y)^4 - 100(x + y)^2$ $(x + y)^2(x + y + 10)(x + y - 10)$

75. $2x^2 - 7xy^2 + 3y^4$ $(2x - y^2)(x - 3y^2)$

76. $3x^2 + 5xy^2 + 2y^4$ $(3x - y^2)(x + 2y^2)$

77. $x^3 - y^3 - x + y$ $(x - y)(x^2 + xy + y^2 - 1)$

78. $x^3 + y^3 + x^2 - y^2$ $(x + y)(x^2 - xy + y^2 + x - y)$

79. $x^6y^3 + x^3 - 8x^3y^3 - 8$ $(xy + 1)(x^2y^2 - xy + 1)(x - 2)(x^2 + 2x + 4)$

80. $x^6y^3 - x^3 + x^3y^3 - 1$
$(xy - 1)(x^2y^2 + xy + 1)(x + 1)(x^2 - x + 1)$

Application Exercises

In Exercises 81–86,

a. *Write an expression for the area of the shaded region.*

b. *Write the expression in factored form.*

81.

a. $x(x + y) - y(x + y)$

b. $(x + y)(x - y)$

82.

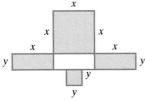

a. $x^2 + xy + xy + y^2 = x^2 + 2xy + y^2$ **b.** $(x + y)^2$

83.

a. $xy + xy + xy + 3x(x) = 3xy + 3x^2$ **b.** $3x(y + x)$

84.

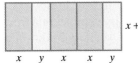

a. $3x(x + y) + 2y(x + y)$
b. $(x + y)(3x + 2y)$

85.

a. $8x^2 - 2\pi x^2$ **b.** $2x^2(4 - \pi)$

86.

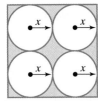

a. $16x^2 - 4\pi x^2$ **b.** $4x^2(4 - \pi)$

Writing in Mathematics

87. Describe a strategy that can be used to factor polynomials.
   87–88. Answers will vary.

88. Describe some of the difficulties in factoring polynomials. What suggestions can you offer to overcome these difficulties?

Technology Exercises

In Exercises 89–92, use a graphing utility to graph the function on each side of the equation in the same viewing rectangle. Use end behavior to show a complete picture of the polynomial on the left side. Do the graphs coincide? If so, the factorization is correct. If not, factor correctly and then use your graphing utility to verify the factorization.   89–92. See graphing answer section.

89. $4x^2 - 12x + 9 = (4x - 3)^2$
   Graphs do not coincide; $4x^2 - 12x + 9 = (2x - 3)^2$.

90. $2x^3 + 10x^2 - 2x - 10 = 2(x + 5)(x^2 + 1)$ Graphs do not coincide; $2x^3 + 10x^2 - 2x - 10 = 2(x + 5)(x + 1)(x - 1)$.

91. $x^4 - 16 = (x^2 + 4)(x + 2)(x - 2)$
   Graphs coincide; factored correctly.

92. $x^3 + 1 = (x + 1)^3$
   Graphs do not coincide; $x^3 + 1 = (x + 1)(x^2 - x + 1)$.

93. Use the $\boxed{\text{TABLE}}$ feature of a graphing utility to verify any two of your complete factorizations in Exercises 15–20.
   Answers will vary.

Critical Thinking Exercises

94. Which one of the following is true?   b

   a. $x^4 - 16$ is factored completely as $(x^2 + 4)(x^2 - 4)$.

   b. The trinomial $x^2 - 4x - 4$ is a prime polynomial.

   c. $x^2 + 36 = (x + 6)^2$

   d. $x^3 - 64 = (x + 4)(x^2 + 4x - 16)$

In Exercises 95–97, factor completely. Assume that variable exponents represent whole numbers.

95. $x^{2n+3} - 10x^{n+3} + 25x^3$ $x^3(x^n - 5)^2$

96. $3x^{n+2} - 13x^{n+1} + 4x^n$ $x^n(3x - 1)(x - 4)$

97. $x^{4n+1} - xy^{4n}$ $x(x^{2n} + y^{2n})(x^n + y^n)(x^n - y^n)$

98. In certain circumstances, the sum of two perfect squares can be factored by adding and subtracting the same perfect square. For example,

$$x^4 + 4 = x^4 + 4x^2 + 4 - 4x^2.$$ Add and subtract $4x^2$.

Use this first step to factor $x^4 + 4$.
   $(x^2 + 2x + 2)(x^2 - 2x + 2)$

99. Express $x^3 + x + 2x^4 + 4x^2 + 2$ as the product of two polynomials of degree 2. $(x^2 + 1)(2x^2 + x + 2)$

Review Exercises

100. Solve: $\dfrac{3x - 1}{5} + \dfrac{2x + 2}{2} = -\dfrac{3}{10}$.
   (Section 1.4, Example 4) -1

101. Simplify: $(4x^3 y^{-1})^2 (2x^{-3} y)^{-1}$.
   (Section 1.6, Example 9) $\dfrac{8x^9}{y^3}$

102. Evaluate: $\begin{vmatrix} 0 & -3 & 2 \\ 1 & 5 & 3 \\ -2 & 1 & 4 \end{vmatrix}$.
   (Section 3.5, Example 3)   52

POLYNOMIAL EQUATIONS AND THEIR APPLICATIONS

SECTION 5.7

Objectives

1 Solve quadratic equations by factoring.

2 Solve higher-degree polynomial equations by factoring.

3 Solve problems using polynomial equations.

If you are a coffee drinker, do you become a tad unnerved if you do not get your caffeine rush at a regular time? We opened the chapter with a polynomial function that models the percentage of coffee drinkers, $P(x)$, who are x years old who become irritable if they do not have coffee at their regular time:

$$P(x) = -0.05x^2 + 4.2x - 26.$$

How can we use this function to find the age at which 36% of coffee drinkers become irritable if they do not have coffee at their regular time? Substitute 36 for $P(x)$ and solve for x:

$$36 = -0.05x^2 + 4.2x - 26.$$

Do you see how this equation differs from a linear equation? The highest exponent on x is 2. Solving such an equation involves finding the numbers that will make the equation a true statement. In this section, we use factoring for solving equations in the form $ax^2 + bx + c = 0$. We also look at applications of these equations.

The Standard Form of a Quadratic Equation We begin by defining a quadratic equation.

> **DEFINITION OF A QUADRATIC EQUATION** A **quadratic equation** in x is an equation that can be written in the **standard form**
>
> $$ax^2 + bx + c = 0,$$
>
> where a, b, and c are real numbers, with $a \neq 0$. A quadratic equation in x is also called a **second-degree polynomial equation** in x.

Here is an example of a quadratic equation in standard form:

$$x^2 - 12x + 27 = 0.$$

$a = 1$ $b = -12$ $c = 27$

1 Solve quadratic equations by factoring.

Solving Quadratic Equations by Factoring We can factor the left side of the quadratic equation $x^2 - 12x + 27 = 0$. We obtain $(x - 3)(x - 9) = 0$. If a quadratic equation has zero on one side and a factored expression on the other side, it can be solved using the **zero-product principle**.

THE ZERO-PRODUCT PRINCIPLE If the product of two algebraic expressions is zero, then at least one of the factors is equal to zero.

$$\text{If } AB = 0, \text{ then } A = 0 \text{ or } B = 0.$$

For example, consider the equation $(x - 3)(x - 9) = 0$. According to the zero-product principle, this product can be zero only if at least one of the factors is zero. We set each individual factor equal to zero and solve the resulting equations for x.

$$(x - 3)(x - 9) = 0$$

$$x - 3 = 0 \quad \text{or} \quad x - 9 = 0$$
$$x = 3 \qquad\qquad x = 9$$

The solutions of the original quadratic equation, $x^2 - 12x + 27 = 0$, are 3 and 9. The solution set is $\{3, 9\}$.

SOLVING A QUADRATIC EQUATION BY FACTORING

1. If necessary, rewrite the equation in the standard form $ax^2 + bx + c = 0$, moving all terms to one side, thereby obtaining zero on the other side.
2. Factor completely.
3. Apply the zero-product principle, setting each factor containing a variable equal to zero.
4. Solve the equations in step 3.
5. Check the solutions in the original equation.

EXAMPLE 1 Solving a Quadratic Equation by Factoring

Solve: $2x^2 - 5x = 12$.

SOLUTION

Step 1. **Move all terms to one side and obtain zero on the other side.** Subtract 12 from both sides and write the equation in standard form.

$$2x^2 - 5x - 12 = 12 - 12$$

$$2x^2 - 5x - 12 = 0$$

Step 2. **Factor.**

$$(2x + 3)(x - 4) = 0$$

Steps 3 and 4. **Set each factor equal to zero and solve the resulting equations.**

$$2x + 3 = 0 \quad \text{or} \quad x - 4 = 0$$
$$2x = -3 \qquad\qquad x = 4$$
$$x = -\frac{3}{2}$$

STUDY TIP

Do not confuse factoring a polynomial with solving a quadratic equation by factoring.

Incorrect!

Factor: ~~$2x^2 - 5x - 12$.~~

~~$(2x + 3)(x - 4)$~~

~~$2x + 3 = 0$ or $x - 4 = 0$~~

~~$x = -\frac{3}{2}$ $x = 4$~~

Step 5. **Check the solutions in the original equation.**

Check $-\dfrac{3}{2}$:	Check 4:
$2x^2 - 5x = 12$	$2x^2 - 5x = 12$
$2\left(-\dfrac{3}{2}\right)^2 - 5\left(-\dfrac{3}{2}\right) \overset{?}{=} 12$	$2(4)^2 - 5(4) \overset{?}{=} 12$
$2\left(\dfrac{9}{4}\right) - 5\left(-\dfrac{3}{2}\right) \overset{?}{=} 12$	$2(16) - 5(4) \overset{?}{=} 12$
$\dfrac{9}{2} + \dfrac{15}{2} \overset{?}{=} 12$	$32 - 20 \overset{?}{=} 12$
$\dfrac{24}{2} \overset{?}{=} 12$	$12 = 12,\ \text{true}$
$12 = 12,\ \text{true}$	

The solutions are $-\frac{3}{2}$ and 4, and the solution set is $\left\{-\frac{3}{2}, 4\right\}$. ∎

✔ **CHECK POINT 1** Solve: $2x^2 - 9x = 5.$ $-\frac{1}{2}$ and 5, or $\left\{-\frac{1}{2}, 5\right\}$

There is an important relationship between a quadratic equation in standard form, such as

$$2x^2 - 5x - 12 = 0$$

and a quadratic function, such as

$$y = 2x^2 - 5x - 12.$$

The solutions of $ax^2 + bx + c = 0$ correspond to the x-intercepts of the graph of the quadratic function $y = ax^2 + bx + c$. For example, you can visualize the solutions of $2x^2 - 5x - 12 = 0$ by looking at the x-intercepts of the graph of the quadratic function $y = 2x^2 - 5x - 12$. The cuplike graph is shown in Figure 5.10. The solutions of the equation $2x^2 - 5x - 12 = 0$, $-\frac{3}{2}$ and 4, appear as the graph's x-intercepts.

x-intercept is $-\frac{3}{2}$. x-intercept is 4.

$[-4, 6, 1]$ by $[-16, 4, 1]$ **FIGURE 5.10**

EXAMPLE 2 Solving Quadratic Equations by Factoring

Solve:

a. $5x^2 = 20x$ **b.** $x^2 + 4 = 8x - 12$ **c.** $(x - 7)(x + 5) = -20.$

FIGURE 5.11 The solution set of $5x^2 = 20x$, or $5x^2 - 20x = 0$, is $\{0, 4\}$.

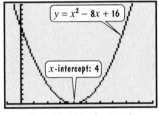

FIGURE 5.12 The solution set of $x^2 + 4 = 8x - 12$, or $x^2 - 8x + 16 = 0$, is $\{4\}$.

SOLUTION

a.

$5x^2 = 20x$	This is the given equation.
$5x^2 - 20x = 0$	Subtract 20x from both sides and write the equation in standard form.
$5x(x - 4) = 0$	Factor.
$5x = 0 \quad$ or $\quad x - 4 = 0$	Set each factor equal to 0.
$x = 0 \qquad\qquad x = 4$	Solve the resulting equations.

Check by substituting 0 and 4 into the given equation. The graph of $y = 5x^2 - 20x$, obtained with a graphing utility, is shown in Figure 5.11. The x-intercepts are 0 and 4. This verifies that the solutions are 0 and 4, and the solution set is $\{0, 4\}$.

b.

$x^2 + 4 = 8x - 12$	This is the given equation.
$x^2 - 8x + 16 = 0$	Write the equation in standard form by subtracting 8x and adding 12 on both sides.
$(x - 4)(x - 4) = 0$	Factor.
$x - 4 = 0 \quad$ or $\quad x - 4 = 0$	Set each factor equal to 0.
$x = 4 \qquad\qquad x = 4$	Solve the resulting equations.

Notice that there is only one solution (or, if you prefer, a repeated solution.) The trinomial $x^2 - 8x + 16$ is a perfect square trinomial that could have been factored as $(x - 4)^2$. The graph of $y = x^2 - 8x + 16$, obtained with a graphing utility, is shown in Figure 5.12. The graph has only one x-intercept at 4. This verifies that the equation's solution is 4, and the solution set is $\{4\}$.

c. Be careful! Although the left side of $(x - 7)(x + 5) = -20$ is factored, we cannot use the zero-product principle. Why not? The right side of the equation is not 0. So we begin by multiplying the factors on the left side of the equation. Then we add 20 to both sides to obtain 0 on the right side.

$(x - 7)(x + 5) = -20$	This is the given equation.
$x^2 - 2x - 35 = -20$	Use the FOIL method to multiply on the left side.
$x^2 - 2x - 15 = 0$	Add 20 to both sides.
$(x + 3)(x - 5) = 0$	Factor.
$x + 3 = 0 \quad$ or $\quad x - 5 = 0$	Set each factor equal to 0.
$x = -3 \qquad\qquad x = 5$	Solve the resulting equations.

Check by substituting -3 and 5 into the given equation. The graph of $y = x^2 - 2x - 15$, obtained with a graphing utility, is shown in Figure 5.13. The x-intercepts are -3 and 5. This verifies that the solutions are -3 and 5, and the solution set is $\{-3, 5\}$.

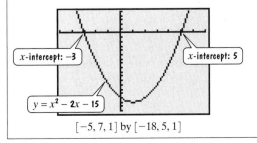

FIGURE 5.13 The solution set of $(x - 7)(x + 5) = -20$, or $x^2 - 2x - 15 = 0$, is $\{-3, 5\}$.

STUDY TIP

Avoid the following errors:

$5x^2 = 20x$

$\dfrac{5x^2}{x} = \dfrac{20x}{x}$ Never divide both sides of an equation by x. Division by zero is undefined

$5x = 20$ and x may be zero. Indeed, the solutions for this equation (Example 2a) are

$x = 4$ 0 and 4. Dividing both sides by x does not permit us to find both solutions.

$(x - 7)(x + 5) = -20$

$x - 7 = -20$ or $x + 5 = -20$ The zero-product principle cannot be used because the

$x = -13$ or $x = -25$ right side of the equation is not equal to 0.

✔ **CHECK POINT 2** Solve:

    a. $3x^2 = 2x$ 0 and $\frac{2}{3}$, or $\left\{0, \frac{2}{3}\right\}$

    b. $x^2 + 7 = 10x - 18$ 5 or $\{5\}$

    c. $(x - 2)(x + 3) = 6.$ -4 and 3, or $\{-4, 3\}$

---

2   Solve higher-degree polynomial equations by factoring.

Polynomial Equations   A **polynomial equation** is the result of setting two polynomials equal to each other. The equation is in **standard form** if one side is 0 and the polynomial on the other side is in standard form, that is, in descending powers of the variable. The **degree of a polynomial equation** is the same as the highest degree of any term in the equation. Here are examples of three polynomial equations:

$$3x + 5 = 14 \qquad 2x^2 + 7x = 4 \qquad x^3 + x^2 = 4x + 4.$$

| This equation is of degree 1 because 1 is the highest degree. | This equation is of degree 2 because 2 is the highest degree. | This equation is of degree 3 because 3 is the highest degree. |

Notice that a polynomial equation of degree 1 is a linear equation. A polynomial equation of degree 2 is a quadratic equation.

    Some polynomial equations of degree 3 or higher can be solved by moving all terms to one side, thereby obtaining 0 on the other side. Once the equation is in standard form, factor and then set each factor equal to 0.

EXAMPLE 3   Solving a Polynomial Equation by Factoring

Solve by factoring: $x^3 + x^2 = 4x + 4.$

SOLUTION

Step 1.   **Move all terms to one side and obtain zero on the other side.** Subtract $4x$ and subtract 4 from both sides.

$$x^3 + x^2 - 4x - 4 = 4x + 4 - 4x - 4$$

$$x^3 + x^2 \underbrace{(- 4x - 4)}_{} = 0$$

A graphing utility's $\boxed{\text{TABLE}}$ feature can be used to numerically verify that $\{-2, -1, 2\}$ is the solution set of

$$x^3 + x^2 = 4x + 4$$

Enter $y_1 = x^3 + x^2$. Enter $y_2 = 4x + 4$.

y_1 and y_2 are equal when $x = -2$, $x = -1$, and $x = 2$.

Step 2. **Factor.** Use factoring by grouping. Group terms that have a common factor.

$$\boxed{x^3 + x^2} + \boxed{-4x - 4} = 0$$

Common factor is x^2.   Common factor is -4.

$x^2(x + 1) - 4(x + 1) = 0$ Factor x^2 from the first two terms and -4 from the last two terms.

$(x + 1)(x^2 - 4) = 0$ Factor out the common binomial, $x + 1$, from each term.

$(x + 1)(x + 2)(x - 2) = 0$ Factor completely by factoring $x^2 - 4$ as the difference of two squares.

Steps 3 and 4. **Set each factor equal to zero and solve the resulting equations.**

$$x + 1 = 0 \quad \text{or} \quad x + 2 = 0 \quad \text{or} \quad x - 2 = 0$$
$$x = -1 \qquad\qquad x = -2 \qquad\qquad x = 2$$

Step 5. **Check the solutions in the original equation.** Check the three solutions, $-1, -2,$ and 2, by substituting them into the original equation. Can you verify that the solutions are $-1, -2,$ and 2, and the solution set is $\{-2, -1, 2\}$? ■

DISCOVER FOR YOURSELF

Suggest a method involving intersecting graphs that can be used with a graphing utility to verify that $\{-2, -1, 2\}$ is the solution set of

$$x^3 + x^2 = 4x + 4.$$

Apply this method to verify the solution set.

x-intercept: -1

x-intercept: -2 x-intercept: 2

$y = x^3 + x^2 - 4x - 4$

$[-5, 5, 1]$ by $[-8, 2, 1]$

You can use a graphing utility to check the solutions to $x^3 + x^2 - 4x - 4 = 0$. Graph $y = x^3 + x^2 - 4x - 4$, as shown on the left. Is the graph complete? Because the degree, 3, is odd, and the leading coefficient, 1, is positive, it should fall to the left and rise to the right. The graph shows this end behavior and is therefore complete. The x-intercepts are $-2, -1,$ and 2, corresponding to the equation's solutions.

 CHECK POINT 3 Solve by factoring: $2x^3 + 3x^2 = 8x + 12$. $-2, -\frac{3}{2},$ and 2, or $\left\{-2, -\frac{3}{2}, 2\right\}$

3 Solve problems using polynomial equations.

Applications of Polynomial Equations   Solving polynomial equations by factoring can be used to answer questions about variables contained in mathematical models.

EXAMPLE 4 Modeling Motion

You throw a ball straight up from a rooftop 384 feet high with an initial speed of 32 feet per second. The function

$$s(t) = -16t^2 + 32t + 384$$

describes the ball's height above the ground, $s(t)$, in feet, t seconds after you throw it. The ball misses the rooftop on its way down and eventually strikes the ground. How long will it take for the ball to hit the ground?

SOLUTION   The ball hits the ground when $s(t)$, its height above the ground, is 0 feet. Thus, we substitute 0 for $s(t)$ in the given function and solve for t.

$$s(t) = -16t^2 + 32t + 384$$ This is the function that models the ball's height.

$$0 = -16t^2 + 32t + 384$$ Substitute 0 for $s(t)$.

$$0 = -16(t^2 - 2t - 24)$$ Factor out -16.

$$0 = -16(t - 6)(t + 4)$$ Factor the trinomial.

> Do not set the constant, -16, equal to zero: $-16 \neq 0$.

$$t - 6 = 0 \quad \text{or} \quad t + 4 = 0$$ Set each variable factor equal to 0.

$$t = 6 \qquad\qquad t = -4$$ Solve for t.

Because we begin describing the ball's height at $t = 0$, we discard the solution $t = -4$. The ball hits the ground after 6 seconds.   ■

Figure 5.14 shows the graph of the function $s(t) = -16t^2 + 32t + 384$. The horizontal axis is labeled t, for the ball's time in motion. The vertical axis is labeled $s(t)$, for the ball's height above the ground at time t. Because time and height are both positive, the function is graphed in quadrant I only.

The graph visually shows what we discovered algebraically: The ball hits the ground after 6 seconds. The graph also reveals that the ball reaches its maximum height, 400 feet, after 1 second. Then the ball begins to fall.

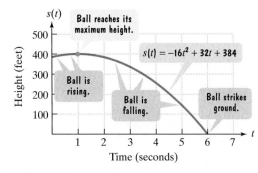

FIGURE 5.14

> ✔ **CHECK POINT 4**   Use the function $s(t) = -16t^2 + 32t + 384$ to determine when the ball's height is 336 feet. Identify your meaningful solution as a point on the graph in Figure 5.14.   *after 3 seconds; (3, 336)*

ENRICHMENT ESSAY

Visualizing Irritability by Age

The polynomial function

$$P(x) = -0.05x^2 + 4.2x - 26$$

models the percentage of coffee drinkers, $P(x)$, who are x years old who become irritable if they do not have coffee at their regular time. Figure 5.15 shows the graph of the function. The graph reveals that 62.2% of 42-year-old coffee drinkers become irritable. This is the maximum percentage for any age, x, in the function's domain.

FIGURE 5.15   *Source:* LMK Associates

In our next example, we use our five-step strategy for solving word problems.

EXAMPLE 5 Solving a Problem Involving Landscape Design

A rectangular garden measures 80 feet by 60 feet. A large path of uniform width is to be added along both shorter sides and one longer side of the garden. The landscape designer doing the work wants to double the garden's area with the addition of this path. How wide should the path be?

SOLUTION

FIGURE 5.16 The garden's area is to be doubled by adding the path.

Step 1. **Let x represent one of the quantities.** We will let

$$x = \text{the width of the path.}$$

The situation is illustrated in Figure 5.16. The figure shows the original 80-by-60 foot rectangular garden and the path of width x added along both shorter sides and one longer side.

Step 2. **Represent other quantities in terms of x.** Because the path is added along both shorter sides and one longer side, Figure 5.16 shows that

$$80 + 2x = \text{the length of the new, expanded rectangle}$$

$$60 + x = \text{the width of the new, expanded rectangle.}$$

Step 3. **Write an equation that describes the conditions.** The area of the rectangle must be doubled by the addition of the path.

The area, or length times width, of the new, expanded rectangle   must be   twice that of   the area of the garden.

$$(80 + 2x)(60 + x) \quad = \quad 2 \;\cdot\; 80 \;\cdot\; 60$$

Step 4. **Solve the equation and answer the question.**

$$(80 + 2x)(60 + x) = 2 \cdot 80 \cdot 60 \qquad \text{This is the equation for the problem's conditions.}$$

$$4800 + 200x + 2x^2 = 9600 \qquad \text{Multiply. Use FOIL on the left side.}$$

$$2x^2 + 200x - 4800 = 0 \qquad \text{Subtract 9600 from both sides and write the equation in standard form.}$$

$$2(x^2 + 100x - 2400) = 0 \qquad \text{Factor out 2, the GCF.}$$

$$2(x - 20)(x + 120) = 0 \qquad \text{Factor the trinomial.}$$

$$x - 20 = 0 \quad \text{or} \quad x + 120 = 0 \qquad \text{Set each variable factor equal to 0.}$$

$$x = 20 \quad \text{or} \qquad x = -120 \qquad \text{Solve for x.}$$

The path cannot have a negative width. Because -120 is geometrically impossible, we use $x = 20$. The width of the path should be 20 feet.

Step 5. **Check the proposed solution in the original wording of the problem.** Has the landscape architect doubled the garden's area with the 20-foot-wide path? The area of the garden is 80 feet times 60 feet, or 4800 square feet. Because $80 + 2x$ and $60 + x$ represent the length and width of the expanded rectangle,

$$80 + 2x = 80 + 2 \cdot 20 = 120 \text{ feet is the expanded rectangle's length.}$$

$$60 + x = 60 + 20 \quad = 80 \text{ feet is the expanded rectangle's width.}$$

The area of the expanded rectangle is 120 feet times 80 feet, or 9600 square feet. This is double the area of the garden, 4800 square feet, as specified by the problem's conditions.

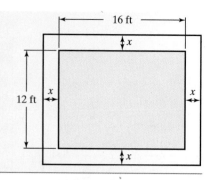

CHECK POINT 5 A rectangular garden measures 16 feet by 12 feet. A path of uniform width is to be added so as to surround the entire garden. The landscape artist doing the work wants the garden and path to cover an area of 320 square feet. How wide should the path be? The path should be 2 feet wide.

The solution to our next problem relies on knowing the **Pythagorean Theorem**. The theorem relates the lengths of the three sides of a **right triangle**, a triangle with one angle measuring 90°. The side opposite the 90° angle is called the **hypotenuse**. The other sides are called **legs**. The legs form the two sides of the right angle.

THE PYTHAGOREAN THEOREM   The sum of the squares of the lengths of the legs of a right triangle equals the square of the length of the hypotenuse.

If the legs have lengths a and b, and the hypotenuse has length c, then

$$a^2 + b^2 = c^2.$$

FIGURE 5.17

EXAMPLE 6   **Using the Pythagorean Theorem to Obtain a Polynomial Equation**

Figure 5.17 shows a tent with wires attached to help stabilize it. The length of each wire is 8 feet greater than the height from the ground where it is attached to the tent. Each wire's location from the tent exceeds this height by 7 feet. Find the length of each wire used to stabilize the tent.

SOLUTION   Figure 5.17 shows a right triangle. The length of the legs are x and $x + 7$. The length of the hypotenuse, $x + 8$, represents the length of the wire. We use the Pythagorean Theorem to find this length.

$$\underbrace{x^2}_{\text{leg}^2} + \underbrace{(x + 7)^2}_{\text{leg}^2} = \underbrace{(x + 8)^2}_{\text{hypotenuse}^2}$$

This is the equation arising from the Pythagorean Theorem.

$$x^2 + x^2 + 14x + 49 = x^2 + 16x + 64$$

Square $x + 7$ and $x + 8$.

$$2x^2 + 14x + 49 = x^2 + 16x + 64$$

Combine like terms: $x^2 + x^2 = 2x^2$.

$$x^2 - 2x - 15 = 0$$

Subtract $x^2 + 16x + 64$ from both sides and write the quadratic equation in standard form.

$$(x - 5)(x + 3) = 0$$

Factor the trinomial.

$$x - 5 = 0 \quad \text{or} \quad x + 3 = 0$$

Set each factor equal to 0.

$$x = 5 \quad \text{or} \quad x = -3$$

Solve for x.

Because x represents the height from the ground where the wire is attached, x cannot be negative. Thus, we only use $x = 5$. Figure 5.17 shows that the length of the wire is $x + 8$ feet. The length of the wire is $5 + 8$ feet, or 13 feet.

We can check to see that the lengths of the three sides of the right triangle, x, $x + 7$, and $x + 8$, satisfy the Pythagorean Theorem when $x = 5$. The lengths are 5 feet, 12 feet, and 13 feet.

$$\underbrace{5^2}_{\text{leg}^2} + \underbrace{12^2}_{\text{leg}^2} \overset{?}{=} \underbrace{13^2}_{\text{hypotenuse}^2}$$

$$25 + 144 \overset{?}{=} 169$$

$$169 = 169, \text{ true}$$

CHECK POINT 6 A guy wire is attached to a tree to help it grow straight. The situation is illustrated in Figure 5.18. The length of the wire is 2 feet greater than the distance from the base of the tree to the stake. Find the length of the wire. *The wire is 5 feet long.*

$x + 1$ $x + 2$

x

FIGURE 5.18

5.7 EXERCISE SET

Student Solutions Manual CD/Video PH Math/Tutor Center MathXL Tutorials on CD MathXL® MyMathLab Interactmath.com

Practice Exercises

In Exercises 1–36, use factoring to solve each quadratic equation. Check by substitution or by using a graphing utility and identifying x-intercepts.

1. $x^2 + x - 12 = 0$ -4 and 3, or $\{-4, 3\}$
2. $x^2 - 2x - 15 = 0$ -3 and 5, or $\{-3, 5\}$
3. $x^2 + 6x = 7$ -7 and 1, or $\{-7, 1\}$
4. $x^2 - 4x = 45$ -5 and 9, or $\{-5, 9\}$
5. $3x^2 + 10x - 8 = 0$ -4 and $\frac{2}{3}$, or $\left\{-4, \frac{2}{3}\right\}$
6. $2x^2 - 5x - 3 = 0$ $-\frac{1}{2}$ and 3, or $\left\{-\frac{1}{2}, 3\right\}$
7. $5x^2 = 8x - 3$ $\frac{3}{5}$ and 1, or $\left\{\frac{3}{5}, 1\right\}$
8. $7x^2 = 30x - 8$ $\frac{2}{7}$ and 4, or $\left\{\frac{2}{7}, 4\right\}$
9. $3x^2 = 2 - 5x$ -2 and $\frac{1}{3}$, or $\left\{-2, \frac{1}{3}\right\}$
10. $5x^2 = 2 + 3x$ $-\frac{2}{5}$ and 1, or $\left\{-\frac{2}{5}, 1\right\}$
11. $x^2 = 8x$ 0 and 8, or $\{0, 8\}$
12. $x^2 = 4x$ 0 and 4, or $\{0, 4\}$
13. $3x^2 = 5x$ 0 and $\frac{5}{3}$, or $\left\{0, \frac{5}{3}\right\}$
14. $2x^2 = 5x$ 0 and $\frac{5}{2}$, or $\left\{0, \frac{5}{2}\right\}$
15. $x^2 + 4x + 4 = 0$ -2 or $\{-2\}$
16. $x^2 + 6x + 9 = 0$ -3 or $\{-3\}$
17. $x^2 = 14x - 49$ 7 or $\{7\}$
18. $x^2 = 12x - 36$ 6 or $\{6\}$

19. $9x^2 = 30x - 25$ $\frac{5}{3}$ or $\left\{\frac{5}{3}\right\}$
20. $4x^2 = 12x - 9$ $\frac{3}{2}$ or $\left\{\frac{3}{2}\right\}$
21. $x^2 - 25 = 0$ -5 and 5, or $\{-5, 5\}$
22. $x^2 - 49 = 0$ -7 and 7, or $\{-7, 7\}$
23. $9x^2 = 100$ $-\frac{10}{3}$ and $\frac{10}{3}$, or $\left\{-\frac{10}{3}, \frac{10}{3}\right\}$
24. $4x^2 = 25$ $-\frac{5}{2}$ and $\frac{5}{2}$, or $\left\{-\frac{5}{2}, \frac{5}{2}\right\}$
25. $x(x - 3) = 18$ -3 and 6, or $\{-3, 6\}$
26. $x(x - 4) = 21$ -3 and 7, or $\{-3, 7\}$
27. $(x - 3)(x + 8) = -30$ -3 and -2, or $\{-3, -2\}$
28. $(x - 1)(x + 4) = 14$ -6 and 3, or $\{-6, 3\}$
29. $x(x + 8) = 16(x - 1)$ 4 or $\{4\}$
30. $x(x + 9) = 4(2x + 5)$ -5 and 4, or $\{-5, 4\}$
31. $(x + 1)^2 - 5(x + 2) = 3x + 7$ -2 and 8, or $\{-2, 8\}$
32. $(x + 1)^2 = 2(x + 5)$ -3 and 3, or $\{-3, 3\}$
33. $x(8x + 1) = 3x^2 - 2x + 2$ -1 and $\frac{2}{5}$, or $\left\{-1, \frac{2}{5}\right\}$
34. $2x(x + 3) = -5x - 15$ -3 and $-\frac{5}{2}$, or $\left\{-3, -\frac{5}{2}\right\}$
35. $\dfrac{x^2}{18} + \dfrac{x}{2} + 1 = 0$ -6 and -3, or $\{-6, -3\}$
36. $\dfrac{x^2}{4} - \dfrac{5x}{2} + 6 = 0$ 4 and 6, or $\{4, 6\}$

In Exercises 37–46, use factoring to solve each polynomial equation. Check by substitution or by using a graphing utility and identifying x-intercepts.

37. $x^3 + 4x^2 - 25x - 100 = 0$ $\quad$ $-5, -4,$ and $5,$ or $\{-5, -4, 5\}$

38. $x^3 - 2x^2 - x + 2 = 0$ $\quad$ $-1, 1,$ and $2,$ or $\{-1, 1, 2\}$

39. $x^3 - x^2 = 25x - 25$ $\quad$ $-5, 1,$ and $5,$ or $\{-5, 1, 5\}$

40. $x^3 + 2x^2 = 16x + 32$ $\quad$ $-4, -2,$ and $4,$ or $\{-4, -2, 4\}$

41. $3x^4 - 48x^2 = 0$ $\quad$ $-4, 0,$ and $4,$ or $\{-4, 0, 4\}$

42. $5x^4 - 20x^2 = 0$ $\quad$ $-2, 0,$ and $2,$ or $\{-2, 0, 2\}$

43. $2x^4 = 16x$ $\quad$ 0 and $2,$ or $\{0, 2\}$

44. $3x^4 = 81x$ $\quad$ 0 and $3,$ or $\{0, 3\}$

45. $2x^3 + 16x^2 + 30x = 0$ $\quad$ $-5, -3,$ and $0,$ or $\{-5, -3, 0\}$

46. $3x^3 - 9x^2 - 30x = 0$ $\quad$ $-2, 0,$ and $5,$ or $\{-2, 0, 5\}$

In Exercises 47–50, determine the x-intercepts of the graph of each quadratic function. Then match the function with its graph, labeled (a)–(d). Each graph is shown in a $[-10, 10, 1]$ by $[-10, 10, 1]$ viewing rectangle.

47. $y = x^2 - 6x + 8$ $\quad$ 2 and $4;$ **d**

48. $y = x^2 - 2x - 8$ $\quad$ -2 and $4;$ **b**

49. $y = x^2 + 6x + 8$ $\quad$ -4 and $-2;$ **c**

50. $y = x^2 + 2x - 8$ $\quad$ -4 and $2;$ **a**

a.

b.

c.

d.

Practice Plus

In Exercises 51–54, solve each polynomial equation.

51. $x(x + 1)^3 - 42(x + 1)^2 = 0$ $\quad$ $-7, -1,$ and $6,$ or $\{-7, -1, 6\}$

52. $x(x - 2)^3 - 35(x - 2)^2 = 0$ $\quad$ $-5, 2,$ and $7,$ or $\left\{-5, 2, 7\right\}$

53. $-4x[x(3x - 2) - 8](25x^2 - 40x + 16) = 0$

$\quad$ $-\frac{4}{3}, 0, \frac{4}{5},$ and $2,$ or $\left\{-\frac{4}{3}, 0, \frac{4}{5}, 2\right\}$

54. $-7x[x(2x - 5) - 12](9x^2 + 30x + 25) = 0$

$\quad$ $-\frac{5}{3}, -\frac{3}{2}, 0,$ and $4,$ or $\left\{-\frac{5}{3}, -\frac{3}{2}, 0, 4\right\}$

In Exercises 55–58, find all values of c satisfying the given conditions.

55. $f(x) = x^2 - 4x - 27$ and $f(c) = 5.$ $\quad$ -4 and 8

56. $f(x) = 5x^2 - 11x + 6$ and $f(c) = 4.$ $\quad$ $\frac{1}{5}$ and 2

57. $f(x) = 2x^3 + x^2 - 8x + 2$ and $f(c) = 6.$ $\quad$ $-2, -\frac{1}{2},$ and 2

58. $f(x) = x^3 + 4x^2 - x + 6$ and $f(c) = 10.$ $\quad$ $-4, -1,$ and 1

In Exercises 59–62, find all numbers satisfying the given conditions.

59. The product of the number decreased by 1 and increased by 4 is 24. $\quad$ -7 and 4

60. The product of the number decreased by 6 and increased by 2 is 20. $\quad$ -4 and 8

61. If 5 is subtracted from 3 times the number, the result is the square of 1 less than the number. $\quad$ 2 and 3

62. If the square of the number is subtracted from 61, the result is the square of 1 more than the number. $\quad$ -6 and 5

In Exercises 63–64, list all numbers that must be excluded from the domain of the given function.

63. $f(x) = \dfrac{3}{x^2 + 4x - 45}$ $\quad$ -9 and 5

64. $f(x) = \dfrac{7}{x^2 - 3x - 28}$ $\quad$ -4 and 7

Application Exercises

A gymnast dismounts the uneven parallel bars at a height of 8 feet with an initial upward velocity of 8 feet per second. The function

$$s(t) = -16t^2 + 8t + 8$$

describes the height of the gymnast's feet above the ground, $s(t)$, in feet, t seconds after dismounting. The graph of the function is shown, with unlabeled tick marks along the horizontal axis. Use the function to solve Exercises 65–66.

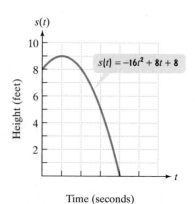

$s(t) = -16t^2 + 8t + 8$

Height (feet)

Time (seconds)

65. How long will it take the gymnast to reach the ground? Use this information to provide a number on each tick mark along the horizontal axis in the figure shown.   1 second; 0.25, 0.5, 0.75, 1

66. When will the gymnast be 8 feet above the ground? Identify the solution(s) as one or more points on the graph.   after 0 seconds, the moment of dismount, and after $\frac{1}{2}$ second; $(0, 8)$ and $\left(\frac{1}{2}, 8\right)$

The Food Stamp Program is America's first line of defense against hunger for millions of families. Over half of all participants are children; one out of six is a low-income older adult. Exercises 67–70 involve the number of participants in the program from 1990 through 2002.
   The function

$$f(x) = -\frac{1}{4}x^2 + 3x + 17$$

models the number of people, $f(x)$, in millions, receiving food stamps x years after 1990. Use the function to solve Exercises 67–68.

67. In which year did 26 million people receive food stamps?   1996

68. In which years did 25 million people receive food stamps?   1994 and 1998

The graph of the quadratic function in Exercises 67–68 is shown. Use the graph to solve Exercises 69–70.

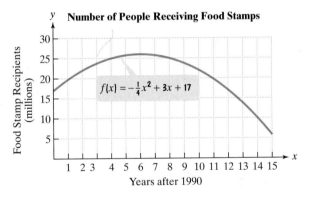

Number of People Receiving Food Stamps

$f(x) = -\frac{1}{4}x^2 + 3x + 17$

Years after 1990

Source: *New York Times*

69. Identify your solution in Exercise 67 as a point on the graph. Describe what is significant about this point.   (6, 26); This is the graph's highest point; During this time period, the greatest number of recipients was 26 million in 1996.

70. Identify your solution in Exercise 68 as one or more points on the graph. Then describe the trend shown by the graph.   (4, 25) and (8, 25); The number of recipients increased each year until 1996 and then began to decrease.

71. The length of a rectangular sign is 3 feet longer than the width. If the sign's area is 54 square feet, find its length and width.   Length is 9 feet and width is 6 feet.

72. A rectangular parking lot has a length that is 3 yards greater than the width. The area of the parking lot is 180 square yards. Find the length and the width.   Length is 15 yards and width is 12 yards.

73. Each side of a square is lengthened by 3 inches. The area of this new, larger square is 64 square inches. Find the length of a side of the original square.   5 inches

74. Each side of a square is lengthened by 2 inches. The area of this new, larger square is 36 square inches. Find the length of a side of the original square.   4 inches

75. A pool measuring 10 meters by 20 meters is surrounded by a path of uniform width, as shown in the figure. If the area of the pool and the path combined is 600 square meters, what is the width of the path?   Width of the path is 5 meters.

76. A vacant rectangular lot is being turned into a community vegetable garden measuring 15 meters by 12 meters. A path of uniform width is to surround the garden. If the area of the lot is 378 square meters, find the width of the path surrounding the garden.   Width of the path is 3 meters.

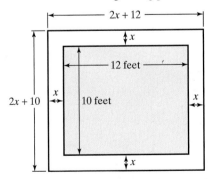

77. As part of a landscaping project, you put in a flower bed measuring 10 feet by 12 feet. You plan to surround the bed with a uniform border of low-growing plants.

a. Write a polynomial that describes the area of the uniform border that surrounds your flower bed. *Hint*: The area of the border is the area of the large rectangle shown in the figure minus the area of the flower bed. $4x^2 + 44x$

b. The low growing plants surrounding the flower bed require 1 square foot each when mature. If you have 168 of these plants, how wide a strip around the flower bed should you prepare for the border?   The border should be 3 feet wide.

78. As part of a landscaping project, you put in a flower bed measuring 20 feet by 30 feet. To finish off the project, you are putting in a uniform border of pine bark around the outside of the rectangular garden. You have enough pine bark to cover 336 square feet. How wide should the border be?   The border should be 3 feet wide.

79. A machine produces open boxes using square sheets of metal. The figure illustrates that the machine cuts equal-sized squares measuring 2 inches on a side from the corners and then shapes the metal into an open box by turning up the sides. If each box must have a volume of 200 cubic inches, find the length and width of the open box.   Length and width are both 10 inches.

80. A machine produces open boxes using square sheets of metal. The machine cuts equal-sized squares measuring 3 inches on a side from the corners and then shapes the metal into an open box by turning up the sides. If each box must have a volume of 75 cubic inches, find the length and width of the open box.   Length and width are both 5 inches.

81. The rectangular floor of a closet is divided into two right triangles by drawing a diagonal, as shown in the figure. One leg of the right triangle is 2 feet more than twice the other leg. The hypotenuse is 13 feet. Determine the closet's length and width.   Length is 12 feet and width is 5 feet.

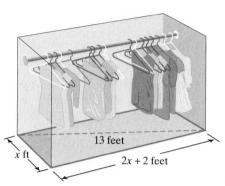

82. A piece of wire measuring 20 feet is attached to a telephone pole as a guy wire. The distance along the ground from the bottom of the pole to the end of the wire is 4 feet greater than the height where the wire is attached to the pole. How far up the pole does the guy wire reach?   The guy wire reaches 12 feet up the pole.

83. A tree is supported by a wire anchored in the ground 15 feet from its base. The wire is 4 feet longer than the height that it reaches on the tree. Find the length of the wire.   The wire is $30\frac{1}{8}$ feet long.

84. A tree is supported by a wire anchored in the ground 5 feet from its base. The wire is 1 foot longer than the height that it reaches on the tree. Find the length of the wire.   The wire is 13 feet long.

Writing in Mathematics

85. What is a quadratic equation?   85–95. Answers will vary.

86. What is the zero-product principle?

87. Explain how to solve $x^2 - x = 6$.

88. Describe the relationship between the solutions of a quadratic equation and the graph of the corresponding quadratic function.

89. What is a polynomial equation? When is it in standard form?

90. What is the degree of a polynomial equation? What are polynomial equations of degree 1 and degree 2, respectively, called?

91. Explain how to solve $x^3 + x^2 = x + 1$.

92. If something is thrown straight up, or possibly dropped, describe a situation in which it is important to know how long it will take the object to hit the ground or possibly the water.

93. A toy rocket is launched vertically upward. Using a quadratic equation, we find that the rocket will reach a height of 220 feet at 2.5 seconds and again at 5.5 seconds. How can this be?

94. Describe a situation in which a landscape designer might use polynomials and polynomial equations.

95. In your own words, state the Pythagorean Theorem.

Technology Exercises

In Exercises 96–99, use a graphing utility with a viewing rectangle large enough to show end behavior to graph each polynomial function. Then use the x-intercepts for the graph to solve the polynomial equation. Check by substitution. 96–99. See graphing answer section.

96. Use the graph of $y = x^2 + 3x - 4$ to solve $x^2 + 3x - 4 = 0$. -4 and 1, or $\{-4, 1\}$

97. Use the graph of $y = x^3 + 3x^2 - x - 3$ to solve $x^3 + 3x^2 - x - 3 = 0$. $-3, -1$, and 1, or $\{-3, -1, 1\}$

98. Use the graph of $y = 2x^3 - 3x^2 - 11x + 6$ to solve $2x^3 - 3x^2 - 11x + 6 = 0$. $-2, \frac{1}{2}$, and 3, or $\{-2, \frac{1}{2}, 3\}$

99. Use the graph of $y = -x^4 + 4x^3 - 4x^2$ to solve $-x^4 + 4x^3 - 4x^2 = 0$. 0 and 2, or $\{0, 2\}$

100. Use the ⌷TABLE⌷ feature of a graphing utility to verify the solution sets for any two equations in Exercises 31–32 or 39–40. Answers will vary.

Critical Thinking Exercises

101. Which one of the following is true? c

   a. Quadratic equations solved by factoring always have two different solutions.

   b. If $4x(x^2 + 49) = 0$, then
   $$4x = 0 \quad \text{or} \quad x^2 + 49 = 0$$
   $$x = 0 \quad \text{or} \qquad\qquad x = 7 \quad \text{or} \quad x = -7.$$

   c. If -4 is a solution of $7y^2 + (2k - 5)y - 20 = 0$, then k must equal 14.

   d. Some quadratic equations have more than two solutions.

102. Write a quadratic equation in standard form whose solutions are -3 and 7. Answers will vary; An example is $x^2 - 4x - 21 = 0$.

103. Solve: $|x^2 + 2x - 36| = 12$. $-8, -6, 4,$ and 6, or $\{-8, -6, 4, 6\}$

Review Exercises

104. Solve: $|3x - 2| = 8$. -2 and $\frac{10}{3}$, or $\{-2, \frac{10}{3}\}$ (Section 4.3, Example 1)

105. Simplify: $3(5 - 7)^2 + \sqrt{16} + 12 \div (-3)$. 12 (Section 1.2, Example 7)

106. You invested $3000 in two accounts paying 5% and 8% annual interest, respectively. If the total interest earned for the year is $189, how much was invested at each rate? (Section 3.2, Example 2) $1700 was invested at 5% and $1300 at 8%.

GROUP PROJECT

CHAPTER 5

Additional group projects can be found in the *Instructor's Resource Manual.*

Divide the group in half. Without looking at any factoring problems in the book, each group should create five factoring problems. Make sure that some of your problems require at least two factoring strategies. Next, exchange problems with the other half of the group. Work to factor the five problems. After completing the factorizations, evaluate the factoring problems that you were given. Are they too easy? Too difficult? Can the polynomials really be factored? Share your responses with the half of the group that wrote the problems. Finally, grade each other's work in factoring the polynomials. Each factoring problem is worth 20 points. You may award partial credit. If you take off points, explain why points are deducted and how you decided to take off a particular number of points for the error(s) that you found.

CHAPTER 5 SUMMARY

Definitions and Concepts	Examples

Section 5.1 Introduction to Polynomials and Polynomial Functions

A polynomial is a single term or the sum of two or more terms containing variables in the numerator with whole-number exponents. A monomial is a polynomial with exactly one term; a binomial has exactly two terms; a trinomial has exactly three terms. If $a \neq 0$, the degree of ax^n is n and the degree of $ax^n y^m$ is $n + m$. The degree of a nonzero constant is 0. The constant 0 has no defined degree. The degree of a polynomial is the greatest of the degrees of its terms. The leading term is the term of greatest degree. Its coefficient is called the leading coefficient.

$$7x^3y \;-\; 4x^5y^4 \;-\; 2x^4y$$

Degree is $3 + 1 = 4$.   Degree is $5 + 4 = 9$.   Degree is $4 + 1 = 5$.

The degree of the polynomial is 9. The leading term is $-4x^5y^4$. The leading coefficient is -4. This polynomial is a trinomial.

In a polynomial function, the expression that defines the function is a polynomial. Polynomial functions have graphs that are smooth and continuous. The behavior of the graph of a polynomial function to the far left or the far right is called its end behavior.

The Leading Coefficient Test

1. Odd-degree polynomial functions have graphs with opposite behavior at each end. If the leading coefficient is positive, the graph falls to the left and rises to the right. If the leading coefficient is negative, the graph rises to the left and falls to the right.
2. Even-degree polynomial functions have graphs with the same behavior at each end. If the leading coefficient is positive, the graph rises to the left and rises to the right. If the leading coefficient is negative, the graph falls to the left and falls to the right.

Describe the end behavior of the graph of each polynomial function:

• $f(x) = -2x^3 + 3x^2 + 11x - 6$

The degree, 3, is odd. The leading coefficient -2 is negative. The graph rises to the left and falls to the right.

• $f(x) = x^2 - 4$

The degree, 2, is even. The leading coefficient, 1, is positive. The graph rises to the left and rises to the right.

To add polynomials, add like terms.

$$(6x^3y + 5x^2y - 7y) + (-9x^3y + x^2y + 6y)$$

$$= (6x^3y - 9x^3y) + (5x^2y + x^2y) + (-7y + 6y)$$

$$= -3x^3y + 6x^2y - y$$

Definitions and Concepts	Examples

Section 5.1 Introduction to Polynomials and Polynomial Functions (continued)

Definitions and Concepts	Examples
To subtract two polynomials, change the sign of every term of the second polynomial. Add this result to the first polynomial.	$(5y^3 - 9y^2 - 4) - (3y^3 - 12y^2 - 5)$ $= (5y^3 - 9y^2 - 4) + (-3y^3 + 12y^2 + 5)$ $= (5y^3 - 3y^3) + (-9y^2 + 12y^2) + (-4 + 5)$ $= 2y^3 + 3y^2 + 1$

Section 5.2 Multiplication of Polynomials

Definitions and Concepts	Examples
To multiply monomials, multiply coefficients and add exponents.	$(-2x^2y^4)(-3x^3y)$ $= (-2)(-3)x^{2+3}y^{4+1} = 6x^5y^5$
To multiply a monomial and a polynomial, multiply each term of the polynomial by the monomial.	$7x^2y(4x^3y^5 - 2xy - 1)$ $= 7x^2y \cdot 4x^3y^5 - 7x^2y \cdot 2xy - 7x^2y \cdot 1$ $= 28x^5y^6 - 14x^3y^2 - 7x^2y$
To multiply polynomials if neither is a monomial, multiply each term of one by each term of the other.	$(x^3 + 2x)(5x^2 - 3x + 4)$ $= x^3(5x^2 - 3x + 4) + 2x(5x^2 - 3x + 4)$ $= 5x^5 - 3x^4 + 4x^3 + 10x^3 - 6x^2 + 8x$ $= 5x^5 - 3x^4 + 14x^3 - 6x^2 + 8x$
The FOIL method may be used when multiplying two binomials: First terms multiplied. Outside terms multiplied. Inside terms multiplied. Last terms multiplied.	$\quad$ F $\quad$ O $\quad$ I $\quad$ L $(5x - 3y)(2x + y) = 5x \cdot 2x + 5x \cdot y + (-3y) \cdot 2x + (-3y) \cdot y$ $= 10x^2 + 5xy - 6xy - 3y^2$ $= 10x^2 - xy - 3y^2$
The Square of a Binomial Sum $(A + B)^2 = A^2 + 2AB + B^2$	$(x^2 + 6)^2 = (x^2)^2 + 2 \cdot x^2 \cdot 6 + 6^2$ $= x^4 + 12x^2 + 36$
The Square of a Binomial Difference $(A - B)^2 = A^2 - 2AB + B^2$	$(4x - 5)^2 = (4x)^2 - 2 \cdot 4x \cdot 5 + 5^2$ $= 16x^2 - 40x + 25$
The Product of the Sum and Difference of Two Terms $(A + B)(A - B) = A^2 - B^2$	$\bullet \ (3x + 7y)(3x - 7y) = (3x)^2 - (7y)^2$ $= 9x^2 - 49y^2$ $\bullet \ [(x + 2) - 4y][(x + 2) + 4y]$ $= (x + 2)^2 - (4y)^2 = x^2 + 4x + 4 - 16y^2$

Definitions and Concepts	Examples

Section 5.3 Greatest Common Factors and Factoring by Grouping

Factoring a polynomial containing the sum of monomials means finding an equivalent expression that is a product. Polynomials that cannot be factored using integer coefficients are called prime polynomials over the integers. The greatest common factor, GCF, is an expression that divides every term of the polynomial. The GCF is the product of the largest common numerical factor and the variable of lowest degree common to every term of the polynomial. To factor a monomial from a polynomial, express each term as the product of the GCF and its other factor. Then use the distributive property to factor out the GCF. When the leading coefficient of a polynomial is negative, it is often desirable to factor out a common factor with a negative coefficient.	Factor: $4x^4y^2 - 12x^2y^3 + 20xy^2$. (GCF is $4xy^2$) $$= 4xy^2 \cdot x^3 - 4xy^2 \cdot 3xy + 4xy^2 \cdot 5$$ $$= 4xy^2(x^3 - 3xy + 5)$$ Factor: $-25x^3 + 10x^2 - 15x$. (Use $-5x$ as a common factor.) $$= -5x(5x^2) - 5x(-2x) - 5x(3)$$ $$= -5x(5x^2 - 2x + 3)$$
To factor by grouping, factor out the GCF from each group. Then factor out the remaining factor.	$xy + 7x - 2y - 14$ $$= x(y + 7) - 2(y + 7)$$ $$= (y + 7)(x - 2)$$

Section 5.4 Factoring Trinomials

To factor a trinomial of the form $x^2 + bx + c$, find two numbers whose product is c and whose sum is b. The factorization is $$(x + \text{one number})(x + \text{other number}).$$	Factor: $x^2 + 9x + 20$. Find two numbers whose product is 20 and whose sum is 9. The numbers are 4 and 5. $$x^2 + 9x + 20 = (x + 4)(x + 5)$$
In some trinomials, the highest power is greater than 2, and the exponent in one of the terms is half that of the other term. Factor by introducing a substitution. Let t equal the variable to the smaller power.	Factor: $x^6 - 7x^3 + 12$. $$= (x^3)^2 - 7x^3 + 12$$ $$= t^2 - 7t + 12 \qquad \text{Let } t = x^3.$$ $$= (t - 4)(t - 3)$$ $$= (x^3 - 4)(x^3 - 3)$$
To factor $ax^2 + bx + c$ by trial and error, try various combinations of factors of ax^2 and c until a middle term of bx is obtained for the sum of the outside and inside products.	Factor: $2x^2 + 7x - 15$. Factors of $2x^2$: $2x, x$ Factors of -15: 1 and -15, -1 and 15, 3 and -5, -3 and 5 $$(2x - 3)(x + 5)$$ Sum of outside and inside products should equal $7x$. $$10x - 3x = 7x$$ Thus, $2x^2 + 7x - 15 = (2x - 3)(x + 5)$.
To factor $ax^2 + bx + c$ by grouping, find the factors of ac whose sum is b. Write bx using these factors. Then factor by grouping.	Factor: $2x^2 + 7x - 15$. Find the factors of $2(-15)$, or -30, whose sum is 7. They are 10 and -3. $2x^2 + 7x - 15$ $$= 2x^2 + 10x - 3x - 15$$ $$= 2x(x + 5) - 3(x + 5) = (x + 5)(2x - 3)$$

Definitions and Concepts	Examples

Section 5.5 Factoring Special Forms

The Difference of Two Squares

$$A^2 - B^2 = (A + B)(A - B)$$

$16x^2 - 9y^2$
$= (4x)^2 - (3y)^2 = (4x + 3y)(4x - 3y)$

Perfect Square Trinomials

$$A^2 + 2AB + B^2 = (A + B)^2$$
$$A^2 - 2AB + B^2 = (A - B)^2$$

$x^2 + 20x + 100 = x^2 + 2 \cdot x \cdot 10 + 10^2 = (x + 10)^2$

$9x^2 - 30x + 25 = (3x)^2 - 2 \cdot 3x \cdot 5 + 5^2 = (3x - 5)^2$

Sum and Difference of Cubes

$$A^3 + B^3 = (A + B)(A^2 - AB + B^2)$$
$$A^3 - B^3 = (A - B)(A^2 + AB + B^2)$$

$125x^3 - 8 = (5x)^3 - 2^3$
$\qquad = (5x - 2)[(5x)^2 + 5x \cdot 2 + 2^2]$
$\qquad = (5x - 2)(25x^2 + 10x + 4)$

When using factoring by grouping, terms can sometimes be grouped to obtain the difference of two squares. One of the squares is a perfect square trinomial.

$x^2 + 18x + 81 - 25y^2$
$= (x + 9)^2 - (5y)^2$
$= (x + 9 + 5y)(x + 9 - 5y)$

Section 5.6 A General Factoring Strategy

A Factoring Strategy

1. Factor out the GCF or a common factor with a negative coefficient.

2. **a.** If two terms, try
   $$A^2 - B^2 = (A + B)(A - B)$$
   $$A^3 + B^3 = (A + B)(A^2 - AB + B^2)$$
   $$A^3 - B^3 = (A - B)(A^2 + AB + B^2).$$

   b. If three terms, try
   $$A^2 + 2AB + B^2 = (A + B)^2$$
   $$A^2 - 2AB + B^2 = (A - B)^2.$$

   If not a perfect square trinomial, try trial and error or grouping.

   c. If four terms, try factoring by grouping.

3. See if any factors can be factored further.

Factor: $3x^4 + 12x^3 - 3x^2 - 12x$.

The GCF is $3x$.

$$3x^4 + 12x^3 - 3x^2 - 12x$$
$$= 3x(x^3 + 4x^2 - x - 4)$$

Four terms: Try grouping.

$$= 3x[x^2(x + 4) - 1(x + 4)]$$
$$= 3x(x + 4)(x^2 - 1)$$

This can be factored further.

$$= 3x(x + 4)(x + 1)(x - 1)$$

Section 5.7 Polynomial Equations and Their Applications

A quadratic equation in x can be written in the standard form
$$ax^2 + bx + c, a \neq 0.$$
A polynomial equation is the result of setting two polynomials equal to each other. The equation is in standard form if one side is 0 and the polynomial on the other side is in standard form, that is in descending powers of the variable. Its degree is the highest degree of any term in the equation. A polynomial equation of degree 1 is a linear equation and of degree 2 a quadratic equation. Some polynomial equations can be solved by writing the equation in standard form, factoring, and then using the zero-product principle: If a product is 0, then at least one of the factors is equal to 0.

Solve: $5x^2 + 7x = 6$.
$$5x^2 + 7x - 6 = 0$$
$$(5x - 3)(x + 2) = 0$$

$5x - 3 = 0 \qquad$ or $\qquad x + 2 = 0$
$\qquad 5x = 3 \qquad\qquad\qquad\qquad x = -2$
$\qquad x = \frac{3}{5}$

The solutions are -2 and $\frac{3}{5}$, and the solution set is $\left\{-2, \frac{3}{5}\right\}$. (The solutions are the x-intercepts of the graph of $y = 5x^2 + 7x - 6$.)

CHAPTER 5   REVIEW EXERCISES

5.1 *In Exercises 1–2, determine the coefficient of each term, the degree of each term, the degree of the polynomial, the leading term, and the leading coefficient of the polynomial.*

1. $-5x^3 + 7x^2 - x + 2$

Term	Coefficient	Degree
$-5x^3$	-5	3
$7x^2$	7	2
$-x$	-1	1
2	2	0

The degree of the polynomial is 3, the leading term is $-5x^3$, and the leading coefficient is -5.

2. $8x^4y^2 - 7xy^6 - x^3y$

Term	Coefficient	Degree
$8x^4y^2$	8	6
$-7xy^6$	-7	7
$-x^3y$	-1	4

The degree of the polynomial is 7, the leading term is $-7xy^6$, and the leading coefficient is -7.

3. If $f(x) = x^3 - 4x^2 + 3x - 1$, find $f(-2)$. -31

4. The polynomial function $f(x) = 0.26x^2 + 2.09x + 16.4$ models the number of people in the United States, $f(x)$, in thousands, waiting for organ transplants x years after 1990. Find and interpret $f(10)$.   63.3; There were 63.3 thousand, or 63,300 people waiting for transplants in the United States in 2000.

In Exercises 5–8, use the Leading Coefficient Test to determine the end behavior of the graph of the given polynomial function. Then use this end behavior to match the polynomial function with its graph. [The graphs are labeled (a) through (d).]

5. $f(x) = -x^3 + x^2 + 2x$ rises to the left and falls to the right; **c**

6. $f(x) = x^6 - 6x^4 + 9x^2$ rises to the left and rises to the right; **b**

7. $f(x) = x^5 - 5x^3 + 4x$ falls to the left and rises to the right; **a**

8. $f(x) = -x^4 + 1$ falls to the left and falls to the right; **d**

a.

b.

c.

d.

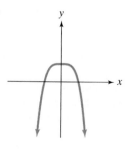

9. The function $f(x) = -0.0013x^3 + 0.78x^2 - 1.43x + 18.1$ models the percentage of U.S. families below the poverty level, $f(x)$, x years after 1960. Use end behavior to explain why the model is valid only for a limited period of time.   The graph of this function falls to the right which indicates that the percentage would eventually be negative, which is not possible.

In Exercises 10–11, add the polynomials.

10. $(-8x^3 + 5x^2 - 7x + 4) + (9x^3 - 11x^2 + 6x - 13)$
$x^3 - 6x^2 - x - 9$

11. $(7x^3y - 13x^2y - 6y) + (5x^3y + 11x^2y - 8y - 17)$
$12x^3y - 2x^2y - 14y - 17$

In Exercises 12–13, subtract the polynomials.

12. $(7x^3 - 6x^2 + 5x - 11) - (-8x^3 + 4x^2 - 6x - 3)$
$15x^3 - 10x^2 + 11x - 8$

13. $(4x^3y^2 - 7x^3y - 4) - (6x^3y^2 - 3x^3y + 4)$
$-2x^3y^2 - 4x^3y - 8$

14. Subtract $-2x^3 - x^2y + xy^2 + 7y^3$ from $x^3 + 4x^2y - y^3$.
$3x^3 + 5x^2y - xy^2 - 8y^3$

5.2 *In Exercises 15–27, multiply the polynomials.*

15. $(4x^2yz^5)(-3x^4yz^2)$ $-12x^6y^2z^7$

16. $6x^3\left(\dfrac{1}{3}x^5 - 4x^2 - 2\right)$ $2x^8 - 24x^5 - 12x^3$

17. $7xy^2(3x^4y^2 - 5xy - 1)$ $21x^5y^4 - 35x^2y^3 - 7xy^2$

18. $(2x + 5)(3x^2 + 7x - 4)$ $6x^3 + 29x^2 + 27x - 20$

19. $(x^2 + x - 1)(x^2 + 3x + 2)$ $x^4 + 4x^3 + 4x^2 - x - 2$

20. $(4x - 1)(3x - 5)$ $12x^2 - 23x + 5$

21. $(3xy - 2)(5xy + 4)$ $15x^2y^2 + 2xy - 8$

22. $(3x + 7y)^2$ $9x^2 + 42xy + 49y^2$

23. $(x^2 - 5y)^2$ $x^4 - 10x^2y + 25y^2$

24. $(2x + 7y)(2x - 7y)$ $4x^2 - 49y^2$

25. $(3xy^2 - 4x)(3xy^2 + 4x)$ $9x^2y^4 - 16x^2$

26. $[(x + 3) + 5y][(x + 3) - 5y]$ $x^2 + 6x + 9 - 25y^2$

27. $(x + y + 4)^2$ $x^2 + 2xy + y^2 + 8x + 8y + 16$

28. Let $f(x) = x - 3$ and $g(x) = 2x + 5$. Find $(fg)(x)$ and $(fg)(-4)$. $2x^2 - x - 15$; 21

29. Let $f(x) = x^2 - 7x + 2$. Find each of the following and simplify:
   a. $f(a - 1)$ $a^2 - 9a + 10$
   b. $f(a + h) - f(a)$. $2ah + h^2 - 7h$

5.3 *In Exercises 30–33, factor the greatest common factor from each polynomial.*

30. $16x^3 + 24x^2$ $8x^2(2x + 3)$

31. $2x - 36x^2$ $2x(1 - 18x)$

32. $21x^2y^2 - 14xy^2 + 7xy$ $7xy(3xy - 2y + 1)$

33. $18x^3y^2 - 27x^2y$ $9x^2y(2xy - 3)$

In Exercises 34–35, factor out a common factor with a negative coefficient.

34. $-12x^2 + 8x - 48$ $-4(3x^2 - 2x + 12)$

35. $-x^2 - 11x + 14$ $-1(x^2 + 11x - 14)$

In Exercises 36–38, factor by grouping.

36. $x^3 - x^2 - 2x + 2$ $(x - 1)(x^2 - 2)$

37. $xy - 3x - 5y + 15$ $(y - 3)(x - 5)$

38. $5ax - 15ay + 2bx - 6by$ $(x - 3y)(5a + 2b)$

5.4 *In Exercises 39–47, factor each trinomial completely, or state that the trinomial is prime.*

39. $x^2 + 8x + 15$ $(x + 5)(x + 3)$

40. $x^2 + 16x - 80$ $(x + 20)(x - 4)$

41. $x^2 + 16xy - 17y^2$ $(x + 17y)(x - y)$

42. $3x^3 - 36x^2 + 33x$ $3x(x - 1)(x - 11)$

43. $3x^2 + 22x + 7$ $(x + 7)(3x + 1)$

44. $6x^2 - 13x + 6$ $(3x - 2)(2x - 3)$

45. $5x^2 - 6xy - 8y^2$ $(5x + 4y)(x - 2y)$

46. $6x^3 + 5x^2 - 4x$ $x(3x + 4)(2x - 1)$

47. $2x^2 + 11x + 15$ $(x + 3)(2x + 5)$

In Exercises 48–51, factor by introducing an appropriate substitution.

48. $x^6 + x^3 - 30$ $(x^3 + 6)(x^3 - 5)$

49. $x^4 - 10x^2 - 39$ $(x^2 + 3)(x^2 - 13)$

50. $(x + 5)^2 + 10(x + 5) + 24$ $(x + 11)(x + 9)$

51. $5x^6 + 17x^3 + 6$ $(5x^3 + 2)(x^3 + 3)$

5.5 *In Exercises 52–55, factor each difference of two squares.*

52. $4x^2 - 25$ $(2x + 5)(2x - 5)$

53. $1 - 81x^2y^2$ $(1 + 9xy)(1 - 9xy)$

54. $x^8 - y^6$ $(x^4 + y^3)(x^4 - y^3)$

55. $(x - 1)^2 - y^2$ $(x - 1 + y)(x - 1 - y)$

In Exercises 56–60, factor any perfect square trinomials, or state that the polynomial is prime.

56. $x^2 + 16x + 64$ $(x + 8)^2$

57. $9x^2 - 6x + 1$ $(3x - 1)^2$

58. $25x^2 + 20xy + 4y^2$ $(5x + 2y)^2$

59. $49x^2 + 7x + 1$ prime

60. $25x^2 - 40xy + 16y^2$ $(5x - 4y)^2$

In Exercises 61–62, factor by grouping to obtain the difference of two squares.

61. $x^2 + 18x + 81 - y^2$ $(x + 9 + y)(x + 9 - y)$

62. $z^2 - 25x^2 + 10x - 1$ $(z + 5x - 1)(z - 5x + 1)$

In Exercises 63–65, factor using the formula for the sum or difference of two cubes.

63. $64x^3 + 27$ $(4x + 3)(16x^2 - 12x + 9)$

64. $125x^3 - 8$ $(5x - 2)(25x^2 + 10x + 4)$

65. $x^3y^3 + 1$ $(xy + 1)(x^2y^2 - xy + 1)$

5.6 *In Exercises 66–90, factor completely, or state that the polynomial is prime.*

66. $15x^2 + 3x$ $3x(5x + 1)$

67. $12x^4 - 3x^2$ $3x^2(2x + 1)(2x - 1)$

68. $20x^4 - 24x^3 + 28x^2 - 12x$ $4x(5x^3 - 6x^2 + 7x - 3)$

69. $x^3 - 15x^2 + 26x$ $x(x - 2)(x - 13)$

70. $-2y^4 + 24y^3 - 54y^2$ $-2y^2(y - 3)(y - 9)$

71. $9x^2 - 30x + 25$ $(3x - 5)^2$

72. $5x^2 - 45$ $5(x + 3)(x - 3)$

73. $2x^3 - x^2 - 18x + 9$ $(2x - 1)(x + 3)(x - 3)$

74. $6x^2 - 23xy + 7y^2$ $(3x - y)(2x - 7y)$

75. $2y^3 + 12y^2 + 18y$ $2y(y + 3)^2$

76. $x^2 + 6x + 9 - 4a^2$ $(x + 3 + 2a)(x + 3 - 2a)$

77. $8x^3 - 27$ $(2x - 3)(4x^2 + 6x + 9)$

78. $x^5 - x$ $x(x^2 + 1)(x + 1)(x - 1)$

79. $x^4 - 6x^2 + 9$ $(x^2 - 3)^2$

80. $x^2 + xy + y^2$ prime

81. $4a^3 + 32$ $4(a + 2)(a^2 - 2a + 4)$

82. $x^4 - 81$ $(x^2 + 9)(x + 3)(x - 3)$

83. $ax + 3bx - ay - 3by$ $(a + 3b)(x - y)$

84. $27x^3 - 125y^3$ $(3x - 5y)(9x^2 + 15xy + 25y^2)$

85. $10x^3y + 22x^2y - 24xy$ $2xy(x + 3)(5x - 4)$

86. $6x^6 + 13x^3 - 5$ $(2x^3 + 5)(3x^3 - 1)$

87. $2x + 10 + x^2y + 5xy$ $(x + 5)(2 + xy)$

88. $y^3 + 2y^2 - 25y - 50$ $(y + 2)(y + 5)(y - 5)$

89. $a^8 - 1$ $(a^4 + 1)(a^2 + 1)(a + 1)(a - 1)$

90. $9(x - 4) + y^2(4 - x)$ $(x - 4)(3 + y)(3 - y)$

In Exercises 91–92,

   a. Write an expression for the area of the shaded region.

   b. Write the expression in factored form.

91.   **a.** $2xy + 2y^2$ **b.** $2y(x + y)$

92.

a. $x^2 - 4y^2$ **b.** $(x + 2y)(x - 2y)$

5.7 *In Exercises 93–97, use factoring to solve each polynomial equation.*

93. $x^2 + 6x + 5 = 0$ −5 and −1, or $\{-5, -1\}$

94. $3x^2 = 22x - 7$ $\frac{1}{3}$ and 7, or $\{\frac{1}{3}, 7\}$

95. $(x + 3)(x - 2) = 50$ −8 and 7, or $\{-8, 7\}$

96. $3x^2 = 12x$ 0 and 4, or $\{0, 4\}$

97. $x^3 + 5x^2 = 9x + 45$ −5, −3, and 3, or $\{-5, -3, 3\}$

98. A model rocket is launched from the top of a cliff 144 feet above sea level. The function

$$s(t) = -16t^2 + 128t + 144$$

describes the rocket's height above the water, $s(t)$, in feet, t seconds after it is launched. The rocket misses the edge of the cliff on its way down and eventually lands in the ocean. How long will it take for the rocket to hit the water?   It will take 9 seconds for the rocket to hit the water.

99. The function

$$f(x) = 40x^2 - 120x + 600$$

models the average amount paid to plaintiffs in personal-injury cases, $f(x)$, in hundreds of thousands of dollars, x years after 1996. In which year will this amount reach 3400 hundred thousand dollars, or $3,400,000?   in 2006

Jury Awards in Personal-Injury Cases

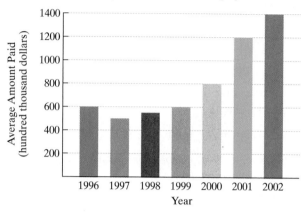

Source: American Lawyer Media

100. The length of a rectangular sign is 3 feet longer than the width. If the sign has space for 54 square feet of advertising, find its length and its width.   Length is 9 feet and width is 6 feet.

101. A painting measuring 10 inches by 16 inches is surrounded by a frame of uniform width. If the combined area of the painting and frame is 280 square inches, determine the width of the frame.   The width of the frame is 2 inches.

102. A lot is in the shape of a right triangle. The longer leg of the triangle is 20 yards longer than twice the length of the shorter leg. The hypotenuse is 30 yards longer than twice the length of the shorter leg. What are the lengths of the three sides?   The shorter leg is 50 yards, the longer leg is 120 yards, and the hypotenuse is 130 yards.

CHAPTER 5 TEST

Remember to use your Chapter Test Prep Video CD to see the worked-out solutions to the test questions you want to review.

In Exercises 1–2, give the degree and the leading coefficient of the polynomial.

1. $7x - 5 + x^2 - 6x^3$ degree: 3; leading coefficient: −6

2. $4xy^3 + 7x^4y^5 - 3xy^4$ degree: 9; leading coefficient: 7

3. If $f(x) = 3x^3 + 5x^2 - x + 6$, find $f(0)$ and $f(-2)$.   6; 4

In Exercises 4–5, use the Leading Coefficient Test to describe the end behavior of the graph of the polynomial function.

4. $f(x) = -16x^2 + 160x$ falls to the left and falls to the right

5. $f(x) = 4x^3 + 12x^2 - x - 3$ falls to the left and rises to the right

In Exercises 6–13, perform the indicated operations.

6. $(4x^3y - 19x^2y - 7y) + (3x^3y + x^2y + 6y - 9)$ $7x^3y - 18x^2y - y - 9$

7. $(6x^2 - 7x - 9) - (-5x^2 + 6x - 3)$ $11x^2 - 13x - 6$

8. $(-7x^3y)(-5x^4y^2)$ $35x^7y^3$

9. $(x - y)(x^2 - 3xy - y^2)$ $x^3 - 4x^2y + 2xy^2 + y^3$

10. $(7x - 9y)(3x + y)$ $21x^2 - 20xy - 9y^2$

11. $(2x - 5y)(2x + 5y)$ $4x^2 - 25y^2$

12. $(4y - 7)^2$ $16y^2 - 56y + 49$

13. $[(x + 2) + 3y][(x + 2) - 3y]$ $x^2 + 4x + 4 - 9y^2$

14. Let $f(x) = x + 2$ and $g(x) = 3x - 5$. Find $(fg)(x)$ and $(fg)(-5)$. $3x^2 + x - 10; 60$

15. Let $f(x) = x^2 - 5x + 3$. Find $f(a + h) - f(a)$ and simplify. $2ah + h^2 - 5h$

In Exercises 16–33, factor completely, or state that the polynomial is prime.

16. $14x^3 - 15x^2$ $x^2(14x - 15)$

17. $81y^2 - 25$ $(9y + 5)(9y - 5)$

18. $x^3 + 3x^2 - 25x - 75$ $(x + 3)(x + 5)(x - 5)$

19. $25x^2 - 30x + 9$ $(5x - 3)^2$

20. $x^2 + 10x + 25 - 9y^2$ $(x + 5 + 3y)(x + 5 - 3y)$

21. $x^4 + 1$ prime

22. $y^2 - 16y - 36$ $(y + 2)(y - 18)$

23. $14x^2 + 41x + 15$ $(2x + 5)(7x + 3)$

24. $5x^3 - 5$ $5(x - 1)(x^2 + x + 1)$

25. $12x^2 - 3y^2$ $3(2x + y)(2x - y)$

26. $12x^2 - 34x + 10$ $2(3x - 1)(2x - 5)$

27. $3x^4 - 3$ $3(x^2 + 1)(x + 1)(x - 1)$

28. $x^8 - y^8$ $(x^4 + y^4)(x^2 + y^2)(x + y)(x - y)$

29. $12x^2y^4 + 8x^3y^2 - 36x^2y$ $4x^2y(3y^3 + 2xy - 9)$

30. $x^6 - 12x^3 - 28$ $(x^3 + 2)(x^3 - 14)$

31. $x^4 - 2x^2 - 24$ $(x^2 - 6)(x^2 + 4)$

32. $12x^2y - 27xy + 6y$ $3y(4x - 1)(x - 2)$

33. $y^4 - 3y^3 + 2y^2 - 6y$ $y(y - 3)(y^2 + 2)$

In Exercises 34–37, solve each polynomial equation.

34. $3x^2 = 5x + 2$ $-\frac{1}{3}$ and 2, or $\left\{-\frac{1}{3}, 2\right\}$

35. $(5x + 4)(x - 1) = 2$ -1 and $\frac{6}{5}$, or $\left\{-1, \frac{6}{5}\right\}$

36. $15x^2 - 5x = 0$ 0 and $\frac{1}{3}$, or $\left\{0, \frac{1}{3}\right\}$

37. $x^3 - 4x^2 - x + 4 = 0$ -1, 1, and 4, or $\{-1, 1, 4\}$

38. A baseball is thrown straight up from a rooftop 448 feet high. The function
$$s(t) = -16t^2 + 48t + 448$$
describes the ball's height above the ground, $s(t)$, in feet, t seconds after it is thrown. How long will it take for the ball to hit the ground?   It will take 7 seconds for the ball to hit the ground.

39. An architect is allowed 15 square yards of floor space to add a small bedroom to a house. Because of the room's design in relationship to the existing structure, the width of the rectangular floor must be 7 yards less than two times the length. Find the length and width of the rectangular floor that the architect is permitted.   Length is 5 yards and width is 3 yards.

40. Find the lengths of the three sides of the right triangle in the figure shown.   One leg has length of 12 units, the other leg has length of 9 units, and the hypotenuse has length of 15 units.

Note to Instructor: The Chapter 5 Test file in the TestGen program provides algorithms specifically matched to these problems so you can easily replicate this chapter test for additional practice or assessment purposes.

CUMULATIVE REVIEW EXERCISES (CHAPTERS 1–5)

In Exercises 1–7, solve each equation, inequality, of system of equations.

1. $8(x + 2) - 3(2 - x) = 4(2x + 6) - 2$ 4 or $\{4\}$

2. $2x + 4y = -6$
$x = 2y - 5$ $\left(-4, \frac{1}{2}\right)$ or $\left\{\left(-4, \frac{1}{2}\right)\right\}$

3. $2x - y + 3z = 0$
$2y + z = 1$
$x + 2y - z = 5$ $(2, 1, -1)$ or $\{(2, 1, -1)\}$

4. $2x + 4 < 10$ and $3x - 1 > 5$ $\{x | 2 < x < 3\}$ or $(2, 3)$

5. $|2x - 5| \geq 9$ $\{x | x \leq -2$ or $x \geq 7\}$ or $(-\infty, -2] \cup [7, \infty)$

6. $2x^2 = 7x - 5$ 1 and $\frac{5}{2}$, or $\left\{1, \frac{5}{2}\right\}$

7. $2x^3 + 6x^2 = 20x$ -5, 0, and 2, or $\{-5, 0, 2\}$

8. Solve for x: $x = \dfrac{ax + b}{c}$. $x = \dfrac{b}{c - a}$

9. Use function notation to write the equation of the line passing through $(-2, -3)$ and $(2, 5)$. $f(x) = 2x + 1$

10. In a campuswide election for student government president, 2800 votes were cast for the two candidates. If the winner had 160 more votes than the loser, how many votes were cast for each candidate?   winner: 1480 votes; loser: 1320 votes

In Exercises 11–13, graph each equation or inequality in a rectangular coordinate system.   **11–13.** See graphing answer section.

11. $f(x) = -\dfrac{1}{3}x + 1$

12. $4x - 5y < 20$

13. $y \leq -1$

14. Simplify: $-\dfrac{8x^3y^6}{16x^9y^{-4}}$. $-\dfrac{y^{10}}{2x^6}$

15. Write in scientific notation: 0.0000706. 7.06×10^{-5}

In Exercises 16–17, perform the indicated operations.

16. $(3x^2 - y)^2$ $9x^4 - 6x^2y + y^2$

17. $(3x^2 - y)(3x^2 + y)$ $9x^4 - y^2$

In Exercises 18–20, factor completely.

18. $x^3 - 3x^2 - 9x + 27$ $(x + 3)(x - 3)^2$

19. $x^6 - x^2$ $x^2(x^2 + 1)(x + 1)(x - 1)$

20. $14x^3y^2 - 28x^4y^2$ $14x^3y^2(1 - 2x)$

During the 1980s, the label "supply-side economics" was applied to the theory that lower tax rates would improve private sector incentives, leading to higher employment, productivity, and output in the economy. George Bush, father of president George W. Bush and an opponent of Ronald Reagan in the 1980 primaries, referred to an extreme version of this theory espoused by Reagan as "voodoo economics." In this version, a cut in tax rates was predicted to result in an increase in tax revenue, and thus not increase the government deficit. George W. Bush ignored his father's opinions about supply-side economics and based his administration's economic policy on supply-side theory.

This discussion is developed algebraically in the Section 6.4 opener and in Exercises 55–58 in Exercise Set 6.4.

U.S. Budget Surplus/Deficit

U.S. Unemployment Rate

Rational Expressions, Functions, and Equations

6.1 Rational Expressions and Functions: Multiplying and Dividing

6.2 Adding and Subtracting Rational Expressions

6.3 Complex Rational Expressions

6.4 Division of Polynomials

6.5 Synthetic Division and the Remainder Theorem

6.6 Rational Equations

6.7 Formulas and Applications of Rational Equations

6.8 Modeling Using Variation

Budget deficits exploded in the 1980s and again in the early 2000s after tax rates were cut in 1981 and 2001. In this chapter, you will see how a function consisting of the quotient of two polynomials models the most controversial version of supply-side economics. Functions involving fractional expressions will give you new insights into phenomena as diverse as the cost of environmental cleanup, the relationship between high production levels and competitively low prices, the percentage of deaths from a disease caused by smoking cigarettes, and even our ongoing processes of learning and forgetting.

SECTION 6.1

Objectives

1 Evaluate rational functions.

2 Find the domain of a rational function.

3 Interpret information given by the graph of a rational function.

4 Simplify rational expressions.

5 Multiply rational expressions.

6 Divide rational expressions.

RATIONAL EXPRESSIONS AND FUNCTIONS: MULTIPLYING AND DIVIDING

Environmental scientists and municipal planners often make decisions using **cost-benefit models**. These mathematical models estimate the cost of removing a pollutant from the atmosphere as a function of the percentage of pollutant removed. What kinds of functions describe the cost of reducing environmental pollution? In this section, we introduce this new category of functions, called *rational functions*.

Rational Expressions   A **rational expression** consists of a polynomial divided by a nonzero polynomial. Here are some examples of rational expressions:

$$\frac{120x}{100 - x}, \quad \frac{2x + 1}{2x^2 - x - 1}, \quad \frac{3x^2 + 12xy - 15y^2}{6x^3 - 6xy^2}.$$

Rational Functions   A **rational function** is a function defined by a formula that is a rational expression.

1   Evaluate rational functions.

EXAMPLE 1   Using a Cost-Benefit Model

The rational function

$$f(x) = \frac{120x}{100 - x}$$

models the cost, $f(x)$, in thousands of dollars, to remove $x\%$ of the pollutants that a city has discharged into a lake. Find and interpret each of the following:

a. $f(20)$ **b.** $f(80)$.

SOLUTION   We use substitution to evaluate a rational function, just as we did to evaluate other functions in Chapter 2.

a. $f(x) = \dfrac{120x}{100 - x}$ This is the given rational function.

$f(20) = \dfrac{120(20)}{100 - 20}$ To find f of 20, replace x with 20.

$= \dfrac{2400}{80} = 30$ Perform the indicated operations.

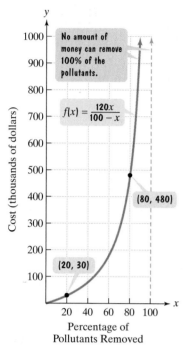

FIGURE 6.1

2 Find the domain of a rational function.

Thus, $f(20) = 30$. This means that the cost to remove 20% of the lake's pollutants is $30 thousand. Figure 6.1 illustrates the solution by the point $(20, 30)$ on the graph of the rational function.

b. $f(x) = \dfrac{120x}{100 - x}$ *This is the given rational function.*

$$f(80) = \frac{120(80)}{100 - 80} = \frac{9600}{20} = 480$$ *To find f of 80, replace x with 80.*

Thus, $f(80) = 480$. This means that the cost to remove 80% of the lake's pollutants is $480 thousand. Figure 6.1 illustrates the solution by the point $(80, 480)$ on the graph of the cost-benefit model. The graph illustrates that costs rise steeply as the percentage of pollutants removed increases.   ∎

✔ **CHECK POINT 1** Use the rational function in Example 1 to find and interpret: **a.** $f(40)$ **b.** $f(60)$. Identify each solution as a point on the graph of the rational function in Figure 6.1.   **a.** 80; The cost to remove 40% of the lake's pollutants is $80 thousand; $(40, 80)$.   **b.** 180; The cost to remove 60% of the lake's pollutants is $180 thousand; $(60, 180)$.

The Domain of a Rational Function   Does the cost-benefit model

$$f(x) = \frac{120x}{100 - x}$$

indicate that the city can clean up its lake completely? To do this, the city must remove 100% of the pollutants. The problem is that the rational function is undefined for $x = 100$.

$$f(x) = \frac{120x}{100 - x}$$ *If x = 100, the value of the denominator is 0.*

Notice how the graph of the rational function in Figure 6.1 approaches, but never touches, the dashed green vertical line whose equation is $x = 100$. The graph continues to rise more and more steeply, visually showing the escalating costs. By never touching the dashed vertical line, the graph illustrates that no amount of money will be enough to remove all pollutants from the lake.

The **domain of a rational function** is the set of all real numbers except those for which the denominator is zero. We can find the domain by determining when the denominator is zero. For the cost-benefit model, the denominator is zero when $x = 100$. Furthermore, for this model, negative values of x and values of x greater than 100 are not meaningful. The domain of the function is $[0, 100)$ and excludes 100.

Inspection can sometimes be used to find a rational function's domain. Here are two examples.

$$f(x) = \frac{4}{x - 2} \qquad\qquad g(x) = \frac{x}{(x + 1)(x - 1)}$$

This denominator would equal zero if x = 2.   *This factor would equal zero if x = −1.*   *This factor would equal zero if x = 1.*

The domain of f can be expressed in set-builder or interval notation:

Domain of $f = \{x \mid x \text{ is a real number and } x \neq 2\}$

Domain of $f = (-\infty, 2) \cup (2, \infty)$.

Likewise, the domain of g can be expressed in set-builder or interval notation:

$$\text{Domain of } g = \{x \mid x \text{ is a real number and } x \neq -1 \text{ and } x \neq 1\}$$
$$\text{Domain of } g = (-\infty, -1) \cup (-1, 1) \cup (1, \infty).$$

Ask your professor if a particular notation is preferred in the course.

EXAMPLE 2 Finding the Domain of a Rational Function

Find the domain of f if

$$f(x) = \frac{2x + 1}{2x^2 - x - 1}.$$

SOLUTION The domain of f is the set of all real numbers except those for which the denominator is zero. We can identify such numbers by setting the denominator equal to zero and solving for x.

$$2x^2 - x - 1 = 0 \qquad \text{Set the denominator equal to 0.}$$
$$(2x + 1)(x - 1) = 0 \qquad \text{Factor.}$$
$$2x + 1 = 0 \quad \text{or} \quad x - 1 = 0 \qquad \text{Set each factor equal to 0.}$$
$$2x = -1 \qquad\qquad x = 1 \qquad \text{Solve the resulting equations.}$$
$$x = -\frac{1}{2}$$

Because $-\frac{1}{2}$ and 1 make the denominator zero, these are the values to exclude. Thus,

$$\text{Domain of } f = \left\{ x \mid x \text{ is a real number and } x \neq -\frac{1}{2} \text{ and } x \neq 1 \right\}$$

or

$$\text{Domain of } f = \left(-\infty, -\frac{1}{2}\right) \cup \left(-\frac{1}{2}, 1\right) \cup (1, \infty).$$

USING TECHNOLOGY

We can use the TABLE feature of a graphing utility (with ΔTbl = .5) to verify our work with

$$f(x) = \frac{2x + 1}{2x^2 - x - 1}.$$

Enter

$$y_1 = (\!(2 \fbox{X} \fbox{+} 1)\!) \fbox{÷}$$
$$(\!(2 \fbox{X} \fbox{$\wedge$} 2 \fbox{$-$} \fbox{X} \fbox{$-$} 1)\!)$$

and press TABLE.

The error messages verify that $-\frac{1}{2}$ and 1 must be excluded from the domain.

 CHECK POINT 2 Find the domain of f if

$$f(x) = \frac{x - 5}{2x^2 + 5x - 3}.$$

Domain of $f = \left\{ x \mid x \text{ is a real number and } x \neq -3 \text{ and } x \neq \frac{1}{2} \right\}$ or $(-\infty, -3) \cup \left(-3, \frac{1}{2}\right) \cup \left(\frac{1}{2}, \infty\right)$

Does every rational function have values to exclude? The answer is no. For example, consider

$$f(x) = \frac{2}{x^2 + 1}.$$

No real-number values of x cause this denominator to equal zero.

The domain of f consists of the set of all real numbers. Recall that this set can be expressed in three different ways:

$$\text{Domain of } f = \{x \mid x \text{ is a real number}\} \text{ or } \mathbb{R} \text{ or } (-\infty, \infty).$$

3 Interpret information given by the graph of a rational function.

What Graphs of Rational Functions Look Like In everyday speech, a continuous process is one that goes on without interruption and without abrupt changes. In mathematics, a continuous function has much the same meaning. The graph of a continuous function does not have interrupting breaks, such as holes, gaps, or jumps. Thus, the graph of a continuous function can be drawn without lifting a pencil off the paper.

Most rational functions are not continuous functions. For example, the graph of the rational function

$$f(x) = \frac{2x + 1}{2x^2 - x - 1}$$

is shown in Figure 6.2. In Example 2, we excluded $-\frac{1}{2}$ and 1 from this function's domain. Unlike the graph of a polynomial function, this graph has two breaks in it—one at each of the excluded values. At $-\frac{1}{2}$, a hole in the graph appears. The graph is composed of two distinct branches. Each branch approaches, but never touches, the dashed vertical line drawn at 1, the other excluded value.

A vertical line that the graph of a function approaches, but does not touch, is said to be a **vertical asymptote** of the graph. In Figure 6.2, the line $x = 1$ is a vertical asymptote of the graph of f. A rational function may have no vertical asymptotes, one vertical asymptote, or several vertical asymptotes. The graph of a rational function never intersects a vertical asymptote.

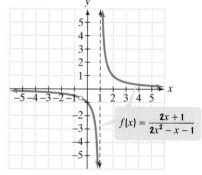

FIGURE 6.2

USING TECHNOLOGY

When using a graphing utility to graph a rational function, you might not be pleased with the quality of the display. The graph of the rational function

$$y = \frac{2x + 1}{2x^2 - x - 1}$$

is shown using the $\boxed{\text{DOT}}$ mode in a $[-6, 6, 1]$ by $[-6, 6, 1]$ viewing rectangle. In this mode, the utility displays unconnected points that have been calculated. Would you agree that the graph's behavior is better illustrated by the hand-drawn graph in Figure 6.2?

Unlike the graph of a polynomial function, the graph of the rational function in Figure 6.2 does not go up or down at each end. At the far left and the far right, the graph is getting closer to, but not actually reaching, the x-axis. This shows that as x increases or decreases without bound, the function values are approaching 0. The line $y = 0$ (that is, the x-axis) is a **horizontal asymptote** of the graph. Many, but not all, rational functions have horizontal asymptotes.

For the remainder of this section, we will focus on the rational expressions that define rational functions. Operations with these expressions should remind you of those performed in arithmetic with fractions.

4 Simplify rational expressions.

Simplifying Rational Expressions A rational expression is **simplified** if its numerator and denominator have no common factors other than 1 or −1. The following procedure can be used to simplify rational expressions:

> ### SIMPLIFYING RATIONAL EXPRESSIONS
>
> **1.** Factor the numerator and the denominator completely.
> **2.** Divide both the numerator and the denominator by any common factors.

EXAMPLE 3 Simplifying a Rational Expression

Simplify: $\dfrac{x^2 + 4x + 3}{x + 1}$.

SOLUTION

$$\frac{x^2 + 4x + 3}{x + 1} = \frac{(x + 1)(x + 3)}{1(x + 1)} \qquad \text{Factor the numerator and denominator.}$$

$$= \frac{\cancel{(x + 1)}(x + 3)}{1\cancel{(x + 1)}} \qquad \text{Divide out the common factor, } x + 1.$$

$$= x + 3$$

Simplifying a rational expression can change the numbers that make it undefined. For example, we just showed that

$$\frac{x^2 + 4x + 3}{x + 1} = x + 3.$$

> This is undefined for $x = -1$.   The simplified form is defined for all real numbers.

Thus, to equate the two expressions, we must restrict the values of x in the simplified expression to exclude −1. We can write

$$\frac{x^2 + 4x + 3}{x + 1} = x + 3, x \neq -1.$$

Without this restriction, the expressions are not equal. The slight difference between them is illustrated using the graphs of

$$f(x) = \frac{x^2 + 4x + 3}{x + 1} \quad \text{and} \quad g(x) = x + 3$$

in Figure 6.3.

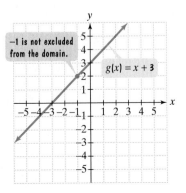

FIGURE 6.3 Visualizing the difference between $\dfrac{x^2 + 4x + 3}{x + 1}$ and $x + 3$

Hereafter, we will assume that a simplified rational expression is equal to the original rational expression for all real numbers except those for which either denominator is 0.

 CHECK POINT 3 Simplify: $\dfrac{x^2 + 7x + 10}{x + 2}$. $x + 5$

EXAMPLE 4 Simplifying Rational Expressions

Simplify:

a. $\dfrac{x^2 - 7x - 18}{2x^2 + 3x - 2}$ **b.** $\dfrac{3x^2 + 12xy - 15y^2}{6x^3 - 6xy^2}$.

SOLUTION

a. $\dfrac{x^2 - 7x - 18}{2x^2 + 3x - 2} = \dfrac{(x - 9)(x + 2)}{(2x - 1)(x + 2)}$ Factor the numerator and denominator.

$= \dfrac{(x - 9)\cancel{(x + 2)}}{(2x - 1)\cancel{(x + 2)}}$ Divide out the common factor, $x + 2$.

$= \dfrac{x - 9}{2x - 1}$

b. $\dfrac{3x^2 + 12xy - 15y^2}{6x^3 - 6xy^2} = \dfrac{3(x^2 + 4xy - 5y^2)}{6x(x^2 - y^2)}$ Factor out the GCF in the numerator and denominator.

$= \dfrac{3(x + 5y)(x - y)}{6x(x + y)(x - y)}$ Factor the numerator and denominator completely.

$= \dfrac{\overset{1}{\cancel{3}}(x + 5y)\cancel{(x - y)}}{\underset{2}{\cancel{6}}x(x + y)\cancel{(x - y)}}$ Divide out the common factor, $3(x - y)$.

$= \dfrac{x + 5y}{2x(x + y)}$ It is not necessary to carry out the multiplication in the denominator.

 CHECK POINT 4 Simplify:

a. $\dfrac{x^2 - 2x - 15}{3x^2 + 8x - 3}$ $\dfrac{x - 5}{3x - 1}$ **b.** $\dfrac{3x^2 + 9xy - 12y^2}{9x^3 - 9xy^2}$. $\dfrac{x + 4y}{3x(x + y)}$

STUDY TIP

When simplifying rational expressions, you can only divide out, or cancel, factors common to the numerator and denominator. **It is incorrect to divide out common terms from the numerator and denominator.**

Study Tip (continued)

The first two expressions, $\dfrac{x+4}{x}$ and $\dfrac{x^2-4}{4}$, have no common factors in their numerators and denominators. Only when expressions are multiplied can they be factors. If you can't factor, then don't try to cancel. The third rational expression shown on the previous page, $\dfrac{x^2-9}{x-3}$, can be simplified as follows:

Correct

$$\frac{x^2-9}{x-3} = \frac{(x+3)\cancel{(x-3)}}{1\cancel{(x-3)}} = x+3.$$

> Divide out the common factor, $x - 3$.

5 Multiply rational expressions.

Multiplying Rational Expressions   The product of two rational expressions is the product of their numerators divided by the product of their denominators. For example,

$$\frac{x^2}{y+3} \cdot \frac{x+5}{y-7} = \frac{x^2(x+5)}{(y+3)(y-7)}.$$

> Multiply numerators.

> Multiply denominators.

Here is a step-by-step procedure for multiplying rational expressions. Before multiplying, divide out any factors common to both a numerator and a denominator.

MULTIPLYING RATIONAL EXPRESSIONS

1. Factor all numerators and denominators completely.
2. Divide numerators and denominators by common factors.
3. Multiply the remaining factors in the numerators and multiply the remaining factors in the denominators.

EXAMPLE 5   Multiplying Rational Expressions

Multiply: $\dfrac{x+3}{x-4} \cdot \dfrac{x^2-2x-8}{x^2-9}.$

SOLUTION

$$\frac{x+3}{x-4} \cdot \frac{x^2-2x-8}{x^2-9}$$

$$= \frac{1(x+3)}{1(x-4)} \cdot \frac{(x-4)(x+2)}{(x+3)(x-3)} \qquad \text{Factor all numerators and denominators completely.}$$

$$= \frac{1\cancel{(x+3)}}{1\cancel{(x-4)}} \cdot \frac{\cancel{(x-4)}(x+2)}{\cancel{(x+3)}(x-3)} \qquad \text{Divide numerators and denominators by common factors.}$$

$$= \frac{x+2}{x-3} \qquad \text{Multiply the remaining factors in the numerators and in the denominators.} \ \blacksquare$$

✔ **CHECK POINT 5** Multiply: $\dfrac{x+4}{x-7} \cdot \dfrac{x^2-4x-21}{x^2-16} \cdot \dfrac{x+3}{x-4}$

Some rational expressions contain factors in the numerator and denominator that are opposites, or additive inverses. Here is an example of such an expression:

$$\frac{(2x + 5)(2x - 5)}{3(5 - 2x)}.$$

> The factors $2x - 5$ and $5 - 2x$ are opposites. They differ only in sign.

Although you can factor out -1 from the numerator or the denominator and then divide out the common factor, there is an even faster way to simplify this rational expression.

SIMPLIFYING RATIONAL EXPRESSIONS WITH OPPOSITE FACTORS IN THE NUMERATOR AND DENOMINATOR The quotient of two polynomials that have opposite signs and are additive inverses is -1.

For example,

$$\frac{(2x + 5)(2x - 5)}{3(5 - 2x)} = \frac{(2x + 5)\overset{(-1)}{\cancel{(2x - 5)}}}{3\cancel{(5 - 2x)}} = \frac{-(2x + 5)}{3} \quad \text{or} \quad -\frac{2x + 5}{3} \quad \text{or} \quad \frac{-2x - 5}{3}.$$

EXAMPLE 6 Multiplying Rational Expressions

Multiply: $\dfrac{5x + 5}{7x - 7x^2} \cdot \dfrac{2x^2 + x - 3}{4x^2 - 9}.$

SOLUTION

$$\frac{5x + 5}{7x - 7x^2} \cdot \frac{2x^2 + x - 3}{4x^2 - 9}$$

$$= \frac{5(x + 1)}{7x(1 - x)} \cdot \frac{(2x + 3)(x - 1)}{(2x + 3)(2x - 3)}$$
Factor all numerators and denominators completely.

$$= \frac{5(x + 1)}{7x\cancel{(1 - x)}} \cdot \frac{\overset{(-1)}{\cancel{(2x + 3)}\cancel{(x - 1)}}}{\cancel{(2x + 3)}(2x - 3)}$$
Divide numerators and denominators by common factors. Because $1 - x$ and $x - 1$ are opposites, their quotient is -1.

$$= \frac{-5(x + 1)}{7x(2x - 3)} \quad \text{or} \quad -\frac{5(x + 1)}{7x(2x - 3)}$$
Multiply the remaining factors in the numerators and in the denominators. ■

✔ **CHECK POINT 6** Multiply: $\dfrac{4x + 8}{6x - 3x^2} \cdot \dfrac{3x^2 - 4x - 4}{9x^2 - 4}.$ $\dfrac{-4(x + 2)}{3x(3x - 2)}$ or $-\dfrac{4(x + 2)}{3x(3x - 2)}$

6 Divide rational expressions.

Dividing Rational Expressions The quotient of two rational expressions is the product of the first expression and the multiplicative inverse, or reciprocal, of the second expression. The reciprocal is found by interchanging the numerator and the denominator of the expression.

DIVIDING RATIONAL EXPRESSIONS If P, Q, R, and S are polynomials, where $Q \neq 0$, $R \neq 0$, and $S \neq 0$, then

$$\frac{P}{Q} \div \frac{R}{S} = \frac{P}{Q} \cdot \frac{S}{R} = \frac{PS}{QR}.$$

Change division to multiplication.

Replace $\dfrac{R}{S}$ with its reciprocal by interchanging its numerator and denominator.

Thus, **we find the quotient of two rational expressions by inverting the divisor and multiplying**. For example,

$$\frac{x}{7} \div \frac{6}{y} = \frac{x}{7} \cdot \frac{y}{6} = \frac{xy}{42}.$$

Change the division to multiplication.

Replace $\frac{6}{y}$ with its reciprocal by interchanging its numerator and denominator.

STUDY TIP

When performing operations with rational expressions, if a rational expression is written without a denominator, it is helpful to write the expression with a denominator of 1. In Example 7(a), we wrote

$$4x^2 - 25 \quad \text{as} \quad \frac{4x^2 - 25}{1}.$$

EXAMPLE 7 Dividing Rational Expressions

Divide:

a. $(4x^2 - 25) \div \dfrac{2x + 5}{14}$ **b.** $\dfrac{x^2 + 3x - 10}{2x} \div \dfrac{x^2 - 5x + 6}{x^2 - 3x}.$

SOLUTION

a. $(4x^2 - 25) \div \dfrac{2x + 5}{14}$

$= \dfrac{4x^2 - 25}{1} \div \dfrac{2x + 5}{14}$ Write $4x^2 - 25$ with a denominator of 1.

$= \dfrac{4x^2 - 25}{1} \cdot \dfrac{14}{2x + 5}$ Invert the divisor and multiply.

$= \dfrac{(2x + 5)(2x - 5)}{1} \cdot \dfrac{14}{1(2x + 5)}$ Factor.

$= \dfrac{\cancel{(2x + 5)}(2x - 5)}{1} \cdot \dfrac{14}{1\cancel{(2x + 5)}}$ Divide the numerator and denominator by the common factor, $2x + 5$.

$= 14(2x - 5)$ Multiply the remaining factors in the numerators and in the denominators.

b. $\dfrac{x^2 + 3x - 10}{2x} \div \dfrac{x^2 - 5x + 6}{x^2 - 3x}$

$= \dfrac{x^2 + 3x - 10}{2x} \cdot \dfrac{x^2 - 3x}{x^2 - 5x + 6}$ Invert the divisor and multiply.

$= \dfrac{(x + 5)(x - 2)}{2x} \cdot \dfrac{x(x - 3)}{(x - 3)(x - 2)}$ Factor.

$= \dfrac{(x + 5)\cancel{(x - 2)}}{2\cancel{x}} \cdot \dfrac{\cancel{x}\cancel{(x - 3)}}{\cancel{(x - 3)}\cancel{(x - 2)}}$ Divide numerators and denominators by common factors.

$= \dfrac{x + 5}{2}$ Multiply the remaining factors in the numerators and in the denominators. ∎

✔ **CHECK POINT 7** Divide:

a. $(9x^2 - 49) \div \dfrac{3x - 7}{9}$ $9(3x + 7)$

b. $\dfrac{x^2 - x - 12}{5x} \div \dfrac{x^2 - 10x + 24}{x^2 - 6x}.$ $\dfrac{x + 3}{5}$

6.1 EXERCISE SET

Student Solutions Manual   CD/Video   PH Math/Tutor Center   MathXL Tutorials on CD   MathXL®   MyMathLab   Interactmath.com

Practice Exercises

In Exercises 1–6, use the given rational function to find the indicated function values. If a function value does not exist, so state.

1. $f(x) = \dfrac{x^2 - 9}{x + 3}$; $f(-2), f(0), f(5)$ $-5; -3; 2$

2. $f(x) = \dfrac{x^2 - 16}{x + 4}$; $f(-2), f(0), f(5)$ $-6; -4; 1$

3. $f(x) = \dfrac{x^2 - 2x - 3}{4 - x}$; $f(-1), f(4), f(6)$ 0; does not exist; $-\dfrac{21}{2}$

4. $f(x) = \dfrac{x^2 - 3x - 4}{3 - x}$; $f(-1), f(3), f(5)$ 0; does not exist; -3

5. $g(t) = \dfrac{2t^3 - 5}{t^2 + 1}$; $g(-1), g(0), g(2)$ $-\dfrac{7}{2}; -5; \dfrac{11}{5}$

6. $g(t) = \dfrac{2t^3 - 1}{t^2 + 4}$; $g(-1), g(0), g(2)$ $-\dfrac{3}{5}; -\dfrac{1}{4}; \dfrac{15}{8}$

In Exercises 7–16, use inspection or set the denominator equal to zero to find the domain of the given rational function. Use the notation of your choice (set-builder or interval) or the notation requested by your professor.

7. $f(x) = \dfrac{x - 2}{x - 5}$ domain of $f = \{x | x$ is a real number and $x \neq 5\}$ or domain of $f = (-\infty, 5) \cup (5, \infty)$

8. $f(x) = \dfrac{x - 3}{x - 6}$ domain of $f = \{x | x$ is a real number and $x \neq 6\}$ or domain of $f = (-\infty, 6) \cup (6, \infty)$

9. $f(x) = \dfrac{x - 4}{(x - 1)(x + 3)}$ domain of $f = \{x | x$ is a real number and $x \neq -3$ and $x \neq 1\}$ or domain of $f = (-\infty, -3) \cup (-3, 1) \cup (1, \infty)$

10. $f(x) = \dfrac{x - 5}{(x - 2)(x + 4)}$ domain of $f = \{x | x$ is a real number and $x \neq -4$ and $x \neq 2\}$ or domain of $f = (-\infty, -4) \cup (-4, 2) \cup (2, \infty)$

11. $f(x) = \dfrac{2x}{(x + 5)^2}$ domain of $f = \{x | x$ is a real number and $x \neq -5\}$ or domain of $f = (-\infty, -5) \cup (-5, \infty)$

12. $f(x) = \dfrac{2x}{(x + 7)^2}$ domain of $f = \{x | x$ is a real number and $x \neq -7\}$ or domain of $f = (-\infty, -7) \cup (-7, \infty)$

13. $f(x) = \dfrac{3x}{x^2 - 8x + 15}$ domain of $f = \{x | x$ is a real number and $x \neq 3$ and $x \neq 5\}$ or domain of $f = (-\infty, 3) \cup (3, 5) \cup (5, \infty)$

14. $f(x) = \dfrac{3x}{x^2 - 13x + 36}$ domain of $f = \{x | x$ is a real number and $x \neq 4$ and $x \neq 9\}$ or domain of $f = (-\infty, 4) \cup (4, 9) \cup (9, \infty)$

15. $f(x) = \dfrac{(x - 1)^2}{3x^2 - 2x - 8}$ domain of $f = \{x | x$ is a real number and $x \neq -\frac{4}{3}$ and $x \neq 2\}$ or domain of $f = \left(-\infty, -\frac{4}{3}\right) \cup \left(-\frac{4}{3}, 2\right) \cup (2, \infty)$

16. $f(x) = \dfrac{(x - 1)^2}{4x^2 - 13x + 3}$ domain of $f = \{x | x$ is a real number and $x \neq \frac{1}{4}$ and $x \neq 3\}$ or domain of $f = \left(-\infty, \frac{1}{4}\right) \cup \left(\frac{1}{4}, 3\right) \cup (3, \infty)$

The graph of a rational function, f, is shown in the figure. Use the graph to solve Exercises 17–26.

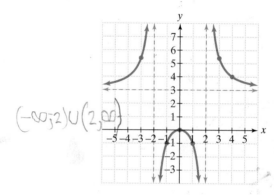

17. Find $f(4)$. 4

18. Find $f(1)$. -1

19. What is the domain of f? What is the range of f?   domain of $f = \{x | x$ is a real number and $x \neq -2$ and $x \neq 2\}$ or $(-\infty, -2) \cup (-2, 2) \cup (2, \infty)$; range of $f = \{y | y \leq 0$ or $y > 3\}$ or $(-\infty, 0] \cup (3, \infty)$

20. What are the vertical asymptotes of the graph of f? $x = -2$ and $x = 2$

21. Describe the end behavior of the graph at the far left. What is the equation of the horizontal asymptote? As x decreases, the function values are approaching 3; $y = 3$.

22. Describe the end behavior of the graph at the far right. What is the equation of the horizontal asymptote? As x increases, the function values are approaching 3; $y = 3$.

23. Explain how the graph shows that $f(-2)$ does not exist. There is no point on the graph with an x-coordinate of -2.

24. Explain how the graph shows that $f(2)$ does not exist. There is no point on the graph with an x-coordinate of 2.

25. How can you tell that this is not the graph of a polynomial function? The graph is not continuous. Furthermore, it neither rises nor falls to the left or the right.

26. List two real numbers that are not function values of f. Answers will vary; any two numbers from the set $\{y | 0 < y \leq 3\}$ comprise a correct answer.

In Exercises 27–50, simplify each rational expression. If the rational expression cannot be simplified, so state.

27. $\dfrac{x^2 - 4}{x - 2}$ $x + 2$

28. $\dfrac{x^2 - 25}{x - 5}$ $x + 5$

29. $\dfrac{x + 2}{x^2 - x - 6}$ $\dfrac{1}{x - 3}$

30. $\dfrac{x + 1}{x^2 - 2x - 3}$ $\dfrac{1}{x - 3}$

31. $\dfrac{4x + 20}{x^2 + 5x}$ $\dfrac{4}{x}$

32. $\dfrac{5x + 30}{x^2 + 6x}$ $\dfrac{5}{x}$

33. $\dfrac{4y - 20}{y^2 - 25}$ $\dfrac{4}{y + 5}$

34. $\dfrac{6y - 42}{y^2 - 49}$ $\dfrac{6}{y + 7}$

35. $\dfrac{3x - 5}{25 - 9x^2}$ $-\dfrac{1}{3x + 5}$

36. $\dfrac{5x - 2}{4 - 25x^2}$ $-\dfrac{1}{5x + 2}$

37. $\dfrac{y^2 - 49}{y^2 - 14y + 49}$ $\dfrac{y + 7}{y - 7}$

38. $\dfrac{y^2 - 9}{y^2 - 6y + 9}$ $\dfrac{y + 3}{y - 3}$

39. $\dfrac{x^2 + 7x - 18}{x^2 - 3x + 2}$ $\dfrac{x + 9}{x - 1}$

40. $\dfrac{x^2 - 4x - 5}{x^2 + 5x + 4}$ $\dfrac{x - 5}{x + 4}$

41. $\dfrac{3x + 7}{3x + 10}$ cannot be simplified

42. $\dfrac{2x + 3}{2x + 5}$ cannot be simplified

43. $\dfrac{x^2 - x - 12}{16 - x^2}$ $-\dfrac{x + 3}{x + 4}$

44. $\dfrac{x^2 - 7x + 12}{9 - x^2}$ $-\dfrac{x - 4}{x + 3}$

45. $\dfrac{x^2 + 3xy - 10y^2}{3x^2 - 7xy + 2y^2}$ $\dfrac{x + 5y}{3x - y}$

46. $\dfrac{x^2 + 2xy - 3y^2}{2x^2 + 5xy - 3y^2}$ $\dfrac{x - y}{2x - y}$

47. $\dfrac{x^3 - 8}{x^2 - 4}$ $\dfrac{x^2 + 2x + 4}{x + 2}$

48. $\dfrac{x^3 - 1}{x^2 - 1}$ $\dfrac{x^2 + x + 1}{x + 1}$

49. $\dfrac{x^3 + 4x^2 - 3x - 12}{x + 4}$ $x^2 - 3$

50. $\dfrac{x^3 - 2x^2 + x - 2}{x - 2}$ $x^2 + 1$

In Exercises 51–72, multiply as indicated.

51. $\dfrac{x - 3}{x + 7} \cdot \dfrac{3x + 21}{2x - 6}$ $\dfrac{3}{2}$

52. $\dfrac{x - 2}{x + 3} \cdot \dfrac{2x + 6}{5x - 10}$ $\dfrac{2}{5}$

53. $\dfrac{x^2 - 49}{x^2 - 4x - 21} \cdot \dfrac{x + 3}{x}$ $\dfrac{x + 7}{x}$

54. $\dfrac{x^2 - 25}{x^2 - 3x - 10} \cdot \dfrac{x + 2}{x}$ $\dfrac{x + 5}{x}$

55. $\dfrac{x^2 - 9}{x^2 - x - 6} \cdot \dfrac{x^2 + 5x + 6}{x^2 + x - 6}$ $\dfrac{x + 3}{x - 2}$

56. $\dfrac{x^2 - 1}{x^2 - 4} \cdot \dfrac{x^2 - 5x + 6}{x^2 - 2x - 3}$ $\dfrac{x - 1}{x + 2}$

57. $\dfrac{x^2 + 4x + 4}{x^2 + 8x + 16} \cdot \dfrac{(x + 4)^3}{(x + 2)^3}$ $\dfrac{x + 4}{x + 2}$

58. $\dfrac{x^2 - 2x + 1}{x^2 - 4x + 4} \cdot \dfrac{(x - 2)^3}{(x - 1)^3}$ $\dfrac{x - 2}{x - 1}$

59. $\dfrac{8y + 2}{y^2 - 9} \cdot \dfrac{3 - y}{4y^2 + y}$ $-\dfrac{2}{y(y + 3)}$

60. $\dfrac{6y + 2}{y^2 - 1} \cdot \dfrac{1 - y}{3y^2 + y}$ $-\dfrac{2}{y(y + 1)}$

61. $\dfrac{y^3 - 8}{y^2 - 4} \cdot \dfrac{y + 2}{2y}$ $\dfrac{y^2 + 2y + 4}{2y}$

62. $\dfrac{y^2 + 6y + 9}{y^3 + 27} \cdot \dfrac{1}{y + 3}$ $\dfrac{1}{y^2 - 3y + 9}$

63. $(x - 3) \cdot \dfrac{x^2 + x + 1}{x^2 - 5x + 6}$ $\dfrac{x^2 + x + 1}{x - 2}$

64. $(x + 1) \cdot \dfrac{x + 2}{x^2 + 7x + 6}$ $\dfrac{x + 2}{x + 6}$

65. $\dfrac{x^2 + xy}{x^2 - y^2} \cdot \dfrac{4x - 4y}{x}$ 4

66. $\dfrac{x^2 - y^2}{x} \cdot \dfrac{x^2 + xy}{x + y}$ $(x + y)(x - y)$

67. $\dfrac{x^2 + 2xy + y^2}{x^2 - 2xy + y^2} \cdot \dfrac{4x - 4y}{3x + 3y}$ $\dfrac{4(x + y)}{3(x - y)}$

68. $\dfrac{2x^2 - 3xy - 2y^2}{3x^2 - 4xy + y^2} \cdot \dfrac{3x^2 - 2xy - y^2}{x^2 + xy - 6y^2}$ $\dfrac{(2x + y)(3x + y)}{(3x - y)(x + 3y)}$

69. $\dfrac{4a^2 + 2ab + b^2}{2a + b} \cdot \dfrac{4a^2 - b^2}{8a^3 - b^3}$ 1

70. $\dfrac{27a^3 - 8b^3}{b^2 - b - 6} \cdot \dfrac{bc - b - 3c + 3}{3ac - 2bc - 3a + 2b}$ $\dfrac{9a^2 + 6ab + 4b^2}{b + 2}$

71. $\dfrac{10z^2 + 13z - 3}{3z^2 - 8z + 5} \cdot \dfrac{2z^2 - 3z - 2z + 3}{25z^2 - 10z + 1} \cdot \dfrac{15z^2 - 28z + 5}{4z^2 - 9}$ 1

72. $\dfrac{2z^2 - 2z - 12}{z^2 - 49} \cdot \dfrac{4z^2 - 1}{2z^2 + 5z + 2} \cdot \dfrac{2z^2 - 13z - 7}{2z^2 - 7z + 3}$ $\dfrac{2(2z + 1)}{z + 7}$

In Exercises 73–90, divide as indicated.

73. $\dfrac{x + 5}{7} \div \dfrac{4x + 20}{9}$ $\dfrac{9}{28}$

74. $\dfrac{x + 1}{3} \div \dfrac{3x + 3}{7}$ $\dfrac{7}{9}$

75. $\dfrac{4}{y - 6} \div \dfrac{40}{7y - 42}$ $\dfrac{7}{10}$

76. $\dfrac{7}{y - 5} \div \dfrac{28}{3y - 15}$ $\dfrac{3}{4}$

77. $\dfrac{x^2 - 2x}{15} \div \dfrac{x - 2}{5}$ $\dfrac{x}{3}$

78. $\dfrac{x^2 - x}{15} \div \dfrac{x - 1}{5}$ $\dfrac{x}{3}$

79. $\dfrac{y^2 - 25}{2y - 2} \div \dfrac{y^2 + 10y + 25}{y^2 + 4y - 5}$ $\dfrac{y - 5}{2}$

80. $\dfrac{y^2 + y}{y^2 - 4} \div \dfrac{y^2 - 1}{y^2 + 5y + 6}$ $\dfrac{y(y + 3)}{(y - 2)(y - 1)}$

81. $(x^2 - 16) \div \dfrac{x^2 + 3x - 4}{x^2 + 4}$ $\dfrac{(x^2 + 4)(x - 4)}{x - 1}$

82. $(x^2 + 4x - 5) \div \dfrac{x^2 - 25}{x + 7}$ $\dfrac{(x + 7)(x - 1)}{x - 5}$

83. $\dfrac{y^2 - 4y - 21}{y^2 - 10y + 25} \div \dfrac{y^2 + 2y - 3}{y^2 - 6y + 5}$ $\dfrac{y - 7}{y - 5}$

84. $\dfrac{y^2 + 4y - 21}{y^2 + 3y - 28} \div \dfrac{y^2 + 14y + 48}{y^2 + 4y - 32}$ $\dfrac{y - 3}{y + 6}$

85. $\dfrac{8x^3 - 1}{4x^2 + 2x + 1} \div \dfrac{x - 1}{(x - 1)^2}$ $(x - 1)(2x - 1)$

86. $\dfrac{x^2 - 9}{x^3 - 27} \div \dfrac{x^2 + 6x + 9}{x^2 + 3x + 9}$ $\dfrac{1}{x + 3}$

87. $\dfrac{x^2 - 4y^2}{x^2 + 3xy + 2y^2} \div \dfrac{x^2 - 4xy + 4y^2}{x + y}$ $\dfrac{1}{x - 2y}$

88. $\dfrac{xy - y^2}{x^2 + 2x + 1} \div \dfrac{2x^2 + xy - 3y^2}{2x^2 + 5xy + 3y^2}$ $\dfrac{y(x + y)}{(x + 1)^2}$

89. $\dfrac{x^4 - y^8}{x^2 + y^4} \div \dfrac{x^2 - y^4}{3x^2}$ $3x^2$

90. $\dfrac{(x - y)^3}{x^3 - y^3} \div \dfrac{x^2 - 2xy + y^2}{x^2 - y^2}$ $\dfrac{(x + y)(x - y)}{x^2 + xy + y^2}$

Practice Plus

In Exercises 91–98, perform the indicated operation or operations.

91. $\dfrac{x^3 - 4x^2 + x - 4}{2x^3 - 8x^2 + x - 4} \cdot \dfrac{2x^3 + 2x^2 + x + 1}{x^4 - x^3 + x^2 - x}$ $\dfrac{x + 1}{x(x - 1)}$

92. $\dfrac{y^3 + y^2 + yz^2 + z^2}{y^3 + y + y^2 + 1} \cdot \dfrac{y^3 + y + y^2z + z}{2y^2 + 2yz - yz^2 - z^3}$ $\dfrac{y^2 + z^2}{2y - z^2}$

93. $\dfrac{ax - ay + 3x - 3y}{x^3 + y^3} \div \dfrac{ab + 3b + ac + 3c}{xy - x^2 - y^2}$ $\dfrac{-(x - y)}{(x + y)(b + c)}$

94. $\dfrac{a^3 + b^3}{ac - ad - bc + bd} \div \dfrac{ab - a^2 - b^2}{ac - ad + bc - bd}$ $\dfrac{-(a + b)^2}{a - b}$ or $\dfrac{(a + b)^2}{b - a}$

95. $\dfrac{a^2b + b}{3a^2 - 4a - 20} \cdot \dfrac{a^2 + 5a}{2a^2 + 11a + 5} \div \dfrac{ab^2}{6a^2 - 17a - 10}$ $\dfrac{a^2 + 1}{b(a + 2)}$

96. $\dfrac{a^2 - 8a + 15}{2a^3 - 10a^2} \cdot \dfrac{2a^2 + 3a}{3a^3 - 27a} \div \dfrac{14a + 21}{a^2 - 6a - 27}$ $\dfrac{a - 9}{42a^2}$

97. $\dfrac{a - b}{4c} \div \left(\dfrac{b - a}{c} \div \dfrac{a - b}{c^2} \right)$ $-\dfrac{a - b}{4c^2}$ or $\dfrac{b - a}{4c^2}$

98. $\left(\dfrac{a - b}{4c} \div \dfrac{b - a}{c} \right) \div \dfrac{a - b}{c^2}$ $\dfrac{-c^2}{4(a - b)}$ or $\dfrac{c^2}{4(b - a)}$

In Exercises 99–102, find $\dfrac{f(a + h) - f(a)}{h}$ *and simplify.*

99. $f(x) = 7x - 4$ 7

100. $f(x) = -3x + 5$ -3

101. $f(x) = x^2 - 5x + 3$ $2a + h - 5$

102. $f(x) = 3x^2 - 4x + 7$ $6a + 3h - 4$

In Exercises 103–104, let

$$f(x) = \dfrac{(x + 2)^2}{1 - 2x} \quad \text{and} \quad g(x) = \dfrac{x + 2}{2x - 1}.$$

103. Find $\left(\dfrac{f}{g} \right)(x)$ and the domain of $\dfrac{f}{g}$. $\left(\dfrac{f}{g} \right)(x) = -x - 2$; $\{x | x$ is a real number and $x \neq -2$ and $x \neq \frac{1}{2}\}$ or $(-\infty, -2) \cup \left(-2, \frac{1}{2}\right) \cup \left(\frac{1}{2}, \infty\right)$

104. Find $\left(\dfrac{g}{f} \right)(x)$ and the domain of $\dfrac{g}{f}$. $\left(\dfrac{g}{f} \right)(x) = -\dfrac{1}{x + 2}$ or $\dfrac{1}{-x - 2}$; $\{x | x$ is a real number and $x \neq -2$ and $x \neq \frac{1}{2}\}$ or $(-\infty, -2) \cup \left(-2, \frac{1}{2}\right) \cup \left(\frac{1}{2}, \infty\right)$

Application Exercises

The rational function

$$f(x) = \dfrac{130x}{100 - x}$$

models the cost, f(x), in millions of dollars, to inoculate x% of the population against a particular strain of flu. The graph of the rational function is shown. Use the function's equation to solve Exercises 105–108.

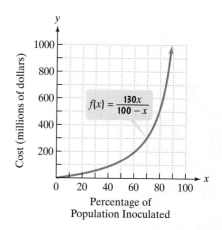

105. Find and interpret $f(60)$. Identify your solution as a point on the graph. 195; The cost to inoculate 60% of the population is $195 million; (60, 195).

106. Find and interpret $f(80)$. Identify your solution as a point on the graph. 520; The cost to inoculate 80% of the population is $520 million; (80, 520).

107. What value of x must be excluded from the rational function's domain? Does the cost model indicate that we can inoculate all of the population against the flu? Explain. 100; This indicates that we cannot inoculate 100% of the population.

108. What happens to the cost as x approaches 100%? How is this shown by the graph? Explain what this means. As x approaches 100%, the cost is increasing. Answers will vary.

The rational function

$$P(x) = \dfrac{72,900}{100x^2 + 729}$$

models the percentage of people in the United States, $P(x)$, with x years of education who are unemployed. The graph of the rational function is shown. Use the function's equation or graph, as specified, to solve Exercises 109–112.

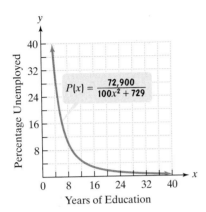

$$P(x) = \frac{72,900}{100x^2 + 729}$$

Source: Bureau of Labor Statistics

109. Use the function's equation to find and interpret $P(10)$. Round to the nearest whole percent. Identify your solution as a point on the graph. 7; About 7% of people in the United States with 10 years of education are unemployed; $(10, 7)$.

110. Use the function's equation to find and interpret $P(20)$. Round to the nearest whole percent. Identify your solution as a point on the graph. 2; About 2% of people in the United States with 20 years of education are unemployed; $(20, 2)$.

111. Describe the end behavior at the far right of the graph. Is there an education level that leads to guaranteed employment? If not, how is this indicated by the horizontal asymptote, the x-axis? The function values are approaching 0; no; The function values never actually reach 0.

112. Describe the behavior of the graph as x approaches 0. What does this mean in terms of the percentage unemployed with decreasing levels of education? The function values are increasing as x approaches 0. The unemployment rate approaches 100% as x approaches 0.

Among all deaths from a particular disease, the percentage that are smoking related (21–39 cigarettes per day) is a function of the disease's **incidence ratio**. *The incidence ratio describes the number of times more likely smokers are than nonsmokers to die from the disease. The following table shows the incidence ratios for heart disease and lung cancer for two age groups.*

Incidence Ratios

	Heart Disease	Lung Cancer
Ages 55–64	1.9	10
Ages 65–74	1.7	9

Source: Alexander M. Walker, *Observations and Inference,* Epidemiology Resources Inc., 1991.

For example, the incidence ratio of 9 in the table means that smokers between the ages of 65 and 74 are 9 times more likely than nonsmokers in the same age group to die from lung cancer. The rational function

$$P(x) = \frac{100(x - 1)}{x}$$

models the percentage of smoking-related deaths among all deaths from a disease, $P(x)$, in terms of the disease's incidence ratio, x. The graph of the rational function is shown. Use this function to solve Exercises 113–116.

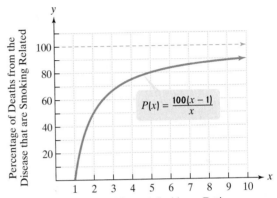

$$P(x) = \frac{100(x - 1)}{x}$$

The Disease's Incidence Ratio:
The number of times more likely smokers are than nonsmokers to die from the disease

113. Find $P(10)$. Describe what this means in terms of the incidence ratio, 10, given in the table. Identify your solution as a point on the graph. 90; Lung cancer has an incidence ratio of 10 between the ages of 55 and 64; 90% of deaths from lung cancer in this group are smoking related; $(10, 90)$.

114. Find $P(9)$. Round to the nearest percent. Describe what this means in terms of the incidence ratio, 9, given in the table. Identify your solution as a point on the graph. 89; Lung cancer has a incidence ratio of 9 between the ages of 65 and 74; 89% of deaths from lung cancer in this group are smoking related; $(9, 89)$.

115. What is the horizontal asymptote of the graph? Describe what this means about the percentage of deaths caused by smoking with increasing incidence ratios. The horizontal asymptote of the graph is $y = 100$; As the incidence ratio increases, the percentage of smoking-related deaths is approaching 100%.

116. According to the model and its graph, is there a disease for which all deaths are caused by smoking? Explain your answer. no; The percentage of smoking-related deaths for a particular disease will approach but never actually reach 100%.

Writing in Mathematics

117. What is a rational expression? Give an example with your explanation. **117–128.** Answers will vary.

118. What is a rational function? Provide an example.

119. What is the domain of a rational function?

120. If you are given the equation of a rational function, explain how to determine its domain.

121. Describe two ways the graph of a rational function differs from the graph of a polynomial function.

122. What is a vertical asymptote?

123. What is a horizontal asymptote?

124. Explain how to simplify a rational expression.

125. Explain how to simplify a rational expression with opposite factors in the numerator and the denominator.

126. Explain how to multiply rational expressions.

127. Explain how to divide rational expressions.

128. Although your friend has a family history of heart disease, he smokes, on average, 25 cigarettes per day. He sees the table showing incidence ratios for heart disease (see Exercises 113–116) and feels comfortable that they are less than 2, compared to 9 and 10 for lung cancer. He claims that all family deaths have been from heart disease and decides not to give up smoking. Use the given function and its graph to describe some additional information not given in the table that might influence his decision.

Technology Exercises

In Exercises 129–132, determine if the multiplication or division has been performed correctly by graphing the function on each side of the equation in the same viewing rectangle. If the graphs do not coincide, correct the expression on the right side and then verify your correction using the graphing utility. 129–132. See graphing answer section.

129. $\dfrac{x^2 + x}{3x} \cdot \dfrac{6x}{x + 1} = 2x$ Graphs coincide; multiplied correctly.

130. $\dfrac{x^3 - 25x}{x^2 - 3x - 10} \cdot \dfrac{x + 2}{x} = x + 5$ Graphs coincide; multiplied correctly.

131. $\dfrac{x^2 - 9}{x + 4} \div \dfrac{x - 3}{x + 4} = x - 3$ Graphs do not coincide; Right side should be $x + 3$.

132. $(x - 5) \div \dfrac{2x^2 - 11x + 5}{4x^2 - 1} = 2x - 1$ Graphs do not coincide; Right side should be $2x + 1$.

133. Use the $\boxed{\text{TABLE}}$ feature of a graphing utility to verify the domains that you determined for any two functions in Exercises 7–16.   Answers will vary.

134. a. Graph $f(x) = \dfrac{x^2 - x - 2}{x - 2}$ and $g(x) = x + 1$ in the same viewing rectangle. What do you observe?   See graphing answer section. The graphs coincide.

b. Simplify the formula in the definition of function f. Do f and g represent exactly the same function? Explain.
$\dfrac{x^2 - x - 2}{x - 2} = x + 1$; no; $g(2) = 3$, but $f(2)$ is undefined.

c. Display the graphs of f and g separately in the utility's viewing rectangle. $\boxed{\text{TRACE}}$ along each of the curves until you get to $x = 2$. What difference do you observe? What does this mean?   When tracing the graph of f, for $x = 2$, no y-value is given; however, on the graph of g, for $x = 2$, the y-value 3 is displayed; f and g are not equal when $x = 2$.

135. The rational function

$$f(x) = \frac{27{,}725(x - 14)}{x^2 + 9} - 5x$$

models the number of arrests, $f(x)$, per 100,000 drivers, for driving under the influence of alcohol, as a function of a driver's age, x.

a. Graph the function in a $[0, 70, 5]$ by $[0, 400, 20]$ viewing rectangle.   See graphing answer section.

b. Describe the trend shown by the graph.   As x increases, y increases and then decreases.

c. Use the $\boxed{\text{ZOOM}}$ and $\boxed{\text{TRACE}}$ features or the maximum function feature of your graphing utility to find the age that corresponds to the greatest number of arrests. How many arrests, per 100,000 drivers, are there for this age group?   25 years old; about 356 arrests per 100,000 drivers

Critical Thinking Exercises

136. Which one of the following is true?   d

a. $\dfrac{x^2 - 25}{x - 5} = x - 5$

b. $\dfrac{x^2 + 7}{7} = x^2 + 1$

c. The domain of $f(x) = \dfrac{7}{x(x - 3) + 5(x - 3)}$ is $(-\infty, 3) \cup (3, \infty)$.

d. None of the above is true.

137. Graph: $f(x) = \dfrac{x^2 - x - 2}{x - 2}$.   See graphing answer section.

138. Simplify: $\dfrac{6x^{3n} + 6x^{2n}y^n}{x^{2n} - y^{2n}} \cdot \dfrac{6x^{2n}}{x^n - y^n}$.

139. Divide: $\dfrac{y^{2n} - 1}{y^{2n} + 3y^n + 2} \div \dfrac{y^{2n} + y^n - 12}{y^{2n} - y^n - 6} \cdot \dfrac{y^n - 1}{y^n + 4}$.

140. Solve for x in terms of a and write the resulting rational expression in simplified form. Include any necessary restrictions on a.

$$a^2(x - 1) = 4(x - 1) + 5a + 10$$

$$x = \frac{a + 3}{a - 2}, a \neq -2, a \neq 2$$

Review Exercises

141. Graph: $4x - 5y \geq 20$. (Section 4.4, Example 1) See graphing answer section.

142. Multiply: $(2x - 5)(x^2 - 3x - 6)$. (Section 5.2, Example 3) $2x^3 - 11x^2 + 3x + 30$

143. Simplify: $\left(\dfrac{ab^{-3}c^{-4}}{4a^5b^{10}c^{-3}}\right)^{-2}$. (Section 1.6, Example 9) $16a^8b^{26}c^2$

SECTION

6.2

ADDING AND SUBTRACTING RATIONAL EXPRESSIONS

Objectives

1 Add rational expressions with the same denominator.

2 Subtract rational expressions with the same denominator.

3 Find the least common denominator.

4 Add and subtract rational expressions with different denominators.

5 Add and subtract rational expressions with opposite denominators.

Did you know that people in California are at far greater risk of injury or death from drunk drivers than from earthquakes? According to the U.S. Bureau of Justice statistics, half the arrests for driving under the influence of alcohol involve drivers ages 25 through 34. The rational function

$$f(x) = \frac{27,725(x - 14)}{x^2 + 9} - 5x$$

models the number of arrests for driving under the influence, $f(x)$, per 100,000 drivers, as a function of a driver's age, x.

The formula for function f involves the subtraction of two expressions. It is possible to perform this subtraction and express the function's formula as a single rational expression. In this section, we will draw on your experience from arithmetic to add and subtract rational expressions.

1 Add rational expressions with the same denominator.

Addition and Subtraction when Denominators Are the Same To add rational numbers having the same denominators, such as $\frac{2}{9}$ and $\frac{5}{9}$, we add the numerators and place the sum over the common denominator:

$$\frac{2}{9} + \frac{5}{9} = \frac{2 + 5}{9} = \frac{7}{9}.$$

We add rational expressions with the same denominator in an identical manner.

ADDING RATIONAL EXPRESSIONS WITH COMMON DENOMINATORS

If $\dfrac{P}{R}$ and $\dfrac{Q}{R}$ are rational expressions, then

$$\frac{P}{R} + \frac{Q}{R} = \frac{P+Q}{R}.$$

To add rational expressions with the same denominator, add numerators and place the sum over the common denominator. If possible, simplify the result.

EXAMPLE 1 Adding Rational Expressions when Denominators Are the Same

Add: $\dfrac{x^2 + 2x - 2}{x^2 + 3x - 10} + \dfrac{5x + 12}{x^2 + 3x - 10}.$

SOLUTION

$$\frac{x^2 + 2x - 2}{x^2 + 3x - 10} + \frac{5x + 12}{x^2 + 3x - 10}$$

$$= \frac{x^2 + 2x - 2 + 5x + 12}{x^2 + 3x - 10}$$

Add numerators. Place this sum over the common denominator.

$$= \frac{x^2 + 7x + 10}{x^2 + 3x - 10}$$

Combine like terms: $2x + 5x = 7x$ and $-2 + 12 = 10$.

$$= \frac{(x + 2)(x + 5)}{(x - 2)(x + 5)}$$

Factor and simplify by dividing out the common factor, $x + 5$.

$$= \frac{x + 2}{x - 2}$$

✔ **CHECK POINT 1** Add: $\dfrac{x^2 - 5x - 15}{x^2 + 5x + 6} + \dfrac{2x + 5}{x^2 + 5x + 6}.$ $\dfrac{x-5}{x+3}$

The following box shows how to subtract rational expressions with the same denominator:

2 Subtract rational expressions with the same denominator.

SUBTRACTING RATIONAL EXPRESSIONS WITH COMMON DENOMINATORS

If $\dfrac{P}{R}$ and $\dfrac{Q}{R}$ are rational expressions, then

$$\frac{P}{R} - \frac{Q}{R} = \frac{P-Q}{R}.$$

To subtract rational expressions with the same denominator, subtract numerators and place the difference over the common denominator. If possible, simplify the result.

EXAMPLE 2 Subtracting Rational Expressions when Denominators Are the Same

Subtract: $\dfrac{3y^3 - 5x^3}{x^2 - y^2} - \dfrac{4y^3 - 6x^3}{x^2 - y^2}$.

SOLUTION

$$\dfrac{3y^3 - 5x^3}{x^2 - y^2} - \dfrac{4y^3 - 6x^3}{x^2 - y^2} = \dfrac{3y^3 - 5x^3 - (4y^3 - 6x^3)}{x^2 - y^2}$$

Subtract numerators and include parentheses to indicate that both terms are subtracted. Place this difference over the common denominator.

$$= \dfrac{3y^3 - 5x^3 - 4y^3 + 6x^3}{x^2 - y^2}$$

Remove parentheses and then distribute the minus to change the sign of each term. Combine like terms: $-5x^3 + 6x^3 = x^3$ and $3y^3 - 4y^3 = -y^3$.

$$= \dfrac{x^3 - y^3}{x^2 - y^2}$$

$$= \dfrac{(x - y)(x^2 + xy + y^2)}{(x + y)(x - y)}$$

Factor and simplify by dividing out the common factor, $x - y$.

$$= \dfrac{x^2 + xy + y^2}{x + y}$$

STUDY TIP

When a numerator is being subtracted, be sure to **subtract every term in that expression**.

The − sign applies to the entire numerator, $4y^3 - 6x^3$.

Insert parentheses to indicate this.

The sign of every term of $4y^3 - 6x^3$ changes.

$$\dfrac{3y^3 - 5x^3}{x^2 - y^2} - \dfrac{4y^3 - 6x^3}{x^2 - y^2} = \dfrac{3y^3 - 5x^3 - (4y^3 - 6x^3)}{x^2 - y^2} = \dfrac{3y^3 - 5x^3 - 4y^3 + 6x^3}{x^2 - y^2}$$

The entire numerator of the second rational expression must be subtracted. Avoid the common error of subtracting only the first term.

Incorrect!

$-6x^3$ must also be subtracted.

$$\dfrac{3y^3 - 5x^3}{x^2 - y^2} - \dfrac{4y^3 - 6x^3}{x^2 - y^2} \quad \dfrac{3y^3 - 5x^3 - 4y^3 - 6x^3}{x^2 - y^2}$$

✔ **CHECK POINT 2** Subtract: $\dfrac{5x - y}{x^2 - y^2} - \dfrac{4x - 2y}{x^2 - y^2}$. $\dfrac{1}{x - y}$

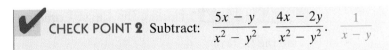

3 Find the least common denominator.

Finding the Least Common Denominator We can gain insight into adding rational expressions with different denominators by looking closely at what we do when adding fractions with different denominators. For example, suppose that we want to add $\frac{1}{2}$ and $\frac{2}{3}$. We must first write the fractions with the same denominator. We look for the smallest number that contains both 2 and 3 as factors. This number, 6, is then used as the *least common denominator*, or LCD.

The **least common denominator** of several rational expressions is a polynomial consisting of the product of all prime factors in the denominators, with each factor raised to the greatest power of its occurrence in any denominator.

FINDING THE LEAST COMMON DENOMINATOR

1. Factor each denominator completely.
2. List the factors of the first denominator.
3. Add to the list in step 2 any factors of the second denominator that do not appear in the list.
4. Form the product of each different factor from the list in step 3. This product is the least common denominator.

EXAMPLE 3 Finding the Least Common Denominator

Find the LCD of

$$\frac{3}{10x^2} \quad \text{and} \quad \frac{7}{15x}.$$

SOLUTION

Step 1. **Factor each denominator completely.**

$$10x^2 = 5 \cdot 2x^2 \quad (\text{or } 5 \cdot 2 \cdot x \cdot x)$$
$$15x = 5 \cdot 3x$$

Step 2. **List the factors of the first denominator.**

$$5, 2, x^2 \quad (\text{or } 5, 2, x, x)$$

Step 3. **Add any unlisted factors from the second denominator.** Two factors from $5 \cdot 3x$ are already in our list. These factors include 5 and x. We add the unlisted factor, 3, to our list. We have

$$3, 5, 2, x^2.$$

Step 4. **The least common denominator is the product of all factors in the final list.** Thus,

$$3 \cdot 5 \cdot 2 \cdot x^2$$

or $30x^2$ is the least common denominator.   ∎

STUDY TIP

After factoring denominators completely, the LCD can be determined by taking each factor to the highest power it appears in any factorization. The *Mathematics Teacher* magazine accused the LCD of keeping up with the Joneses. The LCD wants everything (all of the factors) the other denominators have.

✔ **CHECK POINT 3** Find the LCD of

$$\frac{7}{6x^2} \quad \text{and} \quad \frac{2}{9x}.$$

EXAMPLE 4 Finding the Least Common Denominator

Find the LCD of

$$\frac{9}{7x^2 + 28x} \quad \text{and} \quad \frac{11}{x^2 + 8x + 16}.$$

SOLUTION

Step 1. **Factor each denominator completely.**

$$7x^2 + 28x = 7x(x + 4)$$
$$x^2 + 8x + 16 = (x + 4)^2$$

Step 2. **List the factors of the first denominator.**

$$7, x, (x + 4)$$

Step 3. **Add any unlisted factors from the second denominator.** The second denominator is $(x + 4)^2$, or $(x + 4)(x + 4)$. One factor of $x + 4$ is already in our list, but the other factor is not. We add a second factor of $x + 4$ to the list. We have

$$7, x, (x + 4), (x + 4).$$

Step 4. **The least common denominator is the product of all factors in the final list.** Thus,

$$7x(x + 4)(x + 4), \text{ or } 7x(x + 4)^2,$$

is the least common denominator.   ■

✔ CHECK POINT 4 Find the LCD of

$$\frac{7}{5x^2 + 15x} \text{ and } \frac{9}{x^2 + 6x + 9}. \qquad 5x(x + 3)(x + 3) \text{ or } 5x(x + 3)^2$$

---

4 Add and subtract rational expressions with different denominators.

Addition and Subtraction When Denominators Are Different Finding the least common denominator for two (or more) rational expressions is the first step needed to add or subtract the expressions. For example, to add $\frac{1}{2}$ and $\frac{2}{3}$, we first determine that the LCD is 6. Then we write each fraction in terms of the LCD.

$$\frac{1}{2} + \frac{2}{3} = \frac{1}{2} \cdot \frac{3}{3} + \frac{2}{3} \cdot \frac{2}{2}$$

Multiply the numerator and denominator of each fraction by whatever extra factors are required to form 6, the LCD.

$\frac{3}{3} = 1$ and $\frac{2}{2} = 1$. Multiplying by 1 does not change a fraction's value.

$$= \frac{3}{6} + \frac{4}{6}$$

$$= \frac{3 + 4}{6}$$

Add numerators. Place this sum over the LCD.

$$= \frac{7}{6}$$

We follow the same steps in adding or subtracting rational expressions with different denominators.

ADDING AND SUBTRACTING RATIONAL EXPRESSIONS THAT HAVE DIFFERENT DENOMINATORS

1. Find the LCD of the rational expressions.
2. Rewrite each rational expression as an equivalent expression whose denominator is the LCD. To do so, multiply the numerator and the denominator of each rational expression by any factor(s) needed to convert the denominator into the LCD.
3. Add or subtract numerators, placing the resulting expression over the LCD.
4. If possible, simplify the resulting rational expression.

EXAMPLE 5 Adding Rational Expressions with Different Denominators

Add: $\dfrac{3}{10x^2} + \dfrac{7}{15x}$.

SOLUTION

Step 1. **Find the least common denominator.** In Example 3, we found that the LCD for these rational expressions is $30x^2$.

Step 2. **Write equivalent expressions with the LCD as denominators.** We must rewrite each rational expression with a denominator of $30x^2$.

$$\frac{3}{10x^2} \cdot \frac{3}{3} = \frac{9}{30x^2} \qquad \frac{7}{15x} \cdot \frac{2x}{2x} = \frac{14x}{30x^2}$$

Multiply the numerator and denominator by 3 to get $30x^2$, the LCD.

Multiply the numerator and denominator by $2x$ to get $30x^2$, the LCD.

Because $\frac{3}{3} = 1$ and $\frac{2x}{2x} = 1$, we are not changing the value of either rational expression, only its appearance. In summary, we have

$$\frac{3}{10x^2} + \frac{7}{15x} \qquad \text{The LCD is } 30x^2.$$

$$= \frac{3}{10x^2} \cdot \frac{3}{3} + \frac{7}{15x} \cdot \frac{2x}{2x} \qquad \text{Write equivalent expressions with the LCD.}$$

$$= \frac{9}{30x^2} + \frac{14x}{30x^2}.$$

Steps 3 and 4. **Add numerators, putting this sum over the LCD. Simplify, if possible.**

$$= \frac{9 + 14x}{30x^2} \quad \text{or} \quad \frac{14x + 9}{30x^2}$$

The numerator is prime and further simplification is not possible. ∎

✔ **CHECK POINT 5** Add: $\dfrac{7}{6x^2} + \dfrac{2}{9x}$.

STUDY TIP

It is incorrect to add rational expressions by adding numerators and adding denominators. Avoid this common error.

EXAMPLE 6 Adding Rational Expressions with Different Denominators

Add: $\dfrac{x}{x - 3} + \dfrac{x - 1}{x + 3}$.

SOLUTION

Step 1. **Find the least common denominator.** Begin by factoring the denominators.

$$x - 3 = 1(x - 3)$$

$$x + 3 = 1(x + 3)$$

The factors of the first denominator are 1 and $x - 3$. The only factor from the second denominator that is unlisted is $x + 3$. Thus, the least common denominator is $1(x - 3)(x + 3)$, or $(x - 3)(x + 3)$.

Step 2. **Write equivalent expressions with the LCD as denominators.**

$$\frac{x}{x-3} + \frac{x-1}{x+3}$$

$$= \frac{x(x+3)}{(x-3)(x+3)} + \frac{(x-1)(x-3)}{(x-3)(x+3)}$$

Multiply each numerator and denominator by the extra factor required to form $(x-3)(x+3)$, the LCD.

Steps 3 and 4. **Add numerators, putting this sum over the LCD. Simplify, if possible.**

$$= \frac{x(x+3) + (x-1)(x-3)}{(x-3)(x+3)}$$

$$= \frac{x^2 + 3x + x^2 - 4x + 3}{(x-3)(x+3)}$$

Perform the multiplications using the distributive property and FOIL.

$$= \frac{2x^2 - x + 3}{(x-3)(x+3)}$$

Combine like terms: $x^2 + x^2 = 2x^2$ and $3x - 4x = -x$.

The numerator is prime and further simplification is not possible. ∎

✔ **CHECK POINT 6** Add: $\dfrac{x}{x-4} + \dfrac{x-2}{x+4}$. $\dfrac{2x^2 - 2x + 8}{(x-4)(x+4)}$ or $\dfrac{2(x^2 - x + 4)}{(x-4)(x+4)}$

EXAMPLE 7 Subtracting Rational Expressions with Different Denominators

Subtract: $\dfrac{x-1}{x^2 + x - 6} - \dfrac{x-2}{x^2 + 4x + 3}$.

SOLUTION

Step 1. **Find the least common denominator.** Begin by factoring the denominators.

$$x^2 + x - 6 = (x+3)(x-2)$$

$$x^2 + 4x + 3 = (x+3)(x+1)$$

The factors of the first denominator are $x+3$ and $x-2$. The only factor from the second denominator that is unlisted is $x+1$. Thus, the least common denominator is $(x+3)(x-2)(x+1)$.

Step 2. **Write equivalent expressions with the LCD as denominators.**

$$\frac{x-1}{x^2 + x - 6} - \frac{x-2}{x^2 + 4x + 3}$$

$$= \frac{x-1}{(x+3)(x-2)} - \frac{x-2}{(x+3)(x+1)}$$

Factor denominators. The LCD is $(x+3)(x-2)(x+1)$.

$$= \frac{(x-1)(x+1)}{(x+3)(x-2)(x+1)} - \frac{(x-2)(x-2)}{(x+3)(x-2)(x+1)}$$

Multiply each numerator and denominator by the extra factor required to form $(x+3)(x-2)(x+1)$, the LCD.

Steps 3 and 4. **Subtract numerators, putting this difference over the LCD. Simplify if possible.**

$$= \frac{(x-1)(x+1) - (x-2)(x-2)}{(x+3)(x-2)(x+1)}$$

$$= \frac{x^2 - 1 - (x^2 - 4x + 4)}{(x+3)(x-2)(x+1)}$$

Perform the multiplications in the numerator. Don't forget the parentheses.

$$= \frac{x^2 - 1 - x^2 + 4x - 4}{(x+3)(x-2)(x+1)}$$

Remove parentheses and change the sign of each term in parentheses.

$$= \frac{4x - 5}{(x+3)(x-2)(x+1)}$$

Combine like terms.

The numerator is prime and further simplification is not possible. ∎

✔ **CHECK POINT 7** Subtract: $\dfrac{2x-3}{x^2-5x+6} - \dfrac{x+4}{x^2-2x-3} \cdot \dfrac{x^2-3x+5}{(x-3)(x+1)(x-2)}$

EXAMPLE 8 Adding and Subtracting Rational Expressions with Different Denominators

Perform the indicated operations:

$$\frac{3y+2}{y-5} + \frac{4}{3y+4} - \frac{7y^2+24y+28}{3y^2-11y-20}.$$

SOLUTION

Step 1. **Find the least common denominator.** Begin by factoring the denominators.

$$y - 5 = 1(y - 5)$$

$$3y + 4 = 1(3y + 4)$$

$$3y^2 - 11y - 20 = (3y + 4)(y - 5)$$

The factors of the first denominator are 1 and $y - 5$. The only factor from the second denominator that is unlisted is $3y + 4$. Adding this factor to our list, we have 1, $y - 5$, and $3y + 4$. We have listed all factors from the third denominator. Thus, the least common denominator is $1(y - 5)(3y + 4)$, or $(y - 5)(3y + 4)$.

Step 2. **Write equivalent expressions with the LCD as denominators.**

$$\frac{3y+2}{y-5} + \frac{4}{3y+4} - \frac{7y^2+24y+28}{3y^2-11y-20}$$

$$= \frac{3y+2}{y-5} + \frac{4}{3y+4} - \frac{7y^2+24y+28}{(3y+4)(y-5)}$$

Factor denominators. The LCD is $(y-5)(3y+4)$.

$$= \frac{(3y+2)(3y+4)}{(y-5)(3y+4)} + \frac{4(y-5)}{(y-5)(3y+4)} - \frac{7y^2+24y+28}{(3y+4)(y-5)}$$

Multiply the first two numerators and denominators by the extra factor required to form the LCD.

Step 3. **Add and subtract numerators, putting this result over the LCD. Simplify if possible.**

$$= \frac{(3y+2)(3y+4) + 4(y-5) - (7y^2 + 24y + 28)}{(y-5)(3y+4)}$$

$$= \frac{9y^2 + 18y + 8 + 4y - 20 - 7y^2 - 24y - 28}{(y-5)(3y+4)}$$

Perform multiplications. Remove parentheses and change the sign of each term.

$$= \frac{2y^2 - 2y - 40}{(y-5)(3y+4)}$$

Combine like terms:
$9y^2 - 7y^2 = 2y^2, 18y + 4y - 24y = -2y,$
and $8 - 20 - 28 = -40.$

$$= \frac{2(y^2 - y - 20)}{(y-5)(3y+4)}$$

Factor out the GCF in the numerator.

$$= \frac{2(y+4)\cancel{(y-5)}}{\cancel{(y-5)}(3y+4)}$$

Factor completely and simplify.

$$= \frac{2(y+4)}{3y+4}$$

∎

CHECK POINT 8 Perform the indicated operations:

$$\frac{y-1}{y-2} + \frac{y-6}{y^2-4} - \frac{y+1}{y+2} \cdot \frac{3}{y+2}$$

5 Add and subtract rational expressions with opposite denominators.

In some situations, we need to add or subtract rational expressions with denominators that are opposites, or additive inverses. Multiply the numerator and the denominator of either of the rational expressions by -1. Then they will have the same denominators.

EXAMPLE 9 Adding Rational Expressions when Denominators Are Opposites

Add: $\dfrac{4x - 16y}{x - 5y} + \dfrac{x - 6y}{5y - x}.$

SOLUTION

$$\frac{4x - 16y}{x - 5y} + \frac{x - 6y}{5y - x}$$

The denominators, $x - 5y$ and $5y - x$, are opposites, or additive inverses.

$$= \frac{4x - 16y}{x - 5y} + \frac{(-1)}{(-1)} \cdot \frac{x - 6y}{5y - x}$$

Multiply the numerator and denominator of the second rational expression by -1.

$$= \frac{4x - 16y}{x - 5y} + \frac{-x + 6y}{-5y + x}$$

Perform the multiplications by -1.

$$= \frac{4x - 16y}{x - 5y} + \frac{-x + 6y}{x - 5y}$$

Rewrite $-5y + x$ as $x - 5y$. Both rational expressions have the same denominator.

$$= \frac{4x - 16y - x + 6y}{x - 5y}$$

Add numerators. Place this sum over the common denominator.

$$= \frac{3x - 10y}{x - 5y}$$

Combine like terms.

■

CHECK POINT 9 Add: $\dfrac{4x - 7y}{x - 3y} + \dfrac{x - 2y}{3y - x}.$ $\dfrac{3x - 5y}{x - 3y}$

6.2 EXERCISE SET

 Student Solutions Manual   CD/Video   PH Math/Tutor Center MathXL Tutorials on CD   Math XL MathXL® MyMathLab   MyMathLab Interactmath.com

Practice Exercises

In Exercises 1–16, perform the indicated operations. These exercises involve addition and subtraction when denominators are the same. Simplify the result, if possible.

1. $\dfrac{2}{9x} + \dfrac{4}{9x}$ $\dfrac{2}{3x}$

2. $\dfrac{11}{6x} + \dfrac{4}{6x}$ $\dfrac{5}{2x}$

3. $\dfrac{x}{x - 5} + \dfrac{9x + 3}{x - 5}$ $\dfrac{10x + 3}{x - 5}$

4. $\dfrac{x}{x - 3} + \dfrac{11x + 5}{x - 3}$ $\dfrac{12x + 5}{x - 3}$

5. $\dfrac{x^2 - 2x}{x^2 + 3x} + \dfrac{x^2 + x}{x^2 + 3x}$ $\dfrac{2x - 1}{x + 3}$

6. $\dfrac{x^2 + 7x}{x^2 - 5x} + \dfrac{x^2 - 4x}{x^2 - 5x}$ $\dfrac{2x + 3}{x - 5}$

7. $\dfrac{y^2}{y^2 - 9} + \dfrac{9 - 6y}{y^2 - 9}$ $\dfrac{y - 3}{y + 3}$

8. $\dfrac{y^2}{y^2 - 25} + \dfrac{25 - 10y}{y^2 - 25}$ $\dfrac{y - 5}{y + 5}$

9. $\dfrac{3x}{4x - 3} - \dfrac{2x - 1}{4x - 3}$ $\dfrac{x + 1}{4x - 3}$

10. $\dfrac{3x}{7x - 4} - \dfrac{2x - 1}{7x - 4}$ $\dfrac{x + 1}{7x - 4}$

11. $\dfrac{x^2 - 2}{x^2 + 6x - 7} - \dfrac{19 - 4x}{x^2 + 6x - 7}$ $\dfrac{x - 3}{x - 1}$

12. $\dfrac{x^2 + 6x + 2}{x^2 + x - 6} - \dfrac{2x - 1}{x^2 + x - 6}$ $\dfrac{x + 1}{x - 2}$

13. $\dfrac{20y^2 + 5y + 1}{6y^2 + y - 2} - \dfrac{8y^2 - 12y - 5}{6y^2 + y - 2}$ $\dfrac{4y + 3}{2y - 1}$

14. $\dfrac{y^2 + 3y - 6}{y^2 - 5y + 4} - \dfrac{4y - 4 - 2y^2}{y^2 - 5y + 4}$ $\dfrac{3y + 2}{y - 4}$

15. $\dfrac{2x^3 - 3y^3}{x^2 - y^2} - \dfrac{x^3 - 2y^3}{x^2 - y^2}$ $\dfrac{x^2 + xy + y^2}{x + y}$

16. $\dfrac{4y^3 - 3x^3}{y^2 - x^2} - \dfrac{3y^3 - 2x^3}{y^2 - x^2}$ $\dfrac{y^2 + xy + x^2}{y + x}$

In Exercises 17–28, find the least common denominator of the rational expressions.

17. $\dfrac{11}{25x^2}$ and $\dfrac{14}{35x}$ $175x^2$

18. $\dfrac{7}{15x^2}$ and $\dfrac{9}{24x}$ $120x^2$

19. $\dfrac{2}{x - 5}$ and $\dfrac{3}{x^2 - 25}$ $(x - 5)(x + 5)$

20. $\dfrac{2}{x + 3}$ and $\dfrac{5}{x^2 - 9}$ $(x + 3)(x - 3)$

21. $\dfrac{7}{y^2 - 100}$ and $\dfrac{13}{y(y - 10)}$ $y(y + 10)(y - 10)$

22. $\dfrac{7}{y^2 - 4}$ and $\dfrac{15}{y(y + 2)}$ $y(y + 2)(y - 2)$

23. $\dfrac{8}{x^2 - 16}$ and $\dfrac{x}{x^2 - 8x + 16}$ $(x + 4)(x - 4)(x - 4)$

24. $\dfrac{3}{x^2 - 25}$ and $\dfrac{x}{x^2 - 10x + 25}$ $(x + 5)(x - 5)(x - 5)$

25. $\dfrac{7}{y^2 - 5y - 6}$ and $\dfrac{y}{y^2 - 4y - 5}$ $(y - 5)(y - 6)(y + 1)$

26. $\dfrac{3}{y^2 - y - 20}$ and $\dfrac{y}{2y^2 + 7y - 4}$ $(y - 5)(y + 4)(2y - 1)$

27. $\dfrac{7y}{2y^2 + 7y + 6}, \dfrac{3}{y^2 - 4}$, and $\dfrac{-7y}{2y^2 - 3y - 2}$
$(2y + 3)(y + 2)(y - 2)(2y + 1)$

28. $\dfrac{5y}{y^2 - 9}, \dfrac{8}{y^2 + 6y + 9}$, and $\dfrac{-5y}{2y^2 + 5y - 3}$
$(y + 3)^2(y - 3)(2y - 1)$

In Exercises 29–66, perform the indicated operations. These exercises involve addition and subtraction when denominators are different. Simplify the result, if possible.

29. $\dfrac{3}{5x^2} + \dfrac{10}{x}$ $\dfrac{3 + 50x}{5x^2}$

30. $\dfrac{7}{2x^2} + \dfrac{4}{x}$ $\dfrac{7 + 8x}{2x^2}$

31. $\dfrac{4}{x - 2} + \dfrac{3}{x + 1}$ $\dfrac{7x - 2}{(x - 2)(x + 1)}$

32. $\dfrac{2}{x-3} + \dfrac{7}{x+2}$ $\dfrac{9x-17}{(x-3)(x+2)}$

33. $\dfrac{3x}{x^2+x-2} + \dfrac{2}{x^2-4x+3}$ $\dfrac{3x-4}{(x+2)(x-3)}$

34. $\dfrac{7x}{x^2+2x-8} + \dfrac{3}{x^2-3x+2}$ $\dfrac{7x^2-4x+12}{(x+4)(x-2)(x-1)}$

35. $\dfrac{x-6}{x+5} + \dfrac{x+5}{x-6}$ $\dfrac{2x^2-2x+61}{(x+5)(x-6)}$

36. $\dfrac{x-2}{x+7} + \dfrac{x+7}{x-2}$ $\dfrac{2x^2+10x+53}{(x+7)(x-2)}$

37. $\dfrac{3x}{x^2-25} - \dfrac{4}{x+5}$ $-\dfrac{x-20}{x^2-25}$ or $\dfrac{-x+20}{x^2-25}$

38. $\dfrac{8x}{x^2-16} - \dfrac{5}{x+4}$ $\dfrac{3x+20}{x^2-16}$

39. $\dfrac{3y+7}{y^2-5y+6} - \dfrac{3}{y-3}$ $\dfrac{13}{(y-3)(y-2)}$

40. $\dfrac{2y+9}{y^2-7y+12} - \dfrac{2}{y-3}$ $\dfrac{17}{(y-3)(y-4)}$

41. $\dfrac{x^2-6}{x^2+9x+18} - \dfrac{x-4}{x+6}$ $\dfrac{1}{x+3}$

42. $\dfrac{x^2-39}{x^2+3x-10} - \dfrac{x-7}{x-2}$ $\dfrac{2}{x+5}$

43. $\dfrac{4x+1}{x^2+7x+12} + \dfrac{2x+3}{x^2+5x+4}$ $\dfrac{6x^2+14x+10}{(x+1)(x+4)(x+3)}$

44. $\dfrac{3x-2}{x^2-x-6} + \dfrac{4x-3}{x^2-9}$ $\dfrac{7x^2+12x-12}{(x+3)(x-3)(x+2)}$

45. $\dfrac{x+4}{x^2-x-2} - \dfrac{2x+3}{x^2+2x-8}$ $\dfrac{x^2-3x-13}{(x+4)(x-2)(x+1)}$

46. $\dfrac{2x+1}{x^2-7x+6} - \dfrac{x+3}{x^2-5x-6}$ $\dfrac{x^2+x+4}{(x-6)(x+1)(x-1)}$

47. $4 + \dfrac{1}{x-3}$ $\dfrac{4x-11}{x-3}$

48. $7 + \dfrac{1}{x-5}$ $\dfrac{7x-34}{x-5}$

49. $\dfrac{y-7}{y^2-16} + \dfrac{7-y}{16-y^2}$ $\dfrac{2y-14}{(y+4)(y-4)}$

50. $\dfrac{y-3}{y^2-25} + \dfrac{y-3}{25-y^2}$ 0

51. $\dfrac{x+7}{3x+6} + \dfrac{x}{4-x^2}$ $\dfrac{x^2+2x-14}{3(x+2)(x-2)}$

52. $\dfrac{x+5}{4x+12} + \dfrac{x}{9-x^2}$ $\dfrac{x-5}{4(x-3)}$

53. $\dfrac{2x}{x-4} + \dfrac{64}{x^2-16} - \dfrac{2x}{x+4}$ $\dfrac{16}{x-4}$

54. $\dfrac{x}{x-3} + \dfrac{x+2}{x^2-2x-3} - \dfrac{4}{x+1}$ $\dfrac{x^2-2x+14}{(x+1)(x-3)}$

55. $\dfrac{5x}{x^2-y^2} - \dfrac{7}{y-x}$ $\dfrac{12x+7y}{x^2-y^2}$

56. $\dfrac{9x}{x^2-y^2} - \dfrac{10}{y-x}$ $\dfrac{19x+10y}{x^2-y^2}$

57. $\dfrac{3}{5x+6} - \dfrac{4}{x-2} + \dfrac{x^2-x}{5x^2-4x-12}$ $\dfrac{x^2-18x-30}{(5x+6)(x-2)}$

58. $\dfrac{x-1}{x^2+2x+1} - \dfrac{3}{2x-2} + \dfrac{x}{x^2-1}$ $\dfrac{x^2-8x-1}{2(x+1)(x+1)(x-1)}$

59. $\dfrac{3x-y}{x^2-9xy+20y^2} + \dfrac{2y}{x^2-25y^2}$ $\dfrac{3x^2+16xy-13y^2}{(x+5y)(x-5y)(x-4y)}$

60. $\dfrac{x+2y}{x^2+4xy+4y^2} - \dfrac{2x}{x^2-4y^2}$ $-\dfrac{1}{x-2y}$

61. $\dfrac{3x}{x^2-4} + \dfrac{5x}{x^2+x-2} - \dfrac{3}{x^2-4x+4}$

$\dfrac{8x^3-32x^2+23x+6}{(x+2)(x-2)(x-2)(x-1)}$

62. $\dfrac{1}{x} + \dfrac{4}{x^2-4} - \dfrac{2}{x^2-2x}$ $\dfrac{x+4}{x(x+2)}$

63. $\dfrac{6a+5b}{6a^2+5ab-4b^2} - \dfrac{a+2b}{9a^2-16b^2}$ $\dfrac{16a^2-12ab-18b^2}{(3a+4b)(3a-4b)(2a-b)}$

64. $\dfrac{5a-b}{a^2+ab-2b^2} - \dfrac{3a+2b}{a^2+5ab-6b^2}$ $\dfrac{2a^2+21ab-10b^2}{(a+6b)(a-b)(a+2b)}$

65. $\dfrac{1}{m^2+m-2} - \dfrac{3}{2m^2+3m-2} + \dfrac{2}{2m^2-3m+1}$

$\dfrac{m+6}{(m+2)(m-1)(2m-1)}$

66. $\dfrac{5}{2m^2-5m-3} + \dfrac{3}{2m^2+5m+2} - \dfrac{1}{m^2-m-6}$

$\dfrac{6m}{(2m+1)(m-3)(m+2)}$

Practice Plus

In Exercises 67–74, perform the indicated operations. Simplify the result, if possible.

67. $\left(\dfrac{2x+3}{x+1} \cdot \dfrac{x^2+4x-5}{2x^2+x-3}\right) - \dfrac{2}{x+2}$ $\dfrac{x^2+5x+8}{(x+2)(x+1)}$ ·

68. $\dfrac{1}{x^2-2x-8} \div \left(\dfrac{1}{x-4} - \dfrac{1}{x+2}\right)$ $\dfrac{1}{6}$

69. $\left(2 - \dfrac{6}{x+1}\right)\left(1 + \dfrac{3}{x-2}\right)$ 2

70. $\left(4 - \dfrac{3}{x+2}\right)\left(1 + \dfrac{5}{x-1}\right)$ $\dfrac{(4x+5)(x+4)}{(x+2)(x-1)}$

71. $\left(\dfrac{1}{x+h} - \dfrac{1}{x}\right) \div h$ $-\dfrac{1}{x(x+h)}$

72. $\left(\dfrac{5}{x-5} - \dfrac{2}{x+3}\right) \div (3x+25)$ $\dfrac{1}{(x-5)(x+3)}$

73. $\left(\dfrac{1}{a^3 - b^3} \cdot \dfrac{ac + ad - bc - bd}{1}\right) - \dfrac{c - d}{a^2 + ab + b^2} \quad \dfrac{2d}{a^2 + ab + b^2}$

74. $\dfrac{ab}{a^2 + ab + b^2} + \left(\dfrac{ac - ad - bc + bd}{ac - ad + bc - bd} \div \dfrac{a^3 - b^3}{a^3 + b^3}\right) \quad \dfrac{a^2 + b^2}{a^2 + ab + b^2}$

75. If $f(x) = \dfrac{2x - 3}{x + 5}$ and $g(x) = \dfrac{x^2 - 4x - 19}{x^2 + 8x + 15}$, find $(f - g)(x)$ and the domain of $f - g$. $(f - g)(x) = \dfrac{x + 2}{x + 3}$; $\{x \mid x$ is a real number and $x \neq -5$ and $x \neq -3\}$ or $(-\infty, -5) \cup (-5, -3) \cup (-3, \infty)$

76. If $f(x) = \dfrac{2x - 1}{x^2 + x - 6}$ and $g(x) = \dfrac{x + 2}{x^2 + 5x + 6}$, find $(f - g)(x)$ and the domain of $f - g$. $(f - g)(x) = \dfrac{x + 1}{(x + 3)(x - 2)}$; $\{x \mid x$ is a real number and $x \neq -3, x \neq -2,$ and $x \neq 2\}$ or $(-\infty, -3) \cup (-3, -2) \cup (-2, 2) \cup (2, \infty)$

Application Exercises

You plan to drive from Miami, Florida, to Atlanta, Georgia. Your trip involves approximately 470 miles of travel in Florida and 250 miles in Georgia. The speed limit is 70 miles per hour in Florida and 65 miles per hour in Georgia. If you average x miles per hour over these speed limits, the total driving time, T(x), in hours, is given by the function

$$T(x) = \dfrac{470}{x + 70} + \dfrac{250}{x + 65}.$$

The graph of T is shown in the figure. Use the function's equation to solve Exercises 77–82.

Driving Time from Miami to Atlanta

77. Find and interpret $T(0)$. Round to the nearest hour. Identify your solution as a point on the graph. 11; If you average 0 miles per hour over the speed limits, the total driving time is about 11 hours; (0, 11).

78. Find and interpret $T(5)$. Round to the nearest hour. Identify your solution as a point on the graph. 10; If you average 5 miles per hour over the speed limits, the total driving time is about 10 hours; (5, 10).

79. Find a simplified form of $T(x)$ by adding the rational expressions in the function's formula. Then use this form of the function to find $T(0)$. $\dfrac{720x + 48,050}{(x + 70)(x + 65)}$; $T(0) \approx 11$

80. Find a simplified form of $T(x)$ by adding the rational expressions in the function's formula. Then use this form of the function to find $T(5)$. $\dfrac{720x + 48,050}{(x + 70)(x + 65)}$; $T(5) \approx 10$

81. Use the graph to answer this question. If you want the driving time to be 9 hours, how much over the speed limits do you need to drive? Round to the nearest mile per hour. Does this seem like a realistic driving time for the trip? Explain your answer. You need to drive about 12 miles per hour over the speed limits. Answers will vary.

82. Use the graph to answer this question. If you want the driving time to be 8 hours, how much over the speed limits do you need to drive? Round to the nearest mile per hour. Does this seem like a realistic driving time for the trip? Explain your answer. You need to drive about 22 miles per hour over the speed limits. Answers will vary.

In Exercises 83–84, express the perimeter of each rectangle as a single rational expression.

83. $\dfrac{x}{x + 5}$; $\dfrac{x}{x + 4}$; $\dfrac{2x(2x + 9)}{(x + 4)(x + 5)}$

84. $\dfrac{x}{x + 8}$; $\dfrac{x}{x + 7}$; $\dfrac{2x(2x + 15)}{(x + 7)(x + 8)}$

85. In the section opener, we saw that the rational function

$$f(x) = \dfrac{27,725(x - 14)}{x^2 + 9} - 5x$$

models the number of arrests, $f(x)$, per 100,000 drivers, for driving under the influence of alcohol as a function of a driver's age, x. Perform the subtraction and express the function as a single rational expression. $\dfrac{-5x^3 + 27,680x - 388,150}{x^2 + 9}$

Writing in Mathematics

86. Explain how to add rational expressions when denominators are the same. Give an example with your explanation. 86–90. Answers will vary.

87. Explain how to subtract rational expressions when denominators are the same. Give an example with your explanation.

88. Explain how to find the least common denominator for denominators of $x^2 - 100$ and $x^2 - 20x + 100$.

89. Explain how to add rational expressions that have different denominators. Use $\dfrac{5}{x + 1} + \dfrac{3}{x + 4}$ in your explanation.

90. Explain how to add rational expressions when denominators are opposites. Use an example to support your explanation.

Explain the error in Exercises 91–92. Then rewrite the right side of the equation to correct the error that now exists.

91. $\dfrac{1}{a} + \dfrac{1}{b} = \dfrac{1}{a + b}$ Answers will vary; $\dfrac{b + a}{ab}$

92. $\dfrac{1}{x} + \dfrac{3}{7} = \dfrac{4}{x + 7}$ Answers will vary; $\dfrac{7 + 3x}{7x}$

Technology Exercises

93. Graph the given rational function and the function expressed as a single rational expression in Exercise 85 in a $[0, 70, 10]$ by $[0, 400, 50]$ viewing rectangle. The graphs should be identical. What trend do the graphs show in terms of the relationship between a driver's age and the number of arrests per 100,000 drivers for driving under the influence of alcohol?   See graphing answer section. As age increases, the number of arrests per 100,000 drivers increases until about age 25 and then decreases.

Critical Thinking Exercises

94. Which one of the following is true?   d

a. $\dfrac{a}{b} + \dfrac{a}{c} = \dfrac{a}{b + c}$

b. $6 + \dfrac{1}{x} = \dfrac{7}{x}$

c. $\dfrac{1}{x + 3} + \dfrac{x + 3}{2} = \dfrac{1}{\cancel{(x + 3)}} + \dfrac{\cancel{(x + 3)}}{2} = 1 + \dfrac{1}{2} = \dfrac{3}{2}$

d. None of the above is true.

In Exercises 95–97, perform the indicated operations.

95. $\dfrac{1}{x^n - 1} - \dfrac{1}{x^n + 1} - \dfrac{1}{x^{2n} - 1}$ $\dfrac{1}{x^{2n} - 1}$

96. $\left(1 - \dfrac{1}{x}\right)\left(1 - \dfrac{1}{x + 1}\right)\left(1 - \dfrac{1}{x + 2}\right)\left(1 - \dfrac{1}{x + 3}\right)$ $\dfrac{x - 1}{x + 3}$

97. $(x - y)^{-1} + (x - y)^{-2}$ $\dfrac{x - y + 1}{(x - y)(x - y)}$

Review Exercises

98. Simplify: $\left(\dfrac{3x^2y^{-2}}{y^3}\right)^{-2}$. (Section 1.6, Example 9) $\dfrac{y^{10}}{9x^4}$

99. Solve: $|3x - 1| \le 14$. (Section 4.3, Example 3) $\left\{x \mid -\dfrac{13}{3} \le x \le 5\right\}$ or $\left[-\dfrac{13}{3}, 5\right]$

100. Factor completely: $50x^3 - 18x$. (Section 5.6, Example 2) $2x(5x + 3)(5x - 3)$

SECTION

6.3

COMPLEX RATIONAL EXPRESSIONS

Objectives

1 Simplify complex rational expressions by multiplying by one.

2 Simplify complex rational expressions by dividing.

One area in finance of great interest to us ordinary folks when we buy a new car is the amount of each monthly payment. If P is the principal, or the amount borrowed, i is the monthly interest rate, and n is the number of monthly payments, then the amount, A, of each monthly payment is

$$A = \dfrac{Pi}{1 - \dfrac{1}{(1 + i)^n}}.$$

Do you notice anything unusual about the rational expression for the amount of each payment? It has a separate rational expression in its denominator.

Complex rational expressions, also called **complex fractions**, have numerators or denominators containing one or more rational expressions. Here is another example of such an expression:

In this section, we study two methods for simplifying complex rational expressions.

Simplifying Complex Rational Expressions by Multiplying by One One method for simplifying a complex rational expression is to find the least common denominator of all the rational expressions in its numerator and denominator. Then multiply each term in its numerator and denominator by this least common denominator. Because we are multiplying by a form of 1, we will obtain an equivalent expression that does not contain fractions in its numerator or denominator.

1 Simplify complex rational expressions by multiplying by one.

SIMPLIFYING A COMPLEX RATIONAL EXPRESSION BY MULTIPLYING BY ONE IN THE FORM $\dfrac{\text{LCD}}{\text{LCD}}$

1. Find the LCD of all rational expressions within the complex rational expression.
2. Multiply both the numerator and the denominator of the complex rational expression by this LCD.
3. Use the distributive property and multiply each term in the numerator and denominator by this LCD. Simplify. No fractional expressions should remain.
4. If possible, factor and simplify.

EXAMPLE 1 Simplifying a Complex Rational Expression

Simplify:

$$\frac{\dfrac{1}{x} + \dfrac{y}{x^2}}{\dfrac{1}{y} + \dfrac{x}{y^2}}.$$

SOLUTION The denominators in the complex rational expression are x, x^2, y, and y^2. The LCD is $x^2 y^2$. Multiply both the numerator and the denominator of the complex rational expression by $x^2 y^2$.

$$\frac{\dfrac{1}{x}+\dfrac{y}{x^2}}{\dfrac{1}{y}+\dfrac{x}{y^2}}=\frac{x^2y^2}{x^2y^2}\cdot\frac{\left(\dfrac{1}{x}+\dfrac{y}{x^2}\right)}{\left(\dfrac{1}{y}+\dfrac{x}{y^2}\right)}$$

$$=\frac{\boxed{x}^2y^2\cdot\dfrac{1}{\boxed{x}}+\boxed{x}^2y^2\cdot\dfrac{y}{\boxed{x}^2}}{x^2\boxed{y}^2\cdot\dfrac{1}{\boxed{y}}+x^2\boxed{y}^2\cdot\dfrac{x}{\boxed{y}^2}}$$

Use the distributive property.

> In all four rational expressions, we have divided numerators and denominators by common boxed factors.

$$=\frac{xy^2+y^3}{x^2y+x^3}$$

Simplify.

$$=\frac{y^2(x+y)}{x^2(y+x)}$$

Factor and simplify.

$$=\frac{y^2}{x^2}$$

■

✔ **CHECK POINT 1** Simplify:

$$\frac{\dfrac{x}{y}-1}{\dfrac{x^2}{y^2}-1}. \qquad \frac{y}{x+y}$$

Complex rational expressions are often written with negative exponents. For example,

$$\frac{x^{-1}+x^{-2}y}{y^{-1}+xy^{-2}}\quad\text{means}\quad\frac{\dfrac{1}{x}+\dfrac{y}{x^2}}{\dfrac{1}{y}+\dfrac{x}{y^2}}.$$

This is the expression that we simplified in Example 1. If an expression contains negative exponents, first rewrite it as an equivalent expression with positive exponents. Then simplify by multiplying the numerator and the denominator by the LCD.

EXAMPLE 2 Simplifying a Complex Rational Expression

Simplify:

$$\frac{\dfrac{1}{x+h}-\dfrac{1}{x}}{h}.$$

SOLUTION The denominators in the complex rational expression are $x + h$ and x. The LCD is $x(x + h)$. Multiply both the numerator and the denominator of the complex rational expression by $x(x + h)$.

$$\frac{\dfrac{1}{x + h} - \dfrac{1}{x}}{h} = \frac{x(x + h)}{x(x + h)} \cdot \frac{\left(\dfrac{1}{x + h} - \dfrac{1}{x}\right)}{h}$$

$$= \frac{x(x + h) \cdot \dfrac{1}{x + h} - x(x + h) \cdot \dfrac{1}{x}}{x(x + h)h}$$

Use the distributive property in the numerator.

$$= \frac{x - (x + h)}{x(x + h)h}$$

Simplify: $x \cancel{(x + h)} \cdot \dfrac{1}{\cancel{(x + h)}} = x$ and $\cancel{x}(x + h) \cdot \dfrac{1}{\cancel{x}} = x + h$.

$$= \frac{x - x - h}{x(x + h)h}$$

Remove parentheses and change the sign of each term.

$$= \frac{-h}{x(x + h)h}$$

Simplify.

$$= \frac{-\cancel{h}}{x(x + h)\cancel{h}}$$

Divide the numerator and the denominator by the common factor, h.

$$= -\frac{1}{x(x + h)}$$

∎

✔ **CHECK POINT 2** Simplify:

$$\frac{\dfrac{1}{x + 7} - \dfrac{1}{x}}{7} \cdot -\frac{1}{x(x + 7)}$$

---

2 Simplify complex rational expressions by dividing.

Simplifying Complex Rational Expressions by Dividing A second method for simplifying a complex rational expression is to combine its numerator into a single rational expression and combine its denominator into a single rational expression. Then perform the division by inverting the denominator and multiplying.

SIMPLIFYING A COMPLEX RATIONAL EXPRESSION BY DIVIDING

1. If necessary, add or subtract to get a single rational expression in the numerator.
2. If necessary, add or subtract to get a single rational expression in the denominator.
3. Perform the division indicated by the main fraction bar: Invert the denominator of the complex rational expression and multiply.
4. If possible, simplify.

EXAMPLE 3   Simplifying a Complex Rational Expression

Simplify:

$$\frac{\dfrac{x+1}{x}+\dfrac{x+1}{x-1}}{\dfrac{x+2}{x}-\dfrac{2}{x-1}}.$$

SOLUTION

Step 1.   Add to get a single rational expression in the numerator.

$$\frac{x+1}{x}+\frac{x+1}{x-1}$$

The LCD is $x(x-1)$.

$$=\frac{(x+1)(x-1)}{x(x-1)}+\frac{x(x+1)}{x(x-1)}=\frac{(x+1)(x-1)+x(x+1)}{x(x-1)}=\frac{x^2-1+x^2+x}{x(x-1)}=\frac{2x^2+x-1}{x(x-1)}$$

Step 2.   Subtract to get a single rational expression in the denominator.

$$\frac{x+2}{x}-\frac{2}{x-1}$$

The LCD is $x(x-1)$.

$$=\frac{(x+2)(x-1)}{x(x-1)}-\frac{2x}{x(x-1)}=\frac{(x+2)(x-1)-2x}{x(x-1)}=\frac{x^2+x-2-2x}{x(x-1)}=\frac{x^2-x-2}{x(x-1)}$$

Steps 3 and 4. Perform the division indicated by the main fraction bar: Invert and multiply. If possible, simplify.

$$\frac{\dfrac{x+1}{x}+\dfrac{x+1}{x-1}}{\dfrac{x+2}{x}-\dfrac{2}{x-1}}=\frac{\dfrac{2x^2+x-1}{x(x-1)}}{\dfrac{x^2-x-2}{x(x-1)}}$$

These are the single rational expressions from steps 1 and 2.

$$=\frac{2x^2+x-1}{x(x-1)}\cdot\frac{x(x-1)}{x^2-x-2}=\frac{(2x-1)(x+1)}{x(x-1)}\cdot\frac{x(x-1)}{(x-2)(x+1)}=\frac{2x-1}{x-2}$$

Invert and multiply.

✔ CHECK POINT 3 Simplify:

$$\frac{\dfrac{x+1}{x-1}-\dfrac{x-1}{x+1}}{\dfrac{x-1}{x+1}+\dfrac{x+1}{x-1}}.\quad\frac{2x}{x^2+1}$$

Which of the two methods do you prefer? Let's try them both in Example 4.

EXAMPLE 4 Simplifying a Complex Rational Expression: Comparing Methods

Simplify:

$$\frac{1 - x^{-2}}{1 - 4x^{-1} + 3x^{-2}}.$$

SOLUTION First rewrite the expression without negative exponents.

$$\frac{1 - x^{-2}}{1 - 4x^{-1} + 3x^{-2}} = \frac{1 - \dfrac{1}{x^2}}{1 - \dfrac{4}{x} + \dfrac{3}{x^2}}$$

Method 1   Multiplying by One

$$\frac{1 - \dfrac{1}{x^2}}{1 - \dfrac{4}{x} + \dfrac{3}{x^2}} = \frac{x^2}{x^2} \cdot \frac{\left(1 - \dfrac{1}{x^2}\right)}{\left(1 - \dfrac{4}{x} + \dfrac{3}{x^2}\right)}$$

Multiply the numerator and denominator by x^2, the LCD of all fractions.

$$= \frac{x^2 \cdot 1 - x^2 \cdot \dfrac{1}{x^2}}{x^2 \cdot 1 - x^2 \cdot \dfrac{4}{x} + x^2 \cdot \dfrac{3}{x^2}}$$

Apply the distributive property.

$$= \frac{x^2 - 1}{x^2 - 4x + 3}$$

Simplify.

$$= \frac{(x + 1)\cancel{(x - 1)}}{(x - 3)\cancel{(x - 1)}}$$

Factor and simplify.

$$= \frac{x + 1}{x - 3}$$

Method 2   Dividing

$$\frac{1 - \dfrac{1}{x^2}}{1 - \dfrac{4}{x} + \dfrac{3}{x^2}} = \frac{\dfrac{x^2}{x^2} - \dfrac{1}{x^2}}{\dfrac{x^2}{x^2} - \dfrac{4}{x} \cdot \dfrac{x}{x} + \dfrac{3}{x^2}}$$

Get a single rational expression in the numerator and in the denominator.

$$= \frac{\dfrac{x^2 - 1}{x^2}}{\dfrac{x^2 - 4x + 3}{x^2}}$$

$$= \frac{x^2 - 1}{x^2} \cdot \frac{x^2}{x^2 - 4x + 3}$$

Invert and multiply.

$$= \frac{(x + 1)\cancel{(x - 1)}}{\cancel{x^2}} \cdot \frac{\cancel{x^2}}{\cancel{(x - 1)}(x - 3)}$$

Factor and simplify.

$$= \frac{x + 1}{x - 3}$$

✔ **CHECK POINT 4** Simplify by the method of your choice:

$$\frac{1 - 4x^{-2}}{1 - 7x^{-1} + 10x^{-2}}. \quad \frac{x + 2}{x - 5}$$

6.3 EXERCISE SET

Practice Exercises

In Exercises 1–42, simplify each complex rational expression by the method of your choice.

1. $\dfrac{4 + \dfrac{2}{x}}{1 - \dfrac{3}{x}}$

2. $\dfrac{5 - \dfrac{2}{x}}{3 + \dfrac{1}{x}}$

3. $\dfrac{\dfrac{3}{x} + \dfrac{x}{3}}{\dfrac{x}{3} - \dfrac{3}{x}}$

4. $\dfrac{\dfrac{x}{5} - \dfrac{5}{x}}{\dfrac{1}{5} + \dfrac{1}{x}}$

5. $\dfrac{\dfrac{1}{x} + \dfrac{1}{y}}{\dfrac{1}{x} - \dfrac{1}{y}}$ $\dfrac{y + x}{y - x}$

6. $\dfrac{\dfrac{x}{y} + \dfrac{1}{x}}{\dfrac{y}{x} + \dfrac{1}{x}}$

7. $\dfrac{8x^{-2} - 2x^{-1}}{10x^{-1} - 6x^{-2}}$ $\quad \dfrac{4 - x}{5x - 3}$

8. $\dfrac{12x^{-2} - 3x^{-1}}{15x^{-1} - 9x^{-2}}$ $\quad \dfrac{4 - x}{5x - 3}$

9. $\dfrac{\dfrac{1}{x - 2}}{1 - \dfrac{1}{x - 2}}$ $\quad \dfrac{1}{x - 3}$

10. $\dfrac{\dfrac{1}{x + 2}}{1 + \dfrac{1}{x + 2}}$ $\quad \dfrac{1}{x + 3}$

11. $\dfrac{\dfrac{1}{x + 5} - \dfrac{1}{x}}{5}$ $\quad -\dfrac{1}{x(x + 5)}$

12. $\dfrac{\dfrac{1}{x + 6} - \dfrac{1}{x}}{6}$ $\quad -\dfrac{1}{x(x + 6)}$

13. $\dfrac{\dfrac{4}{x + 4}}{\dfrac{1}{x + 4} - \dfrac{1}{x}}$ $\quad -x$

14. $\dfrac{\dfrac{7}{x + 7}}{\dfrac{1}{x + 7} - \dfrac{1}{x}}$ $\quad -x$

15. $\dfrac{\dfrac{1}{x - 1} + 1}{\dfrac{1}{x + 1} - 1}$ $\quad -\dfrac{x + 1}{x - 1}$

16. $\dfrac{\dfrac{1}{x + 1} - 1}{\dfrac{1}{x - 1} + 1}$ $\quad -\dfrac{x - 1}{x + 1}$

17. $\dfrac{x^{-1} + y^{-1}}{(x + y)^{-1}}$ $\quad \dfrac{(x + y)(x + y)}{xy}$

18. $(x^{-1} + y^{-1})^{-1}$ $\quad \dfrac{xy}{x + y}$

19. $\dfrac{\dfrac{x + 2}{x - 2} - \dfrac{x - 2}{x + 2}}{\dfrac{x - 2}{x + 2} + \dfrac{x + 2}{x - 2}}$ $\quad \dfrac{4x}{x^2 + 4}$

20. $\dfrac{\dfrac{x + 1}{x - 1} - \dfrac{x - 1}{x + 1}}{\dfrac{x - 1}{x + 1} + \dfrac{x + 2}{x - 1}}$ $\quad \dfrac{4x}{2x^2 + x + 3}$

21. $\dfrac{\dfrac{2}{x^3 y} + \dfrac{5}{xy^4}}{\dfrac{5}{x^3 y} - \dfrac{3}{xy}}$ $\quad \dfrac{2y^3 + 5x^2}{y^3(5 - 3x^2)}$

22. $\dfrac{\dfrac{3}{xy^2} + \dfrac{2}{x^2 y}}{\dfrac{1}{x^2 y} + \dfrac{2}{xy^3}}$ $\quad \dfrac{y(3x + 2y)}{y^2 + 2x}$

23. $\dfrac{\dfrac{3}{x + 2} - \dfrac{3}{x - 2}}{\dfrac{5}{x^2 - 4}}$ $\quad -\dfrac{12}{5}$

24. $\dfrac{\dfrac{3}{x + 1} - \dfrac{3}{x - 1}}{\dfrac{5}{x^2 - 1}}$ $\quad -\dfrac{6}{5}$

25. $\dfrac{3a^{-1} + 3b^{-1}}{4a^{-2} - 9b^{-2}}$ $\quad \dfrac{3ab(b + a)}{(2b + 3a)(2b - 3a)}$

26. $\dfrac{5a^{-1} - 2b^{-1}}{25a^{-2} - 4b^{-2}}$ $\quad \dfrac{ab}{2a + 5b}$

27. $\dfrac{\dfrac{4x}{x^2 - 4} - \dfrac{5}{x - 2}}{\dfrac{2}{x - 2} + \dfrac{3}{x + 2}}$ $\quad -\dfrac{x + 10}{5x - 2}$

28. $\dfrac{\dfrac{2}{x + 3} + \dfrac{5x}{x^2 - 9}}{\dfrac{4}{x + 3} + \dfrac{2}{x - 3}}$ $\quad \dfrac{7x - 6}{6(x - 1)}$

29. $\dfrac{\dfrac{2y}{y^2 + 4y + 3}}{\dfrac{1}{y + 3} + \dfrac{2}{y + 1}}$ $\quad \dfrac{2y}{3y + 7}$

30. $\dfrac{\dfrac{5y}{y^2 - 5y + 6}}{\dfrac{3}{y - 3} + \dfrac{2}{y - 2}}$ $\quad \dfrac{5y}{5y - 12}$

31. $\dfrac{\dfrac{2}{a^2} - \dfrac{1}{ab} - \dfrac{1}{b^2}}{\dfrac{1}{a^2} - \dfrac{3}{ab} + \dfrac{2}{b^2}}$ $\quad \dfrac{2b + a}{b - 2a}$

32. $\dfrac{\dfrac{2}{b^2} - \dfrac{5}{ab} - \dfrac{3}{a^2}}{\dfrac{2}{b^2} + \dfrac{7}{ab} + \dfrac{3}{a^2}}$ $\quad \dfrac{a - 3b}{a + 3b}$

33. $\dfrac{\dfrac{2x}{x^2 - 25} + \dfrac{1}{3x - 15}}{\dfrac{5}{x - 5} + \dfrac{3}{4x - 20}}$ $\quad \dfrac{4(7x + 5)}{69(x + 5)}$

34. $\dfrac{\dfrac{7x}{2x - 2} + \dfrac{x}{x^2 - 1}}{\dfrac{4}{x + 1} - \dfrac{1}{3x + 3}}$ $\quad \dfrac{3x(7x + 9)}{22(x - 1)}$

35. $\dfrac{\dfrac{3}{x + 2y} - \dfrac{2y}{x^2 + 2xy}}{\dfrac{3y}{x^2 + 2xy} + \dfrac{5}{x}}$ $\quad \dfrac{3x - 2y}{13y + 5x}$

36. $\dfrac{\dfrac{1}{x^3 - y^3}}{\dfrac{1}{x - y} - \dfrac{1}{x^2 + xy + y^2}}$ $\quad \dfrac{1}{x^2 + xy + y^2 - x + y}$

37. $\dfrac{\dfrac{2}{m^2 - 3m + 2} + \dfrac{2}{m^2 - m - 2}}{\dfrac{2}{m^2 - 1} + \dfrac{2}{m^2 + 4m + 3}}$ $\quad \dfrac{m(m + 3)}{(m - 2)(m + 1)}$

38. $\dfrac{\dfrac{m}{m^2 - 9} - \dfrac{2}{m^2 - 4m + 4}}{\dfrac{3}{m^2 - 5m + 6} + \dfrac{m}{m^2 + m - 6}}$ $\quad \dfrac{m^3 - 6m^2 + 4m + 18}{(m - 2)(m^2 + 9)}$

39. $\dfrac{\dfrac{2}{a^2 + 2a - 8} + \dfrac{1}{a^2 + 5a + 4}}{\dfrac{1}{a^2 - 5a + 6} + \dfrac{2}{a^2 - a - 2}}$ $\quad \dfrac{3a(a - 3)}{(a + 4)(3a - 5)}$

40. $\dfrac{\dfrac{3}{a^2 + 10a + 25} - \dfrac{1}{a^2 - a - 2}}{\dfrac{4}{a^2 + 6a + 5} - \dfrac{2}{a^2 + 3a - 10}}$ $\quad \dfrac{2a^2 - 13a - 31}{2(a + 5)(a - 5)}$

41. $\dfrac{x}{1 - \dfrac{1}{1 + \dfrac{1}{x}}}$ $\quad x(x + 1)$

42. $\dfrac{\dfrac{1}{x + 1}}{x - \dfrac{1}{x + \dfrac{1}{x}}}$ $\quad \dfrac{x^2 + 1}{(x + 1)x^3}$

Practice Plus

In Exercises 43–46, perform the indicated operations. Simplify the result, if possible.

43. $\dfrac{\dfrac{x - 1}{x^2 - 4}}{1 + \dfrac{1}{x - 2}} - \dfrac{1}{x - 2}$ $\quad -\dfrac{4}{(x + 2)(x - 2)}$

44. $\dfrac{\dfrac{x-3}{x^2-16}}{1+\dfrac{1}{x-4}} - \dfrac{1}{x-4} - \dfrac{8}{(x+4)(x-4)}$

45. $\dfrac{3}{1-\dfrac{3}{3+x}} - \dfrac{3}{\dfrac{3}{3-x}-1}$ 6

46. $\dfrac{5}{1-\dfrac{5}{5+x}} - \dfrac{5}{\dfrac{5}{5-x}-1}$ 10

In Exercises 47–48, let $f(x) = \dfrac{1+x}{1-x}$.

47. Find $f\left(\dfrac{1}{x+3}\right)$ and simplify. $\dfrac{x+4}{x+2}$

48. Find $f\left(\dfrac{1}{x-6}\right)$ and simplify. $\dfrac{x-5}{x-7}$

In Exercises 49–50, use the given rational function to find and simplify

$$\dfrac{f(a+h)-f(a)}{h}.$$

49. $f(x) = \dfrac{3}{x}$ $-\dfrac{3}{a(a+h)}$

50. $f(x) = \dfrac{1}{x^2}$ $-\dfrac{2a+h}{a^2(a+h)^2}$

Application Exercises

51. How much are your monthly payments on a loan? If P is the principal, or amount borrowed, i is the monthly interest rate (as a decimal), and n is the number of monthly payments, then the amount, A, of each monthly payment is

$$A = \dfrac{Pi}{1-\dfrac{1}{(1+i)^n}}. \qquad \dfrac{Pi(1+i)^n}{(1+i)^n-1}$$

a. Simplify the complex rational expression for the amount of each payment.

b. You purchase a $20,000 automobile at 1% monthly interest to be paid over 48 months. How much do you pay each month? Use the simplified rational expression from part (a) and a calculator. Round to the nearest dollar. $527 per month

52. The average rate on a round-trip commute having a one-way distance d is given by the complex rational expression

$$\dfrac{2d}{\dfrac{d}{r_1}+\dfrac{d}{r_2}}$$

in which r_1 and r_2 are the rates on the outgoing and return trips, respectively.

a. Simplify the complex rational expression. $\dfrac{2r_1r_2}{r_2+r_1}$

b. Find your average rate if you drive to campus averaging 30 miles per hour and return home on the same route averaging 40 miles per hour. Use the simplified rational expression from part (a).   about 34 miles per hour

53. If three resistors with resistances R_1, R_2, and R_3 are connected in parallel, their combined resistance, R, is given by the formula

$$R = \dfrac{1}{\dfrac{1}{R_1}+\dfrac{1}{R_2}+\dfrac{1}{R_3}}.$$

Simplify the complex rational expression on the right side of the formula. Then find R, to the nearest hundredth of an ohm, when R_1 is 4 ohms, R_2 is 8 ohms, and R_3 is 12 ohms.

$R = \dfrac{R_1R_2R_3}{R_2R_3+R_1R_3+R_1R_2}$; about 2.18 ohms

54. A camera lens has a measurement called its focal length, f. When an object is in focus, its distance from the lens, p, and its image distance from the lens, q, satisfy the formula

$$f = \dfrac{1}{\dfrac{1}{p}+\dfrac{1}{q}}.$$

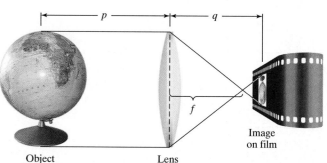

Object   Lens

Simplify the complex rational expression on the right side of the formula. $f = \dfrac{pq}{q+p}$

Writing in Mathematics

55. What is a complex rational expression? Give an example with your explanation.   55–58. Answers will vary.

56. Describe two ways to simplify $\dfrac{\dfrac{2}{x}+\dfrac{2}{y}}{\dfrac{2}{x}-\dfrac{2}{y}}$.

57. Which method do you prefer for simplifying complex rational expressions? Why?

58. Of the four complex rational expressions in Exercises 51–54, which one do you find most useful? Explain how you might use this expression in a practical situation.

Technology Exercises

In Exercises 59–62, use a graphing utility to determine if the simplification is correct by graphing the function on each side of the equation in the same viewing rectangle. If the graphs do not coincide, correct the expression on the right side and then verify your correction using the graphing utility. 59–62. See graphing answer section.

59. $\dfrac{x - \dfrac{1}{2x + 1}}{1 - \dfrac{x}{2x + 1}} = 2x - 1$ Graphs coincide; simplified correctly

60. $\dfrac{\dfrac{1}{x} + 1}{\dfrac{1}{x}} = 2$ Graphs do not coincide; $\dfrac{\dfrac{1}{x} + 1}{\dfrac{1}{x}} = x + 1$

61. $\dfrac{\dfrac{1}{x} + \dfrac{1}{3}}{\dfrac{1}{3x}} = x + \dfrac{1}{3}$ Graphs do not coincide; $\dfrac{\dfrac{1}{x} + \dfrac{1}{3}}{\dfrac{1}{3x}} = x + 3$

62. $\dfrac{\dfrac{x}{3}}{\dfrac{2}{x + 1}} = \dfrac{3(x + 1)}{2}$ Graphs do not coincide; $\dfrac{\dfrac{x}{3}}{\dfrac{2}{x + 1}} = \dfrac{x(x + 1)}{6}$

Critical Thinking Exercises

63. Simplify:

$$\dfrac{\dfrac{x + h}{x + h + 1} - \dfrac{x}{x + 1}}{h} \qquad \dfrac{1}{(x + h + 1)(x + 1)}$$

64. Simplify:

$$x + \dfrac{1}{x + \dfrac{1}{x + \dfrac{1}{x}}} \qquad \dfrac{x^4 + 3x^2 + 1}{x(x^2 + 2)}$$

65. If $f(x) = \dfrac{1}{x + 1}$, find $f(f(a))$ and simplify. $\dfrac{a + 1}{a + 2}$

66. Let x represent the first of two consecutive integers. Find a simplified expression that represents the reciprocal of the sum of the reciprocals of the two integers. $\dfrac{x(x + 1)}{2x + 1}$

Review Exercises

67. Solve: $x^2 + 27 = 12x$. (Section 5.7, Example 2)
3 and 9, or $\{3, 9\}$

68. Multiply: $(4x^2 - y)^2$. (Section 5.2, Example 7)
$16x^4 - 8x^2 y + y^2$

69. Solve: $-4 < 3x - 7 < 8$. (Section 4.2, Example 4)
$\{x \mid 1 < x < 5\}$ or $(1, 5)$

DIVISION OF POLYNOMIALS

Objectives

1 Divide a polynomial by a monomial.

2 Use long division to divide by a polynomial containing more than one term.

During the 1980s, the controversial economist Arthur Laffer promoted the idea that tax *increases* lead to a *reduction* in government revenue. Called supply-side economics, the theory uses rational functions as models. One such function

$$f(x) = \dfrac{80x - 8000}{x - 110}, \qquad 30 \le x \le 100$$

models the government tax revenue, $f(x)$, in tens of billions of dollars, as a function of the tax rate, x. The graph of the rational function is shown in Figure 6.4. The graph shows tax revenue decreasing quite dramatically as the tax rate increases. At a tax rate of (gasp) 100%, the government takes all our money and no one has an incentive to work. With no income earned, zero dollars in tax revenue is generated.

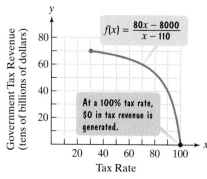

Like all rational functions, the Laffer model consists of the quotient of polynomials. Although the rational expression in the model cannot be simplified, it is possible to perform the division, thereby expressing the function in another form. In this section, you will learn to divide polynomials.

FIGURE 6.4

1 Divide a polynomial by a monomial.

Dividing a Polynomial by a Monomial   We have seen that to divide monomials, we divide the coefficients and subtract the exponents when bases are the same. For example,

$$\frac{25x^{12}}{5x^4} = \frac{25}{5}x^{12-4} = 5x^8 \quad \text{and} \quad \frac{60x^4y^3}{-30x^2y} = \frac{60}{-30}x^{4-2}y^{3-1} = -2x^2y^2.$$

How do we divide a polynomial that is not a monomial by a monomial? We divide each term of the polynomial by the monomial. For example,

$$\underbrace{\frac{10x^8 + 15x^6}{5x^3}}_{\substack{\text{Polynomial}\\\text{dividend}\\\text{Monomial}\\\text{divisor}}} = \underbrace{\frac{10x^8}{5x^3}}_{\substack{\text{Divide the first}\\\text{term by } 5x^3.}} + \underbrace{\frac{15x^6}{5x^3}}_{\substack{\text{Divide the second}\\\text{term by } 5x^3.}} = \frac{10}{5}x^{8-3} + \frac{15}{5}x^{6-3} = 2x^5 + 3x^3.$$

DIVIDING A POLYNOMIAL THAT IS NOT A MONOMIAL BY A MONOMIAL
To divide a polynomial by a monomial, divide each term of the polynomial by the monomial.

EXAMPLE 1   Dividing a Polynomial by a Monomial

Divide: $(15x^3 - 5x^2 + x + 5) \div (5x)$.

SOLUTION

$$\frac{15x^3 - 5x^2 + x + 5}{5x} \qquad \text{Rewrite the division in a vertical format.}$$

$$= \frac{15x^3}{5x} - \frac{5x^2}{5x} + \frac{x}{5x} + \frac{5}{5x} \qquad \text{Divide each term of the polynomial by the monomial.}$$

$$= 3x^2 - x + \frac{1}{5} + \frac{1}{x} \qquad \text{Simplify each quotient.} \qquad \blacksquare$$

✔ **CHECK POINT 1** Divide: $(16x^3 - 32x^2 + 2x + 4) \div 4x.$ $4x^2 - 8x + \frac{1}{2} + \frac{1}{x}$

EXAMPLE 2 Dividing a Polynomial by a Monomial

Divide: $8x^4y^5 - 10x^4y^3 + 12x^2y^3$ by $4x^3y^2$.

SOLUTION

$$\frac{8x^4y^5 - 10x^4y^3 + 12x^2y^3}{4x^3y^2}$$

Express the division in a vertical format.

$$= \frac{8x^4y^5}{4x^3y^2} - \frac{10x^4y^3}{4x^3y^2} + \frac{12x^2y^3}{4x^3y^2}$$

Divide each term of the polynomial by the monomial.

$$= 2xy^3 - \frac{5}{2}xy + \frac{3y}{x}$$

Simplify each quotient. ∎

✔ CHECK POINT 2 Divide: $15x^4y^5 - 5x^3y^4 + 10x^2y^2$ by $5x^2y^3$. $3x^2y^2 - xy + \frac{2}{y}$

Dividing by a Polynomial Containing More Than One Term We now look at division by a polynomial containing more than one term, such as

$$x - 2\overline{)x^2 - 14x + 24}.$$

Divisor has two terms.   Dividend has three terms.

When a divisor has more than one term, the four steps used to divide whole numbers—**divide, multiply, subtract, bring down the next term**—form the repetitive procedure for polynomial long division.

EXAMPLE 3 Dividing a Polynomial by a Binomial

Divide: $x^2 - 14x + 24$ by $x - 2$.

SOLUTION   The following steps illustrate how polynomial division is very similar to numerical division.

$$x - 2\overline{)x^2 - 14x + 24}$$

Arrange the terms of the dividend ($x^2 - 14x + 24$) and the divisor ($x - 2$) in descending powers of x.

$$x - 2\overline{)x^2 - 14x + 24}^{\,x}$$

DIVIDE x^2 (the first term in the dividend) by x (the first term in the divisor): $\frac{x^2}{x} = x$. Align like terms.

$x(x - 2) = x^2 - 2x.$ $x - 2\overline{)x^2 - 14x + 24}^{\,x}$
$\phantom{x - 2\overline{)}}x^2 - 2x$

MULTIPLY each term in the divisor ($x - 2$) by x, aligning terms of the product under like terms in the dividend.

$$x$$
$$x - 2 \overline{)x^2 - 14x + 24}$$
$$\ominus \, x^2 \oplus 2x$$
$$\overline{ -12x}$$

Change signs of the polynomial being subtracted.

Subtract $x^2 - 2x$ from $x^2 - 14x$ by changing the sign of each term in the lower expression and adding.

$$x$$
$$x - 2 \overline{)x^2 - 14x + 24}$$
$$x^2 - 2x \quad \downarrow$$
$$\overline{ -12x + 24}$$

Bring down 24 from the original dividend and add algebraically to form a new dividend.

$$x - 12$$
$$x - 2 \overline{)x^2 - 14x + 24}$$
$$x^2 - 2x$$
$$\overline{ -12x + 24}$$

Find the second term of the quotient. Divide the first term of $-12x + 24$ by x, the first term of the divisor: $\dfrac{-12x}{x} = -12.$

$$-12(x-2) = -12x + 24$$

$$x - 12$$
$$x - 2 \overline{)x^2 - 14x + 24}$$
$$x^2 - 2x$$
$$\overline{ -12x + 24}$$
$$\oplus -12x \ominus 24$$
$$\overline{ 0}$$

Remainder

Multiply the divisor $(x - 2)$ by -12, aligning under like terms in the new dividend. Then subtract to obtain the remainder of 0.

The quotient is $x - 12$ and the remainder is 0. We will not list a remainder of 0 in the answer. Thus,

$$(x^2 - 14x + 24) \div (x - 2) = x - 12.$$

After performing polynomial long division, the answer can be checked. Find the product of the divisor and the quotient, and add the remainder. If the result is the dividend, the answer to the division problem is correct. For example, let's check our work in Example 3.

$$(x^2 - 14x + 24) \div (x - 2) = x - 12$$

Dividend   **Divisor**   **Quotient to be checked**

Multiply the divisor and the quotient, and add the remainder, 0:

$$(x - 2)(x - 12) + 0 = x^2 - 12x - 2x + 24 + 0 = x^2 - 14x + 24.$$

Divisor   **Quotient**   **Remainder**   **This is the dividend.**

Because we obtained the dividend, the quotient is correct.

CHECK POINT 3 Divide: $3x^2 - 14x + 16$ by $x - 2$. $3x - 8$

Before considering additional examples, let's summarize the general procedure for dividing by a polynomial that contains more than one term.

LONG DIVISION OF POLYNOMIALS

1. **Arrange the terms** of both the dividend and the divisor in descending powers of any variable.

2. **Divide** the first term in the dividend by the first term in the divisor. The result is the first term of the quotient.

3. **Multiply** every term in the divisor by the first term in the quotient. Write the resulting product beneath the dividend with like terms lined up.

4. **Subtract** the product from the dividend.

5. **Bring down** the next term in the original dividend and write it next to the remainder to form a new dividend.

6. Use this new expression as the dividend and repeat this process until the remainder can no longer be divided. This will occur when the degree of the remainder (the highest exponent on a variable in the remainder) is less than the degree of the divisor.

In our next long division, we will obtain a nonzero remainder.

EXAMPLE 4 Long Division of Polynomials

Divide: $4 - 5x - x^2 + 6x^3$ by $3x - 2$.

SOLUTION We begin by writing the dividend in descending powers of x.

$$4 - 5x - x^2 + 6x^3 = 6x^3 - x^2 - 5x + 4$$

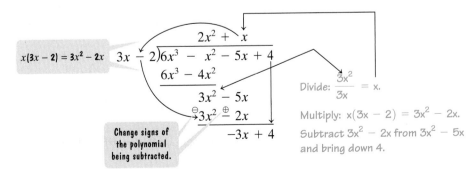

$2x^2(3x - 2) = 6x^3 - 4x^2$

$$\begin{array}{r} 2x^2 \\ 3x - 2 \overline{)6x^3 - x^2 - 5x + 4} \\ 6x^3 \ominus 4x^2 \\ \hline 3x^2 - 5x \end{array}$$

Change signs of the polynomial being subtracted.

Divide: $\dfrac{6x^2}{3x} = 2x^2$.

Multiply: $2x^2(3x - 2) = 6x^3 - 4x^2$.

Subtract $6x^3 - 4x^2$ from $6x^3 - x^2$ and bring down $-5x$.

Now we divide $3x^2$ by $3x$ to obtain x, multiply x and the divisor, and subtract.

$x(3x - 2) = 3x^2 - 2x$

$$\begin{array}{r} 2x^2 + x \\ 3x - 2 \overline{)6x^3 - x^2 - 5x + 4} \\ 6x^3 - 4x^2 \\ \hline 3x^2 - 5x \\ 3x^2 \ominus 2x \\ \hline -3x + 4 \end{array}$$

Change signs of the polynomial being subtracted.

Divide: $\dfrac{3x^2}{3x} = x$.

Multiply: $x(3x - 2) = 3x^2 - 2x$.

Subtract $3x^2 - 2x$ from $3x^2 - 5x$ and bring down 4.

Now we divide $-3x$ by $3x$ to obtain -1, multiply -1 and the divisor, and subtract.

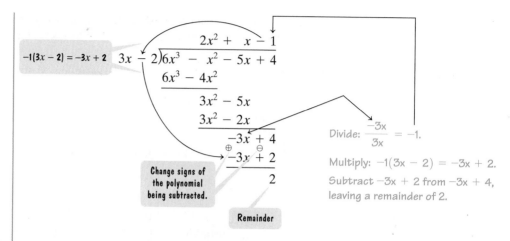

$-1(3x - 2) = -3x + 2$

Divide: $\dfrac{-3x}{3x} = -1.$

Multiply: $-1(3x - 2) = -3x + 2.$

Subtract $-3x + 2$ from $-3x + 4$, leaving a remainder of 2.

Change signs of the polynomial being subtracted.

Remainder

The quotient is $2x^2 + x - 1$ and the remainder is 2. When there is a nonzero remainder, as in this example, list the quotient, plus the remainder above the divisor. Thus,

$$\frac{6x^3 - x^2 - 5x + 4}{3x - 2} = 2x^2 + x - 1 + \frac{2}{3x - 2}.$$

Remainder above divisor

Quotient

Check this result by showing that the product of the divisor and the quotient,

$$(3x - 2)(2x^2 + x - 1),$$

plus the remainder, 2, is the dividend, $6x^3 - x^2 - 5x + 4$. ∎

✔ **CHECK POINT 4** Divide: $-9 + 7x - 4x^2 + 4x^3$ by $2x - 1$. $2x^2 - x + 3 - \dfrac{6}{2x - 1}$

If a power of x is missing in either a dividend or a divisor, add that power of x with a coefficient of 0 and then divide. In this way, like terms will be aligned as you carry out the long division.

EXAMPLE 5 Long Division of Polynomials

Divide: $6x^4 + 5x^3 + 3x - 5$ by $3x^2 - 2x$.

SOLUTION We write the dividend, $6x^4 + 5x^3 + 3x - 5$, as $6x^4 + 5x^3 + 0x^2 + 3x - 5$ to keep all like terms aligned.

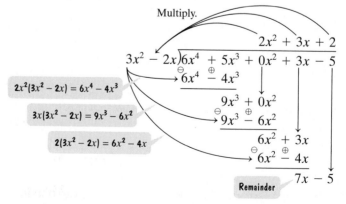

Multiply.

$2x^2(3x^2 - 2x) = 6x^4 - 4x^3$

$3x(3x^2 - 2x) = 9x^3 - 6x^2$

$2(3x^2 - 2x) = 6x^2 - 4x$

$$3x^2 - 2x \overline{)6x^4 + 5x^3 + 0x^2 + 3x - 5} \quad 2x^2 + 3x + 2$$

$6x^4 - 4x^3$

$9x^3 + 0x^2$

$9x^3 - 6x^2$

$6x^2 + 3x$

$6x^2 - 4x$

$7x - 5$

Remainder

The division process is finished because the degree of $7x - 5$, which is 1, is less than the degree of the divisor $3x^2 - 2x$, which is 2. The answer is

$$\frac{6x^4 + 5x^3 + 3x - 5}{3x^2 - 2x} = 2x^2 + 3x + 2 + \frac{7x - 5}{3x^2 - 2x}.$$

 CHECK POINT 5 Divide: $2x^4 + 3x^3 - 7x - 10$ by $x^2 - 2x$. $2x^2 + 7x + 14 + \dfrac{21x - 10}{x^2 - 2x}$

6.4 EXERCISE SET

Student Solutions Manual   CD/Video   PH Math/Tutor Center   MathXL Tutorials on CD   MathXL®   MyMathLab   Interactmath.com

Practice Exercises

In Exercises 1–12, divide the polynomial by the monomial.

1. $\dfrac{25x^7 - 15x^5 + 10x^3}{5x^3}$ $5x^4 - 3x^2 + 2$

2. $\dfrac{49x^7 - 28x^5 + 14x^3}{7x^3}$ $7x^4 - 4x^2 + 2$

3. $\dfrac{18x^3 + 6x^2 - 9x - 6}{3x}$ $6x^2 + 2x - 3 - \dfrac{2}{x}$

4. $\dfrac{25x^3 + 50x^2 - 40x - 10}{5x}$ $5x^2 + 10x - 8 - \dfrac{2}{x}$

5. $(28x^3 - 7x^2 - 16x) \div (4x^2)$ $7x - \dfrac{7}{4} - \dfrac{4}{x}$

6. $(70x^3 - 10x^2 - 14x) \div (7x^2)$ $10x - \dfrac{10}{7} - \dfrac{2}{x}$

7. $(25x^8 - 50x^7 + 3x^6 - 40x^5) \div (-5x^5)$

   $-5x^3 + 10x^2 - \dfrac{3}{5}x + 8$

8. $(18x^7 - 9x^6 + 20x^5 - 10x^4) \div (-2x^4)$

   $-9x^3 + \dfrac{9}{2}x^2 - 10x + 5$

9. $(18a^3b^2 - 9a^2b - 27ab^2) \div (9ab)$ $2a^2b - a - 3b$

10. $(12a^2b^2 + 6a^2b - 15ab^2) \div (3ab)$ $4ab + 2a - 5b$

11. $(36x^4y^3 - 18x^3y^2 - 12x^2y) \div (6x^3y^3)$ $6x - \dfrac{3}{y} - \dfrac{2}{xy^2}$

12. $(40x^4y^3 - 20x^3y^2 - 50x^2y) \div (10x^3y^3)$ $4x - \dfrac{2}{y} - \dfrac{5}{xy^2}$

In Exercises 13–36, divide as indicated. Check at least five of your answers by showing that the product of the divisor and the quotient, plus the remainder, is the dividend.

13. $(x^2 + 8x + 15) \div (x + 5)$ $x + 3$

14. $(x^2 + 3x - 10) \div (x - 2)$ $x + 5$

15. $(x^3 - 2x^2 - 5x + 6) \div (x - 3)$ $x^2 + x - 2$

16. $(x^3 + 5x^2 + 7x + 2) \div (x + 2)$ $x^2 + 3x + 1$

17. $(x^2 - 7x + 12) \div (x - 5)$ $x - 2 + \dfrac{2}{x - 5}$

18. $(2x^2 + x - 9) \div (x - 2)$ $2x + 5 + \dfrac{1}{x - 2}$

19. $(2x^2 + 13x + 5) \div (2x + 3)$ $x + 5 - \dfrac{10}{2x + 3}$

20. $(8x^2 + 6x - 25) \div (4x + 9)$ $2x - 3 + \dfrac{2}{4x + 9}$

21. $(x^3 + 3x^2 + 5x + 4) \div (x + 1)$ $x^2 + 2x + 3 + \dfrac{1}{x + 1}$

22. $(x^3 + 6x^2 - 2x + 3) \div (x - 1)$ $x^2 + 7x + 5 + \dfrac{8}{x - 1}$

23. $(4y^3 + 12y^2 + 7y - 3) \div (2y + 3)$ $2y^2 + 3y - 1$

24. $(6y^3 + 7y^2 + 12y - 5) \div (3y - 1)$ $2y^2 + 3y + 5$

25. $(9x^3 - 3x^2 - 3x + 4) \div (3x + 2)$ $3x^2 - 3x + 1 + \dfrac{2}{3x + 2}$

26. $(2x^3 + 13x^2 + 9x - 6) \div (2x + 3)$ $x^2 + 5x - 3 + \dfrac{3}{2x + 3}$

27. $(4x^3 - 6x - 11) \div (2x - 4)$ $2x^2 + 4x + 5 + \dfrac{9}{2x - 4}$

28. $(2x^3 + 6x - 4) \div (x + 4)$ $2x^2 - 8x + 38 - \dfrac{156}{x + 4}$

29. $(4y^3 - 5y) \div (2y - 1)$ $2y^2 + y - 2 - \dfrac{2}{2y - 1}$

30. $(6y^3 - 5y) \div (2y - 1)$ $3y^2 + \dfrac{3}{2}y - \dfrac{7}{4} - \dfrac{7}{4(2y - 1)}$

31. $(4y^4 - 17y^2 + 14y - 3) \div (2y - 3)$ $2y^3 + 3y^2 - 4y + 1$

32. $(2y^4 - y^3 + 16y^2 - 4) \div (2y - 1)$ $y^3 + 8y + 4$

33. $(4x^4 + 3x^3 + 4x^2 + 9x - 6) \div (x^2 + 3)$ $4x^2 + 3x - 8 + \dfrac{18}{x^2 + 3}$

34. $(3x^5 - x^3 + 4x^2 - 12x - 8) \div (x^2 - 2)$

$3x^3 + 5x + 4 - \dfrac{2x}{x^2 - 2}$

35. $(15x^4 + 3x^3 + 4x^2 + 4) \div (3x^2 - 1)$ $\quad 5x^2 + x + 3 + \dfrac{x + 7}{3x^2 - 1}$

36. $(18x^4 + 9x^3 + 3x^2) \div (3x^2 + 1)$ $\quad 6x^2 + 3x - 1 + \dfrac{1 - 3x}{3x^2 + 1}$

In Exercises 37–40, find a simplified expression for $\left(\dfrac{f}{g}\right)(x).$

37. $f(x) = 8x^3 - 38x^2 + 49x - 10,$

$g(x) = 4x - 1$ $\quad \left(\dfrac{f}{g}\right)(x) = 2x^2 - 9x + 10$

38. $f(x) = 2x^3 - 9x^2 - 17x + 39,$

$g(x) = 2x - 3$ $\quad \left(\dfrac{f}{g}\right)(x) = x^2 - 3x - 13$

39. $f(x) = 2x^4 - 7x^3 + 7x^2 - 9x + 10,$

$g(x) = 2x - 5$ $\quad \left(\dfrac{f}{g}\right)(x) = x^3 - x^2 + x - 2$

40. $f(x) = 4x^4 + 6x^3 + 3x - 1,$

$g(x) = 2x^2 + 1$ $\quad \left(\dfrac{f}{g}\right)(x) = 2x^2 + 3x - 1$

Practice Plus

In Exercises 41–50, divide as indicated.

41. $\dfrac{x^4 + y^4}{x + y}$ $\quad x^3 - x^2y + xy^2 - y^3 + \dfrac{2y^4}{x + y}$

42. $\dfrac{x^5 + y^5}{x + y}$ $\quad x^4 - x^3y + x^2y^2 - xy^3 + y^4$

43. $\dfrac{3x^4 + 5x^3 + 7x^2 + 3x - 2}{x^2 + x + 2}$ $\quad 3x^2 + 2x - 1$

44. $\dfrac{x^4 - x^3 - 7x^2 - 7x - 2}{x^2 - 3x - 2}$ $\quad x^2 + 2x + 1$

45. $\dfrac{4x^3 - 3x^2 + x + 1}{x^2 + x + 1}$ $\quad 4x - 7 + \dfrac{4x + 8}{x^2 + x + 1}$

46. $\dfrac{x^4 - x^2 + 1}{x^2 + x + 1}$ $\quad x^2 - x - 1 + \dfrac{2x + 2}{x^2 + x + 1}$

47. $\dfrac{x^5 - 1}{x^2 - x + 2}$ $\quad x^3 + x^2 - x - 3 + \dfrac{-x + 5}{x^2 - x + 2}$

48. $\dfrac{5x^5 - 7x^4 + 3x^3 - 20x^2 + 28x - 12}{x^3 - 4}$ $\quad 5x^2 - 7x + 3$

49. $\dfrac{4x^3 - 7x^2y - 16xy^2 + 3y^3}{x - 3y}$ $\quad 4x^2 + 5xy - y^2$

50. $\dfrac{12x^3 - 19x^2y + 13xy^2 - 10y^3}{4x - 5y}$ $\quad 3x^2 - xy + 2y^2$

In Exercises 51–52, find $\left(\dfrac{f - g}{h}\right)(x)$ *and the domain of* $\dfrac{f - g}{h}.$

51. $f(x) = 3x^3 + 4x^2 - x - 4,$ $g(x) = -5x^3 + 22x^2 - 28x - 12,$

$h(x) = 4x + 1$ $\quad \left(\dfrac{f - g}{h}\right)(x) = 2x^2 - 5x + 8;$ $\{x | x$ is a

real number and $x \neq -\dfrac{1}{4}\}$ or $\left(-\infty, -\dfrac{1}{4}\right) \cup \left(-\dfrac{1}{4}, \infty\right)$

52. $f(x) = x^3 + 9x^2 - 6x + 25,$ $g(x) = -3x^3 + 2x^2 - 14x + 5,$

$h(x) = 2x + 4$ $\quad \left(\dfrac{f - g}{h}\right)(x) = 2x^2 - \dfrac{1}{2}x + 5;$ $\{x | x$ is a

real number and $x \neq -2\}$ or $(-\infty, -2) \cup (-2, \infty)$

In Exercises 53–54, solve each equation for x in terms of a and simplify.

53. $ax + 2x + 4 = 3a^3 + 10a^2 + 6a$ $\quad x = 3a^2 + 4a - 2$

54. $ax - 3x + 6 = a^3 - 6a^2 + 11a$ $\quad x = a^2 - 3a + 2$

Application Exercises

In the section opener, we saw that

$$f(x) = \frac{80x - 8000}{x - 110}, \quad 30 \leq x \leq 100$$

models the government tax revenue, f(x), in tens of billions of dollars, as a function of the tax rate percentage, x. Use this function to solve Exercises 55–58. Round to the nearest ten billion dollars.

55. Find and interpret $f(30)$. Identify the solution as a point on the graph of the function in Figure 6.4 on page 420. 70; At the tax rate of 30%, the government tax revenue is $70 tens of billions, or $700 billion; (30, 70).

56. Find and interpret $f(40)$. Identify the solution as a point on the graph of the function in Figure 6.4 on page 420. about 69; At the tax rate of 40%, the government tax revenue is about $69 tens of billions, or $690 billion; (40, 69).

57. Rewrite the function by using long division to perform

$$(80x - 8000) \div (x - 110).$$

Then use this new form of the function to find $f(30)$. Do you obtain the same answer as you did in Exercise 55? Which form of the function do you find easier to use? $f(x) = 80 + \dfrac{800}{x - 110};$ $f(30) = 70;$ Answers will vary.

58. Rewrite the function by using long division to perform

$$(80x - 8000) \div (x - 110).$$

Then use this new form of the function to find $f(40)$. Do you obtain the same answer as you did in Exercise 56? Which form of the function do you find easier to use? $f(x) = 80 + \dfrac{800}{x - 110};$ $f(40) \approx 69;$ yes; Answers will vary.

Writing in Mathematics

59. Explain how to divide a polynomial that is not a monomial by a monomial. Give an example.   *59–64.* Answers will vary.

60. In your own words, explain how to divide by a polynomial containing more than one term. Use $\dfrac{x^2 + 4}{x + 2}$ in your explanation.

61. When performing polynomial long division, explain when to stop dividing.

62. After performing polynomial long division, explain how to check the answer.

63. When performing polynomial long division with missing terms, explain the advantage of writing the missing terms with zero coefficients.

64. The idea of supply-side economics is that an increase in the tax rate may actually reduce government revenue. What explanation can you offer for this theory?

Technology Exercises

In Exercises 65–67, use a graphing utility to determine if the division has been performed correctly. Graph the function on each side of the equation in the same viewing rectangle. If the graphs do not coincide, correct the expression on the right side by using polynomial long division. Then verify your correction using the graphing utility.   *65–67.* See graphing answer section.

65. $(6x^2 + 16x + 8) \div (3x + 2) = 2x + 4$ Graphs coincide; division correct.

66. $(4x^3 + 7x^2 + 8x + 20) \div (2x + 4) = 2x^2 - \dfrac{1}{2}x + 3$

Graphs do not coincide; right side should be $2x^2 - \dfrac{1}{2}x + 5$.

67. $(3x^4 + 4x^3 - 32x^2 - 5x - 20) \div (x + 4) = 3x^3 - 8x^2 + 5$
Graphs do not coincide; right side should be $3x^3 - 8x^2 - 5$.

68. Use the $\boxed{\text{TABLE}}$ feature of a graphing utility to verify any two division results that you obtained in Exercises 13–36. Answers will vary.

Critical Thinking Exercises

69. Which one of the following is true?   d

a. All long-division problems can be done by the alternative method of factoring the dividend and canceling identical factors in the dividend and the divisor.

b. Polynomial long division always shows that the quotient of two polynomials is a polynomial.

c. The long division process should be continued until the degree of the remainder is the same as the degree of the divisor.

d. If a polynomial long-division problem results in a remainder that is zero, then the divisor is a factor of the dividend.

In Exercises 70–71, divide as indicated.

70. $(x^{3n} - 4x^{2n} - 2x^n - 12) \div (x^n - 5)$ $x^{2n} + x^n + 3 + \dfrac{3}{x^n - 5}$

71. $(x^{3n} + 1) \div (x^n + 1)$ $x^{2n} - x^n + 1$

72. When $2x^2 - 7x + 9$ is divided by a polynomial, the quotient is $2x - 3$ and the remainder is 3. Find the polynomial. $x - 2$

73. Find k so that the remainder is 0:
$(20x^3 + 23x^2 - 10x + k) \div (4x + 3)$. -12

Review Exercises

74. Solve: $|2x - 3| > 4$. (Section 4.3, Example 4)
$\left\{x \mid x < -\dfrac{1}{2} \text{ or } x > \dfrac{7}{2}\right\}$, or $\left(-\infty, -\dfrac{1}{2}\right) \cup \left(\dfrac{7}{2}, \infty\right)$

75. Write 40,610,000 in scientific notation. (Section 1.7, Example 2) 4.061×10^7

76. Simplify: $2x - 4[x - 3(2x + 1)]$. (Section 1.2, Example 14) $22x + 12$

✔ MID-CHAPTER CHECK POINT

CHAPTER 6

What You Know: We learned that the domain of a rational function is the set of all real numbers except those for which the denominator is zero. We saw that graphs of rational functions have vertical asymptotes or breaks at these excluded values. We learned to simplify rational expressions by dividing the numerator and the denominator by common factors. We performed a variety of operations with rational expressions, including multiplication, division, addition, and subtraction. We used two methods (multiplying by 1 and dividing) to simplify complex rational expressions. Finally, we used long division when dividing by a polynomial with more than one term.

In Exercises 1–19, simplify the given rational expression or perform the indicated operation(s) and, if possible, simplify.

1. $\dfrac{x^2 - x - 6}{x^2 + 3x - 18}$ $\dfrac{x + 2}{x + 6}$

2. $\dfrac{2x^2 - 8x - 11}{x^2 + 3x - 4} + \dfrac{x^2 + 14x - 13}{x^2 + 3x - 4}$ $\dfrac{3(x - 2)}{x - 1}$

3. $\dfrac{x^3 - 27}{4x^2 - 4x} \cdot \dfrac{4x}{x - 3}$ $\dfrac{x^2 + 3x + 9}{x - 1}$

4. $5 + \dfrac{7}{x-2}$ $\dfrac{5x-3}{x-2}$

5. $\dfrac{x - \dfrac{4}{x+6}}{\dfrac{1}{x+6} + x}$ $\dfrac{x^2 + 6x - 4}{x^2 + 6x + 1}$

6. $(2x^4 - 13x^3 + 17x^2 + 18x - 24) \div (x - 4)$
$2x^3 - 5x^2 - 3x + 6$

7. $\dfrac{x^3y - y^3x}{x^2y - xy^2}$ $x + y$

8. $(28x^8y^3 - 14x^6y^2 + 3x^2y^2) \div (7x^2y)$
$4x^6y^2 - 2x^4y + \dfrac{3}{7}y$

9. $\dfrac{2x-1}{x+6} - \dfrac{x+3}{x-2}$ $\dfrac{x^2 - 14x - 16}{(x+6)(x-2)}$

10. $\dfrac{3}{x-2} - \dfrac{2}{x+2} - \dfrac{x}{x^2-4}$ $\dfrac{10}{(x+2)(x-2)}$

11. $\dfrac{3x^2 - 7x - 6}{3x^2 - 13x - 10} \div \dfrac{2x^2 - x - 1}{4x^2 - 18x - 10}$ $\dfrac{2(x-3)}{x-1}$

12. $\dfrac{3}{7-x} + \dfrac{x-2}{x-7}$ $\dfrac{x-5}{x-7}$

13. $(6x^4 - 3x^3 - 11x^2 + 2x + 4) \div (3x^2 - 1)$
$2x^2 - x - 3 + \dfrac{x+1}{3x^2 - 1}$

14. $\dfrac{5 + \dfrac{2}{x}}{3 - \dfrac{1}{x}}$ $\dfrac{5x+2}{3x-1}$

15. $\dfrac{x}{x^2 - 7x + 6} - \dfrac{x}{x^2 - 2x - 24}$ $\dfrac{5x}{(x-6)(x-1)(x+4)}$

16. $\dfrac{\dfrac{3}{x+1} + \dfrac{4}{x}}{\dfrac{4}{x}}$ $\dfrac{7x+4}{4(x+1)}$

17. $\dfrac{x^2 - x - 6}{x+1} \div \left(\dfrac{x^2 - 9}{x^2 - 1} \cdot \dfrac{x-1}{x+3} \right)$ $x + 2$

18. $(64x^3 + 4) \div (4x + 2)$ $16x^2 - 8x + 4 - \dfrac{2}{2x+1}$

19. $\dfrac{x+1}{x^2 + x - 2} - \dfrac{1}{x^2 - 3x + 2} + \dfrac{2x}{x^2 - 4}$ $\dfrac{3x+2}{(x+2)(x-1)}$

20. Find the domain of $f(x) = \dfrac{5x - 10}{x^2 + 5x - 14}$. Then simplify the function's equation.

Domain of $f = \{x \mid x$ is a real number and $x \neq -7$ and $x \neq 2\}$ or $(-\infty, -7) \cup (-7, 2) \cup (2, \infty); f(x) = \dfrac{5}{x+7}$

SECTION

6.5

Objectives

1 Divide polynomials using synthetic division.

2 Evaluate a polynomial function using the Remainder Theorem.

3 Show that a number is a solution of a polynomial equation using the Remainder Theorem.

SYNTHETIC DIVISION AND THE REMAINDER THEOREM

A moth has moved into your closet. She appeared in your bedroom at night, but somehow her relatively stout body escaped your clutches. Within a few weeks, swarms of moths in your tattered wardrobe suggest that Mama Moth was in the family way. There must be at least 200 critters nesting in every crevice of your clothing.

Two hundred plus moth-tykes from one female moth—is this possible? Indeed it is. The number of eggs, $f(x)$, in a female moth is a function of her abdominal width, x, in millimeters, modeled by

$$f(x) = 14x^3 - 17x^2 - 16x + 34, \quad 1.5 \le x \le 3.5.$$

Because there are 200 moths feasting on your favorite sweaters, Mama's abdominal width can be estimated by finding the solutions of the polynomial equation

$$14x^3 - 17x^2 - 16x + 34 = 200.$$

With mathematics present even in your quickly disappearing attire, we move from rags to a shortcut for long division, called *synthetic division*. In the exercise set, you will use this shortcut to find Mama Moth's abdominal width.

1 Divide polynomials using synthetic division.

Dividing Polynomials Using Synthetic Division We can use **synthetic division** to divide polynomials if the divisor is of the form $x - c$. This method provides a quotient more quickly than long division. Let's compare the two methods showing $x^3 + 4x^2 - 5x + 5$ divided by $x - 3$.

Long Division   **Synthetic Division**

Quotient

$$
\begin{array}{r}
x^2 + 7x + 16 \\
x - 3 \overline{)x^3 + 4x^2 - 5x + 5} \\
\end{array}
$$

Dividend

Divisor $x - c$; $c = 3$

$$
\begin{array}{r}
\underline{x^3 - 3x^2} \\
7x^2 - 5x \\
\underline{7x^2 - 21x} \\
16x + 5 \\
\underline{16x - 48} \\
53
\end{array}
$$

Remainder

$$
\begin{array}{r|rrrr}
3 & 1 & 4 & -5 & 5 \\
  &   & 3 & 21 & 48 \\
\hline
  & 1 & 7 & 16 & 53
\end{array}
$$

Notice the relationship between the polynomials in the long division process and the numbers that appear in synthetic division.

These are the coefficients of the dividend $x^3 + 4x^2 - 5x + 5$.

The divisor is $x - 3$. This is 3, or c, in $x - c$.

$$
\begin{array}{r|rrrr}
3 & 1 & 4 & -5 & 5 \\
  &   & 3 & 21 & 48 \\
\hline
  & 1 & 7 & 16 & 53
\end{array}
$$

These are the coefficients of the quotient $x^2 + 7x + 16$.

This is the remainder.

Now let's look at the steps involved in synthetic division.

SYNTHETIC DIVISION   To divide a polynomial by $x - c$:

Example

1. Arrange polynomials in descending powers, with a 0 coefficient for any missing term.

$$x - 3\overline{)x^3 + 4x^2 - 5x + 5}$$

2. Write c for the divisor, $x - c$. To the right, write the coefficients of the dividend.

$$3\rfloor\ 1\quad 4\quad -5\quad 5$$

3. Write the leading coefficient of the dividend on the bottom row.

$$
\begin{array}{c|cccc}
3 & 1 & 4 & -5 & 5 \\
& \downarrow & & \text{Bring down 1}. \\
\hline
& 1
\end{array}
$$

4. Multiply c (in this case, 3) times the value just written on the bottom row. Write the product in the next column in the second row.

$$
\begin{array}{c|cccc}
3 & 1 & 4 & -5 & 5 \\
& & 3 \\
\hline
& 1
\end{array}
$$

Multiply by 3: $3 \cdot 1 = 3$.

5. Add the values in this new column, writing the sum in the bottom row.

$$
\begin{array}{c|cccc}
3 & 1 & 4 & -5 & 5 \\
& & 3 & \text{Add.} \\
\hline
& 1 & 7
\end{array}
$$

6. Repeat this series of multiplications and additions until all columns are filled in.

$$
\begin{array}{c|cccc}
3 & 1 & 4 & -5 & 5 \\
& & 3 & 21 & \text{Add.} \\
\hline
& 1 & 7 & 16
\end{array}
$$

Multiply by 3: $3 \cdot 7 = 21$.

$$
\begin{array}{c|cccc}
3 & 1 & 4 & -5 & 5 \\
& & 3 & 21 & 48 & \text{Add.} \\
\hline
& 1 & 7 & 16 & 53
\end{array}
$$

Multiply by 3: $3 \cdot 16 = 48$.

7. Use the numbers in the last row to write the quotient, plus the remainder above the divisor. **The degree of the first term of the quotient is one less than the degree of the first term of the dividend.** The final value in this row is the remainder.

Written from
1   7   16   53
the last row of the synthetic division

$$1x^2 + 7x + 16 + \dfrac{53}{x - 3}$$
$$x - 3\overline{)x^3 + 4x^2 - 5x + 5}$$

EXAMPLE 1   Using Synthetic Division

Use synthetic division to divide $5x^3 + 6x + 8$ by $x + 2$.

SOLUTION   The divisor must be in the form $x - c$. Thus, we write $x + 2$ as $x - (-2)$. This means that $c = -2$. Writing a 0 coefficient for the missing x^2-term in the dividend, we can express the division as follows:

$$x - (-2)\overline{)5x^3 + 0x^2 + 6x + 8}.$$

Now we are ready to set up the problem so that we can use synthetic division.

Use the coefficients of the dividend
$5x^3 + 0x^2 + 6x + 8$ in descending powers of x.

This is c in
$x - (-2)$.

$$-2 \enclose{verticalstrike}{} \quad 5 \quad 0 \quad 6 \quad 8$$

We begin the synthetic division process by bringing down 5. This is followed by a series of multiplications and additions.

1. Bring down 5.

$$-2 \enclose{verticalstrike}{} \quad 5 \quad 0 \quad 6 \quad 8$$
$$5$$

2. Multiply: $-2(5) = -10$.

$$-2 \enclose{verticalstrike}{} \quad 5 \quad 0 \quad 6 \quad 8$$
$$-10$$
$$5$$
Multiply by -2.

3. Add: $0 + (-10) = -10$.

$$-2 \enclose{verticalstrike}{} \quad 5 \quad 0 \quad 6 \quad 8$$
$$-10 \quad \text{Add.}$$
$$5 \quad -10$$

4. Multiply: $-2(-10) = 20$.

$$-2 \enclose{verticalstrike}{} \quad 5 \quad 0 \quad 6 \quad 8$$
$$-10 \quad 20$$
$$5 \quad -10$$
Multiply by -2.

5. Add: $6 + 20 = 26$.

$$-2 \enclose{verticalstrike}{} \quad 5 \quad 0 \quad 6 \quad 8$$
$$-10 \quad 20 \quad \text{Add.}$$
$$5 \quad -10 \quad 26$$

6. Multiply: $-2(26) = -52$.

$$-2 \enclose{verticalstrike}{} \quad 5 \quad 0 \quad 6 \quad 8$$
$$-10 \quad 20 \quad -52$$
$$5 \quad -10 \quad 26$$
Multiply by -2.

7. Add: $8 + (-52) = -44$.

$$-2 \enclose{verticalstrike}{} \quad 5 \quad 0 \quad 6 \quad 8$$
$$-10 \quad 20 \quad -52 \quad \text{Add.}$$
$$5 \quad -10 \quad 26 \quad -44$$

The numbers in the last row represent the coefficients of the quotient and the remainder. The degree of the first term of the quotient is one less than that of the dividend. Because the degree of the dividend, $5x^3 + 6x + 8$, is 3, the degree of the quotient is 2. This means that the 5 in the last row represents $5x^2$.

$$-2 \enclose{verticalstrike}{} \quad 5 \quad 0 \quad 6 \quad 8$$
$$-10 \quad 20 \quad -52$$
$$5 \quad -10 \quad 26 \quad -44$$

The quotient is $5x^2 - 10x + 26$.

The remainder is -44.

Thus,

$$x + 2 \overline{\smash{)}5x^3 + 6x + 8} = 5x^2 - 10x + 26 - \frac{44}{x + 2}$$

CHECK POINT **1** Use synthetic division to divide

$$x^3 - 7x - 6 \text{ by } x + 2. \qquad x^2 - 2x - 3$$

2 Evaluate a polynomial function using the Remainder Theorem.

The Remainder Theorem   We have seen that the answer to a long division problem can be checked: Find the product of the divisor and the quotient and add the remainder. The result should be the dividend. If the divisor is $x - c$, we can express this idea symbolically:

$$f(x) = (x - c)q(x) + r.$$

| Dividend | Divisor | Quotient | The remainder, r, is a constant when dividing by $x - c$. |

Now let's evaluate f at c.

$$f(c) = (c - c)q(c) + r \qquad \text{Find } f(c) \text{ by letting } x = c \text{ in } f(x) = (x - c)q(x) + r.$$
$$\text{This will give an expression for } r.$$

$$f(c) = 0 \cdot q(c) + r \qquad c - c = 0$$

$$f(c) = r \qquad 0 \cdot q(c) = 0 \text{ and } 0 + r = r.$$

What does this last equation mean? If a polynomial is divided by $x - c$, the remainder is the value of the polynomial at c. This result is called the **Remainder Theorem**.

> **THE REMAINDER THEOREM**   If the polynomial $f(x)$ is divided by $x - c$, then the remainder is $f(c)$.

Example 2 shows how we can use the Remainder Theorem to evaluate a polynomial function at 2. Rather than substituting 2 for x, we divide the function by $x - 2$. The remainder is $f(2)$.

EXAMPLE 2   Using the Remainder Theorem to Evaluate a Polynomial Function

Given $f(x) = x^3 - 4x^2 + 5x + 3$, use the Remainder Theorem to find $f(2)$.

SOLUTION   By the Remainder Theorem, if $f(x)$ is divided by $x - 2$, then the remainder is $f(2)$. We'll use synthetic division to divide.

$$
\begin{array}{r|rrrr}
2 & 1 & -4 & 5 & 3 \\
  &   & 2 & -4 & 2 \\
\hline
  & 1 & -2 & 1 & 5 \\
\end{array}
$$

Remainder

The remainder, 5, is the value of $f(2)$. Thus, $f(2) = 5$. We can verify that this is correct by evaluating $f(2)$ directly. Using $f(x) = x^3 - 4x^2 + 5x + 3$, we obtain

$$f(2) = 2^3 - 4 \cdot 2^2 + 5 \cdot 2 + 3 = 8 - 16 + 10 + 3 = 5.$$

■

 CHECK POINT 2   Given $f(x) = 3x^3 + 4x^2 - 5x + 3$, use the Remainder Theorem to find $f(-4)$. $f(-4) = -105$

3 Show that a number is a solution of a polynomial equation using the Remainder Theorem.

If the polynomial $f(x)$ is divided by $x - c$ and the remainder is zero, then $f(c) = 0$. This means that c is a solution of the polynomial equation $f(x) = 0$.

EXAMPLE 3 Using the Remainder Theorem

Show that 3 is a solution of the equation

$$2x^3 - 3x^2 - 11x + 6 = 0.$$

Then solve the polynomial equation.

SOLUTION One way to show that 3 is a solution is to substitute 3 for x in the equation and obtain 0. An easier way is to use synthetic division and the Remainder Theorem.

Proposed solution
$$
\begin{array}{r|rrrr}
3 & 2 & -3 & -11 & 6 \\
  &   & 6 & 9 & -6 \\
\hline
  & 2 & 3 & -2 & 0
\end{array}
$$

$$
\begin{array}{r}
2x^2 + 3x - 2 \\
x - 3 \overline{)\,2x^3 - 3x^2 - 11x + 6}
\end{array}
$$

Remainder

Equivalently,
$$2x^3 - 3x^2 - 11x + 6 = (x - 3)(2x^2 + 3x - 2).$$

Because the remainder is 0, the polynomial has a value of 0 when $x = 3$. Thus, 3 is a solution of the given equation.

The synthetic division also shows that $x - 3$ divides the polynomial with a zero remainder. Thus, $x - 3$ is a factor of the polynomial, as shown to the right of the synthetic division. The other factor is the quotient found in the last row of the synthetic division. Now we can solve the polynomial equation.

$2x^3 - 3x^2 - 11x + 6 = 0$	This is the given equation.
$(x - 3)(2x^2 + 3x - 2) = 0$	Factor using the result from the synthetic division.
$(x - 3)(2x - 1)(x + 2) = 0$	Factor the trinomial.
$x - 3 = 0$ or $2x - 1 = 0$ or $x + 2 = 0$	Set each factor equal to 0.
$x = 3$ $\qquad x = \frac{1}{2}$ $\qquad x = -2$	Solve for x.

The solutions are -2, $\frac{1}{2}$, and 3, and the solution set is $\left\{-2, \frac{1}{2}, 3\right\}$. ■

 CHECK POINT 3 Use synthetic division to show that -1 is a solution of the equation

$$15x^3 + 14x^2 - 3x - 2 = 0.$$

Then solve the polynomial equation.   The remainder is zero; $-1, -\frac{1}{3}$, and $\frac{2}{5}$, or $\left\{-1, -\frac{1}{3}, \frac{2}{5}\right\}$

USING TECHNOLOGY

Because the solution set of

$$2x^3 - 3x^2 - 11x + 6 = 0$$

is $\left\{-2, \frac{1}{2}, 3\right\}$, this implies that the polynomial function

$$f(x) = 2x^3 - 3x^2 - 11x + 6$$

has x-intercepts at -2, $\frac{1}{2}$, and 3. This is verified by the graph of f.

x-intercept: $\frac{1}{2}$

x-intercept: -2

x-intercept: 3

$[-10, 10, 1]$ by $[-15, 15, 1]$

6.5 EXERCISE SET

Student Solutions Manual CD/Video PH Math/Tutor Center MathXL Tutorials on CD MathXL® MyMathLab Interactmath.com

Practice Exercises

In Exercises 1–18, divide using synthetic division. In the first two exercises, begin the process as shown.

1. $(2x^2 + x - 10) \div (x - 2)$
$$
\begin{array}{r|rrr}
2 & 2 & 1 & -10
\end{array}
$$ $2x + 5$

2. $(x^2 + x - 2) \div (x - 1)$
$$
\begin{array}{r|rrr}
1 & 1 & 1 & -2
\end{array}
$$ $x + 2$

3. $(3x^2 + 7x - 20) \div (x + 5)$ $3x - 8 + \dfrac{20}{x + 5}$

4. $(5x^2 - 12x - 8) \div (x + 3)$ $5x - 27 + \dfrac{73}{x + 3}$

5. $(4x^3 - 3x^2 + 3x - 1) \div (x - 1)$ $4x^2 + x + 4 + \dfrac{3}{x - 1}$

6. $(5x^3 - 6x^2 + 3x + 11) \div (x - 2)$ $5x^2 + 4x + 11 + \dfrac{33}{x - 2}$

7. $(6x^5 - 2x^3 + 4x^2 - 3x + 1) \div (x - 2)$

$6x^4 + 12x^3 + 22x^2 + 48x + 93 + \dfrac{187}{x - 2}$

8. $(x^5 + 4x^4 - 3x^2 + 2x + 3) \div (x - 3)$

$x^4 + 7x^3 + 21x^2 + 60x + 182 + \dfrac{549}{x - 3}$

9. $(x^2 - 5x - 5x^3 + x^4) \div (5 + x)$

$x^3 - 10x^2 + 51x - 260 + \dfrac{1300}{5 + x}$

10. $(x^2 - 6x - 6x^3 + x^4) \div (6 + x)$

$x^3 - 12x^2 + 73x - 444 + \dfrac{2664}{6 + x}$

11. $(3x^3 + 2x^2 - 4x + 1) \div \left(x - \dfrac{1}{3}\right)$ $3x^2 + 3x - 3$

12. $(2x^4 - x^3 + 2x^2 - 3x + 1) \div \left(x - \dfrac{1}{2}\right)$ $2x^3 + 2x - 2$

13. $\dfrac{x^5 + x^3 - 2}{x - 1}$ $x^4 + x^3 + 2x^2 + 2x + 2$

14. $\dfrac{x^7 + x^5 - 10x^3 + 12}{x + 2}$

$x^6 - 2x^5 + 5x^4 - 10x^3 + 10x^2 - 20x + 40 - \dfrac{68}{x + 2}$

15. $\dfrac{x^4 - 256}{x - 4}$ $x^3 + 4x^2 + 16x + 64$

16. $\dfrac{x^7 - 128}{x - 2}$ $x^6 + 2x^5 + 4x^4 + 8x^3 + 16x^2 + 32x + 64$

17. $\dfrac{2x^5 - 3x^4 + x^3 - x^2 + 2x - 1}{x + 2}$

$2x^4 - 7x^3 + 15x^2 - 31x + 64 - \dfrac{129}{x + 2}$

18. $\dfrac{x^5 - 2x^4 - x^3 + 3x^2 - x + 1}{x - 2}$ $x^4 - x^2 + x + 1 + \dfrac{3}{x - 2}$

In Exercises 19–26, use synthetic division and the Remainder Theorem to find the indicated function value.

19. $f(x) = 2x^3 - 11x^2 + 7x - 5;$ $f(4)$ -25

20. $f(x) = x^3 - 7x^2 + 5x - 6;$ $f(3)$ -27

21. $f(x) = 3x^3 - 7x^2 - 2x + 5;$ $f(-3)$ -133

22. $f(x) = 4x^3 + 5x^2 - 6x - 4;$ $f(-2)$ -4

23. $f(x) = x^4 + 5x^3 + 5x^2 - 5x - 6;$ $f(3)$ 240

24. $f(x) = x^4 - 5x^3 + 5x^2 + 5x - 6;$ $f(2)$ 0

25. $f(x) = 2x^4 - 5x^3 - x^2 + 3x + 2;$ $f\left(-\dfrac{1}{2}\right)$ 1

26. $f(x) = 6x^4 + 10x^3 + 5x^2 + x + 1;$ $f\left(-\dfrac{2}{3}\right)$ $\dfrac{7}{9}$

In Exercises 27–32, use synthetic division to show that the number given to the right of each equation is a solution of the equation. Then solve the polynomial equation.

27. $x^3 - 4x^2 + x + 6 = 0;$ -1 The remainder is zero; $-1, 2,$ and 3, or $\{-1, 2, 3\}$

28. $x^3 - 2x^2 - x + 2 = 0;$ -1 The remainder is zero; $-1, 1,$ and 2, or $\{-1, 1, 2\}$

29. $2x^3 - 5x^2 + x + 2 = 0;$ 2 The remainder is zero; $-\dfrac{1}{2}, 1,$ and 2, or $\left\{-\dfrac{1}{2}, 1, 2\right\}$

30. $2x^3 - 3x^2 - 11x + 6 = 0;$ -2 The remainder is zero; $-2, \dfrac{1}{2},$ and 3, or $\left\{-2, \dfrac{1}{2}, 3\right\}$

31. $6x^3 + 25x^2 - 24x + 5 = 0;$ -5 The remainder is zero; $-5, \dfrac{1}{3},$ and $\dfrac{1}{2}$, or $\left\{-5, \dfrac{1}{3}, \dfrac{1}{2}\right\}$

32. $3x^3 + 7x^2 - 22x - 8 = 0;$ -4 The remainder is zero; $-4, -\dfrac{1}{3},$ and 2, or $\left\{-4, -\dfrac{1}{3}, 2\right\}$

Practice Plus

In Exercises 33–36, use the graph or the table to determine a solution of each equation. Use synthetic division to verify that this number is a solution of the equation. Then solve the polynomial equation.

33. $x^3 + 2x^2 - 5x - 6 = 0$

2; The remainder is zero; $-3, -1,$ and 2, or $\{-3, -1, 2\}$

$[0, 4, 1]$ by $[-25, 25, 5]$

34. $2x^3 + x^2 - 13x + 6 = 0$

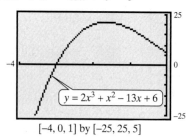

-3; The remainder is zero; $-3, \dfrac{1}{2},$ and 2, or $\left\{-3, \dfrac{1}{2}, 2\right\}$

$[-4, 0, 1]$ by $[-25, 25, 5]$

35. $6x^3 - 11x^2 + 6x - 1 = 0$

1; The remainder is zero; $\dfrac{1}{3}, \dfrac{1}{2},$ and 1, or $\left\{\dfrac{1}{3}, \dfrac{1}{2}, 1\right\}$

36. $2x^3 + 11x^2 - 7x - 6 = 0$

1; The remainder is zero; $-6, -\frac{1}{2}$, and 1, or $\left\{-6, -\frac{1}{2}, 1\right\}$

In Exercises 37–38, perform the given operations.

37. $(22x - 24 + 7x^3 - 29x^2 + 4x^4)(x + 4)^{-1}$ $\quad 4x^3 - 9x^2 + 7x - 6$

38. $(9 - x^2 + 6x + 2x^3)(x + 1)^{-1}$ $\quad 2x^2 - 3x + 9$

In Exercises 39–40, write a polynomial that represents the length of each rectangle.

39.

The width is $x + 0.2$ units.

The area is $0.5x^3 - 0.3x^2 + 0.22x + 0.06$ square units.

$0.5x^2 - 0.4x + 0.3$

40.

The width is $x + \frac{3}{4}$ units.

The area is $8x^3 - 6x^2 - 5x + 3$ square units.

$8x^2 - 12x + 4$

Application Exercises

41. a. Use synthetic division to show that 3 is a solution of the polynomial equation   The remainder is zero.

$$14x^3 - 17x^2 - 16x - 177 = 0.$$

b. Use the solution from part (a) to solve this problem. The number of eggs, $f(x)$, in a female moth is a function of her abdominal width, x, in millimeters, modeled by

$$f(x) = 14x^3 - 17x^2 - 16x + 34.$$

What is the abdominal width when there are 211 eggs?   3 millimeters

42. a. Use synthetic division to show that 2 is a solution of the polynomial equation   The remainder is zero.

$$2h^3 + 14h^2 - 72 = 0.$$

b. Use the solution from part (a) to solve this problem. The width of a rectangular box is twice the height and the length is 7 inches more than the height. If the volume is 72 cubic inches, find the dimensions of the box.

h

$2h$

$h + 7$

2 inches by 4 inches by 9 inches

Writing in Mathematics

43. Explain how to perform synthetic division. Use the division problem

$$(2x^3 - 3x^2 - 11x + 7) \div (x - 3)$$

to support your explanation.   43–46. Answers will vary.

44. State the Remainder Theorem.

45. Explain how the Remainder Theorem can be used to find $f(-6)$ if $f(x) = x^4 + 7x^3 + 8x^2 + 11x + 5$. What advantage is there to using the Remainder Theorem in this situation rather than evaluating $f(-6)$ directly?

46. Explain how the Remainder Theorem and synthetic division can be used to determine whether -4 is a solution of the following equation:

$$5x^3 + 22x^2 + x - 28 = 0.$$

Technology Exercise

47. For each equation that you solved in Exercises 27–32, use a graphing utility to graph the polynomial function on the left side of the equation. Use end behavior to obtain a complete graph. Then use the graph's x-intercepts to verify your solutions.   See graphing answer section.

Critical Thinking Exercises

48. Synthetic division is a process for dividing a polynomial by $x - c$. The coefficient of x is 1. How might synthetic division be used if you are dividing by $2x - 4$? $2x - 4 = 2(x - 2)$; Use synthetic division to divide by $x - 2$ and then divide the coefficients of the dividend by 2.

49. Use synthetic division to show that 5 is a solution of

$$x^4 - 4x^3 - 9x^2 + 16x + 20 = 0.$$

Then solve the polynomial equation. *Hint:* Use factoring by grouping when working with the quotient factor.   The remainder is zero; $-2, -1, 2$, and 5, or $\{-2, -1, 2, 5\}$

Review Exercises

50. Solve: $4x + 3 - 13x - 7 < 2(3 - 4x)$. (Section 4.1, Example 4) $\{x | x > -10\}$ or $(-10, \infty)$

51. Solve: $2x(x + 3) + 6(x - 3) = -28$. (Section 5.7, Example 2) -5 and -1, or $\{-5, -1\}$

52. Solve by Cramer's rule:

$$7x - 6y = 17$$
$$3x + y = 18.$$

(Section 3.5, Example 2) $(5, 3)$ or $\{(5, 3)\}$

Objectives

1 Solve rational equations.

2 Solve problems involving rational functions that model applied situations.

RATIONAL EQUATIONS

The lake is in one of the city's favorite parks and the time has come to clean it up. Voters in the city have committed $80 thousand for the cleanup. We know that

$$f(x) = \frac{120x}{100 - x}$$

models the cost, $f(x)$, in thousands of dollars, to remove $x\%$ of the lake's pollutants. What percentage of the pollutants can be removed for $80 thousand?

To determine the percentage, we use the given cost-benefit model. Voters have committed $80 thousand, so substitute 80 for $f(x)$:

$$80 = \frac{120x}{100 - x}.$$

> This equation contains a rational expression.

Now we have to solve the equation and find the value for x. This variable represents the percentage of the pollutants that can be removed for $80 thousand.

A **rational equation**, also called a **fractional equation**, is an equation containing one or more rational expressions. The equation shown above is an example of a rational equation. Do you see that there is a variable in a denominator? This is a characteristic of many rational equations. In this section, you will learn a procedure for solving such equations.

1 Solve rational equations.

Solving Rational Equations We have seen that the LCD is used to add and subtract rational expressions. By contrast, when solving rational equations, **the LCD is used as a multiplier that clears an equation of fractions**.

EXAMPLE 1 Solving a Rational Equation

Solve: $\dfrac{x + 4}{2x} + \dfrac{x + 20}{3x} = 3$.

SOLUTION Notice that the variable x appears in both denominators. We must avoid any values of the variable that make a denominator zero.

$$\frac{x + 4}{2x} + \frac{x + 20}{3x} = 3$$

> This denominator would equal zero if $x = 0$.

> This denominator would equal zero if $x = 0$.

We see that x cannot equal zero.

The denominators are $2x$ and $3x$. The least common denominator is $6x$. We begin by multiplying both sides of the equation by $6x$. We will also write the restriction that x cannot equal zero to the right of the equation.

$$\frac{x+4}{2x} + \frac{x+20}{3x} = 3, \ x \neq 0 \qquad \text{This is the given equation.}$$

$$6x\left(\frac{x+4}{2x} + \frac{x+20}{3x}\right) = 6x \cdot 3 \qquad \text{Multiply both sides by 6x, the LCD.}$$

$$\frac{6x}{1} \cdot \frac{x+4}{2x} + \frac{6x}{1} \cdot \frac{x+20}{3x} = 18x \qquad \text{Use the distributive property.}$$

$$3(x+4) + 2(x+20) = 18x \qquad \text{Divide out common factors in the multiplications.}$$

Observe that the equation is now cleared of fractions.

$$3x + 12 + 2x + 40 = 18x \qquad \text{Use the distributive property.}$$

$$5x + 52 = 18x \qquad \text{Combine like terms.}$$

$$52 = 13x \qquad \text{Subtract 5x from both sides.}$$

$$4 = x \qquad \text{Divide both sides by 13.}$$

The proposed solution, 4, is not part of the restriction $x \neq 0$. It should check in the original equation.

Check 4:

$$\frac{x+4}{2x} + \frac{x+20}{3x} = 3$$

$$\frac{4+4}{2\cdot 4} + \frac{4+20}{3\cdot 4} \overset{?}{=} 3$$

$$\frac{8}{8} + \frac{24}{12} \overset{?}{=} 3$$

$$1 + 2 \overset{?}{=} 3$$

$$3 = 3, \quad \text{true}$$

This true statement verifies that the solution is 4 and the solution set is $\{4\}$. ▪

 CHECK POINT 1 Solve: $\dfrac{x+6}{2x} + \dfrac{x+24}{5x} = 2.$ 6 or $\{6\}$

The following steps may be used to solve a rational equation:

SOLVING RATIONAL EQUATIONS

1. List restrictions on the variable. Avoid any values of the variable that make a denominator zero.

2. Clear the equation of fractions by multiplying both sides by the LCD of all rational expressions in the equation.

3. Solve the resulting equation.

4. Reject any proposed solution that is in the list of restrictions on the variable. Check other proposed solutions in the original equation.

EXAMPLE 2 Solving a Rational Equation

Solve: $\quad \dfrac{x+1}{x+10} = \dfrac{x-2}{x+4}.$

SOLUTION

Step 1. **List restrictions on the variable.**

$$\frac{x+1}{x+10} = \frac{x-2}{x+4}$$

This denominator would equal zero if $x = -10$.   This denominator would equal zero if $x = -4$.

The restrictions are $x \neq -10$ and $x \neq -4$.

Step 2. **Multiply both sides by the LCD.** The denominators are $x + 10$ and $x + 4$. Thus, the LCD is $(x + 10)(x + 4)$.

$$\frac{x+1}{x+10} = \frac{x-2}{x+4}, \quad x \neq -10, x \neq -4 \qquad \text{This is the given equation.}$$

$$(x+10)(x+4) \cdot \left(\frac{x+1}{x+10}\right) = (x+10)(x+4) \cdot \left(\frac{x-2}{x+4}\right) \qquad \text{Multiply both sides by the LCD.}$$

$$(x+4)(x+1) = (x+10)(x-2) \qquad \text{Simplify.}$$

Step 3. **Solve the resulting equation.**

$$(x+4)(x+1) = (x+10)(x-2) \qquad \text{This is the equation cleared of fractions.}$$

$$x^2 + 5x + 4 = x^2 + 8x - 20 \qquad \text{Use FOIL multiplication on each side.}$$

$$5x + 4 = 8x - 20 \qquad \text{Subtract } x^2 \text{ from both sides.}$$

$$-3x + 4 = -20 \qquad \text{Subtract 8x from both sides.}$$

$$-3x = -24 \qquad \text{Subtract 4 from both sides.}$$

$$x = 8 \qquad \text{Divide both sides by } -3.$$

Step 4. **Check the proposed solution in the original equation.** The proposed solution, 8, is not part of the restriction that $x \neq -10$ and $x \neq -4$. Substitute 8 for x in the given equation. You should obtain the true statement $\frac{1}{2} = \frac{1}{2}$. The solution is 8 and the solution set is $\{8\}$. ∎

CHECK POINT 2 Solve: $\quad \dfrac{x-3}{x+1} = \dfrac{x-2}{x+6}. \qquad$ 4 or $\{4\}$

EXAMPLE 3 Solving a Rational Equation

Solve: $\quad \dfrac{x}{x-3} = \dfrac{3}{x-3} + 9.$

SOLUTION

Step 1. **List restrictions on the variable.**

$$\frac{x}{x-3} = \frac{3}{x-3} + 9$$

These denominators are zero if $x = 3$.

The restriction is $x \neq 3$.

Step 2.   **Multiply both sides by the LCD.** The LCD is $x - 3$.

$$\frac{x}{x - 3} = \frac{3}{x - 3} + 9, \quad x \neq 3$$

This is the given equation.

$$(x - 3) \cdot \frac{x}{x - 3} = (x - 3)\left(\frac{3}{x - 3} + 9\right)$$

Multiply both sides by the LCD.

$$\cancel{(x - 3)} \cdot \frac{x}{\cancel{x - 3}} = \cancel{(x - 3)} \cdot \frac{3}{\cancel{x - 3}} + 9(x - 3)$$

Use the distributive property on the right side.

$$x = 3 + 9(x - 3)$$

Simplify.

Step 3.   **Solve the resulting equation.**

$$x = 3 + 9(x - 3)$$

This is the equation cleared of fractions.

$$x = 3 + 9x - 27$$

Use the distributive property on the right side.

$$x = 9x - 24$$

Combine numerical terms.

$$-8x = -24$$

Subtract 9x from both sides.

$$x = 3$$

Divide both sides by -8.

Step 4.   **Check proposed solutions.** The proposed solution, 3, is *not* a solution because of the restriction that $x \neq 3$. Notice that 3 makes both of the denominators zero in the original equation. There is no solution for this equation. The solution set is $\emptyset$, the empty set. ■

✔ **CHECK POINT 3** Solve: $\dfrac{8x}{x + 1} = 4 - \dfrac{8}{x + 1}$.   no solution or $\emptyset$

Examples 4 and 5 involve rational equations that become quadratic equations after clearing fractions.

EXAMPLE 4   Solving a Rational Equation

Solve: $\dfrac{x}{3} + \dfrac{9}{x} = 4$.

SOLUTION

Step 1.   **List restrictions on the variable.**

$$\frac{x}{3} + \frac{9}{x} = 4$$

This denominator would equal zero if $x = 0$.

The restriction is $x \neq 0$.

Step 2. **Multiply both sides by the LCD.** The denominators are 3 and x. Thus, the LCD is $3x$.

$$\frac{x}{3} + \frac{9}{x} = 4, \quad x \neq 0 \qquad \text{This is the given equation.}$$

$$3x\left(\frac{x}{3} + \frac{9}{x}\right) = 3x \cdot 4 \qquad \text{Multiply both sides by the LCD.}$$

$$3x \cdot \frac{x}{3} + 3x \cdot \frac{9}{x} = 12x \qquad \text{Use the distributive property on the left side.}$$

$$x^2 + 27 = 12x \qquad \text{Simplify.}$$

Step 3. **Solve the resulting equation.** Can you see that we have a quadratic equation? Write the equation in standard form and solve for x.

$$x^2 + 27 = 12x \qquad \text{This is the equation cleared of fractions.}$$

$$x^2 - 12x + 27 = 0 \qquad \text{Subtract 12x from both sides.}$$

$$(x - 9)(x - 3) = 0 \qquad \text{Factor.}$$

$$x - 9 = 0 \quad \text{or} \quad x - 3 = 0 \qquad \text{Set each factor equal to 0.}$$

$$x = 9 \qquad\qquad x = 3 \qquad \text{Solve the resulting equations.}$$

Step 4. **Check proposed solutions in the original equation.** The proposed solutions, 9 and 3, are not part of the restriction that $x \neq 0$. Substitute 9 for x, and then 3 for x, in the given equation. In each case, you should obtain the true statement $4 = 4$. The solutions are 3 and 9, and the solution set is $\{3, 9\}$. ∎

USING TECHNOLOGY

The graphs of

$$y_1 = \frac{x}{3} + \frac{9}{x}$$

and

$$y_2 = 4$$

have two intersection points. The x-coordinates of the points are 3 and 9. This verifies that

$$\frac{x}{3} + \frac{9}{x} = 4$$

has both 3 and 9 as solutions.

$[-2, 12, 1]$ by $[-1, 5, 1]$

✔ **CHECK POINT 4** Solve: $\dfrac{x}{2} + \dfrac{12}{x} = 5$.   4 and 6, or $\{4, 6\}$

EXAMPLE 5   Solving a Rational Equation

Solve: $\dfrac{2x}{x - 3} + \dfrac{6}{x + 3} = -\dfrac{28}{x^2 - 9}$.

SOLUTION

Step 1. **List restrictions on the variable.** By factoring denominators, it makes it easier to see values that make the denominators zero.

$$\frac{2x}{x - 3} + \frac{6}{x + 3} = -\frac{28}{(x + 3)(x - 3)}$$

This denominator is zero if $x = 3$.   This denominator is zero if $x = -3$.   This denominator is zero if $x = -3$ or $x = 3$.

The restrictions are $x \neq -3$ and $x \neq 3$.

Step 2. **Multiply both sides by the LCD.** The LCD is $(x + 3)(x - 3)$.

$$\frac{2x}{x - 3} + \frac{6}{x + 3} = -\frac{28}{(x + 3)(x - 3)}, \quad x \neq -3, x \neq 3$$

This is the given equation with a denominator factored.

$$(x + 3)(x - 3)\left(\frac{2x}{x - 3} + \frac{6}{x + 3}\right) = (x + 3)(x - 3)\left(-\frac{28}{(x + 3)(x - 3)}\right)$$

Multiply both sides by the LCD.

$$(x + 3)(x - 3) \cdot \frac{2x}{x - 3} + (x + 3)(x - 3) \cdot \frac{6}{x + 3} = (x + 3)(x - 3)\left(-\frac{28}{(x + 3)(x - 3)}\right)$$

Use the distributive property on the left side.

$$2x(x + 3) + 6(x - 3) = -28$$

Simplify.

Step 3. **Solve the resulting equation.**

$$2x(x + 3) + 6(x - 3) = -28$$ This is the equation cleared of fractions.

$$2x^2 + 6x + 6x - 18 = -28$$ Use the distributive property twice on the left side.

$$2x^2 + 12x - 18 = -28$$ Combine like terms.

$$2x^2 + 12x + 10 = 0$$ Add 28 to both sides and write the quadratic equation in standard form.

$$2(x^2 + 6x + 5) = 0$$ Factor out the GCF.

$$2(x + 5)(x + 1) = 0$$ Factor the trinomial.

$$x + 5 = 0 \quad \text{or} \quad x + 1 = 0$$ Set each variable factor equal to 0.

$$x = -5 \qquad\qquad x = -1$$ Solve for x.

Step 4. **Check the proposed solutions in the original equation.** The proposed solutions, -5 and -1, are not part of the restriction that $x \neq -3$ and $x \neq 3$. Substitute -5 for x, and then -1 for x, in the given equation. The resulting true statements verify that -5 and -1 are the solutions and that $\{-5, -1\}$ is the solution set. ∎

✔ **CHECK POINT 5** Solve: $\dfrac{3}{x - 3} + \dfrac{5}{x - 4} = \dfrac{x^2 - 20}{x^2 - 7x + 12}$.   1 and 7, or $\{1, 7\}$

2 Solve problems involving rational functions that model applied situations.

Applications of Rational Equations Solving rational equations can be used to answer questions about variables contained in rational functions.

EXAMPLE 6   Using a Cost-Benefit Model

The function

$$f(x) = \frac{120x}{100 - x}$$

models the cost, $f(x)$, in thousands of dollars, to remove $x\%$ of a lake's pollutants. If voters commit $80 thousand for this project, what percentage of the pollutants can be removed?

SOLUTION Substitute 80, the cost in thousands of dollars, for $f(x)$ and solve the resulting rational equation for x.

$$80 = \frac{120x}{100 - x}$$ The LCD is $100 - x$.

$$(100 - x)80 = \cancel{(100 - x)} \cdot \frac{120x}{\cancel{100 - x}}$$ Multiply both sides by the LCD.

$$80(100 - x) = 120x$$ Simplify.

$$8000 - 80x = 120x$$ Use the distributive property on the left side.

$$8000 = 200x$$ Add 80x to both sides.

$$40 = x$$ Divide both sides by 200.

If voters commit $80 thousand, 40% of the lake's pollutants can be removed. ∎

 CHECK POINT 6 Use the cost-benefit model in Example 6 to answer this question: If voters in the city commit $120 thousand for the project, what percentage of the lake's pollutants can be removed? If voters commit $120 thousand, 50% of the lake's pollutants can be removed.

6.6 EXERCISE SET

Student Solutions Manual | CD/Video | PH Math/Tutor Center | MathXL Tutorials on CD | MathXL® | MyMathLab | Interactmath.com

Practice Exercises

In Exercises 1–34, solve each rational equation. If an equation has no solution, so state.

1. $\dfrac{1}{x} + 2 = \dfrac{3}{x}$ 1 or {1}

2. $\dfrac{1}{x} - 3 = \dfrac{4}{x}$ −1 or {−1}

3. $\dfrac{5}{x} + \dfrac{1}{3} = \dfrac{6}{x}$ 3 or {3}

4. $\dfrac{4}{x} + \dfrac{1}{2} = \dfrac{5}{x}$ 2 or {2}

5. $\dfrac{x - 2}{2x} + 1 = \dfrac{x + 1}{x}$ 4 or {4}

6. $\dfrac{7x - 4}{5x} = \dfrac{9}{5} - \dfrac{4}{x}$ 8 or {8}

7. $\dfrac{3}{x + 1} = \dfrac{5}{x - 1}$ −4 or {−4}

8. $\dfrac{6}{x + 3} = \dfrac{4}{x - 3}$ 15 or {15}

9. $\dfrac{x - 6}{x + 5} = \dfrac{x - 3}{x + 1}$ $\dfrac{9}{7}$ or $\left\{\dfrac{9}{7}\right\}$

10. $\dfrac{x + 2}{x + 10} = \dfrac{x - 3}{x + 4}$ 38 or {38}

11. $\dfrac{x + 6}{x + 3} = \dfrac{3}{x + 3} + 2$ no solution or ∅

12. $\dfrac{3x + 1}{x - 4} = \dfrac{6x + 5}{2x - 7}$ no solution or ∅

13. $1 - \dfrac{4}{x + 7} = \dfrac{5}{x + 7}$ 2 or {2}

14. $5 - \dfrac{2}{x - 5} = \dfrac{3}{x - 5}$ 6 or {6}

15. $\dfrac{4x}{x + 2} + \dfrac{2}{x - 1} = 4$ 2 or {2}

16. $\dfrac{3x}{x + 1} + \dfrac{4}{x - 2} = 3$ −10 or {−10}

17. $\dfrac{8}{x^2 - 9} + \dfrac{4}{x + 3} = \dfrac{2}{x - 3}$ 5 or {5}

18. $\dfrac{32}{x^2 - 25} = \dfrac{4}{x + 5} + \dfrac{2}{x - 5}$ 7 or {7}

19. $x + \dfrac{7}{x} = -8$ −7 and −1, or {−7, −1}

20. $x + \dfrac{6}{x} = -7$ −6 and −1, or {−6, −1}

21. $\dfrac{6}{x} - \dfrac{x}{3} = 1$ −6 and 3, or {−6, 3}

22. $\dfrac{x}{2} - \dfrac{12}{x} = 1$ −4 and 6, or {−4, 6}

23. $\dfrac{x + 6}{3x - 12} = \dfrac{5}{x - 4} + \dfrac{2}{3}$ −1 or {−1}

24. $\dfrac{1}{5x + 5} = \dfrac{3}{x + 1} - \dfrac{7}{5}$ 1 or {1}

25. $\dfrac{1}{x-1} + \dfrac{1}{x+1} = \dfrac{2}{x^2-1}$ no solution or $\varnothing$

26. $\dfrac{1}{x-2} + \dfrac{1}{x+2} = \dfrac{4}{x^2-4}$ no solution or $\varnothing$

27. $\dfrac{5}{x+4} + \dfrac{3}{x+3} = \dfrac{12x+19}{x^2+7x+12}$ 2 or $\{2\}$

28. $\dfrac{2x-1}{x^2+2x-8} + \dfrac{2}{x+4} = \dfrac{1}{x-2}$ 3 or $\{3\}$

29. $\dfrac{4x}{x+3} - \dfrac{12}{x-3} = \dfrac{4x^2+36}{x^2-9}$ no solution or $\varnothing$

30. $\dfrac{2}{x+3} - \dfrac{5}{x+1} = \dfrac{3x+5}{x^2+4x+3}$ no solution or $\varnothing$

31. $\dfrac{4}{x^2+3x-10} + \dfrac{1}{x^2+9x+20} = \dfrac{2}{x^2+2x-8}$ $-\dfrac{4}{3}$ or $\left\{-\dfrac{4}{3}\right\}$

32. $\dfrac{4}{x^2+3x-10} - \dfrac{1}{x^2+x-6} = \dfrac{3}{x^2-x-12}$ $\dfrac{1}{7}$ or $\left\{\dfrac{1}{7}\right\}$

33. $\dfrac{3y}{y^2+5y+6} + \dfrac{2}{y^2+y-2} = \dfrac{5y}{y^2+2y-3}$

-6 and $\dfrac{1}{2}$, or $\left\{-6, \dfrac{1}{2}\right\}$

34. $\dfrac{y-1}{y^2-4} + \dfrac{y}{y^2-y-2} = \dfrac{2y-1}{y^2+3y+2}$ $\dfrac{3}{7}$ or $\left\{\dfrac{3}{7}\right\}$

In Exercises 35–38, a rational function g is given. Find all values of a for which g(a) is the indicated value.

35. $g(x) = \dfrac{x}{2} + \dfrac{20}{x}; g(a) = 7$ 4, 10

36. $g(x) = \dfrac{x}{4} + \dfrac{5}{x}; g(a) = 3$ 2, 10

37. $g(x) = \dfrac{5}{x+2} + \dfrac{25}{x^2+4x+4}; g(a) = 20$ $-3, -\dfrac{3}{4}$

38. $g(x) = \dfrac{3x-2}{x+1} + \dfrac{x+2}{x-1}; g(a) = 4$ 4

Practice Plus

In Exercises 39–46, solve or simplify, whichever is appropriate.

39. $\dfrac{x+2}{x^2-x} - \dfrac{6}{x^2-1}$ $\dfrac{x-2}{x(x+1)}$

40. $\dfrac{x+3}{x^2-x} - \dfrac{8}{x^2-1}$ $\dfrac{x-3}{x(x+1)}$

41. $\dfrac{x+2}{x^2-x} - \dfrac{6}{x^2-1} = 0$ 2 or $\{2\}$

42. $\dfrac{x+3}{x^2-x} - \dfrac{8}{x^2-1} = 0$ 3 or $\{3\}$

43. $\dfrac{1}{x^3-8} + \dfrac{3}{(x-2)(x^2+2x+4)} = \dfrac{2}{x^2+2x+4}$ 4 or $\{4\}$

44. $\dfrac{2}{x^3-1} + \dfrac{4}{(x-1)(x^2+x+1)} = -\dfrac{1}{x^2+x+1}$ -5 or $\{-5\}$

45. $\dfrac{1}{x^3-8} + \dfrac{3}{(x-2)(x^2+2x+4)} - \dfrac{2}{x^2+2x+4}$

$\dfrac{-2(x-4)}{(x-2)(x^2+2x+4)}$

46. $\dfrac{2}{x^3-1} + \dfrac{4}{(x-1)(x^2+x+1)} + \dfrac{1}{x^2+x+1}$

$\dfrac{x+5}{(x-1)(x^2+x+1)}$

In Exercises 47–48, find all values of a for which f(a) = g(a) + 1.

47. $f(x) = \dfrac{x+2}{x+3}, g(x) = \dfrac{x+1}{x^2+2x-3}$ 0

48. $f(x) = \dfrac{4}{x-3}, g(x) = \dfrac{10}{x^2+x-12}$ $-3, 6$

In Exercises 49–50, find all values of a for which $(f+g)(a) = h(a)$.

49. $f(x) = \dfrac{5}{x-4}, g(x) = \dfrac{3}{x-3}, h(x) = \dfrac{x^2-20}{x^2-7x+12}$ 1, 7

50. $f(x) = \dfrac{6}{x+3}, g(x) = \dfrac{2x}{x-3}, h(x) = -\dfrac{28}{x^2-9}$ $-5, -1$

Application Exercises

The function

$$f(x) = \dfrac{250x}{100-x}$$

models the cost, $f(x)$, in millions of dollars, to remove $x\%$ of a river's pollutants. Use this function to solve Exercises 51–52.

51. If the government commits \$375 million for this project, what percentage of the pollutants can be removed?
60% of pollutants can be removed.

52. If the government commits \$750 million for this project, what percentage of the pollutants can be removed?
75% of pollutants can be removed.

In an experiment about memory, students in a language class are asked to memorize 40 vocabulary words in Latin, a language with which they are not familiar. After studying the words for one day, students are tested each day thereafter to see how many words they remember. The class average is then found. The function

$$f(x) = \dfrac{5x+30}{x}$$

models the average number of Latin words remembered by the students, $f(x)$, after x days. The graph of the rational function is shown. Use the function to solve Exercises 53–56.

Average Number of Words Remembered Over Time

$f(x) = \dfrac{5x+30}{x}$

Average Number of Words the Group Remembers

Number of Days

(In Exercises 53–56, be sure to refer to the function and the graph on the previous page.)

53. After how many days do the students remember 8 words? Identify your solution as a point on the graph.   after 10 days; (10, 8)

54. After how many days do the students remember 7 words? Identify your solution as a point on the graph.   after 15 days; (15, 7)

55. What is the horizontal asymptote of the graph? Describe what this means about the average number of Latin words remembered by the students over an extended period of time. $y = 5$; On average, the students remembered 5 words over an extended period of time.

56. According to the graph, between which two days do students forget the most? Describe the trend shown by the memory function's graph.   between the first and second days; Answers will vary.

Rational functions can be used to model learning. Many of these functions model the proportion of correct responses as a function of the number of trials of a particular task. One such model, called a learning curve, is

$$f(x) = \frac{0.9x - 0.4}{0.9x + 0.1}$$

where $f(x)$ is the proportion of correct responses after x trials. If $f(x) = 0$, there are no correct responses. If $f(x) = 1$, all responses are correct. The graph of the rational function is shown. Use the function to solve Exercises 57–60.

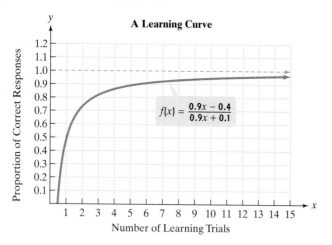

A Learning Curve

Proportion of Correct Responses vs. Number of Learning Trials

$$f(x) = \frac{0.9x - 0.4}{0.9x + 0.1}$$

57. How many learning trials are necessary for 0.95 of the responses to be correct? Identify your solution as a point on the graph.   11 learning trials; (11, 0.95)

58. How many learning trials are necessary for 0.5 of the responses to be correct? Identify your solution as a point on the graph.   1 learning trial; (1, 0.5)

59. Describe the trend shown by the graph in terms of learning new tasks. What happens initially and what happens as time increases?   As the number of learning trials increases, the proportion of correct responses increases; Initially the proportion of correct responses increases rapidly, but slows down as time increases.

60. What is the horizontal asymptote of the graph? Once the performance level approaches peak efficiency, what effect does additional practice have on performance? Describe how this is shown by the graph. $y = 1$; Additional practice has little effect when performance is near peak efficiency; The graph is leveling off.

61. A company wants to increase the 10% peroxide content of its product by adding pure peroxide (100% peroxide). If x liters of pure peroxide are added to 500 liters of its 10% solution, the concentration, $C(x)$, of the new mixture is given by

$$C(x) = \frac{x + 0.1(500)}{x + 500}.$$

How many liters of pure peroxide should be added to produce a new product that is 28% peroxide?   125 liters

62. Suppose that x liters of pure acid are added to 200 liters of a 35% acid solution.

   a. Write a function that gives the concentration, $C(x)$, of the new mixture. (*Hint:* See Exercise 61.) $C(x) = \dfrac{x + 0.35(200)}{x + 200}$

   b. How many liters of pure acid should be added to produce a new mixture that is 74% acid?   300 liters

Writing in Mathematics

63. What is a rational equation?   63–69. Answers will vary.

64. Explain how to solve a rational equation.

65. Explain how to find restrictions on the variable in a rational equation.

66. Why should restrictions on the variable in a rational equation be listed before you begin solving the equation?

67. Describe similarities and differences between the procedures needed to solve the following problems:

$$\text{Add: } \frac{2}{x} + \frac{3}{4}. \qquad \text{Solve: } \frac{2}{x} + \frac{3}{4} = 1.$$

68. Rational functions model learning and forgetting. Use the graphs in Exercises 53–56 and in Exercises 57–60 to describe one similarity and one difference between learning and forgetting over time.

69. Does the graph of the learning curve shown in Exercises 57–60 indicate that practice makes perfect? Explain. Does this have anything to do with what psychologists call "the curse of perfection"?

Technology Exercises

In Exercises 70–74, use a graphing utility to solve each rational equation. Graph each side of the equation in the given viewing rectangle. The solution is the first coordinate of the point(s) of intersection. Check by direct substitution.   70–74. See graphing answer section.

70. $\dfrac{x}{2} + \dfrac{x}{4} = 6$

   $[-5, 10, 1]$ by $[-5, 10, 1]$ 8 or $\{8\}$

71. $\dfrac{50}{x} = 2x$

$[-10, 10, 1]$ by $[-20, 20, 2]$ -5 and 5, or $\{-5, 5\}$

72. $x + \dfrac{6}{x} = -5$

$[-10, 10, 1]$ by $[-7, 10, 1]$ -3 and -2, or $\{-3, -2\}$

73. $\dfrac{2}{x} = x + 1$

$[-5, 5, 1]$ by $[-5, 5, 1]$ -2 and 1, or $\{-2, 1\}$

74. $\dfrac{3}{x} - \dfrac{x + 21}{3x} = \dfrac{5}{3}$

$[-5, 5, 1]$ by $[-5, 5, 1]$ -2 or $\{-2\}$

Critical Thinking Exercises

75. Which one of the following is true?   d

a. Once a rational equation is cleared of fractions, all solutions of the resulting equation are also solutions of the rational equation.

b. We find

$$\dfrac{4}{x} - \dfrac{2}{x + 1}$$

by multiplying each term by the LCD, $x(x + 1)$, thereby clearing fractions.

c. All real numbers satisfy the equation $\dfrac{7}{x} - \dfrac{2}{x} = \dfrac{5}{x}$.

d. None of the above is true.

76. Solve: $\left(\dfrac{1}{x + 1} + \dfrac{x}{1 - x}\right) \div \left(\dfrac{x}{x + 1} - \dfrac{1}{x - 1}\right) = -1$.

no solution or $\varnothing$

77. Solve: $\left|\dfrac{x + 1}{x + 8}\right| = \dfrac{2}{3}$. $-\dfrac{19}{5}$ and 13, or $\left\{-\dfrac{19}{5}, 13\right\}$

78. Write an original rational equation that has no solution.
Answers will vary.

79. Solve

$$\left(\dfrac{4}{x - 1}\right)^2 + 2\left(\dfrac{4}{x - 1}\right) + 1 = 0$$

by introducing the substitution $t = \dfrac{4}{x - 1}$. -3 or $\{-3\}$

Review Exercises

80. Graph the solution set:

$$x + 2y \geq 2$$
$$x - y \geq -4.$$

(Section 4.4, Example 4)   See graphing answer section.

81. Solve:

$$\dfrac{x - 4}{2} - \dfrac{1}{5} = \dfrac{7x + 1}{20}.$$

(Section 1.4, Example 4) 15 or $\{15\}$

82. Solve for F:

$$C = \dfrac{5F - 160}{9}.$$

(Section 1.5, Example 8) $F = \dfrac{9C + 160}{5}$

SECTION **6.7**

FORMULAS AND APPLICATIONS OF RATIONAL EQUATIONS

Objectives

1 Solve a formula with a rational expression for a variable.

2 Solve business problems involving average cost.

3 Solve problems involving time in motion.

4 Solve problems involving work.

The edges between reality and virtual reality are becoming blurred. The people who built your computer talk of massive bandwidth that will give you the visual experience, in high-definition 3-D format, of being in the same room with a person who is actually in another city. It is already possible to "experience" an exciting activity such as skiing down a mountain.

As computers become more powerful, technology is now promising to make the virtual experience even more real and bring it to our home computer. The cost of manufacturing computers with massive bandwidth can be modeled by rational functions. In this section, you will see how high production levels of "tele-immersion" computers can eventually make them affordable for home use. Get ready to feel the wind in your hair, the frost on your eyebrows, and the gentle heat of the sun on your face, as well as the shudder through your ski boots as you race down mountains without ever leaving your home.

1 Solve a formula with a rational expression for a variable.

Solving a Formula for a Variable Formulas and mathematical models frequently contain rational expressions. We solve for a specified variable using the procedure for solving rational equations. The goal is to get the specified variable alone on one side of the equation. To do so, collect all terms with this variable on one side and all other terms on the other side. It is sometimes necessary to factor out the variable you are solving for.

FIGURE 6.5

EXAMPLE 1 Solving for a Variable in a Formula

If you wear glasses, did you know that each lens has a measurement called its focal length, f? When an object is in focus, its distance from the lens, p, and the distance from the lens to your retina, q, satisfy the formula

$$\frac{1}{p} + \frac{1}{q} = \frac{1}{f}.$$

(See Figure 6.5.) Solve this formula for p.

SOLUTION Our goal is to isolate the variable p. We begin by multiplying both sides by the least common denominator, pqf, to clear the equation of fractions.

$$\frac{1}{p} + \frac{1}{q} = \frac{1}{f}$$ This is the given formula.

$$pqf\left(\frac{1}{p} + \frac{1}{q}\right) = pqf\left(\frac{1}{f}\right)$$ Multiply both sides by pqf, the LCD.

$$pqf\left(\frac{1}{p}\right) + pqf\left(\frac{1}{q}\right) = pqf\left(\frac{1}{f}\right)$$ Use the distributive property on the left side.

$$qf + pf = pq$$ Simplify.

Observe that the formula is now cleared of fractions. Collect terms with p, the specified variable, on one side of the equation. To do so, subtract pf from both sides.

$$qf + pf = pq$$ This is the equation cleared of fractions.

$$qf = pq - pf$$ Subtract pf from both sides.

$$qf = p(q - f)$$ Factor out p, the specified variable.

$$\frac{qf}{q - f} = \frac{p(q - f)}{q - f}$$ Divide both sides by $q - f$ and solve for p.

$$\frac{qf}{q - f} = p$$ Simplify.

■

✔ CHECK POINT 1 Solve $\dfrac{1}{x} + \dfrac{1}{y} = \dfrac{1}{z}$ for x. $x = \dfrac{yz}{y - z}$

2 Solve business problems involving average cost.

Business Problems Involving Average Cost   We have seen that the cost function for a business is the sum of its fixed and variable costs:

$$C(x) = (\text{fixed cost}) + cx.$$

> Cost per unit times the number of units produced, x.

The **average cost** per unit for a company to produce x units is the sum of its fixed and variable costs divided by the number of units produced. The **average cost function** is a rational function that is denoted by $\overline{C}$. Thus,

> Cost of producing x units: fixed plus variable costs

$$\overline{C}(x) = \frac{(\text{fixed cost}) + cx}{x}.$$

> Number of units produced

EXAMPLE 2   Average Cost for a Business

The time: the not-too-distant future. A new company is hoping to replace traditional computers and two-dimensional monitors with its virtual reality system. The company's fixed monthly cost will be $600,000 and it will cost $500 to produce each system.

   a. Write the cost function, C, of producing x virtual reality systems.

   b. Write the average cost function, $\overline{C}$, of producing x virtual reality systems.

   c. How many virtual reality systems must be produced each month for the company to reach an average cost of $2000 per system?

SOLUTION

   a. The cost function, C, is the sum of the fixed cost and and the variable costs.

$$C(x) = 600,000 + 500x$$

> Fixed cost is $600,000.   Variable cost: $500 for each system produced

   b. The average cost function, $\overline{C}$, is the sum of fixed and variable costs divided by the number of virtual reality systems produced.

$$\overline{C}(x) = \frac{600,000 + 500x}{x}$$

   c. We are interested in the company's production level that results in an average cost of $2000 per virtual reality system. Substitute 2000, the average cost, for $\overline{C}(x)$ and solve the resulting rational equation for x.

$$2000 = \frac{600,000 + 500x}{x} \qquad \text{Substitute 2000 for } \overline{C}(x).$$

$$2000x = 600,000 + 500x \qquad \text{Multiply both sides by the LCD, } x.$$

$$1500x = 600,000 \qquad \text{Subtract 500x from both sides.}$$

$$x = 400 \qquad \text{Divide both sides by 1500.}$$

The company must produce 400 virtual reality systems each month for an average cost of $2000 per system. ∎

Figure 6.6 shows the graph of the average cost function in Example 2. As the production level increases, the average cost of producing each virtual reality system decreases. The horizontal asymptote, $y = 500$, is also shown in the figure. This means that the more systems produced each month, the closer the average cost per system for the company comes to $500. The least possible cost per virtual reality system is approaching $500. Competitively low prices take place with high production levels, posing a major problem for small businesses.

$$\overline{C}(x) = \frac{600,000 + 500x}{x}$$

$y = 500$

Number of Virtual Reality Systems Produced per Month

FIGURE 6.6

 CHECK POINT 2 A company is planning to manufacture wheelchairs that are light, fast, and beautiful. Fixed monthly cost will be $500,000 and it will cost $400 to produce each radically innovative chair.

a. Write the cost function, C, of producing x wheelchairs. $C(x) = 500,000 + 400x$

b. Write the average cost function, $\overline{C}$, of producing x wheelchairs. $\overline{C}(x) = \dfrac{500,000 + 400x}{x}$

c. How many wheelchairs must be produced each month for the company to reach an average cost of $450 per chair?   10,000 wheelchairs must be produced each month for the company to reach an average cost of $450 per chair.

3 Solve problems involving time in motion.

Problems Involving Motion   We have seen that the distance, d, covered by any moving body is the product of its average rate, r, and its time in motion, t: $d = rt$. Rational expressions appear in motion problems when the conditions of the problem involve the time traveled. We can obtain an expression for t, the time traveled, by dividing both sides of $d = rt$ by r.

$$d = rt \qquad \text{Distance equals rate times time.}$$

$$\frac{d}{r} = \frac{rt}{r} \qquad \text{Divide both sides by } r.$$

$$\frac{d}{r} = t \qquad \text{Simplify.}$$

TIME IN MOTION

$$t = \frac{d}{r}$$

$$\text{Time traveled} = \frac{\text{Distance traveled}}{\text{Rate of travel}}$$

EXAMPLE 3 A Motion Problem Involving Time

You commute to work a distance of 40 miles and return on the same route at the end of the day. Your average rate on the return trip is 30 miles per hour faster than your average rate on the outgoing trip. If the round trip takes 2 hours, what is your average rate on the outgoing trip to work?

SOLUTION

Step 1. **Let x represent one of the quantities.** Let

$$x = \text{the rate on the outgoing trip.}$$

Step 2. **Represent other quantities in terms of x.** Because the average rate on the return trip is 30 miles per hour faster than the average rate on the outgoing trip, let

$$x + 30 = \text{the rate on the return trip.}$$

Step 3. **Write an equation that describes the conditions.** By reading the problem again, we discover that the crucial idea is that the time for the round trip is 2 hours. Thus, the time on the outgoing trip plus the time on the return trip is 2 hours.

	Distance	Rate	Time $= \dfrac{\text{Distance}}{\text{Rate}}$
Outgoing Trip	40	x	$\dfrac{40}{x}$
Return Trip	40	$x + 30$	$\dfrac{40}{x + 30}$

The sum of these times is 2 hours.

We are now ready to write an equation that describes the problem's conditions.

Time on the outgoing trip plus time on the return trip equals 2 hours.

$$\frac{40}{x} + \frac{40}{x + 30} = 2$$

Step 4. **Solve the equation and answer the question.**

$$\frac{40}{x} + \frac{40}{x + 30} = 2 \qquad \text{This is the equation for the problem's conditions.}$$

$$x(x + 30)\left(\frac{40}{x} + \frac{40}{x + 30}\right) = 2x(x + 30) \qquad \text{Multiply both sides by the LCD, } x(x + 30).$$

$$\cancel{x}(x + 30) \cdot \frac{40}{\cancel{x}} + x\cancel{(x + 30)} \cdot \frac{40}{\cancel{x + 30}} = 2x^2 + 60x \qquad \text{Use the distributive property on each side.}$$

$$40(x + 30) + 40x = 2x^2 + 60x \qquad \text{Simplify.}$$

$$40x + 1200 + 40x = 2x^2 + 60x \qquad \text{Use the distributive property.}$$

$$80x + 1200 = 2x^2 + 60x \qquad \text{Combine like terms: } 40x + 40x = 80x.$$

The graph of the rational function for the time for the round trip

$$f(x) = \frac{40}{x} + \frac{40}{x + 30}$$

is shown in a $[0, 60, 30]$ by $[0, 10, 1]$ viewing rectangle.

At an outgoing rate of 30 miles per hour, the trip takes 2 hours.

The graph is falling from left to right. This shows that the time for the round trip decreases as the rate increases.

$$0 = 2x^2 - 20x - 1200 \qquad \text{Subtract } 80x + 1200 \text{ from both sides.}$$

$$0 = 2(x^2 - 10x - 600) \qquad \text{Factor out the GCF.}$$

$$0 = 2(x - 30)(x + 20) \qquad \text{Factor completely.}$$

$$x - 30 = 0 \quad \text{or} \quad x + 20 = 0 \qquad \text{Set each variable factor equal to 0.}$$

$$x = 30 \qquad\qquad x = -20 \qquad \text{Solve for x.}$$

Because x represents the rate on the outgoing trip, we reject the negative value, -20. The rate on the outgoing trip is 30 miles per hour. At an outgoing rate of 30 miles per hour, the trip should take 2 hours.

Step 5. **Check the proposed solution in the original wording of the problem.** Does the round trip take 2 hours? Because the rate on the return trip is 30 miles per hour faster than the rate on the outgoing trip, the rate on the return trip is $30 + 30$, or 60 miles per hour.

$$\text{Time on the outgoing trip} = \frac{\text{Distance}}{\text{Rate}} = \frac{40}{30} = \frac{4}{3} \text{ hours}$$

$$\text{Time on the return trip} = \frac{\text{Distance}}{\text{Rate}} = \frac{40}{60} = \frac{2}{3} \text{ hour}$$

The total time for the round trip is $\frac{4}{3} + \frac{2}{3} = \frac{6}{3}$, or 2 hours. This checks with the original conditions of the problem. ■

 CHECK POINT 3 After riding at a steady speed for 40 miles, a bicyclist had a flat tire and walked 5 miles to a repair shop. The cycling rate was 4 times faster than the walking rate. If the time spent cycling and walking was 5 hours, at what rate was the cyclist riding? *The cyclist was riding at a rate of 12 miles per hour.*

4 Solve problems involving work.

Problems Involving Work You are thinking of designing your own Web site. You estimate that it will take 15 hours to do the job. In 1 hour, $\frac{1}{15}$ of the job is completed. In 2 hours, $\frac{2}{15}$ of the job is completed. In 3 hours, the fractional part of the job done is $\frac{3}{15}$, or $\frac{1}{5}$. In x hours, the fractional part of the job that you can complete is $\frac{x}{15}$.

Your friend, who has experience developing Web sites, took 10 hours working on her own to design an impressive site. You wonder about the possibility of working together. How long would it take both of you to design your Web site?

Problems involving work usually have two (or more) people or machines working together to complete a job. The amount of time it takes each person to do the job working alone is frequently known. The question deals with how long it will take both people working together to complete the job.

In work problems, **the number 1 represents one whole job completed**. For example, the completion of your Web site is represented by 1. Equations in work problems are often based on the following condition:

| Fractional part of the job done by the first person | + | fractional part of the job done by the second person | = | 1 (one whole job completed). |

EXAMPLE 4 Solving a Problem Involving Work

You can design a Web site in 15 hours. Your friend can design the same site in 10 hours. How long will it take to design the Web site if you both work together?

SOLUTION

Step 1. **Let x represent one of the quantities.** Let

$$x = \text{the time, in hours, for you and your friend}$$
$$\text{working together to design the Web site.}$$

Step 2. **Represent other quantities in terms of x.** Because there are no other unknown quantities, we can skip this step.

Step 3. **Write an equation that describes the conditions.** We construct a table to help find the fractional parts of the task completed by you and your friend in x hours.

		Fractional part of job completed in 1 hour	Time working together	Fractional part of job completed in x hours
You can design the site in 15 hours.	**You**	$\dfrac{1}{15}$	x	$\dfrac{x}{15}$
Your friend can design the site in 10 hours.	**Your friend**	$\dfrac{1}{10}$	x	$\dfrac{x}{10}$

Fractional part of the job done by you	$+$	fractional part of the job done by your friend	$=$	one whole job.
$\dfrac{x}{15}$	$+$	$\dfrac{x}{10}$	$=$	1

Step 4. **Solve the equation and answer the question.**

$$\frac{x}{15} + \frac{x}{10} = 1 \qquad \text{This is the equation for the problem's conditions.}$$

$$30\left(\frac{x}{15} + \frac{x}{10}\right) = 30 \cdot 1 \qquad \text{Multiply both sides by 30, the LCD.}$$

$$30 \cdot \frac{x}{15} + 30 \cdot \frac{x}{10} = 30 \qquad \text{Use the distributive property on the left side.}$$

$$2x + 3x = 30 \qquad \text{Simplify: } \frac{\overset{2}{\cancel{30}}}{1} \cdot \frac{x}{\cancel{15}} = 2x \text{ and } \frac{\overset{3}{\cancel{30}}}{1} \cdot \frac{x}{\cancel{10}} = 3x.$$

$$5x = 30 \qquad \text{Combine like terms.}$$

$$x = 6 \qquad \text{Divide both sides by 5.}$$

If you both work together, you can design your Web site in 6 hours.

Step 5. **Check the proposed solution in the original wording of the problem.** Will you both complete the job in 6 hours? Because you can design the site in 15 hours, in 6 hours, you can complete $\frac{6}{15}$, or $\frac{2}{5}$, of the job. Because your friend can design the site in 10 hours, in 6 hours, she can complete $\frac{6}{10}$, or $\frac{3}{5}$, of the job. Notice that $\frac{2}{5} + \frac{3}{5} = 1$, which represents the completion of the entire job, or one whole job. ∎

STUDY TIP

Let

a = the time it takes person A to do a job working alone, and

b = the time it takes person B to do the same job working alone.

If x represents the time it takes for A and B to complete the entire job working together, then the situation can be modeled by the rational equation

$$\frac{x}{a} + \frac{x}{b} = 1.$$

✔ **CHECK POINT 4** A new underwater tunnel is being built using tunnel-boring machines that begin at opposite ends of the tunnel. One tunnel-boring machine can complete the tunnel in 18 months. A faster machine can tunnel to the other side in 9 months. If both machines start at opposite ends and work at the same time, in how many months will the tunnel be finished?   6 months

---

EXAMPLE 5   Solving a Problem Involving Work

After designing your Web site, you and your friend decide to go into business setting up sites for others. With lots of practice, you can now work together and design a modest site in 4 hours. Your friend is still a faster worker. Working alone, you require 6 more hours than she does to design a site for a client. How many hours does it take your friend to design a Web site if she works alone?

SOLUTION

Step 1. **Let x represent one of the quantities.** Let

x = the time, in hours, for your friend to design a Web site working alone.

Step 2. **Represent other quantities in terms of x.** Because you require 6 more hours than your friend to do the job, let

$x + 6$ = the time, in hours, for you to design a Web site working alone.

Step 3. **Write an equation that describes the conditions.** Working together, you and your friend can complete the job in 4 hours. We construct a table to find the fractional part of the task completed by you and your friend in 4 hours.

		Fractional part of job completed in 1 hour	Time working together	Fractional part of job completed in 4 hours
Your friend can design the site in x hours.	**Your friend**	$\dfrac{1}{x}$	4	$\dfrac{4}{x}$
You can design the site in $x + 6$ hours.	**You**	$\dfrac{1}{x + 6}$	4	$\dfrac{4}{x + 6}$

Because you can both complete the job in 4 hours,

| Fractional part of the job done by your friend | + | fractional part of the job done by you | = | one whole job. |

$$\frac{4}{x} + \frac{4}{x+6} = 1.$$

Step 4. Solve the equation and answer the question.

$$\frac{4}{x} + \frac{4}{x+6} = 1 \qquad \text{This is the equation for the problem's conditions.}$$

$$x(x+6)\left(\frac{4}{x} + \frac{4}{x+6}\right) = x(x+6) \cdot 1 \qquad \text{Multiply both sides by } x(x+6), \text{ the LCD.}$$

$$\cancel{x}(x+6) \cdot \frac{4}{\cancel{x}} + x\cancel{(x+6)} \cdot \frac{4}{\cancel{(x+6)}} = x^2 + 6x \qquad \text{Use the distributive property on each side.}$$

$$4(x+6) + 4x = x^2 + 6x \qquad \text{Simplify.}$$

$$4x + 24 + 4x = x^2 + 6x \qquad \text{Use the distributive property.}$$

$$8x + 24 = x^2 + 6x \qquad \text{Combine like terms: } 4x + 4x = 8x.$$

$$0 = x^2 - 2x - 24 \qquad \text{Subtract } 8x + 24 \text{ from both sides and write the quadratic equation in standard form.}$$

$$0 = (x-6)(x+4) \qquad \text{Factor.}$$

$$x - 6 = 0 \quad \text{or} \quad x + 4 = 0 \qquad \text{Set each factor equal to 0.}$$

$$x = 6 \qquad\qquad x = -4 \qquad \text{Solve for x.}$$

Because x represents the time for your friend to design a Web site working alone, we reject the negative value, -4. Your friend can design a Web site in 6 hours.

Step 5. Check the proposed solution in the original wording of the problem. Will you both complete the job in 4 hours? Working alone, your friend takes 6 hours. Because you require 6 more hours than your friend, you require 12 hours to complete the job on your own.

In 4 hours, your friend completes $\frac{4}{6} = \frac{2}{3}$ of the job.

In 4 hours, you complete $\frac{4}{12} = \frac{1}{3}$ of the job.

Notice that $\frac{2}{3} + \frac{1}{3} = 1$, which represents the completion of the entire job, or the design of one Web site. ■

  CHECK POINT **5** An experienced carpenter can panel a room 3 times faster than an apprentice can. Working together, they can panel the room in 6 hours. How long would it take each person working alone to do the job? The experienced carpenter can panel a room in 8 hours and the apprentice in 24 hours, working alone.

6.7 EXERCISE SET

Student Solutions Manual   CD/Video   PH Math/Tutor Center   MathXL Tutorials on CD   MathXL®   MyMathLab   Interactmath.com

Practice Exercises

In Exercises 1–14, solve each formula for the specified variable.

1. $\dfrac{V_1}{V_2} = \dfrac{P_2}{P_1}$ for P_1 (chemistry) $P_1 = \dfrac{P_2 V_2}{V_1}$

2. $\dfrac{V_1}{V_2} = \dfrac{P_2}{P_1}$ for V_2 (chemistry) $V_2 = \dfrac{V_1 P_1}{P_2}$

3. $\dfrac{1}{p} + \dfrac{1}{q} = \dfrac{1}{f}$ for f (optics) $f = \dfrac{pq}{q + p}$

4. $\dfrac{1}{p} + \dfrac{1}{q} = \dfrac{1}{f}$ for q (optics) $q = \dfrac{pf}{p - f}$

5. $P = \dfrac{A}{1 + r}$ for r (investment) $r = \dfrac{A - P}{P}$

6. $S = \dfrac{a}{1 - r}$ for r (mathematics) $r = \dfrac{S - a}{S}$ or $r = -\dfrac{a - S}{S}$

7. $F = \dfrac{G m_1 m_2}{d^2}$ for m_1 (physics) $m_1 = \dfrac{F d^2}{G m_2}$

8. $F = \dfrac{G m_1 m_2}{d^2}$ for m_2 (physics) $m_2 = \dfrac{F d^2}{G m_1}$

9. $z = \dfrac{x - \bar{x}}{s}$ for x (statistics) $x = \bar{x} + zs$

10. $z = \dfrac{x - \bar{x}}{s}$ for s (statistics) $s = \dfrac{x - \bar{x}}{z}$

11. $I = \dfrac{E}{R + r}$ for R (electronics) $R = \dfrac{E - Ir}{I}$

12. $I = \dfrac{E}{R + r}$ for r (electronics) $r = \dfrac{E - IR}{I}$

13. $f = \dfrac{f_1 f_2}{f_1 + f_2}$ for f_1 (optics) $f_1 = \dfrac{f f_2}{f_2 - f}$

14. $f = \dfrac{f_1 f_2}{f_1 + f_2}$ for f_2 (optics) $f_2 = \dfrac{f f_1}{f_1 - f}$

Application Exercises

The figure shows the graph of the average cost function for the company described in Check Point 2 that manufactures wheelchairs. Use the graph to solve Exercises 15–18.

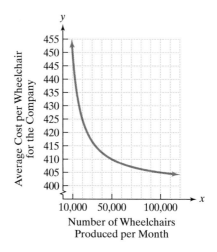

15. How many wheelchairs must be produced each month for the company to have an average cost of $410 per chair? 50,000 wheelchairs must be produced.

16. How many wheelchairs must be produced each month for the company to have an average cost of $425 per chair? 20,000 wheelchairs must be produced.

17. What is the horizontal asymptote of the graph? What is the meaning of the horizontal asymptote as production level increases? $y = 400$; Average cost is nearing $400 as production level increases.

18. Describe the end behavior of the graph at the far right. Is there a production level that results in an average cost of $400 per chair? Explain your answer   Average cost is decreasing and approaching $400 as production per month increases; Average cost will never be $400 per chair.

19. A company is planning to manufacture mountain bikes. Fixed monthly cost will be $100,000 and it will cost $100 to produce each bicycle.

    a. Write the cost function, C, of producing x mountain bikes. $C(x) = 100,000 + 100x$

    b. Write the average cost function, $\bar{C}$, of producing x mountain bikes. $\bar{C}(x) = \dfrac{100,000 + 100x}{x}$

    c. How many mountain bikes must be produced each month for the company to have an average cost of $300 per bike?   500 mountain bikes must be produced each month.

20. A company is planning to manufacture small canoes. Fixed monthly cost will be $20,000 and it will cost $20 to produce each canoe.

    a. Write the cost function, C, of producing x canoes.
    $C(x) = 20,000 + 20x$

    b. Write the average cost function, $\overline{C}$, of producing x canoes.
$$\overline{C}(x) = \frac{20,000 + 20x}{x}$$

    c. How many canoes must be produced each month for the company to have an average cost of $40 per canoe?
    1000 canoes must be produced each month.

It's vacation time. You drive 90 miles along a scenic highway and then take a 5-mile run along a hiking trail. Your driving rate is nine times that of your running rate. The graph shows the total time you spend driving and running, $f(x)$, as a function of your running rate, x. Use the graph of this rational function to solve Exercises 21–24.

21. If the total time for driving and running is 3 hours, what is your running rate?   Running rate is 5 miles per hour.

22. If the total time for driving and running is 5 hours, what is your running rate?   Running rate is 3 miles per hour.

23. Describe the behavior of the graph as x approaches 0. What does this show about the time driving and running as your running rate is close to zero miles per hour (a stupefied crawl)?   The time increases as the running rate is close to zero miles per hour.

24. The graph is falling from left to right. What does this show?   As the running rate increases, the time decreases.

Use a rational equation to solve Exercises 25–36. Each exercise is a problem involving motion.

25. A car can travel 300 miles in the same amount of time it takes a bus to travel 180 miles. If the rate of the bus is 20 miles per hour slower than the rate of the car, find the average rate for each.   car: 50 miles per hour; bus: 30 miles per hour

26. A passenger train can travel 240 miles in the same amount of time it takes a freight train to travel 160 miles. If the rate of the freight train is 20 miles per hour slower than the rate of the passenger train, find the average rate of each.   passenger train: 60 miles per hour; freight train: 40 miles per hour

27. You ride your bike to campus a distance of 5 miles and return home on the same route. Going to campus, you ride mostly downhill and average 9 miles per hour faster than on your return trip home. If the round trip takes one hour and ten minutes—that is $\frac{7}{6}$ hours—what is your average rate on the return trip?   6 miles per hour

28. An engine pulls a train 140 miles. Then a second engine, whose average rate is 5 miles per hour faster than the first engine, takes over and pulls the train 200 miles. The total time required for both engines is 9 hours. Find the average rate of each engine.   first engine: 35 miles per hour; second engine: 40 miles per hour

29. In still water, a boat averages 7 miles per hour. It takes the same amount of time to travel 20 miles downstream, with the current, as 8 miles upstream, against the current. What is the rate of the water's current?   3 miles per hour

	Distance	Rate	Time = $\dfrac{\text{Distance}}{\text{Rate}}$
With the Current	20	$7 + x$	
Against the Current	8	$7 - x$	

These times are equal.

30. In still water, a boat averages 8 miles per hour. It takes the same amount of time to travel 30 miles downstream, with the current, as 18 miles upstream, against the current. What is the rate of the water's current?   2 miles per hour

	Distance	Rate	Time = $\dfrac{\text{Distance}}{\text{Rate}}$
With the Current	30	$8 + x$	
Against the Current	18	$8 - x$	

These times are equal.

31. The rate of the jet stream is 100 miles per hour. Traveling with the jet stream, an airplane can fly 2400 miles in the same amount of time as it takes to fly 1600 miles against the jet stream. What is the airplane's average rate in calm air? 500 miles per hour

32. The wind is blowing at an average rate of 10 miles per hour. Riding with the wind, a bicyclist can cycle 75 miles in the same amount of time it takes to cycle 15 miles against the wind. What is the bicyclist's average rate in calm air? 15 miles per hour

33. A moving sidewalk at an airport glides at a rate of 1.8 feet per second. Walking on the moving sidewalk, you travel 100 feet forward in the same time it takes to travel 40 feet in the opposite direction. Find your walking speed on a nonmoving sidewalk. 4.2 feet per second

34. A moving sidewalk at an airport glides at a rate of 1.8 feet per second. Walking on the moving sidewalk, you travel 105 feet forward in the same time it takes to travel 50 feet in the opposite direction. Find your walking speed on a nonmoving sidewalk. Round to the nearest tenth. about 5.1 feet per second

35. Two runners, one averaging 8 miles per hour and the other 6 miles per hour, start at the same place and run along the same trail. The slower runner arrives at the end of the trail a half hour after the faster runner. How far did each person run? 12 miles

36. Two sailboats, one averaging 20 miles per hour and the other 18 miles per hour, start at the same place and follow the same course. The slower boat arrives at the end of the course $\frac{1}{6}$ of an hour after the faster sailboat. How far did each boat travel? 30 miles

Use a rational equation to solve Exercises 37–48. Each exercise is a problem involving work.

37. You promised your parents that you would wash the family car. You have not started the job and they are due home in 20 minutes. You can wash the car in 45 minutes and your sister claims she can do it in 30 minutes. If you work together, how long will it take to do the job? Will this give you enough time before your parents return? It will take 18 minutes, which is enough time before your parents return.

38. You must leave for campus in half an hour, or you will be late for class. Unfortunately, you are snowed in. You can shovel the driveway in 45 minutes and your brother claims he can do it in 36 minutes. If you shovel together, how long will it take to clear the driveway? Will this give you enough time before you have to leave? It will take 20 minutes, which is enough time before you have to leave.

39. A pool can be filled by one pipe in 6 hours and by a second pipe in 12 hours. How long will it take using both pipes to fill the pool? It will take 4 hours to fill the pool.

40. A pond can be filled by one pipe in 8 hours and by a second pipe in 24 hours. How long will it take using both pipes to fill the pond? It will 6 hours to fill the pond.

41. Working with your cousin, you can refinish a table in 3 hours. Working alone, your cousin can complete the job in 4 hours. How long would it take you to refinish the table working alone? It would take 12 hours working alone.

42. Working with your cousin, you can split a cord of firewood in 5 hours. Working alone, your cousin can complete the job in 7 hours. How long would it take you to split the firewood working alone? It would take 17.5 hours working alone.

43. An earthquake strikes and an isolated area is without food or water. Three crews arrive. One can dispense needed supplies in 20 hours, a second in 30 hours, and a third in 60 hours. How long will it take all three crews working together to dispense food and water? It will take 10 hours for all three crews working together.

44. A hurricane strikes and a rural area is without food or water. Three crews arrive. One can dispense needed supplies in 10 hours, a second in 15 hours, and a third in 20 hours. How long will it take all three crews working together to dispense food and water? It will take about 4.6 hours, or about 4 hours and 37 minutes, for all three crews working together.

45. An office has an old copying machine and a new one. Working together, it takes both machines 6 hours to make all the copies of the annual financial report. Working alone, it takes the old copying machine 5 hours longer than the new one to make all the copies of the report. How long would it take the new copying machine to make all the copies working alone? 10 hours

46. A demolition company wants to build a brick wall to hide from public view the area where they store wrecked cars. Working together, an experienced bricklayer and an apprentice can build the wall in 12 hours. Working alone, it takes the apprentice 10 hours longer than the experienced bricklayer to do the job. How long would it take the experienced bricklayer to build the wall working alone? 20 hours

47. A faucet can fill a sink in 5 minutes. It takes twice that long for the drain to empty the sink. How long will it take to fill the sink if the drain is open and the faucet is on? 10 minutes

48. A pool can be filled by a pipe in 3 hours. It takes 3 times as long for another pipe to empty the pool. How long will it take to fill the pool if both pipes are open? 4.5 hours

Exercises 49–56 contain a variety of problems. Use a rational equation to solve each exercise.

49. What number multiplied by the numerator and added to the denominator of $\frac{4}{5}$ makes the resulting fraction equivalent to $\frac{3}{2}$? 3

50. What number multiplied by the numerator and subtracted from the denominator of $\frac{9}{11}$ makes the resulting fraction equivalent to $-\frac{12}{5}$? −4

51. The sum of 2 times a number and twice its reciprocal is $\frac{20}{3}$. Find the number(s). $\frac{1}{3}$ and 3

52. If 2 times the reciprocal of a number is subtracted from 3 times the number, the difference is 1. Find the number(s). $-\frac{2}{3}$ and 1

53. You have 35 hits in 140 times at bat. Your batting average is $\frac{35}{140}$, or 0.25. How many consecutive hits must you get to increase your batting average to 0.30? 10

54. You have 30 hits in 120 times at bat. Your batting average is $\frac{30}{120}$, or 0.25. How many consecutive hits must you get to increase your batting average to 0.28?   5

55. If one pipe can fill a pool in a hours and a second pipe can fill the pool in b hours, write a formula for the time, x, in terms of a and b, for the number of hours it takes both pipes, working together, to fill the pool. $x = \dfrac{ab}{a + b}$

56. If one pipe can fill a pool in a hours and a second pipe can empty the pool in b hours, write a formula for the time, x, in terms of a and b, for the number of hours it takes to fill the pool with both of these pipes open. $x = \dfrac{ab}{b - a}$

Writing in Mathematics

57. Without showing the details, explain how to solve the formula

$$\frac{1}{R} = \frac{1}{R_1} + \frac{1}{R_2}$$

for R_1. (The formula is used in electronics.)   57–63. Answers will vary.

58. Explain how to find the average cost function for a business.

59. How does the average cost function illustrate a problem for small businesses?

60. What is the relationship among time traveled, distance traveled, and rate of travel?

61. If you know how many hours it takes for you to do a job, explain how to find the fractional part of the job you can complete in x hours.

62. If you can do a job in 6 hours and your friend can do the same job in 3 hours, explain how to find how long it takes to complete the job working together. It is not necessary to solve the problem.

63. When two people work together to complete a job, describe one factor that can result in more or less time than the time given by the rational equations we have been using.

Technology Exercises

64. For Exercises 19–20, use a graphing utility to graph the average cost function described by the problem's conditions. Then $\boxed{\text{TRACE}}$ along the curve and find the point that visually shows the solution in part (c).
See graphing answer section.

65. For Exercises 45–46, use a graphing utility to graph the function representing the sum of the fractional parts of the job done by the two machines or the two people. Then $\boxed{\text{TRACE}}$ along the curve and find the point that visually shows the problem's solution.   See graphing answer section.

66. A boat can travel 10 miles per hour in still water. The boat travels 24 miles upstream, against the current, and then 24 miles downstream, with the current.

   a. Let $x =$ the rate of the current. Write a function in terms of x that models the total time for the boat to travel upstream and downstream. $y = \dfrac{24}{10 + x} + \dfrac{24}{10 - x}$

   b. Use a graphing utility to graph the rational function in part (a).   See graphing answer section.

   c. $\boxed{\text{TRACE}}$ along the curve and determine the current's rate if the trip's time is 5 hours. Then verify this result algebraically.   2 miles per hour

Critical Thinking Exercises

67. Which one of the following is true?   c

   a. As production level increases, the average cost for a company to produce each unit of its product also increases.

   b. To solve $qf + pf = pq$ for p, subtract qf from both sides and then divide by f.

   c. If you can clean the house in 3 hours and your sloppy friend can completely mess it up in 6 hours, then $\dfrac{x}{3} - \dfrac{x}{6} = 1$ can be used to find how long it takes to clean the house if you both "work" together.

   d. None of the above is true.

68. Solve $\dfrac{1}{s} = f + \dfrac{1 - f}{p}$ for f. $f = \dfrac{p - s}{ps - s}$

69. A new schedule for a train requires it to travel 351 miles in $\frac{1}{4}$ hour less time than before. To accomplish this, the rate of the train must be increased by 2 miles per hour. What should the average rate of the train be so that it can keep on the new schedule?   The average rate of the train should be 54 miles per hour.

70. It takes Mr. Todd 4 hours longer to prepare an order of pies than it takes Mrs. Lovett. They bake together for 2 hours when Mrs. Lovett leaves. Mr. Todd takes 7 additional hours to complete the work. Working alone, how long does it take Mrs. Lovett to prepare the pies?   It takes Mrs. Lovett 8 hours to prepare the pies.

Review Exercises

71. Factor: $x^2 + 4x + 4 - 9y^2$. (Section 5.6, Example 4) $(x + 2 + 3y)(x + 2 - 3y)$

72. Solve using matrices:

$$2x + 5y = -5$$
$$x + 2y = -1.$$

(Section 3.4, Example 3) $(5, -3)$ or $\{(5, -3)\}$

73. Solve the system:

$$x + y + z = 4$$
$$2x + 5y \quad\;\; = 1$$
$$x - y - 2z = 0.$$

(Section 3.3, Example 3) $(3, -1, 2)$ or $\{(3, -1, 2)\}$

Objectives

1 Solve direct variation problems.

2 Solve inverse variation problems.

3 Solve combined variation problems.

4 Solve problems involving joint variation.

MODELING USING VARIATION

Have you ever wondered how telecommunication companies estimate the number of phone calls expected per day between two cities? The formula

$$C = \frac{0.02 P_1 P_2}{d^2}$$

shows that the daily number of phone calls, C, increases as the populations of the cities, P_1 and P_2, in thousands, increase and decreases as the distance, d, between the cities increases.

Certain formulas occur so frequently in applied situations that they are given special names. Variation formulas show how one quantity changes in relation to other quantities. Quantities can vary *directly*, *inversely*, or *jointly*. In this section, we look at situations that can be modeled by each of these kinds of variation. And think of this. The next time you get one of those "all-circuits-are-busy" messages, you will be able to use a variation formula to estimate how many other callers you're competing with for those precious 5-cent minutes.

1 Solve direct variation problems.

Direct Variation When you swim underwater, the pressure in your ears depends on the depth at which you are swimming. The formula

$$p = 0.43d$$

describes the water pressure, p, in pounds per square inch, at a depth of d feet. We can use this linear function to determine the pressure in your ears at various depths:

If $d = 20$, $p = 0.43(20) = 8.6$. At a depth of 20 feet, water pressure is 8.6 pounds per square inch.

Doubling the depth doubles the pressure.

If $d = 40$, $p = 0.43(40) = 17.2$. At a depth of 40 feet, water pressure is 17.2 pounds per square inch.

Doubling the depth doubles the pressure.

If $d = 80$, $p = 0.43(80) = 34.4$. At a depth of 80 feet, water pressure is 34.4 pounds per square inch.

The formula $p = 0.43d$ illustrates that water pressure is a constant multiple of your underwater depth. If your depth is doubled, the pressure is doubled; if your depth is tripled, the pressure is triped; and so on. Because of this, the pressure in your ears is said to **vary directly** as your underwater depth. The **equation of variation** is

$$p = 0.43d.$$

Generalizing, we obtain the following statement:

DIRECT VARIATION   If a situation is described by an equation in the form

$$y = kx$$

where k is a constant, we say that **y varies directly as x**. The number k is called the **constant of variation**.

Can you see that the direct variation equation, $y = kx$, is a special case of the linear function $y = mx + b$? When $m = k$ and $b = 0$, $y = mx + b$ becomes $y = kx$. Thus, the slope of a direct variation equation is k, the constant of variation. Because b, the y-intercept, is 0, the graph of a direct variation equation is a line through the origin. This is illustrated in Figure 6.7, which shows the graph of $p = 0.43d$: Water pressure varies directly as depth.

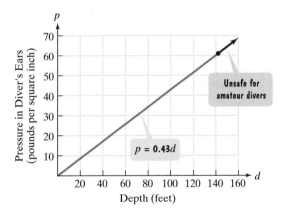

FIGURE 6.7 Water pressure at various depths

Problems involving direct variation can be solved using the following procedure. This procedure applies to direct variation problems, as well as to the other kinds of variation problems that we will discuss.

SOLVING VARIATION PROBLEMS

1. Write an equation that describes the given English statement.
2. Substitute the given pair of values into the equation in step 1 and find the value of k.
3. Substitute the value of k into the equation in step 1.
4. Use the equation from step 3 to answer the problem's question.

EXAMPLE 1 Solving a Direct Variation Problem

Many areas of Northern California depend on the snowpack of the Sierra Nevada mountain range for their water supply. The volume of water produced from melting snow varies directly as the volume of snow. Meteorologists have determined that 250 cubic centimeters of snow will melt to 28 cubic centimeters of water. How much water does 1200 cubic centimeters of melting snow produce?

SOLUTION

Step 1. **Write an equation.** We know that y varies directly as x is expressed as

$$y = kx.$$

By changing letters, we can write an equation that describes the following English statement: Volume of water, W, varies directly as volume of snow, S.

$$W = kS$$

Step 2. **Use the given values to find k.** We are told that 250 cubic centimeters of snow will melt to 28 cubic centimeters of water. Substitute 28 for W and 250 for S in the direct variation equation. Then solve for k.

$W = kS$	Volume of water varies directly as volume of melting snow.
$28 = k(250)$	250 cubic centimeters of snow melt to 28 cubic centimeters of water.
$\dfrac{28}{250} = \dfrac{k(250)}{250}$	Divide both sides by 250.
$0.112 = k$	Simplify.

Step 3. **Substitute the value of k into the equation.**

$W = kS$	This is the equation from step 1.
$W = 0.112S$	Replace k, the constant of variation, with 0.112.

Step 4. **Answer the problem's question.** How much water does 1200 cubic centimeters of melting snow produce? Substitute 1200 for S in $W = 0.112S$ and solve for W.

$W = 0.112S$	Use the equation from step 3.
$W = 0.112(1200)$	Substitute 1200 for S.
$W = 134.4$	Multiply.

A snowpack measuring 1200 cubic centimeters will produce 134.4 cubic centimeters of water. ■

 CHECK POINT 1 The number of gallons of water, W, used when taking a shower varies directly as the time, t, in minutes, in the shower. A shower lasting 5 minutes uses 30 gallons of water. How much water is used in a shower lasting 11 minutes?   66 gallons.

The direct variation equation $y = kx$ is a linear function. If $k > 0$, then the slope of the line is positive. Consequently, as x increases, y also increases.

A direct variation situation can involve variables to higher powers. For example, y can vary directly as x^2 ($y = kx^2$) or as x^3 ($y = kx^3$).

DIRECT VARIATION WITH POWERS **y varies directly as the nth power of x** if there exists some nonzero constant k such that

$$y = kx^n.$$

Direct variation with whole number powers is modeled by polynomial functions. In our next example, the graph of the variation equation is the graph of a quadratic function.

EXAMPLE 2 Solving a Direct Variation Problem

The distance, s, that a body falls from rest varies directly as the square of the time, t, of the fall. If skydivers fall 64 feet in 2 seconds, how far will they fall in 4.5 seconds?

SOLUTION

Step 1. **Write an equation.** We know that y varies directly as the square of x is expressed as

$$y = kx^2.$$

By changing letters, we can write an equation that describes the following English statement: Distance, s, varies directly as the square of time, t, of the fall.

$$s = kt^2$$

Step 2. **Use the given values to find k.** Skydivers fall 64 feet in 2 seconds. Substitute 64 for s and 2 for t in the direct variation equation. Then solve for k.

$s = kt^2$	Distance varies directly as the square of time.
$64 = k \cdot 2^2$	Skydivers fall 64 feet in 2 seconds.
$64 = 4k$	Simplify: $2^2 = 4$.
$\dfrac{64}{4} = \dfrac{4k}{4}$	Divide both sides by 4.
$16 = k$	Simplify.

Step 3. **Substitute the value of k into the equation.**

$s = kt^2$	Use the equation from step 1.
$s = 16t^2$	Replace k, the constant of variation, with 16.

Step 4. **Answer the problem's question.** How far will the skydivers fall in 4.5 seconds? Substitute 4.5 for t in $s = 16t^2$ and solve for s.

$$s = 16(4.5)^2 = 16(20.25) = 324$$

Thus, in 4.5 seconds, the skydivers will fall 324 feet.

Distance Fallen by Skydivers over Time

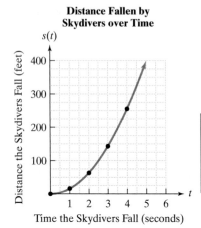

FIGURE 6.8   The graph of $s(t) = 16t^2$

2   Solve inverse variation problems.

We can express the variation equation from Example 2 in function notation, writing

$$s(t) = 16t^2.$$

The distance that a body falls from rest is a function of the time, t, of the fall. The graph of this quadratic function is shown in Figure 6.8. The graph increases rapidly from left to right, showing the effects of the acceleration of gravity.

 CHECK POINT 2   The distance required to stop a car varies directly as the square of its speed. If 200 feet are required to stop a car traveling 60 miles per hour, how many feet are required to stop a car traveling 100 miles per hour?   About 556 feet are required to stop a car traveling 100 miles per hour.

Inverse Variation   The distance from San Francisco to Los Angeles is 420 miles. The time that it takes to drive from San Francisco to Los Angeles depends on the rate at which one drives and is given by

$$\text{Time} = \frac{420}{\text{Rate}}.$$

For example, if you average 30 miles per hour, the time for the drive is

$$\text{Time} = \frac{420}{30} = 14,$$

or 14 hours. If you average 50 miles per hour, the time for the drive is

$$\text{Time} = \frac{420}{50} = 8.4,$$

or 8.4 hours. As your rate (or speed) increases, the time for the trip decreases and vice versa. This is illustrated by the graph in Figure 6.9.

FIGURE 6.9

We can express the time for the San Francisco–Los Angeles trip using t for time and r for rate:

$$t = \frac{420}{r}.$$

This equation is an example of an **inverse variation** equation. Time, t, **varies inversely** as rate, r. When two quantities vary inversely, one quantity increases as the other decreases, and vice versa.

Generalizing, we obtain the following statement:

INVERSE VARIATION   If a situation is described by an equation in the form

$$y = \frac{k}{x}$$

where k is a constant, we say that y **varies inversely as x**. The number k is called the **constant of variation**.

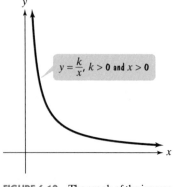

FIGURE 6.10   The graph of the inverse variation equation

Notice that the inverse variation equation

$$y = \frac{k}{x}, \quad \text{or} \quad f(x) = \frac{k}{x},$$

is a rational function. For $k > 0$ and $x > 0$, the graph of the function takes on the shape shown in Figure 6.10.

We use the same procedure to solve inverse variation problems as we did to solve direct variation problems. Example 3 illustrates this procedure.

EXAMPLE 3   Solving an Inverse Variation Problem

P $2P$

$2V$

V

Doubling the pressure halves the volume.

When you use a spray can and press the valve at the top, you decrease the pressure of the gas in the can. This decrease of pressure causes the volume of the gas in the can to increase. Because the gas needs more room than is provided in the can, it expands in spray form through the small hole near the valve. In general, if the temperature is constant, the pressure, P, of a gas in a container varies inversely as the volume, V, of the container. The pressure of a gas sample in a container whose volume is 8 cubic inches is 12 pounds per square inch. If the sample expands to a volume of 22 cubic inches, what is the new pressure of the gas?

SOLUTION

Step 1.   **Write an equation.** We know that y varies inversely as x is expressed as

$$y = \frac{k}{x}.$$

By changing letters, we can write an equation that describes the following English statement: The pressure, P, of a gas in a container varies inversely as the volume, V.

$$P = \frac{k}{V}.$$

Step 2.   **Use the given values to find k.** The pressure of a gas sample in a container whose volume is 8 cubic inches is 12 pounds per square inch. Substitute 12 for P and 8 for V in the inverse variation equation. Then solve for k.

$$P = \frac{k}{V} \qquad \text{Pressure varies inversely as volume.}$$

$$12 = \frac{k}{8} \qquad \text{The pressure in an 8-cubic-inch container is 12 pounds per square inch.}$$

$$12 \cdot 8 = \frac{k}{8} \cdot 8 \qquad \text{Multiply both sides by 8.}$$

$$96 = k \qquad \text{Simplify.}$$

Step 3.   **Substitute the value of k into the equation.**

$$P = \frac{k}{V} \qquad \text{Use the equation from step 1.}$$

$$P = \frac{96}{V} \qquad \text{Replace } k \text{, the constant of variation, with 96.}$$

Step 4.   **Answer the problem's question.** We need to find the pressure when the volume expands to 22 cubic inches. Substitute 22 for V and solve for P.

$$P = \frac{96}{V} = \frac{96}{22} = 4\frac{4}{11}$$

When the volume is 22 cubic inches, the pressure of the gas is $4\frac{4}{11}$ pounds per square inch.

> ✓ CHECK POINT 3 The length of a violin string varies inversely as the frequency of its vibrations. A violin string 8 inches long vibrates at a frequency of 640 cycles per second. What is the frequency of a 10-inch string? 512 cycles per second

3 Solve combined variation problems.

Combined Variation In **combined variation**, direct and inverse variation occur at the same time. For example, as the advertising budget, A, of a company increases, its monthly sales, S, also increase. Monthly sales vary directly as the advertising budget:

$$S = kA.$$

By contrast, as the price of the company's product, P, increases, its monthly sales, S, decrease. Monthly sales vary inversely as the price of the product:

$$S = \frac{k}{P}.$$

We can combine these two variation equations into one combined equation:

$$S = \frac{kA}{P}.$$

Monthly sales , S, vary directly as the advertising budget, A, and inversely as the price of the product, P.

The following example illustrates an application of combined variation.

EXAMPLE 4 Solving a Combined Variation Problem

The owners of Rollerblades Now determine that the monthly sales, S, of its skates vary directly as its advertising budget, A, and inversely as the price of the skates, P. When $60,000 is spent on advertising and the price of the skates is $40, the monthly sales are 12,000 pairs of rollerblades.

a. Write an equation of variation that describes this situation.

b. Determine monthly sales if the amount of the advertising budget is increased to $70,000.

SOLUTION

a. Write an equation.

$$S = \frac{kA}{P}.$$

Translate "sales vary directly as the advertising budget and inversely as the skates' price."

Use the given values to find k.

$$12{,}000 = \frac{k(60{,}000)}{40}$$

When $60,000 is spent on advertising ($A = 60{,}000$) and the price is $40 ($P = 40$), monthly sales are 12,000 units ($S = 12{,}000$).

$$12{,}000 = k \cdot 1500$$ Divide 60,000 by 40.

$$\frac{12{,}000}{1500} = \frac{k \cdot 1500}{1500}$$ Divide both sides of the equation by 1500.

$$8 = k$$ Simplify.

Therefore, the equation of variation that describes monthly sales is

$$S = \frac{8A}{P}.$$ Substitute 8 for k in $S = \frac{kA}{P}$.

b. The advertising budget is increased to $70,000, so $A = 70,000$. The skates' price is still $40, so $P = 40$.

$$S = \frac{8A}{P}$$ This is the equation from part (a).

$$S = \frac{8(70,000)}{40}$$ Substitute 70,000 for A and 40 for P.

$$S = 14,000$$ Simplify.

With a $70,000 advertising budget and $40 price, the company can expect to sell 14,000 pairs of rollerblades in a month (up from 12,000). ■

 CHECK POINT 4 The number of minutes needed to solve an exercise set of variation problems varies directly as the number of problems and inversely as the number of people working to solve the problems. It takes 4 people 32 minutes to solve 16 problems. How many minutes will it take 8 people to solve 24 problems?
It will take 24 minutes for 8 people to solve 24 problems.

4 Solve problems involving joint variation.

Joint Variation   Joint variation is a variation in which a variable varies directly as the product of two or more other variables. Thus, the equation $y = kxz$ is read "y varies jointly as x and z."

Joint variation plays a critical role in Isaac Newton's formula for gravitation:

$$F = G\frac{m_1 m_2}{d^2}.$$

The formula states that the force of gravitation, F, between two bodies varies jointly as the product of their masses, m_1 and m_2, and inversely as the square of the distance between them, d. (G is the gravitational constant.) The formula indicates that gravitational force exists between any two objects in the universe, increasing as the distance between the bodies decreases. One practical result is that the pull of the moon on the oceans is greater on the side of Earth closer to the moon. This gravitational imbalance is what produces tides.

EXAMPLE 5   Modeling Centrifugal Force

The centrifugal force, C, of a body moving in a circle varies jointly with the radius of the circular path, r, and the body's mass, m, and inversely with the square of the time, t, it takes to move about one full circle. A 6-gram body moving in a circle with radius 100 centimeters at a rate of 1 revolution in 2 seconds has a centrifugal force of 6000 dynes. Find the centrifugal force of an 18-gram body moving in a circle with radius 100 centimeters at a rate of 1 revolution in 3 seconds.

SOLUTION

$$C = \frac{krm}{t^2}$$

Translate "Centrifugal force, C, varies jointly with radius, r, and mass, m, and inversely with the square of time, t."

$$6000 = \frac{k(100)(6)}{2^2}$$

If r = 100, m = 6, and t = 2, then C = 6000.

$$6000 = 150k$$

Simplify.

$$40 = k$$

Divide both sides by 150 and solve for k.

$$C = \frac{40rm}{t^2}$$

Substitute 40 for k in the model for centrifugal force.

$$C = \frac{40(100)(18)}{3^2}$$

Find C when r = 100, m = 18, and t = 3.

$$= 8000$$

Simplify.

The centrifugal force is 8000 dynes.

> **CHECK POINT 5** The volume of a cone, V, varies jointly as its height, h, and the square of its radius r. A cone with a radius measuring 6 feet and a height measuring 10 feet has a volume of 120π cubic feet. Find the volume of a cone having a radius of 12 feet and a height of 2 feet. *The volume of a cone having a radius of 12 feet and a height of 2 feet is 96π cubic feet.*

6.8 EXERCISE SET

Student Solutions Manual   CD/Video   PH Math/Tutor Center   MathXL Tutorials on CD   MathXL®   MyMathLab   Interactmath.com

Practice Exercises

Use the four-step procedure for solving variation problems given on page 459 to solve Exercises 1–10.

1. y varies directly as x. $y = 65$ when $x = 5$. Find y when $x = 12$.   *$y = 156$*

2. y varies directly as x. $y = 45$ when $x = 5$. Find y when $x = 13$.   *$y = 117$*

3. y varies inversely as x. $y = 12$ when $x = 5$. Find y when $x = 2$.   *$y = 30$*

4. y varies inversely as x. $y = 6$ when $x = 3$. Find y when $x = 9$.   *$y = 2$*

5. y varies directly as x and inversely as the square of z. $y = 20$ when $x = 50$ and $z = 5$. Find y when $x = 3$ and $z = 6$.   *$y = \frac{5}{6}$*

6. a varies directly as b and inversely as the square of c. $a = 7$ when $b = 9$ and $c = 6$. Find a when $b = 4$ and $c = 8$.   *$a = \frac{7}{4}$*

7. y varies jointly as x and z. $y = 25$ when $x = 2$ and $z = 5$. Find y when $x = 8$ and $z = 12$.   *$y = 240$*

8. C varies jointly as A and T. $C = 175$ when $A = 2100$ and $T = 4$. Find C when $A = 2400$ and $T = 6$.   *$C = 300$*

9. y varies jointly as a and b and inversely as the square root of c. $y = 12$ when $a = 3, b = 2$, and $c = 25$. Find y when $a = 5, b = 3$, and $c = 9$.   *$y = 50$*

10. y varies jointly as m and the square of n and inversely as p. $y = 15$ when $m = 2, n = 1$, and $p = 6$. Find y when $m = 3, n = 4$, and $p = 10$.   *$y = 216$*

Practice Plus

In Exercises 11–20, write an equation that expresses each relationship. Then solve the equation for y.

11. x varies jointly as y and z.   *$x = kyz$; $y = \dfrac{x}{kz}$*

12. x varies jointly as y and the square of z.   *$x = kyz^2$; $y = \dfrac{x}{kz^2}$*

13. x varies directly as the cube of z and inversely as y.   *$x = \dfrac{kz^3}{y}$; $y = \dfrac{kz^3}{x}$*

14. x varies directly as the cube root of z and inversely as y.   *$x = \dfrac{k\sqrt[3]{z}}{y}$; $y = \dfrac{k\sqrt[3]{z}}{x}$*

15. x varies jointly as y and z and inversely as the square root of w. $\quad x = \dfrac{kyz}{\sqrt{w}}; \quad y = \dfrac{x\sqrt{w}}{kz}$

16. x varies jointly as y and z and inversely as the square of w.
$x = \dfrac{kyz}{w^2}; \quad y = \dfrac{xw^2}{kz}$

17. x varies jointly as z and the sum of y and w.
$x = kz(y + w); \quad y = \dfrac{x - kzw}{kz}$

18. x varies jointly as z and the difference between y and w.
$x = kz(y - w); \quad y = \dfrac{x + kzw}{kz}$

19. x varies directly as z and inversely as the difference between y and w. $\quad x = \dfrac{kz}{y - w}; \quad y = \dfrac{xw + kz}{x}$

20. x varies directly as z and inversely as the sum of y and w.
$x = \dfrac{kz}{y + w}; \quad y = \dfrac{kz - xw}{x}$

Application Exercises

Use the four-step procedure for solving variation problems given on page 459 to solve Exercises 21–36.

21. An alligator's tail length, T, varies directly as its body length, B. An alligator with a body length of 4 feet has a tail length of 3.6 feet. What is the tail length of an alligator whose body length is 6 feet?   The tail length is 5.4 feet.

|← —— Body length, B —— →|← —— Tail length, T —— →|

22. An object's weight on the moon, M, varies directly as its weight on Earth, E. Neil Armstrong, the first person to step on the moon on July 20, 1969, weighed 360 pounds on Earth (with all of his equipment on) and 60 pounds on the moon. What is the moon weight of a person who weighs 186 pounds on Earth?   A person who weighs 186 pounds on Earth weighs 31 pounds on the moon.

23. The height that a ball bounces varies directly as the height from which it was dropped. A tennis ball dropped from 12 inches bounces 8.4 inches. From what height was the tennis ball dropped if it bounces 56 inches?   It was dropped from 80 inches.

24. The distance that a spring will stretch varies directly as the force applied to the spring. A force of 12 pounds is needed to stretch a spring 9 inches. What force is required to stretch the spring 15 inches?   A force of 20 pounds is needed.

25. If all men had identical body types, their weight would vary directly as the cube of their height. Shown at the top of the next column is Robert Wadlow, who reached a record height of 8 feet 11 inches (107 inches) before his death at age 22. If a man who is 5 feet 10 inches tall (70 inches) with

the same body type as Mr. Wadlow weighs 170 pounds, what was Robert Wadlow's weight shortly before his death?   Robert Wadlow's weight was about 607 pounds.

26. On a dry asphalt road, a car's stopping distance varies directly as the square of its speed. A car traveling at 45 miles per hour can stop in 67.5 feet. What is the stopping distance for a car traveling at 60 miles per hour?   The stopping distance is 120 feet.

27. The figure shows that a bicyclist tips the cycle when making a turn. The angle B, formed by the vertical direction and the bicycle, is called the banking angle. The banking angle varies inversely as the cycle's turning radius. When the turning radius is 4 feet, the banking angle is 28°. What is the banking angle when the turning radius is 3.5 feet?   The banking angle is 32° when the turning radius is 3.5 feet.

28. The water temperature of the Pacific Ocean varies inversely as the water's depth. At a depth of 1000 meters, the water temperature is 4.4° Celsius. What is the water temperature at a depth of 5000 meters?   The water temperature is 0.88° Celsius at a depth of 5000 meters.

29. Radiation machines, used to treat tumors, produce an intensity of radiation that varies inversely as the square of the distance from the machine. At 3 meters, the radiation intensity is 62.5 milliroentgens per hour. What is the intensity at a distance of 2.5 meters? *The intensity is 90 milliroentgens per hour.*

30. The illumination provided by a car's headlight varies inversely as the square of the distance from the headlight. A car's headlight produces an illumination of 3.75 footcandles at a distance of 40 feet. What is the illumination when the distance is 50 feet? *The illumination is 2.4 footcandles at a distance of 50 feet.*

31. Body-mass index, or BMI, takes both weight and height into account when assessing whether an individual is underweight or overweight. BMI varies directly as one's weight, in pounds, and inversely as the square of one's height, in inches. In adults, normal values for the BMI are between 20 and 25, inclusive. Values below 20 indicate that an individual is underweight and values above 30 indicate that an individual is obese. A person who weighs 180 pounds and is 5 feet, or 60 inches, tall has a BMI of 35.15. What is the BMI, to the nearest tenth, for a 170 pound person who is 5 feet 10 inches tall. Is this person overweight? *This person has a BMI of 24.4 and is not overweight.*

32. One's intelligence quotient, or IQ, varies directly as a person's mental age and inversely as that person's chronological age. A person with a mental age of 25 and a chronological age of 20 has an IQ of 125. What is the chronological age of a person with a mental age of 40 and an IQ of 80? *The chronological age is 50.*

33. The heat loss of a glass window varies jointly as the window's area and the difference between the outside and inside temperatures. A window 3 feet wide by 6 feet long loses 1200 Btu per hour when the temperature outside is 20° colder than the temperature inside. Find the heat loss through a glass window that is 6 feet wide by 9 feet long when the temperature outside is 10° colder than the temperature inside. *The heat loss is 1800 Btu.*

34. Kinetic energy varies jointly as the mass and the square of the velocity. A mass of 8 grams and velocity of 3 centimeters per second has a kinetic energy of 36 ergs. Find the kinetic energy for a mass of 4 grams and velocity of 6 centimeters per second. *A mass of 4 grams and velocity of 6 centimeters per second has a kinetic energy of 72 ergs.*

35. Sound intensity varies inversely as the square of the distance from the sound source. If you are in a movie theater and you change your seat to one that is twice as far from the speakers, how does the new sound intensity compare to that of your original seat? *The sound intensity is $\frac{1}{4}$ of what it was originally.*

36. Many people claim that as they get older, time seems to pass more quickly. Suppose that the perceived length of a period of time is inversely proportional to your age. How long will a year seem to be when you are three times as old as you are now? *A year will seem to be $\frac{1}{3}$ of a year.*

37. The average number of daily phone calls, C, between two cities varies jointly as the product of their populations, P_1 and P_2, and inversely as the square of the distance, d, between them.

 a. Write an equation that expresses this relationship. $C = \dfrac{kP_1P_2}{d^2}$

 b. The distance between San Francisco (population: 777,000) and Los Angeles (population: 3,695,000) is 420 miles. If the average number of daily phone calls between the cities is 326,000, find the value of k to two decimal places and write the equation of variation. $k \approx 0.02;\quad C = \dfrac{0.02\,P_1P_2}{d^2}$

 c. Memphis (population: 650,000) is 400 miles from New Orleans (population: 490,000). Find the average number of daily phone calls, to the nearest whole number, between these cities. *approximately 39,813 daily phone calls*

38. The force of wind blowing on a window positioned at a right angle to the direction of the wind varies jointly as the area of the window and the square of the wind's speed. It is known that a wind of 30 miles per hour blowing on a window measuring 4 feet by 5 feet exerts a force of 150 pounds. During a storm with winds of 60 miles per hour, should hurricane shutters be placed on a window that measures 3 feet by 4 feet and is capable of withstanding 300 pounds of force? *Yes, the wind will exert a force of 360 pounds on the window.*

39. The table shows the values for the current, I, in an electric circuit and the resistance, R, of the circuit.

I (amperes)	0.5	1.0	1.5	2.0	2.5	3.0	4.0	5.0
R (ohms)	12	6.0	4.0	3.0	2.4	2.0	1.5	1.2

 a. Graph the ordered pairs in the table of values, with values of I along the x-axis and values of R along the y-axis. Connect the eight points with a smooth curve. *See graphing answer section.*

 b. Does current vary directly or inversely as resistance? Use your graph and explain how you arrived at your answer. *Current varies inversely as resistance. Answers will vary.*

 c. Write an equation of variation for I and R, using one of the ordered pairs in the table to find the constant of variation. Then use your variation equation to verify the other seven ordered pairs in the table. $R = \dfrac{6}{I}$

Writing in Mathematics

40. What does it mean if two quantities vary directly? *40–44. Answers will vary.*

41. In your own words, explain how to solve a variation problem.

42. What does it mean if two quantities vary inversely?

43. Explain what is meant by combined variation. Give an example with your explanation.

44. Explain what is meant by joint variation. Give an example with your explanation.

In Exercises 45–46, describe in words the variation shown by the given equation.

45. $z = \dfrac{k\sqrt{x}}{y^2}$ *z varies directly as the square root of x and inversely as the square of y.*

46. $z = kx^2\sqrt{y}$ *z* varies jointly as the square of *x* and the square root of *y*.

47. We have seen that the daily number of phone calls between two cities varies jointly as their populations and inversely as the square of the distance between them. This model, used by telecommunication companies to estimate the line capacities needed among various cities, is called the *gravity model*. Compare the model to Newton's formula for gravitation on page 465 and describe why the name *gravity model* is appropriate.   Answers will vary.

Technology Exercise

48. Use a graphing utility to graph any three of the variation equations in Exercises 21–30. Then TRACE along each curve and identify the point that corresponds to the problem's solution.   Answers will vary.

Critical Thinking Exercises

49. In a hurricane, the wind pressure varies directly as the square of the wind velocity. If wind pressure is a measure of a hurricane's destructive capacity, what happens to this destructive power when the wind speed doubles?   If wind speed doubles, the wind pressure is 4 times more destructive.

50. The heat generated by a stove element varies directly as the square of the voltage and inversely as the resistance. If the voltage remains constant, what needs to be done to triple the amount of heat generated?   To triple the amount of heat, the resistance should be divided by 3.

51. Galileo's telescope brought about revolutionary changes in astronomy. A comparable leap in our ability to observe the universe took place as a result of the Hubble Space Telescope. The space telescope can see stars and galaxies whose brightness is $\frac{1}{50}$ of the faintest objects now observable using ground-based telescopes. Use the fact that the brightness of a point source, such as a star, varies inversely as the square of its distance from an observer to show that the space telescope can see about seven times farther than a ground-based telescope.   Distance is increased by $\sqrt{50}$, or about 7.07, for the space telescope.

Review Exercises

52. Evaluate:

$$\begin{vmatrix} -1 & 2 \\ 3 & -4 \end{vmatrix}.$$

(Section 3.5, Example 1)   −2

53. Factor completely:

$$x^2y - 9y - 3x^2 + 27.$$

(Section 5.6, Example 3) $(x + 3)(x - 3)(y - 3)$

54. Find the degree of

$$7xy + x^2y^2 - 5x^3 - 7.$$

(Section 5.1, Example 1)   4

GROUP PROJECT

CHAPTER

6

Additional group projects can be found in the *Instructor's Resource Manual*.

A cost-benefit analysis compares the estimated costs of a project with the benefits that will be achieved. Costs and benefits are given monetary values and compared using a benefit-cost ratio. As shown in the figure, a favorable ratio for a project means that the benefits outweigh the costs and the project is cost-effective. As a group, select an environmental project that was implemented in your area of the country. Research the cost and benefit

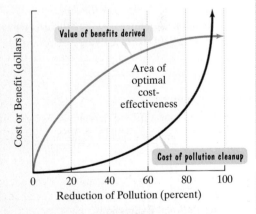

graphs that resulted in the implementation of this project. How were the benefits converted into monetary terms? Is there an equation for either the cost model or the benefit model? Group members may need to interview members of environmental groups and businesses that were part of this project. You may wish to consult an environmental science textbook to find out more about cost-benefit analyses. After doing your research, the group should write or present a report explaining why the cost-benefit analysis resulted in the project's implementation.

CHAPTER 6 SUMMARY

Definitions and Concepts	Examples

Section 6.1 Rational Expressions and Functions: Multiplying and Dividing

A rational expression consists of a polynomial divided by a nonzero polynomial. A rational function is defined by a formula that is a rational expression. The domain of a rational function is the set of all real numbers except those for which the denominator is zero. Graphs of rational functions often contain disconnected branches. The graphs often approach, but do not touch, vertical or horizontal lines, called vertical asymptotes and horizontal asymptotes, respectively.

Let $f(x) = \dfrac{x + 2}{x^2 + x - 6}$. Find the domain of f.

$$x^2 + x - 6 = 0$$

$$(x + 3)(x - 2) = 0$$

$$x + 3 = 0 \quad \text{or} \quad x - 2 = 0$$

$$x = -3 \qquad x = 2$$

These values make the denominator zero. Exclude them from the domain. Domain of $f = (-\infty, -3) \cup (-3, 2) \cup (2, \infty)$. The branches of the graph of f break at -3 and 2.

Simplifying Rational Expressions

1. Factor the numerator and the denominator completely.
2. Divide both the numerator and the denominator by any common factors.

Simplify: $\dfrac{4x + 28}{x^2 - 49}$.

$$\frac{4x + 28}{x^2 - 49} = \frac{4\cancel{(x + 7)}}{\cancel{(x + 7)}(x - 7)} = \frac{4}{x - 7}$$

Multiplying Rational Expressions

1. Factor completely.
2. Divide numerators and denominators by common factors.
3. Multiply remaining factors in the numerators and in the denominators.

$$\frac{x^2 + 3x - 18}{x^2 - 3x} \cdot \frac{x^2}{x^2 - 36}$$

$$= \frac{\cancel{(x + 6)}\cancel{(x - 3)}}{x\cancel{(x - 3)}} \cdot \frac{x \cdot x}{\cancel{(x + 6)}(x - 6)}$$

$$= \frac{x}{x - 6}$$

Dividing Rational Expressions
Invert the divisor and multiply.

$$\frac{5y + 15}{(y + 5)^2} \div \frac{y^2 - 9}{y + 5}$$

$$= \frac{5y + 15}{(y + 5)^2} \cdot \frac{y + 5}{y^2 - 9} = \frac{5\cancel{(y + 3)}}{(y + 5)\cancel{(y + 5)}} \cdot \frac{\cancel{(y + 5)}}{\cancel{(y + 3)}(y - 3)}$$

$$= \frac{5}{(y + 5)(y - 3)}$$

Definitions and Concepts	Examples

Section 6.2 Adding and Subtracting Rational Expressions

Adding or Subtracting Rational Expressions
If the denominators are the same, add or subtract the numerators. Place the result over the common denominator. If the denominators are different, write all rational expressions with the least common denominator. The LCD is a polynomial consisting of the product of all prime factors in the denominators, with each factor raised to the greatest power of occurrence in any denominator. Once all rational expressions are written in terms of the LCD, add or subtract as described above. Simplify the result, if possible.

$$\frac{x+1}{2x-2} - \frac{2x}{x^2+2x-3}$$

$$= \frac{x+1}{2(x-1)} - \frac{2x}{(x-1)(x+3)} \quad \text{LCD is } 2(x-1)(x+3).$$

$$= \frac{(x+1)(x+3)}{2(x-1)(x+3)} - \frac{2x \cdot 2}{2(x-1)(x+3)}$$

$$= \frac{x^2+4x+3-4x}{2(x-1)(x+3)} = \frac{x^2+3}{2(x-1)(x+3)}$$

Section 6.3 Complex Rational Expressions

Complex rational expressions have numerators or denominators containing one or more rational expressions. Complex rational expressions can be simplified by multiplying the numerator and the denominator by the LCD. They can also be simplified by obtaining single expressions in the numerator and denominator and then dividing.

Simplify: $\dfrac{\dfrac{x+3}{x}}{x - \dfrac{9}{x}}$.

Multiplying by the LCD, x:

$$\frac{x \cdot \left(\dfrac{x+3}{x}\right)}{x \cdot \left(x - \dfrac{9}{x}\right)} = \frac{x+3}{x^2-9} = \frac{(x+3)}{(x+3)(x-3)} = \frac{1}{x-3}.$$

Simplifying by dividing:

$$\frac{\dfrac{x+3}{x}}{x - \dfrac{9}{x}} = \frac{\dfrac{x+3}{x}}{\dfrac{x}{1} \cdot \dfrac{x}{x} - \dfrac{9}{x}} = \frac{\dfrac{x+3}{x}}{\dfrac{x^2-9}{x}} = \frac{x+3}{x} \cdot \frac{x}{x^2-9}$$

LCD is x.

$$= \frac{x+3}{x} \cdot \frac{x}{(x+3)(x-3)} = \frac{1}{x-3}.$$

Section 6.4 Division of Polynomials

To divide a polynomial by a monomial, divide each term of the polynomial by the monomial.

$$\frac{9x^3y^4 - 12x^3y^2 - 7x^2y^2}{3xy^2}$$

$$= \frac{9x^3y^4}{3xy^2} - \frac{12x^3y^2}{3xy^2} - \frac{7x^2y^2}{3xy^2} = 3x^2y^2 - 4x^2 - \frac{7x}{3}$$

To divide by a polynomial containing more than one term, use long division. If necessary, arrange the dividend in descending powers of the variable. If a power of a variable is missing, add that power with a coefficient of 0. Repeat the four steps of the long-division process—divide, multiply, subtract, bring down the next term—until the degree of the remainder is less than the degree of the divisor.

Divide: $(2x^3 - x^2 - 7) \div (x - 2)$.

$$\require{enclose}
\begin{array}{r}
2x^2 + 3x + 6 \\
x-2 \enclose{longdiv}{2x^3 - x^2 + 0x - 7} \\
\underline{2x^3 - 4x^2} \\
3x^2 + 0x \\
\underline{3x^2 - 6x} \\
6x - 7 \\
\underline{6x - 12} \\
5
\end{array}$$

The answer is $2x^2 + 3x + 6 + \dfrac{5}{x-2}$.

Definitions and Concepts	Examples

Section 6.5 Synthetic Division and the Remainder Theorem

A shortcut to long division, called synthetic division, can be used to divide a polynomial by a binomial of the form $x - c$.

Here is the division problem shown on page 471 using synthetic division.

$$(2x^3 - x^2 - 7) \div (x - 2)$$

Coefficients of the dividend,
$2x^3 - x^2 + 0x - 7$

This is c in $x - c$.
For $x - 2$, c is 2.

$$\begin{array}{r|rrrr} 2 & 2 & -1 & 0 & -7 \\ & \downarrow & 4 & 6 & 12 \\ \hline & 2 & 3 & 6 & 5 \end{array}$$

Coefficients of quotient   Remainder

The answer is $2x^2 + 3x + 6 + \dfrac{5}{x - 2}$.

The Remainder Theorem
If the polynomial $f(x)$ is divided by $x - c$, then the remainder is $f(c)$. This can be used to evaluate a polynomial function at c. Rather than substituting c for x, divide the function by $x - c$. The remainder is $f(c)$.

If $f(x) = x^4 - 3x^2 - 2x + 5$, use the Remainder Theorem to find $f(-2)$.

$$\begin{array}{r|rrrrr} -2 & 1 & 0 & -3 & -2 & 5 \\ & & -2 & 4 & -2 & 8 \\ \hline & 1 & -2 & 1 & -4 & 13 \end{array}$$

Remainder

Thus, $f(-2) = 13$.

Section 6.6 Rational Equations

A rational equation is an equation containing one or more rational expressions.

Solving Rational Equations

1. List restrictions on the variable.
2. Clear fractions by multiplying both sides by the LCD.
3. Solve the resulting equations.
4. Reject any proposed solution in the list of restrictions. Check other proposed solutions in the original equation.

Solve: $\dfrac{7x}{x^2 - 4} + \dfrac{5}{x - 2} = \dfrac{2x}{x^2 - 4}$.

$$\frac{7x}{(x + 2)(x - 2)} + \frac{5}{x - 2} = \frac{2x}{(x + 2)(x - 2)}$$

Denominators would equal 0 if $x = -2$ or $x = 2$. Restrictions: $x \neq -2$ and $x \neq 2$.

LCD is $(x + 2)(x - 2)$.

$$(x + 2)(x - 2)\left[\frac{7x}{(x + 2)(x - 2)} + \frac{5}{x - 2}\right] =$$

$$(x + 2)(x - 2) \cdot \frac{2x}{(x + 2)(x - 2)}$$

$$7x + 5(x + 2) = 2x$$
$$7x + 5x + 10 = 2x$$
$$12x + 10 = 2x$$
$$10 = -10x$$
$$-1 = x$$

The proposed solution, -1, is not part of the restriction $x \neq -2$ and $x \neq 2$. It checks. The solution is -1 and the solution set is $\{-1\}$.

Definitions and Concepts	Examples

Section 6.7 Formulas and Applications of Rational Equations

To solve a formula for a variable, get the specified variable alone on one side of the formula. When working with formulas containing rational expressions, it is sometimes necessary to factor out the variable you are solving for.

Solve: $\dfrac{e}{E} = \dfrac{r}{r + R}$ for r.

$$E(r + R) \cdot \frac{e}{E} = E(r + R) \cdot \frac{r}{r + R} \quad \text{LCD is } E(r + R).$$

$$e(r + R) = Er$$

$$er + eR = Er$$

$$eR = Er - er$$

$$eR = (E - e)r$$

$$\frac{eR}{E - e} = r$$

The average cost function, $\overline{C}$, for a business is a rational function representing the average cost for the company to produce each unit of its product. The function consists of the sum of fixed and variable costs divided by the number of units produced. As production level increases, the average cost to produce each unit of a product decreases.

A company has a fixed cost of \$80,000 monthly and it costs \$20 to produce each unit of its product. Its cost function, C, of producing x units is

$$C(x) = 80{,}000 + 20x.$$

Its average cost function, $\overline{C}$, is

$$\overline{C}(x) = \frac{80{,}000 + 20x}{x}.$$

Problems involving time in motion and work translate into rational equations. Motion problems involving time are solved using

$$\text{Time traveled} = \frac{\text{Distance traveled}}{\text{Rate of travel}}.$$

Work problems are solved using the following condition:

Fraction of job done by first	+	fraction of job done by second	=	1.

It takes a cyclist who averages 16 miles per hour in still air the same time to travel 48 miles with the wind as 16 miles against the wind. What is the wind's rate?

$$x = \text{wind's rate}$$

$$16 + x = \text{cyclist's rate with wind}$$

$$16 - x = \text{cyclist's rate against wind}$$

	Distance	Rate	Time = $\dfrac{\text{Distance}}{\text{Rate}}$
With wind	48	$16 + x$	$\dfrac{48}{16 + x}$
Against wind	16	$16 - x$	$\dfrac{16}{16 - x}$

These times are equal.

$$(16 + x)(16 - x) \cdot \frac{48}{16 + x} = \frac{16}{16 - x} \cdot (16 + x)(16 - x)$$

$$48(16 - x) = 16(16 + x)$$

Solving this equation, $x = 8$. The wind's rate is 8 miles per hour.

Definitions and Concepts	Examples

Section 6.8 Modeling Using Variation

English Statement	Equation
y varies directly as x.	$y = kx$
y varies directly as x^n.	$y = kx^n$
y varies inversely as x.	$y = \dfrac{k}{x}$
y varies inversely as x^n.	$y = \dfrac{k}{x^n}$
y varies directly as x and inversely as z.	$y = \dfrac{kx}{z}$
y varies jointly as x and z.	$y = kxz$

The time that it takes you to drive a certain distance varies inversely as your driving rate. Averaging 40 miles per hour, it takes you 10 hours to drive the distance. How long would the trip take averaging 50 miles per hour?

1. $t = \dfrac{k}{r}$ Time, t, varies inversely as rate, r.

2. It takes 10 hours at 40 miles per hour.

$$10 = \frac{k}{40}$$

$$k = 10(40) = 400$$

3. $t = \dfrac{400}{r}$

4. How long at 50 miles per hour? Substitute 50 for r.

$$t = \frac{400}{50} = 8$$

It takes 8 hours at 50 miles per hour.

CHAPTER 6 REVIEW EXERCISES

6.1

1. If $f(x) = \dfrac{x^2 + 2x - 3}{x^2 - 4}$

find the following function values. If a function value does not exist, so state.

a. $f(4)$ $\frac{7}{4}$ **b.** $f(0)$ $\frac{3}{4}$

c. $f(-2)$ does not exist **d.** $f(-3)$ 0

In Exercises 2–3, find the domain of the given rational function.

2. $f(x) = \dfrac{x - 6}{(x - 3)(x + 4)}$ Domain of $f = \{x | x$ is a real number and $x \neq -4$ and $x \neq 3\}$ or $(-\infty, -4) \cup (-4, 3) \cup (3, \infty)$

3. $f(x) = \dfrac{x + 2}{x^2 + x - 2}$ Domain of $f = \{x | x$ is a real number and $x \neq -2$ and $x \neq 1\}$ or $(-\infty, -2) \cup (-2, 1) \cup (1, \infty)$

In Exercises 4–8, simplify each rational expression. If the rational expression cannot be simplified, so state.

4. $\dfrac{5x^3 - 35x}{15x^2}$ $\dfrac{x^2 - 7}{3x}$

5. $\dfrac{x^2 + 6x - 7}{x^2 - 49}$ $\dfrac{x - 1}{x - 7}$

6. $\dfrac{6x^2 + 7x + 2}{2x^2 - 9x - 5}$ $\dfrac{3x + 2}{x - 5}$

7. $\dfrac{x^2 + 4}{x^2 - 4}$ cannot be simplified

8. $\dfrac{x^3 - 8}{x^2 - 4}$ $\dfrac{x^2 + 2x + 4}{x + 2}$

In Exercises 9–15, multiply or divide as indicated.

9. $\dfrac{5x^2 - 5}{3x + 12} \cdot \dfrac{x + 4}{x - 1}$ $\dfrac{5(x + 1)}{3}$

10. $\dfrac{2x + 5}{4x^2 + 8x - 5} \cdot \dfrac{4x^2 - 4x + 1}{x + 1}$ $\dfrac{2x - 1}{x + 1}$

11. $\dfrac{x^2 - 9x + 14}{x^3 + 2x^2} \cdot \dfrac{x^2 - 4}{x^2 - 4x + 4}$ $\dfrac{x - 7}{x^2}$

12. $\dfrac{1}{x^2 + 8x + 15} \div \dfrac{3}{x + 5}$ $\dfrac{1}{3(x + 3)}$

13. $\dfrac{x^2 + 16x + 64}{2x^2 - 128} \div \dfrac{x^2 + 10x + 16}{x^2 - 6x - 16}$ $\dfrac{1}{2}$

14. $\dfrac{y^2 - 16}{y^3 - 64} \div \dfrac{y^2 - 3y - 18}{y^2 + 5y + 6}$ $\dfrac{(y + 4)(y + 2)}{(y - 6)(y^2 + 4y + 16)}$

15. $\dfrac{x^2 - 4x + 4 - y^2}{2x^2 - 11x + 15} \cdot \dfrac{x^4 y}{x - 2 + y} \div \dfrac{x^3 y - 2x^2 y - x^2 y^2}{3x - 9}$ $\dfrac{3x^2}{2x - 5}$

16. Deer are placed into a newly acquired habitat. The deer population over time is modeled by a rational function whose graph is shown in the figure. Use the graph to answer each of the following questions.

   a. How many deer were introduced into the habitat?
     50 deer

   b. What is the population after 10 years? The population after 10 years is 150 deer.

   c. What is the equation of the horizontal asymptote shown in the figure? What does this mean in terms of the deer population? $y = 225$; The population will approach 225 deer over time.

Deer Population over Time

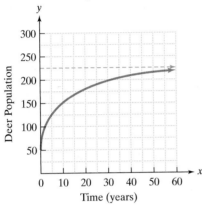

6.2 *In Exercises 17–19, add or subtract as indicated. Simplify the result, if possible.*

17. $\dfrac{4x + 1}{3x - 1} + \dfrac{8x - 5}{3x - 1}$ 4

18. $\dfrac{2x - 7}{x^2 - 9} - \dfrac{x - 4}{x^2 - 9}$ $\dfrac{1}{x + 3}$

19. $\dfrac{4x^2 - 11x + 4}{x - 3} - \dfrac{x^2 - 4x + 10}{x - 3}$ $3x + 2$

In Exercises 20–21, find the least common denominator of the rational expressions.

20. $\dfrac{7}{9x^3}$ and $\dfrac{5}{12x}$ $36x^3$

21. $\dfrac{x + 7}{x^2 + 2x - 35}$ and $\dfrac{x}{x^2 + 9x + 14}$ $(x + 7)(x - 5)(x + 2)$

In Exercises 22–28, perform the indicated operations. Simplify the result, if possible.

22. $\dfrac{1}{x} + \dfrac{2}{x - 5}$ $\dfrac{3x - 5}{x(x - 5)}$

23. $\dfrac{2}{x^2 - 5x + 6} + \dfrac{3}{x^2 - x - 6}$ $\dfrac{5x - 2}{(x - 3)(x + 2)(x - 2)}$

24. $\dfrac{x - 3}{x^2 - 8x + 15} + \dfrac{x + 2}{x^2 - x - 6}$
   $\dfrac{2x - 8}{(x - 3)(x - 5)}$ or $\dfrac{2(x - 4)}{(x - 3)(x - 5)}$

25. $\dfrac{3x^2}{9x^2 - 16} - \dfrac{x}{3x + 4}$ $\dfrac{4x}{(3x + 4)(3x - 4)}$

26. $\dfrac{y}{y^2 + 5y + 6} - \dfrac{2}{y^2 + 3y + 2}$ $\dfrac{y - 3}{(y + 3)(y + 1)}$

27. $\dfrac{x}{x + 3} + \dfrac{x}{x - 3} - \dfrac{9}{x^2 - 9}$ $\dfrac{2x^2 - 9}{(x + 3)(x - 3)}$

28. $\dfrac{3x^2}{x - y} + \dfrac{3y^2}{y - x}$ $3(x + y)$ or $3x + 3y$

6.3 *In Exercises 29–34, simplify each complex rational expression.*

29. $\dfrac{\dfrac{3}{x} - 3}{\dfrac{8}{x} - 8}$ $\dfrac{3}{8}$

30. $\dfrac{\dfrac{5}{x} + 1}{1 - \dfrac{25}{x^2}}$ $\dfrac{x}{x - 5}$

31. $\dfrac{3 - \dfrac{1}{x + 3}}{3 + \dfrac{1}{x + 3}}$ $\dfrac{3x + 8}{3x + 10}$

32. $\dfrac{\dfrac{4}{x + 3}}{\dfrac{2}{x - 2} - \dfrac{1}{x^2 + x - 6}}$ $\dfrac{4(x - 2)}{2x + 5}$

33. $\dfrac{\dfrac{2}{x^2 - x - 6} + \dfrac{1}{x^2 - 4x + 3}}{\dfrac{3}{x^2 + x - 2} - \dfrac{2}{x^2 + 5x + 6}}$ $\dfrac{3x^2 + 9x}{x^2 + 8x - 33}$

34. $\dfrac{x^{-2} + x^{-1}}{x^{-2} - x^{-1}}$ $\dfrac{1 + x}{1 - x}$

6.4 *In Exercises 35–36, divide the polynomial by the monomial.*

35. $(15x^3 - 30x^2 + 10x - 2) \div (5x^2)$ $3x - 6 + \dfrac{2}{x} - \dfrac{2}{5x^2}$

36. $(36x^4y^3 + 12x^2y^3 - 60x^2y^2) \div (6xy^2)$ $6x^3y + 2xy - 10x$

In Exercises 37–40, divide as indicated.

37. $(6x^2 - 5x + 5) \div (2x + 3)$ $3x - 7 + \dfrac{26}{2x + 3}$

38. $(10x^3 - 26x^2 + 17x - 13) \div (5x - 3)$
   $2x^2 - 4x + 1 - \dfrac{10}{5x - 3}$

39. $(x^6 + 3x^5 - 2x^4 + x^2 - 3x + 2) \div (x - 2)$
   $x^5 + 5x^4 + 8x^3 + 16x^2 + 33x + 63 + \dfrac{128}{x - 2}$

40. $(4x^4 + 6x^3 + 3x - 1) \div (2x^2 + 1)$ $2x^2 + 3x - 1$

6.5 *In Exercises 41–43, divide using synthetic division.*

41. $(4x^3 - 3x^2 - 2x + 1) \div (x + 1)$ $4x^2 - 7x + 5 - \dfrac{4}{x + 1}$

42. $(3x^4 - 2x^2 - 10x - 20) \div (x - 2)$ $3x^3 + 6x^2 + 10x + 10$

43. $(x^4 + 16) \div (x + 4)$ $x^3 - 4x^2 + 16x - 64 + \dfrac{272}{x + 4}$

In Exercises 44–45, use synthetic division and the Remainder Theorem to find the indicated function value.

44. $f(x) = 2x^3 - 5x^2 + 4x - 1;\quad f(2)$ 3

45. $f(x) = 3x^4 + 7x^3 + 8x^2 + 2x + 4;\quad f\left(-\frac{1}{3}\right)$ 4

In Exercises 46–47, use synthetic division to determine whether or not the number given to the right of each equation is a solution of the equation.

46. $2x^3 - x^2 - 8x + 4 = 0;\quad -2$ solution

47. $x^4 - x^3 - 7x^2 + x + 6 = 0;\quad 4$ 4 is not a solution.

48. Use synthetic division to show that $\frac{1}{2}$ is a solution of

$$6x^3 + x^2 - 4x + 1 = 0.$$

Then solve the polynomial equation.   The remainder is zero; $-1, \frac{1}{3},$ and $\frac{1}{2},$ or $\left\{-1, \frac{1}{3}, \frac{1}{2}\right\}$

6.6 *In Exercises 49–55, solve each rational equation. If an equation has no solution, so state.*

49. $\dfrac{3}{x} + \dfrac{1}{3} = \dfrac{5}{x}$ 6 or {6} $3 = 60 + 8$

50. $\dfrac{5}{3x + 4} = \dfrac{3}{2x - 8}$ 52 or {52}

51. $\dfrac{1}{x - 5} - \dfrac{3}{x + 5} = \dfrac{6}{x^2 - 25}$ 7 or {7}

52. $\dfrac{x + 5}{x + 1} - \dfrac{x}{x + 2} = \dfrac{4x + 1}{x^2 + 3x + 2}$ $-\dfrac{9}{2}$ or $\left\{-\dfrac{9}{2}\right\}$

53. $\dfrac{2}{3} - \dfrac{5}{3x} = \dfrac{1}{x^2}$ $-\dfrac{1}{2}$ and 3, or $\left\{-\dfrac{1}{2}, 3\right\}$

54. $\dfrac{2}{x - 1} = \dfrac{1}{4} + \dfrac{7}{x + 2}$ -23 and 2, or $\{-23, 2\}$

55. $\dfrac{2x + 7}{x + 5} - \dfrac{x - 8}{x - 4} = \dfrac{x + 18}{x^2 + x - 20}$ -3 and 2, or $\{-3, 2\}$

56. The function

$$f(x) = \frac{4x}{100 - x}$$

models the cost, $f(x)$, in millions of dollars, to remove $x\%$ of pollution from a river due to pesticide runoff from area farms. What percentage of the pollutants can be removed for $16 million?   80% of pollutants can be removed for $16 million.

6.7 *In Exercises 57–61, solve each formula for the specified variable.*

57. $P = \dfrac{R - C}{n}$ for C $C = R - nP$

58. $\dfrac{P_1 V_1}{T_1} = \dfrac{P_2 V_2}{T_2}$ for T_1 $T_1 = \dfrac{P_1 V_1 T_2}{P_2 V_2}$

59. $T = \dfrac{A - P}{Pr}$ for P $P = \dfrac{A}{rT + 1}$

60. $\dfrac{1}{R} = \dfrac{1}{R_1} + \dfrac{1}{R_2}$ for R $R = \dfrac{R_1 R_2}{R_2 + R_1}$

61. $I = \dfrac{nE}{R + nr}$ for n $n = \dfrac{IR}{E - Ir}$

62. A company is planning to manufacture affordable graphing calculators. Fixed monthly cost will be $50,000, and it will cost $25 to produce each calculator.

   a. Write the cost function, C, of producing x graphing calculators. $C(x) = 50,000 + 25x$

   b. Write the average cost function, $\overline{C}$, of producing x graphing calculators. $\overline{C}(x) = \dfrac{50,000 + 25x}{x}$

   c. How many graphing calculators must be produced each month for the company to have an average cost of $35 per graphing calculator?   5000 graphing calculators must be produced each month.

63. After riding at a steady rate for 60 miles, a bicyclist had a flat tire and walked 8 miles to a repair shop. The cycling rate was 3 times faster than the walking rate. If the time spent cycling and walking was 7 hours, at what rate was the cyclist riding?   The cyclist was riding at a rate of 12 miles per hour.

64. The current of a river moves at 3 miles per hour. It takes a boat a total of 3 hours to travel 12 miles upstream against the current, and return the same distance traveling downstream, with the current. What is the boat's rate in still water?   The boat's rate in still water is 9 miles per hour.

65. Working alone, two people can clean their house in 3 hours and 6 hours, respectively. They have agreed to clean together so that they can finish in time to watch a TV program that begins in $1\frac{1}{2}$ hours. How long will it take them to clean the house working together? Can they finish before the program starts?   It will take them 2 hours to clean the house, which is not enough time before the program starts.

66. Working together, two crews can clear snow from the city's streets in 20 hours. Working alone, the faster crew requires 9 hours less time than the slower crew. How many hours would it take each crew to clear the streets working alone?   It would take the faster crew 36 hours and the slower crew 45 hours to clear the streets working alone.

67. An inlet faucet can fill a small pond in 60 minutes. The pond can be emptied by an outlet pipe in 80 minutes. You begin filling the empty pond. By accident, the outlet pipe that empties the pond is left open. Under these conditions, how long will it take for the pond to fill?   It will take 240 minutes, or 4 hours, to fill the pond.

6.8 *Solve the variation problems in Exercises 68–73.*

68. A company's profit varies directly as the number of products it sells. The company makes a profit of $1175 on the sale of 25 products. What is the company's profit when it sells 105 products?   The company's profit is $4935.

69. The distance that a body falls from rest varies directly as the square of the time of the fall. If skydivers fall 144 feet in 3 seconds, how far will they fall in 10 seconds?   Skydivers will fall 1600 feet in 10 seconds.

70. The pitch of a musical tone varies inversely as its wavelength. A tone has a pitch of 660 vibrations per second and a wavelength of 1.6 feet. What is the pitch of a tone that has a wavelength of 2.4 feet?   The tone's pitch is 440 vibrations per second.

71. The loudness of a stereo speaker, measured in decibels, varies inversely as the square of your distance from the speaker. When you are 8 feet from the speaker, the loudness is 28 decibels. What is the loudness when you are 4 feet from the speaker? The loudness at 4 feet from the speaker is 112 decibels.

72. The time required to assemble computers varies directly as the number of computers assembled and inversely as the number of workers. If 30 computers can be assembled by 6 workers in 10 hours, how long would it take 5 workers to assemble 40 computers? It will take 16 hours for 5 workers to assemble 40 computers.

73. The volume of a pyramid varies jointly as its height and the area of its base. A pyramid with a height of 15 feet and a base with an area of 35 square feet has a volume of 175 cubic feet. Find the volume of a pyramid with a height of 20 feet and a base with an area of 120 square feet. The volume of a pyramid with a height of 20 feet and an area of 120 square feet is 800 cubic feet.

CHAPTER 6 TEST

Remember to use your Chapter Test Prep Video CD to see the worked-out solutions to the test questions you want to review.

1. Find the domain of $f(x) = \dfrac{x^2 - 2x}{x^2 - 7x + 10}$. Then simplify the right side of the function's equation Domain of $f = \{x \mid x \text{ is a real number and } x \neq 2 \text{ and } x \neq 5\}$, or $(-\infty, 2) \cup (2, 5) \cup (5, \infty)$; $\dfrac{x}{x - 5}$

In Exercises 2–11, perform the indicated operations. Simplify where possible.

2. $\dfrac{x^2}{x^2 - 16} \cdot \dfrac{x^2 + 7x + 12}{x^2 + 3x}$ $\dfrac{x}{x - 4}$

3. $\dfrac{x^3 + 27}{x^2 - 1} \div \dfrac{x^2 - 3x + 9}{x^2 - 2x + 1}$ $\dfrac{(x + 3)(x - 1)}{x + 1}$

4. $\dfrac{x^2 + 3x - 10}{x^2 + 4x + 3} \cdot \dfrac{x^2 + x - 6}{x^2 + 10x + 25} \cdot \dfrac{x + 1}{x - 2}$ $\dfrac{x - 2}{x + 5}$

5. $\dfrac{x^2 - 6x - 16}{x^3 + 3x^2 + 2x} \cdot (x^2 - 3x - 4) \div \dfrac{x^2 - 7x + 12}{3x}$ $\dfrac{3x - 24}{x - 3}$

6. $\dfrac{x^2 - 5x - 2}{6x^2 - 11x - 35} - \dfrac{x^2 - 7x + 5}{6x^2 - 11x - 35}$ $\dfrac{1}{3x + 5}$

7. $\dfrac{x}{x + 3} + \dfrac{5}{x - 3}$ $\dfrac{x^2 + 2x + 15}{(x + 3)(x - 3)}$

8. $\dfrac{2}{x^2 - 4x + 3} + \dfrac{3x}{x^2 + x - 2}$ $\dfrac{3x - 4}{(x + 2)(x - 3)}$

9. $\dfrac{5x}{x^2 - 4} - \dfrac{2}{x^2 + x - 2}$ $\dfrac{5x^2 - 7x + 4}{(x + 2)(x - 1)(x - 2)}$

10. $\dfrac{x - 4}{x - 5} - \dfrac{3}{x + 5} - \dfrac{10}{x^2 - 25}$ $\dfrac{x + 3}{x + 5}$

11. $\dfrac{1}{10 - x} + \dfrac{x - 1}{x - 10}$ $\dfrac{x - 2}{x - 10}$

In Exercises 12–13, simplify each rational expression.

12. $\dfrac{\dfrac{x}{4} - \dfrac{1}{x}}{1 + \dfrac{x + 4}{x}}$ $\dfrac{x - 2}{8}$

13. $\dfrac{\dfrac{1}{x} - \dfrac{3}{x + 2}}{\dfrac{2}{x^2 + 2x}}$ $1 - x$

In Exercises 14–16, divide as indicated.

14. $(12x^4y^3 + 16x^2y^3 - 10x^2y^2) \div (4x^2y)$ $3x^2y^2 + 4y^2 - \dfrac{5}{2}y$

15. $(9x^3 - 3x^2 - 3x + 4) \div (3x + 2)$ $3x^2 - 3x + 1 + \dfrac{2}{3x + 2}$

16. $(3x^4 + 2x^3 - 8x + 6) \div (x^2 - 1)$ $3x^2 + 2x + 3 + \dfrac{9 - 6x}{x^2 - 1}$

17. Divide using synthetic division: $3x^3 - 4x^2 + 7$
$$(3x^4 + 11x^3 - 20x^2 + 7x + 35) \div (x + 5).$$

18. Given that
$$f(x) = x^4 - 2x^3 - 11x^2 + 5x + 34,$$
use synthetic division and the Remainder Theorem to find $f(-2)$. 12

19. Use synthetic division to decide whether -2 is a solution of $2x^3 - 3x^2 - 11x + 6 = 0$. solution

In Exercises 20–21, solve each rational equation.

20. $\dfrac{x}{x + 4} = \dfrac{11}{x^2 - 16} + 2$ -7 and 3, or $\{-7, 3\}$

21. $\dfrac{x + 1}{x^2 + 2x - 3} - \dfrac{1}{x + 3} = \dfrac{1}{x - 1}$ -1 or $\{-1\}$

22. Park rangers introduce 50 elk into a wildlife preserve. The function
$$f(t) = \dfrac{250(3t + 5)}{t + 25}$$
models the elk population, $f(t)$, after t years. How many years will it take for the population to increase to 125 elk? It will take 3 years for the population to increase to 125 elk.

23. Solve: $R = \dfrac{as}{a + s}$ for a. $\quad a = \dfrac{Rs}{s - R}$ or $-\dfrac{Rs}{R - s}$

24. A company is planning to manufacture pocket-sized televisions. Fixed monthly cost will be $300,000 and it will cost $10 to produce each television.

   a. Write the cost function, $\overline{C}$, of producing x televisions. $C(x) = 300,000 + 10x$

   b. Write the average cost function, C, of producing x televisions. $\overline{C}(x) = \dfrac{300,000 + 10x}{x}$

   c. How many pocket-sized televisions must be produced each month for the company to have an average cost of $25 per television? 20,000 televisions must be produced each month.

Note to Instructor: The Chapter 6 Test file in the TestGen program provides algorithms specifically matched to these problems so you can easily replicate this chapter test for additional practice or assessment purposes.

25. It takes one pipe 3 hours to fill a pool and a second pipe 4 hours to drain the pool. The pool is empty and the first pipe begins to fill it. The second pipe is accidently left open, so the water is also draining out of the pool. Under these conditions, how long will it take to fill the pool? It will take 12 hours to fill the pool.

26. A motorboat averages 20 miles per hour in still water. It takes the boat the same amount of time to travel 3 miles with the current as it does to travel 2 miles against the current. What is the current's rate? The current's rate is 4 miles per hour.

27. The intensity of light received at a source varies inversely as the square of the distance from the source. A particular light has an intensity of 20 foot-candles at 15 feet. What is the light's intensity at 10 feet? The light's intensity at 10 feet is 45 foot-candles.

CUMULATIVE REVIEW EXERCISES (CHAPTERS 1–6)

In Exercises 1–5, solve each equation, inequality, or system.

1. $2x + 5 \le 11$ and $-3x > 18$ $\quad \{x \mid x < -6\}$ or $(-\infty, -6)$

2. $2x^2 = 7x + 4$ $\quad -\frac{1}{2}$ and 4, or $\left\{-\frac{1}{2}, 4\right\}$

3. $4x + 3y + 3z = 4$
$\quad 3x \qquad + 2z = 2$
$\quad 2x - 5y \qquad = -4$ $\quad (-2, 0, 4)$ or $\{(-2, 0, 4)\}$

4. $|3x - 4| \le 10$ $\quad \left\{x \mid -2 \le x \le \dfrac{14}{3}\right\}$ or $\left[-2, \dfrac{14}{3}\right]$

5. $\dfrac{x}{x - 8} + \dfrac{6}{x - 2} = \dfrac{x^2}{x^2 - 10x + 16}$ $\quad$ 12 or $\{12\}$

6. Solve for s: $\quad I = \dfrac{2R}{w + 2s}$. $\quad s = \dfrac{2R - Iw}{2I}$

7. Solve by graphing: $\begin{aligned} 2x - y &= 4 \\ x + y &= 5. \end{aligned}$ $\quad (3, 2)$ or $\{(3, 2)\}$

8. Use function notation to write the equation of the line with slope -3 and passing through the point $(1, -5)$. $\quad f(x) = -3x - 2$

In Exercises 9–11, graph each equation, inequality, or system in a rectangular coordinate system. $\quad$ 9–11. See graphing answer section.

9. $y = |x| + 2$

10. $y \ge 2x - 1$
$\quad x \ge 1$

11. $2x - y < 4$

In Exercises 12–15, perform the indicated operations.

12. $[(x + 2) + 3y][(x + 2) - 3y]$ $\quad x^2 + 4x + 4 - 9y^2$

13. $\dfrac{2x^2 + x - 1}{2x^2 - 9x + 4} \div \dfrac{6x + 15}{3x^2 - 12x}$ $\quad \dfrac{x(x + 1)}{2x + 5}$

14. $\dfrac{3x}{x^2 - 9x + 20} - \dfrac{5}{2x - 8}$ $\quad \dfrac{x + 25}{2(x - 4)(x - 5)}$

15. $(3x^2 + 10x + 10) \div (x + 2)$ $\quad 3x + 4 + \dfrac{2}{x + 2}$

In Exercises 16–17, factor completely.

16. $xy - 6x + 2y - 12$ $\quad (y - 6)(x + 2)$

17. $24x^3y + 16x^2y - 30xy$ $\quad 2xy(2x + 3)(6x - 5)$

18. A baseball is thrown straight up from a height of 64 feet. The function

$$s(t) = -16t^2 + 48t + 64$$

describes the ball's height above the ground, $s(t)$, in feet, t seconds after it is thrown. How long will it take for the ball to hit the ground? It will take 4 seconds for the ball to hit the ground.

19. The local cable television company offers two deals. Basic cable service with one movie channel costs $35 per month. Basic service with two movie channels costs $45 per month. Find the charge for the basic cable service and the charge for each movie channel. The charge for the basic cable service is $25 and the charge for each movie channel is $10.

20. A rectangular garden 10 feet wide and 12 feet long is surrounded by a rock border of uniform width. The area of the garden and rock border combined is 168 square feet. What is the width of the rock border? The width of the rock border is 1 foot.

You enjoy the single life. You have a terrific career, a support group of close friends, and your own home. Still, you feel the pressure to marry. Your parents belong to a generation that believes that you are not complete until you have found your "other half." They see your staying unattached as a "failure to marry." Perhaps you will someday enter into a permanent relationship, but at the moment you are satisfied with your life and career, and plan to stay single into the future.

This discussion is developed algebraically in the Section 7.6 opener and in Example 6 in Section 7.6.

An increase in the median age at which Americans first marry and the increased rate of divorce have contributed to the growth of our single population. Data indicate that the number of Americans living alone increased quite rapidly toward the end of the twentieth century. However, now this growth rate is beginning to slow down. In this chapter, you will see why radical functions with square roots are used to describe phenomena that are continuing to grow, but whose growth is leveling off. By learning about radicals and radical functions, you will have new algebraic tools for describing your world.

Radicals, Radical Functions, and Rational Exponents

7.1 Radical Expressions and Functions

7.2 Rational Exponents

7.3 Multiplying and Simplifying Radical Expressions

7.4 Adding, Subtracting, and Dividing Radical Expressions

7.5 Multiplying with More Than One Term and Rationalizing Denominators

7.6 Radical Equations

7.7 Complex Numbers

SECTION

7.1

Objectives

1 Evaluate square roots.

2 Evaluate square root functions.

3 Find the domain of square root functions.

4 Use models that are square root functions.

5 Simplify expressions of the form $\sqrt{a^2}$.

6 Evaluate cube root functions.

7 Simplify expressions of the form $\sqrt[3]{a^3}$.

8 Find even and odd roots.

9 Simplify expressions of the form $\sqrt[n]{a^n}$.

RADICAL EXPRESSIONS AND FUNCTIONS

S = Sail area

L = Length

D = Displacement

The America's Cup is the supreme event in ocean sailing. Competition is fierce and the costs are huge. Competitors look to mathematics to provide the critical innovation that can make the difference between winning and losing. The basic dimensions of competitors' yachts must satisfy an inequality containing square roots and cube roots:

$$L + 1.25\sqrt{S} - 9.8\sqrt[3]{D} \le 16.296.$$

In the inequality, L is the yacht's length, in meters, S is its sail area, in square meters, and D is its displacement, in cubic meters.

In this section, we introduce a new category of expressions and functions that contain roots. You will see why square root functions are used to describe phenomena that are continuing to grow but whose growth is leveling off.

1 Evaluate square roots.

Square Roots From our earlier work with exponents, we are aware that the square of both 5 and -5 is 25:

$$5^2 = 25 \quad \text{and} \quad (-5)^2 = 25.$$

The reverse operation of squaring a number is finding the *square root* of the number. For example,

- One square root of 25 is 5 because $5^2 = 25$.

- Another square root of 25 is -5 because $(-5)^2 = 25$.

In general, **if $b^2 = a$, then b is a square root of a.**

The symbol $\sqrt{}$ is used to denote the *positive* or *principal square root* of a number. For example,

- $\sqrt{25} = 5$ because $5^2 = 25$ and 5 is positive.

- $\sqrt{100} = 10$ because $10^2 = 100$ and 10 is positive.

The symbol $\sqrt{}$ that we use to denote the principal square root is called a **radical sign**. The number under the radical sign is called the **radicand**. Together we refer to the radical sign and its radicand as a **radical expression**.

$$\underbrace{\text{Radical sign}}\quad \sqrt{a}\quad \underbrace{\text{Radicand}}$$

Radical expression

> **DEFINITION OF THE PRINCIPAL SQUARE ROOT** If a is a nonnegative real number, the nonnegative number b such that $b^2 = a$, denoted by $b = \sqrt{a}$, is the **principal square root** of a.

The symbol $-\sqrt{}$ is used to denote the negative square root of a number. For example,

- $-\sqrt{25} = -5$ because $(-5)^2 = 25$ and -5 is negative.
- $-\sqrt{100} = -10$ because $(-10)^2 = 100$ and -10 is negative.

EXAMPLE 1 Evaluating Square Roots

Evaluate:

a. $\sqrt{81}$ **b.** $-\sqrt{9}$ **c.** $\sqrt{\dfrac{4}{49}}$

d. $\sqrt{0.0064}$ **e.** $\sqrt{36 + 64}$ **f.** $\sqrt{36} + \sqrt{64}$.

SOLUTION

a. $\sqrt{81} = 9$ The principal square root of 81 is 9 because $9^2 = 81$.

b. $-\sqrt{9} = -3$ The negative square root of 9 is -3 because $(-3)^2 = 9$.

c. $\sqrt{\dfrac{4}{49}} = \dfrac{2}{7}$ The principal square root of $\dfrac{4}{49}$ is $\dfrac{2}{7}$ because $\left(\dfrac{2}{7}\right)^2 = \dfrac{4}{49}$.

d. $\sqrt{0.0064} = 0.08$ The principal square root of 0.0064 is 0.08 because $(0.08)^2 = (0.08)(0.08) = 0.0064$.

e. $\sqrt{36 + 64} = \sqrt{100}$ Simplify the radicand.
$\qquad\qquad\quad = 10$ Take the principal square root of 100, which is 10.

f. $\sqrt{36} + \sqrt{64} = 6 + 8$ $\sqrt{36} = 6$ because $6^2 = 36$. $\sqrt{64} = 8$
$\qquad\qquad\qquad = 14$ because $8^2 = 64$.

STUDY TIP

In Example 1, parts (e) and (f), observe that $\sqrt{36 + 64}$ is not equal to $\sqrt{36} + \sqrt{64}$. In general,

$$\sqrt{a + b} \neq \sqrt{a} + \sqrt{b}$$

and

$$\sqrt{a - b} \neq \sqrt{a} - \sqrt{b}.$$

✔ **CHECK POINT 1** Evaluate:

a. $\sqrt{64}$ 8   **b.** $-\sqrt{49}$ -7 **c.** $\sqrt{\dfrac{16}{25}}$ $\dfrac{4}{5}$

c. $\sqrt{0.0081}$ 0.09   **e.** $\sqrt{9 + 16}$ 5   **f.** $\sqrt{9} + \sqrt{16}$.   7

Let's see what happens to the radical expression $\sqrt{x}$ if x is a negative number. Is the square root of a negative number a real number? For example, consider $\sqrt{-25}$. Is there a real number whose square is -25? No. Thus, $\sqrt{-25}$ is not a real number. In general, **a square root of a negative number is not a real number**.

2 Evaluate square root functions.

Square Root Functions Because each nonnegative real number, x, has precisely one principal square root, $\sqrt{x}$, there is a **square root function** defined by

$$f(x) = \sqrt{x}.$$

The domain of this function is $[0, \infty)$. We can graph $f(x) = \sqrt{x}$ by selecting nonnegative real numbers for x. It is easiest to choose perfect squares, numbers that have rational square roots. Table 7.1 shows five such choices for x and the calculations for the corresponding outputs. We plot these ordered pairs as points in the rectangular coordinate system and connect the points with a smooth curve. The graph of $f(x) = \sqrt{x}$ is shown in Figure 7.1.

Table 7.1

x	$f(x) = \sqrt{x}$	(x, y) or $(x, f(x))$
0	$f(0) = \sqrt{0} = 0$	$(0, 0)$
1	$f(1) = \sqrt{1} = 1$	$(1, 1)$
4	$f(4) = \sqrt{4} = 2$	$(4, 2)$
9	$f(9) = \sqrt{9} = 3$	$(9, 3)$
16	$f(16) = \sqrt{16} = 4$	$(16, 4)$

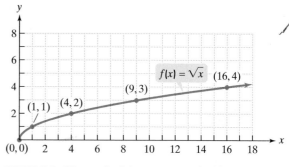

FIGURE 7.1 The graph of the square root function $f(x) = \sqrt{x}$

Is it possible to choose values of x for Table 7.1 that are not squares of integers, or perfect squares? Yes. For example, we can let $x = 3$. Thus, $f(3) = \sqrt{3}$. Because 3 is not a perfect square, $\sqrt{3}$ is an irrational number, one that cannot be expressed as a quotient of integers. We can use a calculator to find a decimal approximation of $\sqrt{3}$.

Many Scientific Calculators	**Many Graphing Calculators**
3 $\boxed{\sqrt{}}$	$\boxed{\sqrt{}}$ 3 $\boxed{\text{ENTER}}$

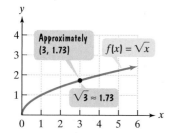

FIGURE 7.2 Visualizing $\sqrt{3}$ as a point on the graph of $f(x) = \sqrt{x}$

Rounding the displayed number to two decimal places, $\sqrt{3} \approx 1.73$. This information is shown visually as a point, approximately $(3, 1.73)$, on the graph of $f(x) = \sqrt{x}$ in Figure 7.2.

To evaluate a square root function, we use substitution, just as we did to evaluate other functions.

EXAMPLE 2 Evaluating Square Root Functions

For each function, find the indicated function value:

a. $f(x) = \sqrt{5x - 6}; f(2)$ **b.** $g(x) = -\sqrt{64 - 8x}; g(-3)$.

SOLUTION

a. $f(2) = \sqrt{5 \cdot 2 - 6}$ Substitute 2 for x in $f(x) = \sqrt{5x - 6}$.

$= \sqrt{4} = 2$ Simplify the radicand and take the square root.

b. $g(-3) = -\sqrt{64 - 8(-3)}$ Substitute -3 for x in $g(x) = -\sqrt{64 - 8x}$.

$= -\sqrt{88} \approx -9.38$ Simplify the radicand:

$64 - 8(-3) = 64 - (-24) = 64 + 24 = 88$.

Then use a calculator to approximate $\sqrt{88}$.

CHECK POINT 2 For each function, find the indicated function value:

a. $f(x) = \sqrt{12x - 20}; f(3)$ $f(3) = 4$
b. $g(x) = -\sqrt{9 - 3x}; g(-5).$ $g(-5) = -\sqrt{24} \approx -4.90$

3 Find the domain of square root functions.

We have seen that the domain of a function f is the largest set of real numbers for which the value of $f(x)$ is a real number. Because only nonnegative numbers have real square roots, the domain of a square root function is the set of real numbers for which the radicand is nonnegative.

EXAMPLE 3 Finding the Domain of a Square Root Function

Find the domain of
$$f(x) = \sqrt{3x + 12}.$$

SOLUTION   The domain is the set of real numbers, x, for which the radicand, $3x + 12$, is nonnegative. We set the radicand greater than or equal to 0 and solve the resulting inequality.
$$3x + 12 \geq 0$$
$$3x \geq -12$$
$$x \geq -4$$
The domain of f is $\{x | x \geq -4\}$ or $[-4, \infty)$.

Figure 7.3 shows the graph of $f(x) = \sqrt{3x + 12}$ in a $[-10, 10, 1]$ by $[-10, 10, 1]$ viewing rectangle. The graph appears only for $x \geq -4$, verifying $[-4, \infty)$ as the domain. Can you see how the graph also illustrates this square root function's range? The graph only appears for nonnegative values of y. Thus, the range is $\{y | y \geq 0\}$ or $[0, \infty)$.

 CHECK POINT 3 Find the domain of
$$f(x) = \sqrt{9x - 27}.$$ Domain of f is $\{x | x \geq 3\}$ or $[3, \infty)$.

$f(x) = \sqrt{3x + 12}$

FIGURE 7.3

4 Use models that are square root functions.

The graph of the square root function $f(x) = \sqrt{x}$ is increasing from left to right. However, the rate of increase is slowing down as the graph moves to the right. This is why square root functions are often used to model growing phenomena with growth that is leveling off.

EXAMPLE 4 Modeling with a Square Root Function

The graph in Figure 7.4 shows the percentage of the U.S. population who used the Internet from 2000 through 2003. Although this percent grew from year to year, the growth was leveling off. The data can be modeled by the function

$$P(x) = 6.7\sqrt{x} + 44.4,$$

where $P(x)$ is the percentage of the U.S. population who used the Internet x years after 2000. According to the model, what percentage used the Internet in 2003? Round to the nearest tenth of a percent. How well does the model describe the actual data?

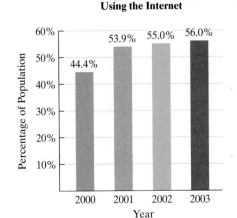

Percentage of U.S. Population Using the Internet

FIGURE 7.4

Source: U.S. Department of Commerce

SOLUTION   Because 2003 is 3 years after 2000, we substitute 3 for x and evaluate the function at 3.

$$P(x) = 6.7\sqrt{x} + 44.4 \qquad \text{Use the given function.}$$

$$P(3) = 6.7\sqrt{3} + 44.4 \qquad \text{Substitute 3 for x.}$$

$$\approx 56.0 \qquad \text{Use a calculator.}$$

The model indicates that approximately 56% of the U.S. population used the Internet in 2003. Figure 7.4 on the previous page shows 56%, so our model provides an excellent description of the actual data value for 2003. ■

CHECK POINT 4   Use the square root function in Example 4 to find the percentage of Internet users in 2002. Round to the nearest tenth of a percent. How well does the model describe the actual data?   Approximately 53.9% used the Internet in 2002. This is the percent given for 2001, not 2002, so the model does not provide a totally accurate description of the data.

5 Simplify expressions of the form $\sqrt{a^2}$.

Simplifying Expressions of the Form $\sqrt{a^2}$ You may think that $\sqrt{a^2} = a$. However, this is not necessarily true. Consider the following examples:

$$\sqrt{4^2} = \sqrt{16} = 4$$
$$\sqrt{(-4)^2} = \sqrt{16} = 4.$$

The result is not −4, but rather the absolute value of −4, or 4.

Here is a rule for simplifying expressions of the form $\sqrt{a^2}$:

SIMPLIFYING $\sqrt{a^2}$ For any real number a,

$$\sqrt{a^2} = |a|.$$

In words, the principal square root of a^2 is the absolute value of a.

EXAMPLE 5   Simplifying Radical Expressions

Simplify each expression:

a. $\sqrt{(-6)^2}$ **b.** $\sqrt{(x + 5)^2}$ **c.** $\sqrt{25x^6}$ **d.** $\sqrt{x^2 - 4x + 4}$.

SOLUTION   The principal square root of an expression squared is the absolute value of that expression. In parts (a) and (b), we are given squared radicands. In parts (c) and (d), it will first be necessary to express the radicand as an expression that is squared.

a. $\sqrt{(-6)^2} = |-6| = 6$

b. $\sqrt{(x + 5)^2} = |x + 5|$

c. To simplify $\sqrt{25x^6}$, first write $25x^6$ as an expression that is squared: $25x^6 = (5x^3)^2$. Then simplify.

$$\sqrt{25x^6} = \sqrt{(5x^3)^2} = |5x^3| \text{ or } 5|x^3|$$

d. To simplify $\sqrt{x^2 - 4x + 4}$, first write $x^2 - 4x + 4$ as an expression that is squared by factoring the perfect square trinomial: $x^2 - 4x + 4 = (x - 2)^2$. Then simplify.

$$\sqrt{x^2 - 4x + 4} = \sqrt{(x - 2)^2} = |x - 2| \qquad ■$$

 CHECK POINT 5 Simplify each expression:

a. $\sqrt{(-7)^2}$ 7

b. $\sqrt{(x + 8)^2}$ $|x + 8|$

c. $\sqrt{49x^{10}}$ $|7x^5|$ or $7|x^5|$

d. $\sqrt{x^2 - 6x + 9}$. $|x - 3|$

In some situations, we are told that no radicands involve negative quantities raised to even powers. When the expression being squared is nonnegative, it is not necessary to use absolute value when simplifying $\sqrt{a^2}$. For example, assuming that no radicands contain negative quantities that are squared,

$$\sqrt{x^8} = \sqrt{(x^4)^2} = x^4$$

$$\sqrt{25x^2 + 10x + 1} = \sqrt{(5x + 1)^2} = 5x + 1.$$

Cube Roots and Cube Root Functions   Finding the square root of a number reverses the process of squaring a number. Similarly, finding the cube root of a number reverses the process of cubing a number. For example, $2^3 = 8$, and so the cube root of 8 is 2. The notation that we use is $\sqrt[3]{8} = 2$.

6   Evaluate cube root functions.

DEFINITION OF THE CUBE ROOT OF A NUMBER   The **cube root** of a real number a is written $\sqrt[3]{a}$.

$$\sqrt[3]{a} = b \quad \text{means that} \quad b^3 = a.$$

For example,

$$\sqrt[3]{64} = 4 \qquad \text{because} \qquad 4^3 = 64.$$

$$\sqrt[3]{-27} = -3 \quad \text{because} \quad (-3)^3 = -27.$$

In contrast to square roots, the cube root of a negative number is a real number. All real numbers have cube roots. The cube root of a positive number is positive. The cube root of a negative number is negative.

Because every real number, x, has precisely one cube root, $\sqrt[3]{x}$, there is a **cube root function** defined by

$$f(x) = \sqrt[3]{x}.$$

The domain of this function is the set of all real numbers. We can graph $f(x) = \sqrt[3]{x}$ by selecting perfect cubes, numbers that have rational cube roots, for x.

Table 7.2 shows five such choices for x and the calculations for the corresponding outputs. We plot these ordered pairs as points in the rectangular coordinate system and connect the points with a smooth curve. The graph of $f(x) = \sqrt[3]{x}$ is shown in Figure 7.5.

Table 7.2

x	$f(x) = \sqrt[3]{x}$	(x, y) or $(x, f(x))$
-8	$f(-8) = \sqrt[3]{-8} = -2$	$(-8, -2)$
-1	$f(-1) = \sqrt[3]{-1} = -1$	$(-1, -1)$
0	$f(0) = \sqrt[3]{0} = 0$	$(0, 0)$
1	$f(1) = \sqrt[3]{1} = 1$	$(1, 1)$
8	$f(8) = \sqrt[3]{8} = 2$	$(8, 2)$

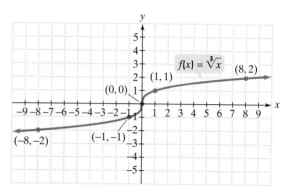

FIGURE 7.5 The graph of the cube root function $f(x) = \sqrt[3]{x}$

Notice that both the domain and the range of $f(x) = \sqrt[3]{x}$ are the set of all real numbers.

EXAMPLE 6 Evaluating Cube Root Functions

For each function, find the indicated function value:

a. $f(x) = \sqrt[3]{x} - 2$; $f(127)$ **b.** $g(x) = \sqrt[3]{8x - 8}$; $g(-7)$.

SOLUTION

a. $f(127) = \sqrt[3]{127} - 2$ Substitute 127 for x in f(x) $= \sqrt[3]{x} - 2$.

$\quad\quad\quad\quad = \sqrt[3]{125}$ Simplify the radicand.

$\quad\quad\quad\quad = 5$ $\sqrt[3]{125} = 5$ because $5^3 = 125$.

b. $g(-7) = \sqrt[3]{8(-7) - 8}$ Substitute -7 for x in g(x) $= \sqrt[3]{8x - 8}$.

$\quad\quad\quad\quad = \sqrt[3]{-64}$ Simplify the radicand: $8(-7) - 8 = -56 - 8 = -64$.

$\quad\quad\quad\quad = -4$ $\sqrt[3]{-64} = -4$ because $(-4)^3 = -64$. ■

 CHECK POINT 6 For each function, find the indicated function value:

a. $f(x) = \sqrt[3]{x - 6}$; $f(33)$ $f(33) = 3$

b. $g(x) = \sqrt[3]{2x + 2}$; $g(-5)$. $g(-5) = -2$

Because the cube root of a positive number is positive and the cube root of a negative number is negative, absolute value is not needed to simplify expressions of the form $\sqrt[3]{a^3}$.

7 Simplify expressions of the form $\sqrt[3]{a^3}$.

SIMPLIFYING $\sqrt[3]{a^3}$ For any real number a,

$$\sqrt[3]{a^3} = a.$$

In words, the cube root of any expression cubed is that expression.

EXAMPLE 7 Simplifying a Cube Root

Simplify: $\sqrt[3]{-64x^3}$.

SOLUTION Begin by expressing the radicand as an expression that is cubed: $-64x^3 = (-4x)^3$. Then simplify.

$$\sqrt[3]{-64x^3} = \sqrt[3]{(-4x)^3} = -4x$$

We can check our answer by cubing $-4x$:

$$(-4x)^3 = (-4)^3 x^3 = -64x^3.$$

By obtaining the original radicand, we know that our simplification is correct. ∎

✔ **CHECK POINT 7** Simplify: $\sqrt[3]{-27x^3}$. $-3x$

---

8 Find even and odd roots.

Even and Odd nth Roots Up to this point, we have focused on square roots and cube roots. Other radical expressions have different roots. For example, the fifth root of a, written $\sqrt[5]{a}$, is the number b for which $b^5 = a$. Thus,

$$\sqrt[5]{32} = 2 \quad \text{because} \quad 2^5 = 2 \cdot 2 \cdot 2 \cdot 2 \cdot 2 = 32.$$

The radical expression $\sqrt[n]{a}$ represents the **nth root** of a. The number n is called the **index**. An index of 2 represents a square root and is not written. An index of 3 represents a cube root.

If the index n in $\sqrt[n]{a}$ is an odd number, a root is said to be an **odd root**. A cube root is an odd root. Other odd roots have the same characteristics as cube roots. Every real number has exactly one real root when n is odd. The (odd) nth root of a, $\sqrt[n]{a}$, is the number b for which $b^n = a$. An odd root of a positive number is positive and an odd root of a negative number is negative. For example,

$$\sqrt[5]{243} = 3 \quad \text{because} \quad 3^5 = 3 \cdot 3 \cdot 3 \cdot 3 \cdot 3 = 243$$

and $\quad \sqrt[5]{-243} = -3 \quad \text{because} \quad (-3)^5 = (-3)(-3)(-3)(-3)(-3) = -243.$

If the index n in $\sqrt[n]{a}$ is an even number, a root is said to be an **even root**. A square root is an even root. Other even roots have the same characteristics as square roots. Every positive real number has two real roots when n is even. One root is positive and one is negative. The positive root, called the **principal nth root** and represented by $\sqrt[n]{a}$, is the nonnegative number b for which $b^n = a$. The symbol $-\sqrt[n]{a}$ is used to denote the negative nth root. **An even root of a negative number is not a real number.**

STUDY TIP

Some higher even and odd roots occur so frequently that you might want to memorize them.

Fourth Roots	Fifth Roots
$\sqrt[4]{1} = 1$	$\sqrt[5]{1} = 1$
$\sqrt[4]{16} = 2$	$\sqrt[5]{32} = 2$
$\sqrt[4]{81} = 3$	$\sqrt[5]{243} = 3$
$\sqrt[4]{256} = 4$	
$\sqrt[4]{625} = 5$	

EXAMPLE 8 Finding Even and Odd Roots

Find the indicated root, or state that the expression is not a real number:

   a. $\sqrt[4]{81}$ **b.** $-\sqrt[4]{81}$ **c.** $\sqrt[4]{-81}$ **d.** $\sqrt[5]{-32}$.

SOLUTION

   a. $\sqrt[4]{81} = 3$ The principal fourth root of 81 is 3 because $3^4 = 3 \cdot 3 \cdot 3 \cdot 3 = 81$.

   b. $-\sqrt[4]{81} = -3$ The negative fourth root of 81 is -3 because $(-3)^4 = (-3)(-3)(-3)(-3) = 81$.

c. $\sqrt[4]{-81}$ is not a real number because the index, 4, is even and the radicand, -81, is negative. No real number can be raised to the fourth power to give a negative result such as -81. Real numbers to even powers can only result in nonnegative numbers.

d. $\sqrt[5]{-32} = -2$ because $(-2)^5 = (-2)(-2)(-2)(-2)(-2) = -32$. An odd root of a negative real number is always negative. ∎

 CHECK POINT 8 Find the indicated root, or state that the expression is not a real number:

a. $\sqrt[4]{16}$ 2

b. $-\sqrt[4]{16}$ -2

c. $\sqrt[4]{-16}$ not a real number

d. $\sqrt[5]{-1}$. -1

9 Simplify expressions of the form $\sqrt[n]{a^n}$.

Simplifying Expressions of the Form $\sqrt[n]{a^n}$ We have seen that

$$\sqrt{a^2} = |a| \quad \text{and} \quad \sqrt[3]{a^3} = a.$$

Expressions of the form $\sqrt[n]{a^n}$ can be simplified in the same manner. Unless a is known to be nonnegative, absolute value notation is needed when n is even. When the index is odd, absolute value bars are not necessary.

SIMPLIFYING $\sqrt[n]{a^n}$ For any real number a,

1. If n is even, $\sqrt[n]{a^n} = |a|$.

2. If n is odd, $\sqrt[n]{a^n} = a$.

EXAMPLE 9 Simplifying Radical Expressions

Simplify:

a. $\sqrt[4]{(x-3)^4}$ **b.** $\sqrt[5]{(2x+7)^5}$ **c.** $\sqrt[6]{(-5)^6}$.

SOLUTION Each expression involves the nth root of a radicand raised to the nth power. Thus, each radical expression can be simplified. Absolute value bars are necessary in parts (a) and (c) because the index, n, is even.

a. $\sqrt[4]{(x-3)^4} = |x-3|$ $\sqrt[n]{a^n} = |a|$ if n is even.

b. $\sqrt[5]{(2x+7)^5} = 2x+7$ $\sqrt[n]{a^n} = a$ if n is odd.

c. $\sqrt[6]{(-5)^6} = |-5| = 5$ $\sqrt[n]{a^n} = |a|$ if n is even. ∎

 CHECK POINT 9 Simplify:

a. $\sqrt[4]{(x+6)^4}$ $|x+6|$

b. $\sqrt[5]{(3x-2)^5}$ $3x-2$

c. $\sqrt[6]{(-8)^6}$. 8

7.1 EXERCISE SET

Student Solutions Manual   CD/Video   PH Math/Tutor Center   MathXL Tutorials on CD   MathXL®   MyMathLab   Interactmath.com

Practice Exercises

In Exercises 1–20, evaluate each expression, or state that the expression is not a real number.

1. $\sqrt{36}$ 6

2. $\sqrt{16}$ 4

3. $-\sqrt{36}$ −6

4. $-\sqrt{16}$ −4

5. $\sqrt{-36}$ not a real number

6. $\sqrt{-16}$ not a real number

7. $\sqrt{\dfrac{1}{25}}$ $\dfrac{1}{5}$

8. $\sqrt{\dfrac{1}{49}}$ $\dfrac{1}{7}$

9. $-\sqrt{\dfrac{9}{16}}$ $-\dfrac{3}{4}$

10. $-\sqrt{\dfrac{4}{25}}$ $-\dfrac{2}{5}$

11. $\sqrt{0.81}$ 0.9

12. $\sqrt{0.49}$ 0.7

13. $-\sqrt{0.04}$ −0.2

14. $-\sqrt{0.64}$ −0.8

15. $\sqrt{25-16}$ 3

16. $\sqrt{144+25}$ 13

17. $\sqrt{25}-\sqrt{16}$ 1

18. $\sqrt{144}+\sqrt{25}$ 17

19. $\sqrt{16-25}$ not a real number

20. $\sqrt{25-144}$ not a real number

In Exercises 21–26, find the indicated function values for each function. If necessary, round to two decimal places. If the function value is not a real number and does not exist, so state.

21. $f(x) = \sqrt{x-2}$; $f(18), f(3), f(2), f(-2)$
4; 1; 0; not a real number

22. $f(x) = \sqrt{x-3}$; $f(28), f(4), f(3), f(-1)$
5; 1; 0; not a real number

23. $g(x) = -\sqrt{2x+3}$; $g(11), g(1), g(-1), g(-2)$
−5; $-\sqrt{5} \approx -2.24$; −1; not a real number

24. $g(x) = -\sqrt{2x+1}$; $g(4), g(1), g\left(-\dfrac{1}{2}\right), g(-1)$
−3; $-\sqrt{3} \approx -1.73$; 0; not a real number

25. $h(x) = \sqrt{(x-1)^2}$; $h(5), h(3), h(0), h(-5)$ 4; 2; 1; 6

26. $h(x) = \sqrt{(x-2)^2}$; $h(5), h(3), h(0), h(-5)$ 3; 1; 2; 7

In Exercises 27–32, find the domain of each square root function. Then use the domain to match the radical function with its graph. [The graphs are labeled (a) through (f) and are shown in $[-10, 10, 1]$ by $[-10, 10, 1]$ viewing rectangles.]

27. $f(x) = \sqrt{x-3}$ Domain of f is $\{x | x \geq 3\} = [3, \infty)$; c

28. $f(x) = \sqrt{x+2}$ Domain of f is $\{x | x \geq -2\} = [-2, \infty)$; f

29. $f(x) = \sqrt{3x+15}$ Domain of f is $\{x | x \geq -5\} = [-5, \infty)$; d

30. $f(x) = \sqrt{3x-15}$ Domain of f is $\{x | x \geq 5\} = [5, \infty)$; a

31. $f(x) = \sqrt{6-2x}$ Domain of f is $\{x | x \leq 3\} = (-\infty, 3]$; e

32. $f(x) = \sqrt{8-2x}$ Domain of f is $\{x | x \leq 4\} = (-\infty, 4]$; b

a.

b.

c.

d.

e.

f.

In Exercises 33–46, simplify each expression.

33. $\sqrt{5^2}$ 5

34. $\sqrt{7^2}$ 7

35. $\sqrt{(-4)^2}$ 4   **36.** $\sqrt{(-10)^2}$ 10

37. $\sqrt{(x-1)^2}$ $|x-1|$ **38.** $\sqrt{(x-2)^2}$ $|x-2|$

39. $\sqrt{36x^4}$ $|6x^2|$ or $6x^2$ **40.** $\sqrt{81x^4}$ $|9x^2|$ or $9x^2$

41. $-\sqrt{100x^6}$ $-|10x^3|$ or $-10|x^3|$

42. $-\sqrt{49x^6}$ $-|7x^3|$ or $-7|x^3|$

43. $\sqrt{x^2+12x+36}$ $|x+6|$

44. $\sqrt{x^2+14x+49}$ $|x+7|$

45. $-\sqrt{x^2-8x+16}$ $-|x-4|$

46. $-\sqrt{x^2-10x+25}$ $-|x-5|$

In Exercises 47–52, find each cube root.

47. $\sqrt[3]{27}$ 3   **48.** $\sqrt[3]{64}$ 4

49. $\sqrt[3]{-27}$ -3 **50.** $\sqrt[3]{-64}$ -4

51. $\sqrt[3]{\dfrac{1}{125}}$ $\dfrac{1}{5}$ **52.** $\sqrt[3]{\dfrac{1}{1000}}$ $\dfrac{1}{10}$

In Exercises 53–56, find the indicated function values for each function.

53. $f(x) = \sqrt[3]{x-1}$; $f(28)$, $f(9)$, $f(0)$, $f(-63)$ $3; 2; -1; -4$

54. $f(x) = \sqrt[3]{x-3}$; $f(30)$, $f(11)$, $f(2)$, $f(-122)$ $3; 2; -1; -5$

55. $g(x) = -\sqrt[3]{8x-8}$; $g(2)$, $g(1)$, $g(0)$ $-2; 0; 2$

56. $g(x) = -\sqrt[3]{2x+1}$; $g(13)$, $g(0)$, $g(-63)$ $-3; -1; 5$

In Exercises 57–74, find the indicated root, or state that the expression is not a real number.

57. $\sqrt[4]{1}$ 1   **58.** $\sqrt[5]{1}$ 1

59. $\sqrt[4]{16}$ 2   **60.** $\sqrt[4]{81}$ 3

61. $-\sqrt[4]{16}$ -2 **62.** $-\sqrt[4]{81}$ -3

63. $\sqrt[4]{-16}$ not a real number   **64.** $\sqrt[4]{-81}$ not a real number

65. $\sqrt[5]{-1}$ -1 **66.** $\sqrt[7]{-1}$ -1

67. $\sqrt[6]{-1}$ not a real number   **68.** $\sqrt[8]{-1}$ not a real number

69. $-\sqrt[4]{256}$ -4 **70.** $-\sqrt[4]{10,000}$ -10

71. $\sqrt[6]{64}$ 2   **72.** $\sqrt[5]{32}$ 2

73. $-\sqrt[5]{32}$ -2 **74.** $-\sqrt[6]{64}$ -2

In Exercises 75–88, simplify each expression. Include absolute value bars where necessary.

75. $\sqrt[3]{x^3}$ x **76.** $\sqrt[5]{x^5}$ x

77. $\sqrt[4]{y^4}$ $|y|$ **78.** $\sqrt[6]{y^6}$ $|y|$

79. $\sqrt[3]{-8x^3}$ $-2x$ **80.** $\sqrt[3]{-125x^3}$ $-5x$

81. $\sqrt[3]{(-5)^3}$ -5 **82.** $\sqrt[3]{(-6)^3}$ -6

83. $\sqrt[4]{(-5)^4}$ 5   **84.** $\sqrt[6]{(-6)^6}$ 6

85. $\sqrt[4]{(x+3)^4}$ $|x+3|$ **86.** $\sqrt[4]{(x+5)^4}$ $|x+5|$

87. $\sqrt[5]{-32(x-1)^5}$ $-2(x-1)$

88. $\sqrt[5]{-32(x-2)^5}$ $-2(x-2)$

Practice Plus

In Exercises 89–92, complete each table and graph the given function. Identify the function's domain and range.   **89–92.** See graphing answer section.

89. $f(x) = \sqrt{x} + 3$

x	$f(x) = \sqrt{x} + 3$
0	3
1	4
4	5
9	6

Domain: $\{x \mid x \geq 0\} = [0, \infty)$; Range: $\{y \mid y \geq 3\} = [3, \infty)$

90. $f(x) = \sqrt{x} - 2$

x	$f(x) = \sqrt{x} - 2$
0	-2
1	-1
4	0
9	1

Domain: $\{x \mid x \geq 0\} = [0, \infty)$; Range: $\{y \mid y \geq -2\} = [-2, \infty)$

91. $f(x) = \sqrt{x-3}$

x	$f(x) = \sqrt{x-3}$
3	0
4	1
7	2
12	3

Domain: $\{x \mid x \geq 3\} = [3, \infty)$; Range: $\{y \mid y \geq 0\} = [0, \infty)$

92. $f(x) = \sqrt{4-x}$

x	$f(x) = \sqrt{4-x}$
-5	3
0	2
3	1
4	0

Domain: $\{x \mid x \leq 4\} = (-\infty, 4]$; Range: $\{y \mid y \geq 0\} = [0, \infty)$

In Exercises 93–96, find the domain of each function.

93. $f(x) = \dfrac{\sqrt[3]{x}}{\sqrt{30-2x}}$ $\{x \mid x < 15\} = (-\infty, 15)$

94. $f(x) = \dfrac{\sqrt[3]{x}}{\sqrt{80-5x}}$ $\{x \mid x < 16\} = (-\infty, 16)$

95. $f(x) = \dfrac{\sqrt{x-1}}{\sqrt{3-x}}$ $\{x \mid 1 \le x < 3\} = [1, 3)$

96. $f(x) = \dfrac{\sqrt{x-2}}{\sqrt{7-x}}$ $\{x \mid 2 \le x < 7\} = [2, 7)$

In Exercises 97–98, evaluate each expression.

97. $\sqrt[3]{\sqrt[4]{16} + \sqrt{625}}$ 3

98. $\sqrt[3]{\sqrt{\sqrt{169} + \sqrt{9}} + \sqrt{\sqrt[3]{1000} + \sqrt[3]{216}}}$ 2

Application Exercises

The table shows the median, or average, heights for boys of various ages in the United States, from birth through 60 months, or five years. The data can be modeled by the radical function

$$f(x) = 2.9\sqrt{x} + 20.1,$$

where $f(x)$ is the median height, in inches, of boys who are x months of age. Use the function to solve Exercises 99–100.

Boys' Median Heights

Age (months)	Height (inches)
0	20.5
6	27.0
12	30.8
18	32.9
24	35.0
36	37.5
48	40.8
60	43.4

Source: The Portable Pediatrician for Parents, by Laura Walther Nathanson, M.D., FAAP

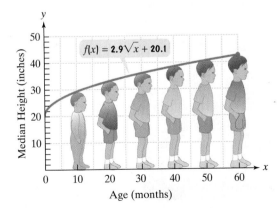

99. According to the model, what is the median height of boys who are 48 months, or four years, old? Use a calculator and round to the nearest tenth of an inch. How well does the model describe the actual median height shown in the table? According to the model, the median height of boys who are 48 months old is approximately 40.2 inches. The model describes actual data well.

100. According to the model, what is the median height of boys who are 60 months, or five years, old? Use a calculator and round to the nearest tenth of an inch. How well does the model describe the actual median height shown in the table? According to the model, the median height of boys who are 60 months old is approximately 42.6 inches. The model describes actual data well.

Police use the function $f(x) = \sqrt{20x}$ to estimate the speed of a car, $f(x)$, in miles per hour, based on the length, x, in feet, of its skid marks upon sudden braking on a dry asphalt road. Use the function to solve Exercises 101–102.

101. A motorist is involved in an accident. A police officer measures the car's skid marks to be 245 feet long. Estimate the speed at which the motorist was traveling before braking. If the posted speed limit is 50 miles per hour and the motorist tells the officer he was not speeding, should the officer believe him? Explain. The motorist was traveling at 70 miles per hour before braking. The officer should not believe the motorist; Answers will vary.

102. A motorist is involved in an accident. A police officer measures the car's skid marks to be 45 feet long. Estimate the speed at which the motorist was traveling before braking. If the posted speed limit is 35 miles per hour and the motorist tells the officer she was not speeding, should the officer believe her? Explain. The motorist was traveling at 30 miles per hour before braking. The officer should believe the motorist; Answers will vary.

Writing in Mathematics

103. What are the square roots of 36? Explain why each of these numbers is a square root.   103–113. Answers will vary.

104. What does the symbol $\sqrt{}$ denote? Which of your answers in Exercise 103 is given by this symbol? Write the symbol needed to obtain the other answer.

105. Explain why $\sqrt{-1}$ is not a real number.

106. Explain how to find the domain of a square root function.

107. Explain how to simplify $\sqrt{a^2}$. Give an example with your explanation.

108. Explain why $\sqrt[3]{8}$ is 2. Then describe what is meant by the cube root of a real number.

109. Describe two differences between odd and even roots.

110. Explain how to simplify $\sqrt[n]{a^n}$ if n is even and if n is odd. Give examples with your explanations.

111. Explain the meaning of the words "radical," "radicand," and "index." Give an example with your explanation.

112. Describe the trend in a boy's growth from birth through five years, shown in the table for Exercises 99–100. Why is a square root function a useful model for the data?

113. The function $f(x) = 6.75\sqrt{x} + 12$ models the amount of new student loans, $f(x)$, in billions of dollars, x years after 1993. Write a word problem using the function. Then solve the problem.

Technology Exercises

114. Use a graphing utility to graph $y_1 = \sqrt{x}$, $y_2 = \sqrt{x + 4}$, and $y_3 = \sqrt{x - 3}$ in the same $[-5, 10, 1]$ by $[0, 6, 1]$ viewing rectangle. Describe one similarity and one difference that you observe among the graphs. Use the word "shift" in your response. *114–117. See graphing answer section.*

115. Use a graphing utility to graph $y = \sqrt{x}$, $y = \sqrt{x + 4}$, and $y = \sqrt{x - 3}$ in the same $[-1, 10, 1]$ by $[-10, 10, 1]$ viewing rectangle. Describe one similarity and one difference that you observe among the graphs.

116. Use a graphing utility to graph $f(x) = \sqrt{x}$, $g(x) = -\sqrt{x}$, $h(x) = \sqrt{-x}$, and $k(x) = -\sqrt{-x}$ in the same $[-10, 10, 1]$ by $[-4, 4, 1]$ viewing rectangle. Use the graphs to describe the domains and the ranges of functions $f, g, h,$ and k. *Domain of f is $\{x | x \geq 0\} = [0, \infty)$; Range of f is $\{y | y \geq 0\} = [0, \infty)$. Domain of g is $\{x | x \geq 0\} = [0, \infty)$; Range of g is $\{y | y \leq 0\} = (-\infty, 0]$. Domain of h is $\{x | x \leq 0\} = (-\infty, 0]$; Range of h is $\{y | y \geq 0\} = [0, \infty)$. Domain of k is $\{x | x \leq 0\} = (-\infty, 0]$; Range of k is $\{y | y \leq 0\} = (-\infty, 0]$.*

117. Use a graphing utility to graph $y_1 = \sqrt{x^2}$ and $y_2 = -x$ in the same viewing rectangle.

    a. For what values of x is $\sqrt{x^2} = -x$? *$\{x | x \leq 0\}$*

    b. For what values of x is $\sqrt{x^2} \neq -x$? *$\{x | x > 0\}$*

Critical Thinking Exercises

118. Which one of the following is true? *d*

    a. The domain of $f(x) = \sqrt[3]{x - 4}$ is $[4, \infty)$.

    b. If n is odd and b is negative, then $\sqrt[n]{b}$ is not a real number.

    c. The expression $\sqrt[n]{4}$ represents increasingly larger numbers for $n = 2, 3, 4, 5, 6,$ and so on.

    d. None of the above is true.

119. Write a function whose domain is $(-\infty, 5]$. *Answers will vary; an example is $f(x) = \sqrt{15 - 3x}$.*

120. Let $f(x) = \sqrt{x - 3}$ and $g(x) = \sqrt{x + 1}$. Find the domain of $f + g$ and $\dfrac{f}{g}$. *Domain of $f + g$ is $\{x | x \geq 3\} = [3, \infty)$ and domain of $\dfrac{f}{g}$ is $\{x | x \geq 3\} = [3, \infty)$.*

121. Simplify: $\sqrt{(2x + 3)^{10}}$. *$|(2x + 3)^5|$*

In Exercises 122–123, graph each function by hand. Then describe the relationship between the function that you graphed and the graph of $f(x) = \sqrt{x}$. *122–123. See graphing answer section.*

122. $g(x) = \sqrt{x} + 2$ *The graph of g is the graph of f shifted up 2 units.*

123. $h(x) = \sqrt{x + 3}$ *The graph of h is the graph of f shifted left 3 units.*

Review Exercises

124. Simplify: $3x - 2[x - 3(x + 5)]$. (Section 1.2, Example 14) *$7x + 30$*

125. Simplify: $(-3x^{-4}y^3)^{-2}$. (Section 1.6, Example 7c) *$\dfrac{x^8}{9y^6}$*

126. Solve: $|3x - 4| > 11$. (Section 4.3, Example 4) *$\{x | x < -\frac{7}{3} \text{ or } x > 5\}$ or $\left(-\infty, -\frac{7}{3}\right) \cup (5, \infty)$*

7.2
RATIONAL EXPONENTS

Objectives

1 Use the definition of $a^{\frac{1}{n}}$.

2 Use the definition of $a^{\frac{m}{n}}$.

3 Use the definition of $a^{-\frac{m}{n}}$.

4 Use models that contain rational exponents.

5 Simplify expressions with rational exponents.

6 Simplify radical expressions using rational exponents.

Marine iguanas of the Galápagos Islands

The Galápagos Islands are a chain of volcanic islands lying 600 miles west of Ecuador. They are famed for over 5000 species of plants and animals, including a rare flightless cormorant, marine iguanas, and giant tortoises weighing more than 600 pounds. Early in 2001, the plants and wildlife that live in the Galápagos were at risk from a massive oil spill that flooded 150,000 gallons of toxic fuel into one of the world's most fragile ecosystems. The long-term danger of the accident is that fuel sinking to the ocean floor will destroy algae that is vital to the food chain. Any imbalance in the food chain could threaten the rare Galápagos plant and animal species that have evolved for thousands of years in isolation with little human intervention.

At risk on these ecologically vulnerable islands are unique flora and fauna that helped to inspire Charles Darwin's theory of evolution. Darwin made an enormous collection of the islands' plant species. The function

$$f(x) = 29x^{\frac{1}{3}}$$

models the number of plant species, $f(x)$, on the various islands of the Galápagos in terms of the area, x, in square miles, of a particular island. But x to the *what* power? How can we interpret the information given by this function? In this section, we turn our attention to rational exponents such as $\frac{1}{3}$ and their relationship to roots of real numbers.

Defining Rational Exponents We define rational exponents so that their properties are the same as the properties for integer exponents. For example, suppose that $x = 7^{\frac{1}{3}}$. We know that exponents are multiplied when an exponential expression is raised to a power. For this to be true,

$$x^3 = \left(7^{\frac{1}{3}}\right)^3 = 7^{\frac{1}{3} \cdot 3} = 7^1 = 7.$$

We see that $x^3 = 7$. This means that x is the number whose cube is 7. Thus, $x = \sqrt[3]{7}$. Remember that we began with $x = 7^{\frac{1}{3}}$. This means that

$$7^{\frac{1}{3}} = \sqrt[3]{7}.$$

We can generalize this idea with the following definition:

1 Use the definition of $a^{\frac{1}{n}}$.

> **THE DEFINITION OF $a^{\frac{1}{n}}$** If $\sqrt[n]{a}$ represents a real number and $n \geq 2$ is an integer, then
>
> $$a^{\frac{1}{n}} = \sqrt[n]{a}.$$
>
> The denominator of the rational exponent is the radical's index.
>
> If a is negative, n must be odd. If a is nonnegative, n can be any index.

EXAMPLE 1   Using the Definition of $a^{\frac{1}{n}}$

Use radical notation to rewrite each expression. Simplify, if possible:

a. $64^{\frac{1}{2}}$ **b.** $(-125)^{\frac{1}{3}}$ **c.** $(6x^2y)^{\frac{1}{5}}$.

SOLUTION

a. $64^{\frac{1}{2}} = \sqrt{64} = 8$

The denominator is the index.

b. $(-125)^{\frac{1}{3}} = \sqrt[3]{-125} = -5$ **c.** $(6x^2y)^{\frac{1}{5}} = \sqrt[5]{6x^2y}$

CHECK POINT 1 Use radical notation to rewrite each expression. Simplify, if possible:

a. $25^{\frac{1}{2}}$ $\sqrt{25} = 5$ **b.** $(-8)^{\frac{1}{3}}$ $\sqrt[3]{-8} = -2$ **c.** $(5xy^2)^{\frac{1}{4}}$. $\sqrt[4]{5xy^2}$

In our next example, we begin with radical notation and rewrite the expression with rational exponents.

The radical's index becomes the exponent's denominator.

$$\sqrt[n]{a} = a^{\frac{1}{n}}$$

The radicand becomes the base.

EXAMPLE 2 Using the Definition of $a^{\frac{1}{n}}$

Rewrite with rational exponents:

a. $\sqrt[5]{13ab}$ **b.** $\sqrt[7]{\dfrac{xy^2}{17}}$.

SOLUTION Parentheses are needed to show that the entire radicand becomes the base.

a. $\sqrt[5]{13ab} = (13ab)^{\frac{1}{5}}$

The index is the exponent's denominator.

b. $\sqrt[7]{\dfrac{xy^2}{17}} = \left(\dfrac{xy^2}{17}\right)^{\frac{1}{7}}$ ■

 CHECK POINT 2 Rewrite with rational exponents:

a. $\sqrt[4]{5xy}$ $(5xy)^{\frac{1}{4}}$ **b.** $\sqrt[5]{\dfrac{a^3b}{2}}$. $\left(\dfrac{a^3b}{2}\right)^{\frac{1}{5}}$

2 Use the definition of $a^{\frac{m}{n}}$.

Can rational exponents have numerators other than 1? The answer is yes. If the numerator is some other integer, we still want to multiply exponents when raising a power to a power. For this reason,

$$a^{\frac{2}{3}} = (a^{\frac{1}{3}})^2 \quad \text{and} \quad a^{\frac{2}{3}} = (a^2)^{\frac{1}{3}}.$$

This means $(\sqrt[3]{a})^2$.   This means $\sqrt[3]{a^2}$.

Thus,

$$a^{\frac{2}{3}} = (\sqrt[3]{a})^2 = \sqrt[3]{a^2}.$$

Do you see that the denominator, 3, of the rational exponent is the same as the index of the radical? The numerator, 2, of the rational exponent serves as an exponent in each of the two radical forms. We generalize these ideas with the following definition:

THE DEFINITION OF $a^{\frac{m}{n}}$ If $\sqrt[n]{a}$ represents a real number, $\dfrac{m}{n}$ is a positive rational number reduced to lowest terms, and $n \geq 2$ is an integer, then

and

$$a^{\frac{m}{n}} = \left(\sqrt[n]{a}\right)^{m} \quad \text{First take the } n\text{th root of } a.$$

$$a^{\frac{m}{n}} = \sqrt[n]{a^{m}}. \quad \text{First raise } a \text{ to the } m \text{ power.}$$

The first form of the definition shown in the box involves taking the root first. This form is often preferable because smaller numbers are involved. Notice that the rational exponent consists of two parts, indicated by the following voice balloons:

The numerator is the exponent.

$$a^{\frac{m}{n}} = \left(\sqrt[n]{a}\right)^{m}.$$

The denominator is the radical's index.

EXAMPLE 3 Using the Definition of $a^{\frac{m}{n}}$

Use radical notation to rewrite each expression and simplify:

    a. $1000^{\frac{2}{3}}$ **b.** $16^{\frac{3}{2}}$ **c.** $-32^{\frac{3}{5}}$.

SOLUTION

    a. $(1000)^{\frac{2}{3}} = \left(\sqrt[3]{1000}\right)^{2} = 10^{2} = 100$

The denominator of $\frac{2}{3}$ is the root and the numerator is the exponent.

    b. $16^{\frac{3}{2}} = \left(\sqrt{16}\right)^{3} = 4^{3} = 64$

    c. $-32^{\frac{3}{5}} = -\left(\sqrt[5]{32}\right)^{3} = -2^{3} = -8$

The base is 32 and the negative sign is not affected by the exponent.

TECHNOLOGY

Here are the calculator key-stroke sequences for $1000^{\frac{2}{3}}$:

Many Scientific Calculators

$1000 \boxed{y^x} \boxed{(} 2 \boxed{\div} 3 \boxed{)} \boxed{=}$

Many Graphing Calculators

$1000 \boxed{\wedge} \boxed{(} 2 \boxed{\div} 3 \boxed{)} \boxed{\text{ENTER}}$

✔ **CHECK POINT 3** Use radical notation to rewrite each expression and simplify:

    a. $8^{\frac{4}{3}}$ $\left(\sqrt[3]{8}\right)^{4} = 16$ **b.** $25^{\frac{3}{2}}$ $\left(\sqrt{25}\right)^{3} = 125$ **c.** $-81^{\frac{3}{4}}$. $-\left(\sqrt[4]{81}\right)^{3} = -27$

In our next example, we begin with radical notation and rewrite the expression with rational exponents. When changing from radical form to exponential form, the index becomes the denominator of the rational exponent.

EXAMPLE 4 Using the Definition of $a^{\frac{m}{n}}$

Rewrite with rational exponents:

    a. $\sqrt[3]{7^{5}}$ **b.** $\left(\sqrt[4]{13xy}\right)^{9}$.

SOLUTION

a. $\sqrt[3]{7^5} = 7^{\frac{5}{3}}$

The index is the exponent's denominator.

b. $\left(\sqrt[4]{13xy}\right)^9 = (13xy)^{\frac{9}{4}}$

 CHECK POINT 4 Rewrite with rational exponents:

a. $\sqrt[3]{6^4}$ $6^{\frac{4}{3}}$ **b.** $\left(\sqrt[5]{2xy}\right)^7$. $(2xy)^{\frac{7}{5}}$

3   Use the definition of $a^{-\frac{m}{n}}$.

Can a rational exponent be negative? Yes. The way that negative rational exponents are defined is similar to the way that negative integer exponents are defined.

THE DEFINITION OF $a^{-\frac{m}{n}}$ If $a^{\frac{m}{n}}$ is a nonzero real number, then

$$a^{-\frac{m}{n}} = \frac{1}{a^{\frac{m}{n}}}.$$

EXAMPLE 5   Using the Definition of $a^{-\frac{m}{n}}$

Rewrite each expression with a positive exponent. Simplify, if possible:

a. $36^{-\frac{1}{2}}$ **b.** $125^{-\frac{1}{3}}$ **c.** $16^{-\frac{3}{4}}$ **d.** $(7xy)^{-\frac{4}{7}}$.

SOLUTION

a. $36^{-\frac{1}{2}} = \frac{1}{36^{\frac{1}{2}}} = \frac{1}{\sqrt{36}} = \frac{1}{6}$

b. $125^{-\frac{1}{3}} = \frac{1}{125^{\frac{1}{3}}} = \frac{1}{\sqrt[3]{125}} = \frac{1}{5}$

c. $16^{-\frac{3}{4}} = \frac{1}{16^{\frac{3}{4}}} = \frac{1}{\left(\sqrt[4]{16}\right)^3} = \frac{1}{2^3} = \frac{1}{8}$

d. $(7xy)^{-\frac{4}{7}} = \frac{1}{(7xy)^{\frac{4}{7}}}$

 CHECK POINT 5 Rewrite each expression with a positive exponent. Simplify, if possible:

a. $100^{-\frac{1}{2}}$ $\frac{1}{100^{\frac{1}{2}}} = \frac{1}{10}$ **b.** $8^{-\frac{1}{3}}$ $\frac{1}{8^{\frac{1}{3}}} = \frac{1}{2}$

c. $32^{-\frac{3}{5}}$ $\frac{1}{32^{\frac{3}{5}}} = \frac{1}{8}$ **d.** $(3xy)^{-\frac{5}{9}}$. $\frac{1}{(3xy)^{\frac{5}{9}}}$

4   Use models that contain rational exponents.

Applications   Now that you know the meaning of rational exponents, you can work with mathematical models that contain these exponents.

EXAMPLE 6 Calculating Windchill

The way that we perceive the temperature on a cold day depends on both air temperature and wind speed. The windchill is what the air temperature would have to be with no wind to achieve the same chilling effect on the skin. In 2002, the National Weather Service issued new windchill temperatures, shown in Table 7.3. (One reason for this new windchill index is that the wind speed is now calculated at 5 feet, the average height of the human body's face, rather than 33 feet, the height of the standard anemometer, an instrument that calculates wind speed.)

Table 7.3	**New Windchill Temperature Index**										

	Air Temperature (°F)										
	30	25	20	15	10	5	0	−5	−10	−15	−20	−25
5	25	19	13	7	1	−5	−11	−16	−22	−28	−34	−40
10	21	15	9	3	−4	−10	−16	−22	−28	−35	−41	−47
15	19	13	6	0	−7	−13	−19	−26	−32	−39	−45	−51
20	17	11	4	−2	−9	−15	−22	−29	−35	−42	−48	−55
25	16	9	3	−4	−11	−17	−24	−31	−37	−44	−51	−58
30	15	8	1	−5	−12	−19	−26	−33	−39	−46	−53	−60
35	14	7	0	−7	−14	−21	−27	−34	−41	−48	−55	−62
40	13	6	−1	−8	−15	−22	−29	−36	−43	−50	−57	−64
45	12	5	−2	−9	−16	−23	−30	−37	−44	−51	−58	−65
50	12	4	−3	−10	−17	−24	−31	−38	−45	−52	−60	−67
55	11	4	−3	−11	−18	−25	−32	−39	−46	−54	−61	−68
60	10	3	−4	−11	−19	−26	−33	−40	−48	−55	−62	−69

Wind Speed (miles per hour)

▨ Frostbite occurs in 15 minutes or less.

Source: National Weather Service

The windchill temperatures shown in Table 7.3 can be calculated using

$$C = 35.74 + 0.6215t - 35.74v^{\frac{4}{25}} + 0.4275tv^{\frac{4}{25}}$$

in which C is the windchill, in degrees Fahrenheit, t is the air temperature, in degrees Fahrenheit, and v is the wind speed, in miles per hour. Use the formula to find the windchill temperature, to the nearest degree, when the air temperature is 25°F and the wind speed is 30 miles per hour.

SOLUTION Because we are interested in the windchill when the air temperature is 25°F and the wind speed is 30 miles per hour, substitute 25 for t and 30 for v. Then calculate C.

$C = 35.74 + 0.6215t - 35.74v^{\frac{4}{25}} + 0.4275tv^{\frac{4}{25}}$ This is the given formula.

$C = 35.74 + 0.6215(25) - 35.74(30)^{\frac{4}{25}} + 0.4275(25)(30)^{\frac{4}{25}}$ Substitute 25 for t and 30 for v.

≈ 8 Use a calculator.

When the air temperature is 25°F and the wind speed is 30 miles per hour, the windchill temperature is 8°. In other words, a 30-mile-per-hour wind makes 25° feel like 8°. ■

 CHECK POINT 6 Use the windchill formula in Example 6 to find the windchill temperature, to the nearest degree, when the air temperature is 35°F and the wind speed is 15 miles per hour. The windchill temperature is 25°.

Properties of Rational Exponents The same properties apply to rational exponents as to integer exponents. The following is a summary of these properties:

> **PROPERTIES OF RATIONAL EXPONENTS** If m and n are rational exponents, and a and b are real numbers for which the following expressions are defined, then
>
> **1.** $b^m \cdot b^n = b^{m+n}$ When multiplying exponential expressions with the same base, add the exponents. Use this sum as the exponent of the common base.
>
> **2.** $\dfrac{b^m}{b^n} = b^{m-n}$ When dividing exponential expressions with the same base, subtract the exponents. Use this difference as the exponent of the common base.
>
> **3.** $(b^m)^n = b^{mn}$ When an exponential expression is raised to a power, multiply the exponents. Place the product of the exponents on the base and remove the parentheses.
>
> **4.** $(ab)^n = a^n b^n$ When a product is raised to a power, raise each factor to that power and multiply.
>
> **5.** $\left(\dfrac{a}{b}\right)^n = \dfrac{a^n}{b^n}$ When a quotient is raised to a power, raise the numerator to that power and divide by the denominator to that power.

5   Simplify expressions with rational exponents.

We can use these properties to simplify exponential expressions with rational exponents. As with integer exponents, an expression with rational exponents is **simplified** when:

- No parentheses appear.
- No powers are raised to powers.
- Each base occurs only once.
- No negative or zero exponents appear.

EXAMPLE 7   Simplifying Expressions with Rational Exponents

Simplify:

a. $6^{\frac{1}{7}} \cdot 6^{\frac{4}{7}}$ **b.** $\dfrac{32x^{\frac{1}{2}}}{16x^{\frac{3}{4}}}$ **c.** $\left(8.3^{\frac{3}{4}}\right)^{\frac{2}{3}}$ **d.** $\left(x^{-\frac{2}{5}}y^{\frac{1}{3}}\right)^{\frac{1}{2}}$.

SOLUTION

a. $6^{\frac{1}{7}} \cdot 6^{\frac{4}{7}} = 6^{\frac{1}{7}+\frac{4}{7}}$ To multiply with the same base, add exponents.

$\qquad\qquad = 6^{\frac{5}{7}}$ Simplify: $\dfrac{1}{7} + \dfrac{4}{7} = \dfrac{5}{7}$.

b. $\dfrac{32x^{\frac{1}{2}}}{16x^{\frac{3}{4}}} = \dfrac{32}{16}x^{\frac{1}{2}-\frac{3}{4}}$ Divide coefficients. To divide with the same base, subtract exponents.

$\qquad\qquad = 2x^{\frac{2}{4}-\frac{3}{4}}$ Write exponents in terms of the LCD, 4.

$\qquad\qquad = 2x^{-\frac{1}{4}}$ Subtract: $\dfrac{2}{4} - \dfrac{3}{4} = -\dfrac{1}{4}$.

$\qquad\qquad = \dfrac{2}{x^{\frac{1}{4}}}$ Rewrite with a positive exponent: $a^{-\frac{m}{n}} = \dfrac{1}{a^{\frac{m}{n}}}$.

c. $\left(8.3^{\frac{3}{4}}\right)^{\frac{2}{3}} = 8.3^{\left(\frac{3}{4}\right)\left(\frac{2}{3}\right)}$ To raise a power to a power, multiply exponents.

$= 8.3^{\frac{1}{2}}$ Multiply: $\dfrac{3}{4} \cdot \dfrac{2}{3} = \dfrac{6}{12} = \dfrac{1}{2}$.

d. $\left(x^{-\frac{2}{5}}y^{\frac{1}{3}}\right)^{\frac{1}{2}} = \left(x^{-\frac{2}{5}}\right)^{\frac{1}{2}}\left(y^{\frac{1}{3}}\right)^{\frac{1}{2}}$ To raise a product to a power, raise each factor to the power.

$= x^{-\frac{1}{5}}y^{\frac{1}{6}}$ Multiply: $-\dfrac{2}{5} \cdot \dfrac{1}{2} = -\dfrac{1}{5}$ and $\dfrac{1}{3} \cdot \dfrac{1}{2} = \dfrac{1}{6}$.

$= \dfrac{y^{\frac{1}{6}}}{x^{\frac{1}{5}}}$ Rewrite with positive exponents.

 CHECK POINT 7 Simplify:

a. $7^{\frac{1}{2}} \cdot 7^{\frac{1}{3}}$ $7^{\frac{5}{6}}$ **b.** $\dfrac{50x^{\frac{1}{3}}}{10x^{\frac{4}{3}}}$ $\dfrac{5}{x}$ **c.** $\left(9.1^{\frac{2}{5}}\right)^{\frac{3}{4}}$ $9.1^{\frac{3}{10}}$ **d.** $\left(x^{-\frac{3}{5}}y^{\frac{1}{4}}\right)^{\frac{1}{3}}$. $\dfrac{y^{\frac{1}{12}}}{x^{\frac{1}{5}}}$

6 Simplify radical expressions using rational exponents.

Using Rational Exponents to Simplify Radical Expressions Some radical expressions can be simplified using rational exponents. We will use the following procedure:

SIMPLIFYING RADICAL EXPRESSIONS USING RATIONAL EXPONENTS

1. Rewrite each radical expression as an exponential expression with a rational exponent.
2. Simplify using properties of rational exponents.
3. Rewrite in radical notation if rational exponents still appear.

EXAMPLE 8   Simplifying Radical Expressions Using Rational Exponents

Use rational exponents to simplify:

a. $\sqrt[10]{x^5}$ **b.** $\sqrt[3]{27a^{15}}$ **c.** $\sqrt[4]{x^6 y^2}$ **d.** $\sqrt{x} \cdot \sqrt[3]{x}$ **e.** $\sqrt[3]{\sqrt{x}}$.

SOLUTION

a. $\sqrt[10]{x^5} = x^{\frac{5}{10}}$ Rewrite as an exponential expression.

$= x^{\frac{1}{2}}$ Simplify the exponent.

$= \sqrt{x}$ Rewrite in radical notation.

b. $\sqrt[3]{27a^{15}} = (27a^{15})^{\frac{1}{3}}$ Rewrite as an exponential expression.

$= 27^{\frac{1}{3}}(a^{15})^{\frac{1}{3}}$ Raise each factor in parentheses to the $\frac{1}{3}$ power.

$= \sqrt[3]{27} \cdot a^{15\left(\frac{1}{3}\right)}$ To raise a power to a power, multiply exponents.

$= 3a^5$ $\sqrt[3]{27} = 3$. Multiply exponents: $15 \cdot \frac{1}{3} = 5$.

c. $\sqrt[4]{x^6 y^2} = (x^6 y^2)^{\frac{1}{4}}$ Rewrite as an exponential expression.

$= (x^6)^{\frac{1}{4}}(y^2)^{\frac{1}{4}}$ Raise each factor in parentheses to the $\frac{1}{4}$ power.

$= x^{\frac{6}{4}} y^{\frac{2}{4}}$ To raise powers to powers, multiply.

$= x^{\frac{3}{2}} y^{\frac{1}{2}}$ Simplify.

$= (x^3 y)^{\frac{1}{2}}$ $a^n b^n = (ab)^n$

$= \sqrt{x^3 y}$ Rewrite in radical notation.

d. $\sqrt{x} \cdot \sqrt[3]{x} = x^{\frac{1}{2}} \cdot x^{\frac{1}{3}}$ Rewrite as exponential expressions.

$= x^{\frac{1}{2} + \frac{1}{3}}$ To multiply with the same base, add exponents.

$= x^{\frac{3}{6} + \frac{2}{6}}$ Write exponents in terms of the LCD, 6.

$= x^{\frac{5}{6}}$ Add: $\frac{3}{6} + \frac{2}{6} = \frac{5}{6}$.

$= \sqrt[6]{x^5}$ Rewrite in radical notation.

e. $\sqrt[3]{\sqrt{x}} = \sqrt[3]{x^{\frac{1}{2}}}$ Write the radicand as an exponential expression.

$= \left(x^{\frac{1}{2}}\right)^{\frac{1}{3}}$ Write the entire expression in exponential form.

$= x^{\frac{1}{6}}$ To raise powers to powers, multiply: $\frac{1}{2} \cdot \frac{1}{3} = \frac{1}{6}$.

$= \sqrt[6]{x}$ Rewrite in radical notation.

✔ **CHECK POINT 8** Use rational exponents to simplify:

a. $\sqrt[6]{x^3}$ $\sqrt{x}$ **b.** $\sqrt[3]{8a^{12}}$ $2a^4$ **c.** $\sqrt[8]{x^4 y^2}$ $\sqrt[4]{x^2 y}$ **d.** $\dfrac{\sqrt{x}}{\sqrt[3]{x}}$ $\sqrt[6]{x}$ **e.** $\sqrt{\sqrt[3]{x}}$. $\sqrt[6]{x}$

7.2 EXERCISE SET

Student Solutions Manual   CD/Video   PH Math/Tutor Center   MathXL Tutorials on CD   MathXL®   MyMathLab   Interactmath.com

Practice Exercises

In Exercises 1–20, use radical notation to rewrite each expression. Simplify, if possible.

1. $49^{\frac{1}{2}}$ $\sqrt{49} = 7$ **2.** $100^{\frac{1}{2}}$ $\sqrt{100} = 10$

3. $(-27)^{\frac{1}{3}}$ $\sqrt[3]{-27} = -3$ **4.** $(-64)^{\frac{1}{3}}$ $\sqrt[3]{-64} = -4$

5. $-16^{\frac{1}{4}}$ $-\sqrt[4]{16} = -2$ **6.** $-81^{\frac{1}{4}}$ $-\sqrt[4]{81} = -3$

7. $(xy)^{\frac{1}{3}}$ $\sqrt[3]{xy}$ **8.** $(xy)^{\frac{1}{4}}$ $\sqrt[4]{xy}$

9. $(2xy^3)^{\frac{1}{5}}$ $\sqrt[5]{2xy^3}$ **10.** $(3xy^4)^{\frac{1}{5}}$ $\sqrt[5]{3xy^4}$

11. $81^{\frac{3}{2}}$ $\left(\sqrt{81}\right)^3 = 729$ **12.** $25^{\frac{3}{2}}$ $\left(\sqrt{25}\right)^3 = 125$

13. $125^{\frac{2}{3}}$ $\left(\sqrt[3]{125}\right)^2 = 25$ **14.** $1000^{\frac{2}{3}}$ $\left(\sqrt[3]{1000}\right)^2 = 100$

15. $(-32)^{\frac{3}{5}}$ $\left(\sqrt[5]{-32}\right)^3 = -8$ **16.** $(-27)^{\frac{2}{3}}$ $\left(\sqrt[3]{-27}\right)^2 = 9$

17. $27^{\frac{2}{3}} + 16^{\frac{3}{4}}$ $\left(\sqrt[3]{27}\right)^2 + \left(\sqrt[4]{16}\right)^3 = 17$

18. $4^{\frac{5}{2}} - 8^{\frac{2}{3}}$ $\left(\sqrt{4}\right)^5 - \left(\sqrt[3]{8}\right)^2 = 28$

19. $(xy)^{\frac{4}{7}}$ $\sqrt[7]{(xy)^4}$

20. $(xy)^{\frac{4}{9}}$ $\sqrt[9]{(xy)^4}$

In Exercises 21–38, rewrite each expression with rational exponents.

21. $\sqrt{7}$ $7^{\frac{1}{2}}$ **22.** $\sqrt{13}$ $13^{\frac{1}{2}}$

23. $\sqrt[3]{5}$ $5^{\frac{1}{3}}$ **24.** $\sqrt[3]{6}$ $6^{\frac{1}{3}}$

25. $\sqrt[5]{11x}$ $(11x)^{\frac{1}{5}}$ **26.** $\sqrt[5]{13x}$ $(13x)^{\frac{1}{5}}$

27. $\sqrt{x^3}$ $x^{\frac{3}{2}}$ **28.** $\sqrt{x^5}$ $x^{\frac{5}{2}}$

29. $\sqrt[5]{x^3}$ $x^{\frac{3}{5}}$ **30.** $\sqrt[7]{x^4}$ $x^{\frac{4}{7}}$

31. $\sqrt[5]{x^2 y}$ $(x^2 y)^{\frac{1}{5}}$ **32.** $\sqrt[7]{xy^3}$ $(xy^3)^{\frac{1}{7}}$

33. $\left(\sqrt{19xy}\right)^3$ $(19xy)^{\frac{3}{2}}$ **34.** $\left(\sqrt{11xy}\right)^3$ $(11xy)^{\frac{3}{2}}$

35. $\left(\sqrt[6]{7xy^2}\right)^5$ $(7xy^2)^{\frac{5}{6}}$ **36.** $\left(\sqrt[6]{9x^2y}\right)^5$ $(9x^2y)^{\frac{5}{6}}$

37. $2x\sqrt[3]{y^2}$ $2xy^{\frac{2}{3}}$ **38.** $4x\sqrt[5]{y^2}$ $4xy^{\frac{2}{5}}$

In Exercises 39–54, rewrite each expression with a positive rational exponent. Simplify, if possible.

39. $49^{-\frac{1}{2}}$ $\dfrac{1}{49^{\frac{1}{2}}}=\dfrac{1}{7}$ **40.** $9^{-\frac{1}{2}}$ $\dfrac{1}{9^{\frac{1}{2}}}=\dfrac{1}{3}$

41. $27^{-\frac{1}{3}}$ $\dfrac{1}{27^{\frac{1}{3}}}=\dfrac{1}{3}$ **42.** $125^{-\frac{1}{3}}$ $\dfrac{1}{125^{\frac{1}{3}}}=\dfrac{1}{5}$

43. $16^{-\frac{3}{4}}$ $\dfrac{1}{16^{\frac{3}{4}}}=\dfrac{1}{8}$ **44.** $81^{-\frac{5}{4}}$ $\dfrac{1}{81^{\frac{5}{4}}}=\dfrac{1}{243}$

45. $8^{-\frac{2}{3}}$ $\dfrac{1}{8^{\frac{2}{3}}}=\dfrac{1}{4}$ **46.** $32^{-\frac{4}{5}}$ $\dfrac{1}{32^{\frac{4}{5}}}=\dfrac{1}{16}$

47. $\left(\dfrac{8}{27}\right)^{-\frac{1}{3}}$ $\left(\dfrac{27}{8}\right)^{\frac{1}{3}}=\dfrac{3}{2}$ **48.** $\left(\dfrac{8}{125}\right)^{-\frac{1}{3}}$ $\left(\dfrac{125}{8}\right)^{\frac{1}{3}}=\dfrac{5}{2}$

49. $(-64)^{-\frac{2}{3}}$ $\dfrac{1}{(-64)^{\frac{2}{3}}}=\dfrac{1}{16}$ **50.** $(-8)^{-\frac{2}{3}}$ $\dfrac{1}{(-8)^{\frac{2}{3}}}=\dfrac{1}{4}$

51. $(2xy)^{-\frac{7}{10}}$ $\dfrac{1}{(2xy)^{\frac{7}{10}}}$ **52.** $(4xy)^{-\frac{4}{7}}$ $\dfrac{1}{(4xy)^{\frac{4}{7}}}$

53. $5xz^{-\frac{1}{3}}$ $\dfrac{5x}{z^{\frac{1}{3}}}$ **54.** $7xz^{-\frac{1}{4}}$ $\dfrac{7x}{z^{\frac{1}{4}}}$

In Exercises 55–78, use properties of rational exponents to simplify each expression. Assume that all variables represent positive numbers.

55. $3^{\frac{3}{4}}\cdot 3^{\frac{1}{4}}$ 3 **56.** $5^{\frac{2}{3}}\cdot 5^{\frac{1}{3}}$ 5

57. $\dfrac{16^{\frac{3}{4}}}{16^{\frac{1}{4}}}$ 4 **58.** $\dfrac{100^{\frac{3}{4}}}{100^{\frac{1}{4}}}$ 10

59. $x^{\frac{1}{2}}\cdot x^{\frac{1}{3}}$ $x^{\frac{5}{6}}$ **60.** $x^{\frac{1}{2}}\cdot x^{\frac{2}{3}}$ $x^{\frac{7}{6}}$

61. $\dfrac{x^{\frac{4}{5}}}{x^{\frac{1}{5}}}$ $x^{\frac{3}{5}}$ **62.** $\dfrac{x^{\frac{3}{7}}}{x^{\frac{1}{7}}}$ $x^{\frac{2}{7}}$

63. $\dfrac{x^{\frac{1}{3}}}{x^{\frac{3}{4}}}$ $\dfrac{1}{x^{\frac{5}{12}}}$ **64.** $\dfrac{x^{\frac{1}{4}}}{x^{\frac{3}{5}}}$ $\dfrac{1}{x^{\frac{7}{20}}}$

65. $\left(5^{\frac{2}{3}}\right)^3$ 25 **66.** $\left(3^{\frac{4}{5}}\right)^5$ 81

67. $\left(y^{-\frac{2}{3}}\right)^{\frac{1}{4}}$ $\dfrac{1}{y^{\frac{1}{6}}}$ **68.** $\left(y^{-\frac{3}{4}}\right)^{\frac{1}{6}}$ $\dfrac{1}{y^{\frac{1}{8}}}$

69. $\left(2x^{\frac{1}{5}}\right)^5$ $32x$ **70.** $\left(2x^{\frac{1}{4}}\right)^4$ $16x$

71. $(25x^4y^6)^{\frac{1}{2}}$ $5x^2y^3$ **72.** $(125x^9y^6)^{\frac{1}{3}}$ $5x^3y^2$

73. $\left(x^{\frac{1}{2}}y^{-\frac{3}{5}}\right)^{\frac{1}{2}}$ $\dfrac{x^{\frac{1}{4}}}{y^{\frac{3}{10}}}$ **74.** $\left(x^{\frac{1}{4}}y^{-\frac{2}{5}}\right)^{\frac{1}{3}}$ $\dfrac{x^{\frac{1}{12}}}{y^{\frac{2}{15}}}$

75. $\dfrac{3^{\frac{1}{2}}\cdot 3^{\frac{3}{4}}}{3^{\frac{1}{4}}}$ 3 **76.** $\dfrac{5^{\frac{3}{4}}\cdot 5^{\frac{1}{2}}}{5^{\frac{1}{4}}}$ 5

77. $\dfrac{\left(3y^{\frac{1}{4}}\right)^3}{y^{\frac{1}{12}}}$ $27y^{\frac{2}{3}}$ **78.** $\dfrac{\left(2y^{\frac{1}{5}}\right)^4}{y^{\frac{3}{10}}}$ $16y^{\frac{1}{2}}$

In Exercises 79–112, use rational exponents to simplify each expression. If rational exponents appear after simplifying, write the answer in radical notation. Assume that all variables represent positive numbers.

79. $\sqrt[8]{x^2}$ $\sqrt[4]{x}$ **80.** $\sqrt[10]{x^2}$ $\sqrt[5]{x}$

81. $\sqrt[3]{8a^6}$ $2a^2$ **82.** $\sqrt[3]{27a^{12}}$ $3a^4$

83. $\sqrt[5]{x^{10}y^{15}}$ x^2y^3 **84.** $\sqrt[5]{x^{15}y^{20}}$ x^3y^4

85. $\left(\sqrt[3]{xy}\right)^{18}$ x^6y^6 **86.** $\left(\sqrt[3]{xy}\right)^{21}$ x^7y^7

87. $\sqrt[10]{(3y)^2}$ $\sqrt[5]{3y}$ **88.** $\sqrt[12]{(3y)^2}$ $\sqrt[6]{3y}$

89. $\left(\sqrt[6]{2a}\right)^4$ $\sqrt[3]{4a^2}$ **90.** $\left(\sqrt[8]{2a}\right)^6$ $\sqrt[4]{8a^3}$

91. $\sqrt[9]{x^6y^3}$ $\sqrt[3]{x^2y}$ **92.** $\sqrt[4]{x^2y^6}$ $\sqrt{xy^3}$

93. $\sqrt{2}\cdot\sqrt[3]{2}$ $\sqrt[6]{2^5}$ or $\sqrt[6]{32}$ **94.** $\sqrt{3}\cdot\sqrt[3]{3}$ $\sqrt[6]{3^5}$ or $\sqrt[6]{243}$

95. $\sqrt[5]{x^2}\cdot\sqrt{x}$ $\sqrt[10]{x^9}$ **96.** $\sqrt[7]{x^2}\cdot\sqrt{x}$ $\sqrt[14]{x^{11}}$

97. $\sqrt[4]{a^2b}\cdot\sqrt[3]{ab}$ $\sqrt[12]{a^{10}b^7}$ **98.** $\sqrt[6]{ab^2}\cdot\sqrt[3]{a^2b}$ $\sqrt[6]{a^5b^4}$

99. $\dfrac{\sqrt[4]{x}}{\sqrt[5]{x}}$ $\sqrt[20]{x}$ **100.** $\dfrac{\sqrt[3]{x}}{\sqrt[4]{x}}$ $\sqrt[12]{x}$

101. $\dfrac{\sqrt[3]{y^2}}{\sqrt[6]{y}}$ $\sqrt{y}$ **102.** $\dfrac{\sqrt[5]{y^2}}{\sqrt[10]{y^3}}$ $\sqrt[10]{y}$

103. $\sqrt[4]{\sqrt{x}}$ $\sqrt[8]{x}$ **104.** $\sqrt[5]{\sqrt{x}}$ $\sqrt[10]{x}$

105. $\sqrt{\sqrt{x^2y}}$ $\sqrt[4]{x^2y}$ **106.** $\sqrt{\sqrt{xy^2}}$ $\sqrt[4]{xy^2}$

107. $\sqrt[4]{\sqrt[3]{2x}}$ $\sqrt[12]{2x}$ **108.** $\sqrt[5]{\sqrt[3]{2x}}$ $\sqrt[15]{2x}$

109. $\left(\sqrt[4]{x^3y^5}\right)^{12}$ x^9y^{15} **110.** $\left(\sqrt[5]{x^4y^2}\right)^{20}$ $x^{16}y^8$

111. $\dfrac{\sqrt[4]{a^5b^5}}{\sqrt{ab}}$ $\sqrt[4]{a^3b^3}$ **112.** $\dfrac{\sqrt[4]{a^3b^3}}{\sqrt{ab}}$ $\sqrt[4]{ab}$

Practice Plus

In Exercises 113–116, use the distributive property or the FOIL method to perform each multiplication.

113. $x^{\frac{1}{3}}\left(x^{\frac{1}{3}}-x^{\frac{2}{3}}\right)$ $x^{\frac{2}{3}}-x$

114. $x^{-\frac{1}{4}}\left(x^{\frac{9}{4}}-x^{\frac{1}{4}}\right)$ x^2-1

115. $\left(x^{\frac{1}{2}}-3\right)\left(x^{\frac{1}{2}}+5\right)$ $x+2x^{\frac{1}{2}}-15$

116. $\left(x^{\frac{1}{3}}-2\right)\left(x^{\frac{1}{3}}+6\right)$ $x^{\frac{2}{3}}+4x^{\frac{1}{3}}-12$

In Exercises 117–120, factor out the greatest common factor from each expression.

117. $6x^{\frac{1}{2}} + 2x^{\frac{3}{2}}$ $2x^{\frac{1}{2}}(3 + x)$

118. $8x^{\frac{1}{4}} + 4x^{\frac{5}{4}}$ $4x^{\frac{1}{4}}(2 + x)$

119. $15x^{\frac{1}{3}} - 60x$ $15x^{\frac{1}{3}}\left(1 - 4x^{\frac{2}{3}}\right)$

120. $7x^{\frac{1}{3}} - 70x$ $7x^{\frac{1}{3}}\left(1 - 10x^{\frac{2}{3}}\right)$

In Exercises 121–124, simplify each expression. Assume that all variables represent positive numbers.

121. $(49x^{-2}y^4)^{-\frac{1}{2}}\left(xy^{\frac{1}{2}}\right)$ $\dfrac{x^2}{7y^{\frac{3}{2}}}$

122. $(8x^{-6}y^3)^{\frac{1}{3}}\left(x^{\frac{5}{6}}y^{-\frac{1}{3}}\right)^6$ $\dfrac{2x^3}{y}$

123. $\left(\dfrac{x^{-\frac{5}{4}}y^{\frac{1}{3}}}{x^{-\frac{3}{4}}}\right)^{-6}$ $\dfrac{x^3}{y^2}$

124. $\left(\dfrac{x^{\frac{1}{2}}y^{-\frac{7}{4}}}{y^{-\frac{5}{4}}}\right)^{-4}$ $\dfrac{y^2}{x^2}$

Application Exercises

The Galápagos Islands, lying 600 miles west of Ecuador, are famed for their extraordinary wildlife. The function

$$f(x) = 29x^{\frac{1}{3}}$$

models the number of plant species, $f(x)$, on the various islands of the Galápagos chain in terms of the area, x, in square miles, of a particular island. Use the function to solve Exercises 125–126.

125. How many species of plants are on a Galápagos island that has an area of 8 square miles?   58 species of plants

126. How many species of plants are on a Galápagos island that has an area of 27 square miles?   87 species of plants

The function

$$f(x) = 70x^{\frac{3}{4}}$$

models the number of calories per day, $f(x)$, a person needs to maintain life in terms of that person's weight, x, in kilograms. (1 kilogram is approximately 2.2 pounds.) Use this model and a calculator to solve Exercises 127–128. Round answers to the nearest calorie.

127. How many calories per day does a person who weighs 80 kilograms (approximately 176 pounds) need to maintain life?   about 1872 calories per day

128. How many calories per day does a person who weighs 70 kilograms (approximately 154 pounds) need to maintain life?   about 1694 calories per day

We have seen that when the air temperature is t degrees Fahrenheit and the wind speed is v miles per hour, the windchill temperature, C, in degrees Fahrenheit, is given by the formula

$$C = 35.74 + 0.6215t - 35.74v^{\frac{4}{25}} + 0.4275tv^{\frac{4}{25}}.$$

Use this formula to solve Exercises 129–130.

129. a. Substitute 0 for t and write a function $C(v)$ that gives the windchill temperature as a function of wind speed for an air temperature of 0 °F.
$C(v) = 35.74 - 35.74v^{\frac{4}{25}}$

   b. Find and interpret $C(25)$. Use a calculator and round to the nearest degree. $C(25) \approx -24$; When the air temperature is 0°F and the wind speed is 25 miles per hour, the windchill temperature is $-24°$.

   c. Identify your solution to part (b) on the graph shown.   the point $(25, -24)$

Air Speed (miles per hour)

Windchill Temperature (°F)

Windchill temperature as a function of air speed for an air temperature of 0°F

$[0, 60, 5]$ by $[-35, 0, 5]$

130. a. Substitute 30 for t and write a function $C(v)$ that gives the windchill temperature as a function of wind speed for an air temperature of 30°F. Simplify the function's formula so that it contains exactly two terms.
$C(v) = 54.385 - 22.915v^{\frac{4}{25}}$

   b. Find and interpret $C(40)$. Use a calculator and round to the nearest degree. $C(40) \approx 13$; When the air temperature is 30°F and the wind speed is 40 miles per hour, the windchill temperature is 13°.

   c. Identify your solution to part (b) on the graph shown.   the point $(40, 13)$

Windchill temperature as a function of air speed for an air temperature of 30°F

Windchill Temperature (°F)

Air Speed (miles per hour)
$[0, 60, 5]$ by $[0, 25, 5]$

Your job is to determine whether or not yachts are eligible for the America's Cup, the supreme event in ocean sailing. The basic dimensions of competitors' yachts must satisfy

$$L + 1.25\sqrt{S} - 9.8\sqrt[3]{D} \leq 16.296,$$

where L is the yacht's length, in meters, S is its sail area, in square meters, and D is its displacement, in cubic meters. Use this information to solve Exercises 131–132.

131. a. Rewrite the inequality using rational exponents.

$$L + 1.25S^{\frac{1}{2}} - 9.8D^{\frac{1}{3}} \leq 16.296$$

b. Use your calculator to determine if a yacht with length 20.85 meters, sail area 276.4 square meters, and displacement 18.55 cubic meters is eligible for the America's Cup. eligible

132. a. Rewrite the inequality using rational exponents.

$$L + 1.25S^{\frac{1}{2}} - 9.8D^{\frac{1}{3}} \leq 16.296$$

b. Use your calculator to determine if a yacht with length 22.85 meters, sail area 312.5 square meters, and displacement 22.34 cubic meters is eligible for the America's Cup. not eligible

Writing in Mathematics

133. What is the meaning of $a^{\frac{1}{n}}$? Give an example to support your explanation. 133–140. Answers will vary.

134. What is the meaning of $a^{\frac{m}{n}}$? Give an example.

135. What is the meaning of $a^{-\frac{m}{n}}$? Give an example.

136. Explain why $a^{\frac{1}{n}}$ is negative when n is odd and a is negative. What happens if n is even and a is negative? Why?

137. In simplifying $36^{\frac{3}{2}}$, is it better to use $a^{\frac{m}{n}} = \sqrt[n]{a^m}$ or $a^{\frac{m}{n}} = (\sqrt[n]{a})^m$? Explain.

138. How can you tell if an expression with rational exponents is simplified?

139. Explain how to simplify $\sqrt[3]{x} \cdot \sqrt{x}$.

140. Explain how to simplify $\sqrt[3]{\sqrt{x}}$.

Technology Exercises

141. Use a scientific or graphing calculator to verify your results in Exercises 15–18.

142. Use a scientific or graphing calculator to verify your results in Exercises 45–50.

Exercises 143–145 show a number of simplifications, not all of which are correct. Enter the left side of each equation as y_1 and the right side as y_2. Then use your graphing utility's TABLE *feature to determine if the simplification is correct. If it is not, correct the right side and use the* TABLE *feature to verify your simplification.* 143–145. See graphing answer section.

143. $x^{\frac{3}{5}} \cdot x^{-\frac{1}{10}} = x^{\frac{1}{2}}$ simplified correctly

144. $\left(x^{-\frac{1}{2}} \cdot x^{\frac{3}{4}}\right)^{-2} = x^{\frac{1}{2}}$ Right side should be $\dfrac{1}{x^{\frac{1}{2}}}$.

145. $\dfrac{x^{\frac{1}{4}}}{x^{\frac{1}{2}} \cdot x^{-\frac{3}{4}}} = \dfrac{1}{x^{\frac{1}{2}}}$ Right side should be $x^{\frac{1}{2}}$.

Critical Thinking Exercises

146. Which one of the following is true? a

a. If n is odd, then $(-b)^{\frac{1}{n}} = -b^{\frac{1}{n}}$.

b. $(a + b)^{\frac{1}{n}} = a^{\frac{1}{n}} + b^{\frac{1}{n}}$

c. $8^{-\frac{2}{3}} = -4$

d. None of the above is true.

147. A mathematics professor recently purchased a birthday cake for her son with the inscription

Happy $\left(2^{\frac{5}{2}} \cdot 2^{\frac{3}{4}} \div 2^{\frac{1}{4}}\right)$th Birthday.

How old is the son? 8 years old

148. The birthday boy in Exercise 147, excited by the inscription on the cake, tried to wolf down the whole thing. Professor Mom, concerned about the possible metamorphosis of her son into a blimp, exclaimed, "Hold on! It is your birthday, so why not take $\dfrac{8^{-\frac{4}{3}} + 2^{-2}}{16^{-\frac{3}{4}} + 2^{-1}}$ of the cake? I'll eat half of what's left over." How much of the cake did the professor eat? $\frac{1}{4}$ of the cake

149. Simplify: $\left[3 + \left(27^{\frac{2}{3}} + 32^{\frac{2}{5}}\right)^{\frac{3}{2}}\right]^{\frac{3}{2}} - 9^{\frac{1}{2}}$. 61

150. Find the domain of $f(x) = (x - 3)^{\frac{1}{2}}(x + 4)^{-\frac{1}{2}}$. Domain of f is $\{x | x \geq 3\}$ or $[3, \infty)$.

Review Exercises

151. Write the equation of the linear function whose graph passes through $(5, 1)$ and $(4, 3)$. (Section 2.4, Example 2) $y = -2x + 11$ or $f(x) = -2x + 11$

152. Graph $y \leq -\dfrac{3}{2}x + 3$. (Section 4.4, Example 2)

See graphing answer section.

153. Solve by Cramer's rule:

$$5x - 3y = 3$$
$$7x + y = 25.$$

(Section 3.5, Example 2) $(3, 4)$ or $\{(3, 4)\}$

Objectives

1 Use the product rule to multiply radicals.

2 Use factoring and the product rule to simplify radicals.

3 Multiply radicals and then simplify.

MULTIPLYING AND SIMPLIFYING RADICAL EXPRESSIONS

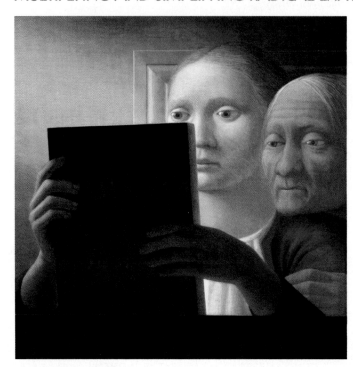

George Tooker (b. 1920) "Mirror II" 1963, egg tempera on gesso panel, 1968.4.

A difficult challenge of middle adulthood is confronting the aging process. Middle-aged adults are forced to acknowledge their mortality as they witness the deaths of parents, colleagues, and friends. In addition, there are a number of physical transformations, including changes in vision that require glasses for reading, the onset of wrinkles and sagging skin, and a decrease in heart response. A change in heart response occurs fairly early; after 20, our hearts become less adept at accelerating in response to exercise. In this section, you will see how a radical function models changes in heart function throughout the aging process, as we turn to multiplying and simplifying radical expressions.

1 Use the product rule to multiply radicals.

The Product Rule for Radicals A rule for multiplying radicals can be generalized by comparing $\sqrt{25} \cdot \sqrt{4}$ and $\sqrt{25 \cdot 4}$. Notice that

$$\sqrt{25} \cdot \sqrt{4} = 5 \cdot 2 = 10 \quad \text{and} \quad \sqrt{25 \cdot 4} = \sqrt{100} = 10.$$

Because we obtain 10 in both situations, the original radical expressions must be equal. That is,

$$\sqrt{25} \cdot \sqrt{4} = \sqrt{25 \cdot 4}.$$

This result is a special case of the **product rule for radicals** that can be generalized as follows:

> **THE PRODUCT RULE FOR RADICALS** If $\sqrt[n]{a}$ and $\sqrt[n]{b}$ are real numbers, then
> $$\sqrt[n]{a} \cdot \sqrt[n]{b} = \sqrt[n]{ab}.$$
> The product of two nth roots is the nth root of the product.

EXAMPLE 1 Using the Product Rule for Radicals

Multiply:

a. $\sqrt{3} \cdot \sqrt{7}$ **b.** $\sqrt{x+7} \cdot \sqrt{x-7}$ **c.** $\sqrt[3]{7} \cdot \sqrt[3]{9}$ **d.** $\sqrt[8]{10x} \cdot \sqrt[8]{8x^4}$.

SOLUTION In each problem, the indices are the same. Thus, we multiply by multiplying the radicands.

a. $\sqrt{3} \cdot \sqrt{7} = \sqrt{3 \cdot 7} = \sqrt{21}$

b. $\sqrt{x+7} \cdot \sqrt{x-7} = \sqrt{(x+7)(x-7)} = \sqrt{x^2 - 49}$

> This is not equal to $\sqrt{x^2} - \sqrt{49}$.

c. $\sqrt[3]{7} \cdot \sqrt[3]{9} = \sqrt[3]{7 \cdot 9} = \sqrt[3]{63}$

d. $\sqrt[8]{10x} \cdot \sqrt[8]{8x^4} = \sqrt[8]{10x \cdot 8x^4} = \sqrt[8]{80x^5}$ ∎

✔ **CHECK POINT 1** Multiply:

a. $\sqrt{5} \cdot \sqrt{11}$ $\quad \sqrt{55}$ **b.** $\sqrt{x+4} \cdot \sqrt{x-4}$ $\quad \sqrt{x^2 - 16}$

c. $\sqrt[3]{6} \cdot \sqrt[3]{10}$ $\quad \sqrt[3]{60}$ **d.** $\sqrt[7]{2x} \cdot \sqrt[7]{6x^3}$. $\quad \sqrt[7]{12x^4}$

---

2 Use factoring and the product rule to simplify radicals.

Using Factoring and the Product Rule to Simplify Radicals A number that is the square of an integer is a **perfect square**. For example, 100 is a perfect square because $100 = 10^2$. A number is a **perfect cube** if it is the cube of an integer. Thus, 125 is a perfect cube because $125 = 5^3$. In general, a number is a **perfect nth power** if it is the nth power of an integer. Thus, p is a perfect nth power if there is an integer q such that $p = q^n$.

A radical of index n is **simplified** when its radicand has no factors other than 1 that are perfect nth powers. For example, $\sqrt{300}$ is not simplified because it can be expressed as $\sqrt{100 \cdot 3}$ and 100 is a perfect square. We can use the product rule in the form

$$\sqrt[n]{ab} = \sqrt[n]{a} \cdot \sqrt[n]{b}$$

to simplify $\sqrt[n]{ab}$ when $\sqrt[n]{a}$ or $\sqrt[n]{b}$ is a perfect nth power. Consider $\sqrt{300}$. To simplify, we factor 300 so that one of its factors is the greatest perfect square possible.

$$\sqrt{300} = \sqrt{100 \cdot 3} \qquad \text{Factor 300. 100 is the greatest perfect square factor.}$$

$$= \sqrt{100} \cdot \sqrt{3} \qquad \text{Use the product rule: } \sqrt[n]{ab} = \sqrt[n]{a} \cdot \sqrt[n]{b}.$$

$$= 10\sqrt{3} \qquad \text{Write } \sqrt{100} \text{ as 10. We read } 10\sqrt{3} \text{ as "ten times the square root of three."}$$

SIMPLIFYING RADICAL EXPRESSIONS BY FACTORING A radical expression whose index is n is **simplified** when its radicand has no factors that are perfect nth powers. To simplify, use the following procedure:

1. Write the radicand as the product of two factors, one of which is the greatest perfect nth power.

2. Use the product rule to take the nth root of each factor.

3. Find the nth root of the perfect nth power.

EXAMPLE 2 Simplifying Radicals by Factoring

Simplify by factoring:

a. $\sqrt{75}$ **b.** $\sqrt[3]{54}$ **c.** $\sqrt[5]{64}$ **d.** $\sqrt{500xy^2}$.

SOLUTION

a.
$$\sqrt{75} = \sqrt{25 \cdot 3}$$
25 is the greatest perfect square that is a factor of 75.
$$= \sqrt{25} \cdot \sqrt{3}$$
Take the square root of each factor: $\sqrt[n]{ab} = \sqrt[n]{a} \cdot \sqrt[n]{b}$.
$$= 5\sqrt{3}$$
Write $\sqrt{25}$ as 5.

b.
$$\sqrt[3]{54} = \sqrt[3]{27 \cdot 2}$$
27 is the greatest perfect cube that is a factor of 54: $27 = 3^3$.
$$= \sqrt[3]{27} \cdot \sqrt[3]{2}$$
Take the cube root of each factor: $\sqrt[n]{ab} = \sqrt[n]{a} \cdot \sqrt[n]{b}$.
$$= 3\sqrt[3]{2}$$
Write $\sqrt[3]{27}$ as 3.

c.
$$\sqrt[5]{64} = \sqrt[5]{32 \cdot 2}$$
32 is the greatest perfect fifth power that is a factor of 64: $32 = 2^5$.
$$= \sqrt[5]{32} \cdot \sqrt[5]{2}$$
Take the fifth root of each factor: $\sqrt[n]{ab} = \sqrt[n]{a} \cdot \sqrt[n]{b}$.
$$= 2\sqrt[5]{2}$$
Write $\sqrt[5]{32}$ as 2.

d.
$$\sqrt{500xy^2} = \sqrt{100y^2 \cdot 5x}$$
$100y^2$ is the greatest perfect square that is a factor of $500xy^2$: $100y^2 = (10y)^2$.
$$= \sqrt{100y^2} \cdot \sqrt{5x}$$
Factor into two radicals.
$$= 10|y|\sqrt{5x}$$
Take the square root of $100y^2$. ∎

✔ **CHECK POINT 2** Simplify by factoring:

a. $\sqrt{80}$ $4\sqrt{5}$ **b.** $\sqrt[3]{40}$ $2\sqrt[3]{5}$

c. $\sqrt[4]{32}$ $2\sqrt[4]{2}$ **d.** $\sqrt{200x^2y}$. $10|x|\sqrt{2y}$

EXAMPLE 3 Simplifying a Radical Function

If
$$f(x) = \sqrt{2x^2 + 4x + 2},$$
express the function, f, in simplified form.

SOLUTION   Begin by factoring the radicand. The GCF is 2. Simplification is possible if we obtain a factor that is a perfect square.

$$f(x) = \sqrt{2x^2 + 4x + 2}$$
This is the given function.
$$= \sqrt{2(x^2 + 2x + 1)}$$
Factor out the GCF.
$$= \sqrt{2(x + 1)^2}$$
Factor the perfect square trinomial: $A^2 + 2AB + B^2 = (A + B)^2$.
$$= \sqrt{2} \cdot \sqrt{(x + 1)^2}$$
Take the square root of each factor. The factor $(x + 1)^2$ is a perfect square.
$$= \sqrt{2}\,|x + 1|$$
Take the square root of $(x + 1)^2$.

In simplified form,
$$f(x) = \sqrt{2}\,|x + 1|.$$

USING TECHNOLOGY

The graphs of

$$f(x) = \sqrt{2x^2 + 4x + 2}, \quad g(x) = \sqrt{2}\,|x + 1|, \quad \text{and} \quad h(x) = \sqrt{2}(x + 1)$$

are shown in three separate $[-5, 5, 1]$ by $[-5, 5, 1]$ viewing rectangles. The graphs in parts (a) and (b) are identical. This verifies that our simplification in Example 3 is correct: $\sqrt{2x^2 + 4x + 2} = \sqrt{2}\,|x + 1|$. Now compare the graphs in parts (a) and (c). Can you see that they are not the same? This illustrates the importance of not leaving out absolute value bars: $\sqrt{2x^2 + 4x + 2} \neq \sqrt{2}(x + 1)$.

$y = \sqrt{2x^2 + 4x + 2}$

(a)

$y = \sqrt{2}\,|x + 1|$

(b)

$y = \sqrt{2}\,(x + 1)$

(c)

 CHECK POINT 3 If $f(x) = \sqrt{3x^2 - 12x + 12}$, express the function, f, in simplified form. $f(x) = \sqrt{3}\,|x - 2|$

For the remainder of this chapter, in situations that do not involve functions, we will **assume that no radicands involve negative quantities raised to even powers. Based upon this assumption, absolute value bars are not necessary when taking even roots.**

SIMPLIFYING WHEN VARIABLES TO EVEN POWERS IN A RADICAND ARE NONNEGATIVE QUANTITIES For any nonnegative real number a,
$$\sqrt[n]{a^n} = a.$$

In simplifying an nth root, how do we find variable factors in the radicand that are perfect nth powers? The **perfect nth powers have exponents that are divisible by n.** Simplification is possible by observation or by using rational exponents. Here are some examples:

- $\sqrt{x^6} = \sqrt{(x^3)^2} = x^3$ or $\sqrt{x^6} = (x^6)^{\frac{1}{2}} = x^3$

  6 is divisible by the index, 2. Thus, x^6 is a perfect square.

- $\sqrt[3]{y^{21}} = \sqrt[3]{(y^7)^3} = y^7$ or $\sqrt[3]{y^{21}} = (y^{21})^{\frac{1}{3}} = y^7$

  21 is divisible by the index, 3. Thus, y^{21} is a perfect cube.

- $\sqrt[6]{z^{24}} = \sqrt[6]{(z^4)^6} = z^4$ or $\sqrt[6]{z^{24}} = (z^{24})^{\frac{1}{6}} = z^4$.

  24 is divisible by the index, 6. Thus, z^{24} is a perfect 6th power.

EXAMPLE 4 Simplifying a Radical by Factoring

Simplify: $\sqrt{x^5 y^{13} z^7}$.

SOLUTION We write the radicand as the product of the greatest perfect square factor and another factor. Because the index is 2, variables that have exponents that are divisible by 2 are part of the perfect square factor. We use the greatest exponents that are divisible by 2.

$$\sqrt{x^5y^{13}z^7} = \sqrt{x^4 \cdot x \cdot y^{12} \cdot y \cdot z^6 \cdot z} \qquad \text{\textit{Use the greatest even power of each variable.}}$$

$$= \sqrt{(x^4y^{12}z^6)(xyz)} \qquad \text{\textit{Group the perfect square factors.}}$$

$$= \sqrt{x^4y^{12}z^6} \cdot \sqrt{xyz} \qquad \text{\textit{Factor into two radicals.}}$$

$$= x^2y^6z^3\sqrt{xyz} \qquad \sqrt{x^4y^{12}z^6} = (x^4y^{12}z^6)^{\frac{1}{2}} = x^2y^6z^3 \quad ■$$

 CHECK POINT 4 Simplify: $\sqrt{x^9y^{11}z^3}$. $x^4y^5z\sqrt{xyz}$

EXAMPLE 5   Simplifying a Radical by Factoring

Simplify: $\sqrt[3]{32x^8y^{16}}$.

SOLUTION   We write the radicand as the product of the greatest perfect cube factor and another factor. Because the index is 3, variables that have exponents that are divisible by 3 are part of the perfect cube factor. We use the greatest exponents that are divisible by 3.

$$\sqrt[3]{32x^8y^{16}} = \sqrt[3]{8 \cdot 4 \cdot x^6 \cdot x^2 \cdot y^{15} \cdot y} \qquad \text{\textit{Identify perfect cube factors.}}$$

$$= \sqrt[3]{(8x^6y^{15})(4x^2y)} \qquad \text{\textit{Group the perfect cube factors.}}$$

$$= \sqrt[3]{8x^6y^{15}} \cdot \sqrt[3]{4x^2y} \qquad \text{\textit{Factor into two radicals.}}$$

$$= 2x^2y^5\sqrt[3]{4x^2y} \qquad \sqrt[3]{8} = 2 \text{ and } \sqrt[3]{x^6y^{15}} = (x^6y^{15})^{\frac{1}{3}} = x^2y^5. \quad ■$$

✔ **CHECK POINT 5** Simplify: $\sqrt[3]{40x^{10}y^{14}}$. $2x^3y^4\sqrt[3]{5xy^2}$

EXAMPLE 6   Simplifying a Radical by Factoring

Simplify: $\sqrt[5]{64x^3y^7z^{29}}$.

SOLUTION   We write the radicand as the product of the greatest perfect 5th power and another factor. Because the index is 5, variables that have exponents that are divisible by 5 are part of the perfect fifth factor. We use the greatest exponents that are divisible by 5.

$$\sqrt[5]{64x^3y^7z^{29}} = \sqrt[5]{32 \cdot 2 \cdot x^3 \cdot y^5 \cdot y^2 \cdot z^{25} \cdot z^4} \qquad \text{\textit{Identify perfect fifth factors.}}$$

$$= \sqrt[5]{(32y^5z^{25})(2x^3y^2z^4)} \qquad \text{\textit{Group the perfect fifth factors.}}$$

$$= \sqrt[5]{32y^5z^{25}} \cdot \sqrt[5]{2x^3y^2z^4} \qquad \text{\textit{Factor into two radicals.}}$$

$$= 2yz^5\sqrt[5]{2x^3y^2z^4} \qquad \sqrt[5]{32} = 2 \text{ and } \sqrt[5]{y^5z^{25}} = (y^5z^{25})^{\frac{1}{5}} = yz^5. \quad ■$$

✔ **CHECK POINT 6** Simplify: $\sqrt[5]{32x^{12}y^2z^8}$. $2x^2z\sqrt[5]{x^2y^2z^3}$

---

3   Multiply radicals and then simplify.

Multiplying and Simplifying Radicals   We have seen how to use the product rule when multiplying radicals with the same index. Sometimes after multiplying, we can simplify the resulting radical.

EXAMPLE 7 Multiplying Radicals and Then Simplifying

Multiply and simplify:

a. $\sqrt{15} \cdot \sqrt{3}$ **b.** $7\sqrt[3]{4} \cdot 5\sqrt[3]{6}$ **c.** $\sqrt[4]{8x^3y^2} \cdot \sqrt[4]{8x^5y^3}$.

SOLUTION

a.
$$\sqrt{15} \cdot \sqrt{3} = \sqrt{15 \cdot 3}$$
Use the product rule.
$$= \sqrt{45} = \sqrt{9 \cdot 5}$$
9 is the greatest perfect square factor of 45.
$$= \sqrt{9} \cdot \sqrt{5} = 3\sqrt{5}$$

b.
$$7\sqrt[3]{4} \cdot 5\sqrt[3]{6} = 35\sqrt[3]{4 \cdot 6}$$
Use the product rule.
$$= 35\sqrt[3]{24} = 35\sqrt[3]{8 \cdot 3}$$
8 is the greatest perfect cube factor of 24.
$$= 35\sqrt[3]{8} \cdot \sqrt[3]{3} = 35 \cdot 2 \cdot \sqrt[3]{3}$$
$$= 70\sqrt[3]{3}$$

c.
$$\sqrt[4]{8x^3y^2} \cdot \sqrt[4]{8x^5y^3} = \sqrt[4]{8x^3y^2 \cdot 8x^5y^3}$$
Use the product rule.
$$= \sqrt[4]{64x^8y^5}$$
Multiply.
$$= \sqrt[4]{16 \cdot 4 \cdot x^8 \cdot y^4 \cdot y}$$
Identify perfect fourth factors.
$$= \sqrt[4]{(16x^8y^4)(4y)}$$
Group the perfect fourth factors.
$$= \sqrt[4]{16x^8y^4} \cdot \sqrt[4]{4y}$$
Factor into two radicals.
$$= 2x^2y\sqrt[4]{4y}$$
$\sqrt[4]{16} = 2$ and
$\sqrt[4]{x^8y^4} = (x^8y^4)^{\frac{1}{4}} = x^2y$. ■

 CHECK POINT **7** Multiply and simplify:

a. $\sqrt{6} \cdot \sqrt{2}$ $2\sqrt{3}$

b. $10\sqrt[3]{16} \cdot 5\sqrt[3]{2}$ $100\sqrt[3]{4}$

c. $\sqrt[4]{4x^2y} \cdot \sqrt[4]{8x^6y^3}$. $2x^2y\sqrt[4]{2}$

7.3 EXERCISE SET

 Student Solutions Manual   CD/Video   PH Math/Tutor Center   MathXL Tutorials on CD   MathXL®   MyMathLab   Interactmath.com

Practice Exercises

In Exercises 1–20, use the product rule to multiply.

1. $\sqrt{3} \cdot \sqrt{5}$ $\sqrt{15}$ **2.** $\sqrt{7} \cdot \sqrt{5}$ $\sqrt{35}$

3. $\sqrt[3]{2} \cdot \sqrt[3]{9}$ $\sqrt[3]{18}$ **4.** $\sqrt[3]{5} \cdot \sqrt[3]{4}$ $\sqrt[3]{20}$

5. $\sqrt[4]{11} \cdot \sqrt[4]{3}$ $\sqrt[4]{33}$ **6.** $\sqrt[5]{9} \cdot \sqrt[5]{3}$ $\sqrt[5]{27}$

7. $\sqrt{3x} \cdot \sqrt{11y}$ $\sqrt{33xy}$ **8.** $\sqrt{5x} \cdot \sqrt{11y}$ $\sqrt{55xy}$

9. $\sqrt[5]{6x^3} \cdot \sqrt[5]{4x}$ $\sqrt[5]{24x^4}$ **10.** $\sqrt[4]{6x^2} \cdot \sqrt[4]{3x}$ $\sqrt[4]{18x^3}$

11. $\sqrt{x+3} \cdot \sqrt{x-3}$ $\sqrt{x^2-9}$

12. $\sqrt{x+6} \cdot \sqrt{x-6}$ $\sqrt{x^2-36}$

13. $\sqrt[6]{x-4} \cdot \sqrt[6]{(x-4)^4}$ $\sqrt[6]{(x-4)^5}$

14. $\sqrt[6]{x-5} \cdot \sqrt[6]{(x-5)^4}$ $\sqrt[6]{(x-5)^5}$

15. $\sqrt{\dfrac{2x}{3}} \cdot \sqrt{\dfrac{3}{2}}$ $\sqrt{x}$ **16.** $\sqrt{\dfrac{2x}{5}} \cdot \sqrt{\dfrac{5}{2}}$ $\sqrt{x}$

17. $\sqrt[4]{\dfrac{x}{7}} \cdot \sqrt[4]{\dfrac{3}{y}}$ $\sqrt[4]{\dfrac{3x}{7y}}$ **18.** $\sqrt[4]{\dfrac{x}{3}} \cdot \sqrt[4]{\dfrac{7}{y}}$ $\sqrt[4]{\dfrac{7x}{3y}}$

19. $\sqrt[7]{7x^2y} \cdot \sqrt[7]{11x^3y^2}$ $\sqrt[7]{77x^5y^3}$

20. $\sqrt[9]{12x^2y^3} \cdot \sqrt[9]{3x^3y^4}$ $\sqrt[9]{36x^5y^7}$

In Exercises 21–32, simplify by factoring.

21. $\sqrt{50}$ $5\sqrt{2}$

22. $\sqrt{27}$ $3\sqrt{3}$

23. $\sqrt{45}$ $3\sqrt{5}$

24. $\sqrt{28}$ $2\sqrt{7}$

25. $\sqrt{75x}$ $5\sqrt{3x}$

26. $\sqrt{40x}$ $2\sqrt{10x}$

27. $\sqrt[3]{16}$ $2\sqrt[3]{2}$

28. $\sqrt[3]{54}$ $3\sqrt[3]{2}$

29. $\sqrt[3]{27x^3}$ $3x$

30. $\sqrt[3]{250x^3}$ $5x\sqrt[3]{2}$

31. $\sqrt[3]{-16x^2y^3}$ $-2y\sqrt[3]{2x^2}$ **32.** $\sqrt[3]{-32x^2y^3}$ $-2y\sqrt[3]{4x^2}$

In Exercises 33–38, express the function, f, in simplified form. Assume that x can be any real number.

33. $f(x) = \sqrt{36(x+2)^2}$ $6|x+2|$

34. $f(x) = \sqrt{81(x-2)^2}$ $9|x-2|$

35. $f(x) = \sqrt[3]{32(x+2)^3}$ $2(x+2)\sqrt[3]{4}$

36. $f(x) = \sqrt[3]{48(x-2)^3}$ $2(x-2)\sqrt[3]{6}$

37. $f(x) = \sqrt{3x^2 - 6x + 3}$ $|x-1|\sqrt{3}$

38. $f(x) = \sqrt{5x^2 - 10x + 5}$ $|x-1|\sqrt{5}$

In Exercises 39–60, simplify by factoring. Assume that all variables in a radicand represent positive real numbers and no radicands involve negative quantities raised to even powers.

39. $\sqrt{x^7}$ $x^3\sqrt{x}$

40. $\sqrt{x^5}$ $x^2\sqrt{x}$

41. $\sqrt{x^8y^9}$ $x^4y^4\sqrt{y}$

42. $\sqrt{x^6y^7}$ $x^3y^3\sqrt{y}$

43. $\sqrt{48x^3}$ $4x\sqrt{3x}$

44. $\sqrt{40x^3}$ $2x\sqrt{10x}$

45. $\sqrt[3]{y^8}$ $y^2\sqrt[3]{y^2}$

46. $\sqrt[3]{y^{11}}$ $y^3\sqrt[3]{y^2}$

47. $\sqrt[3]{x^{14}y^3z}$ $x^4y\sqrt[3]{x^2z}$ **48.** $\sqrt[3]{x^3y^{17}z^2}$ $xy^5\sqrt[3]{y^2z^2}$

49. $\sqrt[3]{81x^8y^6}$ $3x^2y^2\sqrt[3]{3x^2}$ **50.** $\sqrt[3]{32x^9y^{17}}$ $2x^3y^5\sqrt[3]{4y^2}$

51. $\sqrt[3]{(x+y)^5}$ $(x+y)\sqrt[3]{(x+y)^2}$

52. $\sqrt[3]{(x+y)^4}$ $(x+y)\sqrt[3]{x+y}$

53. $\sqrt[5]{y^{17}}$ $y^3\sqrt[5]{y^2}$

54. $\sqrt[5]{y^{18}}$ $y^3\sqrt[5]{y^3}$

55. $\sqrt[5]{64x^6y^{17}}$ $2xy^3\sqrt[5]{2xy^2}$ **56.** $\sqrt[5]{64x^7y^{16}}$ $2xy^3\sqrt[5]{2x^2y}$

57. $\sqrt[4]{80x^{10}}$ $2x^2\sqrt[4]{5x^2}$ **58.** $\sqrt[4]{96x^{11}}$ $2x^2\sqrt[4]{6x^3}$

59. $\sqrt[4]{(x-3)^{10}}$ $(x-3)^2\sqrt[4]{(x-3)^2}$ or $(x-3)^2\sqrt{x-3}$

60. $\sqrt[4]{(x-2)^{14}}$ $(x-2)^3\sqrt[4]{(x-2)^2}$ or $(x-2)^3\sqrt{x-2}$

In Exercises 61–82, multiply and simplify. Assume that all variables in a radicand represent positive real numbers and no radicands involve negative quantities raised to even powers.

61. $\sqrt{12} \cdot \sqrt{2}$ $2\sqrt{6}$

62. $\sqrt{3} \cdot \sqrt{6}$ $3\sqrt{2}$

63. $\sqrt{5x} \cdot \sqrt{10y}$ $5\sqrt{2xy}$

64. $\sqrt{8x} \cdot \sqrt{10y}$ $4\sqrt{5xy}$

65. $\sqrt{12x} \cdot \sqrt{3x}$ $6x$

66. $\sqrt{20x} \cdot \sqrt{5x}$ $10x$

67. $\sqrt{50xy} \cdot \sqrt{4xy^2}$ $10xy\sqrt{2y}$

68. $\sqrt{5xy} \cdot \sqrt{10xy^2}$ $5xy\sqrt{2y}$

69. $2\sqrt{5} \cdot 3\sqrt{40}$ $60\sqrt{2}$ **70.** $3\sqrt{15} \cdot 5\sqrt{6}$ $45\sqrt{10}$

71. $\sqrt[3]{12} \cdot \sqrt[3]{4}$ $2\sqrt[3]{6}$ **72.** $\sqrt[4]{4} \cdot \sqrt[4]{8}$ $2\sqrt[4]{2}$

73. $\sqrt{5x^3} \cdot \sqrt{8x^2}$ $2x^2\sqrt{10x}$

74. $\sqrt{2x^7} \cdot \sqrt{12x^4}$ $2x^5\sqrt{6x}$

75. $\sqrt[3]{25x^4y^2} \cdot \sqrt[3]{5xy^{12}}$ $5xy^4\sqrt[3]{x^2y^2}$

76. $\sqrt[3]{6x^7y} \cdot \sqrt[3]{9x^4y^{12}}$ $3x^3y^4\sqrt[3]{2x^2y}$

77. $\sqrt[4]{8x^2y^3z^6} \cdot \sqrt[4]{2x^4yz}$ $2xyz\sqrt[4]{x^2z^3}$

78. $\sqrt[4]{4x^2y^3z^3} \cdot \sqrt[4]{8x^4yz^6}$ $2xyz^2\sqrt[4]{2x^2z}$

79. $\sqrt[5]{8x^4y^6z^2} \cdot \sqrt[5]{8xy^7z^4}$ $2xy^2z\sqrt[5]{2y^3z}$

80. $\sqrt[5]{8x^4y^3z^3} \cdot \sqrt[5]{8xy^9z^8}$ $2xy^2z^2\sqrt[5]{2y^2z}$

81. $\sqrt[3]{x-y} \cdot \sqrt[3]{(x-y)^7}$ $(x-y)^2\sqrt[3]{(x-y)^2}$

82. $\sqrt[3]{x-6} \cdot \sqrt[3]{(x-6)^7}$ $(x-6)^2\sqrt[3]{(x-6)^2}$

Practice Plus

In Exercises 83–92, simplify each expression. Assume that all variables in a radicand represent positive real numbers and no radicands involve negative quantities raised to even powers.

83. $-2x^2y\left(\sqrt[3]{54x^3y^7z^2}\right)$ $-6x^3y^3\sqrt[3]{2yz^2}$

84. $\dfrac{-x^2y^7}{2}\left(\sqrt[3]{-32x^4y^9z^7}\right)$ $x^3y^{10}z^2\sqrt[3]{4xz}$

85. $-3y\left(\sqrt[5]{64x^3y^6}\right)$ $-6y^2\sqrt[5]{2x^3y}$

86. $-4x^2y^7\left(\sqrt[5]{-32x^{11}y^{17}}\right)$ $8x^4y^{10}\sqrt[5]{xy^2}$

87. $\left(-2xy^2\sqrt{3x}\right)\left(xy\sqrt{6x}\right)$ $-6x^3y^3\sqrt{2}$

88. $\left(-5x^2y^3z\sqrt{2xyz}\right)\left(-x^4z\sqrt{10xz}\right)$ $10x^7y^3z^3\sqrt{5y}$

89. $\left(2x^2y\sqrt[4]{8xy}\right)\left(-3xy^2\sqrt[4]{2x^2y^3}\right)$ $-12x^3y^4\sqrt[4]{x^3}$

90. $\left(5a^2b\sqrt[4]{8a^2b}\right)\left(4ab\sqrt[4]{4a^3b^2}\right)$ $40a^4b^2\sqrt[4]{2ab^3}$

91. $\sqrt[5]{8x^4y^6}\left(\sqrt[5]{2xy^7} + \sqrt[5]{4x^6y^9}\right)$ $xy^2\sqrt[5]{16y^3} + 2x^2y^3$

92. $\sqrt[4]{2x^3y^2}\left(\sqrt[4]{6x^5y^6} + \sqrt[4]{8x^5y^7}\right)$ $x^2y^2\sqrt[4]{12} + 2x^2y^2\sqrt[4]{y}$

Application Exercises

The function

$$d(x) = \sqrt{\frac{3x}{2}}$$

models the distance, $d(x)$, in miles, that a person h feet high can see to the horizon. Use this function to solve Exercises 93–94.

93. The pool deck on a cruise ship is 72 feet above the water. How far can passengers on the pool deck see? Write the answer in simplified radical form. Then use the simplified radical form and a calculator to express the answer to the nearest tenth of a mile. $6\sqrt{3}$ miles; 10.4 miles

94. The captain of a cruise ship is on the star deck, which is 120 feet above the water. How far can the captain see? Write the answer in simplified radical form. Then use the simplified radical form and a calculator to express the answer to the nearest tenth of a mile. $6\sqrt{5}$ miles; 13.4 miles

Paleontologists use the function

$$W(x) = 4\sqrt{2x}$$

to estimate the walking speed of a dinosaur, $W(x)$, in feet per second, where x is the length, in feet, of the dinosaur's leg. The graph of W is shown in the figure. Use this information to solve Exercises 95–96.

Dinosaur Walking Speeds

$W(x) = 4\sqrt{2x}$

Walking Speed (feet per second)

Leg Length (feet)
$[0, 12, 1]$ by $[0, 20, 2]$

·95. What is the walking speed of a dinosaur whose leg length is 6 feet? Use the function's equation and express the answer in simplified radical form. Then use the function's graph to estimate the answer to the nearest foot per second. $8\sqrt{3}$ feet per second; 14 feet per second

96. What is the walking speed of a dinosaur whose leg length is 10 feet? Use the function's equation and express the answer in simplified radical form. Then use the function's graph to estimate the answer to the nearest foot per second. $8\sqrt{5}$ feet feet per second; 18 feet per second

*Your **cardiac index** is your heart's output, in liters of blood per minute, divided by your body's surface area, in square meters. The cardiac index, $C(x)$, can be modeled by*

$$C(x) = \frac{7.644}{\sqrt[4]{x}}, \qquad 10 \leq x \leq 80,$$

where x is an individual's age, in years. The graph of the function is shown. Use the function to solve Exercises 97–98.

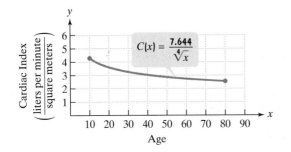

Cardiac Index (liters per minute / square meters)

$C(x) = \dfrac{7.644}{\sqrt[4]{x}}$

Age

97. a. Find the cardiac index of a 32-year-old. Express the denominator in simplified radical form and reduce the fraction. $\dfrac{7.644}{2\sqrt[4]{2}} = \dfrac{3.822}{\sqrt[4]{2}}$

    b. Use the form of the answer in part (a) and a calculator to express the cardiac index to the nearest hundredth. Identify your solution as a point on the graph.   3.21 liters of blood per minute per square meter; (32, 3.21)

98. a. Find the cardiac index of an 80-year-old. Express the denominator in simplified radical form and reduce the fraction. $\dfrac{7.644}{2\sqrt[4]{5}} = \dfrac{3.822}{\sqrt[4]{5}}$

    b. Use the form of the answer in part (a) and a calculator to express the cardiac index to the nearest hundredth. Identify your solution as a point on the graph.   2.56 liters of blood per minute per square meter; (80, 2.56)

Writing in Mathematics

99. What is the product rule for radicals? Give an example to show how it is used.   **99–103.** Answers will vary.

100. Explain why $\sqrt{50}$ is not simplified. What do we mean when we say a radical expression is simplified?

101. In simplifying an nth root, explain how to find variable factors in the radicand that are perfect nth powers.

102. Without showing all the details, explain how to simplify $\sqrt[3]{16x^{14}}$.

103. As you get older, what would you expect to happen to your heart's output? Explain how this is shown in the graph for Exercises 97–98. Is this trend taking place progressively more rapidly or more slowly over the entire interval? What does this mean about this aspect of aging?

Technology Exercises

104. Use a calculator to provide numerical support for your simplifications in Exercises 21–24 and 27–28. In each case, find a decimal approximation for the given expression. Then find a decimal approximation for your simplified expression. The approximations should be the same.

In Exercises 105–108, determine if each simplification is correct by graphing the function on each side of the equation with your graphing utility. Use the given viewing rectangle. The graphs should be the same. If they are not, correct the right side of the equation and then use your graphing utility to verify the simplification.   See graphing answer section.

105. $\sqrt{x^4} = x^2$ Graphs are the same;
$[0, 5, 1]$ by $[0, 20, 1]$ simplification is correct.

106. $\sqrt{8x^2} = 4x\sqrt{2}$ Graphs are not the same;
$[-5, 5, 1]$ by $[-5, 20, 1]$ $\sqrt{8x^2} = 2|x|\sqrt{2}$.

107. $\sqrt{3x^2 - 6x + 3} = (x - 1)\sqrt{3}$ Graphs are not the same;
$[-5, 5, 1]$ by $[-5, 5, 1]$ $\sqrt{3x^2 - 6x + 3} = |x - 1|\sqrt{3}$.

108. $\sqrt[3]{2x} \cdot \sqrt[3]{4x^2} = 4x$ Graphs are not the same;
$[-10, 10, 1]$ by $[-10, 10, 1]$ $\sqrt[3]{2x} \cdot \sqrt[3]{4x^2} = 2x$.

Critical Thinking Exercises

109. Which one of the following is true? d

  a. $2\sqrt{5} \cdot 6\sqrt{5} = 12\sqrt{5}$

  b. $\sqrt[3]{4} \cdot \sqrt[3]{4} = 4$

  c. $\sqrt{12} = 2\sqrt{6}$

  d. $\sqrt[3]{3^{15}} = 243$

110. If a number is tripled, what happens to its square root?   Its square root is multiplied by $\sqrt{3}$.

111. What must be done to a number so that its cube root is tripled?   The number must be multiplied by 27.

112. If $f(x) = \sqrt[3]{2x}$ and $(fg)(x) = 2x$, find $g(x)$. $g(x) = \sqrt[3]{4x^2}$

113. Graph $f(x) = \sqrt{(x-1)^2}$ by hand.   See graphing answer section.

Review Exercises

114. Solve: $2x - 1 \le 21$ and $2x + 2 \ge 12$.
   (Section 4.2, Example 2) $\{x | 5 \le x \le 11\}$ or $[5, 11]$

115. Solve:

$$5x + 2y = 2$$
$$4x + 3y = -4.$$

   (Section 3.1, Example 6) $(2, -4)$ or $\{(2, -4)\}$

116. Factor: $64x^3 - 27$. (Section 5.5, Example 9)
   $(4x - 3)(16x^2 + 12x + 9)$

SECTION

7.4

Objectives

1 Add and subtract radical expressions.

2 Use the quotient rule to simplify radical expressions.

3 Use the quotient rule to divide radical expressions.

ADDING, SUBTRACTING, AND DIVIDING RADICAL EXPRESSIONS

The future is now: You have the opportunity to explore the cosmos in a starship traveling near the speed of light. The experience will enable you to understand the mysteries of the universe first hand, transporting you to unimagined levels of knowing and being. The down side: According to Einstein's theory of relativity, close to the speed of light, your aging rate relative to friends on Earth is nearly zero. You will return from your two-year journey to a futuristic world in which friends and loved ones are long dead. Do you explore space or stay here on Earth? In this section, in addition to learning to perform various operations with radicals, you will see how they model your return to a world where the people you knew have long since departed.

1 Add and subtract radical expressions

Adding and Subtracting Radical Expressions   Two or more radical expressions that have the same indices and radicands are called **like radicals**. Like radicals are combined in exactly the same way that we combine like terms. For example,

$$7\sqrt{11} + 6\sqrt{11} = (7 + 6)\sqrt{11} = 13\sqrt{11}.$$

7 square roots of 11 plus 6 square roots of 11 result in 13 square roots of 11.

EXAMPLE 1 Adding and Subtracting Like Radicals

Simplify (add or subtract) by combining like radical terms:

a. $7\sqrt{2} + 8\sqrt{2}$ **b.** $\sqrt[3]{5} - 4x\sqrt[3]{5} + 8\sqrt[3]{5}$ **c.** $8\sqrt[6]{5x} - 5\sqrt[6]{5x} + 4\sqrt[3]{5x}$.

SOLUTION

a. $7\sqrt{2} + 8\sqrt{2}$

$= (7 + 8)\sqrt{2}$ Apply the distributive property.

$= 15\sqrt{2}$ Simplify.

b. $\sqrt[3]{5} - 4x\sqrt[3]{5} + 8\sqrt[3]{5}$

$= (1 - 4x + 8)\sqrt[3]{5}$ Apply the distributive property.

$= (9 - 4x)\sqrt[3]{5}$ Simplify.

c. $8\sqrt[6]{5x} - 5\sqrt[6]{5x} + 4\sqrt[3]{5x}$

$= (8 - 5)\sqrt[6]{5x} + 4\sqrt[3]{5x}$ Apply the distributive property to the two terms with like radicals.

$= 3\sqrt[6]{5x} + 4\sqrt[3]{5x}$ The indices, 6 and 3, differ. These are not like radicals and cannot be combined. ∎

 CHECK POINT 1 Simplify by combining like radical terms:

a. $8\sqrt{13} + 2\sqrt{13}$ $10\sqrt{13}$

b. $9\sqrt[3]{7} - 6x\sqrt[3]{7} + 12\sqrt[3]{7}$ $(21 - 6x)\sqrt[3]{7}$

c. $7\sqrt[4]{3x} - 2\sqrt[4]{3x} + 2\sqrt[3]{3x}$. $5\sqrt[4]{3x} + 2\sqrt[3]{3x}$

In some cases, radical expressions can be combined once they have been simplified. For example, to add $\sqrt{2}$ and $\sqrt{8}$, we can write $\sqrt{8}$ as $\sqrt{4 \cdot 2}$ because 4 is a perfect square factor of 8.

$$\sqrt{2} + \sqrt{8} = \sqrt{2} + \sqrt{4 \cdot 2} = 1\sqrt{2} + 2\sqrt{2} = (1 + 2)\sqrt{2} = 3\sqrt{2}$$

EXAMPLE 2 Combining Radicals That First Require Simplification

Simplify by combining like radical terms, if possible:

a. $7\sqrt{18} + 5\sqrt{8}$ **b.** $4\sqrt{27x} - 8\sqrt{12x}$ **c.** $7\sqrt{3} - 2\sqrt{5}$.

SOLUTION

a. $7\sqrt{18} + 5\sqrt{8}$

$= 7\sqrt{9 \cdot 2} + 5\sqrt{4 \cdot 2}$ Factor the radicands using the greatest perfect square factors.

$= 7\sqrt{9} \cdot \sqrt{2} + 5\sqrt{4} \cdot \sqrt{2}$ Take the square root of each factor.

$= 7 \cdot 3 \cdot \sqrt{2} + 5 \cdot 2 \cdot \sqrt{2}$ $\sqrt{9} = 3$ and $\sqrt{4} = 2$.

$= 21\sqrt{2} + 10\sqrt{2}$ Multiply.

$= (21 + 10)\sqrt{2}$ Apply the distributive property.

$= 31\sqrt{2}$ Simplify.

b. $4\sqrt{27x} - 8\sqrt{12x}$

$= 4\sqrt{9 \cdot 3x} - 8\sqrt{4 \cdot 3x}$ *Factor the radicands using the greatest perfect square factors.*

$= 4\sqrt{9} \cdot \sqrt{3x} - 8\sqrt{4} \cdot \sqrt{3x}$ *Take the square root of each factor.*

$= 4 \cdot 3 \cdot \sqrt{3x} - 8 \cdot 2 \cdot \sqrt{3x}$ $\sqrt{9} = 3$ *and* $\sqrt{4} = 2$.

$= 12\sqrt{3x} - 16\sqrt{3x}$ *Multiply.*

$= (12 - 16)\sqrt{3x}$ *Apply the distributive property.*

$= -4\sqrt{3x}$ *Simplify.*

c. $7\sqrt{3} - 2\sqrt{5}$ cannot be simplified. The radical expressions have different radicands, namely 3 and 5, and are not like terms. ■

 CHECK POINT 2 Simplify by combining like radical terms, if possible:

a. $3\sqrt{20} + 5\sqrt{45}$ $21\sqrt{5}$

b. $3\sqrt{12x} - 6\sqrt{27x}$ $-12\sqrt{3x}$

c. $8\sqrt{5} - 6\sqrt{2}$.   cannot be simplified

EXAMPLE 3   Adding and Subtracting with Higher Indices

Simplify by combining like radical terms, if possible:

    a. $2\sqrt[3]{16} - 4\sqrt[3]{54}$ **b.** $5\sqrt[3]{xy^2} + \sqrt[3]{8x^4y^5}$.

SOLUTION

a. $2\sqrt[3]{16} - 4\sqrt[3]{54}$

$= 2\sqrt[3]{8 \cdot 2} - 4\sqrt[3]{27 \cdot 2}$ *Factor the radicands using the greatest perfect cube factors.*

$= 2\sqrt[3]{8} \cdot \sqrt[3]{2} - 4\sqrt[3]{27} \cdot \sqrt[3]{2}$ *Take the cube root of each factor.*

$= 2 \cdot 2 \cdot \sqrt[3]{2} - 4 \cdot 3 \cdot \sqrt[3]{2}$ $\sqrt[3]{8} = 2$ *and* $\sqrt[3]{27} = 3$.

$= 4\sqrt[3]{2} - 12\sqrt[3]{2}$ *Multiply.*

$= (4 - 12)\sqrt[3]{2}$ *Apply the distributive property.*

$= -8\sqrt[3]{2}$ *Simplify.*

b. $5\sqrt[3]{xy^2} + \sqrt[3]{8x^4y^5}$

$= 5\sqrt[3]{xy^2} + \sqrt[3]{(8x^3y^3)xy^2}$ *Factor the second radicand using the greatest perfect cube factor.*

$= 5\sqrt[3]{xy^2} + \sqrt[3]{8x^3y^3} \cdot \sqrt[3]{xy^2}$ *Take the cube root of each factor.*

$= 5\sqrt[3]{xy^2} + 2xy\sqrt[3]{xy^2}$ $\sqrt[3]{8} = 2$ *and* $\sqrt[3]{x^3y^3} = (x^3y^3)^{\frac{1}{3}} = xy$.

$= (5 + 2xy)\sqrt[3]{xy^2}$ *Apply the distributive property.* ■

 CHECK POINT 3 Simplify by combining like radical terms, if possible.

a. $3\sqrt[3]{24} - 5\sqrt[3]{81}$ $-9\sqrt[3]{3}$

b. $5\sqrt[3]{x^2y} + \sqrt[3]{27x^5y^4}$. $(5 + 3xy)\sqrt[3]{x^2y}$

Dividing Radical Expressions We have seen that the root of a product is the product of the roots. The root of a quotient can also be expressed as the quotient of roots. Here is an example:

$$\sqrt{\frac{64}{4}} = \sqrt{16} = 4 \quad \text{and} \quad \frac{\sqrt{64}}{\sqrt{4}} = \frac{8}{2} = 4.$$

> This expression is the square root of a quotient.

> This expression is the quotient of two square roots.

The two procedures produce the same result, 4. This is a special case of the **quotient rule for radicals**.

2 Use the quotient rule to simplify radical expressions.

> THE QUOTIENT RULE FOR RADICALS If $\sqrt[n]{a}$ and $\sqrt[n]{b}$ are real numbers and $b \neq 0$, then
>
> $$\sqrt[n]{\frac{a}{b}} = \frac{\sqrt[n]{a}}{\sqrt[n]{b}}.$$
>
> The nth root of a quotient is the quotient of the nth roots.

We know that a radical is simplified when its radicand has no factors other than 1 that are perfect nth powers. The quotient rule can be used to simplify some radicals. Keep in mind that all variables in radicands represent positive real numbers.

EXAMPLE 4 Using the Quotient Rule to Simplify Radicals

Simplify using the quotient rule:

a. $\sqrt[3]{\frac{16}{27}}$ **b.** $\sqrt{\frac{x^2}{25y^6}}$ **c.** $\sqrt[4]{\frac{7y^5}{x^{12}}}$.

SOLUTION We simplify each expression by taking the roots of the numerator and the denominator. Then we use factoring to simplify the resulting radicals, if possible.

a. $\sqrt[3]{\frac{16}{27}} = \frac{\sqrt[3]{16}}{\sqrt[3]{27}} = \frac{\sqrt[3]{8 \cdot 2}}{3} = \frac{\sqrt[3]{8} \cdot \sqrt[3]{2}}{3} = \frac{2\sqrt[3]{2}}{3}$

b. $\sqrt{\frac{x^2}{25y^6}} = \frac{\sqrt{x^2}}{\sqrt{25y^6}} = \frac{x}{5(y^6)^{\frac{1}{2}}} = \frac{x}{5y^3}$

> Try to do this step mentally.

c. $\sqrt[4]{\frac{7y^5}{x^{12}}} = \frac{\sqrt[4]{7y^5}}{\sqrt[4]{x^{12}}} = \frac{\sqrt[4]{y^4 \cdot 7y}}{\sqrt[4]{x^{12}}} = \frac{y\sqrt[4]{7y}}{x^3}$

■

✔ CHECK POINT 4 Simplify using the quotient rule:

a. $\sqrt[3]{\frac{24}{125}}$ $\frac{2\sqrt[3]{3}}{5}$ **b.** $\sqrt{\frac{9x^3}{y^{10}}}$ $\frac{3x\sqrt{x}}{y^5}$ **c.** $\sqrt[3]{\frac{8y^7}{x^{12}}}$. $\frac{2y^2\sqrt[3]{y}}{x^4}$

By reversing the two sides of the quotient rule, we obtain a procedure for dividing radical expressions.

3 Use the quotient rule to divide radical expressions.

> **DIVIDING RADICAL EXPRESSIONS** If $\sqrt[n]{a}$ and $\sqrt[n]{b}$ are real numbers and $b \neq 0$, then
>
> $$\frac{\sqrt[n]{a}}{\sqrt[n]{b}} = \sqrt[n]{\frac{a}{b}}.$$
>
> To divide two radical expressions with the same index, divide the radicands and retain the common index.

EXAMPLE 5 Dividing Radical Expressions

Divide and, if possible, simplify:

a. $\dfrac{\sqrt{48x^3}}{\sqrt{6x}}$ **b.** $\dfrac{\sqrt{45xy}}{2\sqrt{5}}$ **c.** $\dfrac{\sqrt[3]{16x^5y^2}}{\sqrt[3]{2xy^{-1}}}$.

SOLUTION In each part of this problem, the indices in the numerator and the denominator are the same. Perform each division by dividing the radicands and retaining the common index.

a. $\dfrac{\sqrt{48x^3}}{\sqrt{6x}} = \sqrt{\dfrac{48x^3}{6x}} = \sqrt{8x^2} = \sqrt{4x^2 \cdot 2} = \sqrt{4x^2} \cdot \sqrt{2} = 2x\sqrt{2}$

b. $\dfrac{\sqrt{45xy}}{2\sqrt{5}} = \dfrac{1}{2} \cdot \sqrt{\dfrac{45xy}{5}} = \dfrac{1}{2} \cdot \sqrt{9xy} = \dfrac{1}{2} \cdot 3\sqrt{xy}$ or $\dfrac{3\sqrt{xy}}{2}$

c. $\dfrac{\sqrt[3]{16x^5y^2}}{\sqrt[3]{2xy^{-1}}} = \sqrt[3]{\dfrac{16x^5y^2}{2xy^{-1}}}$ Divide the radicands and retain the common index.

$= \sqrt[3]{8x^{5-1}y^{2-(-1)}}$ Divide factors in the radicand. Subtract exponents on common bases.

$= \sqrt[3]{8x^4y^3}$ Simplify.

$= \sqrt[3]{(8x^3y^3)x}$ Factor using the greatest perfect cube factor.

$= \sqrt[3]{8x^3y^3} \cdot \sqrt[3]{x}$ Factor into two radicals.

$= 2xy\sqrt[3]{x}$ Simplify.

CHECK POINT 5 Divide and, if possible, simplify:

a. $\dfrac{\sqrt{40x^5}}{\sqrt{2x}}$ $2x^2\sqrt{5}$ **b.** $\dfrac{\sqrt{50xy}}{2\sqrt{2}}$ $\dfrac{5\sqrt{xy}}{2}$

c. $\dfrac{\sqrt[3]{48x^7y}}{\sqrt[3]{6xy^{-2}}}$. $2x^2y$

7.4 EXERCISE SET

Student Solutions Manual   CD/Video   PH Math/Tutor Center   MathXL Tutorials on CD   MathXL®   MyMathLab   Interactmath.com

Practice Exercises

In this exercise set, assume that all variables represent positive real numbers.

In Exercises 1–10, add or subtract as indicated.

1. $8\sqrt{5} + 3\sqrt{5}$ $11\sqrt{5}$ **2.** $7\sqrt{3} + 2\sqrt{3}$ $9\sqrt{3}$

3. $9\sqrt[3]{6} - 2\sqrt[3]{6}$ $7\sqrt[3]{6}$ **4.** $9\sqrt[3]{7} - 4\sqrt[3]{7}$ $5\sqrt[3]{7}$

5. $4\sqrt[5]{2} + 3\sqrt[5]{2} - 5\sqrt[5]{2}$ $2\sqrt[5]{2}$ **6.** $6\sqrt[5]{3} + 2\sqrt[5]{3} - 3\sqrt[5]{3}$ $5\sqrt[5]{3}$

7. $3\sqrt{13} - 2\sqrt{5} - 2\sqrt{13} + 4\sqrt{5}$ $\sqrt{13} + 2\sqrt{5}$

8. $8\sqrt{17} - 5\sqrt{19} - 6\sqrt{17} + 4\sqrt{19}$ $2\sqrt{17} - \sqrt{19}$

9. $3\sqrt{5} - \sqrt[3]{x} + 4\sqrt{5} + 3\sqrt[3]{x}$ $7\sqrt{5} + 2\sqrt[3]{x}$

10. $6\sqrt{7} - \sqrt[3]{x} + 2\sqrt{7} + 5\sqrt[3]{x}$ $8\sqrt{7} + 4\sqrt[3]{x}$

In Exercises 11–28, add or subtract as indicated. You will need to simplify terms to identify the like radicals.

11. $\sqrt{3} + \sqrt{27}$ $4\sqrt{3}$ **12.** $\sqrt{5} + \sqrt{20}$ $3\sqrt{5}$

13. $7\sqrt{12} + \sqrt{75}$ $19\sqrt{3}$ **14.** $5\sqrt{12} + \sqrt{75}$ $15\sqrt{3}$

15. $3\sqrt{32x} - 2\sqrt{18x}$ $6\sqrt{2x}$ **16.** $5\sqrt{45x} - 2\sqrt{20x}$ $11\sqrt{5x}$

17. $5\sqrt[3]{16} + \sqrt[3]{54}$ $13\sqrt[3]{2}$ **18.** $3\sqrt[3]{24} + \sqrt[3]{81}$ $9\sqrt[3]{3}$

19. $3\sqrt{45x^3} + \sqrt{5x}$ $(9x + 1)\sqrt{5x}$

20. $8\sqrt{45x^3} + \sqrt{5x}$ $(24x + 1)\sqrt{5x}$

21. $\sqrt[3]{54xy^3} + y\sqrt[3]{128x}$ $7y\sqrt[3]{2x}$

22. $\sqrt[3]{24xy^3} + y\sqrt[3]{81x}$ $5y\sqrt[3]{3x}$

23. $\sqrt[3]{54x^4} - \sqrt[3]{16x}$ $(3x - 2)\sqrt[3]{2x}$

24. $\sqrt[3]{81x^4} - \sqrt[3]{24x}$ $(3x - 2)\sqrt[3]{3x}$

25. $\sqrt{9x - 18} + \sqrt{x - 2}$ $4\sqrt{x - 2}$

26. $\sqrt{4x - 12} + \sqrt{x - 3}$ $3\sqrt{x - 3}$

27. $2\sqrt[3]{x^4y^2} + 3x\sqrt[3]{xy^2}$ $5x\sqrt[3]{xy^2}$

28. $4\sqrt[3]{x^4y^2} + 5x\sqrt[3]{xy^2}$ $9x\sqrt[3]{xy^2}$

In Exercises 29–44, simplify using the quotient rule.

29. $\sqrt{\dfrac{11}{4}}$ $\dfrac{\sqrt{11}}{2}$ **30.** $\sqrt{\dfrac{19}{25}}$ $\dfrac{\sqrt{19}}{5}$

31. $\sqrt[3]{\dfrac{19}{27}}$ $\dfrac{\sqrt[3]{19}}{3}$ **32.** $\sqrt[3]{\dfrac{11}{64}}$ $\dfrac{\sqrt[3]{11}}{4}$

33. $\sqrt{\dfrac{x^2}{36y^8}}$ $\dfrac{x}{6y^4}$ **34.** $\sqrt{\dfrac{x^2}{144y^{12}}}$ $\dfrac{x}{12y^6}$

35. $\sqrt{\dfrac{8x^3}{25y^6}}$ $\dfrac{2x\sqrt{2x}}{5y^3}$ **36.** $\sqrt{\dfrac{50x^3}{81y^8}}$ $\dfrac{5x\sqrt{2x}}{9y^4}$

37. $\sqrt[3]{\dfrac{x^4}{8y^3}}$ $\dfrac{x\sqrt[3]{x}}{2y}$ **38.** $\sqrt[3]{\dfrac{x^5}{125y^3}}$ $\dfrac{x\sqrt[3]{x^2}}{5y}$

39. $\sqrt[3]{\dfrac{50x^8}{27y^{12}}}$ $\dfrac{x^2\sqrt[3]{50x^2}}{3y^4}$ **40.** $\sqrt[3]{\dfrac{81x^8}{8y^{15}}}$ $\dfrac{3x^2\sqrt[3]{3x^2}}{2y^5}$

41. $\sqrt[4]{\dfrac{9y^6}{x^8}}$ $\dfrac{y\sqrt[4]{9y^2}}{x^2}$ **42.** $\sqrt[4]{\dfrac{13y^7}{x^{12}}}$ $\dfrac{y\sqrt[4]{13y^3}}{x^3}$

43. $\sqrt[5]{\dfrac{64x^{13}}{y^{20}}}$ $\dfrac{2x^2\sqrt[5]{2x^3}}{y^4}$ **44.** $\sqrt[5]{\dfrac{64x^{14}}{y^{15}}}$ $\dfrac{2x^2\sqrt[5]{2x^4}}{y^3}$

In Exercises 45–62, divide and, if possible, simplify.

45. $\dfrac{\sqrt{40}}{\sqrt{5}}$ $2\sqrt{2}$ **46.** $\dfrac{\sqrt{200}}{\sqrt{10}}$ $2\sqrt{5}$

47. $\dfrac{\sqrt[3]{48}}{\sqrt[3]{6}}$ 2 **48.** $\dfrac{\sqrt[3]{54}}{\sqrt[3]{2}}$ 3

49. $\dfrac{\sqrt{54x^3}}{\sqrt{6x}}$ $3x$ **50.** $\dfrac{\sqrt{72x^3}}{\sqrt{2x}}$ $6x$

51. $\dfrac{\sqrt{x^5y^3}}{\sqrt{xy}}$ x^2y **52.** $\dfrac{\sqrt{x^7y^6}}{\sqrt{x^3y^2}}$ x^2y^2

53. $\dfrac{\sqrt{200x^3}}{\sqrt{10x^{-1}}}$ $2x^2\sqrt{5}$ **54.** $\dfrac{\sqrt{500x^3}}{\sqrt{10x^{-1}}}$ $5x^2\sqrt{2}$

55. $\dfrac{\sqrt{48a^8b^7}}{\sqrt{3a^{-2}b^{-3}}}$ $4a^5b^5$ **56.** $\dfrac{\sqrt{54a^7b^{11}}}{\sqrt{3a^{-4}b^{-2}}}$ $3a^5b^6\sqrt{2ab}$

57. $\dfrac{\sqrt{72xy}}{2\sqrt{2}}$ $3\sqrt{xy}$ **58.** $\dfrac{\sqrt{50xy}}{2\sqrt{2}}$ $\dfrac{5\sqrt{xy}}{2}$

59. $\dfrac{\sqrt[3]{24x^3y^5}}{\sqrt[3]{3y^2}}$ $2xy$ **60.** $\dfrac{\sqrt[3]{250x^5y^3}}{\sqrt[3]{2x^3}}$ $5y\sqrt[3]{x^2}$

61. $\dfrac{\sqrt[4]{32x^{10}y^8}}{\sqrt[4]{2x^2y^{-2}}}$ $2x^2y^2\sqrt[4]{y^2}$ or $2x^2y^2\sqrt{y}$

62. $\dfrac{\sqrt[5]{96x^{12}y^{11}}}{\sqrt[5]{3x^2y^{-2}}}$ $2x^2y^2\sqrt[5]{y^3}$ **63.** $\dfrac{\sqrt[3]{x^2 + 5x + 6}}{\sqrt[3]{x + 2}}$ $\sqrt[3]{x + 3}$

64. $\dfrac{\sqrt[3]{x^2 + 7x + 12}}{\sqrt[3]{x + 3}}$ $\sqrt[3]{x + 4}$ **65.** $\dfrac{\sqrt[3]{a^3 + b^3}}{\sqrt[3]{a + b}}$ $\sqrt[3]{a^2 - ab + b^2}$

66. $\dfrac{\sqrt[3]{a^3 - b^3}}{\sqrt[3]{a - b}}$ $\sqrt[3]{a^2 + ab + b^2}$

Practice Plus

In Exercises 67–76, perform the indicated operations.

67. $\dfrac{\sqrt{32}}{5} + \dfrac{\sqrt{18}}{7}$ $\dfrac{43\sqrt{2}}{35}$ **68.** $\dfrac{\sqrt{27}}{2} + \dfrac{\sqrt{75}}{7}$ $\dfrac{31\sqrt{3}}{14}$

69. $3x\sqrt{8xy^2} - 5y\sqrt{32x^3} + \sqrt{18x^3y^2}$ $-11xy\sqrt{2x}$

70. $6x\sqrt{3xy^2} - 4x^2\sqrt{27xy} - 5\sqrt{75x^5y}$ $6xy\sqrt{3x} - 37x^2\sqrt{3xy}$

71. $5\sqrt{2x^3} + \dfrac{30x^3\sqrt{24x^2}}{3x^2\sqrt{3x}}$ $25x\sqrt{2x}$

72. $7\sqrt{2x^3} + \dfrac{40x^3\sqrt{150x^2}}{5x^2\sqrt{3x}}$ $47x\sqrt{2x}$

73. $2x\sqrt{75xy} - \dfrac{\sqrt{81xy^2}}{\sqrt{3x^{-2}y}}$ $\quad 7x\sqrt{3xy}$

74. $5\sqrt{8x^2y^3} - \dfrac{9x^2\sqrt{64y}}{3x\sqrt{2y^{-2}}}$ $\quad -2xy\sqrt{2y}$

75. $\dfrac{15x^4\sqrt[3]{80x^3y^2}}{5x^3\sqrt[3]{2x^2y}} - \dfrac{75\sqrt[3]{5x^3y}}{25\sqrt[3]{x^{-1}}}$ $\quad 3x\sqrt[3]{5xy}$

76. $\dfrac{16x^4\sqrt[3]{48x^3y^2}}{8x^3\sqrt[3]{3x^2y}} - \dfrac{20\sqrt[3]{2x^3y}}{4\sqrt[3]{x^{-1}}}$ $\quad -x\sqrt[3]{2xy}$

In Exercises 77–80, find $\left(\dfrac{f}{g}\right)(x)$ and the domain of $\left(\dfrac{f}{g}\right)$. Express each quotient function in simplified form.

77. $f(x) = \sqrt{48x^5}$, $g(x) = \sqrt{3x^2}$

$\left(\dfrac{f}{g}\right)(x) = 4x\sqrt{x}$; Domain of $\dfrac{f}{g} = \{x|x > 0\}$ or $(0, \infty)$

78. $f(x) = \sqrt{x^2 - 25}$, $g(x) = \sqrt{x + 5}$

$\left(\dfrac{f}{g}\right)(x) = \sqrt{x - 5}$; Domain of $\dfrac{f}{g} = \{x|x \geq 5\}$ or $[5, \infty)$

79. $f(x) = \sqrt[3]{32x^6}$, $g(x) = \sqrt[3]{2x^2}$ $\left(\dfrac{f}{g}\right)(x) = 2x\sqrt[3]{2x}$;

Domain of $\dfrac{f}{g} = \{x|x \text{ is a real number and } x \neq 0\}$ or $(-\infty, 0) \cup (0, \infty)$

80. $f(x) = \sqrt[3]{2x^6}$, $g(x) = \sqrt[3]{16x}$ $\left(\dfrac{f}{g}\right)(x) = \dfrac{x\sqrt[3]{x^2}}{2}$;

Domain of $\dfrac{f}{g} = \{x|x \text{ is a real number and } x \neq 0\}$ or $(-\infty, 0) \cup (0, \infty)$

Application Exercises

What does travel in space have to do with radicals? Imagine that in the future we will be able to travel in starships at velocities approaching the speed of light (approximately 186,000 miles per second). According to Einstein's theory of relativity, time would pass more quickly on Earth than it would in the moving starship. The radical expression

$$R_f \dfrac{\sqrt{c^2 - v^2}}{\sqrt{c^2}}$$

gives the aging rate of an astronaut relative to the aging rate of a friend, R_f, on Earth. In the expression, v is the astronaut's velocity and c is the speed of light. Use the expression to solve Exercises 81–82. Imagine that you are the astronaut on the starship.

81. a. Use the quotient rule and simplify the expression that shows your aging rate relative to a friend on Earth. Working in a step-by-step manner, express your aging rate as

$$R_f\sqrt{1 - \left(\dfrac{v}{c}\right)^2}.$$

b. You are moving at velocities approaching the speed of light. Substitute c, the speed of light, for v in the simplified expression from part (a). Simplify completely. Close to the speed of light, what is your aging rate relative to a friend on Earth? What does this mean?

82. a. Use the quotient rule and simplify the expression that shows your aging rate relative to a friend on Earth. Working step-by-step, express your aging rate as

$$R_f\sqrt{1 - \left(\dfrac{v}{c}\right)^2}.$$

b. You are moving at 90% of the speed of light. Substitute $0.9c$ for v, your velocity, in the simplified expression from part (a). What is your aging rate, correct to two decimal places, relative to a friend on Earth? If you are gone for 44 weeks, approximately how many weeks have passed for your friend? $0.44R_f$; 100 weeks

81–82.a. $R_f \dfrac{\sqrt{c^2 - v^2}}{\sqrt{c^2}} = R_f\sqrt{\dfrac{c^2 - v^2}{c^2}} = R_f\sqrt{\dfrac{c^2}{c^2} - \dfrac{v^2}{c^2}} =$

$= R_f\sqrt{1 - \dfrac{v^2}{c^2}} = R_f\sqrt{1 - \left(\dfrac{v}{c}\right)^2}$

81.b. $R_f\sqrt{1 - \left(\dfrac{c}{c}\right)^2} = 0$; Close to the speed of light, the aging rate is 0 relative to a friend on Earth. A person moving close to the speed of light does not age relative to a friend on Earth.

In Exercises 83–84, find the perimeter and area of each rectangle. Express answers in simplified radical form.

83.

$2\sqrt{20}$ feet

$\sqrt{125}$ feet

$P = 18\sqrt{5}$ feet
$A = 100$ square feet

84.

$4\sqrt{20}$ feet

$\sqrt{80}$ feet

$P = 24\sqrt{5}$ feet
$A = 160$ square feet

Writing in Mathematics

85. What are like radicals? Give an example with your explanation. 85–90. Answers will vary.

86. Explain how to add like radicals. Give an example with your explanation.

87. If only like radicals can be combined, why is it possible to add $\sqrt{2}$ and $\sqrt{8}$?

88. Explain how to simplify a radical expression using the quotient rule. Provide an example.

89. Explain how to divide radical expressions with the same index.

90. Answer the question posed in the section opener on page 512. What will you do: explore space or stay here on Earth? What are the reasons for your choice?

Technology Exercises

91. Use a calculator to provide numerical support to any four exercises that you worked from Exercises 1–66 that do not contain variables. Begin by finding a decimal approximation for the given expression. Then find a decimal approximation for your answer. The two decimal approximations should be the same.   Answers will vary.

In Exercises 92–94, determine if each operation is performed correctly by graphing the function on each side of the equation with your graphing utility. Use the given viewing rectangle. The graphs should be the same. If they are not, correct the right side of the equation and then use your graphing utility to verify the correction.   92–94. See graphing answer section.

92. $\sqrt{4x} + \sqrt{9x} = 5\sqrt{x}$

[0, 5, 1] by [0, 10, 1]
Graphs are the same; Simplification is correct.

93. $\sqrt{16x} - \sqrt{9x} = \sqrt{7x}$

[0, 5, 1] by [0, 5, 1]
Graphs are not the same; $\sqrt{16x} - \sqrt{9x} = \sqrt{x}$.

94. $x\sqrt{8} + x\sqrt{2} = x\sqrt{10}$

[−5, 5, 1] by [−15, 15, 1]
Graphs are not the same; $x\sqrt{8} + x\sqrt{2} = 3x\sqrt{2}$.

Critical Thinking Exercises

95. Which one of the following is true?   d
  a. $\sqrt{5} + \sqrt{5} = \sqrt{10}$
  b. $4\sqrt{3} + 5\sqrt{3} = 9\sqrt{6}$
  c. If any two radical expressions are completely simplified, they can then be combined through addition or subtraction.
  d. None of the above is true.

96. If an irrational number is decreased by $2\sqrt{18} - \sqrt{50}$, the result is $\sqrt{2}$. What is the number? $2\sqrt{2}$

97. Simplify: $\dfrac{\sqrt{20}}{3} + \dfrac{\sqrt{45}}{4} - \sqrt{80}$. $-\dfrac{31\sqrt{5}}{12}$

98. Simplify: $\dfrac{6\sqrt{49xy}\,\sqrt{ab^2}}{7\sqrt{36x^{-3}y^{-5}}\,\sqrt{a^{-9}b^{-1}}}$. $x^2y^3a^5b\sqrt{b}$

Review Exercises

99. Solve: $2(3x - 1) - 4 = 2x - (6 - x)$.
  (Section 1.4, Example 3)   0 or {0}

100. Factor: $x^2 - 8xy + 12y^2$.
  (Section 5.4, Example 4) $(x - 2y)(x - 6y)$

101. Add: $\dfrac{2}{x^2 + 5x + 6} + \dfrac{3x}{x^2 + 6x + 9}$.
  (Section 6.2, Example 7) $\dfrac{3x^2 + 8x + 6}{(x + 3)^2(x + 2)}$

✔ **MID-CHAPTER CHECK POINT**

CHAPTER 7

What You Know: We learned to find roots of numbers. We saw that the domain of a square root function is the set of real numbers for which the radicand is nonnegative. We learned to simplify radical expressions, using $\sqrt[n]{a^n} = |a|$ if n is even and $\sqrt[n]{a^n} = a$ if n is odd. The definition $a^{\frac{m}{n}} = (\sqrt[n]{a})^m = \sqrt[n]{a^m}$ connected rational exponents and radicals. Finally, we performed various operations with radicals, including multiplication, addition, subtraction, and division.

In Exercises 1–23, simplify the given expression or perform the indicated operation(s) and, if possible, simplify. Assume that all variables represent positive real numbers.

1. $\sqrt{100} - \sqrt[3]{-27}$ 13   **2.** $\sqrt{8x^5y^7}$ $2x^2y^3\sqrt{2xy}$

3. $3\sqrt[3]{4x^2} + 2\sqrt[3]{4x^2}$ $5\sqrt[3]{4x^2}$

4. $\left(3\sqrt[3]{4x^2}\right)\left(2\sqrt[4]{4x^2}\right)$ $12x\sqrt[3]{2x}$

5. $27^{\frac{2}{3}} + (-32)^{\frac{3}{5}}$ 1   **6.** $\left(64x^3y^{\frac{1}{4}}\right)^{\frac{1}{3}}$ $4xy^{\frac{1}{12}}$

7. $5\sqrt{27} - 4\sqrt{48}$ $-\sqrt{3}$ **8.** $\sqrt{\dfrac{500x^3}{4y^4}}$ $\dfrac{5x\sqrt{5x}}{y^2}$

9. $\dfrac{x}{\sqrt[4]{x}}$ $\sqrt[4]{x^3}$ **10.** $\sqrt[3]{54x^5}$ $3x\sqrt[3]{2x^2}$

11. $\dfrac{\sqrt[3]{160}}{\sqrt[3]{2}}$ $2\sqrt[3]{10}$ **12.** $\sqrt[5]{\dfrac{x^{10}}{y^{20}}}$ $\dfrac{x^2}{y^4}$

13. $\dfrac{\left(x^{\frac{2}{3}}\right)^2}{\left(x^{\frac{1}{4}}\right)^3}$ $x^{\frac{7}{12}}$ **14.** $\sqrt[6]{x^6y^4}$ $x\sqrt[3]{y^2}$

15. $\sqrt[7]{(x-2)^3} \cdot \sqrt[7]{(x-2)^6}$ $(x-2)\sqrt[7]{(x-2)^2}$

16. $\sqrt[4]{32x^{11}y^{17}}$ $2x^2y^4\sqrt[4]{2x^3y}$ **17.** $4\sqrt[3]{16} + 2\sqrt[3]{54}$ $14\sqrt[3]{2}$

18. $\dfrac{\sqrt[7]{x^4y^9}}{\sqrt[7]{x^{-5}y^7}}$ $x\sqrt[7]{x^2y^2}$ **19.** $(-125)^{-\frac{2}{3}}$ $\dfrac{1}{25}$

20. $\sqrt{2} \cdot \sqrt[3]{2}$ $\sqrt[6]{32}$ **21.** $\sqrt[3]{\dfrac{32x}{y^4}} \cdot \sqrt[3]{\dfrac{2x^2}{y^2}}$ $\dfrac{4x}{y^2}$

22. $\sqrt{32xy^2} \cdot \sqrt{2x^3y^5}$ $8x^2y^3\sqrt{y}$

23. $4x\sqrt{6x^4y^3} - 7y\sqrt{24x^6y}$ $-10x^3y\sqrt{6y}$

In Exercises 24–25, find the domain of each function.

24. $f(x) = \sqrt{30 - 5x}$ $\{x | x \le 6\} = (-\infty, 6]$

25. $g(x) = \sqrt[3]{3x - 15}$ $\{x | x \text{ is a real number}\} = (-\infty, \infty)$

SECTION

7.5

Objectives

1 Multiply radicals with more than one term.

2 Use polynomial special products to multiply radicals.

3 Rationalize denominators containing one term.

4 Rationalize denominators containing two terms.

5 Rationalize numerators.

MULTIPLYING WITH MORE THAN ONE TERM AND RATIONALIZING DENOMINATORS

PEANUTS reprinted by permission of United Feature Syndicate, Inc.

The late Charles Schulz, creator of the "Peanuts" comic strip, transfixed 350 million readers worldwide with the joys and angst of his hapless Charlie Brown and Snoopy, a romantic self-deluded beagle. In 18,250 comic strips that spanned nearly 50 years, mathematics was often featured. Is the discussion of radicals shown above the real thing, or is it just an algebraic scam? By the time you complete this section on multiplying and dividing radicals, you will be able to decide.

1   Multiply radicals with more than one term.

Multiplying Radicals with More Than One Term   Radical expressions with more than one term are multiplied in much the same way that polynomials with more than one term are multiplied. Example 1 uses the distributive property and the FOIL method to perform multiplications.

EXAMPLE 1   Multiplying Radicals

Multiply:

     a. $\sqrt{7}\left(x + \sqrt{2}\right)$ **b.** $\sqrt[3]{x}\left(\sqrt[3]{6} - \sqrt[3]{x^2}\right)$ **c.** $\left(5\sqrt{2} + 2\sqrt{3}\right)\left(4\sqrt{2} - 3\sqrt{3}\right)$.

SOLUTION

   a. $\sqrt{7}\left(x + \sqrt{2}\right)$

       $= \sqrt{7} \cdot x + \sqrt{7} \cdot \sqrt{2}$ *Use the distributive property.*

       $= x\sqrt{7} + \sqrt{14}$ *Multiply the radicals.*

   b. $\sqrt[3]{x}\left(\sqrt[3]{6} - \sqrt[3]{x^2}\right)$

       $= \sqrt[3]{x} \cdot \sqrt[3]{6} - \sqrt[3]{x} \cdot \sqrt[3]{x^2}$ *Use the distributive property.*

       $= \sqrt[3]{6x} - \sqrt[3]{x^3}$ *Multiply the radicals:* $\sqrt[n]{a} \cdot \sqrt[n]{b} = \sqrt[n]{ab}.$

       $= \sqrt[3]{6x} - x$ *Simplify:* $\sqrt[3]{x^3} = x.$

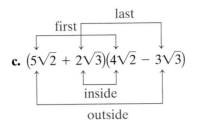

c. $(5\sqrt{2} + 2\sqrt{3})(4\sqrt{2} - 3\sqrt{3})$ Use the FOIL method.

F	O	I	L

$$= (5\sqrt{2})(4\sqrt{2}) + (5\sqrt{2})(-3\sqrt{3}) + (2\sqrt{3})(4\sqrt{2}) + (2\sqrt{3})(-3\sqrt{3})$$

$$= 20 \cdot 2 - 15\sqrt{6} + 8\sqrt{6} - 6 \cdot 3 \qquad \text{Multiply. Note that } \sqrt{2} \cdot \sqrt{2} = \sqrt{4} = 2 \text{ and}$$
$$\sqrt{3} \cdot \sqrt{3} = \sqrt{9} = 3.$$

$$= 40 - 15\sqrt{6} + 8\sqrt{6} - 18 \qquad \text{Complete the multiplications.}$$

$$= (40 - 18) + \left(-15\sqrt{6} + 8\sqrt{6}\right) \qquad \text{Group like terms. Try to do this step mentally.}$$

$$= 22 - 7\sqrt{6} \qquad \text{Combine numerical terms and like radicals. } \blacksquare$$

✓ **CHECK POINT 1** Multiply:

a. $\sqrt{6}(x + \sqrt{10})$ $x\sqrt{6} + 2\sqrt{15}$

b. $\sqrt[3]{y}(\sqrt[3]{y^2} - \sqrt[3]{7})$ $y - \sqrt[3]{7y}$

c. $(6\sqrt{5} + 3\sqrt{2})(2\sqrt{5} - 4\sqrt{2})$. $36 - 18\sqrt{10}$

Some radicals can be multiplied using the special products for multiplying polynomials.

2 Use polynomial special products to multiply radicals.

EXAMPLE 2 Using Special Products to Multiply Radicals

Multiply:

a. $(\sqrt{3} + \sqrt{7})^2$ **b.** $(\sqrt{7} + \sqrt{3})(\sqrt{7} - \sqrt{3})$ **c.** $(\sqrt{a} - \sqrt{b})(\sqrt{a} + \sqrt{b})$.

SOLUTION   Use the special-product formulas shown.

$$(A + B)^2 = A^2 + 2 \cdot A \cdot B + B^2$$

a. $(\sqrt{3} + \sqrt{7})^2 = (\sqrt{3})^2 + 2 \cdot \sqrt{3} \cdot \sqrt{7} + (\sqrt{7})^2$ Use the special product for $(A + B)^2$.

$$= 3 + 2\sqrt{21} + 7 \qquad \text{Multiply the radicals.}$$
$$= 10 + 2\sqrt{21} \qquad \text{Simplify.}$$

$$(A + B) \cdot (A - B) = A^2 - B^2$$

b. $(\sqrt{7} + \sqrt{3})(\sqrt{7} - \sqrt{3}) = (\sqrt{7})^2 - (\sqrt{3})^2$ Use the special product for $(A + B)(A - B)$.

$$= 7 - 3 \qquad \text{Simplify: } (\sqrt{a})^2 = a.$$
$$= 4$$

$$(A - B) \cdot (A + B) = A^2 - B^2$$

c. $(\sqrt{a} - \sqrt{b})(\sqrt{a} + \sqrt{b}) = (\sqrt{a})^2 - (\sqrt{b})^2 = a - b$ $\blacksquare$

STUDY TIP

Just to recap, special products for multiplying polynomials are summarized as follows:

Square of a Binomial:

$$(A + B)^2 = A^2 + 2AB + B^2$$
$$(A - B)^2 = A^2 - 2AB + B^2$$

Product of the Sum and Difference of Two Terms:

$$(A + B)(A - B) = A^2 - B^2.$$

Radical expressions that involve the sum and difference of the same two terms are called **conjugates**. For example,

$$\sqrt{7} + \sqrt{3} \quad \text{and} \quad \sqrt{7} - \sqrt{3}$$

are conjugates of each other. Parts (b) and (c) of Example 2 illustrate that the product of two radical expressions need not be a radical expression:

$$\left(\sqrt{7} + \sqrt{3}\right)\left(\sqrt{7} - \sqrt{3}\right) = 4$$
$$\left(\sqrt{a} - \sqrt{b}\right)\left(\sqrt{a} + \sqrt{b}\right) = a - b.$$

> The product of conjugates does not contain a radical.

Later in this section, we will use conjugates to simplify quotients.

✔ **CHECK POINT 2** Multiply:

a. $\left(\sqrt{5} + \sqrt{6}\right)^2$ $11 + 2\sqrt{30}$ **b.** $\left(\sqrt{6} + \sqrt{5}\right)\left(\sqrt{6} - \sqrt{5}\right)$ 1

c. $\left(\sqrt{a} - \sqrt{7}\right)\left(\sqrt{a} + \sqrt{7}\right)$. $a - 7$

3 Rationalize denominators containing one term.

Rationalizing Denominators Containing One Term You can use a calculator to compare the approximate values for $\dfrac{1}{\sqrt{3}}$ and $\dfrac{\sqrt{3}}{3}$. The two approximations are the same. This is not a coincidence:

$$\frac{1}{\sqrt{3}} = \frac{1}{\sqrt{3}} \cdot \boxed{\frac{\sqrt{3}}{\sqrt{3}}} = \frac{\sqrt{3}}{\sqrt{9}} = \frac{\sqrt{3}}{3}.$$

> Any number divided by itself is 1. Multiplication by 1 does not change the value of $\dfrac{1}{\sqrt{3}}$.

This process involves rewriting a radical expression as an equivalent expression in which the denominator no longer contains any radicals. The process is called **rationalizing the denominator**. When the denominator contains a single radical with an *n*th root, **multiply the numerator and the denominator by a radical of index *n* that produces a perfect *n*th power in the denominator's radicand.**

EXAMPLE 3 Rationalizing Denominators

Rationalize each denominator:

a. $\dfrac{\sqrt{5}}{\sqrt{6}}$ **b.** $\sqrt[3]{\dfrac{7}{16}}$.

SOLUTION

a. If we multiply the numerator and the denominator of $\dfrac{\sqrt{5}}{\sqrt{6}}$ by $\sqrt{6}$, the denominator becomes $\sqrt{6} \cdot \sqrt{6} = \sqrt{36} = 6$. The denominator's radicand, 36, is a perfect square.

The denominator no longer contains a radical. Therefore, we multiply by 1, choosing $\frac{\sqrt{6}}{\sqrt{6}}$ for 1.

$$\frac{\sqrt{5}}{\sqrt{6}} = \frac{\sqrt{5}}{\sqrt{6}} \cdot \frac{\sqrt{6}}{\sqrt{6}}$$ Multiply the numerator and denominator by $\sqrt{6}$ to remove the radical in the denominator.

$$= \frac{\sqrt{30}}{\sqrt{36}}$$ Multiply numerators and multiply denominators. The denominator's radicand, 36, is a perfect square.

$$= \frac{\sqrt{30}}{6}$$ Simplify: $\sqrt{36} = 6$.

b. Using the quotient rule, we can express $\sqrt[3]{\frac{7}{16}}$ as $\frac{\sqrt[3]{7}}{\sqrt[3]{16}}$. We have cube roots, so we want the denominator's radicand to be a perfect cube. Right now, the denominator's radicand is 16 or 4^2. We know that $\sqrt[3]{4^3} = 4$. If we multiply the numerator and the denominator of $\frac{\sqrt[3]{7}}{\sqrt[3]{16}}$ by $\sqrt[3]{4}$, the denominator becomes

$$\sqrt[3]{16} \cdot \sqrt[3]{4} = \sqrt[3]{4^2} \cdot \sqrt[3]{4} = \sqrt[3]{4^3} = 4.$$

The denominator's radicand, 4^3, is a perfect cube. The denominator no longer contains a radical. Therefore, we multiply by 1, choosing $\frac{\sqrt[3]{4}}{\sqrt[3]{4}}$ for 1.

$$\sqrt[3]{\frac{7}{16}} = \frac{\sqrt[3]{7}}{\sqrt[3]{16}}$$ Use the quotient rule and rewrite as the quotient of radicals.

$$= \frac{\sqrt[3]{7}}{\sqrt[3]{4^2}}$$ Write the denominator's radicand as an exponential expression.

$$= \frac{\sqrt[3]{7}}{\sqrt[3]{4^2}} \cdot \frac{\sqrt[3]{4}}{\sqrt[3]{4}}$$ Multiply the numerator and denominator by $\sqrt[3]{4}$ to remove the radical in the denominator.

$$= \frac{\sqrt[3]{28}}{\sqrt[3]{4^3}}$$ Multiply numerators and denominators. The denominator's radicand, 4^3, is a perfect cube.

$$= \frac{\sqrt[3]{28}}{4}$$ Simplify: $\sqrt[3]{4^3} = 4$. ■

✔ CHECK POINT **3** Rationalize each denominator:

a. $\dfrac{\sqrt{3}}{\sqrt{7}}$ $\dfrac{\sqrt{21}}{7}$ **b.** $\sqrt[3]{\dfrac{2}{9}}$ · $\dfrac{\sqrt[3]{6}}{3}$

Example 3 showed that it is helpful to express the denominator's radicand using exponents. In this way, we can find the extra factor or factors needed to produce a perfect nth power. For example, suppose that $\sqrt[5]{8}$ appears in the denominator. We want a perfect fifth power. By expressing $\sqrt[5]{8}$ as $\sqrt[5]{2^3}$, we would multiply the numerator and denominator by $\sqrt[5]{2^2}$ because

$$\sqrt[5]{2^3} \cdot \sqrt[5]{2^2} = \sqrt[5]{2^5} = 2.$$

EXAMPLE 4   Rationalizing Denominators

Rationalize each denominator:

a. $\sqrt{\dfrac{3x}{5y}}$ b. $\dfrac{\sqrt[3]{x}}{\sqrt[3]{36y}}$ c. $\dfrac{10y}{\sqrt[5]{4x^3y}}$.

SOLUTION   By examining each denominator, you can determine how to multiply by 1. Let's first look at the denominators. For the square root, we must produce exponents of 2 in the radicand. For the cube root, we need exponents of 3, and for the fifth root, we want exponents of 5.

- $\sqrt{5y}$

- $\sqrt[3]{36y}$ or $\sqrt[3]{6^2y}$

- $\sqrt[5]{4x^3y}$ or $\sqrt[5]{2^2x^3y}$

> **Multiply by $\sqrt{5y}$:**
> $\sqrt{5y}\cdot\sqrt{5y}=\sqrt{25y^2}=5y.$

> **Multiply by $\sqrt[3]{6y^2}$:**
> $\sqrt[3]{6^2y}\cdot\sqrt[3]{6y^2}=\sqrt[3]{6^3y^3}=6y.$

> **Multiply by $\sqrt[5]{2^3x^2y^4}$:**
> $\sqrt[5]{2^2x^3y}\cdot\sqrt[5]{2^3x^2y^4}=\sqrt[5]{2^5x^5y^5}=2xy.$

a. $\sqrt{\dfrac{3x}{5y}}=\dfrac{\sqrt{3x}}{\sqrt{5y}}=\dfrac{\sqrt{3x}}{\sqrt{5y}}\cdot\dfrac{\sqrt{5y}}{\sqrt{5y}}=\dfrac{\sqrt{15xy}}{\sqrt{25y^2}}=\dfrac{\sqrt{15xy}}{5y}$

> **Multiply by 1.** **$25y^2$ is a perfect square.**

b. $\dfrac{\sqrt[3]{x}}{\sqrt[3]{36y}}=\dfrac{\sqrt[3]{x}}{\sqrt[3]{6^2y}}=\dfrac{\sqrt[3]{x}}{\sqrt[3]{6^2y}}\cdot\dfrac{\sqrt[3]{6y^2}}{\sqrt[3]{6y^2}}=\dfrac{\sqrt[3]{6xy^2}}{\sqrt[3]{6^3y^3}}=\dfrac{\sqrt[3]{6xy^2}}{6y}$

> **Multiply by 1.** **6^3y^3 is a perfect cube.**

c. $\dfrac{10y}{\sqrt[5]{4x^3y}}=\dfrac{10y}{\sqrt[5]{2^2x^3y}}=\dfrac{10y}{\sqrt[5]{2^2x^3y}}\cdot\dfrac{\sqrt[5]{2^3x^2y^4}}{\sqrt[5]{2^3x^2y^4}}$

> **Multiply by 1.**

$=\dfrac{10y\sqrt[5]{2^3x^2y^4}}{\sqrt[5]{2^5x^5y^5}}=\dfrac{10y\sqrt[5]{8x^2y^4}}{2xy}=\dfrac{5\sqrt[5]{8x^2y^4}}{x}$

> **$2^5x^5y^5$ is a perfect 5th power.**   **Simplify: Divide numerator and denominator by $2y$.**

 CHECK POINT 4   Rationalize each denominator:

a. $\sqrt{\dfrac{2x}{7y}}$ $\dfrac{\sqrt{14xy}}{7y}$

b. $\dfrac{\sqrt[3]{x}}{\sqrt[3]{9y}}$ $\dfrac{\sqrt[3]{3xy^2}}{3y}$

c. $\dfrac{6x}{\sqrt[5]{8x^2y^4}}$. $\dfrac{3\sqrt[5]{4x^3y}}{y}$

Rationalizing Denominators Containing Two Terms How can we rationalize a denominator if the denominator contains two terms with one or more square roots? **Multiply the numerator and the denominator by the conjugate of the denominator.** Here are three examples of such expressions:

- $$\frac{8}{3\sqrt{2} + 4}$$

  The conjugate of the denominator is $3\sqrt{2} - 4$.

- $$\frac{2 + \sqrt{5}}{\sqrt{6} - \sqrt{3}}$$

  The conjugate of the denominator is $\sqrt{6} + \sqrt{3}$.

- $$\frac{h}{\sqrt{x + h} - \sqrt{x}}$$

  The conjugate of the denominator is $\sqrt{x + h} + \sqrt{x}$.

The product of the denominator and its conjugate is found using the formula

$$(A + B)(A - B) = A^2 - B^2.$$

The simplified product will not contain a radical.

ENRICHMENT ESSAY

Golden Rectangles

The early Greeks believed that the most pleasing of all rectangles were **golden rectangles**, whose ratio of width to height is

$$\frac{w}{h} = \frac{2}{\sqrt{5} - 1}.$$

The Parthenon at Athens fits into a golden rectangle once the triangular pediment is reconstructed. By working Exercise 128 in the exercise set, you will rationalize the denominator of this rational expression and obtain a better understanding about the ratio of adjacent sides in all golden rectangles.

EXAMPLE 5 Rationalizing a Denominator Containing Two Terms

Rationalize the denominator: $\dfrac{8}{3\sqrt{2} + 4}$.

SOLUTION The conjugate of the denominator is $3\sqrt{2} - 4$. If we multiply the numerator and the denominator by $3\sqrt{2} - 4$, the simplified denominator will not contain a radical. Therefore, we multiply by 1, choosing $\dfrac{3\sqrt{2} - 4}{3\sqrt{2} - 4}$ for 1.

$$\frac{8}{3\sqrt{2} + 4} = \frac{8}{3\sqrt{2} + 4} \cdot \frac{3\sqrt{2} - 4}{3\sqrt{2} - 4} \qquad \text{Multiply by 1.}$$

$$= \frac{8(3\sqrt{2} - 4)}{(3\sqrt{2})^2 - 4^2} \qquad (A + B)(A - B) = A^2 - B^2$$

> Leave the numerator in factored form. This helps simplify, if possible.

$$= \frac{8(3\sqrt{2} - 4)}{18 - 16} \qquad (3\sqrt{2})^2 = 9 \cdot 2 = 18$$

$$= \frac{8(3\sqrt{2} - 4)}{2} \qquad \text{This expression can still be simplified.}$$

$$= \frac{\overset{4}{8}(3\sqrt{2} - 4)}{\underset{1}{2}} \qquad \text{Divide the numerator and denominator by 2.}$$

$$= 4(3\sqrt{2} - 4) \quad \text{or} \quad 12\sqrt{2} - 16 \qquad ■$$

✓ **CHECK POINT 5** Rationalize the denominator: $\dfrac{18}{2\sqrt{3} + 3}$. $\quad 12\sqrt{3} - 18$

EXAMPLE 6 Rationalizing a Denominator Containing Two Terms

Rationalize the denominator: $\dfrac{2 + \sqrt{5}}{\sqrt{6} - \sqrt{3}}$.

SOLUTION Multiplication of both the numerator and denominator by $\sqrt{6} + \sqrt{3}$ will rationalize the denominator.

$$\frac{2 + \sqrt{5}}{\sqrt{6} - \sqrt{3}} = \frac{2 + \sqrt{5}}{\sqrt{6} - \sqrt{3}} \cdot \frac{\sqrt{6} + \sqrt{3}}{\sqrt{6} + \sqrt{3}}$$

Multiply by 1.

F O I L

$$= \frac{2\sqrt{6} + 2\sqrt{3} + \sqrt{5} \cdot \sqrt{6} + \sqrt{5} \cdot \sqrt{3}}{\left(\sqrt{6}\right)^2 - \left(\sqrt{3}\right)^2}$$

Use FOIL in the numerator and $(A - B)(A + B) = A^2 - B^2$ in the denominator.

$$= \frac{2\sqrt{6} + 2\sqrt{3} + \sqrt{30} + \sqrt{15}}{6 - 3}$$

$$= \frac{2\sqrt{6} + 2\sqrt{3} + \sqrt{30} + \sqrt{15}}{3}$$

Further simplification is not possible.

■

✔ **CHECK POINT 6** Rationalize the denominator: $\dfrac{3 + \sqrt{7}}{\sqrt{5} - \sqrt{2}}$.

$\dfrac{3\sqrt{5} + 3\sqrt{2} + \sqrt{35} + \sqrt{14}}{3}$

5 Rationalize numerators.

Rationalizing Numerators We have seen that square root functions are often used to model growing phenomena with growth that is leveling off. Figure 7.6 shows a male's height as a function of his age. The pattern of his growth suggests modeling with a square root function.

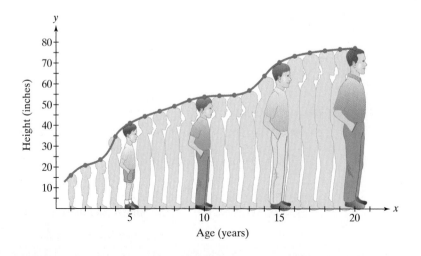

FIGURE 7.6

If we use $f(x) = \sqrt{x}$ to model height, $f(x)$, at age x, the expression

$$\frac{f(a+h) - f(a)}{h} = \frac{\sqrt{a+h} - \sqrt{a}}{h}$$

describes the man's average growth rate from age a to age $a + h$. Can you see that this expression is not defined if $h = 0$? However, to explore the man's average growth rates for successively shorter periods of time, we need to know what happens to the expression as h takes on values that get closer and closer to 0.

What happens to growth near the instant in time that the man is age a? The question is answered by first **rationalizing the numerator**. The procedure is similar to rationalizing the denominator. **To rationalize a numerator, multiply by 1 to eliminate the radical in the *numerator*.**

EXAMPLE 7   Rationalizing a Numerator

Rationalize the numerator:

$$\frac{\sqrt{a+h} - \sqrt{a}}{h}.$$

SOLUTION   The conjugate of the numerator is $\sqrt{a+h} + \sqrt{a}$. If we multiply the numerator and the denominator by $\sqrt{a+h} + \sqrt{a}$, the simplified numerator will not contain radicals. Therefore, we multiply by 1, choosing $\dfrac{\sqrt{a+h} + \sqrt{a}}{\sqrt{a+h} + \sqrt{a}}$ for 1.

$$\frac{\sqrt{a+h} - \sqrt{a}}{h} = \frac{\sqrt{a+h} - \sqrt{a}}{h} \cdot \frac{\sqrt{a+h} + \sqrt{a}}{\sqrt{a+h} + \sqrt{a}} \quad \text{Multiply by 1.}$$

$$= \frac{\left(\sqrt{a+h}\right)^2 - \left(\sqrt{a}\right)^2}{h\left(\sqrt{a+h} + \sqrt{a}\right)} \quad \begin{array}{l}(A - B)(A + B) = A^2 - B^2 \\ \text{Leave the denominator in} \\ \text{factored form.}\end{array}$$

$$= \frac{a + h - a}{h\left(\sqrt{a+h} + \sqrt{a}\right)} \quad \begin{array}{l}\left(\sqrt{a+h}\right)^2 = a + h \\ \text{and } (\sqrt{a})^2 = a.\end{array}$$

$$= \frac{h}{h\left(\sqrt{a+h} + \sqrt{a}\right)} \quad \text{Simplify the numerator.}$$

$$= \frac{1}{\sqrt{a+h} + \sqrt{a}} \quad \begin{array}{l}\text{Simplify by dividing the numerator} \\ \text{and the denominator by } h.\end{array}$$

What happens to the rationalized expression in Example 7 as h is close to 0? The expression is close to

$$\frac{1}{\sqrt{a+0} + \sqrt{a}} = \frac{1}{\sqrt{a} + \sqrt{a}} = \frac{1}{2\sqrt{a}}.$$

The man's growth rate at one instant in time, age a, can be determined with this expression. This expression reveals how the square root function $f(x) = \sqrt{x}$ is changing, instant by instant.

✓ **CHECK POINT 7** Rationalize the numerator: $\dfrac{\sqrt{x+3} - \sqrt{x}}{3}$. $\dfrac{1}{\sqrt{x+3} + \sqrt{x}}$

7.5 EXERCISE SET

Student Solutions Manual   CD/Video   PH Math/Tutor Center   MathXL Tutorials on CD   MathXL®   MyMathLab   Interactmath.com

Practice Exercises

In this exercise set, assume that all variables represent positive real numbers.

In Exercises 1–38, multiply as indicated. If possible, simplify any radical expressions that appear in the product.

1. $\sqrt{2}(x + \sqrt{7})$ $x\sqrt{2} + \sqrt{14}$

2. $\sqrt{5}(x + \sqrt{3})$ $x\sqrt{5} + \sqrt{15}$

3. $\sqrt{6}(7 - \sqrt{6})$ $7\sqrt{6} - 6$

4. $\sqrt{3}(5 - \sqrt{3})$ $5\sqrt{3} - 3$

5. $\sqrt{3}(4\sqrt{6} - 2\sqrt{3})$ $12\sqrt{2} - 6$

6. $\sqrt{6}(4\sqrt{6} - 3\sqrt{2})$ $24 - 6\sqrt{3}$

7. $\sqrt[3]{2}(\sqrt[3]{6} + 4\sqrt[3]{5})$ $\sqrt[3]{12} + 4\sqrt[3]{10}$

8. $\sqrt[3]{3}(\sqrt[3]{6} + 7\sqrt[3]{4})$ $\sqrt[3]{18} + 7\sqrt[3]{12}$

9. $\sqrt[3]{x}(\sqrt[3]{16x^2} - \sqrt[3]{x})$ $2x\sqrt[3]{2} - \sqrt[3]{x^2}$

10. $\sqrt[3]{x}(\sqrt[3]{24x^2} - \sqrt[3]{x})$ $2x\sqrt[3]{3} - \sqrt[3]{x^2}$

11. $(5 + \sqrt{2})(6 + \sqrt{2})$ $32 + 11\sqrt{2}$

12. $(7 + \sqrt{2})(8 + \sqrt{2})$ $58 + 15\sqrt{2}$

13. $(6 + \sqrt{5})(9 - 4\sqrt{5})$ $34 - 15\sqrt{5}$

14. $(4 + \sqrt{5})(10 - 3\sqrt{5})$ $25 - 2\sqrt{5}$

15. $(6 - 3\sqrt{7})(2 - 5\sqrt{7})$ $117 - 36\sqrt{7}$

16. $(7 - 2\sqrt{7})(5 - 3\sqrt{7})$ $77 - 31\sqrt{7}$

17. $(\sqrt{2} + \sqrt{7})(\sqrt{3} + \sqrt{5})$ $\sqrt{6} + \sqrt{10} + \sqrt{21} + \sqrt{35}$

18. $(\sqrt{3} + \sqrt{2})(\sqrt{10} + \sqrt{11})$ $\sqrt{30} + \sqrt{33} + 2\sqrt{5} + \sqrt{22}$

19. $(\sqrt{2} - \sqrt{7})(\sqrt{3} - \sqrt{5})$ $\sqrt{6} - \sqrt{10} - \sqrt{21} + \sqrt{35}$

20. $(\sqrt{3} - \sqrt{2})(\sqrt{10} - \sqrt{11})$ $\sqrt{30} - \sqrt{33} - 2\sqrt{5} + \sqrt{22}$

21. $(3\sqrt{2} - 4\sqrt{3})(2\sqrt{2} + 5\sqrt{3})$ $-48 + 7\sqrt{6}$

22. $(3\sqrt{5} - 2\sqrt{3})(4\sqrt{5} + 5\sqrt{3})$ $30 + 7\sqrt{15}$

23. $(\sqrt{3} + \sqrt{5})^2$ $8 + 2\sqrt{15}$

24. $(\sqrt{2} + \sqrt{7})^2$ $9 + 2\sqrt{14}$

25. $(\sqrt{3x} - \sqrt{y})^2$ $3x - 2\sqrt{3xy} + y$

26. $(\sqrt{2x} - \sqrt{y})^2$ $2x - 2\sqrt{2xy} + y$

27. $(\sqrt{5} + 7)(\sqrt{5} - 7)$ -44

28. $(\sqrt{6} + 2)(\sqrt{6} - 2)$ 2

29. $(2 - 5\sqrt{3})(2 + 5\sqrt{3})$ -71

30. $(3 - 5\sqrt{2})(3 + 5\sqrt{2})$ -41

31. $(3\sqrt{2} + 2\sqrt{3})(3\sqrt{2} - 2\sqrt{3})$ 6

32. $(4\sqrt{3} + 3\sqrt{2})(4\sqrt{3} - 3\sqrt{2})$ 30

33. $(3 - \sqrt{x})(2 - \sqrt{x})$ $6 - 5\sqrt{x} + x$

34. $(4 - \sqrt{x})(3 - \sqrt{x})$ $12 - 7\sqrt{x} + x$

35. $(\sqrt[3]{x} - 4)(\sqrt[3]{x} + 5)$ $\sqrt[3]{x^2} + \sqrt[3]{x} - 20$

36. $(\sqrt[3]{x} - 3)(\sqrt[3]{x} + 7)$ $\sqrt[3]{x^2} + 4\sqrt[3]{x} - 21$

37. $(x + \sqrt[3]{y^2})(2x - \sqrt[3]{y^2})$ $2x^2 + x\sqrt[3]{y^2} - y\sqrt[3]{y}$

38. $(x - \sqrt[5]{y^3})(2x + \sqrt[5]{y^3})$ $2x^2 - x\sqrt[5]{y^3} - y\sqrt[5]{y}$

In Exercises 39–64, rationalize each denominator.

39. $\dfrac{\sqrt{2}}{\sqrt{5}}$ $\dfrac{\sqrt{10}}{5}$

40. $\dfrac{\sqrt{7}}{\sqrt{3}}$ $\dfrac{\sqrt{21}}{3}$

41. $\sqrt{\dfrac{11}{x}}$ $\dfrac{\sqrt{11x}}{x}$

42. $\sqrt{\dfrac{6}{x}}$ $\dfrac{\sqrt{6x}}{x}$

43. $\dfrac{9}{\sqrt{3y}}$ $\dfrac{3\sqrt{3y}}{y}$

44. $\dfrac{12}{\sqrt{3y}}$ $\dfrac{4\sqrt{3y}}{y}$

45. $\dfrac{1}{\sqrt[3]{2}}$ $\dfrac{\sqrt[3]{4}}{2}$

46. $\dfrac{1}{\sqrt[3]{3}}$ $\dfrac{\sqrt[3]{9}}{3}$

47. $\dfrac{6}{\sqrt[3]{4}}$ $3\sqrt[3]{2}$

48. $\dfrac{10}{\sqrt[3]{5}}$ $2\sqrt[3]{25}$

49. $\sqrt[3]{\dfrac{2}{3}}$ $\dfrac{\sqrt[3]{18}}{3}$

50. $\sqrt[3]{\dfrac{3}{4}}$ $\dfrac{\sqrt[3]{6}}{2}$

51. $\dfrac{4}{\sqrt[3]{x}}$ $\dfrac{4\sqrt[3]{x^2}}{x}$

52. $\dfrac{7}{\sqrt[3]{x}}$ $\dfrac{7\sqrt[3]{x^2}}{x}$

53. $\sqrt[3]{\dfrac{2}{y^2}}$ $\dfrac{\sqrt[3]{2y}}{y}$

54. $\sqrt[3]{\dfrac{5}{y^2}}$ $\dfrac{\sqrt[3]{5y}}{y}$

55. $\dfrac{7}{\sqrt[3]{2x^2}}$ $\dfrac{7\sqrt[3]{4x}}{2x}$

56. $\dfrac{10}{\sqrt[3]{4x^2}}$ $\dfrac{5\sqrt[3]{2x}}{x}$

57. $\sqrt[3]{\dfrac{2}{xy^2}}$ $\dfrac{\sqrt[3]{2x^2y}}{xy}$

58. $\sqrt[3]{\dfrac{3}{xy^2}}$ $\dfrac{\sqrt[3]{3x^2y}}{xy}$

59. $\dfrac{3}{\sqrt[4]{x}}$ $\dfrac{3\sqrt[4]{x^3}}{x}$

60. $\dfrac{5}{\sqrt[4]{x}}$ $\dfrac{5\sqrt[4]{x^3}}{x}$

61. $\dfrac{6}{\sqrt[5]{8x^3}}$ $\dfrac{3\sqrt[5]{4x^2}}{x}$

62. $\dfrac{10}{\sqrt[5]{16x^2}}$ $\dfrac{5\sqrt[5]{2x^3}}{x}$

63. $\dfrac{2x^2y}{\sqrt[5]{4x^2y^4}}$ $x\sqrt[5]{8x^3y}$ **64.** $\dfrac{3xy^2}{\sqrt[5]{8xy^3}}$ $\dfrac{3y\sqrt[5]{4x^4y^2}}{2}$

In Exercises 65–74, simplify each radical expression and then rationalize the denominator.

65. $\dfrac{9}{\sqrt{3x^2y}}$ $\dfrac{3\sqrt{3y}}{xy}$ **66.** $\dfrac{25}{\sqrt{5x^2y}}$ $\dfrac{5\sqrt{5y}}{xy}$

67. $-\sqrt{\dfrac{75a^5}{b^3}}$ $-\dfrac{5a^2\sqrt{3ab}}{b^2}$ **68.** $-\sqrt{\dfrac{150a^3}{b^5}}$ $-\dfrac{5a\sqrt{6ab}}{b^3}$

69. $\sqrt{\dfrac{7m^2n^3}{14m^3n^2}}$ $\dfrac{\sqrt{2mn}}{2m}$ **70.** $\sqrt{\dfrac{5m^4n^6}{15m^3n^4}}$ $\dfrac{n\sqrt{3m}}{3}$

71. $\dfrac{3}{\sqrt[4]{x^5y^3}}$ $\dfrac{3\sqrt[4]{x^3y}}{x^2y}$ **72.** $\dfrac{5}{\sqrt[4]{x^2y^7}}$ $\dfrac{5\sqrt[4]{x^2y}}{xy^2}$

73. $\dfrac{12}{\sqrt[3]{-8x^5y^8}}$ $-\dfrac{6\sqrt[3]{xy}}{x^2y^3}$ **74.** $\dfrac{15}{\sqrt[3]{-27x^4y^{11}}}$ $-\dfrac{5\sqrt[3]{x^2y}}{x^2y^4}$

In Exercises 75–92, rationalize each denominator. Simplify, if possible.

75. $\dfrac{8}{\sqrt{5}+2}$ $8\sqrt{5}-16$ **76.** $\dfrac{15}{\sqrt{6}+1}$ $3\sqrt{6}-3$

77. $\dfrac{13}{\sqrt{11}-3}$ $\dfrac{13\sqrt{11}+39}{2}$ **78.** $\dfrac{17}{\sqrt{10}-2}$ $\dfrac{17\sqrt{10}+34}{6}$

79. $\dfrac{6}{\sqrt{5}+\sqrt{3}}$ $3\sqrt{5}-3\sqrt{3}$ **80.** $\dfrac{12}{\sqrt{7}+\sqrt{3}}$ $3\sqrt{7}-3\sqrt{3}$

81. $\dfrac{\sqrt{a}}{\sqrt{a}-\sqrt{b}}$ $\dfrac{a+\sqrt{ab}}{a-b}$ **82.** $\dfrac{\sqrt{b}}{\sqrt{a}-\sqrt{b}}$ $\dfrac{\sqrt{ab}+b}{a-b}$

83. $\dfrac{25}{5\sqrt{2}-3\sqrt{5}}$ $25\sqrt{2}+15\sqrt{5}$

84. $\dfrac{35}{5\sqrt{2}-3\sqrt{5}}$ $35\sqrt{2}+21\sqrt{5}$

85. $\dfrac{\sqrt{5}+\sqrt{3}}{\sqrt{5}-\sqrt{3}}$ $4+\sqrt{15}$ **86.** $\dfrac{\sqrt{11}-\sqrt{5}}{\sqrt{11}+\sqrt{5}}$ $\dfrac{8-\sqrt{55}}{3}$

87. $\dfrac{\sqrt{x}+1}{\sqrt{x}+3}$ $\dfrac{x-2\sqrt{x}-3}{x-9}$ **88.** $\dfrac{\sqrt{x}-2}{\sqrt{x}-5}$ $\dfrac{x+3\sqrt{x}-10}{x-25}$

89. $\dfrac{5\sqrt{3}-3\sqrt{2}}{3\sqrt{2}-2\sqrt{3}}$ $\dfrac{3\sqrt{6}+4}{2}$ **90.** $\dfrac{2\sqrt{6}+\sqrt{5}}{3\sqrt{6}-\sqrt{5}}$ $\dfrac{41+5\sqrt{30}}{49}$

91. $\dfrac{2\sqrt{x}+\sqrt{y}}{\sqrt{y}-2\sqrt{x}}$ $\dfrac{4\sqrt{xy}+4x+y}{y-4x}$

92. $\dfrac{3\sqrt{x}+\sqrt{y}}{\sqrt{y}-3\sqrt{x}}$ $\dfrac{6\sqrt{xy}+9x+y}{y-9x}$

In Exercises 93–104, rationalize each numerator. Simplify, if possible.

93. $\sqrt{\dfrac{3}{2}}$ $\dfrac{3}{\sqrt{6}}$ **94.** $\sqrt{\dfrac{5}{3}}$ $\dfrac{5}{\sqrt{15}}$

95. $\dfrac{\sqrt[3]{4x}}{\sqrt[3]{y}}$ $\dfrac{2x}{\sqrt[3]{2x^2y}}$ **96.** $\dfrac{\sqrt[3]{2x}}{\sqrt[3]{y}}$ $\dfrac{2x}{\sqrt[3]{4x^2y}}$

97. $\dfrac{\sqrt{x}+3}{\sqrt{x}}$ $\dfrac{x-9}{x-3\sqrt{x}}$ **98.** $\dfrac{\sqrt{x}+4}{\sqrt{x}}$ $\dfrac{x-16}{x-4\sqrt{x}}$

99. $\dfrac{\sqrt{a}+\sqrt{b}}{\sqrt{a}-\sqrt{b}}$ $\dfrac{a-b}{a-2\sqrt{ab}+b}$

100. $\dfrac{\sqrt{a}-\sqrt{b}}{\sqrt{a}+\sqrt{b}}$ $\dfrac{a-b}{a+2\sqrt{ab}+b}$

101. $\dfrac{\sqrt{x+5}-\sqrt{x}}{5}$ $\dfrac{1}{\sqrt{x+5}+\sqrt{x}}$

102. $\dfrac{\sqrt{x+7}-\sqrt{x}}{7}$ $\dfrac{1}{\sqrt{x+7}+\sqrt{x}}$

103. $\dfrac{\sqrt{x}+\sqrt{y}}{x^2-y^2}$ $\dfrac{1}{(x+y)(\sqrt{x}-\sqrt{y})}$

104. $\dfrac{\sqrt{x}-\sqrt{y}}{x^2-y^2}$ $\dfrac{1}{(x+y)(\sqrt{x}+\sqrt{y})}$

Practice Plus

In Exercises 105–112, add or subtract as indicated. Begin by rationalizing denominators for all terms in which denominators contain radicals.

105. $\sqrt{2}+\dfrac{1}{\sqrt{2}}$ $\dfrac{3\sqrt{2}}{2}$ **106.** $\sqrt{5}+\dfrac{1}{\sqrt{5}}$ $\dfrac{6\sqrt{5}}{5}$

107. $\sqrt[3]{25}-\dfrac{15}{\sqrt[3]{5}}$ $-2\sqrt[3]{25}$ **108.** $\sqrt[4]{8}-\dfrac{20}{\sqrt[4]{2}}$ $-9\sqrt[4]{8}$

109. $\sqrt{6}-\sqrt{\dfrac{1}{6}}+\sqrt{\dfrac{2}{3}}$ $\dfrac{7\sqrt{6}}{6}$

110. $\sqrt{15}-\sqrt{\dfrac{5}{3}}+\sqrt{\dfrac{3}{5}}$ $\dfrac{13\sqrt{15}}{15}$

111. $\dfrac{2}{\sqrt{2}+\sqrt{3}}+\sqrt{75}-\sqrt{50}$ $7\sqrt{3}-7\sqrt{2}$

112. $\dfrac{5}{\sqrt{2}+\sqrt{7}}-2\sqrt{32}+\sqrt{28}$ $-9\sqrt{2}+3\sqrt{7}$

113. Let $f(x)=x^2-6x-4$. Find $f\big(3-\sqrt{13}\big)$.   0

114. Let $f(x)=x^2+4x-2$. Find $f\big(-2+\sqrt{6}\big)$.   0

115. Let $f(x)=\sqrt{9+x}$. Find $f\big(3\sqrt{5}\big)\cdot f\big(-3\sqrt{5}\big)$.   6

116. Let $f(x)=x^2$. Find $f\big(\sqrt{a+1}-\sqrt{a-1}\big)$. $2a-2\sqrt{a^2-1}$

Application Exercises

The bar graph shows the percentage of U.S. households online from 1997 through 2003. The function

$$P(t) = 15.92\sqrt{t} + 19$$

models the percentage, $P(t)$, of U.S. households online t years after 1997. Use the function to solve Exercises 117–118.

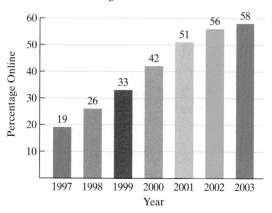

Percentage of U.S. Households Online

Source: U.S. Department of Commerce

117. Find $P(4)$ and describe what this means. Round to the nearest percent. How well does your answer model the actual data shown in the graph? 51; In 2001, approximately 51% of U.S. households were online; Answer models data well.

118. Find $P(6)$ and describe what this means. Use a calculator and round to the nearest percent. How well does your answer model the actual data shown in the graph? 58; In 2003, approximately 58% of U.S. households were online; Answer models data well.

In Exercises 119–120, use the data shown in the bar graph for the percentage of U.S. households online.

119. The average rate of change in the percentage of households online from 1997 through 2003 is

> Change in percent

$$\frac{\text{percentage online in 2003} - \text{percentage online in 1997}}{2003 - 1997}$$

> Change in time

Find the average yearly increase in the percentage of online households during this period. 6.5%

120. The average rate of change in the percentage of households online from 2001 through 2003 is

> Change in percent

$$\frac{\text{percentage online in 2003} - \text{percentage online in 2001}}{2003 - 2001}$$

> Change in time

Find the average yearly increase in the percentage of online households during this period. 3.5%

The algebraic expression

$$15.92\left(\frac{\sqrt{t + h} - \sqrt{t}}{h}\right)$$

models the average rate of change in the percentage of households online from t years after 1997 to $t + h$ years after 1997. Use the expression to solve Exercises 121–124.

121. Let $t = 0$ and $h = 6$. Evaluate the expression to find the average rate of change in the percentage of households online from 1997 through 2003—that is, between 0 years and 6 years after 1997. How well does your answer model the actual yearly increase in the percentage that you found in Exercise 119? The average rate of change in percentage during this period is approximately 6.5%. This answer models the actual yearly percentage increase well.

122. Let $t = 4$ and $h = 2$. Evaluate the expression to find the average rate of change in the percentage of households online from 2001 through 2003—that is, between 4 years and $4 + 2$, or 6, years after 1997. How well does your answer model the actual yearly increase in the percentage that you found in Exercise 120? The average rate of change in percentage during this period is approximately 3.6%. This answer models the actual yearly percentage increase well.

123. a. Rewrite the average rate of change as an equivalent expression by rationalizing the numerator of the rational expression in parentheses. $\dfrac{15.92}{\sqrt{t + h} + \sqrt{t}}$

b. Substitute 0 for h in your expression from part (a) and simplify. The resulting expression gives the rate of change in the percentage of households online precisely t years after 1997. $\dfrac{15.92}{2\sqrt{t}}$ or $\dfrac{7.96}{\sqrt{t}}$

c. Use the expression from part (b) to find the rate of change in the percentage of households online in 2003. Round to the nearest tenth of a percent. The rate of change in percentage in 2003 was 3.2%.

124. Solve parts (a) and (b) of Exercise 123. Then find the rate of change in the percentage of households online in 2001. Round to the nearest tenth of a percent.

a. $\dfrac{15.92}{\sqrt{t + h} + \sqrt{t}}$ **b.** $\dfrac{7.96}{\sqrt{t}}$

The rate of change in percentage in 2001 was 4.0%.

In Exercises 125–126, write expressions for the perimeter and area of each figure. Then simplify these expressions. Assume that all measures are given in inches.

125. $\sqrt{8} + 1$

$\sqrt{8} - 1$

$P = 8\sqrt{2}$ inches
$A = 7$ square inches

126. $2\sqrt{3} + \sqrt{2}$

$2\sqrt{3} + \sqrt{2}$

$P = 8\sqrt{3} + 4\sqrt{2}$ inches
$A = 14 + 4\sqrt{6}$ square inches

127. In the "Peanuts" cartoon shown in the section opener on page 520, Woodstock appears to be working steps mentally. Fill in the missing steps that show how to go from $\dfrac{7\sqrt{2\cdot2\cdot3}}{6}$

to $\dfrac{7}{3}\sqrt{3}$. $\dfrac{7\sqrt{2\cdot2\cdot3}}{6} = \dfrac{7\cdot2\sqrt{3}}{6} = \dfrac{14\sqrt{3}}{6} = \dfrac{7\sqrt{3}}{3} = \dfrac{7}{3}\sqrt{3}$

128. Rationalize the denominator of the golden ratio

$$\frac{2}{\sqrt{5}-1},$$

discussed in the enrichment essay on page 525. Then use a calculator and find the ratio of width to height, correct to the nearest hundredth, in golden rectangles. $\dfrac{\sqrt{5}+1}{2}$; 1.62 to 1

Writing in Mathematics

129. Explain how to perform this multiplication: $\sqrt{2}\,(\sqrt{7}+\sqrt{10})$. 129–137. Answers will vary.

130. Explain how to perform this multiplication: $(2+\sqrt{3})(4+\sqrt{3})$.

131. Explain how to perform this multiplication: $(2+\sqrt{3})^2$.

132. What are conjugates? Give an example with your explanation.

133. Describe how to multiply conjugates.

134. Describe what it means to rationalize a denominator. Use both $\dfrac{1}{\sqrt{5}}$ and $\dfrac{1}{5+\sqrt{5}}$ in your explanation.

135. When a radical expression has its denominator rationalized, we change the denominator so that it no longer contains any radicals. Doesn't this change the value of the radical expression? Explain.

136. Square the real number $\dfrac{2}{\sqrt{3}}$. Observe that the radical is eliminated from the denominator. Explain whether this process is equivalent to rationalizing the denominator.

137. Describe the trend shown in the bar graph in Exercises 117–118 for the percentage of U.S. households online from 1997 through 2003. What explanation can you offer for this trend?

Technology Exercises

In Exercises 138–141, determine if each operation is performed correctly by graphing the function on each side of the equation with your graphing utility. Use the given viewing rectangle. The graphs should be the same. If they are not, correct the right side of the equation and then use your graphing utility to verify the correction. 138–141. See graphing answer section.

138. $(\sqrt{x}-1)(\sqrt{x}-1) = x+1$

[0, 5, 1] by [−1, 2, 1] Graphs are not the same; $(\sqrt{x}-1)(\sqrt{x}-1) = x - 2\sqrt{x} + 1.$

139. $(\sqrt{x}+2)(\sqrt{x}-2) = x^2-4$ for $x \geq 0$

[0, 10, 1] by [−10, 10, 1] Graphs are not the same; $(\sqrt{x}+2)(\sqrt{x}-2) = x - 4$ for $x \geq 0$.

140. $(\sqrt{x}+1)^2 = x+1$

[0, 8, 1] by [0, 15, 1]
Graphs are not the same; $(\sqrt{x}+1)^2 = x + 2\sqrt{x} + 1.$

141. $\dfrac{3}{\sqrt{x+3}-\sqrt{x}} = \sqrt{x+3}+\sqrt{x}$

[0, 8, 1] by [0, 6, 1]
Graphs are the same; Simplification is correct.

Critical Thinking Exercises

142. Which one of the following is true? c

a. $\dfrac{\sqrt{3}+7}{\sqrt{3}-2} = -\dfrac{7}{2}$

b. $\dfrac{4}{\sqrt{x}+y} = \dfrac{4\sqrt{x}-y}{x-y}$

c. $\dfrac{4\sqrt{x}}{\sqrt{x}-y} = \dfrac{4x+4y\sqrt{x}}{x-y^2}$

d. $(\sqrt{x}-7)^2 = x - 49$

143. Solve:

$$7[(2x-5)-(x+1)] = (\sqrt{7}+2)(\sqrt{7}-2). \quad \dfrac{45}{7} \text{ or } \left\{\dfrac{45}{7}\right\}$$

144. Simplify: $(\sqrt{2+\sqrt{3}}+\sqrt{2-\sqrt{3}})^2.$ 6

145. Rationalize the denominator: $\dfrac{1}{\sqrt{2}+\sqrt{3}+\sqrt{4}}.$

$$\dfrac{5\sqrt{2}+3\sqrt{3}-4\sqrt{6}+2}{23}$$

Review Exercises

146. Add: $\dfrac{2}{x-2}+\dfrac{3}{x^2-4}.$

(Section 6.2, Example 6) $\dfrac{2x+7}{x^2-4}$

147. Solve: $\dfrac{2}{x-2}+\dfrac{3}{x^2-4} = 0.$

(Section 6.6, Example 5) $-\dfrac{7}{2}$ or $\left\{-\dfrac{7}{2}\right\}$

148. If $f(x) = x^4 - 3x^2 - 2x + 5$, use synthetic division and the Remainder Theorem to find $f(-2)$. (Section 6.5, Example 2) $f(-2) = 13$

SECTION 7.6

Objectives

1 Solve radical equations.

2 Use models that are radical functions to solve problems.

RADICAL EQUATIONS

A best guess at the look of our nation in the next decades indicates that the number of men and women living alone will increase each year. By 2010, approximately 28 million men and women are projected to be living alone. The function

$$f(x) = 2.6\sqrt{x} + 11$$

models the number of Americans living alone, $f(x)$, in millions, x years after 1970. How can we predict the year when, say, 29 million of us will live alone? Substitute 29 for $f(x)$ and solve for x:

$$29 = 2.6\sqrt{x} + 11.$$

The resulting equation contains a variable in the radicand and is called a *radical equation*. A **radical equation** is an equation in which the variable occurs in a square root, cube root, or any higher root. Some examples of radical equations are

$$\sqrt{2x + 3} = 5, \quad \sqrt{3x + 1} - \sqrt{x + 4} = 1, \quad \text{and} \quad \sqrt[3]{3x - 1} + 4 = 0.$$

Variables occur in radicands.

In this section, you will learn how to solve radical equations. Solving such equations will enable you to solve new kinds of problems using radical functions.

1 Solve radical equations.

Solving Radical Equations   Consider the following radical equation:

$$\sqrt{x} = 9.$$

We solve the equation by squaring both sides:

Squaring both sides eliminates the square root.

$$(\sqrt{x})^2 = 9^2$$
$$x = 81.$$

The proposed solution, 81, can be checked in the original equation, $\sqrt{x} = 9$. Because $\sqrt{81} = 9$, the solution is 81 and the solution set is {81}.

In general, we solve radical equations with square roots by squaring both sides of the equation. We solve radical equations with nth roots by raising both sides of the equation to the nth power. Unfortunately, if n is even, all the solutions of the equation raised to the even power may not be solutions of the original equation. Consider, for example, the equation

$$x = 4.$$

If we square both sides, we obtain

$$x^2 = 16.$$

This new equation has two solutions, -4 and 4. By contrast, only 4 is a solution of the original equation, $x = 4$. For this reason, **when raising both sides of an equation to an even power, always check proposed solutions in the original equation**. Here is a general method for solving radical equations with nth roots:

SOLVING RADICAL EQUATIONS CONTAINING nTH ROOTS

1. If necessary, arrange terms so that one radical is isolated on one side of the equation.
2. Raise both sides of the equation to the nth power to eliminate the nth root.
3. Solve the resulting equation. If this equation still contains radicals, repeat steps 1 and 2.
4. Check all proposed solutions in the original equation.

EXAMPLE 1 Solving a Radical Equation

Solve: $\sqrt{2x + 3} = 5$.

SOLUTION

Step 1. **Isolate a radical on one side.** The radical, $\sqrt{2x + 3}$, is already isolated on the left side of the equation, so we can skip this step.

Step 2. **Raise both sides to the nth power.** Because n, the index, is 2, we square both sides.

$$\sqrt{2x + 3} = 5 \qquad \text{This is the given equation.}$$
$$\left(\sqrt{2x + 3}\right)^2 = 5^2 \qquad \text{Square both sides to eliminate the radical.}$$
$$2x + 3 = 25 \qquad \text{Simplify.}$$

Step 3. **Solve the resulting equation.**

$$2x + 3 = 25 \qquad \text{The resulting equation is a linear equation.}$$
$$2x = 22 \qquad \text{Subtract 3 from both sides.}$$
$$x = 11 \qquad \text{Divide both sides by 2.}$$

Step 4. **Check the proposed, solution in the original equation.** Because both sides were raised to an even power, this check is essential.

Check 11:
$$\sqrt{2x + 3} = 5$$
$$\sqrt{2 \cdot 11 + 3} \overset{?}{=} 5$$
$$\sqrt{25} \overset{?}{=} 5$$
$$5 = 5, \qquad \text{true}$$

The solution is 11 and the solution set is $\{11\}$.

CHECK POINT 1 Solve: $\sqrt{3x + 4} = 8$.   20 or $\{20\}$

EXAMPLE 2 Solving a Radical Equation

Solve: $\sqrt{x-3}+6=5$.

SOLUTION

Step 1. **Isolate a radical on one side.** The radical, $\sqrt{x-3}$, can be isolated by subtracting 6 from both sides. We obtain

$$\sqrt{x-3}=-1.$$

> A principal square root cannot be negative. This equation has no solution. Let's continue the solution procedure to see what happens.

Step 2. **Raise both sides to the nth power.** Because n, the index, is 2, we square both sides.

$$\left(\sqrt{x-3}\right)^2=(-1)^2$$
$$x-3=1 \qquad \text{Simplify.}$$

Step 3. **Solve the resulting equation.**

$$x-3=1 \qquad \text{The resulting equation is a linear equation.}$$
$$x=4 \qquad \text{Add 3 to both sides.}$$

Step 4. **Check the proposed solution in the original equation.**

Check 4:
$$\sqrt{x-3}+6=5$$
$$\sqrt{4-3}+6\overset{?}{=}5$$
$$\sqrt{1}+6\overset{?}{=}5$$
$$1+6\overset{?}{=}5$$
$$7=5, \quad \text{false}$$

This false statement indicates that 4 is not a solution. Thus, the equation has no solution. The solution set is ∅, the empty set. ∎

Example 2 illustrates that extra solutions may be introduced when you raise both sides of a radical equation to an even power. Such solutions, which are not solutions of the given equation, are called **extraneous solutions**. Thus, 4 is an extraneous solution of $\sqrt{x-3}+6=5$.

 CHECK POINT 2 Solve: $\sqrt{x-1}+7=2$. no solution or ∅

EXAMPLE 3 Solving a Radical Equation

Solve: $x+\sqrt{26-11x}=4$.

SOLUTION

Step 1. **Isolate a radical on one side.** We isolate the radical, $\sqrt{26-11x}$, by subtracting x from both sides.

$$x+\sqrt{26-11x}=4 \qquad \text{This is the given equation.}$$
$$\sqrt{26-11x}=4-x. \qquad \text{Subtract } x \text{ from both sides.}$$

Step 2. **Square both sides.**

$$\left(\sqrt{26 - 11x}\right)^2 = (4 - x)^2$$

$$26 - 11x = 16 - 8x + x^2 \quad \text{Simplify. Use the special-product formula}$$
$$(A - B)^2 = A^2 - 2AB + B^2 \text{ to square } 4 - x.$$

Step 3. **Solve the resulting equation.** Because of the x^2-term, the resulting equation is a quadratic equation. We need to write this quadratic equation in standard form. We can obtain zero on the left side by subtracting 26 and adding $11x$ on both sides.

$$26 - 26 - 11x + 11x = 16 - 26 - 8x + 11x + x^2$$
$$0 = x^2 + 3x - 10 \quad \text{Simplify.}$$
$$0 = (x + 5)(x - 2) \quad \text{Factor.}$$
$$x + 5 = 0 \quad \text{or} \quad x - 2 = 0 \quad \text{Set each factor equal to zero.}$$
$$x = -5 \quad\quad\quad x = 2 \quad \text{Solve for x.}$$

Step 4. **Check the proposed solutions in the original equation.**

Check -5:	**Check 2:**

$$x + \sqrt{26 - 11x} = 4$$
$$-5 + \sqrt{26 - 11(-5)} \stackrel{?}{=} 4$$
$$-5 + \sqrt{81} \stackrel{?}{=} 4$$
$$-5 + 9 \stackrel{?}{=} 4$$
$$4 = 4, \quad \text{true}$$

$$x + \sqrt{26 - 11x} = 4$$
$$2 + \sqrt{26 - 11 \cdot 2} \stackrel{?}{=} 4$$
$$2 + \sqrt{4} \stackrel{?}{=} 4$$
$$2 + 2 \stackrel{?}{=} 4$$
$$4 = 4, \quad \text{true}$$

The solutions are -5 and 2, and the solution set is $\{-5, 2\}$. ■

USING TECHNOLOGY

You can use a graphing utility to provide a graphic check that $\{-5, 2\}$ is the solution set of $x + \sqrt{26 - 11x} = 4$.

Use the given equation	Use the equivalent equation
$x + \sqrt{26 - 11x} = 4.$	$x + \sqrt{26 - 11x} - 4 = 0.$

Enter $y_1 = x + \sqrt{26 - 11x}$ in the $\boxed{y=}$ screen.   Enter $y_2 = 4$ in the $\boxed{y=}$ screen.

Enter $y_1 = x + \sqrt{26 - 11x} - 4$ in the $\boxed{y=}$ screen.

Display graphs for y_1 and y_2. The solutions are the x-coordinates of the intersection points. These x-coordinates are -5 and 2. This verifies $\{-5, 2\}$ as the solution set of $x + \sqrt{26 - 11x} = 4$.

Display the graph for y_1. The solutions are the x-intercepts. The x-intercepts are -5 and 2. This verifies $\{-5, 2\}$ as the solution set of $x + \sqrt{26 - 11x} - 4 = 0$.

$[-10, 3, 1]$ by $[-4, 3, 1]$

$y_1 = x + \sqrt{26 - 11x}$

x-coordinate of intersection point is -5.

x-coordinate of intersection point is 2.

$y_2 = 4$

$[-8, 5, 1]$ by $[0, 7, 1]$

✔ CHECK POINT **3** Solve: $\sqrt{6x + 7} - x = 2$. -1 and 3, or $\{-1, 3\}$

The solution of radical equations with two or more square root expressions involves isolating a radical, squaring both sides, and then repeating this process. Let's consider an equation containing two square root expressions.

EXAMPLE 4 Solving an Equation That Has Two Radicals

Solve: $\sqrt{3x + 1} - \sqrt{x + 4} = 1$.

SOLUTION

Step 1. **Isolate a radical on one side.** We can isolate the radical $\sqrt{3x + 1}$ by adding $\sqrt{x + 4}$ to both sides. We obtain

$$\sqrt{3x + 1} = \sqrt{x + 4} + 1.$$

Step 2. **Square both sides.**

$$\left(\sqrt{3x + 1}\right)^2 = \left(\sqrt{x + 4} + 1\right)^2$$

Squaring the expression on the right side of the equation can be a bit tricky. We have to use the formula

$$(A + B)^2 = A^2 + 2AB + B^2.$$

Focusing on just the right side, here is how the squaring is done:

$$
\underset{(A+B)^2}{\left(\sqrt{x+4}+1\right)^2} = \underset{A^2}{\left(\sqrt{x+4}\right)^2} + \underset{2 \cdot A \cdot B}{2 \cdot \sqrt{x+4} \cdot 1} + \underset{B^2}{1^2} = x + 4 + 2\sqrt{x+4} + 1.
$$

Now let's return to squaring both sides.

$$\left(\sqrt{3x + 1}\right)^2 = \left(\sqrt{x + 4} + 1\right)^2 \qquad \text{Square both sides of the equation with an isolated radical.}$$

$$3x + 1 = x + 4 + 2\sqrt{x + 4} + 1 \qquad \left(\sqrt{3x + 1}\right)^2 = 3x + 1; \text{ square the right side using the formula for } (A + B)^2.$$

$$3x + 1 = x + 5 + 2\sqrt{x + 4} \qquad \text{Combine numerical terms on the right side: } 4 + 1 = 5.$$

Can you see that the resulting equation still contains a radical, namely $\sqrt{x + 4}$? Thus, we need to repeat the first two steps.

Repeat Step 1. **Isolate a radical on one side.** We isolate $2\sqrt{x + 4}$, the radical term, by subtracting $x + 5$ from both sides. We obtain

$$3x + 1 = x + 5 + 2\sqrt{x + 4} \qquad \text{This is the equation from our last step.}$$

$$2x - 4 = 2\sqrt{x + 4}. \qquad \text{Subtract } x \text{ and subtract 5 from both sides.}$$

Although we can simplify the equation by dividing both sides by 2, this sort of simplification is not always helpful. Thus, we will work with the equation in this form.

Repeat Step 2. **Square both sides.**

> Be careful in squaring both sides. Use $(A - B)^2 = A^2 - 2AB + B^2$ to square the left side. Use $(AB)^2 = A^2B^2$ to square the right side.

$$(2x - 4)^2 = \left(2\sqrt{x + 4}\right)^2 \quad \text{Square both sides.}$$

$$4x^2 - 16x + 16 = 4(x + 4) \quad \begin{array}{l}\text{Square both 2 and } \sqrt{x + 4} \text{ on}\\\text{the right side.}\end{array}$$

Step 3. **Solve the resulting equation.** We solve this quadratic equation by writing it in standard form.

$$4x^2 - 16x + 16 = 4x + 16 \quad \text{Use the distributive property.}$$
$$4x^2 - 20x = 0 \quad \text{Subtract } 4x + 16 \text{ from both sides.}$$
$$4x(x - 5) = 0 \quad \text{Factor.}$$
$$4x = 0 \quad \text{or} \quad x - 5 = 0 \quad \text{Set each factor equal to zero.}$$
$$x = 0 \qquad\qquad x = 5 \quad \text{Solve for x.}$$

Step 4. **Check the proposed solutions in the original equation.**

Check 0:	**Check 5:**
$\sqrt{3x + 1} - \sqrt{x + 4} = 1$	$\sqrt{3x + 1} - \sqrt{x + 4} = 1$
$\sqrt{3 \cdot 0 + 1} - \sqrt{0 + 4} \overset{?}{=} 1$	$\sqrt{3 \cdot 5 + 1} - \sqrt{5 + 4} \overset{?}{=} 1$
$\sqrt{1} - \sqrt{4} \overset{?}{=} 1$	$\sqrt{16} - \sqrt{9} \overset{?}{=} 1$
$1 - 2 \overset{?}{=} 1$	$4 - 3 \overset{?}{=} 1$
$-1 = 1, \;\;\text{false}$	$1 = 1, \;\;\text{true}$

The check indicates that 0 is not a solution. It is an extraneous solution brought about by squaring each side of the equation. The only solution is 5, and the solution set is $\{5\}$. ∎

 CHECK POINT **4** Solve: $\sqrt{x + 5} - \sqrt{x - 3} = 2$.   4 or $\{4\}$

EXAMPLE 5 Solving a Radical Equation

Solve: $(3x - 1)^{\frac{1}{3}} + 4 = 0$.

SOLUTION Although we can rewrite the equation in radical form

$$\sqrt[3]{3x - 1} + 4 = 0,$$

it is not necessary to do so. Because the equation involves a cube root, we isolate the radical term—that is, the term with the rational exponent—and cube both sides.

$$(3x - 1)^{\frac{1}{3}} + 4 = 0 \quad \text{This is the given equation.}$$

$$(3x - 1)^{\frac{1}{3}} = -4 \quad \begin{array}{l}\text{Subtract 4 from both sides and isolate the term}\\\text{with the rational exponent.}\end{array}$$

$$\left[(3x - 1)^{\frac{1}{3}}\right]^3 = (-4)^3 \quad \text{Cube both sides.}$$

$$3x - 1 = -64 \quad \text{Multiply exponents on the left side and simplify.}$$

$$3x = -63 \quad \text{Add 1 to both sides.}$$

$$x = -21 \quad \text{Divide both sides by 3.}$$

Because both sides were raised to an odd power, it is not essential to check the proposed solution, -21. However, checking is always a good idea. Do so now and verify that -21 is the solution and the solution set is $\{-21\}$. ∎

Example 5 illustrates that a radical equation with rational exponents can be solved by

1. isolating the expression with the rational exponent, and
2. raising both sides of the equation to a power that is the reciprocal of the rational exponent.

Keep in mind that it is essential to check when both sides have been raised to even powers. Thus, equations with rational exponents such as $\frac{1}{2}$ and $\frac{1}{4}$ must be checked.

 CHECK POINT 5 Solve: $(2x - 3)^{\frac{1}{3}} + 3 = 0.$ -12 or $\{-12\}$

---

2 Use models that are radical functions to solve problems.

Applications of Radical Equations   Radical equations can be solved to answer questions about variables contained in radical functions.

EXAMPLE 6   Living Alone

The bar graph in Figure 7.7 shows the number of people in the United States living alone for selected years from 1970 through 2000. The function

$$f(x) = 2.6\sqrt{x} + 11$$

models the number of Americans living alone, $f(x)$, in millions, x years after 1970. According to the model, when will 29 million Americans live alone?

SOLUTION   To find when 29 million Americans will live alone, substitute 29 for $f(x)$ in the given function. Then solve for x, the number of years after 1970.

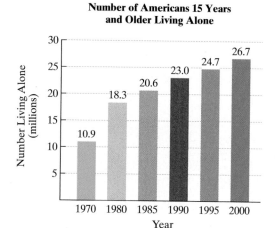

Number of Americans 15 Years and Older Living Alone

FIGURE 7.7

Source: U.S. Census Bureau

$$f(x) = 2.6\sqrt{x} + 11 \qquad \text{This is the given function.}$$

$$29 = 2.6\sqrt{x} + 11 \qquad \text{Substitute 29 for } f(x).$$

$$18 = 2.6\sqrt{x} \qquad \text{Subtract 11 from both sides.}$$

$$\frac{18}{2.6} = \sqrt{x} \qquad \text{Divide both sides by 2.6.}$$

$$\left(\frac{18}{2.6}\right)^2 = (\sqrt{x})^2 \qquad \text{Square both sides.}$$

$$48 \approx x \qquad \text{Use a calculator.}$$

The model indicates that 29 million Americans will live alone approximately 48 years after 1970. Because 1970 + 48 = 2018, this is predicted to occur in 2018.

CHECK POINT 6 The function $f(x) = 2.9\sqrt{x} + 11$ serves as another model for the number of Americans living alone, $f(x)$, in millions, x years after 1970. According to this model, when will 29 million Americans live alone? Use a calculator and round to the nearest year. Does this model predict faster or slower growth in the number of people living alone than the function used in Example 6?   2009; faster

7.6 EXERCISE SET

Student Solutions Manual   CD/Video   PH Math/Tutor Center   MathXL Tutorials on CD   MathXL®   MyMathLab   Interactmath.com

Practice Exercises
In Exercises 1–38, solve each radical equation.

1. $\sqrt{3x - 2} = 4$ 6 or {6}

2. $\sqrt{5x - 1} = 8$ 13 or {13}

3. $\sqrt{5x - 4} - 9 = 0$ 17 or {17}

4. $\sqrt{3x - 2} - 5 = 0$ 9 or {9}

5. $\sqrt{3x + 7} + 10 = 4$ no solution or ∅

6. $\sqrt{2x + 5} + 11 = 6$ no solution or ∅

7. $x = \sqrt{7x + 8}$ 8 or {8}

8. $x = \sqrt{6x + 7}$ 7 or {7}

9. $\sqrt{5x + 1} = x + 1$ 0 and 3, or {0, 3}

10. $\sqrt{2x + 1} = x - 7$ 12 or {12}

11. $x = \sqrt{2x - 2} + 1$ 1 and 3, or {1, 3}

12. $x = \sqrt{3x + 7} - 3$ −2 and −1, or {−2, −1}

13. $x - 2\sqrt{x - 3} = 3$ 3 and 7, or {3, 7}

14. $3x - \sqrt{3x + 7} = -5$ −1 or {−1}

15. $\sqrt{2x - 5} = \sqrt{x + 4}$ 9 or {9}

16. $\sqrt{6x + 2} = \sqrt{5x + 3}$ 1 or {1}

17. $\sqrt[3]{2x + 11} = 3$ 8 or {8}

18. $\sqrt[3]{6x - 3} = 3$ 5 or {5}

19. $\sqrt[3]{2x - 6} - 4 = 0$ 35 or {35}

20. $\sqrt[3]{4x - 3} - 5 = 0$ 32 or {32}

21. $\sqrt{x - 7} = 7 - \sqrt{x}$ 16 or {16}

22. $\sqrt{x - 8} = \sqrt{x} - 2$ 9 or {9}

23. $\sqrt{x + 2} + \sqrt{x - 1} = 3$ 2 or {2}

24. $\sqrt{x - 4} + \sqrt{x + 4} = 4$ 5 or {5}

25. $2\sqrt{4x + 1} - 9 = x - 5$ 2 and 6, or {2, 6}

26. $2\sqrt{x - 3} + 4 = x + 1$ 3 and 7, or {3, 7}

27. $(2x + 3)^{\frac{1}{3}} + 4 = 6$ $\frac{5}{2}$ or $\left\{\frac{5}{2}\right\}$

28. $(3x - 6)^{\frac{1}{3}} + 5 = 8$ 11 or {11}

29. $(3x + 1)^{\frac{1}{4}} + 7 = 9$ 5 or {5}

30. $(2x + 3)^{\frac{1}{4}} + 7 = 10$ 39 or {39}

31. $(x + 2)^{\frac{1}{2}} + 8 = 4$ no solution or ∅

32. $(x - 3)^{\frac{1}{2}} + 8 = 6$ no solution or ∅

33. $\sqrt{2x - 3} - \sqrt{x - 2} = 1$ 2 and 6, or {2, 6}

34. $\sqrt{x + 2} + \sqrt{3x + 7} = 1$ −2 or {−2}

35. $3x^{\frac{1}{3}} = (x^2 + 17x)^{\frac{1}{3}}$ 0 and 10, or {0, 10}

36. $2(x - 1)^{\frac{1}{3}} = (x^2 + 2x)^{\frac{1}{3}}$ 2 and 4, or {2, 4}

37. $(x + 8)^{\frac{1}{4}} = (2x)^{\frac{1}{4}}$ 8 or {8}

38. $(x - 2)^{\frac{1}{4}} = (3x - 8)^{\frac{1}{4}}$ 3 or {3}

Practice Plus

39. If $f(x) = x + \sqrt{x + 5}$, find all values of x for which $f(x) = 7$.   4

40. If $f(x) = x - \sqrt{x - 2}$, find all values of x for which $f(x) = 4$.   6

41. If $f(x) = (5x + 16)^{\frac{1}{3}}$ and $g(x) = (x - 12)^{\frac{1}{3}}$, find all values of x for which $f(x) = g(x)$.   −7

42. If $f(x) = (9x + 2)^{\frac{1}{4}}$ and $g(x) = (5x + 18)^{\frac{1}{4}}$, find all values of x for which $f(x) = g(x)$.   4

In Exercises 43–46, solve each formula for the specified variable.

43. $r = \sqrt{\dfrac{3V}{\pi h}}$ for V $V = \dfrac{\pi r^2 h}{3}$ or $\dfrac{1}{3}\pi r^2 h$

44. $r = \sqrt{\dfrac{A}{4\pi}}$ for A $A = 4\pi r^2$

45. $t = 2\pi\sqrt{\dfrac{l}{32}}$ for l $l = \dfrac{8t^2}{\pi^2}$

46. $v = \sqrt{\dfrac{FR}{m}}$ for m $m = \dfrac{FR}{v^2}$

47. If 5 times a number is decreased by 4, the principal square root of this difference is 2 less than the number. Find the number(s).   8

48. If a number is decreased by 3, the principal square root of this difference is 5 less than the number. Find the number(s).   7

In Exercises 49–50, find the x-intercept(s) of the graph of each function without graphing the function.

49. $f(x) = \sqrt{x + 16} - \sqrt{x} - 2$ 9

50. $f(x) = \sqrt{2x - 3} - \sqrt{2x} + 1$ 2

Application Exercises

In 2002, the average surface temperature on Earth was 57.9°F, approximately 1.4° higher than it was one hundred years ago. Worldwide temperatures have risen only 9°F since the end of the last ice age 12,000 years ago. Most climatologists are convinced that over the next one hundred years, global temperatures will continue to increase, possibly setting off a chain of devastating events beginning with a rise in sea levels worldwide and ending with the destruction of water supplies, forests, and agriculture in many parts of the world. The graph shows global annual average temperatures from 1880 through 2002, with projections from 2002 through 2100.

Global Annual Average Temperatures and Projections through 2100

Source: National Oceanic and Atmospheric Administration

The temperature projections shown in the graph can be modeled by two functions:

$f(x) = 0.083x + 57.9$ Models temperatures at the high end of the range

$g(x) = 0.36\sqrt{x} + 57.9$.   Models temperatures at the low end of the range

In these functions, $f(x)$ and $g(x)$ describe projected global annual average temperatures, in degrees Fahrenheit, x years after 2002, where $0 \le x \le 98$. Use the functions to solve Exercises 51–54.

51. Use f and g to determine the temperatures at the high and low end of the range of projected global average temperatures for 2100. Round to the nearest tenth of a degree.
High: 66.0°; Low: 61.5°

52. Use f and g to determine the temperatures at the high and low end of the range of projected global average temperatures for 2080. Round to the nearest tenth of a degree.
High: 64.4°; Low: 61.1°

53. Use f and g to determine in which years the projected global average temperature will exceed the 2002 average of 57.9° by one degree. Round to the nearest year.   Using f: 2014; Using g: 2010

54. Use f and g to determine in which years the projected global average temperature will exceed the 2002 average of 57.9° by two degrees.   Using f: 2026; Using g: 2033

Out of a group of 50,000 births, the number of people, $f(x)$, surviving to age x is modeled by the function

$$f(x) = 5000\sqrt{100 - x}.$$

The graph of the function is shown. Use the function to solve Exercises 55–56.

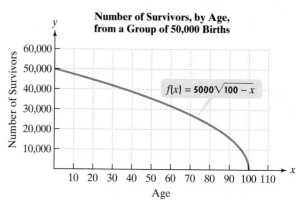

Number of Survivors, by Age, from a Group of 50,000 Births

$f(x) = 5000\sqrt{100 - x}$

55. To what age will 40,000 people in the group survive? Identify the solution as a point on the graph of the function.   40,000 people in the group will survive to age 36; (36, 40,000).

56. To what age will 35,000 people in the group survive? Identify the solution as a point on the graph of the function.   35,000 people in the group will survive to age 51; (51, 35,000).

The function

$$f(x) = 29x^{\frac{1}{3}}$$

models the number of plant species, $f(x)$, on the islands of the Galápagos in terms of the area, x, in square miles, of a particular island. Use the function to solve Exercises 57–58.

57. What is the area of a Galápagos island that has 87 species of plants?   27 square miles.

58. What is the area of a Galápagos island that has 58 species of plants?   8 square miles.

For each planet in our solar system, its year is the time it takes the planet to revolve once around the sun. The function

$$f(x) = 0.2x^{\frac{3}{2}}$$

models the number of Earth days in a planet's year, $f(x)$, where x is the average distance of the planet from the sun, in millions of kilometers. Use the function to solve Exercises 59–60.

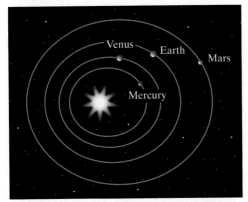

59. We, of course, have 365 Earth days in our year. What is the average distance of Earth from the sun? Use a calculator and round to the nearest million kilometers. The average distance of Earth from the sun is approximately 149 million kilometers.

60. There are approximately 88 Earth days in the year of the planet Mercury. What is the average distance of Mercury from the sun? Use a calculator and round to the nearest million kilometers. The average distance of Mercury from the sun is approximately 58 million kilometers.

Writing in Mathematics

61. What is a radical equation? 61–67. Answers will vary.
62. In solving $\sqrt{2x - 1} + 2 = x$, why is it a good idea to isolate the radical term? What if we don't do this and simply square each side? Describe what happens.
63. What is an extraneous solution to a radical equation?
64. Explain why $\sqrt{x} = -1$ has no solution.
65. Explain how to solve a radical equation with rational exponents.
66. The radical function in Example 6 on page 538 shows a gradual year-by-year increase in the number of Americans living alone. What explanations can you offer for this trend? Describe an event that might occur in the future that could cause this trend to change.
67. Describe the trend shown by the graph of f in Exercises 55–56. When is the rate of decrease most rapid? What does this mean about survival rate by age?

Technology Exercises

In Exercises 68–72, use a graphing utility to solve each radical equation. Graph each side of the equation in the given viewing rectangle. The equation's solution is given by the x-coordinate of the point(s) of intersection. Check by substitution. 68–72. See graphing answer section.

68. $\sqrt{2x + 2} = \sqrt{3x - 5}$
  $[-1, 10, 1]$ by $[-1, 5, 1]$ 7 or {7}

69. $\sqrt{x} + 3 = 5$
  $[-1, 6, 1]$ by $[-1, 6, 1]$ 4 or {4}

70. $\sqrt{x^2 + 3} = x + 1$
  $[-1, 6, 1]$ by $[-1, 6, 1]$ 1 or {1}

71. $4\sqrt{x} = x + 3$
  $[-1, 10, 1]$ by $[-1, 14, 1]$ 1 and 9, or {1, 9}

72. $\sqrt{x} + 4 = 2$
  $[-2, 18, 1]$ by $[0, 10, 1]$ no solution or ∅

Critical Thinking Exercises

73. Which one of the following is true? c
  a. The first step in solving $\sqrt{x + 6} = x + 2$ is to square both sides, obtaining $x + 6 = x^2 + 4$.
  b. The equations $\sqrt{x + 4} = -5$ and $x + 4 = 25$ have the same solution set.
  c. If $T = 2\pi\sqrt{\dfrac{L}{32}}$, then $L = \dfrac{8T^2}{\pi^2}$.
  d. The equation $\sqrt{x^2 + 9x + 3} = -x$ has no solution because a principal square root is always nonnegative.

74. Find the length of the three sides of the right triangle shown in the figure.

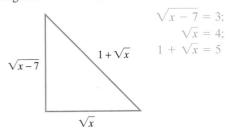

$\sqrt{x - 7} = 3$;
$\sqrt{x} = 4$;
$1 + \sqrt{x} = 5$

In Exercises 75–77, solve each equation.

75. $\sqrt[3]{x\sqrt{x}} = 9$ 81 or {81}
76. $\sqrt{\sqrt{x} + \sqrt{x + 9}} = 3$ 16 or {16}
77. $(x - 4)^{\frac{2}{3}} = 25$ 129 or {129}

Review Exercises

78. Divide using synthetic division:
  $(4x^4 - 3x^3 + 2x^2 - x - 1) \div (x + 3)$.
  (Section 6.5, Example 1) $4x^3 - 15x^2 + 47x - 142 + \dfrac{425}{x + 3}$

79. Divide:
  $$\dfrac{3x^2 - 12}{x^2 + 2x - 8} \div \dfrac{6x + 18}{x + 4}.$$
  (Section 6.1, Example 7) $\dfrac{x + 2}{2(x + 3)}$

80. Factor: $y^2 - 6y + 9 - 25x^2$.
  (Section 5.5, Example 6) $(y - 3 + 5x)(y - 3 - 5x)$

<table>
<tr><td>
SECTION

7.7

Objectives

1 Express square roots of negative numbers in terms of i.

2 Add and subtract complex numbers.

3 Multiply complex numbers.

4 Divide complex numbers.

5 Simplify powers of i.
</td></tr>
</table>

COMPLEX NUMBERS

Who is this kid warning us about our eyeballs turning black if we attempt to find the square root of -9? Don't believe what you hear on the street. Although square roots of negative numbers are not real numbers, they do play a significant role in algebra. In this section, we move beyond the real numbers and discuss square roots with negative radicands.

1 Express square roots of negative numbers in terms of i.

The Imaginary Unit i In Chapter 8, we will study equations whose solutions involve the square roots of negative numbers. Because the square of a real number is never negative, there is no real number x such that $x^2 = -1$. To provide a setting in which such equations have solutions, mathematicians invented an expanded system of numbers, the complex numbers. The *imaginary number i*, defined to be a solution of the equation $x^2 = -1$, is the basis of this new set.

THE IMAGINARY UNIT i   The **imaginary unit i** is defined as

$$i = \sqrt{-1}, \quad \text{where} \quad i^2 = -1.$$

Using the imaginary unit i, we can express the square root of any negative number as a real multiple of i. For example,

$$\sqrt{-25} = \sqrt{25(-1)} = \sqrt{25}\sqrt{-1} = 5i.$$

We can check that $\sqrt{-25} = 5i$ by squaring $5i$ and obtaining -25.

$$(5i)^2 = 5^2 i^2 = 25(-1) = -25$$

> **THE SQUARE ROOT OF A NEGATIVE NUMBER** If b is a positive real number, then
> $$\sqrt{-b} = \sqrt{b(-1)} = \sqrt{b}\sqrt{-1} = \sqrt{b}i \quad \text{or} \quad i\sqrt{b}.$$

EXAMPLE 1 Expressing Square Roots of Negative Numbers as Multiples of i

Write as a multiple of i:

   a. $\sqrt{-9}$ **b.** $\sqrt{-3}$ **c.** $\sqrt{-80}$.

SOLUTION

   a. $\sqrt{-9} = \sqrt{9(-1)} = \sqrt{9}\sqrt{-1} = 3i$

> Be sure not to write i under the radical.

   b. $\sqrt{-3} = \sqrt{3(-1)} = \sqrt{3}\sqrt{-1} = \sqrt{3}\,i$
   c. $\sqrt{-80} = \sqrt{80(-1)} = \sqrt{80}\sqrt{-1} = \sqrt{16 \cdot 5}\sqrt{-1} = 4\sqrt{5}\,i$

✔ **CHECK POINT 1** Write as a multiple of i:

   a. $\sqrt{-64}$ $8i$ **b.** $\sqrt{-11}$ $\sqrt{11}i$ **c.** $\sqrt{-48}$. $4\sqrt{3}i$

A new system of numbers, called *complex numbers*, is based on adding multiples of i, such as $5i$, to the real numbers.

FIGURE 7.8 The complex number system

> **COMPLEX NUMBERS AND IMAGINARY NUMBERS** The set of all numbers in the form
> $$a + bi$$
> with real numbers a and b, and i, the imaginary unit, is called the set of **complex numbers**. The real number a is called the **real part**, and the real number b is called the **imaginary part** of the complex number $a + bi$. If $b \neq 0$, then the complex number is called an **imaginary number** (Figure 7.8).

Here are some examples of complex numbers. Each number can be written in the form $a + bi$.

Can you see that b, the imaginary part, is not zero in the first two complex numbers? Because $b \neq 0$, these complex numbers are imaginary numbers. By contrast, the imaginary part of the complex number on the right is zero. This complex number is not an imaginary number. The number 3, or $3 + 0i$, is a real number.

Adding and Subtracting Complex Numbers The form of a complex number $a + bi$ is like the binomial $a + bx$. Consequently, we can add, subtract, and multiply complex numbers using the same methods we used for binomials, remembering that $i^2 = -1$.

2   Add and subtract complex numbers.

ADDING AND SUBTRACTING COMPLEX NUMBERS

1. $(a + bi) + (c + di) = (a + c) + (b + d)i$
   In words, this says that you add complex numbers by adding their real parts, adding their imaginary parts, and expressing the sum as a complex number.
2. $(a + bi) - (c + di) = (a - c) + (b - d)i$
   In words, this says that you subtract complex numbers by subtracting their real parts, subtracting their imaginary parts, and expressing the difference as a complex number.

EXAMPLE 2 Adding and Subtracting Complex Numbers

Perform the indicated operations, writing the result in the form $a + bi$:

   a. $(5 - 11i) + (7 + 4i)$ **b.** $(-5 + 7i) - (-11 - 6i)$.

SOLUTION

a. $(5 - 11i) + (7 + 4i)$
   $= 5 - 11i + 7 + 4i$ Remove the parentheses.
   $= 5 + 7 - 11i + 4i$ Group real and imaginary terms.
   $= (5 + 7) + (-11 + 4)i$ Add real parts and add imaginary parts.
   $= 12 - 7i$ Simplify.

b. $(-5 + 7i) - (-11 - 6i)$
   $= -5 + 7i + 11 + 6i$ Remove the parentheses. Change signs of real and imaginary parts in the complex number being subtracted.
   $= -5 + 11 + 7i + 6i$ Group real and imaginary terms.
   $= (-5 + 11) + (7 + 6)i$ Add real parts and add imaginary parts.
   $= 6 + 13i$ Simplify.

STUDY TIP

The following examples, using the same integers as in Example 2, show how operations with complex numbers are just like operations with polynomials.

a. $(5 - 11x) + (7 + 4x)$
   $= 12 - 7x$
b. $(-5 + 7x) - (-11 - 6x)$
   $= -5 + 7x + 11 + 6x$
   $= 6 + 13x$

CHECK POINT 2 Add or subtract as indicated:

   a. $(5 - 2i) + (3 + 3i)$ $8 + i$
   b. $(2 + 6i) - (12 - 4i)$. $-10 + 10i$

---

3 Multiply complex numbers.

Multiplying Complex Numbers   Multiplication of complex numbers is performed the same way as multiplication of polynomials, using the distributive property and the FOIL method. After completing the multiplication, we replace i^2 with -1. This idea is illustrated in the next example.

EXAMPLE 3 Multiplying Complex Numbers

Find the products:

   a. $4i(3 - 5i)$ **b.** $(7 - 3i)(-2 - 5i)$.

SOLUTION

a. $4i(3 - 5i)$
   $= 4i \cdot 3 - 4i \cdot 5i$ Distribute $4i$ throughout the parentheses.
   $= 12i - 20i^2$ Multiply.
   $= 12i - 20(-1)$ Replace i^2 with -1.
   $= 20 + 12i$ Simplify to $12i + 20$ and write in $a + bi$ form.

b. $(7 - 3i)(-2 - 5i)$

F   O   I   L

$= -14 - 35i + 6i + 15i^2.$ Use the FOIL method.

$= -14 - 35i + 6i + 15(-1)$ $i^2 = -1$

$= -14 - 15 - 35i + 6i$ Group real and imaginary terms.

$= -29 - 29i$ Combine real and imaginary terms.   ■

 CHECK POINT 3 Find the products:

   a. $7i(2 - 9i)$ $63 + 14i$ **b.** $(5 + 4i)(6 - 7i).$ $58 - 11i$

Consider the multiplication problem

$$5i \cdot 2i = 10i^2 = 10(-1) = -10.$$

This problem can also be given in terms of square roots of negative numbers:

$$\sqrt{-25} \cdot \sqrt{-4}.$$

Because the product rule for radicals only applies to real numbers, multiplying radicands is incorrect. **When performing operations with square roots of negative numbers, begin by expressing all square roots in terms of i.** Then perform the indicated operation.

CORRECT: **INCORRECT!**

$\sqrt{-25} \cdot \sqrt{-4} = \sqrt{25}\sqrt{-1} \cdot \sqrt{4}\sqrt{-1}$ $\cancel{\sqrt{-25} \cdot \sqrt{-4} = \sqrt{(-25)(-4)}}$

$= 5i \cdot 2i$ $\cancel{= \sqrt{100}}$

$= 10i^2 = 10(-1) = -10$ $\cancel{= 10}$

EXAMPLE 4   Multiplying Square Roots of Negative Numbers

Multiply: $\sqrt{-3} \cdot \sqrt{-5}.$

SOLUTION

$$\sqrt{-3} \cdot \sqrt{-5} = \sqrt{3}\sqrt{-1} \cdot \sqrt{5}\sqrt{-1}$$

$$= \sqrt{3}i \cdot \sqrt{5}i \qquad \text{Express square roots in terms of } i.$$

$$= \sqrt{15}i^2 \qquad \sqrt{3} \cdot \sqrt{5} = \sqrt{15} \text{ and } i \cdot i = i^2.$$

$$= \sqrt{15}(-1) \qquad i^2 = -1$$

$$= -\sqrt{15} \qquad\qquad\qquad\qquad\qquad ■$$

 CHECK POINT 4 Multiply: $\sqrt{-5} \cdot \sqrt{-7}.$ $-\sqrt{35}$

Conjugates and Division   It is possible to multiply imaginary numbers and obtain a real number. Here is an example:

4   Divide complex numbers.

F   O   I   L

$$(4 + 7i)(4 - 7i) = 16 - 28i + 28i - 49i^2$$

$$= 16 - 49i^2 = 16 - 49(-1) = 65.$$

Replace i^2 with -1.

You can also perform $(4 + 7i)(4 - 7i)$ using the formula

$$(A + B)(A - B) = A^2 - B^2.$$

A real number is obtained even faster:

$$(4 + 7i)(4 - 7i) = 4^2 - (7i)^2 = 16 - 49i^2 = 16 - 49(-1) = 65.$$

The **conjugate** of the complex number $a + bi$ is $a - bi$. The **conjugate** of the complex number $a - bi$ is $a + bi$. The multiplication problem that we just performed involved conjugates. The multiplication of conjugates always results in a real number:

$$(a + bi)(a - bi) = a^2 - (bi)^2 = a^2 - b^2i^2 = a^2 - b^2(-1) = a^2 + b^2.$$

> The product eliminates i.

Conjugates are used to divide complex numbers. By multiplying the numerator and the denominator of the division by the conjugate of the denominator, you will obtain a real number in the denominator. Here are two examples of such divisions:

$$\bullet \; \frac{7 + 4i}{2 - 5i} \qquad \bullet \; \frac{5i - 4}{3i} \text{ or } \frac{5i - 4}{0 + 3i}.$$

> The conjugate of the denominator is $2 + 5i$.

> The conjugate of the denominator is $0 - 3i$, or $-3i$.

The procedure for dividing complex numbers, illustrated in Examples 5 and 6, should remind you of rationalizing denominators.

EXAMPLE 5 Using Conjugates to Divide Complex Numbers

Divide and simplify to the form $a + bi$:

$$\frac{7 + 4i}{2 - 5i}.$$

SOLUTION The conjugate of the denominator is $2 + 5i$. Multiplication of both the numerator and the denominator by $2 + 5i$ will eliminate i from the denominator.

$$\frac{7 + 4i}{2 - 5i} = \frac{7 + 4i}{2 - 5i} \cdot \frac{2 + 5i}{2 + 5i} \qquad \text{Multiply by 1.}$$

$$= \frac{\overset{F}{14} + \overset{O}{35i} + \overset{I}{8i} + \overset{L}{20i^2}}{2^2 - (5i)^2} \qquad \begin{array}{l}\text{Use FOIL in the numerator and} \\ (A - B)(A + B) = A^2 - B^2 \text{ in} \\ \text{the denominator.}\end{array}$$

$$= \frac{14 + 43i + 20i^2}{4 - 25i^2} \qquad \text{Simplify.}$$

$$= \frac{14 + 43i + 20(-1)}{4 - 25(-1)} \qquad i^2 = -1$$

$$= \frac{14 + 43i - 20}{4 + 25} \qquad \text{Perform the multiplications involving } -1.$$

$$= \frac{-6 + 43i}{29} \qquad \begin{array}{l}\text{Combine real terms in the numerator} \\ \text{and denominator.}\end{array}$$

$$= -\frac{6}{29} + \frac{43}{29}i \qquad \text{Express the answer in the form } a + bi.$$

CHECK POINT 5 Divide and simplify to the form $a + bi$:

$$\frac{6 + 2i}{4 - 3i}. \qquad \frac{18}{25} + \frac{26}{25}i$$

EXAMPLE 6 Using Conjugates to Divide Complex Numbers

Divide and simplify to the form $a + bi$:

$$\frac{5i - 4}{3i}.$$

SOLUTION The conjugate of the denominator, $0 + 3i$, is $0 - 3i$. Multiplication of both the numerator and the denominator by $-3i$ will eliminate i from the denominator.

$$\frac{5i - 4}{3i} = \frac{5i - 4}{3i} \cdot \frac{-3i}{-3i} \qquad \text{Multiply by 1.}$$

$$= \frac{-15i^2 + 12i}{-9i^2} \qquad \text{Multiply. Use the distributive property in the numerator.}$$

$$= \frac{-15(-1) + 12i}{-9(-1)} \qquad i^2 = -1$$

$$= \frac{15 + 12i}{9} \qquad \text{Perform the multiplications involving } -1.$$

$$= \frac{15}{9} + \frac{12}{9}i \qquad \text{Express the division in the form } a + bi.$$

$$= \frac{5}{3} + \frac{4}{3}i \qquad \text{Simplify real and imaginary parts.} \qquad \blacksquare$$

 CHECK POINT 6 Divide and simplify to the form $a + bi$:

$$\frac{3 - 2i}{4i}. \qquad -\frac{1}{2} - \frac{3}{4}i$$

5 Simplify powers of i.

Powers of i Cycles govern many aspects of life—heartbeats, sleep patterns, seasons, and tides all follow regular, predictable cycles. Surprisingly, so do powers of i. To see how this occurs, use the fact that $i^2 = -1$ and express each power of i in terms of i^2:

$$i$$

$$i^2 = -1$$

$$i^3 = i^2 \cdot i = (-1)i = -i$$

$$i^4 = (i^2)^2 = (-1)^2 = 1$$

$$i^5 = i^4 \cdot i = (i^2)^2 \cdot i = (-1)^2 \cdot i = i$$

$$i^6 = (i^2)^3 = (-1)^3 = -1.$$

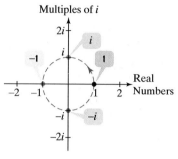

Multiples of i

FIGURE 7.9

Can you see that the powers of i are cycling through the values $i, -1, -i$, and 1? The cycle is illustrated in Figure 7.9. In the figure, tick marks on the horizontal axis represent real numbers and tick marks along the vertical axis represent multiples of i. Using this representation, powers of i are equally spaced at 90° intervals on a circle having radius 1.

Here is a procedure for simplifying powers of i:

SIMPLIFYING POWERS OF i

1. Express the given power of i in terms of i^2.
2. Replace i^2 with -1 and simplify. Use the fact that -1 to an even power is 1 and -1 to an odd power is -1.

EXAMPLE 7 Simplifying Powers of i

Simplify:

    a. i^{12} **b.** i^{39} **c.** i^{50}.

SOLUTION

    a. $i^{12} = (i^2)^6 = (-1)^6 = 1$

    b. $i^{39} = i^{38}i = (i^2)^{19}i = (-1)^{19}i = (-1)i = -i$

    c. $i^{50} = (i^2)^{25} = (-1)^{25} = -1$

✔ CHECK POINT 7 Simplify:

    a. i^{16} 1   **b.** i^{25} i **c.** i^{35}. $-i$

ENRICHMENT ESSAY

The Patterns of Chaos

One of the new frontiers of mathematics suggests that there is an underlying order in things that appear to be random, such as the hiss and crackle of background noises as you tune a radio. Irregularities in the heartbeat, some of them severe enough to cause a heart attack, or irregularities in our sleeping patterns, such as insomnia, are examples of chaotic behavior. Chaos in the mathematical sense does not mean a complete lack of form or arrangement. In mathematics, chaos is used to describe something that appears to be random but is not actually random. The patterns of chaos appear in images like the one on the left, called the Mandelbrot set. Magnified portions of this image yield repetitions of the original structure, as well as new and unexpected patterns. The Mandelbrot set transforms the hidden structure of chaotic events into a source of wonder and inspiration.

The Mandelbrot set is made possible by opening up graphing to include complex numbers in the form $a + bi$. Although the details are beyond the scope of this text, the coordinate system that is used is shown in Figure 7.9. Plot certain complex numbers in this system, add color to the magnified boundary of the graph, and the patterns of chaos begin to appear.

R. F. Voss "29-Fold M-set Seahorse"
computer-generated image

7.7 EXERCISE SET

Student Solutions Manual   CD/Video   PH Math/Tutor Center   MathXL Tutorials on CD   MathXL®   MyMathLab   Interactmath.com

Practice Exercises

In Exercises 1–16, express each number in terms of i and simplify, if possible.

1. $\sqrt{-100}$ $10i$

2. $\sqrt{-49}$ $7i$

3. $\sqrt{-23}$ $\sqrt{23}i$

4. $\sqrt{-21}$ $\sqrt{21}i$

5. $\sqrt{-18}$ $3\sqrt{2}i$

6. $\sqrt{-125}$ $5\sqrt{5}i$

7. $\sqrt{-63}$ $3\sqrt{7}i$

8. $\sqrt{-28}$ $2\sqrt{7}i$

9. $-\sqrt{-108}$ $-6\sqrt{3}i$

10. $-\sqrt{-300}$ $-10\sqrt{3}i$

11. $5 + \sqrt{-36}$ $5 + 6i$

12. $7 + \sqrt{-4}$ $7 + 2i$

13. $15 + \sqrt{-3}$ $15 + \sqrt{3}i$

14. $20 + \sqrt{-5}$ $20 + \sqrt{5}i$

15. $-2 - \sqrt{-18}$ $-2 - 3\sqrt{2}i$

16. $-3 - \sqrt{-27}$ $-3 - 3\sqrt{3}i$

In Exercises 17–32, add or subtract as indicated. Write the result in the form a + bi.

17. $(3 + 2i) + (5 + i)$ $8 + 3i$

18. $(6 + 5i) + (4 + 3i)$ $10 + 8i$

19. $(7 + 2i) + (1 - 4i)$ $8 - 2i$

20. $(-2 + 6i) + (4 - i)$ $2 + 5i$

21. $(10 + 7i) - (5 + 4i)$ $5 + 3i$

22. $(11 + 8i) - (2 + 5i)$ $9 + 3i$

23. $(9 - 4i) - (10 + 3i)$ $-1 - 7i$

24. $(8 - 5i) - (6 + 2i)$ $2 - 7i$

25. $(3 + 2i) - (5 - 7i)$ $-2 + 9i$

26. $(-7 + 5i) - (9 - 11i)$ $-16 + 16i$

27. $(-5 + 4i) - (-13 - 11i)$ $8 + 15i$

28. $(-9 + 2i) - (-17 - 6i)$ $8 + 8i$

29. $8i - (14 - 9i)$ $-14 + 17i$

30. $15i - (12 - 11i)$ $-12 + 26i$

31. $\left(2 + \sqrt{3}i\right) + \left(7 + 4\sqrt{3}i\right)$ $9 + 5\sqrt{3}i$

32. $\left(4 + \sqrt{5}i\right) + \left(8 + 6\sqrt{5}i\right)$ $12 + 7\sqrt{5}i$

In Exercises 33–62, find each product. Write imaginary results in the form a + bi.

33. $2i(5 + 3i)$ $-6 + 10i$

34. $5i(4 + 7i)$ $-35 + 20i$

35. $3i(7i - 5)$ $-21 - 15i$

36. $8i(4i - 3)$ $-32 - 24i$

37. $-7i(2 - 5i)$ $-35 - 14i$

38. $-6i(3 - 5i)$ $-30 - 18i$

39. $(3 + i)(4 + 5i)$ $7 + 19i$

40. $(4 + i)(5 + 6i)$ $14 + 29i$

41. $(7 - 5i)(2 - 3i)$ $-1 - 31i$

42. $(8 - 4i)(3 - 2i)$ $16 - 28i$

43. $(6 - 3i)(-2 + 5i)$ $3 + 36i$

44. $(7 - 2i)(-3 + 6i)$ $-9 + 48i$

45. $(3 + 5i)(3 - 5i)$ 34

46. $(2 + 7i)(2 - 7i)$ 53

47. $(-5 + 3i)(-5 - 3i)$ 34

48. $(-4 + 2i)(-4 - 2i)$ 20

49. $\left(3 - \sqrt{2}i\right)\left(3 + \sqrt{2}i\right)$ 11

50. $\left(5 - \sqrt{3}i\right)\left(5 + \sqrt{3}i\right)$ 28

51. $(2 + 3i)^2$ $-5 + 12i$

52. $(3 + 2i)^2$ $5 + 12i$

53. $(5 - 2i)^2$ $21 - 20i$

54. $(5 - 3i)^2$ $16 - 30i$

55. $\sqrt{-7} \cdot \sqrt{-2}$ $-\sqrt{14}$

56. $\sqrt{-7} \cdot \sqrt{-3}$ $-\sqrt{21}$

57. $\sqrt{-9} \cdot \sqrt{-4}$ -6

58. $\sqrt{-16} \cdot \sqrt{-4}$ -8

59. $\sqrt{-7} \cdot \sqrt{-25}$ $-5\sqrt{7}$

60. $\sqrt{-3} \cdot \sqrt{-36}$ $-6\sqrt{3}$

61. $\sqrt{-8} \cdot \sqrt{-3}$ $-2\sqrt{6}$

62. $\sqrt{-9} \cdot \sqrt{-5}$ $-3\sqrt{5}$

In Exercises 63–84, divide and simplify to the form a + bi.

63. $\dfrac{2}{3 + i}$ $\dfrac{3}{5} - \dfrac{1}{5}i$

64. $\dfrac{3}{4 + i}$ $\dfrac{12}{17} - \dfrac{3}{17}i$

65. $\dfrac{2i}{1 + i}$ $1 + i$

66. $\dfrac{5i}{2 + i}$ $1 + 2i$

67. $\dfrac{7}{4 - 3i}$ $\dfrac{28}{25} + \dfrac{21}{25}i$

68. $\dfrac{9}{1 - 2i}$ $\dfrac{9}{5} + \dfrac{18}{5}i$

69. $\dfrac{6i}{3 - 2i}$ $-\dfrac{12}{13} + \dfrac{18}{13}i$

70. $\dfrac{5i}{2 - 3i}$ $-\dfrac{15}{13} + \dfrac{10}{13}i$

71. $\dfrac{1 + i}{1 - i}$ $0 + i$ or i

72. $\dfrac{1 - i}{1 + i}$ $0 - i$ or $-i$

73. $\dfrac{2 - 3i}{3 + i}$ $\dfrac{3}{10} - \dfrac{11}{10}i$

74. $\dfrac{2 + 3i}{3 - i}$ $\dfrac{3}{10} + \dfrac{11}{10}i$

75. $\dfrac{5 - 2i}{3 + 2i}$ $\dfrac{11}{13} - \dfrac{16}{13}i$

76. $\dfrac{6 - 3i}{4 + 2i}$ $\dfrac{9}{10} - \dfrac{6}{5}i$

77. $\dfrac{4 + 5i}{3 - 7i}$ $-\dfrac{23}{58} + \dfrac{43}{58}i$

78. $\dfrac{5 - i}{3 - 2i}$ $\dfrac{17}{13} + \dfrac{7}{13}i$

79. $\dfrac{7}{3i}$ $0 - \dfrac{7}{3}i$ or $-\dfrac{7}{3}i$

80. $\dfrac{5}{7i}$ $0 - \dfrac{5}{7}i$ or $-\dfrac{5}{7}i$

81. $\dfrac{8 - 5i}{2i}$ $-\dfrac{5}{2} - 4i$

82. $\dfrac{3 + 4i}{5i}$ $\dfrac{4}{5} - \dfrac{3}{5}i$

83. $\dfrac{4 + 7i}{-3i}$ $-\dfrac{7}{3} + \dfrac{4}{3}i$

84. $\dfrac{5 + i}{-4i}$ $-\dfrac{1}{4} + \dfrac{5}{4}i$

In Exercises 85–100, simplify each expression.

85. i^{10} -1

86. i^{14} -1

87. i^{11} $-i$

88. i^{15} $-i$

89. i^{22} -1

90. i^{46} -1

91. i^{200} 1

92. i^{400} 1

93. i^{17} i

94. i^{21} i

95. $(-i)^4$ 1

96. $(-i)^6$ -1

97. $(-i)^9$ $-i$

98. $(-i)^{13}$ $-i$

99. $i^{24} + i^2$ 0

100. $i^{28} + i^{30}$ 0

Practice Plus

In Exercises 101–108, perform the indicated operation(s) and write the result in the form a + bi.

101. $(2 - 3i)(1 - i) - (3 - i)(3 + i)$ $-11 - 5i$

102. $(8 + 9i)(2 - i) - (1 - i)(1 + i)$ $23 + 10i$

103. $(2 + i)^2 - (3 - i)^2$ $-5 + 10i$

104. $(4 - i)^2 - (1 + 2i)^2$ $18 - 12i$

105. $5\sqrt{-16} + 3\sqrt{-81}$ $0 + 47i$ or $47i$

106. $5\sqrt{-8} + 3\sqrt{-18}$ $0 + 19\sqrt{2}i$ or $19\sqrt{2}i$

107. $\dfrac{i^4 + i^{12}}{i^8 - i^7}$ $1 - i$

108. $\dfrac{i^8 + i^{40}}{i^4 + i^3}$ $1 + i$

109. Let $f(x) = x^2 - 2x + 2$. Find $f(1 + i)$. 0

110. Let $f(x) = x^2 - 2x + 5$. Find $f(1 - 2i)$. 0

In Exercises 111–114, simplify each evaluation to the form a + bi.

111. Let $f(x) = x - 3i$ and $g(x) = 4x + 2i$. Find $(fg)(-1)$.
$10 + 10i$

112. Let $f(x) = 12x - i$ and $g(x) = 6x + 3i$. Find $(fg)\left(-\dfrac{1}{3}\right)$.
$11 - 10i$

113. Let $f(x) = \dfrac{x^2 + 19}{2 - x}$. Find $f(3i)$. $\dfrac{20}{13} + \dfrac{30}{13}i$

114. Let $f(x) = \dfrac{x^2 + 11}{3 - x}$. Find $f(4i)$. $-\dfrac{3}{5} - \dfrac{4}{5}i$

Application Exercises

Complex numbers are used in electronics to describe the current in an electric circuit. Ohm's law relates the current in a circuit, I, in amperes, the voltage of the circuit, E, in volts, and the resistance of the circuit, R, in ohms, by the formula E = IR. Use this formula to solve Exercises 115–116.

115. Find E, the voltage of a circuit, if $I = (4 - 5i)$ amperes and $R = (3 + 7i)$ ohms. $(47 + 13i)$ volts

116. Find E, the voltage of a circuit, if $I = (2 - 3i)$ amperes and $R = (3 + 5i)$ ohms. $(21 + i)$ volts

117. The mathematician Girolamo Cardano is credited with the first use (in 1545) of negative square roots in solving the now-famous problem, "Find two numbers whose sum is 10 and whose product is 40." Show that the complex numbers $5 + \sqrt{15}i$ and $5 - \sqrt{15}i$ satisfy the conditions of the problem. (Cardano did not use the symbolism $\sqrt{15}i$ or even $\sqrt{-15}$. He wrote R.m 15 for $\sqrt{-15}$, meaning "radix minus 15." He regarded the numbers $5 +$ R.m 15 and $5 -$ R.m 15 as "fictitious" or "ghost numbers," and considered the problem "manifestly impossible." But in a mathematically adventurous spirit, he exclaimed, "Nevertheless, we will operate.") $\left(5 + \sqrt{15}i\right) + \left(5 - \sqrt{15}i\right) = 10;$
$\left(5 + \sqrt{15}i\right)\left(5 - \sqrt{15}i\right) = 25 - 15i^2 = 25 + 15 = 40$

Writing in Mathematics

118. What is the imaginary unit i? **118–130.** Answers will vary.

119. Explain how to write $\sqrt{-64}$ as a multiple of i.

120. What is a complex number? Explain when a complex number is a real number and when it is an imaginary number. Provide examples with your explanation.

121. Explain how to add complex numbers. Give an example.

122. Explain how to subtract complex numbers. Give an example.

123. Explain how to find the product of $2i$ and $5 + 3i$.

124. Explain how to find the product of $2i + 3$ and $5 + 3i$.

125. Explain how to find the product of $2i + 3$ and $2i - 3$.

126. Explain how to find the product of $\sqrt{-1}$ and $\sqrt{-4}$. Describe a common error in the multiplication that needs to be avoided.

127. What is the conjugate of $2 + 3i$? What happens when you multiply this complex number by its conjugate?

128. Explain how to divide complex numbers. Provide an example with your explanation.

129. Explain each of the three jokes in the cartoon on page 542.

130. A stand-up comedian uses algebra in some jokes, including one about a telephone recording that announces "You have just reached an imaginary number. Please multiply by i and dial again." Explain the joke.

Explain the error in Exercises 131–132.

131. $\sqrt{-9} + \sqrt{-16} = \sqrt{-25} = \sqrt{25}i = 5i$
$\sqrt{-9} + \sqrt{-16} = 3i + 4i = 7i$

132. $\left(\sqrt{-9}\right)^2 = \sqrt{-9} \cdot \sqrt{-9} = \sqrt{81} = 9$
$\left(\sqrt{-9}\right)^2 = \sqrt{-9} \cdot \sqrt{-9} = 3i \cdot 3i = 9i^2 = -9$

Critical Thinking Exercises

133. Which one of the following is true? **d**

    a. Some irrational numbers are not complex numbers.

    b. $(3 + 7i)(3 - 7i)$ is an imaginary number.

    c. $\dfrac{7 + 3i}{5 + 3i} = \dfrac{7}{5}$

    d. In the complex number system, $x^2 + y^2$ (the sum of two squares) can be factored as $(x + yi)(x - yi)$.

In Exercises 134–136, perform the indicated operations and write the result in the form a + bi.

134. $\dfrac{4}{(2 + i)(3 - i)}$ $\dfrac{14}{25} - \dfrac{2}{25}i$

135. $\dfrac{1 + i}{1 + 2i} + \dfrac{1 - i}{1 - 2i}$ $\dfrac{6}{5} + 0i$ or $\dfrac{6}{5}$

136. $\dfrac{8}{1 + \dfrac{2}{i}}$ $\dfrac{8}{5} + \dfrac{16}{5}i$

Review Exercises

137. Simplify:

$$\dfrac{\dfrac{x}{y^2} + \dfrac{1}{y}}{\dfrac{y}{x^2} + \dfrac{1}{x}}.$$

(Section 6.3, Example 1) $\dfrac{x^2}{y^2}$

138. Solve for x: $\dfrac{1}{x} + \dfrac{1}{y} = \dfrac{1}{z}$.

(Section 6.7, Example 1) $x = \dfrac{yz}{y - z}$

139. Solve:

$$2x - \dfrac{x - 3}{8} = \dfrac{1}{2} + \dfrac{x + 5}{2}.$$

(Section 1.4, Example 4) $\dfrac{21}{11}$ or $\left\{\dfrac{21}{11}\right\}$

GROUP PROJECT

Additional group projects can be found in the *Instructor's Resource Manual*.

Group members should consult an almanac, newspaper, magazine, or the Internet and return to the group with as much data as possible that show phenomena that are continuing to grow over time, but whose growth is leveling off. Select the five data sets that you find most intriguing. Let x represent the number of years after the first year in each data set. Model the data by hand using

$$f(x) = a\sqrt{x} + b.$$

Use the first and last data points to find values for a and b. The first data point corresponds to $x = 0$. Its second coordinate gives the value of b. To find a, substitute the second data point into $f(x) = a\sqrt{x} + b$, with the value that you obtained for b. Now solve the equation and find a. Substitute a and b into $f(x) = a\sqrt{x} + b$ to obtain a square root function that models each data set. Then use the function to make predictions about what might occur in the future. Are there circumstances that might affect the accuracy of the predictions? List some of these circumstances.

CHAPTER 7 SUMMARY

Definitions and Concepts	Examples

Section 7.1 Radical Expressions and Functions

Definitions and Concepts	Examples
If $b^2 = a$, then b is a square root of a. The principal square root of a, designated $\sqrt{a}$, is the nonnegative number satisfying $b^2 = a$. The negative square root of a is written $-\sqrt{a}$. A square root of a negative number is not a real number. A radical function in x is a function defined by an expression containing a root of x. The domain of a square root function is the set of real numbers for which the radicand is nonnegative.	Let $f(x) = \sqrt{6 - 2x}$. $f(-15) = \sqrt{6 - 2(-15)} = \sqrt{6 + 30} = \sqrt{36} = 6$ $f(5) = \sqrt{6 - 2 \cdot 5} = \sqrt{6 - 10} = \sqrt{-4}$, not a real number Domain of f: Set the radicand greater than or equal to zero. $$6 - 2x \geq 0$$ $$-2x \geq -6$$ $$x \leq 3$$ Domain of $f = \{x \mid x \leq 3\}$ or $(-\infty, 3]$.
The cube root of a real number a is written $\sqrt[3]{a}$. $$\sqrt[3]{a} = b \quad \text{means that} \quad b^3 = a.$$ The nth root of a real number a is written $\sqrt[n]{a}$. The number n is the index. Every real number has one root when n is odd. The odd nth root of a, $\sqrt[n]{a}$, is the number b for which $b^n = a$. Every positive real number has two real roots when n is even. An even root of a negative number is not a real number. If n is even, then $\sqrt[n]{a^n} = \lvert a \rvert$. If n is odd, then $\sqrt[n]{a^n} = a$.	• $\sqrt[3]{-8} = -2$ because $(-2)^3 = -8$. • $\sqrt[4]{-16}$ is not a real number. • $\sqrt{x^2 - 14x + 49} = \sqrt{(x - 7)^2} = \lvert x - 7 \rvert$ • $\sqrt[3]{125(x + 6)^3} = 5(x + 6)$

Definitions and Concepts	Examples

Section 7.2 Rational Exponents

- $a^{\frac{1}{n}} = \sqrt[n]{a}$

- $a^{\frac{m}{n}} = (\sqrt[n]{a})^m \text{ or } \sqrt[n]{a^m}$

- $a^{-\frac{m}{n}} = \dfrac{1}{a^{\frac{m}{n}}}$

- $121^{\frac{1}{2}} = \sqrt{121} = 11$

- $64^{\frac{1}{3}} = \sqrt[3]{64} = 4$

- $27^{\frac{5}{3}} = \left(\sqrt[3]{27}\right)^5 = 3^5 = 3 \cdot 3 \cdot 3 \cdot 3 \cdot 3 = 243$

- $16^{-\frac{3}{4}} = \dfrac{1}{16^{\frac{3}{4}}} = \dfrac{1}{\left(\sqrt[4]{16}\right)^3} = \dfrac{1}{2^3} = \dfrac{1}{8}$

- $\left(\sqrt[3]{7xy}\right)^4 = (7xy)^{\frac{4}{3}}$

Properties of integer exponents are true for rational exponents. An expression with rational exponents is simplified when no parentheses appear, no powers are raised to powers, each base occurs once, and no negative or zero exponents appear.

Simplify: $\left(8x^{\frac{1}{3}}y^{-\frac{1}{2}}\right)^{\frac{1}{3}}$.

$$= 8^{\frac{1}{3}}\left(x^{\frac{1}{3}}\right)^{\frac{1}{3}}\left(y^{-\frac{1}{2}}\right)^{\frac{1}{3}}$$

$$= 2x^{\frac{1}{9}}y^{-\frac{1}{6}} = \dfrac{2x^{\frac{1}{9}}}{y^{\frac{1}{6}}}$$

Some radical expressions can be simplifed using rational exponents. Rewrite the expression using rational exponents, simplify, and rewrite in radical notation if rational exponents still appear.

- $\sqrt[9]{x^3} = x^{\frac{3}{9}} = x^{\frac{1}{3}} = \sqrt[3]{x}$

- $\sqrt[5]{x^2} \cdot \sqrt[4]{x} = x^{\frac{2}{5}} \cdot x^{\frac{1}{4}} = x^{\frac{2}{5}+\frac{1}{4}}$

$$= x^{\frac{8}{20}+\frac{5}{20}} = x^{\frac{13}{20}} = \sqrt[20]{x^{13}}$$

Section 7.3 Multiplying and Simplifying Radical Expressions

The product rule for radicals can be used to multiply radicals with the same indices:
$$\sqrt[n]{a} \cdot \sqrt[n]{b} = \sqrt[n]{ab}.$$

$\sqrt[3]{7x} \cdot \sqrt[3]{10y^2} = \sqrt[3]{7x \cdot 10y^2} = \sqrt[3]{70xy^2}$

The product rule for radicals can be used to simplify radicals:
$$\sqrt[n]{ab} = \sqrt[n]{a} \cdot \sqrt[n]{b}.$$

A radical expression with index n is simplified when its radicand has no factors that are perfect nth powers. To simplify, write the radicand as the product of two factors, one of which is the greatest perfect nth power. Then use the product rule to take the nth root of each factor. If all variables in a radicand are positive, then
$$\sqrt[n]{a^n} = a.$$

Some radicals can be simplified after the multiplication is performed.

- Simplify: $\sqrt[3]{54x^7y^{11}}$.

$$= \sqrt[3]{27 \cdot 2 \cdot x^6 \cdot x \cdot y^9 \cdot y^2}$$

$$= \sqrt[3]{(27x^6y^9)(2xy^2)}$$

$$= \sqrt[3]{27x^6y^9} \cdot \sqrt[3]{2xy^2} = 3x^2y^3\sqrt[3]{2xy^2}$$

- Assuming positive variables, multiply and

simplify: $\sqrt[4]{4x^2y} \cdot \sqrt[4]{4xy^3}$.

$$= \sqrt[4]{4x^2y \cdot 4xy^3} = \sqrt[4]{16x^3y^4}$$

$$= \sqrt[4]{16y^4} \cdot \sqrt[4]{x^3} = 2y\sqrt[4]{x^3}$$

Definitions and Concepts	Examples

Section 7.4 Adding, Subtracting, and Dividing Radical Expressions

Definitions and Concepts	Examples
Like radicals have the same indices and radicands. Like radicals can be added or subtracted using the distributive property. In some cases, radicals can be combined once they have been simplified.	$4\sqrt{18} - 6\sqrt{50}$ $= 4\sqrt{9 \cdot 2} - 6\sqrt{25 \cdot 2} = 4 \cdot 3\sqrt{2} - 6 \cdot 5\sqrt{2}$ $= 12\sqrt{2} - 30\sqrt{2} = -18\sqrt{2}$
The quotient rule for radicals can be used to simplify radicals: $$\sqrt[n]{\frac{a}{b}} = \frac{\sqrt[n]{a}}{\sqrt[n]{b}}.$$	$\sqrt[3]{-\dfrac{8}{x^{12}}} = \dfrac{\sqrt[3]{-8}}{\sqrt[3]{x^{12}}} = -\dfrac{2}{x^4}$ $\sqrt[3]{x^{12}} = (x^{12})^{\frac{1}{3}} = x^4$
The quotient rule for radicals can be used to divide radicals with the same indices: $$\frac{\sqrt[n]{a}}{\sqrt[n]{b}} = \sqrt[n]{\frac{a}{b}}.$$ Some radicals can be simplified after the division is performed.	Assuming a positive variable, divide and simplify: $$\frac{\sqrt[4]{64x^5}}{\sqrt[4]{2x^{-2}}} = \sqrt[4]{32x^{5-(-2)}} = \sqrt[4]{32x^7}$$ $$= \sqrt[4]{16 \cdot 2 \cdot x^4 \cdot x^3} = \sqrt[4]{16x^4} \cdot \sqrt[4]{2x^3}$$ $$= 2x\sqrt[4]{2x^3}.$$

Section 7.5 Multiplying with More Than One Term and Rationalizing Denominators

Definitions and Concepts	Examples
Radical expressions with more than one term are multiplied in much the same way that polynomials with more than one term are multiplied.	• $\sqrt{5}(2\sqrt{6} - \sqrt{3}) = 2\sqrt{30} - \sqrt{15}$ • $(4\sqrt{3} - 2\sqrt{2})(\sqrt{3} + \sqrt{2})$ $\quad\quad\quad$ F $\quad\quad$ O $\quad\quad$ I $\quad\quad$ L $= 4\sqrt{3} \cdot \sqrt{3} + 4\sqrt{3} \cdot \sqrt{2} - 2\sqrt{2} \cdot \sqrt{3} - 2\sqrt{2} \cdot \sqrt{2}$ $= 4 \cdot 3 + 4\sqrt{6} - 2\sqrt{6} - 2 \cdot 2$ $= 12 + 4\sqrt{6} - 2\sqrt{6} - 4 = 8 + 2\sqrt{6}$
Radical expressions that involve the sum and difference of the same two terms are called conjugates. Use $$(A + B)(A - B) = A^2 - B^2$$ to multiply conjugates.	$(8 + 2\sqrt{5})(8 - 2\sqrt{5})$ $= 8^2 - (2\sqrt{5})^2 = 64 - 4 \cdot 5$ $= 64 - 20 = 44$
The process of rewriting a radical expression as an equivalent expression without any radicals in the denominator is called rationalizing the denominator. When the denominator contains a single radical with an nth root, multiply the numerator and the denominator by a radical of index n that produces a perfect nth power in the denominator's radicand.	Rationalize the denominator: $\dfrac{7}{\sqrt[3]{2x}}$. $= \dfrac{7}{\sqrt[3]{2x}} \cdot \dfrac{\sqrt[3]{4x^2}}{\sqrt[3]{4x^2}} = \dfrac{7\sqrt[3]{4x^2}}{\sqrt[3]{8x^3}} = \dfrac{7\sqrt[3]{4x^2}}{2x}$

Definitions and Concepts	Examples

Section 7.5 Multiplying with More Than One Term and Rationalizing Denominators (continued)

If the denominator contains two terms, rationalize the denominator by multiplying the numerator and the denominator by the conjugate of the denominator.

$$\frac{13}{5 - \sqrt{3}} = \frac{13}{5 - \sqrt{3}} \cdot \frac{5 + \sqrt{3}}{5 + \sqrt{3}}$$

$$= \frac{13(5 + \sqrt{3})}{5^2 - (\sqrt{3})^2}$$

$$= \frac{13(5 + \sqrt{3})}{25 - 3} = \frac{13(5 + \sqrt{3})}{22}$$

Section 7.6 Radical Equations

A radical equation is an equation in which the variable occurs in a radicand.

Solving Radical Equations Containing *n*th Roots

1. Isolate one radical on one side of the equation.
2. Raise both sides to the *n*th power.
3. Solve the resulting equation.
4. Check proposed solutions in the original equation. Solutions of an equation to an even power that is radical-free, but not the original equation, are called extraneous solutions.

Solve: $\sqrt{6x + 13} - 2x = 1.$

$\sqrt{6x + 13} = 2x + 1$	Isolate the radical.
$(\sqrt{6x + 13})^2 = (2x + 1)^2$	Square both sides.
$6x + 13 = 4x^2 + 4x + 1$	
$0 = 4x^2 - 2x - 12$	Subtract $6x + 13$ from both sides.
$0 = 2(2x^2 - x - 6)$	Factor out the GCF.
$0 = 2(2x + 3)(x - 2)$	Factor completely.
$2x + 3 = 0 \quad$ or $\quad x - 2 = 0$	Set variable factors equal to zero.
$2x = -3 \qquad\qquad x = 2$	Solve for x.
$x = -\dfrac{3}{2}$	

Check both proposed solutions. 2 checks, but $-\dfrac{3}{2}$ is extraneous.

The solution is 2 and the solution set is $\{2\}$.

Section 7.7 Complex Numbers

The imaginary unit i is defined as
$$i = \sqrt{-1}, \quad \text{where} \quad i^2 = -1.$$
The set of numbers in the form $a + bi$ is called the set of complex numbers; a is the real part and b is the imaginary part. If $b = 0$, the complex number is a real number. If $b \neq 0$, the complex number is an imaginary number.

- $\sqrt{-81} = \sqrt{81(-1)} = \sqrt{81}\sqrt{-1} = 9i$
- $\sqrt{-75} = \sqrt{75(-1)} = \sqrt{25 \cdot 3}\sqrt{-1} = 5\sqrt{3}i$

To add or subtract complex numbers, add or subtract their real parts and add or subtract their imaginary parts.

$(2 - 4i) - (7 - 10i)$

$= 2 - 4i - 7 + 10i$

$= (2 - 7) + (-4 + 10)i = -5 + 6i$

To multiply complex numbers, multiply as if they are polynomials. After completing the multiplication, replace i^2 with -1. When performing operations with square roots of negative numbers, begin by expressing all square roots in terms of i. Then multiply.

$\qquad\qquad$ **F O I L**

- $(2 - 3i)(4 + 5i) = 8 + 10i - 12i - 15i^2$

$\qquad\qquad\qquad = 8 + 10i - 12i - 15(-1)$

$\qquad\qquad\qquad = 23 - 2i$

- $\sqrt{-36} \cdot \sqrt{-100} = \sqrt{36(-1)} \cdot \sqrt{100(-1)}$

$\qquad = 6i \cdot 10i = 60i^2 = 60(-1) = -60$

Definitions and Concepts	Examples

Section 7.7 Complex Numbers (continued)

The complex numbers $a + bi$ and $a - bi$ are conjugates. Conjugates can be multiplied using the formula $$(A + B)(A - B) = A^2 - B^2.$$ The multiplication of conjugates results in a real number.	$(3 + 5i)(3 - 5i)$ $= 3^2 - (5i)^2$ $= 9 - 25i^2$ $= 9 - 25(-1) = 34$
To divide complex numbers, multiply the numerator and the denominator by the conjugate of the denominator.	$\dfrac{5 + 2i}{4 - i} = \dfrac{5 + 2i}{4 - i} \cdot \dfrac{4 + i}{4 + i} = \dfrac{20 + 5i + 8i + 2i^2}{16 - i^2}$ $= \dfrac{20 + 13i + 2(-1)}{16 - (-1)}$ $= \dfrac{20 + 13i - 2}{16 + 1}$ $= \dfrac{18 + 13i}{17} = \dfrac{18}{17} + \dfrac{13}{17}i$
To simplify powers of i, rewrite the expression in terms of i^2. Then replace i^2 with -1 and simplify.	Simplify: i^{27}. $i^{27} = i^{26} \cdot i = (i^2)^{13}i$ $= (-1)^{13}i = (-1)i = -i$

CHAPTER 7 REVIEW EXERCISES

7.1 *In Exercises 1–5, find the indicated root, or state that the expression is not a real number.*

1. $\sqrt{81}$ 9

2. $-\sqrt{\dfrac{1}{100}}$ $-\dfrac{1}{10}$

3. $\sqrt[3]{-27}$ -3

4. $\sqrt[4]{-16}$ not a real number

5. $\sqrt[5]{-32}$ -2

In Exercises 6–7, find the indicated function values for each function. If necessary, round to two decimal places. If the function value is not a real number and does not exist, so state.

6. $f(x) = \sqrt{2x - 5}$; $f(15), f(4), f\left(\dfrac{5}{2}\right), f(1)$

5; 1.73; 0; not a real number

7. $g(x) = \sqrt[3]{4x - 8}$; $g(4), g(0), g(-14)$ 2; −2; −4

In Exercises 8–9, find the domain of each square root function.

8. $f(x) = \sqrt{x - 2}$ $\{x | x \geq 2\}$ or $[2, \infty)$

9. $g(x) = \sqrt{100 - 4x}$ $\{x | x \leq 25\}$ or $(-\infty, 25]$

In Exercises 10–15, simplify each expression. Assume that each variable can represent any real number, so include absolute value bars where necessary.

10. $\sqrt{25x^2}$ $5|x|$

11. $\sqrt{(x + 14)^2}$ $|x + 14|$

12. $\sqrt{x^2 - 8x + 16}$ $|x - 4|$ **13.** $\sqrt[3]{64x^3}$ $4x$

14. $\sqrt[4]{16x^4}$ $2|x|$ **15.** $\sqrt[5]{-32(x + 7)^5}$ $-2(x + 7)$

7.2 *In Exercises 16–18, use radical notation to rewrite each expression. Simplify, if possible.*

16. $(5xy)^{\frac{1}{3}}$ $\sqrt[3]{5xy}$ **17.** $16^{\frac{3}{2}}$ $\left(\sqrt{16}\right)^3 = 64$

18. $32^{\frac{4}{5}}$ $\left(\sqrt[5]{32}\right)^4 = 16$

In Exercises 19–20, rewrite each expression with rational exponents.

19. $\sqrt{7x}$ $(7x)^{\frac{1}{2}}$ **20.** $\left(\sqrt[3]{19xy}\right)^5$ $(19xy)^{\frac{5}{3}}$

In Exercises 21–22, rewrite each expression with a positive rational exponent. Simplify, if possible.

21. $8^{-\frac{2}{3}}$ $\dfrac{1}{8^{\frac{2}{3}}} = \dfrac{1}{4}$ **22.** $3x(ab)^{-\frac{4}{5}}$ $\dfrac{3x}{(ab)^{\frac{4}{5}}} = \dfrac{3x}{\sqrt[5]{(ab)^4}}$

In Exercises 23–26, use properties of rational exponents to simplify each expression.

23. $x^{\frac{1}{3}} \cdot x^{\frac{1}{4}}$ $x^{\frac{7}{12}}$ **24.** $\dfrac{5^{\frac{1}{2}}}{5^{\frac{1}{3}}}$ $5^{\frac{1}{6}}$

25. $(8x^6y^3)^{\frac{1}{3}}$ $2x^2y$

26. $\left(x^{-\frac{2}{3}}y^4\right)^{\frac{1}{2}}$ $\dfrac{y^{\frac{1}{8}}}{x^{\frac{1}{3}}}$

In Exercises 27–31, use rational exponents to simplify each expression. If rational exponents appear after simplifying, write the answer in radical notation.

27. $\sqrt[3]{x^9y^{12}}$ x^3y^4

28. $\sqrt[9]{x^3y^9}$ $y\sqrt[3]{x}$

29. $\sqrt{x}\cdot\sqrt[3]{x}$ $\sqrt[6]{x^5}$

30. $\dfrac{\sqrt[3]{x^2}}{\sqrt[4]{x^2}}$ $\sqrt[6]{x}$

31. $\sqrt[5]{\sqrt[3]{x}}$ $\sqrt[15]{x}$

32. The function $f(x) = 350x^{\frac{2}{3}}$ models the expenditures, $f(x)$, in millions of dollars, for the U.S. National Park Service x years after 1985. According to this model, what will expenditures be in 2012?   Expenditures will be $3150 million in 2012.

7.3 *In Exercises 33–35, use the product rule to multiply.*

33. $\sqrt{3x}\cdot\sqrt{7y}$ $\sqrt{21xy}$

34. $\sqrt[5]{7x^2}\cdot\sqrt[5]{11x}$ $\sqrt[5]{77x^3}$

35. $\sqrt[6]{x-5}\cdot\sqrt[6]{(x-5)^4}$ $\sqrt[6]{(x-5)^5}$

36. If $f(x) = \sqrt{7x^2 - 14x + 7}$, express the function, f, in simplified form. Assume that x can be any real number. $f(x) = \sqrt{7}|x-1|$

In Exercises 37–39, simplify by factoring. Assume that all variables in a radicand represent positive real numbers.

37. $\sqrt{20x^3}$ $2x\sqrt{5x}$

38. $\sqrt[3]{54x^8y^6}$ $3x^2y^2\sqrt[3]{2x^2}$

39. $\sqrt[4]{32x^3y^{11}z^5}$ $2y^2z\sqrt[4]{2x^3y^3z}$

In Exercises 40–43, multiply and simplify, if possible. Assume that all variables in a radicand represent positive real numbers.

40. $\sqrt{6x^3}\cdot\sqrt{4x^2}$ $2x^2\sqrt{6x}$

41. $\sqrt[3]{4x^2y}\cdot\sqrt[3]{4xy^4}$ $2xy\sqrt[3]{2y^2}$

42. $\sqrt[5]{2x^4y^3z^4}\cdot\sqrt[5]{8xy^6z^7}$ $xyz^2\sqrt[5]{16y^4z}$

43. $\sqrt{x+1}\cdot\sqrt{x-1}$ $\sqrt{x^2-1}$

7.4 *Assume that all variables represent positive real numbers.*

In Exercises 44–47, add or subtract as indicated.

44. $6\sqrt[3]{3} + 2\sqrt[3]{3}$ $8\sqrt[3]{3}$

45. $5\sqrt{18} - 3\sqrt{8}$ $9\sqrt{2}$

46. $\sqrt[3]{27x^4} + \sqrt[3]{xy^6}$ $(3x+y^2)\sqrt[3]{x}$

47. $2\sqrt[3]{6} - 5\sqrt[3]{48}$ $-8\sqrt[3]{6}$

In Exercises 48–50, simplify using the quotient rule.

48. $\sqrt[3]{\dfrac{16}{125}}$ $\dfrac{2}{5}\sqrt[3]{2}$

49. $\sqrt{\dfrac{x^3}{100y^4}}$ $\dfrac{x\sqrt{x}}{10y^2}$

50. $\sqrt[4]{\dfrac{3y^5}{16x^{20}}}$ $\dfrac{y\sqrt[4]{3y}}{2x^5}$

In Exercises 51–54, divide and, if possible, simplify.

51. $\dfrac{\sqrt{48}}{\sqrt{2}}$ $2\sqrt{6}$

52. $\dfrac{\sqrt[3]{32}}{\sqrt[3]{2}}$ $2\sqrt[3]{2}$

53. $\dfrac{\sqrt[4]{64x^7}}{\sqrt[4]{2x^2}}$ $2x\sqrt[4]{2x}$

54. $\dfrac{\sqrt{200x^3y^2}}{\sqrt{2x^{-2}y}}$ $10x^2\sqrt{xy}$

7.5 *Assume that all variables represent positive real numbers.*

In Exercises 55–62, multiply as indicated. If possible, simplify any radical expressions that appear in the product.

55. $\sqrt{3}(2\sqrt{6} + 4\sqrt{15})$ $6\sqrt{2} + 12\sqrt{5}$

56. $\sqrt[3]{5}(\sqrt[3]{50} - \sqrt[3]{2})$ $5\sqrt[3]{2} - \sqrt[3]{10}$

57. $(\sqrt{7} - 3\sqrt{5})(\sqrt{7} + 6\sqrt{5})$ $-83 + 3\sqrt{35}$

58. $(\sqrt{x} - \sqrt{11})(\sqrt{y} - \sqrt{11})$ $\sqrt{xy} - \sqrt{11x} - \sqrt{11y} + 11$

59. $(\sqrt{5} + \sqrt{8})^2$ $13 + 4\sqrt{10}$

60. $(2\sqrt{3} - \sqrt{10})^2$ $22 - 4\sqrt{30}$

61. $(\sqrt{7} + \sqrt{13})(\sqrt{7} - \sqrt{13})$ -6

62. $(7 - 3\sqrt{5})(7 + 3\sqrt{5})$ 4

In Exercises 63–75, rationalize each denominator. Simplify, if possible.

63. $\dfrac{4}{\sqrt{6}}$ $\dfrac{2\sqrt{6}}{3}$

64. $\sqrt{\dfrac{2}{7}}$ $\dfrac{\sqrt{14}}{7}$

65. $\dfrac{12}{\sqrt[3]{9}}$ $4\sqrt[3]{3}$

66. $\sqrt{\dfrac{2x}{5y}}$ $\dfrac{\sqrt{10xy}}{5y}$

67. $\dfrac{14}{\sqrt[3]{2x^2}}$ $\dfrac{7\sqrt[3]{4x}}{x}$

68. $\sqrt[4]{\dfrac{7}{3x}}$ $\dfrac{\sqrt[4]{189x^3}}{3x}$

69. $\dfrac{5}{\sqrt[5]{32x^4y}}$ $\dfrac{5\sqrt[5]{xy^4}}{2xy}$

70. $\dfrac{6}{\sqrt{3} - 1}$ $3\sqrt{3} + 3$

71. $\dfrac{\sqrt{7}}{\sqrt{5} + \sqrt{3}}$ $\dfrac{\sqrt{35} - \sqrt{21}}{2}$

72. $\dfrac{10}{2\sqrt{5} - 3\sqrt{2}}$ $10\sqrt{5} + 15\sqrt{2}$

73. $\dfrac{\sqrt{x} + 5}{\sqrt{x} - 3}$ $\dfrac{x + 8\sqrt{x} + 15}{x - 9}$

74. $\dfrac{\sqrt{7} + \sqrt{3}}{\sqrt{7} - \sqrt{3}}$ $\dfrac{5 + \sqrt{21}}{2}$

75. $\dfrac{2\sqrt{3} + \sqrt{6}}{2\sqrt{6} + \sqrt{3}}$ $\dfrac{3\sqrt{2} + 2}{7}$

In Exercises 76–79, rationalize each numerator. Simplify, if possible.

76. $\sqrt{\dfrac{2}{7}}$ $\dfrac{2}{\sqrt{14}}$

77. $\dfrac{\sqrt[3]{3x}}{\sqrt[3]{y}}$ $\dfrac{3x}{\sqrt[3]{9x^2y}}$

78. $\dfrac{\sqrt{7}}{\sqrt{5} + \sqrt{3}}$ $\dfrac{7}{\sqrt{35} + \sqrt{21}}$

79. $\dfrac{\sqrt{7} + \sqrt{3}}{\sqrt{7} - \sqrt{3}}$ $\dfrac{2}{5 - \sqrt{21}}$

7.6 *In Exercises 80–84, solve each radical equation.*

80. $\sqrt{2x+4}=6$ 16 or {16}

81. $\sqrt{x-5}+9=4$ no solution or ∅

82. $\sqrt{2x-3}+x=3$ 2 or {2}

83. $\sqrt{x-4}+\sqrt{x+1}=5$ 8 or {8}

84. $(x^2+6x)^{\frac{1}{3}}+2=0$ −4 and −2, or {−4, −2}

85. The time, $f(x)$, in seconds, for a free-falling object to fall x feet is modeled by the function

$$f(x)=\sqrt{\frac{x}{16}}.$$

If a worker accidently drops a hammer from a building and it hits the ground after 4 seconds, from what height was the hammer dropped?   The hammer was dropped from 256 feet.

86. Out of a group of 50,000 births, the number of people, $f(x)$, surviving to age x is modeled by the function

$$f(x)=5000\sqrt{100-x}.$$

To what age will 20,000 people in the group survive?   20,000 people in the group will survive to 84 years old.

7.7 *In Exercises 87–89, express each number in terms of i and simplify, if possible.*

87. $\sqrt{-81}$ 9i

88. $\sqrt{-63}$ $3\sqrt{7}i$

89. $-\sqrt{-8}$ $-2\sqrt{2}i$

In Exercises 90–99, perform the indicated operation. Write the result in the form a + bi.

90. $(7+12i)+(5-10i)$ 12 + 2i

91. $(8-3i)-(17-7i)$ −9 + 4i

92. $4i(3i-2)$ −12 − 8i

93. $(7-5i)(2+3i)$ 29 + 11i

94. $(3-4i)^2$ −7 − 24i

95. $(7+8i)(7-8i)$ 113 + 0i or 113

96. $\sqrt{-8}\cdot\sqrt{-3}$ $-2\sqrt{6}+0i$ or $-2\sqrt{6}$

97. $\dfrac{6}{5+i}$ $\dfrac{15}{13}-\dfrac{3}{13}i$

98. $\dfrac{3+4i}{4-2i}$ $\dfrac{1}{5}+\dfrac{11}{10}i$

99. $\dfrac{5+i}{3i}$ $\dfrac{1}{3}-\dfrac{5}{3}i$

In Exercises 100–101, simplify each expression.

100. i^{16} 1

101. i^{23} −i

CHAPTER 7 TEST

Remember to use your Chapter Test Prep Video CD to see the worked-out solutions to the test questions you want to review.

1. Let $f(x)=\sqrt{8-2x}$.

   a. Find $f(-14)$.   6

   b. Find the domain of f. $\{x|x\le 4\}$ or $(-\infty,4]$

2. Evaluate: $27^{-\frac{4}{3}}$. $\dfrac{1}{81}$

3. Simplify: $(25x^{-\frac{1}{2}}y^4)^{\frac{1}{2}}$. $\dfrac{5\sqrt[8]{y}}{\sqrt[4]{x}}$

In Exercises 4–5, use rational exponents to simplify each expression. If rational exponents appear after simplifying, write the answer in radical notation.

4. $\sqrt[8]{x^4}$ $\sqrt{x}$

5. $\sqrt[4]{x}\cdot\sqrt[5]{x}$ $\sqrt[20]{x^9}$

In Exercises 6–9, simplify each expression. Assume that each variable can represent any real number.

6. $\sqrt{75x^2}$ $5|x|\sqrt{3}$

7. $\sqrt{x^2-10x+25}$ $|x-5|$

8. $\sqrt[3]{16x^4y^8}$ $2xy^2\sqrt[3]{2xy^2}$

9. $\sqrt[5]{-\dfrac{32}{x^{10}}}$ $-\dfrac{2}{x^2}$

In Exercises 10–17, perform the indicated operation and, if possible, simplify. Assume that all variables represent positive real numbers.

10. $\sqrt[3]{5x^2}\cdot\sqrt[3]{10y}$ $\sqrt[3]{50x^2y}$

11. $\sqrt[4]{8x^3y}\cdot\sqrt[4]{4xy^2}$ $2x\sqrt[4]{2y^3}$

12. $3\sqrt{18}-4\sqrt{32}$ $-7\sqrt{2}$

13. $\sqrt[3]{8x^4}+\sqrt[3]{xy^6}$ $(2x+y^2)\sqrt[3]{x}$

14. $\dfrac{\sqrt[3]{16x^8}}{\sqrt[3]{2x^4}}$ $2x\sqrt[3]{x}$

15. $\sqrt{3}(4\sqrt{6}-\sqrt{5})$ $12\sqrt{2}-\sqrt{15}$

16. $(5\sqrt{6}-2\sqrt{2})(\sqrt{6}+\sqrt{2})$ $26+6\sqrt{3}$

17. $(7-\sqrt{3})^2$ $52-14\sqrt{3}$

In Exercises 18–20, rationalize each denominator. Simplify, if possible. Assume all variables represent positive real numbers.

18. $\sqrt{\dfrac{5}{x}}$ $\dfrac{\sqrt{5x}}{x}$

19. $\dfrac{5}{\sqrt[3]{5x^2}}$ $\dfrac{\sqrt[3]{25x}}{x}$

20. $\dfrac{\sqrt{2}-\sqrt{3}}{\sqrt{2}+\sqrt{3}}$ $-5+2\sqrt{6}$

In Exercises 21–23, solve each radical equation.

21. $3 + \sqrt{2x - 3} = x$ 6 or {6}

22. $\sqrt{x + 9} - \sqrt{x - 7} = 2$ 16 or {16}

23. $(11x + 6)^{\frac{1}{3}} + 3 = 0$ −3 or {−3}

24. The function

$$f(x) = 2.9\sqrt{x} + 20.1$$

models the average height, $f(x)$, in inches, of boys who are x months of age, $0 \le x \le 60$. Find the age at which the average height of boys is 40.4 inches.   49 months

25. Express in terms of i and simplify: $\sqrt{-75}$. $5\sqrt{3}i$

In Exercises 26–29, perform the indicated operation. Write the result in the form $a + bi$.

26. $(5 - 3i) - (6 - 9i)$ $-1 + 6i$

27. $(3 - 4i)(2 + 5i)$ $26 + 7i$

28. $\sqrt{-9} \cdot \sqrt{-4}$ $-6 + 0i$ or -6

29. $\dfrac{3 + i}{1 - 2i}$ $\dfrac{1}{5} + \dfrac{7}{5}i$

30. Simplify: i^{35}. $-i$

Note to Instructor: The Chapter 7 Test file in the TestGen program provides algorithms specifically matched to these problems so you can easily replicate this chapter test for additional practice or assessment purposes.

CUMULATIVE REVIEW EXERCISES (CHAPTERS 1–7)

In Exercises 1–5, solve each equation, inequality, or system.

1.
$$2x - y + z = -5$$
$$x - 2y - 3z = 6$$
$$x + y - 2z = 1$$ $(-2, -1, -2)$ or $\{(-2, -1, -2)\}$

2. $3x^2 - 11x = 4$ $-\frac{1}{3}$ and 4, or $\left\{-\frac{1}{3}, 4\right\}$

3. $2(x + 4) < 5x + 3(x + 2)$ $\left\{x \mid x > \frac{1}{3}\right\}$ or $\left(\frac{1}{3}, \infty\right)$

4. $\dfrac{1}{x + 2} + \dfrac{15}{x^2 - 4} = \dfrac{5}{x - 2}$ $\frac{3}{4}$ or $\left\{\frac{3}{4}\right\}$

5. $\sqrt{x + 2} - \sqrt{x + 1} = 1$ −1 or {−1}

6. Graph the solution set of the system:

$$x + 2y < 2$$
$$2y - x > 4.$$ See graphing answer section.

In Exercises 7–15, perform the indicated operations.

7. $\dfrac{8x^2}{3x^2 - 12} \div \dfrac{40}{x - 2}$ $\dfrac{x^2}{15(x + 2)}$

8. $\dfrac{x + \dfrac{1}{y}}{y + \dfrac{1}{x}}$ $\dfrac{x}{y}$

9. $(2x - 3)(4x^2 - 5x - 2)$ $8x^3 - 22x^2 + 11x + 6$

10. $\dfrac{7x}{x^2 - 2x - 15} - \dfrac{2}{x - 5}$ $\dfrac{5x - 6}{(x - 5)(x + 3)}$

11. $7(8 - 10)^3 - 7 + 3 \div (-3)$ −64

12. $\sqrt{80x} - 5\sqrt{20x} + 2\sqrt{45x}$ 0

13. $\dfrac{\sqrt{3} - 2}{2\sqrt{3} + 5}$ $-\dfrac{16 - 9\sqrt{3}}{13}$

14. $(2x^3 - 3x^2 + 3x - 4) \div (x - 2)$ $2x^2 + x + 5 + \dfrac{6}{x - 2}$

15. $\left(2\sqrt{3} + 5\sqrt{2}\right)\left(\sqrt{3} - 4\sqrt{2}\right)$ $-34 - 3\sqrt{6}$

In Exercises 16–17, factor completely.

16. $24x^2 + 10x - 4$ $2(3x + 2)(4x - 1)$

17. $16x^4 - 1$ $(4x^2 + 1)(2x + 1)(2x - 1)$

18. The amount of light provided by a light bulb varies inversely as the square of the distance from the bulb. The illumination provided is 120 lumens at a distance of 10 feet. How many lumens are provided at a distance of 15 feet?   About 53 lumens are provided at a distance of 15 feet.

19. You invested $6000 in two accounts paying 7% and 9% annual interest, respectively. At the end of the year, the total interest from these investments was $510. How much was invested at each rate? $1500 was invested at 7% and $4500 at 9%.

20. Although there are 2332 students enrolled in the college, this is 12% fewer students than there were enrolled last year. How many students were enrolled last year?   2650 students were enrolled last year.

Thanks to the efficiency of U.S. farms, most Americans enjoy an abundance of inexpensive foods. Nearly one in five U.S. kids ages 9 through 13 say they've already been on a weight-loss diet. By contrast, worldwide, every three seconds a child dies of hunger. If all the world's undernourished people were gathered into a new country, it would be the third most populous nation, just behind China and India.

Source: The State of Food Insecurity in the World, 2000; 2000 BBC News

The growth and decline in the number of U.S. farms is developed algebraically in the technology box at the conclusion of Section 8.3 and in Exercise 58 of Exercise Set 8.3.

8 CHAPTER

Quadratic Equations and Functions

8.1 The Square Root Property and Completing the Square

8.2 The Quadratic Formula

8.3 Quadratic Functions and Their Graphs

8.4 Equations Quadratic in Form

8.5 Polynomial and Rational Inequalities

Because we tend to take our food abundance for granted, it might surprise you to know that the number of U.S. farms has declined since the 1920s, as individually owned family farms have been swallowed up by huge agribusinesses owned by corporations. Currently, there are about 1.8 acres of cropland to grow food for each American. If the current trends in population growth and loss of farmland continue, there will be only 0.6 acre per American in the year 2050. This is the rate that currently exists worldwide.

Like the number of farms and the amount of cropland in the United States, many phenomena follow trends that involve growth and decline, or vice versa. In this chapter, you will learn to solve equations and graph functions that provide new ways of looking at and understanding these phenomena.

SECTION

8.1

THE SQUARE ROOT PROPERTY AND COMPLETING THE SQUARE

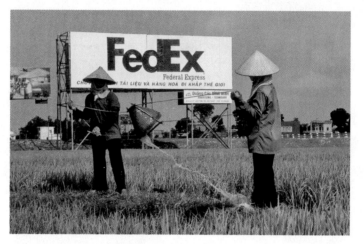

Objectives

1 Solve quadratic equations using the square root property.

2 Complete the square of a binomial.

3 Solve quadratic equations by completing the square.

4 Solve problems using the square root property.

"In the future there will be two kinds of corporations; those that go global, and those that go bankrupt."

C. Michael Armstrong, CEO, AT&T

For better or worse, ours is the era of the multinational corporation. New technology that the multinational corporations control is expanding their power. And their numbers are growing fast. There were approximately 300 multinationals in 1900. By 1970, there were close to 7000, and by 1990 the number had swelled to 30,000. In 2001, more than 65,000 multinational corporations enveloped the world.

In this section, you will learn two new methods for solving quadratic equations. These methods are called the *square root method* and *completing the square*. Using these techniques, you will explore the growth of multinational corporations and make predictions about the number of global corporations in the future.

STUDY TIP

Here is a summary of what we already know about quadratic equations and quadratic functions.

1. A **quadratic equation** in x can be written in the standard form

$$ax^2 + bx + c = 0, \quad a \neq 0.$$

2. Some quadratic equations can be solved by factoring.

Solve: $\quad 2x^2 + 7x - 4 = 0.$

$$(2x - 1)(x + 4) = 0$$

$$2x - 1 = 0 \quad \text{or} \quad x + 4 = 0$$

$$2x = 1 \qquad\qquad x = -4$$

$$x = \tfrac{1}{2}$$

x-intercept is **−4**.

x-intercept is $\tfrac{1}{2}$.

$f(x) = 2x^2 + 7x - 4$

FIGURE 8.1

The solutions are -4 and $\tfrac{1}{2}$, and the solution set is $\left\{-4, \tfrac{1}{2}\right\}$.

3. A polynomial function of the form

$$f(x) = ax^2 + bx + c, \quad a \neq 0$$

is a **quadratic function**. Graphs of quadratic functions have cuplike shapes with the same behavior at each end.

4. The real solutions of $ax^2 + bx + c = 0$ correspond to the x-intercepts for the graph of the quadratic function $f(x) = ax^2 + bx + c$. For example, the solutions of the equation $2x^2 + 7x - 4 = 0$ are -4 and $\tfrac{1}{2}$. Figure 8.1 shows that the solutions appear as x-intercepts on the graph of the quadratic function $f(x) = 2x^2 + 7x - 4$.

Now that we've summarized what we know, let's look at where we go. How do we solve a quadratic equation, $ax^2 + bx + c = 0$, if the trinomial $ax^2 + bx + c$ cannot be factored? Methods other than factoring are needed. In this section, we look at other ways of solving quadratic equations.

1 Solve quadratic equations using the square root property.

The Square Root Property   Let's begin with a relatively simple quadratic equation:

$$x^2 = 9.$$

The value of x must be a number whose square is 9. There are two numbers whose square is 9:

$$x = \sqrt{9} = 3 \quad \text{or} \quad x = -\sqrt{9} = -3.$$

Thus, the solutions of $x^2 = 9$ are 3 and -3. This is an example of the **square root property**.

DISCOVER FOR YOURSELF

Solve $x^2 = 9$, or

$$x^2 - 9 = 0,$$

by factoring. What is the advantage of using the square root property?

THE SQUARE ROOT PROPERTY   If u is an algebraic expression and d is a nonzero real number, then $u^2 = d$ has exactly two solutions:

$$\text{If } u^2 = d, \quad \text{then } u = \sqrt{d} \text{ or } u = -\sqrt{d}.$$

Equivalently,

$$\text{If } u^2 = d, \quad \text{then } u = \pm\sqrt{d}.$$

Notice that $u = \pm\sqrt{d}$ is a shorthand notation to indicate that $u = \sqrt{d}$ or $u = -\sqrt{d}$. Although we usually read $u = \pm\sqrt{d}$ as "u equals plus or minus the square root of d," we actually mean that u is the positive square root of d or the negative square root of d.

EXAMPLE 1   Solving a Quadratic Equation by the Square Root Property

Solve: $3x^2 = 18$.

SOLUTION   To apply the square root property, we need a squared expression by itself on one side of the equation.

$$3x^2 = 18$$

We want x^2 by itself.

We can get x^2 by itself if we divide both sides by 3.

$3x^2 = 18$	This is the original equation.
$\dfrac{3x^2}{3} = \dfrac{18}{3}$	Divide both sides by 3.
$x^2 = 6$	Simplify.
$x = \sqrt{6} \quad \text{or} \quad x = -\sqrt{6}$	Apply the square root property.

Now let's check these proposed solutions in the original equation.

Check $\sqrt{6}$:	**Check** $-\sqrt{6}$:
$3x^2 = 18$	$3x^2 = 18$
$3\left(\sqrt{6}\right)^2 \stackrel{?}{=} 18$	$3\left(-\sqrt{6}\right)^2 \stackrel{?}{=} 18$
$3 \cdot 6 \stackrel{?}{=} 18$	$3 \cdot 6 \stackrel{?}{=} 18$
$18 = 18,$ true	$18 = 18,$ true

The solutions are $-\sqrt{6}$ and $\sqrt{6}$. The solution set is $\left\{-\sqrt{6}, \sqrt{6}\right\}$ or $\left\{\pm\sqrt{6}\right\}$. ∎

 CHECK POINT 1 Solve: $4x^2 = 28$. $-\sqrt{7}$ and $\sqrt{7}$, or $\left\{\pm\sqrt{7}\right\}$

In this section, we will express irrational solutions in simplified radical form, rationalizing denominators when possible.

> **EXAMPLE 2** Solving a Quadratic Equation by the Square Root Property

Solve: $2x^2 - 7 = 0$.

SOLUTION To solve by the square root property, we isolate the squared expression on one side of the equation.

$$2x^2 - 7 = 0$$

We want x^2 by itself.

$2x^2 - 7 = 0$	This is the original equation.
$2x^2 = 7$	Add 7 to both sides.
$x^2 = \dfrac{7}{2}$	Divide both sides by 2.
$x = \sqrt{\dfrac{7}{2}} \quad\text{or}\quad x = -\sqrt{\dfrac{7}{2}}$	Apply the square root property.

Because the proposed solutions are opposites, we can rationalize both denominators at once:

$$\pm\sqrt{\frac{7}{2}} = \pm\frac{\sqrt{7}}{\sqrt{2}} \cdot \frac{\sqrt{2}}{\sqrt{2}} = \pm\frac{\sqrt{14}}{2}.$$

Substitute these values into the original equation and verify that the solutions are $-\dfrac{\sqrt{14}}{2}$ and $\dfrac{\sqrt{14}}{2}$. The solution set is $\left\{-\dfrac{\sqrt{14}}{2}, \dfrac{\sqrt{14}}{2}\right\}$ or $\left\{\pm\dfrac{\sqrt{14}}{2}\right\}$. ∎

 CHECK POINT 2 Solve: $3x^2 - 11 = 0$. $\pm\frac{\sqrt{33}}{3}$ or $\left\{\pm\frac{\sqrt{33}}{3}\right\}$

Some quadratic equations have solutions that are imaginary numbers.

> **EXAMPLE 3** Solving a Quadratic Equation by the Square Root Property

Solve: $9x^2 + 25 = 0$.

SOLUTION

$$9x^2 + 25 = 0$$ This is the original equation.

> We need to isolate x^2.

$$9x^2 = -25$$ Subtract 25 from both sides.

$$x^2 = -\frac{25}{9}$$ Divide both sides by 9.

$$x = \sqrt{-\frac{25}{9}} \quad \text{or} \quad x = -\sqrt{-\frac{25}{9}}$$ Apply the square root property.

$$x = \sqrt{\frac{25}{9}}\sqrt{-1} \qquad x = -\sqrt{\frac{25}{9}}\sqrt{-1}$$

$$x = \frac{5}{3}i \qquad\qquad x = -\frac{5}{3}i$$ $\sqrt{-1} = i$

Because the equation has an x^2-term and no x-term, we can check both proposed solutions, $\pm\frac{5}{3}i$, at once.

Check $\frac{5}{3}i$ and $-\frac{5}{3}i$:

$$9x^2 + 25 = 0$$

$$9\left(\pm\frac{5}{3}i\right)^2 + 25 \stackrel{?}{=} 0$$

$$9\left(\frac{25}{9}i^2\right) + 25 \stackrel{?}{=} 0$$

$$25i^2 + 25 \stackrel{?}{=} 0$$

> $i^2 = -1$

$$25(-1) + 25 \stackrel{?}{=} 0$$

$$0 = 0, \quad \text{true}$$

The solutions are $-\frac{5}{3}i$ and $\frac{5}{3}i$. The solution set is $\left\{-\frac{5}{3}i, \frac{5}{3}i\right\}$ or $\left\{\pm\frac{5}{3}i\right\}$. ∎

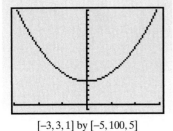
✔ **CHECK POINT 3** Solve: $4x^2 + 9 = 0$. $\pm\frac{3}{2}i$ or $\left\{\pm\frac{3}{2}i\right\}$

Can we solve an equation such as $(x - 1)^2 = 5$ using the square root property? Yes. The equation is in the form $u^2 = d$, where u^2, the squared expression, is by itself on the left side.

$$(x - 1)^2 \qquad = \qquad 5$$

> This is u^2 in $u^2 = d$ with $u = x - 1$.

> This is d in $u^2 = d$ with $d = 5$.

Try solving

$$(x - 1)^2 = 5$$

by writing the equation in standard form and factoring. What problem do you encounter?

EXAMPLE 4 Solving a Quadratic Equation by the Square Root Property

Solve by the square root property: $(x - 1)^2 = 5$.

SOLUTION

$$(x - 1)^2 = 5 \qquad \text{This is the original equation.}$$

$$x - 1 = \sqrt{5} \quad \text{or} \quad x - 1 = -\sqrt{5} \qquad \text{Apply the square root property.}$$

$$x = 1 + \sqrt{5} \qquad x = 1 - \sqrt{5} \qquad \text{Add 1 to both sides in each equation.}$$

Check $1 + \sqrt{5}$:

$$(x - 1)^2 = 5$$
$$\left(1 + \sqrt{5} - 1\right)^2 \overset{?}{=} 5$$
$$\left(\sqrt{5}\right)^2 \overset{?}{=} 5$$
$$5 = 5, \quad \text{true}$$

Check $1 - \sqrt{5}$:

$$(x - 1)^2 = 5$$
$$\left(1 - \sqrt{5} - 1\right)^2 \overset{?}{=} 5$$
$$\left(-\sqrt{5}\right)^2 \overset{?}{=} 5$$
$$5 = 5, \quad \text{true}$$

The solutions are $1 \pm \sqrt{5}$, and the solution set is $\left\{1 + \sqrt{5}, 1 - \sqrt{5}\right\}$ or $\left\{1 \pm \sqrt{5}\right\}$. ■

CHECK POINT 4 Solve: $(x - 3)^2 = 10$. $3 \pm \sqrt{10}$ or $\left\{3 \pm \sqrt{10}\right\}$

---

2 Complete the square of a binomial.

Completing the Square We return to the question that opened this section: How do we solve a quadratic equation, $ax^2 + bx + c = 0$, if the trinomial $ax^2 + bx + c$ cannot be factored? We can convert the equation into an equivalent equation that can be solved using the square root property. This is accomplished by **completing the square**.

> **COMPLETING THE SQUARE** If $x^2 + bx$ is a binomial, then by adding $\left(\dfrac{b}{2}\right)^2$, which is the square of half the coefficient of x, a perfect square trinomial will result. That is,
>
> $$x^2 + bx + \left(\frac{b}{2}\right)^2 = \left(x + \frac{b}{2}\right)^2.$$

EXAMPLE 5 Completing the Square

What term should be added to each binomial so that it becomes a perfect square trinomial? Write and factor the trinomial.

a. $x^2 + 8x$ **b.** $x^2 - 7x$ **c.** $x^2 + \dfrac{3}{5}x$

SOLUTION To complete the square, we must add a term to each binomial. The term that should be added is the square of half the coefficient of x.

$$x^2 + 8x \qquad x^2 - 7x \qquad x^2 + \frac{3}{5}x$$

Add $\left(\frac{8}{2}\right)^2 = 4^2$. Add $\left(\frac{7}{2}\right)^2$, or $\frac{49}{4}$, Add $\left(\frac{1}{2} \cdot \frac{3}{5}\right)^2 = \left(\frac{3}{10}\right)^2$.
Add 16 to complete to complete Add $\frac{9}{100}$ to complete
the square. the square. the square.

Visualizing Completing the Square

This figure, with area $x^2 + 8x$, is not a complete square. The bottom-right corner is missing.

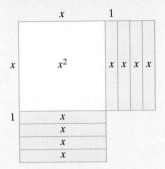

Area: $x^2 + 8x$

Add 16 square units to the missing portion and you, literally, complete the square.

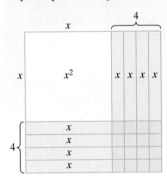

Area: $x^2 + 8x + 16 = (x + 4)^2$

a. The coefficient of the x-term in $x^2 + 8x$ is 8. Half of 8 is 4, and $4^2 = 16$. Add 16. The result is a perfect square trinomial.

$$x^2 + 8x + 16 = (x + 4)^2$$

b. The coefficient of the x-term in $x^2 - 7x$ is -7. Half of -7 is $-\dfrac{7}{2}$, and $\left(-\dfrac{7}{2}\right)^2 = \dfrac{49}{4}$. Add $\dfrac{49}{4}$. The result is a perfect square trinomial.

$$x^2 - 7x + \frac{49}{4} = \left(x - \frac{7}{2}\right)^2$$

c. The coefficient of the x-term in $x^2 + \dfrac{3}{5}x$ is $\dfrac{3}{5}$. Half of $\dfrac{3}{5}$ is $\dfrac{1}{2} \cdot \dfrac{3}{5}$, or $\dfrac{3}{10}$, and $\left(\dfrac{3}{10}\right)^2 = \dfrac{9}{100}$. Add $\dfrac{9}{100}$. The result is a perfect square trinomial.

$$x^2 + \frac{3}{5}x + \frac{9}{100} = \left(x + \frac{3}{10}\right)^2$$

STUDY TIP

You may not be accustomed to factoring perfect square trinomials in which fractions are involved. The constant in the factorization is always half the coefficient of x.

$$x^2 - 7x + \frac{49}{4} = \left(x - \frac{7}{2}\right)^2 \qquad x^2 + \frac{3}{5}x + \frac{9}{100} = \left(x + \frac{3}{10}\right)^2$$

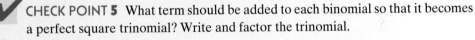

Half the coefficient of x, -7, is $-\frac{7}{2}$.   Half the coefficient of x, $\frac{3}{5}$, is $\frac{3}{10}$.

✓ **CHECK POINT 5** What term should be added to each binomial so that it becomes a perfect square trinomial? Write and factor the trinomial.

a. $x^2 + 10x$ 25; $x^2 + 10x + 25 = (x + 5)^2$

b. $x^2 - 3x$ $\dfrac{9}{4}$; $x^2 - 3x + \dfrac{9}{4} = \left(x - \dfrac{3}{2}\right)^2$

c. $x^2 + \dfrac{3}{4}x$. $\dfrac{9}{64}$; $x^2 + \dfrac{3}{4}x + \dfrac{9}{64} = \left(x + \dfrac{3}{8}\right)^2$

3 Solve quadratic equations by completing the square.

Solving Quadratic Equations by Completing the Square We can solve any quadratic equation by completing the square. If the coefficient of the x^2-term is one, we add the square of half the coefficient of x to both sides of the equation. **When you add a constant term to one side of the equation to complete the square, be certain to add the same constant to the other side of the equation.** These ideas are illustrated in Example 6.

EXAMPLE 6 Solving a Quadratic Equation by Completing the Square

Solve by completing the square: $x^2 - 6x + 4 = 0$.

SOLUTION We begin by subtracting 4 from both sides. This is done to isolate the binomial $x^2 - 6x$ so that we can complete the square.

$$x^2 - 6x + 4 = 0 \qquad \text{This is the original equation.}$$
$$x^2 - 6x = -4 \qquad \text{Subtract 4 from both sides.}$$

Next, we work with $x^2 - 6x = -4$ and complete the square. Find half the coefficient of the x-term and square it. The coefficient of the x-term is -6. Half of -6 is -3 and $(-3)^2 = 9$. Thus, we add 9 to both sides of the equation.

$$x^2 - 6x + 9 = -4 + 9 \qquad \text{Add 9 to both sides to complete the square.}$$

$$(x - 3)^2 = 5 \qquad \text{Factor and simplify.}$$

$$x - 3 = \sqrt{5} \quad \text{or} \quad x - 3 = -\sqrt{5} \qquad \text{Apply the square root property.}$$

$$x = 3 + \sqrt{5} \qquad x = 3 - \sqrt{5} \qquad \text{Add 3 to both sides in each equation.}$$

The solutions are $3 \pm \sqrt{5}$, and the solution set is $\{3 + \sqrt{5}, 3 - \sqrt{5}\}$ or $\{3 \pm \sqrt{5}\}$. ∎

If you solve a quadratic equation by completing the square and the solutions are rational numbers, the equation can also be solved by factoring. By contrast, quadratic equations with irrational solutions cannot be solved by factoring. However, all quadratic equations can be solved by completing the square.

CHECK POINT 6 Solve by completing the square: $x^2 + 4x - 1 = 0$.
$-2 \pm \sqrt{5}$ or $\{-2 \pm \sqrt{5}\}$

If the coefficient of the x^2-term in a quadratic equation is not 1, you must divide each side of the equation by this coefficient before completing the square. For example, to solve $9x^2 - 6x - 4 = 0$ by completing the square, first divide every term by 9:

$$\frac{9x^2}{9} - \frac{6x}{9} - \frac{4}{9} = \frac{0}{9}$$

$$x^2 - \frac{6}{9}x - \frac{4}{9} = 0$$

$$x^2 - \frac{2}{3}x - \frac{4}{9} = 0.$$

Now that the coefficient of the x^2-term is 1, we can solve by completing the square.

EXAMPLE 7 Solving a Quadratic Equation by Completing the Square

Solve by completing the square: $9x^2 - 6x - 4 = 0$.

SOLUTION

$$9x^2 - 6x - 4 = 0 \qquad \text{This is the original equation.}$$

$$x^2 - \frac{2}{3}x - \frac{4}{9} = 0 \qquad \text{Divide both sides by 9.}$$

$$x^2 - \frac{2}{3}x = \frac{4}{9} \qquad \text{Add } \tfrac{4}{9} \text{ to both sides to isolate the binomial.}$$

$$x^2 - \frac{2}{3}x + \frac{1}{9} = \frac{4}{9} + \frac{1}{9} \qquad \text{Complete the square: Half of } -\tfrac{2}{3} \text{ is } -\tfrac{2}{6}, \text{ or } -\tfrac{1}{3}, \text{ and } \left(-\tfrac{1}{3}\right)^2 = \tfrac{1}{9}.$$

$$\left(x - \frac{1}{3}\right)^2 = \frac{5}{9} \qquad \text{Factor and simplify.}$$

Obtain a decimal approximation for each solution of
$9x^2 - 6x - 4 = 0$, the equation in Example 7.

$$\frac{1 + \sqrt{5}}{3} \approx 1.1$$

$$\frac{1 - \sqrt{5}}{3} \approx -0.4$$

$[-2, 2, 1]$ by $[-10, 10, 1]$

The x-intercepts of
$f(x) = 9x^2 - 6x - 4$ verify
the solutions.

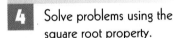

4 Solve problems using the square root property.

$$x - \frac{1}{3} = \sqrt{\frac{5}{9}} \quad \text{or} \quad x - \frac{1}{3} = -\sqrt{\frac{5}{9}} \qquad \text{Apply the square root property.}$$

$$x - \frac{1}{3} = \frac{\sqrt{5}}{3} \qquad x - \frac{1}{3} = -\frac{\sqrt{5}}{3} \qquad \sqrt{\frac{5}{9}} = \frac{\sqrt{5}}{\sqrt{9}} = \frac{\sqrt{5}}{3}$$

$$x = \frac{1}{3} + \frac{\sqrt{5}}{3} \qquad x = \frac{1}{3} - \frac{\sqrt{5}}{3} \qquad \text{Add } \tfrac{1}{3} \text{ to both sides and solve for } x.$$

$$x = \frac{1 + \sqrt{5}}{3} \qquad x = \frac{1 - \sqrt{5}}{3} \qquad \text{Express solutions with a common denominator.}$$

The solutions are $\dfrac{1 \pm \sqrt{5}}{3}$, and the solution set is $\left\{\dfrac{1 \pm \sqrt{5}}{3}\right\}$. ■

✔ **CHECK POINT 7** Solve by completing the square: $2x^2 + 3x - 4 = 0$.

$$\frac{-3 \pm \sqrt{41}}{4} \quad \text{or} \quad \left\{\frac{-3 \pm \sqrt{41}}{4}\right\}$$

Applications We all want a wonderful life with fulfilling work, good health, and loving relationships. And let's be honest: Financial security wouldn't hurt! Achieving this goal depends on understanding how money in a savings account grows in remarkable ways as a result of *compound interest*. **Compound interest** is interest computed on your original investment as well as on any accumulated interest. For example, suppose you deposit $1000, the principal, in a savings account at a rate of 5%. Table 8.1 shows how the investment grows if the interest earned is automatically added on to the principal.

Table 8.1	Compound Interest on $1000		
Year	**Starting Balance**	**Interest Earned:** $I = Pr$	**New Balance**
1	$1000	$1000 × 0.05 = $50	$1050
2	$1050	$1050 × 0.05 = $52.50	$1102.50
3	$1102.50	$1102.50 × 0.05 ≈ $55.13	$1157.63

A faster way to determine the amount, A, in an account subject to compound interest is to use the following formula:

A FORMULA FOR COMPOUND INTEREST Suppose that an amount of money, P, is invested at interest rate r, compounded annually. In t years, the amount, A, or balance, in the account is given by the formula

$$A = P(1 + r)^t.$$

Some compound interest problems can be solved using quadratic equations.

EXAMPLE 8 Solving a Compound Interest Problem

You invested $1000 in an account whose interest is compounded annually. After 2 years, the amount, or balance, in the account is $1210. Find the annual interest rate.

SOLUTION We are given that

$$P \quad \text{(the amount invested)} = \$1000$$

$$t \quad \text{(the time of the investment)} = 2 \text{ years}$$

$$A \quad \text{(the amount, or balance, in the account)} = \$1210.$$

We are asked to find the annual interest rate, r. We substitute the three given values into the compound interest formula and solve for r.

$A = P(1 + r)^t$	*Use the compound interest formula.*
$1210 = 1000(1 + r)^2$	*Substitute the given values.*
$\dfrac{1210}{1000} = (1 + r)^2$	*Divide both sides by 1000.*
$\dfrac{121}{100} = (1 + r)^2$	*Simplify the fraction.*
$1 + r = \sqrt{\dfrac{121}{100}} \quad \text{or} \quad 1 + r = -\sqrt{\dfrac{121}{100}}$	*Apply the square root property.*
$1 + r = \dfrac{11}{10} \qquad 1 + r = -\dfrac{11}{10}$	$\sqrt{\dfrac{121}{100}} = \dfrac{\sqrt{121}}{\sqrt{100}} = \dfrac{11}{10}$
$r = \dfrac{11}{10} - 1 \qquad r = -\dfrac{11}{10} - 1$	*Subtract 1 from both sides and solve for r.*
$r = \dfrac{1}{10} \qquad\quad r = -\dfrac{21}{10}$	$\dfrac{11}{10} - 1 = \dfrac{11}{10} - \dfrac{10}{10} = \dfrac{1}{10}$ and $-\dfrac{11}{10} - 1 = -\dfrac{11}{10} - \dfrac{10}{10} = -\dfrac{21}{10}.$

Because the interest rate cannot be negative, we reject $-\dfrac{21}{10}$. Thus, the annual interest rate is $\dfrac{1}{10} = 0.10 = 10\%$.

We can check this answer using the formula $A = P(1 + r)^t$. If \$1000 is invested for 2 years at 10% interest, compounded annually, the balance in the account is

$$A = \$1000(1 + 0.10)^2 = \$1000(1.10)^2 = \$1210.$$

Because this is precisely the amount given by the problem's conditions, the annual interest rate is, indeed, 10% compounded annually. ∎

 CHECK POINT 8 You invested \$3000 in an account whose interest is compounded annually. After 2 years, the amount, or balance, in the account is \$4320. Find the annual interest rate. *The annual interest rate is 20%.*

In Chapter 5, we solved problems using the Pythagorean Theorem. Recall that in a right triangle, the side opposite the 90° angle is the hypotenuse and the other sides are legs. The Pythagorean Theorem states that the sum of the squares of the lengths of the legs equals the square of the length of the hypotenuse. Some problems that involve the Pythagorean Theorem can be solved using the square root property.

80 feet

80 feet

FIGURE 8.2

| EXAMPLE 9 | Using the Pythagorean Theorem and the Square Root Property |

A supporting wire is to be attached to the top of an 80-foot antenna, as shown in Figure 8.2. Because of surrounding trees, the wire must be anchored 80 feet from the base of the antenna. What length of wire is required?

SOLUTION Let w = the wire's length. We can find w in the right triangle using the Pythagorean Theorem.

| (Leg)² | plus | (Leg)² | equals | (Hypotenuse)². |

$$80^2 \quad + \quad 80^2 \quad = \quad w^2.$$

We solve this equation using the square root property.

$$6400 + 6400 = w^2 \qquad \text{Square 80.}$$

$$12{,}800 = w^2 \qquad \text{Add.}$$

$$w = \sqrt{12{,}800} \quad \text{or} \quad w = -\sqrt{12{,}800} \qquad \text{Apply the square root property.}$$

$$w = \sqrt{6400}\,\sqrt{2} \quad \text{or} \quad w = -\sqrt{6400}\,\sqrt{2} \qquad \text{6400 is the greatest perfect square factor of 12,800.}$$

$$w = 80\sqrt{2} \quad \text{or} \quad w = -80\sqrt{2}$$

Because w represents the wire's length, we reject the negative value. Thus, a wire of length $80\sqrt{2}$ feet is needed to support the antenna. Using a calculator, this is approximately 113.1 feet of wire. ■

Take another look at the right triangle in Figure 8.2. Do you see that both legs are the same length? If the lengths of both legs of a right triangle are the same, the triangle is called an **isosceles right triangle**. Figure 8.3 shows such a triangle. In Example 9, the wire's length was the length of a leg, 80 feet, times $\sqrt{2}$. We can use Figure 8.3 to show that the length of the hypotenuse of any isosceles right triangle is the length of a leg times $\sqrt{2}$.

45°

a

45°

a

FIGURE 8.3 An isosceles right triangle has legs that are the same length and acute angles each measuring 45°.

$$c^2 = a^2 + a^2 \qquad \text{Apply the Pythagorean Theorem to the isosceles right triangle in Figure 8.3.}$$

$$c^2 = 2a^2 \qquad \text{Combine like terms.}$$

$$c = \sqrt{2a^2} \quad \text{or} \quad c = -\sqrt{2a^2} \qquad \text{Apply the square root property.}$$

Reject because c cannot be negative.

$$c = \sqrt{2}\,a \quad \text{or} \quad a\sqrt{2} \qquad \text{Simplify: } \sqrt{2a^2} = \sqrt{2}\,\sqrt{a^2} = \sqrt{2}\,a.$$

LENGTHS WITHIN ISOSCELES RIGHT TRIANGLES The length of the hypotenuse of an isosceles right triangle is the length of a leg times $\sqrt{2}$.

a

$a\sqrt{2}$

a

CHECK POINT 9 A 50-foot supporting wire is to be attached to an antenna. The wire is anchored 20 feet from the base of the antenna. How high up the antenna is the wire attached? Express the answer in simplified radical form. Then find a decimal approximation to the nearest tenth of a foot.   The wire is attached $10\sqrt{21}$ feet, or about 45.8 feet, up the antenna.

50 feet

x feet

20 feet

8.1 EXERCISE SET

 Student Solutions Manual   CD/Video   PH Math/Tutor Center   MathXL Tutorials on CD   Math XL MathXL®   MyMathLab MyMathLab   Interactmath.com

Practice Exercises

In Exercises 1–22, solve each equation by the square root property. If possible, simplify radicals or rationalize denominators. Express imaginary solutions in the form a + bi.

1. $3x^2 = 75$ ± 5 or $\{\pm 5\}$

2. $5x^2 = 20$ ± 2 or $\{\pm 2\}$

3. $7x^2 = 42$ $\pm\sqrt{6}$ or $\{\pm\sqrt{6}\}$

4. $8x^2 = 40$ $\pm\sqrt{5}$ or $\{\pm\sqrt{5}\}$

5. $16x^2 = 25$ $\pm\frac{5}{4}$ or $\{\pm\frac{5}{4}\}$

6. $4x^2 = 49$ $\pm\frac{7}{2}$ or $\{\pm\frac{7}{2}\}$

7. $3x^2 - 2 = 0$ $\pm\dfrac{\sqrt{6}}{3}$ or $\left\{\pm\dfrac{\sqrt{6}}{3}\right\}$

8. $3x^2 - 5 = 0$ $\pm\dfrac{\sqrt{15}}{3}$ or $\left\{\pm\dfrac{\sqrt{15}}{3}\right\}$

9. $25x^2 + 16 = 0$ $\pm\frac{4}{5}i$ or $\{\pm\frac{4}{5}i\}$

10. $4x^2 + 49 = 0$ $\pm\frac{7}{2}i$ or $\{\pm\frac{7}{2}i\}$

11. $(x + 7)^2 = 9$ -10 and -4, or $\{-10, -4\}$

12. $(x + 3)^2 = 64$ -11 and 5, or $\{-11, 5\}$

13. $(x - 3)^2 = 5$ $3 \pm \sqrt{5}$ or $\{3 \pm \sqrt{5}\}$

14. $(x - 4)^2 = 3$ $4 \pm \sqrt{3}$ or $\{4 \pm \sqrt{3}\}$

15. $2(x + 2)^2 = 16$ $-2 \pm 2\sqrt{2}$ or $\{-2 \pm 2\sqrt{2}\}$

16. $3(x + 2)^2 = 36$ $-2 \pm 2\sqrt{3}$ or $\{-2 \pm 2\sqrt{3}\}$

17. $(x - 5)^2 = -9$ $5 \pm 3i$ or $\{5 \pm 3i\}$

18. $(x - 5)^2 = -4$ $5 \pm 2i$ or $\{5 \pm 2i\}$

19. $\left(x + \dfrac{3}{4}\right)^2 = \dfrac{11}{16}$ $\dfrac{-3 \pm \sqrt{11}}{4}$ or $\left\{\dfrac{-3 \pm \sqrt{11}}{4}\right\}$

20. $\left(x + \dfrac{2}{5}\right)^2 = \dfrac{7}{25}$ $\dfrac{-2 \pm \sqrt{7}}{5}$ or $\left\{\dfrac{-2 \pm \sqrt{7}}{5}\right\}$

21. $x^2 - 6x + 9 = 36$ -3 and 9, or $\{-3, 9\}$

22. $x^2 - 6x + 9 = 49$ -4 and 10, or $\{-4, 10\}$

In Exercises 23–34, determine the constant that should be added to the binomial so that it becomes a perfect square trinomial. Then write and factor the trinomial.

23. $x^2 + 2x$ $1; x^2 + 2x + 1 = (x + 1)^2$

24. $x^2 + 4x$ $4; x^2 + 4x + 4 = (x + 2)^2$

25. $x^2 - 14x$ $49; x^2 - 14x + 49 = (x - 7)^2$

26. $x^2 - 10x$ $25; x^2 - 10x + 25 = (x - 5)^2$

27. $x^2 + 7x$ $\frac{49}{4}; x^2 + 7x + \frac{49}{4} = \left(x + \frac{7}{2}\right)^2$

28. $x^2 + 9x$ $\frac{81}{4}; x^2 + 9x + \frac{81}{4} = \left(x + \frac{9}{2}\right)^2$

29. $x^2 - \dfrac{1}{2}x$ $\dfrac{1}{16}; x^2 - \dfrac{1}{2}x + \dfrac{1}{16} = \left(x - \dfrac{1}{4}\right)^2$

30. $x^2 - \dfrac{1}{3}x$ $\dfrac{1}{36}; x^2 - \dfrac{1}{3}x + \dfrac{1}{36} = \left(x - \dfrac{1}{6}\right)^2$

31. $x^2 + \dfrac{4}{3}x$ $\dfrac{4}{9}; x^2 + \dfrac{4}{3}x + \dfrac{4}{9} = \left(x + \dfrac{2}{3}\right)^2$

32. $x^2 + \dfrac{4}{5}x$ $\dfrac{4}{25}; x^2 + \dfrac{4}{5}x + \dfrac{4}{25} = \left(x + \dfrac{2}{5}\right)^2$

33. $x^2 - \dfrac{9}{4}x$ $\dfrac{81}{64}; x^2 - \dfrac{9}{4}x + \dfrac{81}{64} = \left(x - \dfrac{9}{8}\right)^2$

34. $x^2 - \dfrac{9}{5}x \quad \dfrac{81}{100}; x^2 - \dfrac{9}{5}x + \dfrac{81}{100} = \left(x - \dfrac{9}{10}\right)^2$

In Exercises 35–56, solve each quadratic equation by completing the square.

35. $x^2 + 4x = 32$ -8 and 4, or $\{-8, 4\}$

36. $x^2 + 6x = 7$ -7 and 1, or $\{-7, 1\}$

37. $x^2 + 6x = -2$ $-3 \pm \sqrt{7}$ or $\{-3 \pm \sqrt{7}\}$

38. $x^2 + 2x = 5$ $-1 \pm \sqrt{6}$ or $\{-1 \pm \sqrt{6}\}$

39. $x^2 - 8x + 1 = 0$ $4 \pm \sqrt{15}$ or $\{4 \pm \sqrt{15}\}$

40. $x^2 + 8x - 5 = 0$ $-4 \pm \sqrt{21}$ or $\{-4 \pm \sqrt{21}\}$

41. $x^2 + 2x + 2 = 0$ $-1 \pm i$ or $\{-1 \pm i\}$

42. $x^2 - 4x + 8 = 0$ $2 \pm 2i$ or $\{2 \pm 2i\}$

43. $x^2 + 3x - 1 = 0$ $\dfrac{-3 \pm \sqrt{13}}{2}$ or $\left\{\dfrac{-3 \pm \sqrt{13}}{2}\right\}$

44. $x^2 - 3x - 5 = 0$ $\dfrac{3 \pm \sqrt{29}}{2}$ or $\left\{\dfrac{3 \pm \sqrt{29}}{2}\right\}$

45. $x^2 + \dfrac{4}{7}x + \dfrac{3}{49} = 0$ $-\dfrac{3}{7}$ and $-\dfrac{1}{7}$, or $\left\{-\dfrac{3}{7}, -\dfrac{1}{7}\right\}$

46. $x^2 + \dfrac{6}{5}x + \dfrac{8}{25} = 0$ $-\dfrac{4}{5}$ and $-\dfrac{2}{5}$, or $\left\{-\dfrac{4}{5}, -\dfrac{2}{5}\right\}$

47. $x^2 + x - 1 = 0$ $\dfrac{-1 \pm \sqrt{5}}{2}$ or $\left\{\dfrac{-1 \pm \sqrt{5}}{2}\right\}$

48. $x^2 + 3x - 1 = 0$ $\dfrac{-3 \pm \sqrt{13}}{2}$ or $\left\{\dfrac{-3 \pm \sqrt{13}}{2}\right\}$

49. $2x^2 + 3x - 5 = 0$ $-\dfrac{5}{2}$ and 1, or $\left\{-\dfrac{5}{2}, 1\right\}$

50. $2x^2 + 5x - 3 = 0$ -3 and $\dfrac{1}{2}$, or $\left\{-3, \dfrac{1}{2}\right\}$

51. $3x^2 + 6x + 1 = 0$ $\dfrac{-3 \pm \sqrt{6}}{3}$ or $\left\{\dfrac{-3 \pm \sqrt{6}}{3}\right\}$

52. $3x^2 - 6x + 2 = 0$ $\dfrac{3 \pm \sqrt{3}}{3}$ or $\left\{\dfrac{3 \pm \sqrt{3}}{3}\right\}$

53. $3x^2 - 8x + 1 = 0$ $\dfrac{4 \pm \sqrt{13}}{3}$ or $\left\{\dfrac{4 \pm \sqrt{13}}{3}\right\}$

54. $2x^2 + 3x - 4 = 0$ $\dfrac{-3 \pm \sqrt{41}}{4}$ or $\left\{\dfrac{-3 \pm \sqrt{41}}{4}\right\}$

55. $8x^2 - 4x + 1 = 0$ $\dfrac{1}{4} \pm \dfrac{1}{4}i$ or $\left\{\dfrac{1}{4} \pm \dfrac{1}{4}i\right\}$

56. $9x^2 - 6x + 5 = 0$ $\dfrac{1}{3} \pm \dfrac{2}{3}i$ or $\left\{\dfrac{1}{3} \pm \dfrac{2}{3}i\right\}$

57. If $f(x) = (x - 1)^2$, find all values of x for which $f(x) = 36$. $-5, 7$

58. If $f(x) = (x + 2)^2$, find all values of x for which $f(x) = 25$. $-7, 3$

59. If $g(x) = \left(x - \dfrac{2}{5}\right)^2$, find all values of x for which $g(x) = \dfrac{9}{25}$. $-\dfrac{1}{5}, 1$

60. If $g(x) = \left(x + \dfrac{1}{3}\right)^2$, find all values of x for which $g(x) = \dfrac{4}{9}$. $-1, \dfrac{1}{3}$

61. If $h(x) = 5(x + 2)^2$, find all values of x for which $h(x) = -125$. $-2 \pm 5i$

62. If $h(x) = 3(x - 4)^2$, find all values of x for which $h(x) = -12$. $4 \pm 2i$

Practice Plus

63. Three times the square of the difference between a number and 2 is -12. Find the number(s). $2 + 2i$ and $2 - 2i$

64. Three times the square of the difference between a number and 9 is -27. Find the number(s). $9 + 3i$ and $9 - 3i$

In Exercises 65–68, solve the formula for the specified variable. Because each variable is nonnegative, list only the principal square root. If possible, simplify radicals or rationalize denominators.

65. $h = \dfrac{v^2}{2g}$ for v $v = \sqrt{2gh}$

66. $s = \dfrac{kwd^2}{l}$ for d $d = \dfrac{\sqrt{slkw}}{kw}$

67. $A = P(1 + r)^2$ for r $r = \dfrac{\sqrt{AP}}{P} - 1$

68. $C = \dfrac{kP_1 P_2}{d^2}$ for d $d = \dfrac{\sqrt{kP_1 P_2 C}}{C}$

In Exercises 69–72, solve each quadratic equation by completing the square.

69. $\dfrac{x^2}{3} + \dfrac{x}{9} - \dfrac{1}{6} = 0$ $\dfrac{-1 \pm \sqrt{19}}{6}$ or $\left\{\dfrac{-1 \pm \sqrt{19}}{6}\right\}$

70. $\dfrac{x^2}{2} - \dfrac{x}{6} - \dfrac{3}{4} = 0$ $\dfrac{1 \pm \sqrt{55}}{6}$ or $\left\{\dfrac{1 \pm \sqrt{55}}{6}\right\}$

71. $x^2 - bx = 2b^2$ $-b$ and $2b$, or $\{-b, 2b\}$

72. $x^2 - bx = 6b^2$ $-2b$ and $3b$, or $\{-2b, 3b\}$

Application Exercises

In Exercises 73–76, use the compound interest formula

$$A = P(1 + r)^t$$

to find the annual interest rate, r.

73. In 2 years, an investment of $2000 grows to $2880.   20%

74. In 2 years, an investment of $2000 grows to $2420.   10%

75. In 2 years, an investment of $1280 grows to $1445.   6.25%

76. In 2 years, an investment of $80,000 grows to $101,250.   12.5%

Of the one hundred largest economies in the world, 53 are multinational corporations. In 1970, there were approximately 7000 multinational corporations. By 2001, more than 65,000 corporations enveloped the world. The graph shows this rapid growth from 1970 through 2001, including the starting dates of some notable corporations.

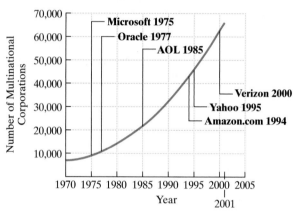

Number of Multinational Corporations in the World

Source: Medard Gabel, *Global Inc.*, The New Press, 2003

The data shown can be modeled by the function

$$f(x) = 62.2x^2 + 7000,$$

where $f(x)$ represents the number of multinational corporations in the world x years after 1970. Use this function and the square root property to solve Exercises 77–78.

77. In which year will there be 92,000 multinational corporations? In approximately 2007, there will be 92,000 corporations.

78. In which year were there 46,000 multinational corporations? How well does the function model the actual number of corporations for that year shown in the graph? In approximately 1995, there were 46,000 corporations. The function models the actual number quite well.

The function $s(t) = 16t^2$ models the distance, $s(t)$, in feet, that an object falls in t seconds. Use this function and the square root property to solve Exercises 79–80. Express answers in simplified radical form. Then use your calculator and find a decimal approximation to the nearest tenth of a second.

79. A sky diver jumps from an airplane and falls for 4800 feet before opening a parachute. For how many seconds was the diver in a free fall? The diver was in a free fall $10\sqrt{3}$ seconds, or about 17.3 seconds.

80. A sky diver jumps from an airplane and falls for 3200 feet before opening a parachute. For how many seconds was the diver in a free fall? The diver was in a free fall $10\sqrt{2}$ seconds, or about 14.1 seconds.

Use the Pythagorean Theorem and the square root property to solve Exercises 81–86. Express answers in simplified radical form. Then find a decimal approximation to the nearest tenth.

81. A rectangular park is 6 miles long and 3 miles wide. How long is a pedestrian route that runs diagonally across the park? The route is $3\sqrt{5}$ miles, or about 6.7 miles long.

82. A rectangular park is 4 miles long and 2 miles wide. How long is a pedestrian route that runs diagonally across the park? The route is $2\sqrt{5}$ miles, or about 4.5 miles long.

83. The base of a 30-foot ladder is 10 feet from the building. If the ladder reaches the flat roof, how tall is the building?

The building is $20\sqrt{2}$ feet tall, or about 28.3 feet.

84. The doorway into a room is 4 feet wide and 8 feet high. What is the length of the longest rectangular panel that can be taken through this doorway diagonally?

The length is $4\sqrt{5}$ feet, or about 8.9 feet.

85. A supporting wire is to be attached to the top of a 50-foot antenna. If the wire must be anchored 50 feet from the base of the antenna, what length of wire is required? The wire needs to be $50\sqrt{2}$ feet, or about 70.7 feet.

86. A supporting wire is to be attached to the top of a 70-foot antenna. If the wire must be anchored 70 feet from the base of the antenna, what length of wire is required? The wire needs to be $70\sqrt{2}$ feet, or about 99.0 feet.

87. A square flower bed is to be enlarged by adding 2 meters on each side. If the larger square has an area of 196 square meters, what is the length of a side of the original square?   10 meters

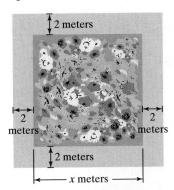

2 meters

2 meters

2 meters

2 meters

x meters

88. A square flower bed is to be enlarged by adding 4 feet on each side. If the larger square has an area of 225 square feet, what is the length of a side of the original square?   7 feet

Writing in Mathematics

89. What is the square root property?   89–96. Answers will vary.

90. Explain how to solve $(x - 1)^2 = 16$ using the square root property.

91. Explain how to complete the square for a binomial. Use $x^2 + 6x$ to illustrate your explanation.

92. Explain how to solve $x^2 + 6x + 8 = 0$ by completing the square.

93. What is compound interest?

94. In your own words, describe the compound interest formula

$$A = P(1 + r)^t.$$

95. What is an isosceles right triangle? What is the relationship between the length of its hypotenuse and the length of a leg?

96. A basketball player's hang time is the time spent in the air when shooting a basket. The function $f(x) = 4x^2$ models the player's vertical leap, $f(x)$, in feet, in terms of the hang time, x, measured in seconds. Use this function to write a problem that can be solved by the square root property. Then solve the problem. Be realistic: It is unlikely that hang time can exceed 1 second.

Technology Exercises

97. Use a graphing utility to solve $4 - (x + 1)^2 = 0$. Graph $y = 4 - (x + 1)^2$ in a $[-5, 5, 1]$ by $[-5, 5, 1]$ viewing rectangle. The equation's solutions are the graph's x-intercepts. Check by substitution in the given equation. -3 and 1, or $\{-3, 1\}$ **97–98.** See graphing answer section.

98. Use a graphing utility to solve $(x - 1)^2 - 9 = 0$. Graph $y = (x - 1)^2 - 9$ in a $[-5, 5, 1]$ by $[-9, 3, 1]$ viewing rectangle. The equation's solutions are the graph's x-intercepts. Check by substitution in the given equation. -2 and 4, or $\{-2, 4\}$

99. Use a graphing utility and x-intercepts to verify any of the real solutions that you obtained for five of the quadratic equations in Exercises 35–56.   Answers will vary.

Critical Thinking Exercises

100. Which one of the following is true?   d

a. The equation $(x - 5)^2 = 12$ is equivalent to $x - 5 = 2\sqrt{3}$.

b. In completing the square for $2x^2 - 6x = 5$, we should add 9 to both sides.

c. Although not every quadratic equation can be solved by completing the square, they can all be solved by factoring.

d. The graph of $y = (x - 2)^2 + 3$ cannot have x-intercepts.

101. Solve for y: $\dfrac{x^2}{a^2} + \dfrac{y^2}{b^2} = 1$. $y = \pm\dfrac{b\sqrt{a^2 - x^2}}{a}$

102. Solve by completing the square:

$x^2 + x + c = 0$. $\dfrac{-1 \pm \sqrt{1 - 4c}}{2}$ or $\left\{\dfrac{-1 \pm \sqrt{1 - 4c}}{2}\right\}$

103. Solve by completing the square:

$x^2 + bx + c = 0$. $\dfrac{-b \pm \sqrt{b^2 - 4c}}{2}$ or $\left\{\dfrac{-b \pm \sqrt{b^2 - 4c}}{2}\right\}$

104. Solve: $x^4 - 8x^2 + 15 = 0$. $\pm\sqrt{5}$ and $\pm\sqrt{3}$, or $\left\{\pm\sqrt{5}, \pm\sqrt{3}\right\}$

Review Exercises

105. Simplify: $4x - 2 - 3[4 - 2(3 - x)]$. (Section 1.2, Example 14) $4 - 2x$

106. Factor: $1 - 8x^3$. (Section 5.5, Example 9) $(1 - 2x)(1 + 2x + 4x^2)$

107. Divide: $(x^4 - 5x^3 + 2x^2 - 6) \div (x - 3)$. (Section 6.5, Example 1) $x^3 - 2x^2 - 4x - 12 - \dfrac{42}{x - 3}$

SECTION

8.2

Objectives

1 Solve quadratic equations using the quadratic formula.

2 Use the discriminant to determine number and type of solutions.

3 Determine the most efficient method to use when solving a quadratic equation.

4 Write quadratic equations from solutions.

5 Use the quadratic formula to solve problems.

THE QUADRATIC FORMULA

America is a nation of immigrants. Since 1820, more than 40 million people have immigrated to the United States from all over the world. They chose to come for various reasons, such as to live in freedom, to practice religion freely, to escape poverty or oppression, and to make better lives for themselves and their children. As a result, a substantial percentage of the U.S. population is foreign-born. The graph in Figure 8.4 shows the percentage of foreign-born Americans from 1930 through 2002.

Percentage of U.S. Population That Was Foreign-Born, 1930-2002

FIGURE 8.4

Source: U.S. Census Bureau

The data shown in Figure 8.4 can be modeled by the quadratic function

$$f(x) = 0.0049x^2 - 0.361x + 11.79,$$

where $f(x)$ is the percentage of the U.S. population that was foreign-born x years after 1930. If the trend shown by the data continues, in which year will 15% of the U.S. population be foreign-born? To answer the question, it is necessary to substitute 15 for $f(x)$ in the given function and solve for x, the number of years after 1930:

$$15 = 0.0049x^2 - 0.361x + 11.79.$$

In this section, we will derive a formula that will enable you to solve quadratic equations more quickly than the method of completing the square. Using this formula, we will work with the function that models the percentage of foreign-born Americans and determine the year in which 15% of our population will be foreign-born.

1	Solve quadratic equations using the quadratic formula.

Solving Quadratic Equations Using the Quadratic Formula We can use the method of completing the square to derive a formula that can be used to solve all quadratic equations. The derivation given here also shows a particular quadratic equation, $3x^2 - 2x - 4 = 0$, to specifically illustrate each of the steps.

Deriving the Quadratic Formula

Standard Form of a Quadratic Equation	Comment	A Specific Example
$ax^2 + bx + c = 0, a > 0$	This is the given equation.	$3x^2 - 2x - 4 = 0$
$x^2 + \dfrac{b}{a}x + \dfrac{c}{a} = 0$	Divide both sides by the coefficient of x^2.	$x^2 - \dfrac{2}{3}x - \dfrac{4}{3} = 0$
$x^2 + \dfrac{b}{a}x = -\dfrac{c}{a}$	Isolate the binomial by adding $-\dfrac{c}{a}$ on both sides.	$x^2 - \dfrac{2}{3}x = \dfrac{4}{3}$
$x^2 + \dfrac{b}{a}x + \left(\dfrac{b}{2a}\right)^2 = -\dfrac{c}{a} + \left(\dfrac{b}{2a}\right)^2$ $\underset{(\text{half})^2}{\underbrace{\qquad}}$	Complete the square. Add the square of half the coefficient of x to both sides.	$x^2 - \dfrac{2}{3}x + \left(-\dfrac{1}{3}\right)^2 = \dfrac{4}{3} + \left(-\dfrac{1}{3}\right)^2$ $\underset{(\text{half})^2}{\underbrace{\qquad}}$
$x^2 + \dfrac{b}{a}x + \dfrac{b^2}{4a^2} = -\dfrac{c}{a} + \dfrac{b^2}{4a^2}$		$x^2 - \dfrac{2}{3}x + \dfrac{1}{9} = \dfrac{4}{3} + \dfrac{1}{9}$
$\left(x + \dfrac{b}{2a}\right)^2 = -\dfrac{c}{a}\cdot\dfrac{4a}{4a} + \dfrac{b^2}{4a^2}$	Factor on the left side and obtain a common denominator on the right side.	$\left(x - \dfrac{1}{3}\right)^2 = \dfrac{4}{3}\cdot\dfrac{3}{3} + \dfrac{1}{9}$
$\left(x + \dfrac{b}{2a}\right)^2 = \dfrac{-4ac + b^2}{4a^2}$	Add fractions on the right side.	$\left(x - \dfrac{1}{3}\right)^2 = \dfrac{12 + 1}{9}$
$\left(x + \dfrac{b}{2a}\right)^2 = \dfrac{b^2 - 4ac}{4a^2}$		$\left(x - \dfrac{1}{3}\right)^2 = \dfrac{13}{9}$
$x + \dfrac{b}{2a} = \pm\sqrt{\dfrac{b^2 - 4ac}{4a^2}}$	Apply the square root property.	$x - \dfrac{1}{3} = \pm\sqrt{\dfrac{13}{9}}$
$x + \dfrac{b}{2a} = \pm\dfrac{\sqrt{b^2 - 4ac}}{2a}$	Take the square root of the quotient, simplifying the denominator.	$x - \dfrac{1}{3} = \pm\dfrac{\sqrt{13}}{3}$
$x = \dfrac{-b}{2a} \pm \dfrac{\sqrt{b^2 - 4ac}}{2a}$	Solve for x by subtracting $\dfrac{b}{2a}$ from both sides.	$x = \dfrac{1}{3} \pm \dfrac{\sqrt{13}}{3}$
$x = \dfrac{-b \pm \sqrt{b^2 - 4ac}}{2a}$	Combine fractions on the right side.	$x = \dfrac{1 \pm \sqrt{13}}{3}$

The formula shown at the bottom of the left column is called the *quadratic formula*. A similar proof shows that the same formula can be used to solve quadratic equations if a, the coefficient of the x^2-term, is negative.

THE QUADRATIC FORMULA The solutions of a quadratic equation in standard form $ax^2 + bx + c = 0$, with $a \neq 0$, are given by the **quadratic formula**

$$x = \dfrac{-b \pm \sqrt{b^2 - 4ac}}{2a}.$$

x equals negative b plus or minus the square root of $b^2 - 4ac$, all divided by 2a.

To use the quadratic formula, write the quadratic equation in standard form if necessary. Then determine the numerical values for a (the coefficient of the x^2-term), b (the coefficient of the x-term), and c (the constant term). Substitute the values of a, b, and c into the quadratic formula and evaluate the expression. The $\pm$ sign indicates that there are two solutions of the equation.

EXAMPLE 1 Solving a Quadratic Equation Using the Quadratic Formula

Solve using the quadratic formula: $8x^2 + 2x - 1 = 0$.

SOLUTION The given equation is in standard form. Begin by identifying the values for a, b, and c.

$$8x^2 + 2x - 1 = 0$$

$a = 8 \qquad b = 2 \qquad c = -1$

Substituting these values into the quadratic formula and simplifying gives the equation's solutions.

$$x = \frac{-b \pm \sqrt{b^2 - 4ac}}{2a}$$ Use the quadratic formula.

$$x = \frac{-2 \pm \sqrt{2^2 - 4(8)(-1)}}{2(8)}$$ Substitute the values for a, b, and c: $a = 8$, $b = 2$, and $c = -1$.

$$= \frac{-2 \pm \sqrt{4 - (-32)}}{16}$$ $2^2 - 4(8)(-1) = 4 - (-32)$

$$= \frac{-2 \pm \sqrt{36}}{16}$$ $4 - (-32) = 4 + 32 = 36$

$$= \frac{-2 \pm 6}{16}$$ $\sqrt{36} = 6$

Now we will evaluate this expression in two different ways to obtain the two solutions. On the left, we will *add* 6 to -2. On the right, we will *subtract* 6 from -2.

$$x = \frac{-2 + 6}{16} \quad \text{or} \quad x = \frac{-2 - 6}{16}$$

$$= \frac{4}{16} = \frac{1}{4} \qquad\qquad = \frac{-8}{16} = -\frac{1}{2}$$

The solutions are $-\frac{1}{2}$ and $\frac{1}{4}$, and the solution set is $\left\{-\frac{1}{2}, \frac{1}{4}\right\}$. ∎

In Example 1, the solutions of $8x^2 + 2x - 1 = 0$ are rational numbers. This means that the equation can also be solved by factoring. The reason that the solutions are rational numbers is that $b^2 - 4ac$, the radicand in the quadratic formula, is 36, which is a perfect square. If a, b, and c are rational numbers, all quadratic equations for which $b^2 - 4ac$ is a perfect square have rational solutions.

USING TECHNOLOGY

The graph of the quadratic function

$$y = 8x^2 + 2x - 1$$

has x-intercepts at $-\frac{1}{2}$ and $\frac{1}{4}$. This verifies that $\left\{-\frac{1}{2}, \frac{1}{4}\right\}$ is the solution set of the quadratic equation

$$8x^2 + 2x - 1 = 0.$$

$y = 8x^2 + 2x - 1$

x-intercept is $-\frac{1}{2}$. x-intercept is $\frac{1}{4}$.

$[-2, 2, 1]$ by $[-3, 10, 1]$

✔ **CHECK POINT 1** Solve using the quadratic formula: $2x^2 + 9x - 5 = 0$.

-5 and $\frac{1}{2}$, or $\left\{-5, \frac{1}{2}\right\}$

EXAMPLE 2 Solving a Quadratic Equation Using the Quadratic Formula

Solve using the quadratic formula:

$$2x^2 = 4x + 1.$$

SOLUTION The quadratic equation must be in standard form to identify the values for a, b, and c. To move all terms to one side and obtain zero on the right, we subtract $4x + 1$ from both sides. Then we can identify the values for a, b, and c.

$$2x^2 = 4x + 1 \qquad \text{This is the given equation.}$$

$$2x^2 - 4x - 1 = 0 \qquad \text{Subtract } 4x + 1 \text{ from both sides.}$$

$$a = 2 \qquad b = -4 \qquad c = -1$$

Substituting these values into the quadratic formula and simplifying gives the equation's solutions.

$$x = \frac{-b \pm \sqrt{b^2 - 4ac}}{2a} \qquad \text{Use the quadratic formula.}$$

$$x = \frac{-(-4) \pm \sqrt{(-4)^2 - 4(2)(-1)}}{2(2)} \qquad \begin{array}{l} \text{Substitute the values for } a, b, \text{ and } c: \\ a = 2, b = -4, \text{ and } c = -1. \end{array}$$

$$= \frac{4 \pm \sqrt{16 - (-8)}}{4} \qquad (-4)^2 - 4(2)(-1) = 16 - (-8)$$

$$= \frac{4 \pm \sqrt{24}}{4} \qquad 16 - (-8) = 16 + 8 = 24$$

$$= \frac{4 \pm 2\sqrt{6}}{4} \qquad \sqrt{24} = \sqrt{4 \cdot 6} = \sqrt{4} \cdot \sqrt{6} = 2\sqrt{6}$$

$$= \frac{2(2 \pm \sqrt{6})}{4} \qquad \text{Factor out 2 from the numerator.}$$

$$= \frac{2 \pm \sqrt{6}}{2} \qquad \text{Divide the numerator and denominator by 2.}$$

The solutions are $\dfrac{2 \pm \sqrt{6}}{2}$, and the solution set is $\left\{ \dfrac{2 + \sqrt{6}}{2}, \dfrac{2 - \sqrt{6}}{2} \right\}$ or $\left\{ \dfrac{2 \pm \sqrt{6}}{2} \right\}$.

USING TECHNOLOGY

You can use a graphing utility to verify that the solutions of $2x^2 - 4x - 1 = 0$ are $\dfrac{2 \pm \sqrt{6}}{2}$. Begin by entering $y_1 = 2x^2 - 4x - 1$ in the $\boxed{Y=}$ screen. Then evaluate this function at each of the proposed solutions.

```
Y₁((2+√(6))/2)
                    0
Y₁((2-√(6))/2)
                    0
■
```

In each case, the function value is 0, verifying that the solutions satisfy $2x^2 - 4x - 1 = 0$.

In Example 2, the solutions of $2x^2 = 4x + 1$ are irrational numbers. This means that the equation cannot be solved by factoring. The reason that the solutions are irrational numbers is that $b^2 - 4ac$, the radicand in the quadratic formula, is 24, which is not a perfect square. Notice, too, that the solutions, $\dfrac{2 + \sqrt{6}}{2}$ and $\dfrac{2 - \sqrt{6}}{2}$, are conjugates.

STUDY TIP

Many students use the quadratic formula correctly until the last step, where they make an error in simplifying the solutions. Be sure to factor the numerator before dividing the numerator and the denominator by the greatest common factor.

$$\frac{4 \pm 2\sqrt{6}}{4} = \frac{2(2 \pm \sqrt{6})}{4} = \frac{\overset{1}{\cancel{2}}(2 \pm \sqrt{6})}{\underset{2}{\cancel{4}}} = \frac{2 \pm \sqrt{6}}{2}$$

You cannot divide just one term in the numerator and the denominator by their greatest common factor.

Incorrect!

$$\frac{\overset{1}{\cancel{4}} \pm 2\sqrt{6}}{\underset{1}{\cancel{4}}} = 1 \pm 2\sqrt{6} \qquad\qquad \frac{4 \pm \overset{1}{\cancel{2}}\sqrt{6}}{\underset{2}{\cancel{4}}} = \frac{4 \pm \sqrt{6}}{2}$$

Can all irrational solutions of quadratic equations be simplified? No. The following solutions cannot be simplified:

$$\frac{5 \pm 2\sqrt{7}}{2} \qquad\qquad \frac{-4 \pm 3\sqrt{7}}{2}.$$

Other than 1, terms in each numerator have no common factor.

 CHECK POINT 2 Solve using the quadratic formula: $2x^2 = 6x - 1$.

$$\frac{3 \pm \sqrt{7}}{2} \text{ or } \left\{ \frac{3 \pm \sqrt{7}}{2} \right\}$$

EXAMPLE 3 Solving a Quadratic Equation Using the Quadratic Formula

Solve using the quadratic formula:

$$3x^2 + 2 = -4x.$$

SOLUTION Begin by writing the quadratic equation in standard form.

$$3x^2 + 2 = -4x \qquad \text{This is the given equation.}$$

$$3x^2 + 4x + 2 = 0 \qquad \text{Add 4x to both sides.}$$

$$\boxed{a = 3} \quad \boxed{b = 4} \quad \boxed{c = 2}$$

Substituting these values into the quadratic formula and simplifying gives the equation's solutions.

$$x = \frac{-b \pm \sqrt{b^2 - 4ac}}{2a}$$ Use the quadratic formula.

$$x = \frac{-4 \pm \sqrt{4^2 - 4 \cdot 3 \cdot 2}}{2 \cdot 3}$$ Substitute the values for a, b, and c: $a = 3$, $b = 4$, and $c = 2$.

$$= \frac{-4 \pm \sqrt{16 - 24}}{6}$$ Multiply under the radical.

$$= \frac{-4 \pm \sqrt{-8}}{6}$$ Subtract under the radical.

$$= \frac{-4 \pm 2\sqrt{2}i}{6}$$ $\sqrt{-8} = \sqrt{8(-1)} = \sqrt{8}\sqrt{-1} = \sqrt{8}i$
$= \sqrt{4 \cdot 2}i = 2\sqrt{2}i$

$$= \frac{2(-2 \pm \sqrt{2}i)}{6}$$ Factor out 2 from the numerator.

$$= \frac{-2 \pm \sqrt{2}i}{3}$$ Divide the numerator and denominator by 2.

$$= -\frac{2}{3} \pm \frac{\sqrt{2}}{3}i$$ Write in the form $a + bi$.

The solutions are $-\dfrac{2}{3} \pm \dfrac{\sqrt{2}}{3}i$, and the solution set is $\left\{ -\dfrac{2}{3} + \dfrac{\sqrt{2}}{3}i, -\dfrac{2}{3} - \dfrac{\sqrt{2}}{3}i \right\}$ or $\left\{ -\dfrac{2}{3} \pm \dfrac{\sqrt{2}}{3}i \right\}$.

USING TECHNOLOGY

The graph of the quadratic function

$$y = 3x^2 + 4x + 2$$

has no x-intercepts. This verifies that the equation in Example 3

$$3x^2 + 2 = -4x, \quad \text{or}$$
$$3x^2 + 4x + 2 = 0$$

has imaginary solutions.

$y = 3x^2 + 4x + 2$

$[-3, 3, 1]$ by $[-2, 10, 1]$

In Example 3, the solutions of $3x^2 + 2 = -4x$ are imaginary numbers. This means that the equation cannot be solved using factoring. The reason that the solutions are imaginary numbers is that $b^2 - 4ac$, the radicand in the quadratic formula, is -8, which is negative. Notice, too, that the solutions are complex conjugates.

✔ **CHECK POINT 3** Solve using the quadratic formula: $3x^2 + 5 = -6x$.

$-1 \pm \dfrac{\sqrt{6}}{3}i$ or $\left\{ -1 \pm \dfrac{\sqrt{6}}{3}i \right\}$

Some rational equations can be solved using the quadratic formula. For example, consider the equation

$$3 + \frac{4}{x} = -\frac{2}{x^2}.$$

The denominators are x and x^2. The least common denominator is x^2. We clear fractions by multiplying both sides of the equation by x^2. Notice that x cannot equal zero.

$$x^2\left(3 + \frac{4}{x}\right) = x^2\left(-\frac{2}{x^2}\right), \quad x \neq 0$$

$$3x^2 + \frac{4}{x} \cdot x^2 = x^2\left(-\frac{2}{x^2}\right)$$ Use the distributive property.

$$3x^2 + 4x = -2$$ Simplify.

By adding 2 to both sides of $3x^2 + 4x = -2$, we obtain the standard form of the quadratic equation:

$$3x^2 + 4x + 2 = 0.$$

This is the equation that we solved in Example 3. The two imaginary solutions are not part of the restriction that $x \neq 0$.

2 Use the discriminant to determine the number and type of solutions.

The Discriminant The quantity $b^2 - 4ac$, which appears under the radical sign in the quadratic formula, is called the **discriminant**. The discriminant of the quadratic equation $ax^2 + bx + c = 0$ determines the number and type of solutions. In Example 1, the discriminant was 36, a positive number that is a perfect square. The equation had two solutions that were rational numbers. In Example 2, the discriminant was 24, a positive number that is not a perfect square. The equation had two solutions that were irrational numbers. Finally, in Example 3, the discriminant was -8, a negative number. The equation had solutions that were imaginary numbers involving i. In this case, the graph of the corresponding quadratic function had no x-intercepts.

These observations are generalized in Table 8.2.

STUDY TIP

Checking irrational and imaginary solutions can be time-consuming. The solutions given by the quadratic formula are always correct, unless you have made a careless error. Checking for computational errors or errors in simplification is sufficient.

Table 8.2 **The Discriminant and the Kinds of Solutions to $ax^2 + bx + c = 0$**

Discriminant $b^2 - 4ac$	Kinds of Solutions to $ax^2 + bx + c = 0$	Graph of $y = ax^2 + bx + c$
$b^2 - 4ac > 0$	**Two unequal real solutions;** If $a, b,$ and c are rational numbers and the discriminant is a perfect square, the solutions are rational. If the discriminant is not a perfect square, the solutions are irrational conjugates.	 Two x-intercepts
$b^2 - 4ac = 0$	**One real solution (a repeated solution);** If $a, b,$ and c are rational numbers, the repeated solution is also a rational number.	 One x-intercept
$b^2 - 4ac < 0$	**No real solution; two imaginary solutions;** The solutions are complex conjugates.	 No x-intercepts

EXAMPLE 4 Using the Discriminant

For each equation, compute the discriminant. Then determine the number and type of solutions:

a. $3x^2 + 4x - 5 = 0$ **b.** $9x^2 - 6x + 1 = 0$ **c.** $3x^2 - 8x + 7 = 0.$

SOLUTION Begin by identifying the values for a, b, and c in each equation. Then compute $b^2 - 4ac$, the discriminant.

a. $3x^2 + 4x - 5 = 0$

$a = 3$ $b = 4$ $c = -5$

Substitute and compute the discriminant:

$$b^2 - 4ac = 4^2 - 4 \cdot 3(-5) = 16 - (-60) = 16 + 60 = 76.$$

The discriminant, 76, is a positive number that is not a perfect square. Thus, there are two irrational solutions. (These solutions are conjugates of each other.)

b. $9x^2 - 6x + 1 = 0$

$a = 9$ $b = -6$ $c = 1$

Substitute and compute the discriminant:

$$b^2 - 4ac = (-6)^2 - 4 \cdot 9 \cdot 1 = 36 - 36 = 0.$$

The discriminant, 0, shows that there is only one real solution. This real solution is a rational number.

c. $3x^2 - 8x + 7 = 0$

$a = 3$ $b = -8$ $c = 7$

$$b^2 - 4ac = (-8)^2 - 4 \cdot 3 \cdot 7 = 64 - 84 = -20$$

The negative discriminant, -20, shows that there are two imaginary solutions. (These solutions are complex conjugates of each other.) ■

 CHECK POINT 4 For each equation, compute the discriminant. Then determine the number and type of solutions:

a. $x^2 + 6x + 9 = 0$ 0; one real solution

b. $2x^2 - 7x - 4 = 0$ 81; two rational solutions

c. $3x^2 - 2x + 4 = 0.$ -44; two imaginary solutions that are complex conjugates

3 Determine the most efficient method to use when solving a quadratic equation.

Determining Which Method to Use All quadratic equations can be solved by the quadratic formula. However, if an equation is in the form $u^2 = d$, such as $x^2 = 5$ or $(2x + 3)^2 = 8$, it is faster to use the square root property, taking the square root of both sides. If the equation is not in the form $u^2 = d$, write the quadratic equation in standard form $(ax^2 + bx + c = 0)$. Try to solve the equation by factoring. If $ax^2 + bx + c$ cannot be factored, then solve the quadratic equation by the quadratic formula.

Because we used the method of completing the square to derive the quadratic formula, we no longer need it for solving quadratic equations. However, we will use completing the square in Chapter 10 to help graph certain kinds of equations.

Table 8.3 summarizes our observations about which technique to use when solving a quadratic equation.

Table 8.3	**Determining the Most Efficient Technique to Use when Solving a Quadratic Equation**	
Description and Form of the Quadratic Equation	**Most Efficient Solution Method**	**Example**
$ax^2 + bx + c = 0$ and $ax^2 + bx + c$ can be factored easily.	Factor and use the zero-product principle.	$3x^2 + 5x - 2 = 0$ $(3x - 1)(x + 2) = 0$ $3x - 1 = 0$ or $x + 2 = 0$ $x = \dfrac{1}{3}$ $\quad x = -2$
$ax^2 + c = 0$ The quadratic equation has no x-term. $(b = 0)$	Solve for x^2 and apply the square root property.	$4x^2 - 7 = 0$ $4x^2 = 7$ $x^2 = \dfrac{7}{4}$ $x = \pm\dfrac{\sqrt{7}}{2}$
$u^2 = d$; u is a first-degree polynomial.	Use the square root property.	$(x + 4)^2 = 5$ $x + 4 = \pm\sqrt{5}$ $x = -4 \pm \sqrt{5}$
$ax^2 + bx + c = 0$ and $ax^2 + bx + c$ cannot be factored or the factoring is too difficult.	Use the quadratic formula: $x = \dfrac{-b \pm \sqrt{b^2 - 4ac}}{2a}.$	$x^2 - 2x - 6 = 0$ $\boxed{a = 1} \quad \boxed{b = -2} \quad \boxed{c = -6}$ $x = \dfrac{-(-2) \pm \sqrt{(-2)^2 - 4(1)(-6)}}{2(1)}$ $= \dfrac{2 \pm \sqrt{4 - 4(1)(-6)}}{2(1)}$ $= \dfrac{2 \pm \sqrt{28}}{2} = \dfrac{2 \pm \sqrt{4}\sqrt{7}}{2}$ $= \dfrac{2 \pm 2\sqrt{7}}{2} = \dfrac{2(1 \pm \sqrt{7})}{2}$ $= 1 \pm \sqrt{7}$

4 Write quadratic equations from solutions.

Writing Quadratic Equations from Solutions   Using the zero-product principle, the equation $(x - 3)(x + 5) = 0$ has two solutions, 3 and -5. By applying the zero-product principle in reverse, we can find a quadratic equation that has two given numbers as its solutions.

THE ZERO-PRODUCT PRINCIPLE IN REVERSE   If $A = 0$ or $B = 0$, then $AB = 0$.

Jasper Johns, "Zero". © Jasper Johns/Licensed by VAGA, New York, NY.

The special properties of zero make it possible to write a quadratic equation from its solutions.

EXAMPLE 5 Writing Equations from Solutions

Write a quadratic equation with the given solution set:

a. $\left\{-\dfrac{5}{3}, \dfrac{1}{2}\right\}$ **b.** $\{-5i, 5i\}$.

SOLUTION

a. Because the solution set is $\left\{-\dfrac{5}{3}, \dfrac{1}{2}\right\}$, then

$$x = -\frac{5}{3} \quad \text{or} \quad x = \frac{1}{2}.$$

$$x + \frac{5}{3} = 0 \quad \text{or} \quad x - \frac{1}{2} = 0 \qquad \textit{Obtain zero on one side of each equation.}$$

$$3x + 5 = 0 \quad \text{or} \quad 2x - 1 = 0 \qquad \textit{Clear fractions, multiplying by 3 and 2, respectively.}$$

$$(3x + 5)(2x - 1) = 0 \qquad \textit{Use the zero-product principle in reverse: If A = 0 or B = 0, then AB = 0.}$$

$$6x^2 - 3x + 10x - 5 = 0 \qquad \textit{Use the FOIL method to multiply.}$$

$$6x^2 + 7x - 5 = 0. \qquad \textit{Combine like terms.}$$

Thus, one equation is $6x^2 + 7x - 5 = 0$. Many other quadratic equations have $\left\{-\dfrac{5}{3}, \dfrac{1}{2}\right\}$ for their solution sets. These equations can be obtained by multiplying both sides of $6x^2 + 7x - 5 = 0$ by any nonzero real number.

b. Because the solution set is $\{-5i, 5i\}$, then

$$x = -5i \quad \text{or} \quad x = 5i$$

$$x + 5i = 0 \quad \text{or} \quad x - 5i = 0 \qquad \textit{Obtain zero on one side of each equation.}$$

$$(x + 5i)(x - 5i) = 0 \qquad \textit{Use the zero-product principle in reverse: If A = 0 or B = 0, then AB = 0.}$$

$$x^2 - (5i)^2 = 0 \qquad \textit{Multiply conjugates using } (A + B)(A - B) = A^2 - B^2.$$

$$x^2 - 25i^2 = 0 \qquad (5i)^2 = 5^2 i^2 = 25 i^2$$

$$x^2 - 25(-1) = 0 \qquad i^2 = -1$$

$$x^2 + 25 = 0. \qquad \textit{This is the required equation.} \qquad ■$$

 CHECK POINT 5 Write a quadratic equation with the given solution set:

a. $\left\{-\dfrac{3}{5}, \dfrac{1}{4}\right\}$ $20x^2 + 7x - 3 = 0$ **b.** $\{-7i, 7i\}$. $x^2 + 49 = 0$

Applications   Quadratic equations can be solved to answer questions about variables contained in quadratic functions.

5 Use the quadratic formula to solve problems.

EXAMPLE 6 Making Predictions about the U.S. Foreign-Born Population

The quadratic function

$$f(x) = 0.0049x^2 - 0.361x + 11.79$$

models the percentage of the U.S. population, $f(x)$, that was foreign-born x years after 1930. According to this model, in which year will 15% of the U.S. population be foreign-born?

SOLUTION Because we are interested in when the foreign-born percentage will reach 15%, we substitute 15 for $f(x)$ in the given function. Then we solve for x, the number of years after 1930.

$$f(x) = 0.0049x^2 - 0.361x + 11.79 \qquad \text{This is the given function.}$$

$$15 = 0.0049x^2 - 0.361x + 11.79 \qquad \text{Substitute 15 for f(x).}$$

$$0 = 0.0049x^2 - 0.361x - 3.21 \qquad \text{Subtract 15 from both sides and write the quadratic equation in standard form.}$$

$$\boxed{a = 0.0049} \quad \boxed{b = -0.361} \quad \boxed{c = -3.21}$$

Because the trinomial on the right side of the equation is not easily factored, if it can be factored at all, we solve using the quadratic formula.

$$x = \frac{-b \pm \sqrt{b^2 - 4ac}}{2a} \qquad \text{Use the quadratic formula.}$$

$$= \frac{-(-0.361) \pm \sqrt{(-0.361)^2 - 4(0.0049)(-3.21)}}{2(0.0049)} \qquad \text{Substitute the values for a, b, and c: } a = 0.0049, b = -0.361, c = -3.21.$$

$$= \frac{0.361 \pm \sqrt{0.193237}}{0.0098} \qquad \text{Perform the indicated computations. Use a calculator to simplify the radicand.}$$

$$x = \frac{0.361 + \sqrt{0.193237}}{0.0098} \quad \text{or} \quad x = \frac{0.361 - \sqrt{0.193237}}{0.0098}$$

$$x \approx 82 \qquad\qquad x \approx -8$$

Reject this solution. The model applies to years **after**, not **before**, 1930.

According to the model, approximately 82 years after 1930, in 2012, 15% of the U.S. population will be foreign-born. ∎

USING TECHNOLOGY

A graphing utility's TABLE feature can be used to verify the solution to Example 6.

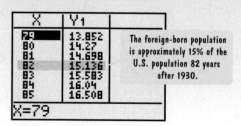

CHECK POINT 6 According to the model in Example 6, in which year(s) was 5.4% of the U.S. population foreign-born? How well does the function describe the situation for that (these) year(s) shown in Figure 8.4 on page 574? 1960 and 1974; very well

8.2 EXERCISE SET

Student Solutions Manual   CD/Video   PH Math/Tutor Center   MathXL Tutorials on CD   MathXL®   MyMathLab   Interactmath.com

Practice Exercises

In Exercises 1–16, solve each equation using the quadratic formula. Simplify solutions, if possible.

1. $x^2 + 8x + 12 = 0$ -6 and -2, or $\{-6, -2\}$

2. $x^2 + 8x + 15 = 0$ -5 and -3, or $\{-5, -3\}$

3. $2x^2 - 7x = -5$ 1 and $\frac{5}{2}$, or $\left\{1, \frac{5}{2}\right\}$

4. $5x^2 + 8x = -3$ -1 and $-\frac{3}{5}$, or $\left\{-1, -\frac{3}{5}\right\}$

5. $x^2 + 3x - 20 = 0$ $\dfrac{-3 \pm \sqrt{89}}{2}$ or $\left\{\dfrac{-3 \pm \sqrt{89}}{2}\right\}$

6. $x^2 + 5x - 10 = 0$ $\dfrac{-5 \pm \sqrt{65}}{2}$ or $\left\{\dfrac{-5 \pm \sqrt{65}}{2}\right\}$

7. $3x^2 - 7x = 3$ $\dfrac{7 \pm \sqrt{85}}{6}$ or $\left\{\dfrac{7 \pm \sqrt{85}}{6}\right\}$

8. $4x^2 + 3x = 2$ $\dfrac{-3 \pm \sqrt{41}}{8}$ or $\left\{\dfrac{-3 \pm \sqrt{41}}{8}\right\}$

9. $6x^2 = 2x + 1$ $\dfrac{1 \pm \sqrt{7}}{6}$ or $\left\{\dfrac{1 \pm \sqrt{7}}{6}\right\}$

10. $2x^2 = -4x + 5$ $\dfrac{-2 \pm \sqrt{14}}{2}$ or $\left\{\dfrac{-2 \pm \sqrt{14}}{2}\right\}$

11. $4x^2 - 3x = -6$ $\dfrac{3}{8} \pm \dfrac{\sqrt{87}}{8}i$ or $\left\{\dfrac{3}{8} \pm \dfrac{\sqrt{87}}{8}i\right\}$

12. $9x^2 + x = -2$ $-\dfrac{1}{18} \pm \dfrac{\sqrt{71}}{18}i$ or $\left\{-\dfrac{1}{18} \pm \dfrac{\sqrt{71}}{18}i\right\}$

13. $x^2 - 4x + 8 = 0$ $2 \pm 2i$ or $\{2 \pm 2i\}$

14. $x^2 + 6x + 13 = 0$ $-3 \pm 2i$ or $\{-3 \pm 2i\}$

15. $3x^2 = 8x - 7$ $\dfrac{4}{3} \pm \dfrac{\sqrt{5}}{3}i$ or $\left\{\dfrac{4}{3} \pm \dfrac{\sqrt{5}}{3}i\right\}$

16. $3x^2 = 4x - 6$ $\dfrac{2}{3} \pm \dfrac{\sqrt{14}}{3}i$ or $\left\{\dfrac{2}{3} \pm \dfrac{\sqrt{14}}{3}i\right\}$

17. $2x(x - 2) = x + 12$ $-\frac{3}{2}$ and 4, or $\left\{-\frac{3}{2}, 4\right\}$

18. $2x(x + 4) = 3x - 3$ $-\frac{3}{2}$ and -1, or $\left\{-\frac{3}{2}, -1\right\}$

In Exercises 19–30, compute the discriminant. Then determine the number and type of solutions for the given equation.

19. $x^2 + 8x + 3 = 0$ 52; two irrational solutions

20. $x^2 + 7x + 4 = 0$ 33; two irrational solutions

21. $x^2 + 6x + 8 = 0$ 4; two rational solutions

22. $x^2 + 2x - 3 = 0$ 16; two rational solutions

23. $2x^2 + x + 3 = 0$ -23; two imaginary solutions

24. $2x^2 - 4x + 3 = 0$ -8; two imaginary solutions

25. $2x^2 + 6x = 0$ 36; two rational solutions

26. $3x^2 - 5x = 0$ 25; two rational solutions

27. $5x^2 + 3 = 0$ -60; two imaginary solutions

28. $5x^2 + 4 = 0$ -80; two imaginary solutions

29. $9x^2 = 12x - 4$ 0; one (repeated) rational solution

30. $4x^2 = 20x - 25$ 0; one (repeated) rational solution

In Exercises 31–46, solve each equation by the method of your choice. Simplify solutions, if possible.

31. $3x^2 - 4x = 4$ $-\frac{2}{3}$ and 2, or $\left\{-\frac{2}{3}, 2\right\}$

32. $2x^2 - x = 1$ $-\frac{1}{2}$ and 1, or $\left\{-\frac{1}{2}, 1\right\}$

33. $x^2 - 2x = 1$ $1 \pm \sqrt{2}$ or $\{1 \pm \sqrt{2}\}$

34. $2x^2 + 3x = 1$ $\dfrac{-3 \pm \sqrt{17}}{4}$ or $\left\{\dfrac{-3 \pm \sqrt{17}}{4}\right\}$

35. $(2x - 5)(x + 1) = 2$ $\dfrac{3 \pm \sqrt{65}}{4}$ or $\left\{\dfrac{3 \pm \sqrt{65}}{4}\right\}$

36. $(2x + 3)(x + 4) = 1$ $\dfrac{-11 \pm \sqrt{33}}{4}$ or $\left\{\dfrac{-11 \pm \sqrt{33}}{4}\right\}$

37. $(3x - 4)^2 = 16$ 0 and $\frac{8}{3}$, or $\left\{0, \frac{8}{3}\right\}$

38. $(2x + 7)^2 = 25$ -6 and -1, or $\{-6, -1\}$

39. $\dfrac{x^2}{2} + 2x + \dfrac{2}{3} = 0$ $\dfrac{-6 \pm 2\sqrt{6}}{3}$ or $\left\{\dfrac{-6 \pm 2\sqrt{6}}{3}\right\}$

40. $\dfrac{x^2}{3} - x - \dfrac{1}{6} = 0$ $\dfrac{3 \pm \sqrt{11}}{2}$ or $\left\{\dfrac{3 \pm \sqrt{11}}{2}\right\}$

41. $(3x - 2)^2 = 10$ $\dfrac{2 \pm \sqrt{10}}{3}$ or $\left\{\dfrac{2 \pm \sqrt{10}}{3}\right\}$

42. $(4x - 1)^2 = 15$ $\dfrac{1 \pm \sqrt{15}}{4}$ or $\left\{\dfrac{1 \pm \sqrt{15}}{4}\right\}$

43. $\dfrac{1}{x} + \dfrac{1}{x + 2} = \dfrac{1}{3}$ $2 \pm \sqrt{10}$ or $\{2 \pm \sqrt{10}\}$

44. $\dfrac{1}{x} + \dfrac{1}{x + 3} = \dfrac{1}{4}$ $\dfrac{5 \pm \sqrt{73}}{2}$ or $\left\{\dfrac{5 \pm \sqrt{73}}{2}\right\}$

45. $(2x - 6)(x + 2) = 5(x - 1) - 12$ 1 and $\frac{5}{2}$, or $\left\{1, \frac{5}{2}\right\}$

46. $7x(x - 2) = 3 - 2(x + 4)$ $\frac{5}{7}$ and 1, or $\left\{\frac{5}{7}, 1\right\}$

In Exercises 47–60, write a quadratic equation in standard form with the given solution set.

47. $\{-3, 5\}$ $x^2 - 2x - 15 = 0$

48. $\{-2, 6\}$ $x^2 - 4x - 12 = 0$

49. $\left\{-\dfrac{2}{3}, \dfrac{1}{4}\right\}$ $12x^2 + 5x - 2 = 0$

50. $\left\{-\dfrac{5}{6}, \dfrac{1}{3}\right\}$ $18x^2 + 9x - 5 = 0$

51. $\{-6i, 6i\}$ $x^2 + 36 = 0$

52. $\{-8i, 8i\}$ $x^2 + 64 = 0$

53. $\left\{-\sqrt{2}, \sqrt{2}\right\}$ $x^2 - 2 = 0$

54. $\left\{-\sqrt{3}, \sqrt{3}\right\}$ $x^2 - 3 = 0$

55. $\left\{-2\sqrt{5}, 2\sqrt{5}\right\}$ $x^2 - 20 = 0$

56. $\left\{-3\sqrt{5}, 3\sqrt{5}\right\}$ $x^2 - 45 = 0$

57. $\left\{1 + i, 1 - i\right\}$ $x^2 - 2x + 2 = 0$

58. $\left\{2 + i, 2 - i\right\}$ $x^2 - 4x + 5 = 0$

59. $\left\{1 + \sqrt{2}, 1 - \sqrt{2}\right\}$ $x^2 - 2x - 1 = 0$

60. $\left\{1 + \sqrt{3}, 1 - \sqrt{3}\right\}$ $x^2 - 2x - 2 = 0$

Practice Plus

Exercises 61–64 describe quadratic equations. Match each description with the graph of the corresponding qudratic function. Each graph is shown in a $[-10, 10, 1]$ *by* $[-10, 10, 1]$ *viewing rectangle.*

61. A quadratic equation whose solution set contains imaginary numbers   b

62. A quadratic equation whose discriminant is 0   c

63. A quadratic equation whose solution set is $\{3 \pm \sqrt{2}\}$ a

64. A quadratic equation whose solution set contains integers   d

a.

b.

c.

d.

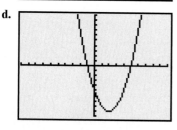

65. When the sum of 6 and twice a positive number is subtracted from the square of the number, 0 results. Find the number. $1 + \sqrt{7}$

66. When the sum of 1 and twice a negative number is subtracted from twice the square of the number, 0 results. Find the number. $\dfrac{1 - \sqrt{3}}{2}$

In Exercises 67–72, solve each equation by the method of your choice.

67. $\dfrac{1}{x^2 - 3x + 2} = \dfrac{1}{x + 2} + \dfrac{5}{x^2 - 4}$ $\dfrac{-1 \pm \sqrt{21}}{2}$ or $\left\{\dfrac{-1 \pm \sqrt{21}}{2}\right\}$

68. $\dfrac{x - 1}{x - 2} + \dfrac{x}{x - 3} = \dfrac{1}{x^2 - 5x + 6}$ $\dfrac{3 \pm \sqrt{5}}{2}$ or $\left\{\dfrac{3 \pm \sqrt{5}}{2}\right\}$

69. $\sqrt{2}x^2 + 3x - 2\sqrt{2} = 0$ $-2\sqrt{2}$ and $\dfrac{\sqrt{2}}{2}$, or $\left\{-2\sqrt{2}, \dfrac{\sqrt{2}}{2}\right\}$

70. $\sqrt{3}x^2 + 6x + 7\sqrt{3} = 0$ $-\sqrt{3} \pm 2i$ or $\left\{-\sqrt{3} \pm 2i\right\}$

71. $\left|x^2 + 2x\right| = 3$ $-3, 1$, and $-1 \pm \sqrt{2}i$, or $\left\{-3, 1, -1 \pm \sqrt{2}i\right\}$

72. $\left|x^2 + 3x\right| = 2$ $-2, -1$, and $\dfrac{-3 \pm \sqrt{17}}{2}$, or $\left\{-2, -1, \dfrac{-3 \pm \sqrt{17}}{2}\right\}$

Application Exercises

A driver's age has something to do with his or her chance of getting into a fatal car crash. The bar graph shows the number of fatal vehicle crashes per 100 million miles driven for drivers of various age groups. For example, 25-year-old drivers are involved in 4.1 fatal crashes per 100 million miles driven. Thus, when a group of 25-year-old Americans have driven a total of 100 million miles, approximately 4 have been in accidents in which someone died.

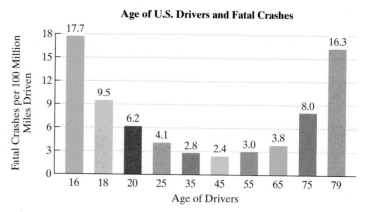

Age of U.S. Drivers and Fatal Crashes

Source: Insurance Institute for Highway Safety

The number of fatal vehicle crashes per 100 million miles, $f(x)$, for drivers of age x can be modeled by the quadratic function

$$f(x) = 0.013x^2 - 1.19x + 28.24.$$

Use the function to solve Exercises 73–74.

73. What age groups are expected to be involved in 3 fatal crashes per 100 million miles driven? How well does the function model the trend in the actual data shown in the bar graph on the previous page?   33-year-olds and 58-year-olds; The function models the actual data well.

74. What age groups are expected to be involved in 10 fatal crashes per 100 million miles driven? How well does the function model the trend in the actual data shown in the bar graph on the previous page?   19-year-olds and 72-year-olds; Answers will vary.

Throwing events in track and field include the shot put, the discus throw, the hammer throw, and the javelin throw. The distance that an athlete can achieve depends on the initial velocity of the object thrown and the angle above the horizontal at which the object leaves the hand.

In Exercises 75–76, an athlete whose event is the shot put releases the shot with the same initial velocity, but at different angles.

75. When the shot is released at an angle of 35°, its path can be modeled by the function

$$f(x) = -0.01x^2 + 0.7x + 6.1$$

in which x is the shot's horizontal distance, in feet, and $f(x)$ is its height, in feet. This function is shown by one of the graphs, (a) or (b), in the figure. Use the function to determine the shot's maximum distance. Use a calculator and round to the nearest tenth of a foot. Which graph, (a) or (b), shows the shot's path?   77.8 feet; (b)

Horizontal Distance
[0, 80, 10] by [0, 40, 10]

76. When the shot is released at an angle of 65°, its path can be modeled by the function

$$f(x) = -0.04x^2 + 2.1x + 6.1$$

in which x is the shot's horizontal distance, in feet, and $f(x)$ is its height, in feet. This function is shown by one of the graphs, (a) or (b), in the figure at the bottom of the previous column. Use the function to determine the shot's maximum distance. Use a calculator and round to the nearest tenth of a foot. Which graph, (a) or (b), shows the shot's path?   55.3 feet; (a)

77. The length of a rectangle is 4 meters longer than the width. If the area is 8 square meters, find the rectangle's dimensions. Round to the nearest tenth of a meter.   1.5 meters by 5.5 meters

78. The length of a rectangle exceeds twice its width by 3 inches. If the area is 10 square inches, find the rectangle's dimensions. Round to the nearest tenth of an inch.   1.6 inches by 6.2 inches

79. One leg of a right triangle exceeds the shorter leg by 1 inch, and the hypotenuse exceeds the longer leg by 7 inches. Find the lengths of the legs. Round to the nearest tenth of a inch.   17.6 inches and 18.6 inches

80. The hypotenuse of a right triangle is 6 feet long. One leg is 2 feet shorter than the other. Find the lengths of the legs. Round to the nearest tenth of a foot.   5.1 feet and 3.1 feet

81. A rain gutter is made from sheets of aluminum that are 20 inches wide. As shown in the figure, the edges are turned up to form right angles. Determine the depth of the gutter that will allow a cross-sectional area of 13 square inches. Show that there are two different solutions to the problem. Round to the nearest tenth of an inch.

9.3 inches and 0.7 inch

82. A piece of wire is 8 inches long. The wire is cut into two pieces and then each piece is bent into a square. Find the length of each piece if the sum of the areas of these squares is to be 2 square inches.

4 inches and 4 inches

83. Working together, two people can mow a large lawn in 4 hours. One person can do the job alone 1 hour faster than the other person. How long does it take each person working alone to mow the lawn? Round to the nearest tenth of an hour.   7.5 hours and 8.5 hours

84. A pool has an inlet pipe to fill it and an outlet pipe to empty it. It takes 2 hours longer to empty the pool than it does to fill it. The inlet pipe is turned on to fill the pool, but the outlet pipe is accidentally open. Despite this, the pool fills in 8 hours. How long does it take the outlet pipe to empty the pool? Round to the nearest tenth of an hour.   5.1 hours

Writing in Mathematics

85. What is the quadratic formula and why is it useful?
85–92. Answers will vary.

86. Without going into specific details for every step, describe how the quadratic formula is derived.

87. Explain how to solve $x^2 + 6x + 8 = 0$ using the quadratic formula.

88. If a quadratic equation has imaginary solutions, how is this shown on the graph of the corresponding quadratic function?

89. What is the discriminant and what information does it provide about a quadratic equation?

90. If you are given a quadratic equation, how do you determine which method to use to solve it?

91. Explain how to write a quadratic equation from its solution set. Give an example with your explanation.

92. Describe the trend shown by the data for the percentage of the U.S. population that was foreign-born from 1930 through 2002. (See Figure 8.4 on page 574.) Do you believe that this trend is likely to continue or might something occur that would make it impossible to extend the model in Example 6 on page 583 into the future? Explain your answer.

Technology Exercises

93. Use a graphing utility to graph the quadratic function related to any five of the quadratic equations in Exercises 19–30. How does each graph illustrate what you determined algebraically using the discriminant?   Answers will vary.

94. Reread Exercise 81. The cross-sectional area of the gutter is given by the quadratic function

$$f(x) = x(20 - 2x).$$

Graph the function in a $[0, 10, 1]$ by $[0, 60, 5]$ viewing rectangle. Then $\boxed{\text{TRACE}}$ along the curve or use the maximum function feature to determine the depth of the gutter that will maximize its cross-sectional area and allow the greatest amount of water to flow. What is the maximum area? Does the situation described in Exercise 81 take full advantage of the sheets of aluminum?   See graphing answer section. Depth: 5 inches; maximum area: 50 square inches; Answers will vary.

Critical Thinking Exercises

95. Which one of the following is true?   d
  a. The quadratic formula is developed by applying factoring and the zero-product principle to the quadratic equation $ax^2 + bx + c = 0$.
  b. In using the quadratic formula to solve the quadratic equation $5x^2 = 2x - 7$, we have $a = 5, b = 2$, and $c = -7$.
  c. The quadratic formula cannot be used to solve the equation $x^2 - 9 = 0$.
  d. Any quadratic equation that can be solved by completing the square can be solved by the quadratic formula.

96. Solve for t: $s = -16t^2 + v_0 t$. $t = \dfrac{v_0 \pm \sqrt{v_0^2 - 64s}}{32}$

97. A rectangular swimming pool is 12 meters long and 8 meters wide. A tile border of uniform width is to be built around the pool using 120 square meters of tile. The tile is from a discontinued stock (so no additional materials are available) and all 120 square meters are to be used. How wide should the border be? Round to the nearest tenth of a meter. If zoning laws require at least a 2-meter-wide border around the pool, can this be done with the available tile?   The border should be about 2.4 meters wide. Yes, the available tile meets the zoning laws.

98. The area of the shaded region outside the rectangle and inside the triangle is 10 square yards. Find the triangle's height, represented by $2x$. Round to the nearest tenth of a yard.

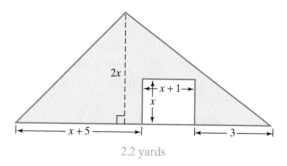

2.2 yards

Review Exercises

99. Solve: $|5x + 2| = |4 - 3x|$. (Section 4.3, Example 2)
  -3 and $\frac{1}{4}$, or $\left\{-3, \frac{1}{4}\right\}$

100. Solve: $\sqrt{2x - 5} - \sqrt{x - 3} = 1$. (Section 7.6, Example 4)
  3 and 7, or $\{3, 7\}$

101. Rationalize the denominator: $\dfrac{5}{\sqrt{3} + x}$. (Section 7.5, Example 5)
  $\dfrac{5\sqrt{3} - 5x}{3 - x^2}$

SECTION 8.3

Objectives

1 Recognize characteristics of parabolas.

2 Graph parabolas in the form $f(x) = a(x - h)^2 + k$.

3 Graph parabolas in the form $f(x) = ax^2 + bx + c$.

4 Solve problems involving minimizing or maximizing quadratic functions.

QUADRATIC FUNCTIONS AND THEIR GRAPHS

We have a long history of throwing things. Before 400 B.C., the Greeks competed in games that included discus throwing. In the seventeenth century, English soldiers organized cannonball-throwing competitions. In 1827, a Yale University student, disappointed over failing an exam, took out his frustrations at the passing of a collection plate in chapel. Seizing the monetary tray, he flung it in the direction of a large open space on campus. Yale students see this act of frustration as the origin of the Frisbee.

In this section, we study quadratic functions and their graphs. By graphing functions that model the paths of the things we throw, you will be able to determine both the maximum height and the distance of these objects.

1 Recognize characteristics of parabolas.

Graphs of Quadratic Functions The graph of the quadratic function

$$f(x) = ax^2 + bx + c, \quad a \neq 0,$$

is called a **parabola**. Parabolas are shaped like cups, as shown in Figure 8.5. If the coefficient of x^2 (the value of a in $ax^2 + bx + c$) is positive, the parabola opens upward. If the coefficient of x^2 is negative, the parabola opens downward. The **vertex** (or turning point) of the parabola is the lowest point on the graph when it opens upward and the highest point on the graph when it opens downward.

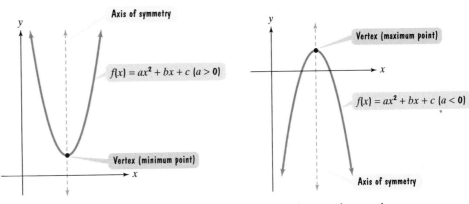

$a > 0$: Parabola opens upward. $a < 0$: Parabola opens downward.

FIGURE 8.5 Characteristics of graphs of quadratic functions

Look at the unusual image of the word "mirror" shown below. The artist, Scott Kim, has created the image so that the two halves of the whole are mirror images of each other. A parabola shares this kind of symmetry, in which a line through the vertex divides the figure in half. Parabolas are symmetric with respect to this line, called the **axis of symmetry**. If a parabola is folded along its axis of symmetry, the two halves match exactly.

Graphing Quadratic Functions in the Form $f(x) = a(x - h)^2 + k$ One way to obtain the graph of a quadratic function is to use point plotting. Let's begin by graphing the functions $f(x) = x^2$, $g(x) = 2x^2$, and $h(x) = \frac{1}{2}x^2$ in the same rectangular coordinate system. Select integers for x, starting with -3 and ending with 3. A partial table of coordinates for each function is shown below. The three parabolas are shown in Figure 8.6.

2 Graph parabolas in the form
$$f(x) = a(x - h)^2 + k.$$

x	$f(x) = x^2$	(x, y) or $(x, f(x))$
-3	$f(-3) = (-3)^2 = 9$	$(-3, 9)$
-2	$f(-2) = (-2)^2 = 4$	$(-2, 4)$
-1	$f(-1) = (-1)^2 = 1$	$(-1, 1)$
0	$f(0) = 0^2 = 0$	$(0, 0)$
1	$f(1) = 1^2 = 1$	$(1, 1)$
2	$f(2) = 2^2 = 4$	$(2, 4)$
3	$f(3) = 3^2 = 9$	$(3, 9)$

x	$g(x) = 2x^2$	(x, y) or $(x, g(x))$
-3	$g(-3) = 2(-3)^2 = 18$	$(-3, 18)$
-2	$g(-2) = 2(-2)^2 = 8$	$(-2, 8)$
-1	$g(-1) = 2(-1)^2 = 2$	$(-1, 2)$
0	$g(0) = 2 \cdot 0^2 = 0$	$(0, 0)$
1	$g(1) = 2 \cdot 1^2 = 2$	$(1, 2)$
2	$g(2) = 2 \cdot 2^2 = 8$	$(2, 8)$
3	$g(3) = 2 \cdot 3^2 = 18$	$(3, 18)$

x	$h(x) = \dfrac{1}{2}x^2$	(x, y) or $(x, h(x))$
-3	$h(-3) = \dfrac{1}{2}(-3)^2 = \dfrac{9}{2}$	$\left(-3, \dfrac{9}{2}\right)$
-2	$h(-2) = \dfrac{1}{2}(-2)^2 = 2$	$(-2, 2)$
-1	$h(-1) = \dfrac{1}{2}(-1)^2 = \dfrac{1}{2}$	$\left(-1, \dfrac{1}{2}\right)$
0	$h(0) = \dfrac{1}{2} \cdot 0^2 = 0$	$(0, 0)$
1	$h(1) = \dfrac{1}{2} \cdot 1^2 = \dfrac{1}{2}$	$\left(1, \dfrac{1}{2}\right)$
2	$h(2) = \dfrac{1}{2} \cdot 2^2 = 2$	$(2, 2)$
3	$h(3) = \dfrac{1}{2} \cdot 3^2 = \dfrac{9}{2}$	$\left(3, \dfrac{9}{2}\right)$

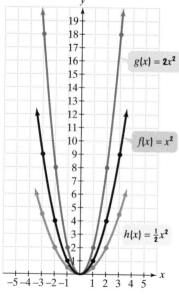

FIGURE 8.6

Can you see that the graphs of f, g, and h all have the same vertex, $(0, 0)$? They also have the same axis of symmetry, the y-axis, or $x = 0$. This is true for all graphs of the form $f(x) = ax^2$. However, the blue graph of $g(x) = 2x^2$ is a narrower parabola than the red graph of $f(x) = x^2$. By contrast, the green graph of $h(x) = \frac{1}{2}x^2$ is a flatter parabola than the red graph of $f(x) = x^2$.

Is there a more efficient method than point plotting to obtain the graph of a quadratic function? The answer is yes. The method is based on comparing graphs of the form $g(x) = a(x - h)^2 + k$ to those of the form $f(x) = ax^2$.

In Figure 8.7(a), the graph of $f(x) = ax^2$ for $a > 0$ is shown in black. The parabola's vertex is $(0, 0)$ and it opens upward. In Figure 8.7(b), the graph of $f(x) = ax^2$ for $a < 0$ is shown in black. The parabola's vertex is $(0, 0)$ and it opens downward.

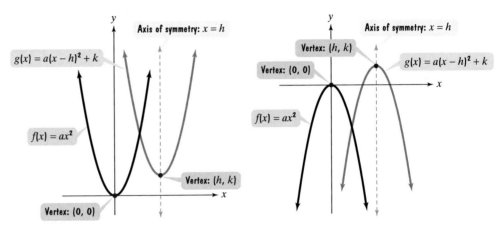

(a) $a > 0$: Parabola opens upward.   **(b)** $a < 0$: Parabola opens downward.

FIGURE 8.7 Moving, or shifting, the graph of $f(x) = ax^2$

Figure 8.7 also shows the graph of $g(x) = a(x - h)^2 + k$ in blue. Compare these graphs to those of $f(x) = ax^2$. Observe that h causes a horizontal move, or shift, and k causes a vertical shift of the graph of $f(x) = ax^2$. Consequently, the vertex $(0, 0)$ on the graph of $f(x) = ax^2$ moves to the point (h, k) on the graph of $g(x) = a(x - h)^2 + k$. The axis of symmetry is the vertical line whose equation is $x = h$.

The form of the expression for g is convenient because it immediately identifies the vertex of the parabola as (h, k). The sign of a in $g(x) = a(x - h)^2 + k$ determines whether the parabola opens upward or downward. Furthermore, if $|a|$ is small, the parabola opens more flatly than if $|a|$ is large.

GRAPHING QUADRATIC FUNCTIONS WITH EQUATIONS IN THE FORM $f(x) = a(x - h)^2 + k$ To graph $f(x) = a(x - h)^2 + k$,

1. Determine whether the parabola opens upward or downward. If $a > 0$, it opens upward. If $a < 0$, it opens downward.

2. Determine the vertex of the parabola. The vertex is (h, k).

3. Find any x-intercepts by replacing $f(x)$ with 0. Solve the resulting quadratic equation for x.

4. Find the y-intercept by replacing x with 0.

5. Plot the intercepts and vertex and additional points as necessary. Connect these points with a smooth curve that is shaped like a cup.

EXAMPLE 1 Graphing a Quadratic Function in the Form $f(x) = a(x - h)^2 + k$

Graph the quadratic function $f(x) = -2(x - 3)^2 + 8$.

SOLUTION We can graph this function by following the steps in the preceding box. We begin by identifying values for a, h, and k.

$$f(x) = a(x - h)^2 + k$$

$$\boxed{a = -2} \quad \boxed{h = 3} \quad \boxed{k = 8}$$

$$f(x) = -2(x - 3)^2 + 8$$

Step 1. **Determine how the parabola opens.** Note that a, the coefficient of x^2, is -2. Thus, $a < 0$; this negative value tells us that the parabola opens downward.

Step 2. **Find the vertex.** The vertex of the parabola is at (h, k). Because $h = 3$ and $k = 8$, the parabola has its vertex at $(3, 8)$.

Step 3. **Find the x-intercepts.** Replace $f(x)$ with 0 in $f(x) = -2(x - 3)^2 + 8$.

$$0 = -2(x - 3)^2 + 8 \qquad \text{Find x-intercepts, setting f(x) equal to 0.}$$
$$2(x - 3)^2 = 8 \qquad \text{Solve for x. Add } 2(x - 3)^2 \text{ to both sides of the equation.}$$
$$(x - 3)^2 = 4 \qquad \text{Divide both sides by 2.}$$
$$x - 3 = \sqrt{4} \quad \text{or} \quad x - 3 = -\sqrt{4} \qquad \text{Apply the square root property.}$$
$$x - 3 = 2 \qquad\qquad x - 3 = -2 \qquad \sqrt{4} = 2$$
$$x = 5 \qquad\qquad x = 1 \qquad \text{Add 3 to both sides in each equation.}$$

The x-intercepts are 5 and 1. The parabola passes through $(5, 0)$ and $(1, 0)$.

Step 4. **Find the y-intercept.** Replace x with 0 in $f(x) = -2(x - 3)^2 + 8$.

$$f(0) = -2(0 - 3)^2 + 8 = -2(-3)^2 + 8 = -2(9) + 8 = -10$$

The y-intercept is -10. The parabola passes through $(0, -10)$.

Step 5. **Graph the parabola.** With a vertex at $(3, 8)$, x-intercepts at 5 and 1, and a y-intercept at -10, the graph of f is shown in Figure 8.8. The axis of symmetry is the vertical line whose equation is $x = 3$. ■

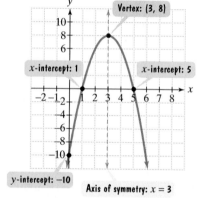

FIGURE 8.8 The graph of $f(x) = -2(x - 3)^2 + 8$

✔ **CHECK POINT 1** Graph the quadratic function $f(x) = -(x - 1)^2 + 4$.
See graphing answer section.

EXAMPLE 2 Graphing a Quadratic Function in the Form $f(x) = a(x - h)^2 + k$

Graph the quadratic function $f(x) = (x + 3)^2 + 1$.

SOLUTION We begin by finding values for a, h, and k.

$$f(x) = a(x - h)^2 + k \qquad \text{Form of quadratic function}$$
$$f(x) = (x + 3)^2 + 1 \qquad \text{Given function}$$
or
$$f(x) = 1(x - (-3))^2 + 1$$

$$\boxed{a = 1} \quad \boxed{h = -3} \quad \boxed{k = 1}$$

Axis of symmetry: $x = -3$

y-intercept: 10

Vertex: $(-3, 1)$

FIGURE 8.9 The graph of $f(x) = (x + 3)^2 + 1$

Step 1. **Determine how the parabola opens.** Note that a, the coefficient of x^2, is 1. Thus, $a > 0$; this positive value tells us that the parabola opens upward.

Step 2. **Find the vertex.** The vertex of the parabola is at $(h, k.)$ Because $h = -3$ and $k = 1$, the parabola has its vertex at $(-3, 1)$.

Step 3. **Find the x-intercepts.** Replace $f(x)$ with 0 in $f(x) = (x + 3)^2 + 1$. Because the vertex is at $(-3, 1)$, which lies above the x-axis, and the parabola opens upward, it appears that this parabola has no x-intercepts. We can verify this observation algebraically.

$$0 = (x + 3)^2 + 1$$ Find possible x-intercepts, setting $f(x)$ equal to 0.

$$-1 = (x + 3)^2$$ Solve for x. Subtract 1 from both sides.

$x + 3 = \sqrt{-1}$ or $x + 3 = -\sqrt{-1}$ Apply the square root property.

$x + 3 = i$ $x + 3 = -i$ $\sqrt{-1} = i$

$x = -3 + i$ $x = -3 - i$ The solutions are $-3 \pm i$.

Because this equation has no real solutions, the parabola has no x-intercepts.

Step 4. **Find the y-intercept.** Replace x with 0 in $f(x) = (x + 3)^2 + 1$.

$$f(0) = (0 + 3)^2 + 1 = 3^2 + 1 = 9 + 1 = 10$$

The y-intercept is 10. The parabola passes through $(0, 10)$.

Step 5. **Graph the parabola.** With a vertex at $(-3, 1)$, no x-intercepts, and a y-intercept at 10, the graph of f is shown in Figure 8.9. The axis of symmetry is the vertical line whose equation is $x = -3$. ∎

 CHECK POINT 2 Graph the quadratic function $f(x) = (x - 2)^2 + 1$.
See graphing answer section.

3 Graph parabolas in the form $f(x) = ax^2 + bx + c$.

Graphing Quadratic Functions in the Form $f(x) = ax^2 + bx + c$ Quadratic functions are frequently expressed in the form $f(x) = ax^2 + bx + c$. How can we identify the vertex of a parabola whose equation is in this form? Completing the square provides the answer to this question.

$$f(x) = ax^2 + bx + c$$

$$= a\left(x^2 + \frac{b}{a}x\right) + c$$ Factor out a from $ax^2 + bx$.

$$= a\left(x^2 + \frac{b}{a}x + \frac{b^2}{4a^2}\right) + c - a\left(\frac{b^2}{4a^2}\right)$$

Complete the square by adding the square of half the coefficient of x.

By completing the square, we added $a \cdot \frac{b^2}{4a^2}$. To avoid changing the function's equation, we must subtract this term.

$$= a\left(x + \frac{b}{2a}\right)^2 + c - \frac{b^2}{4a}$$

Write the trinomial as the square of a binomial and simplify the constant term.

Now let's compare the form of the equation on the bottom of the previous page with a quadratic function in the form $f(x) = a(x - h)^2 + k$.

The form we know how to graph → $f(x) = a(x - h)^2 + k$

$$h = -\frac{b}{2a} \qquad k = c - \frac{b^2}{4a}$$

Equation under discussion → $f(x) = a\left(x - \left(-\frac{b}{2a}\right)\right)^2 + c - \frac{b^2}{4a}$

The important part of this observation is that h, the x-coordinate of the vertex, is $-\frac{b}{2a}$. The y-coordinate can be found by evaluating the function at $-\frac{b}{2a}$.

THE VERTEX OF A PARABOLA WHOSE EQUATION IS $f(x) = ax^2 + bx + c$

Consider the parabola defined by the quadratic function $f(x) = ax^2 + bx + c$. The parabola's vertex is $\left(-\frac{b}{2a}, f\left(-\frac{b}{2a}\right)\right)$.

We can apply our five-step procedure and graph parabolas in the form $f(x) = ax^2 + bx + c$. The only step that is different is how we determine the vertex.

EXAMPLE 3 Graphing a Quadratic Function in the Form $f(x) = ax^2 + bx + c$

Graph the quadratic function $f(x) = -x^2 - 2x + 1$. Use the graph to identify the function's domain and its range.

SOLUTION

Step 1. **Determine how the parabola opens.** Note that a, the coefficient of x^2, is -1. Thus, $a < 0$; this negative value tells us that the parabola opens downward.

Step 2. **Find the vertex.** We know that the x-coordinate of the vertex is $x = -\frac{b}{2a}$. We identify $a, b,$ and c in $f(x) = ax^2 + bx + c$.

$$f(x) = -x^2 - 2x + 1$$

$$a = -1 \qquad b = -2 \qquad c = 1$$

Substitute the values of a and b into the equation for the x-coordinate:

$$x = -\frac{b}{2a} = -\frac{-2}{2(-1)} = -\left(\frac{-2}{-2}\right) = -1.$$

The x-coordinate of the vertex is -1. We substitute -1 for x in the equation of the function to find the y-coordinate:

$$f(-1) = -(-1)^2 - 2(-1) + 1 = -1 + 2 + 1 = 2.$$

The vertex is $(-1, 2)$.

Step 3. **Find the x-intercepts.** Replace $f(x)$ with 0 in $f(x) = -x^2 - 2x + 1$. We obtain $0 = -x^2 - 2x + 1$. This equation cannot be solved by factoring. We will use the quadratic formula to solve it.

$$-x^2 - 2x + 1 = 0$$

$$a = -1 \qquad b = -2 \qquad c = 1$$

$$x = \frac{-b + \sqrt{b^2 - 4ac}}{2a} = \frac{-(-2) \pm \sqrt{(-2)^2 - 4(-1)(1)}}{2(-1)} = \frac{2 \pm \sqrt{4 - (-4)}}{-2}$$

To locate the x-intercepts, we need decimal approximations. Thus, there is no need to simplify the radical form of the solutions.

$$x = \frac{2 + \sqrt{8}}{-2} \approx -2.4 \qquad \text{or} \qquad x = \frac{2 - \sqrt{8}}{-2} \approx 0.4$$

The x-intercepts are approximately -2.4 and 0.4. The parabola passes through $(-2.4, 0)$ and $(0.4, 0)$.

Step 4. **Find the y-intercept.** Replace x with 0 in $f(x) = -x^2 - 2x + 1$.

$$f(0) = -0^2 - 2 \cdot 0 + 1 = 1$$

The y-intercept is 1. The parabola passes through $(0, 1)$.

Step 5. **Graph the parabola.** With a vertex at $(-1, 2)$, x-intercepts at -2.4 and 0.4, and a y-intercept at 1, the graph of f is shown in Figure 8.10(a). The axis of symmetry is the vertical line whose equation is $x = -1$.

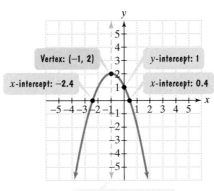

Vertex: $(-1, 2)$
y-intercept: 1
x-intercept: -2.4
x-intercept: 0.4
Axis of symmetry: $x = -1$

Domain: Inputs on x-axis include all real numbers.
Range: Outputs on y-axis fall at or below 2.

FIGURE 8.10(a) The graph of $f(x) = -x^2 - 2x + 1$

FIGURE 8.10(b) Determining the domain and range of $f(x) = -x^2 - 2x + 1$

STUDY TIP

The domain of any quadratic function includes all real numbers. If the vertex is the graph's highest point, the range includes all real numbers at or below the y-coordinate of the vertex. If the vertex is the graph's lowest point, the range includes all real numbers at or above the y-coordinate of the vertex.

Now we are ready to determine the domain and range of $f(x) = -x^2 - 2x + 1$. We can use the parabola, shown again in Figure 8.10(b), to do so. To find the domain, look for all the inputs on the x-axis that correspond to points on the graph. As the graph widens and continues to fall at both ends, can you see that these inputs include all real numbers?

Domain of f is $\{x \mid x \text{ is a real number}\}$ or $(-\infty, \infty)$.

To find the range, look for all the outputs on the y-axis that correspond to points on the graph. Figure 8.10(b) shows that the parabola's vertex, $(-1, 2)$, is the highest point on the graph. Because the y-coordinate of the vertex is 2, outputs on the y-axis fall at or below 2.

Range of f is $\{y \mid y \leq 2\}$ or $(-\infty, 2]$.

> ✓ **CHECK POINT 3** Graph the quadratic function $f(x) = -x^2 + 4x + 1$. Use the graph to identify the function's domain and its range. See graphing answer section.
> Domain = $\{x \mid x \text{ is a real number}\}$ or $(-\infty, \infty)$, range = $\{y \mid y \le 5\}$ or $(-\infty, 5]$.

4 Solve problems involving minimizing or maximizing quadratic functions.

Applications of Quadratic Functions When do the things we throw reach their maximum height and what is that height? How many hours of sleep per night minimize the risk of death? How many units of a product should a business manufacture to maximize its profits? The answers to these questions involve finding the maximum or minimum value of quadratic functions.

Consider the quadratic function $f(x) = ax^2 + bx + c$. If $a > 0$, the parabola opens upward and the vertex is its lowest point. If $a < 0$, the parabola opens downward and the vertex is its highest point. The x-coordinate of the vertex is $-\dfrac{b}{2a}$. Thus, we can find the minimum or maximum value of f by evaluating the quadratic function at $x = -\dfrac{b}{2a}$.

MINIMUM AND MAXIMUM: QUADRATIC FUNCTIONS Consider $f(x) = ax^2 + bx + c$.

1. If $a > 0$, then f has a minimum that occurs at $x = -\dfrac{b}{2a}$.
This minimum value is $f\left(-\dfrac{b}{2a}\right)$.

2. If $a < 0$, then f has a maximum that occurs at $x = -\dfrac{b}{2a}$.
This maximum value is $f\left(-\dfrac{b}{2a}\right)$.

EXAMPLE 4 Parabolic Paths of a Shot Put

An athlete whose event is the shot put releases the shot with the same initial velocity, but at different angles. Figure 8.11 shows the parabolic paths for shots released at angles of 35° and 65°.

FIGURE 8.11 Two paths of a shot put

Throwing to Achieve Maximum Distance

Although Example 4 focuses on maximum height of a propelled object, the goal of the shot put throw is to achieve the greatest possible distance. If an object is propelled from ground level, an angle of 45° produces the maximum distance for a given initial velocity. An angle slightly less than 45° is better when an object is thrown from above the ground. For the shot in Example 4, released from 6.1 feet above the ground, an angle of approximately 43° produces the maximum distance.

When the shot is released at an angle of 35°, its path can be modeled by the function

$$f(x) = -0.01x^2 + 0.7x + 6.1$$

in which x is the shot's horizontal distance, in feet, and $f(x)$ is its height, in feet. What is the maximum height of this shot's path?

SOLUTION The quadratic function is in the form $f(x) = ax^2 + bx + c$, with $a = -0.01$ and $b = 0.7$. Because $a < 0$, the function has a maximum that occurs at $x = -\dfrac{b}{2a}$.

$$x = -\frac{b}{2a} = -\frac{0.7}{2(-0.01)} = -(-35) = 35$$

This means that the shot's maximum height occurs when its horizontal distance is 35 feet. Can you see how this is shown by the blue graph of f in Figure 8.11? The maximum height of this path is

$$f(35) = -0.01(35)^2 + 0.7(35) + 6.1 = 18.35$$

or 18.35 feet. ∎

 CHECK POINT 4 Use function g, whose equation and graph are shown in Figure 8.11, to find the maximum height, to the nearest tenth of a foot, when the shot is released at an angle of 65°.   33.7 feet

Quadratic functions can also be formed from verbal conditions. Once we have obtained a quadratic function, we can then use the x-coordinate of the vertex to determine its maximum or minimum value. Here is a step-by-step strategy for solving these kinds of problems:

STRATEGY FOR SOLVING PROBLEMS INVOLVING MAXIMIZING OR MINIMIZING QUADRATIC FUNCTIONS

1. Read the problem carefully and decide which quantity is to be maximized or minimized.
2. Use the conditions of the problem to express the quantity as a function in one variable.
3. Rewrite the function in the form $f(x) = ax^2 + bx + c$.
4. Calculate $-\dfrac{b}{2a}$. If $a > 0$, f has a minimum at $x = -\dfrac{b}{2a}$. This minimum value is $f\left(-\dfrac{b}{2a}\right)$. If $a < 0$, f has a maximum at $x = -\dfrac{b}{2a}$. This maximum value is $f\left(-\dfrac{b}{2a}\right)$.
5. Answer the question posed in the problem.

EXAMPLE 5 Minimizing a Product

Among all pairs of numbers whose difference is 10, find a pair whose product is as small as possible. What is the minimum product?

SOLUTION

Step 1. **Decide what must be maximized or minimized.** We must minimize the product of two numbers. Calling the numbers x and y, and calling the product P, we must minimize

$$P = xy.$$

Step 2. **Express this quantity as a function in one variable.** In the formula $P = xy$, P is expressed in terms of two variables, x and y. However, because the difference of the numbers is 10, we can write

$$x - y = 10.$$

We can solve this equation for y in terms of x (or vice versa), substitute the result into $P = xy$, and obtain P as a function of one variable.

$$-y = -x + 10 \qquad \text{Subtract } x \text{ from both sides of } x - y = 10.$$

$$y = x - 10 \qquad \text{Multiply both sides of the equation by } -1 \text{ and solve for } y$$

Now we substitute $x - 10$ for y in $P = xy$.

$$P = xy = x(x - 10).$$

Because P is now a function of x, we can write

$$P(x) = x(x - 10).$$

Step 3. **Write the function in the form $f(x) = ax^2 + bx + c$.** We apply the distributive property to obtain

$$P(x) = x(x - 10) = x^2 - 10x$$

$a = 1$ $b = -10$

Step 4. **Calculate $-\dfrac{b}{2a}$. If $a > 0$, the function has a minimum at this value.** The voice balloons show that $a = 1$ and $b = -10$.

$$x = -\frac{b}{2a} = -\frac{-10}{2(1)} = -(-5) = 5$$

This means that the product, P, of two numbers whose difference is 10 is a minimum when one of the numbers, x, is 5.

Step 5. **Answer the question posed by the problem.** The problem asks for the two numbers and the minimum product. We found that one of the numbers, x, is 5. Now we must find the second number, y.

$$y = x - 10 = 5 - 10 = -5.$$

The number pair whose difference is 10 and whose product is as small as possible is 5, −5. The minimum product is 5(−5), or −25. ∎

 CHECK POINT 5 Among all pairs of numbers whose difference is 8, find a pair whose product is as small as possible. What is the minimum product? 4, −4; −16

USING TECHNOLOGY

The $\boxed{\text{TABLE}}$ feature of a graphing utility can be used to verify our work in Example 5.

Enter $y_1 = x^2 - 10x$, the function for the product, when one of the numbers is x.

X	Y1
2	-16
3	-21
4	-24
5	-25
6	-24
7	-21
8	-16

X=2

The product is a minimum, −25, when one of the numbers is 5.

EXAMPLE 6 Maximizing Area

You have 100 yards of fencing to enclose a rectangular region. Find the dimensions of the rectangle that maximize the enclosed area. What is the maximum area?

SOLUTION

Step 1. **Decide what must be maximized or minimized.** We must maximize area. What we do not know are the rectangle's dimensions, x and y.

Step 2. **Express this quantity as a function in one variable.** Because we must maximize area, we have $A = xy$. We need to transform this into a function in which A is represented by one variable. Because you have 100 yards of fencing, the perimeter of the rectangle is 100 yards. This means that

$$2x + 2y = 100.$$

We can solve this equation for y in terms of x, substitute the result into $A = xy$, and obtain A as a function in one variable. We begin by solving for y.

$$2y = 100 - 2x \quad \text{Subtract 2x from both sides.}$$

$$y = \frac{100 - 2x}{2} \quad \text{Divide both sides by 2.}$$

$$y = 50 - x \quad \text{Divide each term in the numerator by 2.}$$

Now we substitute $50 - x$ for y in $A = xy$.

$$A = xy = x(50 - x)$$

The rectangle and its dimensions are illustrated in Figure 8.12. Because A is now a function of x, we can write

$$A(x) = x(50 - x).$$

This function models the area, $A(x)$, of any rectangle whose perimeter is 100 yards in terms of one of its dimensions, x.

Step 3. **Write the function in the form $f(x) = ax^2 + bx + c$.** We apply the distributive property to obtain

$$A(x) = x(50 - x) = 50x - x^2 = -x^2 + 50x.$$

$a = -1$ $b = 50$

FIGURE 8.12 What value of x will maximize the rectangle's area?

$50 - x$

x

USING TECHNOLOGY

The graph of the area function

$$A(x) = x(50 - x)$$

was obtained with a graphing utility using a $[0, 50, 2]$ by $[0, 700, 25]$ viewing rectangle. The maximum function feature verifies that a maximum area of 625 square yards occurs when one of the dimensions is 25 yards.

Step 4. **Calculate** $-\dfrac{b}{2a}$**. If** $a < 0$**, the function has a maximum at this value.** The voice balloons on the previous page show that $a = -1$ and $b = 50$.

$$x = -\frac{b}{2a} = -\frac{50}{2(-1)} = 25$$

This means that the area, $A(x)$, of a rectangle with perimeter 100 yards is a maximum when one of the rectangle's dimensions, x, is 25 yards.

FIGURE 8.12 (repeated)

Step 5. **Answer the question posed by the problem.** We found that $x = 25$. Figure 8.12 shows that the rectangle's other dimension is $50 - x = 50 - 25 = 25$. The dimensions of the rectangle that maximize the enclosed area are 25 yards by 25 yards. The rectangle that gives the maximum area is actually a square with an area of 25 yards · 25 yards, or 625 square yards. ∎

CHECK POINT 6 You have 120 feet of fencing to enclose a rectangular region. Find the dimensions of the rectangle that maximize the enclosed area. What is the maximum area?   The dimensions are 30 feet by 30 feet with an area of 900 square feet.

USING TECHNOLOGY

We've come a long way from the small nation of "embattled farmers" who launched the American Revolution. In the early days of our Republic, 95% of the population was involved in farming. The graph in Figure 8.13 shows the number of farms in the United States from 1850 through 2010 (projected). Because the graph is shaped like a cup, with an increasing number of farms from 1850 to 1910 and a decreasing number of farms from 1910 to 2010, a quadratic function is an appropriate model for the data. You can use the statistical menu of a graphing utility to enter the data in Figure 8.13. We entered the data using (number of decades after 1850, millions of U.S. farms). The data are shown to the right of Figure 8.13.

Number of U. S. Farms, 1850–2010

FIGURE 8.13

Source: U. S. Bureau of the Census

Data:

$(0, 2.3), (2, 3.3), (4, 5.1),$
$(6, 6.7), (8, 6.4), (10, 5.8),$
$(12, 3.6), (14, 2.9), (16, 2.3)$

```
QuadReg
 y=ax²+bx+c
 a=-.0643668831
 b=.9873701299
 c=2.203636364
```

Upon entering the QUADratic REGression program, we obtain the results shown in the screen. Thus, the quadratic function of best fit is approximately

$$f(x) = -0.064x^2 + 0.99x + 2.2,$$

where x represents the number of decades after 1850 and $f(x)$ represents the number of U.S. farms, in millions.

8.3 EXERCISE SET

Student Solutions Manual   CD/Video   PH Math/Tutor Center   MathXL Tutorials on CD   MathXL®   MyMathLab   Interactmath.com

Practice Exercises

In Exercises 1–4, the graph of a quadratic function is given. Write the function's equation, selecting from the following options:

$$f(x) = (x + 1)^2 - 1, g(x) = (x + 1)^2 + 1,$$
$$h(x) = (x - 1)^2 + 1, j(x) = (x - 1)^2 - 1.$$

1. $h(x) = (x - 1)^2 + 1$

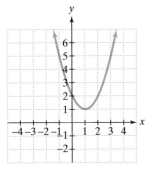

2. $g(x) = (x + 1)^2 + 1$

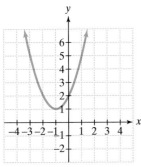

3. $j(x) = (x - 1)^2 - 1$

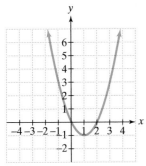

4. $f(x) = (x + 1)^2 - 1$

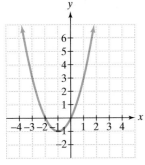

In Exercises 5–8, the graph of a quadratic function is given. Write the function's equation, selecting from the following options:

$$f(x) = x^2 + 2x + 1, g(x) = x^2 - 2x + 1,$$
$$h(x) = x^2 - 1, j(x) = -x^2 - 1.$$

5. $h(x) = x^2 - 1$

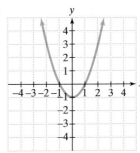

6. $f(x) = x^2 + 2x + 1$

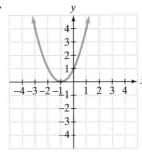

7. $g(x) = x^2 - 2x + 1$

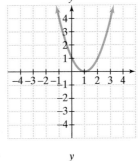

8. $j(x) = -x^2 - 1$

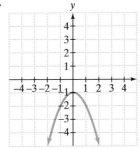

In Exercises 9–16, find the coordinates of the vertex for the parabola defined by the given quadratic function.

9. $f(x) = 2(x - 3)^2 + 1$ $(3, 1)$

10. $f(x) = -3(x - 2)^2 + 12$ $(2, 12)$

11. $f(x) = -2(x + 1)^2 + 5$ $(-1, 5)$

12. $f(x) = -2(x + 4)^2 - 8$ $(-4, -8)$

13. $f(x) = 2x^2 - 8x + 3$ $(2, -5)$

14. $f(x) = 3x^2 - 12x + 1$ $(2, -11)$

15. $f(x) = -x^2 - 2x + 8$ $(-1, 9)$

16. $f(x) = -2x^2 + 8x - 1$ $(2, 7)$

In Exercises 17–34, use the vertex and intercepts to sketch the graph of each quadratic function. Give the equation for the parabola's axis of symmetry. Use the parabola to identify the function's range.   **17–34.** See graphing answer section.

17. $f(x) = (x - 4)^2 - 1$ $x = 4; \{y|y \geq -1\}$ or $[-1, \infty)$

18. $f(x) = (x - 1)^2 - 2$ $x = 1; \{y|y \geq -2\}$ or $[-2, \infty)$

19. $f(x) = (x - 1)^2 + 2$ $x = 1; \{y|y \geq 2\}$ or $[2, \infty)$

20. $f(x) = (x - 3)^2 + 2$ $x = 3; \{y|y \geq 2\}$ or $[2, \infty)$

21. $y - 1 = (x - 3)^2$ $x = 3; \{y|y \geq 1\}$ or $[1, \infty)$

22. $y - 3 = (x - 1)^2$ $x = 1; \{y|y \geq 3\}$ or $[3, \infty)$

23. $f(x) = 2(x + 2)^2 - 1$ $x = -2; \{y|y \geq -1\}$ or $[-1, \infty)$

24. $f(x) = \dfrac{5}{4} - \left(x - \dfrac{1}{2}\right)^2$ $x = \dfrac{1}{2}; \{y|y \leq \frac{5}{4}\}$ or $\left(-\infty, \frac{5}{4}\right]$

25. $f(x) = 4 - (x - 1)^2$ $x = 1; \{y|y \leq 4\}$ or $(-\infty, 4]$

26. $f(x) = 1 - (x - 3)^2$ $x = 3; \{y|y \leq 1\}$ or $(-\infty, 1]$

27. $f(x) = x^2 - 2x - 3$ $x = 1; \{y|y \geq -4\}$ or $[-4, \infty)$

28. $f(x) = x^2 - 2x - 15$ $x = 1; \{y|y \geq -16\}$ or $[-16, \infty)$

29. $f(x) = x^2 + 3x - 10$ $x = -\frac{3}{2}; \{y|y \geq -\frac{49}{4}\}$ or $\left[-\frac{49}{4}, \infty\right)$

30. $f(x) = 2x^2 - 7x - 4$ $x = \frac{7}{4}; \{y|y \geq -\frac{81}{8}\}$ or $\left[-\frac{81}{8}, \infty\right)$

31. $f(x) = 2x - x^2 + 3$ $x = 1; \{y|y \leq 4\}$ or $(-\infty, 4]$

32. $f(x) = 5 - 4x - x^2$ $x = -2; \{y|y \leq 9\}$ or $(-\infty, 9]$

33. $f(x) = 2x - x^2 - 2$ $x = 1; \{y|y \leq -1\}$ or $(-\infty, -1]$

34. $f(x) = 6 - 4x + x^2$ $x = 2; \{y|y \geq 2\}$ or $[2, \infty)$

In Exercises 35–40, determine, without graphing, whether the given quadratic function has a minimum value or a maximum value. Then find the coordinates of the minimum or the maximum point.

35. $f(x) = 3x^2 - 12x - 1$ minimum; $(2, -13)$

36. $f(x) = 2x^2 - 8x - 3$ minimum; $(2, -11)$

37. $f(x) = -4x^2 + 8x - 3$ maximum; $(1, 1)$

38. $f(x) = -2x^2 - 12x + 3$ maximum; $(-3, 21)$

39. $f(x) = 5x^2 - 5x$ minimum; $\left(\frac{1}{2}, -\frac{5}{4}\right)$

40. $f(x) = 6x^2 - 6x$ minimum; $\left(\frac{1}{2}, -\frac{3}{2}\right)$

Practice Plus

In Exercises 41–44, give the domain and the range of each quadratic function whose graph is described.

41. The vertex is $(-1, -2)$ and the parabola opens up.
domain = $\{x|x$ is a real number$\}$ or $(-\infty, \infty)$; range = $\{y|y \geq -2\}$ or $[-2, \infty)$

42. The vertex is $(-3, -4)$ and the parabola opens down.
domain = $\{x|x$ is a real number$\}$ or $(-\infty, \infty)$; range = $\{y|y \leq -4\}$ or $(-\infty, -4]$

43. The maximum point on the parabola is $(10, 3)$.
domain = $\{x|x$ is a real number$\}$ or $(-\infty, \infty)$; range = $\{y|y \leq 3\}$ or $(-\infty, 3]$

44. The minimum point on the parabola is $(5, 12)$.
domain = $\{x|x$ is a real number$\}$ or $(-\infty, \infty)$; range = $\{y|y \geq 12\}$ or $[12, \infty)$

In Exercises 45–48, write an equation of the parabola that has the same shape as the graph of $f(x) = 2x^2$, but with the given point as the vertex.

45. $(5, 3)$ $f(x) = 2(x - 5)^2 + 3$

46. $(7, 4)$ $f(x) = 2(x - 7)^2 + 4$

47. $(-10, -5)$ $f(x) = 2(x + 10)^2 - 5$

48. $(-8, -6)$ $f(x) = 2(x + 8)^2 - 6$

In Exercises 49–52, write an equation of the parabola that has the same shape as the graph of $f(x) = 3x^2$ or $g(x) = -3x^2$, but with the given point as a maximum or minimum point.

49. maximum: $(-2, 4)$ $f(x) = -3(x + 2)^2 + 4$

50. maximum: $(5, -7)$ $f(x) = -3(x - 5)^2 - 7$

51. minimum: $(11, 0)$ $f(x) = 3(x - 11)^2$

52. minimum: $(9, 0)$ $f(x) = 3(x - 9)^2$

Application Exercises

In the United States, HCV, or hepatitis C virus, is four times as widespread as HIV. Few of the nation's three to four million carriers have any idea they are infected. The graph shows the projected mortality, in thousands, from the virus. Use the information in the graph to solve Exercises 53–54.

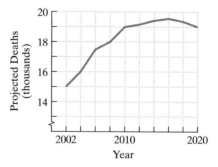

Projected Hepatitis C Deaths in the U. S.

Source: American Journal of Public Health

53. Suppose that a quadratic function is used to model the data shown with ordered pairs representing (number of years after 2002, thousands of hepatitis C deaths). Determine, without obtaining an actual quadratic function that models the data, the approximate coordinates of the vertex for the function's graph. Describe what this means in practical terms. Use the word "maximum" in your description. $(14, 19.5)$; In 2016, the projected mortality will reach a maximum of 19,500.

54. Why is a quadratic function an appropriate model for the data shown in the graph on the previous page?   The graph has the shape of a parabola.

55. A person standing close to the edge on the top of a 160-foot building throws a baseball vertically upward. The quadratic function

$$s(t) = -16t^2 + 64t + 160$$

models the ball's height above the ground, $s(t)$, in feet, t seconds after it was thrown.

a. After how many seconds does the ball reach its maximum height? What is the maximum height?   2 seconds; 224 feet

b. How many seconds does it take until the ball finally hits the ground? Round to the nearest tenth of a second. 5.7 seconds

c. Find $s(0)$ and describe what this means.   160; 160 feet is the height of the building.

d. Use your results from parts (a) through (c) to graph the quadratic function. Begin the graph with $t = 0$ and end with the value of t for which the ball hits the ground. See graphing answer section.

56. A person standing close to the edge on the top of a 200-foot building throws a baseball vertically upward. The quadratic function

$$s(t) = -16t^2 + 64t + 200$$

models the ball's height above the ground, $s(t)$, in feet, t seconds after it was thrown.

a. After how many seconds does the ball reach its maximum height? What is the maximum height?   2 seconds; 264 feet

b. How many seconds does it take until the ball finally hits the ground? Round to the nearest tenth of a second. 6.1 seconds

c. Find $s(0)$ and describe what this means.   200; 200 feet is the height of the building.

d. Use your results from parts (a) through (c) to graph the quadratic function. Begin the graph with $t = 0$ and end with the value of t for which the ball hits the ground. See graphing answer section.

57. The function

$$f(x) = 104.5x^2 - 1501.5x + 6016$$

models the death rate per year per 100,000 males, $f(x)$, for U.S. men who average x hours of sleep each night. How many hours of sleep, to the nearest tenth of an hour, corresponds to the minimum death rate? What is this minimum death rate, to the nearest whole number?   7.2 hours; 622 per 100,000 males

58. The function

$$f(x) = -0.064x^2 + 0.99x + 2.2$$

models the number of U.S. farms, $f(x)$, in millions, x decades after 1850. How many decades after 1850, to the nearest decade, did the number of U.S. farms reach a maximum? How many farms, to the nearest tenth of a million, were there in that year? How well does this model the actual data shown in Figure 8.13 on page 600?   8 decades; 6.0 million farms; Answers will vary.

59. Among all pairs of numbers whose sum is 16, find a pair whose product is as large as possible. What is the maximum product?   The maximum product is 64 when the numbers are 8 and 8.

60. Among all pairs of numbers whose sum is 20, find a pair whose product is as large as possible. What is the maximum product?   The maximum product is 100 when the numbers are 10 and 10.

61. Among all pairs of numbers whose difference is 16, find a pair whose product is as small as possible. What is the minimum product?   The minimum product is -64 when the numbers are 8 and -8.

62. Among all pairs of numbers whose difference is 24, find a pair whose product is as small as possible. What is the minimum product?   The minimum product is -144 when the numbers are 12 and -12.

63. You have 600 feet of fencing to enclose a rectangular plot that borders on a river. If you do not fence the side along the river, find the length and width of the plot that will maximize the area. What is the largest area that can be enclosed?

The maximum area is 45,000 square feet when the length is 300 feet and the width is 150 feet.

64. You have 200 feet of fencing to enclose a rectangular plot that borders on a river. If you do not fence the side along the river, find the length and width of the plot that will maximize the area. What is the largest area that can be enclosed?

The maximum area is 5000 square feet when the length is 100 feet and the width is 50 feet.

65. You have 50 yards of fencing to enclose a rectangular region. Find the dimensions of the rectangle that maximize the enclosed area. What is the maximum area? The maximum area is 156.25 square yards when the dimensions are 12.5 yards by 12.5 yards.

66. You have 80 yards of fencing to enclose a rectangular region. Find the dimensions of the rectangle that maximize the enclosed area. What is the maximum area? The maximum area is 400 square yards when the dimensions are 20 yards by 20 yards.

67. A rain gutter is made from sheets of aluminum that are 20 inches wide by turning up the edges to form right angles. Determine the depth of the gutter that will maximize its cross-sectional area and allow the greatest amount of water to flow. What is the maximum cross-sectional area?

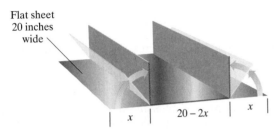

Flat sheet
20 inches
wide

x $20 - 2x$ x

The maximum cross-sectional area is 50 square inches when the depth is 5 inches.

68. A rain gutter is made from sheets of aluminum that are 12 inches wide by turning up the edges to form right angles. Determine the depth of the gutter that will maximize its cross-sectional area and allow the greatest amount of water to flow. What is the maximum cross-sectional area? The maximum cross-sectional area is 18 square inches when the depth is 3 inches.

In Chapter 3, we saw that the profit, P(x), generated after producing and selling x units of a product is given by the function

$$P(x) = R(x) - C(x),$$

where R and C are the revenue and cost functions, respectively. Use these functions to solve Exercises 69–70.

69. Hunky Beef, a local sandwich store, has a fixed weekly cost of $525.00, and variable costs for making a roast beef sandwich are $0.55.

  a. Let x represent the number of roast beef sandwiches made and sold each week. Write the weekly cost function, C, for Hunky Beef. $C(x) = 525 + 0.55x$

  b. The function $R(x) = -0.001x^2 + 3x$ describes the money that Hunky Beef takes in each week from the sale of x roast beef sandwiches. Use this revenue function and the cost function from part (a) to write the store's weekly profit function, P. $P(x) = -0.001x^2 + 3x - (525 + 0.55x)$
$= -0.001x^2 + 2.45x - 525$

  c. Use the store's profit function to determine the number of roast beef sandwiches it should make and sell each week to maximize profit. What is the maximum weekly profit? The maximum weekly profit is $975.63 when 1225 sandwiches are made and sold each week.

70. Virtual Fido is a company that makes electronic virtual pets. The fixed weekly cost is $3000, and variable costs for each pet are $20.

  a. Let x represent the number of virtual pets made and sold each week. Write the weekly cost function, C, for Virtual Fido. $C(x) = 3000 + 20x$

  b. The function $R(x) = -x^2 + 1000x$ describes the money that Virtual Fido takes in each week from the sale of x virtual pets. Use this revenue function and the cost function from part (a) to write the weekly profit function, P. $P(x) = -x^2 + 1000x - (3000 + 20x) = -x^2 + 980x - 3000$

  c. Use the profit function to determine the number of virtual pets that should be made and sold each week to maximize profit. What is the maximum weekly profit? The maximum weekly profit is $237,100 when 490 virtual pets are made and sold each week.

Writing in Mathematics

71. What is a parabola? Describe its shape. 71–75. Answers will vary.

72. Explain how to decide whether a parabola opens upward or downward.

73. Describe how to find a parabola's vertex if its equation is in the form $f(x) = a(x - h)^2 + k$. Give an example.

74. Describe how to find a parabola's vertex if its equation is in the form $f(x) = ax^2 + bx + c$. Use $f(x) = x^2 - 6x + 8$ as an example.

75. A parabola that opens upward has its vertex at $(1, 2)$. Describe as much as you can about the parabola based on this information. Include in your discussion the number of x-intercepts (if any) for the parabola.

Technology Exercises

76. Use a graphing utility to verify any five of your hand-drawn graphs in Exercises 17–34. Answers will vary.

77. a. Use a graphing utility to graph $y = 2x^2 - 82x + 720$ in a standard viewing rectangle. What do you observe?

  b. Find the coordinates of the vertex for the given quadratic function. $(20.5, -120.5)$

  c. The answer to part (b) is $(20.5, -120.5)$. Because the leading coefficient, 2, of the given function is positive, the vertex is a minimum point on the graph. Use this fact to help find a viewing rectangle that will give a relatively complete picture of the parabola. With an axis of symmetry at $x = 20.5$, the setting for x should extend past this, so try Xmin = 0 and Xmax = 30. The setting for y should include (and probably go below) the y-coordinate of the graph's minimum point, so try Ymin = -130. Experiment with Ymax until your utility shows the parabola's major features. See graphing answer section.

  d. In general, explain how knowing the coordinates of a parabola's vertex can help determine a reasonable viewing rectangle on a graphing utility for obtaining a complete picture of the parabola. Answer will vary.

In Exercises 78–81, find the vertex for each parabola. Then determine a reasonable viewing rectangle on your graphing utility and use it to graph the parabola. See graphing answer section

78. $y = -0.25x^2 + 40x$ Vertex = $(80, 1600)$

79. $y = -4x^2 + 20x + 160$ Vertex = $(2.5, 185)$

80. $y = 5x^2 + 40x + 600$ Vertex = $(-4, 520)$

81. $y = 0.01x^2 + 0.6x + 100$ Vertex = $(-30, 91)$

82. The following data show fuel efficiency, in miles per gallon, for all U.S. automobiles in the indicated year.

x (Years after 1940)	y (Average Number of Miles per Gallon for U.S. Automobiles)
1940: 0	14.8
1950: 10	13.9
1960: 20	13.4
1970: 30	13.5
1980: 40	15.9
1990: 50	20.2
2000: 60	22.0

Source: U.S. Department of Transportation

a. Use a graphing utility to draw a scatter plot of the data. Explain why a quadratic function is appropriate for modeling these data.   See graphing answer section. The values of y decrease, then increase.

b. Use the quadratic regression feature to find the quadratic function that best fits the data.
$y = 0.005x^2 - 0.170x + 14.817$

c. Use the equation in part (b) to determine the worst year for automobile fuel efficiency. What was the average number of miles per gallon for that year?   The worst year for fuel efficiency was 1957 with about 13.372 miles per gallon.

d. Use a graphing utility to draw a scatter plot of the data and graph the quadratic function of best fit on the scatter plot.   See graphing answer section.

Critical Thinking Exercises

83. Which one of the following is true?   a

a. No quadratic functions have a range of $(-\infty, \infty)$.

b. The vertex of the parabola described by $f(x) = 2(x - 5)^2 - 1$ is at $(5, 1)$.

c. The graph of $f(x) = -2(x + 4)^2 - 8$ has one y-intercept and two x-intercepts.

d. The maximum value of y for the quadratic function $f(x) = -x^2 + x + 1$ is 1.

In Exercises 84–85, find the axis of symmetry for each parabola whose equation is given. Use the axis of symmetry to find a second point on the parabola whose y-coordinate is the same as the given point.

84. $f(x) = 3(x + 2)^2 - 5$; $(-1, -2)$ $x = -2$; $(-3, -2)$

85. $f(x) = (x - 3)^2 + 2$; $(6, 11)$ $x = 3$; $(0, 11)$

In Exercises 86–87, write the equation of each parabola in $f(x) = a(x - h)^2 + k$ form.

86. Vertex: $(-3, -4)$; The graph passes through the point $(1, 4)$. $f(x) = \frac{1}{2}(x + 3)^2 - 4$

87. Vertex: $(-3, -1)$; The graph passes through the point $(-2, -3)$. $f(x) = -2(x + 3)^2 - 1$

88. A rancher has 1000 feet of fencing to construct six corrals, as shown in the figure. Find the dimensions that maximize the enclosed area. What is the maximum area?

125 feet by about 166.7 feet: about 20,833 square feet

89. The annual yield per lemon tree is fairly constant at 320 pounds when the number of trees per acre is 50 or fewer. For each additional tree over 50, the annual yield per tree for all trees on the acre decreases by 4 pounds due to overcrowding. Find the number of trees that should be planted on an acre to produce the maximum yield. How many pounds is the maximum yield?   65 trees; 16,900 pounds

Review Exercises

90. Solve: $\dfrac{2}{x + 5} + \dfrac{1}{x - 5} = \dfrac{16}{x^2 - 25}$. (Section 6.6, Example 5)
7 or $\{7\}$

91. Simplify: $\dfrac{1 + \dfrac{2}{x}}{1 - \dfrac{4}{x^2}}$. (Section 6.3, Example 1) $\dfrac{x}{x - 2}$

92. Solve using determinants (Cramer's Rule):
$$2x + 3y = 6$$
$$x - 4y = 14.$$
(Section 3.5, Example 2) $(6, -2)$ or $\{(6, -2)\}$

✔ MID-CHAPTER CHECK POINT

CHAPTER 8

What You Know: We saw that not all quadratic equations can be solved by factoring. We learned three new methods for solving these equations - the square root property, completing the square, and the quadratic formula. We saw that the discriminant of $ax^2 + bx + c$, namely $b^2 - 4ac$, determines the number and type of the equation's solutions. We graphed quadratic functions using vertices, intercepts, and additional points, as necessary. We learned that the vertex of $f(x) = a(x - h)^2 + k$ is (h, k) and the vertex of $f(x) = ax^2 + bx + c$ is $\left(-\dfrac{b}{2a}, f\left(-\dfrac{b}{2a}\right)\right)$. We used the vertex to solve problems that involved minimizing or maximizing quadratic functions.

In Exercises 1–13, solve each equation by the method of your choice. Simplify solutions, if possible.

1. $(3x - 5)^2 = 36$ $\quad -\frac{1}{3}$ and $\frac{11}{3}$, or $\left\{-\frac{1}{3}, \frac{11}{3}\right\}$

2. $5x^2 - 2x = 7$ $\quad -1$ and $\frac{7}{5}$, or $\left\{-1, \frac{7}{5}\right\}$

3. $3x^2 - 6x - 2 = 0$ $\quad \dfrac{3 \pm \sqrt{15}}{3}$ or $\left\{\dfrac{3 \pm \sqrt{15}}{3}\right\}$

4. $x^2 + 6x = -2$ $\quad -3 \pm \sqrt{7}$ or $\{-3 \pm \sqrt{7}\}$

5. $5x^2 + 1 = 37$ $\quad \pm\dfrac{6\sqrt{5}}{5}$ or $\left\{\pm\dfrac{6\sqrt{5}}{5}\right\}$

6. $x^2 - 5x + 8 = 0$ $\quad \dfrac{5}{2} \pm \dfrac{\sqrt{7}}{2}i$ or $\left\{\dfrac{5}{2} \pm \dfrac{\sqrt{7}}{2}i\right\}$

7. $2x^2 + 26 = 0$ $\quad \pm\sqrt{13}i$ or $\left\{\pm\sqrt{13}i\right\}$

8. $(2x + 3)(x + 2) = 10$ $\quad -4$ and $\frac{1}{2}$, or $\left\{-4, \frac{1}{2}\right\}$

9. $(x + 3)^2 = 24$ $\quad -3 \pm 2\sqrt{6}$ or $\{-3 \pm 2\sqrt{6}\}$

10. $\dfrac{1}{x^2} - \dfrac{4}{x} + 1 = 0$ $\quad 2 \pm \sqrt{3}$ or $\{2 \pm \sqrt{3}\}$

11. $x(2x - 3) = -4$ $\quad \dfrac{3}{4} \pm \dfrac{\sqrt{23}}{4}i$ or $\left\{\dfrac{3}{4} \pm \dfrac{\sqrt{23}}{4}i\right\}$

12. $\dfrac{x^2}{3} + \dfrac{x}{2} = \dfrac{2}{3}$ $\quad \dfrac{-3 \pm \sqrt{41}}{4}$ or $\left\{\dfrac{-3 \pm \sqrt{41}}{4}\right\}$

13. $\dfrac{2x}{x^2 + 6x + 8} = \dfrac{x}{x + 4} - \dfrac{2}{x + 2}$ $\quad -2$ and 4, or $\{-2, 4\}$

14. Solve by completing the square: $x^2 + 10x - 3 = 0$. $\quad -5 \pm 2\sqrt{7}$ or $\{-5 \pm 2\sqrt{7}\}$

In Exercises 15–18, graph the given quadratic function. Give each function's domain and range.

15–18. See graphing answer section.

15. $f(x) = (x - 3)^2 - 4$ $\quad$ domain: $\{x|x$ is a real number$\}$ or $(-\infty, \infty)$; range: $\{y|y \geq -4\}$ or $[-4, \infty)$

16. $g(x) = 5 - (x + 2)^2$ $\quad$ domain: $\{x|x$ is a real number$\}$ or $(-\infty, \infty)$; range: $\{y|y \leq 5\}$ or $(-\infty, 5]$

17. $h(x) = -x^2 - 4x + 5$ $\quad$ domain: $\{x|x$ is a real number$\}$ or $(-\infty, \infty)$; range: $\{y|y \leq 9\}$ or $(-\infty, 9]$

18. $f(x) = 3x^2 - 6x + 1$ $\quad$ domain: $\{x|x$ is a real number$\}$ or $(-\infty, \infty)$; range: $\{y|y \geq -2\}$ or $[-2, \infty)$

In Exercises 19–20, without solving the equation, determine the number and type of solutions.

19. $2x^2 + 5x + 4 = 0$ $\quad$ two imaginary solutions

20. $10x(x + 4) = 15x - 15$ $\quad$ two rational solutions

In Exercises 21–22, write a quadratic equation in standard form with the given solution set.

21. $\left\{-\dfrac{1}{2}, \dfrac{3}{4}\right\}$ $\quad 8x^2 - 2x - 3 = 0$

22. $\left\{-2\sqrt{3}, 2\sqrt{3}\right\}$ $\quad x^2 - 12 = 0$

23. A company manufactures and sells bath cabinets. The function

$$P(x) = -x^2 + 150x - 4425$$

models the company's daily profit, $P(x)$, when x cabinets are manufactured and sold per day. How many cabinets should be manufactured and sold per day to maximize the company's profit? What is the maximum daily profit? $\quad$ 75 cabinets per day; $1200

24. Among all pairs of numbers whose sum is -18, find a pair whose product is as large as possible. What is the maximum product? $\quad -9$ and -9; 81

25. The base of a triangle measures 40 inches minus twice the measure of its height. For what measure of the height does the triangle have a maximum area? What is the maximum area? $\quad$ 10 inches; 100 square inches

SECTION

8.4

Objective

1 Solve equations that are quadratic in form.

EQUATIONS QUADRATIC IN FORM

"My husband asked me if we have any cheese puffs. Like he can't go and lift the couch cushion up himself."—Roseanne

Photo by Michael Lutzky. © 2000, *The Washington Post.* Reprinted with permission.

How important is it for you to have a clean house? The percentage of people who find this to be quite important varies by age. In the exercise set, you will work with a function that models this phenomenon. Your work will be based on equations that are not quadratic, but that can be written as quadratic equations using an appropriate substitution. Here are some examples:

Given Equation	Substitution	New Equation
$x^4 - 10x^2 + 9 = 0$ or $(x^2)^2 - 10x^2 + 9 = 0$	$t = x^2$	$t^2 - 10t + 9 = 0$
$5x^{\frac{2}{3}} + 11x^{\frac{1}{3}} + 2 = 0$ or $5\left(x^{\frac{1}{3}}\right)^2 + 11x^{\frac{1}{3}} + 2 = 0$	$t = x^{\frac{1}{3}}$	$5t^2 + 11t + 2 = 0$

An equation that is **quadratic in form** is one that can be expressed as a quadratic equation using an appropriate substitution. Both of the preceding given equations are quadratic in form.

Equations that are quadratic in form contain an expression to a power, the same expression to that power squared, and a constant term. By letting t equal the expression to the power, a quadratic equation in t will result. Now it's easy. Solve this quadratic equation for t. Finally, use your substitution to find the values for the variable in the given equation. Example 1 shows how this is done.

1 Solve equations that are quadratic in form.

EXAMPLE 1 Solving an Equation Quadratic in Form

Solve: $x^4 - 10x^2 + 9 = 0$.

SOLUTION Notice that the equation contains an expression to a power, x^2, the same expression to that power squared, x^4 or $(x^2)^2$, and a constant term, 9. We let t equal the expression to the power. Thus,

$$\text{let } t = x^2.$$

The graph of

$$y = x^4 - 10x^2 + 9$$

has x-intercepts at $-3, -1, 1,$ and 3. This verifies that the solution set of

$$x^4 - 10x^2 + 9 = 0$$

is $\{-3, -1, 1, 3\}$.

$[-5, 5, 1]$ by $[-20, 10, 5]$

Now we write the given equation as a quadratic equation in t and solve for t.

$x^4 - 10x^2 + 9 = 0$	This is the given equation.
$(x^2)^2 - 10x^2 + 9 = 0$	The given equation contains x^2 and x^2 squared.
$t^2 - 10t + 9 = 0$	Let $t = x^2$. Replace x^2 with t.
$(t - 9)(t - 1) = 0$	Factor.
$t - 9 = 0 \quad$ or $\quad t - 1 = 0$	Set each factor equal to zero.
$t = 9 \qquad\qquad\quad t = 1$	Solve for t.

We're not done! Why not? We were asked to solve for x and we have values for t. We use the original substitution, $t = x^2$, to solve for x. Replace t with x^2 in each equation shown.

$x^2 = 9 \quad$ or $\quad x^2 = 1$	
$x = \pm\sqrt{9} \qquad\quad x = \pm\sqrt{1}$	Apply the square root property.
$x = \pm 3 \qquad\qquad x = \pm 1$	

Substitute these values into the given equation and verify that the solutions are $-3, 3,$ $-1,$ and 1. The solution set is $\{-3, 3, -1, 1\}$. You may prefer to express the solution set in numerical order as $\{-3, -1, 1, 3\}$. ■

✔ **CHECK POINT 1** Solve: $x^4 - 17x^2 + 16 = 0$. $\quad -4, 4, -1,$ and 1, or $\{-4, 4, -1, 1\}$

If checking proposed solutions is not overly cumbersome, you should do so either algebraically or with a graphing utility. The Using Technology box shows a check of the four solutions in Example 1. Are there situations when solving equations quadratic in form where a check is essential? Yes. **If at any point in the solution process both sides of an equation are raised to an even power, a check is required.** Extraneous solutions that are not solutions of the given equation may have been introduced.

EXAMPLE 2 Solving an Equation Quadratic in Form

Solve: $2x - \sqrt{x} - 10 = 0$.

SOLUTION To identify exponents on the terms, let's rewrite $\sqrt{x}$ as $x^{\frac{1}{2}}$. The equation can be expressed as

$$2x^1 - x^{\frac{1}{2}} - 10 = 0.$$

Notice that the equation contains an expression to a power, $x^{\frac{1}{2}}$, the same expression to that power squared, x^1 or $\left(x^{\frac{1}{2}}\right)^2$, and a constant term, -10. We let t equal the expression to the power. Thus,

$$\text{let } t = x^{\frac{1}{2}}.$$

Now we write the given equation as a quadratic equation in t and solve for t.

$2x^1 - x^{\frac{1}{2}} - 10 = 0$	This is the given equation in exponential form.
$2\left(x^{\frac{1}{2}}\right)^2 - x^{\frac{1}{2}} - 10 = 0$	The equation contains $x^{\frac{1}{2}}$ and $x^{\frac{1}{2}}$ squared.
$2t^2 - t - 10 = 0$	Let $t = x^{\frac{1}{2}}$. Replace $x^{\frac{1}{2}}$ with t.
$(2t - 5)(t + 2) = 0$	Factor.
$2t - 5 = 0 \quad$ or $\quad t + 2 = 0$	Set each factor equal to zero.
$t = \dfrac{5}{2} \qquad\qquad t = -2$	Solve for t.

Use the original substitution, $t = x^{\frac{1}{2}}$, to solve for x. Replace t with $x^{\frac{1}{2}}$ in each of the preceding equations.

$$x^{\frac{1}{2}} = \frac{5}{2} \quad \text{or} \quad x^{\frac{1}{2}} = -2 \qquad \text{Replace } t \text{ with } x^{\frac{1}{2}}.$$

$$\left(x^{\frac{1}{2}}\right)^2 = \left(\frac{5}{2}\right)^2 \qquad \left(x^{\frac{1}{2}}\right)^2 = (-2)^2 \qquad \text{Solve for } x \text{ by squaring both sides of each equation.}$$

> Both sides are raised to even powers. We must check.

$$x = \frac{25}{4} \qquad\qquad x = 4 \qquad \text{Square } \frac{5}{2} \text{ and } -2.$$

It is essential to check both proposed solutions in the original equation.

<table>
<tr><td align="center">**Check** $\frac{25}{4}$**:**</td><td align="center">**Check 4:**</td></tr>
<tr><td align="center">$2x - \sqrt{x} - 10 = 0$</td><td align="center">$2x - \sqrt{x} - 10 = 0$</td></tr>
<tr><td align="center">$2 \cdot \frac{25}{4} - \sqrt{\frac{25}{4}} - 10 \overset{?}{=} 0$</td><td align="center">$2 \cdot 4 - \sqrt{4} - 10 \overset{?}{=} 0$</td></tr>
<tr><td align="center">$\frac{25}{2} - \frac{5}{2} - 10 \overset{?}{=} 0$</td><td align="center">$8 - 2 - 10 \overset{?}{=} 0$</td></tr>
<tr><td align="center">$\frac{20}{2} - 10 \overset{?}{=} 0$</td><td align="center">$6 - 10 \overset{?}{=} 0$</td></tr>
<tr><td align="center">$0 = 0, \quad$ true</td><td align="center">$-4 = 0, \quad$ false</td></tr>
</table>

The check indicates that 4 is not a solution. It is an extraneous solution brought about by squaring each side of the equation. The only solution is $\frac{25}{4}$, and the solution set is $\left\{\frac{25}{4}\right\}$. ∎

✔ **CHECK POINT 2** Solve: $x - 2\sqrt{x} - 8 = 0$.   16 or {16}

The equations in Examples 1 and 2 can be solved by methods other than using substitutions.

$$x^4 - 10x^2 + 9 = 0 \qquad\qquad 2x - \sqrt{x} - 10 = 0$$

> This equation can be solved directly by factoring: $(x^2 - 9)(x^2 - 1) = 0$.

> This equation can be solved by isolating the radical term: $2x - 10 = \sqrt{x}$. Then square both sides.

In the examples that follow, solving the equations by methods other than first introducing a substitution becomes increasingly difficult.

EXAMPLE 3 Solving an Equation Quadratic in Form

Solve: $(x^2 - 5)^2 + 3(x^2 - 5) - 10 = 0$.

SOLUTION This equation contains $x^2 - 5$ and $x^2 - 5$ squared. We let $t = x^2 - 5$.

$(x^2 - 5)^2 + 3(x^2 - 5) - 10 = 0$	This is the given equation.
$t^2 + 3t - 10 = 0$	Let $t = x^2 - 5$.
$(t + 5)(t - 2) = 0$	Factor.
$t + 5 = 0$ or $t - 2 = 0$	Set each factor equal to zero.
$t = -5$ $\qquad$ $t = 2$	Solve for t.

Use the original substitution, $t = x^2 - 5$, to solve for x. Replace t with $x^2 - 5$ in each of the preceding equations.

$x^2 - 5 = -5$ or $x^2 - 5 = 2$	Replace t with $x^2 - 5$.
$x^2 = 0$ $\qquad$ $x^2 = 7$	Solve for x by isolating x^2.
$x = 0$ $\qquad$ $x = \pm\sqrt{7}$	Apply the square root property.

Although we did not raise both sides of an equation to an even power, checking the three proposed solutions in the original equation is a good idea. Do this now and verify that the solutions are $-\sqrt{7}$, 0, and $\sqrt{7}$, and the solution set is $\{-\sqrt{7}, 0, \sqrt{7}\}$. ∎

 CHECK POINT 3 Solve: $(x^2 - 4)^2 + (x^2 - 4) - 6 = 0$.
$-\sqrt{6}, -1, 1, \sqrt{6}$ or $\{-\sqrt{6}, -1, 1, \sqrt{6}\}$

EXAMPLE 4 Solving an Equation Quadratic in Form

Solve: $10x^{-2} + 7x^{-1} + 1 = 0$.

SOLUTION Notice that the equation contains an expression to a power, x^{-1}, the same expression to that power squared, x^{-2} or $(x^{-1})^2$, and a constant term, 1. We let t equal the expression to the power. Thus,

$$\text{let } t = x^{-1}.$$

Now we write the given equation as a quadratic equation in t and solve for t.

$10x^{-2} + 7x^{-1} + 1 = 0$	This is the given equation.
$10(x^{-1})^2 + 7x^{-1} + 1 = 0$	The equation contains x^{-1} and x^{-1} squared.
$10t^2 + 7t + 1 = 0$	Let $t = x^{-1}$.
$(5t + 1)(2t + 1) = 0$	Factor.
$5t + 1 = 0$ or $2t + 1 = 0$	Set each factor equal to zero.
$5t = -1$ $\qquad$ $2t = -1$	Solve each equation for t.
$t = -\dfrac{1}{5}$ $\qquad$ $t = -\dfrac{1}{2}$	

Use the original substitution, $t = x^{-1}$, to solve for x. Replace t with x^{-1} in each of the preceding equations.

$$x^{-1} = -\frac{1}{5} \quad \text{or} \quad x^{-1} = -\frac{1}{2} \qquad \text{Replace } t \text{ with } x^{-1}.$$

$$(x^{-1})^{-1} = \left(-\frac{1}{5}\right)^{-1} \qquad (x^{-1})^{-1} = \left(-\frac{1}{2}\right)^{-1} \qquad \text{Solve for } x \text{ by raising both sides of each equation to the } -1 \text{ power.}$$

$$x = -5 \qquad\qquad x = -2$$

$$\left(-\tfrac{1}{5}\right)^{-1} = \frac{1}{-\frac{1}{5}} = -5 \qquad \left(-\tfrac{1}{2}\right)^{-1} = \frac{1}{-\frac{1}{2}} = -2$$

We did not raise both sides of an equation to an even power. A check will show that both -5 and -2 are solutions of the original equation. The solution set is $\{-5, -2\}$. ∎

 CHECK POINT 4 Solve: $2x^{-2} + x^{-1} - 1 = 0$. $\quad -1 \text{ and } 2, \text{ or } \{-1, 2\}$

EXAMPLE 5 Solving an Equation Quadratic in Form

Solve: $5x^{\frac{2}{3}} + 11x^{\frac{1}{3}} + 2 = 0$.

SOLUTION Notice that the equation contains an expression to a power, $x^{\frac{1}{3}}$, the same expression to that power squared, $x^{\frac{2}{3}}$ or $\left(x^{\frac{1}{3}}\right)^2$, and a constant term, 2. We let t equal the expression to the power. Thus,

$$\text{let } t = x^{\frac{1}{3}}.$$

Now we write the given equation as a quadratic equation in t and solve for t.

$$5x^{\frac{2}{3}} + 11x^{\frac{1}{3}} + 2 = 0 \qquad \text{This is the given equation.}$$

$$5\left(x^{\frac{1}{3}}\right)^2 + 11\left(x^{\frac{1}{3}}\right) + 2 = 0 \qquad \text{The given equation contains } x^{\frac{1}{3}} \text{ and } x^{\frac{1}{3}} \text{ squared.}$$

$$5t^2 + 11t + 2 = 0 \qquad \text{Let } t = x^{\frac{1}{3}}.$$

$$(5t + 1)(t + 2) = 0 \qquad \text{Factor.}$$

$$5t + 1 = 0 \quad \text{or} \quad t + 2 = 0 \qquad \text{Set each factor equal to 0.}$$

$$t = -\tfrac{1}{5} \qquad\qquad t = -2 \qquad \text{Solve for } t.$$

Use the original substitution, $t = x^{\frac{1}{3}}$, to solve for x. Replace t with $x^{\frac{1}{3}}$ in each of the preceding equations.

$$x^{\frac{1}{3}} = -\frac{1}{5} \quad \text{or} \quad x^{\frac{1}{3}} = -2 \qquad \text{Replace } t \text{ with } x^{\frac{1}{3}}.$$

$$\left(x^{\frac{1}{3}}\right)^3 = \left(-\frac{1}{5}\right)^3 \qquad \left(x^{\frac{1}{3}}\right)^3 = (-2)^3 \qquad \text{Solve for } x \text{ by cubing both sides of each equation.}$$

$$x = -\frac{1}{125} \qquad\qquad x = -8$$

We did not raise both sides of an equation to an even power. A check will show that both -8 and $-\frac{1}{125}$ are solutions of the original equation. The solution set is $\left\{-8, -\frac{1}{125}\right\}$. ∎

 CHECK POINT 5 Solve: $3x^{\frac{2}{3}} - 11x^{\frac{1}{3}} - 4 = 0$. $\quad -\frac{1}{27} \text{ and } 64, \text{ or } \left\{-\frac{1}{27}, 64\right\}$

8.4 EXERCISE SET

Student Solutions Manual CD/Video PH Math/Tutor Center MathXL Tutorials on CD MathXL® MyMathLab Interactmath.com

Practice Exercises

In Exercises 1–32, solve each equation by making an appropriate substitution. If at any point in the solution process both sides of an equation are raised to an even power, a check is required.

1. $x^4 - 5x^2 + 4 = 0$ $-2, 2, -1,$ and $1,$ or $\{-2, 2, -1, 1\}$

2. $x^4 - 13x^2 + 36 = 0$ $-3, 3, -2,$ and $2,$ or $\{-3, 3, -2, 2\}$

3. $x^4 - 11x^2 + 18 = 0$
$-3, 3, -\sqrt{2},$ and $\sqrt{2},$ or $\left\{-3, 3, -\sqrt{2}, \sqrt{2}\right\}$

4. $x^4 - 9x^2 + 20 = 0$
$-\sqrt{5}, \sqrt{5}, -2,$ and $2,$ or $\left\{-\sqrt{5}, \sqrt{5}, -2, 2\right\}$

5. $x^4 + 2x^2 = 8$
$-2i, 2i, -\sqrt{2},$ and $\sqrt{2},$ or $\left\{-2i, 2i, -\sqrt{2}, \sqrt{2}\right\}$

6. $x^4 + 4x^2 = 5$
$-\sqrt{5}i, \sqrt{5}i, -1,$ and $1,$ or $\left\{-\sqrt{5}i, \sqrt{5}i, -1, 1\right\}$

7. $x + \sqrt{x} - 2 = 0$ 1 or $\{1\}$

8. $x + \sqrt{x} - 6 = 0$ 4 or $\{4\}$

9. $x - 4x^{\frac{1}{2}} - 21 = 0$ 49 or $\{49\}$

10. $x - 6x^{\frac{1}{2}} + 8 = 0$ 4 and $16,$ or $\{4, 16\}$

11. $x - 13\sqrt{x} + 40 = 0$ 25 and $64,$ or $\{25, 64\}$

12. $2x - 7\sqrt{x} - 30 = 0$ 36 or $\{36\}$

13. $(x - 5)^2 - 4(x - 5) - 21 = 0$ 2 and $12,$ or $\{2, 12\}$

14. $(x + 3)^2 + 7(x + 3) - 18 = 0$ -12 and $-1,$ or $\{-12, -1\}$

15. $(x^2 - 1)^2 - (x^2 - 1) = 2$
$-\sqrt{3}, 0,$ and $\sqrt{3},$ or $\left\{-\sqrt{3}, 0, \sqrt{3}\right\}$

16. $(x^2 - 2)^2 - (x^2 - 2) = 6$
$-\sqrt{5}, 0,$ and $\sqrt{5},$ or $\left\{-\sqrt{5}, 0, \sqrt{5}\right\}$

17. $(x^2 + 3x)^2 - 8(x^2 + 3x) - 20 = 0$
$-5, -2, -1,$ and $2,$ or $\{-5, -2, -1, 2\}$

18. $(x^2 - 2x)^2 - 11(x^2 - 2x) + 24 = 0$
$-2, -1, 3,$ and $4,$ or $\{-2, -1, 3, 4\}$

19. $x^{-2} - x^{-1} - 20 = 0$ $-\frac{1}{4}$ and $\frac{1}{5},$ or $\left\{-\frac{1}{4}, \frac{1}{5}\right\}$

20. $x^{-2} - x^{-1} - 6 = 0$ $-\frac{1}{2}$ and $\frac{1}{3},$ or $\left\{-\frac{1}{2}, \frac{1}{3}\right\}$

21. $2x^{-2} - 7x^{-1} + 3 = 0$ $\frac{1}{3}$ and $2,$ or $\left\{\frac{1}{3}, 2\right\}$

22. $20x^{-2} + 9x^{-1} + 1 = 0$ -5 and $-4,$ or $\{-5, -4\}$

23. $x^{-2} - 4x^{-1} = 3$ $\dfrac{-2 \pm \sqrt{7}}{3}$ or $\left\{\dfrac{-2 \pm \sqrt{7}}{3}\right\}$

24. $x^{-2} - 6x^{-1} = -4$ $\dfrac{3 \pm \sqrt{5}}{4}$ or $\left\{\dfrac{3 \pm \sqrt{5}}{4}\right\}$

25. $x^{\frac{2}{3}} - x^{\frac{1}{3}} - 6 = 0$ -8 and $27,$ or $\{-8, 27\}$

26. $x^{\frac{2}{3}} + 2x^{\frac{1}{3}} - 3 = 0$ -27 and $1,$ or $\{-27, 1\}$

27. $x^{\frac{2}{5}} + x^{\frac{1}{5}} - 6 = 0$ -243 and $32,$ or $\{-243, 32\}$

28. $x^{\frac{2}{5}} + x^{\frac{1}{5}} - 2 = 0$ -32 and $1,$ or $\{-32, 1\}$

29. $2x^{\frac{1}{2}} - x^{\frac{1}{4}} = 1$ 1 or $\{1\}$

30. $2x^{\frac{1}{2}} - 5x^{\frac{1}{4}} = 3$ 81 or $\{81\}$

31. $\left(x - \dfrac{8}{x}\right)^2 + 5\left(x - \dfrac{8}{x}\right) - 14 = 0$
$-8, -2, 1,$ and $4,$ or $\{-8, -2, 1, 4\}$

32. $\left(x - \dfrac{10}{x}\right)^2 + 6\left(x - \dfrac{10}{x}\right) - 27 = 0$
$-10, -2, 1,$ and $5,$ or $\{-10, -2, 1, 5\}$

In Exercises 33–38, find the x-intercepts of the given function, f. Then use the x-intercepts to match each function with its graph. [The graphs are labeled (a) through (f).]

33. $f(x) = x^4 - 5x^2 + 4$ $-2, 2, -1,$ and $1;$ **c**

34. $f(x) = x^4 - 13x^2 + 36$ $-3, 3, -2,$ and $2;$ **a**

35. $f(x) = x^{\frac{1}{3}} + 2x^{\frac{1}{6}} - 3$ $1;$ **e**

36. $f(x) = x^{-2} - x^{-1} - 6$ $-\frac{1}{2}$ and $\frac{1}{3};$ **b**

37. $f(x) = (x + 2)^2 - 9(x + 2) + 20$ 2 and $3;$ **f**

38. $f(x) = 2(x + 2)^2 + 5(x + 2) - 3$ -5 and $-\frac{3}{2};$ **d**

a.

$[-5, 5, 1]$ by $[-10, 40, 5]$

b.

$[-3, 3, 1]$ by $[-10, 10, 1]$

c.

$[-5, 5, 1]$ by $[-4, 10, 1]$

d.

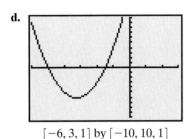

$[-6, 3, 1]$ by $[-10, 10, 1]$

e.

$[-1, 10, 1]$ by $[-3, 3, 1]$

f.

$[-1, 6, 1]$ by $[-1, 10, 1]$

Application Exercises

How important is it for you to have a clean house? The bar graph indicates that the percentage of people who find this to be quite important varies by age. The percentage, $P(x)$, who find having a clean house very important can be modeled by the function

$$P(x) = 0.04(x + 40)^2 - 3(x + 40) + 104,$$

where x is the number of years a person's age is above or below 40. Thus, x is positive for people over 40 and negative for people under 40. Use the function to solve Exercises 47–48.

The Importance of Having a Clean House, by Age

Source: Soap and Detergent Association

Practice Plus

39. Let $f(x) = (x^2 + 3x - 2)^2 - 10(x^2 + 3x - 2)$. Find all x such that $f(x) = -16$. $-5, -4, 1,$ and 2

40. Let $f(x) = (x^2 + 2x - 2)^2 - 7(x^2 + 2x - 2)$. Find all x such that $f(x) = -6$. $-4, -3, 1,$ and 2

41. Let $f(x) = 3\left(\dfrac{1}{x} + 1\right)^2 + 5\left(\dfrac{1}{x} + 1\right)$. Find all x such that $f(x) = 2$. $-\frac{3}{2}$ and $-\frac{1}{3}$

42. Let $f(x) = 2x^{\frac{2}{3}} + 3x^{\frac{1}{3}}$. Find all x such that $f(x) = 2$. -8 and $\frac{1}{8}$

43. Let $f(x) = \dfrac{x}{x - 4}$ and $g(x) = 13\sqrt{\dfrac{x}{x - 4}} - 36$. Find all x such that $f(x) = g(x)$. $\frac{81}{20}$ and $\frac{64}{15}$

44. Let $f(x) = \dfrac{x}{x - 2} + 10$ and $g(x) = -11\sqrt{\dfrac{x}{x - 2}}$. Find all x such that $f(x) = g(x)$.   No values of x satisfy $f(x) = g(x)$.

45. Let $f(x) = 3(x - 4)^{-2}$ and $g(x) = 16(x - 4)^{-1}$. Find all x such that $f(x)$ exceeds $g(x)$ by 12. $\frac{5}{2}$ and $\frac{25}{6}$

46. Let $f(x) = 6\left(\dfrac{2x}{x - 3}\right)^2$ and $g(x) = 5\left(\dfrac{2x}{x - 3}\right)$. Find all x such that $f(x)$ exceeds $g(x)$ by 6. -9 and $\frac{3}{4}$

47. According to the model, at which ages do 60% of us feel that having a clean house is very important? Substitute 60 for $P(x)$ and solve the quadratic-in-form equation. How well does the function model the data shown in the bar graph?
ages 20 and 55; The function models the data well.

48. According to the model, at which ages do 50% of us feel that having a clean house is very important? Substitute 50 for $P(x)$ and solve the quadratic-in-form equation. How well does the function model the data shown in the bar graph?
ages 30 and 45; The function models the data well.

Writing in Mathematics

49. Explain how to recognize an equation that is quadratic in form. Provide two original examples with your explanation.
49–51. Answers will vary.

50. Describe two methods for solving this equation:

$$x - 5\sqrt{x} + 4 = 0.$$

51. In the twenty-first century, collecting and presenting data are big business. Because statisticians both record and influence our behavior, you should always ask yourself if the person or group presenting the data has any special case to make for or against the displayed information. Using this criterion, describe your impressions of the data in the bar graph for Exercises 47–48.

Technology Exercises

52. Use a graphing utility to verify the solutions of any five equations in Exercises 1-32 that you solved algebraically. The real solutions should appear as x-intercepts on the graph of the function related to the given equation.   Answers will vary.

Use a graphing utility to solve the equations in Exercises 53–60. Check by direct substitution.

53. $x^6 - 7x^3 - 8 = 0$ -1 and 2, or $\{-1, 2\}$

54. $3(x - 2)^{-2} - 4(x - 2)^{-1} + 1 = 0$ 3 and 5, or $\{3, 5\}$

55. $x^4 - 10x^2 + 9 = 0$ $-1, 1, -3,$ and 3, or $\{-1, 1, -3, 3\}$

56. $2x + 6\sqrt{x} = 8$ 1 or $\{1\}$

57. $2(x + 1)^2 = 5(x + 1) + 3$ $-\frac{3}{2}$ and 2, or $\left\{-\frac{3}{2}, 2\right\}$

58. $(x^2 - 3x)^2 + 2(x^2 - 3x) - 24 = 0$ -1 and 4, or $\{-1, 4\}$

59. $x^{\frac{1}{2}} + 4x^{\frac{1}{4}} = 5$ 1 or $\{1\}$

60. $x^{\frac{2}{3}} - 3x^{\frac{1}{3}} + 2 = 0$ 1 and 8, or $\{1, 8\}$

Critical Thinking Exercises

61. Which one of the following is true?   b

a. If an equation is quadratic in form, there is only one method that can be used to obtain its solution.

b. An equation that is quadratic in form must have a variable factor in one term that is the square of the variable factor in another term.

c. Because x^6 is the square of x^3, the equation $x^6 - 5x^3 + 6x = 0$ is quadratic in form.

d. To solve $x - 9\sqrt{x} + 14 = 0$, we let $\sqrt{t} = x$.

In Exercises 62–64, use a substitution to solve each equation.

62. $x^4 - 5x^2 - 2 = 0$ $\pm\dfrac{\sqrt{10 + 2\sqrt{33}}}{2}$ and $\pm\dfrac{\sqrt{10 - 2\sqrt{33}}}{2}$

or $\left\{\pm\dfrac{\sqrt{10 + 2\sqrt{33}}}{2}, \pm\dfrac{\sqrt{10 - 2\sqrt{33}}}{2}\right\}$

63. $5x^6 + x^3 = 18$ $\sqrt[3]{-2}$ and $\dfrac{\sqrt[3]{225}}{5}$, or $\left\{\sqrt[3]{-2}, \dfrac{\sqrt[3]{225}}{5}\right\}$

64. $\sqrt{\dfrac{x + 4}{x - 1}} + \sqrt{\dfrac{x - 1}{x + 4}} = \dfrac{5}{2}\left(\text{Let } t = \sqrt{\dfrac{x + 4}{x - 1}}.\right)$

$-\dfrac{17}{3}$ and $\dfrac{8}{3}$, or $\left\{-\dfrac{17}{3}, \dfrac{8}{3}\right\}$

Review Exercises

65. Simplify:

$$\frac{2x^2}{10x^3 - 2x^2}.$$

(Section 6.1, Example 4) $\dfrac{1}{5x - 1}$

66. Divide: $\dfrac{2 + i}{1 - i}$. (Section 7.7, Example 5) $\dfrac{1}{2} + \dfrac{3}{2}i$

67. Solve using matrices:

$$2x + y = 6$$
$$x - 2y = 8.$$

(Section 3.4, Example 3) $(4, -2)$ or $\{(4, -2)\}$

POLYNOMIAL AND RATIONAL INEQUALITIES

Objectives

1 Solve polynomial inequalities.

2 Solve rational inequalities.

3 Solve problems modeled by polynomial or rational inequalities.

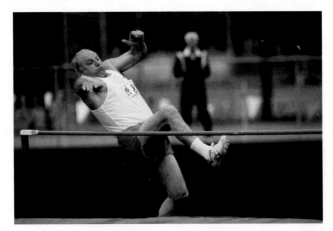

People are going to live longer in the twenty-first century. This will put added pressure on the Social Security and Medicare systems. The bar graph in Figure 8.14 shows the cost of Medicare, in billions of dollars, through 2005.

Medicare Spending

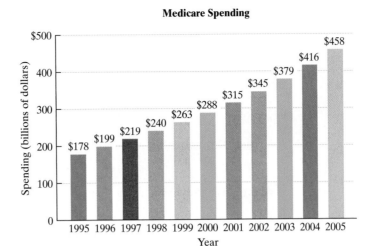

FIGURE 8.14

Source: Congressional Budget Office

Medicare spending, $f(x)$, in billions of dollars, x years after 1995 can be modeled by the quadratic function

$$f(x) = 1.2x^2 + 15.2x + 181.4.$$

To determine in which years Medicare spending will exceed $500 billion, we must solve the inequality

$$1.2x^2 + 15.2x + 181.4 > 500.$$

Medicare spending   exceeds $500 billion.

We begin by subtracting 500 from both sides. This will give us zero on the right:

$$1.2x^2 + 15.2x + 181.4 - 500 > 500 - 500$$
$$1.2x^2 + 15.2x - 318.6 > 0.$$

The form of this inequality is $ax^2 + bx + c > 0$. Such an inequality is called a *polynomial inequality*.

> **DEFINITION OF A POLYNOMIAL INEQUALITY**   A polynomial inequality is any inequality that can be put in one of the forms
>
> $$f(x) < 0, \quad f(x) > 0, \quad f(x) \le 0, \quad \text{or} \quad f(x) \ge 0,$$
>
> where f is a polynomial function.

In this section, we establish the basic techniques for solving polynomial inequalities. We will use these techniques to solve inequalities involving rational functions.

Solving Polynomial Inequalities   Graphs can help us visualize the solutions of polynomial inequalities. For example, the graph of $f(x) = x^2 - 7x + 10$ is shown in Figure 8.15. The x-intercepts, 2 and 5, are **boundary points** between where the graph lies above the x-axis, shown in blue, and where the graph lies below the x-axis, shown in red.

Locating the x-intercepts of a polynomial function, f, is an important step in finding the solution set for polynomial inequalities in the form $f(x) < 0$ or $f(x) > 0$.

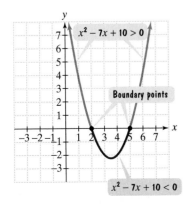

FIGURE 8.15

We use the x-intercepts of f as boundary points that divide the real number line into intervals. On each interval, the graph of f is either above the x-axis $[f(x) > 0]$ or below the x-axis $[f(x) < 0]$. For this reason, x-intercepts play a fundamental role in solving polynomial inequalities. The x-intercepts are found by solving the equation $f(x) = 0$.

1 Solve polynomial inequalities.

PROCEDURE FOR SOLVING POLYNOMIAL INEQUALITIES

1. Express the inequality in the form

$$f(x) < 0 \quad \text{or} \quad f(x) > 0,$$

where f is a polynomial function

2. Solve the equation $f(x) = 0$. The real solutions are the **boundary points**.

3. Locate these boundary points on a number line, thereby dividing the number line into intervals.

4. Choose one representative number within each interval and evaluate f at that number.

   a. If the value of f is positive, then $f(x) > 0$ for all numbers, x, in the interval.

   b. If the value of f is negative, then $f(x) < 0$ for all numbers, x, in the interval.

5. Write the solution set, selecting the interval(s) that satisfy the given inequality.

This procedure is valid if $<$ is replaced by $\leq$ or $>$ is replaced by $\geq$. However, if the inequality involves $\leq$ or $\geq$, include the boundary points [the solutions of $f(x) = 0$] in the solution set.

EXAMPLE 1 Solving a Polynomial Inequality

Solve and graph the solution set on a real number line: $2x^2 + x > 15$.

SOLUTION

Step 1. **Express the inequality in the form $f(x) < 0$ or $f(x) > 0$.** We begin by rewriting the inequality so that 0 is on the right side.

$$2x^2 + x > 15 \qquad \text{This is the given inequality.}$$
$$2x^2 + x - 15 > 15 - 15 \qquad \text{Subtract 15 from both sides.}$$
$$2x^2 + x - 15 > 0 \qquad \text{Simplify.}$$

This inequality is equivalent to the one we wish to solve. It is in the form $f(x) > 0$, where $f(x) = 2x^2 + x - 15$.

Step 2. **Solve the equation $f(x) = 0$.** We find the x-intercepts of $f(x) = 2x^2 + x - 15$ by solving the equation $2x^2 + x - 15 = 0$.

$$2x^2 + x - 15 = 0 \qquad \text{This polynomial equation is a quadratic equation.}$$
$$(2x - 5)(x + 3) = 0 \qquad \text{Factor.}$$
$$2x - 5 = 0 \quad \text{or} \quad x + 3 = 0 \qquad \text{Set each factor equal to 0.}$$
$$x = \tfrac{5}{2} \qquad\qquad x = -3 \qquad \text{Solve for x.}$$

The x-intercepts of f are -3 and $\tfrac{5}{2}$. We will use these x-intercepts as boundary points on a number line.

Step 3. **Locate the boundary points on a number line and separate the line into intervals.** The number line with the boundary points is shown as follows:

The boundary points divide the number line into three intervals:

$$(-\infty, -3) \quad \left(-3, \tfrac{5}{2}\right) \quad \left(\tfrac{5}{2}, \infty\right).$$

Step 4. **Choose one representative number within each interval and evaluate f at that number.**

Interval	Representative Number	Substitute into $f(x) = 2x^2 + x - 15$	Conclusion
$(-\infty, -3)$	-4	$f(-4) = 2(-4)^2 + (-4) - 15$ $= 13,$ positive	$f(x) > 0$ for all x in $(-\infty, -3)$.
$\left(-3, \dfrac{5}{2}\right)$	0	$f(0) = 2 \cdot 0^2 + 0 - 15$ $= -15,$ negative	$f(x) < 0$ for all x in $\left(-3, \dfrac{5}{2}\right)$.
$\left(\dfrac{5}{2}, \infty\right)$	3	$f(3) = 2 \cdot 3^2 + 3 - 15$ $= 6,$ positive	$f(x) > 0$ for all x in $\left(\dfrac{5}{2}, \infty\right)$.

Step 5. **Write the solution set, selecting the interval(s) that satisfy the given inequality.** Based on our work in step 4, we see that $f(x) > 0$ for all x in $(-\infty, -3)$ or $\left(\tfrac{5}{2}, \infty\right)$. Thus, the solution set of the given inequality, $2x^2 + x > 15$, or, equivalently, $2x^2 + x - 15 > 0$, is

$$(-\infty, -3) \cup \left(\tfrac{5}{2}, \infty\right) \text{ or } \left\{x \mid x < -3 \text{ or } x > \tfrac{5}{2}\right\}.$$

The graph of the solution set on a number line is shown as follows:

 CHECK POINT 1 Solve and graph the solution set: $x^2 - x > 20$.
See graphing answer section. $(-\infty, -4) \cup (5, \infty)$ or $\{x \mid x < -4 \text{ or } x > 5\}$

EXAMPLE 2   Solving a Polynomial Inequality

Solve and graph the solution set on a real number line: $x^3 + x^2 \le 4x + 4$.

SOLUTION

Step 1. **Express the inequality in the form $f(x) \le 0$ or $f(x) \ge 0$.** We begin by rewriting the inequality so that 0 is on the right side.

$x^3 + x^2 \le 4x + 4$	This is the given inequality.
$x^3 + x^2 - 4x - 4 \le 4x + 4 - 4x - 4$	Subtract $4x + 4$ from both sides.
$x^3 + x^2 - 4x - 4 \le 0$	Simplify.

This inequality is equivalent to the one we wish to solve. It is in the form $f(x) \le 0$, where $f(x) = x^3 + x^2 - 4x - 4$.

Step 2. **Solve the equation $f(x) = 0$.** We find the x-intercepts of $f(x) = x^3 + x^2 - 4x - 4$ by solving the equation $x^3 + x^2 - 4x - 4 = 0$.

$x^3 + x^2 - 4x - 4 = 0$	*This polynomial equation is of degree 3.*
$x^2(x + 1) - 4(x + 1) = 0$	*Factor x^2 from the first two terms and -4 from the last two terms.*
$(x + 1)(x^2 - 4) = 0$	*A common factor of $x + 1$ is factored from the expression.*
$x + 1 = 0$ or $x^2 - 4 = 0$	*Set each factor equal to 0.*
$x = -1 \qquad x^2 = 4$	*Solve for x.*
$x = \pm 2$	*Use the square root property.*

The x-intercepts of f are $-2, -1,$ and 2. We will use these x-intercepts as boundary points on a number line.

Step 3. **Locate the boundary points on a number line and separate the line into intervals.** The number line with the boundary points is shown as follows:

The boundary points divide the number line into four intervals:

$$(-\infty, -2) \quad (-2, -1) \quad (-1, 2) \quad (2, \infty).$$

Step 4. **Choose one representative number within each interval and evaluate f at that number.**

Interval	Representative Number	Substitute into $f(x) = x^3 + x^2 - 4x - 4$	Conclusion
$(-\infty, -2)$	-3	$f(-3) = (-3)^3 + (-3)^2 - 4(-3) - 4$ $= -10$, negative	$f(x) < 0$ for all x in $(-\infty, -2)$.
$(-2, -1)$	-1.5	$f(-1.5) = (-1.5)^3 + (-1.5)^2 - 4(-1.5) - 4$ $= 0.875$, positive	$f(x) > 0$ for all x in $(-2, -1)$.
$(-1, 2)$	0	$f(0) = 0^3 + 0^2 - 4 \cdot 0 - 4$ $= -4$, negative	$f(x) < 0$ for all x in $(-1, 2)$.
$(2, \infty)$	3	$f(3) = 3^3 + 3^2 - 4 \cdot 3 - 4$ $= 20$, positive	$f(x) > 0$ for all x in $(2, \infty)$.

Step 5. **Write the solution set, selecting the interval(s) that satisfy the given inequality.** Based on our work in step 4, we see that $f(x) < 0$ for all x in $(-\infty, -2)$ or $(-1, 2)$. However, because the inequality involves $\le$ (less than or *equal to*), we must also include the solutions of $x^3 + x^2 - 4x - 4 = 0$, namely $-2, -1,$ and 2, in the solution set. Thus, the solution set of the given inequality, $x^3 + x^2 \le 4x + 4$, or, equivalently, $x^3 + x^2 - 4x - 4 \le 0$, is

$$(-\infty, -2] \cup [-1, 2]$$

or $\{x \mid x \le -2 \text{ or } -1 \le x \le 2\}$.

The graph of the solution set on a number line is shown as follows:

✔ **CHECK POINT 2** Solve and graph the solution set on a real number line: $x^3 + 3x^2 \leq x + 3$.   See graphing answer section.
$(-\infty, -3] \cup [-1, 1]$ or $\{x \mid x \leq -3 \text{ or } -1 \leq x \leq 1\}$

2 Solve rational inequalities.

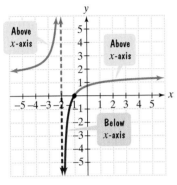

Above x-axis
Above x-axis
Below x-axis

FIGURE 8.16 The graph of $f(x) = \dfrac{3x + 3}{2x + 4}$

Solving Rational Inequalities A **rational inequality** is any inequality that can be put in one of the forms

$$f(x) < 0, \quad f(x) > 0, \quad f(x) \leq 0, \quad \text{or} \quad f(x) \geq 0,$$

where f is a rational function. An example of a rational inequality is

$$\frac{3x + 3}{2x + 4} > 0.$$

This inequality is in the form $f(x) > 0$, where f is the rational function given by

$$f(x) = \frac{3x + 3}{2x + 4}.$$

The graph of f is shown in Figure 8.16.

We can find the x-intercept of f by setting the numerator equal to 0:

$$3x + 3 = 0$$
$$3x = -3 \quad \boxed{f \text{ has an } x\text{-intercept at } -1 \text{ and passes through } (-1, 0).}$$
$$x = -1.$$

We can determine where f is undefined by setting the denominator equal to 0:

$$2x + 4 = 0$$
$$2x = -4 \quad \boxed{f \text{ is undefined at } -2. \text{ Figure 8.16 shows that the function's vertical asymptote is } x = -2.}$$
$$x = -2.$$

By setting the numerator and the denominator of f equal to 0, we obtained the solutions -2 and -1. These numbers separate the x-axis into three intervals: $(-\infty, -2), (-2, -1),$ and $(-1, \infty)$. On each interval, the graph of f is either above the x-axis $[f(x) > 0]$ or below the x-axis $[f(x) < 0]$.

Examine the graph in Figure 8.16 carefully. Can you see that it is above the x-axis for all x in $(-\infty, -2)$ or $(-1, \infty)$, shown in blue? Thus, the solution set of $\dfrac{3x + 3}{2x + 4} > 0$ is $(-\infty, -2) \cup (-1, \infty)$. By contrast, the graph of f lies below the x-axis for all x in $(-2, -1)$, shown in red. Thus, the solution set of $\dfrac{3x + 3}{2x + 4} < 0$ is $(-2, -1)$.

The first step in solving a rational inequality is to bring all terms to one side, obtaining zero on the other side. Then express the rational function on the nonzero side as a single quotient. The second step is to set the numerator and the denominator of f equal to zero. The solutions of these equations serve as boundary points that separate the real number line into intervals. At this point, the procedure is the same as the one we used for solving polynomial inequalities.

EXAMPLE 3 Solving a Rational Inequality

Solve and graph the solution set: $\dfrac{x + 1}{x + 3} \geq 2.$

SOLUTION

Step 1. **Express the inequality so that one side is zero and the other side is a single quotient.** We subtract 2 from both sides to obtain zero on the right.

$$\frac{x + 1}{x + 3} \geq 2 \qquad \text{This is the given inequality.}$$

$$\frac{x + 1}{x + 3} - 2 \geq 0 \qquad \text{Subtract 2 from both sides, obtaining 0 on the right.}$$

$$\frac{x + 1}{x + 3} - \frac{2(x + 3)}{x + 3} \geq 0 \qquad \text{The least common denominator is } x + 3. \text{ Express 2 in terms of this denominator.}$$

$$\frac{x + 1 - 2(x + 3)}{x + 3} \geq 0 \qquad \text{Subtract rational expressions.}$$

$$\frac{x + 1 - 2x - 6}{x + 3} \geq 0 \qquad \text{Apply the distributive property.}$$

$$\frac{-x - 5}{x + 3} \geq 0 \qquad \text{Simplify.}$$

This inequality is equivalent to the one we wish to solve. It is in the form $f(x) \geq 0$, where $f(x) = \dfrac{-x - 5}{x + 3}$.

Step 2. **Set the numerator and the denominator of f equal to zero.** The real solutions are the boundary points.

$$-x - 5 = 0 \qquad x + 3 = 0 \qquad \text{Set the numerator and denominator equal to 0. These are the values that make the previous quotient zero or undefined.}$$

$$x = -5 \qquad x = -3 \qquad \text{Solve for x.}$$

We will use these solutions as boundary points on a number line.

Step 3. **Locate the boundary points on a number line and separate the line into intervals.** The number line with the boundary points is shown as follows:

The boundary points divide the number line into three intervals:

$$(-\infty, -5) \quad (-5, -3) \quad (-3, \infty).$$

Step 4. **Choose one representative number within each interval and evaluate f at that number.**

Interval	Representative Number	Substitute into $f(x) = \dfrac{-x - 5}{x + 3}$	Conclusion
$(-\infty, -5)$	-6	$f(-6) = \dfrac{-(-6) - 5}{-6 + 3}$ $= -\frac{1}{3}$, negative	$f(x) < 0$ for all x in $(-\infty, -5)$.
$(-5, -3)$	-4	$f(-4) = \dfrac{-(-4) - 5}{-4 + 3}$ $= 1$, positive	$f(x) > 0$ for all x in $(-5, -3)$.
$(-3, \infty)$	0	$f(0) = \dfrac{-0 - 5}{0 + 3}$ $= -\frac{5}{3}$, negative	$f(x) < 0$ for all x in $(-3, \infty)$.

Step 5. **Write the solution set, selecting the interval(s) that satisfy the given inequality.** Based on our work in step 4, we see that $f(x) > 0$ for all x in $(-5, -3)$. However, because the inequality involves $\geq$ (greater than or *equal to*), we must also include the solution of $f(x) = 0$, namely the value that we obtained when we set the numerator of f equal to zero. Thus, we must include -5 in the solution set. The solution set of the given inequality is

$$[-5, -3) \text{ or } \{x \mid -5 \leq x < -3\}.$$

The graph of the solution set on a number line is shown as follows:

USING TECHNOLOGY

The solution set for

$$\frac{x + 1}{x + 3} \geq 2$$

or, equivalently,

$$\frac{-x - 5}{x + 3} \geq 0$$

can be verified with a graphing utility. The graph of $f(x) = \dfrac{-x - 5}{x + 3}$ lies on or above the x-axis, representing $\geq$, for all x in $[-5, -3)$.

Graph lies on or above the x-axis.

$x = -5$ $x = -3$

$[-8, 8, 1]$ by $[-3, 3, 1]$

DISCOVER FOR YOURSELF

Because $(x + 3)^2$ is positive, it is possible so solve

$$\frac{x + 1}{x + 3} \geq 2$$

by first multiplying both sides by $(x + 3)^2$ (where $x \neq -3$). This will not reverse the sense of the inequality and will clear the fraction. Try using this solution method and compare it to the one shown on pages 619–621.

 CHECK POINT **3** Solve and graph the solution set: $\dfrac{2x}{x + 1} \geq 1.$

See graphing answer section. $(-\infty, -1) \cup [1, \infty)$ or $\{x \mid x < -1 \text{ or } x \geq 1\}$

3 Solve problems modeled by polynomial or rational inequalities.

Applications Polynomial inequalities can be solved to answer questions about variables contained in polynomial functions.

EXAMPLE 4 Modeling the Position of a Free-Falling Object

A ball is thrown vertically upward from the top of the Leaning Tower of Pisa (190 feet high) with an initial velocity of 96 feet per second (Figure 8.17). The function

$$s(t) = -16t^2 + 96t + 190$$

models the ball's height above the ground, $s(t)$, in feet, t seconds after it was thrown. During which time period will the ball's height exceed that of the tower?

SOLUTION Using the problem's question and the given model for the ball's height, $s(t) = -16t^2 + 96t + 190$, we obtain a polynomial inequality.

$$-16t^2 + 96t + 190 > 190$$

When will the ball's height   exceed that   of the tower?

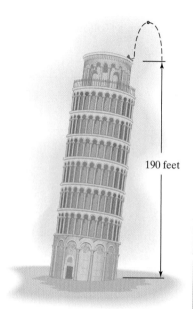

190 feet

FIGURE 8.17 Throwing a ball from the top of the Leaning Tower of Pisa

$-16t^2 + 96t + 190 > 190$ This is the inequality that models the problem's question. We must find t.

$-16t^2 + 96t > 0$ Subtract 190 from both sides. This inequality is in the form $f(t) > 0$, where $f(t) = -16t^2 + 96t$.

$-16t^2 + 96t = 0$ Solve the equation $f(t) = 0$.

$-16t(t - 6) = 0$ Factor.

$-16t = 0$ or $t - 6 = 0$ Set each factor equal to 0.

$t = 0$ $t = 6$ Solve for t. The boundary points are 0 and 6.

number line: -2 -1 0 1 2 3 4 5 6 7 8

Locate these values on a number line, with $t \geq 0$.

The intervals are $(-\infty, 0)$, $(0, 6)$ and $(6, \infty)$. For our purposes, the mathematical model is useful only from $t = 0$ until the ball hits the ground. (By setting $-16t^2 + 96t + 190$ equal to zero, we find $t \approx 7.57$; the ball hits the ground after approximately 7.57 seconds.) Thus, we use $(0, 6)$ and $(6, 7.57)$ for our intervals.

Interval	Representative Number	Substitute into $f(t) = -16t^2 + 96t$	Conclusion
$(0, 6)$	1	$f(1) = -16 \cdot 1^2 + 96 \cdot 1$ $= 80$, positive	$f(t) > 0$ for all t in $(0, 6)$.
$(6, 7.57)$	7	$f(7) = -16 \cdot 7^2 + 96 \cdot 7$ $= -112$, negative	$f(t) < 0$ for all t in $(6, 7.57)$.

We see that $f(t) > 0$ for all t in $(0, 6)$. This means that the ball's height exceeds that of the tower between 0 and 6 seconds.

USING TECHNOLOGY

The graphs of

$$y_1 = -16x^2 + 96x + 190$$

and

$$y_2 = 190$$

are shown in a

$$[0, 8, 1] \text{ by } [0, 360, 36]$$

seconds in motion

height, in feet

viewing rectangle. The graphs show that the ball's height exceeds that of the tower between 0 and 6 seconds.

Height of Ball:
$y_1 = -16x^2 + 96x + 190$

360

Height (feet)

Height of Tower:
$y_2 = 190$

Ball hits ground after 7.57 seconds.

0   6   8

Seconds in Motion
$[0, 8, 1]$ by $[0, 360, 36]$

 CHECK POINT 4 An object is propelled straight up from ground level with an initial velocity of 80 feet per second. Its height at time t is modeled by

$$s(t) = -16t^2 + 80t,$$

where the height, $s(t)$, is measured in feet and the time, t, is measured in seconds. In which time interval will the object be more than 64 feet above the ground? The object will be more than 64 feet above ground between 1 and 4 seconds, excluding $t = 1$ and $t = 4$.

8.5 EXERCISE SET

    Math XL MyMathLab

Student Solutions Manual   CD/Video   PH Math/Tutor Center   MathXL Tutorials on CD   MathXL®   MyMathLab   Interactmath.com

Practice Exercises

Solve each polynomial inequality in Exercises 1–36, and graph the solution set on a real number line.

1–36. See graphing answer section.

1. $(x - 4)(x + 2) > 0$
  $\{x | x < -2 \text{ or } x > 4\}$ or $(-\infty, -2) \cup (4, \infty)$

2. $(x + 3)(x - 5) > 0$
  $\{x | x < -3 \text{ or } x > 5\}$ or $(-\infty, -3) \cup (5, \infty)$

3. $(x - 7)(x + 3) \leq 0$ $\{x | -3 \leq x \leq 7\}$ or $[-3, 7]$

4. $(x + 1)(x - 7) \leq 0$ $\{x | -1 \leq x \leq 7\}$ or $[-1, 7]$

5. $x^2 - 5x + 4 > 0$
  $\{x | x < 1 \text{ or } x > 4\}$ or $(-\infty, -1) \cup (4, \infty)$

6. $x^2 - 4x + 3 < 0$ $\{x | 1 < x < 3\}$ or $(1, 3)$

7. $x^2 + 5x + 4 > 0$
  $\{x | x < -4 \text{ or } x > -1\}$ or $(-\infty, -4) \cup (-1, \infty)$

8. $x^2 + x - 6 > 0$
  $\{x | x < -3 \text{ or } x > 2\}$ or $(-\infty, -3) \cup (2, \infty)$

9. $x^2 - 6x + 8 \leq 0$ $\{x | 2 \leq x \leq 4\}$ or $[2, 4]$

10. $x^2 - 2x - 3 \geq 0$
$\{x|x \leq -1 \text{ or } x \geq 3\}$ or $(-\infty, -1] \cup [3, \infty)$

11. $3x^2 + 10x - 8 \leq 0$ $\{x|-4 \leq x \leq \frac{2}{3}\}$ or $\left[-4, \frac{2}{3}\right]$

12. $9x^2 + 3x - 2 \geq 0$
$\{x|x \leq -\frac{2}{3} \text{ or } x \geq \frac{1}{3}\}$ or $\left(-\infty, -\frac{2}{3}\right] \cup \left[\frac{1}{3}, \infty\right)$

13. $2x^2 + x < 15$ $\{x|-3 < x < \frac{5}{2}\}$ or $\left(-3, \frac{5}{2}\right)$

14. $6x^2 + x > 1$
$\{x|x < -\frac{1}{2} \text{ or } x > \frac{1}{3}\}$ or $\left(-\infty, -\frac{1}{2}\right) \cup \left(\frac{1}{3}, \infty\right)$

15. $4x^2 + 7x < -3$ $\{x|-1 < x < -\frac{3}{4}\}$ or $\left(-1, -\frac{3}{4}\right)$

16. $3x^2 + 16x < -5$ $\{x|-5 < x < -\frac{1}{3}\}$ or $\left(-5, -\frac{1}{3}\right)$

17. $x^2 - 4x \geq 0$ $\{x|x \leq 0 \text{ or } x \geq 4\}$ or $(-\infty, 0] \cup [4, \infty)$

18. $x^2 + 2x < 0$ $\{x|-2 < x < 0\}$ or $(-2, 0)$

19. $2x^2 + 3x > 0$
$\{x|x < -\frac{3}{2} \text{ or } x > 0\}$ or $\left(-\infty, -\frac{3}{2}\right) \cup (0, \infty)$

20. $3x^2 - 5x \leq 0$ $\{x|0 \leq x \leq \frac{5}{3}\}$ or $\left[0, \frac{5}{3}\right]$

21. $-x^2 + x \geq 0$ $\{x|0 \leq x \leq 1\}$ or $[0, 1]$

22. $-x^2 + 2x \geq 0$ $\{x|0 \leq x \leq 2\}$ or $[0, 2]$

23. $x^2 \leq 4x - 2$
$\{x|2 - \sqrt{2} \leq x \leq 2 + \sqrt{2}\}$ or $\left[2 - \sqrt{2}, 2 + \sqrt{2}\right]$

24. $x^2 \leq 2x + 2$
$\{x|1 - \sqrt{3} \leq x \leq 1 + \sqrt{3}\}$ or $\left[1 - \sqrt{3}, 1 + \sqrt{3}\right]$

25. $x^2 - 6x + 9 < 0$ no solution or $\emptyset$

26. $4x^2 - 4x + 1 \geq 0$
$\{x|x \text{ is a real number}\}$ or $\mathbb{R}$ or $(-\infty, \infty)$

27. $(x - 1)(x - 2)(x - 3) \geq 0$
$\{x|1 \leq x \leq 2 \text{ or } x \geq 3\}$ or $[1, 2] \cup [3, \infty)$

28. $(x + 1)(x + 2)(x + 3) \geq 0$
$\{x|-3 \leq x \leq -2 \text{ or } x \geq -1\}$ or $[-3, -2] \cup [-1, \infty)$

29. $x^3 + 2x^2 - x - 2 \geq 0$
$\{x|-2 \leq x \leq -1 \text{ or } x \geq 1\}$ or $[-2, -1] \cup [1, \infty)$

30. $x^3 + 2x^2 - 4x - 8 \geq 0$
$\{x|x = -2 \text{ or } x \geq 2\}$ or $\{-2\} \cup [2, \infty)$

31. $x^3 - 3x^2 - 9x + 27 < 0$ $\{x|x < -3\}$ or $(-\infty, -3)$

32. $x^3 + 7x^2 - x - 7 < 0$
$\{x|x < -7 \text{ or } -1 < x < 1\}$ or $(-\infty, -7) \cup (-1, 1)$

33. $x^3 + x^2 + 4x + 4 > 0$ $\{x|x > -1\}$ or $(-1, \infty)$

34. $x^3 - x^2 + 9x - 9 > 0$ $\{x|x > 1\}$ or $(1, \infty)$

35. $x^3 \geq 9x^2$ $\{x|x = 0 \text{ or } x \geq 9\}$ or $\{0\} \cup [9, \infty)$

36. $x^3 \leq 4x^2$ $\{x|x \leq 4\}$ or $(-\infty, 4]$

Solve each rational inequality in Exercises 37–52 and graph the solution set on a real number line.   **37–52.** See graphing answer section.

37. $\dfrac{x - 4}{x + 3} > 0$ $\{x|x < -3 \text{ or } x > 4\}$ or $(-\infty, -3) \cup (4, \infty)$

38. $\dfrac{x + 5}{x - 2} > 0$ $\{x|x < -5 \text{ or } x > 2\}$ or $(-\infty, -5) \cup (2, \infty)$

39. $\dfrac{x + 3}{x + 4} < 0$ $\{x|-4 < x < -3\}$ or $(-4, -3)$

40. $\dfrac{x + 5}{x + 2} < 0$ $\{x|-5 < x < -2\}$ or $(-5, -2)$

41. $\dfrac{-x + 2}{x - 4} \geq 0$ $\{x|2 \leq x < 4\}$ or $[2, 4)$

42. $\dfrac{-x - 3}{x + 2} \leq 0$ $\{x|x \leq -3 \text{ or } x > -2\}$ or $(-\infty, -3] \cup (-2, \infty)$

43. $\dfrac{4 - 2x}{3x + 4} \leq 0$ $\{x|x < -\frac{4}{3} \text{ or } x \geq 2\}$ or $\left(-\infty, -\frac{4}{3}\right) \cup [2, \infty)$

44. $\dfrac{3x + 5}{6 - 2x} \geq 0$ $\{x|-\frac{5}{3} \leq x < 3\}$ or $\left[-\frac{5}{3}, 3\right)$

45. $\dfrac{x}{x - 3} > 0$ $\{x|x < 0 \text{ or } x > 3\}$ or $(-\infty, 0) \cup (3, \infty)$

46. $\dfrac{x + 4}{x} > 0$ $\{x|x < -4 \text{ or } x > 0\}$ or $(-\infty, -4) \cup (0, \infty)$

47. $\dfrac{x + 1}{x + 3} < 2$ $\{x|x < -5 \text{ or } x > -3\}$ or $(-\infty, -5) \cup (-3, \infty)$

48. $\dfrac{x}{x - 1} > 2$ $\{x|1 < x < 2\}$ or $(1, 2)$

49. $\dfrac{x + 4}{2x - 1} \leq 3$ $\left\{x|x < \frac{1}{2} \text{ or } x \geq \frac{7}{5}\right\}$ or $\left(-\infty, \frac{1}{2}\right) \cup \left[\frac{7}{5}, \infty\right)$

50. $\dfrac{1}{x - 3} < 1$ $\{x|x < 3 \text{ or } x > 4\}$ or $(-\infty, 3) \cup (4, \infty)$

51. $\dfrac{x - 2}{x + 2} \leq 2$ $\{x|x \leq -6 \text{ or } x > -2\}$ or $(-\infty, -6] \cup (-2, \infty)$

52. $\dfrac{x}{x + 2} \geq 2$ $\{x|-4 \leq x < -2\}$ or $[-4, -2)$

In Exercises 53–56, use the given functions to find all values of x that satisfy the required inequality.

53. $f(x) = 2x^2, g(x) = 5x - 2; f(x) \geq g(x)$
$\left\{x|x \leq \frac{1}{2} \text{ or } x \geq 2\right\}$ or $\left(-\infty, \frac{1}{2}\right] \cup [2, \infty)$

54. $f(x) = 4x^2, g(x) = 9x - 2; f(x) < g(x)$
$\left\{x|\frac{1}{4} < x < 2\right\}$ or $\left(\frac{1}{4}, 2\right)$

55. $f(x) = \dfrac{2x}{x + 1}, g(x) = 1; f(x) < g(x)$
$\{x|-1 < x < 1\}$ or $(-1, 1)$

56. $f(x) = \dfrac{x}{2x - 1}, g(x) = 1; f(x) \geq g(x)$
$\left\{x|\frac{1}{2} < x \leq 1\right\}$ or $\left(\frac{1}{2}, 1\right]$

Practice Plus

Solve each inequality in Exercises 57–62 and graph the solution set on a real number line. **57–62.** See graphing answer section.

57. $|x^2 + 2x - 36| > 12$ $\quad \{x | x < -8 \text{ or } -6 < x < 4 \text{ or } x > 6\}$
or $(-\infty, -8) \cup (-6, 4) \cup (6, \infty)$

58. $|x^2 + 6x + 1| > 8$
$\{x | x < -7 \text{ or } x > 1\}$ or $(-\infty, -7) \cup (1, \infty)$

59. $\dfrac{3}{x + 3} > \dfrac{3}{x - 2}$ $\quad \{x | -3 < x < 2\}$ or $(-3, 2)$

60. $\dfrac{1}{x + 1} > \dfrac{2}{x - 1}$
$\{x | x < -3 \text{ or } -1 < x < 1\}$ or $(-\infty, -3) \cup (-1, 1)$

61. $\dfrac{x^2 - x - 2}{x^2 - 4x + 3} > 0$ $\quad \{x | x < -1 \text{ or } 1 < x < 2 \text{ or } x > 3\}$
or $(-\infty, -1) \cup (1, 2) \cup (3, \infty)$

62. $\dfrac{x^2 - 3x + 2}{x^2 - 2x - 3} > 0$ $\quad \{x | x < -1 \text{ or } 1 < x < 2 \text{ or } x > 3\}$
or $(-\infty, -1) \cup (1, 2) \cup (3, \infty)$

In Exercises 63–64, use the graph of the polynomial function to solve each inequality.

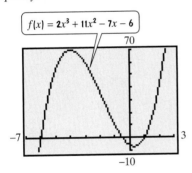

$f(x) = 2x^3 + 11x^2 - 7x - 6$

63. $2x^3 + 11x^2 \geq 7x + 6$
$\{x | -6 \leq x \leq -\frac{1}{2} \text{ or } x \geq 1\}$ or $[-6, -\frac{1}{2}] \cup [1, \infty)$

64. $2x^3 + 11x^2 < 7x + 6$
$\{x | x < -6 \text{ or } -\frac{1}{2} < x < 1\}$ or $(-\infty, -6) \cup (\frac{1}{2}, 1)$

In Exercises 65–66, use the graph of the rational function to solve each inequality.

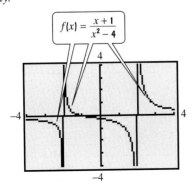

$f(x) = \dfrac{x + 1}{x^2 - 4}$

65. $\dfrac{1}{4(x + 2)} \leq -\dfrac{3}{4(x - 2)}$
$\{x | x < -2 \text{ or } -1 \leq x < 2\}$ or $(-\infty, -2) \cup [-1, 2)$

66. $\dfrac{1}{4(x + 2)} > -\dfrac{3}{4(x - 2)}$
$\{x | -2 < x < -1 \text{ or } x > 2\}$ or $(-2, -1) \cup (2, \infty)$

Application Exercises

67. You throw a ball straight up from a rooftop 160 feet high with an initial speed of 48 feet per second. The function

$$s(t) = -16t^2 + 48t + 160$$

models the ball's height above the ground, $s(t)$, in feet, t seconds after it was thrown. During which time period will the ball's height exceed that of the rooftop? 0 and 3 seconds, excluding $t = 0$ and $t = 3$

68. Divers in Acapulco, Mexico, dive headfirst from the top of a cliff 87 feet above the Pacific Ocean. The function

$$s(t) = -16t^2 + 8t + 87$$

models a diver's height above the ocean, $s(t)$, in feet, t seconds after leaping. During which time period will the diver's height exceed that of the cliff? 0 and $\frac{1}{2}$ second, excluding $t = 0$ and $t = \frac{1}{2}$

The bar graph in Figure 8.14 on page 615 shows the cost of Medicare, in billions of dollars, through 2005. Using the regression feature of a graphing utility, these data can be modeled by

a linear function, $f(x) = 27x + 163$;
a quadratic function, $g(x) = 1.2x^2 + 15.2x + 181.4$.

In each function, x represents the number of years after 1995. Use these functions to solve Exercises 69–72.

69. The graph indicates that Medicare spending reached $379 billion in 2003. Find the amount predicted by each of the functions, f and g, for that year. How well do the functions model the value in the graph? f: $379 billion; g: $379.8 billion; The functions model the value in the graph well.

70. The graph indicates that Medicare spending reached $458 billion in 2005. Find the amount predicted by each of the functions, f and g, for that year. How well do the functions model the value in the graph? Which function serves as a better model for that year? f: $433 billion; g: $453.4 billion; g serves as a better model for 2005.

71. For which years does the quadratic model indicate that Medicare spending will exceed $536.6 billion? years after 2007

72. For which years does the quadratic model indicate that Medicare spending will exceed $629.4 billion? years after 2009

A company manufactures wheelchairs. The average cost function, $\overline{C}$, of producing x wheelchairs per month is given by

$$\overline{C}(x) = \frac{500,000 + 400x}{x}.$$

The graph of the rational function is shown. Use the function to solve Exercises 73–74.

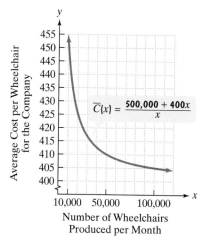

73. Describe the company's production level so that the average cost of producing each wheelchair does not exceed $425. Use a rational inequality to solve the problem. Then explain how your solution is shown on the graph.   The company's production level must be at least 20,000 wheelchairs per month. For values of x greater than or equal to 20,000, the graph lies on or below the line $y = 425$.

74. Describe the company's production level so that the average cost of producing each wheelchair does not exceed $410. Use a rational inequality to solve the problem. Then explain how your solution is shown on the graph.   The company's production level must be at least 50,000 wheelchairs per month. For values of x greater than or equal to 50,000, the graph lies on or below the line $y = 410$.

75. The perimeter of a rectangle is 50 feet. Describe the possible length of a side if the area of the rectangle is not to exceed 114 square feet.   The length of the shorter side cannot exceed 6 feet.

76. The perimeter of a rectangle is 180 feet. Describe the possible lengths of a side if the area of the rectangle is not to exceed 800 square feet.   The length of the shorter side cannot exceed 10 feet.

Writing in Mathematics

77. What is a polynomial inequality?   **77–79.** Answers will vary.

78. What is a rational inequality?

79. Describe similarities and differences between the solutions of

$$(x - 2)(x + 5) \geq 0 \quad \text{and} \quad \frac{x - 2}{x + 5} \geq 0.$$

Technology Exercises

Solve each inequality in Exercises 80–85 using a graphing utility.

80. $x^2 + 3x - 10 > 0$ $\{x | x < -5 \text{ or } x > 2\}$ or $(-\infty, -5) \cup (2, \infty)$

81. $2x^2 + 5x - 3 \leq 0$ $\{x | -3 \leq x \leq \frac{1}{2}\}$ or $\left[-3, \frac{1}{2}\right]$

82. $\frac{x - 4}{x - 1} \leq 0$ $\{x | 1 < x \leq 4\}$ or $(1, 4]$

83. $\frac{x + 2}{x - 3} \leq 2$ $\{x | x < 3 \text{ or } x \geq 8\}$ or $(-\infty, 3) \cup [8, \infty)$

84. $\frac{1}{x + 1} \leq \frac{2}{x + 4}$
$\{x | -4 < x < -1 \text{ or } x \geq 2\}$ or $(-4, -1) \cup [2, \infty)$

85. $x^3 + 2x^2 - 5x - 6 > 0$
$\{x | -3 < x < -1 \text{ or } x > 2\}$ or $(-3, -1) \cup (2, \infty)$

Critical Thinking Exercises

86. Which one of the following is true?   **d**

   a. The solution set of $x^2 > 25$ is $(5, \infty)$.

   b. The inequality $\frac{x - 2}{x + 3} < 2$ can be solved by multiplying both sides by $x + 3$, resulting in the equivalent inequality $x - 2 < 2(x + 3)$.

   c. $(x + 3)(x - 1) \geq 0$ and $\frac{x + 3}{x - 1} \geq 0$ have the same solution set.

   d. None of the above statements is true.

87. Write a quadratic inequality whose solution set is $[-3, 5]$. Answers will vary; An example is $x^2 - 2x - 15 \leq 0$.

88. Write a rational inequality whose solution set is $(-\infty, -4) \cup [3, \infty)$.   Answers will vary; An example is $\frac{x - 3}{x + 4} \geq 0$.

In Exercises 89–92, use inspection to describe each inequality's solution set. Do not solve any of the inequalities.

89. $(x - 2)^2 > 0$ $\{x | x < 2 \text{ or } x > 2\}$ or $(-\infty, 2) \cup (2, \infty)$

90. $(x - 2)^2 \leq 0$ $\{2\}$

91. $(x - 2)^2 < -1$ ∅

92. $\frac{1}{(x - 2)^2} > 0$ $\{x | x < 2 \text{ or } x > 2\}$ or $(-\infty, 2) \cup (2, \infty)$

93. The graphing calculator screen shows the graph of $y = 4x^2 - 8x + 7$.

[-2, 6, 1] by [-2, 8, 1]

a. Use the graph to describe the solution set for $4x^2 - 8x + 7 > 0$.   all real numbers

b. Use the graph to describe the solution set for $4x^2 - 8x + 7 < 0$.   empty set

c. Use an algebraic approach to verify each of your descriptions in parts (a) and (b).

94. The graphing calculator screen shows the graph of $y = \sqrt{27 - 3x^2}$. Write and solve a quadratic inequality that explains why the graph only appears for $-3 \le x \le 3$.

$[-5, 5, 1]$ by $[0, 6, 1]$

$27 - 3x^2 \ge 0; \{x \mid -3 \le x \le 3\}$ or $[-3, 3]$

Review Exercises

95. Solve: $\left| \dfrac{x - 5}{3} \right| < 8$. (Section 4.3, Example 3)

$\{x \mid -19 < x < 29\}$ or $(-19, 29)$

96. Divide:

$$\frac{2x + 6}{x^2 + 8x + 16} \div \frac{x^2 - 9}{x^2 + 3x - 4}.$$

(Section 6.1, Example 7) $\dfrac{2(x - 1)}{(x + 4)(x - 3)}$

97. Factor completely: $x^4 - 16y^4$. (Section 5.5, Example 3)

$(x^2 + 4y^2)(x + 2y)(x - 2y)$

GROUP PROJECT

CHAPTER 8

Additional group projects can be found in the *Instructor's Resource Manual*.

Throughout the chapter, we have considered functions that model the position of free-falling objects. Any object that is falling, or vertically projected into the air, has its height above the ground, $s(t)$, in feet, after t seconds in motion, modeled by the quadratic function

$$s(t) = -16t^2 + v_o t + s_o,$$

where v_o is the original velocity (initial velocity) of the object, in feet per second, and s_o is the original height (initial height) of the object, in feet, above the ground. In this exercise, group members will be working with this position function. The exercise is appropriate for groups of three to five people.

a. Drop a ball from a height of 3 feet, 6 feet, and 12 feet. Record the number of seconds it takes for the ball to hit the ground.

b. For each of the three initial positions, use the position function to determine the time required for the ball to hit the ground.

c. What factors might result in differences between the times that you recorded and the times indicated by the function?

d. What appears to be happening to the time required for a free-falling object to hit the ground as its initial height is doubled? Verify this observation algebraically and with a graphing utility.

e. Repeat part (a) using a sheet of paper rather than a ball. What differences do you observe? What factor seems to be ignored in the position function?

f. What is meant by the acceleration of gravity and how does this number appear in the position function for a free-falling object?

CHAPTER 8 SUMMARY

Definitions and Concepts	Examples

Section 8.1 The Square Root Property and Completing the Square

The Square Root Property

If u is an algebraic expression and d is a real number, then

If $u^2 = d$, then $u = \sqrt{d}$ or $u = -\sqrt{d}$.

Equivalently,

If $u^2 = d$, then $u = \pm\sqrt{d}$.

Solve: $(x - 6)^2 = 50$.

$$x - 6 = \pm\sqrt{50}$$
$$x - 6 = \pm\sqrt{25 \cdot 2}$$
$$x - 6 = \pm 5\sqrt{2}$$
$$x = 6 \pm 5\sqrt{2}$$

The solutions are $6 \pm 5\sqrt{2}$ and the solution set is $\left\{6 \pm 5\sqrt{2}\right\}$.

Completing the Square

If $x^2 + bx$ is a binomial, then by adding $\left(\dfrac{b}{2}\right)^2$, the square of half the coefficient of x, a perfect square trinomial will result. That is,

$$x^2 + bx + \left(\frac{b}{2}\right)^2 = \left(x + \frac{b}{2}\right)^2.$$

Complete the square:

$$x^2 + \frac{2}{7}x.$$

Half of $\frac{2}{7}$ is $\frac{1}{2} \cdot \frac{2}{7} = \frac{1}{7}$ and $\left(\frac{1}{7}\right)^2 = \frac{1}{49}$.

$$x^2 + \frac{2}{7}x + \frac{1}{49} = \left(x + \frac{1}{7}\right)^2$$

Solving Quadratic Equations by Completing the Square

1. If the coefficient of x^2 is not 1, divide both sides by this coefficient.

2. Isolate variable terms on one side.

3. Complete the square by adding the square of half the coefficient of x to both sides.

4. Factor the perfect square trinomial.

5. Solve by applying the square root property.

Solve by completing the square.

$$2x^2 + 16x - 6 = 0.$$

$$\frac{2x^2}{2} + \frac{16x}{2} - \frac{6}{2} = \frac{0}{2} \qquad \text{Divide by 2.}$$

$$x^2 + 8x - 3 = 0 \qquad \text{Simplify.}$$

$$x^2 + 8x = 3 \qquad \text{Add 3.}$$

The coefficient of x is 8. Half of 8 is 4 and $4^2 = 16$. Add 16 to both sides.

$$x^2 + 8x + 16 = 3 + 16$$
$$(x + 4)^2 = 19$$
$$x + 4 = \pm\sqrt{19}$$
$$x = -4 \pm \sqrt{19}$$

Section 8.2 The Quadratic Formula

The solutions of a quadratic equation in standard form

$$ax^2 + bx + c = 0, \quad a \neq 0,$$

are given by the quadratic formula

$$x = \frac{-b \pm \sqrt{b^2 - 4ac}}{2a}.$$

Solve using the quadratic formula:

$$2x^2 = 6x - 3.$$

First write the equation in standard form by subtracting $6x$ and adding 3 on both sides.

$$2x^2 - 6x + 3 = 0$$

$$\boxed{a = 2} \quad \boxed{b = -6} \quad \boxed{c = 3}$$

$$x = \frac{-(-6) \pm \sqrt{(-6)^2 - 4 \cdot 2 \cdot 3}}{2 \cdot 2} = \frac{6 \pm \sqrt{36 - 24}}{4}$$

$$= \frac{6 \pm \sqrt{12}}{4} = \frac{6 \pm \sqrt{4 \cdot 3}}{4} = \frac{6 \pm 2\sqrt{3}}{4}$$

$$= \frac{2(3 \pm \sqrt{3})}{2 \cdot 2} = \frac{3 \pm \sqrt{3}}{2}$$

Definitions and Concepts	Examples

Section 8.2 The Quadratic Formula (continued)

The Discriminant
The discriminant, $b^2 - 4ac$, of the quadratic equation $ax^2 + bx + c = 0$ determines the number and type of solutions.

Discriminant	Solutions
Positive perfect square, with $a, b,$ and c rational numbers	2 rational solutions
Positive and not a perfect square	2 irrational solutions
Zero, with $a, b,$ and c rational numbers	1 rational solution
Negative	2 imaginary solutions

• $2x^2 - 7x - 4 = 0$

$$\boxed{a = 2} \quad \boxed{b = -7} \quad \boxed{c = -4}$$

$$b^2 - 4ac = (-7)^2 - 4(2)(-4)$$
$$= 49 - (-32) = 49 + 32 = 81$$

$\boxed{\text{Positive perfect square}}$

The equation has 2 rational solutions.

Writing Quadratic Equations from Solutions
The zero-product principle in reverse makes it possible to write a quadratic equation from solutions:

$$\text{If } A = 0 \quad \text{or} \quad B = 0, \quad \text{then} \quad AB = 0.$$

Write a quadratic equation with the solution set $\{-2\sqrt{3}, 2\sqrt{3}\}$.

$$x = -2\sqrt{3} \quad \text{or} \qquad x = 2\sqrt{3}$$
$$x + 2\sqrt{3} = 0 \qquad \text{or} \quad x - 2\sqrt{3} = 0$$
$$\left(x + 2\sqrt{3}\right)\left(x - 2\sqrt{3}\right) = 0$$
$$x^2 - \left(2\sqrt{3}\right)^2 = 0$$
$$x^2 - 12 = 0$$

Section 8.3 Quadratic Functions and Their Graphs

The graph of the quadratic function
$$f(x) = a(x - h)^2 + k, \quad a \neq 0,$$
is called a parabola. The vertex, or turning point, is (h, k). The graph opens upward if a is positive and downward if a is negative. The axis of symmetry is a vertical line passing through the vertex. The graph can be obtained using the vertex, x-intercepts, if any, (set $f(x)$ equal to zero), and the y-intercept (set $x = 0$).

Graph: $f(x) = -(x + 3)^2 + 1$.

$$f(x) = -1(x - (-3))^2 + 1$$

$$\boxed{a = -1} \quad \boxed{h = -3} \quad \boxed{k = 1}$$

• Vertex: $(h, k) = (-3, 1)$.
• Opens downward because $a < 0$
• x-intercepts: Set $f(x) = 0$.
$$0 = -(x + 3)^2 + 1$$
$$(x + 3)^2 = 1$$
$$x + 3 = \pm\sqrt{1}$$
$$x + 3 = 1 \quad \text{or} \quad x + 3 = -1$$
$$x = -2 \qquad\qquad x = -4$$

• y-intercept: Set $x = 0$.
$$f(0) = -(0 + 3)^2 + 1 = -9 + 1 = -8$$

Definitions and Concepts	Examples

Section 8.3 Quadratic Functions and Their Graphs (continued)

A parabola whose equation is in the form

$$f(x) = ax^2 + bx + c, \quad a \neq 0,$$

has its vertex at

$$\left(-\frac{b}{2a}, f\left(-\frac{b}{2a}\right)\right).$$

The parabola is graphed as described in the left column on the previous page. The only difference is how we determine the vertex. If $a > 0$, then f has a minimum that occurs at $x = -\dfrac{b}{2a}$. This minimum value is $f\left(-\dfrac{b}{2a}\right)$. If $a < 0$, then f has a maximum that occurs at $x = -\dfrac{b}{2a}$. This maximum value is $f\left(-\dfrac{b}{2a}\right)$.

Graph: $f(x) = x^2 - 6x + 5$.

$$a = 1 \quad b = -6 \quad c = 5$$

- Vertex: $x = -\dfrac{b}{2a} = -\dfrac{-6}{2 \cdot 1} = 3$

$$f(3) = 3^2 - 6 \cdot 3 + 5 = -4$$

Vertex is at $(3, -4)$.

- Opens upward because $a > 0$.
- x-intercepts: Set $f(x) = 0$.

$$x^2 - 6x + 5 = 0$$
$$(x - 1)(x - 5) = 0$$
$$x = 1 \quad \text{or} \quad x = 5$$

- y-intercept: Set $x = 0$.

$$f(0) = 0^2 - 6 \cdot 0 + 5 = 5$$

Section 8.4 Equations Quadratic in Form

An equation that is quadratic in form is one that can be expressed as a quadratic equation using an appropriate substitution. These equations contain an expression to a power, the same expression to that power squared, and a constant term: Let $t =$ the expression to the power. If at any point in the solution process both sides of an equation are raised to an even power, a check is required.

Solve: $\qquad x^{\frac{2}{3}} - 3x^{\frac{1}{3}} + 2 = 0.$

$$\left(x^{\frac{1}{3}}\right)^2 - 3x^{\frac{1}{3}} + 2 = 0$$

Let $t = x^{\frac{1}{3}}$.

$$t^2 - 3t + 2 = 0$$
$$(t - 1)(t - 2) = 0$$
$$t - 1 = 0 \quad \text{or} \quad t - 2 = 0$$
$$t = 1 \qquad\qquad t = 2$$
$$x^{\frac{1}{3}} = 1 \qquad\qquad x^{\frac{1}{3}} = 2$$
$$\left(x^{\frac{1}{3}}\right)^3 = 1^3 \qquad \left(x^{\frac{1}{3}}\right)^3 = 2^3$$
$$x = 1 \qquad\qquad x = 8$$

The solutions are 1 and 8, and the solution set is $\{1, 8\}$.

Definitions and Concepts	Examples

Section 8.5 Polynomial and Rational Inequalities

Solving Polynomial Inequalities

1. Express the inequality in the form
$$f(x) < 0 \quad \text{or} \quad f(x) > 0,$$
where f is a polynomial function.

2. Solve the equation $f(x) = 0$. The real solutions are the boundary points.

3. Locate these boundary points on a number line, thereby dividing the number line into intervals.

4. Choose one representative number within each interval and evaluate f at that number.

   a. If the value of f is positive, then $f(x) > 0$ for all x in the interval.

   b. If the value of f is negative, then $f(x) < 0$ for all x in the interval.

5. Write the solution set, selecting the interval(s) that satisfy the given inequality.

This procedure is valid if $<$ is replaced by $\leq$ and $>$ is replaced by $\geq$. In these cases, include the boundary points in the solution set.

Solve: $2x^2 + x - 6 > 0$.
The form of the inequality is $f(x) > 0$ with $f(x) = 2x^2 + x - 6$. Solve $f(x) = 0$.

$$2x^2 + x - 6 = 0$$
$$(2x - 3)(x + 2) = 0$$
$$2x - 3 = 0 \quad \text{or} \quad x + 2 = 0$$
$$x = \frac{3}{2} \qquad\qquad x = -2$$

$$f(-3) = 2(-3)^2 + (-3) - 6 = 9, \text{positive}$$
$$f(x) > 0 \text{ for all } x \text{ in } (-\infty, -2).$$
$$f(0) = 2 \cdot 0^2 + 0 - 6 = -6, \text{negative}$$
$$f(x) < 0 \text{ for all } x \text{ in } \left(-2, \frac{3}{2}\right).$$
$$f(2) = 2 \cdot 2^2 + 2 - 6 = 4, \text{positive}$$
$$f(x) > 0 \text{ for all } x \text{ in } \left(\frac{3}{2}, \infty\right).$$

The solution set is $\left\{x \mid x < -2 \text{ or } x > \frac{3}{2}\right\}$ or $(-\infty, -2) \cup \left(\frac{3}{2}, \infty\right)$.

Solving Rational Inequalities

1. Express the inequality in the form
$$f(x) < 0 \quad \text{or} \quad f(x) > 0,$$
where f is a rational function.

2. Set the numerator and the denominator of f equal to zero. The real zeros are the boundary points.

3. Locate these boundary points on a number line, thereby dividing the number line into intervals.

4. Choose one representative number within each interval and evaluate f at that number.

   a. If the value of f is positive, then $f(x) > 0$ for all x in the interval.

   b. If the value of f is negative, then $f(x) < 0$ for all x in the interval.

5. Write the solution set, selecting the interval(s) that satisfy the given inequality.

This procedure is valid if $<$ is replaced by $\leq$ and $>$ is replaced by $\geq$. In these cases, include any values that make the numerator of f zero. Always exclude any values that make the denominator zero.

Solve: $\dfrac{x}{x + 4} \geq 2$.

$$\frac{x}{x + 4} - \frac{2(x + 4)}{x + 4} \geq 0$$
$$\frac{-x - 8}{x + 4} \geq 0$$

The form of the inequality is $f(x) \geq 0$ with $f(x) = \dfrac{-x - 8}{x + 4}$.

Set the numerator and the denominator equal to zero.

$$-x - 8 = 0 \qquad x + 4 = 0$$
$$-8 = x \qquad\qquad x = -4$$

$$f(-9) = \frac{-(-9) - 8}{-9 + 4} = \frac{1}{-5}, \text{negative}$$
$$f(x) < 0 \text{ for all } x \text{ in } (-\infty, -8).$$
$$f(-7) = \frac{-(-7) - 8}{-7 + 4} = \frac{-1}{-3} = \frac{1}{3}, \text{positive}$$
$$f(x) > 0 \text{ for all } x \text{ in } (-8, -4)$$
$$f(-3) = \frac{-(-3) - 8}{-3 + 4} = \frac{-5}{1} = -5, \text{negative}$$
$$f(x) < 0 \text{ for all } x \text{ in } (-4, \infty).$$

Because of $\geq$, include -8, the value that makes the numerator zero, in the solution set.
The solution set is $\{x \mid -8 \leq x < -4\}$ or $[-8, -4)$.

CHAPTER 8 REVIEW EXERCISES

8.1 *In Exercises 1–5, solve each equation by the square root property. If possible, simplify radicals or rationalize denominators. Express imaginary solutions in the form a + bi.*

1. $2x^2 - 3 = 125$ ± 8 or $\{\pm 8\}$

2. $3x^2 - 150 = 0$ $\pm 5\sqrt{2}$ or $\{\pm 5\sqrt{2}\}$

3. $3x^2 - 2 = 0$ $\pm\dfrac{\sqrt{6}}{3}$ or $\left\{\pm\dfrac{\sqrt{6}}{3}\right\}$

4. $(x - 4)^2 = 18$ $4 \pm 3\sqrt{2}$ or $\{4 \pm 3\sqrt{2}\}$

5. $(x + 7)^2 = -36$ $-7 \pm 6i$ or $\{-7 \pm 6i\}$

In Exercises 6–7, determine the constant that should be added to the binomial so that it becomes a perfect square trinomial. Then write and factor the trinomial.

6. $x^2 + 20x$ 100; $x^2 + 20x + 100 = (x + 10)^2$

7. $x^2 - 3x$ $\dfrac{9}{4}$; $x^2 - 3x + \dfrac{9}{4} = \left(x - \dfrac{3}{2}\right)^2$

In Exercises 8–10, solve each quadratic equation by completing the square.

8. $x^2 - 12x + 27 = 0$ 3 and 9, or $\{3, 9\}$

9. $x^2 - 7x - 1 = 0$ $\dfrac{7 \pm \sqrt{53}}{2}$, or $\left\{\dfrac{7 \pm \sqrt{53}}{2}\right\}$

10. $2x^2 + 3x - 4 = 0$ $\dfrac{-3 \pm \sqrt{41}}{4}$, or $\left\{\dfrac{-3 \pm \sqrt{41}}{4}\right\}$

11. In 2 years, an investment of \$2500 grows to \$2916. Use the compound interest formula
$$A = P(1 + r)^t$$
to find the annual interest rate, r.   The annual interest rate is 8%.

12. The function $W(t) = 3t^2$ models the weight of a human fetus, $W(t)$, in grams, after t weeks, where $0 \leq t \leq 39$. After how many weeks does the fetus weigh 588 grams?   After 14 weeks, the fetus weighs 588 grams.

13. A building casts a shadow that is double the length of the building's height. If the distance from the end of the shadow to the top of the building is 300 meters, how high is the building? Express the answer in simplified radical form. Then find a decimal approximation to the nearest tenth of a meter.   The building is $60\sqrt{5}$ meters, or about 134.2 meters high.

8.2 *In Exercises 14–16, solve each equation using the quadratic formula. Simplify solutions, if possible.*

14. $x^2 = 2x + 4$ $1 \pm \sqrt{5}$ or $\{1 \pm \sqrt{5}\}$

15. $x^2 - 2x + 19 = 0$ $1 \pm 3\sqrt{2}i$ or $\{1 \pm 3\sqrt{2}i\}$

16. $2x^2 = 3 - 4x$ $\dfrac{-2 \pm \sqrt{10}}{2}$ or $\left\{\dfrac{-2 \pm \sqrt{10}}{2}\right\}$

In Exercises 17–19, without solving the given quadratic equation, determine the number and type of solutions.

17. $x^2 - 4x + 13 = 0$ two imaginary solutions

18. $9x^2 = 2 - 3x$ two rational solutions

19. $2x^2 + 4x = 3$ two irrational solutions

In Exercises 20–26, solve each equation by the method of your choice. Simplify solutions, if possible.

20. $3x^2 - 10x - 8 = 0$ $-\frac{2}{3}$ and 4, or $\left\{-\frac{2}{3}, 4\right\}$

21. $(2x - 3)(x + 2) = x^2 - 2x + 4$ -5 and 2, or $\{-5, 2\}$

22. $5x^2 - x - 1 = 0$ $\dfrac{1 \pm \sqrt{21}}{10}$ or $\left\{\dfrac{1 \pm \sqrt{21}}{10}\right\}$

23. $x^2 - 16 = 0$ -4 and 4, or $\{-4, 4\}$

24. $(x - 3)^2 - 8 = 0$ $3 \pm 2\sqrt{2}$ or $\{3 \pm 2\sqrt{2}\}$

25. $3x^2 - x + 2 = 0$ $\dfrac{1}{6} \pm \dfrac{\sqrt{23}}{6}i$ or $\left\{\dfrac{1}{6} \pm \dfrac{\sqrt{23}}{6}i\right\}$

26. $\dfrac{5}{x + 1} + \dfrac{x - 1}{4} = 2$ $4 \pm \sqrt{5}$ or $\{4 \pm \sqrt{5}\}$

In Exercises 27–29, write a quadratic equation in standard form with the given solution set.

27. $\left\{-\dfrac{1}{3}, \dfrac{3}{5}\right\}$ $15x^2 - 4x - 3 = 0$

28. $\{-9i, 9i\}$ $x^2 + 81 = 0$

29. $\{-4\sqrt{3}, 4\sqrt{3}\}$ $x^2 - 48 = 0$

30. The bar graph shows the number of convictions of police officers throughout the United States for seven years. The number of police officers convicted of felonies, $f(x)$, x years after 1990 can be modeled by the quadratic function
$$f(x) = 23x^2 - 259x + 816.$$

Convictions of Police Officers

Source: F.B.I.

In which year after 1990 were 1020 police officers convicted of felonies?   2002

31. A baseball is hit by a batter. The function
$$s(t) = -16t^2 + 140t + 3$$
models the ball's height above the ground, $s(t)$, in feet, t seconds after it is hit. How long will it take for the ball to strike the ground? Round to the nearest tenth of a second.   It will take about 8.8 seconds for the ball to strike the ground.

8.3 *In Exercises 32–35, use the vertex and intercepts to sketch the graph of each quadratic function. Give the equation for the parabola's axis of symmetry.*

32–35. See graphing answer section.

32. $f(x) = -(x + 1)^2 + 4$ $x = -1$

33. $f(x) = (x + 4)^2 - 2$ $x = -4$

34. $f(x) = -x^2 + 2x + 3$ $x = 1$

35. $f(x) = 2x^2 - 4x - 6$ $x = 1$

36. The function

$$f(x) = -0.02x^2 + x + 1$$

models the yearly growth of a young redwood tree, $f(x)$, in inches, with x inches of rainfall per year. How many inches of rainfall per year result in maximum tree growth? What is the maximum yearly growth?   The maximum yearly growth is 13.5 inches with 25 inches of rainfall per year.

37. A model rocket is launched upward from a platform 40 feet above the ground. The quadratic function

$$s(t) = -16t^2 + 400t + 40$$

models the rocket's height above the ground, $s(t)$, in feet, t seconds after it was launched. After how many seconds does the rocket reach its maximum height? What is the maximum height?   The maximum height is 2540 feet after 12.5 seconds.

38. Suppose that a quadratic function is used to model the data shown in the graph using

(number of years after 1960, divorce rate per 1000 population).

U. S. Divorce Rate

Source: National Center for Health Statistics

Determine, without obtaining an actual quadratic function that models the data, the approximate coordinates of the vertex for the function's graph. Describe what this means in practical terms. $(25, 5)$; The maximum divorce rate of 5 per 1000 population was in 1985.

39. A field bordering a straight stream is to be enclosed. The side bordering the stream is not to be fenced. If 1000 yards of fencing material is to be used, what are the dimensions of the largest rectangular field that can be fenced? What is the maximum area?

$1000 - 2x$

The maximum area is 125,000 square yards when the dimensions are 250 by 500 yards.

40. Among all pairs of numbers whose difference is 14, find a pair whose product is as small as possible. What is the minimum product?   The minimum product is -49 when the numbers are -7 and 7.

8.4 *In Exercises 41–46, solve each equation by making an appropriate substitution. When necessary, check proposed solutions.*

41. $x^4 - 6x^2 + 8 = 0$ $-\sqrt{2}, \sqrt{2}, -2,$ and $2,$ or $\{-\sqrt{2}, \sqrt{2}, -2, 2\}$

42. $x + 7\sqrt{x} - 8 = 0$ 1 or $\{1\}$

43. $(x^2 + 2x)^2 - 14(x^2 + 2x) = 15$ $-5, -1,$ and $3,$ or $\{-5, -1, 3\}$

44. $x^{-2} + x^{-1} - 56 = 0$ $-\frac{1}{8}$ and $\frac{1}{7}$, or $\{-\frac{1}{8}, \frac{1}{7}\}$

45. $x^{\frac{2}{3}} - x^{\frac{1}{3}} - 12 = 0$ -27 and 64, or $\{-27, 64\}$

46. $x^{\frac{1}{2}} + 3x^{\frac{1}{4}} - 10 = 0$ 16 or $\{16\}$

8.5 *In Exercises 47–51, solve each inequality and graph the solution set on a real number line.*   47–51. See graphing answer section.

47. $2x^2 + 5x - 3 < 0$ $\{x|-3 < x < \frac{1}{2}\}$ or $(-3, \frac{1}{2})$

48. $2x^2 + 9x + 4 \geq 0$ $\{x|x \leq -4$ or $x \geq -\frac{1}{2}\}$ or $(-\infty, -4] \cup [-\frac{1}{2}, \infty)$

49. $x^3 + 2x^2 > 3x$ $\{x|-3 < x < 0$ or $x > 1\}$ or $(-3, 0) \cup (1, \infty)$

50. $\dfrac{x - 6}{x + 2} > 0$ $\{x|x < -2$ or $x > 6\}$ or $(-\infty, -2) \cup (6, \infty)$

51. $\dfrac{x + 3}{x - 4} \leq 5$ $\{x|x < 4$ or $x \geq \frac{23}{4}\}$ or or $(-\infty, 4) \cup [\frac{23}{4}, \infty)$

52. A model rocket is launched from ground level. The function

$$s(t) = -16t^2 + 48t$$

models the rocket's height above the ground, $s(t)$, in feet, t seconds after it was launched. During which time period will the rocket's height exceed 32 feet?   The rocket's height will exceed 32 feet between 1 and 2 seconds, excluding $t = 1$ and $t = 2$.

53. The function

$$H(x) = \frac{15}{8}x^2 - 30x + 200$$

models heart rate, $H(x)$, in beats per minute, x minutes after a strenuous workout.

a. What is the heart rate immediately following the workout? The heart rate is 200 beats per minute immediately following the workout.

b. According to the model, during which intervals of time after the strenuous workout does the heart rate exceed 110 beats per minute? For which of these intervals has model breakdown occurred? Which interval provides a more realistic answer? How did you determine this? The heart rate exceeds 110 beats per minute between 0 and 4 minutes after the workout and more than 12 minutes after the workout. Between 0 and 4 minutes provides a more realistic answer; Answers will vary.

CHAPTER 8 TEST

Remember to use your Chapter Test Prep Video CD to see the worked-out solutions to the test questions you want to review.

Express solutions to all equations in simplified form. Rationalize denominators, if possible.

In Exercises 1–2, solve each equation by the square root property.

1. $2x^2 - 5 = 0$ $\pm\dfrac{\sqrt{10}}{2}$ or $\left\{\pm\dfrac{\sqrt{10}}{2}\right\}$

2. $(x - 3)^2 = 20$ $3 \pm 2\sqrt{5}$ or $\left\{3 \pm 2\sqrt{5}\right\}$

In Exercises 3–4, determine the constant that should be added to the binomial so that it becomes a perfect square trinomial. Then write and factor the trinomial.

3. $x^2 - 16x$ $64; x^2 - 16x + 64 = (x - 8)^2$

4. $x^2 + \dfrac{2}{5}x$ $\dfrac{1}{25}; x^2 + \dfrac{2}{5}x + \dfrac{1}{25} = \left(x + \dfrac{1}{5}\right)^2$

5. Solve by completing the square: $x^2 - 6x + 7 = 0$.
$3 \pm \sqrt{2}$ or $\left\{3 \pm \sqrt{2}\right\}$

6. Use the measurements determined by the surveyor to find the width of the pond. Express the answer in simplified radical form.

50 feet   50 feet

The width of the pond is $50\sqrt{2}$ feet.

In Exercises 7–8, without solving the given quadratic equation, determine the number and type of solutions.

7. $3x^2 + 4x - 2 = 0$ two irrational solutions

8. $x^2 = 4x - 8$ two imaginary solutions

In Exercises 9–12, solve each equation by the method of your choice.

9. $2x^2 + 9x = 5$ -5 and $\frac{1}{2}$, or $\left\{-5, \frac{1}{2}\right\}$

10. $x^2 + 8x + 5 = 0$ $-4 \pm \sqrt{11}$ or $\left\{-4 \pm \sqrt{11}\right\}$

11. $(x + 2)^2 + 25 = 0$ $-2 \pm 5i$ or $\{-2 \pm 5i\}$

12. $2x^2 - 6x + 5 = 0$ $\frac{3}{2} \pm \frac{1}{2}i$ or $\left\{\frac{3}{2} \pm \frac{1}{2}i\right\}$

In Exercises 13–14, write a quadratic equation in standard form with the given solution set.

13. $\{-3, 7\}$ $x^2 - 4x - 21 = 0$

14. $\{-10i, 10i\}$ $x^2 + 100 = 0$

15. The function

$$f(x) = -0.5x^2 + 4x + 19$$

models the number of people in the United States, $f(x)$, in millions, receiving food stamps x years after 1990. In which year(s) were 20 million people receiving food stamps? Use a calculator and round to the nearest year(s). 20 million people were receiving food stamps in 1990 and 1998.

In Exercises 16–17, use the vertex and intercepts to sketch the graph of each quadratic function. Give the equation for the parabola's axis of symmetry. **16–17.** See graphing answer section.

16. $f(x) = (x + 1)^2 + 4$ $x = -1$

17. $f(x) = x^2 - 2x - 3$ $x = 1$

A baseball player hits a pop fly into the air. The function

$$s(t) = -16t^2 + 64t + 5$$

models the ball's height above the ground, $s(t)$, in feet, t seconds after it is hit. Use the function to solve Exercises 18–19.

18. When does the baseball reach its maximum height? What is that height? The baseball reaches its maximum height of 69 feet after 2 seconds.

19. After how many seconds does the baseball hit the ground? Round to the nearest tenth of a second. The baseball hits the ground after about 4.1 seconds.

20. The function $f(x) = -x^2 + 46x - 360$ models the daily profit, $f(x)$, in hundreds of dollars, for a company that manufactures x computers daily. How many computers should be manufactured each day to maximize profit? What is the maximum daily profit? The maximum daily profit is $169 hundreds, or $16,900, when 23 computers are manufactured each day.

In Exercises 21–23, solve each equation by making an appropriate substitution. When necessary, check proposed solutions.

21. $(2x - 5)^2 + 4(2x - 5) + 3 = 0$ 1 and 2, or $\{1, 2\}$

22. $x^4 - 13x^2 + 36 = 0$ $-3, 3, -2,$ and 2, or $\{-3, 3, -2, 2\}$

23. $x^{\frac{2}{3}} - 9x^{\frac{1}{3}} + 8 = 0$ 1 and 512, or $\{1, 512\}$

In Exercises 24–25, solve each inequality and graph the solution set on a real number line. **24–25.** See graphing answer section.

24. $x^2 - x - 12 < 0$ $\{x \mid -3 < x < 4\}$ or $(-3, 4)$

25. $\dfrac{2x + 1}{x - 3} \leq 3$ $\{x \mid x < 3 \text{ or } x \geq 10\}$ or $(-\infty, 3) \cup [10, \infty)$

Note to Instructor: The Chapter 8 Test file in the TestGen program provides algorithms specifically matched to these problems so you can easily replicate this chapter test for additional practice or assessment purposes.

CUMULATIVE REVIEW EXERCISES (CHAPTERS 1–8)

In Exercises 1–7, solve each equation, inequality, or system.

1. $8 - (4x - 5) = x - 7$ 4 or $\{4\}$

2. $\begin{aligned} 5x + 4y &= 22 \\ 3x - 8y &= -18 \end{aligned}$ $(2, 3)$ or $\{(2, 3)\}$

3. $\begin{aligned} -3x + 2y + 4z &= 6 \\ 7x - y + 3z &= 23 \\ 2x + 3y + z &= 7 \end{aligned}$ $(2, 0, 3)$ or $\{(2, 0, 3)\}$

4. $|x - 1| > 3$ $\{x \mid x < -2 \text{ or } x > 4\}$ or $(-\infty, -2) \cup (4, \infty)$

5. $\sqrt{x + 4} - \sqrt{x - 4} = 2$ 5 or $\{5\}$

6. $x - 4 \geq 0$ and $-3x \leq -6$ $\{x \mid x \geq 4\}$ or $[4, \infty)$

7. $2x^2 = 3x - 2$ $\dfrac{3}{4} \pm \dfrac{\sqrt{7}}{4} i$ or $\left\{ \dfrac{3}{4} \pm \dfrac{\sqrt{7}}{4} i \right\}$

In Exercises 8–12, graph each function, equation, or inequality in a rectangular coordinate system. **8–12.** See graphing answer section.

8. $3x = 15 + 5y$

9. $2x - 3y > 6$

10. $f(x) = -\dfrac{1}{2}x + 1$

11. $f(x) = x^2 + 6x + 8$

12. $f(x) = (x - 3)^2 - 4$

13. Evaluate:
$$\begin{vmatrix} 3 & 1 & 0 \\ 0 & 5 & -6 \\ -2 & -1 & 0 \end{vmatrix}. \quad -6$$

14. Solve for c: $A = \dfrac{cd}{c + d}$. $c = \dfrac{Ad}{d - A}$

15. Write the equation of the linear function whose graph contains the point $(-2, 4)$ and is perpendicular to the line whose equation is $2x + y = 10$. $y = \frac{1}{2}x + 5$ or $f(x) = \frac{1}{2}x + 5$

In Exercises 16–20, perform the indicated operations and simplify, if possible.

16. $\dfrac{-5x^3 y^7}{15x^4 y^{-2}} - \dfrac{y^9}{3x}$

17. $(4x^2 - 5y)^2$ $16x^4 - 40x^2 y + 25y^2$

18. $(5x^3 - 24x^2 + 9) \div (5x + 1)$ $x^2 - 5x + 1 + \dfrac{8}{5x + 1}$

19. $\dfrac{\sqrt[3]{32xy^{10}}}{\sqrt[3]{2xy^2}}$ $2y^2 \sqrt[3]{2y^2}$

20. $\dfrac{x + 2}{x^2 - 6x + 8} + \dfrac{3x - 8}{x^2 - 5x + 6}$ $\dfrac{4x - 13}{(x - 3)(x - 4)}$

In Exercises 21–22, factor completely.

21. $x^4 - 4x^3 + 8x - 32$ $(x - 4)(x + 2)(x^2 - 2x + 4)$

22. $2x^2 + 12xy + 18y^2$ $2(x + 3y)^2$

23. The length of a rectangular carpet is 4 feet greater than twice its width. If the area is 48 square feet, find the carpet's length and width. The carpet's length is 12 feet and its width is 4 feet.

24. Working alone, you can mow the lawn in 2 hours and your sister can do it in 3 hours. How long will it take you to do the job if you work together? It will take $\frac{6}{5}$ hours, or 1 hour and 12 minutes, to do the job if you work together.

25. Your motorboat can travel 15 miles per hour in still water. Traveling with the river's current, the boat can cover 20 miles in the same time it takes to go 10 miles against the current. Find the rate of the current. The rate of the current is 5 miles per hour.

DISTANCE AND MIDPOINT FORMULAS; CIRCLES

Objectives

1 Find the distance between two points.

2 Find the midpoint of a line segment.

3 Write the standard form of a circle's equation.

4 Give the center and radius of a circle whose equation is in standard form.

5 Convert the general form of a circle's equation to standard form.

The Distance Formula Using the Pythagorean Theorem, we can find the distance between the two points $P_1(x_1, y_1)$ and $P_2(x_2, y_2)$ in the rectangular coordinate system. The two points are illustrated in Figure A.1.

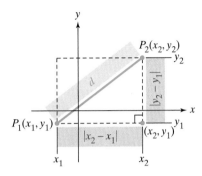

FIGURE A.1

The distance that we want to find is represented by d and shown in blue. Notice that the distance between two points on the dashed horizontal line is the absolute value of the difference between the x-coordinates of the two points. This distance, $|x_2 - x_1|$, is shown in pink. Similarly, the distance between two points on the dashed vertical line is the absolute value of the difference between the y-coordinates of the two points. This distance, $|y_2 - y_1|$, is also shown in pink.

Because the dashed lines are horizontal and vertical, a right triangle is formed. Thus, we can use the Pythagorean Theorem to find the distance d. Squaring the lengths of the triangle's sides results in positive numbers, so absolute value notation is not necessary.

$$d^2 = (x_2 - x_1)^2 + (y_2 - y_1)^2$$
Apply the Pythagorean Theorem to the right triangle in Figure A.1.

$$d = \pm\sqrt{(x_2 - x_1)^2 + (y_2 - y_1)^2}$$
Apply the square root property.

$$d = \sqrt{(x_2 - x_1)^2 + (y_2 - y_1)^2}$$
Because distance is nonnegative, write only the principal square root.

This result is called the **distance formula**.

1 Find the distance between two points.

THE DISTANCE FORMULA The distance, d, between the points (x_1, y_1) and (x_2, y_2) in the rectangular coordinate system is

$$d = \sqrt{(x_2 - x_1)^2 + (y_2 - y_1)^2}.$$

When using the distance formula, it does not matter which point you call (x_1, y_1) and which you call (x_2, y_2).

EXAMPLE 1 Using the Distance Formula

Find the distance between $(-1, -3)$ and $(2, 3)$.

SOLUTION Letting $(x_1, y_1) = (-1, -3)$ and $(x_2, y_2) = (2, 3)$, we obtain

$$d = \sqrt{(x_2 - x_1)^2 + (y_2 - y_1)^2} \qquad \text{Use the distance formula.}$$

$$= \sqrt{[2 - (-1)]^2 + [3 - (-3)]^2} \qquad \text{Substitute the given values.}$$

$$= \sqrt{(2 + 1)^2 + (3 + 3)^2} \qquad \text{Perform subtractions within the grouping symbols.}$$

$$= \sqrt{3^2 + 6^2} \qquad \text{Perform the resulting additions.}$$

$$= \sqrt{9 + 36} \qquad \text{Square 3 and 6.}$$

$$= \sqrt{45} \qquad \text{Add.}$$

$$= 3\sqrt{5} \approx 6.71. \qquad \sqrt{45} = \sqrt{9 \cdot 5} = \sqrt{9}\,\sqrt{5} = 3\sqrt{5}$$

The distance between the given points is $3\sqrt{5}$ units, or approximately 6.71 units. The situation is illustrated in Figure A.2. ∎

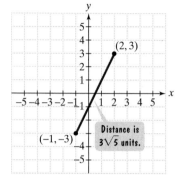

FIGURE A.2 Finding the distance between two points

✔ **CHECK POINT 1** Find the distance between $(2, -2)$ and $(5, 2)$. 5 units

2 Find the midpoint of a line segment.

The Midpoint Formula The distance formula can be used to derive a formula for finding the midpoint of a line segment between two given points. The formula is given as follows:

> **THE MIDPOINT FORMULA** Consider a line segment whose endpoints are (x_1, y_1) and (x_2, y_2). The coordinates of the segment's midpoint are
> $$\left(\frac{x_1 + x_2}{2}, \frac{y_1 + y_2}{2} \right).$$
> To find the midpoint, take the average of the two x-coordinates and the average of the two y-coordinates.

EXAMPLE 2 Using the Midpoint Formula

Find the midpoint of the line segment with endpoints $(1, -6)$ and $(-8, -4)$.

SOLUTION To find the coordinates of the midpoint, we average the coordinates of the endpoints.

$$\text{Midpoint} = \left(\frac{1 + (-8)}{2}, \frac{-6 + (-4)}{2} \right) = \left(\frac{-7}{2}, \frac{-10}{2} \right) = \left(-\frac{7}{2}, -5 \right)$$

Average the x-coordinates.   Average the y-coordinates.

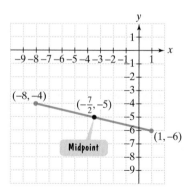

FIGURE A.3 Finding a line segment's midpoint

Figure A.3 illustrates that the point $\left(-\frac{7}{2}, -5 \right)$ is midway between the points $(1, -6)$ and $(-8, -4)$. ∎

✔ CHECK POINT **2** Find the midpoint of the line segment with endpoints $(1, 2)$ and $(7, -3)$. $\left(4, -\frac{1}{2}\right)$

Circles Our goal is to translate a circle's geometric definition into an equation. We begin with this geometric definition.

DEFINITION OF A CIRCLE A **circle** is the set of all points in a plane that are equidistant from a fixed point, called the **center**. The fixed distance from the circle's center to any point on the circle is called the **radius**.

Figure A.4 is our starting point for obtaining a circle's equation. We've placed the circle into a rectangular coordinate system. The circle's center is (h, k) and its radius is r. We let (x, y) represent the coordinates of any point on the circle.

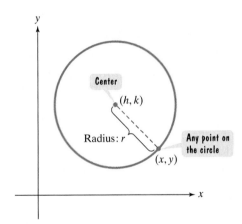

FIGURE A.4 A circle centered at (h, k) with radius r

What does the geometric definition of a circle tell us about the point (x, y) in Figure A.4? The point is on the circle if and only if its distance from the center is r. We can use the distance formula to express this idea algebraically:

The distance between (x, y) and (h, k) is always r.

$$\sqrt{(x - h)^2 + (y - k)^2} = r.$$

Squaring both sides of this equation yields the *standard form of the equation of a circle*.

THE STANDARD FORM OF THE EQUATION OF A CIRCLE The **standard form of the equation of a circle** with center (h, k) and radius r is

$$(x - h)^2 + (y - k)^2 = r^2.$$

3 Write the standard form of a circle's equation.

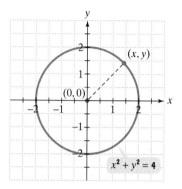

FIGURE A.5 The graph of $x^2 + y^2 = 4$

EXAMPLE 3   Finding the Standard Form of a Circle's Equation

Write the standard form of the equation of the circle with center $(0, 0)$ and radius 2. Graph the circle.

SOLUTION   The center is $(0, 0)$. Because the center is represented as (h, k) in the standard form of the equation, $h = 0$ and $k = 0$. The radius is 2, so we will let $r = 2$ in the equation.

$$(x - h)^2 + (y - k)^2 = r^2 \qquad \text{This is the standard form of a circle's equation.}$$
$$(x - 0)^2 + (y - 0)^2 = 2^2 \qquad \text{Substitute 0 for } h, \text{ 0 for } k, \text{ and 2 for } r.$$
$$x^2 + y^2 = 4 \qquad \text{Simplify.}$$

The standard form of the equation of the circle is $x^2 + y^2 = 4$. Figure A.5 shows the graph. ■

✔ **CHECK POINT 3** Write the standard form of the equation of the circle with center $(0, 0)$ and radius 4. $x^2 + y^2 = 16$

Example 3 and Check Point 3 involved circles centered at the origin. The standard form of the equation of all such circles is $x^2 + y^2 = r^2$, where r is the circle's radius. Now, let's consider a circle whose center is not at the origin.

EXAMPLE 4   Finding the Standard Form of a Circle's Equation

Write the standard form of the equation of the circle with center $(-2, 3)$ and radius 4.

SOLUTION   The center is $(-2, 3)$. Because the center is represented as (h, k) in the standard form of the equation, $h = -2$ and $k = 3$. The radius is 4, so we will let $r = 4$ in the equation.

$$(x - h)^2 + (y - k)^2 = r^2 \qquad \text{This is the standard form of a circle's equation.}$$
$$[x - (-2)]^2 + (y - 3)^2 = 4^2 \qquad \text{Substitute } -2 \text{ for } h, \text{ 3 for } k, \text{ and 4 for } r.$$
$$(x + 2)^2 + (y - 3)^2 = 16 \qquad \text{Simplify.}$$

The standard form of the equation of the circle is $(x + 2)^2 + (y - 3)^2 = 16$. ■

✔ **CHECK POINT 4** Write the standard form of the equation of the circle with center $(5, -6)$ and radius 10. $(x - 5)^2 + (y + 6)^2 = 100$

4 Give the center and radius of a circle whose equation is in standard form.

EXAMPLE 5   Using the Standard Form of a Circle's Equation to Graph the Circle

Find the center and radius of the circle whose equation is

$$(x - 2)^2 + (y + 4)^2 = 9$$

and graph the equation.

SOLUTION To graph the circle, we have to know its center, (h, k), and its radius, r. We can find the values for $h, k,$ and r by comparing the given equation to the standard form of the equation of a circle, $(x - h)^2 + (y - k)^2 = r^2$.

$$(x - 2)^2 + (y + 4)^2 = 9$$

$$(x - 2)^2 + (y - (-4))^2 = 3^2$$

This is $(x - h)^2$, with $h = 2$.   This is $(y - k)^2$, with $k = -4$.   This is r^2, with $r = 3$.

We see that $h = 2, k = -4,$ and $r = 3$. Thus, the circle has center $(h, k) = (2, -4)$ and a radius of 3 units. To graph this circle, first plot the center $(2, -4)$. Because the radius is 3, you can locate at least four points on the circle by going out three units to the right, to the left, up, and down from the center.

The points three units to the right and to the left of $(2, -4)$ are $(5, -4)$ and $(-1, -4)$, respectively. The points three units up and down from $(2, -4)$ are $(2, -1)$ and $(2, -7)$, respectively.

Using these points, we obtain the graph in Figure A.6. ■

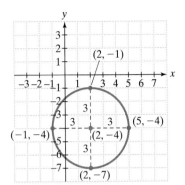

FIGURE A.6 The graph of $(x - 2)^2 + (y + 4)^2 = 9$

✔ **CHECK POINT 5** Find the center and radius of the circle whose equation is

$$(x + 3)^2 + (y - 1)^2 = 4$$

and graph the equation.   The center is $(-3, 1)$ and the radius is 2 units. See graphing answer section.

If we square $x - 2$ and $y + 4$ in the standard form of the equation of Example 5, we obtain another form for the circle's equation.

$$(x - 2)^2 + (y + 4)^2 = 9 \qquad \text{This is the standard form of the equation from Example 5.}$$

$$x^2 - 4x + 4 + y^2 + 8y + 16 = 9 \qquad \text{Square } x - 2 \text{ and } y + 4.$$

$$x^2 + y^2 - 4x + 8y + 20 = 9 \qquad \text{Combine numerical terms and rearrange terms.}$$

$$x^2 + y^2 - 4x + 8y + 11 = 0 \qquad \text{Subtract 9 from both sides.}$$

This result suggests that an equation in the form $x^2 + y^2 + Dx + Ey + F = 0$ can represent a circle. This is called the *general form of the equation of a circle*.

THE GENERAL FORM OF THE EQUATION OF A CIRCLE The **general form of the equation of a circle** is

$$x^2 + y^2 + Dx + Ey + F = 0.$$

5 Convert the general form of a circle's equation to standard form.

We can convert the general form of the equation of a circle to the standard form $(x - h)^2 + (y - k)^2 = r^2$. We do so by completing the square on x and y. Let's see how this is done.

EXAMPLE 6 Converting the General Form of a Circle's Equation to Standard Form and Graphing the Circle

Write in standard form and graph: $x^2 + y^2 + 4x - 6y - 23 = 0$.

SOLUTION Because we plan to complete the square on both x and y, let's rearrange the terms so that x-terms are arranged in descending order, y-terms are arranged in descending order, and the constant term appears on the right.

$$x^2 + y^2 + 4x - 6y - 23 = 0$$

This is the given equation.

$$(x^2 + 4x \quad) + (y^2 - 6y \quad) = 23$$

Rewrite in anticipation of completing the square.

$$(x^2 + 4x + 4) + (y^2 - 6y + 9) = 23 + 4 + 9$$

Complete the square on x: $\frac{1}{2} \cdot 4 = 2$ and $2^2 = 4$, so add 4 to both sides. Complete the square on y: $\frac{1}{2}(-6) = -3$ and $(-3)^2 = 9$, so add 9 to both sides.

> Remember that numbers added on the left side must also be added on the right side.

$$(x + 2)^2 + (y - 3)^2 = 36$$

Factor on the left and add on the right.

This last equation is in standard form. We can identify the circle's center and radius by comparing this equation to the standard form of the equation of a circle, $(x - h)^2 + (y - k)^2 = r^2$.

$$(x + 2)^2 + (y - 3)^2 = 36$$

$$(x - (-2))^2 + (y - 3)^2 = 6^2$$

| This is $(x - h)^2$, with $h = -2$. | This is $(y - k)^2$, with $k = 3$. | This is r^2, with $r = 6$. |

We use the center, $(h, k) = (-2, 3)$, and the radius, $r = 6$, to graph the circle. The graph is shown in Figure A.7. ∎

FIGURE A.7 The graph of $(x + 2)^2 + (y - 3)^2 = 36$

✔ **CHECK POINT 6** Write in standard form and graph:
$$x^2 + y^2 + 4x - 4y - 1 = 0.$$
$(x + 2)^2 + (y - 2)^2 = 9$; See graphing answer section.

A EXERCISE SET

Student Solutions Manual   CD/Video   PH Math/Tutor Center   MathXL Tutorials on CD   MathXL®   MyMathLab   Interactmath.com

Practice Exercises

In Exercises 1–18, find the distance between each pair of points. If necessary, round answers to two decimal places.

1. $(2, 3)$ and $(14, 8)$ 13 units

2. $(5, 1)$ and $(8, 5)$ 5 units

3. $(4, 1)$ and $(6, 3)$ $2\sqrt{2}$ or 2.83 units

4. $(2, 3)$ and $(3, 5)$ $\sqrt{5}$ or 2.24 units

5. $(0, 0)$ and $(-3, 4)$ 5 units

6. $(0, 0)$ and $(3, -4)$ 5 units

7. $(-2, -6)$ and $(3, -4)$ $\sqrt{29}$ or 5.39 units

8. $(-4, -1)$ and $(2, -3)$ $2\sqrt{10}$ or 6.32 units

9. $(0, -3)$ and $(4, 1)$ $4\sqrt{2}$ or 5.66 units

10. $(0, -2)$ and $(4, 3)$ $\sqrt{41}$ or 6.40 units

11. $(3.5, 8.2)$ and $(-0.5, 6.2)$ $2\sqrt{5}$ or 4.47 units

12. $(2.6, 1.3)$ and $(1.6, -5.7)$ $5\sqrt{2}$ or 7.07 units

13. $\left(0, -\sqrt{3}\right)$ and $\left(\sqrt{5}, 0\right)$ $2\sqrt{2}$ or 2.83 units

14. $\left(0, -\sqrt{2}\right)$ and $\left(\sqrt{7}, 0\right)$ 3 units

15. $\left(3\sqrt{3}, \sqrt{5}\right)$ and $\left(-\sqrt{3}, 4\sqrt{5}\right)$ $\sqrt{93}$ or 9.64 units

16. $\left(2\sqrt{3}, \sqrt{6}\right)$ and $\left(-\sqrt{3}, 5\sqrt{6}\right)$ $\sqrt{123}$ or 11.09 units

17. $\left(\dfrac{7}{3}, \dfrac{1}{5}\right)$ and $\left(\dfrac{1}{3}, \dfrac{6}{5}\right)$ $\sqrt{5}$ or 2.24 units

18. $\left(-\dfrac{1}{4}, -\dfrac{1}{7}\right)$ and $\left(\dfrac{3}{4}, \dfrac{6}{7}\right)$ $\sqrt{2}$ or 1.41 units

In Exercises 19–30, find the midpoint of the line segment with the given endpoints.

19. $(6, 8)$ and $(2, 4)$ $(4, 6)$

20. $(10, 4)$ and $(2, 6)$ $(6, 5)$

21. $(-2, -8)$ and $(-6, -2)$ $(-4, -5)$

22. $(-4, -7)$ and $(-1, -3)$ $\left(-\frac{5}{2}, -5\right)$

23. $(-3, -4)$ and $(6, -8)$ $\left(\frac{3}{2}, -6\right)$

24. $(-2, -1)$ and $(-8, 6)$ $\left(-5, \frac{5}{2}\right)$

25. $\left(-\dfrac{7}{2}, \dfrac{3}{2}\right)$ and $\left(-\dfrac{5}{2}, -\dfrac{11}{2}\right)$ $(-3, -2)$

26. $\left(-\dfrac{2}{5}, \dfrac{7}{15}\right)$ and $\left(-\dfrac{2}{5}, -\dfrac{4}{15}\right)$ $\left(-\dfrac{2}{5}, \dfrac{1}{10}\right)$

27. $\left(8, 3\sqrt{5}\right)$ and $\left(-6, 7\sqrt{5}\right)$ $\left(1, 5\sqrt{5}\right)$

28. $\left(7\sqrt{3}, -6\right)$ and $\left(3\sqrt{3}, -2\right)$ $\left(5\sqrt{3}, -4\right)$

29. $\left(\sqrt{18}, -4\right)$ and $\left(\sqrt{2}, 4\right)$ $\left(2\sqrt{2}, 0\right)$

30. $\left(\sqrt{50}, -6\right)$ and $\left(\sqrt{2}, 6\right)$ $\left(3\sqrt{2}, 0\right)$

In Exercises 31–40, write the standard form of the equation of the circle with the given center and radius.

31. Center $(0, 0)$, $r = 7$ $x^2 + y^2 = 49$

32. Center $(0, 0)$, $r = 8$ $x^2 + y^2 = 64$

33. Center $(3, 2)$, $r = 5$ $(x - 3)^2 + (y - 2)^2 = 25$

34. Center $(2, -1)$, $r = 4$ $(x - 2)^2 + (y + 1)^2 = 16$

35. Center $(-1, 4)$, $r = 2$ $(x + 1)^2 + (y - 4)^2 = 4$

36. Center $(-3, 5)$, $r = 3$ $(x + 3)^2 + (y - 5)^2 = 9$

37. Center $(-3, -1)$, $r = \sqrt{3}$ $(x + 3)^2 + (y + 1)^2 = 3$

38. Center $(-5, -3)$, $r = \sqrt{5}$ $(x + 5)^2 + (y + 3)^2 = 5$

39. Center $(-4, 0)$, $r = 10$ $(x + 4)^2 + y^2 = 100$

40. Center $(-2, 0)$, $r = 6$ $(x + 2)^2 + y^2 = 36$

In Exercises 41–48, give the center and radius of the circle described by the equation and graph each equation.   **41–48.** See graphing answer section.

41. $x^2 + y^2 = 16$ center: $(0, 0)$; $r = 4$

42. $x^2 + y^2 = 49$ center: $(0, 0)$; $r = 7$

43. $(x - 3)^2 + (y - 1)^2 = 36$ center: $(3, 1)$; $r = 6$

44. $(x - 2)^2 + (y - 3)^2 = 16$ center: $(2, 3)$; $r = 4$

45. $(x + 3)^2 + (y - 2)^2 = 4$ center: $(-3, 2)$; $r = 2$

46. $(x + 1)^2 + (y - 4)^2 = 25$ center: $(-1, 4)$; $r = 5$

47. $(x + 2)^2 + (y + 2)^2 = 4$ center: $(-2, -2)$; $r = 2$

48. $(x + 4)^2 + (y + 5)^2 = 36$ center: $(-4, -5)$; $r = 6$

In Exercises 49–56, complete the square and write the equation in standard form. Then give the center and radius of each circle and graph the equation.   **49–56.** See graphing answer section.

49. $x^2 + y^2 + 6x + 2y + 6 = 0$
$(x + 3)^2 + (y + 1)^2 = 4$; center: $(-3, -1)$; $r = 2$

50. $x^2 + y^2 + 8x + 4y + 16 = 0$
$(x + 4)^2 + (y + 2)^2 = 4$; center: $(-4, -2)$; $r = 2$

51. $x^2 + y^2 - 10x - 6y - 30 = 0$
$(x - 5)^2 + (y - 3)^2 = 64$; center: $(5, 3)$; $r = 8$

52. $x^2 + y^2 - 4x - 12y - 9 = 0$
$(x - 2)^2 + (y - 6)^2 = 49$; center: $(2, 6)$; $r = 7$

53. $x^2 + y^2 + 8x - 2y - 8 = 0$
$(x + 4)^2 + (y - 1)^2 = 25$; center: $(-4, 1)$; $r = 5$

54. $x^2 + y^2 + 12x - 6y - 4 = 0$
$(x + 6)^2 + (y - 3)^2 = 49$; center: $(-6, 3)$; $r = 7$

55. $x^2 - 2x + y^2 - 15 = 0$
$(x - 1)^2 + y^2 = 16$; center: $(1, 0)$; $r = 4$

56. $x^2 + y^2 - 6y - 7 = 0$
$x^2 + (y - 3)^2 = 16$; center: $(0, 3)$; $r = 4$

1 Use summation notation.

SUMMATION NOTATION AND THE BINOMIAL THEOREM

Summation Notation   There is a compact notation for expressing a sum of terms.

> SUMMATION NOTATION   The sum $a_1 + a_2 + a_3 + a_4 + \cdots + a_n$ is represented by the **summation notation**
>
> $$\sum_{i=1}^{n} a_i = a_1 + a_2 + a_3 + a_4 + \cdots + a_n,$$
>
> where i is the **index of summation**, n is the **upper limit of summation**, and 1 is the **lower limit of summation.**

Any letter can be used for the index of summation. The letters $i, j,$ and k are used commonly. Furthermore, the lower limit of summation can be an integer other than 1.

When we write out a sum that is given in summation notation, we are **expanding the summation notation**. Example 1 shows how to do this.

EXAMPLE 1   Using Summation Notation

Expand and evaluate the sum:

a. $\displaystyle\sum_{i=1}^{6} (i^2 + 1)$ **b.** $\displaystyle\sum_{k=4}^{7} [(-2)^k - 5]$ **c.** $\displaystyle\sum_{i=1}^{5} 3.$

SOLUTION

a. We must replace i in the expression $i^2 + 1$ with all consecutive integers from 1 to 6, inclusive. Then we add.

$$\sum_{i=1}^{6} (i^2 + 1) = (1^2 + 1) + (2^2 + 1) + (3^2 + 1) + (4^2 + 1)$$
$$+ (5^2 + 1) + (6^2 + 1)$$
$$= 2 + 5 + 10 + 17 + 26 + 37$$
$$= 97$$

b. This time the index of summation is k. First we evaluate $(-2)^k - 5$ for all consecutive integers from 4 through 7, inclusive. Then we add.

$$\sum_{k=4}^{7} [(-2)^k - 5] = [(-2)^4 - 5] + [(-2)^5 - 5]$$
$$+ [(-2)^6 - 5] + [(-2)^7 - 5]$$
$$= (16 - 5) + (-32 - 5) + (64 - 5) + (-128 - 5)$$
$$= 11 + (-37) + 59 + (-133)$$
$$= -100$$

c. To find $\displaystyle\sum_{i=1}^{5} 3$, we observe that every term of the sum is 3. The notation $i = 1$ through 5 indicates that we must add the first five terms in which every term is 3.

$$\sum_{i=1}^{5} 3 = 3 + 3 + 3 + 3 + 3 = 15$$

✔ CHECK POINT **1** Expand and evaluate the sum:

a. $\displaystyle\sum_{i=1}^{6} 2i^2$ $2(1)^2 + 2(2)^2 + 2(3)^2 + 2(4)^2 + 2(5)^2 + 2(6)^2 = 182$

b. $\displaystyle\sum_{k=3}^{5} (2^k - 3)$ $(2^3 - 3) + (2^4 - 3) + (2^5 - 3) = 47$

c. $\displaystyle\sum_{i=1}^{5} 4.$ $4 + 4 + 4 + 4 + 4 = 20$

2   Use factorial notation.

Factorial Notation Products of consecutive positive integers occur in the Binomial Theorem. These products can be expressed in a special notation, called **factorial notation**.

> **FACTORIAL NOTATION**   If n is a positive integer, the notation $n!$ (read "n factorial") is the product of all positive integers from n down through 1.
>
> $$n! = n(n - 1)(n - 2) \cdots (3)(2)(1)$$
>
> 0! (zero factorial), by definition, is 1.
>
> $$0! = 1$$

The values of $n!$ for the first six positive integers are

$$1! = 1$$
$$2! = 2 \cdot 1 = 2$$
$$3! = 3 \cdot 2 \cdot 1 = 6$$
$$4! = 4 \cdot 3 \cdot 2 \cdot 1 = 24$$
$$5! = 5 \cdot 4 \cdot 3 \cdot 2 \cdot 1 = 120$$
$$6! = 6 \cdot 5 \cdot 4 \cdot 3 \cdot 2 \cdot 1 = 720.$$

EXAMPLE 2   Using Summation and Factorial Notations

Expand and evaluate the sum:

$$\sum_{j=3}^{6} 4(j - 3)!.$$

SOLUTION   We replace j in the expression $4(j - 3)!$ with all consecutive integers from 3 to 6, inclusive. Then we use factorial notation to evaluate terms. Finally, we add the terms.

$$\sum_{j=3}^{6} 4(j - 3)!$$

$$= 4(3 - 3)! + 4(4 - 3)! + 4(5 - 3)! + 4(6 - 3)!$$
$$= 4 \cdot 0! \quad + 4 \cdot 1! \quad + 4 \cdot 2! \quad + 4 \cdot 3!$$
$$= 4 \cdot 1 \quad + 4 \cdot 1 \quad + 4 \cdot 2 \cdot 1 \quad + 4 \cdot 3 \cdot 2 \cdot 1$$
$$= 4 \quad + 4 \quad + 8 \quad + 24$$
$$= 40$$

✔ CHECK POINT **2** Expand and evaluate the sum:

$$\sum_{k=2}^{5} k^2(k - 1)!.\quad \text{718}$$

3 Evaluate a binomial coefficient.

Binomial Coefficients Before turning to the Binomial Theorem, we introduce a special notation that uses factorials.

> **DEFINITION OF A BINOMIAL COEFFICIENT** $\binom{n}{r}$ For nonnegative integers n and r, with $n \geq r$, the expression $\binom{n}{r}$ (read "n above r") is called a **binomial coefficient** and is defined by
>
> $$\binom{n}{r} = \frac{n!}{r!(n - r)!}.$$

The symbol $_nC_r$ is often used in place of $\binom{n}{r}$ to denote binomial coefficients.

Can you see that the definition of a binomial coefficient involves a fraction with factorials in the numerator and the denominator? When evaluating such an expression, try to reduce the fraction before performing the multiplications. For example, consider $\frac{26!}{21!}$. Rather than writing out 26! as the product of all integers from 26 down to 1, we can express 26! as

$$26! = 26 \cdot 25 \cdot 24 \cdot 23 \cdot 22 \cdot 21!.$$

In this way, we can divide both the numerator and the denominator by the common factor, 21!.

$$\frac{26!}{21!} = \frac{26 \cdot 25 \cdot 24 \cdot 23 \cdot 22 \cdot 21!}{21!} = 26 \cdot 25 \cdot 24 \cdot 23 \cdot 22 = 7{,}893{,}600$$

EXAMPLE 3 Evaluating Binomial Coefficients

Evaluate: **a.** $\binom{6}{2}$ **b.** $\binom{3}{0}$ **c.** $\binom{9}{3}$ **d.** $\binom{4}{4}$.

SOLUTION In each case, we apply the definition of the binomial coefficient.

a. $\binom{6}{2} = \frac{6!}{2!(6 - 2)!} = \frac{6!}{2!4!} = \frac{6 \cdot 5 \cdot 4!}{2 \cdot 1 \cdot 4!} = 15$

b. $\binom{3}{0} = \frac{3!}{0!(3 - 0)!} = \frac{3!}{0!3!} = \frac{1}{1} = 1$

Remember that 0! = 1.

c. $\binom{9}{3} = \frac{9!}{3!(9 - 3)!} = \frac{9!}{3!6!} = \frac{9 \cdot 8 \cdot 7 \cdot 6!}{3 \cdot 2 \cdot 1 \cdot 6!} = 84$

d. $\binom{4}{4} = \frac{4!}{4!(4 - 4)!} = \frac{4!}{4!0!} = \frac{1}{1} = 1$

> **CHECK POINT 3** Evaluate:
>
> **a.** $\binom{6}{3}$ 20   **b.** $\binom{6}{0}$ 1   **c.** $\binom{8}{2}$ 28   **d.** $\binom{3}{3}$.   1

4   Expand a binomial raised to a power.

The Binomial Theorem When we write out the *binomial expression* $(a + b)^n$, where n is a positive integer, several patterns begin to appear.

$$(a + b)^1 = a + b$$
$$(a + b)^2 = a^2 + 2ab + b^2$$
$$(a + b)^3 = a^3 + 3a^2b + 3ab^2 + b^3$$
$$(a + b)^4 = a^4 + 4a^3b + 6a^2b^2 + 4ab^3 + b^4$$
$$(a + b)^5 = a^5 + 5a^4b + 10a^3b^2 + 10a^2b^3 + 5ab^4 + b^5$$

Observe the following patterns:

1. The first term is a^n. The exponents on a decrease by 1 in each successive term.
2. The exponents on b increase by 1 in each successive term. In the first term, the exponent on b is 0. (Because $b^0 = 1$, b is not shown in the first term.) The last term is b^n.
3. The sum of the exponents on the variables in any term is equal to n, the exponent on $(a + b)^n$.
4. There is one more term in the polynomial expansion than there is in the power of the binomial, n. There are $n + 1$ terms in the expanded form of $(a + b)^n$.

Using these observations, the variable parts of the expansion of $(a + b)^6$ are

$$a^6, \quad a^5b, \quad a^4b^2, \quad a^3b^3, \quad a^2b^4, \quad ab^5, \quad b^6.$$

The first term is a^6, with the exponents on a decreasing by 1 in each successive term. The exponents on b increase from 0 to 6, with the last term being b^6. The sum of the exponents in each term is equal to 6.

We can generalize from these observations to obtain the variable parts of the expansion of $(a + b)^n$. They are

$$a^n, \quad a^{n-1}b, \quad a^{n-2}b^2, \quad a^{n-3}b^3, \ldots, \quad ab^{n-1}, \quad b^n.$$

Exponents on a are decreasing by 1. Exponents on b are increasing by 1.

Sum of exponents: $n - 1 + 1 = n$ Sum of exponents: $n - 3 + 3 = n$ Sum of exponents: $1 + n - 1 = n$

If we use binomial coefficients and the pattern for the variable part of each term, a formula called the **Binomial Theorem** can be used to expand any positive integral power of a binomial.

A FORMULA FOR EXPANDING BINOMIALS: THE BINOMIAL THEOREM
For any positive integer n,

$$(a + b)^n = \binom{n}{0}a^n + \binom{n}{1}a^{n-1}b + \binom{n}{2}a^{n-2}b^2 + \binom{n}{3}a^{n-3}b^3 + \cdots + \binom{n}{n}b^n$$

$$= \sum_{r=0}^{n} \binom{n}{r}a^{n-r}b^r$$

EXAMPLE 4   Using the Binomial Theorem

Expand: $(x + 2)^4$.

SOLUTION   We use the Binomial Theorem

$$(a + b)^n = \binom{n}{0}a^n + \binom{n}{1}a^{n-1}b + \binom{n}{2}a^{n-2}b^2 + \binom{n}{3}a^{n-3}b^3 + \cdots + \binom{n}{n}b^n$$

to expand $(x + 2)^4$. In $(x + 2)^4$, $a = x$, $b = 2$, and $n = 4$. In the expansion, powers of x are in descending order, starting with x^4. Powers of 2 are in ascending order, starting with 2^0. (Because $2^0 = 1$, a 2 is not shown in the first term.) The sum of the exponents on x and 2 in each term is equal to 4, the exponent in the expression $(x + 2)^4$.

$$(x + 2)^4 = \binom{4}{0}x^4 + \binom{4}{1}x^3 \cdot 2 + \binom{4}{2}x^2 \cdot 2^2 + \binom{4}{3}x \cdot 2^3 + \binom{4}{4}2^4$$

> These binomial coefficients are evaluated using $\binom{n}{r} = \frac{n!}{r!(n-r)!}$.

$$= \frac{4!}{0!4!}x^4 + \frac{4!}{1!3!}x^3 \cdot 2 + \frac{4!}{2!2!}x^2 \cdot 4 + \frac{4!}{3!1!}x \cdot 8 + \frac{4!}{4!0!} \cdot 16$$

> $\frac{4!}{2!2!} = \frac{4 \cdot 3 \cdot 2!}{2! \cdot 2 \cdot 1} = \frac{12}{2} = 6$
> Take a few minutes to verify the other factorial evaluations.

$$= 1 \cdot x^4 + 4x^3 \cdot 2 + 6x^2 \cdot 4 + 4x \cdot 8 + 1 \cdot 16$$

$$= x^4 + 8x^3 + 24x^2 + 32x + 16 \qquad \blacksquare$$

✔ **CHECK POINT 4** Expand: $(x + 1)^4$. $x^4 + 4x^3 + 6x^2 + 4x + 1$

EXAMPLE 5   Using the Binomial Theorem

Expand: $(2x - y)^5$.

SOLUTION   Because the Binomial Theorem involves the addition of two terms raised to a power, we rewrite $(2x - y)^5$ as $[2x + (-y)]^5$. We use the Binomial Theorem

$$(a + b)^n = \binom{n}{0}a^n + \binom{n}{1}a^{n-1}b + \binom{n}{2}a^{n-2}b^2 + \binom{n}{3}a^{n-3}b^3 + \cdots + \binom{n}{n}b^n$$

to expand $[2x + (-y)]^5$. In $[2x + (-y)]^5$, $a = 2x$, $b = -y$, and $n = 5$. In the expansion, powers of $2x$ are in descending order, starting with $(2x)^5$. Powers of $-y$ are in ascending order, starting with $(-y)^0$. [Because $(-y)^0 = 1$, a $-y$ is not shown in the first term.] The sum of the exponents on $2x$ and $-y$ in each term is equal to 5, the exponent in the expression $(2x - y)^5$.

$$(2x - y)^5 = [2x + (-y)]^5$$

$$= \binom{5}{0}(2x)^5 + \binom{5}{1}(2x)^4(-y) + \binom{5}{2}(2x)^3(-y)^2 + \binom{5}{3}(2x)^2(-y)^3 + \binom{5}{4}(2x)(-y)^4 + \binom{5}{5}(-y)^5$$

> Evaluate binomial coefficients using $\binom{n}{r} = \frac{n!}{r!(n-r)!}$.

$$= \frac{5!}{0!5!}(2x)^5 + \frac{5!}{1!4!}(2x)^4(-y) + \frac{5!}{2!3!}(2x)^3(-y)^2 + \frac{5!}{3!2!}(2x)^2(-y)^3 + \frac{5!}{4!1!}(2x)(-y)^4 + \frac{5!}{5!0!}(-y)^5$$

> $\frac{5!}{2!3!} = \frac{5 \cdot 4 \cdot 3!}{2 \cdot 1 \cdot 3!} = 10$
> Take a few minutes to verify the other factorial evaluations.

$$= 1(2x)^5 + 5(2x)^4(-y) + 10(2x)^3(-y)^2 + 10(2x)^2(-y)^3 + 5(2x)(-y)^4 + 1(-y)^5$$

> Raise both factors in these parentheses to the indicated powers.

$$= 1(32x^5) + 5(16x^4)(-y) + 10(8x^3)(-y)^2 + 10(4x^2)(-y)^3 + 5(2x)(-y)^4 + 1(-y)^5$$

> Now raise $-y$ to the indicated powers.

$$= 1(32x^5) + 5(16x^4)(-y) + 10(8x^3)y^2 + 10(4x^2)(-y^3) + 5(2x)y^4 + 1(-y^5)$$

Multiplying factors in each of the six terms gives us the desired expansion:
$$(2x - y)^5 = 32x^5 - 80x^4y + 80x^3y^2 - 40x^2y^3 + 10xy^4 - y^5.$$

✔ **CHECK POINT 5** Expand: $(x - 2y)^5$. $\quad x^5 - 10x^4y + 40x^3y^2 - 80x^2y^3 + 80xy^4 - 32y^5$

B EXERCISE SET

Student Solutions Manual   CD/Video   PH Math/Tutor Center   MathXL Tutorials on CD   MathXL®   MyMathLab   Interactmath.com

In Exercises 1–14, find each indicated sum.

1. $\sum_{i=1}^{6} 5i$ 105

2. $\sum_{i=1}^{6} 7i$ 147

3. $\sum_{i=1}^{4} 2i^2$ 60

4. $\sum_{i=1}^{5} i^3$ 225

5. $\sum_{k=1}^{5} k(k+4)$ 115

6. $\sum_{k=1}^{4} (k-3)(k+2)$ -4

7. $\sum_{i=1}^{4} \left(-\frac{1}{2}\right)^i$ $-\frac{5}{16}$

8. $\sum_{i=2}^{4} \left(-\frac{1}{3}\right)^i$ $\frac{7}{81}$

9. $\sum_{i=5}^{9} 11$ 55

10. $\sum_{i=3}^{7} 12$ 60

11. $\sum_{i=0}^{4} \frac{(-1)^i}{i!}$ $\frac{3}{8}$

12. $\sum_{i=0}^{4} \frac{(-1)^{i+1}}{(i+1)!}$ $-\frac{19}{30}$

13. $\sum_{i=1}^{5} \frac{i!}{(i-1)!}$ 15

14. $\sum_{i=1}^{5} \frac{(i+2)!}{i!}$ 110

In Exercises 15–22, evaluate the given binomial coefficient.

15. $\binom{8}{3}$ 56

16. $\binom{7}{2}$ 21

17. $\binom{12}{1}$ 12

18. $\binom{11}{1}$ 11

19. $\binom{6}{6}$ 1

20. $\binom{15}{2}$ 105

21. $\binom{100}{2}$ 4950

22. $\binom{100}{98}$ 4950

In Exercises 23–44, use the Binomial Theorem to expand each binomial and express the result in simplified form.

23. $(x + 2)^3$ $x^3 + 6x^2 + 12x + 8$

24. $(x + 4)^3$ $x^3 + 12x^2 + 48x + 64$

25. $(3x + y)^3$ $27x^3 + 27x^2y + 9xy^2 + y^3$

26. $(x + 3y)^3$ $x^3 + 9x^2y + 27xy^2 + 27y^3$

27. $(5x - 1)^3$ $125x^3 - 75x^2 + 15x - 1$

28. $(4x - 1)^3$ $64x^3 - 48x^2 + 12x - 1$

29. $(2x + 1)^4$ $16x^4 + 32x^3 + 24x^2 + 8x + 1$

30. $(3x + 1)^4$ $81x^4 + 108x^3 + 54x^2 + 12x + 1$

31. $(x^2 + 2y)^4$ $x^8 + 8x^6y + 24x^4y^2 + 32x^2y^3 + 16y^4$

32. $(x^2 + y)^4$ $x^8 + 4x^6y + 6x^4y^2 + 4x^2y^3 + y^4$

33. $(y - 3)^4$ $y^4 - 12y^3 + 54y^2 - 108y + 81$

34. $(y - 4)^4$ $y^4 - 16y^3 + 96y^2 - 256y + 256$

35. $(2x^3 - 1)^4$ $16x^{12} - 32x^9 + 24x^6 - 8x^3 + 1$

36. $(2x^5 - 1)^4$ $16x^{20} - 32x^{15} + 24x^{10} - 8x^5 + 1$

37. $(c + 2)^5$ $c^5 + 10c^4 + 40c^3 + 80c^2 + 80c + 32$

38. $(c + 3)^5$ $c^5 + 15c^4 + 90c^3 + 270c^2 + 405c + 243$

39. $(x - 1)^5$ $x^5 - 5x^4 + 10x^3 - 10x^2 + 5x - 1$

40. $(x - 2)^5$ $x^5 - 10x^4 + 40x^3 - 80x^2 + 80x - 32$

41. $(3x - y)^5$ $243x^5 - 405x^4y + 270x^3y^2 - 90x^2y^3 + 15xy^4 - y^5$

42. $(x - 3y)^5$ $x^5 - 15x^4y + 90x^3y^2 - 270x^2y^3 + 405xy^4 - 243y^5$

43. $(2a + b)^6$
$64a^6 + 192a^5b + 240a^4b^2 + 160a^3b^3 + 60a^2b^4 + 12ab^5 + b^6$

44. $(a + 2b)^6$
$a^6 + 12a^5b + 60a^4b^2 + 160a^3b^3 + 240a^2b^4 + 192ab^5 + 64b^6$

Answers to Selected Exercises

CHAPTER 1

Section 1.1

Check Point Exercises

1. a. $8x + 5$ **b.** $\dfrac{x}{7} - 2x$ **2.** 51; In 1950, approximately 51% of American adults smoked cigarettes.   **3.** 608
4. 2311; The formula models the data quite well.   **5. a.** true   **b.** true   **6. a.** -8 is less than -2.; true   **b.** 7 is greater than -3.; true
c. -1 is less than or equal to -4.; false   **d.** 5 is greater than or equal to 5.; true   **e.** 2 is greater than or equal to -14.; true

Exercise Set 1.1

1. $x + 5$ **3.** $x - 4$ **5.** $4x$ **7.** $2x + 10$ **9.** $6 - \dfrac{1}{2}x$ **11.** $\dfrac{4}{x} - 2$ **13.** $\dfrac{3}{5 - x}$ **15.** 57   **17.** 10   **19.** 88   **21.** 10   **23.** 44
25. 46   **27.** $\{1, 2, 3, 4\}$ **29.** $\{-7, -6, -5, -4\}$ **31.** $\{8, 9, 10, \dots\}$ **33.** $\{1, 3, 5, 7, 9\}$ **35.** true   **37.** true   **39.** false   **41.** true
43. false   **45.** true   **47.** false   **49.** -6 is less than -2.; true   **51.** 5 is less than -7.; true   **53.** 0 is less than -4.; false
55. -4 is less than or equal to 1.; true   **57.** -2 is less than or equal to -6.; false   **59.** -2 is less than or equal to -2.; true
61. -2 is greater than or equal to -2.; true   **63.** 2 is less than or equal to $-\dfrac{1}{2}$.; false   **65.** true   **67.** false   **69.** true   **71.** false
73. false   **75.** \$4664; underestimates by \$8   **77.** 32 million; fairly well   **79.** $10°C$ **81.** 60 ft   **101.** c   **103.** $(2 \cdot 3 + 3) \cdot 5 = 45$
105. -6 and -5

Section 1.2

Check Point Exercises

1. a. 6   **b.** 4.5   **c.** 0   **2. a.** -28 **b.** 0.7   **c.** $-\dfrac{1}{10}$ **3. a.** 8   **b.** $-\dfrac{1}{3}$ **4. a.** -3 **b.** 10.5   **c.** -6 **5. a.** 25   **b.** -25
c. -64 **d.** 81   **6. a.** -8 **b.** $\dfrac{8}{15}$ **7.** 74   **8.** -4 **9.** addition: $9 + 4x$; multiplication: $x \cdot 4 + 9$ **10. a.** $(6 + 12) + x = 18 + x$
b. $(-7 \cdot 4)x = -28x$ **11.** $-28x - 8$ **12.** $14x + 15x^2$ or $15x^2 + 14x$ **13.** $12x - 40$ **14.** $42 - 4x$

Exercise Set 1.2

1. 7   **3.** 4   **5.** 7.6   **7.** $\dfrac{\pi}{2}$ **9.** $\sqrt{2}$ **11.** $-\dfrac{2}{5}$ **13.** -11 **15.** -4 **17.** -4.5 **19.** $\dfrac{2}{15}$ **21.** $-\dfrac{35}{36}$ **23.** -8.2 **25.** -12.4
27. 0   **29.** -11 **31.** 5   **33.** 0   **35.** -12 **37.** 18   **39.** -15 **41.** $-\dfrac{1}{4}$ **43.** 5.5   **45.** $\sqrt{2}$ **47.** -90 **49.** 33   **51.** $-\dfrac{15}{13}$
53. 0   **55.** -8 **57.** 48   **59.** 100   **61.** -100 **63.** -8 **65.** 1   **67.** -1 **69.** $\dfrac{1}{8}$ **71.** -3 **73.** 45   **75.** 0   **77.** undefined
79. $\dfrac{9}{14}$ **81.** -15 **83.** -2 **85.** -24 **87.** 45   **89.** $\dfrac{1}{121}$ **91.** 14   **93.** $-\dfrac{8}{3}$ **95.** $-\dfrac{1}{2}$ **97.** 31   **99.** 37
101. addition: $10 + 4x$; multiplication: $x \cdot 4 + 10$ **103.** addition: $-5 + 7x$; multiplication: $x \cdot 7 - 5$ **105.** $(4 + 6) + x = 10 + x$
107. $(-7 \cdot 3)x = -21x$ **109.** $\left(-\dfrac{1}{3} \cdot -3\right)y = y$ **111.** $6x + 15$ **113.** $-14x - 21$ **115.** $-3x + 6$ **117.** $12x$ **119.** $5x^2$
121. $10x + 12x^2$ **123.** $18x - 40$ **125.** $8y - 12$ **127.** $16y - 25$ **129.** $12x^2 + 11$ **131.** $x - (x + 4); -4$ **133.** $6(-5x); -30x$
135. $5x - 2x; 3x$ **137.** $8x - (3x + 6); 5x - 6$ **139.** Hong Kong is 16 hours ahead.   **141.** Los Angeles is 3 hours behind.   **143.** 8 points
145. 126 points   **147. a.** $1200 - 0.07x$ **b.** \$780   **167.** d   **169.** $\left(2 \cdot 5 - \dfrac{1}{2} \cdot 10\right) \cdot 9 = 45$ **171.** $\dfrac{10}{x} - 4x$ **172.** 42   **173.** true

Section 1.3
Check Point Exercises

1.

2.

3.

4. a. 0 to 3 hr   **b.** 3 to 13 hr   **c.** 0.05 mg per 100 ml; after 3 hr   **d.** None of the drug is left in the body.
5. minimum x-value: -100; maximum x-value: 100;
distance between tick marks on x-axis: 50;
minimum y-value: -100; maximum y-value: 100;
distance between tick marks on y-axis: 10

Exercise Set 1.3

1–9.   **11.** $y = x^2 - 4$ **13.**   **15.** $y = 2x + 1$

17. $y = -\frac{1}{2}x$ **19.** $y = |x| + 1$ **21.** $y = 2|x|$ **23.** $y = -x^2$ **25.** $y = x^3$

27. c   **29.** b   **31.** c   **33.** no   **35.** $(2, 0)$ **37.** $(-2, 4)$ and $(1, 1)$ **39.** $y = 2x + 4$

41. $y = 3 - x^2$ **43.**   **45.**

47. 0 to 4 years   **49.** 4th year; about 8.3%   **51.** 1980; about 72%   **53.** about 3%   **55.** a   **57.** b   **59.** b   **61.** about 181 lb
63. about 135 lb   **75.** c;   **77.** c;   **79.** d   **81.** 8.8 sec   **82.** 14.3
83. 3   **84.** $-14x - 25$

Section 1.4

Check Point Exercises

1. 6 or $\{6\}$ **2.** -1 or $\{-1\}$ **3.** 6 or $\{6\}$ **4.** 1 or $\{1\}$ **5.** no solution; inconsistent   **6.** all real numbers; identity   **7.** 2007

Exercise Set 1.4

1. 3   **3.** 11   **5.** 11   **7.** 7   **9.** 13   **11.** 2   **13.** -4 **15.** 9   **17.** -5 **19.** 6   **21.** 19   **23.** $\frac{5}{2}$ **25.** 12   **27.** 24
29. -15 **31.** 5   **33.** 13   **35.** -12 **37.** $\frac{46}{5}$ **39.** all real numbers; identity   **41.** no solution; inconsistent   **43.** 0; conditional
45. -10; conditional   **47.** no solution; inconsistent   **49.** 0; conditional   **51.** $3(x - 4) = 3(2 - 2x)$; 2   **53.** $-3(x - 3) = 5(2 - x)$; 0.5

55. 2 **57.** −7 **59.** −2 **61.** Ø or no solution **63.** 10 **65.** −2 **67.** 2006 **69.** 40 years old; approximately (40, 117) on female graph

83. 5

85. −5

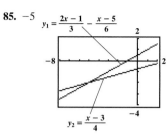

87. $x = \dfrac{c - b}{a}$ **89.** Answers will vary. **91.** $\dfrac{3}{10}$

92. −60 **93.**

Mid-Chapter 1 Check Point

1. $3x + 10$ **2.** 6 **3.** −15 **4.** −7 **5.** 3 **6.** $13x - 23$ **7.** 0 **8.** $-7x - 34$ **9.** 7
10. no solution or Ø or inconsistent equation **11.** 3 **12.** −4 **13.** $\{x \mid x$ is a real number$\}$ or $\mathbb{R}$ or identity **14.** 2
15. **16.** **17.** **18.** true **19.** false **20.** false **21.** true

Section 1.5

Check Point Exercises

1. English: 57; French: 33; Arabic: 23 **2.** 2012 **3.** 300 min **4.** $1200 **5.** 50 ft by 94 ft **6.** $w = \dfrac{P - 2l}{2}$ **7.** $h = \dfrac{V}{lw}$
8. $W = 106 + 6H$ or $W = 6H + 106$ **9.** $C = \dfrac{P}{1 + M}$

Exercise Set 1.5

1. 6 **3.** 25 **5.** 120 **7.** 320 **9.** 19 and 45 **11.** U.S.: 169 million; Japan: 56 million; China: 46 million **13.** 94°, 47°, 39°
15. 59°, 60°, 61° **17.** 2050 **19.** 2010 **21.** after 7 years **23.** after 5 months; $165 **25.** 30 times **27. a.** 2014; 22,300 students
b. $y_1 = 13300 + 1000x; y_2 = 26800 - 500x$ **29.** $420 **31.** $150 **33.** $36,000 **35.** $467.20 **37.** 50 yd by 100 yd **39.** 36 ft by 78 ft
41. 2 in **43.** 11 min **45.** $l = \dfrac{A}{w}$ **47.** $b = \dfrac{2A}{h}$ **49.** $P = \dfrac{I}{rt}$ **51.** $p = \dfrac{T - D}{m}$ **53.** $a = \dfrac{2A}{h} - b$ or $a = \dfrac{2A - hb}{h}$ **55.** $h = \dfrac{3V}{\pi r^2}$
57. $m = \dfrac{y - y_1}{x - x_1}$ **59.** $d_1 = Vt + d_2$ **61.** $x = \dfrac{C - By}{A}$ **63.** $v = \dfrac{2s - at^2}{2t}$ **65.** $n = \dfrac{L - a}{d} + 1$ **67.** $l = \dfrac{A - 2wh}{2w + 2h}$ **69.** $I = \dfrac{E}{R + r}$
79.

81. $200 **83.** Mrs. Ricardo: $4000; boy: $8000; girl: $2000 **85.** $C = \dfrac{LV - SN}{L - N}$
86. −6 is less than or equal to −6.; true **87.** $\dfrac{7}{3}$ **88.** −8

Section 1.6

Check Point Exercises

1. a. b^{11} **b.** $40x^5y^{10}$ **2. a.** b^{12} **b.** $9x^{11}y^3$ **3. a.** 1 **b.** 1 **c.** −1 **d.** 10 **e.** 1 **4. a.** $\dfrac{1}{25}$ **b.** $-\dfrac{1}{27}$ **c.** 16 **d.** $\dfrac{3y^4}{x^6}$
5. a. $\dfrac{4^3}{7^2} = \dfrac{64}{49}$ **b.** $\dfrac{x^2}{5}$ **6. a.** x^{15} **b.** $\dfrac{1}{y^{14}}$ **c.** b^{12} **7. a.** $16x^4$ **b.** $-27y^6$ **c.** $\dfrac{y^2}{16x^{10}}$ **8. a.** $\dfrac{x^{15}}{64}$ **b.** $\dfrac{16}{x^{12}y^8}$ **c.** $x^{15}y^{20}$
9. a. $-12y^9$ **b.** $\dfrac{4y^{14}}{x^6}$ **c.** $\dfrac{64}{x^9y^{15}}$

Exercise Set 1.6

1. b^{11} **3.** x^4 **5.** 32 **7.** $6x^6$ **9.** $20y^{12}$ **11.** $100x^{10}y^{12}$ **13.** $21x^5yz^4$ **15.** b^9 **17.** $5x^5$ **19.** x^5y^5 **21.** $10xy^3$

23. $-8a^{11}b^8c^4$ **25.** 1 **27.** 1 **29.** −1 **31.** 13 **33.** 1 **35.** $\dfrac{1}{3^2} = \dfrac{1}{9}$ **37.** $\dfrac{1}{(-5)^2} = \dfrac{1}{25}$ **39.** $-\dfrac{1}{5^2} = -\dfrac{1}{25}$ **41.** $\dfrac{x^2}{y^3}$ **43.** $\dfrac{8y^3}{x^7}$

45. $5^3 = 125$ **47.** $(-3)^4 = 81$ **49.** $\dfrac{y^5}{x^2}$ **51.** $\dfrac{b^7c^3}{a^4}$ **53.** x^{60} **55.** $\dfrac{1}{b^{12}}$ **57.** 7^{20} **59.** $64x^3$ **61.** $9x^{14}$ **63.** $8x^3y^6$

65. $9x^4y^{10}$ **67.** $-\dfrac{x^6}{27}$ **69.** $\dfrac{y^8}{25x^6}$ **71.** $\dfrac{x^{20}}{16y^{16}z^8}$ **73.** $\dfrac{16}{x^4}$ **75.** $\dfrac{x^6}{25}$ **77.** $\dfrac{81x^4}{y^4}$ **79.** $\dfrac{x^{24}}{y^{12}}$ **81.** x^9y^{12} **83.** a^8b^{12} **85.** $\dfrac{1}{x^6}$

87. $-\dfrac{4}{x}$ **89.** $\dfrac{2}{x^7}$ **91.** $\dfrac{10a^2}{b^5}$ **93.** $\dfrac{1}{x^9}$ **95.** $-\dfrac{6}{a^2}$ **97.** $\dfrac{3y^4z^3}{x^7}$ **99.** $3x^{10}$ **101.** $\dfrac{1}{x^{10}}$ **103.** $-\dfrac{5y^8}{x^6}$ **105.** $\dfrac{8a^9c^{12}}{b}$

107. x^{16} **109.** $-\dfrac{27b^{15}}{a^{18}}$ **111.** 1 **113.** $\dfrac{81x^{20}}{y^{32}}$ **115.** $\dfrac{1}{100a^4b^{12}c^8}$ **117.** $10x^2y^4$ **119.** $\dfrac{8}{3xy^{10}}$ **121.** $\dfrac{y^5}{8x^{14}}$ **123.** $\dfrac{1}{128x^7y^{16}}$

125. 125 kg; not safe   **127.** 0.55 astronomical unit   **129.** 1; Earth is 1 astronomical unit from the sun.

139. about 100 years or 2086

141. d   **143.** x^{9n} **145.** $x^{3n}y^{6n+3}$ **146.**

147. $y = \dfrac{C - Ax}{B}$ **148.** 40 m by 75 m

Section 1.7

Check Point Exercises

1. a. $-2{,}600{,}000{,}000$ **b.** 0.000003017 **2. a.** 5.21×10^9 **b.** -6.893×10^{-8} **3.** 4.1×10^9 **4. a.** 3.55×10^{-1} **b.** 4×10^8
5. \$7448   **6.** 3.1×10^7 mi, or 31 million mi

Exercise Set 1.7

1. 380   **3.** 0.0006   **5.** $-7{,}160{,}000$ **7.** 1.4   **9.** 0.79   **11.** -0.00415 **13.** $-60{,}000{,}100{,}000$ **15.** 3.2×10^4 **17.** 6.38×10^{17}
19. -3.17×10^2 **21.** -5.716×10^3 **23.** 2.7×10^{-3} **25.** -5.04×10^{-9} **27.** 7×10^{-3} **29.** 3.14159×10^0 **31.** 6.3×10^7
33. 6.4×10^4 **35.** 1.22×10^{-11} **37.** 2.67×10^{13} **39.** 2.1×10^3 **41.** 4×10^5 **43.** 2×10^{-8} **45.** 5×10^3 **47.** 4×10^{15}
49. 9×10^{-3} **51.** 6×10^{13} **53.** -6.2×10^3 **55.** 1.63×10^{19} **57.** -3.6×10^5 **59.** 6.26×10^7 people   **61.** 9.63×10^7 people
63. approximately 69 hot dogs per person   **65.** $2.5 \times 10^2 = 250$ chickens   **67.** 5.2×10^5 mi   **69.** 1.06×10^{-18} g   **71.** 3.1536×10^7
79. d   **81.** 1.25×10^{-15} **83.** Answers will vary.   **84.** $85x - 26$ **85.** 4   **86.** $\dfrac{y^6}{64x^8}$

Review Exercises

1. $2x - 10$ **2.** $6x + 4$ **3.** $\dfrac{9}{x} + \dfrac{1}{2}x$ **4.** 34   **5.** 60   **6.** 15   **7.** $\{1, 2\}$ **8.** $\{-3, -2, -1, 0, 1\}$ **9.** false   **10.** true   **11.** true
12. -5 is less than 2.; true   **13.** -7 is greater than or equal to -3.; false   **14.** -7 is less than or equal to -7.; true   **15.** 124 ft   **16.** 9.7
17. 5.003   **18.** 0   **19.** -7.6 **20.** -4.4 **21.** 13   **22.** 60   **23.** $-\dfrac{1}{10}$ **24.** $-\dfrac{3}{35}$ **25.** -240 **26.** 16   **27.** -32 **28.** $-\dfrac{5}{12}$
29. 7   **30.** -9.1 **31.** 7   **32.** 9   **33.** -2 **34.** -18 **35.** 55   **36.** 1   **37.** -4 **38.** -13 **39.** $17x - 15$ **40.** $9x^2 + x$
41. $5y - 17$ **42.** $10x$ **43.** $-3x - 8$ **44–46.**

47.

48.

49.

50.

51. minimum x-value: -20;
maximum x-value: 40;
distance between tick marks on x-axis: 10;
minimum y-value: -5;
maximum y-value: 5;
distance between tick marks on y-axis: 1

52. 20%   **53.** 85 years old   **54.** Answers will vary.   **55.** c   **56.** 6   **57.** −10   **58.** 5   **59.** −13   **60.** −3   **61.** −1   **62.** 2

63. 2   **64.** $\dfrac{72}{11}$ **65.** −12   **66.** $\dfrac{77}{15}$ **67.** no solution; inconsistent   **68.** all real numbers; identity   **69.** 0; conditional   **70.** $\dfrac{3}{2}$; conditional

71. no solution; inconsistent   **72.** 2007   **73.** Chicken Caesar: 495; Express Taco: 620; Mandarin Chicken: 590   **74.** 25°, 35°, 120°   **75.** 2013

76. 500 min   **77.** $60   **78.** $10,000 in sales   **79.** 44 yd by 126 yd   **80. a.** $14,100 + 1500x = 41,700 − 800x$ **b.** 2017; 32,100

81. $h = \dfrac{3V}{B}$ **82.** $x = \dfrac{y - y_1}{m} + x_1$ **83.** $R = \dfrac{E}{I} - r$ or $R = \dfrac{E - Ir}{I}$ **84.** $F = \dfrac{9C + 160}{5}$ or $F = \dfrac{9}{5}C + 32$ **85.** $g = \dfrac{s - vt}{t^2}$

86. $g = \dfrac{T}{r + vt}$ **87.** $15x^{13}$ **88.** $\dfrac{x^2}{y^5}$ **89.** $\dfrac{x^4 y^7}{9}$ **90.** $\dfrac{1}{x^{18}}$ **91.** $49x^6 y^2$ **92.** $-\dfrac{8}{y^7}$ **93.** $-\dfrac{12}{x^7}$ **94.** $3x^{10}$ **95.** $-\dfrac{a^8}{2b^5}$ **96.** $-24x^7 y^4$

97. $\dfrac{3}{4}$ **98.** $\dfrac{y^{12}}{125x^6}$ **99.** $-\dfrac{6x^9}{y^5}$ **100.** $\dfrac{9x^8 y^{14}}{25}$ **101.** $-\dfrac{x^{21}}{8y^{27}}$ **102.** 7,160,000   **103.** 0.000107   **104.** -4.1×10^{13} **105.** 8.09×10^{-3}

106. 1.26×10^8 **107.** 2.5×10^1 **108.** $\$4.35 \times 10^{10}$

Chapter 1 Test

1. $4x - 5$ **2.** 170   **3.** $\{-4, -3, -2, -1\}$ **4.** true   **5.** −3 is greater than −1.; false   **6.** 616 convictions   **7.** 17.9   **8.** −7.6

9. $\dfrac{1}{4}$ **10.** −60   **11.** $\dfrac{1}{8}$ **12.** 3.1   **13.** −3   **14.** 6   **15.** −4   **16.** $-5x - 18$ **17.** $6y - 27$ **18.** $17x - 22$

19.   **20.**   **21.** 2   **22.** −6   **23.** no solution; inconsistent   **24.** 23 and 49   **25.** 5 yr

26. 20 prints; $3.80   **27.** $50   **28.** 120 yd by 380 yd   **29.** $h = \dfrac{3V}{lw}$

30. $y = \dfrac{C - Ax}{B}$ **31.** $-\dfrac{14}{x^5}$ **32.** $\dfrac{40}{x^3 y^8}$ **33.** $\dfrac{x^6}{4y^3}$ **34.** $\dfrac{x^{15}}{64y^6}$ **35.** $\dfrac{x^{16}}{9y^{10}}$

36. 0.0000038   **37.** 4.07×10^{11} **38.** 5×10^3 **39.** 1.26×10^{10}

CHAPTER 2

Section 2.1

Check Point Exercises

1. domain: $\{5, 10, 15, 20, 25\}$; range: $\{12.8, 16.2, 18.9, 20.7, 21.8\}$ **2. a.** not a function   **b.** function   **3. a.** 29   **b.** 65   **c.** 46
d. $6a + 6h + 9$ **4.**   ; The graph of g is the graph of f shifted down by 3 units.   **5. a.** function
b. function
c. not a function
6. a. 16
b. 8

7. a. Domain = $\{x | -2 \le x \le 1\}$; Range = $\{y | 0 \le y \le 3\}$ **b.** Domain = $\{x | -2 < x \le 1\}$; Range = $\{y | -1 \le y < 2\}$
c. Domain = $\{x | -3 \le x < 0\}$; Range = $\{y | y = -3, -2, -1\}$

Exercise Set 2.1

1. function; domain: $\{1, 3, 5\}$; range: $\{2, 4, 5\}$ **3.** not a function; domain: $\{3, 4\}$; range: $\{4, 5\}$ **5.** function; domain: $\{-3, -2, -1, 0\}$;

range: $\{-3, -2, -1, 0\}$ **7.** not a function; domain: $\{1\}$; range: $\{4, 5, 6\}$ **9. a.** 1   **b.** 6   **c.** −7   **d.** $2a + 1$ **e.** $a + 3$

11. a. −2   **b.** −17   **c.** 0   **d.** $12b - 2$ **e.** $3b + 10$ **13. a.** 5   **b.** 8   **c.** 53   **d.** 32   **e.** $48b^2 + 5$

15. a. −1   **b.** 26   **c.** 19   **d.** $2b^2 + 3b - 1$ **e.** $50a^2 + 15a - 1$ **17. a.** $\dfrac{3}{4}$ **b.** −3   **c.** $\dfrac{11}{8}$ **d.** $\dfrac{13}{9}$ **e.** $\dfrac{2a + 2h - 3}{a + h - 4}$

f. Denominator would be zero.

19.   **21.**   **23.**

The graph of g is the graph of f shifted up by 3 units.

The graph of g is the graph of f shifted down by 1 unit.

The graph of g is the graph of f shifted up by 1 unit.

25.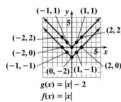

The graph of g is the graph of f shifted down by 2 units.

27.

The graph of g is the graph of f shifted up by 2 units.

29. function   **31.** not a function
33. function   **35.** not a function
37. -4 **39.** 4   **41.** 0   **43.** 2
45. 2   **47.** -2
49. Domain $= \{x | 0 \le x < 5\}$; Range $= \{y | -1 \le y < 5\}$
51. Domain $= \{x | x \ge 0\}$; Range $= \{y | y \ge 1\}$
53. Domain $= \{x | -2 \le x \le 6\}$; Range $= \{y | -2 \le y \le 6\}$
55. Domain $= \{x | x \text{ is a real number}\}$; Range $= \{y | y \le -2\}$
57. Domain $= \{x | x = -5, -2, 0, 1, 3\}$; Range $= \{y | y = 2\}$

59. $-2; 10$ **61.** -38 **63.** $-2x^3 - 2x$ **65. a.** -1 **b.** 7   **c.** 19   **d.** 112   **67.** Domain $= \{x | x \ne 1\}$; Range $= \{y | y \ne 0\}$
69. a. $\{(\text{EL}, 1\%), (\text{L}, 7\%), (\text{SL}, 11\%), (\text{M}, 52\%), (\text{SC}, 13\%), (\text{C}, 13\%), (\text{EC}, 3\%)\}$ **b.** Yes; Each ideology corresponds to exactly one percentage.
c. $\{(1\%, \text{EL}), (7\%, \text{L}), (11\%, \text{SL}), (52\%, \text{M}), (13\%, \text{SC}), (13\%, \text{C}), (3\%, \text{EC})\}$ **d.** No; 13% in the domain corresponds to two ideologies in the range, SC and C   **71.** 5.22; In 2000, there were 5.22 million women enrolled in U.S. colleges; (2000, 5.22).   **73.** 1.4; In 2004, there were 1.4 million more women than men enrolled in U.S. colleges.   **75.** 440; For 20-year-old drivers, there are 440 accidents per 50 million miles driven; (20,440).
77. $x = 45$; $y = 190$; The minimum number of accidents is 190 per 50 million miles driven and is attributed to 45-year-old drivers.
79. 3.1; In 1960, Jewish Americans made up about 3.1% of the U.S. population.   **81.** 19 and 64; In 1919 and in 1964, Jewish Americans made up about 3% of the U.S. population.   **83.** 1940; 3.7%   **85.** Each year corresponds to only one percentage.   **87.** 0.83; It costs $0.83 to mail a 3-ounce first-class letter.   **89.** $0.60   **101.** d   **103.** Answers will vary; an example is $\{(1, 1), (2, 1)\}$.   **105.** 0   **106.** $\dfrac{y^{10}}{9x^4}$ **107.** -15

Section 2.2

Check Point Exercises

1. a. $\{x | x \text{ is a real number}\}$ **b.** $\{x | x \text{ is a real number and } x \ne -5\}$ **2. a.** $3x^2 + 6x + 6$ **b.** 78   **3. a.** $\dfrac{5}{x} + \dfrac{7}{x-8}$
b. $\{x | x \text{ is a real number and } x \ne 0 \text{ and } x \ne 8\}$ **4. a.** 23   **b.** $x^2 - 3x - 3; 1$ **c.** $\dfrac{x^2 - 2x}{x+3}; \dfrac{7}{2}$ **d.** -24

Exercise Set 2.2

1. $\{x | x \text{ is a real number}\}$ **3.** $\{x | x \text{ is a real number and } x \ne -4\}$ **5.** $\{x | x \text{ is a real number and } x \ne 3\}$ **7.** $\{x | x \text{ is a real number and } x \ne 5\}$
9. $\{x | x \text{ is a real number and } x \ne -7 \text{ and } x \ne 9\}$ **11. a.** $5x - 5$ **b.** 20   **13. a.** $3x^2 + x - 5$ **b.** 75   **15. a.** $2x^2 - 2$ **b.** 48
17. $\{x | x \text{ is a real number}\}$ **19.** $\{x | x \text{ is a real number and } x \ne 5\}$ **21.** $\{x | x \text{ is a real number and } x \ne 0 \text{ and } x \ne 5\}$
23. $\{x | x \text{ is a real number and } x \ne 2 \text{ and } x \ne -3\}$ **25.** $\{x | x \text{ is a real number and } x \ne 2\}$ **27.** $\{x | x \text{ is a real number}\}$
29. $x^2 + 3x + 2; 20$ **31.** 0   **33.** $x^2 + 5x - 2; 48$ **35.** -8 **37.** -16 **39.** -135 **41.** $\dfrac{x^2 + 4x}{2 - x}; 5$ **43.** -1
45. $\{x | x \text{ is a real number}\}$ **47.** $\{x | x \text{ is a real number and } x \ne 2\}$ **49.** 5   **51.** -1 **53.** $\{x | -4 \le x \le 3\}$
55.

57. -4 **59.** -4 **61.** $\{x | x = 0, 1, 2, \ldots, 7\}$ **63. a.** $(B - D)(x) = 4565x + 1,578,296$; change in U.S. population
b. $(B - D)(6) = 1,605,686$; In 2001, U.S. population increased by 1,605,686.   **c.** 1,609,508; fairly well
65. $h(x)$ or total world population in year x **67.** 5.9 billion   **69.** $-200,000; 0; 200,000$; The company has a loss of $200,000, breaks even, and a profit of $200,000 when 20,000, 30,000, and 40,000 radios are produced, respectively.
75. Per capita costs of Medicare are increasing more and more rapidly over time.

77.

79.

81. d   **83.** $b = \dfrac{R - 3a}{3}$ or $b = \dfrac{R}{3} - a$ **84.** 7   **85.** $6b + 8$

Mid-Chapter 2 Check Point

1. not a function; Domain: $\{1, 2\}$; Range: $\{-6, 4, 6\}$ **2.** function; Domain: $\{0, 2, 3\}$; Range: $\{1, 4\}$ **3.** function; Domain: $\{x | -2 \le x < 2\}$; Range: $\{y | 0 \le y \le 3\}$ **4.** not a function; Domain: $\{x | -3 < x \le 4\}$; Range: $\{y | -1 \le y \le 2\}$ **5.** not a function; Domain: $\{-2, -1, 0, 1, 2\}$; Range: $\{-2, -1, 1, 3\}$ **6.** function; Domain: $\{x | x \le 1\}$; Range: $\{y | y \ge -1\}$ **7.** No vertical line intersects the graph of f more than once.
8. 3   **9.** -2 **10.** -6 and 2   **11.** $\{x | x \text{ is a real number}\}$ **12.** $\{y | y \le 4\}$ **13.** $\{x | x \text{ is a real number}\}$
14. $\{x | x \text{ is a real number and } x \ne -2 \text{ and } x \ne 2\}$ **15.** 23   **16.** 23   **17.** $a^2 - 5a - 3$ **18.** $x^2 - 5x + 3; 17$ **19.** $x^2 - x + 13; 33$

20. -36 **21.** $\dfrac{x^2 - 3x + 8}{-2x - 5}; 12$ **22.** $\left\{ x \middle| x \text{ is a real number and } x \ne -\dfrac{5}{2} \right\}$

Section 2.3

Check Point Exercises

1. $3x - 2y = 6$ **2. a.** 6   **b.** $-\dfrac{7}{5}$ **3.** $2; -5$ **4.** $y = 4x - 3$

5. $f(x) = -\dfrac{2}{3}x$ **6.** $x = -3$ **7.** 0.01 mg per 100 ml per hr   **8.** $S(x) = 0.22x + 1$

Exercise Set 2.3

1. $x + y = 4$ **3.** $x + 3y = 6$ **5.** $6x - 2y = 12$ **7.** $3x - y = 6$

9. $x - 3y = 9$ **11.** $2x = 3y + 6$ **13.** $6x - 3y = 15$ **15.** $m = 4$; rises   **17.** $m = \dfrac{1}{3}$; rises

19. $m = 0$; horizontal   **21.** $m = -\dfrac{4}{3}$; falls   **23.** undefined slope; vertical   **25.** $L_1: \dfrac{2}{3}; L_2: -2; L_3: -\dfrac{1}{2}$

27. $m = 2; b = 1;$ **29.** $m = -2; b = 1;$ **31.** $m = \dfrac{3}{4}; b = -2;$ **33.** $m = -\dfrac{3}{5}; b = 7;$ **35.** $m = -\dfrac{1}{2}; b = 0;$

37. a. $y = -2x$ **39. a.** $y = \dfrac{4}{5}x$ **41. a.** $y = -3x + 2$ **43. a.** $y = -\dfrac{5}{3}x + 5$

b. $m = -2; b = 0$ **b.** $m = \dfrac{4}{5}; b = 0$ **b.** $m = -3; b = 2$ **b.** $m = -\dfrac{5}{3}; b = 5$

c.   **c.**   **c.**   **c.**

45. $y = 3$ **47.** $f(x) = -2$ **49.** $3y = 18$ **51.** $f(x) = 2$

53.

55.

57.

59. $m = -\dfrac{a}{b}$; falls

61. undefined slope; vertical

63. $m = -\dfrac{A}{B}$; $b = \dfrac{C}{B}$ **65.** -2

67.

69. 5   **71.** m_1, m_3, m_2, m_4

73. $m = -0.4$; The percentage of U.S. men smoking has decreased at a rate of 0.4% per year after 1980.

75. $m = 0.01$; The global average temperature of Earth has increased at a rate of 0.01°F per year since 1995.

77. $m = -1.56\%$; Average decrease of 1.56% of total sales per year   **79. a.** $A(x) = 2x + 33$ **b.** 55 million

81. Predictions will vary. $S(x) = 1700x + 49{,}100$; average salary, $S(x)$, x years after 1995

83. Predictions will vary. $A(x) = -0.28x + 1.7$; percentage of audited taxpayers, $A(x)$, x years after 1996

85. Predictions will vary. $f(x) = 98$; percentage of U.S. households with televisions, $f(x)$, x years after 1980

87. Answers may vary;

107.

$m = -3$

109.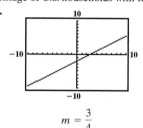

$m = \dfrac{3}{4}$

111. coefficient of x: -6; coefficient of y: 3   **113. a.** $mx_1 + mx_2 + b$
b. $mx_1 + mx_2 + 2b$ **c.** no
114. $16x^4y^6$ **115.** 3.2×10^{-3}
116. $3x + 17$

Section 2.4
Check Point Exercises

1. $y + 3 = -2(x - 4)$; $y = -2x + 5$ or $f(x) = -2x + 5$ **2. a.** $y + 3 = -2(x - 6)$ or $y - 5 = -2(x - 2)$ **b.** $y = -2x + 9$ or $f(x) = -2x + 9$
3. $y = 2.35x + 179.5$ or $f(x) = 2.35x + 179.5$; 297 million   **4.** $y - 5 = 3(x + 2)$; $y = 3x + 11$ or $f(x) = 3x + 11$ **5.** 3

Exercise Set 2.4

1. $y - 5 = 3(x - 2)$; $f(x) = 3x - 1$ **3.** $y - 6 = 5(x + 2)$; $f(x) = 5x + 16$ **5.** $y + 2 = -4(x + 3)$; $f(x) = -4x - 14$

7. $y - 0 = -5(x + 2)$; $f(x) = -5x - 10$ **9.** $y + \dfrac{1}{2} = -1(x + 2)$; $f(x) = -x - \dfrac{5}{2}$ **11.** $y - 0 = \dfrac{1}{4}(x - 0)$; $f(x) = \dfrac{1}{4}x$

13. $y + 4 = -\dfrac{2}{3}(x - 6)$; $f(x) = -\dfrac{2}{3}x$ **15.** $y - 3 = 1(x - 6)$ or $y - 2 = 1(x - 5)$; $f(x) = x - 3$

17. $y - 0 = 2(x + 2)$ or $y - 4 = 2(x - 0)$; $f(x) = 2x + 4$ **19.** $y - 13 = -2(x + 6)$ or $y - 5 = -2(x + 2)$; $f(x) = -2x + 1$

21. $y - 9 = -\dfrac{11}{3}(x - 1)$ or $y + 2 = -\dfrac{11}{3}(x - 4)$; $f(x) = -\dfrac{11}{3}x + \dfrac{38}{3}$ **23.** $y + 5 = 0(x + 2)$ or $y + 5 = 0(x - 3)$; $f(x) = -5$

25. $y - 8 = 2(x - 7)$ or $y - 0 = 2(x - 3)$; $f(x) = 2x - 6$ **27.** $y - 0 = \dfrac{1}{2}(x - 2)$ or $y + 1 = \dfrac{1}{2}(x - 0)$; $f(x) = \dfrac{1}{2}x - 1$

29. a. 5   **b.** $-\dfrac{1}{5}$ **31. a.** -7 **b.** $\dfrac{1}{7}$ **33. a.** $\dfrac{1}{2}$ **b.** -2 **35. a.** $-\dfrac{2}{5}$ **b.** $\dfrac{5}{2}$ **37. a.** -4 **b.** $\dfrac{1}{4}$ **39. a.** $-\dfrac{1}{2}$ **b.** 2

41. a. $\dfrac{2}{3}$ **b.** $-\dfrac{3}{2}$ **43. a.** undefined   **b.** 0   **45.** $y - 2 = 2(x - 4)$; $y = 2x - 6$ or $f(x) = 2x - 6$

47. $y - 4 = -\dfrac{1}{2}(x - 2)$; $y = -\dfrac{1}{2}x + 5$ or $f(x) = -\dfrac{1}{2}x + 5$ **49.** $y + 10 = -4(x + 8)$; $y = -4x - 42$ or $f(x) = -4x - 42$

51. $y + 3 = -5(x - 2)$; $y = -5x + 7$ or $f(x) = -5x + 7$ **53.** $y - 2 = \dfrac{2}{3}(x + 2)$; $y = \dfrac{2}{3}x + \dfrac{10}{3}$ or $f(x) = \dfrac{2}{3}x + \dfrac{10}{3}$

55. $y + 7 = -2(x - 4)$; $y = -2x + 1$ or $f(x) = -2x + 1$ **57.** $f(x) = 5$ **59.** $f(x) = -\dfrac{1}{2}x + 1$ **61.** $f(x) = -\dfrac{2}{3}x - 2$ **63.** $f(x) = 4x - 5$

65. $-\dfrac{A}{B}$ **67. a.** $y - 16 = -0.55(x - 10)$ or $y - 12.7 = -0.55(x - 16)$ **b.** $f(x) = -0.55x + 21.5$ **c.** 10.5%

69. a.

b.

$y - 12.1 = -0.12(x - 20)$
or $y - 10.9 = -0.12(x - 30)$;
$f(x) = -0.12x + 14.5$

c. 9.7%   **71. a.** $m = 1.7$; Stimulant use has increased at a rate of 1.7 per thousand children per year.
b. $m = 1.4$; Antidepressant use has increased at a rate of 1.4 per thousand children per year.
c. no; Stimulant use has increased at a faster rate than antidepressant use.

81. a. 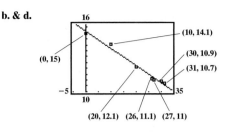   **b. & d.**   **c.**   **83.** -4 **85.** $(-40, 74)$ and $(97, -200)$ **87.** 33   **88.** -56 **89.** $40°, 60°,$ and $80°$

Review Exercises

1. function; domain: $\{3, 4, 5\}$; range: $\{10\}$ **2.** function; domain: $\{1, 2, 3, 4\}$; range: $\{12, 100, \pi, -6\}$

3. not a function; domain: $\{13, 15\}$; range: $\{14, 16, 17\}$ **4. a.** -5 **b.** 16   **c.** -75 **d.** $14a - 5$ **e.** $7a + 9$

5. a. 2   **b.** 52   **c.** 70   **d.** $3b^2 - 5b + 2$ **e.** $48a^2 - 20a + 2$

6.   **7.**   **8.** not a function   **9.** function   **10.** function   **11.** not a function

12. not a function   **13.** function   **14.** -3 **15.** -2 **16.** 3

17. $\{x \mid -3 \le x < 5\}$ **18.** $\{y \mid -5 \le y \le 0\}$

19. a. For each time, there is only one height.
    b. 0; The vulture was on the ground after 15 seconds.
    c. 45 m
    d. 7 and 22; After 7 and 22 seconds, the vulture's height is 20 meters.
    e. Answers will vary.   **20.** $\{x \mid x$ is a real number$\}$

The graph of g is the graph of f shifted down by 1 unit.

The graph of g is the graph of f shifted up by 2 units.

21. $\{x \mid x$ is a real number and $x \ne -8\}$ **22.** $\{x \mid x$ is a real number and $x \ne 5\}$ **23. a.** $6x - 4$ **b.** 14   **24. a.** $5x^2 + 1$ **b.** 46

25. $\{x \mid x$ is a real number and $x \ne 4\}$ **26.** $\{x \mid x$ is a real number and $x \ne -6$ and $x \ne -1\}$ **27.** $x^2 - x - 5; 1$ **28.** 1

29. $x^2 - 3x + 5; 3$ **30.** 9   **31.** -120 **32.** $\dfrac{x^2 - 2x}{x - 5}; -8$ **33.** $\{x \mid x$ is a real number$\}$ **34.** $\{x \mid x$ is a real number and $x \ne 5\}$

35.   **36.**   **37.**   **38.** 2; rises   **39.** $-\dfrac{2}{3}$; falls

40. undefined; vertical

41. 0; horizontal

42. $m = 2; b = -1;$ **43.** $m = -\dfrac{1}{2}; b = 4;$ **44.** $m = \dfrac{2}{3}; b = 0;$

45. $y = -2x + 4; m = -2; b = 4$ **46.** $y = -\dfrac{5}{3}x; m = -\dfrac{5}{3}; b = 0$ **47.** $y = -\dfrac{5}{3}x + 2; m = -\dfrac{5}{3}; b = 2$

48.   **49.**   **50.**   **51.**   **52.**

53. -0.27; Record time has been decreasing at a rate of 0.27 second per year since 1900.

54. $m \approx -3.4$; average decrease of approximately \$3.4 billion per year   **55. a.** $F = \dfrac{9}{5}C + 32$ **b.** $86°$

56. $y - 2 = -6(x + 3); y = -6x - 16$ or $f(x) = -6x - 16$ **57.** $y - 6 = 2(x - 1)$ or $y - 2 = 2(x + 1); y = 2x + 4$ or $f(x) = 2x + 4$

58. $y + 7 = -3(x - 4); y = -3x + 5$ or $f(x) = -3x + 5$ **59.** $y - 6 = -3(x + 2); y = -3x$ or $f(x) = -3x$

60. a. $y - 4.85 = -0.12(x - 20)$ or $y - 4.49 = -0.12(x - 5)$ **b.** $f(x) = -0.12x + 5.09$ **c.** \$3.89

Chapter 2 Test

1. function; domain: $\{1, 3, 5, 6\}$; range: $\{2, 4, 6\}$ **2.** not a function; domain: $\{2, 4, 6\}$; range: $\{1, 3, 5, 6\}$ **3.** $3a + 10$ **4.** 28

5.

The graph of g is the graph of f shifted up by 2 units.

6. function   **7.** not a function   **8.** -3 **9.** -2 and 3
10. $\{x|x \text{ is a real number}\}$ **11.** $\{y|y \le 3\}$ **12.** $\{x|x \text{ is a real number and } x \ne 10\}$
13. $x^2 + 5x + 2$; 26   **14.** $x^2 + 3x - 2$; -4 **15.** -15

16. $\dfrac{x^2 + 4x}{x + 2}$; 3   **17.** $\{x|x \text{ is a real number and } x \ne -2\}$

18.   **19.**

20.
21. $-\dfrac{1}{2}$; falls   **22.** undefined; vertical   **23.** 176; In 2005, the number of Super Bowl viewers was 176 million.

24. 3.6; The number of Super Bowl viewers is increasing at a rate of 3.6 million per year.

25. $y + 3 = 1(x + 1)$ or $y - 2 = 1(x - 4)$; $y = x - 2$ or $f(x) = x - 2$

26. $y - 3 = 2(x + 2)$; $y = 2x + 7$ or $f(x) = 2x + 7$

27. $y + 4 = -\dfrac{1}{2}(x - 6)$; $y = -\dfrac{1}{2}x - 1$ or $f(x) = -\dfrac{1}{2}x - 1$ **28. a.** $y - 2.3 = 0.3(x - 1)$ or $y - 3.2 = 0.3(x - 4)$
b. $f(x) = 0.3x + 2$ **c.** 4.4 million

Cumulative Review Exercises (Chapters 1–2)

1. $\{0, 1, 2, 3\}$ **2.** false   **3.** 7   **4.** 15   **5.** $3x + 7$ **6.** -4 **7.** all real numbers; identity   **8.** -6 **9.** \$2250

10. $t = \dfrac{A - p}{pr}$ **11.** $\dfrac{y^{10}}{9x^8}$ **12.** $\dfrac{9x^{10}}{y^{12}}$ **13.** 2.1×10^{-5} **14.** function; domain: $\{1, 2, 3, 4, 6\}$; range: $\{5\}$

15.   **16.** $\{x|x \text{ is a real number and } x \ne 15\}$ **17.** $2x^2 + x + 5$; 6

18. $f(x) = -2x + 4$ **19.** $x - 2y = 6$ **20.** $y + 5 = 4(x - 3)$;
$y = 4x - 17$ or
$f(x) = 4x - 17$

The graph of g is the graph of f shifted up by 3 units.

CHAPTER 3

Section 3.1

Check Point Exercises

1. solution   **2.** $(1, 4)$ or $\{(1, 4)\}$ **3.** $(6, 11)$ or $\{(6, 11)\}$ **4.** $(-2, 5)$ or $\{(-2, 5)\}$ **5.** $\left(-\dfrac{1}{2}, 2\right)$ or $\left\{\left(-\dfrac{1}{2}, 2\right)\right\}$ **6.** $(2, -1)$ or $\{(2, -1)\}$

7. $\left(\dfrac{37}{7}, \dfrac{19}{7}\right)$ or $\left\{\left(\dfrac{37}{7}, \dfrac{19}{7}\right)\right\}$ **8.** no solution or $\varnothing$ **9.** $\{(x, y)|x = 4y - 8\}$ or $\{(x, y)|5x - 20y = -40\}$

Exercise Set 3.1

1. solution   **3.** not a solution   **5.** solution   **7.** $(3, 1)$ **9.** $\left(\dfrac{1}{2}, 3\right)$ **11.** $(4, 3)$ **13.** $\{(x, y)|2x + 3y = 6\}$ or $\{(x, y)|4x = -6y + 12\}$
15. $(1, 0)$ **17.** $\varnothing$ **19.** $(1, 2)$ **21.** $(3, 1)$ **23.** $\varnothing$ **25.** $(2, 4)$ **27.** $(3, 1)$ **29.** $(2, 1)$ **31.** $(2, -3)$ **33.** $\varnothing$ **35.** $(3, -2)$

37. $\varnothing$ **39.** $(-5, -1)$ **41.** $\left\{(x, y)|y = \dfrac{2}{5}x - 2\right\}$ or $\{(x, y)|2x - 5y = 10\}$ **43.** $(5, 2)$ **45.** $(2, -3)$ **47.** $(-1, 1)$ **49.** $(-2, -7)$

51. $(7, 2)$ **53.** $(4, -1)$ **55.** $\{(x, y)|2x + 6y = 8\}$ or $\{(x, y)|3x + 9y = 12\}$ **57.** $\left(\dfrac{29}{22}, -\dfrac{5}{11}\right)$ **59.** $(-2, -1)$ **61.** $(1, -3)$

63. $(1, -3)$ **65.** $(4, 3)$ **67.** $\{(x, y)|x = 3y - 1\}$ or $\{(x, y)|2x - 6y = -2\}$ **69.** $\varnothing$ **71.** $(5, 1)$ **73.** $\left(\dfrac{32}{7}, -\dfrac{20}{7}\right)$

75. $(-5, 7)$ **77.** $\varnothing$ **79.** $\{(x, y)|x + 2y - 3 = 0\}$ or $\{(x, y)|12 = 8y + 4x\}$ **81.** $(0, 0)$ **83.** $(6, -1)$ **85.** $\left(\dfrac{1}{a}, 3\right)$

87. $m = -4, b = 3$ **89.** $y = x - 4;\ y = -\dfrac{1}{3}x + 4$ **91.** 2063; 87 years; Answers will vary.

93. a. $y = -0.75x + 38$ **b.** $y = -x + 40$ **c.** 1993; 32% **95. a.** 150 sold; 300 supplied **b.** $100; 250

107. chicken: 0.94; red meat: -0.74; Chicken consumption is increasing; red meat consumption is decreasing.; The lines are not parallel.;

Chicken consumption is greater than red meat consumption. **109.** d **111.** Answers will vary. **113.** $\dfrac{10}{9}$ **114.** $-128x^{19}y^8$ **115.** 11

Section 3.2

Check Point Exercises

1. Snickers: 35 grams; Think Thin: 0 grams **2.** $3150 at 9%; $1850 at 11% **3.** 12% solution: 100 oz; 20% solution: 60 oz
4. boat: 35 mph; current: 7 mph **5. a.** $C(x) = 300,000 + 30x$ **b.** $R(x) = 80x$ **c.** $(6000, 480,000)$; The company will break even when it
produces and sells 6000 pairs of shoes. At this level, both revenue and cost are $480,000. **6.** $P(x) = 200x - 500,000$

Exercise Set 3.2

1. 3 and 4 **3.** first number: 2; second number: 5 **5. a.** 1500 units; $48,000 **b.** $P(x) = 17x - 25,500$ **7. a.** 500 units; $122,500 **b.** $P(x) = 140x - 70,000$
9. Bachelor's or higher: 26%; Associate: 10% **11.** 22 computers and 14 hard drives **13.** $2000 at 6% and $5000 at 8%
15. first fund: $8000; second fund: $6000 **17.** $17,000 at 12%; $3000 at a 5% loss **19.** California: 100 gal; French: 100 gal
21. 18-karat gold: 96 grams; 12-karat gold: 204 grams **23.** cheaper candy: 30 lb; more expensive candy: 45 lb **25.** 8 nickels and 7 dimes
27. plane: 130 mph; wind: 30 mph **29.** crew: 6 km/hr; current: 2 km/hr **31.** in still water: 4.5 mph; current: 1.5 mph **33.** 86 and 74
35. $80°, 50°, 50°$ **37.** 70 ft by 40 ft **39.** two-seat table: 6; four-seat table: 11 **41.** 500 radios **43.** -6000; When the company produces and
sells 200 radios, the loss is $6000. **45. a.** $P(x) = 20x - 10,000$ **b.** $190,000 **47. a.** $C(x) = 18,000 + 20x$ **b.** $R(x) = 80x$
c. $(300, 24,000)$; When 300 canoes are produced and sold, both revenue and cost are $24,000. **49. a.** $C(x) = 30,000 + 2500x$ **b.** $R(x) = 3125x$
c. $(48, 150,000)$; For 48 sold-out performances, both cost and revenue are $150,000.

59. $(44, 4070)$ **61.** yes, 8 hexagons and 4 squares
63. 95
65. $45 for mangos and $22 for avocados
66. $y - 5 = -2(x + 2)$ or $y - 13 = -2(x + 6)$; $y = -2x + 1$ or $f(x) = -2x + 1$
67. $y - 0 = 1(x + 3)$; $y = x + 3$ or $f(x) = x + 3$
68. $\{x|x$ is a real number and $x \neq 3\}$

Section 3.3

Check Point Exercises

1. $(-1) - 2(-4) + 3(5) = 22;\ 2(-1) - 3(-4) - 5 = 5;\ 3(-1) + (-4) - 5(5) = -32$ **2.** $(1, 4, -3)$ or $\{(1, 4, -3)\}$
3. $(4, 5, 3)$ or $\{(4, 5, 3)\}$ **4.** $y = 3x^2 - 12x + 13$ or $f(x) = 3x^2 - 12x + 13$

Exercise Set 3.3

1. not a solution **3.** solution **5.** $(2, 3, 3)$ **7.** $(2, -1, 1)$ **9.** $(1, 2, 3)$ **11.** $(3, 1, 5)$ **13.** $(1, 0, -3)$ **15.** $(1, -5, -6)$ **17.** no solution or $\varnothing$

19. infinitely many solutions; dependent equations **21.** $\left(\dfrac{1}{2}, \dfrac{1}{3}, -1\right)$ **23.** $y = 2x^2 - x + 3$ **25.** $y = 2x^2 + x - 5$

27. 7, 4, and 5 **29.** $(4, 8, 6)$ **31.** $y = -\dfrac{3}{4}x^2 + 6x - 11$ **33.** $\left(\dfrac{8}{a}, -\dfrac{3}{b}, -\dfrac{5}{c}\right)$ **35. a.** $(0, 13.6), (70, 4.7), (102, 11.5)$

b. $0a + 0b + c = 13.6;\ 4900a + 70b + c = 4.7;\ 10,404a + 102b + c = 11.5$ **37. a.** $y = -16x^2 + 40x + 200$ **b.** 0; After 5 seconds, the ball hits the ground.

39. Carnegie: $100 billion; Vanderbilt: $96 billion; Gates: $60 billion **41.** $1200 at 8%; $2000 at 10%; $3500 at 12%

43. 200 $8 tickets; 150 $10 tickets; 50 $12 tickets **45.** 4 oz of food A; 0.5 oz of food B; 1 oz of food C

55. c **57.** $60°, 55°, 65°$ **59.** $f(x) = -\dfrac{3}{4}x + 3$ **60.** $-2x + y = 6$ **61.** $f(x) = -5$

Mid-Chapter 3 Check Point

1. $(-1, 2)$ **2.** $(1, -2)$ **3.** $(6, 10)$ **4.** $\{(x, y)|y = 4x - 5\}$ or $\{(x, y)|8x - 2y = 10\}$ **5.** $\left(\dfrac{11}{19}, \dfrac{7}{19}\right)$ **6.** $\varnothing$ **7.** $(-1, 2, -2)$

8. $(4, -2, 3)$ **9.** $(3, 2)$ **10.** $(-2, 1)$ **11. a.** $C(x) = 400,000 + 20x$ **b.** $R(x) = 100x$ **c.** $P(x) = 80x - 400,000$

d. $(5000, 500,000)$; The company will break even when it produces and sells 5000 PDAs. At this level, both revenue and cost are $500,000.

12. 6 roses and 14 carnations   **13.** $6300 at 5% and $8700 at 6%   **14.** 13% nitrogen: 20 gal; 18% nitrogen: 30 gal   **15.** rowing rate in still

water: 3 mph; current: 1.5 mph   **16.** $4500 at 2% and $3500 at 5%   **17.** $y = -x^2 + 2x + 3$ **18.** 8 nickels, 6 dimes, 12 quarters

Section 3.4

Check Point Exercises

1. $(4, -3, 1)$ or $\{(4, -3, 1)\}$ **2. a.** $\begin{bmatrix} 1 & 6 & -3 & | & 7 \\ 4 & 12 & -20 & | & 8 \\ -3 & -2 & 1 & | & -9 \end{bmatrix}$ **b.** $\begin{bmatrix} 1 & 3 & -5 & | & 2 \\ 1 & 6 & -3 & | & 7 \\ -3 & -2 & 1 & | & -9 \end{bmatrix}$ **c.** $\begin{bmatrix} 4 & 12 & -20 & | & 8 \\ 1 & 6 & -3 & | & 7 \\ 0 & 16 & -8 & | & 12 \end{bmatrix}$

3. $(-1, 2)$ or $\{(-1, 2)\}$ **4.** $(5, 2, 3)$ or $\{(5, 2, 3)\}$

Exercise Set 3.4

1. $x - 3y = 11$; $y = -3$; $(2, -3)$ **3.** $x - 3y = 1$; $y = -1$; $(-2, -1)$ **5.** $x - 4z = 5$; $y - 12z = 13$; $z = -\dfrac{1}{2}$; $\left(3, 7, -\dfrac{1}{2}\right)$

7. $x + \dfrac{1}{2}y + z = \dfrac{11}{2}$; $y + \dfrac{3}{2}z = 7$; $z = 4$; $(1, 1, 4)$ **9.** $\begin{bmatrix} 1 & -\frac{3}{2} & | & 5 \\ 2 & 2 & | & 5 \end{bmatrix}$ **11.** $\begin{bmatrix} 1 & -\frac{4}{3} & | & 2 \\ 3 & 5 & | & -2 \end{bmatrix}$ **13.** $\begin{bmatrix} 1 & -3 & | & 5 \\ 0 & 12 & | & -6 \end{bmatrix}$ **15.** $\begin{bmatrix} 1 & -\frac{3}{2} & | & \frac{7}{2} \\ 0 & \frac{17}{2} & | & -\frac{17}{2} \end{bmatrix}$

17. $\begin{bmatrix} 1 & -3 & 2 & | & 5 \\ 1 & 5 & -5 & | & 0 \\ 3 & 0 & 4 & | & 7 \end{bmatrix}$ **19.** $\begin{bmatrix} 1 & -3 & 2 & | & 0 \\ 0 & 10 & -7 & | & 7 \\ 2 & -2 & 1 & | & 3 \end{bmatrix}$ **21.** $\begin{bmatrix} 1 & 1 & -1 & | & 6 \\ 0 & -3 & 3 & | & -15 \\ 0 & -4 & 2 & | & -14 \end{bmatrix}$ **23.** $(4, 2)$ **25.** $(3, -3)$ **27.** $(2, -5)$ **29.** no solution or $\varnothing$

31. infinitely many solutions; dependent equations   **33.** $(1, -1, 2)$ **35.** $(3, -1, -1)$ **37.** $(1, 2, -1)$ **39.** $(1, 2, 3)$ **41.** no solution or $\varnothing$

43. infinitely many solutions; dependent equations   **45.** $(-1, 2, -2)$ **47.** $w - x + y + z = 3$; $x - 2y - z = 0$; $y + 6z = 17$; $z = 3$; $(2, 1, -1, 3)$

49. $\begin{bmatrix} 1 & -1 & 1 & 1 & | & 3 \\ 0 & 1 & -2 & -1 & | & 0 \\ 0 & 2 & 1 & 2 & | & 5 \\ 0 & 6 & -3 & -1 & | & -9 \end{bmatrix}$ **51.** $(1, 2, 3, -2)$ **53. a.** $y = 2x^2 + 22x + 320$ **b.** 2780 thousand, or 2,780,000, inmates

c. Answers will vary.   **55.** yes: 34%; no: 61%; not sure: 5%   **65.** d   **67.** $-6a + 13$ **68.** 15   **69.** $-\dfrac{x^{11}}{3y^{36}}$

Section 3.5

Check Point Exercises

1. a. -4 **b.** -17 **2.** $(4, -2)$ or $\{(4, -2)\}$ **3.** 80   **4.** $(2, -3, 4)$ or $\{(2, -3, 4)\}$

Exercise Set 3.5

1. 1   **3.** -29 **5.** 0   **7.** 33   **9.** $-\dfrac{7}{16}$ **11.** $(5, 2)$ **13.** $(2, -3)$ **15.** $(3, -1)$ **17.** $(4, 0)$ **19.** $(4, 2)$ **21.** $(7, 4)$

23. inconsistent; no solution or $\varnothing$ **25.** dependent equations; infinitely many solutions   **27.** 72   **29.** -75 **31.** 0   **33.** $(-5, -2, 7)$

35. $(2, -3, 4)$ **37.** $(3, -1, 2)$ **39.** $(2, 3, 1)$ **41.** -42 **43.** $2x - 4y = 8$; $3x + 5y = 10$ **45.** -11 **47.** 4   **49.** 28 sq units   **51.** yes

53. $\begin{vmatrix} x & y & 1 \\ 3 & -5 & 1 \\ -2 & 6 & 1 \end{vmatrix} = 0$; $y = -\dfrac{11}{5}x + \dfrac{8}{5}$ **65.** d   **69.** all real numbers or $\{x|x \text{ is a real number}\}$ or $\mathbb{R}$ **70.** $y = \dfrac{2x + 7}{3}$ **71.** 0 or $\{0\}$

Review Exercises

1. not a solution   **2.** solution   **3.** $(2, 3)$ **4.** $\{(x, y)|3x - 2y = 6\}$ or $\{(x, y)|6x - 4y = 12\}$ **5.** $(-5, -6)$ **6.** $\varnothing$ **7.** $(3, 4)$

8. $(23, -43)$ **9.** $(-4, 2)$ **10.** $\left(3, \dfrac{1}{2}\right)$ **11.** $\{(x, y)|y = 4 - x\}$ or $\{(x, y)|3x + 3y = 12\}$ **12.** $\left(3, \dfrac{8}{3}\right)$ **13.** $\varnothing$ **14.** TV: $350; stereo: $370

15. $2500 at 4%; $6500 at 7%   **16.** 10 ml of 34%; 90 ml of 4%   **17.** plane: 630 mph; wind: 90 mph   **18.** 12 ft by 5 ft   **19.** loss of $4500
20. (500, 42,500); When 500 calculators are produced and sold, both cost and revenue are $42,500.   **21.** $P(x) = 45x - 22,500$
22. a. $C(x) = 60,000 + 200x$ **b.** $R(x) = 450x$ **c.** (240, 108,000); When 240 desks are produced and sold, both cost and revenue are $108,000.
23. no   **24.** (0, 1, 2)   **25.** (2, 1, −1)   **26.** infinitely many solutions; dependent equations   **27.** $y = 3x^2 - 4x + 5$
28. per dog: $129; per horse: $97; per cat: $81   **29.** $2x + 3y = -10$; $y = -6$; (4, −6)   **30.** $x + y + 3z = 12$; $y - 2z = -4$; $z = 3$; (1, 2, 3)

31. $\begin{bmatrix} 1 & -8 & | & 3 \\ 0 & 1 & | & -2 \end{bmatrix}$ **32.** $\begin{bmatrix} 1 & -3 & | & 1 \\ 0 & 7 & | & -7 \end{bmatrix}$ **33.** $\begin{bmatrix} 1 & -1 & \frac{1}{2} & | & -\frac{1}{2} \\ 1 & 2 & -1 & | & 2 \\ 6 & 4 & 3 & | & 5 \end{bmatrix}$ **34.** $\begin{bmatrix} 1 & 2 & 2 & | & 2 \\ 0 & 1 & -1 & | & 2 \\ 0 & 0 & 9 & | & -9 \end{bmatrix}$ **35.** (−5, 3)   **36.** inconsistent; no solution or Ø

37. (1, 3, −4)   **38.** (−2, −1, 0)   **39.** 17   **40.** 4   **41.** −86   **42.** −236   **43.** $\left(\frac{7}{4}, -\frac{25}{8}\right)$ **44.** (2, −7)   **45.** (23, −12, 3)   **46.** (−3, 2, 1)

47. $y = \frac{5}{8}x^2 - 50x + 1150$; 30-year-old drivers are involved in 212.5 accidents daily and 50-year-old drivers are involved in 212.5 accidents daily.

Chapter 3 Test
1. (2, 4)   **2.** (6, −5)   **3.** (1, −3)   **4.** $\{(x, y) | 4x = 2y + 6\}$ or $\{(x, y) | y = 2x - 3\}$ **5.** one-bedroom: 15 units; two-bedroom: 35 units
6. $2000 at 6% and $7000 at 7%   **7.** 6% solution: 12 oz; 9% solution: 24 oz   **8.** boat: 14 mph; current: 2 mph   **9.** $C(x) = 360,000 + 850x$
10. $R(x) = 1150x$ **11.** (1200, 1,380,000); When 1200 computers are produced and sold, both cost and revenue are $1,380,000.

12. $P(x) = 85x - 350,000$ **13.** (1, 3, 2)   **14.** $\begin{bmatrix} 1 & 0 & -4 & | & 5 \\ 0 & -1 & 26 & | & -20 \\ 2 & -1 & 4 & | & -3 \end{bmatrix}$ **15.** (4, −2)   **16.** (−1, 2, 2)   **17.** 17   **18.** −10

19. (−1, −6)   **20.** (4, −3, 3)

Cumulative Review Exercises (Chapters 1–3)
1. 1   **2.** $15x - 7$ **3.** −6   **4.** −15   **5.** no solution or Ø   **6.** $2000   **7.** $-\dfrac{x^8}{4y^{30}}$

8. $-4a - 3$ **9.** $\{x | x$ is a real number and $x \neq -3\}$ **10.** $x^2 - 3x - 1$; −1
11. $f(x) = -\frac{2}{3}x + 2$

12. $2x - y = 6$

13. $y - 4 = -3(x - 2)$ or $y + 2 = -3(x - 4)$; $y = -3x + 10$ or $f(x) = -3x + 10$
14. $y - 0 = -3(x + 1)$; $y = -3x - 3$ or $f(x) = -3x - 3$ **15.** $\left(7, \frac{1}{3}\right)$
16. (3, 2, 4)   **17.** pad: $0.80; pen: $0.20   **18.** 23   **19.** (3, −2, 1)   **20.** (−3, 2)

CHAPTER 4

Section 4.1

Check Point Exercises

1. a.

    b. (−6 to 2 number line)   **c.** (−4 to 2 number line)

2. a. $\{x | -2 \leq x < 5\}$
(number line −5 to 5)
    b. $\{x | 1 \leq x \leq 3.5\}$
(number line −5 to 5)
    c. $\{x | x < -1\}$
(number line −5 to 5)

3. $\{x | x > -5\}$ or $(-5, \infty)$
(number line −8 to 2)
    4. $\{x | x < 4\}$ or $(-\infty, 4)$
(number line −3 to 6)
    5. $\{x | x \geq 13\}$ or $[13, \infty)$
(number line 6 to 16)

6. a. $\{x | x$ is a real number$\}$ or $\mathbb{R}$ or $(-\infty, \infty)$ **b.** Ø   **7.** more than 720 miles

Exercise Set 4.1

1.

    3. (number line −7 to 1)   **5.** (number line −6 to 4)

7. (number line −3 to 6)   **9.** (number line −4 to 6)   **11.** (number line −3 to 6)

13. $\{x | 1 < x \le 6\}$

15. $\{x | -5 \le x < 2\}$

17. $\{x | -3 \le x \le 1\}$

19. $\{x | x > 2\}$

21. $\{x | x \ge -3\}$

23. $\{x | x < 3\}$

25. $\{x | x < 5.5\}$

27. $\{x | x < 3\}$ or $(-\infty, 3)$

29. $\{x | x \ge 7\}$ or $[7, \infty)$

31. $\{x | x \le -4\}$ or $(-\infty, -4]$

33. $\left\{ x \,\middle|\, x \le -\dfrac{2}{5} \right\}$ or $\left(-\infty, -\dfrac{2}{5} \right]$

35. $\{x | x \ge 0\}$ or $[0, \infty)$

37. $\{x | x < 1\}$ or $(-\infty, 1)$

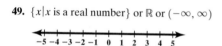

39. $\{x | x \ge 6\}$ or $[6, \infty)$

41. $\{x | x \ge -6\}$ or $[-6, \infty)$

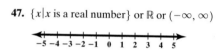

43. $\{x | x < -6\}$ or $(-\infty, -6)$

45. $\{x | x \ge 13\}$ or $[13, \infty)$

47. $\{x | x \text{ is a real number}\}$ or $\mathbb{R}$ or $(-\infty, \infty)$

49. $\{x | x \text{ is a real number}\}$ or $\mathbb{R}$ or $(-\infty, \infty)$

51. $\varnothing$

53. $\{y | y > -6\}$ or $(-6, \infty)$

55. $\{x | x \ge -1\}$ or $[-1, \infty)$

57. $\{y | y < -2\}$ or $(-\infty, -2)$

59. $\{x | x < 5\}$ or $(-\infty, 5)$

61. $\left\{ x \,\middle|\, x \ge -\dfrac{4}{7} \right\}$ or $\left[-\dfrac{4}{7}, \infty \right)$

63. $\{x | x < 2\}$ or $(-\infty, 2)$

65. $x < \dfrac{c - b}{a}$ **67.** $\{x | x \le -3\}$ or $(-\infty, -3]$ **69.** $\{x | x > -1.4\}$ or $(-1.4, \infty)$ **71.** $(0, 4)$

73. intimacy $\ge$ passion or passion $\le$ intimacy   **75.** commitment $>$ passion or passion $<$ commitment   **77.** 9; after 3 years

79. voting years after 2006   **81.** $\{t | t > 175\}$ or $(175, \infty)$; The women's speed skating times will be less than the men's after the year 2075.

83. more than 100 miles per day   **85.** greater than \$32,000   **87.** more than 6250 tapes   **89.** 40 bags or fewer

99.

$\{x | x < -3\}$ or $(-\infty, -3)$

101. $\varnothing$

103. a. plan A: $4 + 0.10x$; plan B: $2 + 0.15x$

b.

c–d. more than 40 checks per month

105. Since $x > y$, $y - x < 0$. Thus, when both sides were multiplied by $y - x$, the sense of the inequality should have been changed.

107. 29   **108.** $(-1, -1, 2)$ **109.** $\dfrac{x^9}{8y^{15}}$

Section 4.2

Check Point Exercises

1. $\{3, 7\}$ **2.** $\{x | x < 1\}$ or $(-\infty, 1)$ **3.** $\varnothing$ **4.** $\{x | -1 \le x < 4\}$ or $[-1, 4)$;

5. $\{3, 4, 5, 6, 7, 8, 9\}$ **6.** $\{x | x \le 1 \text{ or } x > 3\}$ or $(-\infty, 1] \cup (3, \infty)$ **7.** $\{x | x \text{ is a real number}\}$ or $\mathbb{R}$ or $(-\infty, \infty)$

Exercise Set 4.2

1. $\{2, 4\}$ **3.** $\varnothing$ **5.** $\varnothing$

7. $\{x|x > 6\}; (6, \infty)$

9. $\{x|x \le 1\}; (-\infty, 1]$

11. $\{x|-1 \le x < 2\}; [-1, 2)$

13. $\varnothing$

15. $\{x|-6 < x < -4\}; (-6, -4)$

17. $\{x|-3 < x \le 6\}; (-3, 6]$

19. $\{x|2 < x < 5\}; (2, 5)$

21. $\varnothing$

23. $\{x|0 \le x < 2\}; [0, 2)$

25. $\{x|3 < x < 5\}; (3, 5)$

27. $\{x|-1 \le x < 3\}; [-1, 3)$

29. $\{x|-5 < x \le -2\}; (-5, -2]$

31. $\{x|3 \le x < 6\}; [3, 6)$

33. $\{1, 2, 3, 4, 5\}$

35. $\{1, 2, 3, 4, 5, 6, 7, 8, 10\}$

37. $\{a, e, i, o, u\}$

39. $\{x|x > 3\}; (3, \infty)$

41. $\{x|x \le 5\}; (-\infty, 5]$

43. $\{x|x \text{ is a real number}\}; (-\infty, \infty)$

45. $\{x|x < -1 \text{ or } x \ge 2\}; (-\infty, -1) \cup [2, \infty)$

47. $\{x|x < -3 \text{ or } x > 4\}; (-\infty, -3) \cup (4, \infty)$

49. $\{x|x \le 1 \text{ or } x \ge 3\}; (-\infty, 1] \cup [3, \infty)$

51. $\{x|x \text{ is a real number}\}; (-\infty, \infty)$

53. $\{x|x < 2\}; (-\infty, 2)$

55. $\{x|x > 4\} \text{ or } (4, \infty)$ **57.** $\{x|x < 0 \text{ or } x > 6\} \text{ or } (-\infty, 0) \cup (6, \infty)$

59. $\dfrac{b-c}{a} < x < \dfrac{b+c}{a}$ **61.** $\{x|-1 \le x \le 3\} \text{ or } [-1, 3]$ **63.** $\{x|-1 < x < 3\} \text{ or } (-1, 3)$

65. $\{x|-1 \le x < 2\} \text{ or } [-1, 2)$ **67.** $\{-3, -2, -1\}$ **69.** spatial-temporal toys, sports equipment, and toy cars and trucks **71.** dollhouses, spatial-temporal toys, sports equipment, and toy cars and trucks **73.** none of the toys **75.** between 80 and 110 minutes, inclusive

77. $[76, 126)$; If the highest grade is 100, then $[76, 100]$.   **79.** more than 3 and less than 15 crossings per 3-month period

89. $\{x|-6 < x < 2\}$ or $(-6, 2)$

91. $\{x|-3 \le x \le 2\}$ or $[-3, 2]$

93. $(-1, 5]$ **95.** $[-1, \infty)$ **97.** $(-1, 4]$ **99.** $-x^2 + 5x - 9; -15$ **100.** $f(x) = -\dfrac{1}{2}x + 4$ **101.** $17 - 2x$

Section 4.3

Check Point Exercises

1. -2 and 3, or $\{-2, 3\}$ **2.** $\dfrac{4}{3}$ and 10, or $\left\{\dfrac{4}{3}, 10\right\}$

3. $\{x|-3 < x < 7\}$ or $(-3, 7)$

[number line: $-3\ -2\ -1\ 0\ 1\ 2\ 3\ 4\ 5\ 6\ 7$]

4. $\{x|x \le 1 \text{ or } x \ge 4\}$ or $(-\infty, 1] \cup [4, \infty)$

[number line: $-5\ -4\ -3\ -2\ -1\ 0\ 1\ 2\ 3\ 4\ 5$]

5. $\{x|8.1 \le x \le 13.9\}$ or $[8.1, 13.9]$; The percentage of children in the population who think that not being able to do everything they want is a bad thing is between a low of 8.1% and a high of 13.9%.

Exercise Set 4.3

1. $\{-8, 8\}$ **3.** $\{-5, 9\}$ **5.** $\{-3, 4\}$ **7.** $\{-1, 2\}$ **9.** $\emptyset$ **11.** $\{-3\}$ **13.** $\{-11, -1\}$ **15.** $\{-3, 4\}$ **17.** $\left\{-\dfrac{13}{3}, 5\right\}$

19. $\left\{-\dfrac{2}{5}, \dfrac{2}{5}\right\}$ **21.** $\emptyset$ **23.** $\emptyset$ **25.** $\left\{\dfrac{1}{2}\right\}$ **27.** $\left\{\dfrac{3}{4}, 5\right\}$ **29.** $\left\{\dfrac{5}{3}, 3\right\}$ **31.** $\{0\}$ **33.** $\{4\}$ **35.** $\{4\}$ **37.** $\{-1, 15\}$

39. $\{x|-3 < x < 3\}$ or $(-3, 3)$

[number line: $-5\ -4\ -3\ -2\ -1\ 0\ 1\ 2\ 3\ 4\ 5$]

41. $\{x|1 < x < 3\}$ or $(1, 3)$

[number line: $-5\ -4\ -3\ -2\ -1\ 0\ 1\ 2\ 3\ 4\ 5$]

43. $\{x|-3 \le x \le -1\}$ or $[-3, -1]$

[number line: $-5\ -4\ -3\ -2\ -1\ 0\ 1\ 2\ 3\ 4\ 5$]

45. $\{x|-1 < x < 7\}$ or $(-1, 7)$

[number line: $-2\ -1\ 0\ 1\ 2\ 3\ 4\ 5\ 6\ 7\ 8$]

47. $\{x|x < -3 \text{ or } x > 3\}$ or $(-\infty, -3) \cup (3, \infty)$

[number line: $-5\ -4\ -3\ -2\ -1\ 0\ 1\ 2\ 3\ 4\ 5$]

49. $\{x|x < -4 \text{ or } x > -2\}$ or $(-\infty, -4) \cup (-2, \infty)$

[number line: $-5\ -4\ -3\ -2\ -1\ 0\ 1\ 2\ 3\ 4\ 5$]

51. $\{x|x \le 2 \text{ or } x \ge 6\}$ or $(-\infty, 2] \cup [6, \infty)$

[number line: $-10\ -8\ -6\ -4\ -2\ 0\ 2\ 4\ 6\ 8\ 10$]

53. $\left\{x\middle|x < \dfrac{1}{3} \text{ or } x > 5\right\}$ or $\left(-\infty, \dfrac{1}{3}\right) \cup (5, \infty)$

[number line: $-2\ -1\ 0\ 1\ 2\ 3\ 4\ 5\ 6\ 7\ 8$]

55. $\{x|-5 \le x \le 3\}$ or $[-5, 3]$

[number line: $-6\ -5\ -4\ -3\ -2\ -1\ 0\ 1\ 2\ 3\ 4$]

57. $\{x|-6 < x < 0\}$ or $(-6, 0)$

[number line: $-8\ -7\ -6\ -5\ -4\ -3\ -2\ -1\ 0\ 1\ 2$]

59. $\{x|x \le -5 \text{ or } x \ge 3\}$ or $(-\infty, -5] \cup [3, \infty)$

[number line: $-5\ -4\ -3\ -2\ -1\ 0\ 1\ 2\ 3\ 4\ 5$]

61. $\{x|x < -3 \text{ or } x > 12\}$ or $(-\infty, -3) \cup (12, \infty)$

[number line: $-4\ -2\ 0\ 2\ 4\ 6\ 8\ 10\ 12\ 14\ 16$]

63. $\emptyset$

65. $\{x|x \text{ is a real number}\}$ or $\mathbb{R}$ or $(-\infty, \infty)$

[number line: $-5\ -4\ -3\ -2\ -1\ 0\ 1\ 2\ 3\ 4\ 5$]

67. $\{x|-9 \le x \le 5\}$ or $[-9, 5]$

[number line: $-10\ -8\ -6\ -4\ -2\ 0\ 2\ 4\ 6\ 8\ 10$]

69. $\{x|x < 1 \text{ or } x > 2\}$ or $(-\infty, 1) \cup (2, \infty)$

[number line: $-5\ -4\ -3\ -2\ -1\ 0\ 1\ 2\ 3\ 4\ 5$]

71. $-\dfrac{3}{2}$ and 4 **73.** -7 and -2 **75.** $\left\{x\middle|-\dfrac{7}{3} \le x \le 1\right\}$ or $\left[-\dfrac{7}{3}, 1\right]$ **77.** $\{x|x < -1 \text{ or } x > 4\}$ or $(-\infty, -1) \cup (4, \infty)$

79. $\left(-\infty, -\dfrac{1}{3}\right] \cup [3, \infty)$ **81.** $\left\{x\middle|\dfrac{-c - b}{a} < x < \dfrac{c - b}{a}\right\}$ **83.** 3 and 5 **85.** $\{x|-2 \le x \le 1\}$ or $[-2, 1]$

87. $\{x | 58.6 \leq x \leq 61.8\}$ or $[58.6, 61.8]$; The percentage of the U.S. population that watched M*A*S*H is between a low of 58.6% and a high of 61.8%.; 1.6% **89.** $\{T | 50 \leq T \leq 64\}$ or $[50, 64]$; The monthly average temperature for San Francisco, CA is between a low of 50°F and a high of 64°F.
91. $\{x | 8.59 \leq x \leq 8.61\}$ or $[8.59, 8.61]$; A machine part that is supposed to be 8.6 centimeters is acceptable between a low of 8.59 and a high of 8.61 centimeters.   **93.** If the number of outcomes that result in heads is 41 or less or 59 or more, then the coin is unfair.
103. -8 and 0 **105.** $\{x | -4 < x < 1\}$ or $(-4, 1)$ **107.** $\varnothing$ **109.** $\{x | x < 2 \text{ or } x > 6\}$ or $(-\infty, 2) \cup (6, \infty)$

113. a. $|x - 4| < 3$ **b.** $|x - 4| \geq 3$ **115.** 1
116.   **117.**   **118.**

Mid-Chapter 4 Check Point

1. $\{x | x \leq -4\}$; $(-\infty, -4]$ **2.** $\{x | 3 \leq x < 5\}$; $[3, 5)$ **3.** $\left\{\dfrac{1}{2}, 3\right\}$ **4.** $\{x | x < -1\}$; $(-\infty, -1)$

5. $\{x | x < -9 \text{ or } x > -5\}$; $(-\infty, -9) \cup (-5, \infty)$ **6.** $\left\{x \left| -\dfrac{2}{3} \leq x \leq 2\right.\right\}$; $\left[-\dfrac{2}{3}, 2\right]$ **7.** $\left\{\dfrac{1}{2}, \dfrac{13}{4}\right\}$ **8.** $\{x | -4 < x \leq -2\}$; $(-4, -2]$

9. $\{x | x \text{ is real number}\}$; $(-\infty, \infty)$ **10.** $\{x | x \leq -3\}$; $(-\infty, -3]$ **11.** $\{x | x < -3\}$; $(-\infty, -3)$

12. $\left\{x \left| x < -1 \text{ or } x > -\dfrac{1}{5}\right.\right\}$; $(-\infty, -1) \cup \left(-\dfrac{1}{5}, \infty\right)$ **13.** $\{x | x \leq -10 \text{ or } x \geq 2\}$; $(-\infty, -10] \cup [2, \infty)$ **14.** $\varnothing$

15. $\left\{x \left| x < -\dfrac{5}{3}\right.\right\}$; $\left(-\infty, -\dfrac{5}{3}\right)$ **16.** $\{x | x > 4\}$; $(4, \infty)$ **17.** $\{-2, 7\}$ **18.** $\varnothing$ **19.** no more than 80 miles per day   **20.** $[49\%, 99\%)$
21. at least \$120,000   **22.** at least 750,000 discs each month

Section 4.4

Check Point Exercises

1.   **2.**   **3. a.**   **b.**

4.   **5.**   **6.** $B = (60, 20)$; Using $T = 60$ and $P = 20$, each of the three inequalities for grasslands is true: $60 \geq 35$, true; $5(60) - 7(20) \geq 70$, true; $3(60) - 35(20) \leq -140$, true.

Exercise Set 4.4

1. $x + y \geq 3$ **3.** $x - y < 5$ **5.** $x + 2y > 4$ **7.** $3x - y \leq 6$ **9.** $\dfrac{x}{2} + \dfrac{y}{3} < 1$

11.

13.

15.

17.

19.

21.

23.

25.

27.

29.

31.

33.

35.

37. $\emptyset$ **39.**

41.

43.

45.

47. $y \geq -2x + 4$

49. $x + y \leq 4$ and $3x + y \leq 6$

51. $-2 \leq x \leq 2; -3 \leq y \leq 3$

53.

55. no solution   **57.** infinitely many solutions

59. a. $A = (20, 150)$; A 20-year-old with a heart rate of 150 beats per minute is within the target range.
  b. $10 \leq 20 \leq 70$, true; $150 \geq 0.7(220 - 20)$, true; $150 \leq 0.8(220 - 20)$, true
61. $10 \leq a \leq 70; H \geq 0.6(220 - a); H \leq 0.7(220 - a)$

63. a. $y \geq 0; x + y \geq 5; x \geq 1; 200x + 100y \leq 700$

b.

c. 2 nights   **75.**   **77.**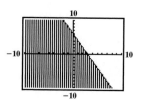

83. $x \geq -2; y > -1$ **85.** Answers will vary.   **87.** $(3, 1)$ or $\{(3, 1)\}$ **88.** $(2, 4)$ or $\{(2, 4)\}$ **89.** 165

Section 4.5

Check Point Exercises

1. $z = 25x + 55y$ **2.** $x + y \leq 80$ **3.** $30 \leq x \leq 80, 10 \leq y \leq 30; z = 25x + 55y, x + y \leq 80, 30 \leq x \leq 80, 10 \leq y \leq 30$
4. 50 bookshelves and 30 desks; $2900   **5.** 30

Exercise Set 4.5

1. $(1, 2)$: 17; $(2, 10)$: 70; $(7, 5)$: 65; $(8, 3)$: 58; maximum: 70; minimum: 17   **3.** $(0, 0)$: 0; $(0, 8)$: 400; $(4, 9)$: 610; $(8, 0)$: 320; maximum: 610; minimum: 0

5. a.

b. $(0, 4)$: 8; $(0, 8)$: 16; $(4, 0)$: 12

c. maximum: 16; at $(0, 8)$

7. a.

b. $(0, 3)$: 3; $(0, 4)$: 4; $(6, 0)$: 24; $(3, 0)$: 12

c. maximum: 24; at $(6, 0)$

9. a.

b. $(1, 2)$: -1; $(1, 4)$: -5; $(5, 8)$: -1; $(5, 2)$: 11

c. maximum: 11; at $(5, 2)$

11. a.

b. $(0, 2)$: 4; $(0, 4)$: 8; $\left(\dfrac{12}{5}, \dfrac{12}{5}\right)$: $\dfrac{72}{5} = 14.4$; $(4, 0)$: 16; $(2, 0)$: 8

c. maximum: 16; at $(4, 0)$

13. a.

b. $(0, 0)$: 0; $(0, 6)$: 72; $(3, 4)$: 78; $(5, 0)$: 50

c. maximum: 78; at $(3, 4)$

15. a. $z = 125x + 200y$

b. $x \le 450$; $y \le 200$; $600x + 900y \le 360{,}000$

c.

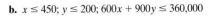

d. 0; 40,000; 77,500; 76,250; 56,250

e. 300; 200; 77,500

17. 40 of model A and 0 of model B   **19.** 300 boxes of food and 200 boxes of clothing   **21.** 100 parents and 50 students   **23.** 10 Boeing 727s and 42 Falcon 20s   **29.** $5000 in stocks and $5000 in bonds   **31.** $54x^7y^{15}$ **32.** $L = \dfrac{12P + W}{2}$ **33.** 10

Review Exercises

1.

2.

3.

4. $\{x|-2 < x \le 3\}$

5. $\{x|-1.5 \le x \le 2\}$

6. $\{x|x > -1\}$

7. $\{x|x \ge -2\}$; $[-2, \infty)$

8. $\left\{x \,\middle|\, x \ge \dfrac{3}{5}\right\}$; $\left[\dfrac{3}{5}, \infty\right)$

9. $\left\{x \,\middle|\, x < -\dfrac{21}{2}\right\}$; $\left(-\infty, -\dfrac{21}{2}\right)$

10. $\{x|x > -3\}$; $(-3, \infty)$

11. $\{x|x \le -2\}$; $(-\infty, -2]$

12. $\varnothing$ **13.** more than 50 checks   **14.** more than $13,500 in sales   **15.** $\{a, c\}$ **16.** $\{a\}$ **17.** $\{a, b, c, d, e\}$ **18.** $\{a, b, c, d, f, g\}$

19. $\{x|x \le 3\}$; $(-\infty, 3]$

20. $\{x|x < 6\}$; $(-\infty, 6)$

21. $\{x|6 < x < 8\}$; $(6, 8)$

22. $\{x|x \le 1\}$; $(-\infty, 1]$

23. $\varnothing$ **24.** $\{x|x < 1 \text{ or } x > 2\}$; $(-\infty, 1) \cup (2, \infty)$ **25.** $\{x|x \le -4 \text{ or } x > 2\}$; $(-\infty, -4] \cup (2, \infty)$

26. $\{x|x < -2\}$; $(-\infty, -2)$

27. $\{x|x \text{ is a real number}\}$; $(-\infty, \infty)$

28. $\{x|-5 < x \le 2\}$; $(-5, 2]$

29. $\left\{x \mid -\dfrac{3}{4} \le x \le 1\right\}; \left[-\dfrac{3}{4}, 1\right]$

30. $[81, 131)$; If the highest grade is 100, then $[81, 100]$.

31. $\{-4, 3\}$ **32.** $\varnothing$ **33.** $\left\{-\dfrac{11}{2}, \dfrac{23}{2}\right\}$ **34.** $\left\{-4, -\dfrac{6}{11}\right\}$

35. $\{x \mid -9 \le x \le 6\}; [-9, 6]$

36. $\{x \mid x < -6 \text{ or } x > 0\}; (-\infty, -6) \cup (0, \infty)$

37. $\{x \mid -3 < x < -2\}; (-3, -2)$

38. $\varnothing$ **39.** Approximately 90% of the population sleeps between 5.5 hours and 7.5 hours daily, inclusive.

40. $3x - 4y > 12$

41. $x - 3y \le 6$

42. $y \le -\dfrac{1}{2}x + 2$

43. $y > \dfrac{3}{5}x$

44. $x \le 2$

45. $y > -3$

46. $2x - y \le 4$
$x + y \ge 5$

47. $y < -x + 4$
$y > x - 4$

48. $-3 \le x < 5$

49. $-2 < y \le 6$

50. $x \ge 3$
$y \le 0$

51. $2x - y > -4$
$x \ge 0$

52. $x + y \le 6$
$y \ge 2x - 3$

53. $3x + 2y \ge 4$
$x - y \le 3$
$x \ge 0, y \ge 0$

54. $\varnothing$

55. $\left(\dfrac{1}{2}, \dfrac{1}{2}\right): \dfrac{5}{2}$;
$(2, 2): 10; (4, 0): 8; (1, 0): 2$;

maximum: 10; minimum: 2

56.

maximum: 24

57.

maximum: 33

58.

maximum: 44

59. a. $z = 500x + 350y$ **b.** $x + y \le 200; x \ge 10; y \ge 80$

c.

d. $(10, 80): 33,000; (10, 190): 71,500; (120, 80): 88,000$

e. $120; 80; 88,000$ **60.** 480 of model A and 240 of model B

Chapter 4 Test

1. $\{x \mid -3 \le x < 2\}$

2. $\{x \mid x \le -1\}$

3. $\{x \mid x \le 12\}; (-\infty, 12]$

4. $\left\{x \mid x \ge \dfrac{21}{8}\right\}; \left[\dfrac{21}{8}, \infty\right)$

5. more than 200 calls   **6.** $\{4, 6\}$

7. $\{2, 4, 6, 8, 10, 12, 14\}$

8. $\{x \mid -2 < x < -1\}; (-2, -1)$

9. $\{x | x \geq -2\}; [-2, \infty)$

10. $\{x | x < 4\}; (-\infty, 4)$

11. $\{x | x \leq -4 \text{ or } x > 2\}; (-\infty, -4] \cup (2, \infty)$

12. $\left\{x \middle| -7 \leq x < \dfrac{13}{2}\right\}; \left[-7, \dfrac{13}{2}\right)$

13. $\left\{-2, \dfrac{4}{5}\right\}$

14. $\left\{-\dfrac{8}{5}, 7\right\}$

15. $\{x | -3 < x < 4\}; (-3, 4)$

16. $\{x | x \leq -1 \text{ or } x \geq 4\}; (-\infty, -1] \cup [4, \infty)$

17. $\{b | b < 90.6 \text{ or } b > 106.6\}$ or $(-\infty, 90.6) \cup (106.6, \infty)$; Hypothermia: Body temperature below 90.6°F; Hyperthermia: Body temperature above 106.6°F

18. $3x - 2y < 6$

19. $y \geq \dfrac{1}{2}x - 1$

20. $y \leq -1$

21. $x + y \geq 2$
$x - y \geq 4$

22. $3x + y \leq 9$
$2x + 3y \geq 6$
$x \geq 0$
$y \geq 0$

23. $-2 < x \leq 4$

24. maximum: 26

25. 50 regular and 100 deluxe; $35,000

Cumulative Review Exercises (Chapters 1–4)

1. -1
2. 8
3. $-\dfrac{2y^7}{3x^5}$
4. $22; 4a^2 - 6a + 4$
5. $2x^2 + x + 2; 12$
6. $f(x) = -\dfrac{1}{2}x + 4$

7. $f(x) = 2x + 1$

8. $y > 2x$

9. $2x - y \geq 6$

10. $f(x) = -1$

11. $(-4, 2, -1)$
12. $(-1, 2)$
13. -17
14. 46 rooms with kitchen facilities and 14 without kitchen facilities
15. a. and b. are functions.

16. $\{x | x \geq -7\}; [-7, \infty)$

17. $\{x | x < -6\}; (-\infty, -6)$

18. $\{x | x \leq 3 \text{ or } x \geq 5\}; (-\infty, 3] \cup [5, \infty)$

19. $\{x | -10 \leq x \leq 7\}; [-10, 7]$

20. $\left\{x \middle| x < \dfrac{1}{3} \text{ or } x > 5\right\}; \left(-\infty, \dfrac{1}{3}\right) \cup (5, \infty)$

CHAPTER 5

Section 5.1

Check Point Exercises

1.

Term	Coefficient	Degree
$8x^4y^5$	8	9
$-7x^3y^2$	-7	5
$-x^2y$	-1	3
$-5x$	-5	1
11	11	0

The degree of the polynomial is 9, the leading term is $8x^4y^5$, and the leading coefficient is 8.

2. 16
3. The graph rises to the left and to the right.
4. This would not be appropriate over long time periods. Since the graph falls to the right, at some point the ratio would be negative, which is not possible.
5. The graph does not show the end behavior of the function. The graph should fall to the left.
6. $-3x^3 + 10x^2 - 10$
7. $9xy^3 + 3xy^2 - 15y - 9$
8. $10x^3 - 2x^2 + 8x - 10$
9. $13x^2y^5 + 2xy^3 - 10$

Exercise Set 5.1

1.

Term	Coefficient	Degree
$-x^4$	-1	4
x^2	1	2

The degree of the polynomial is 4, the leading term is $-x^4$, and the leading coefficient is -1.

3.

Term	Coefficient	Degree
$5x^3$	5	3
$7x^2$	7	2
$-x$	-1	1
9	9	0

The degree of the polynomial is 3, the leading term is $5x^3$, and the leading coefficient is 5.

5.

Term	Coefficient	Degree
$3x^2$	3	2
$-7x^4$	-7	4
$-x$	-1	1
6	6	0

The degree of the polynomial is 4, the leading term is $-7x^4$, and the leading coefficient is -7.

7.

Term	Coefficient	Degree
x^3y^2	1	5
$-5x^2y^7$	-5	9
$6y^2$	6	2
-3	-3	0

The degree of the polynomial is 9, the leading term is $-5x^2y^7$, and the leading coefficient is -5.

9.

Term	Coefficient	Degree
x^5	1	5
$3x^2y^4$	3	6
$7xy$	7	2
$9x$	9	1
-2	-2	0

The degree of the polynomial is 6, the leading term is $3x^2y^4$, and the leading coefficient is 3.

11. 0 **13.** 12 **15.** 56
17. -29 **19.** -1
21–23. Graph #23 is not that of a polynomial function.
25. falls to the left and falls to the right; graph (b)
27. rises to the left and rises to the right; graph (a)
29. $11x^3 + 7x^2 - 12x - 4$

31. $-\frac{2}{5}x^4 + x^3 + \frac{3}{8}x^2$ **33.** $9x^2y - 6xy$ **35.** $2x^2y + 15xy + 15$ **37.** $-9x^4y^2 - 6x^2y^2 - 5x^2y + 2xy$ **39.** $5x^{2n} - 2x^n - 6$
41. $12x^3 + 4x^2 + 12x - 14$ **43.** $22y^5 + 9y^4 + 7y^3 - 13y^2 + 3y - 5$ **45.** $-5x^3 + 8xy - 9y^2$ **47.** $x^4y^2 + 8x^3y + y - 6x$
49. $y^{2n} + 2y^n - 3$ **51.** $8a^2b^4 + 3ab^2 + 8ab$ **53.** $5x^3 + 3x^2y - xy^2 - 4y^3$ **55.** $5x^4 - x^3 + 5x^2 - 5x + 2$
57. $-10x^2y^2 + 4x^2 + 3$ **59.** $-4x^3 - x^2 + 4x + 8; 7$ **61.** $-8x^2 - 2x - 1; -29$ **63.** $-9x^3 - x^2 + 13x + 20$
65. $f(10) = 462{,}446$; In 2000, the cumulative number of AIDS deaths in the U.S. was 462,446. **67.** $(10, 462{,}446)$
69. $f(10) = 462{,}446; g(10) = 465{,}325; f$ **71.** falls to the right; no; Cumulative number of deaths cannot decrease.
73. falls to the right; The viral particles eventually decrease as the days increase. **75.** falls to the right; Eventually the elk population will be extinct.
89. Answers will vary; an example is $f(x) = -x^2 - x + 1$. **91.** Answers will vary; an example is $f(x) = x^3 + x + 1$.

93.

95.

97.

99. $3x^{2n} - x^n + 4$ **101.** $9x^2 - 3x + 5$ **102.** $\frac{2}{3}$ **103.**

$2x - 3y < -6$
104. $y - 5 = 3(x + 2); y = 3x + 11$ or $f(x) = 3x + 11$

Section 5.2

Check Point Exercises

1. a. $-18x^7y^{11}$ **b.** $30x^{10}y^6z^8$ **2. a.** $12x^9 - 18x^6 + 24x^4$ **b.** $10x^5y^9 - 8x^7y^7 - 10x^4y^3$ **3.** $6x^3 - 2x^2 - x + 2$
4. $12x^2y^6 - 8x^2y^4 + 6xy^5 + 2y^2$ **5. a.** $x^2 + 8x + 15$ **b.** $14x^2 + xy - 4y^2$ **c.** $4x^6 - 12x^4 - 5x^3 + 15x$ **6. a.** $x^2 + 16x + 64$
b. $16x^2 + 40xy + 25y^2$ **7. a.** $x^2 - 10x + 25$ **b.** $4x^2 - 24xy^4 + 36y^8$ **8. a.** $x^2 - 9$ **b.** $25x^2 - 49y^2$ **c.** $25a^2b^4 - 16a^2$
9. a. $9x^2 + 12x + 4 - 25y^2$ **b.** $4x^2 + 4xy + y^2 + 12x + 6y + 9$ **10. a.** $x^2 - 10x + 21$ **b.** 5 **11. a.** $a^2 + a - 2$ **b.** $2ah + h^2 - 5h$

Exercise Set 5.2

1. $15x^6$ **3.** $15x^3y^{11}$ **5.** $-6x^2y^9z^9$ **7.** $2x^{3n}y^{n-2}$ **9.** $12x^3 + 8x^2$ **11.** $2y^3 - 10y^2$ **13.** $10x^8 - 20x^5 + 45x^3$ **15.** $28x^2y + 12xy^2$
17. $18a^3b^5 + 15a^2b^3$ **19.** $-12x^6y^3 + 28x^3y^4 - 24x^2y$ **21.** $-12x^{3n} + 20x^{2n} - 2x^{n+1}$ **23.** $x^3 - x^2 - x - 15$ **25.** $x^3 - 1$
27. $a^3 - b^3$ **29.** $x^4 + 5x^3 + x^2 - 11x + 4$ **31.** $x^3 - 4x^2y + 4xy^2 - y^3$ **33.** $x^3y^3 + 8$ **35.** $x^2 + 11x + 28$ **37.** $y^2 - y - 30$
39. $10x^2 + 11x + 3$ **41.** $6y^2 - 11y + 4$ **43.** $15x^2 - 22x + 8$ **45.** $2x^2 + xy - 21y^2$ **47.** $14x^2y^2 - 19xy - 3$
49. $x^3 - 4x^2 - 5x + 20$ **51.** $8x^5 - 40x^3 + 3x^2 - 15$ **53.** $3x^{2n} + 5x^ny^n - 2y^{2n}$ **55.** $x^2 + 6x + 9$
57. $y^2 - 10y + 25$ **59.** $4x^2 + 4xy + y^2$ **61.** $25x^2 - 30xy + 9y^2$ **63.** $4x^4 + 12x^2y + 9y^2$ **65.** $16x^2y^4 - 8x^2y^3 + x^2y^2$
67. $a^{2n} + 8a^nb^n + 16b^{2n}$ **69.** $x^2 - 16$ **71.** $25x^2 - 9$ **73.** $16x^2 - 49y^2$ **75.** $y^6 - 4$ **77.** $1 - y^{10}$ **79.** $49x^2y^4 - 100y^2$
81. $25a^{2n} - 49$ **83.** $4x^2 + 12x + 9 - 16y^2$ **85.** $x^2 + 2xy + y^2 - 9$ **87.** $25x^2 + 70xy + 49y^2 - 4$ **89.** $25y^2 - 4x^2 - 12x - 9$
91. $x^2 + 2xy + y^2 + 2x + 2y + 1$ **93.** $x^4 - 1$ **95. a.** $x^2 + 4x - 12$ **b.** -15 **c.** -12 **97. a.** $x^3 - 27$ **b.** -35 **c.** -27

99. a. $a^2 + a + 5$ **b.** $2ah + h^2 - 3h$ **101. a.** $3a^2 + 14a + 15$ **b.** $6ah + 3h^2 + 2h$ **103.** $48xy$ **105.** $-9x^2 + 3x + 9$ **107.** $16x^4 - 625$
109. $x^3 - 3x^2 + 3x - 1$ **111.** $(2x - 7)^2 = 4x^2 - 28x + 49$ **113. a.** $x^2 + 6x + 4x + 24$ **b.** $(x + 6)(x + 4) = x^2 + 10x + 24$
115. a. $x^2 + 12x + 27$ **b.** $x^2 + 6x + 5$ **c.** $6x + 22$ **117. a.** $4x^2 - 36x + 80$ **b.** $4x^3 - 36x^2 + 80x$
129.

conclusion: $y_1 = y_2$

131.

conclusion: $y_1 = y_2$

133. d   **135.** $x^3 + 7x^2 - 3x$ **137.** $6y^n - 13$

139. $\left\{ x \,\middle|\, x \le -\dfrac{14}{3} \text{ or } x \ge 2 \right\}$ or $\left(-\infty, -\dfrac{14}{3} \right) \cup [2, \infty)$

140. $\{x | x \ge -3\}$ or $[-3, \infty)$ **141.** 8.034×10^9

Section 5.3

Check Point Exercises

1. $10x(2x + 3)$ **2. a.** $3x^2(3x^2 + 7)$ **b.** $5x^3y^2(3 - 5xy)$ **c.** $4x^2y^3(4x^2y^2 - 2xy + 1)$ **3.** $-2x(x^2 - 5x + 3)$
4. a. $(x - 4)(3 + 7a)$ **b.** $(a + b)(7x - 1)$ **5.** $(x - 4)(x^2 + 5)$ **6.** $(x + 5)(4x - 3y)$

Exercise Set 5.3

1. $2x(5x + 2)$ **3.** $y(y - 4)$ **5.** $x^2(x + 5)$ **7.** $4x^2(3x^2 - 2)$ **9.** $2x^2(16x^2 + x + 4)$ **11.** $2xy(2xy^2 + 3)$ **13.** $10xy^2(3xy - 1)$
15. $2x(6y - 3z + 2w)$ **17.** $3x^2y^4(5xy^2 - 3x^2 + 4y)$ **19.** $5x^2y^4z^2(5xy^2 - 3x^2z^2 + 5yz)$ **21.** $5x^n(3x^n - 5)$ **23.** $-4(x - 3)$ **25.** $-8(x + 6)$
27. $-2(x^2 - 3x + 7)$ **29.** $-5(y^2 - 8x)$ **31.** $-4x(x^2 - 8x + 5)$ **33.** $-1(x^2 + 7x - 5)$ **35.** $(x + 3)(4 + a)$ **37.** $(y - 6)(x - 7)$
39. $(x + y)(3x - 1)$ **41.** $(3x - 1)(4x^2 + 1)$ **43.** $(x + 3)(2x + 1)$ **45.** $(x + 3)(x + 5)$ **47.** $(x + 7)(x - 4)$ **49.** $(x - 3)(x^2 + 4)$
51. $(y - 6)(x + 2)$ **53.** $(y + 1)(x - 7)$ **55.** $(5x - 6y)(2x + 7y)$ **57.** $(4x - 1)(x^2 - 3)$ **59.** $(x - a)(x - b)$ **61.** $(x - 3)(x^2 + 4)$
63. $(a - b)(y - x)$ **65.** $(a + 2b)(y^2 - 3x)$ **67.** $(x^n + 1)(y^n + 3)$ **69.** $(a + 1)(b - c)$ **71.** $(x^3 - 5)(1 + 4y)$ **73.** $y^6(3x - 1)^4(6xy - 2y - 7)$
75. $(x^2 + 5x - 2)(a + b)$ **77.** $(x + y + z)(a - b + c)$ **79. a.** 16; The ball is 16 feet above the ground after 2 seconds.   **b.** 0; The ball is on the ground
after 2.5 seconds.   **c.** $-8t(2t - 5)$; $f(t) = -8t(2t - 5)$ **d.** 16; 0; yes; no; Answers will vary.   **81. a.** $(x - 0.4x)(1 - 0.4) = (0.6x)(0.6) = 0.36x$
b. no; 36%   **83.** $A = P + Pr + (P + Pr)r = P(1 + r) + Pr(1 + r) = (1 + r)(P + Pr) = P(1 + r)(1 + r) = P(1 + r)^2$ **85.** $A = r(\pi r + 2l)$
93.

Graphs coincide; factored correctly.

95.

Graphs do not coincide;
$x^2 + 2x + x + 2 = (x + 2)(x + 1)$.

97. c

99. $3x^{2m}y^m(x^m - 2y^m)$ **101.** Answers will vary; an example is $-4x^4 + 8x^3 - 16x^2$.

103. $\left(\dfrac{20}{11}, -\dfrac{14}{11} \right)$ or $\left\{ \left(\dfrac{20}{11}, -\dfrac{14}{11} \right) \right\}$ **104. a.** function   **b.** not a function   **105.** length: 8 ft; width: 3 ft

Section 5.4

Check Point Exercises

1. $(x + 4)(x + 2)$ or $(x + 2)(x + 4)$ **2.** $(x - 5)(x - 4)$ **3.** $(y + 22)(y - 3)$ **4.** $(x - 3y)(x - 2y)$ **5.** $3x(x - 7)(x + 2)$
6. $(x^3 - 5)(x^3 - 2)$ **7.** $(3x - 14)(x - 2)$ **8.** $x^4(3x - 1)(2x + 7)$ **9.** $(2x - y)(x - 3y)$ **10.** $(3y^2 - 2)(y^2 + 4)$ **11.** $(2x - 5)(4x - 1)$

Exercise Set 5.4

1. $(x + 3)(x + 2)$ **3.** $(x + 6)(x + 2)$ **5.** $(x + 5)(x + 4)$ **7.** $(y + 8)(y + 2)$ **9.** $(x - 3)(x - 5)$ **11.** $(y - 2)(y - 10)$
13. $(a + 7)(a - 2)$ **15.** $(x + 6)(x - 5)$ **17.** $(x + 4)(x - 7)$ **19.** $(y + 4)(y - 9)$ **21.** prime   **23.** $(x - 2y)(x - 7y)$
25. $(x + 5y)(x - 6y)$ **27.** prime   **29.** $(a - 10b)(a - 8b)$ **31.** $3(x + 3)(x - 2)$ **33.** $2x(x - 3)(x - 4)$ **35.** $3y(y - 2)(y - 3)$
37. $2x^2(x + 3)(x - 16)$ **39.** $(x^3 + 2)(x^3 - 3)$ **41.** $(x^2 - 6)(x^2 + 1)$ **43.** $(x + 6)(x + 2)$ **45.** $(3x + 5)(x + 1)$
47. $(x + 11)(5x + 1)$ **49.** $(y + 8)(3y - 2)$ **51.** $(y + 2)(4y + 1)$ **53.** $(2x + 3)(5x + 2)$ **55.** $(4x - 3)(2x - 3)$
57. $(y - 3)(6y - 5)$ **59.** prime   **61.** $(x + y)(3x + y)$ **63.** $(2x + y)(3x - 5y)$ **65.** $(5x - 2y)(3x - 5y)$ **67.** $(3a - 7b)(a + 2b)$
69. $5x(3x - 2)(x - 1)$ **71.** $2x^2(3x + 2)(4x - 1)$ **73.** $y^3(5y + 1)(3y - 1)$ **75.** $3(8x + 9y)(x - y)$ **77.** $2b(a + 3)(3a - 10)$
79. $2y(2x - y)(3x - 7y)$ **81.** $13x^3y(y + 4)(y - 1)$ **83.** $(2x^2 - 3)(x^2 + 1)$ **85.** $(2x^3 + 5)(x^3 + 3)$ **87.** $(2y^5 + 1)(y^5 + 3)$

89. $(5x + 12)(x + 2)$ **91.** $(2x - 13)(x - 2)$ **93.** $(x - 0.3)(x - 0.2)$ **95.** $\left(x + \dfrac{3}{7} \right)\left(x - \dfrac{1}{7} \right)$ **97.** $(ax - b)(cx + d)$

99. $-x^3y^2(4x - 3y)(x - y)$ **101.** $f(x) = 3x - 13$ and $g(x) = x - 3$, or vice versa   **103.** $2x + 1$ by $x + 3$
105. a. 32; The diver is 32 feet above the water after 1 second.   **b.** 0; The diver hits the water after 2 seconds.

c. $-16(t - 2)(t + 1); f(t) = -16(t - 2)(t + 1)$ **d.** $32; 0$ **107. a.** $x^2 + x + x + x + 1 + 1 = x^2 + 3x + 2$
b. $(x + 2)(x + 1)$ **c.** Answers will vary.   **117.**

119.

Graphs coincide; factored correctly.   Graphs coincide; factored correctly.

123. $-3, 0, 3$ **125.** $(9x^n - 8)(x^n + 1)$ **127.** $a^2(a^n - 3)(a^n + 2)$ **129.** $c^n(3c - 1)(c - 3)$
131. $\{x|x > 5\}$ or $(5, \infty)$ **132.** $(2, -1, 3)$ or $\{(2, -1, 3)\}$ **133.** $x - y \geq -4$
$x + 2y \geq 2$

Mid-Chapter 5 Check Point

1. $-x^3 + 4x^2 + 6x + 17$ **2.** $-2x^7y^3z^5$ **3.** $30x^5y^3 - 35x^3y^2 - 2x^2y$ **4.** $3x^3 + 4x^2 - 39x + 40$ **5.** $2x^4 - x^3 - 8x^2 + 11x - 4$
6. $-x^2 - 5x + 5$ **7.** $-4x^3y - 6x^2y - 7y - 5$ **8.** $8x^2 + 18x - 5$ **9.** $10x^2y^2 - 11xy - 6$ **10.** $9x^2 - 4y^2$
11. $6x^3y - 9xy^2 + 2x^2 - 3y$ **12.** $49x^6y^2 - 25x^2$ **13.** $6xh + 3h^2 - 2h$ **14.** $x^4 - 6x^2 + 9$ **15.** $x^5 + 2x^3 + 2x^2 - 15x - 6$
16. $4x^2 + 20xy + 25y^2$ **17.** $x^2 + 12x + 36 - 9y^2$ **18.** $x^2 + 2xy + y^2 + 10x + 10y + 25$ **19.** $(x - 8)(x + 3)$ **20.** $(5x + 2)(3y + 1)$
21. $(5x - 2)(x + 2)$ **22.** $5(7x^2 + 2x - 10)$ **23.** $9(x - 2)(x + 1)$ **24.** $5x^2y(2xy - 4y + 7)$ **25.** $(3x + 1)(6x + 5)$
26. $(4x - 3y)(3x - 4)$ **27.** $(3x - 4)(3x - 1)$ **28.** $(3x^3 + 5)(x^3 + 2)$ **29.** $x(5x - 2)(5x + 7)$ **30.** $(x^3 - 3)(2x - y)$

Section 5.5

Check Point Exercises

1. a. $(4x + 5)(4x - 5)$ **b.** $(10y^3 + 3x^2)(10y^3 - 3x^2)$ **2.** $6y(1 + xy^3)(1 - xy^3)$ **3.** $(4x^2 + 9)(2x + 3)(2x - 3)$
4. $(x + 7)(x + 2)(x - 2)$ **5. a.** $(x + 3)^2$ **b.** $(4x + 5y)^2$ **c.** $(2y^2 - 5)^2$ **6.** $(x + 5 + y)(x + 5 - y)$ **7.** $(a + b - 2)(a - b + 2)$
8. a. $(x + 3)(x^2 - 3x + 9)$ **b.** $(x^2 + 10y)(x^4 - 10x^2y + 100y^2)$ **9. a.** $(x - 2)(x^2 + 2x + 4)$ **b.** $(1 - 3xy)(1 + 3xy + 9x^2y^2)$

Exercise Set 5.5

1. $(x + 2)(x - 2)$ **3.** $(3x + 5)(3x - 5)$ **5.** $(3 + 5y)(3 - 5y)$ **7.** $(6x + 7y)(6x - 7y)$ **9.** $(xy + 1)(xy - 1)$
11. $(3x^2 + 5y^3)(3x^2 - 5y^3)$ **13.** $(x^7 + y^2)(x^7 - y^2)$ **15.** $(x - 3 + y)(x - 3 - y)$ **17.** $(a + b - 2)(a - b + 2)$
19. $(x^n + 5)(x^n - 5)$ **21.** $(1 + a^n)(1 - a^n)$ **23.** $2x(x + 2)(x - 2)$ **25.** $2(5 + y)(5 - y)$ **27.** $8(x + y)(x - y)$
29. $2xy(x + 3)(x - 3)$ **31.** $a(ab + 7c)(ab - 7c)$ **33.** $5y(1 + xy^3)(1 - xy^3)$ **35.** $8(x^2 + y^2)$ **37.** prime
39. $(x^2 + 4)(x + 2)(x - 2)$ **41.** $(9x^2 + 1)(3x + 1)(3x - 1)$ **43.** $2x(x^2 + y^2)(x + y)(x - y)$ **45.** $(x + 3)(x + 2)(x - 2)$
47. $(x - 7)(x + 1)(x - 1)$ **49.** $(x + 2)^2$ **51.** $(x - 5)^2$ **53.** $(x^2 - 2)^2$ **55.** $(3y + 1)^2$ **57.** $(8y - 1)^2$ **59.** $(x - 6y)^2$
61. prime   **63.** $(3x + 8y)^2$ **65.** $(x - 3 + y)(x - 3 - y)$ **67.** $(x + 10 + x^2)(x + 10 - x^2)$ **69.** $(3x - 5 + 6y)(3x - 5 - 6y)$
71. $(x^2 + x + 1)(x^2 - x - 1)$ **73.** $(z + x - 2y)(z - x + 2y)$ **75.** $(x + 4)(x^2 - 4x + 16)$ **77.** $(x - 3)(x^2 + 3x + 9)$
79. $(2y + 1)(4y^2 - 2y + 1)$ **81.** $(5x - 2)(25x^2 + 10x + 4)$ **83.** $(xy + 3)(x^2y^2 - 3xy + 9)$ **85.** $x(4 - x)(16 + 4x + x^2)$
87. $(x^2 + 3y)(x^4 - 3x^2y + 9y^2)$ **89.** $(5x^2 - 4y^2)(25x^4 + 20x^2y^2 + 16y^4)$ **91.** $(x + 1)(x^2 - x + 1)(x^6 - x^3 + 1)$

93. $(x - 2y)(x^2 - xy + y^2)$ **95.** $(0.2x + 0.3)^2$ or $\dfrac{1}{100}(2x + 3)^2$ **97.** $x\left(2x - \dfrac{1}{2}\right)\left(4x^2 + x + \dfrac{1}{4}\right)$

99. $(x - 1)(x^2 + x + 1)(x - 2)(x^2 + 2x + 4)$ **101.** $(x^4 + 1)(x^2 + 4)(x + 2)(x - 2)$ **103.** $(x + 1)(x - 1)(x - 2)(x^2 + 2x + 4)$
105. $x^2 - 16 = (x + 4)(x - 4)$ **107.** $9x^2 - 16 = (3x + 4)(3x - 4)$ **109.** $4a^3 - 4ab^2 = 4a(a + b)(a - b)$
115.

117.

119.

121.

Graphs coincide; factored correctly.   Graphs coincide; factored correctly.   Graphs coincide; factored correctly.   Graphs do not coincide;
$x^3 - 1 = (x - 1)(x^2 + x + 1)$.

125. $(y + x)(y^2 - xy + x^2 + 1)$ **127.** $(x^n + y^{4n})(x^{2n} - x^ny^{4n} + y^{8n})$
129. $x^6 - y^6 = (x^3 + y^3)(x^3 - y^3) = (x + y)(x^2 - xy + y^2)(x - y)(x^2 + xy + y^2)$;
$x^6 - y^6 = (x^2 - y^2)(x^4 + x^2y^2 + y^4) = (x + y)(x - y)(x^4 + x^2y^2 + y^4)$; $x^4 + x^2y^2 + y^4 = (x^2 - xy + y^2)(x^2 + xy + y^2)$

131. 1 **132.** $\{x|5 \leq x \leq 11\}$ or $[5, 11]$ **133.** $(-2, 1)$ or $\{(-2, 1)\}$ **134.** $\dfrac{y^6}{16x^5}$

Section 5.6

Check Point Exercises

1. $3x(x - 5)^2$ **2.** $3y(x + 2)(x - 6)$ **3.** $(x + y)(4a + 5)(4a - 5)$
4. $(x + 10 + 6a)(x + 10 - 6a)$ **5.** $x(x + 2)(x^2 - 2x + 4)(x^6 - 8x^3 + 64)$

Exercise Set 5.6

1. $x(x + 4)(x - 4)$ **3.** $3(x + 3)^2$ **5.** $3(3x - 1)(9x^2 + 3x + 1)$ **7.** $(x + 4)(x - 4)(y - 2)$ **9.** $2b(2a + 5)(a - 3)$
11. $(y + 2)(y - 2)(a - 4)$ **13.** $11x(x^2 + y)(x^2 - y)$ **15.** $4x(x^2 + 4)(x + 2)(x - 2)$ **17.** $(x - 4)(x + 3)(x - 3)$
19. $2x^2(x + 3)(x^2 - 3x + 9)$ **21.** $3y(x^2 + 4y^2)(x + 2y)(x - 2y)$ **23.** $3x(2x + 3y)^2$ **25.** $(x - 6 + 7y)(x - 6 - 7y)$ **27.** prime
29. $12xy(x + y)(x - y)$ **31.** $6b(x^2 + y^2)$ **33.** $(x + y)(x - y)(x^2 + xy + y^2)$ **35.** $(x + 6 + 2a)(x + 6 - 2a)$ **37.** $(x^3 - 2)(x^3 + 7)$
39. $(2x - y)(2 + x^2)$ **41.** $2(3x - 2y)(9x^2 + 6xy + 4y^2)$ **43.** $(x + 5 + y)(x + 5 - y)$ **45.** $(x^4 + y^4)(x^2 + y^2)(x + y)(x - y)$
47. $xy(x + 4y)(x - 4y)$ **49.** $x(1 + 2x)(1 - 2x + 4x^2)$ **51.** $2(4y + 1)(2y - 1)$ **53.** $y(14y^2 + 7y - 10)$ **55.** $3(3x + 2y)^2$
57. $3x(4x^2 + y^2)$ **59.** $x^3y^3(xy - 1)(x^2y^2 + xy + 1)$ **61.** $2(x + 5)(x - 5)$ **63.** $(x - y)(a + 2)(a - 2)$
65. $(c + d - 1)(c + d + 1)[(c + d)^2 + 1]$ **67.** $(p + q)^2(p - q)$ **69.** $(x + 2y)(x - 2y)(x + y)(x - y)$
71. $(x + y + 3)^2$ **73.** $(x - y)^2(x - y + 2)(x - y - 2)$ **75.** $(2x - y^2)(x - 3y^2)$ **77.** $(x - y)(x^2 + xy + y^2 - 1)$
79. $(xy + 1)(x^2y^2 - xy + 1)(x - 2)(x^2 + 2x + 4)$ **81. a.** $x(x + y) - y(x + y)$ **b.** $(x + y)(x - y)$
83. a. $xy + xy + xy + 3x(x) = 3xy + 3x^2$ **b.** $3x(y + x)$ **85. a.** $8x^2 - 2\pi x^2$ **b.** $2x^2(4 - \pi)$
89. **91.** **95.** $x^3(x^n - 5)^2$
97. $x(x^{2n} + y^{2n})(x^n + y^n)(x^n - y^n)$
99. $(x^2 + 1)(2x^2 + x + 2)$

100. -1 or $\{-1\}$ **101.** $\dfrac{8x^9}{y^3}$ **102.** 52

Graphs do not coincide; Graphs coincide; factored
$4x^2 - 12x + 9 = (2x - 3)^2$. correctly.

Section 5.7

Check Point Exercises

1. $-\dfrac{1}{2}$ and 5, or $\left\{-\dfrac{1}{2}, 5\right\}$ **2. a.** 0 and $\dfrac{2}{3}$, or $\left\{0, \dfrac{2}{3}\right\}$ **b.** 5 or $\{5\}$ **c.** -4 and 3, or $\{-4, 3\}$

3. $-2, -\dfrac{3}{2}$, and 2, or $\left\{-2, -\dfrac{3}{2}, 2\right\}$ **4.** after 3 seconds; $(3, 336)$ **5.** 2 ft **6.** 5 ft

Exercise Set 5.7

1. -4 and 3, or $\{-4, 3\}$ **3.** -7 and 1, or $\{-7, 1\}$ **5.** -4 and $\dfrac{2}{3}$, or $\left\{-4, \dfrac{2}{3}\right\}$ **7.** $\dfrac{3}{5}$ and 1, or $\left\{\dfrac{3}{5}, 1\right\}$ **9.** -2 and $\dfrac{1}{3}$, or $\left\{-2, \dfrac{1}{3}\right\}$

11. 0 and 8, or $\{0, 8\}$ **13.** 0 and $\dfrac{5}{3}$, or $\left\{0, \dfrac{5}{3}\right\}$ **15.** -2 or $\{-2\}$ **17.** 7 or $\{7\}$ **19.** $\dfrac{5}{3}$ or $\left\{\dfrac{5}{3}\right\}$ **21.** -5 and 5, or $\{-5, 5\}$

23. $-\dfrac{10}{3}$ and $\dfrac{10}{3}$, or $\left\{-\dfrac{10}{3}, \dfrac{10}{3}\right\}$ **25.** -3 and 6, or $\{-3, 6\}$ **27.** -3 and -2, or $\{-3, -2\}$ **29.** 4 or $\{4\}$ **31.** -2 and 8, or $\{-2, 8\}$

33. -1 and $\dfrac{2}{5}$, or $\left\{-1, \dfrac{2}{5}\right\}$ **35.** -6 and -3, or $\{-6, -3\}$ **37.** $-5, -4$, and 5, or $\{-5, -4, 5\}$ **39.** $-5, 1$, and 5, or $\{-5, 1, 5\}$
41. $-4, 0$, and 4, or $\{-4, 0, 4\}$ **43.** 0 and 2, or $\{0, 2\}$ **45.** $-5, -3$, and 0, or $\{-5, -3, 0\}$ **47.** 2 and 4; d **49.** -4 and -2; c
51. $-7, -1$, and 6, or $\{-7, -1, 6\}$ **53.** $-\dfrac{4}{3}, 0, \dfrac{4}{5}$, and 2, or $\left\{-\dfrac{4}{3}, 0, \dfrac{4}{5}, 2\right\}$ **55.** -4 and 8 **57.** $-2, -\dfrac{1}{2}$, and 2 **59.** -7 and 4

63. -9 and 5 **65.** 1 second; 0.25, 0.5, 0.75, 1 **67.** 1996 **69.** $(6, 26)$; This is the highest point on the graph. During this time period,
the greatest number of recipients was 26 million in 1996. **71.** length: 9 ft; width: 6 ft **73.** 5 in. **75.** 5 m

77. a. $4x^2 + 44x$ **b.** 3 ft **79.** length: 10 in.; width: 10 in. **81.** length: 12 ft; width: 5 ft **83.** $30\dfrac{1}{8}$ ft

97. **99.** **101.** c **103.** $-8, -6, 4$, and 6, or $\{-8, -6, 4, 6\}$

104. -2 and $\dfrac{10}{3}$, or $\left\{-2, \dfrac{10}{3}\right\}$ **105.** 12

106. \$1700 at 5% and \$1300 at 8%

$-3, -1$, and 1, or $\{-3, -1, 1\}$ 0 and 2, or $\{0, 2\}$

Review Exercises

1.

Term	Coefficient	Degree
$-5x^3$	-5	3
$7x^2$	7	2
$-x$	-1	1
2	2	0

The degree of the polynomial is 3, the leading term is $-5x^3$, and the leading coefficient is -5.

2.

Term	Coefficient	Degree
$8x^4y^2$	8	6
$-7xy^6$	-7	7
$-x^3y$	-1	4

The degree of the polynomial is 7, the leading term is $-7xy^6$, and the leading coefficient is -7.

3. -31
4. 63.3; There were 63.3 thousand, or 63,300, people waiting for transplants in the United States in 2000.
5. rises to the left and falls to the right; c
6. rises to the left and rises to the right; b
7. falls to the left and rises to the right; a
8. falls to the left and falls to the right; d

9. The graph of this function falls to the right which indicates that the percentage would eventually be negative, which is not possible.
10. $x^3 - 6x^2 - x - 9$ **11.** $12x^3y - 2x^2y - 14y - 17$ **12.** $15x^3 - 10x^2 + 11x - 8$ **13.** $-2x^3y^2 - 4x^3y - 8$
14. $3x^3 + 5x^2y - xy^2 - 8y^3$ **15.** $-12x^6y^2z^7$ **16.** $2x^8 - 24x^5 - 12x^3$ **17.** $21x^5y^4 - 35x^2y^3 - 7xy^2$ **18.** $6x^3 + 29x^2 + 27x - 20$
19. $x^4 + 4x^3 + 4x^2 - x - 2$ **20.** $12x^2 - 23x + 5$ **21.** $15x^2y^2 + 2xy - 8$ **22.** $9x^2 + 42xy + 49y^2$ **23.** $x^4 - 10x^2y + 25y^2$
24. $4x^2 - 49y^2$ **25.** $9x^2y^4 - 16x^2$ **26.** $x^2 + 6x + 9 - 25y^2$ **27.** $x^2 + 2xy + y^2 + 8x + 8y + 16$ **28.** $2x^2 - x - 15; 21$
29. a. $a^2 - 9a + 10$ **b.** $2ah + h^2 - 7h$ **30.** $8x^2(2x + 3)$ **31.** $2x(1 - 18x)$ **32.** $7xy(3xy - 2y + 1)$ **33.** $9x^2y(2xy - 3)$
34. $-4(3x^2 - 2x + 12)$ **35.** $-1(x^2 + 11x - 14)$ **36.** $(x - 1)(x^2 - 2)$ **37.** $(y - 3)(x - 5)$ **38.** $(x - 3y)(5a + 2b)$
39. $(x + 5)(x + 3)$ **40.** $(x + 20)(x - 4)$ **41.** $(x + 17y)(x - y)$ **42.** $3x(x - 1)(x - 11)$ **43.** $(x + 7)(3x + 1)$
44. $(3x - 2)(2x - 3)$ **45.** $(5x + 4y)(x - 2y)$ **46.** $x(3x + 4)(2x - 1)$ **47.** $(x + 3)(2x + 5)$ **48.** $(x^3 + 6)(x^3 - 5)$
49. $(x^2 + 3)(x^2 - 13)$ **50.** $(x + 11)(x + 9)$ **51.** $(5x^3 + 2)(x^3 + 3)$ **52.** $(2x + 5)(2x - 5)$ **53.** $(1 + 9xy)(1 - 9xy)$
54. $(x^4 + y^3)(x^4 - y^3)$ **55.** $(x - 1 + y)(x - 1 - y)$ **56.** $(x + 8)^2$ **57.** $(3x - 1)^2$ **58.** $(5x + 2y)^2$ **59.** prime **60.** $(5x - 4y)^2$
61. $(x + 9 + y)(x + 9 - y)$ **62.** $(z + 5x + 1)(z - 5x + 1)$ **63.** $(4x + 3)(16x^2 - 12x + 9)$ **64.** $(5x - 2)(25x^2 + 10x + 4)$
65. $(xy + 1)(x^2y^2 - xy + 1)$ **66.** $3x(5x + 1)$ **67.** $3x^2(2x + 1)(2x - 1)$ **68.** $4x(5x^3 - 6x^2 + 7x - 3)$ **69.** $x(x - 2)(x - 13)$
70. $-2y^2(y - 3)(y - 9)$ **71.** $(3x - 5)^2$ **72.** $5(x + 3)(x - 3)$ **73.** $(2x - 1)(x + 3)(x - 3)$ **74.** $(3x - y)(2x - 7y)$
75. $2y(y + 3)^2$ **76.** $(x + 3 + 2a)(x + 3 - 2a)$ **77.** $(2x - 3)(4x^2 + 6x + 9)$ **78.** $x(x^2 + 1)(x + 1)(x - 1)$ **79.** $(x^2 - 3)^2$
80. prime **81.** $4(a + 2)(a^2 - 2a + 4)$ **82.** $(x^2 + 9)(x + 3)(x - 3)$ **83.** $(a + 3b)(x - y)$ **84.** $(3x - 5y)(9x^2 + 15xy + 25y^2)$
85. $2xy(x + 3)(5x - 4)$ **86.** $(2x^3 + 5)(3x^3 - 1)$ **87.** $(x + 5)(2 + xy)$ **88.** $(y + 2)(y + 5)(y - 5)$ **89.** $(a^4 + 1)(a^2 + 1)(a + 1)(a - 1)$
90. $(x - 4)(3 + y)(3 - y)$ **91. a.** $2xy + 2y^2$ **b.** $2y(x + y)$ **92. a.** $x^2 - 4y^2$ **b.** $(x + 2y)(x - 2y)$ **93.** -5 and -1, or $\{-5, -1\}$
94. $\frac{1}{3}$ and 7, or $\left\{\frac{1}{3}, 7\right\}$ **95.** -8 and 7, or $\{-8, 7\}$ **96.** 0 and 4, or $\{0, 4\}$ **97.** $-5, -3$, and 3, or $\{-5, -3, 3\}$ **98.** 9 seconds
99. 2006 **100.** length: 9 ft; width: 6 ft **101.** 2 in. **102.** 50 yd, 120 yd, and 130 yd

Chapter 5 Test

1. degree: 3; leading coefficient: -6 **2.** degree: 9; leading coefficient: 7 **3.** 6; 4 **4.** falls to the left and falls to the right **5.** falls to the left and rises to the right **6.** $7x^3y - 18x^2y - y - 9$ **7.** $11x^2 - 13x - 6$ **8.** $35x^7y^3$ **9.** $x^3 - 4x^2y + 2xy^2 + y^3$ **10.** $21x^2 - 20xy - 9y^2$
11. $4x^2 - 25y^2$ **12.** $16y^2 - 56y + 49$ **13.** $x^2 + 4x + 4 - 9y^2$ **14.** $3x^2 + x - 10; 60$ **15.** $2ah + h^2 - 5h$ **16.** $x^2(14x - 15)$
17. $(9y + 5)(9y - 5)$ **18.** $(x + 3)(x + 5)(x - 5)$ **19.** $(5x - 3)^2$ **20.** $(x + 5 + 3y)(x + 5 - 3y)$ **21.** prime **22.** $(y + 2)(y - 18)$
23. $(2x + 5)(7x + 3)$ **24.** $5(x - 1)(x^2 + x + 1)$ **25.** $3(2x + y)(2x - y)$ **26.** $2(3x - 1)(2x - 5)$ **27.** $3(x^2 + 1)(x + 1)(x - 1)$
28. $(x^4 + y^4)(x^2 + y^2)(x + y)(x - y)$ **29.** $4x^2y(3y^3 + 2xy - 9)$ **30.** $(x^3 + 2)(x^3 - 14)$ **31.** $(x^2 - 6)(x^2 + 4)$
32. $3y(4x - 1)(x - 2)$ **33.** $y(y - 3)(y^2 + 2)$ **34.** $-\frac{1}{3}$ and 2, or $\left\{-\frac{1}{3}, 2\right\}$ **35.** -1 and $\frac{6}{5}$, or $\left\{-1, \frac{6}{5}\right\}$ **36.** 0 and $\frac{1}{3}$, or $\left\{0, \frac{1}{3}\right\}$
37. $-1, 1$, and 4, or $\{-1, 1, 4\}$ **38.** 7 seconds **39.** length: 5 yd; width: 3 yd **40.** 12 units, 9 units, and 15 units

Cumulative Review Exercises (Chapters 1–5)

1. 4 or $\{4\}$ **2.** $\left(-4, \frac{1}{2}\right)$ or $\left\{\left(-4, \frac{1}{2}\right)\right\}$ **3.** $(2, 1, -1)$ or $\{(2, 1, -1)\}$ **4.** $\{x | 2 < x < 3\}$ or $(2, 3)$

5. $\{x | x \leq -2 \text{ or } x \geq 7\}$ or $(-\infty, -2] \cup [7, \infty)$ **6.** 1 and $\frac{5}{2}$, or $\left\{1, \frac{5}{2}\right\}$ **7.** $-5, 0$, and 2, or $\{-5, 0, 2\}$ **8.** $x = \frac{b}{c - a}$ **9.** $f(x) = 2x + 1$

10. winner: 1480 votes; loser: 1320 votes **11.** $f(x) = -\frac{1}{3}x + 1$ **12.** $4x - 5y < 20$ **13.** $y \leq -1$

14. $-\dfrac{y^{10}}{2x^6}$ **15.** 7.06×10^{-5} **16.** $9x^4 - 6x^2y + y^2$ **17.** $9x^4 - y^2$ **18.** $(x + 3)(x - 3)^2$
19. $x^2(x^2 + 1)(x + 1)(x - 1)$ **20.** $14x^3y^2(1 - 2x)$

CHAPTER 6

Section 6.1

Check Point Exercises

1. a. 80; The cost to remove 40% of the lake's pollutants is $80 thousand.; (40, 80)　　**b.** 180; The cost to remove 60% of the lake's pollutants is $180 thousand.; (60, 180)　　**2.** $\left\{x \mid x \text{ is a real number and } x \neq -3 \text{ and } x \neq \frac{1}{2}\right\}$ or $(-\infty, -3) \cup \left(-3, \frac{1}{2}\right) \cup \left(\frac{1}{2}, \infty\right)$　　**3.** $x + 5$
4. a. $\dfrac{x - 5}{3x - 1}$　　**b.** $\dfrac{x + 4y}{3x(x + y)}$　　**5.** $\dfrac{x + 3}{x - 4}$　　**6.** $\dfrac{-4(x + 2)}{3x(x - 2)}$ or $-\dfrac{4(x + 2)}{3x(3x - 2)}$　　**7. a.** $9(3x + 7)$　　**b.** $\dfrac{x + 3}{5}$

Exercise Set 6.1

1. $-5; -3; 2$　　**3.** 0; does not exist; $-\dfrac{21}{2}$　　**5.** $-\dfrac{7}{2}; -5; \dfrac{11}{5}$　　**7.** $\{x \mid x \text{ is a real number and } x \neq 5\}$ or $(-\infty, 5) \cup (5, \infty)$　　**9.** $\{x \mid x \text{ is a real number and }$
$x \neq -3 \text{ and } x \neq 1\}$ or $(-\infty, -3) \cup (-3, 1) \cup (1, \infty)$　　**11.** $\{x \mid x \text{ is a real number and } x \neq -5\}$ or $(-\infty, -5) \cup (-5, \infty)$

13. $\{x \mid x \text{ is a real number and } x \neq 3 \text{ and } x \neq 5\}$ or $(-\infty, 3) \cup (3, 5) \cup (5, \infty)$　　**15.** $\{x \mid x \text{ is a real number and } x \neq -\dfrac{4}{3} \text{ and } x \neq 2\}$ or
$\left(-\infty, -\dfrac{4}{3}\right) \cup \left(-\dfrac{4}{3}, 2\right) \cup (2, \infty)$　　**17.** 4　　**19.** domain of $f = \{x \mid x \text{ is a real number and } x \neq -2 \text{ and } x \neq 2\}$ or
$(-\infty, -2) \cup (-2, 2) \cup (2, \infty)$; range of $f = \{y \mid y \leq 0 \text{ or } y > 3\}$ or $(-\infty, 0] \cup (3, \infty)$
21. As x decreases, the function values are approaching 3.; $y = 3$　　**23.** There is no point on the graph with x-coordinate -2.

25. The graph is not continuous. Furthermore, it neither rises nor falls to either the left or the right.　　**27.** $x + 2$　　**29.** $\dfrac{1}{x - 3}$　　**31.** $\dfrac{4}{x}$

33. $\dfrac{4}{y + 5}$　　**35.** $-\dfrac{1}{3x + 5}$　　**37.** $\dfrac{y + 7}{y - 7}$　　**39.** $\dfrac{x + 9}{x - 1}$　　**41.** cannot be simplified　　**43.** $-\dfrac{x + 3}{x + 4}$　　**45.** $\dfrac{x + 5y}{3x - y}$　　**47.** $\dfrac{x^2 + 2x + 4}{x + 2}$

49. $x^2 - 3$　　**51.** $\dfrac{3}{2}$　　**53.** $\dfrac{x + 7}{x}$　　**55.** $\dfrac{x + 3}{x - 2}$　　**57.** $\dfrac{x + 4}{x + 2}$　　**59.** $-\dfrac{2}{y(y + 3)}$　　**61.** $\dfrac{y^2 + 2y + 4}{2y}$　　**63.** $\dfrac{x^2 + x + 1}{x - 2}$　　**65.** 4

67. $\dfrac{4(x + y)}{3(x - y)}$　　**69.** 1　　**71.** 1　　**73.** $\dfrac{9}{28}$　　**75.** $\dfrac{7}{10}$　　**77.** $\dfrac{x}{3}$　　**79.** $\dfrac{y - 5}{2}$　　**81.** $\dfrac{(x^2 + 4)(x - 4)}{x - 1}$　　**83.** $\dfrac{y - 7}{y - 5}$

85. $(x - 1)(2x - 1)$　　**87.** $\dfrac{1}{x - 2y}$　　**89.** $3x^2$　　**91.** $\dfrac{x + 1}{x(x - 1)}$　　**93.** $\dfrac{-(x - y)}{(x + y)(b + c)}$　　**95.** $\dfrac{a^2 + 1}{b(a + 2)}$　　**97.** $-\dfrac{a - b}{4c^2}$ or $\dfrac{b - a}{4c^2}$　　**99.** 7

101. $2a + h - 5$　　**103.** $\left(\dfrac{f}{g}\right)(x) = -x - 2; \left\{x \mid x \text{ is a real number and } x \neq -2 \text{ and } x \neq \dfrac{1}{2}\right\}$ or $(-\infty, -2) \cup \left(-2, \dfrac{1}{2}\right) \cup \left(\dfrac{1}{2}, \infty\right)$

105. 195; The cost to inoculate 60% of the population is $195 million.; (60, 195)　　**107.** 100; This indicates that we cannot inoculate 100% of the population.
109. 7; About 7% of people with 10 years of education are unemployed.; (10, 7)　　**111.** The function values are approaching 0.; no; The function values never actually reach 0.　　**113.** 90; Lung cancer has an incidence ratio of 10 between the ages of 55 and 64. 90% of deaths from lung cancer in this group are smoking related.; (10, 90)　　**115.** $y = 100$; As the incidence ratio increases, the percentage of smoking-related deaths is approaching 100%.

129.

Graphs coincide; multiplied correctly.

131.

Graphs do not coincide; right side should be $x + 3$.

135. a.

$y = \dfrac{27{,}725(x - 14)}{x^2 + 9} - 5x$

b. As x increases, y increases and then decreases.
c. 25 years old; about 356 arrests per 100,000 drivers

137. $f(x) = \dfrac{x^2 - x - 2}{x - 2}$

139. $\dfrac{y^n - 1}{y^n + 4}$　　**141.** $4x - 5y \geq 20$

142. $2x^3 - 11x^2 + 3x + 30$
143. $16a^8 b^{26} c^2$

Section 6.2

Check Point Exercises

1. $\dfrac{x-5}{x+3}$ **2.** $\dfrac{1}{x-y}$ **3.** $18x^2$ **4.** $5x(x+3)(x+3)$ or $5x(x+3)^2$ **5.** $\dfrac{21+4x}{18x^2}$ or $\dfrac{4x+21}{18x^2}$

6. $\dfrac{2x^2-2x+8}{(x-4)(x+4)}$ or $\dfrac{2(x^2-x+4)}{(x-4)(x+4)}$ **7.** $\dfrac{x^2-3x+5}{(x-3)(x+1)(x-2)}$ **8.** $\dfrac{3}{y+2}$ **9.** $\dfrac{3x-5y}{x-3y}$

Exercise Set 6.2

1. $\dfrac{2}{3x}$ **3.** $\dfrac{10x+3}{x-5}$ **5.** $\dfrac{2x-1}{x+3}$ **7.** $\dfrac{y-3}{y+3}$ **9.** $\dfrac{x+1}{4x-3}$ **11.** $\dfrac{x-3}{x-1}$ **13.** $\dfrac{4y+3}{2y-1}$ **15.** $\dfrac{x^2+xy+y^2}{x+y}$ **17.** $175x^2$

19. $(x-5)(x+5)$ **21.** $y(y+10)(y-10)$ **23.** $(x+4)(x-4)(x-4)$ **25.** $(y-5)(y-6)(y+1)$

27. $(y+2)(2y+3)(y-2)(2y+1)$ **29.** $\dfrac{3+50x}{5x^2}$ **31.** $\dfrac{7x-2}{(x-2)(x+1)}$ **33.** $\dfrac{3x-4}{(x+2)(x-3)}$ **35.** $\dfrac{2x^2-2x+61}{(x+5)(x-6)}$

37. $\dfrac{20-x}{(x-5)(x+5)}$ **39.** $\dfrac{13}{(y-3)(y-2)}$ **41.** $\dfrac{1}{x+3}$ **43.** $\dfrac{6x^2+14x+10}{(x+1)(x+4)(x+3)}$ **45.** $-\dfrac{x^2-3x-13}{(x+4)(x-2)(x+1)}$ **47.** $\dfrac{4x-11}{x-3}$

49. $\dfrac{2y-14}{(y-4)(y+4)}$ **51.** $\dfrac{x^2+2x-14}{3(x+2)(x-2)}$ **53.** $\dfrac{16}{x-4}$ **55.** $\dfrac{12x+7y}{(x-y)(x+y)}$ **57.** $\dfrac{x^2-18x-30}{(5x+6)(x-2)}$ **59.** $\dfrac{3x^2+16xy-13y^2}{(x+5y)(x-5y)(x-4y)}$

61. $\dfrac{8x^3-32x^2+23x+6}{(x+2)(x-2)(x-2)(x-1)}$ **63.** $\dfrac{16a^2-12ab-18b^2}{(3a+4b)(3a-4b)(2a-b)}$ **65.** $\dfrac{m+6}{(m-1)(m+2)(2m-1)}$ **67.** $\dfrac{x^2+5x+8}{(x+2)(x+1)}$ **69.** 2

71. $-\dfrac{1}{x(x+h)}$ **73.** $\dfrac{2d}{a^2+ab+b^2}$ **75.** $(f-g)(x)=\dfrac{x+2}{x+3}$; $\{x\,|\,x$ is a real number and $x\neq -5$ and $x\neq -3\}$ or $(-\infty,-5)\cup(-5,-3)\cup(-3,\infty)$

77. 11; If you average 0 miles per hour over the speed limits, the total driving time is about 11 hours; (0, 11). **79.** $\dfrac{720x+48{,}050}{(x+70)(x+65)}$; 11

81. 12 mph; Answers will vary. **83.** $\dfrac{2x(2x+9)}{(x+4)(x+5)}$ **85.** $\dfrac{-5x^3+27{,}680x-388{,}150}{x^2+9}$ **91.** Answers will vary.; $\dfrac{b+a}{ab}$

93.

As age increases, the number of arrests per 100,000 drivers increases until about age 25 and then decreases.

95. $\dfrac{1}{x^{2n}-1}$

97. $\dfrac{x-y+1}{(x-y)(x-y)}$

98. $\dfrac{y^{10}}{9x^4}$

99. $\left\{x\left|-\dfrac{13}{3}\le x\le 5\right.\right\}$ or $\left[-\dfrac{13}{3},5\right]$

100. $2x(5x+3)(5x-3)$

Section 6.3

Check Point Exercises

1. $\dfrac{y}{x+y}$ **2.** $-\dfrac{1}{x(x+7)}$ **3.** $\dfrac{2x}{x^2+1}$ **4.** $\dfrac{x+2}{x-5}$

Exercise Set 6.3

1. $\dfrac{4x+2}{x-3}$ **3.** $\dfrac{x^2+9}{(x-3)(x+3)}$ **5.** $\dfrac{y+x}{y-x}$ **7.** $\dfrac{4-x}{5x-3}$ **9.** $\dfrac{1}{x-3}$ **11.** $-\dfrac{1}{x(x+5)}$ **13.** $-x$ **15.** $-\dfrac{x+1}{x-1}$

17. $\dfrac{(x+y)(x+y)}{xy}$ **19.** $\dfrac{4x}{x^2+4}$ **21.** $\dfrac{2y^3+5x^2}{y^3(5-3x^2)}$ **23.** $-\dfrac{12}{5}$ **25.** $\dfrac{3ab(b+a)}{(2b+3a)(2b-3a)}$ **27.** $-\dfrac{x+10}{5x-2}$

29. $\dfrac{2y}{3y+7}$ **31.** $\dfrac{2b+a}{b-2a}$ **33.** $\dfrac{4(7x+5)}{69(x+5)}$ **35.** $\dfrac{3x-2y}{13y+5x}$ **37.** $\dfrac{m(m+3)}{(m-2)(m+1)}$ **39.** $\dfrac{3a(a-3)}{(a+4)(3a-5)}$

41. $x(x+1)$ **43.** $-\dfrac{4}{(x+2)(x-2)}$ **45.** 6 **47.** $\dfrac{x+4}{x+2}$ **49.** $-\dfrac{3}{a(a+h)}$

51. a. $\dfrac{Pi(1+i)^n}{(1+i)^n-1}$ **b.** \$527 per month **53.** $R=\dfrac{R_1R_2R_3}{R_2R_3+R_1R_3+R_1R_2}$; about 2.18 ohms

59.

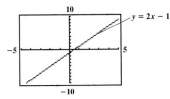

Graphs coincide; simplified correctly.

61.

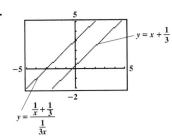

Graphs do not coincide; $\dfrac{\dfrac{1}{x}+\dfrac{1}{3}}{\dfrac{1}{3x}} = x + 3.$

63. $\dfrac{1}{(x + h + 1)(x + 1)}$

65. $\dfrac{a + 1}{a + 2}$

67. 3 and 9, or $\{3, 9\}$

68. $16x^4 - 8x^2y + y^2$

69. $\{x | 1 < x < 5\}$ or $(1, 5)$

Section 6.4

Check Point Exercises

1. $4x^2 - 8x + \dfrac{1}{2} + \dfrac{1}{x}$　　**2.** $3x^2y^2 - xy + \dfrac{2}{y}$　　**3.** $3x - 8$　　**4.** $2x^2 - x + 3 - \dfrac{6}{2x - 1}$　　**5.** $2x^2 + 7x + 14 + \dfrac{21x - 10}{x^2 - 2x}$

Exercise Set 6.4

1. $5x^4 - 3x^2 + 2$　　**3.** $6x^2 + 2x - 3 - \dfrac{2}{x}$　　**5.** $7x - \dfrac{7}{4} - \dfrac{4}{x}$　　**7.** $-5x^3 + 10x^2 - \dfrac{3}{5}x + 8$　　**9.** $2a^2b - a - 3b$　　**11.** $6x - \dfrac{3}{y} - \dfrac{2}{xy^2}$

13. $x + 3$　　**15.** $x^2 + x - 2$　　**17.** $x - 2 + \dfrac{2}{x - 5}$　　**19.** $x + 5 - \dfrac{10}{2x + 3}$　　**21.** $x^2 + 2x + 3 + \dfrac{1}{x + 1}$　　**23.** $2y^2 + 3y - 1$

25. $3x^2 - 3x + 1 + \dfrac{2}{3x + 2}$　　**27.** $2x^2 + 4x + 5 + \dfrac{9}{2x - 4}$　　**29.** $2y^2 + y - 2 - \dfrac{2}{2y - 1}$　　**31.** $2y^3 + 3y^2 - 4y + 1$　　**33.** $4x^2 + 3x - 8 + \dfrac{18}{x^2 + 3}$

35. $5x^2 + x + 3 + \dfrac{x + 7}{3x^2 - 1}$　　**37.** $\left(\dfrac{f}{g}\right)(x) = 2x^2 - 9x + 10$　　**39.** $\left(\dfrac{f}{g}\right)(x) = x^3 - x^2 + x - 2$　　**41.** $x^3 - x^2y + xy^2 - y^3 + \dfrac{2y^4}{x + y}$

43. $3x^2 + 2x - 1$　　**45.** $4x - 7 + \dfrac{4x + 8}{x^2 + x + 1}$　　**47.** $x^3 + x^2 - x - 3 + \dfrac{-x + 5}{x^2 - x + 2}$　　**49.** $4x^2 + 5xy - y^2$

51. $\left(\dfrac{f - g}{h}\right)(x) = 2x^2 - 5x + 8;\ \left\{x | x \text{ is a real number and } x \neq -\dfrac{1}{4}\right\}$ or $\left(-\infty, -\dfrac{1}{4}\right) \cup \left(-\dfrac{1}{4}, \infty\right)$

53. $x = 3a^2 + 4a - 2$　　**55.** 70; At the tax rate of 30%, the government tax revenue is \$70 tens of billions, or \$700 billion.; $(30, 70)$

57. $f(x) = 80 + \dfrac{800}{x - 110}$; 70; yes; Answers will vary.

65.

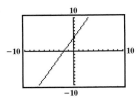

Graphs coincide; division correct.

67.

Graphs do not coincide; right side should be $3x^3 - 8x^2 - 5$.

69. d　　**71.** $x^{2n} - x^n + 1$

73. -12

74. $\left\{x \middle| x < -\dfrac{1}{2} \text{ or } x > \dfrac{7}{2}\right\}$ or $\left(-\infty, -\dfrac{1}{2}\right) \cup \left(\dfrac{7}{2}, \infty\right)$

75. 4.061×10^7

76. $22x + 12$

Mid-Chapter 6 Check Point

1. $\dfrac{x + 2}{x + 6}$　　**2.** $\dfrac{3(x - 2)}{x - 1}$　　**3.** $\dfrac{x^2 + 3x + 9}{x - 1}$　　**4.** $\dfrac{5x - 3}{x - 2}$　　**5.** $\dfrac{x^2 + 6x - 4}{x^2 + 6x + 1}$　　**6.** $2x^3 - 5x^2 - 3x + 6$　　**7.** $x + y$

8. $4x^6y^2 - 2x^4y + \dfrac{3}{7}y$　　**9.** $\dfrac{x^2 - 14x - 16}{(x + 6)(x - 2)}$　　**10.** $\dfrac{10}{(x + 2)(x - 2)}$　　**11.** $\dfrac{2(x - 3)}{x - 1}$　　**12.** $\dfrac{x - 5}{x - 7}$　　**13.** $2x^2 - x - 3 + \dfrac{x + 1}{3x^2 - 1}$

14. $\dfrac{5x + 2}{3x - 1}$　　**15.** $\dfrac{5x}{(x - 6)(x - 1)(x + 4)}$　　**16.** $\dfrac{7x + 4}{4(x + 1)}$　　**17.** $x + 2$　　**18.** $16x^2 - 8x + 4 - \dfrac{2}{2x + 1}$　　**19.** $\dfrac{3x + 2}{(x + 2)(x - 1)}$

20. Domain of $f = \{x | x \text{ is a real number and } x \neq -7 \text{ and } x \neq 2\}$ or $(-\infty, -7) \cup (-7, 2) \cup (2, \infty); f(x) = \dfrac{5}{x + 7}$

Section 6.5

Check Point Exercises

1. $x^2 - 2x - 3$ **2.** -105 **3.** The remainder is zero.; $-1, -\dfrac{1}{3}$, and $\dfrac{2}{5}$, or $\left\{-1, -\dfrac{1}{3}, \dfrac{2}{5}\right\}$

Exercise Set 6.5

1. $2x + 5$ **3.** $3x - 8 + \dfrac{20}{x + 5}$ **5.** $4x^2 + x + 4 + \dfrac{3}{x - 1}$ **7.** $6x^4 + 12x^3 + 22x^2 + 48x + 93 + \dfrac{187}{x - 2}$ **9.** $x^3 - 10x^2 + 51x - 260 + \dfrac{1300}{5 + x}$

11. $3x^2 + 3x - 3$ **13.** $x^4 + x^3 + 2x^2 + 2x + 2$ **15.** $x^3 + 4x^2 + 16x + 64$ **17.** $2x^4 - 7x^3 + 15x^2 - 31x + 64 - \dfrac{129}{x + 2}$ **19.** -25 **21.** -133

23. 240 **25.** 1 **27.** The remainder is 0.; $-1, 2$, and 3, or $\{-1, 2, 3\}$ **29.** The remainder is 0.; $-\dfrac{1}{2}, 1$, and 2, or $\left\{-\dfrac{1}{2}, 1, 2\right\}$ **31.** The remainder is 0.;

$-5, \dfrac{1}{3}$, and $\dfrac{1}{2}$, or $\left\{-5, \dfrac{1}{3}, \dfrac{1}{2}\right\}$ **33.** 2; The remainder is zero; $-3, -1$, and 2, or $\{-3, -1, 2\}$ **35.** 1; The remainder is zero; $\dfrac{1}{3}, \dfrac{1}{2}$, and 1, or $\left\{\dfrac{1}{3}, \dfrac{1}{2}, 1\right\}$

37. $4x^3 - 9x^2 + 7x - 6$ **39.** $0.5x^2 - 0.4x + 0.3$ **41. a.** The remainder is 0. **b.** 3 mm

47. a. For Exercise 27:

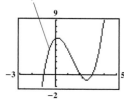
$y = x^3 - 4x^2 + x + 6$

b. For Exercise 29:

$y = 2x^3 - 5x^2 + x + 2$

c. For Exercise 31:

$y = 6x^3 + 25x^2 - 24x + 5$

49. The remainder is 0.; $-2, -1, 2$, and 5, or $\{-2, -1, 2, 5\}$

50. $\{x \mid x > -10\}$ or $(-10, \infty)$

51. -5 and -1, or $\{-5, -1\}$

52. $(5, 3)$ or $\{(5, 3)\}$

Section 6.6

Check Point Exercises

1. 6 or $\{6\}$ **2.** 4 or $\{4\}$ **3.** no solution or $\varnothing$ **4.** 4 and 6, or $\{4, 6\}$ **5.** 1 and 7, or $\{1, 7\}$ **6.** 50%

Exercise Set 6.6

1. 1 or $\{1\}$ **3.** 3 or $\{3\}$ **5.** 4 or $\{4\}$ **7.** -4 or $\{-4\}$ **9.** $\dfrac{9}{7}$ or $\left\{\dfrac{9}{7}\right\}$ **11.** no solution or $\varnothing$ **13.** 2 or $\{2\}$

15. 2 or $\{2\}$ **17.** 5 or $\{5\}$ **19.** -7 and -1, or $\{-7, -1\}$ **21.** -6 and 3, or $\{-6, 3\}$ **23.** -1 or $\{-1\}$

25. no solution or $\varnothing$ **27.** 2 or $\{2\}$ **29.** no solution or $\varnothing$ **31.** $-\dfrac{4}{3}$ or $\left\{-\dfrac{4}{3}\right\}$ **33.** -6 and $\dfrac{1}{2}$, or $\left\{-6, \dfrac{1}{2}\right\}$

35. 4, 10 **37.** $-3, -\dfrac{3}{4}$ **39.** $\dfrac{x - 2}{x(x + 1)}$ **41.** 2 or $\{2\}$ **43.** 4 or $\{4\}$ **45.** $\dfrac{-2(x - 4)}{(x - 2)(x^2 + 2x + 4)}$ **47.** 0

49. 1,7   **51.** 60%   **53.** 10 days; $(10, 8)$ **55.** $y = 5$; On average, the students remembered 5 words over an extended period of time.

57. 11 learning trials; $(11, 0.95)$ **59.** As the number of learning trials increases, the proportion of correct responses increases.; Initially, the proportion of correct responses increases rapidly, but slows down as time increases.   **61.** 125 liters

71. 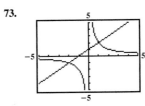 -5 and 5, or $\{-5, 5\}$ **73.** -2 and 1, or $\{-2, 1\}$

75. d   **77.** $-\dfrac{19}{5}$ and 13, or $\left\{-\dfrac{19}{5}, 13\right\}$ **79.** -3 or $\{-3\}$ **80.** $\begin{array}{l} x + 2y \geq 2 \\ x - y \geq -4 \end{array}$ **81.** 15 or $\{15\}$ **82.** $F = \dfrac{9C + 160}{5}$

Section 6.7

Check Point Exercises

1. $x = \dfrac{yz}{y - z}$ **2. a.** $C(x) = 500{,}000 + 400x$ **b.** $\overline{C}(x) = \dfrac{500{,}000 + 400x}{x}$ **c.** 10,000 wheelchairs

3. 12 mph   **4.** 6 months   **5.** experienced carpenter: 8 hr; apprentice: 24 hr

Exercise Set 6.7

1. $P_1 = \dfrac{P_2 V_2}{V_1}$ **3.** $f = \dfrac{pq}{q + p}$ **5.** $r = \dfrac{A - P}{P}$ **7.** $m_1 = \dfrac{Fd^2}{Gm_2}$ **9.** $x = \overline{x} + zs$ **11.** $R = \dfrac{E - Ir}{I}$ **13.** $f_1 = \dfrac{ff_2}{f_2 - f}$

15. 50,000 wheelchairs   **17.** $y = 400$; Average cost is nearing \$400 as production level increases.   **19. a.** $C(x) = 100{,}000 + 100x$

b. $\overline{C}(x) = \dfrac{100{,}000 + 100x}{x}$ **c.** 500 bikes   **21.** 5 mph   **23.** The time increases as the running rate is close to zero miles per hour.

25. car: 50 mph; bus: 30 mph   **27.** 6 mph   **29.** 3 mph   **31.** 500 mph   **33.** 4.2 ft per sec   **35.** 12 miles   **37.** 18 min; yes   **39.** 4 hr

41. 12 hr   **43.** 10 hr   **45.** 10 hr   **47.** 10 min   **49.** 3   **51.** $\dfrac{1}{3}$ and 3   **53.** 10   **55.** $x = \dfrac{ab}{a + b}$

65. For Exercise 45:

```
10
Y1=6/X+6/(X+5)
0  X=10      Y=1      20
0
```

67. c   **69.** 54 mph

71. $(x + 2 + 3y)(x + 2 - 3y)$

72. $(5, -3)$ or $\{(5, -3)\}$

73. $(3, -1, 2)$ or $\{(3, -1, 2)\}$

Section 6.8

Check Point Exercises

1. 66 gal   **2.** about 556 ft   **3.** 512 cycles per second   **4.** 24 min   **5.** 96π cubic feet

Exercise Set 6.8

1. 156   **3.** 30   **5.** $\dfrac{5}{6}$ **7.** 240   **9.** 50   **11.** $x = kyz$; $y = \dfrac{x}{kz}$ **13.** $x = \dfrac{kz^3}{y}$; $y = \dfrac{kz^3}{x}$

15. $x = \dfrac{kyz}{\sqrt{w}}$; $y = \dfrac{x\sqrt{w}}{kz}$ **17.** $x = kz(y + w)$; $y = \dfrac{x - kzw}{kz}$ **19.** $x = \dfrac{kz}{y - w}$; $y = \dfrac{xw + kz}{x}$ **21.** 5.4 ft   **23.** 80 in.

25. about 607 lb   **27.** $32°$ **29.** 90 milliroentgens per hour   **31.** This person has a BMI of 24.4 and is not overweight.   **33.** 1800 Btu

35. $\dfrac{1}{4}$ of what it was originally   **37. a.** $C = \dfrac{kP_1 P_2}{d^2}$ **b.** $k \approx 0.02$; $C = \dfrac{0.02\, P_1 P_2}{d^2}$ **c.** approximately 39,813 daily phone calls

39. a.

b. Current varies inversely as resistance. **c.** $R = \dfrac{6}{I}$

45. z varies directly as the square root of x and inversely as the square of y.

49. The wind pressure is 4 times more destructive.

51. Distance is increased by $\sqrt{50}$, or about 7.07, for the space telescope.

52. -2 **53.** $(x + 3)(x - 3)(y - 3)$ **54.** 4

Review Exercises

1. a. $\dfrac{7}{4}$ **b.** $\dfrac{3}{4}$ **c.** does not exist **d.** 0 **2.** $\{x \mid x$ is a real number and $x \neq -4$ and $x \neq 3\}$ or $(-\infty, -4) \cup (-4, 3) \cup (3, \infty)$

3. $\{x \mid x$ is a real number and $x \neq -2$ and $x \neq 1\}$ or $(-\infty, -2) \cup (-2, 1) \cup (1, \infty)$ **4.** $\dfrac{x^2 - 7}{3x}$ **5.** $\dfrac{x - 1}{x - 7}$ **6.** $\dfrac{3x + 2}{x - 5}$

7. cannot be simplified **8.** $\dfrac{x^2 + 2x + 4}{x + 2}$ **9.** $\dfrac{5(x + 1)}{3}$ **10.** $\dfrac{2x - 1}{x + 1}$ **11.** $\dfrac{x - 7}{x^2}$ **12.** $\dfrac{1}{3(x + 3)}$ **13.** $\dfrac{1}{2}$ **14.** $\dfrac{(y + 4)(y + 2)}{(y - 6)(y^2 + 4y + 16)}$

15. $\dfrac{3x^2}{2x - 5}$ **16. a.** 50 deer **b.** 150 deer **c.** $y = 225$; The population will approach 225 deer over time. **17.** 4

18. $\dfrac{1}{x + 3}$ **19.** $3x + 2$ **20.** $36x^3$ **21.** $(x + 7)(x - 5)(x + 2)$ **22.** $\dfrac{3x - 5}{x(x - 5)}$ **23.** $\dfrac{5x - 2}{(x - 3)(x + 2)(x - 2)}$

24. $\dfrac{2x - 8}{(x - 3)(x - 5)}$ or $\dfrac{2(x - 4)}{(x - 3)(x - 5)}$ **25.** $\dfrac{4x}{(3x + 4)(x - 4)}$ **26.** $\dfrac{y - 3}{(y + 1)(y + 3)}$ **27.** $\dfrac{2x^2 - 9}{(x + 3)(x - 3)}$ **28.** $3(x + y)$ or $3x + 3y$

29. $\dfrac{3}{8}$ **30.** $\dfrac{x}{x - 5}$ **31.** $\dfrac{3x + 8}{3x + 10}$ **32.** $\dfrac{4(x - 2)}{2x + 5}$ **33.** $\dfrac{3x^2 + 9x}{x^2 + 8x - 33}$ **34.** $\dfrac{1 + x}{1 - x}$ **35.** $3x - 6 + \dfrac{2}{x} - \dfrac{2}{5x^2}$ **36.** $6x^3y + 2xy - 10x$

37. $3x - 7 + \dfrac{26}{2x + 3}$ **38.** $2x^2 - 4x + 1 - \dfrac{10}{5x - 3}$ **39.** $x^5 + 5x^4 + 8x^3 + 16x^2 + 33x + 63 + \dfrac{128}{x - 2}$ **40.** $2x^2 + 3x - 1$

41. $4x^2 - 7x + 5 - \dfrac{4}{x + 1}$ **42.** $3x^3 + 6x^2 + 10x + 10$ **43.** $x^3 - 4x^2 + 16x - 64 + \dfrac{272}{x + 4}$ **44.** 3 **45.** 4 **46.** solution

47. not a solution **48.** The remainder is 0.; $-1, \dfrac{1}{3}$, and $\dfrac{1}{2}$, or $\left\{-1, \dfrac{1}{3}, \dfrac{1}{2}\right\}$ **49.** 6 or $\{6\}$ **50.** 52 or $\{52\}$ **51.** 7 or $\{7\}$

52. $-\dfrac{9}{2}$ or $\left\{-\dfrac{9}{2}\right\}$ **53.** $-\dfrac{1}{2}$ and 3, or $\left\{-\dfrac{1}{2}, 3\right\}$ **54.** -23 and 2, or $\{-23, 2\}$ **55.** -3 and 2, or $\{-3, 2\}$ **56.** 80%

57. $C = R - nP$ **58.** $T_1 = \dfrac{P_1 V_1 T_2}{P_2 V_2}$ **59.** $P = \dfrac{A}{rT + 1}$ **60.** $R = \dfrac{R_1 R_2}{R_2 + R_1}$ **61.** $n = \dfrac{IR}{E - Ir}$ **62. a.** $C(x) = 50{,}000 + 25x$

b. $\overline{C}(x) = \dfrac{50{,}000 + 25x}{x}$ **c.** 5000 graphing calculators **63.** 12 mph **64.** 9 mph **65.** 2 hr; no **66.** faster crew: 36 hr; slower crew: 45 hr

67. 240 min or 4 hr **68.** \$4935 **69.** 1600 ft **70.** 440 vibrations per second **71.** 112 decibels **72.** 16 hr **73.** 800 cubic feet

Chapter 6 Test

1. $\{x \mid x$ is a real number and $x \neq 2$ and $x \neq 5\}$ or $(-\infty, 2) \cup (2, 5) \cup (5, \infty)$; $\dfrac{x}{x - 5}$ **2.** $\dfrac{x}{x - 4}$ **3.** $\dfrac{(x + 3)(x - 1)}{x + 1}$ **4.** $\dfrac{x - 2}{x + 5}$

5. $\dfrac{3x - 24}{x - 3}$ **6.** $\dfrac{1}{3x + 5}$ **7.** $\dfrac{x^2 + 2x + 15}{(x + 3)(x - 3)}$ **8.** $\dfrac{3x - 4}{(x + 2)(x - 3)}$ **9.** $\dfrac{5x^2 - 7x + 4}{(x + 2)(x - 1)(x - 2)}$ **10.** $\dfrac{x + 3}{x + 5}$ **11.** $\dfrac{x - 2}{x - 10}$

12. $\dfrac{x - 2}{8}$ **13.** $1 - x$ **14.** $3x^2y^2 + 4y^2 - \dfrac{5}{2}y$ **15.** $3x^2 - 3x + 1 + \dfrac{2}{3x + 2}$ **16.** $3x^2 + 2x + 3 + \dfrac{9 - 6x}{x^2 - 1}$ **17.** $3x^3 - 4x^2 + 7$

18. 12 **19.** solution **20.** -7 and 3, or $\{-7, 3\}$ **21.** -1 or $\{-1\}$ **22.** 3 years **23.** $a = \dfrac{Rs}{s - R}$ or $a = -\dfrac{Rs}{R - s}$

24. a. $C(x) = 300{,}000 + 10x$ **b.** $\overline{C}(x) = \dfrac{300{,}000 + 10x}{x}$ **c.** 20,000 televisions **25.** 12 hr **26.** 4 mph **27.** 45 foot-candles

Cumulative Review Exercises (Chapters 1–6)

1. $\{x \mid x < -6\}$ or $(-\infty, -6)$ **2.** $-\dfrac{1}{2}$ and 4, or $\left\{-\dfrac{1}{2}, 4\right\}$ **3.** $(-2, 0, 4)$ or $\{(-2, 0, 4)\}$ **4.** $\left\{x \mid -2 \leq x \leq \dfrac{14}{3}\right\}$ or $\left[-2, \dfrac{14}{3}\right]$

5. 12 or $\{12\}$ **6.** $s = \dfrac{2R - Iw}{2I}$ **7.** $(3, 2)$ or $\{(3, 2)\}$ **8.** $f(x) = -3x - 2$ **9.**

$y = |x| + 2$

10. $y \geq 2x - 1$
$x \geq 1$

11. $2x - y < 4$

12. $x^2 + 4x + 4 - 9y^2$ **13.** $\dfrac{x(x + 1)}{2x + 5}$ **14.** $\dfrac{x + 25}{2(x - 4)(x - 5)}$ **15.** $3x + 4 + \dfrac{2}{x + 2}$

16. $(y - 6)(x + 2)$ **17.** $2xy(2x + 3)(6x - 5)$ **18.** 4 sec

19. basic cable service: $25; each movie channel: $10   **20.** 1 ft

CHAPTER 7

Section 7.1

Check Point Exercises

1. a. 8   **b.** -7 **c.** $\dfrac{4}{5}$ **d.** 0.09   **e.** 5   **f.** 7   **2. a.** 4   **b.** $-\sqrt{24} \approx -4.90$ **3.** $\{x \mid x \geq 3\}$ or $[3, \infty)$
4. 53.9%; This is the percent given for 2001, not 2002, so the model does not provide a totally accurate description of the data.
5. a. 7   **b.** $|x + 8|$ **c.** $|7x^5|$ or $7|x^5|$ **d.** $|x - 3|$ **6. a.** 3   **b.** -2 **7.** $-3x$ **8. a.** 2   **b.** -2 **c.** not a real number   **d.** -1
9. a. $|x + 6|$ **b.** $3x - 2$ **c.** 8

Exercise Set 7.1

1. 6   **3.** -6 **5.** not a real number   **7.** $\dfrac{1}{5}$ **9.** $-\dfrac{3}{4}$ **11.** 0.9   **13.** -0.2 **15.** 3   **17.** 1   **19.** not a real number
21. 4; 1; 0; not a real number   **23.** $-5; -\sqrt{5} \approx -2.24; -1;$ not a real number   **25.** 4; 2; 1; 6   **27.** $\{x \mid x \geq 3\}$ or $[3, \infty)$; c
29. $\{x \mid x \geq -5\}$ or $[-5, \infty)$; d   **31.** $\{x \mid x \leq 3\}$ or $(-\infty, 3]$; e   **33.** 5   **35.** 4   **37.** $|x - 1|$ **39.** $|6x^2|$ or $6x^2$
41. $-|10x^3|$ or $-10|x^3|$ **43.** $|x + 6|$ **45.** $-|x - 4|$ **47.** 3   **49.** -3 **51.** $\dfrac{1}{5}$ **53.** 3; 2; $-1; -4$ **55.** $-2; 0; 2$
57. 1   **59.** 2   **61.** -2 **63.** not a real number   **65.** -1 **67.** not a real number   **69.** -4 **71.** 2   **73.** -2
75. x **77.** $|y|$ **79.** $-2x$ **81.** -5 **83.** 5   **85.** $|x + 3|$ **87.** $-2(x - 1)$

89.

y↑ $f(x) = \sqrt{x} + 3$

(4, 5)
(9, 6)
(0, 3) (1, 4)

Domain: $\{x \mid x \geq 0\} = [0, \infty)$;
Range: $\{y \mid y \geq 3\} = [3, \infty)$

91.

y↑ $f(x) = \sqrt{x - 3}$

(12, 3)
(4, 1) (7, 2)
(3, 0)

Domain: $\{x \mid x \geq 3\} = [3, \infty)$;
Range: $\{y \mid y \geq 0\} = [0, \infty)$

93. $\{x \mid x < 15\} = (-\infty, 15)$
95. $\{x \mid 1 \leq x < 3\} = [1, 3)$
97. 3
99. 40.2 in.; The model describes actual data well.
101. 70 mph; The officer should not believe the motorist.; Answers will vary.

115.

; Answers will vary.

117.

a. $\{x \mid x \leq 0\}$ **b.** $\{x \mid x > 0\}$
119. Answers will vary; an example is $f(x) = \sqrt{15 - 3x}$.
121. $|(2x + 3)^5|$
123. The graph of h is the graph of f shifted left by 3 units.

y↑ $h(x) = \sqrt{x + 3}$

(6, 3)
(-2, 1) (1, 2)
(-3, 0)

124. $7x + 30$ **125.** $\dfrac{x^8}{9y^6}$ **126.** $\left\{ x \mid x < -\dfrac{7}{3} \text{ or } x > 5 \right\}$ or $\left(-\infty, -\dfrac{7}{3}\right) \cup (5, \infty)$

Section 7.2

Check Point Exercises

1. a. 5   **b.** -2 **c.** $\sqrt[4]{5xy^2}$ **2. a.** $(5xy)^{1/4}$ **b.** $\left(\dfrac{a^3b}{2}\right)^{1/5}$ **3. a.** $(\sqrt[3]{8})^4 = 16$ **b.** $(\sqrt{25})^3 = 125$ **c.** $-(\sqrt[4]{81})^3 = -27$ **4. a.** $6^{4/3}$ **b.** $(2xy)^{7/5}$

5. a. $\dfrac{1}{100^{1/2}} = \dfrac{1}{10}$ **b.** $\dfrac{1}{8^{1/3}} = \dfrac{1}{2}$ **c.** $\dfrac{1}{32^{3/5}} = \dfrac{1}{8}$ **d.** $\dfrac{1}{(3xy)^{5/9}}$ **6.** 25° **7. a.** $7^{5/6}$ **b.** $\dfrac{5}{x}$ **c.** $9.1^{3/10}$ **d.** $\dfrac{y^{1/12}}{x^{1/5}}$

8. a. $\sqrt{x}$ **b.** $2a^4$ **c.** $\sqrt[4]{x^2y}$ **d.** $\sqrt[6]{x}$ **e.** $\sqrt[6]{x}$

Exercise Set 7.2

1. $\sqrt{49} = 7$ **3.** $\sqrt[3]{-27} = -3$ **5.** $-\sqrt[4]{16} = -2$ **7.** $\sqrt[3]{xy}$ **9.** $\sqrt[5]{2xy^3}$ **11.** $(\sqrt{81})^3 = 729$ **13.** $(\sqrt[3]{125})^2 = 25$ **15.** $(\sqrt[5]{-32})^3 = -8$

17. $(\sqrt[3]{27})^2 + (\sqrt[4]{16})^3 = 17$ **19.** $\sqrt[7]{(xy)^4}$ **21.** $7^{1/2}$ **23.** $5^{1/3}$ **25.** $(11x)^{1/5}$ **27.** $x^{3/2}$ **29.** $x^{3/5}$ **31.** $(x^2y)^{1/5}$ **33.** $(19xy)^{3/2}$

35. $(7xy^2)^{5/6}$ **37.** $2xy^{2/3}$ **39.** $\dfrac{1}{49^{1/2}} = \dfrac{1}{7}$ **41.** $\dfrac{1}{27^{1/3}} = \dfrac{1}{3}$ **43.** $\dfrac{1}{16^{3/4}} = \dfrac{1}{8}$ **45.** $\dfrac{1}{8^{2/3}} = \dfrac{1}{4}$ **47.** $\left(\dfrac{27}{8}\right)^{1/3} = \dfrac{3}{2}$ **49.** $\dfrac{1}{(-64)^{2/3}} = \dfrac{1}{16}$

51. $\dfrac{1}{(2xy)^{7/10}}$ **53.** $\dfrac{5x}{z^{1/3}}$ **55.** 3 **57.** 4 **59.** $x^{5/6}$ **61.** $x^{3/5}$ **63.** $\dfrac{1}{x^{5/12}}$ **65.** 25 **67.** $\dfrac{1}{y^{1/6}}$ **69.** $32x$ **71.** $5x^2y^3$ **73.** $\dfrac{x^{1/4}}{y^{3/10}}$

75. 3 **77.** $27y^{2/3}$ **79.** $\sqrt[4]{x}$ **81.** $2a^2$ **83.** x^2y^3 **85.** x^6y^6 **87.** $\sqrt[5]{3y}$ **89.** $\sqrt[3]{4a^2}$ **91.** $\sqrt[3]{x^2y}$ **93.** $\sqrt[6]{2^5}$ or $\sqrt[6]{32}$ **95.** $\sqrt[10]{x^9}$

97. $\sqrt[12]{a^{10}b^7}$ **99.** $\sqrt[20]{x}$ **101.** $\sqrt{y}$ **103.** $\sqrt[3]{x}$ **105.** $\sqrt[4]{x^2y}$ **107.** $\sqrt[12]{2x}$ **109.** x^9y^{15} **111.** $\sqrt[4]{a^3b^3}$ **113.** $x^{2/3} - x$ **115.** $x + 2x^{1/2} - 15$

117. $2x^{1/2}(3 + x)$ **119.** $15x^{1/3}(1 - 4x^{2/3})$ **121.** $\dfrac{x^2}{7y^{3/2}}$ **123.** $\dfrac{x^3}{y^2}$ **125.** 58 species of plants **127.** about 1872 calories per day

129. a. $C(v) = 35.74 - 35.74v^{4/25}$ **b.** $C(25) \approx -24$; When the air temperature is 0°F and the wind speed is 25 miles per hour, the windchill temperature is $-24°$. **c.** the point $(25, -24)$ **131. a.** $L + 1.25S^{1/2} - 9.8D^{1/3} \le 16.296$ **b.** eligible

143. simplified correctly

X	Y₁	Y₂
1	1	1
2	1.4142	1.4142
3	1.7321	1.7321
4	2	2
5	2.2361	2.2361
6	2.4495	2.4495
7	2.6458	2.6458

X=5

145. Right side should be $x^{1/2}$.

X	Y₁	Y₂
1	1	1
2	1.4142	.70711
3	1.7321	.57735
4	2	.5
5	2.2361	.44721
6	2.4494	.40824
7	2.6457	.37796

X=1

147. 8 years old **149.** 61 **151.** $y = -2x + 11$ or $f(x) = -2x + 11$ **152.**

$y \le -\dfrac{3}{2}x + 3$

153. $(3, 4)$ or $\{(3, 4)\}$

Section 7.3

Check Point Exercises

1. a. $\sqrt{55}$ **b.** $\sqrt{x^2 - 16}$ **c.** $\sqrt[3]{60}$ **d.** $\sqrt[4]{12x^4}$ **2. a.** $4\sqrt{5}$ **b.** $2\sqrt[3]{5}$ **c.** $2\sqrt[4]{2}$ **d.** $10|x|\sqrt{2y}$
3. $f(x) = \sqrt{3}|x - 2|$ **4.** $x^4y^5z\sqrt{xyz}$ **5.** $2x^3y^4\sqrt[3]{5xy^2}$ **6.** $2x^2z\sqrt[5]{x^2y^2z^3}$ **7. a.** $2\sqrt{3}$ **b.** $100\sqrt[3]{4}$ **c.** $2x^2y\sqrt[4]{2}$

Exercise Set 7.3

1. $\sqrt{15}$ **3.** $\sqrt[3]{18}$ **5.** $\sqrt[4]{33}$ **7.** $\sqrt{33xy}$ **9.** $\sqrt[5]{24x^4}$ **11.** $\sqrt{x^2 - 9}$ **13.** $\sqrt[6]{(x - 4)^5}$ **15.** $\sqrt{x}$ **17.** $\sqrt[4]{\dfrac{3x}{7y}}$ **19.** $\sqrt[7]{77x^5y^3}$

21. $5\sqrt{2}$ **23.** $3\sqrt{5}$ **25.** $5\sqrt{3x}$ **27.** $2\sqrt[3]{2}$ **29.** $3x$ **31.** $-2y\sqrt[3]{2x^2}$ **33.** $6|x + 2|$ **35.** $2(x + 2)\sqrt[3]{4}$ **37.** $|x - 1|\sqrt{3}$

39. $x^3\sqrt{x}$ **41.** $x^4y^4\sqrt{y}$ **43.** $4x\sqrt{3x}$ **45.** $y^2\sqrt[3]{y^2}$ **47.** $x^4y\sqrt[3]{x^2z}$ **49.** $3x^2y^2\sqrt[3]{3x^2}$ **51.** $(x + y)\sqrt[5]{(x + y)^2}$ **53.** $y^3\sqrt[5]{y^2}$

55. $2xy^3\sqrt[5]{2xy^2}$ **57.** $2x^2\sqrt[4]{5x^2}$ **59.** $(x - 3)^2\sqrt[4]{(x - 3)^2}$ or $(x - 3)^2\sqrt{x - 3}$ **61.** $2\sqrt{6}$ **63.** $5\sqrt{2xy}$ **65.** $6x$ **67.** $10xy\sqrt{2y}$ **69.** $60\sqrt{2}$

71. $2\sqrt[3]{6}$ **73.** $2x^2\sqrt{10x}$ **75.** $5xy^4\sqrt[3]{x^2y^2}$ **77.** $2xyz\sqrt[4]{x^2z^3}$ **79.** $2xy^2z\sqrt[5]{2y^3z}$ **81.** $(x - y)^2\sqrt[3]{(x - y)^2}$ **83.** $-6x^3y^3\sqrt[3]{2yz^2}$

85. $-6y^2\sqrt[3]{2x^3y}$ **87.** $-6x^3y^3\sqrt{2}$ **89.** $-12x^3y^4\sqrt[4]{x^3}$ **91.** $xy^2\sqrt[5]{16y^3} + 2x^2y^3$ **93.** $6\sqrt{3}$ miles; 10.4 miles **95.** $8\sqrt{3}$ ft per sec; 14 ft per sec

97. a. $\dfrac{7.644}{2\sqrt{2}} = \dfrac{3.822}{\sqrt[4]{2}}$ **b.** 3.21 liters of blood per minute per square meter; (32, 3.21)

105. Graphs are the same; simplification is correct. **107.** Graphs are not the same; $\sqrt{3x^2 - 6x + 3} = |x - 1|\sqrt{3}$. **109.** d

111. The number must be multiplied by 27.

113.
$f(x) = \sqrt{(x-1)^2}$

114. $\{x | 5 \le x \le 11\}$ or $[5, 11]$

115. $(2, -4)$ or $\{(2, -4)\}$

116. $(4x - 3)(16x^2 + 12x + 9)$

Section 7.4

Check Point Exercises

1. a. $10\sqrt{13}$ **b.** $(21 - 6x)\sqrt[3]{7}$ **c.** $5\sqrt[4]{3x} + 2\sqrt[3]{3x}$ **2. a.** $21\sqrt{5}$ **b.** $-12\sqrt{3x}$ **c.** cannot be simplified

3. a. $-9\sqrt[3]{3}$ **b.** $(5 + 3xy)\sqrt[3]{x^2 y}$ **4. a.** $\dfrac{2\sqrt[3]{3}}{5}$ **b.** $\dfrac{3x\sqrt{x}}{y^5}$ **c.** $\dfrac{2y^2\sqrt[3]{y}}{x^4}$ **5. a.** $2x^2\sqrt{5}$ **b.** $\dfrac{5\sqrt{xy}}{2}$ **c.** $2x^2 y$

Exercise Set 7.4

1. $11\sqrt{5}$ **3.** $7\sqrt[3]{6}$ **5.** $2\sqrt[5]{2}$ **7.** $\sqrt{13} + 2\sqrt{5}$ **9.** $7\sqrt{5} + 2\sqrt[3]{x}$ **11.** $4\sqrt{3}$ **13.** $19\sqrt{3}$ **15.** $6\sqrt{2x}$ **17.** $13\sqrt[3]{2}$ **19.** $(9x + 1)\sqrt{5x}$

21. $7y\sqrt[3]{2x}$ **23.** $(3x - 2)\sqrt[3]{2x}$ **25.** $4\sqrt{x - 2}$ **27.** $5x\sqrt[3]{xy^2}$ **29.** $\dfrac{\sqrt{11}}{2}$ **31.** $\dfrac{\sqrt[3]{19}}{3}$ **33.** $\dfrac{x}{6y^4}$ **35.** $\dfrac{2x\sqrt{2x}}{5y^3}$ **37.** $\dfrac{x\sqrt[3]{x}}{2y}$ **39.** $\dfrac{x^2\sqrt[3]{50x^2}}{3y^4}$

41. $\dfrac{y\sqrt[4]{9y^2}}{x^2}$ **43.** $\dfrac{2x^2\sqrt[5]{2x^3}}{y^4}$ **45.** $2\sqrt{2}$ **47.** 2 **49.** $3x$ **51.** $x^2 y$ **53.** $2x^2\sqrt{5}$ **55.** $4a^5 b^5$ **57.** $3\sqrt{xy}$ **59.** $2xy$ **61.** $2x^2 y^2\sqrt[4]{y^2}$ or $2x^2 y^2\sqrt{y}$

63. $\sqrt[3]{x + 3}$ **65.** $\sqrt[3]{a^2 - ab + b^2}$ **67.** $\dfrac{43\sqrt{2}}{35}$ **69.** $-11xy\sqrt{2x}$ **71.** $25x\sqrt{2x}$ **73.** $7x\sqrt{3xy}$ **75.** $3x\sqrt[3]{5xy}$ **77.** $\left(\dfrac{f}{g}\right)(x) = 4x\sqrt{x}$;

Domain of $\dfrac{f}{g} = \{x | x > 0\}$ or $(0, \infty)$ **79.** $\left(\dfrac{f}{g}\right)(x) = 2x\sqrt[3]{2x}$; Domain of $\dfrac{f}{g} = \{x | x \text{ is a real number and } x \ne 0\}$ or $(-\infty, 0) \cup (0, \infty)$

81. a. $R_f \dfrac{\sqrt{c^2 - v^2}}{\sqrt{c^2}} = R_f\sqrt{\dfrac{c^2 - v^2}{c^2}} = R_f\sqrt{\dfrac{c^2}{c^2} - \dfrac{v^2}{c^2}} = R_f\sqrt{1 - \dfrac{v^2}{c^2}} = R_f\sqrt{1 - \left(\dfrac{v}{c}\right)^2}$ **b.** $R_f\sqrt{1 - \left(\dfrac{c}{c}\right)^2} = 0$; Aging rate is 0 relative to a friend on Earth;

A person moving at the speed of light does not age relative to a friend on Earth. **83.** $P = 18\sqrt{5}$ ft ; $A = 100$ sq ft

93. Graphs are not the same.; $\sqrt{16x} - \sqrt{9x} = \sqrt{x}$.

$y = \sqrt{7x}$ $y = \sqrt{16x} - \sqrt{9x}$

95. d **97.** $-\dfrac{31\sqrt{5}}{12}$

99. 0 or $\{0\}$

100. $(x - 2y)(x - 6y)$

101. $\dfrac{3x^2 + 8x + 6}{(x + 3)^2(x + 2)}$

Mid-Chapter 7 Check Point

1. 13 **2.** $2x^2 y^3\sqrt{2xy}$ **3.** $5\sqrt[4]{4x^2}$ **4.** $12x\sqrt[3]{2x}$ **5.** 1 **6.** $4xy^{1/12}$ **7.** $-\sqrt{3}$ **8.** $\dfrac{5x\sqrt{5x}}{y^2}$ **9.** $\sqrt[4]{x^3}$ **10.** $3x\sqrt[3]{2x^2}$ **11.** $2\sqrt[3]{10}$

12. $\dfrac{x^2}{y^4}$ **13.** $x^{7/12}$ **14.** $x\sqrt[6]{y^2}$ **15.** $(x - 2)\sqrt[4]{(x - 2)^2}$ **16.** $2x^2 y^4\sqrt[4]{2x^3 y}$ **17.** $14\sqrt[3]{2}$ **18.** $x\sqrt[6]{x^2 y^2}$ **19.** $\dfrac{1}{25}$ **20.** $\sqrt[6]{32}$

21. $\dfrac{4x}{y^2}$ **22.** $8x^2 y^3\sqrt{y}$ **23.** $-10x^3 y\sqrt{6y}$ **24.** $\{x | x \le 6\} = (-\infty, 6]$ **25.** $\{x | x \text{ is a real number}\} = (-\infty, \infty)$

Section 7.5

Check Point Exercises

1. a. $x\sqrt{6} + 2\sqrt{15}$ **b.** $y - \sqrt[3]{7y}$ **c.** $36 - 18\sqrt{10}$ **2. a.** $11 + 2\sqrt{30}$ **b.** 1 **c.** $a - 7$ **3. a.** $\dfrac{\sqrt{21}}{7}$ **b.** $\dfrac{\sqrt[3]{6}}{3}$

4. a. $\dfrac{\sqrt{14xy}}{7y}$ **b.** $\dfrac{\sqrt[3]{3xy^2}}{3y}$ **c.** $\dfrac{3\sqrt[5]{4x^3 y}}{y}$ **5.** $12\sqrt{3} - 18$ **6.** $\dfrac{3\sqrt{5} + 3\sqrt{2} + \sqrt{35} + \sqrt{14}}{3}$ **7.** $\dfrac{1}{\sqrt{x + 3} + \sqrt{x}}$

Exercise Set 7.5

1. $x\sqrt{2} + \sqrt{14}$ **3.** $7\sqrt{6} - 6$ **5.** $12\sqrt{2} - 6$ **7.** $\sqrt[3]{12} + 4\sqrt[3]{10}$ **9.** $2x\sqrt[3]{2} - \sqrt[3]{x^2}$ **11.** $32 + 11\sqrt{2}$ **13.** $34 - 15\sqrt{5}$

15. $117 - 36\sqrt{7}$ **17.** $\sqrt{6} + \sqrt{10} + \sqrt{21} + \sqrt{35}$ **19.** $\sqrt{6} - \sqrt{10} - \sqrt{21} + \sqrt{35}$ **21.** $-48 + 7\sqrt{6}$ **23.** $8 + 2\sqrt{15}$

25. $3x - 2\sqrt{3xy} + y$ **27.** -44 **29.** -71 **31.** 6 **33.** $6 - 5\sqrt{x} + x$ **35.** $\sqrt[3]{x^2} + \sqrt[3]{x} - 20$ **37.** $2x^2 + x\sqrt[3]{y^2} - y\sqrt[3]{y}$ **39.** $\dfrac{\sqrt{10}}{5}$

41. $\dfrac{\sqrt{11x}}{x}$ **43.** $\dfrac{3\sqrt{3y}}{y}$ **45.** $\dfrac{\sqrt[3]{4}}{2}$ **47.** $3\sqrt[3]{2}$ **49.** $\dfrac{\sqrt[3]{18}}{3}$ **51.** $\dfrac{4\sqrt[3]{x^2}}{x}$ **53.** $\dfrac{\sqrt[3]{2y}}{y}$ **55.** $\dfrac{7\sqrt[3]{4x}}{2x}$ **57.** $\dfrac{\sqrt[3]{2x^2y}}{xy}$ **59.** $\dfrac{3\sqrt[4]{x^3}}{x}$

61. $\dfrac{3\sqrt[5]{4x^2}}{x}$ **63.** $x\sqrt[5]{8x^3y}$ **65.** $\dfrac{3\sqrt[3]{3y}}{xy}$ **67.** $-\dfrac{5a^2\sqrt{3ab}}{b^2}$ **69.** $\dfrac{\sqrt{2mn}}{2m}$ **71.** $\dfrac{3\sqrt[4]{x^3y}}{x^2y}$ **73.** $-\dfrac{6\sqrt[3]{xy}}{x^2y^3}$ **75.** $8\sqrt{5} - 16$ **77.** $\dfrac{13\sqrt{11} + 39}{2}$

79. $3\sqrt{5} - 3\sqrt{3}$ **81.** $\dfrac{a + \sqrt{ab}}{a - b}$ **83.** $25\sqrt{2} + 15\sqrt{5}$ **85.** $4 + \sqrt{15}$ **87.** $\dfrac{x - 2\sqrt{x} - 3}{x - 9}$ **89.** $\dfrac{3\sqrt{6} + 4}{2}$ **91.** $\dfrac{4\sqrt{xy} + 4x + y}{y - 4x}$ **93.** $\dfrac{3}{\sqrt{6}}$

95. $\dfrac{2x}{\sqrt[3]{2x^2y}}$ **97.** $\dfrac{x - 9}{x - 3\sqrt{x}}$ **99.** $\dfrac{a - b}{a - 2\sqrt{ab} + b}$ **101.** $\dfrac{1}{\sqrt{x + 5} + \sqrt{x}}$ **103.** $\dfrac{1}{(x + y)(\sqrt{x} - \sqrt{y})}$ **105.** $\dfrac{3\sqrt{2}}{2}$ **107.** $-2\sqrt[3]{25}$ **109.** $\dfrac{7\sqrt{6}}{6}$

111. $7\sqrt{3} - 7\sqrt{2}$ **113.** 0 **115.** 6 **117.** 51; In 2001, approximately 51% of U.S. households were online.; Answer models data well. **119.** 6.5%

121. approximately 6.5%; This answer models the actual yearly percentage increase well. **123. a.** $\dfrac{15.92}{\sqrt{t + h} + \sqrt{t}}$ **b.** $\dfrac{15.92}{2\sqrt{t}}$ or $\dfrac{7.96}{\sqrt{t}}$ **c.** 3.2%

125. $P = 8\sqrt{2}$ in.; $A = 7$ sq in. **127.** $\dfrac{7\sqrt{2 \cdot 2 \cdot 3}}{6} = \dfrac{7 \cdot 2\sqrt{3}}{6} = \dfrac{14\sqrt{3}}{6} = \dfrac{7\sqrt{3}}{3} = \dfrac{7}{3}\sqrt{3}$

139. $y = x^2 - 4$ $y = (\sqrt{x} + 2)(\sqrt{x} - 2)$

141. $y = \sqrt{x + 3} + \sqrt{x}$

143. $\dfrac{45}{7}$ or $\left\{\dfrac{45}{7}\right\}$

145. $\dfrac{5\sqrt{2} + 3\sqrt{3} - 4\sqrt{6} + 2}{23}$

146. $\dfrac{2x + 7}{x^2 - 4}$

147. $-\dfrac{7}{2}$ or $\left\{-\dfrac{7}{2}\right\}$

148. 13

Graphs are not the same;
$(\sqrt{x} + 2)(\sqrt{x} - 2) = x - 4$ for $x \geq 0$.

Graphs are the same; Simplification is correct.

Section 7.6

Check Point Exercises

1. 20 or $\{20\}$ **2.** no solution or $\varnothing$ **3.** -1 and 3, or $\{-1, 3\}$ **4.** 4 or $\{4\}$ **5.** -12 or $\{-12\}$ **6.** 2009; faster

Exercise Set 7.6

1. 6 or $\{6\}$ **3.** 17 or $\{17\}$ **5.** no solution or $\varnothing$ **7.** 8 or $\{8\}$ **9.** 0 and 3, or $\{0, 3\}$ **11.** 1 and 3, or $\{1, 3\}$ **13.** 3 and 7, or $\{3, 7\}$ **15.** 9 or $\{9\}$

17. 8 or $\{8\}$ **19.** 35 or $\{35\}$ **21.** 16 or $\{16\}$ **23.** 2 or $\{2\}$ **25.** 2 and 6, or $\{2, 6\}$ **27.** $\dfrac{5}{2}$ or $\left\{\dfrac{5}{2}\right\}$ **29.** 5 or $\{5\}$ **31.** no solution or $\varnothing$

33. 2 and 6, or $\{2, 6\}$ **35.** 0 and 10, or $\{0, 10\}$ **37.** 8 or $\{8\}$ **39.** 4 **41.** -7 **43.** $V = \dfrac{\pi r^2 h}{3}$ or $\dfrac{1}{3}\pi r^2 h$ **45.** $l = \dfrac{8t^2}{\pi^2}$ **47.** 8 **49.** 9

51. High: $66.0°$; Low: $61.5°$ **53.** Using f: 2014; Using g: 2010 **55.** 36 years old; (36, 40,000) **57.** 27 sq mi **59.** 149 million km

69. 4 or $\{4\}$ **71.** 1 and 9, or $\{1, 9\}$ **73.** c **75.** 81 or $\{81\}$ **77.** 129 or $\{129\}$

78. $4x^3 - 15x^2 + 47x - 142 + \dfrac{425}{x + 3}$

79. $\dfrac{x + 2}{2(x + 3)}$

80. $(y - 3 + 5x)(y - 3 - 5x)$

Section 7.7

Check Point Exercises

1. a. $8i$ **b.** $\sqrt{11}i$ **c.** $4\sqrt{3}i$ **2. a.** $8 + i$ **b.** $-10 + 10i$ **3. a.** $63 + 14i$ **b.** $58 - 11i$

4. $-\sqrt{35}$ **5.** $\dfrac{18}{25} + \dfrac{26}{25}i$ **6.** $-\dfrac{1}{2} - \dfrac{3}{4}i$ **7. a.** 1 **b.** i **c.** $-i$

Exercise Set 7.7

1. $10i$ **3.** $\sqrt{23}i$ **5.** $3\sqrt{2}i$ **7.** $3\sqrt{7}i$ **9.** $-6\sqrt{3}i$ **11.** $5 + 6i$ **13.** $15 + \sqrt{3}i$ **15.** $-2 - 3\sqrt{2}i$ **17.** $8 + 3i$ **19.** $8 - 2i$
21. $5 + 3i$ **23.** $-1 - 7i$ **25.** $-2 + 9i$ **27.** $8 + 15i$ **29.** $-14 + 17i$ **31.** $9 + 5\sqrt{3}i$ **33.** $-6 + 10i$ **35.** $-21 - 15i$
37. $-35 - 14i$ **39.** $7 + 19i$ **41.** $-1 - 31i$ **43.** $3 + 36i$ **45.** 34 **47.** 34 **49.** 11 **51.** $-5 + 12i$ **53.** $21 - 20i$ **55.** $-\sqrt{14}$
57. -6 **59.** $-5\sqrt{7}$ **61.** $-2\sqrt{6}$ **63.** $\frac{3}{5} - \frac{1}{5}i$ **65.** $1 + i$ **67.** $\frac{28}{25} + \frac{21}{25}i$ **69.** $-\frac{12}{13} + \frac{18}{13}i$ **71.** $0 + i$ or i **73.** $\frac{3}{10} - \frac{11}{10}i$
75. $\frac{11}{13} - \frac{16}{13}i$ **77.** $-\frac{23}{58} + \frac{43}{58}i$ **79.** $0 - \frac{7}{3}i$ or $-\frac{7}{3}i$ **81.** $-\frac{5}{2} - 4i$ **83.** $-\frac{7}{3} + \frac{4}{3}i$ **85.** -1 **87.** $-i$ **89.** -1 **91.** 1
93. i **95.** 1 **97.** $-i$ **99.** 0 **101.** $-11 - 5i$ **103.** $-5 + 10i$ **105.** $0 + 47i$ or $47i$ **107.** $1 - i$ **109.** 0 **111.** $10 + 10i$
113. $\frac{20}{13} + \frac{30}{13}i$ **115.** $(47 + 13i)$ volts **117.** $(5 + \sqrt{15}i) + (5 - \sqrt{15}i) = 10$; $(5 + \sqrt{15}i)(5 - \sqrt{15}i) = 25 - 15i^2 = 25 + 15 = 40$
131. $\sqrt{-9} + \sqrt{-16} = 3i + 4i = 7i$ **133.** d **135.** $\frac{6}{5} + 0i$ or $\frac{6}{5}$ **137.** $\frac{x^2}{y^2}$ **138.** $x = \frac{yz}{y - z}$ **139.** $\frac{21}{11}$ or $\left\{\frac{21}{11}\right\}$

Review Exercises

1. 9 **2.** $-\frac{1}{10}$ **3.** -3 **4.** not a real number **5.** -2 **6.** $5; 1.73; 0;$ not a real number **7.** $2; -2; -4$ **8.** $\{x|x \geq 2\}$ or $[2, \infty)$
9. $\{x|x \leq 25\}$ or $(-\infty, 25]$ **10.** $5|x|$ **11.** $|x + 14|$ **12.** $|x - 4|$ **13.** $4x$ **14.** $2|x|$ **15.** $-2(x + 7)$ **16.** $\sqrt[3]{5xy}$
17. $(\sqrt{16})^3 = 64$ **18.** $(\sqrt[5]{32})^4 = 16$ **19.** $(7x)^{1/2}$ **20.** $(19xy)^{5/3}$ **21.** $\frac{1}{8^{2/3}} = \frac{1}{4}$ **22.** $\frac{3x}{(ab)^{4/5}} = \frac{3x}{\sqrt[5]{(ab)^4}}$ **23.** $x^{7/12}$ **24.** $5^{1/6}$
25. $2x^2y$ **26.** $\frac{y^{1/8}}{x^{1/3}}$ **27.** x^3y^4 **28.** $y\sqrt[3]{x}$ **29.** $\sqrt[6]{x^5}$ **30.** $\sqrt[6]{x}$ **31.** $\sqrt[15]{x}$ **32.** $\$3150$ million **33.** $\sqrt{21xy}$ **34.** $\sqrt[5]{77x^3}$
35. $\sqrt[6]{(x - 5)^5}$ **36.** $f(x) = \sqrt{7}|x - 1|$ **37.** $2x\sqrt{5x}$ **38.** $3x^2y^2\sqrt[3]{2x^2}$ **39.** $2y^2z\sqrt[4]{2x^3y^3z}$ **40.** $2x^2\sqrt{6x}$ **41.** $2xy\sqrt[3]{2y^2}$
42. $xyz^2\sqrt[5]{16y^4z}$ **43.** $\sqrt{x^2 - 1}$ **44.** $8\sqrt[3]{3}$ **45.** $9\sqrt{2}$ **46.** $(3x + y^2)\sqrt[3]{x}$ **47.** $-8\sqrt[3]{6}$ **48.** $\frac{2}{5}\sqrt[3]{2}$ **49.** $\frac{x\sqrt{x}}{10y^2}$ **50.** $\frac{y\sqrt[4]{3y}}{2x^5}$
51. $2\sqrt{6}$ **52.** $2\sqrt[3]{2}$ **53.** $2x\sqrt[4]{2x}$ **54.** $10x^2\sqrt{xy}$ **55.** $6\sqrt{2} + 12\sqrt{5}$ **56.** $5\sqrt[3]{2} - \sqrt[3]{10}$ **57.** $-83 + 3\sqrt{35}$
58. $\sqrt{xy} - \sqrt{11x} - \sqrt{11y} + 11$ **59.** $13 + 4\sqrt{10}$ **60.** $22 - 4\sqrt{30}$ **61.** -6 **62.** 4 **63.** $\frac{2\sqrt{6}}{3}$ **64.** $\frac{\sqrt{14}}{7}$ **65.** $4\sqrt[3]{3}$
66. $\frac{\sqrt{10xy}}{5y}$ **67.** $\frac{7\sqrt[3]{4x}}{x}$ **68.** $\frac{\sqrt[4]{189x^3}}{3x}$ **69.** $\frac{5\sqrt[5]{xy^4}}{2xy}$ **70.** $3\sqrt{3} + 3$ **71.** $\frac{\sqrt{35} - \sqrt{21}}{2}$ **72.** $10\sqrt{5} + 15\sqrt{2}$ **73.** $\frac{x + 8\sqrt{x} + 15}{x - 9}$
74. $\frac{5 + \sqrt{21}}{2}$ **75.** $\frac{3\sqrt{2} + 2}{7}$ **76.** $\frac{2}{\sqrt{14}}$ **77.** $\frac{3x}{\sqrt[3]{9x^2y}}$ **78.** $\frac{7}{\sqrt{35} + \sqrt{21}}$ **79.** $\frac{2}{5 - \sqrt{21}}$ **80.** 16 or $\{16\}$ **81.** no solution or $\varnothing$
82. 2 or $\{2\}$ **83.** 8 or $\{8\}$ **84.** -4 and -2, or $\{-4, -2\}$ **85.** 256 feet **86.** 84 years old **87.** $9i$ **88.** $3\sqrt{7}i$ **89.** $-2\sqrt{2}i$
90. $12 + 2i$ **91.** $-9 + 4i$ **92.** $-12 - 8i$ **93.** $29 + 11i$ **94.** $-7 - 24i$ **95.** $113 + 0i$ or 113 **96.** $-2\sqrt{6} + 0i$ or $-2\sqrt{6}$
97. $\frac{15}{13} - \frac{3}{13}i$ **98.** $\frac{1}{5} + \frac{11}{10}i$ **99.** $\frac{1}{3} - \frac{5}{3}i$ **100.** 1 **101.** $-i$

Chapter 7 Test

1. a. 6 **b.** $\{x|x \leq 4\}$ or $(-\infty, 4]$ **2.** $\frac{1}{81}$ **3.** $\frac{5\sqrt[8]{y}}{\sqrt[4]{x}}$ **4.** $\sqrt{x}$ **5.** $\sqrt[20]{x^9}$ **6.** $5|x|\sqrt{3}$ **7.** $|x - 5|$ **8.** $2xy^2\sqrt[3]{xy^2}$ **9.** $-\frac{2}{x^2}$
10. $\sqrt[3]{50x^2y}$ **11.** $2x\sqrt[4]{2y^3}$ **12.** $-7\sqrt{2}$ **13.** $(2x + y^2)\sqrt[3]{x}$ **14.** $2x\sqrt[3]{x}$ **15.** $12\sqrt{2} - \sqrt{15}$ **16.** $26 + 6\sqrt{3}$ **17.** $52 - 14\sqrt{3}$
18. $\frac{\sqrt{5x}}{x}$ **19.** $\frac{\sqrt[3]{25x}}{x}$ **20.** $-5 + 2\sqrt{6}$ **21.** 6 or $\{6\}$ **22.** 16 or $\{16\}$ **23.** -3 or $\{-3\}$ **24.** 49 months **25.** $5\sqrt{3}i$ **26.** $-1 + 6i$
27. $26 + 7i$ **28.** $-6 + 0i$ or -6 **29.** $\frac{1}{5} + \frac{7}{5}i$ **30.** $-i$

Cumulative Review Exercises (Chapters 1–7)

1. $(-2, -1, -2)$ or $\{(-2, -1, -2)\}$ **2.** $-\frac{1}{3}$ and 4, or $\left\{-\frac{1}{3}, 4\right\}$ **3.** $\left\{x \middle| x > \frac{1}{3}\right\}$ or $\left(\frac{1}{3}, \infty\right)$ **4.** $\frac{3}{4}$ or $\left\{\frac{3}{4}\right\}$ **5.** -1 or $\{-1\}$
6.
$x + 2y < 2$
$2y - x > 4$
7. $\frac{x^2}{15(x + 2)}$ **8.** $\frac{x}{y}$ **9.** $8x^3 - 22x^2 + 11x + 6$ **10.** $\frac{5x - 6}{(x - 5)(x + 3)}$ **11.** -64 **12.** 0 **13.** $-\frac{16 - 9\sqrt{3}}{13}$
14. $2x^2 + x + 5 + \frac{6}{x - 2}$ **15.** $-34 - 3\sqrt{6}$ **16.** $2(3x + 2)(4x - 1)$ **17.** $(4x^2 + 1)(2x + 1)(2x - 1)$
18. about 53 lumens **19.** $\$1500$ at 7% and $\$4500$ at 9% **20.** 2650 students

CHAPTER 8

Section 8.1

Check Point Exercises

1. $\pm\sqrt{7}$ or $\{\pm\sqrt{7}\}$ **2.** $\pm\dfrac{\sqrt{33}}{3}$ or $\left\{\pm\dfrac{\sqrt{33}}{3}\right\}$ **3.** $\pm\dfrac{3}{2}i$ or $\left\{\pm\dfrac{3}{2}i\right\}$ **4.** $3 \pm \sqrt{10}$ or $\{3 \pm \sqrt{10}\}$ **5. a.** 25; $x^2 + 10x + 25 = (x + 5)^2$

b. $\dfrac{9}{4}$; $x^2 - 3x + \dfrac{9}{4} = \left(x - \dfrac{3}{2}\right)^2$ **c.** $\dfrac{9}{64}$; $x^2 + \dfrac{3}{4}x + \dfrac{9}{64} = \left(x + \dfrac{3}{8}\right)^2$ **6.** $-2 \pm \sqrt{5}$ or $\{-2 \pm \sqrt{5}\}$ **7.** $\dfrac{-3 \pm \sqrt{41}}{4}$ or $\left\{\dfrac{-3 \pm \sqrt{41}}{4}\right\}$

8. 20% **9.** $10\sqrt{21}$ ft; 45.8 ft

Exercise Set 8.1

1. ± 5 or $\{\pm 5\}$ **3.** $\pm\sqrt{6}$ or $\{\pm\sqrt{6}\}$ **5.** $\pm\dfrac{5}{4}$ or $\left\{\pm\dfrac{5}{4}\right\}$ **7.** $\pm\dfrac{\sqrt{6}}{3}$ or $\left\{\pm\dfrac{\sqrt{6}}{3}\right\}$ **9.** $\pm\dfrac{4}{5}i$ or $\left\{\pm\dfrac{4}{5}i\right\}$ **11.** -10 and -4, or $\{-10, -4\}$

13. $3 \pm \sqrt{5}$ or $\{3 \pm \sqrt{5}\}$ **15.** $-2 \pm 2\sqrt{2}$ or $\{-2 \pm 2\sqrt{2}\}$ **17.** $5 \pm 3i$ or $\{5 \pm 3i\}$ **19.** $\dfrac{-3 \pm \sqrt{11}}{4}$ or $\left\{\dfrac{-3 \pm \sqrt{11}}{4}\right\}$

21. -3 and 9, or $\{-3, 9\}$ **23.** 1; $x^2 + 2x + 1 = (x + 1)^2$ **25.** 49; $x^2 - 14x + 49 = (x - 7)^2$ **27.** $\dfrac{49}{4}$; $x^2 + 7x + \dfrac{49}{4} = \left(x + \dfrac{7}{2}\right)^2$

29. $\dfrac{1}{16}$; $x^2 - \dfrac{1}{2}x + \dfrac{1}{16} = \left(x - \dfrac{1}{4}\right)^2$ **31.** $\dfrac{4}{9}$; $x^2 + \dfrac{4}{3}x + \dfrac{4}{9} = \left(x + \dfrac{2}{3}\right)^2$ **33.** $\dfrac{81}{64}$; $x^2 - \dfrac{9}{4}x + \dfrac{81}{64} = \left(x - \dfrac{9}{8}\right)^2$ **35.** -8 and 4, or $\{-8, 4\}$

37. $-3 \pm \sqrt{7}$ or $\{-3 \pm \sqrt{7}\}$ **39.** $4 \pm \sqrt{15}$ or $\{4 \pm \sqrt{15}\}$ **41.** $-1 \pm i$ or $\{-1 \pm i\}$ **43.** $\dfrac{-3 \pm \sqrt{13}}{2}$ or $\left\{\dfrac{-3 \pm \sqrt{13}}{2}\right\}$

45. $-\dfrac{3}{7}$ and $-\dfrac{1}{7}$, or $\left\{-\dfrac{3}{7}, -\dfrac{1}{7}\right\}$ **47.** $\dfrac{-1 \pm \sqrt{5}}{2}$ or $\left\{\dfrac{-1 \pm \sqrt{5}}{2}\right\}$ **49.** $-\dfrac{5}{2}$ and 1, or $\left\{-\dfrac{5}{2}, 1\right\}$ **51.** $\dfrac{-3 \pm \sqrt{6}}{3}$ or $\left\{\dfrac{-3 \pm \sqrt{6}}{3}\right\}$

53. $\dfrac{4 \pm \sqrt{13}}{3}$ or $\left\{\dfrac{4 \pm \sqrt{13}}{3}\right\}$ **55.** $\dfrac{1}{4} \pm \dfrac{1}{4}i$ or $\left\{\dfrac{1}{4} \pm \dfrac{1}{4}i\right\}$ **57.** $-5, 7$ **59.** $-\dfrac{1}{5}, 1$ **61.** $-2 \pm 5i$ **63.** $2 \pm 2i$ **65.** $v = \sqrt{2gh}$

67. $r = \dfrac{\sqrt{AP}}{P} - 1$ **69.** $\dfrac{-1 \pm \sqrt{19}}{6}$ or $\left\{\dfrac{-1 \pm \sqrt{19}}{6}\right\}$ **71.** $-b$ and $2b$, or $\{-b, 2b\}$ **73.** 20% **75.** 6.25% **77.** approximately 2007

79. $10\sqrt{3}$ sec; 17.3 sec **81.** $3\sqrt{5}$ mi; 6.7 mi **83.** $20\sqrt{2}$ ft, or 28.3 ft **85.** $50\sqrt{2}$ ft; 70.7 ft **87.** 10 m

97.

-3 and 1, or $\{-3, 1\}$

101. $y = \pm\dfrac{b\sqrt{a^2 - x^2}}{a}$ **103.** $\dfrac{-b \pm \sqrt{b^2 - 4c}}{2}$ or $\left\{\dfrac{-b \pm \sqrt{b^2 - 4c}}{2}\right\}$

105. $4 - 2x$ **106.** $(1 - 2x)(1 + 2x + 4x^2)$

107. $x^3 - 2x^2 - 4x - 12 - \dfrac{42}{x - 3}$

Section 8.2

Check Point Exercises

1. -5 and $\dfrac{1}{2}$, or $\left\{-5, \dfrac{1}{2}\right\}$ **2.** $\dfrac{3 \pm \sqrt{7}}{2}$ or $\left\{\dfrac{3 \pm \sqrt{7}}{2}\right\}$ **3.** $-1 \pm \dfrac{\sqrt{6}}{3}i$ or $\left\{-1 \pm \dfrac{\sqrt{6}}{3}i\right\}$

4. a. 0; one real solution **b.** 81; two rational solutions **c.** -44; two imaginary solutions that are complex conjugates

5. a. $20x^2 + 7x - 3 = 0$ **b.** $x^2 + 49 = 0$ **6.** 1960 and 1974; very well

Exercise Set 8.2

1. -6 and -2, or $\{-6, -2\}$ **3.** 1 and $\dfrac{5}{2}$, or $\left\{1, \dfrac{5}{2}\right\}$ **5.** $\dfrac{-3 \pm \sqrt{89}}{2}$ or $\left\{\dfrac{-3 \pm \sqrt{89}}{2}\right\}$ **7.** $\dfrac{7 \pm \sqrt{85}}{6}$ or $\left\{\dfrac{7 \pm \sqrt{85}}{6}\right\}$

9. $\dfrac{1 \pm \sqrt{7}}{6}$ or $\left\{\dfrac{1 \pm \sqrt{7}}{6}\right\}$ **11.** $\dfrac{3}{8} \pm \dfrac{\sqrt{87}}{8}i$ or $\left\{\dfrac{3}{8} \pm \dfrac{\sqrt{87}}{8}i\right\}$ **13.** $2 \pm 2i$ or $\{2 \pm 2i\}$ **15.** $\dfrac{4}{3} \pm \dfrac{\sqrt{5}}{3}i$ or $\left\{\dfrac{4}{3} \pm \dfrac{\sqrt{5}}{3}i\right\}$

17. $-\dfrac{3}{2}$ and 4, or $\left\{-\dfrac{3}{2}, 4\right\}$ **19.** 52; two irrational solutions   **21.** 4; two rational solutions   **23.** -23; two imaginary solutions

25. 36; two rational solutions   **27.** -60; two imaginary solutions   **29.** 0; one (repeated) rational solution   **31.** $-\dfrac{2}{3}$ and 2, or $\left\{-\dfrac{2}{3}, 2\right\}$

33. $1 \pm \sqrt{2}$ or $\{1 \pm \sqrt{2}\}$ **35.** $\dfrac{3 \pm \sqrt{65}}{4}$ or $\left\{\dfrac{3 \pm \sqrt{65}}{4}\right\}$ **37.** 0 and $\dfrac{8}{3}$, or $\left\{0, \dfrac{8}{3}\right\}$ **39.** $\dfrac{-6 \pm 2\sqrt{6}}{3}$ or $\left\{\dfrac{-6 \pm 2\sqrt{6}}{3}\right\}$

41. $\dfrac{2 \pm \sqrt{10}}{3}$ or $\left\{\dfrac{2 \pm \sqrt{10}}{3}\right\}$ **43.** $2 \pm \sqrt{10}$ or $\{2 \pm \sqrt{10}\}$ **45.** 1 and $\dfrac{5}{2}$, or $\left\{1, \dfrac{5}{2}\right\}$ **47.** $x^2 - 2x - 15 = 0$ **49.** $12x^2 + 5x - 2 = 0$

51. $x^2 + 36 = 0$ **53.** $x^2 - 2 = 0$ **55.** $x^2 - 20 = 0$ **57.** $x^2 - 2x + 2 = 0$ **59.** $x^2 - 2x - 1 = 0$ **61.** b   **63.** a   **65.** $1 + \sqrt{7}$

67. $\dfrac{-1 \pm \sqrt{21}}{2}$ or $\left\{\dfrac{-1 \pm \sqrt{21}}{2}\right\}$ **69.** $-2\sqrt{2}$ and $\dfrac{\sqrt{2}}{2}$, or $\left\{-2\sqrt{2}, \dfrac{\sqrt{2}}{2}\right\}$ **71.** $-3, 1,$ and $-1 \pm \sqrt{2}i$, or $\{-3, 1, -1 \pm \sqrt{2}i\}$

73. 33-year-olds and 58-year-olds; The function models the actual data well.   **75.** 77.8 ft; (b)   **77.** 5.5 m by 1.5 m   **79.** 17.6 in. and 18.6 in.

81. 9.3 in. and 0.7 in.   **83.** 7.5 hr and 8.5 hr   **95.** d   **97.** 2.4 m; yes   **99.** -3 and $\dfrac{1}{4}$, or $\left\{-3, \dfrac{1}{4}\right\}$

100. 3 and 7, or $\{3, 7\}$ **101.** $\dfrac{5\sqrt{3} - 5x}{3 - x^2}$

Section 8.3

Check Point Exercises

1.

2.

3. Domain = $\{x \mid x \text{ is a real number}\}$ or $(-\infty, \infty)$;
Range = $\{y \mid y \le 5\}$ or $(-\infty, 5]$

4. 33.7 ft   **5.** $4, -4; -16$
6. 30 ft by 30 ft; 900 sq ft

Exercise Set 8.3

1. $h(x) = (x - 1)^2 + 1$ **3.** $j(x) = (x - 1)^2 - 1$ **5.** $h(x) = x^2 - 1$ **7.** $g(x) = x^2 - 2x + 1$

9. $(3, 1)$ **11.** $(-1, 5)$ **13.** $(2, -5)$ **15.** $(-1, 9)$

17.
$x = 4$; $\{y \mid y \ge -1\}$ or $[-1, \infty)$

19.
$x = 1$; $\{y \mid y \ge 2\}$ or $[2, \infty)$

21.
$x = 3$; $\{y \mid y \ge 1\}$ or $[1, \infty)$

23.
$x = -2$; $\{y \mid y \ge -1\}$ or $[-1, \infty)$

25.
$x = 1$; $\{y \mid y \le 4\}$ or $(-\infty, 4]$

27.
$x = 1$; $\{y \mid y \ge -4\}$ or $[-4, \infty)$

29.
$x = -\dfrac{3}{2}$; $\left\{y \mid y \ge -\dfrac{49}{4}\right\}$ or $\left[-\dfrac{49}{4}, \infty\right)$

31.
$x = 1$; $\{y \mid y \le 4\}$ or $(-\infty, 4]$

33.
$x = 1$; $\{y \mid y \le -1\}$ or $(-\infty, -1]$

35. minimum; $(2, -13)$ **37.** maximum; $(1, 1)$ **39.** minimum; $\left(\dfrac{1}{2}, -\dfrac{5}{4}\right)$

41. Domain $= \{x \mid x \text{ is a real number}\}$ or $(-\infty, \infty)$; Range $= \{y \mid y \geq -2\}$ or $[-2, \infty)$

43. Domain $= \{x \mid x \text{ is a real number}\}$ or $(-\infty, \infty)$; Range $= \{y \mid y \leq 3\}$ or $(-\infty, 3]$ **45.** $f(x) = 2(x - 5)^2 + 3$ **47.** $f(x) = 2(x + 10)^2 - 5$

49. $f(x) = -3(x + 2)^2 + 4$ **51.** $f(x) = 3(x - 11)^2$ **53.** $(14, 19.5)$; In 2016, the projected mortality will reach a maximum of 19,500.

55. a. 2 sec; 224 ft

b. 5.7 sec

c. 160; 160 feet is the height of the building.

d.

77. a.

b. $(20.5, -120.5)$

c.

d. Answers will vary.

57. 7.2 hr; 622 per 100,000 males

59. 8 and 8; 64

61. 8, −8; −64

63. length: 300 ft; width: 150 ft; maximum area: 45,000 sq ft

65. 12.5 yd by 12.5 yd; 156.25 sq yd

67. 5 in.; 50 sq in.

69. a. $C(x) = 525 + 0.55x$

b. $P(x) = -0.001x^2 + 2.45x - 525$

c. 1225 sandwiches; $975.63

79. $(2.5, 185)$

81. $(-30, 91)$

83. a   **85.** $x = 3$; $(0, 11)$

87. $f(x) = -2(x + 3)^2 - 1$

89. 65 trees; 16,900 lb

90. 7 or $\{7\}$

91. $\dfrac{x}{x - 2}$

92. $(6, -2)$ or $\{(6, -2)\}$

Mid-Chapter 8 Check Point

1. $-\dfrac{1}{3}$ and $\dfrac{11}{3}$, or $\left\{-\dfrac{1}{3}, \dfrac{11}{3}\right\}$ **2.** -1 and $\dfrac{7}{5}$, or $\left\{-1, \dfrac{7}{5}\right\}$ **3.** $\dfrac{3 \pm \sqrt{15}}{3}$ or $\left\{\dfrac{3 \pm \sqrt{15}}{3}\right\}$ **4.** $-3 \pm \sqrt{7}$ or $\{-3 \pm \sqrt{7}\}$ **5.** $\pm \dfrac{6\sqrt{5}}{5}$ or $\left\{\pm \dfrac{6\sqrt{5}}{5}\right\}$

6. $\dfrac{5}{2} \pm \dfrac{\sqrt{7}}{2}i$ or $\left\{\dfrac{5}{2} \pm \dfrac{\sqrt{7}}{2}i\right\}$ **7.** $\pm\sqrt{13}i$ or $\{\pm\sqrt{13}i\}$ **8.** -4 and $\dfrac{1}{2}$, or $\left\{-4, \dfrac{1}{2}\right\}$ **9.** $-3 \pm 2\sqrt{6}$ or $\{-3 \pm 2\sqrt{6}\}$ **10.** $2 \pm \sqrt{3}$ or $\{2 \pm \sqrt{3}\}$

11. $\dfrac{3}{4} \pm \dfrac{\sqrt{23}}{4}i$ or $\left\{\dfrac{3}{4} \pm \dfrac{\sqrt{23}}{4}i\right\}$ **12.** $\dfrac{-3 \pm \sqrt{41}}{4}$ or $\left\{\dfrac{-3 \pm \sqrt{41}}{4}\right\}$ **13.** -2 and 4, or $\{-2, 4\}$ **14.** $-5 \pm 2\sqrt{7}$ or $\{-5 \pm 2\sqrt{7}\}$

15.

$f(x) = (x - 3)^2 - 4$
$(0, 5)$
$(1, 0)$ $(5, 0)$
$(3, -4)$

Domain $= \{x \mid x \text{ is a real number}\}$ or $(-\infty, \infty)$; Range $= \{y \mid y \geq -4\}$ or $[-4, \infty)$

16.

$g(x) = 5 - (x + 2)^2$
$(-2, 5)$
$(0, 1)$
$(-2-\sqrt{5}, 0)$
$(-2+\sqrt{5}, 0)$

Domain $= \{x \mid x \text{ is a real number}\}$ or $(-\infty, \infty)$; Range $= \{y \mid y \leq 5\}$ or $(-\infty, 5]$

17.

$h(x) = -x^2 - 4x + 5$
$(-2, 9)$
$(0, 5)$
$(-5, 0)$
$(1, 0)$

Domain $= \{x \mid x \text{ is a real number}\}$ or $(-\infty, \infty)$; Range $= \{y \mid y \leq 9\}$ or $(-\infty, 9]$

18.

$f(x) = 3x^2 - 6x + 1$
$\left(\dfrac{3+\sqrt{6}}{3}, 0\right)$
$(0, 1)$
$\left(\dfrac{3-\sqrt{6}}{3}, 0\right)$
$(1, -2)$

Domain $= \{x \mid x \text{ is a real number}\}$ or $(-\infty, \infty)$; Range $= \{y \mid y \geq -2\}$ or $[-2, \infty)$

19. two imaginary solutions   **20.** two rational solutions   **21.** $8x^2 - 2x - 3 = 0$ **22.** $x^2 - 12 = 0$

23. 75 cabinets per day; $1200   **24.** −9, −9; 81   **25.** 10 in.; 100 sq in.

Section 8.4

Check Point Exercises

1. $-4, 4, -1,$ and $1,$ or $\{-4, 4, -1, 1\}$ **2.** 16 or $\{16\}$ **3.** $-\sqrt{6}, -1, 1,$ and $\sqrt{6},$ or $\{-\sqrt{6}, -1, 1, \sqrt{6}\}$

4. -1 and $2,$ or $\{-1, 2\}$ **5.** $-\dfrac{1}{27}$ and $64,$ or $\left\{-\dfrac{1}{27}, 64\right\}$

Exercise Set 8.4

1. $-2, 2, -1,$ and $1,$ or $\{-2, 2, -1, 1\}$ **3.** $-3, 3, -\sqrt{2},$ and $\sqrt{2},$ or $\{-3, 3, -\sqrt{2}, \sqrt{2}\}$ **5.** $-2i, 2i, -\sqrt{2},$ and $\sqrt{2},$ or $\{-2i, 2i, -\sqrt{2}, \sqrt{2}\}$

7. 1 or $\{1\}$ **9.** 49 or $\{49\}$ **11.** 25 and 64, or $\{25, 64\}$ **13.** 2 and 12, or $\{2, 12\}$ **15.** $-\sqrt{3}, 0,$ and $\sqrt{3},$ or $\{-\sqrt{3}, 0, \sqrt{3}\}$

17. $-5, -2, -1,$ and $2,$ or $\{-5, -2, -1, 2\}$ **19.** $-\dfrac{1}{4}$ and $\dfrac{1}{5},$ or $\left\{-\dfrac{1}{4}, \dfrac{1}{5}\right\}$ **21.** $\dfrac{1}{3}$ and $2,$ or $\left\{\dfrac{1}{3}, 2\right\}$ **23.** $\dfrac{-2 \pm \sqrt{7}}{3}$ or $\left\{\dfrac{-2 \pm \sqrt{7}}{3}\right\}$

25. -8 and $27,$ or $\{-8, 27\}$ **27.** -243 and $32,$ or $\{-243, 32\}$ **29.** 1 or $\{1\}$ **31.** $-8, -2, 1,$ and $4,$ or $\{-8, -2, 1, 4\}$ **33.** $-2, 2, -1,$ and $1;$ c

35. 1; e   **37.** 2 and 3; f   **39.** $-5, -4, 1,$ and 2   **41.** $-\dfrac{3}{2}$ and $-\dfrac{1}{3}$ **43.** $\dfrac{64}{15}$ and $\dfrac{81}{20}$ **45.** $\dfrac{5}{2}$ and $\dfrac{25}{6}$ **47.** ages 20 and 55;

The function models the data well.   **53.** -1 and $2,$ or $\{-1, 2\}$ **55.** $-1, 1, -3,$ and $3,$ or $\{-1, 1, -3, 3\}$

57. $-\dfrac{3}{2}$ and $2,$ or $\left\{-\dfrac{3}{2}, 2\right\}$ **59.** 1 or $\{1\}$ **61.** b   **63.** $\sqrt[3]{-2}$ and $\dfrac{\sqrt[3]{225}}{5},$ or $\left\{\sqrt[3]{-2}, \dfrac{\sqrt[3]{225}}{5}\right\}$ **65.** $\dfrac{1}{5x-1}$ **66.** $\dfrac{1}{2} + \dfrac{3}{2}i$ **67.** $(4, -2)$ or $\{(4, -2)\}$

Section 8.5

Check Point Exercises

1. $\{x | x < -4 \text{ or } x > 5\}$ or $(-\infty, -4) \cup (5, \infty)$

3. $\{x | x < -1 \text{ or } x \geq 1\}$ or $(-\infty, -1) \cup [1, \infty)$

2. $\{x | x \leq -3 \text{ or } -1 \leq x \leq 1\}$ or $(-\infty, -3] \cup [-1, 1]$

4. between 1 and 4 seconds, excluding $t = 1$ and $t = 4$

Exercise Set 8.5

1. $\{x | x < -2 \text{ or } x > 4\}$ or $(-\infty, -2) \cup (4, \infty)$

5. $\{x | x < 1 \text{ or } x > 4\}$ or $(-\infty, 1) \cup (4, \infty)$

9. $\{x | 2 \leq x \leq 4\}$ or $[2, 4]$

13. $\left\{x \middle| -3 < x < \dfrac{5}{2}\right\}$ or $\left(-3, \dfrac{5}{2}\right)$

17. $\{x | x \leq 0 \text{ or } x \geq 4\}$ or $(-\infty, 0] \cup [4, \infty)$

21. $\{x | 0 \leq x \leq 1\}$ or $[0, 1]$

3. $\{x | -3 \leq x \leq 7\}$ or $[-3, 7]$

7. $\{x | x < -4 \text{ or } x > -1\}$ or $(-\infty, -4) \cup (-1, \infty)$

11. $\left\{x \middle| -4 \leq x \leq \dfrac{2}{3}\right\}$ or $\left[-4, \dfrac{2}{3}\right]$

15. $\left\{x \middle| -1 < x < -\dfrac{3}{4}\right\}$ or $\left(-1, -\dfrac{3}{4}\right)$

19. $\left\{x \middle| x < -\dfrac{3}{2} \text{ or } x > 0\right\}$ or $\left(-\infty, -\dfrac{3}{2}\right) \cup (0, \infty)$

23. $\{x | 2 - \sqrt{2} \leq x \leq 2 + \sqrt{2}\}$ or $[2 - \sqrt{2}, 2 + \sqrt{2}]$

25. no solution or ∅

27. $\{x|1 \le x \le 2 \text{ or } x \ge 3\}$ or $[1, 2] \cup [3, \infty)$

29. $\{x|-2 \le x \le -1 \text{ or } x \ge 1\}$ or $[-2, -1] \cup [1, \infty)$

31. $\{x|x < -3\}$ or $(-\infty, -3)$

33. $\{x|x > -1\}$ or $(-1, \infty)$

35. $\{x|x = 0 \text{ or } x \ge 9\}$ or $\{0\} \cup [9, \infty)$

37. $\{x|x < -3 \text{ or } x > 4\}$ or $(-\infty, -3) \cup (4, \infty)$

39. $\{x|-4 < x < -3\}$ or $(-4, -3)$

41. $\{x|2 \le x < 4\}$ or $[2, 4)$

43. $\left\{x \middle| x < -\dfrac{4}{3} \text{ or } x \ge 2\right\}$ or $\left(-\infty, -\dfrac{4}{3}\right) \cup [2, \infty)$

45. $\{x|x < 0 \text{ or } x > 3\}$ or $(-\infty, 0) \cup (3, \infty)$

47. $\{x|x < -5 \text{ or } x > -3\}$ or $(-\infty, -5) \cup (-3, \infty)$

49. $\left\{x \middle| x < \dfrac{1}{2} \text{ or } x \ge \dfrac{7}{5}\right\}$ or $\left(-\infty, \dfrac{1}{2}\right) \cup \left[\dfrac{7}{5}, \infty\right)$

51. $\{x|x \le -6 \text{ or } x > -2\}$ or $(-\infty, -6] \cup (-2, \infty)$

53. $\left\{x \middle| x \le \dfrac{1}{2} \text{ or } x \ge 2\right\}$ or $\left(-\infty, \dfrac{1}{2}\right] \cup [2, \infty)$

55. $\{x|-1 < x < 1\}$ or $(-1, 1)$

57. $\{x|x < -8 \text{ or } -6 < x < 4 \text{ or } x > 6\}$ or

$(-\infty, -8) \cup (-6, 4) \cup (6, \infty)$

59. $\{x|-3 < x < 2\}$ or $(-3, 2)$

61. $\{x|x < -1 \text{ or } 1 < x < 2 \text{ or } x > 3\}$ or $(-\infty, -1) \cup (1, 2) \cup (3, \infty)$

63. $\left\{x \middle| -6 \le x \le -\dfrac{1}{2} \text{ or } x \ge 1\right\}$ or $\left[-6, -\dfrac{1}{2}\right] \cup [1, \infty)$

65. $\{x|x < 2 \text{ or } -1 \le x < 2\}$ or $(-\infty, 2) \cup [-1, 2)$ **67.** between 0 and 3 seconds, excluding $t = 0$ and $t = 3$ **69.** f: \$379 billion; g: \$379.8 billion; The functions model the value in the graph well.   **71.** after 2007   **73.** The company's production level must be at least 20,000 wheelchairs per month. For values of x greater than or equal to 20,000, the graph lies on or below the line $y = 425$.   **75.** The length of the shorter side cannot exceed 6 ft.

81. $\left\{x \middle| -3 \le x \le \dfrac{1}{2}\right\}$ or $\left[-3, \dfrac{1}{2}\right]$ **83.** $\{x|x < 3 \text{ or } x \ge 8\}$ or $(-\infty, 3) \cup [8, \infty)$ **85.** $\{x|-3 < x < -1 \text{ or } x > 2\}$ or $(-3, -1) \cup (2, \infty)$

87. Answers will vary; example: $x^2 - 2x - 15 \le 0$.   **89.** $\{x|x < 2 \text{ or } x > 2\}$ or $(-\infty, 2) \cup (2, \infty)$ **91.** ∅   **93. a.** all real numbers   **b.** empty set

95. $\{x|-19 < x < 29\}$ or $(-19, 29)$ **96.** $\dfrac{2(x - 1)}{(x + 4)(x - 3)}$ **97.** $(x^2 + 4y^2)(x + 2y)(x - 2y)$

Review Exercises

1. ± 8 or $\{\pm 8\}$ **2.** $\pm 5\sqrt{2}$ or $\{\pm 5\sqrt{2}\}$ **3.** $\pm\dfrac{\sqrt{6}}{3}$ or $\left\{\pm\dfrac{\sqrt{6}}{3}\right\}$ **4.** $4 \pm 3\sqrt{2}$ or $\{4 \pm 3\sqrt{2}\}$ **5.** $-7 \pm 6i$ or $\{-7 \pm 6i\}$

6. 100; $x^2 + 20x + 100 = (x + 10)^2$ **7.** $\dfrac{9}{4}$; $x^2 - 3x + \dfrac{9}{4} = \left(x - \dfrac{3}{2}\right)^2$ **8.** 3 and 9, or $\{3, 9\}$ **9.** $\dfrac{7 \pm \sqrt{53}}{2}$ or $\left\{\dfrac{7 \pm \sqrt{53}}{2}\right\}$

10. $\dfrac{-3 \pm \sqrt{41}}{4}$ or $\left\{\dfrac{-3 \pm \sqrt{41}}{4}\right\}$ **11.** 8%   **12.** 14 weeks   **13.** $60\sqrt{5}$ m; 134.2 m   **14.** $1 \pm \sqrt{5}$ or $\{1 \pm \sqrt{5}\}$

15. $1 \pm 3\sqrt{2}i$ or $\{1 \pm 3\sqrt{2}i\}$ **16.** $\dfrac{-2 \pm \sqrt{10}}{2}$ or $\left\{\dfrac{-2 \pm \sqrt{10}}{2}\right\}$ **17.** two imaginary solutions   **18.** two rational solutions

19. two irrational solutions   **20.** $-\dfrac{2}{3}$ and 4, or $\left\{-\dfrac{2}{3}, 4\right\}$ **21.** -5 and 2, or $\{-5, 2\}$ **22.** $\dfrac{1 \pm \sqrt{21}}{10}$ or $\left\{\dfrac{1 \pm \sqrt{21}}{10}\right\}$

23. -4 and 4, or $\{-4, 4\}$ **24.** $3 \pm 2\sqrt{2}$ or $\{3 \pm 2\sqrt{2}\}$ **25.** $\dfrac{1}{6} \pm \dfrac{\sqrt{23}}{6}i$ or $\left\{\dfrac{1}{6} \pm \dfrac{\sqrt{23}}{6}i\right\}$ **26.** $4 \pm \sqrt{5}$ or $\{4 \pm \sqrt{5}\}$

27. $15x^2 - 4x - 3 = 0$ **28.** $x^2 + 81 = 0$ **29.** $x^2 - 48 = 0$ **30.** 2002   **31.** 8.8 sec

32.

33.

34.

35.

36. 25 in. of rainfall per year; 13.5 in. of growth   **37.** 12.5 sec; 2540 feet   **38.** $(25, 5)$; The maximum divorce rate of about 5 per 1000 population was in 1985.   **39.** 250 yd by 500 yd; 125,000 sq yard   **40.** -7 and 7; -49 **41.** $-\sqrt{2}, \sqrt{2}, -2,$ and 2, or $\{-\sqrt{2}, \sqrt{2}, -2, 2\}$

42. 1 or $\{1\}$ **43.** $-5, -1,$ and 3, or $\{-5, -1, 3\}$ **44.** $-\dfrac{1}{8}$ and $\dfrac{1}{7}$, or $\left\{-\dfrac{1}{8}, \dfrac{1}{7}\right\}$ **45.** -27 and 64, or $\{-27, 64\}$ **46.** 16 or $\{16\}$

47. $\left\{x \mid -3 < x < \dfrac{1}{2}\right\}$ or $\left(-3, \dfrac{1}{2}\right)$;

48. $\left\{x \mid x \leq -4 \text{ or } x \geq -\dfrac{1}{2}\right\}$ or $(-\infty, -4] \cup \left[-\dfrac{1}{2}, \infty\right)$;

49. $\{x \mid -3 < x < 0 \text{ or } x > 1\}$ or $(-3, 0) \cup (1, \infty)$;

50. $\{x \mid x < -2 \text{ or } x > 6\}$ or $(-\infty, -2) \cup (6, \infty)$;

51. $\left\{x \mid x < 4 \text{ or } x \geq \dfrac{23}{4}\right\}$ or $(-\infty, 4) \cup \left[\dfrac{23}{4}, \infty\right)$;

52. between 1 and 2 seconds, excluding $t = 1$ and $t = 2$
53. a. 200 beats per minute   **b.** between 0 and 4 minutes and more than 12 minutes after the workout; between 0 and 4 minutes; Answers will vary.

Chapter 8 Test

1. $\pm\dfrac{\sqrt{10}}{2}$ or $\left\{\pm\dfrac{\sqrt{10}}{2}\right\}$ **2.** $3 \pm 2\sqrt{5}$ or $\{3 \pm 2\sqrt{5}\}$ **3.** 64; $x^2 - 16x + 64 = (x - 8)^2$ **4.** $\dfrac{1}{25}$; $x^2 + \dfrac{2}{5}x + \dfrac{1}{25} = \left(x + \dfrac{1}{5}\right)^2$ **5.** $3 \pm \sqrt{2}$ or $\{3 \pm \sqrt{2}\}$

6. $50\sqrt{2}$ ft   **7.** two irrational solutions   **8.** two imaginary solutions   **9.** -5 and $\dfrac{1}{2}$, or $\left\{-5, \dfrac{1}{2}\right\}$ **10.** $-4 \pm \sqrt{11}$ or $\{-4 \pm \sqrt{11}\}$

11. $-2 \pm 5i$ or $\{-2 \pm 5i\}$ **12.** $\dfrac{3}{2} \pm \dfrac{1}{2}i$ or $\left\{\dfrac{3}{2} \pm \dfrac{1}{2}i\right\}$ **13.** $x^2 - 4x - 21 = 0$ **14.** $x^2 + 100 = 0$ **15.** 1990 and 1998

16.

17.

18. after 2 sec; 69 ft   **19.** 4.1 sec   **20.** 23 computers; $169 hundreds or $16,900
21. 1 and 2, or $\{1, 2\}$ **22.** $-3, 3, -2,$ and 2, or $\{-3, 3, -2, 2\}$
23. 1 and 512, or $\{1, 512\}$
24. $\{x \mid -3 < x < 4\}$ or $(-3, 4)$

25. $\{x \mid x < 3 \text{ or } x \geq 10\}$ or $(-\infty, 3) \cup [10, \infty)$

Cumulative Review Exercises (Chapters 1–8)

1. 4 or $\{4\}$ **2.** $(2, 3)$ or $\{(2, 3)\}$ **3.** $(2, 0, 3)$ or $\{(2, 0, 3)\}$ **4.** $\{x \mid x < -2 \text{ or } x > 4\}$ or $(-\infty, -2) \cup (4, \infty)$ **5.** 5 or $\{5\}$

6. $\{x \mid x \geq 4\}$ or $[4, \infty)$ **7.** $\dfrac{3}{4} \pm \dfrac{\sqrt{7}}{4}i$ or $\left\{\dfrac{3}{4} \pm \dfrac{\sqrt{7}}{4}i\right\}$

8.

9.

10.

11.

12.

$f(x) = (x - 3)^2 - 4$

13. -6 **14.** $c = \dfrac{Ad}{d - A}$ **15.** $y = \dfrac{1}{2}x + 5$ or $f(x) = \dfrac{1}{2}x + 5$ **16.** $-\dfrac{y^9}{3x}$ **17.** $16x^4 - 40x^2y + 25y^2$

18. $x^2 - 5x + 1 + \dfrac{8}{5x + 1}$ **19.** $2y^2\sqrt[3]{2y^2}$ **20.** $\dfrac{4x - 13}{(x - 3)(x - 4)}$ **21.** $(x - 4)(x + 2)(x^2 - 2x + 4)$

22. $2(x + 3y)^2$ **23.** length: 12 ft; width: 4 ft   **24.** $\dfrac{6}{5}$ hr or 1 hr and 12 min   **25.** 5 mph

APPENDIX A

Check Point Exercises

1. 5 units   **2.** $\left(4, -\dfrac{1}{2}\right)$ **3.** $x^2 + y^2 = 16$ **4.** $(x - 5)^2 + (y + 6)^2 = 100$

5. center: $(-3, 1)$; radius: 2 units

$(x + 3)^2 + (y - 1)^2 = 4$

6. $(x + 2)^2 + (y - 2)^2 = 9$

$(x + 2)^2 + (y - 2)^2 = 9$

Appendix A Exercises

1. 13 units   **3.** $2\sqrt{2}$ or 2.83 units   **5.** 5 units   **7.** $\sqrt{29}$ or 5.39 units   **9.** $4\sqrt{2}$ or 5.66 units   **11.** $2\sqrt{5}$ or 4.47 units

13. $2\sqrt{2}$ or 2.83 units   **15.** $\sqrt{93}$ or 9.64 units   **17.** $\sqrt{5}$ or 2.24 units   **19.** $(4, 6)$ **21.** $(-4, -5)$ **23.** $\left(\dfrac{3}{2}, -6\right)$

25. $(-3, -2)$ **27.** $(1, 5\sqrt{5})$ **29.** $(2\sqrt{2}, 0)$ **31.** $x^2 + y^2 = 49$ **33.** $(x - 3)^2 + (y - 2)^2 = 25$ **35.** $(x + 1)^2 + (y - 4)^2 = 4$

37. $(x + 3)^2 + (y + 1)^2 = 3$ **39.** $(x + 4)^2 + y^2 = 100$

41. center: $(0, 0)$; $r = 4$ **43.** center: $(3, 1)$; $r = 6$ **45.** center: $(-3, 2)$; $r = 2$ **47.** center: $(-2, -2)$; $r = 2$

$x^2 + y^2 = 16$ $(x - 3)^2 + (y - 1)^2 = 36$ $(x + 3)^2 + (y - 2)^2 = 4$ $(x + 2)^2 + (y + 2)^2 = 4$

49. $(x + 3)^2 + (y + 1)^2 = 4$; center: $(-3, -1)$; $r = 2$

$x^2 + y^2 + 6x + 2y + 6 = 0$

51. $(x - 5)^2 + (y - 3)^2 = 64$; center: $(5, 3)$; $r = 8$

$x^2 + y^2 - 10x - 6y - 30 = 0$

53. $(x + 4)^2 + (y - 1)^2 = 25$; center: $(-4, 1)$; $r = 5$

$x^2 + y^2 + 8x - 2y - 8 = 0$

55. $(x - 1)^2 + y^2 = 16$; center: $(1, 0)$; $r = 4$

$x^2 - 2x + y^2 - 15 = 0$

APPENDIX B

Check Point Exercises

1. a. $2(1)^2 + 2(2)^2 + 2(3)^2 + 2(4)^2 + 2(5)^2 + 2(6)^2 = 182$ **b.** $(2^3 - 3) + (2^4 - 3) + (2^5 - 3) = 47$ **c.** $4 + 4 + 4 + 4 + 4 = 20$
2. $2^2(2 - 1)! + 3^2(3 - 1)! + 4^2(4 - 1)! + 5^2(5 - 1)! = 718$ **3. a.** 20   **b.** 1   **c.** 28   **d.** 1   **4.** $x^4 + 4x^3 + 6x^2 + 4x + 1$
5. $x^5 - 10x^4y + 40x^3y^2 - 80x^2y^3 + 80xy^4 - 32y^5$

Appendix B Exercises

1 105   **3.** 60   **5.** 115   **7.** $-\dfrac{5}{16}$ **9.** 55   **11.** $\dfrac{3}{8}$ **13.** 15   **15.** 56   **17.** 12   **19.** 1   **21.** 4950   **23.** $x^3 + 6x^2 + 12x + 8$
25. $27x^3 + 27x^2y + 9xy^2 + y^3$ **27.** $125x^3 - 75x^2 + 15x - 1$ **29.** $16x^4 + 32x^3 + 24x^2 + 8x + 1$
31. $x^8 + 8x^6y + 24x^4y^2 + 32x^2y^3 + 16y^4$ **33.** $y^4 - 12y^3 + 54y^2 - 108y + 81$ **35.** $16x^{12} - 32x^9 + 24x^6 - 8x^3 + 1$
37. $c^5 + 10c^4 + 40c^3 + 80c^2 + 80c + 32$ **39.** $x^5 - 5x^4 + 10x^3 - 10x^2 + 5x - 1$ **41.** $243x^5 - 405x^4y + 270x^3y^2 - 90x^2y^3 + 15xy^4 - y^5$
43. $64a^6 + 192a^5b + 240a^4b^2 + 160a^3b^3 + 60a^2b^4 + 12ab^5 + b^6$

Graphing Answer Section

CHAPTER 1

Section 1.3

Check Point Exercises

1.

2.

3.

5.
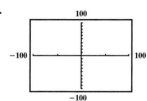

Exercise Set 1.3

1–10.

11.

12.

13.

14.

15.

16.

17.

18.

19.

20.

21.

22.

23.

24.

25.

26.

39.

40.

41.

42.

$y = x^2 + 2$

43.

$y = 5$

44.

$y = -1$

45.

$y = \dfrac{1}{x}$

46.

$y = -\dfrac{1}{x}$

74.

75. c.

76. c.

77. c.

78. c.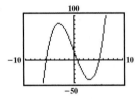

Exercise Set 1.4

82.

$y_2 = 3x + 10$

$y_1 = 5x + 2(x - 1)$

83.

$y_2 = 4x - 7$

$y_1 = 2x + 3(x - 4)$

84.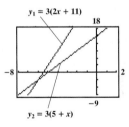

$y_1 = 3(2x + 11)$

$y_2 = 3(5 + x)$

85.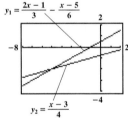

$y_1 = \dfrac{2x - 1}{3} - \dfrac{x - 5}{6}$

$y_2 = \dfrac{x - 3}{4}$

93.

$y = x^2 - 4$

Mid-Chapter Check Point

15.

$y = 2x - 1$

16.

$y = 1 - |x|$

17.

$y = x^2 + 2$

Exercise Set 1.5

78.

79.

Exercise Set 1.6

139.

146.

Review Exercises

44–46.

47.

48.

49.

50.

51.

Chapter 1 Test

19.

20.

CHAPTER **2**

Section 2.1

Check Point Exercises

4. $f(x) = 2x$
$g(x) = 2x - 3$

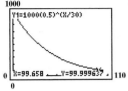

Exercise Set 2.1

19.

20.

21.

22. $f(x) = -2x$
$g(x) = -2x + 3$
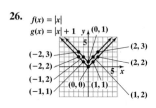

23. $g(x) = x^2 + 1$
$f(x) = x^2$

24. $f(x) = x^2$
$g(x) = x^2 - 2$

25.

$g(x) = |x| - 2$
$f(x) = |x|$

26. $f(x) = |x|$
$g(x) = |x| + 1$

27. $f(x) = x^3$
$g(x) = x^3 + 2$

28. $f(x) = x^3$
$g(x) = x^3 - 1$

100.
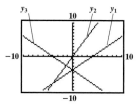

Exercise Set 2.2

55.

56.

76.

77.

78.

79.

82.
Population of Less Developed Regions

Section 2.3

Check Point Exercises

1. $3x - 2y = 6$

4. $y = 4x - 3$

5. $f(x) = -\dfrac{2}{3}x$

6. $x = -3$

Exercise Set 2.3

1. $x + y = 4$

2. $x + y = 2$

3. $x + 3y = 6$

4. $2x + y = 4$

5. $6x - 2y = 12$

6.

7. $3x - y = 6$

8. $x - 4y = 8$

9. $x - 3y = 9$

10. $2x - y = 5$

11. $2x = 3y + 6$

12. $3x = 5y - 15$

13. $6x - 3y = 15$

14. $8x - 2y = 12$

27. $y = 2x + 1$

28. $y = 3x + 2$

29. $y = -2x + 1$

30. $y = -3x + 2$

31. $f(x) = \frac{3}{4}x - 2$

32. $f(x) = \frac{3}{4}x - 3$

33. $f(x) = -\frac{3}{5}x + 7$

34. $f(x) = -\frac{2}{5}x + 6$

35. $y = -\frac{1}{2}x$

36. $y = -\frac{1}{3}x$

37c. $y = -2x$

38c. $y = -3x$

39c. $y = \frac{4}{5}x$

40c. $y = \frac{3}{4}x$

41c. $y = -3x + 2$

42c. $y = -2x + 4$

43c. $y = -\frac{5}{3}x + 5$

44c. $y = -\frac{7}{2}x + 7$

45. $y = 3$

46. $y = 5$

47. $f(x) = -2$

48. $f(x) = -4$

49. $3y = 18$

50. $5y = -30$

51. $f(x) = 2$

52. $f(x) = 1$

53.

54.

55.

56.

57.

58.

67.

68.

87.

88.

106.

107.

108.

109.

Exercise Set 2.4

69. a.

b.

70. a.

b.

80. a.

b.

81. a.

b.

c. LinReg
y=ax+b
a=-.1483291859
b=15.17991468

82. a.

b.

c. LinReg
y=ax+b
a=6.446721311
b=19.84836066

Review Exercises

6. $f(x) = x^2$
$g(x) = x^2 - 1$

7.
$f(x) = |x|$
$g(x) = |x| + 2$

35. $x + 2y = 4$

36. $2x - 3y = 12$

37. $4x = 8 - 2y$

42. $y = 2x - 1$

43. $f(x) = -\dfrac{1}{2}x + 4$

44. $y = \dfrac{2}{3}x$

48. $y = 2$

49. $7y = -21$

50. $f(x) = -4$

51. $x = 3$

52. $2x = -10$

Chapter 2 Test

5. $g(x) = x^2 + 1$
$f(x) = x^2 - 1$

18. $4x - 3y = 12$

19. $f(x) = -\dfrac{1}{3}x + 2$

20. $f(x) = 4$

Cumulative Review Exercises (Chapters 1–2)

15.
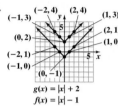

$g(x) = |x| + 2$
$f(x) = |x| - 1$

18. $f(x) = -2x + 4$

19. $x - 2y = 6$

CHAPTER 3

Exercise Set 3.2

58.

59.

Exercise Set 3.3

59. $f(x) = -\frac{3}{4}x + 3$

60. $-2x + y = 6$

61. $f(x) = -5$

Cumulative Review Exercises (Chapters 1–3)

11. $f(x) = -\frac{2}{3}x + 2$

12. $2x - y = 6$

CHAPTER 4

Section 4.1

Check Point Exercises

1. a.

b.

c.

2. a.

b.

c.

3.

4.

5.

Exercise Set 4.1

1.
0 1 2 3 4 5 6 7 8

2.
−7 −6 −5 −4 −3 −2 −1 0 1 2

3.
−7 −6 −5 −4 −3 −2 −1 0 1

4.
−5 −4 −3 −2 −1 0 1 2 3 4 5

5.
−6 −5 −4 −3 −2 −1 0 1 2 3 4

6.
−6 −5 −4 −3 −2 −1 0 1 2 3

7.
−3 −2 −1 0 1 2 3 4 5 6

8.
−1 0 1 2 3 4 5 6 7 8

9.
−4 −3 −2 −1 0 1 2 3 4 5 6

10.
−3 −2 −1 0 1 2 3 4 5 6 7

11.
−3 −2 −1 0 1 2 3 4 5 6

12.
−8 −7 −6 −5 −4 −3 −2 −1 0 1 2

13.
−2 −1 0 1 2 3 4 5 6 7 8

14.
−5 −4 −3 −2 −1 0 1 2 3 4 5

15.
−5 −4 −3 −2 −1 0 1 2 3 4 5

16.
−5 −4 −3 −2 −1 0 1 2 3 4 5

17.
−5 −4 −3 −2 −1 0 1 2 3 4 5

18.
−5 −4 −3 −2 −1 0 1 2 3 4 5

19.
−5 −4 −3 −2 −1 0 1 2 3 4 5

20.
−5 −4 −3 −2 −1 0 1 2 3 4 5

21.
−5 −4 −3 −2 −1 0 1 2 3 4 5

22.
−8 −7 −6 −5 −4 −3 −2 −1 0 1 2

23.
−5 −4 −3 −2 −1 0 1 2 3 4 5

24.
−5 −4 −3 −2 −1 0 1 2 3 4 5

25.
−2 −1 0 1 2 3 4 5 6 7 8

26.
−2 −1 0 1 2 3 4 5 6 7 8

27.
−2 −1 0 1 2 3 4 5 6 7 8

28.
−2 −1 0 1 2 3 4 5 6 7 8

29.
−2 −1 0 1 2 3 4 5 6 7 8

30.
−5 −4 −3 −2 −1 0 1 2 3 4 5

31.
−8 −7 −6 −5 −4 −3 −2 −1 0 1 2

32.
−7 −6 −5 −4 −3 −2 −1 0 1 2

33.
−5 −4 −3 −2 −1 0 1 2 3 4 5

34.
−10 −9 −8 −7 −6 −5 −4 −3 −2 −1 0

35.
−5 −4 −3 −2 −1 0 1 2 3 4 5

36.
−2 −1 0 1 2 3 4 5 6 7 8

37.
−5 −4 −3 −2 −1 0 1 2 3 4 5

38.
−8 −7 −6 −5 −4 −3 −2 −1 0 1 2

39.
−2 −1 0 1 2 3 4 5 6 7 8

40.
−5 −4 −3 −2 −1 0 1 2 3 4 5

41.
−8 −7 −6 −5 −4 −3 −2 −1 0 1 2

42.
−5 −4 −3 −2 −1 0 1 2 3 4 5

43.
−8 −7 −6 −5 −4 −3 −2 −1 0 1 2

44.
0 1 2 3 4 5 6 7 8 9 10

45.
6 7 8 9 10 11 12 13 14 15 16

46.
−5 −4 −3 −2 −1 0 1 2 3 4 5

47.
−5 −4 −3 −2 −1 0 1 2 3 4 5

48.
−5 −4 −3 −2 −1 0 1 2 3 4 5

49.
−5 −4 −3 −2 −1 0 1 2 3 4 5

50.
−5 −4 −3 −2 −1 0 1 2 3 4 5

53.
−7 −6 −5 −4 −3 −2 −1 0 1 2 3

54.
−5 −4 −3 −2 −1 0 1 2 3 4 5

55.
−7 −6 −5 −4 −3 −2 −1 0 1 2 3

56.
−8 −7 −6 −5 −4 −3 −2 −1 0 1 2

57.
−4 −3 −2 −1 −0 1 2 3 4 5 6

58.
−7 −6 −5 −4 −3 −2 −1 0 1 2 3

63.
−3 −2 −1 0 1 2 3 4 5 6 7

64.
−3 −2 −1 0 1 2 3 4 5 6

98.
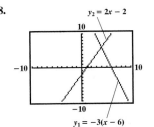
$y_2 = 2x - 2$
$y_1 = -3(x - 6)$

99.

$y_1 = -2(x + 4)$
$y_2 = 6x + 16$

100. a. 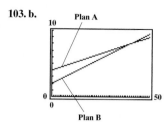 **b.** **103. b.**

Section 4.2

Check Point Exercises

4.

Exercise Set 4.2

7.

8.

9.

10.

11.

12.

13.

14.

15.

16.

17.

18.

19.

20.

21.

22.

23.

24.

25.

26.

27.

28.

29.

30.

31.

32.

39.

40.

41.

42.

43.

44.

45.

46.

47.

48.

49.

50.

51.

52.

53.

54.

88. **89.** **90.** **91.**

Section 4.3

Check Point Exercises

3.

4.

Exercise Set 4.3

39.

40.

41.

42.

43.

44.

45.

46.

47.

48.

49.

50.

51.

52.

53.

54.

55.

56.

57.

58.

59.

60.

61.

62.

65.

66.

67.

68.

69.

70.

102.

103.

104.

105.

106.

107.

108.

109.

110.

116. $3x - 5y = 15$

117. $f(x) = -\frac{2}{3}x$

118. $f(x) = -2$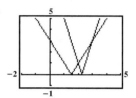

Section 4.4

Check Point Exercises

1. $4x - 2y \geq 8$

2. $y > -\frac{3}{4}x$

3. a. $y > 1$

b. $x \leq -2$

4. $x - 3y < 6$
$2x + 3y \geq -6$

5. $x + y < 2$
$-2 \leq x < 1$
$y > -3$

Exercise Set 4.4

1. $x + y \geq 3$

2. $x + y \geq 2$

3. $x - y < 5$

4. $x - y < 6$

5. $x + 2y > 4$

6. $2x + y > 6$

7. $3x - y \leq 6$

8. $x - 3y \leq 6$

9. $\dfrac{x}{2} + \dfrac{y}{3} < 1$

10. $\dfrac{x}{4} + \dfrac{y}{2} < 1$

11. $y > \dfrac{1}{3}x$

12. $y > \dfrac{1}{4}x$

13. $y \leq 3x + 2$

14. $y \leq 2x - 1$

15. $y < -\dfrac{1}{4}x$

16. $y < -\dfrac{1}{3}x$

17. $x \leq 2$

18. $x \leq -4$

19. $y > -4$

20. $y > -2$

21. $y \geq 0$

22. $x \leq 0$

23. $3x + 6y \leq 6$
$2x + y \leq 8$

24. $x - y \geq 4$
$x + y \leq 6$

25. $2x - 5y \leq 10$
$3x - 2y > 6$

26. $2x - y \leq 4$
$3x + 2y > -6$

27. $y > 2x - 3$
$y < -x + 6$

28. $y < -2x + 4$
$y < x - 4$

29. $x + 2y \leq 4$
$y \geq x - 3$

30. $x + y \leq 4$
$y \geq 2x - 4$

31. $x \leq 2$
$y \geq -1$

32. $x \leq 3$
$y \leq -1$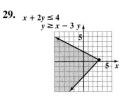

33. $-2 \leq x < 5$

34. $-2 < y \leq 5$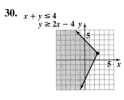

35. $x - y \leq 1$
$x \geq 2$

36. $4x - 5y \geq -20$
$x \geq -3$

39. $x + y > 4$
$x + y > -1$

40. $x + y > 3$
$x + y > -2$

41. $x - y \leq 2$
$x \geq -2$
$y \leq 3$

42. $3x + y \leq 6$
$x \geq -2$
$y \leq 4$

43.
$$x \geq 0$$
$$y \geq 0$$
$$2x + 5y \leq 10$$
$$3x + 4y \leq 12$$

44.
$$x \geq 0$$
$$y \geq 0$$
$$2x + y \leq 4$$
$$2x - 3y \leq 6$$

45.
$$3x + y \leq 6$$
$$2x - y \leq -1$$
$$x \geq -2$$
$$y \leq 4$$

46.
$$2x + y \leq 6$$
$$x + y \geq 2$$
$$1 \leq x \leq 2$$
$$y \leq 3$$

47. $y \geq -2x + 4$

48. $y \geq -3x + 2$

49.
$$x + y \leq 4$$
$$3x + y \leq 6$$

50.
$$x + y \leq 3$$
$$4x + y \leq 6$$

51.
$$-2 \leq x \leq 2$$
$$-3 \leq y \leq 3$$

52.
$$-1 \leq x \leq 1$$
$$-2 \leq y \leq 2$$

53. $y > \frac{3}{2}x - 2 \text{ or } y < 4$

54. $x - y \geq -1 \text{ or } 5x - 2y \leq 10$

63. b.
$$y \geq 0$$
$$x + y \geq 5$$
$$x \geq 1$$
$$200x + 100y \leq 700$$

64. b.
$$x \geq 0, y \geq 0$$
$$80x + 160y > 2000$$

75.

76.

77.

78.
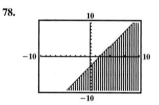

86.
$$y \geq nx + b$$
$$y \leq mx + b$$
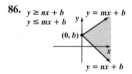

Exercise Set 4.5

5. a.

6. a.

7. a.

8. a.

9. a.

10. a.

11. a.

12. a.

13a.

14a.

15c.

16c.

Review Exercises

1.

2.

3.

4.

5.

6.

7.

8.

9.

10.

11.

19.

20.

21.

22.

24.

25.

26.

27.

28.

29.

35.

36.

37.

40. $3x - 4y > 12$

41. $x - 3y \le 6$

42. $y \le -\frac{1}{2}x + 2$

43. $y > \frac{3}{5}x$

44. $x \le 2$

45. $y > -3$

46. $2x - y \le 4$
$x + y \ge 5$

47. $y < -x + 4$
$y > x - 4$

48. $-3 \le x < 5$

49. $-2 < y \le 6$

50. $x \ge 3$
$y \le 0$

51. $2x - y > -4$
$x \ge 0$

52. $x + y \le 6$
$y \ge 2x - 3$

53. $3x + 2y \ge 4$
$x - y \le 3$
$x \ge 0, y \ge 0$

56.

57.

58.

59c.

Chapter 4 Test

1.

2.

3.

4.

8.

9.

10.

11.

12.

15.

16.

18. $3x - 2y < 6$

19. $y \geq \dfrac{1}{2}x - 1$

20. $y \leq -1$

21. $x + y \geq 2$
$x - y \geq 4$

22. $3x + y \leq 9$
$2x + 3y \geq 6$
$x \geq 0$
$y \geq 0$

23. $-2 < x \leq 4$

Cumulative Review Exercises (Chapters 1–4)

7. $f(x) = 2x + 1$

8. $y > 2x$

9. $2x - y \geq 6$

10. $f(x) = -1$

16.

17.

18.

19.

20.

CHAPTER 5

Exercise Set 5.1

92.

93.

94.

95.

96.

97.

103. $2x - 3y < -6$

Exercise Set 5.2

128.

129.

130.

131.

132.

Exercise Set 5.3

93.

94.

95.

96.

Exercise Set 5.4

117.

118.

119.

120.

133. $x - y \geq -4$
$x + 2y \geq 2$

Exercise Set 5.5

115.

116.

117.

118.

$y = (x + 7)(x - 3)$

$y = 25 - (x^2 + 4x + 4)$

119.

120.

121.

122.

Exercise Set 5.6

89.

90.

91.

92.
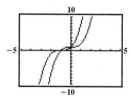

Exercise Set 5.7

96.

97.

98.

99.

Cumulative Review Exercises (Chapters 1–5)

11. $f(x) = -\frac{1}{3}x + 1$

12. $4x - 5y < 20$

13. $y \leq -1$

CHAPTER **6**

Exercise Set 6.1

129.

130.

131.

132.

134. a.

135. a.

$$y = \frac{27{,}725\,(x - 14)}{x^2 + 9} - 5x$$

137. $f(x) = \dfrac{x^2 - x - 2}{x - 2}$

141. $4x - 5y \geq 20$

Exercise Set 6.2

93.

Exercise Set 6.3

59.

$y = 2x - 1$

60.
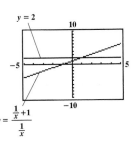

$y = 2$

$y = \dfrac{\frac{1}{x} + 1}{\frac{1}{x}}$

61.
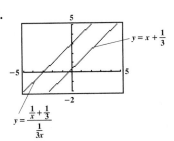

$y = x + \dfrac{1}{3}$

$y = \dfrac{\frac{1}{x} + \frac{1}{3}}{\frac{1}{3x}}$

62.
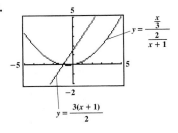

$y = \dfrac{\frac{x}{3}}{\frac{2}{x + 1}}$

$y = \dfrac{3(x + 1)}{2}$

Exercise Set 6.4

65.

66.

67.

Exercise Set 6.5

47. a. For Exercise 27:

$y = x^3 - 4x^2 + x + 6$

b. For Exercise 28:

$y = x^3 - 2x^2 - x + 2$

c. For Exercise 29:

$y = 2x^3 - 5x^2 + x + 2$

d. For Exercise 30:

$y = 2x^3 - 3x^2 - 11x + 6$

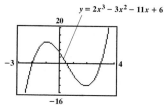

e. For Exercise 31:

$y = 6x^3 + 25x^2 - 24x + 5$

f. For Exercise 32:

$y = 3x^3 + 7x^2 - 22x - 8$

Exercise Set 6.6

70.

71.

72.

73.

74.

80. $x + 2y \geq 2$
$x - y \geq -4$

Exercise Set 6.7

64. For Exercise 19:

For Exercise 20:

65. For Exercise 45:

For Exercise 46:

66. b.

Exercise Set 6.8

39a.

Cumulative Review Exercises (Chapters 1–6)

9.

$y = |x| + 2$

10. $y \geq 2x - 1$
       $x \geq 1$

11. $2x - y < 4$

CHAPTER 7

Exercise Set 7.1

89. $f(x) = \sqrt{x} + 3$

(0, 3) (1, 4) (4, 5) (9, 6)

90. $f(x) = \sqrt{x} - 2$

(0, −2) (1, −1) (4, 0) (9, 1)

91. $f(x) = \sqrt{x - 3}$

(3, 0) (4, 1) (7, 2) (12, 3)

92. $f(x) = \sqrt{4 - x}$

(−5, 3) (0, 2) (3, 1) (4, 0)

114.

115.

116.

117.

122. $g(x) = \sqrt{x} + 2$

(0, 2) (1, 3) (4, 4) (9, 5)

123. $h(x) = \sqrt{x + 3}$

(−3, 0) (−2, 1) (1, 2) (6, 3)

Exercise Set 7.2

143.

X	Y₁	Y₂
1	1	1
2	1.4142	1.4142
3	1.7321	1.7321
4	2	2
5	2.2361	2.2361
6	2.4495	2.4495
7	2.6458	2.6458

X=5

144.

X	Y₁	Y₂
1	1	1.4142
2	.70711	1.7321
3	.57735	2
4	.5	2.2361
5	.44721	2.4495
6	.40825	2.6458
7	.37796	

X=1

145.

X	Y₁	Y₂
1	1	1
2	1.4142	.70711
3	1.7321	.57735
4	2	.5
5	2.2361	.44721
6	2.4494	.40824
7	2.6457	.37796

X=1

152. $y \leq -\dfrac{3}{2}x + 3$

Exercise Set 7.3

105.

$y = x^2$

106.

$y = \sqrt{8x^2}$ $y = 4x\sqrt{2}$

107.

$y = \sqrt{3x^2 - 6x + 3}$ $y = (x - 1)\sqrt{3}$

108.

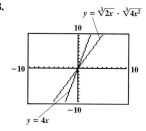

$y = \sqrt[3]{2x} \cdot \sqrt[3]{4x^2}$ $y = 4x$

113.

$f(x) = \sqrt{(x - 1)^2}$

Exercise Set 7.4

92.

$y = 5\sqrt{x}$

93.

$y = \sqrt{7x}$ $y = \sqrt{16x} - \sqrt{9x}$

94.

$y = x\sqrt{8} + x\sqrt{2}$ $y = x\sqrt{10}$

Exercise Set 7.5

138.

$y = x + 1$ $y = (\sqrt{x} - 1)^2$

139.

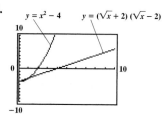

$y = x^2 - 4$ $y = (\sqrt{x} + 2)(\sqrt{x} - 2)$

140.

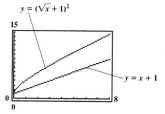

$y = (\sqrt{x} + 1)^2$ $y = x + 1$

141.

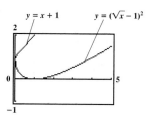

$y = \sqrt{x + 3} + \sqrt{x}$

Exercise Set 7.6

68.

69.

70.

71.

72.

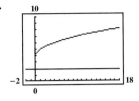

Cumulative Review Exercises (Chapters 1–7)

6. $x + 2y < 2$
$2y - x > 4$

CHAPTER 8

Exercise Set 8.1

97.

98.

Exercise Set 8.2

94.

Section 8.3

Check Point Exercises

1.

$f(x) = -(x - 1)^2 + 4$

2.

3.

Exercise Set 8.3

17.

18.

19.

20.

21.

22.

23.

24.

25.

26.

27.

28.

29.

30.

31.

32.

33.

34.

55. d.

56. d.

77. a.

c.

78.

79.

80.

81.

82. a & d.

Mid-Chapter 8 Check Point

15. $f(x) = (x - 3)^2 - 4$

(0, 5)
(1, 0)
(5, 0)
(3, −4)

16. $g(x) = 5 - (x + 2)^2$

(−2, 5)
(0, 1)
$(-2-\sqrt{5}, 0)$
$(-2+\sqrt{5}, 0)$

17. $h(x) = -x^2 - 4x + 5$

(−2, 9)
(0, 5)
(−5, 0)
(1, 0)

18. $f(x) = 3x^2 - 6x + 1$

$\left(\dfrac{3+\sqrt{6}}{3}, 0\right)$
(0, 1)
$\left(\dfrac{3-\sqrt{6}}{3}, 0\right)$
(1, −2)

Section 8.5

Check Point Exercises

1.

2.

3.

Exercise Set 8.5

1.

2.

3.

4.

5.

6.

7.

8.

9.

10.

11.

12.

13.

14.

15.

16.

17.

18.

19.

20.

21.

22.

23.

24.

25.

26.

27.

28.

29.

30.

31.

32.

33.

34.

35.

36.

37.

38.

39.

40.

41.

42.

43.

44.

45.

46.

47.

48.

49.

50.

51.

52.

57.

58.

59.

60.

61.

62.

Review Exercises

32.
$f(x) = -(x + 1)^2 + 4$
$(-1, 4)$, $(0, 3)$, $(-3, 0)$, $(1, 0)$, $x = -1$

33.
$f(x) = (x + 4)^2 - 2$
$(0, 14)$, $(-5.4, 0)$, $(-2.6, 0)$, $(-4, -2)$, $x = -4$

34.
$f(x) = -x^2 + 2x + 3$
$(1, 4)$, $(3, 0)$, $(0, 3)$, $(-1, 0)$, $x = 1$

35.
$f(x) = 2x^2 - 4x - 6$
$(-1, 0)$, $(3, 0)$, $(0, -6)$, $(1, -8)$, $x = 1$

47.

48.

49.

50.

51.

Chapter 8 Test

16.
$f(x) = (x + 1)^2 + 4$
$(-1, 4)$, $(0, 5)$, $x = -1$

17.
$f(x) = x^2 - 2x - 3$
$(-1, 0)$, $(3, 0)$, $(0, -3)$, $(1, -4)$, $x = 1$

24.

25.

Cumulative Review Exercises (Chapters 1–8)

8.
$3x = 15 + 5y$

9.
$2x - 3y > 6$

10.
$f(x) = -\frac{1}{2}x + 1$

11. $f(x) = x^2 + 6x + 8$

12. $f(x) = (x - 3)^2 - 4$

APPENDIX A

Check Point Exercises

5. $(x + 3)^2 + (y - 1)^2 = 4$

6. $(x + 2)^2 + (y - 2)^2 = 9$

Appendix A Exercises

41. 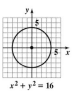 $x^2 + y^2 = 16$

42. $x^2 + y^2 = 49$

43. $(x - 3)^2 + (y - 1)^2 = 36$

44. $(x - 2)^2 + (y - 3)^2 = 16$

45. $(x + 3)^2 + (y - 2)^2 = 4$

46. $(x + 1)^2 + (y - 4)^2 = 25$

47. $(x + 2)^2 + (y + 2)^2 = 4$

48. $(x + 4)^2 + (y + 5)^2 = 36$

49. $x^2 + y^2 + 6x + 2y + 6 = 0$

50. $x^2 + y^2 + 8x + 4y + 16 = 0$

51. $x^2 + y^2 - 10x - 6y - 30 = 0$

52. $x^2 + y^2 - 4x - 12y - 9 = 0$

53. $x^2 + y^2 + 8x - 2y - 8 = 0$

54. $x^2 + y^2 + 12x - 6y - 4 = 0$

55. $x^2 - 2x + y^2 - 15 = 0$

56. $x^2 + y^2 - 6y - 7 = 0$

APPLICATIONS INDEX

A

Age of driver and accidents per day, 235 (Exercise 47)
Aging process, 504
AIDS, 203 (Exercise 52)
   cases diagnosed (U.S.), 298–99, 301, 302
   deaths in U.S. from, 309–10 (Exercises 65–72)
   deaths worldwide from, 160 (Exercise 28)
Airplane flight, 37 (Exercises 59–60)
Airplane's weight/volume constraints, 283–85
Alcohol and risk of accident, arrests and drunk driving, 398 (Exercise 135), 399, 410 (Exercise 85), 411 (Exercise 93)
Alcohol content of wine blend, 190 (Exercise 19)
Alligator's tail length given body length, 467 (Exercise 21)
Alzheimer's prevalence (U.S.), 91–92 (Exercises 52–54)
America's Cup, 480, 503 (Exercises 131–132)
Angle measures, 191 (Exercises 35–36), 204 (Exercise 57)
Animal protein consumption, 84 (Exercises 63–64)
Antifreeze solution, 193 (Exercise 62)
Area
   maximum, 599–600, 603–4 (Exercises 63–66), 605 (Exercise 88), 606 (Exercise 25), 633 (Exercise 39)
   of rectangle, 345 (Exercise 107), 518 (Exercises 83–84), 530 (Exercise 125)
   of square, 530 (Exercise 126)
   of triangle, 226 (Exercises 49–50)
Arrests and drunk driving, 398 (Exercise 135), 399, 410 (Exercise 85), 411 (Exercise 93)
Arts, empty set in, 46
Astronomical units, 77 (Exercises 127–130)
Automobile repair estimate, 261 (Exercise 80)
Average age of whites, blacks, and Hispanics (U.S.), 28

Average cost function, 447–48, 454 (Exercises 19–20), 457 (Exercise 64), 473, 476 (Exercise 62), 478 (Exercise 24)
Average number of paid vacation days for full-time workers, 97, 103–4
Average rate of change, 135–36

B

Babies born out of wedlock, 63 (Exercises 19–20), 65 (Exercise 79)
Ball's height above ground, 202 (Exercise 37), 216 (Exercise 54), 331 (Exercise 79), 367–68, 383 (Exercise 38), 622–23, 625 (Exercise 67), 632 (Exercise 31), 634 (Exercises 18–19)
   baseball, 603 (Exercises 55–56)
   bounce height, 467 (Exercise 23)
   football, 12 (Exercises 81–82)
   time to drop to ground, 478 (Exercise 18)
Bank and credit union charges, 251 (Exercise 86)
Banking angle and turning radius of bicycle, 467 (Exercise 27)
Baseball's height above ground, 603 (Exercises 55–56)
Basketball court dimensions, 58
Basketball player's hang time, 573 (Exercise 96)
Batting average, 456–57 (Exercises 53–54)
Berlin Airlift, 282, 289 (Exercise 20)
Bird purchase, 12 (Exercise 102)
Birthday cake, 503 (Exercises 147–148)
Births and deaths in U.S. from 1995 through 2002, 120–21 (Exercises 61–64)
Boat's rate in still water, 476 (Exercise 64)
Body-mass index, 468 (Exercise 31)
Body temperature, 32–33
Bouquet, mixture of flowers in, 205 (Exercise 12)
Box area/volume, 323 (Exercises 117–118), 345 (Exercise 106)
Box dimensions, 435 (Exercise 42)
Boys and girls in family, 193 (Exercise 64)

Break-even analysis, 186–88
Bridge coupon book/toll passes, 63 (Exercise 26), 261 (Exercise 79)
Building's shadow, 632 (Exercise 13)
Bus fares, 63 (Exercise 25)
Business analysis, 227
Business ventures, 178, 186–89, 192 (Exercises 47–50)
Butterfat in cream, 191 (Exercise 22)

C

Cable service, 478 (Exercise 19)
Calories
   calculations, 205–6
   caloric intake, 502 (Exercises 127–128)
   daily, in U.S. food supply, 140 (Exercise 88)
   in fast-food salads, 92 (Exercise 73)
Campus blood drive, support for, 253–54, 257
Campus election, 383 (Exercise 10)
Candy mix, 191 (Exercise 23)
Canoe manufacturing, 455 (Exercise 20)
Car and SUV sales, 112–13, 114, 136–37
Car depreciation, 63 (Exercises 21–22), 94 (Exercise 25), 140 (Exercise 84)
Cardiac index, 511 (Exercise 103), 511 (Exercises 97–98)
Car loan monthly payments, 411, 418 (Exercise 51)
Carnegie, Andrew, 202 (Exercise 39)
Carpenters, work done by, 453
Carpet length and width, 635 (Exercise 23)
Car rentals, 248–49, 271 (Exercise 19)
Celebrity Jeopardy's five all-time highest earners, 96
Cellular phone plan, 261 (Exercise 75)
Centrifugal force, 465–66
Chaos, patterns of, 548
Checking accounts, 252 (Exercise 103), 293 (Exercise 13)
Chernobyl nuclear power plant accident, 66, 77 (Exercise 139), 77 (Exercises 125–126)
Chicken(s)
   eaten per year, 84 (Exercise 66)
   raised for food, 84 (Exercise 65)
   and red meat consumption, 177 (Exercises 105–6)

Christmas tree sales, 261 (Exercise 87)

Cigarette use in U.S. by ethnicity, 162, 176–77 (Exercises 93–94)

Circular cylinder volume, 59

Clean house, importance of having, 607

Closet's length and width, 374 (Exercise 81)

Coffee drinking and irritability, 297, 362, 368

Coin mix, 191 (Exercises 25–26), 205 (Exercise 18)

Coin tosses, 269 (Exercise 93)

Cold virus, 310 (Exercise 73)

College education, cost of, 39, 40, 48, 49–50 (Exercises 67–68), 50 (Exercise 81)

College enrollment
and gender, 110 (Exercises 71–74)
projected, 63 (Exercise 27), 93 (Exercise 80)

Collinear points, 226 (Exercises 51–52)

Compact disc manufacturing, 271 (Exercise 22)

Compound interest, 331 (Exercises 83–84), 567–68, 571 (Exercises 73–76), 632 (Exercise 11)

Computer and hard drive sales, 190 (Exercise 11)

Computers
price before reduction, 56
sale, 190 (Exercise 11), 331 (Exercise 81)
student to computer ratio in U.S. public schools, 304
in use for U.S. and the world, 202–3 (Exercise 40)

Condominium sales, 235 (Exercise 5)

Cone volume, 466

Constraints, 283–84, 288 (Exercise 15), 288 (Exercise 16), 289 (Exercises 19–21), 289 (Exercise 23)

Cost and revenue functions, 188, 189, 192 (Exercises 41–46), 236 (Exercise 12)

Cost and revenue functions/break-even points, 192 (Exercises 47–50), 192 (Exercise 41), 205 (Exercise 11), 234 (Exercises 19–21), 236 (Exercises 9–11)

Cost before added tax, 63 (Exercises 31–32)

Cost-benefit analysis, 469

Cost function, 476 (Exercise 62), 478 (Exercise 24), 604 (Exercises 69–70). *See also* Cost and revenue functions

Cyclist's rate of travel, 476 (Exercise 63)

with/without wind, 473

Cylindrical tin can surface area, 332 (Exercise 86)

D

Daily calories in the U.S. food supply, 140 (Exercise 88)

Death rate and hours of sleep, 193, 199, 603 (Exercise 57)

Deaths
from AIDS, 160 (Exercise 28), 309–10 (Exercises 65–72)
hepatitis C, in U.S., 602 (Exercises 53–54)

Defective products, 270 (Exercise 114)

Defense budget, countries with largest, 62 (Exercise 12)

Desk manufacturing, 203 (Exercise 46)

Dieting American women and men, 150 (Exercises 67–68)

Digital camera sale, 56

Dinosaur walking speeds, 511 (Exercises 95–96)

Distance to horizon from specific height, 510 (Exercises 93–94)

Distance traveled, 27 (Exercise 148)

Diver's height above water, 345 (Exercise 105), 625 (Exercise 68)

Divorce rates (U.S.), 202 (Exercise 36), 633 (Exercise 38)

Divorces by number of years of marriage, 36 (Exercises 47–50), 38 (Exercise 72)

Doorway's diagonal length, 572 (Exercise 84)

Driver's age
and accidents, 110–11 (Exercises 75–78)
and fatal crashes, 586–87 (Exercises 73–74)

Driving rate and time for trip, 462, 474

Driving time from Miami to Atlanta, 410 (Exercises 77–82)

Drugs
concentration, 33, 135–36
prescription costs, 139 (Exercise 74)
to treat mental disorders in children, 151 (Exercise 71)

Drug use by U.S. high school seniors, 36 (Exercises 51–54)

Dual investments, 180–82, 184, 190 (Exercises 13–18), 205 (Exercise 13), 205 (Exercise 16), 229, 234 (Exercise 15), 236 (Exercise 6), 375 (Exercise 106), 558 (Exercise 19)

E

Earthquake relief, 282–83, 284–85

Education level of Americans by their early thirties, 190 (Exercise 9)

Electrical resistance, 418 (Exercise 53), 468 (Exercise 39)

Elevator capacity, 252 (Exercises 89–90), 280 (Exercise 64)

Elk population, 310 (Exercise 75), 477 (Exercise 22)

Endangered species, 1
in U.S., 92 (Exercise 72), 93 (Exercise 75)

Energy efficiency of building materials, 179–80

Energy formula, Einstein's, 61

Exam grades, 261 (Exercises 77–78), 271 (Exercise 20), 294 (Exercise 30)

Expressway speeds and safe distances, 91 (Exercise 15)

F

Fahrenheit/Celsius temperature interconversions, 12 (Exercises 79–80), 60, 158 (Exercise 55), 261 (Exercise 76)

Families below poverty level, 380 (Exercise 9)

Families with four or more children (U.S.), 140 (Exercise 86)

Farmland loss (U.S.), 559, 600

Farms in U.S., maximum number of, 603

Federal Express aircraft purchase decisions, 289 (Exercise 23)

Federal minimum wage, 132–33
adjusted for inflation, 159 (Exercise 60)

Fencing, maximum area inside, 600, 605 (Exercise 88), 633 (Exercise 39)

Fetal weight, 632 (Exercise 12)

Field's dimensions, 93 (Exercise 79), 94 (Exercise 28)

Film, costs to develop, 94 (Exercise 26)

Floor length and width, 383 (Exercise 39)

Flower bed enlargement, 573 (Exercises 87–88)

Flu inoculation costs, 396 (Exercises 105–108)

Focal length
for camera lens, 418 (Exercise 54)
for eyeglass lenses, 446

Food Stamps, 373 (Exercises 67–70)
recipients, 634 (Exercise 15)

Football field dimensions, 57–58

Football's height above ground, 12 (Exercises 81–82)

Force required to stretch spring, 467 (Exercise 24)

Foreign-born population in U.S., 201 (Exercise 35), 574, 583–84, 588 (Exercise 92)

Frame dimensions, 64 (Exercise 41)

Free-falling object's position, 622–23, 625 (Exercises 67–68), 627, 632 (Exercise 31), 634 (Exercises 18–19)

Free fall time, 572 (Exercises 79–80)

Fuel efficiency, 605 (Exercise 82)

Fundraiser attendance, 190 (Exercise 12)

G

Galápagos Islands plant and animal species, 492–93, 502 (Exercises 125–126), 540 (Exercises 57–58)

Gardens
area of, 323 (Exercise 133)
border width, 478 (Exercise 20)

Gas pressure in can, 463

Gay service members, 38 (Exercise 74)

Gold
in alloys, 190 (Exercise 20)
different karats mixed, 190 (Exercise 21)

Golden rectangle, 525, 531 (Exercise 128)

Granola-raisin mixture, 191 (Exercise 24)

Gravitational force, 465

Gravity model, 469 (Exercise 47)

Gymnast's height above ground, 372–73 (Exercises 65–66)

H

Happiness of Americans in early thirties, 190 (Exercise 10)

Health care expenditures in U.S., per capita, 11 (Exercises 75–76)

Health club membership fees, 63 (Exercise 23)

Heart beats over lifetime, 85 (Exercise 82)

Heart function throughout aging process, 504
cardiac index, 511 (Exercise 103), 511 (Exercises 97–98)

Heart rate, 238, 280 (Exercises 59–62)
model, 634 (Exercise 53)

Heat generated by stove, 469 (Exercise 50)

Heat loss of a glass window, 468 (Exercise 33)

Height. *See also* Ball's height above ground

of ball dropped, 38 (Exercises 80–81)

of debris after explosion, 331 (Exercise 80)

diver's height above water, 345 (Exercise 105), 625 (Exercise 68)

as a function of age, 135, 526–27

gymnast's height above ground, 372–73 (Exercises 65–66)

maximum, 603 (Exercises 55–56), 634 (Exercise 18)

median height of boys, 491 (Exercise 112), 491 (Exercises 99–100)

of object, 202 (Exercise 37), 216 (Exercise 54), 375 (Exercise 92), 557 (Exercise 85)

rocket's height above ground/water, 375 (Exercise 93), 382 (Exercise 98), 633 (Exercise 37), 633 (Exercise 52)

weight and height recommendations/calculations, 61, 467 (Exercise 25)

Hepatitis C deaths in U.S., projected, 602 (Exercises 53–54)

Higher education costs, 39, 40, 48, 49–50 (Exercises 67–68), 50 (Exercise 81)

High school drop out rate (U.S.), 150 (Exercise 69)

HIV/AIDS cases, global, 139 (Exercise 79). *See also* AIDS

Homes with wireless telephones (U.S.), 151 (Exercise 70)

Honeycombs, 347

Hotel room types, 192 (Exercise 40)

Households online (U.S.), 530 (Exercises 117–124)

Households watching final episode of M*A*S*H, 262, 269 (Exercise 87)

Households with televisions (U.S.), 140 (Exercise 85)

House sales prices, 139 (Exercise 80)

Hubble Space Telescope, 469 (Exercise 51)

Hurricane relief, 289 (Exercise 19)

I

Illumination intensity, 468 (Exercise 30), 478 (Exercise 27), 558 (Exercise 18)

Imaginary number joke, 550 (Exercise 130)

Income by level of education, 63–64 (Exercises 33–34)

Inmates in U.S. prisons, 215 (Exercise 53)

Instantaneous change, roller coasters and, 527

Intelligence quotient (IQ) and mental/chronological age, 468 (Exercise 32)

Internet users, countries with the most, 62 (Exercise 11)

Internet users in U.S., 140 (Exercise 82)
percentage of population, 483–84

Investment(s)
amounts invested per rate, 203 (Exercises 41–42)
compound interest, 567–68, 632 (Exercise 11)
for desired return, 271 (Exercise 21)
dual, 180–82, 184, 190 (Exercises 13–18), 205 (Exercise 13), 205 (Exercise 16), 229, 234 (Exercise 15), 236 (Exercise 6), 375 (Exercise 106), 558 (Exercise 19)
and interest rates, 27 (Exercise 147)
maximizing expected returns, 289 (Exercise 29)
simple annual interest rate, 61

J

Jury awards in personal-injury cases, 382 (Exercise 99)

K

Kinetic energy, 468 (Exercise 34)

L

Ladder's reach, 572 (Exercise 83)

Lamp manufacture, 229

Landscape design, 369–70, 373–74 (Exercises 77–78)

Lawn mowing time, 588 (Exercise 83), 635 (Exercise 24)

Learning curve, 444 (Exercises 57–60)

Leg lengths of right triangle, 587 (Exercises 79–80)

Lemonade business, 178, 186, 189

Length of side of square, 373 (Exercises 73–74)

Length of sides of triangle, 382 (Exercise 102), 383 (Exercise 40)

Length of violin string and frequency, 464

Life expectancy, 62 (Exercises 17–18), 65 (Exercise 78), 141, 144–45, 176 (Exercise 91)

Literacy and child mortality, 123, 128, 151 (Exercise 72)

Living alone, number of Americans, 479, 532, 538–39, 541 (Exercise 66)

Loan payments, 411, 418 (Exercise 51)

Long-distance carriers/plans, 54–55, 56, 93 (Exercise 76), 239, 242, 251 (Exercise 84)
Long-distance telephone charges, 64 (Exercise 43)
Lot's dimensions, 191 (Exercises 37–38)
Love over time, course of, 251 (Exercises 71–78)
Lunch menus, 288 (Exercise 18)

M

Mailing costs, 111 (Exercises 87–90)
Manufacturing constraints, 283, 284
Marching band, 193 (Exercise 61)
Margin of error, 269 (Exercises 87–88)
Markup, 64 (Exercises 35–36)
Marriage decisions, 216 (Exercises 55–56)
Mass of hydrogen atoms, 85 (Exercise 70)
Mass of oxygen molecules, 85 (Exercise 69)
Maximum area, 599–600, 603–4 (Exercises 63–66), 605 (Exercise 88), 606 (Exercise 25), 633 (Exercise 39)
Maximum height, 603 (Exercises 55–56), 634 (Exercise 18)
Maximum product, 603 (Exercises 59–60), 606 (Exercise 24)
Maximum scores, 289 (Exercise 22)
Maximum tree growth, 633 (Exercise 36)
Maximum yield, 605 (Exercise 89)
Median height of boys, 491 (Exercise 112), 491 (Exercises 99–100)
Medicare costs/spending, 121 (Exercise 70), 121 (Exercise 75), 614–15, 625 (Exercises 69–72)
Memory retention, 443 (Exercises 53–56)
Men smokers, 139 (Exercise 73)
Miles per gallon, 605 (Exercise 82)
Military spending (U.S.), 158
Minimum product, 597–98, 603 (Exercises 61–62), 633 (Exercise 40)
Minimum wage, 132–33
    adjusted for inflation, 159 (Exercise 60)
Mixture problems, 182–84, 190 (Exercises 19–26), 193 (Exercise 62), 205 (Exercise 14), 205 (Exercise 18), 234 (Exercise 16), 236 (Exercise 7), 444 (Exercises 61–62)
Modernistic painting consisting of geometric figures, 204 (Exercise 56)
Moon weight of person given Earth weight, 467 (Exercise 22)

Motel rooms, 296 (Exercise 14)
Moth eggs and abdominal width, 428–29, 435 (Exercise 41)
Motion problems, 184–85, 191 (Exercises 27–32), 202 (Exercise 38), 205 (Exercise 15), 234 (Exercise 17), 236 (Exercise 8), 448–50, 455–56 (Exercises 25–36), 457 (Exercise 66), 457 (Exercise 69), 635 (Exercise 25)
    distance covered by any moving body, 82–83
Mountain bike manufacturing, 454 (Exercise 19)
Movie spending, 93 (Exercise 108)
Multinational corporations, growth in number of, 560, 572 (Exercises 77–78)

N

National Park Service expenditures, 556 (Exercise 32)
Negative square roots, 550 (Exercise 117)
Newspaper reading, 139 (Exercise 76)
New York City Marathon, 84 (Exercise 67)
Nickels in change machine, 262 (Exercise 98)
Nutritional content, 203 (Exercise 45)

O

Obesity rate, 140 (Exercise 87)
Official languages spoken in greatest number of countries, 53
Ohm's law, 550 (Exercises 115–116)
Open box lengths and widths, 374 (Exercises 79–80)
Organ transplants, people waiting for, 380 (Exercise 4)

P

Painting frame width, 382 (Exercise 382)
Parking lot length and width, 373 (Exercise 72)
Path around swimming pool, dimensions of, 64 (Exercise 42), 373 (Exercises 75–76)
"Peanuts" cartoon, 520, 531 (Exercise 127)
Pedestrian route length, 572 (Exercises 81–82)
Perceived length of time period and age, 468 (Exercise 36)

Percentage of teens in U.S. who say they really need various technologies, 110 (Exercise 70)
Perimeter
    of rectangle, 410 (Exercises 83–84), 518 (Exercises 83–84), 530 (Exercise 125)
    of square, 530 (Exercise 126)
Phone calls between cities, 458, 468 (Exercise 37), 469 (Exercise 47)
Physician visits, 112 (Exercise 100)
Pitch of tone and wavelength, 476 (Exercise 70)
Plane figure area, 324 (Exercise 134)
Plane's weight/volume constraints, 283–85
Planets, years for, 541 (Exercises 59–60)
Plant food mixture, 205 (Exercise 14)
Playing field dimensions, 77 (Exercise 148)
Police officers convicted of felonies, 94 (Exercise 6), 632 (Exercise 30)
Political ideologies in United States, 109–10 (Exercise 69)
Poll's margin of error, 266–67
Pollution removal costs, 385–86, 436, 441–42, 443 (Exercises 51–52), 476 (Exercise 56)
Pond fill time, 476 (Exercise 67)
Pond width, 634 (Exercise 6)
Pool dimensions, 64 (Exercise 38)
    path around, 64 (Exercise 42), 373 (Exercises 75–76)
Pool emptying time, 588 (Exercise 84)
Pool fill time, 457 (Exercises 55–56), 478 (Exercise 25)
Population
    over age 65 (U.S.), 84 (Exercises 59–62)
    deer, 475 (Exercise 16)
    foreign-born (U.S.), 201 (Exercise 35), 574, 583–84, 588 (Exercise 92)
    gray wolf, 2, 4, 5
    Jewish Americans, 111 (Exercises 79–86)
    reindeer, 310 (Exercise 87)
    in scientific notation, 80–81
    single, 479, 532, 538–39, 541 (Exercise 66)
    world, 94 (Exercise 39), 121 (Exercises 65–68)
Population projections, 63 (Exercise 28)
Poverty level, Americans living below, 12 (Exercises 77–78)
Precipitation and temperature, regions resulting from ranges in, 278

Price per acre for Alaska purchase, 85 (Exercise 68)

Price reductions, 63 (Exercises 29–30), 65 (Exercise 81), 89, 93 (Exercise 77), 94 (Exercise 27), 160 (Exercise 9), 324

Prison inmates, 215 (Exercise 53)

Problem solving time, 465

Profit function, 188–89, 192 (Exercises 45–46), 604 (Exercises 69–70)

Profits, 476 (Exercise 68)
  and loss, 193 (Exercise 65)
  maximizing, 288 (Exercise 17), 295 (Exercise 60), 296 (Exercise 25), 604 (Exercises 69–70), 606 (Exercise 23), 635 (Exercise 20)
  maximum daily, 285
  maximum monthly, 288 (Exercise 15)
  on newsprint/writing paper, 295 (Exercise 59)
  production and sales for gains in, 251–52 (Exercises 87–88)
  total monthly, 288 (Exercise 15)

Pyramid volume, 477 (Exercise 73)

R

Radiation intensity and distance of radiation machine, 468 (Exercise 29)

Radio sales, 121 (Exercise 69)

Radios produced and sold, 192 (Exercises 41–46)

Rain gutter cross-sectional area, 587 (Exercise 81), 588 (Exercise 94), 604 (Exercises 67–68)

Rates of travel for cars on round-trip commute, 418 (Exercise 52)

Razor blades sold, 203 (Exercise 44)

Real-estate sales and prices (U.S.), 139 (Exercise 80)
  condominiums, 235 (Exercise 5)

Recorded music sales, 139 (Exercises 77–78)

Record time for women's 400-meter run, 158 (Exercise 53)

Rectangle
  area of, 345 (Exercise 107), 518 (Exercises 83–84), 530 (Exercise 125)
  dimensions of, 587 (Exercises 77–78), 626 (Exercises 75–76)
  golden, 525, 531 (Exercise 128)
  perimeter of, 410 (Exercises 83–84), 518 (Exercises 83–84), 530 (Exercise 125)

Rectangular box dimensions, 435 (Exercise 42)

Rectangular solid volume, 59

Reindeer population, 310 (Exercise 87)

Restaurant tables and maximum occupancy, 192 (Exercise 39)

Revenue functions, 604 (Exercises 69–70). *See also* Cost and revenue functions

River's current rate, 635 (Exercise 25)

Rocket's height above ground/water, 375 (Exercise 93), 382 (Exercise 98), 633 (Exercise 37), 633 (Exercise 52)

Roller coasters and instantaneous change, 527

Running rate/time driving and hiking, 455 (Exercises 21–24)

S

Salaries
  college professor, 140 (Exercise 81)
  gross amount per paycheck, 64 (Exercise 44)
  salesperson's earnings/commissions, 93 (Exercise 78), 293 (Exercise 14)
  summer job comparison, 236 (Exercise 6)
  total weekly earnings, 288 (Exercise 16)

Sales
  car and SUV, 112–13, 114, 136–37
  Christmas tree, 261 (Exercise 87)
  computer, 190 (Exercise 11), 331 (Exercise 81)
  condominium, 235 (Exercise 50)
  digital camera, 56
  monthly, 464–65
  radio, 121 (Exercise 69)
  real estate, 139 (Exercise 80)
  recorded music, 139 (Exercises 77–78)
  television, 331 (Exercise 82)
  theater ticket, 203 (Exercise 43)

Salesperson's earnings, 93 (Exercise 78), 293 (Exercise 14)

SAT scores, 26–27 (Exercises 143–46)

Scatter plots for data, 144, 145, 152

Shaded region areas, 354 (Exercises 105–108), 360 (Exercises 81–86), 381–82 (Exercises 91–92)

Shot put
  parabolic paths of, 596–97
  path of, given angle, 587 (Exercises 75–76)

Sign's length and width, 373 (Exercise 71), 382 (Exercise 100)

Skating rink area, 332 (Exercise 85)

Skid mark measurements and speed of car, 491 (Exercises 101–102)

Skydiver's fall, 461–62, 476 (Exercise 69)

Sleeping time, 294 (Exercise 39)

Smokers
  men, 139 (Exercise 73)
  percentage of, 155

Smoking-related deaths and disease's incidence ratios, 397 (Exercises 113–116), 398 (Exercise 128)

Snow fall, 36–37 (Exercises 55–58)

Soccer field dimension, 64 (Exercise 37)

Sound intensity, 468 (Exercise 35)

Space exploration and relativity theory, 512, 518 (Exercises 81–82)

Speed skating times, 251 (Exercises 81–82)

Square
  area of, 530 (Exercise 126)
  length of side of, 373 (Exercises 73–74)
  perimeter of, 530 (Exercise 126)

Stationery item costs, 237 (Exercise 17)

Stereo speaker loudness, 477 (Exercise 71)

Stolen plants, 65 (Exercise 84)

Stopping distance required for car, 462, 467 (Exercise 26)

Student enrollments, 558 (Exercise 20)

Student loans, 491 (Exercise 113)

Students to computers ratio in U.S. public schools, 304

Super Bowl viewers, 160 (Exercises 23–24)

Supply and demand, 177 (Exercises 95–96)

Supply-side economics, 384, 419–20

Survivors by age, given number of births, 540 (Exercises 55–56), 541 (Exercise 67), 557 (Exercise 86)

Swimming pool dimensions, 64 (Exercise 38), 64 (Exercise 40)
  path around pool, 64 (Exercise 42), 373 (Exercises 75–76)

Swimming pool emptying time, 588 (Exercise 84)

Swimming pool tile border, 588 (Exercise 97)

T

Table height, 204 (Exercise 58)

Table top dimensions, 234 (Exercise 18)

Target heart rate, 50 (Exercises 69–70)

Target heart rate for exercise, 280 (Exercises 59–62)

Taxes
  bills, 251 (Exercise 85)
  burden, per capita, 82
  tax rate percentage and revenue,
    419–20, 426 (Exercises 55–58)
  U.S. population and total tax
    collections, 78
Taxpayers audited by the IRS, 140
    (Exercise 83)
Telephone calling plans, 295 (Exercise 5)
Television programs viewed by greatest
    percentage of U.S. households, 269
    (Exercises 87–88, 101)
Television sale, 331 (Exercise 82)
Temperature
  body, in hyperthermia or
    hypothermia, 295 (Exercise 17)
  and depth of water, 467 (Exercise 28)
  Fahrenheit-Celsius interconversions,
    12 (Exercises 79–80), 60, 158
    (Exercise 55), 261 (Exercise 76)
  global average, 139 (Exercise 75), 540
    (Exercises 51–54)
  monthly average, 269
    (Exercises 89–90)
  regions resulting from range of, 278
  windchill, 497, 502
    (Exercises 129–130)
Tennis court dimensions, 64
    (Exercise 39)
Test scores, 193 (Exercise 63)
  difference between and mean of, 191
    (Exercises 33–34)
Theater attendance, maximizing revenue
    from, 289 (Exercise 21)
Theater ticket sales, 203
    (Exercise 43)
Thefts in U.S., 310 (Exercise 74)
Thrown objects, path of, 589
Time in motion, 448–50
Time zones, world's, 13, 26
    (Exercises 139–142)

Tolerance limits for machine parts, 269
    (Exercises 91–92)
Toy preferences, 253, 260–61
    (Exercises 69–74)
Track-and-field records, 51, 53–54
Train passengers, 92 (Exercise 55)
Triangle
  angle measures, 62 (Exercises 13–16),
    92 (Exercise 74), 152 (Exercise 89),
    191 (Exercises 35–36), 204
    (Exercise 57)
  area, 226 (Exercises 49–50)
  height, 588 (Exercise 98)
  length of sides of, 382 (Exercise 102),
    383 (Exercise 40)
Truck rental costs, 251 (Exercise 83)
Tunnel-boring machines, work done
    by, 452
TV and stereo prices, 234
    (Exercise 14)

U

Unemployment and years of education,
    397 (Exercises 109–112)

V

Vacation lodgings, 280 (Exercise 63)
Veterinary care costs, 234 (Exercise 28)
Video-game rentals, 63 (Exercise 25)
Violin string length and frequency, 464
Virtual reality, 445–46
Vitamin content, 216 (Exercise 66)
Volume(s)
  of cone, 466
  of figures, 324 (Exercises 135–136)
  for given regions, 355
    (Exercises 109–110)
Voting systems usage, 251
    (Exercises 79–80)
Vulture's flight, 157 (Exercise 19)

W

Water current rate, 478 (Exercise 26)
Water pressure and depth, 458–59
Water supply produced by snowpack, 460
Water temperature and depth, 467
    (Exercise 28)
Water used in a shower, 460
Wealth of richest people, 202 (Exercise 39)
Web site design, 450–53
Weight
  chart for men, 37 (Exercises 61–64)
  and height
    recommendations/calculations, 61,
    467 (Exercise 25)
  moon weight of person given Earth
    weight, 467 (Exercise 22)
Weight-loss diets, percentage of
    American adults on, 150
    (Exercises 67–68)
Wheelchair business, 189
Wheelchair manufacturing, 454
    (Exercises 15–18), 626
    (Exercises 73–74)
Will distribution, 65 (Exercise 83)
Windchill temperature, 497, 502
    (Exercises 129–130)
Wind force, 468 (Exercise 38)
Wind pressure, 469 (Exercise 49)
Wire length, 370–71, 374 (Exercises
    82–84), 569, 570, 572 (Exercises
    85–86), 587 (Exercise 82)
Women
  and men in workforce, 163, 176
    (Exercise 92)
  and right to vote, 52–53
Work problems, 450–53, 456 (Exercises
    37–48), 457 (Exercises 55–56), 457
    (Exercise 65), 457 (Exercise 70),
    476 (Exercises 65–67), 477
    (Exercise 72), 478 (Exercise 25),
    635 (Exercise 24)

SUBJECT INDEX

A

Absolute value, 13, 15–16, 87
　definition of, 15
　finding, 13–14
Absolute value bars, 291, 507
　rewriting absolute value equation
　　without, 263
　rewriting absolute value equation
　　with two absolute values
　　without, 264
　rewriting inequalities without, 265
Absolute value equations, 291
　solving, 263–64
Absolute value inequalities, 291–92
　problem solving using, 266–67
　solving, 264–66
　with unusual solution sets, 266
Absolute value symbols, 19
Addition
　associative property of, 21
　commutative property of, 20
　of complex numbers, 543–44, 554
　of functions, 115–16
　inverse property of, 15
　of like radicals, 512–13
　of polynomials, 305–6, 376
　of radical expressions, 553
　of rational expressions, 471
　of rational expressions when
　　denominators are opposites,
　　407–8
　of rational expressions with
　　common denominators, 399–400
　of rational expressions with
　　different denominators, 403–5,
　　406–7
　of real numbers, 14–16, 87
Addition method
　and dual investment problems, 181
　and motion problem, 185
　systems of linear equations in three
　　variables solved by, 195, 230
　systems of linear equations in two
　　variables solved by, 168–71,
　　172, 228
　variables eliminated using, 168–71,
　　172, 195, 228, 230
Addition property of equality, 40
Addition property of inequality, 243

Additive identity, 15
Additive inverses, 15, 392
　adding or subtracting rational
　　expressions with denominators
　　that are, 407–8
Aging process, 504
AIDS, 298–99, 301, 302, 304
Algebraic expressions, 86–87
　defined, 2
　English phrases translated to, 2–3
　equivalent, 20, 21, 22
　evaluating, 3–4, 86
　simplifying, 23–24, 88
　terms of, 22
Algebra of functions, 114–22, 153–54
　difference, 117, 154
　product, 117, 154, 319–21
　quotient, 117, 154
　sum, 115–16, 117, 154
　using, 117–18, 319
American Journal of Public
　　Health, 162
American Revolution, 600
America's Cup, 480
"Annie G. Cantering, Saddled"
　　(photo), 333
Area
　formulas for, 57
　maximizing, 599–600
Arithmetic operations, English phrases
　　involving, 2
Armstrong, C. Michael, 560
Arrests for driving under the
　　influence, 399
Arrows, 153
　in graph, 103
Associative property, 87
　of addition, 21
　and English language, 21
　of multiplication, 21
Asymptotes of graph, horizontal and
　　vertical, 388, 470
Augmented matrix, 206, 207, 209, 211,
　　213, 231
Average cost, business problems
　　involving, 447–48
Average cost function, 447, 448, 473
Average rate of change, 135, 155
　slope as, 134–36, 155

Axis of symmetry, 590, 629
　of parabola, 590

B

Babbage, Charles, 217
Bannister, Roger, 53
Berlin Airlift, 282
Binomial(s), 299, 376
　completing the square of, 564–65
　dividing polynomial by, 421–22
　expanding, Binomial Theorem for,
　　647–49
　multiplying trinomial and,
　　313–14
　product of two, 314–15, 377
　squaring, 316–17
Binomial coefficients
　definition of, 646
　evaluating, 646–47
Binomial difference, square of,
　　316–17, 377
Binomial expression, patterns in
　　writing out, 647
Binomial sum, square of, 316, 377
Binomial Theorem, 647–49
　factorial notation for products of
　　consecutive positive integers
　　in, 645
　using, 648–49
Boundary points, 615
　locating on number line, 616, 617,
　　618, 620, 631
Braces, 6, 19
Brackets, 19, 240
Break-even analysis, 186–89, 229
Break-even point, 187–88, 229
Budget deficits, 384
Bush, George, 384
Bush, George W., 384
Business
　average cost for, 447–48
　functions, 186–89, 227, 229

C

Cage, John, 46
Calculators, exponential expressions
　　evaluated on, 4

Calculus, 320
  and change studies, 137
  and rate of change, 311
Cardano, Girolamo, 550
Cardiac index, 511
Cardona, Carlos, 28
Cartesian coordinate system, 29. *See also* Rectangular coordinate system
Center, circle, 639
Centrifugal force, modeling, 465–66
Challenger space shuttle, 636
Chaos, patterns of, 548
Chernobyl nuclear power plant explosion, 66
Circles, 639–42
  area and perimeter formulas for, 57
  center, 639
  definition, 639
  general form of equation of, 641–42
  radius, 639
  standard form of equation of, 639–41
Circular cylinder, volume formula for, 57, 59
"Clearing equation of fractions," 44
Closed dots, 103, 153
Coefficients, 22, 299
  correlation, 152
  leading, 299, 376
  Leading Coefficient Test, 302–5, 376
  negative, 327, 379
Coffee drinkers who become irritable, 297, 362, 368
College education, cost of, 39, 48
Combined variation, 464–65
Combining like terms, 22
Commutative property, 87
  of addition, 20
  of multiplication, 20
Completing the square, 564–67, 628
  graphing quadratic functions in form $f(x) = ax^2 + bx + c$, 593
  quadratic equations solved by, 565–67, 628
  visualizing, 565
Complex fractions. *See* Complex rational expressions
Complex numbers, 543
  adding and subtracting, 543–44, 554
  dividing, 545–47, 555
  multiplying, 544–45, 554
  on postage stamp, 546
Complex rational expressions, 411–19, 471

simplifying by multiplying by one, 412–14, 416
Compound inequalities, 253–62, 290–91
  examples of, 253
  finding those involving "and," 253–54
  finding those involving "or," 257
  solving those involving "and," 254–57
  solving those involving "or," 257–59
Compound interest, 331, 567–68
  formulas for, 567
  and problem solving, 567–68
Computer, history behind, 217
Conditional equation, 45, 46, 47, 89
Cone, volume formula for, 57
Conic sections
  circles, 57
  parabolas, 589, 590–96, 629, 630
Conjugates, 522, 553, 555, 579
  of complex number, 546
  to divide complex numbers, using, 546–47, 555
  multiplying, 522, 555
Consistent systems, 173, 198, 229
Constant, 2
Constant function, 133
Constant of variation, 459, 462
Constraints, in linear programming, 283–84, 293
Continuous function, 388
Continuous graph, 302, 376
Contradiction, 45
Correlation coefficient, 152
Cost-benefit model, 385, 441–42
  using, 385–86
Cost function, 186–87, 229, 447
Cramer, Gabriel, 217
Cramer's rule, 217
  with inconsistent and dependent systems, 225
  system of linear equations in three variables solved using, 222–24, 232–33
  system of linear equations in two variables solved using, 219–20, 232
Crespo, Luigi, 28
Cube
  perfect, 505
  volume formula for, 57
Cube root functions, 485–86
  evaluating, 486
Cube roots, 480, 551
  defined, 485

simplifying, 486–87
Cubes, sum and difference of, 379
  factoring, 351–52, 379
Curve fitting, 199, 231
Cycles, and simplifying powers of i, 547–48

D

Darwin, Charles, evolution theory of, 493
Data, modeling, 5, 86, 144–45, 199
  with slope-intercept form, 136–37
Decimal notation, 90
  converting from scientific notation to, 79
  converting to scientific notation from, 80
Decimals, fractions expressed as, 7
Degree
  of ax^n, 299
  of nonzero constant, 299, 376
  of polynomial, 299, 300, 376
  of polynomial equation, 366
  of term $ax^n y^m$, 299
De La Hoya, Oscar, 28
Denominators. *See also* Least common denominator (LCD)
  adding rational expressions with the same, 399–400
  negative exponents in, 69
  rationalizing, 522–26, 553–54, 562
  rationalizing those containing one term, 522–24, 553
  rationalizing those containing two terms, 525–26, 554
  simplifying rational expressions with opposite factors in, 392
  subtracting rational expressions with the same, 400–401
Dependent equations, 173, 198, 229, 230, 231, 232, 233
Dependent systems, identifying
  determinants used for, 225
  matrices used for, 213
Dependent variables, 99, 155
Descartes, René, 29
Determinants, 217, 232–33
  inconsistent and dependent systems identified with, 225
  second-order, 217–18, 232
  solving systems of linear equations in two variables using, 218–20, 232

systems of linear equations in three variables solved using, 222–24, 232–33

third-order, 221–22, 232

Difference
binomial, square of, 316–17, 377
of cubes, factoring, 351–52, 379
of functions, 117, 154
of two terms, product of, 317–18, 377

Difference engine, 217

Difference of two squares, 348–51
factoring, 348–49, 379
grouping used for obtaining, 350–51

Direct variation, 458–62
in combined variation, 464–65
with powers, 461
and problem solving, 460–62

Discriminant
for determining number/type of solutions, 580, 629
using, 580–81

Distance formula, 637–38
midpoint formula derived from, 638–39

Distributive property, 21–22, 87
and addition of polynomials, 305
for factoring out Greatest Common Factor, 325, 327, 378
for multiplying complex numbers, 544–45
for multiplying monomial and polynomial that is not monomial, 312
for multiplying radicals with more than one term, 520–21
other forms of, 22

Division
of complex numbers, 545–47, 555
long, 421–25, 429, 471
with numbers in scientific notation, 81
of polynomial by binomial, 421–22
of polynomial by monomial, 420–21
of polynomials, 419–27, 471
of radical expressions, 515–16, 553
of rational expressions, 392–93, 470
of real numbers, 18–19, 87
simplifying complex rational expressions by, 414–15, 416
synthetic, 428–31, 472
by zero, 19

Domain, 98
of any quadratic function, 595
of cube root functions, 485
determining, 116
of function, 113–14, 153
identifying, from function's graph, 104–6
of rational function, 386–88, 470
of relation, 96–97, 153
of square root function, 482, 483, 551

E

Einstein, Albert, 61, 512

Elements, 86
of matrix, 206
of set, 6

Ellipsis, 6

Empty set, 46, 89, 247, 277

Endangered Species Act of 1973, 2

Endangered species formula, 1

End behavior, 357, 376
of polynomial functions, 302–5, 320

Energy efficiency of building materials, 179–80

English language, associative property and, 21

English phrases, 10
translating into algebraic expressions, 2–3, 86

English sentences, and inequalities, 248

Equals sign, 5, 272, 273, 274

Equation(s), 5, 86. See also Linear equations; Polynomial equations
absolute value, 263–64, 291
of circle, general form of, 641–42
of circle, standard form of, 639–41
conditional, 45, 46, 47, 89
containing fractions, 89
dependent, 173, 198, 229, 230, 231, 232, 233
equivalent, 40
functions as, 99
graphing, 28–38, 88
graphing, using point-plotting method, 30–31
inconsistent, 45, 46, 47, 89
inverse variation, 462
quadratic in form, 607–14, 630
radical, 532–41, 554
rational, 436–42, 472, 473
in two variables, 30
types of, 45–47, 89
of variation, 459
zeros of, 40

Equation of line
horizontal lines, 132–33
parallel to given line, 145–46

point-slope form of, 141–52, 156
slope-intercept form of, 128–32, 142–44, 156
standard form of, 124, 144
various forms of, finding, 144
vertical lines, 134

Equivalent algebraic expressions, 20
and associative properties, 21
and distributive property, 22

Equivalent equations, 40

Equivalent inequalities, 242, 243

Evaluating an algebraic expression, 3–4

Evaluating the function, 99

Even-degree polynomial functions, graphs of, 305, 320, 376

Even-degree polynomials, and Leading Coefficient Test, 303

Even-degree quadratic function, graph of, 335

Even roots, 487–88, 551

Expanding binomials, 647–49

Expanding the summation notation, 644

Expansion by minors, 221

Exponential expression, 4
evaluating, 17–18
simplifying, 73–75, 90, 498–99

Exponents, 4, 86. See also Rational exponents
and large numbers, 75
negative, 69–70, 90
negative integers as, 68–70
power rule for, 70–71
products-to-powers rule for, 71–72, 90
quotient-to-power rule for, 72–73, 90
zero, 68, 90

Extraneous solutions, 534, 554

F

Factorial notation, 645–46

Factoring, 325, 378
difference of two squares, 348–49, 379
general strategy for, 379
monomial from polynomial, 325, 378
perfect square trinomials, 349–50, 379, 565
polynomial equations solved by, 366–67
polynomials, 356–59
quadratic equations solved by, 362–66, 560, 582

Factoring, (*cont.*)
radicals simplified by, 505–8
special forms, 347–55, 379
by substitution, 338–39, 342
sum and difference of two cubes,
351–52
trinomials, 378
trinomials by grouping, 342–43
trinomials in two variables, 337–38,
341–43
trinomials whose leading coefficient
is not 1, 339–42
trinomial with leading coefficient of
1, 333–39
Factoring by grouping, 327–29, 378
to obtain difference of two squares,
350–51
trinomials, 342–43, 378
Factoring completely, 338, 349
Factoring out Greatest Common
Factor, 325–27, 338
Factors, 325
of term, 22
Farmland losses (U.S.), 559, 600
First terms, in binomial, 314
FOIL method
and factoring by grouping, 329
and factoring by substitution, 339
and factoring completely, 338
and factoring trinomials, 333, 334,
335, 337, 340
and factoring trinomials by
grouping, 342
for multiplying complex numbers,
544–45
for multiplying polynomials, 314–15
for multiplying radicals with more
than one term, 520–21
for multiplying sum and difference
of two terms, 317–18
for multiplying two binomials, 377
and square of binomial, 316
Formulas, 5, 86
and applications of rational
equations, 473
for area, perimeter, and volume, 57
compound interest, 567
distance, 637–38
and mathematical models, 5
midpoint, 638–39
for motion, 83
problem solving using, 89
quadratic, 575
solving for variable in, 58–61,
446, 473

special-product, 316, 317, 318
time in motion, 448
variation, 458
Fraction(s)
complex. *See* Complex rational
expressions
equations containing, 89
expressing as decimals, 7
linear equations with, 44–45
linear inequalities containing,
246–47
Fractional equation. *See* Rational
equations
Fraction bars, 19
Free-falling object, modeling position
of, 622–23
Function(s), 95–161. *See also*
Exponential functions;
Logarithmic functions;
Polynomial functions; Quadratic
functions; Rational functions
algebra of, 114–22
analyzing graphs of, 103–4
average cost, 447, 448, 473
business, 186–89, 227, 229
constant, 133
continuous, 388
cost, 186–87, 229, 447
defined, 98
determining domain, 116–17
domain of, 104–6
as equations, 99
evaluating, 99
graphing, 100–106
linear, 101, 124–41
objective, 282–83, 285, 286
polynomial multiplication for
evaluating, 320–21
profit, 188–89, 229
quadratic, 199, 320, 335, 589–600,
629–30
radical, 506, 538–39, 551
rational, 385–88, 419–20, 470
relations as, 97–98
revenue, 186–87, 229
square root, 482–84
sum of (addition of), 115–16
vertical line test for, 102, 153
Function machine, 99
Function notation, 99, 100, 153

G

Galápagos Islands, 492–93
Gates, Bill, 202

Gauss, Carl Friedrich, 209
postage stamp honoring work of, 546
Gaussian elimination, 209
Gender stereotypes, 253
Geometric figures, formulas for area,
perimeter, and volume, 57
Gershwin, Ira, 46
Golden rectangles, 525
Graphing calculators/graphing utilities
checking solution of radical
equation on, 535
checking tables and graphs on, 101
converting from decimal to
scientific notation on, 80
determinant of matrix evaluated
on, 222
equations graphed on, 33–35
even-degree quadratic
functions, 335
exponential expressions evaluated
on, 4
functions evaluated on, 99
keystroke sequences for rational
exponents, 495, 496
linear equations graphed on, 126
maximum function feature, 600
minimum function feature, 199
models for scatter plots obtained
on, 145
numeric or graphic check on, 55
polynomial equations, 367
polynomial functions graphed on,
301, 304
QUADratic REGression
program, 600
rational functions graphed with, 388
row-echelon form on, 213
SHADE feature, 274
solutions to linear equations in one
variable on, 44, 47
solving systems on, 165
square root functions on, 482
TABLE feature on, 367, 387,
584, 599
verifying solution set for rational
inequality, 621
verifying solution set of absolute
value equation, 263
verifying solution set of linear
inequality, 245
verifying solution to quadratic
equation, 577
Graphs/graphing
asymptotes of, horizontal and
vertical, 388, 470

of circle, using standard form of circle's equation to, 640–42
equations, 28–38, 88
even-degree polynomial functions, 305, 320, 376
functions, 100–106
inequalities, 239–40
information obtained from, 103–4
linear inequalities in two variables, 272–75
lines, 29
odd-degree polynomial functions, 304, 357, 376
polynomial functions, 304, 305, 320, 357, 376
quadratic functions, 335, 560, 589–96, 629–30
rational functions, 388, 470
real numbers, 8
smooth and continuous, 302, 376
stories told with, 32–33
systems of inequalities, 292
systems of linear equations in two variables solved by, 164–65, 172, 228
systems of linear inequalities, 276–78
using intercepts, 124–26
vertical lines, 134, 155
Gravitation, Newton's formula for, 465
Greater than or equal to symbol, 9, 87, 239
absolute value inequalities using, 265–66
Greatest common binomial factor, factoring out, 328
Greatest Common Factor (GCF), 325, 378
factoring out, 325–27, 338
factoring out, then factoring difference of two squares, 348
greatest common binomial factor, 328
Grouping/grouping method, factoring by, 327–29, 378
special forms used when, 350–51
trinomials, 342–43, 378
Grouping symbols, 19
Group projects
business functions, 227
cost-benefit analysis, 469
factoring problems, 375
free-falling objects, 627
growth phenomena that is leveling off, 551
pricing options, 290

scatter plots for data, 152
word problems, 85
Growth patterns, modeling those leveling off, 479, 483

H

Half-planes, 272–74, 292
Heart rate during aerobic activities, 238
Hexagons, in honeycombs, 347
Hispanic population, in United States, 28
Horizontal asymptotes, 388, 470
Horizontal lines, 144
equations of, 132–33
graphs of, 155
Hypotenuse, of right triangle, 370
isosceles right triangle, 569

I

IBM, 217
Identity, 45, 46, 47, 89
Identity element of addition, 15
i (imaginary unit), 542–43
simplifying powers of, 547–48, 555
Imaginary numbers, 543
as solutions to quadratic equation, 562–63, 579, 580
Imaginary part, of complex number, 543
Incidence ratio, 397
Inconsistent equation, 45, 46, 47, 89
Inconsistent systems, 173, 229, 230, 231, 233
determinants used for identifying, 225
geometric possibilities for, 198
matrices used for identifying, 213
Independent variable, 99, 155
Index, 487
Index of summation, 644
Inequalities. *See also* Compound inequalities
absolute value, 264–67, 291–92
and English sentences, 248
equivalent, 242, 243
graphs of, 239–40
with infinitely many solutions, 247
and intervals, 241
isolating x in middle of, 256
with no solution, 247
polynomial, 615–19, 622–23, 631
rational, 619–23
solving, 242–47

with unusual solution sets, 247–48
Inequality symbols, 9–10, 87, 239, 272, 273, 274
reversal of direction of, 242
Infinity symbol, 241
Innumeracy, 324–25
Innumeracy (Paulos), 324
Inside terms, in binomial, 314
Instantaneous change, roller coasters and, 527
Integers, 6, 86
Integral exponents, properties of, 90
Intercepts
for graphing linear equations, 124–26, 154
graphing using, 124–26
Interest
compound. *See* Compound interest
simple, dual investments with, 180–82, 184
Intersection of sets, 254, 290
Interval notation, 240, 241, 244, 245, 246–47
domain of rational function expressed in, 386–87
Intervals
and inequalities, 241
on real number line, 241
satisfying polynomial inequality, 616, 617, 618, 621, 631
satisfying rational inequalities, 631
Inverse
additive, 15, 392, 407–8
multiplicative (reciprocal), 18, 392
Inverse function of exponential function. *See* Logarithmic functions
Inverse variation, 462–64
in combined variation, 464–65
equations, 462
problem solving, 463
Investments
and compound interest, 567–68
and simple interest, 180–82, 184
Irrational number, 8, 86
as solutions to quadratic equations, 577–78
Isolating x in the middle of inequality, 256
Isosceles right triangle, 569

J

Johns, Jasper, 324, 583
Joint variation, 465–66

K

Kim, Scott, 590
Korean War, 262

L

Laffer, Arthur, 419, 420
Landscape design, problem solving
    with, 369–70
Large numbers
    and exponents, 75
    names of, 79
Last terms, in binomial, 314
Leading coefficient, 299, 376
Leading Coefficient Test, 302–5, 376
Leading term, 299, 376
Leaning Tower of Pisa, 622–23
Least common denominator (LCD), 44
    finding, 401–3, 402, 404, 405, 406
    in solving rational equations, 436,
        438, 439
Legs, of right triangle, 370
Leguizamo, John, 28
Less than or equal to symbol, 9, 87, 239
Less than symbol, 239
    and absolute value inequalities, 265
Like radicals, adding and subtracting,
    512–13
Like terms, 22
Limitations, inequalities to describe,
    283–84
Limits of summation, upper and
    lower, 644
Line(s)
    equations of, 144. *See also* Equation
        of line
    regression, 123, 154
    slope of, 126–28
Linear equations, 379
    algebraic word problems solved
        using, 52–57
    defined, 40
    with fractions, 44–45
    intercepts used for graphing, 125–26
    in one variable, 40–43
    properties of equality for solving, 41
    solving, 39–51, 88–89
Linear equations in three variables, 194
Linear functions, 101, 124–41,
        154, 320
    data modeled with, 144–45
    graphing in slope-intercept form,
        130–32
    and slope, 155

Linear inequalities
    containing fractions, 246–47
    problem solving with, 248–49
    properties of, 243
    solving, 242–47, 290
    with unusual solution sets, 247–48
Linear inequalities in one variable,
    239, 290
Linear inequalities in two variables,
    271–81, 292
    graphing, 272–75
Linear programming, 282–89, 293
    constraints in, 283–84, 293
    objective functions in, 282–83, 285,
        286, 293
    problem solving with, 284–86
Linear systems. *See* Systems of linear
    equations
Line graphs, 29
Living alone, 532, 538–39
Long distance phone service, 239, 242
Long division, 429, 471
    polynomial, 421–25
Lopez, Jennifer, 28
Lower limit of summation, 644

M

Main diagonal, 207
Mandelbrot set, 548
Margin of error, 263, 266–67
Martin, Ricky, 28
M*A*S*H, 262
Mathematical illiteracy, 324
Mathematical models, 86
    applied problems solved using, 48
    and formulas, 5
Matrix (matrices), 206, 231
    augmented, 206, 207, 209, 211, 213, 231
    determinant of 2×2, 217–18, 232
    determinant of 3×3, 221–22, 232
    inconsistent and dependent systems
        identified with, 213
    linear systems solved using, 205–17,
        231
    square, 217, 232
Matrix row operations, 207–9
Maximized quantity, objective function
    describing, 282–83, 285
Maximizing quadratic functions,
    596–97, 599–600, 630
Medicare, 614–15
Midpoint formula, 638–39
Minimized quantity, objective function
    describing, 282–83, 285

Minimizing quadratic functions, 596,
    597–98, 630
Minimum function feature, on
    graphing utility, 199
Minor, 221, 232
Mixture problems, 182–83, 184
Model breakdown, 5
Modeling, with variation, 458–69, 474
Monomials, 299, 376
    factoring from polynomial, 325, 378
    multiplying, 311–12, 377
    multiplying polynomial and,
        312–13, 377
    multiplying polynomial that is not
        monomial and, 312–13
    polynomials divided by, 420–21
Motion, 333
    formula, 83
    modeling, 367–68
Motion problems, 184–85, 473
    involving time, 448–50, 473
Multinational corporations, 560
Multiplication
    associative property of, 21
    of binomial and trinomial, 313–14
    commutative property of, 20
    of complex numbers, 544–45, 554
    of conjugates, 522, 555
    of monomial and polynomial,
        312–13, 377
    of monomials, 311–12, 377
    with numbers in scientific
        notation, 81
    of polynomial functions, 319–21
    of polynomials, 313, 377
    of radical expressions, 508–9, 552
    of radical expressions using special
        products, 521
    of radicals with more than one term,
        520–21, 553
    of rational expressions, 391–92, 470
    of real numbers, 17, 87
    of sum and difference of two terms,
        317–18, 377
Multiplication property of equality, 40
Multiplication property of
    inequality, 243
Multiplicative inverse (or reciprocal),
    18, 392
Muybridge, Eadweard, 333

N

Natural numbers, 6, 86
Negative coefficient

common factor used with, 327
factoring out GCF or common factor with, 379
Negative-exponent rule, 68–69
Negative exponents, 69–70, 90
complex rational expressions written with, 413
Negative integers, as exponents, 68–70
Negative multiplication property of inequality, 243
Negative numbers
cube root of, 485, 486
even root of, 487
square root of, 481
square root of, as multiples of i, 542–43
square root of, multiplying, 545
Negative reciprocal, 147, 156
Negative slope, 128, 155
Negative square root, 481
Negative units, 8
Newton, Isaac, gravitation formula of, 465
n factorial $(n!)$, 645
nth roots
even and odd, 487–88
of real numbers, 551
solving radical equations containing, 533
Nuclear accidents, 66
Numbers, sets of, 6
irrational numbers, 8
rational numbers, 7
real numbers, 6–8, 86
Numerators
negative exponents in, 69
rationalizing, 526–27
simplifying rational expressions with opposite factors in, 392
Numerical coefficient, 22

O

Objective functions, in linear programming, 282–83, 285, 286, 293
O'Brien, Soledad, 28
Odd-degree polynomial functions, graphs of, 304, 357, 376
Odd-degree polynomials, and Leading Coefficient Test, 303
Odd roots, 487–88, 551
One-to-one correspondence, 8
Open dots, 103, 153
Opposites (additive inverses), 15, 392

adding or subtracting rational expressions with denominators that are, 407–8
Ordered pairs, 29, 88
as solutions of systems, 163–64
Ordered triples, as solution of system of linear equation in three variables, 194
Order of operations, 4, 19–20, 87
Origin, 8, 29, 88
Outside terms, in binomial, 314

P

Parabolas, 589
and axis of symmetry, 590
downward opening, 591, 592, 595, 596, 629
in form $f(x) = ax^2 + bx + c$, 593–96
in form $f(x) = a(x - h)^2 + k$, 590–93
upward opening, 591, 593, 596, 629, 630
Parallel lines, 145–46, 247
and slope, 145, 146, 156
Parentheses, 19
and distributive property, 22
in interval notation, 240
and simplifying algebraic expressions, 23
Parthenon (Athens), and golden rectangle, 525
Paulos, John Allen, 324
"Peanuts" comic strip, 520
Percents, 324
Perfect cube, 505
Perfect nth power, 505, 507
with exponents divisible by n, 507
Perfect square, 505
Perfect square trinomials, 351, 365, 564
factoring, 349–50, 379, 565
visualizing factoring of, 350
Perimeter, formulas for, 57
Perpendicular lines, 147
and slope, 147, 156
Plotting points in rectangular system, 29–30
Point-plotting method, graphing equation using, 30–31
Point Reyes National Seashore, 271
Points, plotting in rectangular coordinate system, 29–30
Point-slope form of equation of line, 141–52, 156

to write line's equation, 142–44
Polynomial(s), 297–324
adding, 305–6, 376
defined, 299, 376
degree of, 299, 300, 376
degree of four terms for, 299
dividing, 419–27, 471
dividing by monomials, 420–21
dividing by those containing more than one term, 421–25
dividing using synthetic division, 429–31
factoring, 356–59
factoring completely, 338
long division of, 421–25
monomials factored from, 325, 378
prime, 325, 339, 341, 378
strategy for factoring, 356
subtracting, 306–7, 377
vocabulary of, 299–300
Polynomial equations, 366, 379
applications with, 367–71, 379
degree of, 366
Pythagorean Theorem used for obtaining, 370–71
Remainder Theorem and solving, 433
solving by factoring, 366–67
in standard form, 366
Polynomial functions, 300–305, 320
end behavior of, 302–5
evaluating, 300–301
graphs of, 304, 305, 320, 357, 376
multiplication of, 319–21
Remainder Theorem used for evaluating, 432–33
Polynomial inequality
definition of, 615
solving, 615–19, 631
solving problems modeled by, 622–23
Polynomial multiplication, 313, 377
for evaluating functions, 320–21
FOIL method used in, 314–15
multiplying monomial and polynomial that is not monomial, 312–13
visualizing, 315
when neither is monomial, 313–14
Population growth
foreign-born (U.S.), 574, 583–84
loss of farmland and, 559
Porgy and Bess, 46
Positive multiplication property of inequality, 243

Positive number, cube root of, 485, 486
Positive (or principal) square root, 480, 551
    definition of, 481
Positive slope, 128, 155
Positive units, 8
Power rule, 90
    for exponents (powers to powers), 70–71
Powers of *i*, simplifying, 547–48, 555
Prime polynomials, 339, 341
    over the set of integers, 325, 378
Prime trinomials, 334
Principal, 180
Principal *n*th root, 487
Principal square root, 480, 551
    definition of, 481
Problem solving
    with absolute value inequalities, 266–67
    involving maximizing/minimizing quadratic functions, 596–600
    with linear inequalities, 248–49
    with linear programming, 284–86
    quadratic formula used for, 583–84
    with scientific notation, 82–83
    with square root property, 567–70
    success in, 356
    with systems of equations, 229
    using formulas, 89
Product(s)
    of functions, 117, 154, 319–21
    minimizing, 597–98
    special, 318, 521
    special-product formula, 316, 317, 318
    of sum and difference of two terms, 317–18, 377
    of two binomials, 314–15, 377
Product rule, 66–67, 90
Product rule for radicals, 504–5, 552
    for multiplying radicals with same indices, 552
    for simplifying radicals, 552
    simplifying radicals using factoring and, 505–8
Products-to-powers rule, 71–72, 90
Profit function, 188–89, 229
    gain or loss, 189
Pythagorean Theorem, 370
    distance formula and, 637–38
    polynomial equation obtained using, 370–71
    and square root property, 568–70

Q

Quadrants, 29
Quadratic equations, 379, 560
    compound interest problems solved using, 567–68
    defined, 362
    determining most efficient method for solving, 581–82
    discriminant of, 580–81
    irrational solutions to, 577–78
    rational solutions to, 576
    solving by completing the square, 565–67, 628
    solving by factoring, 362–66, 560, 582
    solving by square root property, 561–64, 582
    solving using quadratic formula, 575–80, 582, 628
    in standard form, 362, 379, 560
    writing from solutions, 582–83, 629
Quadratic formula, 574–88, 581
    deriving, 575
    for problem solving, 583–84
    quadratic equations solved with, 575–80, 582, 628
Quadratic functions, 199, 320
    applications of, 596–600
    graphs of, 335, 560, 589–96, 629–30
    minimizing or maximizing, 596–600, 630
Quadratic in form, solving equations, 607–14, 630
Quotient, of functions, 117, 154
Quotient rule, 67, 90
    for radicals, 515–16, 553
Quotients, 18
    of two rational expressions, 392–93
Quotients-to-powers rule, 72–73, 90

R

Radical equations, 532–41, 554
    applications of, 538–39
    solving, 532–38
    solving those containing *n*th roots, 554
Radical expressions, 481, 551
    adding and subtracting, 512–14, 553
    adding and subtracting with higher indices, 514
    combining those requiring simplification, 513–14
    dividing, 515–16, 553

    multiplying, special products used for, 521
    multiplying, then simplifying, 508–9, 552
    multiplying those with more than one term, 520–21, 553
    product rule for, 504–5, 552
    quotient rule for, 515–16, 553
    simplifying, 484–85, 488
    simplifying by factoring, 505–8
    simplifying using rational exponents, 499–500
    simplifying when variables to even powers in radicand are nonnegative quantities, 507
Radical functions, 551
    problem solving using models that are, 538–39
    simplifying, 506
Radical sign, 481
Radical symbols, 19, 481
Radicand, 481
Radius of circle, 639
Range, 98
    identifying, from function's graph, 104–6
    of relation, 96–97, 153
Rate of change
    average, 134–36, 155
    and calculus, 311
Rational equations, 436
    applications of, 441–42, 473
    formulas and applications of, 473
    solving, 436–41, 472
Rational exponents, 492–503, 552
    applications with, 496–97
    defining, 493
    exponential expressions simplified using, 498–99
    negative, 496
    with numerators other than one, 494–96
    properties of, 498
    radical expressions simplified using, 499–500
    and windchill, 497
Rational expressions, 385, 470
    adding, 471
    addition of, when denominators are opposites, 407–8
    addition of those with common denominators, 399–400
    addition of those with different denominators, 403–5, 406–7
    complex, 411–19, 471

dividing, 392–93, 470
in motion problems, 448
multiplying, 391–92, 470
simplifying, 389–91, 470
subtracting, 471
subtraction of, when denominators
   are opposites, 407–8
subtraction of those with common
   denominators, 400–401
subtraction of those with different
   denominators, 405–7
Rational functions, 385–86, 419–20, 470
   domain of, 386–88, 470
   graphs of, 388, 470
Rational inequalities, 619
   solving, 619–21, 631
   solving problems modeled by, 622–23
Rationalizing denominators, 553–54
   containing one term, 522–24, 553
   containing two terms, 525–26, 554
   in solving quadratic equation by
      square root property, 562
Rationalizing numerators, 526–27
Rational numbers, 7, 86
Reagan, Ronald, 384
Real number line, 8, 241
Real numbers, 8
   adding, 14–16, 87
   dividing, 18–19, 87
   multiplying, 17, 87
   ordering, 8–10
   set of, 6–8, 86
   subtracting, 16, 87
Real part, of complex number, 543
Reciprocal (or multiplicative inverse),
   18, 392
   negative, 147, 156
Rectangle, area and perimeter
   formulas for, 57
Rectangular coordinate system, 29, 88
   graphing equations in, 30–33
   points plotted in, 29–30
Rectangular solid, volume formula for,
   57, 59
Regression line, 123, 154
Reinhardt, Ad, 46
Relations, 96–97, 153
   defined, 96
   as functions, 97–98
Relativity theory, 512
Remainder Theorem, 472
   to evaluate polynomial function,
      432–33
   using to show number as solution of
      polynomial equation, 433

Repeated factorization, 349
Repeating decimals, 7
Representative numbers
   solving polynomial inequalities at,
      616, 617, 618, 631
   solving rational inequality at,
      620, 631
Revenue, 186
Revenue function, 186–87, 229
Right triangle, 370
   isosceles, 569
Rise, 126
Rockefeller, John D., 202
Roller coasters, and instantaneous
   change, 527
Roots of equations, 40
Roseanne, 607
Roster method, 6, 86
Row-echelon form, 207, 213
Row equivalent, 207
Row operations, 231
   on augmented matrix, 207
Run, 126

S

Satisfying the equation, 30, 40
Satisfying the inequality, 239, 272
Scatter plots, 123, 154
Schulz, Charles, 520
Scientific calculators
   converting from decimal to
      scientific notation on, 80
   exponential expressions evaluated
      on, 4
   keystroke sequences for rational
      exponents, 495, 496
   polynomial functions on, 301
   square root functions on, 482
Scientific notation, 78–85, 90
   computations with, 81–82, 90
   converting to decimal notation, 79
   defined, 79
   problem solving with, 82–83
Second-degree polynomial
   equation, 362
Second-order determinants, evaluating,
   217–18, 232
Sense of the inequality, changing,
   242
Set-builder notation, 6, 86, 240, 244,
   245, 246
   domain of rational function
      expressed in, 386–87
   intersection of sets expressed in, 254

to represent function's domain and
   range, 104–5
Sets, 6, 86
   intersection of, 254, 290
   of irrational numbers, 8
   of numbers, 6
   of rational numbers, 7
   of real numbers, 6–8, 86
   union of, 257, 290
Simple interest, dual investments with,
   180–82, 184
Simplifying algebraic expressions,
   23–24, 88
Simplifying complex rational
   expressions
   by dividing, 414–15
   by multiplying by one, 412–14, 416
Simplifying exponential expressions,
   73–75
   common errors in, 74
   with rational exponents, 498–99
Simplifying radical expressions,
   484–85, 488
   cube root, 486–87
   by factoring, 505–8
Simplifying rational expressions,
   389–91, 470
   solving rational equations
      contrasted with, 438
Single population, 479
Slope, 155
   as average rate of change,
      134–36, 155
   defined, 126, 134
   interpreting, 134
   of line, 126–28
   negative, 128, 155
   and parallel lines, 145, 146
   and perpendicular lines, 147
   positive, 128, 155
   zero, 128, 155
Slope-intercept equation of nonvertical
   line, 155
Slope-intercept form of equation of
   line, 128–32, 144, 156
   linear functions graphed in, 130–32
   modeling data with, 136–37
   writing, 142–44
Smooth, continuous graphs, 302, 376
Social Security, 614
Solution(s)
   of compound inequality, 254, 257
   of equation in two variables, 30
   extraneous, 534, 554
   of inequality, 239

Solution(s) (*cont.*)
    of inequality in two variables, 272
    of linear equation in one variable, 40
    of system of linear equations, 163–64
    of system of linear equations in three variables, 194
    of system of linear inequalities, 276
Solution set
    of inequality, 239, 240–41
    of linear equation in one variable, 40
    of system of linear equations in three variables, 194
    of system of linear inequalities, 276
Solving a formula for a variable, 58–61
Solving an inequality, 239
Solving linear equations, 40–43
Special forms
    difference of two squares, 348–51, 379
    factoring, 347–55, 379
    perfect square trinomials, 349–50, 351, 365, 379, 564, 565
    sum and difference of cubes, 379
Special-product formula, 316, 317, 318
    visualizing, 316
Special products
    for multiplying radicals, 521
    using, 318
Sphere, volume formula for, 57
Square
    area and perimeter formulas for, 57
    perfect, 505
Square matrix, 217, 232
Square of binomial difference, 316–17, 377
Square of binomial sum, 316, 377
Square root function, 482
    domain of, 482, 483, 551
    evaluating, 482–83
    modeling with, 483–84
Square root property, 628
    problem solving with, 567–70
    and Pythagorean Theorem, 568–70
    quadratic equations solved by, 561–64, 582
Square roots, 8, 551
    evaluating, 480–81
Square root signs, 19, 480, 481
Square roots of negative numbers, 481
    as multiples of i, 542–43
    multiplying, 545

Standard form
    of equation of circle, 639–41
    of equation of circle, converting general form to, 641–42
    of equation of line, 124, 144
    of polynomial, 299
    polynomial equation in, 366
    quadratic equations in, 362, 379, 560
Standard viewing rectangle, 34
Substitution
    and dual investment problem, 181
    for eliminating variables, 165–67, 195
    factoring by, 338–39, 342
    and mixture problem, 183
    systems of linear equations in two variables solved by, 165–67, 172, 228
Subtraction
    of complex numbers, 543–44, 554
    definition of, 16
    of like radicals, 512–13
    of polynomials, 306–7, 377
    of radical expressions, 512–14, 553
    of rational expressions, 471
    of rational expressions when denominators are opposites, 407
    of rational expressions with common denominators, 400–401
    of rational expressions with different denominators, 405–7
    of real numbers, 16, 87
Sum
    binomial, square of, 316, 377
    of cubes, factoring, 351–52, 379
    of functions, 115–16, 117, 154
    of two terms, product of, 317–18, 377
Summation
    index of, 644
    upper and lower limits of, 644
Summation notation, 644–45
Supercomputers, 213
Supply-side economics, 384, 419–20
Symbols
    approximation, 7–8
    for binomial coefficients, 646
    for elements in sets, 6–7
    empty set, 46, 247
    greater than, 239
    greater than or equal to, 9, 87, 239, 265–66
    grouping, 19
    inequality, 9–10, 87, 239, 242, 272, 273, 274
    infinity, 241

    less than, 239, 265
    less than or equal to, 9, 87, 239
    for opposites (or additive inverses), 15
    radical, 19, 481
    square root, 19, 480, 481
Synthetic division, 428–31, 472
    polynomials divided using, 429–31
Systems of equations. *See* Systems of linear equations
Systems of inequalities. *See* Systems of linear inequalities
Systems of linear equations, 163
    matrix solutions to, 205–17, 231
    problem solving and business applications using, 178–93, 229
Systems of linear equations in three variables, 193–204, 230–31
    inconsistent and dependent systems, 198
    problem solving with, 200
    solution of, 194
    solving by eliminating variables, 194–97, 230
    solving by using matrices, 206–7, 211–13
    solving those with missing terms, 196–97
    solving using determinants and Cramer's rule, 222–24, 232–33
Systems of linear equations in two variables, 163–78, 228–29
    comparing solution methods, 172
    determining if ordered pair is solution of, 163–64
    with infinitely many solutions, 173–74
    with no solutions, 173
    number of solutions to, 172
    solving, using determinants and Cramer's rule, 218–20, 232
    solving by addition method, 168–71, 172, 228
    solving by graphing, 164–65, 172, 228
    solving by substitution method, 165–67, 172, 228
Systems of linear inequalities
    applications of, 278, 282–86
    graphing, 276–78, 292

T

Tax rates, and supply-side economics, 384
"Tele-immersion" computers, 446

Television program viewing, 262
Temperature
    conversions, 60
    windchill, 497
Terminating decimals, 7
Terms
    of algebraic expressions, 22
    leading, 299, 376
    like, 22
    multiplying sum and difference of,
        317–18
    outside, in binomial, 314
Third-order determinants
    defined, 221
    evaluating, 221–22, 232
Time, motion problems involving,
    448–50, 473
Time zone system, 13
Trapezoid, area and perimeter
    formulas for, 57
Triangle
    area and perimeter formulas for, 57
    isosceles right, 569
    right, 370
Trinomials, 299, 376
    factoring, 378
    factoring by grouping, 342–43, 378
    factoring those whose leading
        coefficient is not 1, 339–42
    factoring those with leading
        coefficient of 1, 333–39
    multiplying binomial and, 313–14
    perfect square, 349–50, 351, 365, 379,
        564, 565
    prime, 334
Trinomials in two variables, factoring,
    337–38, 341–43
"29-Fold M-set Seahorse" (Voss), 548

U

Undefined slope, 128, 155
Union of sets, 290
    defined, 257

Unit distance, 8
U.S. Bureau of Justice, 399
Upper limit of summation, 644

V

Value, of second-order determinant,
    217
Vanderbilt, Cornelius, 202
Variable cost, 186
Variables, 2, 86
    dependent, 99, 155
    independent, 99, 155
    solving for, in formulas, 58–61,
        446, 473
Variables, eliminating
    addition method of, 168–71, 172,
        195, 228, 230
    solving linear system in three
        variables by, 194–97, 230
    solving linear systems in two
        variables by, 165–71, 195
    substitution method of, 165–67, 172,
        195, 228
Variation
    combined, 464–65
    constant of, 459, 462
    direct, 458–62
    equations of, 459
    formulas, 458
    inverse, 462–64
    joint, 465–66
    modeling using, 458–69, 474
Variation problems, solving, 459
Vertex (vertices), 284, 286
    of parabola, 589, 591, 592, 593,
        594, 629
Vertical asymptotes, 388, 470
Vertical lines, 144
    equations of, 134
    graphs of, 134, 155
Vertical line test, for functions, 102, 153
Viewing rectangle
    on graphing utility, 34

understanding, 34
Virtual reality, 445–46, 447
Volume, formulas for, 57
Voss, R. F, 548

W

Whole numbers, 6, 86
Windchill, rational exponents
    and, 497
Wolf population, 2, 5
Word problems, solving, 178–85
    strategy for, 52, 89
Work problems, solving, 450–53, 473

X

x-axis, 29, 88, 124
x-coordinate, 29, 88, 125
x-intercept, 124, 125, 154
    solving polynomial inequalities
        and, 616

Y

y-axis, 29, 88
y-coordinate, 29, 88, 124
y-intercept, 125, 154
    graphing using slope and, 130–32

Z

Zero, division by, 19
Zero-exponent rule, 68, 90
Zero factorial (0!), 645
Zero (Johns), 583
Zero-Product Principle, 362–63, 365,
    366, 379
    in reverse, 582, 629
Zero slope, 128, 155
Zeros of equation, 40
0 Through 9 (Johns), 324
Zoom in/zoom out, on graphing
    utilities, 34

PHOTO CREDITS

Robert F. Blitzer
Essentials of Intermediate Algebra for College Students, Chapter Test Prep Video CD
0-13-186611-7
© 2006 Pearson Education, Inc.
Pearson Prentice Hall
Pearson Education, Inc.
Upper Saddle River, NJ 07458
Pearson Prentice Hall™ is a trademark of Pearson Education, Inc.

ESM Media Development
Higher Education Division
Pearson Education, Inc.
1 Lake Street
Upper Saddle River, NJ 07458
Should you have any questions concerning technical support, you
may write to:
New Media Production
Higher Education Division
Pearson Education, Inc.
1 Lake Street
Upper Saddle River, NJ 07458
YOU ACKNOWLEDGE THAT YOU HAVE READ THIS AGREEMENT, UNDERSTAND IT, AND AGREE TO BE
BOUND BY ITS TERMS AND CONDITIONS. YOU FURTHER AGREE THAT IT IS THE COMPLETE AND
EXCLUSIVE STATEMENT OF THE AGREEMENT BETWEEN US THAT SUPERSEDES ANY PROPOSAL OR
PRIOR AGREEMENT, ORAL OR WRITTEN, AND ANY OTHER COMMUNICATIONS BETWEEN US RELATING
TO THE SUBJECT MATTER OF THIS AGREEMENT.

System Requirements
- Windows:
Pentium II 300 MHz processor
Windows 98, NT, 2000, ME, or XP
64 MB RAM (128 MB RAM required for Windows XP)
4.3 available hard drive space (optional—for minimum QuickTime installation)
800 × 600 resolution
8x or faster CD-ROM drive
QuickTime 6.x
Sound Card
- Macintosh:
Power PC G3 233 MHz or better
Mac OS 9.x or 10.x
64 MB RAM
10 MB available hard drive space for Mac OS 9, 19 MB on OS X (optional—if QuickTime installation is needed)
800 x 600 resolution
8x or faster CD-ROM drive
QuickTime 6.x

Support Information
If you are having problems with this software, call (800) 677-6337 between 8:00 a.m. and 8:00 p.m. EST, Monday through Friday,
and 5:00 p.m. through Midnight EST on Sundays. You can also get support by filling out the web form located at :
http://247.prenhall.com/mediaform
Our technical staff will need to know certain things about your system in order to help us solve your problems more quickly and
efficiently. If possible, please be at your computer when you call for support. You should have the following information ready:

- Textbook ISBN
- CD-Rom/Diskette ISBN
- corresponding product and title
- computer make and model
- Operating System (Windows or Macintosh) and Version
- RAM available
- hard disk space available
- Sound card? Yes or No
- printer make and model
- network connection
- detailed description of the problem, including the exact wording of any error messages.

NOTE: Pearson does not support and/or assist with the following:

- third-party software (i.e. Microsoft including Microsoft Office Suite, Apple, Borland, etc.)
- homework assistance
- Textbooks and CD-Rom's purchased used are not supported and are non-replaceable. To purchase a new CD-Rom contact
  Pearson Individual Order Copies at 1-800-282-0693

Definitions, Rules, and Formulas

The Real Numbers

Natural Numbers: $\{1, 2, 3, \dots\}$

Whole Numbers: $\{0, 1, 2, 3, \dots\}$

Integers: $\{\dots, -3, -2, -1, 0, 1, 2, 3, \dots\}$

Rational Numbers: $\{\frac{a}{b} \mid a$ and b are integers, $b \neq 0\}$

Irrational Numbers: $\{x \mid x$ is real and not rational$\}$

Basic Rules of Algebra

Commutative: $a + b = b + a;\ ab = ba$

Associative: $(a + b) + c = a + (b + c);$
$\qquad\qquad (ab)c = a(bc)$

Distributive: $a(b + c) = ab + ac;\ a(b - c) = ab - ac$

Identity: $a + 0 = a;\ a \cdot 1 = a$

Inverse: $a + (-a) = 0;\ a \cdot \frac{1}{a} = 1(a \neq 0)$

Multiplication Properties: $(-1)a = -a;$
$(-1)(-a) = a;\ a \cdot 0 = 0;\ (-a)(b) = (a)(-b) = -ab;$
$(-a)(-b) = ab$

Set-Builder Notation, Interval Notation, and Graphs

$(a, b) = \{x \mid a < x < b\}$

$[a, b) = \{x \mid a \leq x < b\}$

$(a, b] = \{x \mid a < x \leq b\}$

$[a, b] = \{x \mid a \leq x \leq b\}$

$(-\infty, b) = \{x \mid x < b\}$

$(-\infty, b] = \{x \mid x \leq b\}$

$(a, \infty) = \{x \mid x > a\}$

$[a, \infty) = \{x \mid x \geq a\}$

$(-\infty, \infty) = \{x \mid x$ is a real number$\} = \{x \mid x \in R\}$

Slope Formula

$$\text{slope } (m) = \frac{\text{change in } y}{\text{change in } x} = \frac{y_2 - y_1}{x_2 - x_1} \quad (x_1 \neq x_2)$$

Equations of Lines

1. Slope-intercept form: $y = mx + b$
 m is the line's slope and b is its y-intercept.

2. Standard form: $Ax + By = C$

3. Point-slope form: $y - y_1 = m(x - x_1)$
 m is the line's slope and (x_1, y_1) is a fixed point on the line.

4. Horizontal line parallel to the x-axis: $y = b$

5. Vertical line parallel to the y-axis: $x = a$

Systems of Equations

One solution: consistent

No solution: inconsistent

Infinitely many solutions: dependent and consistent

A system of linear equations may be solved: (a) graphically, (b) by the substitution method, (c) by the addition or elimination method, (d) by matrices, or (e) by determinants.

$$\begin{vmatrix} a_1 & b_1 \\ a_2 & b_2 \end{vmatrix} = a_1 b_2 - a_2 b_1$$

Cramer's Rule:

Given a system of a equations of the form

$\begin{aligned} a_1 x + b_1 y &= c_1 \\ a_2 x + b_2 y &= c_2 \end{aligned}$, then $x = \dfrac{\begin{vmatrix} c_1 & b_1 \\ c_2 & b_2 \end{vmatrix}}{\begin{vmatrix} a_1 & b_1 \\ a_2 & b_2 \end{vmatrix}}$ and $y = \dfrac{\begin{vmatrix} a_1 & c_1 \\ a_2 & c_2 \end{vmatrix}}{\begin{vmatrix} a_1 & b_1 \\ a_2 & b_2 \end{vmatrix}}$.

Absolute Value

1. $|x| = \begin{cases} x & \text{if } x \geq 0 \\ -x & \text{if } x < 0 \end{cases}$

2. If $|x| = c$, then $x = c$ or $x = -c$. $(c > 0)$

3. If $|x| < c$, then $-c < x < c$. $(c > 0)$

4. If $|x| > c$, then $x < -c$ or $x > c$. $(c > 0)$

Special Factorizations

1. Difference of two squares:
$$A^2 - B^2 = (A + B)(A - B)$$

2. Perfect square trinomials:
$$A^2 + 2AB + B^2 = (A + B)^2$$
$$A^2 - 2AB + B^2 = (A - B)^2$$

3. Sum of two cubes:
$$A^3 + B^3 = (A + B)(A^2 - AB + B^2)$$

4. Difference of two cubes:
$$A^3 - B^3 = (A - B)(A^2 + AB + B^2)$$

Variation

English Statement	Equation
y varies directly as x.	$y = kx$
y varies directly as x^n.	$y = kx^n$
y varies inversely as x.	$y = \dfrac{k}{x}$
y varies inversely as x^n.	$y = \dfrac{k}{x^n}$
y varies jointly as x and z.	$y = kxz$

Exponents
Definitions of Rational Exponents

1. $a^{\frac{1}{n}} = \sqrt[n]{a}$

2. $a^{\frac{m}{n}} = \left(\sqrt[n]{a}\right)^m$ or $\sqrt[n]{a^m}$

3. $a^{-\frac{m}{n}} = \dfrac{1}{a^{\frac{m}{n}}}$

Properties of Rational Exponents
If m and n are rational exponents, and a and b are real numbers for which the following expressions are defined, then

1. $b^m \cdot b^n = b^{m+n}$

2. $\dfrac{b^m}{b^n} = b^{m-n}$

3. $(b^m)^n = b^{mn}$

4. $(ab)^n = a^n b^n$

5. $\left(\dfrac{a}{b}\right)^n = \dfrac{a^n}{b^n}$

Radicals

1. If n is even, then $\sqrt[n]{a^n} = |a|$.

2. If n is odd, then $\sqrt[n]{a^n} = a$.

3. The product rule: $\sqrt[n]{a} \cdot \sqrt[n]{b} = \sqrt[n]{ab}$

4. The quotient rule: $\dfrac{\sqrt[n]{a}}{\sqrt[n]{b}} = \sqrt[n]{\dfrac{a}{b}}$

Complex Numbers

1. The imaginary unit i is defined as
$$i = \sqrt{-1}, \quad \text{where} \quad i^2 = -1.$$

The set of numbers in the form $a + bi$ is called the set of complex numbers. If $b = 0$, the complex number is a real number. If $b \neq 0$, the complex number is an imaginary number.

2. The complex numbers $a + bi$ and $a - bi$ are conjugates. Conjugates can be multiplied using the formula
$$(A + B)(A - B) = A^2 - B^2.$$
The multiplication of conjugates results in a real number.

3. To simplify powers of i, rewrite the expression in terms of i^2. Then replace i^2 with -1 and simplify.

Quadratic Equations and Functions

1. The solutions of a quadratic equation in standard form
$$ax^2 + bx + c = 0, \quad a \neq 0,$$
are given by the quadratic formula
$$x = \frac{-b \pm \sqrt{b^2 - 4ac}}{2a}.$$

2. The discriminant, $b^2 - 4ac$, of the quadratic equation $ax^2 + bx + c = 0$ determines the number and type of solutions.

Discriminant	Solutions
Positive perfect square with a, b, and c rational numbers	2 rational solutions
Positive and not a perfect square	2 irrational solutions
Zero, with a, b, and c rational numbers	1 rational solution
Negative	2 imaginary solutions

3. The graph of the quadratic function
$$f(x) = a(x - h)^2 + k, \quad a \neq 0,$$
is called a parabola. The vertex, or turning point, is (h, k). The graph opens upward if a is positive and downward if a negative. The axis of symmetry is a vertical line passing through the vertex. The graph can be obtained using the vertex, x-intercepts, if any, [set $f(x)$ equal to zero], and the y-intercept (set $x = 0$).

4. A parabola whose equation is in the form
$$f(x) = ax^2 + bx + c, \quad a \neq 0,$$
has its vertex at
$$\left(-\frac{b}{2a}, f\left(-\frac{b}{2a}\right)\right).$$

If $a > 0$, then f has a minimum that occurs at $x = -\dfrac{b}{2a}$. If $a < 0$, then f has a maximum that occurs at $x = -\dfrac{b}{2a}$.

Definitions, Rules, and Formulas (continued)

Exponential and Logarithmic Functions

1. Exponential Function: $f(x) = b^x, b > 0, b \neq 1$
Graphs:

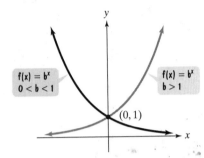

2. Logarithmic Function: $f(x) = \log_b x, b > 0, b \neq 1$
$y = \log_b x$ is equivalent to $x = b^y$.

Graphs:

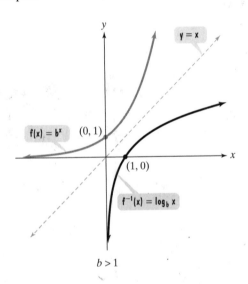

$b > 1$

3. Properties of Logarithms

a. $\log_b(MN) = \log_b M + \log_b N$

b. $\log_b\left(\dfrac{M}{N}\right) = \log_b M - \log_b N$

c. $\log_b M^p = p \log_b M$

d. $\log_b M = \dfrac{\log_a M}{\log_a b} = \dfrac{\ln M}{\ln b} = \dfrac{\log M}{\log b}$

e. $\log_b b^x = x;\ \log 10^x = x;\ \ln e^x = x$

f. $b^{\log_b x} = x;\ 10^{\log x} = x;\ e^{\ln x} = x$

Distance and Midpoint Formulas

1. The distance from (x_1, y_1) to (x_2, y_2) is
$$\sqrt{(x_2 - x_1)^2 + (y_2 - y_1)^2}.$$

2. The midpoint of the line segment with endpoints (x_1, y_1) and (x_2, y_2) is
$$\left(\frac{x_1 + x_2}{2}, \frac{y_1 + y_2}{2}\right).$$

Conic Sections
Circle

Ellipse

Hyperbola

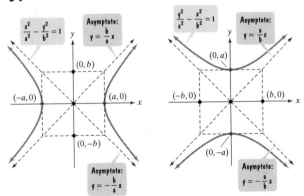